Beginning and Intermediate
ALGEBRA

FIFTH EDITION

SHERRI MESSERSMITH
College of DuPage

NATHALIE M. VEGA-RHODES
Lone Star College—Kingwood

ROBERT S. FELDMAN
University of Massachusetts Amherst

With contributions from William C. Mulford, *The McGraw-Hill Companies*

BEGINNING & INTERMEDIATE ALGEBRA

Published by McGraw-Hill Education, 2 Penn Plaza, New York, NY 10121. Copyright ©2021 by McGraw-Hill Education. All rights reserved. Printed in the United States of America. No part of this publication may be reproduced or distributed in any form or by any means, or stored in a database or retrieval system, without the prior written consent of McGraw-Hill Education, including, but not limited to, in any network or other electronic storage or transmission, or broadcast for distance learning.

Some ancillaries, including electronic and print components, may not be available to customers outside the United States.

This book is printed on acid-free paper.

1 2 3 4 5 6 7 8 9 LWI 24 23 22 21 20

ISBN 978-1-260-57067-0
MHID 1-260-57067-3

Cover Image: ©*Olha Polishchuk/Shutterstock*

The Internet addresses listed in the text were accurate at the time of publication. The inclusion of a website does not indicate an endorsement by the authors or McGraw-Hill Education, and McGraw-Hill Education does not guarantee the accuracy of the information presented at these sites.

About the Authors

© Phil Messersmith

Sherri Messersmith
Professor of Mathematics, College of DuPage

Sherri Messersmith began teaching at the College of DuPage in Glen Ellyn, Illinois, in 1994, where she served as the Chair of the Developmental Math Committee for several years. She has over 25 years of experience teaching many different courses from developmental mathematics through calculus. She earned a Bachelor of Science degree in the Teaching of Mathematics at the University of Illinois at Urbana-Champaign and taught high school for two years. Sherri returned to UIUC and earned a Master of Science degree in Applied Mathematics and stayed on at the university to teach and coordinate large sections of undergraduate math courses as well as teach in the Summer Bridge program for at-risk students. She is the author of many McGraw-Hill Education texts.

Sherri and her husband, Phil, are empty-nesters who recently relocated to the East Bay when her husband took at position at the University of California—Berkeley as a professor of Materials Science and Bioengineering. Sherri enjoys attending conferences and meeting with faculty, and in her free time she loves to read, hang out with her dogs, study French and Italian, and travel.

© Nathalie M. Vega-Rhodes

Nathalie M. Vega-Rhodes
Professor of Mathematics, Lone Star College—Kingwood

Nathalie Vega-Rhodes' career in higher education began 18 years ago and has encompassed a number of student-focused positions. For nearly a decade, she has taught mathematics ranging from developmental courses to calculus, as well as student success courses. She holds a Bachelor of Arts in Mathematics from the University of Houston and a Master of Science in Mathematics from the University of Houston—Clear Lake. In addition to teaching, as both Lead Faculty and Online Faculty Fellow at Lone Star College–Kingwood, she assists math faculty with pedagogical and technology-related implementation strategies. Most recently, she led a team in designing and implementing algebraic corequisite courses as part of a state-required redesign. Her earliest work in higher education focused on academic support, first as a tutor and supplemental instruction (SI) leader, and then as a coordinator for a math tutoring and SI program. These early roles shaped her teaching and led to a sharp focus on developing students' academic mindset in her courses. In her free time, Nathalie enjoys scuba diving and traveling with her husband, hanging out with her dog, and reading.

Robert S. Feldman
Deputy Chancellor and Professor of Psychological and Brain Sciences, University of Massachusetts—Amherst

© Robert S. Feldman

Bob Feldman still remembers those moments of being overwhelmed when he started college at Wesleyan University. "I wondered whether I was up to the challenges that faced me," he recalls, "and although I never would have admitted it then, I really had no idea what it took to be successful at college."

That experience, along with his encounters with many students during his own teaching career, led to a life-long interest in helping students navigate the critical transition that they face at the start of their own college careers. Bob, who went on to graduate with High Honors from Wesleyan and receive a Masters and Doctorate in Psychology from the University of Wisconsin—Madison, teaches at the University of Massachusetts—Amherst, where he is the Senior Advisor to the Chancellor and Professor of Psychological and Brain Sciences. He is founding director of the first-year experience course for incoming students at UMass and is Senior Fellow of the Center for Student Success Research.

Bob is a Fellow of the American Psychological Association, the American Association for the Advancement of Science, and the Association for Psychological Science. He has written and edited more than 250 scientific articles, book chapters, and books, including *P.O.W.E.R. Learning: Strategies for Success in College and Life,* 8e; *Understanding Psychology,* 14e; *The First Year of College: Research, Theory, and Practice on Improving the Student Experience and Increasing Retention,* and *Learning Science: Theory, Research, and Practice.* He is past president of the FABBS Foundation, an umbrella group of societies promoting the behavioral and brain sciences, and he is President of the Board of Directors of New England Public Radio. Bob loves travel, music, and cooking. He and his wife live in a home overlooking the Holyoke mountain range in western Massachusetts.

What is  ?

P.O.W.E.R. is an approach to systematically completing tasks based on five practical steps

P Prepare — The most critical facet of preparation is setting goals. Goal-setting improves student performance by increasing student focus, confidence, motivation, and persistence.

O Organize — Students have to identify and organize the intellectual tools necessary to accomplish their goals. Organizing refers not only to considering how they must apply the most appropriate academic strategies, but also to maintaining and applying good habits outside of class in order to manage their many responsibilities.

W Work — Doing the work—reading the materials, taking good notes in class, and doing in-class exercises—may seem like the most obvious step, but it is an area where students often falter. Using P.O.W.E.R. will improve your students' motivation and help them view success as a product of their hard work and effort.

E Evaluate — In math, concepts build on each other, so student success depends on reaching a level of mastery in each section before progressing. P.O.W.E.R. helps students understand that their work is not complete until they have assessed their progress and identified where they are struggling.

R Rethink — Too often in developmental math, students do not stop to assess their overall performance until after an exam, at which point it may be too late. P.O.W.E.R. prompts students after each section to honestly assess how they are doing and where they may need to change their strategy or ask for help.

The Story of the P.O.W.E.R. Textbooks

In the classroom, I viewed the textbook as merely a *guide* and did many other things on my own to better meet the needs of my students. For example, I taught in bite-sized pieces because developmental students in particular learn better when material is presented in more manageable chunks. As students' basic skills deteriorated, I created several Basic Skills Worksheets to help improve those skills. The students did them in class, and it only took up two or three minutes of precious class time. When McGraw-Hill Education saw some of the materials that I had made for my own classroom, they asked if I would write a textbook. So in 2004 I began writing my first book, *Beginning & Intermediate Algebra,* now in its 5th edition. The material for the book was written to align with how students learn best, and these strategies and activities were included in both the book and its accompanying supplements. Twenty-five years of teaching and writing that first book revealed that *developmental students don't want to fail; they just don't know how to succeed.*

Over the years, the books have evolved to include everything possible to give students *and* instructors the tools they need for success. Explicit student success skills were not included as part of the first several books. But it became more and more clear over time that students needed direct instruction in skills like how to *effectively* do homework, how to read a math textbook, and how to manage their time. I had been addressing all of this in my own classroom, and after meeting Bob Feldman, I began doing it more formally by using his research-based P.O.W.E.R. Learning framework. I asked myself, *"Why not incorporate P.O.W.E.R. into the textbooks to both teach study skills and organize the material according to the way research says students learn best?"* Happily, Bob agreed to come on as a coauthor to give us the math textbooks with P.O.W.E.R. Learning that we have today.

Even with this major evolution in our approach, *we made sure not to compromise the math whatsoever*! The math content and its level of rigor remain the same as in the books written before we added the P.O.W.E.R. framework—so the fact that our books contain student success skills does *not* mean they are lacking in rigor. These books are light-years ahead of the earlier ones because we have now addressed a huge weakness of many college students: *knowing how to learn.*

With this 5th edition of *Beginning and Intermediate Algebra with P.O.W.E.R. Learning,* we have evolved even more. To each chapter we have added a new *Get Ready* chapter opener that contains prerequisite/corequisite lessons, with exercises, to prepare students to learn the concepts coming up in the chapter. Also new to this edition are the two academic mindset topics of developing a growth mindset and developing grit. The power (no pun intended) of the Internet is that it allows us to enhance the learning and teaching experience for students and instructors. We can do things that we couldn't dream of even 10 years ago! That's where Nathalie Vega-Rhodes comes in. A long-time P.O.W.E.R. math textbook user and outstanding instructor, Nathalie has a lot of experience using digital tools. We share the same philosophy of teaching, love for students, and belief in addressing *all* of our students' needs in the classroom, so why not bring her on board as a coauthor with digital expertise? So that's what we did. An immediate benefit was the creation of the Integrated Video & Study Guides, or IVSGs. It was a perfect collaboration; I wrote the IVSGs, and Nathalie made the new online videos.

Our team is complete, but this isn't the end; our story is not finished. The books continue to evolve and improve because of engaged faculty like you. We are so happy that you have chosen our textbook and want to continue the conversation. We would love to hear from you. Tell us your stories, and share your suggestions. Be a part of our story and part of the evolution of the P.O.W.E.R. math textbooks.

Sherri Messersmith (sherri.messersmith@gmail.com)
Nathalie M. Vega-Rhodes (nvegarhodes@gmail.com)
Bob Feldman (feldman@chancellor.umass.edu)

Why should you use *Beginning and Intermediate Algebra with P.O.W.E.R. Learning?*

- *NEW!* A *Get Ready chapter opener* **accompanies each chapter** and provides a just-in-time prerequisite lesson and exercises specific to what is needed for the chapter.
- **Solid, time-tested math content** with the amount of rigor needed to succeed in college-level courses
- **Written with friendly, conversational, non-intimidating language,** making it easier to read than most books while using all of the necessary mathematical language our students need
- **Written in bite-sized pieces** to make it easier to learn "complicated" material
- **Engaging applications** written with students and their interests in mind
- **Rewritten, easy-to-use, research-based student success materials** in every chapter *in the book* that instructors can use at their discretion
- **Research-based pedagogical features throughout the books**
  - **24-hr problems:** Marked in each exercise set, these are indicated by an icon. Even if students do not complete their *entire* homework assignment immediately, they are encouraged to do the 24-hr problems within 24 hours of leaving class so that they better retain what they have learned.
  - **"You Try" exercises:** After almost every example in the book, students can do a You Try problem to work out a problem that is similar to what was presented.
  - **P.O.W.E.R. framework:** Since the first editions of our math with P.O.W.E.R. Learning textbooks, this research-based framework has informed the organization of chapters and sections to match the way that research shows students learn best.
  - **Rethink questions:** This is a *crucial* step in the learning process, and the "R" in P.O.W.E.R., yet it is the one that students overlook the most (or don't even realize exists). Every exercise set is followed by a set of Rethink questions that require students to reflect upon what they have just done.
  - **Student success skills, emPOWERme, and Study Strategies:** Every chapter has a student success theme with an emPOWERme survey and a Study Strategies page *in the book* so that instructors can address students' weaknesses in this area. They are based upon Bob Feldman's research and can be done in any order. If instructors do not want to include them in their courses, they can be skipped easily.
- *Time-saving supplements, especially for adjunct instructors:* Many adjuncts teach at more than one school, and they don't always have time to make materials specific to their classes and their students' needs. We offer many author-created supplements with every textbook.
  - **Basic Skills Worksheets**—Help students improve their basic skills in class while using only 2 or 3 minutes of class time. Their confidence improves, too, when they see their basic skills improve. They start to believe that they *can* learn math!
  - **Section Worksheets**
  - **Worksheets to Tie Multiple Concepts Together**
  - **Guided Student Notes**—These ready-made lessons follow the book exactly and provide students with a framework for taking notes.
  - **Integrated Video & Study Guides (IVSGs)**—Great for flipped classrooms or any class, these require students to actively engage with the videos not only for doing examples but also for filling in information for elements such as definitions and procedures. Students *must* pay attention to complete these guides. The IVSGs are written for every *objective* in every section and designed so that most videos are 3 or 4 minutes long. The videos, and example from the video guides aligned to them, are assignable in ConnectMath Hosted by ALEKS.
  - **PowerPoints**
  - **Instructor Resource Manual**
  - *NEW! Get Ready* chapter opener as well as *Get Ready* exercises at the beginning of selected exercise sets

- **Enhanced exercise sets** with more conceptual questions
- **Videos for selected homework exercises**
- **Putting It All Together** sections in chapters where it is appropriate
- **Group Activities** in every chapter
- **ConnectMath Hosted by ALEKS and ALEKS**
- **ALEKS Prep for Beginning and Intermediate Algebra with P.O.W.E.R. Learning**: This preparatory tool focuses on prerequisite and introductory material for this text. The Prep products can be used during the first 6 weeks of a traditional course or in a corequisite course where students need to quickly narrow the gap in their skill and concept base. ALEKS Prep course products feature artificial intelligence that targets gaps in individual students' knowledge, assessment and learning directed toward individual student needs, an open response environment with realistic input tools, and unlimited online access on both PCs and MACs.

Table of Contents

Instructor P O W E R Tool Kit

Find these resources in your P.O.W.E.R. Tool Kit

In the textbook

- Work Hints*
- *Get Ready* chapter openers*
- In-Class Examples*
- You Trys*
- 24-hr Problems
- Enhanced Exercise Sets
- Putting It All Together sections*
- Group Activities
- Study Strategies and emPOWERme in each chapter*

Supplements

- Basic Skills Worksheets*
- Section Worksheets*
- Worksheets to Tie Multiple Concepts Together*
- Guided Student Notes*
- Comprehensive video package, including Integrated Video & Study Guides and videos for selected exercises*
- Power Points
- Instructor and Student Solution Manuals
- Computerized Test Bank and files
- Instructor Resource Manual**
- ConnectMath Hosted by ALEKS and ALEKS*

* Descriptions of these resources are included in the following pages.
**Additional information about how to use the resources can be found in the Instructor Resource Manual.

WORK HINTS highlight important steps in working out a problem, point out places of common student errors, or give a study tip for learning. Pulling these tips out of the main text makes them more noticeable to students.

 **Hint**
Notice that *like terms* are always lined up in the same columns.

$$
\begin{array}{r}
3x + 4 \\
x + 5\overline{)3x^2 + 19x + 20} \\
-(3x^2 + 15x) \\
\hline
4x + 20 \\
-(4x + 20) \\
\hline
0
\end{array}
$$

1) By what do we multiply x to get $4x$? 4
   Write +4 above +20.
2) Multiply 4 by $(x + 5)$: $4(x + 5) = 4x + 20$.
3) Subtract $(4x + 20) - (4x + 20) = 0$.
4) There are no more terms. The remainder is 0.

A **GET READY** chapter opener accompanies each chapter and provides a just-in-time pre/corequisite lesson, with exercises, specific to what is needed for the chapter. To decide which topics to include in this new feature, the authors broke down the material in the chapter into its component parts and pinpointed the pre/corequisite skills needed to learn the new concepts in the chapter. The topics identified as pre- or corequisites have been in use in many classrooms *before* being included in the book.

Get Ready

In this chapter, we will learn about logarithms. The properties of logarithms and the rules of exponents are related, so let's review some of the rules here.

1) Recall that an **exponent** (or **power**) is used to represent repeated multiplication. For example, $2 \cdot 2 \cdot 2 = 2^3$ where the *exponent*, 3, tells us that the *base*, 2, is multiplied by itself three times. (It is necessary to know the powers of integers listed in Section 1.2.)

2) Here are some of the **rules of exponents** that help us simplify expressions.

   Examples: Simplify using the rules of exponents. Assume all variables represent nonzero real numbers. The answer should contain only positive exponents.

   a) $p^7 \cdot p^3 = p^{7+3} = p^{10}$ b) $(k^2)^6 = k^{2 \cdot 6} = k^{12}$

   c) $(3c)^4 = 3^4 \cdot c^4 = 81c^4$ d) $\dfrac{n^8}{n^3} = n^{8-3} = n^5$

   e) $7^0 = 1$ f) $t^{-2} = \dfrac{1}{t^2}$

3) Remember the **relationship between radical and fractional exponent notations.**

   If n is a positive integer greater than 1 and $\sqrt[n]{a}$ is a real number, then $\sqrt[n]{a} = a^{1/n}$.

   (*The denominator of the fractional exponent is the index of the radical.*)

   Examples: Write

   a)

4) If an equation co

   Examples: Solve

In this chapter, use
WS2 Powers and W

Get Ready Exercises

1) Write each product in exponential form.
   a) $8 \cdot 8 \cdot 8 \cdot 8 \cdot 8 \cdot 8$ b) $5 \cdot 5 \cdot 5 \cdot 5$

2) Evaluate.
   a) 9^2 b) 2^5 c) 4^3 d) 10^4 e) 3^3

For Exercises 3–8, simplify using the rules of exponents. Assume all variables represent nonzero real numbers. The answer should contain only positive exponents.

3) $(2n^4)^5$ 4) $\dfrac{h^5}{h}$ 5) w^{-2} 6) 10^{-3} 7) 9^0 8) $k^7 \cdot k^2$

9) Write 64 as a power of 2.   10) Write 1000 as a power of 10.

11) Write $\dfrac{1}{125}$ as a power of $\dfrac{1}{5}$.   12) Write $\dfrac{1}{49}$ as a power of 7.

Write each expression using a fractional exponent.

13) $\sqrt[3]{5}$ 14) $\sqrt{10}$ 15) $\sqrt{6}$ 16) $\sqrt[5]{32}$

Solve each equation for y.

17) $x = 8y + 5$ 18) $x = -\dfrac{2}{3}y - 4$

Answers

1) a) 8^6 b) 5^4 2) a) 81 b) 32 c) 64 d) 10,000 e) 27   3) $32n^{20}$ 4) h^4 5) $\dfrac{1}{w^2}$ 6) $\dfrac{1}{1000}$ 7) 1 8) k^9 9) 2^6 10) 10^3 11) $\left(\dfrac{1}{5}\right)^3$ 12) 7^{-2} 13) $5^{1/3}$ 14) $10^{1/2}$ 15) $6^{1/2}$ 16) $32^{1/5}$ 17) $y = \dfrac{1}{8}x - \dfrac{5}{8}$ 18) $y = -\dfrac{3}{2}x - 6$

The IN-CLASS EXAMPLES exactly mirror the examples in the book, giving instructors additional problems to use while teaching in class. They are available only in the Annotated Instructor Edition, and they align with the examples in the **Guided Student Notes**.

EXAMPLE 4

In-Class Example 4

Factor completely.
a) $36m - 9m^3$ b) $h^4 - 81$
c) $8p^2 + 800$

Answer: a) $9m(2 + m)(2 - m)$
b) $(h^2 + 9)(h + 3)(h - 3)$
c) $8(p^2 + 100)$

Factor completely.

a) $128t - 2t^3$ b) $5x^2 + 45$ c) $h^4 - 16$

Solution

a) Ask yourself, *"Can I take out a common factor?"* Yes. Factor out $2t$.

$$128t - 2t^3 = 2t(64 - t^2)$$

Now ask yourself, *"Can I factor again?"* Yes. $64 - t^2$ is the difference of two squares. Identify a and b.

YOU TRY problems follow almost every example in the book and exactly mirror the examples. After working through problems in class, the instructor can have the students do a **You Try** to practice on their own to reinforce the lesson. The **Examples**, **In-Class Examples**, and **You Try** problems are the same *types* of problems containing different numbers, providing consistency for students *and* instructors.

YOU TRY 5

Solve each equation.

a) $\sqrt{2y + 1} - \sqrt{y} = 1$ b) $\sqrt{3t + 4} + \sqrt{t + 2} = 2$

The PUTTING IT ALL TOGETHER sections tie multiple concepts together and take students through the thought processes necessary for problem recognition. For example, after learning and practicing four methods for solving quadratic equations *individually,* the ***Putting It All Together*** section presents students with a group of quadratic equations in different forms and explains, step by step, how to decide which method to use to solve a particular equation.

Putting It All Together

 **Prepare**    **Organize**

What is your objective?	How can you accomplish the objective?
1 Decide Which Method to Use to Solve a Quadratic Equation	• While following Example 1 on your own, make a chart that will help you identify when to use each of the four methods. • Complete the given example on your own. • Complete You Try 1.

 **Work**   **Read the explanations, follow the example, take notes, and complete the You Try.**

We have learned four methods for solving quadratic equations.

Methods for Solving Quadratic Equations
1) Factoring
2) Square root property
3) Completing the square
4) Quadratic formula

Video Package: Integrated Video and Study Guides as well as Exercise Videos

Would you like to "flip" your classroom or offer students a structured learning environment outside the classroom? The **Integrated Video and Study Guides**, IVSGs, allow instructors to do just that. Updated for this edition of the textbook, the authors have created one **Integrated Video and Study Guide** for *each objective* in the book. Because each video and guide covers only one objective, most videos are just 3 to 5 minutes long. Students are required to watch the video and fill out the **IVSG** as they go along, stopping along the way to do the similar *Your Turn* problems and *Think About It* conclusion questions designed for deeper conceptual thinking. Each companion video follows the procedures and examples, with detailed explanations aligning perfectly with the textbook. In addition to IVSG companion videos, hundreds of 3–5 minute **Exercise Video** clips show students how to solve various exercises from the textbook. All videos are assignable in ConnectMath Hosted by ALEKS.

▶ *Play.* Often, we have to take out a greatest common factor before applying other factoring techniques. Let's summarize how to factor by grouping and look at another example.

Procedure   Steps for Factoring by Grouping

1) Before trying to factor by grouping, look at each term in the polynomial and ask yourself,

   "_____?" If so, factor out the GCF

   from all of the terms.

2) Make two groups of two terms so that each group has a common factor.

3) Take out the common factor in each group of terms.

4) Factor out the common binomial factor using the distributive property.

5) Check the answer by multiplying the factors.

Example 4: Factor $12t^3 + 12t^2 - 3t^2u - 3tu$ completely.

⏸ *Pause and do Your Turn 4:* Factor $8n^5 + 24n^4 - 16n^3 - 48n^2$ completely.

⏸ *Think About It:* Ms. Szymanski asks her students to factor $2x^3 + 10x^2 + 3x + 15$. Haeshin's answer is $(x+5)(2x+3)$, and Ilhan's answer is $(2x+3)(x+5)$. Who is right, and why?

Example 4: Factor $12t^3 + 12t^2 - 3t^2u - 3tu$ completely.

$$12t^3 + 12t^2 - 3t^2u - 3tu = 3t(4t^2 + 4t - tu - u) \quad \text{Factor out the GCF, } 3t$$

$$3t(\underbrace{4t^2 + 4t} - \underbrace{tu - u}) \quad \text{Take out the common factor in each group.}$$

$$3t[4t(t+1) - u(t+1)] \quad \text{Factor out } t+1 \text{ using the distributive property.}$$

$$= 3t(t+1)(4t-u)$$

Check by multiplying: $3t(t+1)(4t-u) = 3t(4t^2 + 4t - tu - u)$

$$= 12t^3 + 12t^2 - 3t^2u - 3tu$$

⏸ **Your Turn 4:** Factor $8n^5 + 24n^4 - 16n^3 - 48n^2$ completely.

common factor in each group. Factor out t+1 using the distributive property. Check by multiplying. Now it's your turn. Pause the video

Student Success Skills in Every Chapter

Start the student success discussion by having your students do the emPOWERme activity. Found in *every chapter* before the Chapter Summary, most **emPOWERme** activities take the form of a survey so that students can learn something about themselves with respect to the student success skill of the chapter. Follow this up with the **Study Strategies** found at the beginning of the chapter to help your students use the P.O.W.E.R. framework to acquire skills such as developing a growth mindset and grit, reading a math textbook, doing their homework effectively, or managing their time.

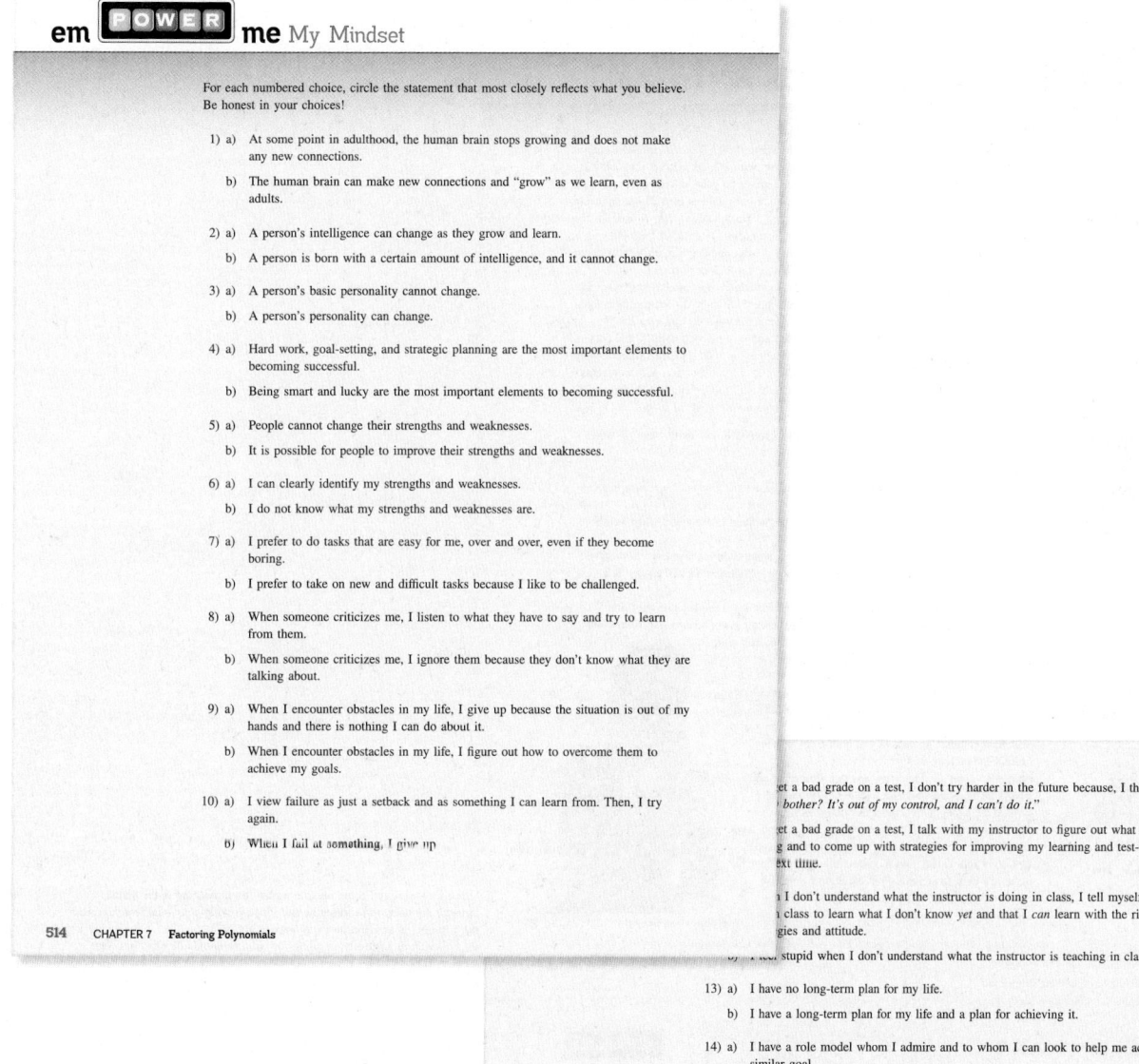

em **POWER** me My Mindset

For each numbered choice, circle the statement that most closely reflects what you believe. Be honest in your choices!

1) a) At some point in adulthood, the human brain stops growing and does not make any new connections.
   b) The human brain can make new connections and "grow" as we learn, even as adults.

2) a) A person's intelligence can change as they grow and learn.
   b) A person is born with a certain amount of intelligence, and it cannot change.

3) a) A person's basic personality cannot change.
   b) A person's personality can change.

4) a) Hard work, goal-setting, and strategic planning are the most important elements to becoming successful.
   b) Being smart and lucky are the most important elements to becoming successful.

5) a) People cannot change their strengths and weaknesses.
   b) It is possible for people to improve their strengths and weaknesses.

6) a) I can clearly identify my strengths and weaknesses.
   b) I do not know what my strengths and weaknesses are.

7) a) I prefer to do tasks that are easy for me, over and over, even if they become boring.
   b) I prefer to take on new and difficult tasks because I like to be challenged.

8) a) When someone criticizes me, I listen to what they have to say and try to learn from them.
   b) When someone criticizes me, I ignore them because they don't know what they are talking about.

9) a) When I encounter obstacles in my life, I give up because the situation is out of my hands and there is nothing I can do about it.
   b) When I encounter obstacles in my life, I figure out how to overcome them to achieve my goals.

10) a) I view failure as just a setback and as something I can learn from. Then, I try again.
    b) When I fail at something, I give up

...et a bad grade on a test, I don't try harder in the future because, I think, ...bother? It's out of my control, and I can't do it."

...et a bad grade on a test, I talk with my instructor to figure out what I did ...g and to come up with strategies for improving my learning and test-taking ...ext time.

...I don't understand what the instructor is doing in class, I tell myself that I ...class to learn what I don't know *yet* and that I *can* learn with the right ...gies and attitude.

...stupid when I don't understand what the instructor is teaching in class.

13) a) I have no long-term plan for my life.
    b) I have a long-term plan for my life and a plan for achieving it.

14) a) I have a role model whom I admire and to whom I can look to help me achieve a similar goal.
    b) I have a role model, but I know I can never achieve anything close to what that person has achieved.

In the *odd*-numbered statements, #1 – #13, part a) indicates a *fixed mindset* while b) indicates a *growth mindset*. In the *even*-numbered statements, #2 – #14, part a) indicates a *growth mindset* while part b) indicates a *fixed mindset*. Next to each of #1 – #14, write an *F* or a *G* to note whether you circled the fixed mindset or growth mindset statement, and count the number of each.

Do your answers show a tendency toward a fixed or a growth mindset? (Most people have some of both.) What, exactly, are these mindsets, and what can they tell you about yourself?

A **mindset** is a person's set of beliefs. People who think that their beliefs and qualities cannot be changed have a **fixed mindset.** On the other hand, people with a **growth mindset** believe that their beliefs, abilities, and personalities can change by using strategies and working hard. In fact, research shows that the brain itself actually changes as you learn new things, no matter your age! So, adopting a growth mindset is important for learning and succeeding in school and in life. (For the original version, see Dweck, Carol S. [2016]. *Mindset: The New Psychology of Success.* New York: Random House.)

Every Chapter Has a Student Success Theme

Find the STUDY STRATEGIES on the second page of every chapter. After doing the **emPOWERme** activity found before the Chapter Summary, continue the discussion of the chapter's student success skill by assigning or discussing the **Study Strategies.** Based on Bob Feldman's research, the **Study Strategies** explain, in easy-to-understand language, how to use each step of the P.O.W.E.R. framework to acquire student success skills such as developing a growth mindset and grit, how to read a math textbook, and how to be a good time manager. The student success topic of each chapter is listed in the Table of Contents, and the topics can be done in any order.

POWER Study Strategy — Developing a Growth Mindset

A **mindset** is a person's set of beliefs. People who think that their beliefs and qualities *cannot* be changed have a **fixed mindset**. For example, a student with a fixed mindset may tell himself, "I'm not a math person. No matter what I do, I'll never be good at math." People with a **growth mindset** believe that their beliefs, abilities, and personalities *can* change by using strategies and working hard. Such a person would say, "I haven't been good at math in the past, but if I get help from my instructor and try new strategies, I can do it." People with a fixed mindset tend to believe that events are out of their control and they can do nothing about them, while those with a growth mindset feel they have control over outcomes.

Carol Dweck is a psychology professor and expert in mindsets. Her research has shown that people's attitudes and strongly held beliefs greatly influence their ability to achieve their goals, and that the *"view you adopt for yourself* profoundly affects the way you lead your life."*

Her research has also revealed that, with a growth mindset, a person's abilities and personalities can change. Among other things, this involves coming up with a plan to succeed; your mindset can hold you back from achieving your goals, or it can put you on a path to success.

So how can *you* adopt (or maintain) a growth mindset? Let's use the P.O.W.E.R. framework to see how to develop a growth mindset to help you achieve your goals in school *and* in life.

P Prepare
- I will develop a growth mindset to help me to succeed in my math class, in other courses, and in life.

O Organize
- Complete the emPOWERme survey that appears before the Chapter Summary to learn whether you have more of a fixed or a growth mindset.
- Write down some examples of when you have had a fixed mindset and when you have had a growth mindset.
- Understand that your brain can actually grow new connections as you learn new things. *Truly believe* that core traits and beliefs *can* change with hard work and a plan. Tell yourself that you can succeed!
- Know that most successful people do not become that way "just because they are smart." The fact is, most successful people achieve their goals because they know their strengths and weaknesses, establish clear goals, work hard and smart, and ask for help from others when they need it.

W Work
- Write a few sentences describing how you feel about learning math. What did you write about that indicates a fixed mindset, and what points to a growth mindset? How can your fixed mindset items hold you back from learning?
- Make a list of your strengths and weaknesses. (For example, *I come to class every day* is a strength. *I don't always do my homework* is a weakness.) Be honest!
- Write down a long-term goal for this course. Also write down short-term goals that could help you achieve your long-term goal. Tell yourself that you *can* achieve those goals!
- Identify a role model who has achieved a goal similar to what you want to achieve. Find out how that person came to be successful, and make a list of what *you* can do to achieve your goal.

Image Source/Getty Images

*Dweck, Carol S. (2016). *Mindset: The New Psychology of Success*. New York: Random House.

448   CHAPTER 7   **Factoring Polynomials**

- ... your instructor and come up with a plan for succeeding in the course.
- ... strategies for using your strengths and for improving your weaknesses. Accept ... ive criticism as an opportunity to learn.
- ... *d and smart* to achieve your goals. Try something new if what you were ... s not working.
- ... poorly on a test, talk with your instructor and devise new strategies for success. Don't give up!

E Evaluate
- Take the emPOWERme survey again. Answer honestly! Did you have more answers that corresponded to a growth mindset than to a fixed mindset?
- Are you achieving your short-term and long-term goals in the course?
- Do you have more confidence in your ability to succeed at difficult tasks than you did before? Do you feel a stronger sense of control over the outcomes?

R Rethink
- Do you have more of a growth mindset than you did before? Why or why not?
- List some strategies that helped you achieve your goals as well as strategies that did not work well.
- How can you apply what you have learned about using a growth mindset to succeed in a math class to achieving goals outside the classroom?

Supplements Include a Suite of Ready-Made Worksheets

Want materials to use in class but don't have a lot of time to make them? Our package includes three types of author-created worksheets: **Basic Skills Worksheets, Section Worksheets,** and **Worksheets to Tie Multiple Concepts Together.** All are available in the student version (without answers) and instructor version (with answers), and *all worksheets are available as PDF or Word files so that instructors can download them and edit them as if they were their own Word documents.*

Worksheet 3C   Name: _____
Messersmith/Vega-Rhodes/Feldman
Beginning & Intermediate Algebra with P.O.W.E.R. Learning, 5th ed.

Find two numbers that...

MULTIPLY TO	and ADD TO	ANSWER
−27	−6	−9 and 3
72	18	
24	−11	
−4	3	
10	−7	
121	22	
−54	−3	
54	29	
16	−10	
30	17	
9	−6	
−8	−2	
21	10	
60	−19	
56	15	
−28	3	
−72	−6	
100	25	
−40	6	
11	−12	
20	12	
−35	−2	
77	18	
108	21	
−3	−2	

BASIC SKILLS WORKSHEETS

The **Basic Skills Worksheets** help students improve their basic skills while using only a few minutes, sometimes only 2 or 3, of class time. Their confidence improves, too, when they see their basic skills improve. Use these *before* reaching the topic where the basic skill is needed so that students can strengthen their weaknesses and be better prepared to learn new content. Each type of **Basic Skills Worksheet** comes in six different versions, incorporating more difficult concepts as they move from version A to version F.

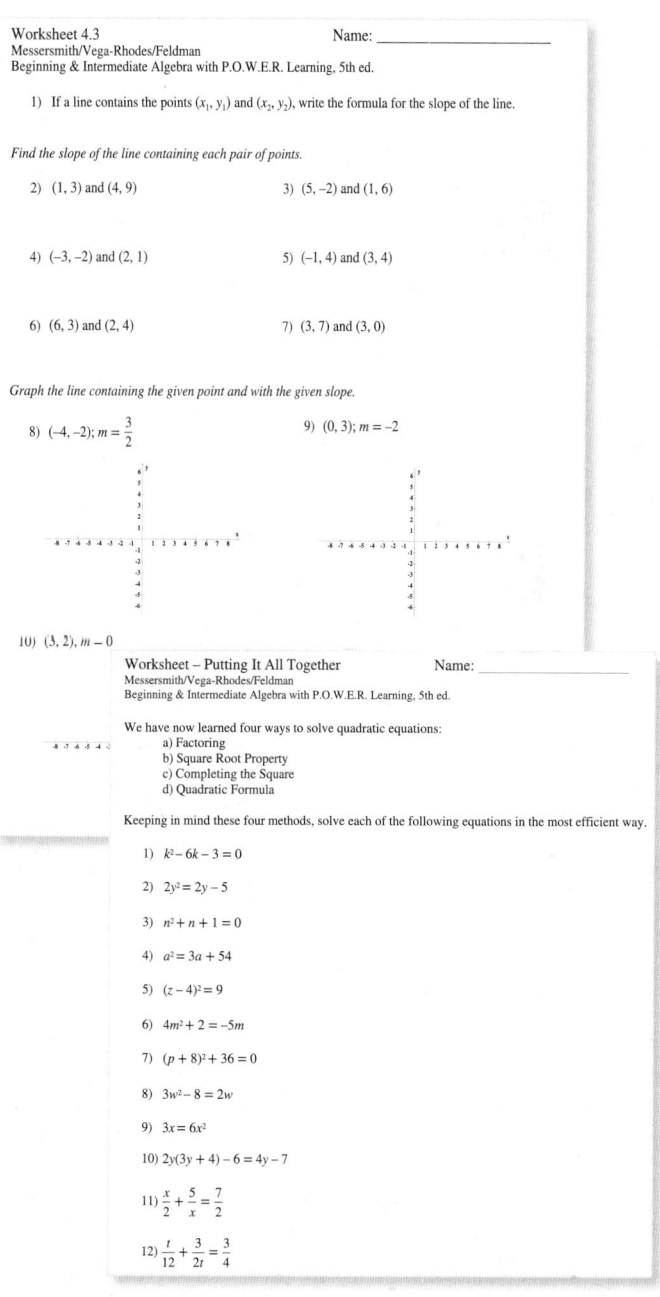

Worksheet 4.3   Name: _____
Messersmith/Vega-Rhodes/Feldman
Beginning & Intermediate Algebra with P.O.W.E.R. Learning, 5th ed.

1) If a line contains the points (x_1, y_1) and (x_2, y_2), write the formula for the slope of the line.

Find the slope of the line containing each pair of points.

2) (1, 3) and (4, 9)   3) (5, −2) and (1, 6)

4) (−3, −2) and (2, 1)   5) (−1, 4) and (3, 4)

6) (6, 3) and (2, 4)   7) (3, 7) and (3, 0)

Graph the line containing the given point and with the given slope.

8) $(-4, -2); m = \frac{3}{2}$ 9) $(0, 3); m = -2$

10) $(3, 2); m = 0$

Worksheet – Putting It All Together   Name: _____
Messersmith/Vega-Rhodes/Feldman
Beginning & Intermediate Algebra with P.O.W.E.R. Learning, 5th ed.

We have now learned four ways to solve quadratic equations:
   a) Factoring
   b) Square Root Property
   c) Completing the Square
   d) Quadratic Formula

Keeping in mind these four methods, solve each of the following equations in the most efficient way.

1) $k^2 - 6k - 3 = 0$

2) $2y^2 = 2y - 5$

3) $n^2 + n + 1 = 0$

4) $a^2 = 3a + 54$

5) $(z - 4)^2 = 9$

6) $4m^2 + 2 = -5m$

7) $(p + 8)^2 + 36 = 0$

8) $3w^2 - 8 = 2w$

9) $3x = 6x^2$

10) $2y(3y + 4) - 6 = 4y - 7$

11) $\frac{x}{2} + \frac{5}{x} = \frac{7}{2}$

12) $\frac{t}{12} + \frac{3}{2t} = \frac{3}{4}$

These author-created, ready-made **Section Worksheets** and **Worksheets to Tie Multiple Concepts Together** can be used for extra practice in class, with students working individually or in groups, or they can be distributed for students to take home. They come in student versions (without answers) and instructor versions (with answers).

SECTION WORKSHEETS

Every section of *every book* comes with at least one **Section Worksheet** to help instructors teach new content or to give students extra practice problems. The **Section Worksheets** *exactly match* the material in the section and are a great way to standardize instruction across a department.

WORKSHEETS THAT TIE MULTIPLE CONCEPTS TOGETHER

When appropriate, there are **Worksheets That Tie Multiple Concepts Together** to help students with problem recognition and differentiation. These worksheets allow students to practice multiple concepts together after having learned them individually.

Help Students Learn How to Take Good Notes

GUIDED STUDENT NOTES

Use the **Guided Student Notes** to help your students become better note-takers. Every section of the book has a ready-made corresponding **Guided Student Note** that mirrors the material in the section. These are "skeleton note outlines" that help students learn how to structure their notes. Because the structure is already given to students, they don't have to write down *everything* that the instructors write in class, so students can concentrate better on what they are learning.

The **Guided Student Notes** save instructors time and help standardize the material that is taught across a math department. They come in both a P.O.W.E.R. format and a standard format and include answer keys. *The Guided Student Notes are available as PDF or Word files so that instructors can download them and edit them as if they were their own Word documents.*

Guided Student Notes
Messersmith/Vega-Rhodes/Feldman
Beginning & Intermediate Algebra with P.O.W.E.R. Learning, 5th ed.

3) $\dfrac{1}{12}x + \dfrac{1}{4}y = \dfrac{3}{2}$

   $\dfrac{2}{5}x - \dfrac{1}{2}y = -3$

5) $2x + y = -3$

   $6x + 3y = 0$

4) $0.3x + 0.1y = 3$

   $0.01x - 0.05y = -0.06$

Guided Student Notes   Name:_____
Messersmith/Vega-Rhodes/Feldman
Beginning & Intermediate Algebra with P.O.W.E.R. Learning, 5th ed.

5.2 Solving Systems by Substitution

Prepare   **What are my goals for this section?**

_____

_____

Organize   **What am I going to do to accomplish these goals?**

_____

_____

Work

Steps for Solving a System by Substitution

Solve each system using substitution.

1) $y = 4x - 1$

   $6x - 5y = -2$

2) $x + 3y = 5$

   $-2x + 5y = 23$

Acknowledgments

Manuscript Reviewers and Focus Group Participants

Thank you to all of the dedicated instructors who reviewed manuscript, participated in focus groups, and provided thoughtful feedback throughout the development of the *P.O.W.E.R.* series.

Darla Aguilar, *Pima Community College;* Scott Albert, *College of DuPage;* Bhagirathi Anand, *Long Beach City College;* Raul Arana, *Lee College;* Jan Archibald, *Ventura College;* Morgan Arnold, *Central Georgia Technical College;* Christy Babu, *Laredo Community College;* Michele Bach, *Kansas City Kansas Community College;* Kelly Bails, *Parkland College;* Vince Bander, *Pierce College, Pullallup;* Kim Banks, *Florence Darlington Technical College;* Michael Bartlett, *University of Wisconsin–Marinette;* Sarah Baxter, *Gloucester County College;* Michelle Beard, *Ventura College;* Annette Benbow, *Tarrant County College, Northwest;* Abraham Biggs, *Broward College;* Leslie Bolinger Horton, *Quinsigamond Community College;* Jessica Bosworth, *Nassau Community College;* Joseph Brenkert, *Front Range Community College;* Michelle Briles, *Gloucester County College;* Kelly Brooks, *Daytona State College (and Pierce);* Connie Buller, *Metropolitan Community College;* Rebecca Burkala, *Rose State College;* Gail Burkett, *Palm Beach State College;* Gale Burtch, *Ivy Tech Community College;* Jennifer Caldwell, *Mesa Community College;* Edie Carter, *Amarillo College;* Allison Cath, *Ivy Tech Community College of Indiana, Indianapolis;* Dawn Chapman, *Columbus Tech College;* Chris Chappa, *Tyler Junior College;* Charles Choo, *University of Pittsburgh at Titusville;* Patricia Clark, *Sinclair Community College;* Judy Kim Clark, *Wayne Community College;* Karen Cliffe, *Southwestern College;* Sherry Clune, *Front Range Community College;* Ela Coronado, *Front Range Community College;* Heather Cotharp, *West Kentucky Community & Tech College;* Danny Cowan, *Tarrant County College, Northwest;* Susanna Crawford, *Solano College;* George Daugavietis, *Solano Community College;* Joseph De Guzman, *Norco College;* Michaelle Downey, *Ivy Tech Community College;* Dale Duke, *Oklahoma City Community College;* Rhonda Duncan, *Midlands Technical College;* Marcial Echenique, *Broward College;* Sarah Ellis, *Dona Ana Community College;* Onunwor Enyinda, *Stark State College;* Chana Epstein, *Sullivan County Community College;* Karen Ernst, *Hawkeye Community College;* Stephen Ester, *St. Petersburg College;* Rosemary Farrar, *Southern West Virginia Community & Technical College;* John Fay, *Chaffey College;* Stephanie Fernandes, *Lewis and Clark Community College;* James Fiebiger, *Front Range Community College;* Angela Fipps, *Durham Technical Community College;* Jennifer Fisher, *Caldwell Community College & Technical Institute;* Elaine Fitt, *Bucks County Community College;* Carol Fletcher, *Hinds Community College;* Claude Fortune, *Atlantic Cape Community College;* Marilyn Frydrych, *Pikes Peak Community College;* Robert Fusco, *Broward College;* Jennifer Ganowsky, *Southern Utah University;* Jared Ganson, *Nassau Community College;* Kristine Glasener, *Cuyahoga Community College;* Ernest Gobert, *Oklahoma City Community College;* Linda Golovin, *Caldwell College;* Suzette Goss, *Lone Star College Kingwood;* Sharon Graber, *Lee College;* Susan Grody, *Broward College;* Leonard Groeneveld, *Springfield Tech Community College;* Joseph Guiciardi, *Community College of Allegheny County;* Susanna Gunther, *Solano College;* Lucy Gurrola, *Dona Ana Community College;* Frederick Hageman, *Delaware Technical & Community College;* Tamela Hanebrink, *Southeast Missouri State University;* Deborah Hanus, *Brookhaven College;* John Hargraves, *St. John's River State College;* Suzanne Harris-Smith, *Central New Mexico Community College;* Michael Helinger, *Clinton Community College;* Mary Hill, *College of DuPage;* Jody Hinson, *Cape Fear Community College;* Kayana Hoagland, *South Puget Sound Community College;* Tracey Hollister, *Casper College;* Wendy Houston, *Everett Community College;* Mary Howard, *Thomas Nelson Community College;* Lisa Hugdahl, *Milwaukee Area Tech College–Milwaukee;* Larry Huntzinger, *Western Oklahoma State College;* Manoj Illickal, *Nassau Community College;* Sharon Jackson, *Brookhaven College;* Lisa Jackson, *Black River Technical College;* Christina Jacobs, *Washington State University;* Gretta Johnson, *Amarillo College;* Lisa Juliano, *El Paso Community College, Northwest Campus;* Elias M. Jureidini, *Lamar State College/Orange;* Ismail Karahouni, *Lamar University;* Cliffe Karen, *Southwestern College;* David Kater, *San Diego City College;* Joe Kemble, *Lamar University;* Joanne Kendall, *Lone Star College–CyFair;* Esmarie Kennedy, *San Antonio College;* Ahmed Khago, *Lamar University;* Michael Kirby, *Tidewater Community College VA Beach Campus;* Corrine Kirkbride, *Solano Community College;* Mary Ann Klicka, *Bucks County Community College;* Alex Kolesnik, *Ventura College;* Tatyana Kravchuk, *Northern Virginia Community College;* Randa Kress, *Idaho State University;* Julianne Labbiento, *Lehigh Carbon Community College;* Robert Leifson, *Pierce College;* Greg Liano, *Brookdale Community College;* Charyl Link, *Kansas City Kansas Community College;* Cassondra Lochard, *California State University–Dominguez Hills;* Wanda Long, *Saint Charles County Community College;* Lorraine Lopez, *San Antonio College;* Luke Mannion, *St. John's University;* Shakir Manshad, *New Mexico State University;* Robert Marinelli, *Durham Technical Community College;* Lydia Matthews-Morales, *Ventura College;* Melvin Mays, *Metropolitan Community College (Omaha NE);* Carrie McCammon, *Ivy Tech Community College;* Milisa Mcilwain, *Meridian Community College;* Valerie Melvin, *Cape Fear Community College;* Christopher Merlo, *Nassau Community College;* Leslie Meyer, *Ivy Tech Community College/Central Indiana;* Beverly Meyers, *Jefferson College;* Laura Middaugh, *McHenry County College;* Karen Mifflin, *Palomar College;* Kris Mudunuri, *Long Beach City College;* Sharon Muehlbacher, *California State Polytechnic University–Pomona;* Donald Munsey, *Louisiana Delta Community College;* Randall Nichols, *Delta College;* Joshua Niemczyk, *Saint Charles County Community College;* David Stumpf, *Lakeland Community College;* Katherine Ocker Stone, *Tusculum College;* Karen Orr, *Roane State;* Staci Osborn, *Cuyahoga Community College;* Steven Ottmann, *Southeast Community College, Lincoln Nebraska;* William Parker, *Greenville Technical College;* Joanne Peeples, *El Paso Community College;* Paul Peery, *Lee College;* Betty Peterson, *Mercer County Community College;* Carol Ann Poore, *Hinds Community College;* Hadley Pridgen, *Gulf Coast State College;* William Radulovich, *Florida State College @ Jacksonville;* Lakshminarayan Rajaram, *St. Petersburg College;* Kumars Ranjbaran, *Mountain View College;* Darian Ransom,

Southeastern Community College; Nimisha Raval, Central Georgia Technical College; Amy Riipinen, Hibbing Community College; Janet Roads, Moberly Area Community College; Marianne Roarty, Metropolitan Community College; Jennifer Robb, Scott Community College; Marie Robison, McHenry County College; Daphne Anne Rossiter, Mesa Community College; Anna Roth, Gloucester County College; Daria Santerre, Norwalk Community College; Kala Sathappan, College of Southern Nevada; Patricia Schubert, Saddleback College; William H. Shaw, Coppin State University; Azzam Shihabi, Long Beach City College; Jed Soifer, Atlantic Cape Community College; Lee Ann Spahr, Durham Technical Community College; Marie St. James, Saint Clair County Community College; Mike Stack, College of DuPage; Ann Starkey, Stark State College of Technology; Thomas Steinmann, Lewis and Clark Community College; Claudia Stewart, Casper College; Kim Taylor, Florence Darlington Technical College; Laura Taylor, Cape Fear Community College; Janet Teeguarden, Ivy Tech Community College; Janine Termine, Bucks County Community College; Yan Tian, Palomar College; Lisa Tolliver, Brown Mackie South Bend; David Usinski, Erie Community College; Hien Van Eaton, Liberty University; Diane Veneziale, Rowan College at Burlington County; Theresa Vecchiarelli, Nassau Community College; Val Villegas, Southwestern College; David Walker, Hinds Community College; Ursula Walsh, Minneapolis Community & Tech College; Dottie Walton, Cuyahoga Community College; LuAnn Walton, San Juan College; Thomas Wells, Delta College; Kathryn Wetzel, Amarillo College; Marjorie Whitmore, North West Arkansas Community College; Sandra Wildfeuer, University of Alaska–Fairbanks; Ross Wiliams, Stark State College of Technology; Gerald Williams, San Juan College; Michelle Wolcott, Pierce College, Puyallup; Mary Young, Brookdale Community College; Loris Zucca, Lone Star College, Kingwood; Michael Zwilling, University of Mount Union

Digital Contributors

Special thanks go to the faculty members who contributed their time and expertise to the digital offerings with *P.O.W.E.R.*

Jennifer Caldwell, *Mesa Community College*
Chris Chappa, *Tyler Junior College*
Tim Chappell, *MCC Penn Valley Community College*
Kim Cozean, *Saddleback College*
Katy Cryer
Cindy Cummins, *Ozarks Technical Community College*
Rob Fusco, *Bergen Community College*
Vicki Garringer, *College of DuPage*
Amy Hoherz, *Lone Star College*
Brian Huyvaert, *University of Oregon*
Sharon Jackson, *Brookhaven College*

Kelly Jackson, *Camden County College*
Theresa Killebrew, *Mesa Community College*
Corrine Kirkbride, *Solano Community College*
Brianna Ashley, *Daytona State College*
Jamie Manche, *Southwestern Illinois College*
Amy Naughten
Christy Peterson, *College of DuPage*
Melissa Rossi, *Southwestern Illinois College*
Lisa Rombes, *Washtenaw Community College*
Janine Termine, *Bucks County Community College*
Linda Schott, *Ozarks Technical Community College*

From the Authors

The authors would like to thank everyone at McGraw-Hill Education who helped make the publication of this book a reality.

In addition, words cannot adequately express our appreciation to Vicki Garringer and Amy Jo Hoherz. You are an integral part of our team, and your ability to turn work around quickly and accurately is nothing short of marvelous. You are truly rock stars! Special thanks also go out to Tim Chappell and Lisa Rombes for the patience and attention to detail you each demonstrated while helping us develop the digital side of our books. Working with both of you is always a pleasure.

From Sherri Messersmith: Thank you to my husband, Phil, daughters, Alex and Cailen, and son-in-law, Justin, for their support and for providing inspiration for applications throughout the books over the years. Shout out to the baristas at Philz, my new hometown coffee shop, for your excellent coffee-making skills and for letting me sit and work in the groovy Berkeley atmosphere for hours on end. Bob and Nathalie, I am extremely grateful that you agreed to become my coauthors; your expertise is essential to what these books have become. Bill Mulford (best student ever), thank you for all that you have done over the years and for introducing Bob and me in the first place. Working with our team of four has been a joy, and I look forward to continuing our commitment of creating the best possible resources for students and instructors in the future.

From Nathalie Vega-Rhodes: First, to my husband, Rob—there is no way I can adequately express in a few sentences how grateful I am for you. Thank you for your unconditional love and support over the years and, more recently, for making sure that I regularly eat (delicious food), sleep, and work out. I love you. And second, to Sherri—thank you. I've seen the impact these books have had on my students each semester. Being a part of helping students and instructors on a larger scale means the world to me. Thank you for our collaboration and also for the many talks we've had about teaching and life.

From Bob Feldman: I am grateful to my children, Jonathan, Joshua, and Sarah; my daughters-in-law Leigh and Julie and son-in-law Jeffrey; my extraordinarily smart, cute, and talented grandchildren Alex, Miles, Naomi, Marina, Lilia, and Rose; and most of all to my wife, Katherine (who no longer can claim to be the sole mathematician in our family). I thank them all, with great love.

ALEKS

Create More Lightbulb Moments.

Every student has different needs and enters your course with varied levels of preparation. ALEKS® pinpoints what students already know, what they don't and, most importantly, what they're ready to learn next. Optimize your class engagement by aligning your course objectives to ALEKS® topics and layer on our textbook as an additional resource for students.

ALEKS® Creates a Personalized and Dynamic Learning Path

ALEKS® creates an optimized path with an ongoing cycle of learning and assessment, celebrating students' small wins along the way with positive real-time feedback. Rooted in research and analytics, ALEKS® improves student outcomes by fostering better preparation, increased motivation and knowledge retention.

*visit **bit.ly/whatmakesALEKSunique** to learn more about the science behind the most powerful adaptive learning tool in education!

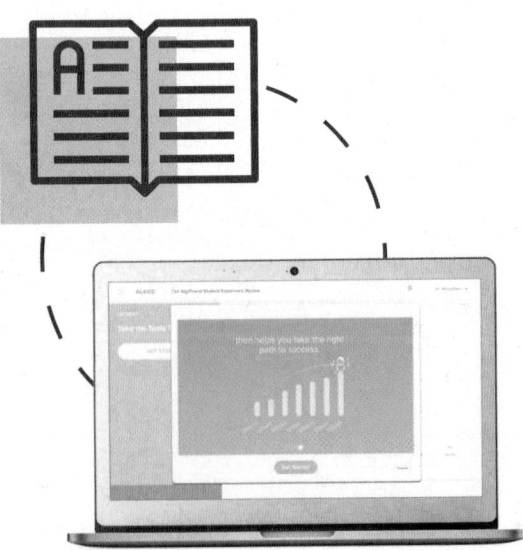

Preparation & Retention

The more prepared your students are, the more effective your instruction is. Because ALEKS® understands the prerequisite skills necessary for mastery, students are better prepared when a topic is presented to them. ALEKS® provides personalized practice and guides students to what they need to learn next to achieve mastery. ALEKS® improves knowledge and student retention through periodic knowledge checks and personalized learning paths. This cycle of learning and assessment ensures that students remember topics they have learned, are better prepared for exams, and are ready to learn new content as they continue into their next course.

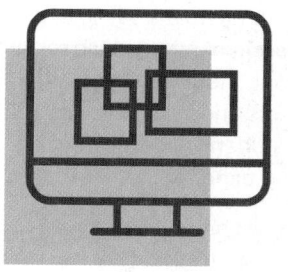

Flexible Implementation: Your Class Your Way!

ALEKS® enables you to structure your course regardless of your instruction style and format. From a traditional classroom, to various co-requisite models, to an online prep course before the start of the term, ALEKS® can supplement your instruction or play a lead role in delivering the content.

*visit **bit.ly/ALEKScasestudies** to see how your peers are delivering better outcomes across various course models!

Outcomes & Efficacy

Our commitment to improve student outcomes services a wide variety of implementation models and best practices, from lecture-based to labs and co-reqs to summer prep courses. Our case studies illustrate our commitment to help you reach your course goals and our research demonstrates our drive to support all students, regardless of their math background and preparation level.

*visit **bit.ly/outcomesandefficacy** to review empirical data from ALEKS® users around the country

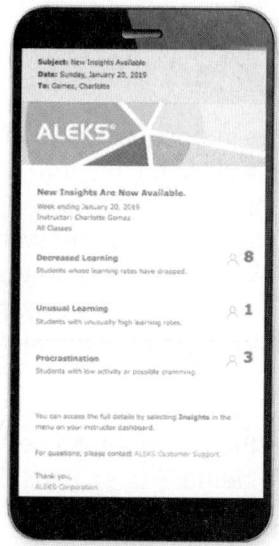

Turn Data Into Actionable Insights

ALEKS® Reports are designed to inform your instruction and create more meaningful interactions with your students when they need it the most. ALEKS® Insights alert you when students might be at risk of falling behind so that you can take immediate action. Insights summarize students exhibiting at least one of four negative behaviors that may require intervention including Failed Topics, Decreased Learning, Unusual Learning and Procrastination & Cramming.

Winner of 2019 Digital Edge 50 Award for Data Analytics!

bit.ly/ALEKS_MHE

connect MATH
McGraw Hill
HOSTED BY ALEKS

Effective, Efficient, Engaging

ConnectMath is the complete homework solution for you and your students. Developed by instructors for instructors, ConnectMath offers access to author-developed, text-specific assignments, learning resources, videos and adaptive learning modules. The platform delivers easy-to-read reports and learns the strengths and weaknesses of each student, allowing you to create a more meaningful learning experience.

New Enhanced Experience

Upgrade to the latest suite of tools! The new Enhanced Experience offers free-response graphing functionality, improved statistical tools, more detailed instant feedback for students, and a more intuitive palette and answer-entry experience. The replacement of flash-based content means improved accessibility and mobile capabilities.

*visit **bit.ly/CHBAe2** to view some brief demo videos for the new content available through the Enhanced Experience.

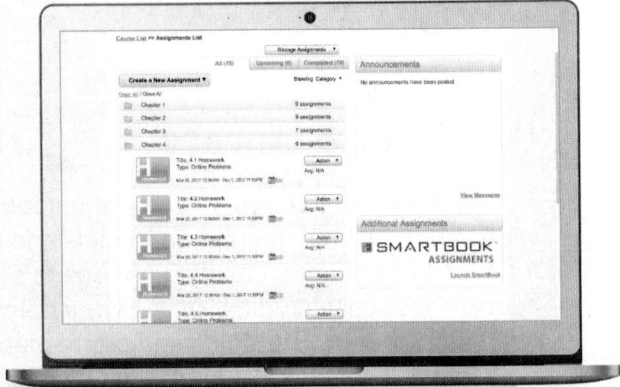

Save Time. Study Smarter.

By housing a wealth of resources in one, easy-to-use platform, you and your students will save time and be able to spend it more efficiently by setting up and working through valuable learning paths. ConnectMath® is your one-stop-shop for homework, quizzes, and tests, conceptual learning, classroom activities, self-study, and more all delivered through highly engaging content, videos, and interactives.

*visit **bit.ly/demoCHBA** to explore the platform yourself through an interactive, clickable demo.

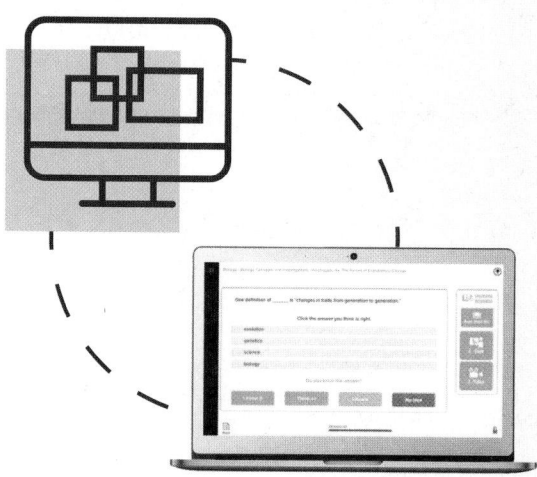

SmartBook® Adaptive Reading

As part of ConnectMath®, your students have access to SmartBook®. SmartBook actively tailors content to the individual needs of each student. It creates a personalized reading experience by focusing on the most impactful concepts a student needs to learn at that moment in time. SmartBook® helps students better prioritize, engage with the content and come to class ready to participate by prompting them with questions based on the material they are studying. By assessing individual answers, SmartBook® then learns what each student knows and identifies which topics they need to practice. This adaptive technology gives each student a personalized learning experience and path to success.

*visit **bit.ly/MHsmartbook** to learn more and to request a SmartBook® demo.

Seamless LMS Integration

You can easily integrate ConnectMath® with your current learning management system across courses, sections and your institution. Simply assign roles and responsibilities, which can be used for current and future terms. Integration provides single sign-on capabilities, quick registration, gradebook synchronization and access to assignments.

Reliability and Technical Support

ConnectMath® is highly reliable with industry leading 99.97% uptime. You'll spend your time teaching, not troubleshooting tech issues. Should there be an issue, tech support is available to you and your students by phone, online or chat.

*visit **supportateverystep.com** to learn more about how we are prioritizing YOU and your students in all that we do!

www.mheducation.com

The Real Number System and Geometry

Get Ready

To succeed in an algebra course, there are certain arithmetic skills we need to know well. In Section 1.1, we will review operations with fractions. Right now, we will review operations with decimals.

Recall that a **decimal** is a number (containing a *decimal point*) that is another way to represent a fraction with a denominator that is a power of 10: $0.7 = \dfrac{7}{10}$, $0.209 = \dfrac{209}{1000}$

To **add or subtract decimals,** write the numbers vertically and *line up the decimal points.* If there are any missing digits to the right of the decimal point, insert zeros. Place the decimal point in the answer *directly below* the decimal point in the problem.

$51.43 + 8.25$

	5	1 .	4	3
+		8 .	2	5
	5	9 .	6	8

$9.7 - 2.18$

	9 .	7	0
−	2 .	1	8
	7 .	5	2

When **multiplying decimals,** we do *not* need to line up the decimal points. Line up the factors (the numbers being multiplied) on the right, and multiply them just as you would multiply whole numbers. *The number of decimal places in the answer is the **total** number of decimal places in the factors.*

13.81×6.7

	1	3 .	8	1
×			6 .	7
	9	6	6	7
8	2	8	6	
9	2 .	5	2	7

When **dividing a decimal by a whole number,** like $0.78 \div 3$, write the decimal point in the quotient directly above the decimal point in the dividend. Divide as if the numbers were whole numbers.

$0.78 \div 3 = 0.26$

```
     0.2 6
 3)0.7 8
    -6
     1 8
    -1 8
        0
```

When **dividing a number by a decimal,** like $1.2\overline{)5.88}$, move the decimal point to the **right** end of the *divisor*, 1.2, to make it a whole number. Move the decimal point in the *dividend*, 5.88, the same number of places to the right. Then, divide as in the previous example.

Change $1.2\overline{)5.88}$ to $12\overline{)58.8}$, then divide.

```
        4.9
 1 2)5 8.8
    -4 8
      1 0 8
     -1 0 8
          0
```

In this chapter, use the following *Basic Skills Worksheets* **to improve the arithmetic skills needed in this, and future, chapters: WS1 Multiplication Facts, WS2 Powers, WS4 Divisibility Rules.**

Perform the indicated operation.

1) $4.1 + 53.8$

2) $12.5 - 7.9$

3) $6 - 1.62$

4) $8.4 + 19.23$

5) $5.7 \cdot 4$

6) 7.65×2.3

7) 0.04×6

8) 0.8×0.02

9) $0.92 \div 4$

10) $\dfrac{1.7}{5}$

11) $21.6 \div 0.12$

12) $72.18 \div 0.9$

Answers

1) 57.9 2) 4.6 3) 4.38 4) 27.63 5) 22.8 6) 17.595
7) 0.24 8) 0.016 9) 0.23 10) 0.34 11) 180 12) 80.2

Study Strategies

An Introduction to the
P.O.W.E.R. Framework

Research shows that successful goal-achievers, and successful learners, do five things to achieve their goals: **P**repare, **O**rganize, **W**ork, **E**valuate, and **R**ethink. If we take the first five letters of those words, we get **P.O.W.E.R.**, as in the P.O.W.E.R. Learning Framework.

We can use this framework to do any task or achieve any goal, whether it is running a marathon, completing a project at work, or learning math! Let's learn what each step of P.O.W.E.R. means and apply it to an example to help you understand it.

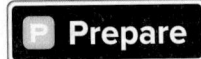

- *Prepare means to explicitly state a goal.* Be very specific.
- Let's do an example: *"I will train for and run a marathon one year from now."*
- Keep in mind that there are long-term goals and short-term goals. Running in a marathon is the long-term goal, but we can set short-term goals along the way to help us achieve our ultimate goal of running in the race. We should train before running a marathon, so a short-term goal might be to increase the number of miles I am able to run every two weeks.

- *Organize means to **organize** the physical and mental tools you need to achieve your goal.*
- A *mental tool* for running a marathon is planning: Where can I get information about training for a marathon? How will I fit the training into my schedule? Is there an app I can use to keep track of my progress? Do I need to change my diet so that I am more physically ready to run a race? Is there another person or a running group with whom I can train?
- I must gather or buy the *physical tools* needed such as new running shoes and clothes, a training schedule, healthy food that will give my body the nutrition it needs, and a group of other people training for the marathon.

- *Work means to **do the work** that needs to be done to achieve the goal.*
- Time to train! I eat healthy foods and consult my training schedule. Some days I work out at the gym and run on the treadmill, and other days I run outside. I keep track of the number of miles I run. Every Saturday, I train with other runners.
- A year passes, and it is time for the race! I grab all of my gear, get to the starting line, and run the race.

 **Evaluate**

- *Evaluate means you should* **evaluate** *what you have done.* Did you achieve your goal or not?
- Look at our short-term and long-term goals. As I am training, I should think about whether I am achieving the short-term goal of increasing the number of miles I run every two weeks so that adjustments can be made, if necessary.
- The ultimate and long-term goal is running the marathon. Did I complete the race?

R Rethink

- **Rethink** *means to* **rethink** *and reflect upon your goal.* If you did *not* achieve your goal, ask yourself, *"Why not?"* If you did achieve your goal, ask yourself what you did *right.*
- First, reflect upon the *short-term goal* along the way. If I *did* increase the number miles I ran every two weeks during training, I should continue to train as I have but also think about ways that I can improve. If I did *not* increase the number of miles I could run, I should ask myself, *"Why not?"* Did I fail to stick to my training schedule? Was I cheating on my training diet? Were my mileage goals unrealistic, and do they need to be adjusted? Continually tweak the training process so that it is the best it can be.
- Now, reflect upon the ultimate goal of running in the marathon. If I completed the race, I should think about how I trained so that I can use some of the same strategies in the future. If I did *not* complete the race, I should ask myself why: Did I have an appropriate training schedule, and did I follow it? Did I have the right equipment to prepare for and run the race? If I plan to run this race again, I should think about my preparation and make adjustments.

You can apply these same steps to help you be a successful student! Throughout the book we will use the P.O.W.E.R. Learning Framework to learn math *and* to acquire study skills to help us succeed in *any* course. And, of course, you can use P.O.W.E.R. outside of school to help you achieve *any* goal!

Andrea Laurita/Getty Images

Study Strategies

Using P.O.W.E.R. to Succeed
in Your Course

Now we know that **P.O.W.E.R.** stands for **P**repare, **O**rganize, **W**ork, **E**valuate, and **R**ethink. We have learned how to apply the steps to a real-life situation, *training for and running a marathon*, but how do we apply the steps to learning math? **Let's apply the P.O.W.E.R. framework to help a fictional student, Derek, succeed in this course.**

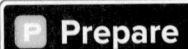

- *Prepare means to explicitly state a goal.* Remember, be specific.
- Derek's goal is **"I will make at least a B in this course."**

- *Organize means to **organize** the physical and mental tools you need to achieve your goal.*
- Some *physical tools* Derek will need are a backpack, a book, a notebook, regular pencils, colored pencils (for taking good notes), and a quiet place to study.
- Some *mental tools* he will need are a positive attitude, a commitment to go to class every day and do his homework, good basic skills such as knowing the multiplication facts from 1 to 12, and good time management.

- *Work means to **do the work** that needs to be done to achieve the goal.*
- Derek now has to **do the work** *so that he can make at least a B in the course!* That work includes going to class every day, putting away his cell phone, reviewing multiplication facts that he has trouble remembering, taking good notes, asking questions in class, studying in a place without distractions, and finishing his homework on time.

- *Evaluate means you **evaluate** what you have done.* Did you achieve your goal or not?
- Derek should *evaluate* how he is doing *as the term goes along* so that he can make adjustments, if necessary. He should ask himself, *"Am I making the kinds of grades that I need to make on the quizzes and tests?"*
- He should *evaluate* how he has done at the *end of the term.* He should ask himself, *"Did I make at least a B in the course?"*

- **Rethink** *means to **rethink** and reflect upon your goal.* If you did *not* achieve your goal, ask yourself, *"Why not?"* If you did achieve your goal, ask yourself what you did *right.*
- First, Derek should **rethink** *throughout the term.* If he is doing well on his quizzes and tests, he should ask himself, *"What have I been doing right so far?"* If he is not making a B at this point, he should ask himself, *"Why not?"* as well as questions such as *"Did I miss too many classes? Did I put off doing my homework until the last minute? Should I go in for help more often? What can I do differently to improve before the end of the term?"* Then, he can make adjustments so that he still has time to improve his grade.
- Derek should **rethink** *at the end of the term.* If he made an A or a B in the course, then he has achieved his goal! *Still,* he should ask himself what he did to be successful so that he can apply similar strategies in the future. If he did *not* make at least a B in the course, he should ask himself, *"Why not?"* as well as ask himself the questions in the previous bullet. This way, he can do things differently in the future.

This is just one example of how you can apply the **P.O.W.E.R.** framework to learning math (or anything)! In the following chapters we will learn more *specific* skills to help you learn math and acquire study skills that will help you in any of your courses and in life.

1.1 Review of Fractions

[P] Prepare

What are your objectives for Section 1.1?

		[O] Organize — How can you accomplish each objective?

What are your objectives for Section 1.1?	**How can you accomplish each objective?**
1 Understand What a Fraction Represents	• Write your own definition of a **fraction,** and include the words *numerator* and *denominator.* • Complete Example 1 on your own. • Complete You Try 1.
2 Write Fractions in Lowest Terms	• Know how to write the prime factorization of a number. • In your own words, write a procedure for writing a fraction in lowest terms. • Complete the examples on your own. • Complete You Trys 2–6.
3 Multiply and Divide Fractions	• In your own words, write a procedure for multiplying fractions and mixed numbers. • Be able to find the reciprocal of a number. • In your own words, write a procedure for dividing fractions and mixed numbers. • Complete the examples on your own. • Complete You Trys 7 and 8.
4 Add and Subtract Fractions	• Understand that to add or subtract fractions, the fractions must have a common denominator. • Know how to find a least common denominator for a group of fractions. • Be able to rewrite a fraction with a different denominator. • In your own words, write a procedure for **Adding and Subtracting Fractions with Unlike Denominators.** • Complete the examples on your own. • Complete You Trys 9–13.

Read the explanations, follow the examples, take notes, and complete the You Trys.

Why review fractions and arithmetic skills? Because the manipulations done in arithmetic and with fractions are precisely the same skills needed to learn algebra.

Let's begin by defining some numbers used in arithmetic:

Natural numbers: 1, 2, 3, 4, 5, …

Whole numbers: 0, 1, 2, 3, 4, 5, …

Natural numbers are often thought of as the counting numbers. **Whole numbers** consist of the natural numbers and zero.

Natural and whole numbers are used to represent complete quantities. To represent a part of a quantity, we can use a fraction.

1 Understand What a Fraction Represents

What is a fraction?

Definition

A **fraction** is a number in the form $\frac{a}{b}$, where $b \neq 0$. a is called the **numerator,** and b is the **denominator.**

Note

1) A fraction describes a part of a whole quantity.

2) $\frac{a}{b}$ means $a \div b$.

EXAMPLE 1

What part of the figure is shaded?

Solution

The whole figure is divided into three equal parts. Two of the parts are shaded. Therefore, the part of the figure that is shaded is $\frac{2}{3}$.

$$\frac{2}{3} \begin{array}{l} \rightarrow \text{ Number of shaded parts} \\ \rightarrow \text{ Total number of equal parts in the figure} \end{array}$$

YOU TRY 1

What part of the figure is shaded?

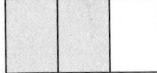

---

2 Write Fractions in Lowest Terms

A fraction is in **lowest terms** when the numerator and denominator have no common factors except 1. Before discussing how to write a fraction in lowest terms, we need to know about factors.

Consider the number 12.

3 and 4 are **factors** of 12. (When we use the term *factors,* we mean natural numbers.) Multiplying 3 and 4 results in 12. 12 is the **product.**

Does 12 have any other factors?

EXAMPLE 2   Find all factors of 12.

Solution

$$12 = 3 \cdot 4 \qquad \text{Factors are 3 and 4.}$$
$$12 = 2 \cdot 6 \qquad \text{Factors are 2 and 6.}$$
$$12 = 1 \cdot 12 \qquad \text{Factors are 1 and 12.}$$

These are all of the ways to write 12 as the product of two factors. The factors of 12 are 1, 2, 3, 4, 6, and 12.

[YOU TRY 2]   Find all factors of 30.

We can also write 12 as a product of *prime numbers.*

Definition

A **prime number** is a natural number whose only two *different* factors are 1 and itself. (The factors are natural numbers.)

EXAMPLE 3   Is 7 a prime number?

Solution

Yes. The only way to write 7 as a product of natural numbers is $1 \cdot 7$.

[YOU TRY 3]   Is 19 a prime number?

Definition

A **composite number** is a natural number with factors other than 1 and itself. Therefore, if a natural number is not prime, it is composite (with the exception of 0 and 1).

Note
The numbers 0 and 1 are neither prime nor composite.

[YOU TRY 4]
a)   What are the first six prime numbers?
b)   What are the first six composite numbers?

To perform various operations in arithmetic and algebra, it is helpful to write a number as the product of its **prime factors.** This is called finding the **prime factorization** of a number. We can use a **factor tree** to help us find the prime factorization of a number.

EXAMPLE 4

Write 120 as the product of its prime factors.

Solution

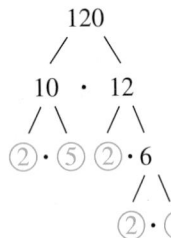

Think of *any* two natural numbers that multiply to 120.

10 and 12 are not prime, so write them as the product of two factors. Circle the primes.
6 is not prime, so write it as the product of two factors. The factors are primes. Circle them.

W Hint

Write a procedure for using a factor tree to write the prime factorization of a number.

Prime factorization: $120 = 2 \cdot 2 \cdot 2 \cdot 3 \cdot 5$.

[**YOU TRY 5**]

Use a factor tree to write each number as the product of its prime factors.

a) 30   b) 175

Let's return to writing a fraction in lowest terms.

EXAMPLE 5

Write $\dfrac{48}{42}$ in lowest terms.

Solution

$\dfrac{48}{42}$ is an **improper fraction.** A fraction is *improper* if its numerator is greater than or equal to its denominator. We will use two methods to express this fraction in lowest terms.

W Hint

Describe two ways to write a fraction in lowest terms.

Method 1

Using a factor tree to get the prime factorizations of 48 and 42 and then dividing out common factors, we have

$$\frac{48}{42} = \frac{\overset{1}{\cancel{2}} \cdot 2 \cdot 2 \cdot 2 \cdot \overset{1}{\cancel{2}}}{\underset{1}{\cancel{2}} \cdot \underset{1}{\cancel{2}} \cdot 7} = \frac{2 \cdot 2 \cdot 2}{7} = \frac{8}{7} \text{ or } 1\frac{1}{7}$$

The answer may be expressed as an improper fraction, $\dfrac{8}{7}$, or as a **mixed number,** $1\dfrac{1}{7}$, as long as each is in lowest terms.

Method 2

48 and 42 are each divisible by 6, so we can divide each by 6.

$$\frac{48}{42} = \frac{48 \div 6}{42 \div 6} = \frac{8}{7} \text{ or } 1\frac{1}{7}$$

[YOU TRY 6]
Write each fraction in lowest terms.

a) $\dfrac{8}{14}$ b) $\dfrac{63}{36}$

3 Multiply and Divide Fractions

Procedure Multiplying Fractions

To multiply fractions, $\dfrac{a}{b} \cdot \dfrac{c}{d}$, we multiply the numerators and multiply the denominators. That is,

$$\frac{a}{b} \cdot \frac{c}{d} = \frac{a \cdot c}{b \cdot d} \quad \text{if } b \neq 0 \text{ and } d \neq 0.$$

EXAMPLE 6

Multiply. Write each answer in lowest terms.

a) $\dfrac{3}{8} \cdot \dfrac{7}{4}$ b) $\dfrac{10}{21} \cdot \dfrac{21}{25}$ c) $4\dfrac{2}{5} \cdot 1\dfrac{7}{8}$

Solution

a) $\dfrac{3}{8} \cdot \dfrac{7}{4} = \dfrac{3 \cdot 7}{8 \cdot 4}$ Multiply numerators; multiply denominators.

$= \dfrac{21}{32}$ 21 and 32 contain no common factors, so $\dfrac{21}{32}$ is in lowest terms.

b) $\dfrac{10}{21} \cdot \dfrac{21}{25}$

We can take out the common factors before we multiply.

> **W Hint**
> Write out the example as you are reading it.

5 is the greatest common factor of 10 and 25. Divide 10 and 25 by 5.

$$\frac{\overset{2}{\cancel{10}}}{\cancel{21}} \cdot \frac{\overset{1}{\cancel{21}}}{\cancel{25}} = \frac{2}{1} \cdot \frac{1}{5} = \frac{2 \cdot 1}{1 \cdot 5} = \boxed{\frac{2}{5}}$$

21 is the greatest common factor of 21 and 21. Divide each 21 by 21.

Note

Usually, it is easier to remove the common factors before multiplying rather than after finding the product.

c) $4\dfrac{2}{5} \cdot 1\dfrac{7}{8}$

Before multiplying mixed numbers, we must change them to improper fractions. Recall that $4\dfrac{2}{5}$ is the same as $4 + \dfrac{2}{5}$. Here is one way to rewrite $4\dfrac{2}{5}$ as an improper fraction:

W Hint

Do you prefer to divide out common factors in the numerators and denominators *before* or *after* multiplying?

1) Multiply the denominator and the whole number: $5 \cdot 4 = 20$.

2) Add the numerator: $20 + 2 = 22$.

3) Put the sum over the denominator: $\dfrac{22}{5}$

To summarize, $4\dfrac{2}{5} = \dfrac{(5 \cdot 4) + 2}{5} = \dfrac{20 + 2}{5} = \dfrac{22}{5}$.

Then, $1\dfrac{7}{8} = \dfrac{(8 \cdot 1) + 7}{8} = \dfrac{8 + 7}{8} = \dfrac{15}{8}$.

$$4\dfrac{2}{5} \cdot 1\dfrac{7}{8} = \dfrac{22}{5} \cdot \dfrac{15}{8}$$

$$= \dfrac{\overset{11}{\cancel{22}}}{\underset{1}{\cancel{5}}} \cdot \dfrac{\overset{3}{\cancel{15}}}{\underset{4}{\cancel{8}}} \qquad \text{5 and 15 each divide by 5.} \qquad \text{8 and 22 each divide by 2.}$$

$$= \dfrac{11}{1} \cdot \dfrac{3}{4}$$

$$= \dfrac{33}{4} \text{ or } 8\dfrac{1}{4} \qquad \text{Write the result as an improper fraction or as a mixed number.}$$

[YOU TRY 7] Multiply. Write the answer in lowest terms.

a) $\dfrac{1}{5} \cdot \dfrac{4}{9}$ b) $\dfrac{8}{25} \cdot \dfrac{15}{32}$ c) $3\dfrac{3}{4} \cdot 2\dfrac{2}{3}$

To divide fractions, we must define a reciprocal.

Definition

The **reciprocal** of a number, $\dfrac{a}{b}$, is $\dfrac{b}{a}$ since $\dfrac{a}{b} \cdot \dfrac{b}{a} = 1$. That is, a nonzero number times its reciprocal equals 1. Notice that $a \neq 0$ and $b \neq 0$.

For example, the reciprocal of $\dfrac{5}{9}$ is $\dfrac{9}{5}$ since $\dfrac{\cancel{5}}{\cancel{9}} \cdot \dfrac{\cancel{9}}{\cancel{5}} = \dfrac{1}{1} = 1$.

Procedure Dividing Fractions

Division of fractions: Let a, b, c, and d represent numbers so that b, c, and d do not equal zero. Then,

$$\frac{a}{b} \div \frac{c}{d} = \frac{a}{b} \cdot \frac{d}{c}.$$

Note

To perform division involving fractions, multiply the first fraction by the reciprocal of the second.

EXAMPLE 7

Divide. Write the answer in lowest terms.

a) $\dfrac{3}{8} \div \dfrac{10}{11}$ b) $\dfrac{3}{2} \div 9$ c) $5\dfrac{1}{4} \div 1\dfrac{1}{13}$

Solution

a) $\dfrac{3}{8} \div \dfrac{10}{11} = \dfrac{3}{8} \cdot \dfrac{11}{10}$ Multiply $\dfrac{3}{8}$ by the reciprocal of $\dfrac{10}{11}$.

$= \dfrac{33}{80}$ Multiply.

b) $\dfrac{3}{2} \div 9 = \dfrac{3}{2} \cdot \dfrac{1}{9}$ The reciprocal of 9 is $\dfrac{1}{9}$.

$= \dfrac{\overset{1}{3}}{2} \cdot \dfrac{1}{\underset{3}{9}}$ Divide out a common factor of 3.

$= \dfrac{1}{6}$ Multiply.

W Hint

In your own words, write a procedure for dividing fractions and mixed numbers.

c) $5\dfrac{1}{4} \div 1\dfrac{1}{13} = \dfrac{21}{4} \div \dfrac{14}{13}$ Change the mixed numbers to improper fractions.

$= \dfrac{21}{4} \cdot \dfrac{13}{14}$ Multiply $\dfrac{21}{4}$ by the reciprocal of $\dfrac{14}{13}$.

$= \dfrac{\overset{3}{21}}{4} \cdot \dfrac{13}{\underset{2}{14}}$ Divide out a common factor of 7.

$= \dfrac{39}{8}$ or $4\dfrac{7}{8}$ Express the answer as an improper fraction or mixed number.

YOU TRY 8

Divide. Write the answer in lowest terms.

a) $\dfrac{2}{7} \div \dfrac{3}{5}$ b) $\dfrac{3}{10} \div \dfrac{9}{16}$ c) $9\dfrac{1}{6} \div 5$

4 Add and Subtract Fractions

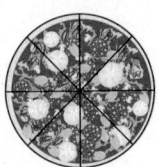

The top pizza is cut into eight equal slices. If you eat two pieces and your friend eats three pieces, what fraction of the pizza was eaten?

Five out of the eight pieces were eaten. As a fraction, we can say that you and your friend ate $\frac{5}{8}$ of the pizza.

Let's set up this problem as the sum of two fractions.

Fraction you ate + Fraction your friend ate = Fraction of the pizza eaten

$$\frac{2}{8} \quad + \quad \frac{3}{8} \quad = \quad \frac{5}{8}$$

To add $\frac{2}{8} + \frac{3}{8}$, we added the numerators and kept the denominator the same. Notice that these fractions have the same denominator.

Procedure Adding and Subtracting Fractions

Let a, b, and c be numbers such that $c \neq 0$.

$$\frac{a}{c} + \frac{b}{c} = \frac{a+b}{c} \quad \text{and} \quad \frac{a}{c} - \frac{b}{c} = \frac{a-b}{c}$$

To add or subtract fractions, the denominators must be the same. (This is called a **common denominator.**) Then, add (or subtract) the numerators and keep the same denominator.

EXAMPLE 8

Perform the operation and simplify.

a) $\frac{3}{11} + \frac{5}{11}$ b) $\frac{17}{30} - \frac{13}{30}$

Solution

a) $\frac{3}{11} + \frac{5}{11} = \frac{3+5}{11}$ Add the numerators, and keep the denominator the same.

$$= \frac{8}{11}$$

b) $\frac{17}{30} - \frac{13}{30} = \frac{17-13}{30}$ Subtract the numerators, and keep the denominator the same.

$$= \frac{4}{30}$$ This is not in lowest terms, so simplify it.

$$= \frac{2}{15}$$ Simplify.

[**YOU TRY 9**] Perform the operation and simplify.

a) $\frac{5}{9} + \frac{2}{9}$ b) $\frac{19}{20} - \frac{7}{20}$

When adding or subtracting mixed numbers, either work with them as mixed numbers or change them to improper fractions first.

EXAMPLE 9

Add $2\dfrac{4}{15} + 1\dfrac{7}{15}$.

Solution

Method 1

To add these numbers while keeping them in mixed number form, add the whole number parts and add the fractional parts.

$$2\dfrac{4}{15} + 1\dfrac{7}{15} = (2 + 1) + \left(\dfrac{4}{15} + \dfrac{7}{15}\right) = 3\dfrac{11}{15}$$

Method 2

Change each mixed number to an improper fraction, then add.

$$2\dfrac{4}{15} + 1\dfrac{7}{15} = \dfrac{34}{15} + \dfrac{22}{15} = \dfrac{34 + 22}{15} = \dfrac{56}{15} \text{ or } 3\dfrac{11}{15}$$

[YOU TRY 10]

Add $4\dfrac{3}{7} + 5\dfrac{1}{7}$.

The examples given so far contain common denominators. How do we add or subtract fractions that do not have common denominators? We find the least common denominator for the fractions and rewrite each fraction with this denominator.

The **least common denominator (LCD)** of two fractions is the least common multiple of the numbers in the denominators.

EXAMPLE 10

Find the LCD for $\dfrac{3}{4}$ and $\dfrac{1}{6}$.

Solution

Method 1

List some multiples of 4 and 6.

4: 4, 8, 12, 16, 20, 24, ...

6: 6, 12, 18, 24, 30, ...

Although 24 is a multiple of 6 and of 4, the *least* common multiple, and therefore the least common denominator, is 12.

Method 2

We can also use the prime factorizations of 4 and 6 to find the LCD.

   To find the LCD:

1) Find the prime factorization of each number.

2) The least common denominator will include each different factor appearing in the factorizations.

W Hint

In your own words, explain how to find the LCD of a group of fractions.

3) If a factor appears more than once in any prime factorization, use it in the LCD the *maximum number of times* it appears in any single factorization. Multiply the factors.

$$4 = 2 \cdot 2$$
$$6 = 2 \cdot 3$$

The least common multiple of 4 and 6 is

$$\underbrace{2 \cdot 2}_{\substack{\text{2 appears at} \\ \text{most twice in} \\ \text{any single} \\ \text{factorization.}}} \quad \cdot \quad \underbrace{3}_{\substack{\text{3 appears} \\ \text{once in a} \\ \text{factorization.}}} \quad = \quad 12$$

The LCD of $\dfrac{3}{4}$ and $\dfrac{1}{6}$ is 12.

[YOU TRY 11] Find the LCD for $\dfrac{5}{6}$ and $\dfrac{4}{9}$.

To add or subtract fractions with unlike denominators, begin by identifying the least common denominator. Then, we must rewrite each fraction with this LCD. This will not change the value of the fraction; we will obtain an *equivalent* fraction.

EXAMPLE 11 Rewrite $\dfrac{3}{4}$ with a denominator of 12.

Solution

We want to find a fraction that is equivalent to $\dfrac{3}{4}$ so that $\dfrac{3}{4} = \dfrac{?}{12}$.

To obtain the new denominator of 12, the "old" denominator, 4, must be multiplied by 3. But, if the denominator is multiplied by 3, the numerator must be multiplied by 3 as well. When we multiply $\dfrac{3}{4}$ by $\dfrac{3}{3}$, we have multiplied by 1 since $\dfrac{3}{3} = 1$. This is why the fractions are equivalent.

$$\frac{3}{4} \cdot \frac{3}{3} = \frac{9}{12} \qquad \text{So, } \frac{3}{4} = \frac{9}{12}.$$

[YOU TRY 12] Rewrite $\dfrac{5}{6}$ with a denominator of 42.

Now we are ready to add and subtract fractions with different denominators.

Procedure Adding or Subtracting Fractions with Unlike Denominators

To add or subtract fractions with unlike denominators:

1) Determine, and write down, the least common denominator (LCD).

2) Rewrite each fraction with the LCD.

3) Add or subtract.

4) Express the answer in lowest terms.

EXAMPLE 12 Add or subtract.

a) $\dfrac{2}{9} + \dfrac{1}{6}$ b) $6\dfrac{7}{8} - 3\dfrac{1}{2}$

Solution

a) $\dfrac{2}{9} + \dfrac{1}{6}$ $\text{LCD} = 18$ Identify the least common denominator.

$\dfrac{2}{9} \cdot \dfrac{2}{2} = \dfrac{4}{18}$ $\dfrac{1}{6} \cdot \dfrac{3}{3} = \dfrac{3}{18}$ Rewrite each fraction with a denominator of 18.

$\dfrac{2}{9} + \dfrac{1}{6} = \dfrac{4}{18} + \dfrac{3}{18} = \dfrac{7}{18}$

W Hint

Whenever you multiply, divide, add, or subtract fractions, you should *always* look at the result and ask yourself, "Is it in lowest terms?" If not, write the answer in lowest terms.

b) $6\dfrac{7}{8} - 3\dfrac{1}{2}$

Method 1

Keep the numbers in mixed number form. Subtract the whole number parts and subtract the fractional parts. Get a common denominator for the fractional parts.

For $6\dfrac{7}{8}$ and $3\dfrac{1}{2}$, the LCD is 8.   Identify the least common denominator.

$6\dfrac{7}{8}$: $\dfrac{7}{8}$ has the LCD of 8.

$3\dfrac{1}{2}$: $\dfrac{1}{2} \cdot \dfrac{4}{4} = \dfrac{4}{8}$.   So, $3\dfrac{1}{2} = 3\dfrac{4}{8}$.   Rewrite $\dfrac{1}{2}$ with a denominator of 8.

$6\dfrac{7}{8} - 3\dfrac{1}{2} = 6\dfrac{7}{8} - 3\dfrac{4}{8}$

$= 3\dfrac{3}{8}$ Subtract whole number parts and subtract fractional parts.

Method 2

Rewrite each mixed number as an improper fraction, get a common denominator, then subtract.

$6\dfrac{7}{8} - 3\dfrac{1}{2} = \dfrac{55}{8} - \dfrac{7}{2}$ $\text{LCD} = 8$ $\dfrac{55}{8}$ already has a denominator of 8.

$\dfrac{7}{2} \cdot \dfrac{4}{4} = \dfrac{28}{8}$ Rewrite $\dfrac{7}{2}$ with a denominator of 8.

$6\dfrac{7}{8} - 3\dfrac{1}{2} = \dfrac{55}{8} - \dfrac{7}{2} = \dfrac{55}{8} - \dfrac{28}{8} = \dfrac{27}{8}$ or $3\dfrac{3}{8}$

Perform the operations and simplify.

a) $\dfrac{11}{12} - \dfrac{5}{8}$ b) $\dfrac{1}{3} + \dfrac{5}{6} + \dfrac{3}{4}$ c) $4\dfrac{2}{5} + 1\dfrac{7}{15}$

ANSWERS TO [YOU TRY] EXERCISES

1) $\dfrac{3}{5}$ 2) 1, 2, 3, 5, 6, 10, 15, 30 3) yes 4) a) 2, 3, 5, 7, 11, 13 b) 4, 6, 8, 9, 10, 12

5) a) $2 \cdot 3 \cdot 5$ b) $5 \cdot 5 \cdot 7$ 6) a) $\dfrac{4}{7}$ b) $\dfrac{7}{4}$ or $1\dfrac{3}{4}$

7) a) $\dfrac{4}{45}$ b) $\dfrac{3}{20}$ c) 10 8) a) $\dfrac{10}{21}$ b) $\dfrac{8}{15}$ c) $\dfrac{11}{6}$ or $1\dfrac{5}{6}$ 9) a) $\dfrac{7}{9}$ b) $\dfrac{3}{5}$ 10) $9\dfrac{4}{7}$

11) 18 12) $\dfrac{35}{42}$ 13) a) $\dfrac{7}{24}$ b) $\dfrac{23}{12}$ or $1\dfrac{11}{12}$ c) $\dfrac{88}{15}$ or $5\dfrac{13}{15}$

E Evaluate 1.1 Exercises Do the exercises, and check your work.

Objective 1: Understand What a Fraction Represents

1) What fraction of each figure is shaded? If the fraction is not in lowest terms, reduce it.

a)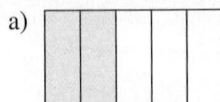

b)

c)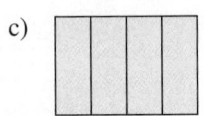

2) What fraction of each figure is *not* shaded? If the fraction is not in lowest terms, reduce it.

a) 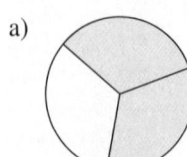 b)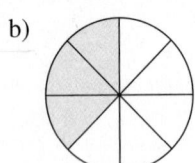

c)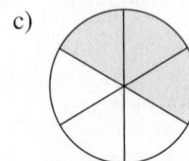

3) Draw a rectangle divided into 8 equal parts. Shade in $\dfrac{4}{8}$ of the rectangle. Write another fraction to represent how much of the rectangle is shaded.

4) Draw a rectangle divided into 6 equal parts. Shade in $\dfrac{2}{6}$ of the rectangle. Write another fraction to represent how much of the rectangle is shaded.

Objective 2: Write Fractions in Lowest Terms

5) Find all factors of each number.

a) 18

b) 40

c) 23

6) Find all factors of each number.

a) 20

b) 17

c) 60

7) Identify each number as prime or composite.

a) 27

b) 34

c) 11

8) Identify each number as prime or composite.

a) 2

b) 57

c) 90

9) Is 3072 prime or composite? Explain your answer.

10) Is 4185 prime or composite? Explain your answer.

11) Use a factor tree to find the prime factorization of each number.

    a) 18   b) 54

    c) 42   d) 150

12) Explain, in words, how to use a factor tree to find the prime factorization of 72.

13) Write each fraction in lowest terms.

    a) $\dfrac{9}{12}$ b) $\dfrac{54}{72}$

    c) $\dfrac{84}{35}$ d) $\dfrac{120}{280}$

14) Write each fraction in lowest terms.

    a) $\dfrac{21}{35}$ b) $\dfrac{48}{80}$

    c) $\dfrac{125}{500}$ d) $\dfrac{900}{450}$

Objective 3: Multiply and Divide Fractions

15) Multiply. Write the answer in lowest terms.

    a) $\dfrac{2}{7} \cdot \dfrac{3}{5}$ b) $\dfrac{15}{26} \cdot \dfrac{4}{9}$

    c) $\dfrac{1}{2} \cdot \dfrac{14}{15}$ d) $\dfrac{42}{55} \cdot \dfrac{22}{35}$

    e) $4 \cdot \dfrac{1}{8}$ f) $6\dfrac{1}{8} \cdot \dfrac{2}{7}$

16) Multiply. Write the answer in lowest terms.

    a) $\dfrac{1}{6} \cdot \dfrac{5}{9}$ b) $\dfrac{9}{20} \cdot \dfrac{6}{7}$

    c) $\dfrac{12}{25} \cdot \dfrac{25}{36}$ d) $\dfrac{30}{49} \cdot \dfrac{21}{100}$

    e) $\dfrac{7}{15} \cdot 10$ f) $7\dfrac{5}{7} \cdot 1\dfrac{5}{9}$

17) When Elizabeth multiplies $5\dfrac{1}{2} \cdot 2\dfrac{1}{3}$, she gets $10\dfrac{1}{6}$. What was her mistake? What is the correct answer?

18) Explain how to multiply mixed numbers.

19) Divide. Write the answer in lowest terms.

    a) $\dfrac{1}{42} \div \dfrac{2}{7}$ b) $\dfrac{3}{11} \div \dfrac{4}{5}$

    c) $\dfrac{18}{35} \div \dfrac{9}{10}$ d) $\dfrac{14}{15} \div \dfrac{2}{15}$

    e) $6\dfrac{2}{5} \div 1\dfrac{13}{15}$ f) $\dfrac{4}{7} \div 8$

20) Explain how to divide mixed numbers.

Objective 4: Add and Subtract Fractions

21) Find the least common multiple of 10 and 15.

22) Find the least common multiple of 12 and 9.

23) Find the least common denominator for each group of fractions.

    a) $\dfrac{9}{10}, \dfrac{11}{30}$ b) $\dfrac{7}{8}, \dfrac{5}{12}$

    c) $\dfrac{4}{9}, \dfrac{1}{6}, \dfrac{3}{4}$

24) Find the least common denominator for each group of fractions.

    a) $\dfrac{3}{14}, \dfrac{2}{7}$ b) $\dfrac{17}{25}, \dfrac{3}{10}$

    c) $\dfrac{29}{30}, \dfrac{3}{4}, \dfrac{9}{20}$

25) Add or subtract. Write the answer in lowest terms.

    a) $\dfrac{6}{11} + \dfrac{2}{11}$ b) $\dfrac{19}{20} - \dfrac{7}{20}$

    c) $\dfrac{4}{25} + \dfrac{2}{25} + \dfrac{9}{25}$ d) $\dfrac{2}{9} + \dfrac{1}{6}$

    e) $\dfrac{3}{5} + \dfrac{11}{30}$ f) $\dfrac{13}{18} - \dfrac{2}{3}$

    g) $\dfrac{4}{7} + \dfrac{5}{9}$ h) $\dfrac{5}{6} - \dfrac{1}{4}$

    i) $\dfrac{3}{10} + \dfrac{7}{20} + \dfrac{3}{4}$ j) $\dfrac{1}{6} + \dfrac{2}{9} + \dfrac{10}{27}$

26) Add or subtract. Write the answer in lowest terms.

    a) $\dfrac{8}{9} - \dfrac{5}{9}$ b) $\dfrac{14}{15} - \dfrac{2}{15}$

    c) $\dfrac{11}{36} + \dfrac{13}{36}$ d) $\dfrac{16}{45} + \dfrac{8}{45} + \dfrac{11}{45}$

    e) $\dfrac{15}{16} - \dfrac{3}{4}$ f) $\dfrac{1}{8} + \dfrac{1}{6}$

    g) $\dfrac{5}{8} - \dfrac{2}{9}$ h) $\dfrac{23}{30} - \dfrac{19}{90}$

    i) $\dfrac{1}{6} + \dfrac{1}{4} + \dfrac{2}{3}$ j) $\dfrac{3}{10} + \dfrac{2}{5} + \dfrac{4}{15}$

27) Add or subtract. Write the answer in lowest terms.

a) $8\frac{5}{11} + 6\frac{2}{11}$ b) $2\frac{1}{10} + 9\frac{3}{10}$

c) $7\frac{11}{12} - 1\frac{5}{12}$ d) $3\frac{1}{5} + 2\frac{1}{4}$

e) $5\frac{2}{3} - 4\frac{4}{15}$ f) $9\frac{5}{8} - 5\frac{3}{10}$

g) $4\frac{3}{7} + 6\frac{3}{4}$ h) $7\frac{13}{20} + \frac{4}{5}$

28) Add or subtract. Write the answer in lowest terms.

a) $3\frac{2}{7} + 1\frac{3}{7}$ b) $8\frac{5}{16} + 7\frac{3}{16}$

c) $5\frac{13}{20} - 3\frac{5}{20}$ d) $10\frac{8}{9} - 2\frac{1}{3}$

e) $1\frac{5}{12} + 2\frac{3}{8}$ f) $4\frac{1}{9} + 7\frac{2}{5}$

g) $1\frac{5}{6} + 4\frac{11}{18}$ h) $3\frac{7}{8} + 4\frac{2}{5}$

Mixed Exercises: Objectives 3 and 4

29) For Valentine's Day, Alex wants to sew teddy bears for her friends. Each bear requires $1\frac{2}{3}$ yd of fabric. If she has 7 yd of material, how many bears can Alex make? How much fabric will be left over?

Creative Crop/ Getty Images

30) A chocolate chip cookie recipe that makes 24 cookies uses $\frac{3}{4}$ cup of brown sugar. If Raphael wants to make 48 cookies, how much brown sugar does he need?

31) During the 2017 Major League Baseball regular season, Cody Bellinger of the Los Angeles Dodgers had been up to bat 480 times. He got a hit $\frac{4}{15}$ of the time. How many hits did he have? (www.espn.com)

32) When all children are present, Ms. Yamoto has 30 children in her fifth-grade class. One day during flu season, $\frac{3}{5}$ of them were absent. How many children were absent on this day?

33) Mr. Burnett plans to have a picture measuring $18\frac{3}{8}$ in. by $12\frac{1}{4}$ in. custom framed. The frame he chose is $2\frac{1}{8}$ in. wide. What will be the new length and width of the picture plus the frame?

34) Andre is building a table in his workshop. For the legs, he bought wood that is 30 in. long. If the legs are to be $26\frac{3}{4}$ in. tall, how many inches must he cut off to get the desired height?

35) When Rosa opens the kitchen cabinet, she finds three partially filled bags of flour. One contains $\frac{2}{3}$ cup, another contains $1\frac{1}{4}$ cups, and the third contains $1\frac{1}{2}$ cups. How much flour does she have all together?

36) Tamika takes the same route to school every day. (See the figure.) How far does she walk to school?

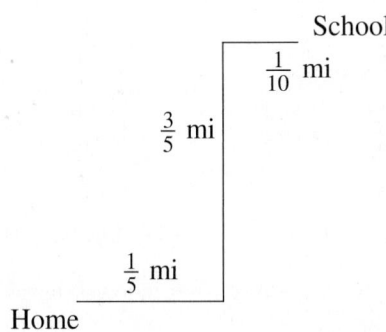

37) The gas tank of Jenny's car holds $11\frac{3}{5}$ gal, while Scott's car holds $16\frac{3}{4}$ gal. How much more gasoline does Scott's car hold?

38) Mr. Johnston is building a brick wall along his driveway. He estimates that one row of brick plus mortar will be $4\frac{1}{4}$ in. high. How many rows will he need to construct a wall that is 34 in. high?

39) For homework, Bill's math teacher assigned 42 problems. Bill finished $\frac{5}{6}$ of them. How many problems did he do?

40) Clarice's parents tell her that she must deposit $\frac{1}{3}$ of the money she earns from babysitting into her savings account, but she can keep the rest. If she earns $117 in one week during the summer, how much does she deposit, and how much does she keep?

41) A welder must construct a beam with a total length of $32\frac{7}{8}$ in. If he has already joined a $14\frac{1}{6}$-in. beam with a $10\frac{3}{4}$-in. beam, find the length of a third beam needed to reach the total length.

42) A survey of 1500 high school students revealed that $\frac{2}{3}$ of them use their phones while they are in school.

How many students surveyed use their phones while school is in session?

Hill Street Studios/Blend Images

R Rethink

R1) Were there any problems you could not answer or that you got wrong? If so, write them down or circle them. Look back in the section to see whether you can figure out how to do them, or ask your instructor for help.

R2) Can you think of a situation where you used fractions in the last week? What was that situation?

R3) Write your own application problem involving fractions, and give it to another student to solve.

1.2 Exponents and Order of Operations

P Prepare

O Organize

What are your objectives for Section 1.2?	How can you accomplish each objective?
1 Use Exponents	• Understand the prime factorization of a number, exponent, and exponential expression. • Memorize the table of commonly used powers. • Complete the given examples on your own. • Complete You Trys 1 and 2.
2 Use the Order of Operations	• Write **The Order of Operations** in your own words. • Understand how the phrase **P**lease **E**xcuse **M**y **D**ear **A**unt **S**ally can help you master the order of operations. • Complete the given examples on your own. • Complete You Trys 3–5.

 **W Work**   **Read the explanations, follow the examples, take notes, and complete the You Trys.**

1 Use Exponents

In Section 1.1, we discussed the prime factorization of a number. The prime factorization of 8 is $8 = 2 \cdot 2 \cdot 2$.

We can write $2 \cdot 2 \cdot 2$ another way, by using an *exponent*.

$$2 \cdot 2 \cdot 2 = 2^3 \leftarrow \text{exponent (or power)}$$
$$\uparrow$$
$$\text{base}$$

W Hint

Notice that the exponent tells you how many times the base is multiplied by itself.

2 is the *base*. 2 is a *factor* that appears three times. 3 is the *exponent* or *power*. An **exponent** represents repeated multiplication. We read 2^3 as "2 to the third power" or "2 cubed." 2^3 is called an **exponential expression.**

EXAMPLE 1

Rewrite each product in exponential form.

a) $9 \cdot 9 \cdot 9 \cdot 9$ b) $7 \cdot 7$

Solution

a) $9 \cdot 9 \cdot 9 \cdot 9 = 9^4$ 9 is the base. It appears as a factor 4 times. So, 4 is the exponent.

b) $7 \cdot 7 = 7^2$ 7 is the base. 2 is the exponent.
    This is read as "7 squared."

[YOU TRY 1]

Rewrite each product in exponential form.

a) $8 \cdot 8 \cdot 8 \cdot 8 \cdot 8$ b) $\dfrac{3}{2} \cdot \dfrac{3}{2} \cdot \dfrac{3}{2} \cdot \dfrac{3}{2}$

We can also *evaluate* an exponential expression.

EXAMPLE 2

Evaluate.

a) 2^5 b) 5^3 c) $\left(\dfrac{4}{7}\right)^2$ d) 8^1 e) 1^4

Solution

a) $2^5 = 2 \cdot 2 \cdot 2 \cdot 2 \cdot 2 = 32$ 2 appears as a factor 5 times.

b) $5^3 = 5 \cdot 5 \cdot 5 = 125$ 5 appears as a factor 3 times.

c) $\left(\dfrac{4}{7}\right)^2 = \dfrac{4}{7} \cdot \dfrac{4}{7} = \dfrac{16}{49}$ $\dfrac{4}{7}$ appears as a factor 2 times.

d) $8^1 = 8$ 8 is a factor only once.

e) $1^4 = 1 \cdot 1 \cdot 1 \cdot 1 = 1$ 1 appears as a factor 4 times.

Note

1 raised to any natural number power is 1 since 1 multiplied by itself equals 1.

Evaluate.

a) 3^4
b) 2^5
c) $\left(\dfrac{8}{11}\right)^2$
d) 9^1
e) 1^3

It is generally agreed that there are some skills in arithmetic that everyone should have in order to be able to acquire other math skills. Knowing the basic multiplication facts, for example, is essential for learning how to add, subtract, multiply, and divide fractions as well as how to perform many other operations in arithmetic and algebra. Similarly, memorizing powers of certain bases is necessary for learning how to apply the rules of exponents (Chapter 2) and for working with radicals (Chapter 10). Therefore, the powers listed here must be memorized in order to be successful in these and other topics. Throughout this book, it is assumed that students know these powers:

W Hint

Write down any patterns you see in the "Powers to Memorize" table.

Powers to Memorize						
$2^1 = 2$	$3^1 = 3$	$4^1 = 4$	$5^1 = 5$	$6^1 = 6$	$8^1 = 8$	$10^1 = 10$
$2^2 = 4$	$3^2 = 9$	$4^2 = 16$	$5^2 = 25$	$6^2 = 36$	$8^2 = 64$	$10^2 = 100$
$2^3 = 8$	$3^3 = 27$	$4^3 = 64$	$5^3 = 125$			$10^3 = 1000$
$2^4 = 16$	$3^4 = 81$					
$2^5 = 32$				$7^1 = 7$	$9^1 = 9$	$11^1 = 11$
$2^6 = 64$				$7^2 = 49$	$9^2 = 81$	$11^2 = 121$
						$12^1 = 12$
						$12^2 = 144$
						$13^1 = 13$
						$13^2 = 169$

(Hint: Making flash cards might help you learn these facts.)

2 Use the Order of Operations

We will begin this topic with a problem for the student:

[YOU TRY 3]

Evaluate $40 - 24 \div 8 + (5 - 3)^2$.

What answer did you get? 41? or 6? or 33? Or, did you get another result?

Most likely you obtained one of the three answers just given. Only one is correct, however. If we do not have rules to guide us in evaluating expressions, it is easy to get the incorrect answer.

Therefore, here are the rules we follow. This is called the **order of operations.**

Procedure The Order of Operations

Simplify expressions in the following order:

1) If parentheses or other grouping symbols appear in an expression, simplify what is in these grouping symbols first.

2) Simplify expressions with exponents.

3) Perform multiplication and division from left to right.

4) Perform addition and subtraction from left to right.

Think about the "You Try" problem. Did you evaluate it using the order of operations? Let's look at that expression.

EXAMPLE 3

Evaluate $40 - 24 \div 8 + (5 - 3)^2$.

Solution

$40 - 24 \div 8 + (5 - 3)^2$ First, perform the operation in the parentheses.

$40 - 24 \div 8 + 2^2$ Exponents are done before division, addition, and subtraction.

$40 - 24 \div 8 + 4$ Perform division before addition and subtraction.

$40 - 3 + 4$ When an expression contains only addition and subtraction,
perform the operations starting at the left and moving to the right.

$37 + 4$

41

W Hint

When practicing these problems, be sure to write out every step.

[YOU TRY 4]

Evaluate $12 \cdot 3 - (2 + 1)^2 \div 9$.

A good way to remember the order of operations is to remember the sentence, "**P**lease **E**xcuse **M**y **D**ear **A**unt **S**ally" (**P**arentheses, **E**xponents, **M**ultiplication, and **D**ivision from left to right, **A**ddition and **S**ubtraction from left to right). Don't forget that multiplication and division are at the same "level" in the process of performing operations and that addition and subtraction are at the same "level."

EXAMPLE 4

Evaluate.

a) $9 + 20 - 5 \cdot 3$ b) $4[3 + (10 \div 2)] - 11$ c) $\dfrac{(9 - 6)^3 \cdot 2}{26 - 4 \cdot 5}$

Solution

a) $9 + 20 - 5 \cdot 3 = 9 + 20 - 15$ Perform multiplication before addition and subtraction.

$= 29 - 15$ When an expression contains only addition and subtraction, work from left to right.

$= 14$ Subtract.

b) $4[3 + (10 \div 2)] - 11$

This expression contains two sets of grouping symbols: **brackets** [ ] and **parentheses** ( ). Perform the operation in the **innermost** grouping symbol first—the parentheses in this case.

$4[3 + (10 \div 2)] - 11 = 4[3 + 5] - 11$ Innermost grouping symbol

$= 4[8] - 11$ Brackets

$= 32 - 11$ Perform multiplication before subtraction.

$= 21$ Subtract.

c) $\dfrac{(9 - 6)^3 \cdot 2}{26 - 4 \cdot 5}$

The fraction bar in this expression acts as a grouping symbol. Therefore, simplify the numerator, simplify the denominator, then simplify the resulting fraction, if possible.

$\dfrac{(9 - 6)^3 \cdot 2}{26 - 4 \cdot 5} = \dfrac{3^3 \cdot 2}{26 - 20}$ Parentheses / Multiply.

$= \dfrac{27 \cdot 2}{6}$ Exponent / Subtract.

$= \dfrac{54}{6}$ Multiply.

$= 9$ Divide.

$\left[\text{YOU TRY 5}\right]$ Evaluate:

a) $35 - 2 \cdot 6 + 1$

b) $3 \cdot 12 - (7 - 4)^3 \div 9$

c) $9 + 2[23 - 4(1 + 2)]$

d) $\dfrac{11^2 - 7 \cdot 3}{20(9 - 4)}$

Using Technology

We can use a graphing calculator or smartphone to check our answer when we evaluate an expression by hand. For example, evaluate the expression

$\dfrac{2(3 + 7)}{13 - 2 \cdot 4}$. To evaluate the expression using a graphing calculator or smartphone, enter the following on the home screen: $(2(3+7))/(13-2\times4)$ and then press ENTER or =. The result is 4, as shown on the screen.

```
(2(3+7))/(13-2*4
)
              4
```

Notice that it is important to enclose the numerator and denominator in parentheses since the fraction bar acts as both a division and a grouping symbol.

Evaluate each expression by hand, and then verify your answer using a graphing calculator or smartphone.

1) $45 - 3 \cdot 2 + 7$

2) $24 \div \dfrac{6}{7} - 5 \cdot 4$

3) $5 + 2(9 - 6)^2$

4) $3 + 2[37 - (4 + 1)^2 - 2 \cdot 6]$

5) $\dfrac{5(7 - 3)}{50 - 3^2 \cdot 4}$

6) $\dfrac{25 - (1 + 3)^2}{6 + 14 \div 2 - 8}$

ANSWERS TO [YOU TRY] EXERCISES

1) a) 8^5 b) $\left(\dfrac{3}{2}\right)^4$ 2) a) 81 b) 32 c) $\dfrac{64}{121}$ d) 9 e) 1   3) 41   4) 35   5) a) 24 b) 33   c) 31 d) 1

ANSWERS TO TECHNOLOGY EXERCISES

1) 46   2) 8   3) 23   4) 3   5) $\dfrac{10}{7}$ or $1\dfrac{3}{7}$ 6) $\dfrac{9}{5}$ or $1\dfrac{4}{5}$

E Evaluate 1.2 Exercises   Do the exercises, and check your work.

Objective 1: Use Exponents

1) Identify the base and the exponent.

a) 6^4

b) 2^3

c) $\left(\dfrac{9}{8}\right)^5$

2) Identify the base and the exponent.

a) 5^1

b) 1^8

c) $\left(\dfrac{3}{7}\right)^2$

SECTION 1.2   **Exponents and Order of Operations**   23

3) Write in exponential form.

   a) $9 \cdot 9 \cdot 9 \cdot 9$

   b) $2 \cdot 2 \cdot 2 \cdot 2 \cdot 2 \cdot 2 \cdot 2 \cdot 2$

   c) $\dfrac{1}{4} \cdot \dfrac{1}{4} \cdot \dfrac{1}{4}$

4) Explain, in words, why $7 \cdot 7 \cdot 7 \cdot 7 \cdot 7 = 7^5$.

5) Evaluate.

   a) 8^2 b) 11^2

   c) 2^4 d) 5^3

   e) 3^4 f) 12^2

   g) 1^2 h) $\left(\dfrac{3}{10}\right)^2$

   i) $\left(\dfrac{1}{2}\right)^6$ j) $(0.3)^2$

6) Evaluate.

   a) 9^2 b) 13^2

   c) 3^3 d) 2^5

   e) 4^3 f) 1^4

   g) 6^2 h) $\left(\dfrac{7}{5}\right)^2$

   i) $\left(\dfrac{2}{3}\right)^4$ j) $(0.02)^2$

7) Evaluate $(0.5)^2$ two different ways.

8) Explain why $1^{200} = 1$.

For Exercises 9–11, fill in the blank with *always*, *sometimes*, or *never*.

9) Raising a positive decimal number that is less than 1 to a natural number power _____ has a result that is less than that number.

10) Raising a positive proper fraction to a natural number power _____ has a result that is less than that fraction.

11) Raising a positive improper fraction (*not equivalent to* 1) to a natural number power _____ has a result that is less than that fraction.

12) Which two numbers, when raised to a natural number power, have a result equal to themselves?

Objective 2: Use the Order of Operations

 13) In your own words, summarize the order of operations.

Evaluate.

14) $20 + 12 - 5$ 15) $17 - 2 + 4$

16) $51 - 18 + 2 - 11$ 17) $48 \div 2 + 14$

18) $15 \cdot 2 - 1$ 19) $20 - 3 \cdot 2 + 9$

20) $28 + 21 \div 7 - 4$ 21) $8 + 12 \cdot \dfrac{3}{4}$

22) $27 \div \dfrac{9}{5} - 1$ 23) $\dfrac{2}{5} \cdot \dfrac{1}{8} + \dfrac{2}{3} \cdot \dfrac{9}{10}$

24) $\dfrac{4}{9} \cdot \dfrac{5}{6} - \dfrac{1}{6} \cdot \dfrac{2}{3}$ 25) $2 \cdot \dfrac{3}{4} - \left(\dfrac{2}{3}\right)^2$

26) $\left(\dfrac{3}{2}\right)^2 - \left(\dfrac{5}{4}\right)^2$ 27) $25 - 11 \cdot 2 + 1$

28) $2 + 16 + 14 \div 2$

29) $39 - 3(9 - 7)^3$

30) $1 + 2(7 - 1)^2$

31) $60 \div 15 + 5 \cdot 3$

32) $27 \div (10 - 7)^2 + 8 \cdot 3$

33) $7[45 \div (19 - 10)] + 2$

34) $6[3 + (14 - 12)^3] - 10$

35) $1 + 2[(3 + 2)^3 \div (11 - 6)^2]$

36) $(4 + 7)^2 - 3[5(6 + 2) - 4^2]$

37) $\dfrac{4(7 - 2)^2}{12^2 - 8 \cdot 3}$ 38) $\dfrac{(8 + 4)^2 - 2^6}{7 \cdot 8 - 6 \cdot 9}$

39) $\dfrac{4(9 - 6)^3}{2^2 + 3 \cdot 8}$ 40) $\dfrac{7 + 3(10 - 8)^4}{6 + 10 \div 2 + 11}$

R Rethink

R1) Where have you encountered exponents other than in this math course?

R2) Where in a 12 × 12 multiplication table can you rewrite the products using an exponent?

R3) Which objective is the most difficult for you, and why?

1.3 Geometry Review

What are your objectives for Section 1.3?	How can you accomplish each objective?
1 Identify Angles and Parallel and Perpendicular Lines	• Learn and draw examples of *acute, right, obtuse, straight, complementary,* and *supplementary angles.* • Complete the given example on your own. • Complete You Try 1. • Understand the relationship between *vertical angles* along with *parallel* and *perpendicular lines.*
2 Identify Triangles	• Know how to identify a triangle by its angles: *acute, obtuse,* and *right triangles.* • Know how to identify a triangle by its sides: *equilateral, isosceles,* and *scalene triangles.* • Complete the given example on your own. • Complete You Try 2.
3 Use Area, Perimeter, and Circumference Formulas	• Understand and memorize the formulas to determine the perimeter and area of common figures. • Take notes on the terms required to determine the circumference and area of a circle. • Complete the given examples on your own. • Complete You Trys 3–5.
4 Use Volume Formulas	• Understand and memorize the formulas to determine the volume of common figures. • Complete the given example on your own. • Complete You Trys 6–7.

Read the explanations, follow the examples, take notes, and complete the You Trys.

We often use geometry to solve algebraic problems. We must review some basic geometric concepts that we will need beginning in Chapter 3.

Let's begin by looking at angles. An angle can be measured in **degrees.** For example, $45°$ is read as "45 degrees."

1 Identify Angles and Parallel and Perpendicular Lines

Definitions

An **acute angle** is an angle whose measure is greater than 0° and less than 90°.

A **right angle** is an angle whose measure is 90°, indicated by the ⌐ symbol.

An **obtuse angle** is an angle whose measure is greater than 90° and less than 180°.

A **straight angle** is an angle whose measure is 180°.

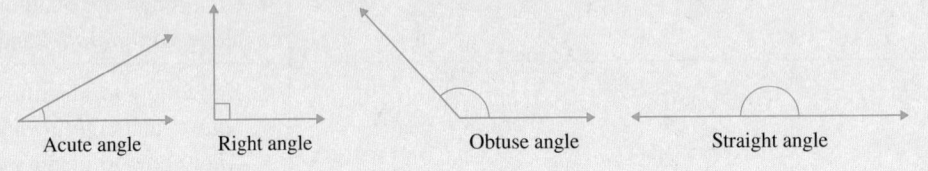

| Acute angle | Right angle | Obtuse angle | Straight angle |

Two angles are **complementary** if their measures add to 90°.

Two angles are **supplementary** if their measures add to 180°.

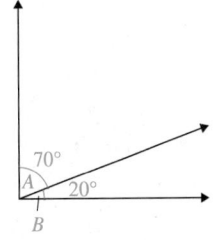

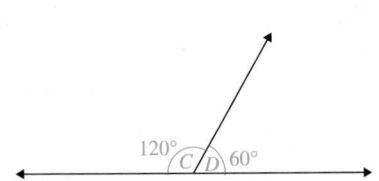

A and B are **complementary angles** since $m\angle A + m\angle B = 70° + 20° = 90°$.

C and D are **supplementary angles** since $m\angle C + m\angle D = 120° + 60° = 180°$.

Note

The measure of angle A is denoted by $m\angle A$.

EXAMPLE 1 $m\angle A = 41°$. Find its complement.

Solution

$$\text{Complement} = 90° - 41° = 49°$$

Since the sum of two complementary angles is 90°, if one angle measures 41°, its complement has a measure of $90° - 41° = 49°$.

[YOU TRY 1] $m\angle A = 62°$. Find its supplement.

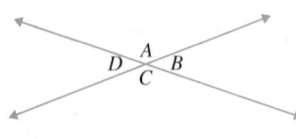

Figure 1.1

When two lines intersect, four angles are formed (see Figure 1.1). The pair of opposite angles are called **vertical angles**. Angles A and C are *vertical angles*, and angles B and D are *vertical angles*. *The measures of vertical angles are equal*. Therefore, $m\angle A = m\angle C$ and $m\angle B = m\angle D$.

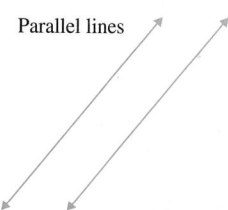

Parallel lines

Figure 1.2

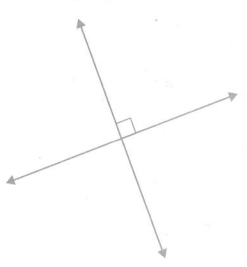

Perpendicular lines

Figure 1.3

Parallel lines are lines in the same plane that do not intersect (Figure 1.2). **Perpendicular lines** are lines that intersect at right angles (Figure 1.3).

2 Identify Triangles

We can classify triangles by their angles and by their sides.

Definitions

An **acute triangle** is one in which all three angles are acute.

An **obtuse triangle** contains one obtuse angle.

A **right triangle** contains one right angle.

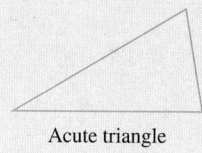

Acute triangle

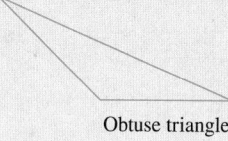

Obtuse triangle

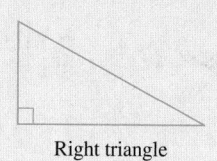

Right triangle

Property

The sum of the measures of the angles of any triangle is 180°.

> **W Hint**
> Use the acronym "EIS" or make up your own way to help you remember these three different types of triangles.

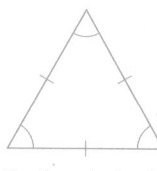

Equilateral triangle

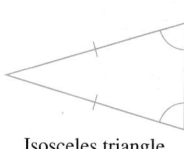

Isosceles triangle

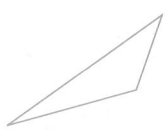

Scalene triangle

If a triangle has three sides of equal length, it is an **equilateral triangle.** (Each angle measure of an equilateral triangle is 60°.)

If a triangle has two sides of equal length, it is an **isosceles triangle.** (The angles opposite the equal sides have the same measure.)

If a triangle has no sides of equal length, it is a **scalene triangle.** (No angles have the same measure.)

EXAMPLE 2

Find the measures of angles A and B in this isosceles triangle.

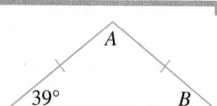

Solution

The single hash marks on the two sides of the triangle mean that those sides are of equal length.

$$m\angle B = 39° \qquad \text{Angle measures opposite sides of equal length are the same.}$$

$$39° + m\angle B = 39° + 39° = 78°$$

We have found that the sum of two of the angles is 78°. Since all of the angle measures add up to 180°,

$$m\angle A = 180° - 78° = 102°$$

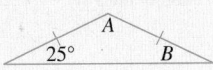

3 Use Area, Perimeter, and Circumference Formulas

The **perimeter** of a figure is the distance around the figure, while the **area** of a figure is the number of square units enclosed within the figure. For some familiar shapes, we have the following formulas:

Figure		Perimeter	Area
Rectangle:		$P = 2l + 2w$	$A = lw$
Square:		$P = 4s$	$A = s^2$
Triangle: h = height		$P = a + b + c$	$A = \dfrac{1}{2}bh$
Parallelogram: h = height		$P = 2a + 2b$	$A = bh$
Trapezoid: h = height		$P = a + c + b_1 + b_2$	$A = \dfrac{1}{2}h(b_1 + b_2)$

The perimeter of a circle is called the **circumference.** The **radius,** r, is the distance from the center of the circle to a point on the circle. A line segment that passes through the center of the circle and has its endpoints on the circle is called a **diameter.**

Pi, π, is the ratio of the circumference of any circle to its diameter. $\pi \approx 3.14159265 \ldots$, but we will use 3.14 as an approximation for π. The symbol $\approx$ is read as "is approximately equal to."

	Circumference	Area
	$C = 2\pi r$	$A = \pi r^2$

EXAMPLE 3 Find the perimeter and area of each figure.

a)

b)

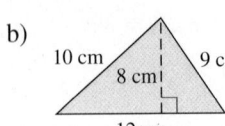

a) This figure is a rectangle.

Perimeter: $P = 2l + 2w$

$P = 2(9 \text{ in.}) + 2(7 \text{ in.})$ Substitute the values.

$P = 18 \text{ in.} + 14 \text{ in.}$

$P = 32 \text{ in.}$

Area: $A = lw$

$A = (9 \text{ in.})(7 \text{ in.})$ Substitute the values.

$A = 63 \text{ in}^2$ or 63 square inches

W Hint

Notice that perimeter is always expressed using one dimension, while area is expressed in two dimensions or square units.

b) This figure is a triangle.

Perimeter: $P = a + b + c$

$P = 9 \text{ cm} + 12 \text{ cm} + 10 \text{ cm}$ Substitute the values.

$P = 31 \text{ cm}$

Area: $A = \dfrac{1}{2}bh$

$A = \dfrac{1}{2}(12 \text{ cm})(8 \text{ cm})$ Substitute the values.

$A = 48 \text{ cm}^2$ or 48 square centimeters

[YOU TRY 3]

Find the perimeter and area of the figure.

8 cm

11 cm

EXAMPLE 4

Find (a) the circumference and (b) the area of the circle shown below right. Give an exact answer for each, and give an approximation using 3.14 for π.

Solution

a) The formula for the circumference of a circle is $C = 2\pi r$.
The radius of the given circle is 4 cm. Replace r with 4 cm.

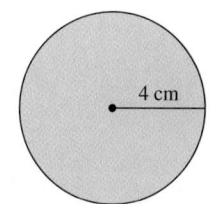

4 cm

$C = 2\pi r$

$= 2\pi(4 \text{ cm})$ Replace r with 4 cm.

$= 8\pi \text{ cm}$ Multiply.

Leaving the answer in terms of π gives us the exact circumference of the circle, 8π cm.

To find an approximation for the circumference, substitute 3.14 for π and simplify.

$C = 8\pi \text{ cm}$

$\approx 8(3.14) \text{ cm} = 25.12 \text{ cm}$

b) The formula for the area of a circle is $A = \pi r^2$. Replace r with 4 cm.

$A = \pi r^2$

$= \pi(4 \text{ cm})^2$ Replace r with 4 cm.

$= 16\pi \text{ cm}^2$ $4^2 = 16$

Leaving the answer in terms of π gives us the exact area of the circle, 16π cm^2.
To find an approximation for the area, substitute 3.14 for π and simplify.

$$A = 16\pi \text{ cm}^2$$
$$\approx 16(3.14) \text{ cm}^2$$
$$= 50.24 \text{ cm}^2$$

[YOU TRY 4] Find (a) the circumference and (b) the area of the circle. Give an exact answer for each, and give an approximation using 3.14 for π.

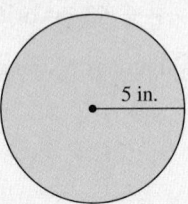

We can find the area and perimeter of some figures by combining formulas for more than one figure.

EXAMPLE 5 Find the perimeter and area of the figure shown here.

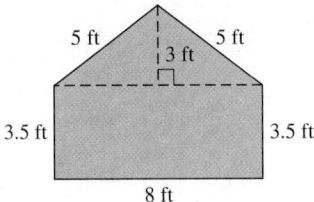

Solution

Perimeter: The perimeter is the distance around the figure.

$$P = 5 \text{ ft} + 5 \text{ ft} + 3.5 \text{ ft} + 8 \text{ ft} + 3.5 \text{ ft}$$
$$P = 25 \text{ ft}$$

Area: To find the area of this figure, think of it as two regions: a triangle and a rectangle.

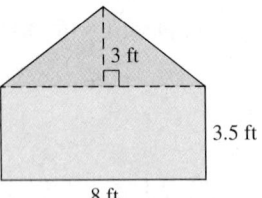

Total area = Area of triangle + Area of rectangle

$$= \frac{1}{2} bh + lw$$

$$= \frac{1}{2} (8 \text{ ft})(3 \text{ ft}) + (8 \text{ ft})(3.5 \text{ ft})$$

$$= 12 \text{ ft}^2 + 28 \text{ ft}^2$$

$$= 40 \text{ ft}^2$$

[YOU TRY 5] Find the perimeter and area of the figure.

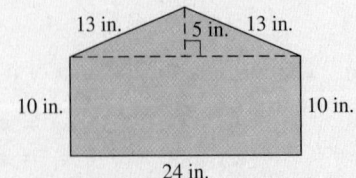

4 Use Volume Formulas

The **volume** of a three-dimensional object is the amount of space occupied by the object. Volume is measured in cubic units such as cubic inches (in³), cubic centimeters (cm³), cubic feet (ft³), and so on. Volume also describes the amount of a substance that can be enclosed within a three-dimensional object. Therefore, volume can also be measured in quarts, liters, gallons, and so on. In the figures, l = length, w = width, h = height, s = length of a side, and r = radius.

Volumes of Three-Dimensional Figures		
Rectangular solid		$V = lwh$
Cube		$V = s^3$
Right circular cylinder		$V = \pi r^2 h$
Sphere		$V = \dfrac{4}{3}\pi r^3$
Right circular cone		$V = \dfrac{1}{3}\pi r^2 h$

EXAMPLE 6   Find the volume of each. In (b), give the answer in terms of π.

a)

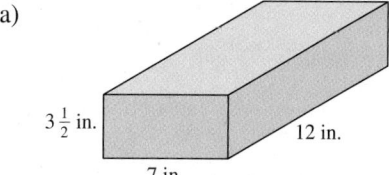

$3\frac{1}{2}$ in.   12 in.   7 in.

b)

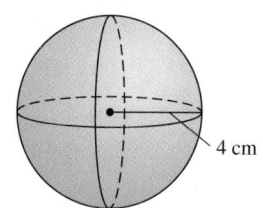

4 cm

Solution

a) $V = lwh$ Volume of a rectangular solid

$\quad = (12 \text{ in.})(7 \text{ in.})\left(3\frac{1}{2}\text{ in.}\right)$ Substitute values.

$\quad = (12 \text{ in.})(7 \text{ in.})\left(\frac{7}{2}\text{ in.}\right)$ Change to an improper fraction.

$\quad = \left(84 \cdot \frac{7}{2}\right) \text{ in}^3$ Multiply.

$\quad = 294 \text{ in}^3 \text{ or } 294 \text{ cubic inches}$

W Hint

Notice that volume is always expressed using three dimensions or cubic units.

b) $V = \dfrac{4}{3}\pi r^3$ Volume of a sphere

      $= \dfrac{4}{3}\pi (4\text{ cm})^3$ Replace r with 4 cm.

      $= \dfrac{4}{3}\pi (64\text{ cm}^3)$ $4^3 = 64$

      $= \dfrac{256}{3}\pi\text{ cm}^3$ Multiply.

$\left[\textbf{YOU TRY 6}\right]$ Find the volume of each figure. In (b), give the answer in terms of π.

     a)   A box with length = 3 ft, width = 2 ft, and height = 1.5 ft

     b)   A sphere with radius = 3 in.

EXAMPLE 7

A large truck has a fuel tank in the shape of a right circular cylinder. Its radius is 1 ft, and it is 4 ft long. (See Figure 1.4.)

     a)   How many cubic feet of diesel fuel will the tank hold? (Use 3.14 for π.)

     b)   How many gallons will it hold? Round to the nearest gallon. (1 ft^3 ≈ 7.48 gal)

     c)   If diesel fuel costs $2.50 per gallon, how much will it cost to fill the tank?

Solution

     a)   We're asked to determine how much fuel the tank will hold. We must find the *volume* of the tank.

$$\text{Volume of a cylinder} = \pi r^2 h$$
$$\approx (3.14)(1\text{ ft})^2 (4\text{ ft})$$
$$= 12.56\text{ ft}^3$$

The tank will hold 12.56 ft^3 of diesel fuel.

     b)   We must convert 12.56 ft^3 to gallons. Since 1 ft^3 ≈ 7.48 gal, we can change units by multiplying:

$$12.56\text{ ft}^3 \cdot \left(\dfrac{7.48\text{ gal}}{1\text{ ft}^3}\right) = 93.9488\text{ gal}$$
$$\approx 94\text{ gal}$$

We can divide out units in fractions the same way we can divide out common factors.

The tank will hold approximately 94 gal.

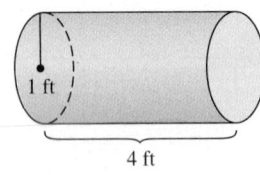

1 ft

4 ft

Figure 1.4

     c)   Diesel fuel costs $2.50 per gallon. We can figure out the total cost of the fuel the same way we did in (b).

$2.50 *per* gallon
↓

$$94\text{ gal} \cdot \left(\dfrac{\$2.50}{\text{gal}}\right) = \$235.00$$ Divide out the units of gallons.

It will cost about $235.00 to fill the tank.

YOU TRY 7

A large truck has a fuel tank in the shape of a right circular cylinder. Its radius is 1 ft, and it is 3 ft long.

a) How many cubic feet of diesel fuel will the tank hold? (Use 3.14 for π.)

b) How many gallons of fuel will it hold? Round to the nearest gallon. (1 ft^3 ≈ 7.48 gal)

c) If diesel fuel costs $2.50 per gallon, how much will it cost to fill the tank?

ANSWERS TO YOU TRY EXERCISES

1) 118°   2) $m\angle A = 130°$; $m\angle B = 25°$ 3) $P = 38$ cm; $A = 88$ cm^2 4) a) $C = 10\pi$ in.;
$C ≈ 31.4$ in.   b) $A = 25\pi$ in^2; $A ≈ 78.5$ in^2 5) $P = 70$ in.; $A = 300$ in^2 6) a) 9 ft^3 b) 36π in^3
7) a) 9.42 ft^3 b) 70 gal   c) $175.00

E Evaluate 1.3 Exercises

Do the exercises, and check your work.

Objective 1: Identify Angles and Parallel and Perpendicular Lines

1) An angle whose measure is between 0° and 90° is a(n) _____ angle.

2) An angle whose measure is 90° is a(n) _____ angle.

3) An angle whose measure is 180° is a(n) _____ angle.

4) An angle whose measure is between 90° and 180° is a(n) _____ angle.

5) If the sum of two angles is 180°, the angles are _____. If the sum of two angles is 90°, the angles are _____.

6) If two angles are supplementary, can both of them be obtuse? Explain.

Find the complement of each angle.

7) 59°   8) 84°

9) 12°   10) 40°

Find the supplement of each angle.

11) 143°   12) 62°

13) 38°   14) 155°

Find the measure of the missing angles.

15)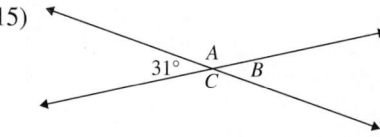

16)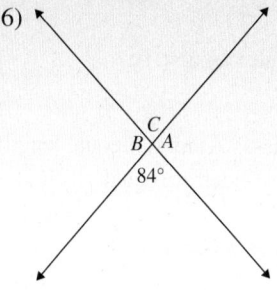

Objective 2: Identify Triangles

17) The sum of the angles in a triangle is _____ degrees.

Find the missing angle and classify each triangle as acute, obtuse, or right.

18)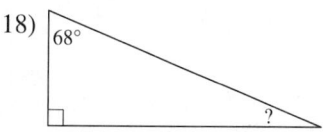

19)

20)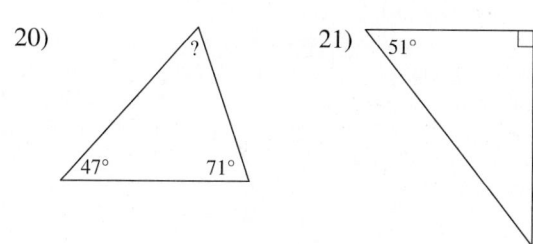

21)

22) Can a triangle contain more than one obtuse angle? Explain.

Classify each triangle as equilateral, isosceles, or scalene.

23)
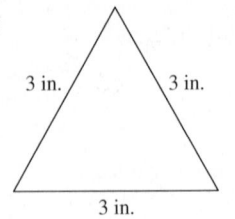
3 in.   3 in.

3 in.

24)
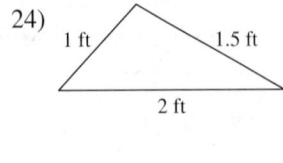
1 ft   1.5 ft
2 ft

25)
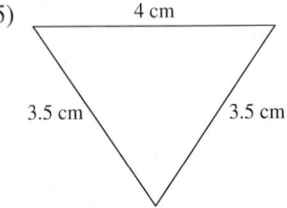
4 cm
3.5 cm   3.5 cm

26) What can you say about the measures of the angles in an equilateral triangle?

27) True or False: A right triangle can also be isosceles.

28) True or False: If a triangle has two sides of equal length, then the angles opposite these sides are equal.

Objective 3: Use Area, Perimeter, and Circumference Formulas

Find the area and perimeter of each figure. Include the correct units.

29)
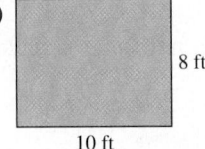
8 ft
10 ft

30)
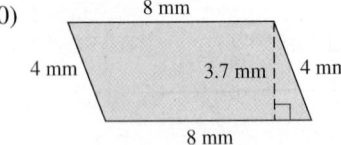
8 mm
4 mm   3.7 mm   4 mm
8 mm

31)
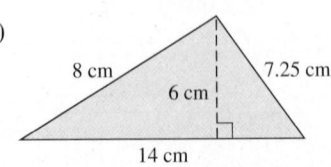
8 cm   7.25 cm
6 cm
14 cm

32)
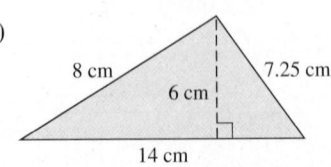
13 in.   7.8 in.
5 in.
18 in.

33)
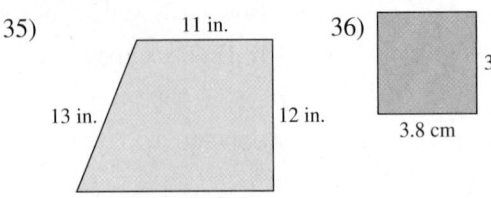
6.5 mi
6.5 mi

34)
$2\frac{1}{2}$ ft
$7\frac{2}{3}$ ft

35)
11 in.
13 in.   12 in.
16 in.

36)
3.8 cm
3.8 cm

Find (a) the area and (b) the circumference of the circle. Give an exact answer for each, and give an approximation using 3.14 for π. Include the correct units.

37)
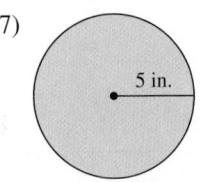
5 in.

38)
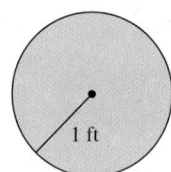
1 ft

39)
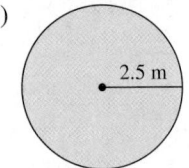
2.5 m

40)
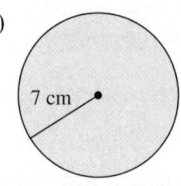
7 cm

Find the exact area and circumference of the circle in terms of π. Include the correct units.

41)
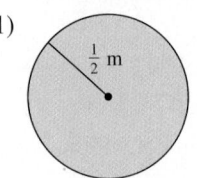
$\frac{1}{2}$ m

42)
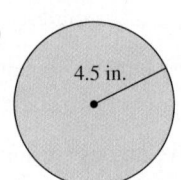
4.5 in.

43)
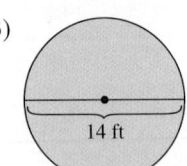
14 ft

44)
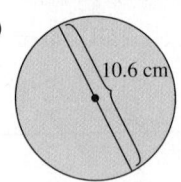
10.6 cm

Find the area and perimeter of each figure. Include the correct units.

45)
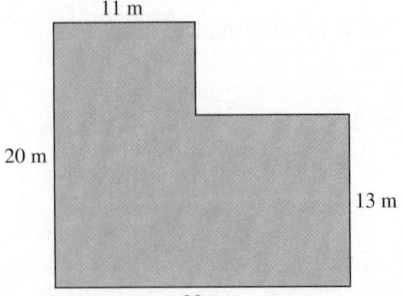
11 m
20 m
13 m
23 m

46)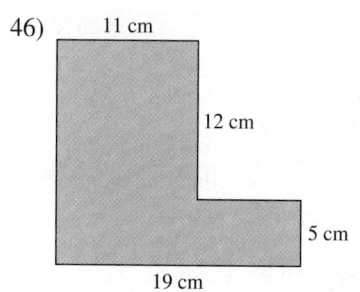
11 cm
12 cm
5 cm
19 cm

47)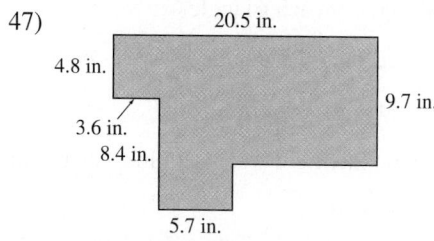
20.5 in.
4.8 in.
9.7 in.
3.6 in.
8.4 in.
5.7 in.

48)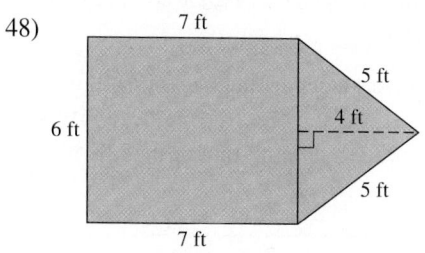
7 ft
5 ft
4 ft
6 ft
5 ft
7 ft

Find the area of the shaded region. Use 3.14 for π. Include the correct units.

49)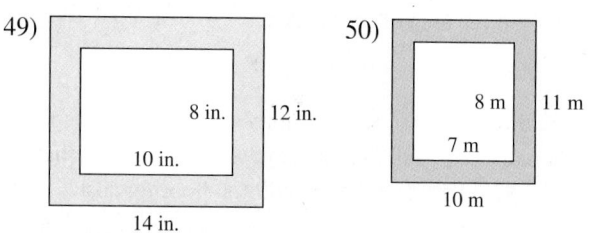
8 in.
12 in.
10 in.
14 in.

50)
8 m
11 m
7 m
10 m

51)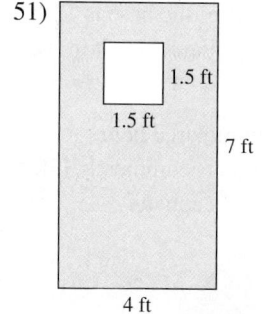
1.5 ft
1.5 ft
7 ft
4 ft

52)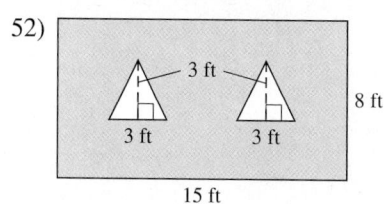
3 ft
8 ft
3 ft
3 ft
15 ft

53)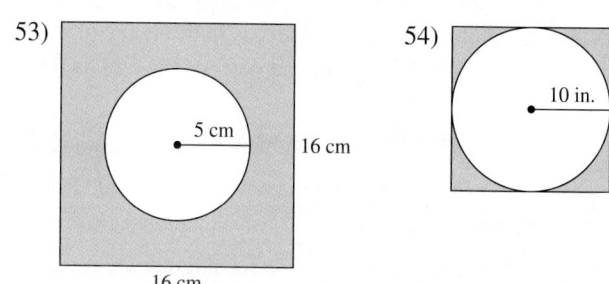
5 cm
16 cm
16 cm

54)
10 in.

Objective 4: Use Volume Formulas

Find the volume of each figure. Where appropriate, give the answer in terms of π. Include the correct units.

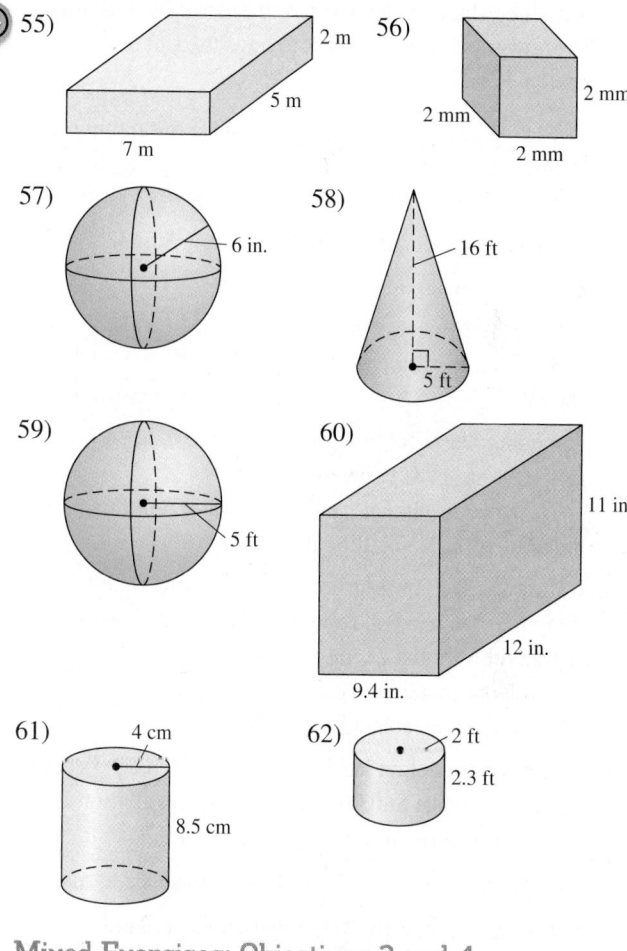

55)
2 m
5 m
7 m

56)
2 mm
2 mm
2 mm

57)
6 in.

58)
16 ft
5 ft

59)
5 ft

60)
11 in.
12 in.
9.4 in.

61)
4 cm
8.5 cm

62)
2 ft
2.3 ft

Mixed Exercises: Objectives 3 and 4

Applications of Perimeter, Area, and Volume: Use 3.14 for π, and include the correct units.

63) To lower her energy costs, Yun would like to replace her rectangular storefront window with low-emissivity (low-e) glass that costs $20.00/ft². The window measures 9 ft by 6.5 ft, and she can spend at most $900.

a) How much glass does she need?

b) Can she afford the low-e glass for this window?

64) An insulated rectangular cooler is 15 in. long, 10 in. wide, and 13.6 in. high. What is the volume of the cooler?

65) A fermentation tank at a winery is in the shape of a right circular cylinder. The diameter of the tank is 6 ft, and it is 8 ft tall.

a) How many cubic feet of wine will the tank hold?

b) How many gallons of wine will the tank hold? Round to the nearest gallon. ($1 \text{ ft}^3 \approx 7.48$ gallons)

66) Yessenia wants a custom-made rectangular area rug measuring 5 ft by 8 ft. She has budgeted $500. She likes the Alhambra carpet sample that costs $9.80/ft² and the Sahara pattern that costs $12.20/ft². Can she afford either of these fabrics to make the area rug, or does she have to choose the cheaper one to remain within her budget? Support your answer by determining how much it would cost to have the rug made in each pattern.

67) The lazy Susan on a table in a Chinese restaurant has a 10-inch radius. (A lazy Susan is a rotating tray used to serve food.)

a) What is the circumference of the lazy Susan?

b) What is its area?

68) Find the perimeter of home plate given the dimensions shown.

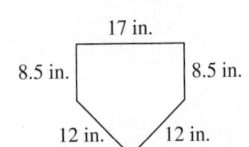

69) A rectangular reflecting pool is 30 ft long, 19 ft wide, and 1.5 ft deep. How many gallons of water will this pool hold? ($1 \text{ ft}^3 \approx 7.48$ gallons)

70) Ralph wants to childproof his house now that his daughter has learned to walk. The round, glass-top coffee table in his living room has a diameter of 36 in. How much soft padding does Ralph need to cover the edges around the table?

71) Nadia is remodeling her kitchen and her favorite granite countertop costs $80.00/ft², including installation. The layout of the countertop is shown, where the counter has a uniform width of $2\frac{1}{4}$ ft. If she can spend at most $2500.00, can she afford her first-choice granite?

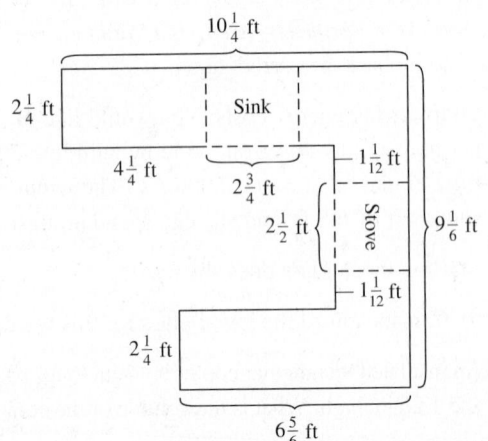

72) A container of lip balm is in the shape of a right circular cylinder with a radius of 1.5 cm and a height of 2 cm. How much lip balm will the container hold?

73) Jaden is making decorations for the bulletin board in his fifth-grade classroom. Each equilateral triangle has a height of 15.6 in. and sides of length 18 in.

a) Find the area of each triangle.

b) Find the perimeter of each triangle.

74) The chamber of a rectangular laboratory water bath measures 6 in. $\times\ 11\frac{3}{4}$ in. $\times\ 5\frac{1}{2}$ in.

a) How many cubic inches of water will the water bath hold?

b) How many liters of water will the water bath hold? ($1 \text{ in}^3 \approx 0.016$ L)

75) A town's public works department will install a flower garden in the shape of a trapezoid. It will be enclosed by decorative fencing that costs $23.50/ft.

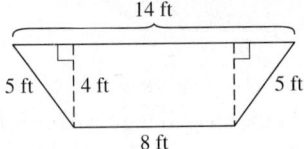

a) Find the area of the garden.

b) Find the cost of the fence.

76) The dimensions of Riyad's home office are 10 ft $\times$ 12 ft. He plans to install laminated hardwood flooring that costs $2.69/ft². How much will the flooring cost?

77) Salt used to melt road ice in winter is piled in the shape of a right circular cone. The radius of the base is 12 ft, and the pile is 8 ft high. Find the volume of salt in the pile.

78) Find the volume of the ice cream pictured here. Assume that the right circular cone is completely filled and that the scoop on top is half of a sphere.

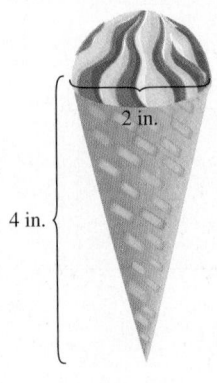

R1) Were there certain types of geometry formulas that were more difficult for you to work with than others? Which were they? What was difficult about them?

R2) Can you think of any jobs that would require computing an area? What are they?

R3) Write about a situation where you have to calculate the area of something to determine its cost.

1.4 Sets of Numbers and Absolute Value

P Prepare

O Organize

What are your objectives for Section 1.4?	How can you accomplish each objective?
1 Identify and Graph Numbers on a Number Line	• Know the definitions of *natural numbers, whole numbers,* and *integers,* and be able to compare the sets of numbers. • Be able to identify and graph any of these numbers on a number line. • Write the definitions of a *rational number,* an *irrational number,* and the *real numbers* in your own words. • Complete the given examples on your own. • Complete You Trys 1–5.
2 Compare Numbers Using Inequality Symbols	• Learn the inequality symbols by using each to compare two numbers in your notes. • Be able to use signed numbers to represent increases or decreases. • Complete the given examples on your own. • Complete You Trys 6–7.
3 Find the Additive Inverse and Absolute Value of a Number	• Write the definitions of *additive inverse* and *absolute value* in your own words. • Complete the given examples on your own. • Complete You Trys 8 and 9.

W Work

Read the explanations, follow the examples, take notes, and complete the You Trys.

1 Identify and Graph Numbers on a Number Line

In Section 1.1, we defined the following sets of numbers:

Natural numbers: {1, 2, 3, 4, ...}

Whole numbers: {0, 1, 2, 3, 4, ...}

We will begin this section by discussing other sets of numbers.

On a **number line,** positive numbers are to the right of zero and negative numbers are to the left of zero.

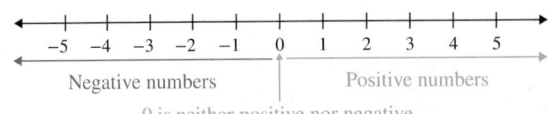

Negative numbers   Positive numbers

0 is neither positive nor negative.

The natural numbers, their negatives, and 0 form the set of numbers called *integers*.

Definition

The set of **integers** includes the set of natural numbers, their negatives, and zero. The set of *integers* is $\{\ldots, -3, -2, -1, 0, 1, 2, 3, \ldots\}$.

EXAMPLE 1

Graph each number on a number line.

$$4, 1, -6, 0, -3$$

Solution

4 and 1 are to the right of zero since they are positive.

-3 is three units to the left of zero, and -6 is six units to the left of zero.

[YOU TRY 1] Graph each number on a number line. $2, -4, 5, -1, -2$

Positive and negative numbers are also called **signed numbers.**

EXAMPLE 2

Given the set of numbers $\left\{4, -7, 0, \dfrac{3}{4}, -6, 10, -3\right\}$, list the

a) whole numbers   b) natural numbers   c) integers

Solution

a) whole numbers: 0, 4, 10   b) natural numbers: 4, 10

c) integers: $-7, -6, -3, 0, 4, 10$

[YOU TRY 2]

Given the set of numbers $\left\{-1, 5, \dfrac{2}{7}, 8, 0, -12\right\}$, list the

a) whole numbers   b) natural numbers   c) integers

Notice in Example 2 that $\dfrac{3}{4}$ did not belong to any of these sets. That is because the whole numbers, natural numbers, and integers do not contain any fractional parts. $\dfrac{3}{4}$ is a *rational number.*

Definition

A **rational number** is any number of the form $\frac{p}{q}$, where p and q are integers and $q \neq 0$.

That is, a rational number is any number that can be written as a fraction where the numerator and denominator are integers and the denominator does not equal zero.

Rational numbers include much more than numbers like $\frac{3}{4}$, which are already in fractional form.

EXAMPLE 3

Explain why each of the following numbers is rational.

a) 7 b) 0.8 c) −5 d) $6\frac{1}{4}$ e) $0.\overline{3}$ f) $\sqrt{4}$

Solution

Rational Number	Reason
7	7 can be written as $\frac{7}{1}$.
0.8	0.8 can be written as $\frac{8}{10}$.
−5	−5 can be written as $\frac{-5}{1}$.
$6\frac{1}{4}$	$6\frac{1}{4}$ can be written as $\frac{25}{4}$.
$0.\overline{3}$	$0.\overline{3}$ can be written as $\frac{1}{3}$.
$\sqrt{4}$	$\sqrt{4} = 2$ and $2 = \frac{2}{1}$.

$\sqrt{4}$ is read as "the square root of 4." This means, "What number times itself equals 4?" That number is 2.

[YOU TRY 3]

Explain why each of the following numbers is rational.

a) 12 b) 0.7 c) −8 d) $2\frac{3}{4}$ e) $0.\overline{6}$ f) $\sqrt{100}$

W Hint

Notice that square roots of perfect squares are always rational numbers!

To summarize, the set of rational numbers includes

1) integers, whole numbers, and natural numbers.

2) repeating decimals.

3) terminating decimals.

4) fractions and mixed numbers.

The set of rational numbers does *not* include nonrepeating, nonterminating decimals. These decimals cannot be written as the quotient of two integers. Numbers such as these are called *irrational numbers*.

Definition

The set of numbers that cannot be written as the quotient of two integers is called the set of **irrational numbers**. Written in decimal form, an *irrational number* is a nonrepeating, nonterminating decimal.

EXAMPLE 4   Explain why each of the following numbers is irrational.

a) 0.827136...   b) π c) $\sqrt{3}$

Solution

Irrational Number	Reason
0.827136...	It is a nonrepeating, nonterminating decimal.
π	$\pi \approx 3.14159265...$ It is a nonrepeating, nonterminating decimal.
$\sqrt{3}$	3 is not a perfect square, and the decimal equivalent of the square root of a nonperfect square is a nonrepeating, nonterminating decimal. Here, $\sqrt{3} \approx 1.73205...$

[YOU TRY 4]   Explain why each of the following numbers is irrational.

a) 2.41895...   b) $\sqrt{2}$

If we put together the sets of numbers we have discussed up to this point, we get the *real numbers*.

Definition

The set of **real numbers** consists of the rational and irrational numbers.

W Hint

In your notes and in your own words, summarize the definitions of the different sets of numbers. Give examples.

We summarize the information next with examples of the different sets of numbers:

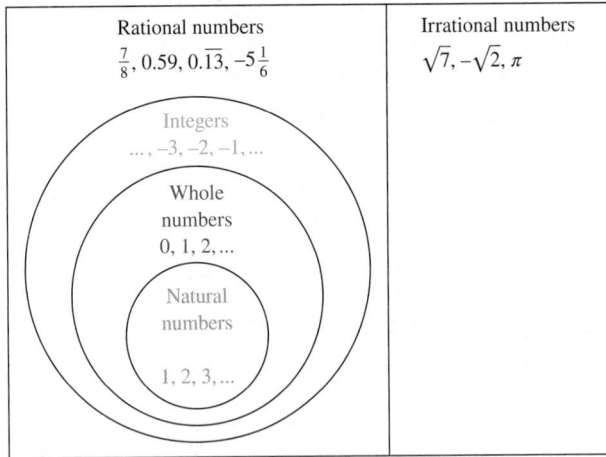

Real Numbers

Rational numbers
$\frac{7}{8}, 0.59, 0.\overline{13}, -5\frac{1}{6}$

Irrational numbers
$\sqrt{7}, -\sqrt{2}, \pi$

Integers
..., −3, −2, −1, ...

Whole numbers
0, 1, 2, ...

Natural numbers
1, 2, 3, ...

From the figure we can see, for example, that all whole numbers {0, 1, 2, 3...} are integers, but not all integers are whole numbers (−3, for example).

EXAMPLE 5   Given the set of numbers $\left\{ -16, 3.82, 0, 29, 0.\overline{6}, -\frac{11}{12}, \sqrt{10}, 5.302981... \right\}$, list the

a) integers   b) natural numbers   c) whole numbers

d) rational numbers   e) irrational numbers   f) real numbers

a)   integers: $-16, 0, 29$

b)   natural numbers: 29

c)   whole numbers: $0, 29$

d)   rational numbers: $-16, 3.82, 0, 29, 0.\overline{6}, -\dfrac{11}{12}$. Each of these numbers can be written as the quotient of two integers.

e)   irrational numbers: $\sqrt{10}, 5.302981\ldots$

f)   real numbers: All of the numbers in this set are real.

$$\left\{ -16, 3.82, 0, 29, 0.\overline{6}, -\frac{11}{12}, \sqrt{10}, 5.302981\ldots \right\}$$

[YOU TRY 5]

Given the set of numbers $\left\{ \dfrac{9}{8}, \sqrt{14}, 34, -41, 6.59, 0.\overline{2}, 0, 7.412835\ldots \right\}$, list the

a)   whole numbers   b)   integers

c)   rational numbers   d)   irrational numbers

---

2   Compare Numbers Using Inequality Symbols

Let's review the inequality symbols.

W Hint

As we move to the *left* on the number line, the numbers get smaller. As we move to the *right* on the number line, the numbers get larger.

$<$ less than	$\le$ less than or equal to
$>$ greater than	$\ge$ greater than or equal to
$\ne$ not equal to	$\approx$ approximately equal to

We use these symbols to compare numbers as in $5 > 2$, $6 \le 17$, $4 \ne 9$, and so on. How do we compare negative numbers?

EXAMPLE 6

Insert $>$ or $<$ to make the statement true. Look at the number line, if necessary.

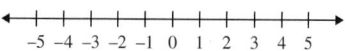

a) 4 __ 2 b) 2 __ 4 c) -3 __ 1 d) -2 __ -5 e) -4 __ -1

Solution

a) $4 \underline{\ >\ } 2$ 4 is to the right of 2.

b) $2 \underline{\ <\ } 4$ 2 is to the left of 4.

c) $-3 \underline{\ <\ } 1$ -3 is to the left of 1.

d) $-2 \underline{\ >\ } -5$ -2 is to the right of -5.

e) $-4 \underline{\ <\ } -1$ -4 is to the left of -1.

[YOU TRY 6]

Insert $>$ or $<$ to make the statement true.

a) 5 ___ 3 b) 3 ___ 5 c) -4 ___ 1 d) -1 ___ -3 e) -5 ___ -2

<table>
<tr><td>

W Hint

Think about why a "is less than" symbol points to the left and why a "is greater than" symbol points to the right.

</td></tr>
</table>

Note

The inequality sign always points to the number farther *left* on the number line.

Look at a) and b) in Example 6, and you will notice that the same numbers, 2 and 4, are being compared but the order and symbol are reversed. We can rewrite the statements by reversing the inequality symbol and interchanging the numbers being compared.

Signed numbers are used in many different ways.

EXAMPLE 7

Use a signed number to represent the change in each situation.

a) In 2016, the percentage of American adults who smoked was 25.8% less than in 2005. (www.cdc.gov)

b) Domestic box office revenues increased by approximately $202,000,000 in June 2018 compared to the same month the previous year. (www.hollywoodreporter.com)

Solution

a) −25.8%   The negative number represents a decrease in the percentage of American adults who smoked.

b) $202,000,000   The positive number represents an increase in box office revenue.

[YOU TRY 7]

Use a signed number to represent the change.
After getting off the highway, Huda decreased his car's speed by 25 mph.

3 Find the Additive Inverse and Absolute Value of a Number

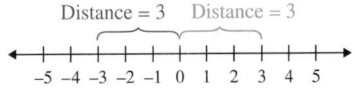

Notice that both −3 and 3 are a distance of 3 units from 0 but are on opposite sides of 0. We say that 3 and −3 are *additive inverses*.

W Hint

Notice that the additive inverse of a negative number is always positive. The additive inverse of a positive number is always negative.

Definition

Two numbers are **additive inverses** if they are the same distance from 0 on the number line but on opposite sides of 0. Therefore, if a is any real number, then $-a$ is its additive inverse.

Furthermore, $-(-a) = a$. We can see this on the number line.

EXAMPLE 8

Find $-(-2)$.

Solution

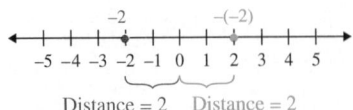

So, beginning with −2, the number on the opposite side of zero and 2 units away from zero is 2.
$-(-2) = 2$

Find −(−3).

We can explain *distance from zero* in another way: *absolute value.*

Definition

The **absolute value** of a number is the distance between that number and 0 on the number line. Furthermore, if a is any real number, then the **absolute value of** a, denoted by $|a|$, is

$$\text{i)}\ a \text{ if } a \geq 0$$
$$\text{ii)}\ -a \text{ if } a < 0$$

Remember, $|a|$ is never negative.

W Hint

The absolute value of a number describes its *distance* from zero and *not* what side of zero the number is on. So, the absolute value of a number is always positive or zero.

EXAMPLE 9

Evaluate each.

a) $|6|$ b) $|-5|$ c) $|0|$ d) $-|12|$ e) $|14 - 5|$

Solution

a) $|6| = 6$ 6 is 6 units from 0.

b) $|-5| = 5$ −5 is 5 units from 0.

c) $|0| = 0$

d) $-|12| = -12$ First, evaluate $|12|$: $|12| = 12$. Then, apply the negative symbol to get −12.

e) $|14 - 5| = |9|$ The absolute value symbols work like parentheses. First, evaluate what is inside: $14 - 5 = 9$.

$ = 9$ Find the absolute value.

[YOU TRY 9]

Evaluate each.

a) $|19|$ b) $|-8|$ c) $-|7|$ d) $|20 - 9|$

ANSWERS TO [YOU TRY] **EXERCISES**

1)
2) a) 0, 5, 8 b) 5, 8 c) −12, −1, 0, 5, 8

3) a) $12 = \dfrac{12}{1}$ b) $0.7 = \dfrac{7}{10}$ c) $-8 = \dfrac{-8}{1}$ d) $2\dfrac{3}{4} = \dfrac{11}{4}$ c) $0.\overline{6} = \dfrac{2}{3}$ f) $\sqrt{100} = 10$ and $10 = \dfrac{10}{1}$

4) a) It is a nonrepeating, nonterminating decimal. b) 2 is not a perfect square, so the decimal equivalent of $\sqrt{2}$ is a nonrepeating, nonterminating decimal.   5) a) 34, 0   b) 34, −41, 0

c) $\dfrac{9}{8}$, 34, −41, 6.59, 0.$\overline{2}$, 0   d) $\sqrt{14}$, 7.412835...   6) a) > b) < c) < d) > e) <

7) −25 mph   8) 3   9) a) 19 b) 8 c) −7 d) 11

Objective 1: Identify and Graph Numbers on a Number Line

1) In your own words, explain the difference between the set of rational numbers and the set of irrational numbers. Give two examples of each type of number.

2) In your own words, explain the difference between the set of whole numbers and the set of natural numbers. Give two examples of each type of number.

In Exercises 3 and 4, given each set of numbers, list the

    a) natural numbers   b) whole numbers

    c) integers   d) rational numbers

    e) irrational numbers   f) real numbers

3) 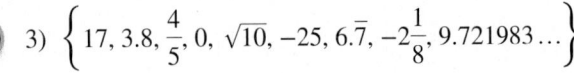 $\left\{ 17, 3.8, \dfrac{4}{5}, 0, \sqrt{10}, -25, 6.\overline{7}, -2\dfrac{1}{8}, 9.721983\ldots \right\}$

4) $\left\{ -6, \sqrt{23}, 21, 5.\overline{62}, 0.4, 3\dfrac{2}{9}, 0, -\dfrac{7}{8}, 2.074816\ldots \right\}$

Fill in the blank with *always*, *sometimes*, or *never* to make the statement true.

5) A whole number is _____ a real number.

6) A real number is _____ an integer.

7) A rational number is _____ a whole number.

8) A whole number is _____ an integer.

9) A natural number is _____ an irrational number.

10) An irrational number is _____ a whole number.

Graph the numbers on a number line. Label each.

11) $5, -2, \dfrac{3}{2}, -3\dfrac{1}{2}, 0$

12) $-4, 3, \dfrac{7}{8}, 4\dfrac{1}{3}, -2\dfrac{1}{4}$

13) $-6.8, -\dfrac{3}{8}, 0.2, 1\dfrac{8}{9}, -4\dfrac{1}{3}$

14) $-3.25, \dfrac{2}{3}, 2, -1\dfrac{3}{8}, 4.1$

Objective 2: Compare Numbers Using Inequality Symbols

Insert > or < to make the statement true.

15) 7 _____ 4   16) 3 _____ 11

17) −4 _____ −1   18) −2 _____ −3

19) 9 _____ −2   20) −3 _____ 5

21) 0 _____ −2   22) 8 _____ 10

23) $-\dfrac{1}{2}$ _____ $-\dfrac{3}{4}$ 24) $-\dfrac{3}{5}$ _____ $-\dfrac{7}{10}$

Write each statement with the inequality symbol reversed.

25) $6 < 11$ 26) $-5 < -1$

27) $4 > -8$ 28) $3 > -12$

Objective 3: Find the Additive Inverse and Absolute Value of a Number

29) What does the absolute value of a number represent?

30) If a is a real number and if $|a|$ is not a positive number, then what is the value of a?

Find the additive inverse of each.

31) 8   32) 6

33) −15   34) −1

35) $-\dfrac{3}{4}$ 36) 4.7

Evaluate.

37) $|-10|$ 38) $|9|$

39) $\left| \dfrac{9}{4} \right|$ 40) $\left| -\dfrac{5}{6} \right|$

41) $-|-14|$ 42) $-|27|$

43) $|17 - 4|$ 44) $-|10 - 6|$

45) $-\left| -4\dfrac{1}{7} \right|$ 46) $|-9.6|$

Determine whether each statement is true or false.

47) The absolute value of the difference between any two real numbers represents the distance between the two numbers on a number line.

48) The additive inverse of the absolute value of any nonzero real number is always negative.

Write each group of numbers from smallest to largest.

49) $7, -2, 3.8, -10, 0, \dfrac{9}{10}$

50) $2.6, 2.06, -1, -5\dfrac{3}{8}, 3, \dfrac{7}{4}$

51) $7\dfrac{5}{6}, -5, -6.5, -6.51, 7\dfrac{1}{3}, 2$

52) $-\dfrac{3}{4}, 0, -0.5, 4, -1, \dfrac{15}{2}$

Mixed Exercises: Objectives 2 and 3
Decide whether each statement is true or false.

53) $16 \geq -11$

54) $-19 < -18$

55) $\dfrac{7}{11} \leq \dfrac{5}{9}$

56) $-1.7 \geq -1.6$

57) $-|-28| = 28$

58) $-|13| = -13$

59) $-5\dfrac{3}{10} < -5\dfrac{3}{4}$

60) $\dfrac{3}{2} \leq \dfrac{3}{4}$

Use a signed number to represent the change in each situation.

61) During the 2010–2011 regular season, Jonathan Toews had 233 shots on goal for the Chicago Blackhawks. During the 2017–2018 regular season, he had 211 shots on goal for a decrease of 22 shots on goal. (www.nhl.com)

62) In 2016, the average price for a gallon of gas in Los Angeles was $2.85. In 2017, the average price was $3.10, an increase of $0.25 per gallon. (www.eia.gov)

63) In the second quarter of 2017, there were 326 million active Twitter users. That number increased to 335 million active users in the second quarter of 2018. The number of Twitter users increased by 9 million from the second quarter of 2017 to 2018. (www.statista.com)

64) The population of Michigan increased by approximately 34,000 from 2016 to 2017. (www.census.gov)

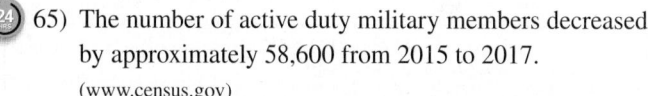 65) The number of active duty military members decreased by approximately 58,600 from 2015 to 2017. (www.census.gov)

66) From 2017 to 2018, the number of new housing starts decreased by approximately 17,000. (www.census.gov)

R Rethink

R1) If the absolute value of a number is greater than the absolute value of another, what does that mean in terms of their locations on a number line?

R2) In a computer programming language, what do you think "abs(−24)" means?

R3) Why is the additive inverse of a negative number always positive?

1.5 Addition and Subtraction of Real Numbers

P Prepare

What are your objectives for Section 1.5?

O Organize

How can you accomplish each objective?

What are your objectives for Section 1.5?	How can you accomplish each objective?
1 Add Integers Using a Number Line	• Complete the given examples on your own. • Complete You Try 1.
2 Add Real Numbers with the Same Sign	• Write the procedure for **Adding Numbers with the Same Sign** in your own words. • Complete the given example on your own. • Complete You Try 2.
3 Add Real Numbers with Different Signs	• Write the procedure for **Adding Numbers with Different Signs** in your own words. • Complete the given example on your own. • Complete You Try 3.
4 Subtract Real Numbers	• Understand that $a - b = a + (-b)$, and write a procedure that will help you subtract numbers. • Use the procedure you developed to complete the given example on your own. • Complete You Try 4.
5 Solve Applied Problems	• Complete the given example on your own. • Complete You Try 5.
6 Apply the Order of Operations to Real Numbers	• Add any additional steps to the Order of Operations you learned in Section 1.1. • Complete the given example on your own. • Complete You Try 6.
7 Translate English Expressions into Mathematical Expressions	• Take notes on the common English expressions, and be able to identify whether an expression refers to addition or subtraction. • Complete the given example on your own. • Complete You Try 7.

 **Work**   **Read the explanations, follow the examples, take notes, and complete the You Trys.**

In Section 1.4, we defined real numbers. In this section, we will discuss adding and subtracting real numbers.

1 Add Integers Using a Number Line

Let's use a number line to add numbers.

EXAMPLE 1   Use a number line to add each pair of numbers.

a) $2 + 5$ 　　　 b) $-1 + (-4)$ 　　　 c) $2 + (-5)$ 　　　 d) $-8 + 12$

Solution

a) 2 + 5: Start at 2 and move 5 units to the right.

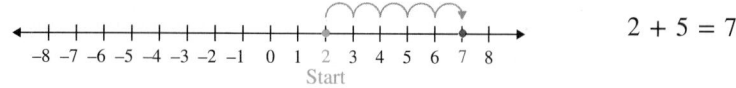

$$2 + 5 = 7$$

b) −1 + (−4): Start at −1 and move 4 units to the left. (Move to the left when adding a negative.)

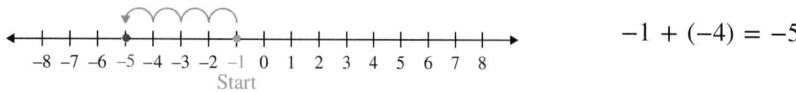

$$-1 + (-4) = -5$$

c) 2 + (−5): Start at 2 and move 5 units to the left.

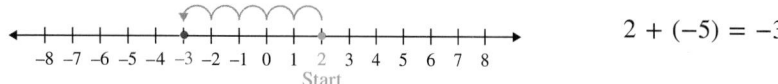

$$2 + (-5) = -3$$

d) −8 + 12: Start at −8 and move 12 units to the right.

$$-8 + 12 = 4$$

Note

When a negative number follows an operation symbol such as +, −, · , or ÷, we put the number in parentheses as in parts b) and c) of this example: −1 + (−4) and 2 + (−5).

[YOU TRY 1]

Use a number line to add each pair of numbers.

a) 1 + 3   b) −3 + (−2)   c) 8 + (−6)   d) −10 + 7

2 Add Real Numbers with the Same Sign

We found that

$$2 + 5 = 7, \quad -1 + (-4) = -5, \quad 2 + (-5) = -3, \quad -8 + 12 = 4.$$

Notice that when we add two numbers with the same sign, the result has the same sign as the numbers being added.

Procedure Adding Numbers with the Same Sign

To add numbers with the same sign, find the absolute value of each number and add them. The sum will have the same sign as the numbers being added.

Apply this rule to −1 + (−4).

The result will be negative
↓
$$-1 + (-4) = -(|-1| + |-4|) = -(1 + 4) = -5$$

Add the absolute value of each number.

Hint

Did you notice that some answers are positive and some are negative? What patterns do you see?

EXAMPLE 2

Add $-23 + (-41)$.

Solution

$-23 + (-41) = -(|-23| + |-41|) = -(23 + 41) = -64$

[YOU TRY 2] Add.

a) $-6 + (-10)$ b) $-38 + (-56)$ c) $-\dfrac{3}{10} + \left(-\dfrac{1}{6}\right)$

3 Add Real Numbers with Different Signs

W Hint

Try writing this procedure using steps.

Step 1: _____

Step 2: _____

In Example 1, we found that $2 + (-5) = -3$ and $-8 + 12 = 4$.

Procedure Adding Numbers with Different Signs

To add two numbers with different signs, find the absolute value of each number. Subtract the smaller absolute value from the larger. The sum will have the sign of the number with the larger absolute value.

EXAMPLE 3

Add.

a) $-17 + 5$ b) $9.8 + (-6.3)$ c) $\dfrac{1}{5} + \left(-\dfrac{2}{3}\right)$ d) $-8 + 8$

Solution

a) $-17 + 5 = -12$ The sum will be negative since the number with the larger absolute value, $|-17|$, is negative.

b) $9.8 + (-6.3) = 3.5$ The sum will be positive since the number with the larger absolute value, $|9.8|$, is positive.

c) $\dfrac{1}{5} + \left(-\dfrac{2}{3}\right) = \dfrac{3}{15} + \left(-\dfrac{10}{15}\right)$ Get a common denominator.
The sum will be negative since the number with the larger

$= -\dfrac{7}{15}$ absolute value, $\left|-\dfrac{10}{15}\right|$, is negative.

d) $-8 + 8 = 0$

Note

The sum of a number and its additive inverse is always 0. That is, if a is a real number, then a + (−a) = 0. Notice in part d) of Example 3 that −8 and 8 are additive inverses.

[YOU TRY 3] Add.

a) $-14 + 2$ b) $20 + (-19)$ c) $-\dfrac{3}{7} + \dfrac{1}{4}$ d) $7.2 + (-7.2)$

4 Subtract Real Numbers

We can use the additive inverse to subtract numbers. Let's start with a basic subtraction problem and use a number line to find $8 - 5$.

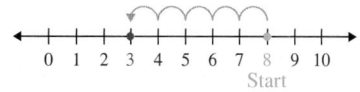

Start at 8. Then to subtract 5, move 5 units to the left to get 3.

$$8 - 5 = 3$$

We use the same procedure to find $8 + (-5)$. This leads us to a definition of subtraction:

Definition

If a and b are real numbers, then $a - b = a + (-b)$.

The definition tells us that to subtract $a - b$,

1) change subtraction to addition.
2) find the additive inverse of b.
3) add a and the additive inverse of b.

EXAMPLE 4

Subtract.

a) $4 - 9$ b) $-10 - 8$ c) $6 - (-25)$

Solution

a) $4 - 9 = 4 + (-9) = -5$

  Change to addition.   Additive inverse of 9

W Hint

Notice the use of parentheses when a negative number follows an addition or subtraction sign.

b) $-10 - 8 = -10 + (-8) = -18$

  Change to addition.   Additive inverse of 8

c) $6 - (-25) = 6 + 25 = 31$

  Change to addition.   Additive inverse of -25

[YOU TRY 4]

Subtract.

a) $31 - 14$ b) $8 - 21$ c) $-9 - 13$ d) $5.2 - (-7.6)$ e) $\dfrac{5}{8} - \dfrac{2}{3}$

In part c) of Example 4, $6 - (-25)$ changed to $6 + 25$. This illustrates that *subtracting a negative number is equivalent to adding a positive number.* Therefore, $-7 - (-15) = -7 + 15 = 8$.

5 Solve Applied Problems

We can use signed numbers to solve real-life problems.

EXAMPLE 5

According to the National Weather Service, the coldest temperature ever recorded in Wyoming was −66°F on February 9, 1944. The record high was 115°F on August 8, 1983. What is the difference between these two temperatures? (www.ncdc.noaa.gov)

Solution

$$\text{Difference} = \text{Highest temperature} - \text{Lowest temperature}$$
$$= \quad 115 \quad - \quad (-66)$$
$$= 115 + 66$$
$$= 181$$

The difference between the temperatures is 181°F.

[YOU TRY 5]

The best score in a golf tournament was −16, and the worst score was +9. What is the difference between these two scores?

6 Apply the Order of Operations to Real Numbers

We discussed the order of operations in Section 1.2. Let's explore it further with the real numbers.

EXAMPLE 6

Simplify.

a) $(10 - 18) + (-4 + 6)$ b) $|-31 - 4| - 7|9 - 4|$

Solution

a) $(10 - 18) + (-4 + 6) = -8 + 2$ First, perform the operations in parentheses.
$$= -6 \qquad\qquad\qquad \text{Add.}$$

b) $|-31 - 4| - 7|9 - 4| = |-31 + (-4)| - 7|9 - 4|$
$$= |-35| - 7|5| \qquad \text{Perform the operations in the absolute values.}$$
$$= 35 - 7(5) \qquad \text{Evaluate the absolute values.}$$
$$= 35 - 35$$
$$= 0$$

[YOU TRY 6]

Simplify.

a) $[12 + (-5)] - [-16 + (-8)]$ b) $-\dfrac{4}{9} + \left(\dfrac{1}{6} - \dfrac{2}{3}\right)$ c) $-|7 - 15| - |4 - 2|$

7 Translate English Expressions into Mathematical Expressions

Knowing how to translate English expressions into mathematical expressions is a skill students need to learn algebra. Here, we will discuss how to "translate" from English into mathematics. Let's look at some key words and phrases you may encounter.

English Expression	Mathematical Operation
sum, more than, increased by	addition
difference between, less than, decreased by	subtraction

EXAMPLE 7

Write a mathematical expression for each and simplify.

a)   9 more than −2   b)   10 less than 41   c)   −8 decreased by 17

d)   the sum of 13 and −4   e)   8 less than the sum of −11 and −3

Solution

a)   9 more than −2

9 more than a quantity means we *add* 9 to the quantity, in this case, −2.

$$-2 + 9 = 7$$

b)   10 less than 41

10 less than a quantity means we *subtract 10 from* that quantity, in this case, 41.

$$41 - 10 = 31$$

c)   −8 decreased by 17

If −8 is being *decreased by 17*, then we *subtract* 17 *from* −8.

$$-8 - 17 = -8 + (-17)$$
$$= -25$$

d)   the sum of 13 and −4

Sum means *add*. $13 + (-4) = 9$

e)   8 less than the sum of −11 and −3

8 less than means we are subtracting 8 *from* something. From what? From the *sum of −11 and −3.*

Sum means add, so we must find the sum of −11 and −3 and subtract 8 from it.

$$[-11 + (-3)] - 8 = -14 - 8 \qquad \text{First, perform the operation in the brackets.}$$
$$= -14 + (-8) \qquad \text{Change to addition.}$$
$$= -22 \qquad \text{Add.}$$

W Hint

Review inequalities in Section 1.4, and notice the difference between "10 less than 41" and "10 is less than 41."

[YOU TRY 7]

Write a mathematical expression for each and simplify.

a)   the sum of 5 and 7   b)   −14 increased by 6   c)   27 less than 15

d)   20 decreased by 9   e)   8 less than the sum of 24 and −16

ANSWERS TO [YOU TRY] EXERCISES

1) a) 4   b) −5   c) 2   d) −3   2) a) −16   b) −94   c) $-\dfrac{7}{15}$ 3) a) −12   b) 1   c) $-\dfrac{5}{28}$ d) 0

4) a) 17   b) −13   c) −22   d) 12.8   e) $-\dfrac{1}{24}$ 5) 25   6) a) 31   b) $-\dfrac{17}{18}$ c) −10

7) a) 5 + 7; 12   b) −14 + 6; −8   c) 15 − 27; −12   d) 20 − 9; 11   e) [24 + (−16)] − 8; 0

Mixed Exercises: Objectives 1–4 and 6

1) Explain, in your own words, how to subtract two negative numbers.

2) Explain, in your own words, how to add two negative numbers.

3) Explain, in your own words, how to add a positive and a negative number.

Use a number line to represent each sum or difference.

4) $-8 + 5$

5) $6 - 11$

6) $-1 - 5$

7) $-2 + (-7)$

8) $10 + (-6)$

Add or subtract as indicated.

9) $8 + (-15)$

10) $-12 + (-6)$

11) $-3 - 11$

12) $-7 + 13$

13) $-31 + 54$

14) $19 - (-14)$

15) $-26 - (-15)$

16) $-20 - (-30)$

17) $-352 - 498$

18) $217 + (-521)$

19) $-\dfrac{7}{12} + \dfrac{3}{4}$

20) $\dfrac{3}{10} - \dfrac{11}{15}$

21) $-\dfrac{1}{6} - \dfrac{7}{8}$

22) $\dfrac{2}{9} - \left(-\dfrac{2}{5}\right)$

23) $-\dfrac{4}{9} - \left(-\dfrac{4}{15}\right)$

24) $-\dfrac{1}{8} + \left(-\dfrac{3}{4}\right)$

25) $19.4 + (-16.7)$

26) $-31.3 - (-19.82)$

27) $-25.8 - (-16.57)$

28) $7.3 - 21.9$

29) $9 - (5 - 11)$

30) $-2 + (3 - 8)$

31) $-1 + (-6 - 4)$

32) $14 - (-10 - 2)$

33) $(-3 - 1) - (-8 + 6)$

34) $[14 + (-9)] + (1 - 8)$

35) $-16 + 4 + 3 - 10$

36) $8 - 28 + 3 - 7$

37) $5 - (-30) - 14 + 2$

38) $-17 - (-9) + 1 - 10$

39) $\dfrac{4}{9} - \left(\dfrac{2}{3} + \dfrac{5}{6}\right)$

40) $-\dfrac{1}{2} + \left(\dfrac{3}{5} - \dfrac{3}{10}\right)$

41) $\left(\dfrac{1}{8} - \dfrac{1}{2}\right) + \left(\dfrac{3}{4} - \dfrac{1}{6}\right)$

42) $\dfrac{11}{12} - \left(\dfrac{3}{8} - \dfrac{2}{3}\right)$

43) $(2.7 + 3.8) - (1.4 - 6.9)$

44) $-9.7 - (-5.5 + 1.1)$

45) $|7 - 11| + |6 + (-13)|$

46) $|8 - (-1)| - |3 + 12|$

47) $-|-2 - (-3)| - 2|-5 + 8|$

48) $|-6 + 7| + 5|-20 - (-11)|$

Fill in the blank with *always*, *sometimes*, or *never* to make the statement true.

49) The sum of a positive number and a negative number is _____ negative.

50) The sum of any two negative numbers is _____ positive.

Determine whether each statement is true or false. For any real numbers a and b,

51) $|a + b| = |a| + |b|$

52) $|a - b| = |b - a|$

53) $|a + b| = a + b$

54) $|a| + |b| = a + b$

55) $-b - (-b) = 0$

56) $a + (-a) = 0$

Objective 5: Solve Applied Problems

Write an expression for each and simplify. Answer the question with a complete sentence.

57) Tiger Woods won his first Masters championship in 1997 at age 21 with a score of -18. When he won the championship in 2019, his score was 5 strokes higher. What was Tiger's score when he won the Masters in 2019? (www.masters.com)

58) In 2014, Abraham Lincoln's Birthplace National Historical Park recorded 241,264 visits, while in 2015 there were 148,605 visits. What was the difference in the number of visits from 2014 to 2015? (irma.nps.gov)

59) In 2016, China's carbon emissions were 10.43 billion metric tons and the carbon emissions of the United States totaled 5.01 billion metric tons. By how much did China's carbon emissions exceed those of the United States? (http://edgar.jrc.ec.europa.eu)

60) In January 2016, San Francisco had 5.79 in. of rain. This was 3.63 in. less than in January 2017. How much rain fell in San Francisco in January 2017? (www.usclimatedata.com)

61) From 2016 to 2017, the number of flights going through O'Hare Airport in Chicago decreased by 586. There were 867,049 flights in 2017. How many flights went through O'Hare in 2016? (www.chicagotribune.com)

62) The lowest temperature ever recorded in Minnesota was −60°F while the highest temperature on record was 175° greater than that. What was the warmest temperature ever recorded in Minnesota? (www.ncdc.noaa.gov)

63) The bar graph shows the total number of daily newspapers in the United States in various years. Use a signed number to represent the change in the number of dailies over the given years. (www.naa.org)

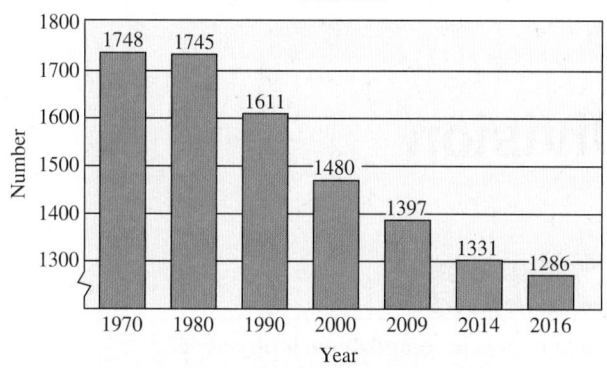

Number of Daily Newspapers in the U.S.

a) 1970–1980   b) 1980–1990

c) 1990–2009   d) 2000–2016

64) The bar graph shows the TV ratings for the World Series over a 5-year period. Each ratings number represents the percentage of people watching TV at the time of the World Series who were tuned into the games. Use a signed number to represent the change in ratings over the given years. (www.baseball-almanac.com and www.forbes.com)

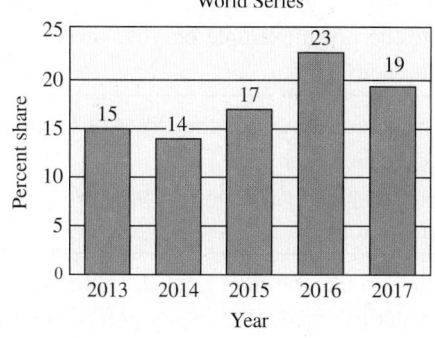

TV Ratings for the World Series

a) 2013–2014   b) 2015–2016

c) 2014–2015   d) 2016–2017

65) The bar graph shows the average number of days a woman was in the hospital for childbirth. Use a signed number to represent the change in hospitalization time over the given years. (www.cdc.gov, www.businessinsider.com)

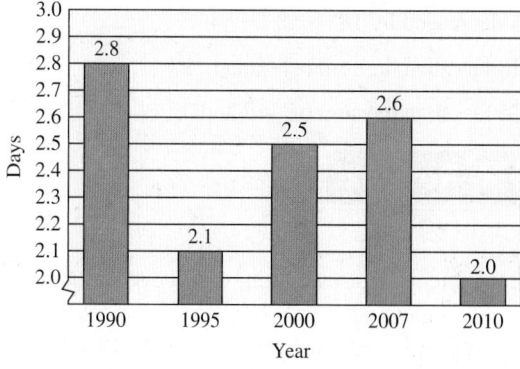

Average Hospital Stay

a) 1990–1995   b) 1995–2000

c) 2000–2010   d) 1990–2010

66) The bar graph shows snowfall totals for different seasons in Syracuse, NY. Use a signed number to represent the difference in snowfall totals over different years. (www.erh.noaa.gov)

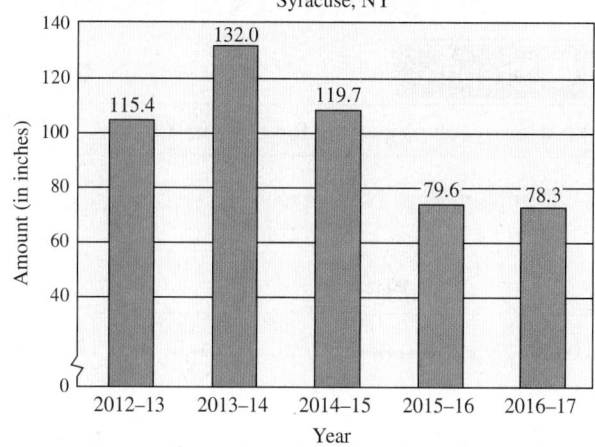

Snowfall Totals for Syracuse, NY

a) 2014–15 to 2015–16   b) 2015–16 to 2016–17

c) 2012–13 to 2013–14   d) 2013–14 to 2014–15

Objective 7: Translate English Expressions into Mathematical Expressions

Write a mathematical expression for each and simplify.

67) 7 more than 5

68) 3 more than 11

69) 16 less than 10

70) 15 less than 4

71) −8 less than 9

72) −25 less than −19

73) The sum of −21 and 13

74) The sum of −7 and 20

75) −20 increased by 30

76) −37 increased by 22

77) 23 decreased by 19

78) 8 decreased by 18

79) 18 less than the sum of −5 and 11

80) 35 less than the sum of −17 and 3

Write a statement, in English, that translates to the given expression.

81) 19 + (−4)

82) −12 − 15

83) (2 + 7) − 6

84) (−8 + 1) + 10

 **R Rethink**

R1) Which exercises in this section do you find most challenging?

R2) Which is easier for you, 7 − 9 or 7 + (−9)? Why?

R3) Where have you used the addition of signed numbers outside of your math class?

R4) If the absolute value of a number is greater than the absolute value of another, how do you determine whether their difference is positive or negative?

1.6 Multiplication and Division of Real Numbers

P Prepare

O Organize

What are your objectives for Section 1.6?	How can you accomplish each objective?
1 Multiply Real Numbers	• Write the procedure for **Multiplying Real Numbers** in your own words. • Complete the given examples on your own. • Complete You Trys 1 and 2.
2 Evaluate Exponential Expressions	• Create a procedure to help you quickly determine whether an exponential expression will give a positive or negative result. • Complete the given examples on your own. • Complete You Trys 3 and 4.
3 Divide Real Numbers	• Write the procedure for **Dividing Real Numbers** in your own words. • Complete the given example on your own. • Complete You Try 5.
4 Apply the Order of Operations	• Add any additional steps to the Order of Operations you updated in Section 1.4. • Complete the given example on your own. • Complete You Try 6.
5 Translate English Expressions into Mathematical Expressions	• Take notes on the common English expressions, and be able to identify whether an expression refers to multiplication or division. • Complete the given example on your own. • Complete You Try 7.

1 Multiply Real Numbers

What is the meaning of $4 \cdot 5$? It is repeated addition.

$$4 \cdot 5 = 4 + 4 + 4 + 4 + 4 = 20$$

So, what is the meaning of $-4 \cdot 5$? It, too, represents repeated addition.

$$-4 \cdot 5 = -4 + (-4) + (-4) + (-4) + (-4) = -20$$

Let's make a table of some products:

×	5	4	③	2	1	0	−1	−2	−3	−4	−5
④	20	16	12	8	4	0	−4	−8	−12	−16	−20

$4 \cdot 3 = 12$

The bottom row represents the product of 4 and the number above it ($4 \cdot 3 = 12$). Notice that as the numbers in the first row decrease by 1, the numbers in the bottom row decrease by 4. Therefore, once we get to $4 \cdot (-1)$, the product is negative. From the table we can see that the product of a positive number and a negative number is negative.

> **Note**
> The product of a positive number and a negative number is negative.

EXAMPLE 1

Multiply.

a) $-6 \cdot 9$ b) $\dfrac{3}{8} \cdot (-12)$ c) $-5 \cdot 0$

Solution

a) $-6 \cdot 9 = -54$ b) $\dfrac{3}{8} \cdot (-12) = \dfrac{3}{8} \cdot \left(-\dfrac{\overset{3}{\cancel{12}}}{1} \right) = -\dfrac{9}{2}$

c) $-5 \cdot 0 = 0$ The product of zero and any real number is zero.

[YOU TRY 1]

Multiply.

a) $-7 \cdot 3$ b) $\dfrac{8}{15} \times (-10)$

What is the sign of the product of two negative numbers? Again, we'll make a table.

×	3	2	1	0	−1	−2	−3
−4	−12	−8	−4	0	4	8	12

As the numbers in the top row decrease by 1, the numbers in the bottom row *increase* by 4. When we reach $-4 \cdot (-1)$, our product is a positive number, 4. The table illustrates that the product of two negative numbers is positive.

We can summarize our findings this way:

> **Procedure** Multiplying Real Numbers
>
> 1) The product of two positive numbers is positive.
> 2) The product of two negative numbers is positive.
> 3) The product of a positive number and a negative number is negative.
> 4) The product of any real number and zero is zero.

EXAMPLE 2

Multiply.

a) $-8 \cdot (-5)$ b) $-1.5 \cdot 6$ c) $-\dfrac{3}{8} \cdot \left(-\dfrac{4}{5}\right)$ d) $-5 \cdot (-2) \cdot (-3)$

Solution

a) $-8 \cdot (-5) = 40$ The product of two negative numbers is positive.

b) $-1.5 \cdot 6 = -9$ The product of a negative number and a positive number is negative.

c) $-\dfrac{3}{8} \cdot \left(-\dfrac{4}{5}\right) = -\dfrac{3}{\overset{}{8}} \cdot \left(-\dfrac{\overset{1}{4}}{5}\right) = \dfrac{3}{10}$ The product of two negatives is positive.

d) $\underbrace{-5 \cdot (-2)}_{10} \cdot (-3) = 10 \cdot (-3) = -30$ Order of operations—multiply from left to right.

[YOU TRY 2]

Multiply.

a) $-6 \cdot 7$ b) $-2.9 \times (-5)$ c) $-4 \cdot (-1) \cdot (-5) \cdot (-2)$

> **Note**
> It is helpful to know that
> 1) an **even number** of negative factors in a product gives a positive result.
> $$-3 \cdot 1 \cdot (-2) \cdot (-1) \cdot (-4) = 24 \qquad \text{Four negative factors}$$
> 2) an **odd number** of negative factors in a product gives a negative result.
> $$5 \cdot (-3) \cdot (-1) \cdot (-2) \cdot (3) = -90 \qquad \text{Three negative factors}$$

2 Evaluate Exponential Expressions

In Section 1.2, we discussed exponential expressions. Recall that exponential notation is a shorthand way to represent repeated multiplication:

$$2^4 = 2 \cdot 2 \cdot 2 \cdot 2 = 16$$

Now we will discuss exponents and negative numbers. Consider a base of -2 raised to different powers. (The -2 is in parentheses to indicate that it is the base.)

$$(-2)^1 = -2$$
$$(-2)^2 = -2 \cdot (-2) = 4$$
$$(-2)^3 = -2 \cdot (-2) \cdot (-2) = -8$$
$$(-2)^4 = -2 \cdot (-2) \cdot (-2) \cdot (-2) = 16$$
$$(-2)^5 = -2 \cdot (-2) \cdot (-2) \cdot (-2) \cdot (-2) = -32$$
$$(-2)^6 = -2 \cdot (-2) \cdot (-2) \cdot (-2) \cdot (-2) \cdot (-2) = 64$$

 **Hint**

A negative number raised to an *odd* power will give a *negative* result. A negative number raised to an *even* power will give a *positive* result.

Do you notice that

1) -2 raised to an *odd* power gives a negative result?

and

2) -2 raised to an *even* power gives a positive result?

This will always be true.

EXAMPLE 3

Evaluate.

a) $(-6)^2$ b) $(-10)^3$

Solution

a) $(-6)^2 = 36$ b) $(-10)^3 = -1000$

YOU TRY 3

Evaluate.

a) $(-9)^2$ b) $(-5)^3$

 **Hint**

The order of operations tells us to evaluate exponents before multiplying. Therefore, to evaluate -2^4, first we evaluate the base using the exponent, then we multiply by -1.

How do $(-2)^4$ and -2^4 differ? Let's identify their bases and evaluate each.

$(-2)^4$:   Base $= -2$ $(-2)^4 = 16$

-2^4:   Since there are no parentheses,

-2^4 is equivalent to $-1 \cdot 2^4$. Therefore, the base is 2.
$$-2^4 = -1 \cdot 2^4$$
$$= -1 \cdot 2 \cdot 2 \cdot 2 \cdot 2$$
$$= -16$$

So, $(-2)^4 = 16$ and $-2^4 = -16$.

 **BE CAREFUL**

When working with exponential expressions, be able to identify the base.

EXAMPLE 4

Evaluate.

a) $(-5)^3$ b) -9^2 c) $\left(-\dfrac{1}{7}\right)^2$

Solution

a) $(-5)^3$: Base $= -5$ $(-5)^3 = -5 \cdot (-5) \cdot (-5) = -125$

b) -9^2: Base $= 9$ $-9^2 = -1 \cdot 9^2$
$$= -1 \cdot 9 \cdot 9$$
$$= -81$$

c) $\left(-\dfrac{1}{7}\right)^2$: Base $= -\dfrac{1}{7}$ $\left(-\dfrac{1}{7}\right)^2 = -\dfrac{1}{7} \cdot \left(-\dfrac{1}{7}\right) = \dfrac{1}{49}$

[YOU TRY 4] Evaluate.

a) -3^4 b) $(-11)^2$ c) -8^2 d) $-\left(-\dfrac{2}{3}\right)^3$

---

3 Divide Real Numbers

Here are the rules for dividing signed numbers:

> **Procedure** Dividing Signed Numbers
>
> 1) The quotient of two positive numbers is a positive number.
> 2) The quotient of two negative numbers is a positive number.
> 3) The quotient of a positive and a negative number is a negative number.

EXAMPLE 5

Divide.

a) $-36 \div 9$ b) $-\dfrac{1}{10} \div \left(-\dfrac{3}{5}\right)$ c) $\dfrac{-8}{-1}$ d) $\dfrac{-24}{42}$

Solution

a) $-36 \div 9 = -4$

b) $-\dfrac{1}{10} \div \left(-\dfrac{3}{5}\right) = -\dfrac{1}{10} \cdot \left(-\dfrac{5}{3}\right)$ When dividing by a fraction, multiply by the reciprocal.

$$= -\dfrac{1}{\overset{}{10}} \cdot \left(-\dfrac{\overset{1}{5}}{3}\right) = \dfrac{1}{6}$$
$$\underset{2}{}$$

W Hint
Remember, write all
answers in lowest terms.

c) $\dfrac{-8}{-1} = 8$ The quotient of two negative numbers is positive, and $\dfrac{8}{1}$ simplifies to 8.

d) $\dfrac{-24}{42} = -\dfrac{24}{42}$ The quotient of a negative number and a positive number is negative, so simplify $\dfrac{24}{42}$.

$$= -\dfrac{4}{7}$$ 24 and 42 each divide by 6.

It is important to note here in part d) that there are three ways to write the answer: $-\frac{4}{7}, \frac{-4}{7},$ or $\frac{4}{-7}$. These are equivalent. However, we usually write the negative sign in front of the entire fraction as in $-\frac{4}{7}$.

[YOU TRY 5]   Divide.

a) $-42 \div 6$ b) $-\frac{8}{5} \div \left(-\frac{6}{5}\right)$ c) $\frac{-30}{-10}$ d) $\frac{21}{-56}$

---

4 Apply the Order of Operations

EXAMPLE 6

Simplify.

a) $-24 \div 12 - 2^2$ b) $-5(-3) - 4(2-3)$

Solution

a) $-24 \div 12 - 2^2 = -24 \div 12 - 4$ Simplify exponent first.
                     $= -2 - 4$ Perform division before subtraction.
                     $= -6$

b) $-5(-3) - 4(2-3) = -5(-3) - 4(-1)$ Simplify the difference in parentheses.
                          $= 15 - (-4)$ Find the products.
                          $= 15 + 4$ Change subtraction to addition.
                          $= 19$

[YOU TRY 6]   Simplify.

a) $-18 + 45 \div (-3)^2$ b) $(-10)^2 + 2[8 - 5(4)]$

---

5 Translate English Expressions into Mathematical Expressions

Here are some words and phrases you may encounter and how they would translate into mathematical expressions:

English Expression	Mathematical Operation
times, product of	multiplication
divided by, quotient of	division

EXAMPLE 7

Write a mathematical expression for each and simplify.

a) The quotient of -56 and 7

b) The product of 4 and the sum of 15 and -6

c) Twice the difference of -10 and -3

d) Half of the sum of -8 and 3

Solution

a) The quotient of −56 and 7:
   Quotient means division with −56 in the numerator and 7 in the denominator.
   The expression is $\dfrac{-56}{7} = -8$.

b) The product of 4 and the sum of 15 and −6:
   The *sum of* 15 *and* −6 means we must add the two numbers. *Product* means multiply.

$$\underbrace{4\overbrace{[15 + (-6)]}^{\substack{\text{Sum of 15}\\\text{and }-6}}}_{\substack{\text{Product of 4}\\\text{and the sum}}} = 4(9) = 36$$

W Hint

In these examples, notice how the word *and* is being used and notice how the number in front of the word *and* is placed into the expression.

c) Twice the difference of −10 and −3:
   The *difference of* −10 *and* −3 will be in parentheses with −3 being subtracted from −10. *Twice* means "two times."

$$2[-10 - (-3)] = 2(-10 + 3) = 2(-7) = -14$$

d) Half of the sum of −8 and 3:
   The *sum of* −8 *and* 3 means that we will add the two numbers. They will be in parentheses. *Half of* means multiply by $\dfrac{1}{2}$.

$$\frac{1}{2}(-8 + 3) = \frac{1}{2}(-5) = -\frac{5}{2}$$

[YOU TRY 7] Write a mathematical expression for each and simplify.

a) The quotient of 18 and 3

b) 12 less than the product of −7 and 4

c) The sum of −41 and −23, divided by the square of −2

ANSWERS TO [YOU TRY] EXERCISES

1) a) −21 b) $-\dfrac{16}{3}$ 2) a) −42 b) 14.5 c) 40   3) a) 81 b) −125

4) a) −81 b) 121 c) −64 d) $\dfrac{8}{27}$ 5) a) −7 b) $\dfrac{4}{3}$ c) 3 d) $-\dfrac{3}{8}$ 6) a) −13 b) 76

7) a) $\dfrac{18}{3}$; 6 b) $(-7) \cdot 4 - 12$; −40 c) $\dfrac{-14 + (-23)}{(-2)^2}$; −16

E Evaluate **1.6** Exercises   Do the exercises, and check your work.

Objective 1: Multiply Real Numbers

Fill in the blank with *positive* or *negative*.

1) The product of a positive number and a negative number is _____.

2) The product of two negative numbers is _____.

Multiply.

3) −8 · 7

4) 4 · (−9)

5) −15 · (−3)

6) −23 · (−48)

7) −4 · 3 · (−7)

8) −5 · (−1) · (−11)

9) $\dfrac{4}{33} \cdot \left(-\dfrac{11}{10}\right)$

10) $-\dfrac{14}{27} \cdot \left(-\dfrac{15}{28}\right)$

11) $(-0.5)(-2.8)$

12) $(-6.1)(5.7)$

13) $-9 \cdot (-5) \cdot (-1) \cdot (-3)$

14) $-1 \cdot (-6) \cdot (4) \cdot (-2) \cdot (3)$

15) $\dfrac{3}{10} \cdot (-7) \cdot (8) \cdot (-1) \cdot (-5)$

16) $-\dfrac{5}{6} \cdot (-4) \cdot 0 \cdot 3$

Objective 2: Evaluate Exponential Expressions

17) For what values of k is k^5 a negative quantity?

18) For what values of k is k^5 a positive quantity?

19) For what values of k is $-k^2$ a negative quantity?

20) Explain the difference between how you would evaluate $(-8)^2$ and -8^2. Then, evaluate each.

Evaluate.

21) $(-6)^2$

22) -6^2

23) -5^3

24) $(-2)^4$

25) $(-3)^2$

26) $(-1)^5$

27) -7^2

28) -4^3

29) -2^5

30) $(-12)^2$

Objective 3: Divide Real Numbers

Fill in the blank with *positive* or *negative*.

31) The quotient of two negative numbers is _____.

32) The quotient of a negative number and a positive number is _____.

Divide.

33) $-50 \div (-5)$

34) $-84 \div 12$

35) $\dfrac{64}{-16}$

36) $\dfrac{-54}{-9}$

37) $\dfrac{-2.4}{0.3}$

38) $\dfrac{16}{-0.5}$

39) $-\dfrac{12}{13} \div \left(-\dfrac{6}{5}\right)$

40) $20 \div \left(-\dfrac{15}{7}\right)$

41) $-\dfrac{0}{7}$

42) $\dfrac{0}{-6}$

43) $\dfrac{270}{-180}$

44) $\dfrac{-64}{-320}$

Objective 4: Apply the Order of Operations

Use the order of operations to simplify.

45) $7 + 8(-5)$

46) $-40 \div 2 - 10$

47) $(9 - 14)^2 - (-3)(6)$

48) $-23 - 6^2 \div 4$

49) $10 - 2(1 - 4)^3 \div 9$

50) $-7(4) + (-8 + 6)^4 + 5$

51) $\left(-\dfrac{3}{4}\right)(8) - 2[7 - (-3)(-6)]$

52) $-2^5 - (-3)(4) + 5[(-9 + 30) \div 7]$

53) $\dfrac{-46 - 3(-12)}{(-5)(-2)(-4)}$

54) $\dfrac{(8)(-6) + 10 - 7}{(-5 + 1)^2 - 12 + 5}$

Objective 5: Translate English Expressions into Mathematical Expressions

Write a mathematical expression for each and simplify.

55) The product of -12 and 6

56) The quotient of -80 and -4

57) 9 more than the product of -7 and -5

58) The product of -10 and 2 increased by 11

59) The quotient of 63 and -9 increased by 7

60) 8 more than the quotient of 54 and -6

61) 19 less than the product of -4 and -8

62) The product of -16 and -3 decreased by 20

63) The quotient of -100 and 4 decreased by the sum of -7 and 2

64) The quotient of -35 and 5 increased by the product of -11 and -2

65) Twice the sum of 18 and -31

66) Twice the difference of -5 and -15

67) Two-thirds of -27

68) Half of -30

69) The product of 12 and -5 increased by half of 36

70) One-third of -18 decreased by half the sum of -21 and -5

Write a statement, in English, that translates to the given expression.

71) $\dfrac{34}{17}$

72) $-8 \cdot 7$

73) $6 \cdot (-10) - 3$

74) $\dfrac{72}{-12} + 11$

R1) What are the similarities between multiplying and dividing signed numbers?

R2) In Objective 2, why is the use of parentheses so important?

R3) In Objective 5, think about the order in which the integer values are given. How does this affect their placement into the mathematical expression?

1.7 Algebraic Expressions and Properties of Real Numbers

P Prepare

O Organize

What are your objectives for Section 1.7?	How can you accomplish each objective?
1 Identify the Terms and Coefficients in an Expression	• Write the definitions of *term, constant term,* and *coefficient* in your own words. • Complete the given example on your own. • Complete You Try 1.
2 Evaluate Algebraic Expressions	• Write the definitions of *variable* and *algebraic expression* in your own words. • Complete the given examples on your own. • Complete You Trys 2 and 3.
3 Identify Like Terms	• Write the definition of *like terms* in your own words. • Complete the given example on your own. • Complete You Try 4.
4 Use the Commutative Properties	• Learn the **commutative properties** for addition and multiplication. • Complete the given example on your own. • Complete You Try 5.
5 Use the Associative Properties	• Learn the **associative properties** for addition and multiplication. • Complete the given examples on your own. • Complete You Trys 6 and 7.
6 Use the Identity and Inverse Properties	• Know that 0 is the additive identity and 1 is the multiplicative identity, and use that information to develop the **identity properties** in your notes. • Know how to find an additive inverse and a reciprocal or multiplicative inverse, and use that information to develop the **inverse properties** in your notes. • Complete the given example on your own. • Complete You Try 8.

What are your objectives for Section 1.7?	How can you accomplish each objective?
7 Use the Distributive Property	• Learn the **distributive properties.** • Complete the given examples on your own. • Complete You Trys 9 and 10.
8 Combine Like Terms	• Use the **distributive property** to combine *like terms*. • Complete the given examples on your own. • Complete You Trys 11 and 12.
9 Translate English Expressions into Mathematical Expressions	• Use the tools you learned from the previous sections to create a procedure from Example 13. • Complete You Try 13.

 **Work**   Read the explanations, follow the examples, take notes, and complete the You Trys.

1 Identify the Terms and Coefficients in an Expression

Here is an algebraic expression: $8x^3 - 5x^2 + \dfrac{2}{7}x + 4$

x is the *variable*. A **variable** is a symbol, usually a letter, used to represent an unknown number. This expression contains four terms. A **term** is a number or a variable or a product or quotient of numbers and variables. 4 is the **constant** or **constant term.** The value of a constant does not change. Each term has a **coefficient.**

Term	Coefficient
$8x^3$	8
$-5x^2$	-5
$\dfrac{2}{7}x$	$\dfrac{2}{7}$
4	4

Definition

An **algebraic expression** is a collection of numbers, variables, and grouping symbols connected by operation symbols such as $+$, $-$, $\times$, and $\div$.

Examples of expressions: $3c + 4,$ $9(p^2 - 7p - 2),$ $-4a^2b^2 + 5ab - 8a + 1$

EXAMPLE 1   List the terms and coefficients of $4x^2y + 7xy - x + \dfrac{y}{9} - 12.$

Solution

Term	Coefficient
$4x^2y$	4
$7xy$	7
$-x$	-1
$\dfrac{y}{9}$	$\dfrac{1}{9}$
-12	-12

The minus sign indicates a negative coefficient.

$\dfrac{y}{9}$ can be rewritten as $\dfrac{1}{9}y$.

-12 is also called the *constant.*

2 Evaluate Algebraic Expressions

We can **evaluate** an algebraic expression by substituting a value for a variable and simplifying. The value of an algebraic expression changes depending on the value that is substituted.

EXAMPLE 2

Evaluate $2x + 5$ when a) $x = 7$ and b) $x = -3$.

Solution

a) $2x + 5$ when $x = 7$ Substitute 7 for x.

$= 2(7) + 5$ Use parentheses when substituting a value for a variable.

$= 14 + 5$ Multiply.

$= 19$ Add.

Ⓦ Hint

A good strategy for reading a math book is to write out the example as you are reading it.

b) $2x + 5$ when $x = -3$ Substitute -3 for x.

$= 2(-3) + 5$ Use parentheses when substituting a value for a variable.

$= -6 + 5$ Multiply.

$= -1$ Add.

[**YOU TRY 2**] Evaluate $8t + 1$ when a) $t = 4$ and b) $t = -5$.

EXAMPLE 3

Evaluate $5x^2 - xy - 11$ when $x = -2$ and $y = 3$.

Solution

$5x^2 - xy - 11 = 5(-2)^2 - (-2)(3) - 11$ Substitute -2 for x and 3 for y.

$= 5 \cdot 4 - (-6) - 11$ Evaluate exponent; multiply.

$= 20 + 6 - 11$ Multiply.

$= 15$

[**YOU TRY 3**] Evaluate $6u^2 - 5uv + 4$ when $u = -1$ and $v = 4$.

3 Identify Like Terms

In the expression $15a + 11a - 8a + 3a$, there are four **terms:** $15a$, $11a$, $-8a$, $3a$. In fact, they are **like terms.** *Like terms contain the same variables with the same exponents.*

EXAMPLE 4

Determine whether the following groups of terms are like terms.

a) $4y^2, -9y^2, \dfrac{2}{3}y^2$

b) $-5x^6, 0.8x^9, 3x^4$

c) $6a^2b^3, a^2b^3, -\dfrac{5}{8}a^2b^3$

d) $9c, 4d$

Solution

a) $4y^2, -9y^2, \dfrac{2}{3}y^2$

Yes. Each contains the variable y with an exponent of 2. They are y^2-terms.

b) $-5x^6, 0.8x^9, 3x^4$

No. Although each contains the variable x, the exponents are not the same.

c) $6a^2b^3, a^2b^3, -\dfrac{5}{8}a^2b^3$

Yes. Each term contains the product of a^2 and b^3.

d) $9c, 4d$

No. The terms contain different variables.

YOU TRY 4

Determine whether the following groups of terms are like terms.

a) $2k^2, -9k^2, \dfrac{1}{5}k^2$

b) $-xy^2, 8xy^2, 7xy^2$

c) $3r^3s^2, -10r^2s^3$

Next, we will learn about the properties of real numbers so that we can use them to work with algebraic expressions.

Like the order of operations, the properties of real numbers guide us in our work with numbers and variables. We begin with the commutative properties of real numbers.

4 Use the Commutative Properties

True or false?

1) $7 + 3 = 3 + 7$ *True:* $7 + 3 = 10$ and $3 + 7 = 10$

2) $8 - 2 = 2 - 8$ *False:* $8 - 2 = 6$ but $2 - 8 = -6$

3) $(-6)(5) = (5)(-6)$ *True:* $(-6)(5) = -30$ and $(5)(-6) = -30$

In 1), we see that adding 7 and 3 in any order still equals 10. The third equation shows that multiplying $(-6)(5)$ and $(5)(-6)$ both equal -30. But, 2) illustrates that changing the order in which numbers are subtracted does *not* necessarily give the same result: $8 - 2 \neq 2 - 8$.

Therefore, subtraction is **not commutative,** while the addition and multiplication of real numbers **are commutative.** This gives us our first properties of real numbers.

Property Commutative Properties

If a and b are real numbers, then

   1) $a + b = b + a$ Commutative property of addition

   2) $ab = ba$ Commutative property of multiplication

We have already shown that subtraction is not commutative. Is division commutative? No. For example,

$$20 \div 4 \overset{?}{=} 4 \div 20$$

$$5 \neq \frac{1}{5}$$

EXAMPLE 5   Use the commutative property to rewrite each expression.

  a) $12 + 5$ b) $k \cdot 3$

Solution

  a) $12 + 5 = 5 + 12$ b) $k \cdot 3 = 3 \cdot k$ or $3k$

[YOU TRY 5]   Use the commutative property to rewrite each expression.

  a) $1 + 16$ b) $n \cdot 6$

5   Use the Associative Properties

Another important property involves the use of grouping symbols. Let's determine whether these two statements are true:

$$(9 + 4) + 2 \overset{?}{=} 9 + (4 + 2) \qquad\qquad (2 \cdot 3)4 \overset{?}{=} 2(3 \cdot 4)$$
$$13 + 2 \overset{?}{=} 9 + 6 \qquad \text{and} \qquad (6)4 \overset{?}{=} 2(12)$$
$$15 = 15 \qquad\qquad\qquad\qquad 24 = 24$$
$$\text{TRUE} \qquad\qquad\qquad\qquad \text{TRUE}$$

We can generalize and say that when adding or multiplying real numbers, the way in which we group them to evaluate them will not affect the result. Notice that the *order* in which the numbers are written does not change.

Property Associative Properties

If a, b, and c are real numbers, then

   1) $(a + b) + c = a + (b + c)$ Associative property of addition

   2) $(ab)c = a(bc)$ Associative property of multiplication

Sometimes, applying the associative property can simplify calculations.

EXAMPLE 6

Apply the associative property to simplify $\left(7 \cdot \dfrac{2}{5}\right)5$.

Solution

By the associative property, $\left(7 \cdot \dfrac{2}{5}\right)5 = 7 \cdot \left(\dfrac{2}{\cancel{5}} \cdot \cancel{5}\right) = 7 \cdot 2 = 14$

[YOU TRY 6]

Apply the associative property to simplify $\left(9 \cdot \dfrac{4}{3}\right)3$.

EXAMPLE 7

Use the associative property to simplify each expression.

a) $-6 + (10 + y)$ b) $\left(-\dfrac{3}{11} \cdot \dfrac{8}{5}\right)\dfrac{5}{8}$

Solution

a) $-6 + (10 + y) = (-6 + 10) + y = 4 + y$

b) $\left(-\dfrac{3}{11} \cdot \dfrac{8}{5}\right)\dfrac{5}{8} = -\dfrac{3}{11}\left(\dfrac{8}{5} \cdot \dfrac{5}{8}\right) = -\dfrac{3}{11}(1) = -\dfrac{3}{11}$ A number times its reciprocal equals 1.

[YOU TRY 7]

Use the associative property to simplify each expression.

a) $(k + 3) + 9$ b) $\left(-\dfrac{9}{7} \cdot \dfrac{8}{5}\right)\dfrac{5}{8}$

The identity properties of addition and multiplication are also ones we need to know.

6 Use the Identity and Inverse Properties

For addition we know that, for example,

$$5 + 0 = 5, \qquad 0 + \dfrac{2}{3} = \dfrac{2}{3}, \qquad -14 + 0 = -14$$

When zero is added to a number, the value of the number is unchanged. *Zero* is the **identity element for addition** (also called the **additive identity**).
   What is the identity element for multiplication?

$$-4(1) = -4 \qquad 1(3.82) = 3.82 \qquad \dfrac{9}{2}(1) = \dfrac{9}{2}$$

When a number is multiplied by 1, the value of the number is unchanged. *One* is the **identity element for multiplication** (also called the **multiplicative identity**).

W Hint

Notice that the identity property of addition uses the number 0, and that the identity property of multiplication uses the number 1.

Property Identity Properties

If a is a real number, then

    1) $a + 0 = 0 + a = a$ Identity property of addition

    2) $a \cdot 1 = 1 \cdot a = a$ Identity property of multiplication

The next properties we will discuss give us the additive and multiplicative identities as results. In Section 1.4, we introduced an **additive inverse.**

Number	Additive Inverse
3	−3
−11	11
$-\dfrac{7}{9}$	$\dfrac{7}{9}$

Let's add each number and its additive inverse:

$$3 + (-3) = 0, \qquad\qquad -11 + 11 = 0, \qquad\qquad -\frac{7}{9} + \frac{7}{9} = 0$$

Given a number such as $\dfrac{3}{5}$, we know that its **reciprocal** (or **multiplicative inverse**) is $\dfrac{5}{3}$. We have also established the fact that the product of a number and its reciprocal is 1 as in $\dfrac{3}{5} \cdot \dfrac{5}{3} = 1$.

Therefore, multiplying a number b by its reciprocal (multiplicative inverse) $\dfrac{1}{b}$ gives us the identity element for multiplication, 1. That is,

$$b \cdot \frac{1}{b} = \frac{1}{b} \cdot b = 1$$

Property Inverse Properties

If a is any real number and b is a real number not equal to 0, then

    1) $a + (-a) = -a + a = 0$ Inverse property of addition

    2) $b \cdot \dfrac{1}{b} = \dfrac{1}{b} \cdot b = 1$ Inverse property of multiplication

W Hint

Notice that the inverse property for addition has a result of 0, and the inverse property for multiplication gives a result of 1.

EXAMPLE 8   Which property is illustrated by each statement?

    a) $0 + 12 = 12$ b) $-9.4 + 9.4 = 0$

    c) $\dfrac{1}{7} \cdot 7 = 1$ d) $2(1) = 2$

Solution

a) $0 + 12 = 12$ Identity property of addition

b) $-9.4 + 9.4 = 0$ Inverse property of addition

c) $\dfrac{1}{7} \cdot 7 = 1$ Inverse property of multiplication

d) $2(1) = 2$ Identity property of multiplication

[YOU TRY 8] Which property is illustrated by each statement?

a) $5 \cdot \dfrac{1}{5} = 1$ b) $-26 + 26 = 0$ c) $2.7(1) = 2.7$ d) $-4 + 0 = -4$

---

7 Use the Distributive Property

The last property we will discuss is the **distributive property.** It involves both multiplication and addition or multiplication and subtraction.

> **Property** Distributive Properties
>
> If a, b, and c are real numbers, then
>
> 1) $a(b + c) = ab + ac$ and $(b + c)a = ba + ca$
> 2) $a(b - c) = ab - ac$ and $(b - c)a = ba - ca$

EXAMPLE 9   Evaluate using the distributive property.

a) $3(2 + 8)$ b) $-(6 + 3)$

Solution

a) $3(2 + 8) = 3 \cdot 2 + 3 \cdot 8$ Apply distributive property

$= 6 + 24$

$= 30$

Note that we would get the same result if we would apply the order of operations:

$$3(2 + 8) = 3(10) = 30$$

[W] Hint

Write out each example as you are reading it!

b) $-(6 + 3) = -1(6 + 3)$

$= -1 \cdot 6 + (-1)(3)$ Apply distributive property.

$= -6 + (-3)$

$= -9$

A negative sign in front of parentheses is the same as multiplying by -1.

[YOU TRY 9] Evaluate using the distributive property.

a) $2(11 - 5)$ b) $-5(3 - 7)$ c) $-(4 + 9)$

The distributive property can be applied when there are more than two terms in parentheses and when there are variables.

EXAMPLE 10

Use the distributive property to rewrite each expression. Simplify if possible.

a) $7(x + 4)$ b) $-(-5c + 4d - 6)$

Solution

a) $7(x + 4) = 7x + 7 \cdot 4$ Apply distributive property.
$$= 7x + 28$$

b) $-(-5c + 4d - 6) = -1(-5c + 4d - 6)$
$$= -1(-5c) + (-1)(4d) - (-1)(6) \quad \text{Apply distributive property.}$$
$$= 5c + (-4d) - (-6) \quad\quad\quad\quad \text{Multiply.}$$
$$= 5c - 4d + 6$$

[YOU TRY 10]

Use the distributive property to rewrite each expression. Simplify if possible.

a) $6(a + 2)$ b) $5(2x - 7y - 4z)$ c) $-(-r + 4s - 9)$

The properties stated previously are summarized next.

Summary Properties of Real Numbers

If a, b, and c are real numbers, then

Commutative properties:	$a + b = b + a$ and $ab = ba$
Associative properties:	$(a + b) + c = a + (b + c)$ and $(ab)c = a(bc)$
Identity properties:	$a + 0 = 0 + a = a$
	$a \cdot 1 = 1 \cdot a = a$
Inverse properties:	$a + (-a) = -a + a = 0$
	$b \cdot \dfrac{1}{b} = \dfrac{1}{b} \cdot b = 1 \; (b \neq 0)$
Distributive properties:	$a(b + c) = ab + ac$ and $(b + c)a = ba + ca$
	$a(b - c) = ab - ac$ and $(b - c)a = ba - ca$

8 Combine Like Terms

To simplify an expression like $15a + 11a - 8a + 3a$, we combine like terms using the distributive property.

$$15a + 11a - 8a + 3a = (15 + 11 - 8 + 3)a \quad \text{Distributive property}$$
$$= (26 - 8 + 3)a \quad\quad \text{Order of operations}$$
$$= (18 + 3)a \quad\quad\quad \text{Order of operations}$$
$$= 21a$$

We can add and subtract only those terms that are like terms.

EXAMPLE 11   Combine like terms.

a) $-9k + 2k$ b) $n + 8 - 4n + 3$ c) $\dfrac{3}{5}t^2 + \dfrac{1}{4}t^2$

d) $10x^2 + 6x - 2x^2 + 5x$

Solution

a) We can use the distributive property to combine like terms.

$$-9k + 2k = (-9 + 2)k = -7k$$

Notice that using the distributive property to combine like terms is the same as combining the coefficients of the terms and leaving the variable and its exponent the same.

b) $n + 8 - 4n + 3 = n - 4n + 8 + 3$ Rewrite like terms together.

$\qquad\qquad\qquad\quad = -3n + 11$ Remember, n is the same as $1n$.

c) $\dfrac{3}{5}t^2 + \dfrac{1}{4}t^2 = \dfrac{12}{20}t^2 + \dfrac{5}{20}t^2$ Get a common denominator.

$\qquad\qquad\quad = \dfrac{17}{20}t^2$

d) $10x^2 + 6x - 2x^2 + 5x = 10x^2 - 2x^2 + 6x + 5x$ Rewrite like terms together.

$\qquad\qquad\qquad\qquad\quad = 8x^2 + 11x$

$8x^2 + 11x$ cannot be simplified more because the terms are *not* like terms.

YOU TRY 11   Combine like terms.

a) $6z + 5z$ b) $q - 9 - 4q + 11$ c) $\dfrac{5}{6}c^2 - \dfrac{2}{3}c^2$

d) $2y^2 + 8y + y^2 - 3y$

If an expression contains parentheses, we use the distributive property to clear the parentheses, and then combine like terms.

EXAMPLE 12   Combine like terms.

a) $5(2c + 3) - 3c + 4$ b) $3(2n + 1) - (6n - 11)$

c) $\dfrac{3}{8}(8 - 4p) + \dfrac{5}{6}(2p - 6)$

Solution

a) $5(2c + 3) - 3c + 4 = 10c + 15 - 3c + 4$ Distributive property

$\qquad\qquad\qquad\qquad = 10c - 3c + 15 + 4$ Rewrite like terms together.

$\qquad\qquad\qquad\qquad = 7c + 19$

b) $3(2n + 1) - (6n - 11) = 3(2n + 1) -1\,(6n - 11)$ Remember, $-(6n - 11)$ is the same as $-1(6n - 11)$.

$\qquad\qquad\qquad\qquad\qquad = 6n + 3 - 6n + 11$ Distributive property

$\qquad\qquad\qquad\qquad\qquad = 6n - 6n + 3 + 11$ Rewrite like terms together.

$\qquad\qquad\qquad\qquad\qquad = 0n + 14$ $0n = 0$

$\qquad\qquad\qquad\qquad\qquad = 14$

c) $\dfrac{3}{8}(8-4p)+\dfrac{5}{6}(2p-6)=\dfrac{3}{8}(8)-\dfrac{3}{8}(4p)+\dfrac{5}{6}(2p)-\dfrac{5}{6}(6)$ Distributive property

$$=3-\dfrac{3}{2}p+\dfrac{5}{3}p-5 \qquad \text{Multiply.}$$

$$=-\dfrac{3}{2}p+\dfrac{5}{3}p+3-5 \qquad \text{Rewrite like terms together.}$$

$$=-\dfrac{9}{6}p+\dfrac{10}{6}p+3-5 \qquad \text{Get a common denominator.}$$

$$=\dfrac{1}{6}p-2 \qquad \text{Combine like terms.}$$

[**YOU TRY 12**]

Combine like terms.

a) $9d^2-7+2d^2+3$ b) $10-3(2k+5)+k-6$

9   Translate English Expressions into Mathematical Expressions

Translating from English into a mathematical expression is a skill that is necessary to solve applied problems. Let's practice writing mathematical expressions.

Read the phrase carefully, choose a variable to represent the unknown quantity, then translate the phrase into a mathematical expression.

EXAMPLE 13

Write a mathematical expression for each and simplify. Define the unknown with a variable.

a) Seven more than twice a number

b) The sum of a number and four times the same number

Solution

a) Seven more than twice a number
   i) **Define the unknown.** This means that you should clearly state on your paper what the variable represents.

$$\text{Let } x = \text{the number.}$$

   ii) **Slowly, break down the phrase.** How do you write an expression for "seven more than" something?

$$+\,7$$

   iii) **What does "twice a number" mean?** It means two times the number. Since our number is represented by x, "twice a number" is $2x$.

   iv) **Put the information together:**

Seven more than twice a number

$2x \qquad + \qquad 7$

The expression is $2x+7$.

b) The sum of a number and four times the same number

   i) **Define the unknown.**

$$\text{Let } y = \text{the number.}$$

   ii) **Slowly, break down the phrase.** What does *sum* mean? **Add.** So, we have to add a number and four times the same number:

$$\text{Number} + 4(\text{Number})$$

   iii) Since y represents the number, *four times the number* is $4y$.

   iv) Therefore, to translate from English into a mathematical expression, we know that we must add the number, y, to four times the number, $4y$. Our expression is $y + 4y$. It simplifies to $5y$.

[YOU TRY 13]

Write a mathematical expression for each and simplify. Let x equal the unknown number.

a) Five less than twice a number

b) The sum of a number and two times the same number

Using Technology

A graphing calculator can be used to evaluate an algebraic expression. This is especially valuable when evaluating expressions for several values of the given variables.

We will evaluate the expression $\dfrac{x^2 - 2xy}{3x + y}$ when $x = -3$ and $y = 8$.

Method 1

Substitute the values for the variables and evaluate the arithmetic expression on the home screen. Each value substituted for a variable should be enclosed in parentheses to guarantee a correct answer. For example, $(-3)^2$ gives the result 9, whereas -3^2 gives the result -9. Be careful to press the negative key [(−)] when entering a negative sign and the minus key [−] when entering the minus operator.

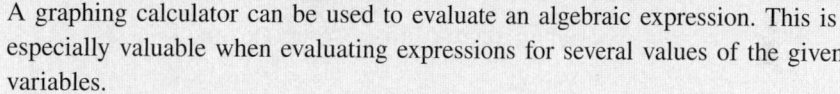

Method 2

Store the given values in the variables and evaluate the algebraic expression on the home screen.

To store -3 in the variable x, press [(−)] [3] [STO>] [x, T, θ, n] [ENTER].

To store 8 in the variable y, press [8] [STO>] [ALPHA] [1] [ENTER].

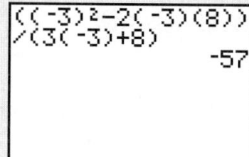

Enter $\dfrac{x^2 - 2xy}{3x + y}$ on the home screen.

The advantage of Method 2 is that we can easily store two different values in x and y. For example, store 5 in x and -2 in y. It is not necessary to enter the expression again because the calculator can recall previous entries.

Press [2nd] [ENTER] three times; then press [ENTER].

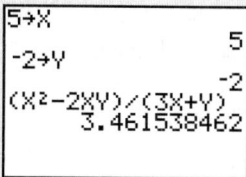

To convert this decimal to a fraction, press $\boxed{\text{MATH}}$ $\boxed{\text{ENTER}}$ $\boxed{\text{ENTER}}$.

Evaluate each expression when $x = -5$ and $y = 2$.

1. $3y - 4x$ 2. $2xy - 5y$ 3. $y^3 - 2x^2$

4. $\dfrac{x - y}{4x}$ 5. $\dfrac{2x + 5y}{x - y}$ 6. $\dfrac{x - y^2}{2x}$

```
Ans▶Frac
                45/13
```

ANSWERS TO $\boxed{\text{YOU TRY}}$ EXERCISES

1)

Term	Coeff.
$-15r^3$	-15
r^2	1
$-4r$	-4
8	8

2) a) 33 b) -39 3) 30   4) a) yes b) yes c) no

5) a) $16 + 1$ b) $6n$ 6) 36   7) a) $k + 12$ b) $-\dfrac{9}{7}$ or $-1\dfrac{2}{7}$

8) a) inverse property of multiplication b) inverse property of addition c) identity property of multiplication d) identity property of addition   9) a) 12 b) 20 c) -13

10) a) $6a + 12$ b) $10x - 35y - 20z$ c) $r - 4s + 9$

11) a) $11z$ b) $-3q + 2$ c) $\dfrac{1}{6}c^2$ d) $3y^2 + 5y$ 12) a) $11d^2 - 4$

b) $-5k - 11$ 13) a) $2x - 5$ b) $x + 2x; 3x$

ANSWERS TO TECHNOLOGY EXERCISES

1. 26   2. -30 3. -42 4. $\dfrac{7}{20}$ 5. 0   6. $\dfrac{9}{10}$

$\boxed{\text{E Evaluate}}$ **1.7** Exercises   Do the exercises, and check your work.

Objective 1: Identify the Terms and Coefficients in an Expression

For each expression, list the terms and their coefficients. Also, identify the constant.

1) $7p^2 - 6p + 4$ 2) $-8z + \dfrac{5}{6}$

3) $x^2y^2 + 2xy - y + 11$ 4) $w^3 - w^2 + 9w - 5$

5) $-2g^5 + \dfrac{g^4}{5} + 3.8g^2 + g - 1$

6) $121c^2 - d^2$

Objective 2: Evaluate Algebraic Expressions

7) What is an algebraic expression?

8) Is the value of an algebraic expression always the same? Explain.

Evaluate $4m + 3$ for each value of m.

9) $m = 5$ 10) $m = 9$

11) $m = -6$ 12) $m = -7$

Evaluate $-6d + 11$ for each value of d.

13) $d = 2$ 14) $d = -9$

15) $d = \dfrac{2}{3}$ 16) $d = \dfrac{5}{6}$

17) $d = 3.4$ 18) $d = 4.7$

Evaluate each expression for the given values.

19) $3r - t + 4$ when $r = -7$ and $t = -4$

20) $5u - v + 9$ when $u = -6$ and $v = -10$

21) $\dfrac{2d - c}{c + 5d + 1}$ when $c = 1$ and $d = -4$

22) $\dfrac{4k - h}{6h + 2k - 1}$ when $h = 5$ and $k = -1$

23) $m^2 + 4mn + n^2$ when $m = -3$ and $n = 2$

24) $-8g^2 - gh + 14$ when $g = -1$ and $h = -6$

Evaluate each expression when $x = 2$, $y = -3$, and $z = 5$.

25) $\dfrac{y^3}{x - z^2 + 2}$

26) $\dfrac{x^4 + 2y}{y + z^2 - 10}$

Objective 3: Identify Like Terms

27) Are $9k$ and $9k^2$ *like* terms? Why or why not?

28) Are $\dfrac{3}{4}n$ and $8n$ *like* terms? Why or why not?

29) Are a^3b and $-7a^3b$ *like* terms? Why or why not?

30) Write three *like* terms that are x^2-terms.

Mixed Exercises: Objectives 4–7

31) What is the identity element for multiplication?

32) What is the identity element for addition?

33) What is the additive inverse of 5?

34) What is the multiplicative inverse of 8?

Which property of real numbers is illustrated by each example? Choose from the commutative, associative, identity, inverse, or distributive property.

35) $9(2 + 8) = 9 \cdot 2 + 9 \cdot 8$

36) $(-16 + 7) + 3 = -16 + (7 + 3)$

37) $14 \cdot 1 = 14$

38) $\left(\dfrac{9}{2}\right)\left(\dfrac{2}{9}\right) = 1$

39) $5(2 \cdot 3) = (5 \cdot 2) \cdot 3$

40) $11 \cdot 7 = 7 \cdot 11$

41) $7(4w) \quad 7(w) = 7(4w - w)$

42) $8(p - q) = 8p - 8q$

Rewrite each expression using the indicated property.

43) $p + 19$; commutative

44) $5(m + n)$; distributive

45) $8 + (1 + 9)$; associative

46) $-2c + 0$; identity

47) $y + 0$; identity

48) $\left(4 \cdot \dfrac{2}{7}\right) \cdot 7$; associative

Use the commutative property to simplify each expression.

49) $-\dfrac{5}{2} + \dfrac{4}{9} + \dfrac{11}{2} + \dfrac{5}{9}$

50) $-\dfrac{8}{3} + \dfrac{2}{5} + \dfrac{2}{3} + \dfrac{8}{5}$

51) $\dfrac{1}{6} \cdot \dfrac{11}{7} \cdot \dfrac{6}{5}$

52) $\dfrac{3}{4} \cdot \dfrac{2}{9} \cdot \dfrac{4}{3}$

53) Is $2a - 7$ equivalent to $7 - 2a$? Why or why not?

54) Is $6 + t$ equivalent to $t + 6$? Why or why not?

Rewrite each expression using the distributive property. Simplify if possible.

55) $2(1 + 9)$

56) $-5(3 + 7)$

57) $9 \cdot a + 9 \cdot b$

58) $4 \cdot c - 4 \cdot d$

59) $8 \cdot 3 - 8 \cdot 10$

60) $12 \cdot 2 + 12 \cdot 5$

61) $-(10 - 4)$

62) $-(3 + 9)$

63) $8(y + 3)$

64) $4(k + 11)$

65) $-\dfrac{2}{3}(z + 6)$

66) $-\dfrac{3}{5}(m + 10)$

67) $-3(x - 4y - 6)$

68) $6(2a - 5b + 1)$

69) $-(-8c + 9d - 14)$

70) $-(x - 10y - 4z)$

Objective 8: Combine Like Terms
Combine like terms and simplify.

71) $10p + 9 + 14p - 2$

72) $11 - k^2 + 12k^2 - 3 + 6k^2$

73) $-18y^2 - 2y^2 + 19 + y^2 - 2 + 13$

74) $-7x - 3x - 1 + 9x + 6 - 2x$

75) $\dfrac{4}{9} + 3r - \dfrac{2}{3} + \dfrac{1}{5}r$

76) $6a - \dfrac{3}{8}a + 2 + \dfrac{1}{4} - \dfrac{3}{4}a$

77) $h^2 - 11h + 5 + 7h^2 - h + 9$

78) $6p^2 + p - 10 - 4p^2 - 5p + 3$

79) $4c^3 + c^2 - 8c - 1 - 3c^3 + 6c^2 + 8c - 5$

80) $-9y^3 - 7y^2 + 2y + 13 + y^3 + 7y^2 + 4y - 4$

81) $2(3w + 5) + w$

82) $9 - 4(3 - x) - 4x + 3$

83) $3[2(5x + 7) - 11] + 4(7 - x)$

84) $22 - [6 + 5(2w - 3)] - (7w + 16)$

 85) $5k^2 + 8k + 4 - (4k^2 + 3k - 2)$

86) $m^2 - 6m + 1 - (9m^2 - m + 10)$

87) $4(4n^2 - 3n + 2) + 2(n^2 + 6n - 9)$

88) $5(4x^2 + 7x + 3) + 4(-5x^2 - 8x + 1)$

89) $\dfrac{4}{5}(2z + 10) - \dfrac{1}{2}(z + 3)$

90) $\dfrac{2}{3}(6c - 7) + \dfrac{5}{12}(2c + 5)$

91) $1 + \dfrac{3}{4}(10t - 3) + \dfrac{5}{8}\left(t + \dfrac{1}{10}\right)$

92) $\dfrac{7}{15} - \dfrac{9}{10}(2y + 1) - \dfrac{2}{5}(4y - 3)$

93) $2.5(x - 4) - 1.2(3x + 8)$

94) $9.4 - 3.8(2a + 5) + 0.6 + 1.9a$

Objective 9: Translate English Expressions into Mathematical Expressions

Write a mathematical expression for each phrase, and combine like terms if possible. Let x represent the unknown quantity.

95) Eighteen more than a number

96) Eleven more than a number

97) Six subtracted from a number

98) Eight subtracted from a number

99) The product of eight and a number

100) The product of five and a number

101) The quotient of a number and seven

102) The quotient of a number and twelve

103) Three less than a number

104) Fourteen less than a number

105) The sum of twelve and twice a number

106) The sum of nine and twice a number

107) One less than half a number

108) Ten less than one-third of a number

109) Seven less than the sum of three and twice a number

110) The sum of -8 and twice a number increased by three

Write a statement, in English, that translates to the given expression.

111) $10x$

112) $x + 7$

113) $13 - x$

114) $\dfrac{x}{-4}$

115) $\dfrac{1}{2}x + 8$

116) $2x - 9$

R Rethink

R1) Describe a situation in your everyday life when you combine like terms.

R2) Think about how a cashier counts money in the cash register at the end of a shift. How does the cashier use like terms to count the money?

R3) Which exercises do you need help mastering?

Group Activity – Magic Squares

A magic square is a table with the same number of rows as columns. The numbers in each row, column, and diagonal add up to the same number.

1) Complete the magic square if the numbers in each row, column, and diagonal add up to 15.

8		6
	9	

2) Complete the magic square if the numbers in each row, column, and diagonal add up to 5/6.

$\frac{1}{9}$		
		$\frac{1}{6}$
$\frac{1}{3}$		

3) Find the answer to each problem. Use your answers to complete the magic square. The numbers in each row, column, and diagonal add up to 83.

Find the perimeter: 3 3 2 8 9		Find the area: 4 8	
	Evaluate: $\dfrac{2 - 4(5)}{2 - 3}$	Evaluate: $2\left(\dfrac{6 + 10 \div 2 + 11}{6 + 2^4}\right)$	Evaluate: $17 - (3 - 6)^3$
	Evaluate: $(6 - 11)^2 - (-2)(9)$	Find the area of the unshaded region: 4 2 2 6	
Evaluate: $7 + 2[(1 + 4)^3 \div 5^2]$		Find the volume: 3 5 3	

As you probably know by now, success in college takes a great deal of time, effort, and dedication. In order to stay motivated, it is important to focus on your goals for attending college. For many people, college is an essential step toward achieving career aspirations. Other people go to college in order to build a better life for their families. Have you thought about why *you* are attending college?

Place a 1, 2, or 3 by the three *most* important reasons that you have for attending college:

_____ The job I want requires a college degree.

_____ I'm not sure what I want to do as a career, but I believe that college is important to get a good job.

_____ I want to make my family proud.

_____ My parents said I have to go to college.

_____ I couldn't find a job.

_____ I want to try something new.

_____ I want to get ahead at my current job.

_____ I want to improve my reading and critical thinking skills.

_____ I want to become a more cultured person.

_____ I want to make more money.

_____ I want to learn more about things that interest me.

_____ A mentor or role model encouraged me to go.

_____ I want to prove to others that I can succeed.

Now think about the following:

- What do your answers tell you about yourself?
- What reasons besides these did you think about when you were applying to college?
- How do you think your reasons compare with those of other students who are starting college with you?
- How do you think your reasons will affect your approach to college?
- How can you use your understanding of why you are in college to help you achieve college success?

Chapter 1: Summary

Definition/Procedure	Example

1.1 Review of Fractions

Writing Fractions in Lowest Terms

A fraction is in **lowest terms** when the numerator and denominator have no common factors other than 1.

Write $\dfrac{36}{48}$ in lowest terms.

$$\dfrac{36}{48} = \dfrac{36 \div 12}{48 \div 12} = \dfrac{3}{4}$$

Multiplying Fractions

To multiply fractions, multiply the numerators and multiply the denominators. Common factors can be divided out either before or after multiplying.

Multiply $\dfrac{21}{45} \cdot \dfrac{9}{14}$.

$$\dfrac{\overset{3}{\cancel{21}}}{\underset{5}{\cancel{45}}} \cdot \dfrac{\overset{1}{\cancel{9}}}{\underset{2}{\cancel{14}}}$$

↙ 9 and 45 each divide by 9.

↖ 21 and 14 each divide by 7.

$$= \dfrac{3}{5} \cdot \dfrac{1}{2} = \dfrac{3}{10}$$

Dividing Fractions

To divide fractions, multiply the first fraction by the reciprocal of the second.

Divide $\dfrac{7}{5} \div \dfrac{4}{3}$. $\dfrac{7}{5} \div \dfrac{4}{3} = \dfrac{7}{5} \cdot \dfrac{3}{4} = \dfrac{21}{20}$ or $1\dfrac{1}{20}$

Adding and Subtracting Fractions

To add or subtract fractions,

1) Identify the least common denominator (LCD).
2) Write each fraction as an equivalent fraction using the LCD.
3) Add or subtract.
4) Express the answer in lowest terms.

Add $\dfrac{5}{11} + \dfrac{2}{11}$. $\dfrac{5}{11} + \dfrac{2}{11} = \dfrac{7}{11}$

Subtract $\dfrac{8}{9} - \dfrac{3}{4}$. $\dfrac{8}{9} - \dfrac{3}{4} = \dfrac{32}{36} - \dfrac{27}{36} = \dfrac{5}{36}$

1.2 Exponents and Order of Operations

Exponents

An **exponent** represents repeated multiplication.

Write $9 \cdot 9 \cdot 9 \cdot 9 \cdot 9$ in exponential form.

$$9 \cdot 9 \cdot 9 \cdot 9 \cdot 9 = 9^5$$

Evaluate 2^4. $2^4 = 2 \cdot 2 \cdot 2 \cdot 2 = 16$

Order of Operations

Parentheses, **E**xponents, **M**ultiplication and **D**ivision, **A**ddition and **S**ubtraction.

Evaluate $8 + (5 - 1)^2 - 6 \cdot 3$.

$$8 + (5 - 1)^2 \quad 6 \cdot 3$$

$= 8 + 4^2 - 6 \cdot 3$	Parentheses
$= 8 + 16 - 6 \cdot 3$	Exponents
$= 8 + 16 - 18$	Multiply.
$= 24 - 18$	Add.
$= 6$	Subtract.

1.3 Geometry Review

Important Angles

The definitions for an acute angle, an obtuse angle, and a right angle can be found in the section.

Two angles are **complementary** if the sum of their measures is 90°.

Two angles are **supplementary** if the sum of their measures is 180°.

The measure of an angle is 73°. Find the measure of its complement and its supplement.

The measure of its complement is 17° since $90° - 73° = 17°$.

The measure of its supplement is 107° since $180° - 73° = 107°$.

Definition/Procedure	Example
Triangle Properties The sum of the measures of the angles of any triangle is 180°. An **equilateral triangle** has three sides of equal length. Each angle measures 60°. An **isosceles triangle** has two sides of equal length. The angles opposite the sides with equal measure have the same measure. A **scalene triangle** has no sides of equal length. No angles have the same measure.	Find the measure of $\angle C$. $m\angle A + m\angle B = 63° + 94° = 157°$ $m\angle C = 180° - 157° = 23°$ 
Perimeter and Area The formulas for the perimeter and area of a rectangle, square, triangle, parallelogram, and trapezoid can be found in the section.	Find the area and perimeter of this rectangle. 6 in. 8 in. Area = (Length)(Width) Perimeter = 2(Length) + 2(Width) = (8 in.)(6 in.) = 2(8 in.) + 2(6 in.) = 48 in^2 = 16 in. + 12 in. = 28 in.
Volume The formulas for the volume of a rectangular solid, cube, right circular cylinder, sphere, and right circular cone can be found in the section.	Find the volume of the cylinder pictured here. 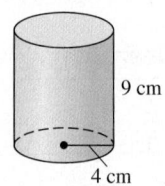 9 cm 4 cm Give an exact answer and give an approximation using 3.14 for π. $V = \pi r^2 h$ $V = 144\pi \text{ cm}^3$ $= \pi(4 \text{ cm})^2(9 \text{ cm})$ $\approx 144(3.14) \text{ cm}^3$ $= \pi(16 \text{ cm}^2)(9 \text{ cm})$ $\approx 452.16 \text{ cm}^3$ $= 144\pi \text{ cm}^3$

1.4 Sets of Numbers and Absolute Value

Natural numbers: $\{1, 2, 3, 4, \ldots\}$ **Whole numbers:** $\{0, 1, 2, 3, 4, \ldots\}$ **Integers:** $\{\ldots, -3, -2, -1, 0, 1, 2, 3, \ldots\}$	
A **rational number** is any number of the form $\dfrac{p}{q}$, where p and q are integers and $q \neq 0$.	The following numbers are rational: $-3, 10, \dfrac{5}{8}, 7.4, 2.\overline{3}$
An **irrational number** cannot be written as the quotient of two integers.	The following numbers are irrational: $\sqrt{6}, 9.2731\ldots$
The set of **real numbers** includes the rational and irrational numbers.	Any number that can be represented on the number line is a real number.
The **additive inverse** of a is $-a$.	The additive inverse of 4 is -4.
Absolute Value $|a|$ is the distance of a from zero.	$|-6| = 6$

Definition/Procedure	Example

1.5 Addition and Subtraction of Real Numbers

Adding Real Numbers

To add numbers with the **same sign,** add the absolute values of the numbers. The sum will have the same sign as the numbers being added.

$-3 + (-9) = -12$

To add two numbers with **different signs,** subtract the smaller absolute value from the larger. The sum will have the sign of the number with the larger absolute value.

$-20 + 15 = -5$

Subtracting Real Numbers

To subtract $a - b$, change subtraction to addition, and add the additive inverse of b: $a - b = a + (-b)$.

$2 - 11 = 2 + (-11) = -9$

$-17 - (-7) = -17 + 7 = -10$

1.6 Multiplication and Division of Real Numbers

Multiplying Real Numbers

The product of two real numbers with the *same* sign is positive.

$8 \cdot 3 = 24 \qquad -7 \cdot (-8) = 56$

The product of a positive number and a negative number is *negative*.

$-2 \cdot 5 = -10 \qquad 9 \cdot (-1) = -9$

An *even number* of negative factors in a product gives a *positive* result.

$\underbrace{(-1)(-6)(-3)(2)(-4)}_{\text{4 negative factors}} = 144$

An *odd number* of negative factors in a product gives a *negative* result.

$\underbrace{(5)(-2)(-3)(1)(-1)}_{\text{3 negative factors}} = -30$

Evaluating Exponential Expressions

Exponential notation is a shorthand way to represent repeated multiplication.

Evaluate $(-3)^4$.   The base is -3.
$(-3)^4 = (-3)(-3)(-3)(-3) = 81$

Evaluate -3^4.   The base is 3.
$-3^4 = -1 \cdot 3^4 = -1 \cdot 3 \cdot 3 \cdot 3 \cdot 3 = -81$

Dividing Real Numbers

The quotient of two numbers with the *same* sign is positive. The quotient of two numbers with *different* signs is negative.

$\dfrac{40}{2} = 20 \qquad -18 \div (-3) = 6$

$\dfrac{-56}{8} = -7 \qquad 48 \div (-4) = -12$

1.7 Algebraic Expressions and Properties of Real Numbers

A **variable** is a symbol, usually a letter, used to represent an unknown number.

In the expression $9c + 2$, c is the variable.

An **algebraic expression** is a collection of numbers, variables, and grouping symbols connected by operation symbols such as $+$, $-$, $\times$, and $\div$.

$4y^2 - 7y + \dfrac{3}{5}$

A **term** in an expression is a number or a variable or a product or quotient of numbers and variables.

The value of a **constant** does not change.

The **coefficient** of a variable is the number that the variable is multiplied by.

List the terms and coefficients of $7x^3 - x^2 - \dfrac{2}{5}x + 8$.

Term	Coefficient
$7x^3$	7
$-x^2$	-1
$-\dfrac{2}{5}x$	$-\dfrac{2}{5}$
8	8

8 is the constant.

Definition/Procedure	Example
Evaluating Expressions We can evaluate expressions for different values of the variables.	Evaluate $2xy - 5y + 1$ when $x = -3$ and $y = 4$. Substitute -3 for x and 4 for y and simplify. $\begin{aligned} 2xy - 5y + 1 &= 2(-3)(4) - 5(4) + 1 \\ &= -24 - 20 + 1 \\ &= -24 + (-20) + 1 \\ &= -43 \end{aligned}$
Like Terms **Like terms** contain the same variables with the same exponents.	In the group of terms $5k^2, -8k, -4k^2, \frac{1}{3}k,$ $5k^2$ and $-4k^2$ are like terms and $-8k$ and $\frac{1}{3}k$ are like terms.
Properties of Real Numbers If a, b, and c are real numbers, then the following properties hold. **Commutative Properties:** $a + b = b + a$ $ab = ba$	 $10 + 3 = 3 + 10$ $(-6)(5) = (5)(-6)$
Associative Properties: $(a + b) + c = a + (b + c)$ $(ab)c = a(bc)$	$(9 + 4) + 2 = 9 + (4 + 2)$ $(5 \cdot 2) \cdot 8 = 5 \cdot (2 \cdot 8)$
Identity Properties: $a + 0 = 0 + a = a$ $a \cdot 1 = 1 \cdot a = a$	$7 + 0 = 7 \qquad \frac{2}{3} \cdot 1 = \frac{2}{3}$
Inverse Properties: $a + (-a) = -a + a = 0$ $b \cdot \dfrac{1}{b} = \dfrac{1}{b} \cdot b = 1$, for $b \neq 0$.	$11 + (-11) = 0 \qquad 5 \cdot \dfrac{1}{5} = 1$
Distributive Properties: $a(b + c) = ab + ac$ and $(b + c)a = ba + ca$ $a(b - c) = ab - ac$ and $(b - c)a = ba - ca$	$\begin{aligned} 6(5 + 8) &= 6 \cdot 5 + 6 \cdot 8 \\ &= 30 + 48 \\ &= 78 \end{aligned}$ $\begin{aligned} 9(w - 2) &= 9w - 9 \cdot 2 \\ &= 9w - 18 \end{aligned}$
Combining Like Terms We can simplify expressions by combining like terms.	Combine like terms and simplify. $\begin{aligned} &4n^2 - 3n + 1 - 2(6n^2 - 5n + 7) \\ &= 4n^2 - 3n + 1 - 12n^2 + 10n - 14 \qquad \text{Distributive property} \\ &= -8n^2 + 7n - 13 \qquad\qquad\qquad\quad \text{Combine like terms.} \end{aligned}$
Writing Mathematical Expressions	Write a mathematical expression for the following: *Sixteen more than twice a number* Let $x = $ the number. $\underbrace{\text{Sixteen more than}}_{+16} \qquad \underbrace{\text{twice a number}}_{2x}$ $2x + 16$

Chapter 1: Review Exercises

(1.1)

1) Find all factors of each number.

   a) 16   b) 37

2) Find the prime factorization of each number.

   a) 28   b) 66

3) Write each fraction in lowest terms.

   a) $\dfrac{12}{30}$ b) $\dfrac{414}{702}$

Perform the indicated operation. Write the answer in lowest terms.

4) $\dfrac{4}{11} \cdot \dfrac{3}{5}$

5) $\dfrac{45}{64} \cdot \dfrac{32}{75}$

6) $\dfrac{5}{8} \div \dfrac{3}{10}$

7) $35 \div \dfrac{7}{8}$

8) $4\dfrac{2}{3} \cdot 1\dfrac{1}{8}$

9) $\dfrac{30}{49} \div 2\dfrac{6}{7}$

10) $\dfrac{2}{9} + \dfrac{4}{9}$

11) $\dfrac{2}{3} + \dfrac{1}{4}$

12) $\dfrac{9}{40} + \dfrac{7}{16}$

13) $\dfrac{1}{5} + \dfrac{1}{3} + \dfrac{1}{6}$

14) $\dfrac{21}{25} - \dfrac{11}{25}$

15) $\dfrac{5}{8} - \dfrac{2}{7}$

16) $3\dfrac{2}{9} + 5\dfrac{3}{8}$

17) $9\dfrac{3}{8} - 2\dfrac{5}{6}$

18) A pattern for a skirt calls for $1\dfrac{7}{8}$ yd of fabric. If Mary Kate wants to make one skirt for herself and one for her twin, how much fabric will she need?

(1.2) Evaluate.

19) 3^4

20) 2^6

21) $\left(\dfrac{3}{4}\right)^3$

22) $(0.6)^2$

23) $13 - 7 + 4$

24) $8 \cdot 3 + 20 \div 4$

25) $\dfrac{12 - 56 \div 8}{(1 + 5)^2 - 2^4}$

(1.3)

26) The complement of 51° is _____.

27) The supplement of 78° is _____.

28) Is this triangle acute, obtuse, or right? Find the missing angle.

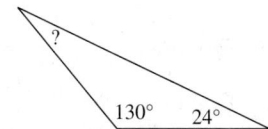

Find the area and perimeter of each figure. Include the correct units.

29)

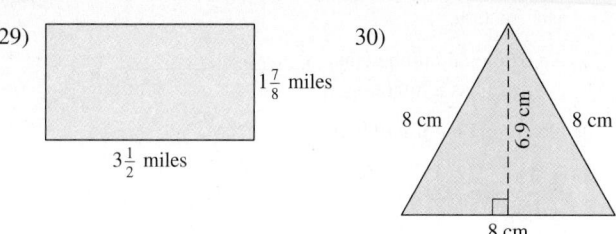

30)

31)

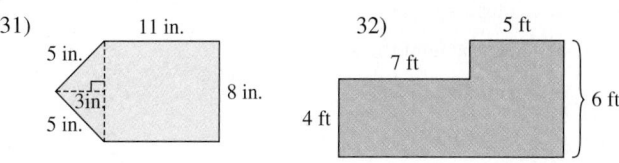

32)

Find a) the area and b) the circumference of each circle. Give an exact answer for each and give an approximation using 3.14 for π. Include the correct units.

33)

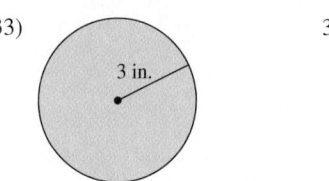

34)

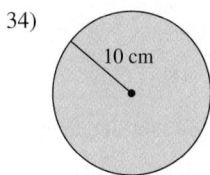

Find the area of the shaded region. Include the correct units.

35)

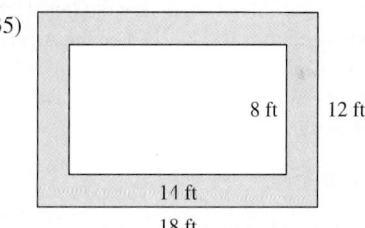

Find the volume of each figure. Where appropriate, give the answer in terms of π. Include the correct units.

36)

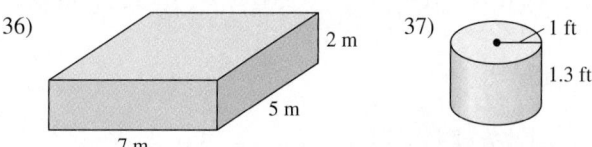

37)

38)

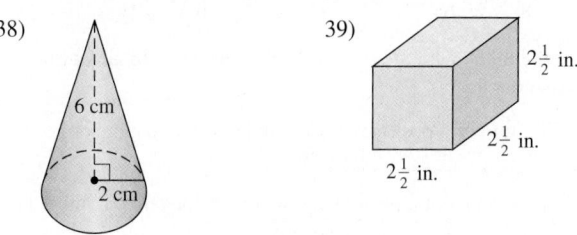

39)

40) The radius of a basketball is approximately 4.7 inches. Find its circumference to the nearest tenth of an inch.

Fill in the blank with *always*, *sometimes*, or *never* to make the statement true.

41) A rational number can _____ be expressed as a fraction.

42) An irrational number can _____ be expressed as a fraction.

43) Given this set of numbers,

$$\left\{ \frac{7}{15}, -16, 0, 3.\overline{2}, 8.5, \sqrt{31}, 4, 6.01832 \dots \right\}$$

list the

a) integers

b) rational numbers

c) natural numbers

d) whole numbers

e) irrational numbers

44) Graph and label these numbers on a number line.

$$-3.5, 4, \frac{9}{10}, 2\frac{1}{3}, -\frac{3}{4}, -5$$

45) Evaluate.

a) $|-18|$ b) $-|7|$

(1.5) Add or subtract as indicated.

46) $-38 + 13$ 47) $-21 - (-40)$

48) $-1.9 + 2.3$ 49) $\dfrac{5}{12} - \dfrac{5}{8}$

50) The lowest temperature on record in the country of Greenland is $-87°F$. The coldest temperature ever reached on the African continent was in Morocco and is 76° higher than Greenland's record low. What is the lowest temperature ever recorded in Africa? (www.ncdc.noaa.gov)

(1.6) Multiply or divide as indicated.

51) $\left(-\dfrac{3}{2}\right)(8)$ 52) $(-4.9)(-3.6)$

53) $(-4)(3)(-2)(-1)(-3)$ 54) $\left(-\dfrac{2}{3}\right)(-5)(2)(-6)$

55) $-108 \div 9$ 56) $\dfrac{56}{-84}$

57) $-3\dfrac{1}{8} \div \left(-\dfrac{5}{6}\right)$ 58) $-\dfrac{9}{10} \div 12$

Fill in the blank with *always*, *sometimes*, or *never* to make the statement true.

59) The sum of a positive number and a negative number is _____ positive.

60) The product of a positive number and a negative number is _____ negative.

61) The absolute value of a number is _____ 0 or positive.

62) The absolute value of a number _____ tells its distance from 0.

63) A negative number raised to an odd power is _____ negative.

64) A negative number raised to an even power is _____ negative.

Evaluate.

65) -6^2 66) $(-6)^2$

67) $(-2)^6$ 68) -1^{10}

69) 3^3 70) $(-5)^3$

Use the order of operations to simplify.

71) $56 \div (-7) - 1$

72) $15 - (2 - 5)^3$

73) $-11 + 4 \cdot 3 + (-8 + 6)^5$

74) $\dfrac{1 + 6(7 - 3)}{2[3 - 2(8 - 1)] - 3}$

Write a mathematical expression for each and simplify.

75) The quotient of -120 and -3

76) Twice the sum of 22 and -10

77) 15 less than the product of -4 and 7

78) 11 more than half of -18

(1.7)

79) Evaluate $9x - 4y$ when $x = -3$ and $y = 7$.

80) Evaluate $\dfrac{2a + b}{a^3 - b^2}$ when $a = -3$ and $b = 5$.

Which property of real numbers is illustrated by each example? Choose from the commutative, associative, identity, inverse, or distributive property.

81) $12 + (5 + 3) = (12 + 5) + 3$

82) $\left(\dfrac{2}{5}\right)\left(\dfrac{5}{2}\right) = 1$

83) $0 + 19 = 19$

84) $-4(7 + 2) = -4(7) + (-4)(2)$

85) $8 \cdot 3 = 3 \cdot 8$

Rewrite each expression using the distributive property. Simplify if possible.

86) $7(3 - 9)$

87) $(10 + 4)5$

88) $-(15 - 3)$

89) $-6(9p - 4q + 1)$

90) List the terms and coefficients of

$$5z^4 - 8z^3 + \frac{3}{5}z^2 - z + 14$$

Combine like terms and simplify.

91) $9m - 14 + 3m + 4$

92) $-5c + d - 2c + 8d$

93) $15y^2 + 8y - 4 + 2y^2 - 11y + 1$

94) $7t + 10 - 3(2t + 3)$

95) $\frac{3}{2}(5n - 4) + \frac{1}{4}(n + 6)$

96) $1.4(a + 5) - (a + 2)$

Write a mathematical expression for each phrase. Let x represent the unknown quantity.

97) Fifteen more than a number

98) A number decreased by 12

99) The product of eight and a number

100) Eleven more than twice a number

Chapter 1:   Test

1) Find the prime factorization of 210.

2) Write in lowest terms:

   a) $\frac{45}{72}$ b) $\frac{420}{560}$

Perform the indicated operations. Write all answers in lowest terms.

3) $\frac{7}{16} \cdot \frac{10}{21}$

4) $\frac{5}{12} + \frac{2}{9}$

5) $10\frac{2}{3} - 3\frac{1}{4}$

6) $\frac{4}{9} \div 12$

7) $\frac{3}{5} - \frac{17}{20}$

8) $-31 - (-14)$

9) $16 + 8 \div 2$

10) $\frac{1}{8} \cdot \left(-\frac{2}{3}\right)$

11) $-15 \cdot (-4)$

12) $-9.5 + 5.8$

13) $23 - 6[-4 + (9 - 11)^4]$

14) $\dfrac{7 \cdot 2 - 4}{48 \div 3 - 8^0}$

15) An extreme sports athlete has reached an altitude of 14,693 ft while ice climbing and has dived to a depth of 518 ft below sea level. What is the difference between these two elevations?

16) Evaluate.

   a) 5^3

   b) $|-43|$

   c) $-|18 - 40| - 3|9 - 4|$

17) Write $(-2)^4$ and -2^4 without exponents. Evaluate each.

18) The supplement of $31°$ is _____.

19) Find the missing angle, and classify the triangle as acute, obtuse, or right.

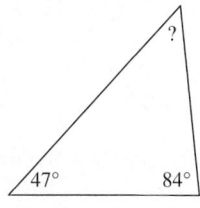

20) Find the area and perimeter of each figure. Include the correct units.

a)

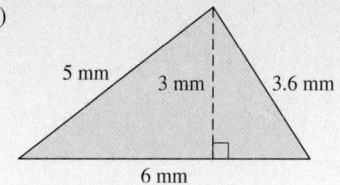

b)

c)

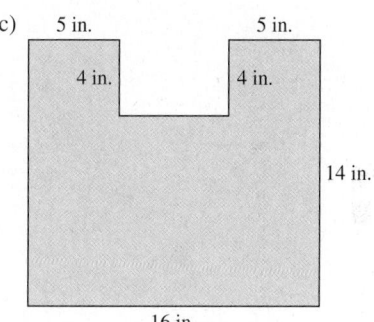

21) Find the volume of this figure:

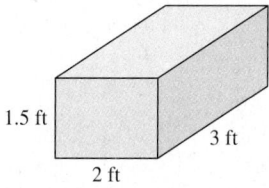

22) Given this set of numbers,

   $\left\{3\frac{1}{5}, 22, -7, \sqrt{43}, 0, 1.\overline{5}, 8.0934 \ldots\right\}$ list the

   a) whole numbers

   b) natural numbers

   c) irrational numbers

   d) integers

   e) rational numbers

23) Graph the numbers on a number line. Label each.

$$4, -5, \frac{2}{3}, -3\frac{1}{2}, -\frac{5}{6}, 2.2$$

24) Write a mathematical expression for each and simplify.

   a) The sum of -4 and 27

   b) The product of 5 and -6 subtracted from 17

25) List the terms and coefficients of

   $$4p^3 - p^2 + \frac{1}{3}p - 10$$

26) Evaluate $\dfrac{x^2 - y^2}{6y + x}$ when $x = 3$ and $y = -4$.

27) Which property of real numbers is illustrated by each example? Choose from the commutative, associative, identity, inverse, or distributive property.

   a) $9 \cdot 5 = 5 \cdot 9$

   b) $16 + (4 + 7) = (16 + 4) + 7$

   c) $\left(\dfrac{10}{3}\right)\left(\dfrac{3}{10}\right) = 1$

   d) $8(1 - 4) = 8 \cdot 1 - 8 \cdot 4$

28) Rewrite each expression using the distributive property. Simplify if possible.

   a) $-4(2 + 7)$

   b) $3(8m - 3n + 11)$

29) Combine like terms and simplify.

   a) $-8k^2 + 3k - 5 + 2k^2 + k - 9$

   b) $\dfrac{4}{3}(6c - 5) - \dfrac{1}{2}(4c + 3)$

30) Write a mathematical expression for "nine less than twice a number." Let x represent the number.

Rules of Exponents

Get Ready

To use the rules of exponents, we need to be able to add, subtract, multiply, and divide real numbers without making "careless" mistakes. We need to know the powers of integers, too. Let's review this material here.

1) The **sum** of *two positive real numbers* is positive: $6 + 11 = 17$

2) The **sum** of *two negative real numbers* is negative:
$-4.5 + (-1.8) = -6.3$

3) The **sum** of *a positive real number and a negative real number* is sometimes positive and sometimes negative: $-5 + 9 = 4$, $1 + (-7) = -6$

4) The **difference** of *two real numbers* is sometimes negative and sometimes positive: $3 - 8 = 3 + (-8) = -5$, $-2.4 - (-9.7) = -2.4 + 9.7 = 7.3$

5) When *multiplying* real numbers, the **product** of *two real numbers with the same sign is positive.* The **product** of *a positive number and a negative number is negative:*

$$7 \cdot 4 = 28, \qquad -3.5 \times (-2.1) = 7.35, \qquad -\frac{4}{9} \cdot \frac{5}{3} = -\frac{20}{27}$$

6) When *dividing* real numbers, the **quotient** of *two real numbers with the same sign is positive.* The **quotient** of *a positive number and a negative number is negative:*

$$50 \div 5 = 10, \qquad \frac{-42}{-6} = 7, \quad 3.6 \div (-4) = -0.9$$

7) It is important to know some powers of integers *by heart* in order learn the algebra in this course. (See Section 1.2 for the table of powers you need to know.) It is important to understand how to work with negative signs, too.

$$5^3 = 125, \qquad 8^2 = 64, \qquad (-8)^2 = 64, \qquad -8^2 = -64$$

In this chapter, use the following *Basic Skills Worksheets* to improve the arithmetic skills needed in this, and future, chapters: WS2 Powers, WS3 Prefactoring.

Perform the indicated operations. Write each answer in lowest terms.

1) $-9 \cdot 6$

2) $5 - (-8)$

3) $-4 + (-10)$

4) $\dfrac{7}{10} \cdot \dfrac{2}{3}$

5) $-9.6 \div (-1.2)$

6) $12 + (-3)$

7) $-11 - (-6)$

8) $4.8 \cdot (-7.2)$

9) $-\dfrac{8}{9} \cdot \left(-\dfrac{1}{12}\right)$

10) $-10 - 4$

11) $-11 + 9$

12) $\dfrac{-60}{-5}$

Evaluate.

13) 11^2

14) 3^4

15) $(-2)^5$

16) -9^2

17) -10^3

18) $(-2)^6$

19) $15 \cdot \dfrac{1}{6} \cdot \dfrac{2}{3}$

20) $8 \cdot \left(\dfrac{1}{4}\right)^2$

Answers
1) -54 2) 13 3) -14 4) $\dfrac{7}{15}$ 5) 8 6) 9 7) -5 8) -34.56 9) $\dfrac{2}{27}$ 10) -14 11) -2 12) 12
13) 121 14) 81 15) -32 16) -81 17) -1000 18) 64 19) $\dfrac{5}{3}$ 20) $\dfrac{1}{2}$

 **Study Strategies** How to Read Math (and Other) Textbooks

Some textbooks seem intimidating to read, especially math books! With some strategies, however, you can learn to read and use your textbook effectively.

- *Prepare means to explicitly state a goal.* **I will learn how to read and use my math textbook.**

- *Organize means to* **organize** *the physical and mental tools you need to achieve your goal.*
- Read the preface and/or the introduction to learn about the features and structure of the book.
- Be aware of additional materials online, such as review sections for prerequisite skills, an online homework system, videos, and other supplements.
- Gather the *physical tools* you may need: the book, a computer, pencils, a notebook, a folder, and a highlighter. Locate a quiet place to study.

- *Work means to* **do the work** *that needs to be done to achieve the goal.*
- **The best way to read a math textbook is to** *write while you read*! Write out the examples in your notebook as you are reading them so that you are actually *doing* the problems.
- Use different colored pencils to write out the steps just like color is used in the book. This helps you to see the step performed as you go from one line to the next.
- Highlight and underline no more than about 10% of what you read.
- In this book, do the You Trys (and check the answers at the end of the section) after you have worked through the examples.
- When you read through definitions and procedures, jot down notes next to them explaining them in your own words.

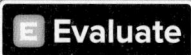 **Evaluate**

- *Evaluate* means you should *evaluate* what you have done. Did you achieve your goal?
- Did you understand what you read, or not? Could you explain major concepts, procedures, and definitions *in your own words* to a classmate? Could you do the You Trys on your own after reading the section? Could you do the homework exercises after reading?

R Rethink

- **Rethink** *means to rethink and reflect upon your goal.* If you did *not* achieve your goal, ask yourself, "Why not?" If you did achieve your goal, ask yourself what you did *right*.
- If you did not understand a section or could not do the You Trys or exercises, ask yourself, "Why not?" Did you read in a place that was full of distractions? Did you write out the examples as you read them, or did you skip that part?
- If you did understand a section after reading it, think about what you did that led to that success. Did you read in a quiet place? Was writing out the examples using different colored pencils helpful to you? Be aware of what worked so that you can do it again.

Rido/123RF

2.1A Basic Rules of Exponents: The Product Rule and Power Rules

P Prepare   **O Organize**

What are your objectives for Section 2.1A?	How can you accomplish each objective?
1 Evaluate Exponential Expressions	• Write the definition of *exponential expression* in your own words, using the words *base* and *exponent*. • Complete the given example on your own. • Complete You Try 1.
2 Use the Product Rule for Exponents	• Learn the property for the **Product Rule,** and write an example in your notes. • Complete the given example on your own. • Complete You Try 2.
3 Use the Power Rule $(a^m)^n = a^{mn}$	• Learn the property for the **Basic Power Rule,** and write an example in your notes. • Complete the given example on your own. • Complete You Try 3.
4 Use the Power Rule $(ab)^n = a^n b^n$	• Learn the property for the **Power Rule for a Product,** and write an example in your notes. • Complete the given example on your own. • Complete You Try 4.

SECTION 2.1A *Basic Rules of Exponents: The Product Rule and Power Rules*   **89**

What are your objectives for Section 2.1A?	How can you accomplish each objective?
5 Use the Power Rule $\left(\dfrac{a}{b}\right)^n = \dfrac{a^n}{b^n}$ where $b \neq 0$	• Learn the property for the **Power Rule for a Quotient,** and write an example in your notes. • Complete the given example on your own. • Complete You Try 5. • Write the summary for the product and power rules of exponents in your notes, and be able to apply them.

 **Work**

Read the explanations, follow the examples, take notes, and complete the You Trys.

1 Evaluate Exponential Expressions

Recall from Chapter 1 that exponential notation is used as a shorthand way to represent a multiplication problem. For example, $3 \cdot 3 \cdot 3 \cdot 3 \cdot 3$ can be written as 3^5.

Definition

An **exponential expression** of the form a^n, where a is any real number and n is a positive integer, is equivalent to $\underbrace{a \cdot a \cdot a \cdot \cdots \cdot a}_{n \text{ factors of } a}$. We say that a is the **base** and n is the **exponent.**

We can also evaluate an exponential expression.

EXAMPLE 1

Identify the base and the exponent in each expression and evaluate.

a) 2^4 b) $(-2)^4$ c) -2^4

Solution

a) 2^4 2 is the base, 4 is the exponent. Therefore, $2^4 = 2 \cdot 2 \cdot 2 \cdot 2 = 16$.

b) $(-2)^4$ -2 is the base, 4 is the exponent. Therefore,
$(-2)^4 = (-2) \cdot (-2) \cdot (-2) \cdot (-2) = 16$.

c) -2^4 It may be very tempting to say that the base is -2. However, there are no parentheses in this expression. Therefore, 2 is the base, and 4 is the exponent. To evaluate,

$$-2^4 = -1 \cdot 2^4 = -1 \cdot 2 \cdot 2 \cdot 2 \cdot 2$$
$$= -16$$

W Hint
In part c), you must follow the *order of operations* and evaluate the exponential expression before multiplying by −1.

BE CAREFUL
The expressions $(-a)^n$ and $-a^n$ are not always equivalent:

$$(-a)^n = \underbrace{(-a) \cdot (-a) \cdot (-a) \cdot \cdots \cdot (-a)}_{n \text{ factors of } -a}$$

$$-a^n = -1 \cdot \underbrace{a \cdot a \cdot a \cdot \cdots \cdot a}_{n \text{ factors of } a}$$

[YOU TRY 1] Identify the base and exponent in each expression and evaluate.

a) 5^3 b) $(-3)^4$ c) -3^4

---

2 Use the Product Rule for Exponents

Is there a rule to help us *multiply* exponential expressions? Let's rewrite each of the following products as a single power of the base using what we already know:

1) $2^3 \cdot 2^2 = \overset{3 \text{ factors of } 2}{\overbrace{2 \cdot 2 \cdot 2}} \cdot \overset{2 \text{ factors of } 2}{\overbrace{2 \cdot 2}} = 2^5$
     2) $5^4 \cdot 5^3 = \overset{4 \text{ factors of } 5}{\overbrace{5 \cdot 5 \cdot 5 \cdot 5}} \cdot \overset{3 \text{ factors of } 5}{\overbrace{5 \cdot 5 \cdot 5}} = 5^7$

Let's summarize: $2^3 \cdot 2^2 = 2^5$, $5^4 \cdot 5^3 = 5^7$

Do you notice a pattern? *When you multiply expressions with the same base, keep the same base and add the exponents.* This is called the **product rule** for exponents.

> **Property** Product Rule
>
> Let a be any real number, and let m and n be positive integers. Then,
> $$a^m \cdot a^n = a^{m+n}$$

EXAMPLE 2 Find each product.

a) $2^2 \cdot 2^4$ b) $x^9 \cdot x^6$ c) $5c^3 \cdot 7c^9$ d) $(-k)^8 \cdot (-k) \cdot (-k)^{11}$

Solution

a) $2^2 \cdot 2^4 = 2^{2+4} = 2^6 = 64$ Since the bases are the same, add the exponents.

b) $x^9 \cdot x^6 = x^{9+6} = x^{15}$

c) $5c^3 \cdot 7c^9 = (5 \cdot 7)(c^3 \cdot c^9)$ Use the associative and commutative properties.
       $= 35c^{12}$

W Hint
If you need to review the powers of whole numbers, refer to the table in Section 1.2.

d) $(-k)^8 \cdot (-k) \cdot (-k)^{11} = (-k)^{8+1+11} = (-k)^{20}$ Product rule

---

[YOU TRY 2] Find each product.

a) $3 \cdot 3^2$ b) $y^{10} \cdot y^4$ c) $-6m^5 \cdot 9m^{11}$ d) $h^4 \cdot h^6 \cdot h^4$ e) $(-3)^2 \cdot (-3)^2$

BE CAREFUL

Can the product rule be applied to $4^3 \cdot 5^2$? **No!** The bases are not the same, so we cannot add the exponents. To evaluate $4^3 \cdot 5^2$, we would evaluate $4^3 = 64$ and $5^2 = 25$, then multiply:

$$4^3 \cdot 5^2 = 64 \cdot 25 = 1600$$

3 Use the Power Rule $(a^m)^n = a^{mn}$

What does $(2^2)^3$ mean? We can rewrite $(2^2)^3$ first as $2^2 \cdot 2^2 \cdot 2^2$.

$$2^2 \cdot 2^2 \cdot 2^2 = 2^{2+2+2} = 2^6 = 64$$

Notice that $(2^2)^3 = 2^{2+2+2}$, or $2^{2 \cdot 3}$. This leads us to the basic power rule for exponents: *When you raise a power to another power, keep the base and multiply the exponents.*

Property Basic Power Rule

Let a be any real number, and let m and n be positive integers. Then,

$$(a^m)^n = a^{mn}$$

EXAMPLE 3   Simplify using the power rule.

a) $(3^8)^4$ b) $(n^3)^7$ c) $((-f)^4)^3$

Solution

a) $(3^8)^4 = 3^{8 \cdot 4} = 3^{32}$ b) $(n^3)^7 = n^{3 \cdot 7} = n^{21}$ c) $((-f)^4)^3 = (-f)^{4 \cdot 3} = (-f)^{12}$

[YOU TRY 3] Simplify using the power rule.

a) $(5^4)^3$ b) $(j^6)^5$ c) $((-2)^3)^2$

4 Use the Power Rule $(ab)^n = a^n b^n$

W Hint

This property demonstrates how you can distribute an exponent to the bases if the bases are being multiplied.

We can use another power rule to simplify an expression such as $(5c)^3$. We can rewrite and simplify $(5c)^3$ as $5c \cdot 5c \cdot 5c = 5 \cdot 5 \cdot 5 \cdot c \cdot c \cdot c = 5^3 c^3 = 125c^3$. *To raise a product to a power, raise each factor to that power.*

> **Property** Power Rule for a Product
>
> Let a and b be real numbers, and let n be a positive integer. Then,
> $$(ab)^n = a^n b^n$$

BE CAREFUL   Notice that $(ab)^n = a^n b^n$ is different from $(a + b)^n$. $(a + b)^n \neq a^n + b^n$. We will study this in Chapter 6.

EXAMPLE 4

Simplify each expression.

a) $(9y)^2$ b) $\left(\dfrac{1}{4}t\right)^3$ c) $(5c^2)^3$ d) $3(6ab)^2$

Solution

a) $(9y)^2 = 9^2 y^2 = 81y^2$ b) $\left(\dfrac{1}{4}t\right)^3 = \left(\dfrac{1}{4}\right)^3 \cdot t^3 = \dfrac{1}{64}t^3$

c) $(5c^2)^3 = 5^3 \cdot (c^2)^3 = 125c^{2 \cdot 3} = 125c^6$

d) $3(6ab)^2 = 3[6^2 \cdot (a)^2 \cdot (b)^2]$ The 3 is not in parentheses; therefore, it will not be squared.
$= 3(36a^2 b^2)$
$= 108a^2 b^2$

YOU TRY 4

Simplify.

a) $(5a)^3$ b) $\left(\dfrac{1}{2}x\right)^6$ c) $(-r^2 s^8)^3$ d) $-4(3tu)^2$

---

5 Use the Power Rule $\left(\dfrac{a}{b}\right)^n = \dfrac{a^n}{b^n}$ where $b \neq 0$

Another power rule allows us to simplify an expression like $\left(\dfrac{2}{x}\right)^4$. We can rewrite and simplify $\left(\dfrac{2}{x}\right)^4$ as $\dfrac{2}{x} \cdot \dfrac{2}{x} \cdot \dfrac{2}{x} \cdot \dfrac{2}{x} = \dfrac{2 \cdot 2 \cdot 2 \cdot 2}{x \cdot x \cdot x \cdot x} = \dfrac{2^4}{x^4} = \dfrac{16}{x^4}$. *To raise a quotient to a power, raise both the numerator and denominator to that power.*

Property Power Rule for a Quotient

Let a and b be real numbers, and let n be a positive integer. Then,

$$\left(\frac{a}{b}\right)^n = \frac{a^n}{b^n}, \text{ where } b \neq 0$$

EXAMPLE 5

Simplify using the power rule for quotients.

a) $\left(\frac{3}{8}\right)^2$ b) $\left(\frac{5}{x}\right)^3$ c) $\left(\frac{t}{u}\right)^9$

Solution

a) $\left(\frac{3}{8}\right)^2 = \frac{3^2}{8^2} = \frac{9}{64}$ b) $\left(\frac{5}{x}\right)^3 = \frac{5^3}{x^3} = \frac{125}{x^3}$ c) $\left(\frac{t}{u}\right)^9 = \frac{t^9}{u^9}$

[YOU TRY 5]

Simplify using the power rule for quotients.

a) $\left(\frac{5}{12}\right)^2$ b) $\left(\frac{2}{d}\right)^5$ c) $\left(\frac{u}{v}\right)^6$

Let's summarize the rules of exponents we have learned in this section:

Summary The Product and Power Rules of Exponents

In the following rules, a and b are any real numbers, and m and n are positive integers.

	Rule	Example
Product Rule	$a^m \cdot a^n = a^{m+n}$	$p^4 \cdot p^{11} = p^{4+11} = p^{15}$
Basic Power Rule	$(a^m)^n = a^{mn}$	$(c^8)^3 = c^{8 \cdot 3} = c^{24}$
Power Rule for a Product	$(ab)^n = a^n b^n$	$(3z)^4 = 3^4 \cdot z^4 = 81z^4$
Power Rule for a Quotient	$\left(\frac{a}{b}\right)^n = \frac{a^n}{b^n}, \ (b \neq 0)$	$\left(\frac{w}{2}\right)^4 = \frac{w^4}{2^4} = \frac{w^4}{16}$

W Hint

Summarize these properties in your notes.

ANSWERS TO [YOU TRY] EXERCISES

1) a) base: 5; exponent: 3; $5^3 = 125$ b) base: -3; exponent: 4; 81   c) base: 3; exponent: 4; -81

2) a) 27   b) y^{14} c) $-54m^{16}$ d) h^{14} e) 81   3) a) 5^{12} b) j^{30} c) 64

4) a) $125a^3$ b) $\frac{1}{64}x^6$ c) $-r^6 s^{24}$ d) $-36t^2 u^2$ 5) a) $\frac{25}{144}$ b) $\frac{32}{d^5}$ c) $\frac{u^6}{v^6}$

Objective 1: Evaluate Exponential Expressions

Rewrite each expression using exponents.

 1) $9 \cdot 9 \cdot 9 \cdot 9 \cdot 9 \cdot 9$

2) $4 \cdot 4 \cdot 4 \cdot 4 \cdot 4 \cdot 4 \cdot 4$

3) $\left(\dfrac{1}{7}\right)\left(\dfrac{1}{7}\right)\left(\dfrac{1}{7}\right)\left(\dfrac{1}{7}\right)$

4) $(0.8)(0.8)(0.8)$

5) $(-5)(-5)(-5)(-5)(-5)(-5)(-5)$

6) $(-c)(-c)(-c)(-c)(-c)$

7) $(-3y)(-3y)(-3y)(-3y)(-3y)(-3y)(-3y)(-3y)$

8) $\left(-\dfrac{5}{4}t\right)\left(-\dfrac{5}{4}t\right)\left(-\dfrac{5}{4}t\right)\left(-\dfrac{5}{4}t\right)$

Identify the base and the exponent in each.

9) 6^8

10) 9^4

11) $(0.05)^7$

12) $(0.3)^{10}$

13) $(-8)^5$

14) $(-7)^6$

15) $(9x)^8$

16) $(13k)^3$

17) $(-11a)^2$

18) $(-2w)^9$

19) $5p^4$

20) $-3m^5$

21) $-\dfrac{3}{8}y^2$

22) $\dfrac{5}{9}t^7$

23) Evaluate $(3+4)^2$ and $3^2 + 4^2$. Are they equivalent? Why or why not?

24) Evaluate $(7-3)^2$ and $7^2 - 3^2$. Are they equivalent? Why or why not?

25) For any values of a and b, does $(a+b)^2 = a^2 + b^2$? Why or why not?

26) Does $-2^4 = (-2)^4$? Why or why not?

27) Are $3t^4$ and $(3t)^4$ equivalent? Why or why not?

28) Is there any value of a for which $(-a)^2 = -a^2$? Support your answer with an example.

Evaluate.

29) 2^5

30) 9^2

31) $(11)^2$

32) 4^3

33) $(-2)^4$

34) $(-5)^3$

35) -3^4

36) -6^2

37) -2^3

38) -8^2

39) $\left(\dfrac{1}{5}\right)^3$

40) $\left(\dfrac{3}{2}\right)^4$

For Exercises 41–44, fill in the blank with *always, sometimes,* or *never* to make the statement true.

41) Raising a negative base to an even exponent power will _____ give a negative result.

42) If the base of an exponential expression is 1, the result will _____ be 1.

43) If b is any integer value except zero, then the exponential expression $(-b)^3$ will _____ give a negative result.

44) If a is any integer value except zero, then the exponential expression $-a^4$ will _____ give a positive result.

Objective 2: Use the Product Rule for Exponents

Evaluate the expression using the product rule, where applicable.

45) $2^2 \cdot 2^3$

46) $5^2 \cdot 5$

47) $3^2 \cdot 3^2$

48) $2^3 \cdot 2^3$

49) $5^2 \cdot 2^3$

50) $4^3 \cdot 3^2$

51) $\left(\dfrac{1}{2}\right)^4 \cdot \left(\dfrac{1}{2}\right)^2$

52) $\left(\dfrac{4}{3}\right) \cdot \left(\dfrac{4}{3}\right)^2$

Simplify the expression using the product rule. Leave your answer in exponential form.

53) $8^3 \cdot 8^9$

54) $6^4 \cdot 6^3$

55) $5^2 \cdot 5^4 \cdot 5^5$

56) $12^4 \cdot 12 \cdot 12^2$

57) $(-7)^2 \cdot (-7)^3 \cdot (-7)^3$

58) $(-3)^5 \cdot (-3) \cdot (-3)^6$

59) $b^2 \cdot b^4$

60) $x^4 \cdot x^3$

61) $k \cdot k^2 \cdot k^3$

62) $n^6 \cdot n^5 \cdot n^2$

63) $8y^3 \cdot y^2$

64) $10c^8 \cdot c^2 \cdot c$

65) $(9m^4)(6m^{11})$

66) $(-10p^8)(-3p)$

67) $(-6r)(7r^4)$

68) $(8h^5)(-5h^2)$

69) $(-7t^6)(t^3)(-4t^7)$

70) $(3k^2)(-4k^5)(2k^4)$

71) $\left(\dfrac{5}{3}x^2\right)(12x)(-2x^3)$

72) $\left(\dfrac{7}{10}y^9\right)(-2y^4)(3y^2)$

73) $\left(\dfrac{8}{21}b\right)(-6b^8)\left(-\dfrac{7}{2}b^6\right)$

74) $(12c^3)\left(\dfrac{14}{15}c^2\right)\left(\dfrac{5}{7}c^6\right)$

Mixed Exercises: Objectives 3–5

Simplify the expression using one of the power rules.

75) $(y^3)^4$

76) $(x^5)^8$

77) $(w^{11})^7$

78) $(a^3)^2$

79) $(3^3)^2$

80) $(2^2)^2$

81) $((-5)^3)^2$

82) $((-4)^5)^3$

83) $\left(\dfrac{1}{3}\right)^4$

84) $\left(\dfrac{5}{2}\right)^3$

85) $\left(\dfrac{6}{a}\right)^2$

86) $\left(\dfrac{v}{4}\right)^3$

87) $\left(\dfrac{m}{n}\right)^5$

88) $\left(\dfrac{t}{u}\right)^{12}$

89) $(10y)^4$

90) $(7w)^2$

91) $(-3p)^4$

92) $(2m)^5$

93) $(-4ab)^3$

94) $(-2cd)^4$

95) $6(xy)^3$

96) $-8(mn)^5$

97) $-9(tu)^4$

98) $2(ab)^6$

Mixed Exercises: Objectives 2–5

99) Find the area and perimeter of each rectangle.

a)

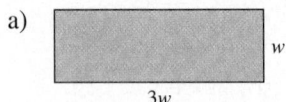

b)

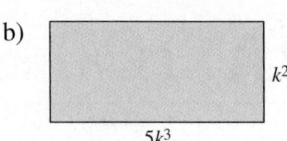

100) Find the area.

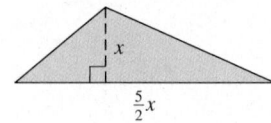

101) Find the area.

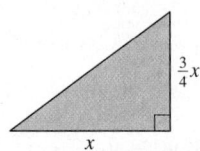

102) Here are the shape and dimensions of the Millers' family room. They will have wall-to-wall carpeting installed, and the carpet they have chosen costs $2.50/ft^2.

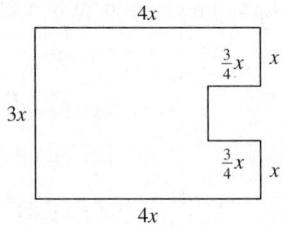

a) Write an expression for the amount of carpet they will need. (Include the correct units.)

b) Write an expression for the cost of carpeting the family room. (Include the correct units.)

2.1B Basic Rules of Exponents: Combining the Rules

P Prepare

O Organize

What is your objective for Section 2.1B?	How can you accomplish the objective?
1 Combine the Product Rule and Power Rules of Exponents	• Be able to follow the order of operations correctly, and complete the given example on your own. • Complete You Try 1.

W Work   Read the explanations, follow the examples, take notes, and complete the You Try.

1 Combine the Product Rule and Power Rules of Exponents

When we combine the rules of exponents, we follow the order of operations.

EXAMPLE 1   Simplify.

a) $(2c)^3(3c^8)^2$ b) $2(5k^4m^3)^3$ c) $\dfrac{(6t^5)^2}{(2u^4)^3}$

Solution

a) $(2c)^3(3c^8)^2$

Because evaluating exponents comes before multiplying in the order of operations, *evaluate the exponents first.*

$$(2c)^3(3c^8)^2 = (2^3c^3)(3^2)(c^8)^2 \qquad \text{Use the power rule.}$$
$$= (8c^3)(9c^{16}) \qquad \text{Use the power rule, and evaluate exponents.}$$
$$= 72c^{19} \qquad \text{Product rule}$$

W Hint

Notice how the parentheses are used in these examples.

b) $2(5k^4m^3)^3$

Which operation should be performed first, multiplying $2 \cdot 5$ or simplifying $(5k^4m^3)^3$? In the order of operations, we evaluate exponents before multiplying, so *we will begin by simplifying $(5k^4m^3)^3$.*

$$2(5k^4m^3)^3 = 2 \cdot (5)^3(k^4)^3(m^3)^3 \qquad \text{Order of operations and power rule}$$
$$= 2 \cdot 125k^{12}m^9 \qquad \text{Power rule}$$
$$= 250k^{12}m^9 \qquad \text{Multiply.}$$

c) $\dfrac{(6t^5)^2}{(2u^4)^3}$

What comes first in the order of operations, dividing or evaluating exponents? *Evaluating exponents.*

$$\frac{(6t^5)^2}{(2u^4)^3} = \frac{36t^{10}}{8u^{12}} \qquad \text{Power rule}$$
$$= \frac{\overset{9}{\cancel{36}}t^{10}}{\underset{2}{\cancel{8}}u^{12}} \qquad \text{Divide out the common factor of 4.}$$
$$= \frac{9t^{10}}{2u^{12}}$$

 **BE CAREFUL**

When simplifying the expression in Example 1c, $\dfrac{(6t^5)^2}{(2u^4)^3}$, it may be tempting to simplify before applying the product rule, like this:

$$\dfrac{(\overset{3}{6}t^5)^2}{(2u^4)^3} \neq \dfrac{(3t^5)^2}{(u^4)^3} = \dfrac{9t^{10}}{u^{12}} \quad \textbf{Wrong!}$$
$$\underset{1}{}$$

You can see, however, that because we did not follow the rules for the order of operations, we did *not* get the correct answer.

[YOU TRY 1] Simplify.

a) $-4(2a^9b^6)^4$

b) $(7x^{10}y)^2(-x^4y^5)^4$

c) $\dfrac{10(m^2n^3)^5}{(5p^4)^2}$

d) $\left(\dfrac{1}{6}w^7\right)^2(3w^{11})^3$

ANSWERS TO [YOU TRY] EXERCISE

1) a) $-64a^{36}b^{24}$ b) $49x^{36}y^{22}$ c) $\dfrac{2m^{10}n^{15}}{5p^8}$ d) $\dfrac{3}{4}w^{47}$

E Evaluate **2.1B** Exercises   Do the exercises, and check your work.

Objective 1: Combine the Product Rule and Power Rules of Exponents

1) When evaluating expressions involving exponents, always keep in mind the order of _____.

2) The first step in evaluating $(9 - 3)^2$ is _____.

Simplify.

3) $(k^9)^2(k^3)^2$

4) $(d^5)^3(d^2)^4$

5) $(5z^4)^2(2z^6)^3$

6) $(3r)^2(6r^8)^2$

7) $6ab(-a^{10}b^2)^3$

8) $-5pq^4(-p^4q)^4$

9) $(9 + 2)^2$

10) $(8 - 5)^3$

11) $(-4t^6u^2)^3(u^4)^5$

12) $(-m^2)^6(-2m^9)^4$

13) $8(6k^7l^2)^2$

14) $5(-7c^4d)^2$

15) $\left(\dfrac{3}{g^5}\right)^3\left(\dfrac{1}{6}\right)^2$

16) $\left(-\dfrac{2}{5}z^5\right)^3(10z)^2$

17) $\left(\dfrac{7}{8}n^2\right)^2(-4n^9)^2$

18) $\left(\dfrac{2}{3}d^8\right)^4\left(\dfrac{9}{2}d^3\right)^2$

19) $h^4(10h^3)^2(-3h^9)^2$

20) $-v^6(-2v^5)^5(-v^4)^3$

21) $3w^{11}(7w^2)^2(-w^6)^5$

22) $5z^3(-4z)^2(2z^3)^2$

23) $\dfrac{(12x^3)^2}{(10y^5)^2}$

24) $\dfrac{(-3a^4)^3}{(6b)^2}$

25) $\dfrac{(4d^9)^2}{(-2c^5)^6}$

26) $\dfrac{(-5m^7)^3}{(5n^{12})^2}$

27) $\dfrac{8(a^4b^7)^9}{(6c)^2}$

28) $\dfrac{(3x^5)^3}{21(yz^2)^6}$

29) $\dfrac{r^4(r^5)^7}{2t(11t^2)^2}$

30) $\dfrac{k^5(k^2)^3}{7m^{10}(2m^3)^2}$

31) $\left(\dfrac{4}{9}x^3y\right)^2\left(\dfrac{3}{2}x^6y^4\right)^3$

32) $(6s^8t^3)^2\left(-\dfrac{10}{3}st^4\right)^2$

33) $\left(-\dfrac{2}{5}c^9d^2\right)^3\left(\dfrac{5}{4}cd^6\right)^2$

34) $-\dfrac{11}{12}\left(\dfrac{3}{2}m^3n^{10}\right)^2$

35) $\left(\dfrac{5x^5y^2}{z^4}\right)^3$

36) $\left(-\dfrac{7a^4b}{8c^6}\right)^2$

37) $\left(-\dfrac{3t^4u^9}{2v^7}\right)^4$

38) $\left(\dfrac{2pr^8}{q^{11}}\right)^5$

39) $\left(\dfrac{12w^5}{4x^3y^6}\right)^2$

40) $\left(\dfrac{10b^3c^5}{15a}\right)^2$

98   CHAPTER 2   **Rules of Exponents**

Determine whether the equation is *true* or *false*.

41) $(2k^2 + 2k^2) = (2k^2)^2$

42) $(4c^3 - 2c^2) = 2c$

For Exercises 43 and 44, fill in the blank with *always, sometimes,* or *never* to make the statement true.

43) If a and b are any integer values except zero, then the exponential expression $-b^2(-a)^4$ will _____ give a positive result.

44) If a and b are any integer values except zero, then the exponential expression $\left(\dfrac{-a}{b}\right)^3$ will _____ give a positive result.

45) The length of a side of a square is $5l^2$ units.

   a) Write an expression for its perimeter.

   b) Write an expression for its area.

46) The width of a rectangle is $2w$ units, and the length of the rectangle is $7w$ units.

   a) Write an expression for its area.

   b) Write an expression for its perimeter.

 47) The length of a rectangle is x units, and the width of the rectangle is $\dfrac{3}{8}x$ units.

   a) Write an expression for its area.

   b) Write an expression for its perimeter.

48) The width of a rectangle is $4y^3$ units, and the length of the rectangle is $\dfrac{13}{2}y^3$ units.

   a) Write an expression for its perimeter.

   b) Write an expression for its area.

49) Write an exponential expression that uses the product rule to simplify to k^{12}.

50) Write an exponential expression that uses the power rule for quotients to simplify to $\dfrac{c^7}{d^7}$.

51) Write an exponential expression that uses a combination of the rules of exponents to simplify to $\dfrac{n^{15}}{m^{24}}$.

52) Write an exponential expression that uses a combination of the rules of exponents to simplify to $16a^{36}$.

R Rethink

R1) How are the parentheses being used in these problems?

R2) Which exercises do you need to come back to and try again? How will this help you prepare for an exam?

2.2A Integer Exponents: Real-Number Bases

P Prepare

O Organize

What are your objectives for Section 2.2A?	How can you accomplish each objective?
1 Use 0 as an Exponent	• Understand the definition of *zero as an exponent*, and write it in your notes. • Complete the given example on your own. • Complete You Try 1.
2 Use Negative Integers as Exponents	• Understand the definition of *negative exponent*, and write it in your notes. • Complete the given example on your own. • Complete You Try 2.

W Work   **Read the explanations, follow the examples, take notes, and complete the You Trys.**

So far, we have defined an exponential expression such as 2^3. The exponent of 3 indicates that $2^3 = 2 \cdot 2 \cdot 2$ (3 factors of 2) so that $2^3 = 2 \cdot 2 \cdot 2 = 8$. Is it possible to have an exponent of zero or a negative exponent? If so, what do they mean?

1 Use 0 as an Exponent

Definition

Zero as an Exponent: If $a \neq 0$, then $a^0 = 1$.

How can this be possible?

Let's evaluate $2^0 \cdot 2^3$. Using the product rule, we get:

$$2^0 \cdot 2^3 = 2^{0+3} = 2^3 = 8$$

But we know that $2^3 = 8$. Therefore, if $2^0 \cdot 2^3 = 8$, then $2^0 = 1$. This is one way to understand that $a^0 = 1$.

EXAMPLE 1

Evaluate each expression.

a) 5^0 b) -8^0 c) $(-7)^0$ d) $-3(2^0)$

Solution

a) $5^0 = 1$ b) $-8^0 = -1 \cdot 8^0 = -1 \cdot 1 = -1$

c) $(-7)^0 = 1$ d) $-3(2^0) = -3(1) = -3$

[YOU TRY 1]   Evaluate.

a) 9^0 b) -2^0 c) $(-5)^0$ d) $3^0(-2)$

2 Use Negative Integers as Exponents

So far, we have worked with exponents that are zero or positive. What does a negative exponent mean?

Let's use the product rule to find $2^3 \cdot 2^{-3}$: $2^3 \cdot 2^{-3} = 2^{3+(-3)} = 2^0 = 1$

Remember that a number multiplied by its reciprocal is 1, and here we have that a quantity, 2^3, times another quantity, 2^{-3}, is 1. Therefore, 2^3 and 2^{-3} are reciprocals!

This leads to the definition of a negative exponent.

W Hint

Did you notice that the signs of the bases do NOT change?

Definition

Negative Exponent: If n is any integer and a and b are not equal to zero, then

$$a^{-n} = \left(\frac{1}{a}\right)^n = \frac{1}{a^n} \quad \text{and} \quad \left(\frac{a}{b}\right)^{-n} = \left(\frac{b}{a}\right)^n.$$

Therefore, to rewrite an expression of the form a^{-n} with a positive exponent, *take the reciprocal of the base and make the exponent positive.*

EXAMPLE 2   Evaluate each expression.

a) 2^{-3} b) $\left(\dfrac{3}{2}\right)^{-4}$ c) $\left(\dfrac{1}{5}\right)^{-3}$ d) $(-7)^{-2}$

Solution

a) 2^{-3}: The reciprocal of 2 is $\dfrac{1}{2}$, so $2^{-3} = \left(\dfrac{1}{2}\right)^{3} = \dfrac{1^3}{2^3} = \dfrac{1}{8}$.

b) $\left(\dfrac{3}{2}\right)^{-4}$: The reciprocal of $\dfrac{3}{2}$ is $\dfrac{2}{3}$, so $\left(\dfrac{3}{2}\right)^{-4} = \left(\dfrac{2}{3}\right)^{4} = \dfrac{2^4}{3^4} = \dfrac{16}{81}$.

 **BE CAREFUL**   Notice that a negative exponent does not make the answer negative!

W Hint

Before working out these examples, try to determine by inspection whether the answer is going to be positive or negative.

c) $\left(\dfrac{1}{5}\right)^{-3}$: The reciprocal of $\dfrac{1}{5}$ is 5, so $\left(\dfrac{1}{5}\right)^{-3} = 5^3 = 125$.

d) $(-7)^{-2}$: The reciprocal of -7 is $-\dfrac{1}{7}$, so

$$(-7)^{-2} = \left(-\dfrac{1}{7}\right)^{2} = \left(-1 \cdot \dfrac{1}{7}\right)^{2} = (-1)^2 \left(\dfrac{1}{7}\right)^{2} = 1 \cdot \dfrac{1^2}{7^2} = \dfrac{1}{49}$$

[YOU TRY 2]   Evaluate.

a) $(10)^{-2}$ b) $\left(\dfrac{1}{4}\right)^{-2}$ c) $\left(\dfrac{2}{3}\right)^{-3}$ d) $(-5)^{-2}$

ANSWERS TO [YOU TRY] EXERCISES

1) a) 1 b) −1 c) 1 d) −2   2) a) $\dfrac{1}{100}$ b) 16 c) $\dfrac{27}{8}$ d) $\dfrac{1}{25}$

E Evaluate **2.2A** Exercises   Do the exercises, and check your work.

Mixed Exercises: Objectives 1 and 2

 1) True or False: Raising a positive base to a negative exponent will give a negative result. (Example: 2^{-4})

2) True or False: $8^0 = 1$.

3) True or False: The reciprocal of 4 is $\dfrac{1}{4}$.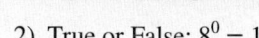

4) True or False: $3^{-2} - 2^{-2} = 1^{-2}$.

Evaluate.

5) 2^0 6) $(-4)^0$

 7) -5^0 8) -1^0

9) 0^8 10) $-(-9)^0$

11) $(5)^0 + (-5)^0$

12) $\left(\dfrac{4}{7}\right)^0 - \left(\dfrac{7}{4}\right)^0$

13) 6^{-2}

14) 9^{-2}

15) 2^{-4}

16) 11^{-2}

17) 5^{-3}

18) 2^{-5}

19) $\left(\dfrac{1}{8}\right)^{-2}$

20) $\left(\dfrac{1}{10}\right)^{-3}$

21) $\left(\dfrac{1}{2}\right)^{-5}$

22) $\left(\dfrac{1}{4}\right)^{-2}$

23) $\left(\dfrac{4}{3}\right)^{-3}$

24) $\left(\dfrac{2}{5}\right)^{-3}$

SECTION 2.2A   Integer Exponents: Real-Number Bases   101

25) $\left(\dfrac{9}{7}\right)^{-2}$

26) $\left(\dfrac{10}{3}\right)^{-2}$

27) $\left(-\dfrac{1}{4}\right)^{-3}$

28) $\left(-\dfrac{1}{12}\right)^{-2}$

 29) $\left(-\dfrac{3}{8}\right)^{-2}$

30) $\left(-\dfrac{5}{2}\right)^{-3}$

31) -2^{-6}

32) -4^{-3}

33) -1^{-5}

34) -9^{-2}

35) $2^{-3} - 4^{-2}$

36) $5^{-2} + 2^{-2}$

37) $2^{-2} + 3^{-2}$

38) $4^{-1} - 6^{-2}$

39) $-9^{-2} + 3^{-3} + (-7)^0$

40) $6^0 - 9^{-1} + 4^0 + 3^{-2}$

41) Write an exponential expression containing a negative exponent that simplifies to $\dfrac{1}{13}$.

42) Write an exponential expression containing a negative exponent that simplifies to $\dfrac{1}{10}$.

Write an exponential expression containing a negative exponent, other than –1, that simplifies to the given expression.

43) $\dfrac{1}{100}$

44) $\dfrac{1}{25}$

45) $\dfrac{125}{27}$

46) $\dfrac{121}{64}$

47) $-\dfrac{4}{49}$

48) $-\dfrac{81}{16}$

R Rethink

R1) When evaluating an expression with negative exponents, when do you get a negative answer?

R2) Explain the importance of fully understanding and using the order of operations in this chapter.

2.2B Integer Exponents: Variable Bases

P Prepare

What are your objectives for Section 2.2B?

O Organize

How can you accomplish each objective?

What are your objectives for Section 2.2B?	How can you accomplish each objective?
1 Use 0 as an Exponent	• Use the same definition of *zero as an exponent* used in the previous section. • Complete the given example on your own. • Complete You Try 1.
2 Rewrite an Exponential Expression with Positive Exponents	• Follow the explanation to understand how to rewrite an expression with positive exponents. • Learn and apply the definition of $\dfrac{a^{-m}}{b^{-n}} = \dfrac{b^n}{a^m}$. • Complete the given examples on your own. • Complete You Trys 2 and 3.

W Work

Read the explanations, follow the examples, take notes, and complete the You Trys.

 Use 0 as an Exponent

We can apply 0 as an exponent to bases containing variables.

EXAMPLE 1

Evaluate each expression. Assume that the variable does not equal zero.

a) t^0 b) $(-k)^0$ c) $-(11p)^0$

Solution

a) $t^0 = 1$ b) $(-k)^0 = 1$ c) $-(11p)^0 = -1 \cdot (11p)^0 = -1 \cdot 1 = -1$

[YOU TRY 1]

Evaluate. Assume that the variable does not equal zero.

a) p^0 b) $(-10x)^0$ c) $-(7s)^0$

2 Rewrite an Exponential Expression with Positive Exponents

Next, let's apply the definition of a negative exponent to bases containing variables. As in Example 1, we will assume that the variable does not equal zero since having zero in the denominator of a fraction will make the fraction undefined.

Recall that $2^{-4} = \left(\dfrac{1}{2}\right)^4 = \dfrac{1}{16}$. That is, to rewrite the expression with a positive exponent, we take the reciprocal of the base.

What is the reciprocal of x? The reciprocal is $\dfrac{1}{x}$.

EXAMPLE 2

Rewrite the expression with positive exponents. Assume that the variable does not equal zero.

a) x^{-6} b) $\left(\dfrac{2}{n}\right)^{-6}$ c) $3a^{-2}$

Solution

a) $x^{-6} = \left(\dfrac{1}{x}\right)^6 = \dfrac{1^6}{x^6} = \dfrac{1}{x^6}$

b) $\left(\dfrac{2}{n}\right)^{-6} = \left(\dfrac{n}{2}\right)^6$ The reciprocal of $\dfrac{2}{n}$ is $\dfrac{n}{2}$.

$= \dfrac{n^6}{2^6} = \dfrac{n^6}{64}$

W Hint

Before working out these examples, be sure you correctly identify the base.

c) $3a^{-2} = 3 \cdot \left(\dfrac{1}{a}\right)^2$ Remember, the base is a, *not* $3a$, since there are no parentheses. Therefore, the exponent of -2 applies only to a.

$= 3 \cdot \dfrac{1}{a^2} = \dfrac{3}{a^2}$

[YOU TRY 2]

Rewrite the expression with positive exponents. Assume that the variable does not equal zero.

a) m^{-4} b) $\left(\dfrac{1}{z}\right)^{-7}$ c) $\left(\dfrac{a}{2}\right)^{-5}$ d) $-2y^{-3}$

How could we rewrite $\dfrac{x^{-2}}{y^{-2}}$ with only positive exponents? One way would be to apply the power rule for exponents: $\dfrac{x^{-2}}{y^{-2}} = \left(\dfrac{x}{y}\right)^{-2} = \left(\dfrac{y}{x}\right)^2 = \dfrac{y^2}{x^2}$

Notice that to rewrite the original expression with only positive exponents, the terms with the negative exponents "switch" their positions in the fraction. We can generalize this way:

Definition

If m and n are any integers and a and b are real numbers not equal to zero, then

$$\frac{a^{-m}}{b^{-n}} = \frac{b^n}{a^m}$$

EXAMPLE 3

Rewrite the expression with positive exponents. Assume that the variables do not equal zero.

a) $\dfrac{c^{-8}}{d^{-3}}$ b) $\dfrac{5p^{-6}}{q^7}$ c) $t^{-2}u^{-1}$ d) $\dfrac{2xy^{-3}}{3z^{-2}}$ e) $\left(\dfrac{ab}{4c}\right)^{-3}$

Solution

a) $\dfrac{c^{-8}}{d^{-3}} = \dfrac{d^3}{c^8}$ To make the exponents positive, "switch" the positions of the terms in the fraction.

b) $\dfrac{5p^{-6}}{q^7} = \dfrac{5}{p^6q^7}$ Since the exponent on q is positive, we do not change its position in the expression.

c) $t^{-2}u^{-1} = \dfrac{t^{-2}u^{-1}}{1}$

$= \dfrac{1}{t^2u^1} = \dfrac{1}{t^2u}$ Move $t^{-2}u^{-1}$ to the denominator to write with positive exponents.

W Hint

If the exponent is already positive, do not change the position of the expression in the fraction.

d) $\dfrac{2xy^{-3}}{3z^{-2}} = \dfrac{2xz^2}{3y^3}$ To make the exponents positive, "switch" the positions of the factors with negative exponents in the fraction.

e) $\left(\dfrac{ab}{4c}\right)^{-3} = \left(\dfrac{4c}{ab}\right)^3$ To make the exponent positive, use the reciprocal of the base.

$= \dfrac{4^3c^3}{a^3b^3}$ Power rule

$= \dfrac{64c^3}{a^3b^3}$ Simplify.

[YOU TRY 3]

Rewrite the expression with positive exponents. Assume that the variables do not equal zero.

a) $\dfrac{n^{-6}}{y^{-2}}$ b) $\dfrac{z^{-9}}{3k^{-4}}$ c) $u^{-1}v^{-5}$ d) $\dfrac{8d^{-4}}{6m^2n^{-1}}$ e) $\left(\dfrac{3xz}{y}\right)^{-2}$

ANSWERS TO [YOU TRY] EXERCISES

1) a) 1 b) 1 c) −1   2) a) $\dfrac{1}{m^4}$ b) z^7 c) $\dfrac{32}{a^5}$ d) $-\dfrac{2}{y^3}$ 3) a) $\dfrac{y^2}{n^6}$ b) $\dfrac{k^4}{3z^9}$

c) $\dfrac{1}{uv^5}$ d) $\dfrac{4n}{3m^2d^4}$ e) $\dfrac{y^2}{9x^2z^2}$

Objective 1: Use 0 as an Exponent

1) Identify the base in each expression.

a) w^0 b) $-3n^{-5}$

c) $(2p)^{-3}$ d) $4c^0$

2) True or False: $6^0 - 4^0 = (6-4)^0$

Evaluate. Assume that the variables do not equal zero.

3) r^0 4) $(5m)^0$

5) $-2k^0$ 6) $-z^0$

7) $x^0 + (2x)^0$ 8) $\left(\dfrac{7}{8}\right)^0 - \left(\dfrac{3}{5}\right)^0$

Objective 2: Rewrite an Exponential Expression with Positive Exponents

Rewrite each expression with only positive exponents. Assume that the variables do not equal zero.

9) d^{-3} 10) y^{-7}

11) p^{-1} 12) a^{-5}

13) $\dfrac{a^{-10}}{b^{-3}}$ 14) $\dfrac{h^{-2}}{k^{-1}}$

15) $\dfrac{y^{-8}}{x^{-5}}$ 16) $\dfrac{v^{-2}}{w^{-7}}$

17) $\dfrac{t^5}{8u^{-3}}$ 18) $\dfrac{9x^{-4}}{y^5}$

19) $5m^6 n^{-2}$ 20) $\dfrac{1}{9}a^{-4}b^3$

21) $\dfrac{2}{t^{-11}u^{-5}}$ 22) $\dfrac{7r}{2t^{-9}u^2}$

23) $\dfrac{8a^6 b^{-1}}{5c^{-10}d}$ 24) $\dfrac{17k^{-8}h^5}{20m^{-7}n^{-2}}$

25) $\dfrac{2z^4}{x^{-7}y^{-6}}$ 26) $\dfrac{1}{a^{-2}b^{-2}c^{-1}}$

27) $\left(\dfrac{a}{6}\right)^{-2}$ 28) $\left(\dfrac{3}{y}\right)^{-4}$

29) $\left(\dfrac{2n}{q}\right)^{-5}$ 30) $\left(\dfrac{w}{5v}\right)^{-3}$

31) $\left(\dfrac{12b}{cd}\right)^{-2}$ 32) $\left(\dfrac{2tu}{v}\right)^{-6}$

33) $-9k^{-2}$ 34) $3g^{-5}$

35) $3t^{-3}$ 36) $8h^{-4}$

37) $-m^{-9}$ 38) $-d^{-5}$

39) $\left(\dfrac{1}{z}\right)^{-10}$ 40) $\left(\dfrac{1}{k}\right)^{-6}$

41) $\left(\dfrac{1}{j}\right)^{-1}$ 42) $\left(\dfrac{1}{c}\right)^{-7}$

43) $5\left(\dfrac{1}{n}\right)^{-2}$ 44) $7\left(\dfrac{1}{t}\right)^{-8}$

45) $c\left(\dfrac{1}{d}\right)^{-3}$ 46) $x^2\left(\dfrac{1}{y}\right)^{-2}$

For Exercises 47–50, fill in the blank with *always, sometimes,* or *never* to make the statement true.

47) If a is any integer value except zero, then the exponential expression $-a^{-2}$ will _____ give a negative result.

48) If b is any integer value except zero, then the exponential expression $(-b)^{-3}$ will _____ give a positive result.

49) If a and b are any integer values except zero, then the exponential expression $a^0 - b^0$ will _____ equal zero.

50) If a and b are any integer values except zero, then the exponential expression $(a^0 b^0)^{-2}$ will _____ equal zero.

Determine whether the equation is *true* or *false.*

51) $2 \div 2t^5 = t^{-5}$

52) $\dfrac{6x^5}{7} \div \dfrac{6x^8}{7} = \dfrac{7}{6}x^{-3}$

53) $(p^{-1} \div 3q^{-1})^{-2} = \left(\dfrac{q}{3p}\right)^2$

54) $(h^{-2} \div 4k^{-2})^2 = \left(\dfrac{2h}{k}\right)^{-4}$

R1) If a variable is raised to the power zero, the answer will always be what number? Assume the variable does not equal zero.

R2) Why is it useful to write expressions without negative exponents?

R3) Were there any problems you were unable to do? If so, write them down or circle them and ask your instructor for help.

2.3 The Quotient Rule

P Prepare

O Organize

What is your objective for Section 2.3?	How can you accomplish the objective?
1 Use the Quotient Rule for Exponents	• Learn the **Quotient Rule for Exponents,** and write an example in your notes. • Complete the given examples on your own. • Complete You Trys 1 and 2.

W Work

Read the explanations, follow the examples, take notes, and complete the You Trys.

1 Use the Quotient Rule for Exponents

In this section, we will discuss how to simplify the quotient of two exponential expressions with the same base. Let's begin by simplifying $\dfrac{8^6}{8^4}$. One way to simplify this expression is to write the numerator and denominator without exponents:

$$\frac{8^6}{8^4} = \frac{8 \cdot 8 \cdot 8 \cdot 8 \cdot 8 \cdot 8}{8 \cdot 8 \cdot 8 \cdot 8} \qquad \text{Divide out common factors.}$$

$$= 8 \cdot 8 = 8^2 = 64$$

Therefore, $\dfrac{8^6}{8^4} = 8^2 = 64$

Do you notice a relationship between the exponents in the original expression and the exponent we get when we simplify?

$$\frac{8^6}{8^4} = 8^{6-4} = 8^2 = 64$$

That's right. We *subtracted* the exponents.

Property Quotient Rule for Exponents

If m and n are any integers and $a \neq 0$, then

$$\frac{a^m}{a^n} = a^{m-n}$$

To divide expressions with the same base, keep the base the same and subtract the denominator's exponent from the numerator's exponent.

EXAMPLE 1

Simplify. Assume that the variables do not equal zero.

a) $\dfrac{2^9}{2^3}$ b) $\dfrac{t^{10}}{t^4}$ c) $\dfrac{3}{3^{-2}}$ d) $\dfrac{n^5}{n^7}$ e) $\dfrac{3^2}{2^4}$

Solution

a) $\dfrac{2^9}{2^3} = 2^{9-3} = 2^6 = 64$ Since the bases are the same, subtract the exponents.

b) $\dfrac{t^{10}}{t^4} = t^{10-4} = t^6$ Because the bases are the same, subtract the exponents.

W Hint

Be careful when you subtract exponents, especially when you are working with negative numbers!

c) $\dfrac{3}{3^{-2}} = \dfrac{3^1}{3^{-2}} = 3^{1-(-2)}$ Since the bases are the same, subtract the exponents.

$\phantom{\dfrac{3}{3^{-2}}} = 3^3 = 27$ Be careful when subtracting the negative exponent!

d) $\dfrac{n^5}{n^7} = n^{5-7} = n^{-2}$ Same base; subtract the exponents.

$\phantom{\dfrac{n^5}{n^7}} = \left(\dfrac{1}{n}\right)^2 = \dfrac{1}{n^2}$ Write with a positive exponent.

e) $\dfrac{3^2}{2^4} = \dfrac{9}{16}$ Because the bases are not the same, we cannot apply the quotient rule. Evaluate the numerator and denominator separately.

[YOU TRY 1]

Simplify. Assume that the variables do not equal zero.

a) $\dfrac{5^7}{5^4}$ b) $\dfrac{c^4}{c^{-1}}$ c) $\dfrac{k^2}{k^{10}}$ d) $\dfrac{2^3}{7^2}$

We can apply the quotient rule to expressions containing more than one variable. Here are more examples.

EXAMPLE 2

Simplify. Assume that the variables do not equal zero.

a) $\dfrac{x^8 y^7}{x^3 y^4}$ b) $\dfrac{12a^{-5}b^{10}}{8a^{-3}b^2}$

Solution

a) $\dfrac{x^8 y^7}{x^3 y^4} = x^{8-3}y^{7-4}$ Subtract the exponents.

$\phantom{\dfrac{x^8 y^7}{x^3 y^4}} = x^5 y^3$

b) $\dfrac{12a^{-5}b^{10}}{8a^{-3}b^2}$

We will simplify $\dfrac{12}{8}$ in addition to applying the quotient rule.

$$\dfrac{\overset{3}{\cancel{12}}a^{-5}b^{10}}{\underset{2}{\cancel{8}}a^{-3}b^2} = \dfrac{3}{2}a^{-5-(-3)}b^{10-2}$$ Subtract the exponents.

$$= \dfrac{3}{2}a^{-5+3}b^8 = \dfrac{3}{2}a^{-2}b^8 = \dfrac{3b^8}{2a^2}$$

[YOU TRY 2] Simplify. Assume that the variables do not equal zero.

a) $\dfrac{r^4 s^{10}}{rs^3}$ b) $\dfrac{30m^6 n^{-8}}{42m^4 n^{-3}}$

ANSWERS TO [YOU TRY] EXERCISES

1) a) 125   b) c^5 c) $\dfrac{1}{k^8}$ d) $\dfrac{8}{49}$ 2) a) $r^3 s^7$ b) $\dfrac{5m^2}{7n^5}$

E Evaluate **2.3** **Exercises**   Do the exercises, and check your work.

Get Ready

1) $7 - 12$

2) $-3 - 4$

3) $-8 - (-5)$

4) $1 - 9$

5) $6 - (-1)$

6) $-2 - (-2)$

Objective 1: Use the Quotient Rule for Exponents

State what is wrong with the following steps and then simplify correctly.

7) $\dfrac{a^5}{a^3} = a^{3-5} = a^{-2} = \dfrac{1}{a^2}$

8) $\dfrac{4^3}{2^6} = \left(\dfrac{4}{2}\right)^{3-6} = 2^{-3} = \dfrac{1}{2^3} = \dfrac{1}{8}$

Determine whether the equation is *true* or *false*.

9) $3^{-3} \div 3^{-4} = \dfrac{3^3}{3^4}$

10) $2^{-4} \div 2^{-3} = \dfrac{2^3}{2^4}$

11) $t^7 \div t^{-5} = t^2$

12) $m^{10} \div m^{-3} = m^7$

13) $r^{-6} \div r^{-2} = r^{-4}$

14) $k^{-5} \div k^{-10} = k^{-5}$

Simplify using the quotient rule. Assume that the variables do not equal zero.

15) $\dfrac{d^{10}}{d^5}$

16) $\dfrac{z^{11}}{z^7}$

(24) 17) $\dfrac{m^9}{m^5}$

18) $\dfrac{a^6}{a}$

19) $\dfrac{8t^{15}}{t^8}$

20) $\dfrac{4k^4}{k^2}$

21) $\dfrac{6^{12}}{6^{10}}$

22) $\dfrac{4^4}{4}$

23) $\dfrac{3^{12}}{3^8}$

24) $\dfrac{2^7}{2^4}$

25) $\dfrac{2^5}{2^9}$

26) $\dfrac{9^5}{9^7}$

27) $\dfrac{5^6}{5^9}$

28) $\dfrac{8^4}{8^6}$

29) $\dfrac{10d^4}{d^2}$

30) $\dfrac{3x^6}{x^2}$

31) $\dfrac{20c^{11}}{30c^6}$

32) $\dfrac{35t^7}{56t^2}$

(24) 33) $\dfrac{y^3}{y^8}$

34) $\dfrac{m^4}{m^{10}}$

35) $\dfrac{x^{-3}}{x^6}$

36) $\dfrac{u^{-20}}{u^{-9}}$

37) $\dfrac{t^{-6}}{t^{-3}}$

38) $\dfrac{y^8}{y^{15}}$

39) $\dfrac{a^{-1}}{a^9}$

40) $\dfrac{m^{-9}}{m^{-3}}$

41) $\dfrac{t^4}{t}$

42) $\dfrac{c^7}{c^{-1}}$

43) $\dfrac{15w^2}{w^{10}}$

44) $\dfrac{-7p^3}{p^{12}}$

45) $\dfrac{-6k}{k^4}$

46) $\dfrac{21h^3}{h^7}$

 47) $\dfrac{a^4b^9}{ab^2}$

48) $\dfrac{p^5q^7}{p^2q^3}$

49) $\dfrac{10k^{-2}l^{-6}}{15k^{-5}l^2}$

50) $\dfrac{28tu^{-2}}{14t^5u^{-9}}$

51) $\dfrac{300x^7y^3}{30x^{12}y^8}$

52) $\dfrac{63a^{-3}b^2}{7a^7b^8}$

53) $\dfrac{6v^{-1}w}{54v^2w^{-5}}$

54) $\dfrac{3a^2b^{-11}}{18a^{-10}b^6}$

55) $\dfrac{3c^5d^{-2}}{8cd^{-3}}$

56) $\dfrac{9x^{-5}y^2}{4x^{-2}y^6}$

57) $\dfrac{(x+y)^9}{(x+y)^2}$

58) $\dfrac{(a+b)^9}{(a+b)^4}$

59) $\dfrac{(c+d)^{-5}}{(c+d)^{-11}}$

60) $\dfrac{(a+2b)^{-3}}{(a+2b)^{-4}}$

Write an exponential expression that simplifies to the given expression when the quotient rule for exponents is applied.

61) n^6

62) w^3

63) a^4b

64) xy^9

65) $\dfrac{4}{k^5}$

66) $\dfrac{7}{c^8}$

R Rethink

R1) How could you change a positive exponent to a negative exponent?

R2) When do you add exponents?

R3) When do you subtract exponents?

R4) When do you multiply exponents?

R5) In what other courses have you seen exponents used?

Putting It All Together

P Prepare

O Organize

What is your objective?	How can you accomplish the objective?
1 Combine the Rules of Exponents	• Understand all of the **Rules of Exponents,** and summarize them in your notes. • Complete the given examples on your own. • Complete You Trys 1 and 2.

W Work

Read the explanations, follow the examples, take notes, and complete the You Try.

W Hint

Summarize all of the rules of exponents in your notes.

1 Combine the Rules of Exponents

Let's see how we can combine the rules of exponents to simplify expressions.

EXAMPLE 1

Simplify using the rules of exponents. Assume all variables represent nonzero real numbers.

a) $(2t^{-6})^3(3t^2)^2$

b) $\left(\dfrac{7c^{10}d^7}{c^4d^2}\right)^2$

c) $\dfrac{w^{-3} \cdot w^4}{w^6}$

d) $\left(\dfrac{12a^{-2}b^9}{30ab^{-2}}\right)^{-3}$

Solution

a) $(2t^{-6})^3 \ (3t^2)^2$ We must follow the order of operations. Therefore, evaluate the exponents first.

$$(2t^{-6})^3 \cdot (3t^2)^2 = 2^3t^{(-6)(3)} \cdot 3^2t^{(2)(2)}$$ Apply the power rule.
$$= 8t^{-18} \cdot 9t^4$$ Simplify.
$$= 72t^{-18+4}$$ Multiply $8 \cdot 9$ and add the exponents.
$$= 72t^{-14}$$
$$= \dfrac{72}{t^{14}}$$ Write the answer using a positive exponent.

> **W Hint**
>
> Write out the examples as you read them. Try writing with colors as done here; it will help you understand the steps better.

b) $\left(\dfrac{7c^{10}d^7}{c^4d^2}\right)^2$ How can we begin this problem? We can use the quotient rule to simplify the expression before squaring it.

$$\left(\dfrac{7c^{10}d^7}{c^4d^2}\right)^2 = (7c^{10-4}d^{7-2})^2$$ Apply the quotient rule in the parentheses.

$$= (7c^6d^5)^2$$ Simplify.
$$= 7^2c^{(6)(2)}d^{(5)(2)}$$ Apply the power rule.
$$= 49c^{12}d^{10}$$

c) $\dfrac{w^{-3} \cdot w^4}{w^6}$ Let's begin by simplifying the numerator:

$$\dfrac{w^{-3} \cdot w^4}{w^6} = \dfrac{w^{-3+4}}{w^6}$$ Add the exponents in the numerator.

$$= \dfrac{w^1}{w^6}$$

Now, we can apply the quotient rule:

$$= w^{1-6} = w^{-5}$$ Subtract the exponents.

$$= \dfrac{1}{w^5}$$ Write the answer using a positive exponent.

d) $\left(\dfrac{12a^{-2}b^9}{30ab^{-2}}\right)^{-3}$ Eliminate the negative exponent *outside* the parentheses by taking the reciprocal of the base. Notice that we have *not* eliminated the negatives on the exponents *inside* the parentheses.

$$\left(\dfrac{12a^{-2}b^9}{30ab^{-2}}\right)^{-3} = \left(\dfrac{30ab^{-2}}{12a^{-2}b^9}\right)^3$$

Next, we *could* apply the exponent of 3 to the quantity inside the parentheses, but we could also simplify $\dfrac{30}{12}$ first and apply the quotient rule before cubing the quantity.

$$\left(\dfrac{30ab^{-2}}{12a^{-2}b^9}\right)^3 = \left(\dfrac{5}{2}a^{1-(-2)}b^{-2-9}\right)^3$$ Simplify $\dfrac{30}{12}$ and subtract the exponents.

$$= \left(\dfrac{5}{2}a^3b^{-11}\right)^3$$

$$= \dfrac{125}{8}a^9b^{-33}$$ Apply the power rule.

$$= \dfrac{125a^9}{8b^{33}}$$ Write the answer using positive exponents.

Simplify using the rules of exponents.

a) $\left(\dfrac{m^{12}n^3}{m^4 n}\right)^4$ b) $(-p^{-5})^4(6p^7)^2$ c) $\dfrac{5a^8 \cdot a^{-3}}{a^{-2}}$ d) $\left(\dfrac{9x^4 y^{-5}}{54x^3 y}\right)^{-2}$

It is possible for variables to appear in exponents. The same rules apply.

EXAMPLE 2   Simplify using the rules of exponents. Assume that the variables represent nonzero integers. Write your final answer so that the exponents have positive coefficients.

a) $c^{4x} \cdot c^{2x}$ b) $\dfrac{x^{5y}}{x^{9y}}$

Solution

a) $c^{4x} \cdot c^{2x} = c^{4x+2x} = c^{6x}$ The bases are the same, so apply the product rule. Add the exponents.

b) $\dfrac{x^{5y}}{x^{9y}} = x^{5y-9y}$ The bases are the same, so apply the quotient rule. Subtract the exponents.

$= x^{-4y}$

$= \dfrac{1}{x^{4y}}$ Write the answer with a positive coefficient in the exponent.

[YOU TRY 2]   Simplify using the rules of exponents. Assume that the variables represent nonzero integers. Write your final answer so that the exponents have positive coefficients.

a) $8^{2k} \cdot 8^k \cdot 8^{10k}$ b) $(w^3)^{-2p}$

ANSWERS TO [YOU TRY] **EXERCISES**

1) a) $m^{32}n^8$ b) $\dfrac{36}{p^6}$ c) $5a^7$ d) $\dfrac{36y^{12}}{x^2}$ 2) a) 8^{13k} b) $\dfrac{1}{w^{6p}}$

Putting It All Together Exercises    Do the exercises, and check your work.

Objective 1: Combine the Rules of Exponents

Use the rules of exponents to evaluate.

1) $\left(\dfrac{2}{3}\right)^4$ 2) $(2^2)^3$

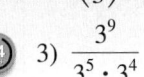 3) $\dfrac{3^9}{3^5 \cdot 3^4}$ 4) $\dfrac{(-5)^6 \cdot (-5)^2}{(-5)^5}$

5) $\left(\dfrac{10}{3}\right)^{-2}$ 6) $\left(\dfrac{3}{7}\right)^{-2}$

7) $(9-6)^2$ 8) $(3-8)^3$

9) 10^{-2} 10) 2^{-3}

11) $\dfrac{2^7}{2^{12}}$ 12) $\dfrac{3^{19}}{3^{15}}$

13) $\left(-\dfrac{5}{3}\right)^{-7} \cdot \left(-\dfrac{5}{3}\right)^4$ 14) $\left(\dfrac{1}{8}\right)^{-2}$

15) $3^{-2} - 12^{-1}$ 16) $2^{-2} + 3^{-2}$

Simplify. Assume that all variables represent nonzero real numbers. The final answer should not contain negative exponents.

(24) 17) $-10(-3g^4)^3$ 18) $7(2d^3)^3$

19) $\dfrac{33s}{s^{12}}$ 20) $\dfrac{c^{-7}}{c^{-2}}$

21) $\left(\dfrac{2xy^4}{3x^{-9}y^{-2}}\right)^4$ 22) $\left(\dfrac{a^6 b^5}{10a^3}\right)^3$

(24) 23) $\left(\dfrac{9m^8}{n^3}\right)^{-2}$ 24) $\left(\dfrac{3s^{-6}}{r^2}\right)^{-4}$

25) $(-b^5)^3$ 26) $(h^{11})^8$

27) $(-3m^5n^2)^3$

28) $(13a^6b)^2$

29) $\left(-\dfrac{9}{4}z^5\right)\left(\dfrac{8}{3}z^{-2}\right)$

30) $(15w^3)\left(-\dfrac{3}{5}w^6\right)$

31) $\left(\dfrac{s^7}{t^3}\right)^{-6}$

32) $\dfrac{m^{-3}}{n^{14}}$

33) $(-ab^3c^5)^2\left(\dfrac{a^4}{bc}\right)^3$

34) $\dfrac{(4v^3)^2}{(6v^8)^2}$

(24) 35) $\left(\dfrac{48u^{-7}v^2}{36u^3v^{-5}}\right)^{-3}$

36) $\left(\dfrac{xy^5}{9x^{-2}y}\right)^{-2}$

37) $\left(\dfrac{-3t^4u}{t^2u^{-4}}\right)^3$

38) $\left(\dfrac{k^7m^7}{12k^{-1}m^6}\right)^2$

39) $(h^{-3})^6$

40) $(-d^4)^{-5}$

41) $\left(\dfrac{h}{2}\right)^4$

42) $13f^{-2}$

43) $-7c^4(-2c^2)^3$

44) $5p^3(4p^6)^2$

45) $(12a^7)^{-1}(6a)^2$

46) $(9r^2s^2)^{-1}$

47) $\left(\dfrac{9}{20}r^4\right)(4r^{-3})\left(\dfrac{2}{33}r^9\right)$

48) $\left(\dfrac{f^8 \cdot f^{-3}}{f^2 \cdot f^9}\right)^6$

49) $\dfrac{(a^2b^{-5}c)^{-3}}{(a^4b^{-3}c)^{-2}}$

50) $\dfrac{(x^{-1}y^7z^4)^3}{(x^4yz^{-5})^{-3}}$

51) $\dfrac{(2mn^{-2})^3(5m^2n^{-3})^{-1}}{(3m^{-3}n^3)^{-2}}$

52) $\dfrac{(4s^3t^{-1})^2(5s^2t^{-3})^{-2}}{(4s^3t^{-1})^3}$

53) $\left(\dfrac{4n^{-3}m}{n^8m^2}\right)^0$

54) $\left(\dfrac{7qr^4}{37r^{-19}}\right)^0$

55) $\left(\dfrac{49c^4d^8}{21c^4d^5}\right)^{-2}$

56) $\dfrac{(2x^4y)^{-2}}{(5xy^3)^2}$

Simplify. Assume that the variables represent nonzero integers. Write your final answer so that the exponents have positive coefficients.

57) $(p^{2c})^6$

58) $(5d^{4t})^2$

59) $y^m \cdot y^{3m}$

60) $x^{-5c} \cdot x^{9c}$

61) $t^{5b} \cdot t^{-8b}$

62) $a^{-4y} \cdot a^{-3y}$

63) $\dfrac{25c^{2x}}{40c^{9x}}$

64) $-\dfrac{3y^{-10a}}{8y^{-2a}}$

R Rethink

R1) Was it more difficult to simplify the exponential expressions in this section when all of the rules of exponents were combined? Why or why not?

R2) Which problems were easy for you to do and which were more difficult? What do you think gave you trouble in the harder problems?

2.4 Scientific Notation

P Prepare

O Organize

What are your objectives for Section 2.4?	How can you accomplish each objective?
1 Multiply a Number by a Power of 10	• Learn how to multiply a number by a positive power of 10. • Learn how to multiply a number by a negative power of 10. • Complete the given examples on your own. • Complete You Trys 1 and 2.
2 Understand Scientific Notation	• Write the definition of *scientific notation* in your own words, and write an example. • Complete the given example on your own. • Complete You Try 3.
3 Write a Number in Scientific Notation	• Write the procedure for **Writing a Number in Scientific Notation** in your own words. • Complete the given example on your own. • Complete You Try 4.
4 Perform Operations with Numbers in Scientific Notation	• Use the rules of exponents and the order of operations to complete the given example on your own. • Complete You Try 5.

112 CHAPTER 2 **Rules of Exponents**

Read the explanations, follow the examples, take notes, and complete the You Trys.

The distance from Earth to the sun is approximately 150,000,000 km. A single rhinovirus (cause of the common cold) measures 0.00002 mm across. Performing operations on very large or very small numbers like these can be difficult. This is why scientists and economists, for example, often work with such numbers in a shorthand form called *scientific notation*. Writing numbers in scientific notation together with applying rules of exponents can simplify calculations with very large and very small numbers.

1 Multiply a Number by a Power of 10

Before discussing scientific notation further, we need to understand some principles behind the notation. Let's look at multiplying numbers by positive powers of 10.

EXAMPLE 1

Multiply.

a) 3.4×10^1 b) $0.0857 \cdot 10^3$

Solution

a) $3.4 \times 10^1 = 3.4 \times 10 = 34$

b) $0.0857 \cdot 10^3 = 0.0857 \cdot 1000 = 85.7$

Notice that when we multiply each of these numbers by a positive power of 10, the result is *larger* than the original number. In fact, the exponent determines how many places to the *right* the decimal point is moved.

$$3.40 \times 10^1 = 3.4 \times 10^1 = 34 \qquad 0.0857 \cdot 10^3 = 85.7$$

1 place to the right   3 places to the right

[YOU TRY 1]

Multiply by moving the decimal point the appropriate number of places.

a) 6.2×10^2 b) $5.31 \cdot 10^5$

What happens to a number when we multiply by a *negative* power of 10?

EXAMPLE 2

Multiply.

a) $41 \cdot 10^{-2}$ b) 367×10^{-4}

Solution

a) $41 \cdot 10^{-2} = 41 \cdot \dfrac{1}{100} = \dfrac{41}{100} = 0.41$

b) $367 \times 10^{-4} = 367 \times \dfrac{1}{10,000} = \dfrac{367}{10,000} = 0.0367$

When we multiply each of these numbers by a negative power of 10, the result is *smaller* than the original number. The exponent determines how many places to the *left* the decimal point is moved:

$$41 \cdot 10^{-2} = 41. \cdot 10^{-2} = 0.41 \qquad 367 \times 10^{-4} = 0367. \times 10^{-4} = 0.0367$$

2 places to the left   4 places to the left

[YOU TRY 2] Multiply.

a) $83 \cdot 10^{-2}$ b) 45×10^{-3}

It is important to understand the previous concepts to understand how to use scientific notation.

2 Understand Scientific Notation

Definition

A number is in **scientific notation** if it is written in the form $a \times 10^n$, where $1 \le |a| < 10$ and n is an integer.

Note

Multiplying $|a|$ by a *positive* power of 10 will result in a number that is *larger* than $|a|$. Multiplying $|a|$ by a *negative* power of 10 will result in a number that is *smaller* than $|a|$. The double inequality $1 \le |a| < 10$ means that a is a number that has *one* nonzero digit to the left of the decimal point.

In other words, a number in scientific notation has one digit to the left of the decimal point and the number is multiplied by a power of 10.

Here are some examples of numbers written in scientific notation:

$$3.82 \times 10^{-5}, \quad 1.2 \cdot 10^3, \quad \text{and } 7 \cdot 10^{-2}$$

The following numbers are *not* in scientific notation:

$$51.94 \times 10^4 \qquad 0.61 \cdot 10^{-3} \qquad 300 \cdot 10^6$$

2 digits to left of decimal point   Zero is to left of decimal point   3 digits to left of decimal point

Now let's convert a number written in scientific notation to a number without exponents.

EXAMPLE 3 Rewrite without exponents.

a) $5.923 \cdot 10^4$ b) 7.4×10^{-3} c) 1.8875×10^3

Solution

a) $5.923 \cdot 10^4 \to 5.9230 = 59{,}230$ Remember, multiplying by a positive power of 10 will make the result *larger* than 5.923.
4 places to the right

W Hint
How does this process compare to Examples 1 and 2?

b) $7.4 \times 10^{-3} \to 007.4 = 0.0074$ Multiplying by a negative power of 10 will make the result *smaller* than 7.4.
3 places to the left

c) $1.8875 \times 10^3 \to 1.8875 = 1887.5$
3 places to the right

3 Write a Number in Scientific Notation

Let's write the number 48,000 in scientific notation. First locate its decimal point: 48,000.

Next, determine where the decimal point will be when the number is in scientific notation:

48,000.
∧
Decimal point will be here.

Therefore, $48,000 = 4.8 \times 10^n$, where n is an integer. Will n be positive or negative? We can see that 4.8 must be multiplied by a *positive* power of 10 to make it larger, 48,000.

48000.
∧   ↖
Decimal point   Decimal point
will be here.   starts here.

Now we count four places between the original and the final decimal place locations.

48000.
1 2 3 4

We use the number of spaces, 4, as the exponent of 10.

$$48,000 = 4.8 \times 10^4$$

EXAMPLE 4   Write each number in scientific notation.

Solution

a) The distance from Earth to the sun is approximately 150,000,000 km.

150,000,000.   150,000,000.   Move decimal point 8 places.
∧   ↖
Decimal point   Decimal point
will be here.   is here.

$150,000,000 \text{ km} = 1.5 \times 10^8 \text{ km}$

b) A single rhinovirus measures 0.00002 mm across.

0.00002 mm $0.00002 \text{ mm} = 2 \times 10^{-5} \text{ mm}$
∧
Decimal point
will be here.

Procedure How to Write a Number in Scientific Notation

1) Locate the decimal point in the original number.

2) Determine where the decimal point will be when converting to scientific notation. Remember, there will be *one* nonzero digit to the left of the decimal point.

3) Count how many places you must move the decimal point to take it from its original place to its position for scientific notation.

4) If the absolute value of the resulting number is *smaller* than the absolute value of the original number, you will multiply the result by a *positive* power of 10.

Example: $350.9 = 3.509 \times 10^2$

If the absolute value of the resulting number is *larger* than the absolute value of the original number, you will multiply the result by a *negative* power of 10.

Example: $0.0000068 = 6.8 \times 10^{-6}$

YOU TRY 4 Write each number in scientific notation.

a) The USDA projects that corn production in the United States in the 2024–2025 season will be 14,655,000,000 bushels. (www.ers.usda.gov)

b) The diameter of a human hair is approximately 0.001 in.

---

4 Perform Operations with Numbers in Scientific Notation

We use the rules of exponents to perform operations with numbers in scientific notation.

EXAMPLE 5

Simplify $\dfrac{3 \times 10^3}{4 \times 10^5}$.

Solution

$$\frac{3 \times 10^3}{4 \times 10^5} = \frac{3}{4} \times \frac{10^3}{10^5}$$
$$= 0.75 \times 10^{-2} \qquad \text{Write } \tfrac{3}{4} \text{ in decimal form.}$$
$$= 7.5 \times 10^{-3} \qquad \text{Use scientific notation.}$$
$$\text{or } 0.0075$$

YOU TRY 5 Perform the operations, and simplify.

a) $(2.6 \cdot 10^2)(5 \cdot 10^4)$

b) $\dfrac{7.2 \times 10^{-9}}{6 \times 10^{-5}}$

We can use a graphing calculator to convert a very large or very small number to scientific notation, or to convert a number in scientific notation to a number written without an exponent. Suppose we are given a very large number such as 35,000,000,000. If you enter any number with more than 10 digits on the home screen on your calculator and press ENTER, the number will automatically be displayed in scientific notation as shown on screen (a). A small number with more than two zeros to the right of the decimal point (such as .000123) will automatically be displayed in scientific notation as shown.

The E shown in the screen refers to a power of 10, so 3.5E10 is the number 3.5×10^{10} in scientific notation. 1.23E-4 is the number 1.23×10^{-4} in scientific notation.

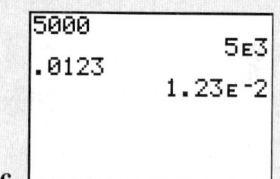

a.

If a large number has 10 or fewer digits, or if a small number has fewer than three zeros to the right of the decimal point, then the number will not automatically be displayed in scientific notation. To display the number using scientific notation, press MODE, select SCI, and press ENTER. When you return to the home screen, all numbers will be displayed in scientific notation as shown on screens (b) and (c).

b.

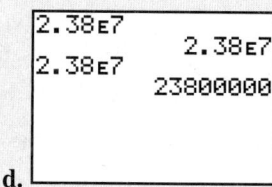
c.

A number written in scientific notation can be entered directly into your calculator. For example, the number 2.38×10^7 can be entered directly on the home screen by typing 2.38 followed by 2nd , 7 ENTER as shown on screen (d). If you wish to display this number without an exponent, change the mode back to NORMAL and enter the number on the home screen as shown.

d.

Write each number without an exponent, using a graphing calculator.

1. 3.4×10^5 2. 9.3×10^7 3. 1.38×10^{-3}

Write each number in scientific notation, using a graphing calculator.

4. 186,000   5. 5280   6. 0.0469

ANSWERS TO [YOU TRY] EXERCISES

1) a) 620 b) 531,000   2) a) 0.83 b) 0.045   3) a) 30,500 b) 0.000083 c) 6918.53
4) a) 1.4655×10^{10} bushels b) 1.0×10^{-3} in.   5) a) 13,000,000 b) 0.00012

ANSWERS TO TECHNOLOGY EXERCISES

1) 314,000   2) 93,000,000   3) 0.00138   4) 1.86×10^5 5) 5.28×10^3 6) 4.69×10^{-2}

Mixed Exercises: Objectives 1 and 2

Determine whether each number is in scientific notation.

1) 7.23×10^5 2) 24.0×10^{-3}

3) $0.16 \cdot 10^{-4}$ 4) $-2.8 \cdot 10^4$

5) -37×10^{-2} 6) 0.9×10^{-1}

7) $-5 \cdot 10^6$ 8) $7.5 \cdot 2^{-10}$

9) Explain, in your own words, how to determine whether a number is expressed in scientific notation.

10) Explain, in your own words, how to write 4.1×10^{-3} without an exponent.

Multiply.

11) $980.2 \cdot 10^4$ 12) $71.765 \cdot 10^2$

13) $0.1502 \cdot 10^8$ 14) $40.6 \cdot 10^{-3}$

15) 0.0674×10^{-1} 16) $1,200,006 \times 10^{-7}$

Objective 1: Multiply a Number by a Power of 10

Write each number without an exponent.

17) 1.92×10^6 18) -6.8×10^{-5}

19) $2.03449 \cdot 10^3$ 20) $-5.26 \cdot 10^4$

21) $-7 \cdot 10^{-4}$ 22) $8 \cdot 10^{-6}$

23) -9.5×10^{-3} 24) 6.021967×10^5

25) 6×10^4 26) 3×10^6

27) $-9.815 \cdot 10^{-2}$ 28) $-7.44 \cdot 10^{-4}$

Write the following quantities without an exponent.

29) A dog's nose has approximately $3 \cdot 10^8$ scent receptors. (www.newscientist.com)

30) As of June 30, 2018, Facebook had $2.23 \cdot 10^9$ monthly active users worldwide. (newsroom.fb.com)

31) The radius of one hydrogen atom is about 2.5×10^{-11} m.

32) The length of a household ant is 2.54×10^{-3} m.

Objective 3: Write a Number in Scientific Notation

Write each number in scientific notation.

33) 2110.5 34) 38.25

35) 0.000096 36) 0.00418

37) $-7,000,000$ 38) $62,000$

39) 3400 40) $-145,000$

41) 0.0008 42) -0.00000022

43) -0.076 44) 990

45) 6000 46) $-500,000$

Write each number in scientific notation.

47) The total weight of the Golden Gate Bridge is 380,800,000 kg. (www.goldengatebridge.com)

Fotosearch/Getty Images

48) A flash drive may hold approximately 256,000,000,000 bytes of data.

49) The diameter of an atom is about 0.00000001 cm.

50) The oxygen-hydrogen bond length in a water molecule is 0.000000001 mm.

Objective 4: Perform Operations with Numbers in Scientific Notation

Perform the operation as indicated. Write the final answer without an exponent.

51) $\dfrac{6 \cdot 10^9}{2 \cdot 10^5}$ 52) $(7 \cdot 10^2)(2 \cdot 10^4)$

53) $(2.3 \times 10^3)(3 \times 10^2)$ 54) $\dfrac{8 \times 10^7}{4 \times 10^4}$

55) $\dfrac{8.4 \times 10^{12}}{-7 \times 10^9}$ 56) $\dfrac{-4.5 \times 10^{-6}}{-1.5 \times 10^{-8}}$

57) $(-1.5 \cdot 10^{-8})(4 \cdot 10^6)$

58) $(-3 \cdot 10^{-2})(-2.6 \cdot 10^{-3})$

59) $\dfrac{-3 \cdot 10^5}{6 \cdot 10^8}$ 60) $\dfrac{2 \cdot 10^1}{5 \cdot 10^4}$

61) $(9.75 \times 10^4) + (6.25 \times 10^4)$

62) $(4.7 \times 10^{-3}) + (8.8 \times 10^{-3})$

63) $(3.19 \cdot 10^{-5}) + (9.2 \cdot 10^{-5})$

64) $(2 \cdot 10^2) + (9.7 \cdot 10^2)$

For each problem, express each number in scientific notation, then solve the problem.

65) Humans shed about $1.44 \cdot 10^7$ particles of skin every day. How many particles would be shed in a year? (Assume 365 days in a year.)

66) Scientists send a lunar probe to land on the moon and send back data. How long will it take for pictures to reach Earth if the distance between Earth and the moon is 360,000 km and if the speed of light is $3 \cdot 10^5$ km/sec?

67) In the United States in 2017, approximately 9,400,000 cows produced $2.15 \cdot 10^{11}$ lb of milk. On average, how much milk did each cow produce? (www.ers.usda.gov)

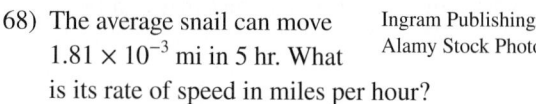

Ingram Publishing/ Alamy Stock Photo

68) The average snail can move 1.81×10^{-3} mi in 5 hr. What is its rate of speed in miles per hour?

69) A photo printer delivers approximately 1.1×10^6 droplets of ink per square inch. How many droplets of ink would a 4 in. × 6 in. photo contain?

70) One year, 3,500,000,000,000 prescription drug orders were filled in the United States. If the average price of each prescription was roughly $65.00, how much did consumers pay for prescription drugs? (www.ncsl.org)

71) During a certain year, Americans spent approximately $1.5 \cdot 10^{12}$ dollars on food. If the population of the United States was about 320,000,000, how much money was spent on food, per person? (www.census.gov)

72) Find the population density of Australia if the estimated population in April 2017 was about 24,000,000 people and the country encompasses about 2,900,000 sq mi. (www.abs.gov.au)

73) Approximately 111,300,000 people watched Super Bowl LI in 2017 between the New England Patriots and the Atlanta Falcons. The Hass Avocado Board estimated that about 105,000,000 lb of avocados were eaten on Super Bowl Sunday, mostly in the form of guacamole. On average, how many pounds of avocados did each viewer eat while watching the Super Bowl? (www.fortune.com, abcnews.go.com)

74) In 2016, the United States produced about 6.51×10^9 metric tons of carbon emissions. The U.S. population that year was about 323 million. Find the amount of carbon emissions produced per person that year. (www.epa.gov, U.S. Census Bureau)

R Rethink

R1) When would you want to write a number in scientific notation?

R2) In which college courses, other than math, do you think scientific notation is used?

Group Activity – Exponent BINGO

1) Make a BINGO card with an asterisk in the middle space. You will need markers (like small pieces of paper or candy) to cover the numbers on the BINGO card.

B	I	N	G	O
		*		

2) Choose 24 numbers from this list, and write one number in each box on the BINGO card. Be sure to choose some positive *and* some negative numbers.

1	-1	4	-4	8	-8	9	-9
16	-16	25	-25	27	-27	32	-32
36	-36	49	-49	64	-64	81	-81
100	-100	121	-121	125	-125	144	-144

3) Choose one person in the class or in the group as the *caller*. The caller will write one of the following expressions on the board or on a piece of paper to show to the group. The expressions should be set aside as they are called so that they can be checked when someone calls BINGO.

1^2	-1^2	$(-1)^2$	2^2	-2^2	2^3	-2^3
2^4	$(-2)^4$	2^5	-2^5	2^6	-2^6	3^2
-3^2	$(-3)^2$	3^3	$(-3)^3$	3^4	-3^4	4^2
-4^2	4^3	$(-4)^3$	5^2	$(-5)^2$	5^3	-5^3
6^2	-6^2	7^2	-7^2	8^2	$(-8)^2$	9^2
-9^2	10^2	$(-10)^2$	11^2	$(-11)^2$	12^2	-12^2

4) As the caller shows each expression, the players simplify the expression and put a marker on the answer if it is on the card. The caller continues to give the players one expression at a time until someone gets BINGO. A player has BINGO when there are markers on five consecutive numbers across, vertical, or diagonal through the middle. The player who calls BINGO must check with the caller to determine whether the answers are correct.

emPOWERme Discover Your Reading Style

Not all people read the same way. Some people take a big-picture approach when they are reading, trying to understand the material as a whole. Other people read with great attention to detail, analyzing each point as they come to it. Use the following questions to learn how *you* read—that is, your characteristic reading style. Rate how well each statement below describes you. Use this rating scale:

1 = Doesn't describe me at all
2 = Describes me only slightly
3 = Describes me fairly well
4 = Describes me very well

	1	2	3	4
1. I often reread passages in books that I particularly like.				
2. I often read good passages aloud to whomever is around.				
3. I often stop while reading to check that I understood what I just read.				
4. If I come across a long, unfamiliar name, I try to sound it out and pronounce it correctly.				
5. If there's a word I don't understand, I look it up in a dictionary right away or mark it to look it up later.				
6. Before I start reading a textbook or other serious book or article, I look for clues about how it is organized.				
7. I often question what I'm reading and "argue" with the author.				
8. I often try to guess what the chapter I'm about to read will cover.				
9. I often write comments or make notes in books that I own.				
10. I'm always finding typographical errors in books and articles I read.				

Reading styles range from a big-picture, noncritical style to a very analytic, critical style. Add up the points you assigned yourself. Use this informal scale to find your reading style:

10–12 = Very broad reading style
13–20 = Mostly broad reading style
21–28 = Mostly analytic reading style
29–40 = Very analytic reading style

How do you think your reading style affects the way you learn material in textbooks? Is your style related to the kinds of subjects you prefer? Do you think your reading style affects your leisure reading?

Chapter 2: Summary

Definition/Procedure	Example

2.1A The Product Rule and Power Rules

Exponential Expression: $$a^n = \underbrace{a \cdot a \cdot a \cdot \cdots \cdot a}_{n \text{ factors of } a}$$ a is the **base,** n is the exponent.	$5^4 = 5 \cdot 5 \cdot 5 \cdot 5$ 5 is the **base,** 4 is the exponent.
Product Rule: $a^m \cdot a^n = a^{m+n}$	$x^8 \cdot x^2 = x^{10}$
Basic Power Rule: $(a^m)^n = a^{mn}$	$(t^3)^5 = t^{15}$
Power Rule for a Product: $(ab)^n = a^n b^n$	$(2c)^4 = 2^4 c^4 = 16c^4$
Power Rule for a Quotient: $$\left(\frac{a}{b}\right)^n = \frac{a^n}{b^n}, \text{ where } b \neq 0.$$	$$\left(\frac{w}{5}\right)^3 = \frac{w^3}{5^3} = \frac{w^3}{125}$$

2.1B Combining the Rules

Remember to follow the order of operations.	Simplify $(3y^4)^2(2y^9)^3$. $= 9y^8 \cdot 8y^{27}$ Exponents come before multiplication. $= 72y^{35}$ Use the product rule, and multiply coefficients.

2.2A Real-Number Bases

Zero Exponent: If $a \neq 0$, then $a^0 = 1$.	$(-9)^0 = 1$
Negative Exponent: For $a \neq 0$, $a^{-n} = \left(\frac{1}{a}\right)^n = \frac{1}{a^n}$.	Evaluate. $\left(\frac{5}{2}\right)^{-3} = \left(\frac{2}{5}\right)^3 = \frac{2^3}{5^3} = \frac{8}{125}$

2.2B Variable Bases

If $a \neq 0$ and $b \neq 0$, then $\left(\frac{a}{b}\right)^{-m} = \left(\frac{b}{a}\right)^m$.	Rewrite p^{-10} with a positive exponent (assume $p \neq 0$). $p^{-10} = \left(\frac{1}{p}\right)^{10} = \frac{1}{p^{10}}$
If $a \neq 0$ and $b \neq 0$, then $\dfrac{a^{-m}}{b^{-n}} = \dfrac{b^n}{a^m}$.	Rewrite each expression with positive exponents. Assume that the variables represent nonzero real numbers. a) $\dfrac{x^{-3}}{y^{-7}} = \dfrac{y^7}{x^3}$ b) $\dfrac{14m^{-6}}{n^{-1}} = \dfrac{14n}{m^6}$

2.3 The Quotient Rule

Quotient Rule: If $a \neq 0$, then $\dfrac{a^m}{a^n} = a^{m-n}$.	Simplify. $$\frac{4^9}{4^6} = 4^{9-6} = 4^3 = 64$$

Putting It All Together

Combine the Rules of Exponents	Simplify. $$\left(\frac{a^4}{2a^7}\right)^{-5} = \left(\frac{2a^7}{a^4}\right)^5 = (2a^3)^5 = 32a^{15}$$

Definition/Procedure	Example

2.4 Scientific Notation

Scientific Notation A number is in **scientific notation** if it is written in the form $a \times 10^n$, where $1 \leq |a| < 10$ and n is an integer. That is, a is a number that has one nonzero digit to the left of the decimal point.	Write in scientific notation. a) $78{,}000 \rightarrow 78{,}000 \rightarrow 7.8 \times 10^4$ b) $0.00293 \rightarrow 0.00293 \rightarrow 2.93 \times 10^{-3}$
Converting from Scientific Notation	Write without exponents. a) $5 \times 10^{-4} \rightarrow 0005. \rightarrow 0.0005$ b) $1.7 \cdot 10^6 = 1.700000 \rightarrow 1{,}700{,}000$
Performing Operations	Multiply $(4 \times 10^2)(2 \times 10^4)$. $\quad = (4 \times 2)(10^2 \times 10^4)$ $\quad = 8 \times 10^6$ $\quad = 8{,}000{,}000$

Chapter 2: Review Exercises

(2.1A)

1) Write in exponential form.

   a) $8 \cdot 8 \cdot 8 \cdot 8 \cdot 8 \cdot 8$

   b) $(-7)(-7)(-7)(-7)$

2) Identify the base and the exponent.

   a) -6^5 b) $(4t)^3$

   c) $4t^3$ d) $-4t^3$

3) Use the rules of exponents to simplify.

   a) $2^3 \cdot 2^2$ b) $\left(\dfrac{1}{3}\right)^2 \cdot \left(\dfrac{1}{3}\right)$

   c) $(7^3)^4$ d) $(k^5)^6$

4) Use the rules of exponents to simplify.

   a) $(3^2)^2$ b) $8^3 \cdot 8^7$

   c) $(m^4)^9$ d) $p^9 \cdot p^7$

5) Simplify using the rules of exponents.

   a) $(5y)^3$ b) $(-7m^4)(2m^{12})$

   c) $\left(\dfrac{a}{b}\right)^6$ d) $6(xy)^2$

   e) $\left(\dfrac{10}{9}c^4\right)(2c)\left(\dfrac{15}{4}c^3\right)$

6) Simplify using the rules of exponents.

   a) $\left(\dfrac{x}{y}\right)^{10}$ b) $(-2z)^5$

   c) $(6t^7)\left(-\dfrac{5}{8}t^5\right)\left(\dfrac{2}{3}t^2\right)$ d) $-3(ab)^4$

   e) $(10j^6)(4j)$

(2.1B)

7) Simplify using the rules of exponents.

   a) $(z^5)^2(z^3)^4$ b) $-2(3c^5d^8)^2$

   c) $(9-4)^3$ d) $\dfrac{(10t^3)^2}{(2u^7)^3}$

8) Simplify using the rules of exponents.

   a) $\left(\dfrac{-20d^4c}{5b^3}\right)^3$ b) $(-2y^8z)^3(3yz^2)^2$

   c) $\dfrac{x^7 \cdot (x^2)^5}{(2y^3)^4}$ d) $(6-8)^2$

(2.2A)

9) Evaluate.

   a) 8^0 b) -3^0

   c) 9^{-1} d) $3^{-2} - 2^{-2}$

   e) $\left(\dfrac{4}{5}\right)^{-3}$

10) Evaluate.

   a) $(-12)^0$ b) $5^0 + 4^0$

   c) -6^{-2} d) 2^{-4}

   e) $\left(\dfrac{10}{3}\right)^{-2}$

(2.2B)

11) Rewrite the expression with positive exponents. Assume that the variables do not equal zero.

   a) v^{-9} b) $\left(\dfrac{9}{c}\right)^{-2}$

   c) $\left(\dfrac{1}{y}\right)^{-8}$ d) $-7k^{-9}$

e) $\dfrac{19z^{-4}}{a^{-1}}$

f) $20m^{-6}n^5$

g) $\left(\dfrac{2j}{k}\right)^{-5}$

12) Rewrite the expression with positive exponents. Assume that the variables do not equal zero.

a) $\left(\dfrac{1}{x}\right)^{-5}$

b) $3p^{-4}$

c) $a^{-8}b^{-3}$

d) $\dfrac{12k^{-3}r^5}{16mn^{-6}}$

e) $\dfrac{c^{-1}d^{-1}}{15}$

f) $\left(-\dfrac{m}{4n}\right)^{-3}$

g) $\dfrac{10b^4}{a^{-9}}$

(2.3) In Exercises 13–16, assume that the variables represent nonzero real numbers. The answers should not contain negative exponents.

13) Simplify using the rules of exponents.

a) $\dfrac{3^8}{3^6}$

b) $\dfrac{r^{11}}{r^3}$

c) $\dfrac{48t^{-2}}{32t^3}$

d) $\dfrac{21xy^2}{35x^{-6}y^3}$

14) Simplify using the rules of exponents.

a) $\dfrac{2^9}{2^{15}}$

b) $\dfrac{d^4}{d^{-10}}$

c) $\dfrac{m^{-5}n^3}{mn^8}$

d) $\dfrac{100a^8b^{-1}}{25a^7b^{-4}}$

15) Simplify by applying one or more of the rules of exponents.

a) $(-3s^4t^5)^4$

b) $\dfrac{(2a^6)^5}{(4a^7)^2}$

c) $\left(\dfrac{z^4}{y^3}\right)^{-6}$

d) $(-x^3y)^5(6x^{-2}y^3)^2$

e) $\left(\dfrac{cd^{-4}}{c^8d^{-9}}\right)^5$

f) $\left(\dfrac{14m^5n^5}{7m^4n}\right)^3$

g) $\left(\dfrac{3k^{-1}t}{5k^{-7}t^4}\right)^{-3}$

h) $\left(\dfrac{40}{21}x^{10}\right)(3x^{-12})\left(\dfrac{49}{20}x^2\right)$

16) Simplify by applying one or more of the rules of exponents.

a) $\left(\dfrac{4}{3}\right)^8\left(\dfrac{4}{3}\right)^{-2}\left(\dfrac{4}{3}\right)^{-3}$

b) $\left(\dfrac{k^{10}}{k^4}\right)^3$

c) $\left(\dfrac{x^{-4}y^{11}}{xy^2}\right)^{-2}$

d) $(-9z^5)^{-2}$

e) $\left(\dfrac{g^2\cdot g^{-1}}{g^{-7}}\right)^{-4}$

f) $(12p^{-3})\left(\dfrac{10}{3}p^5\right)\left(\dfrac{1}{4}p^2\right)^2$

g) $\left(\dfrac{30u^2v^{-3}}{40u^7v^{-7}}\right)^{-2}$

h) $-5(3h^4k^9)^2$

Simplify. Assume that the variables represent nonzero integers. Write your final answer so that the exponents have positive coefficients.

17) $y^{3k}\cdot y^{7k}$

18) $\dfrac{z^{12c}}{z^{5c}}$

(2.4) Write each number without an exponent.

19) 9.38×10^5

20) -4.185×10^2

21) $9\cdot 10^3$

22) $6.7\cdot 10^{-4}$

23) 1.05×10^{-6}

24) 2×10^4

Write each number in scientific notation.

25) 0.0000575

26) $36,940$

27) $32,000,000$

28) 0.0000004

29) 0.0009315

30) 66

Write the number without exponents.

31) Before 2010, golfer Tiger Woods earned over 7×10^7 dollars per year in product endorsements. (www.forbes.com)

Perform the operation as indicated. Write the final answer without an exponent.

32) $\dfrac{8\cdot 10^6}{2\cdot 10^{13}}$

33) $\dfrac{-1\cdot 10^9}{5\cdot 10^{12}}$

34) $(9\times 10^{-8})(4\times 10^7)$

35) $(5\cdot 10^3)(3.8\cdot 10^{-8})$

36) $\dfrac{-3\times 10^{10}}{-4\times 10^6}$

37) $(-4.2\times 10^2)(3.1\times 10^3)$

For each problem, write each of the numbers in scientific notation, then solve the problem. Write the answer without exponents.

38) Eight porcupines have a total of about $2.4\cdot 10^5$ quills on their bodies. How many quills would one porcupine have?

39) In 2015, Nebraska had approximately 4.5×10^7 acres of farmland and about 48,000 farms. What was the average size of a Nebraska farm in 2015? (www.nda.nebraska.gov)

40) One molecule of water has a mass of 2.99×10^{-23} g. Find the mass of 100,000,000 molecules.

Digital Vision/
Getty Images

41) In 2017, users were spending a total of approximately $6.141\cdot 10^9$ min per day using Snapchat. On average, each user spent 34.5 min per day on Snapchat. How many people used Snapchat every day? (www.cnbc.com)

42) When the polls closed on the west coast on November 4, 2008, and Barack Obama was declared the new president, there were about 143,000 visits per second to news websites. If the visits continued at that rate for 3 min, how many visits did the news websites receive during that time? (www.xconomy.com)

Chapter 2: Test

Write in exponential form.

1) $(-3)(-3)(-3)$

2) $x \cdot x \cdot x \cdot x \cdot x$

Use the rules of exponents to simplify.

3) $5^2 \cdot 5$

4) $\left(\dfrac{1}{x}\right)^5 \cdot \left(\dfrac{1}{x}\right)^2$

5) $(8^3)^{12}$

6) $p^7 \cdot p^{-2}$

Evaluate.

7) 3^4

8) 8^0

9) 2^{-5}

10) $4^{-2} + 2^{-3}$

11) $\left(-\dfrac{3}{4}\right)^3$

12) $\left(\dfrac{10}{7}\right)^{-2}$

Simplify using the rules of exponents. Assume that all variables represent nonzero real numbers. The final answer should not contain negative exponents.

13) $(5n^6)^3$

14) $(-3p^4)(10p^8)$

15) $\dfrac{m^{10}}{m^4}$

16) $\dfrac{a^9 b}{a^5 b^7}$

17) $\left(\dfrac{-12t^{-6}u^8}{4t^5 u^{-1}}\right)^{-3}$

18) $(2y^{-4})^6 \left(\dfrac{1}{2}y^5\right)^3$

19) $\left(\dfrac{(9x^2 y^{-2})^3}{4xy}\right)^0$

20) $\dfrac{(2m + n)^3}{(2m + n)^2}$

21) $\dfrac{12a^4 b^{-3}}{20c^{-2}d^3}$

22) $\left(\dfrac{y^{-7} \cdot y^3}{y^5}\right)^{-2}$

23) Simplify $t^{10k} \cdot t^{3k}$. Assume that the variables represent nonzero integers.

24) Rewrite $7.283 \cdot 10^5$ without exponents.

25) Write 0.000165 in scientific notation.

26) Divide. Write the answer without exponents. $\dfrac{-7.5 \times 10^{12}}{1.5 \times 10^8}$

27) Write the number without an exponent: In 2018, the population of Texas was about $2.83 \cdot 10^7$. (www.census.gov)

28) An electron is a subatomic particle with a mass of 9.1×10^{-28} g. What is the mass of 2,000,000,000 electrons? Write the answer without exponents.

Chapter 2: Cumulative Review for Chapters 1–2

1) Write $\dfrac{90}{150}$ in lowest terms.

Perform the indicated operations. Write the answer in lowest terms.

2) $\dfrac{2}{15} + \dfrac{1}{10} + \dfrac{7}{20}$

3) $\dfrac{4}{15} \div \dfrac{20}{21}$

4) $-144 \div (-12)$

5) $-26 + 5 - 7$

6) -9^2

7) $(-1)^5$

8) $(5 + 1)^2 - 2[17 + 5(10 - 14)]$

9) Glen Crest High School is building a new football field. The dimensions of a regulation-size field are $53\dfrac{1}{3}$ yd by 120 yd. (There are 10 yd of end zone on each end.) The sod for the field will cost \$1.80/yd^2.

   a) Find the perimeter of the field.

   b) How much will it cost to sod the field?

10) Evaluate $2p^2 - 11q$ when $p = 3$ and $q = -4$.

11) State the formula for the volume of a sphere.

12) Given this set of numbers $\left\{3, -4, -2.1\overline{3}, \sqrt{11}, 2\dfrac{2}{3}\right\}$, list the

   a) integers

   b) irrational numbers

   c) natural numbers

   d) rational numbers

   e) whole numbers

13) Evaluate $4x^3 + 2x - 3$ when $x = 4$.

14) Rewrite $\dfrac{3}{4}(6m - 20n + 7)$ using the distributive property.

15) Combine like terms and simplify: $5(t^2 + 7t - 3) - 2(4t^2 - t + 5)$

16) Let x represent the unknown quantity, and write a mathematical expression for "thirteen less than half of a number."

Simplify using the rules of exponents. The answer should not contain negative exponents. Assume the variables represent nonzero real numbers.

17) $4^3 \cdot 4^7$

18) $\left(\dfrac{x}{y}\right)^{-3}$

19) $\left(\dfrac{32x^3}{8x^{-2}}\right)^{-1}$

20) $-(3rt^{-3})^4$

21) $(4z^3)(-7z^5)$

22) $\dfrac{n^2}{n^9}$

23) $(-2a^{-6}b)^5$

24) Write 0.000729 in scientific notation.

25) Perform the indicated operation. Write the final answer without an exponent. $(6.2 \times 10^5)(9.4 \times 10^{-2})$

Linear Equations and Inequalities in One Variable

Get Ready

We need many different skills to solve equations and applications in this chapter. Let's review some of them here.

1) When simplifying expressions, we must keep the **distributive property** and **order of operations** in mind.

   Example: Simplify $8 - 3(2c - 5) + 7c$.

   $8 - 3(2c - 5) + 7c = 8 - 6c + 15 + 7c$ Distribute.
   $= c + 23$ Combine like terms.

2) To **change a percent to a decimal,** remove the percent symbol then move the decimal point two places to the *left.* If a whole number does not contain a decimal point, recall that the point can be written at the end of the number.

   $34.2\% = 0.342$ $67\% = 0.67$ $3\% = 0.03\%$

3) To change a **decimal to a percent,** move the decimal point two places to the *right.* Put the percent symbol at the end of the number.

   $0.49 = 49\%$ $0.08 = 8\%$ $1 = 1. = 100\%$

4) It is important to know and be able to use **basic geometry facts.** (See Section 1.3 for a geometry review.)

   Example: Find the perimeter of a rectangle with length 5 in. and width 3 in.

   Use the formula $P = 2l + 2w$:
   $P = 2(5 \text{ in.}) + 2(3 \text{ in.}) = 10 \text{ in.} + 6 \text{ in.} = 16 \text{ in.}$

In this chapter, use the following *Basic Skills Worksheets* to prepare students for future chapters: WS2 Powers (to prepare for operations with polynomials in Chapter 6) and WS3 Prefactoring (so that students can practice the mental arithmetic skills needed to factor polynomials in Chapter 7).

Get Ready Exercises

Simplify each expression.

1) $-9 + 4(a + 3) + a$

2) $5(7k + 1) - 2(6 - k)$

3) $y - (8y - 5) + 3(y + 10)$

4) $4(2p + 9) + 10 - 7(p - 4)$

Change to a decimal.

5) 81%

6) 5%

7) 0.9%

8) 130%

Change to a percent.

9) 0.4

10) 0.197

11) 2

12) 0.06

Solve each problem involving geometry.

13) $m\angle A = 52°$. What is the measure of its supplement?

14) What is the sum of the angles of a triangle?

15) Find the area and perimeter of a rectangle with length 9 cm and width 4 cm.

16) Find the area of a triangle that is 6 in. high and has a base of length 5 in.

Answers
1) $5a + 3$ 2) $37k - 7$ 3) $-4y + 35$ 4) $p + 74$ 5) 0.81 6) 0.05 7) 0.009 8) 1.3 9) 40% 10) 19.7% 11) 200% 12) 6% 13) $128°$ 14) $180°$ 15) $Area = 36$ cm^2; $Perimeter = 26$ cm 16) 15 $in.^2$

Study Strategies   Time Management

Students have to juggle many different things, such as school, work, activites, and family responsibilities. How well students juggle everything affects how well they do in school. So how can you get it all done and get it done *well*? Here's how you can use the P.O.W.E.R. framework to become a better time manager.

- *Prepare means to explicitly state a goal.* **I will get better at managing my time.**

- *Organize means to* **organize** *the physical and mental tools you need to achieve your goal.*
- Complete the emPOWERme survey that appears before the Chapter Summary.
- Create a *weekly master calendar* that displays your class schedule, due dates for assignments and tests, work schedule, and other responsibilities.
- Make a daily *to-do list* as well as one for the week. Rank the items on the lists in order of importance, and plan to complete the most important and urgent items first.

- *Work means to* **do the work** *that needs to be done to achieve the goal.*
- Using the schedules you've created, look at your open blocks of time to decide when you can do your homework, study for a test, meet for a group project, schedule doctor appointments, and so on. Write them on your master calendar.

Insy Shah/Getty Images

- Look at your to-do lists regularly, and continue to prioritize the items on the list. Check off items as you complete them, and add new ones as they arise.
- Avoid distractions. For example, turn off your phone while studying.

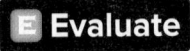

- *Evaluate* means you should **evaluate** what you have done. Did you achieve your goal?
- Look at your daily to-do list. Did you check off the items that you needed to finish?
- Consult your weekly to-do list and master calendar, and ask yourself whether you are on track to complete the rest of the items on the list. At the end of the week, did you finish everything you needed to finish?

- **Rethink** *means to* **rethink** *and reflect upon your goal.*
- Reflect upon the day or week, and ask yourself how you felt. Did you feel rushed or stressed? If necessary, adjust your master calendar. Reassess your priorities.
- If you completed all of your tasks, ask yourself which strategies helped you manage your time well so that you can do them again. Are there other things that you could try in the future to make your use of time even *more* efficient?
- If you did *not* finish what you needed to get done, ask yourself, "Why not?" Did you let distractions get the best of you? Were you overscheduled? What can you do differently in the future?

3.1 Solving Linear Equations Part I

P Prepare

O Organize

What are your objectives for Section 3.1?	How can you accomplish each objective?
1 Define a Linear Equation in One Variable	• Write the definition of a *linear equation in one variable* in your own words, and write an example. • Know the differences between expressions and equations and what solving an equation means.
2 Use the Addition and Subtraction Properties of Equality	• Follow the explanation to understand and then learn the **Addition and Subtraction Properties of Equality.** • Complete the given examples on your own. • Complete You Try 1.
3 Use the Multiplication and Division Properties of Equality	• Learn the **Multiplication and Division Properties of Equality.** • Complete the given examples on your own. • Complete You Try 2.
4 Solve Equations of the Form $ax + b = c$	• Be able to combine the properties learned in Objectives 2 and 3. • Complete the given examples on your own. • Complete You Trys 3 and 4.
5 Solve Equations with Variables on One Side of the Equal Sign by Combining Like Terms	• Follow the procedure for solving equations, and notice the subtle differences between the examples. • Complete the given examples on your own. • Complete You Trys 5–7.

 **Read the explanations, follow the examples, take notes, and complete the You Trys.**

1 Define a Linear Equation in One Variable

What is an equation? It is a mathematical statement that two expressions are equal. For example, $4 + 3 = 7$ is an equation.

> **Note**
>
> An equation contains an "=" sign, and an expression does not.

$$3x + 5 = 17 \text{ is an } equation.$$
$$3x + 5x \text{ is an } expression.$$

We can **solve** equations, and we can **simplify** expressions.

There are many different types of algebraic equations, and in Sections 3.1 and 3.2, we will learn how to solve *linear* equations. Here are some examples of linear equations in one variable:

<aside>

W Hint

Notice that there are no squared variable terms in a linear equation.

</aside>

$$p - 1 = 4 \qquad 3x + 5 = 17 \qquad 8(n + 1) - 7 = 2n + 3 \qquad -\frac{5}{6}y + \frac{1}{3} = y - 2$$

> ## Definition
>
> A **linear equation in one variable** is an equation that can be written in the form $ax + b = 0$, where a and b are real numbers and $a \neq 0$.

The exponent of the variable, x, in a linear equation like $ax + b = 0$ is 1. For this reason, linear equations are also known as first-degree equations. Equations like $k^2 - 13k + 36 = 0$ and $\sqrt{w - 3} = 2$ are not linear equations and are presented later in the text.

To **solve an equation** means to find the value or values of the variable that make the equation true. For example, the solution of the equation $p - 1 = 4$ is $p = 5$ since substituting 5 for the variable makes the equation true.

$$p - 1 = 4$$
$$5 - 1 = 4 \quad \text{True}$$

Usually, we use set notation to list all the solutions of an equation. The **solution set** of an equation is the set of all numbers that make the equation true. Therefore, {5} is the solution set of the equation $p - 1 = 4$. We also say that 5 *satisfies* the equation $p - 1 = 4$.

2 Use the Addition and Subtraction Properties of Equality

Begin with the true statement $8 = 8$. What happens if we add the same number, say 2, to each side? Is the statement still true? Yes.

$$8 = 8$$
$$8 + 2 = 8 + 2$$
$$10 = 10 \quad \text{True}$$

The statement would also be true if we subtracted 2 from both sides of $8 = 8$.

$$8 = 8 \text{ and } 8 + 2 = 8 + 2 \text{ are } equivalent\ equations.$$
$$8 = 8 \text{ and } 8 - 2 = 8 - 2 \text{ are } equivalent\ equations \text{ as well.}$$

We can use these principles to solve an algebraic equation because doing so will not change the equation's solution.

Property Addition and Subtraction Properties of Equality

Let a, b, and c be expressions representing real numbers. Then,

1) If $a = b$, then $a + c = b + c$.   Addition property of equality

2) If $a = b$, then $a - c = b - c$.   Subtraction property of equality

---

EXAMPLE 1

Solve each equation, and check the solution.

a) $x - 8 = 3$ b) $-5 = m + 4$

Solution

a) Remember, to solve the equation means to find the value of the variable that makes the statement true. To do this, we want to get the variable on a side by itself. We call this **isolating the variable.**

   On the left side of the equal sign, the 8 is being **subtracted from** the x. To isolate x, we perform the "opposite" operation—that is, we **add 8** to each side.

$$x - 8 = 3$$
$$x - 8 + 8 = 3 + 8 \quad \text{Add 8 to each side.}$$
$$x = 11 \quad \text{Simplify.}$$

Check: Substitute 11 for x in the original equation.

$$x - 8 = 3$$
$$11 - 8 = 3$$
$$3 = 3 \quad \checkmark$$

The solution set is $\{11\}$.

b) Notice that the 4 is being **added to** the variable, m. We will **subtract 4** from each side to isolate the variable.

$$-5 = m + 4$$
$$-5 - 4 = m + 4 - 4 \quad \text{Subtract 4 from each side.}$$
$$-9 = m \quad \text{Simplify.}$$

Check: Substitute -9 for m in the original equation.

$$-5 = m + 4$$
$$-5 = -9 + 4$$
$$-5 = -5 \quad \checkmark$$

 **Hint**
The variable can be on the left or right side of the equal sign.

The solution set is $\{-9\}$.

YOU TRY 1

Solve each equation, and check the solution.

a) $b - 5 = 9$ b) $-2 = y + 6$

---

3 Use the Multiplication and Division Properties of Equality

It is also true that if we multiply both sides of an equation by the same nonzero number or divide both sides of an equation by the same nonzero number, then we will obtain an equivalent equation.

> **W Hint**
> Did you notice that dividing both sides of an equation by c is the same as multiplying both sides by $1/c$?

Property Multiplication and Division Properties of Equality

Let a, b, and c be expressions representing real numbers where $c \neq 0$. Then,

1) If $a = b$, then $ac = bc$.   Multiplication property of equality

2) If $a = b$, then $\dfrac{a}{c} = \dfrac{b}{c}$.   Division property of equality

---

EXAMPLE 2

Solve each equation, and check the solution.

a) $3k = -9.6$ b) $-m = 19$ c) $\dfrac{x}{4} = 5$ d) $\dfrac{3}{8}y = 12$

Solution

a) On the left-hand side of the equation, the k is being **multiplied** by 3. So, we will perform the "opposite" operation and **divide** each side by 3.

$$3k = -9.6$$
$$\frac{3k}{3} = \frac{-9.6}{3} \qquad \text{Divide each side by 3.}$$
$$k = -3.2 \qquad \text{Simplify.}$$

Check: Substitute -3.2 for k in the original equation.

$$3k = -9.6$$
$$3(-3.2) = -9.6$$
$$-9.6 = -9.6 \ \checkmark$$

The solution set is $\{-3.2\}$.

b) The negative sign in front of the m tells us that the coefficient of m is -1. Since m is being **multiplied** by -1, we will **divide** each side by -1.

$$-m = 19$$
$$\frac{-1m}{-1} = \frac{19}{-1} \qquad \text{Rewrite } -m \text{ as } -1m; \text{ divide each side by } -1.$$
$$m = -19 \qquad \text{Simplify.}$$

The check is left to the student. The solution set is $\{-19\}$.

c) The x in $\frac{x}{4} = 5$ is being **divided** by 4. Therefore, we will **multiply** each side by 4 to get the x on a side by itself.

$$\frac{x}{4} = 5$$

$$4 \cdot \frac{x}{4} = 4 \cdot 5 \qquad \text{Multiply each side by 4.}$$

$$1x = 20 \qquad\qquad \text{Simplify.}$$

$$x = 20$$

The check is left to the student. The solution set is $\{20\}$.

W Hint

Which property from Section 1.7 are you using to complete this example?

d) On the left-hand side of $\frac{3}{8}y = 12$, the y is being **multiplied** by $\frac{3}{8}$. So, we could divide each side by $\frac{3}{8}$. However, recall that dividing a quantity by a fraction is the same as **multiplying by the reciprocal** of the fraction. Therefore, we will multiply each side by the reciprocal of $\frac{3}{8}$.

$$\frac{3}{8}y = 12$$

$$\frac{8}{3} \cdot \frac{3}{8}y = \frac{8}{3} \cdot 12 \qquad \text{The reciprocal of } \frac{3}{8} \text{ is } \frac{8}{3}. \text{ Multiply each side by } \frac{8}{3}.$$

$$1y = \frac{8}{\cancel{3}} \cdot \cancel{12}^{4} \qquad \text{Perform the multiplication.}$$

$$y = 32 \qquad\qquad \text{Simplify.}$$

The check is left to the student. The solution set is $\{32\}$.

[YOU TRY 2] Solve each equation, and check the solution.

a) $-8w = 42.4$ b) $-a = -25$ c) $\frac{h}{7} = 8$ d) $-\frac{5}{9}c = 20$

---

4 Solve Equations of the Form $ax + b = c$

Next, we will combine the properties of addition, subtraction, multiplication, and division to solve equations like $3p + 7 = 31$ and $4x + 9 - 6x + 2 = 17$.

EXAMPLE 3   Solve $3p + 7 = 31$.

Solution

In this equation, there is a number, 7, being **added** to the term containing the variable, and the variable is being multiplied by a number, 3. **In general, we first eliminate the number being added to or subtracted from the variable.** Then we eliminate the coefficient.

$$3p + 7 = 31$$
$$3p + 7 - 7 = 31 - 7 \qquad \text{Subtract 7 from each side.}$$
$$3p = 24 \qquad \text{Combine like terms.}$$
$$\frac{3p}{3} = \frac{24}{3} \qquad \text{Divide by 3.}$$
$$p = 8 \qquad \text{Simplify.}$$

$$\text{Check:} \quad 3p + 7 = 31$$
$$3(8) + 7 = 31$$
$$24 + 7 = 31$$
$$31 = 31 \quad \checkmark$$

The solution set is $\{8\}$.

YOU TRY 3   Solve $2n + 9 = 15$.

EXAMPLE 4   Solve.

a) $-\dfrac{6}{5}c - 1 = 13$ b) $-8.85 = 2.1y - 5.49$

Solution

a) On the left-hand side, the c is being multiplied by $-\dfrac{6}{5}$, and 1 is being subtracted from the c-term. To solve the equation, begin by eliminating the number being subtracted from the c-term.

W Hint

Write down all of the steps in the example on your paper as you are reading it.

$$-\frac{6}{5}c - 1 = 13$$
$$-\frac{6}{5}c - 1 + 1 = 13 + 1 \qquad \text{Add 1 to each side.}$$
$$-\frac{6}{5}c = 14 \qquad \text{Combine like terms.}$$
$$-\frac{5}{6} \cdot \left(-\frac{6}{5}c\right) = -\frac{5}{6} \cdot 14 \qquad \text{Multiply each side by the reciprocal of } -\frac{6}{5}.$$
$$1c = -\frac{5}{\overset{}{\underset{3}{6}}} \cdot \overset{7}{14} \qquad \text{Simplify.}$$
$$c = -\frac{35}{3}$$

The check is left to the student. The solution set is $\left\{-\dfrac{35}{3}\right\}$.

b) The variable is on the right-hand side of the equation. First, we will add 5.49 to each side, then we will divide by 2.1.

$$-8.85 = 2.1y - 5.49$$
$$-8.85 + 5.49 = 2.1y - 5.49 + 5.49 \qquad \text{Add 5.49 to each side.}$$
$$-3.36 = 2.1y \qquad \text{Combine like terms.}$$
$$\frac{-3.36}{2.1} = \frac{2.1y}{2.1} \qquad \text{Divide each side by 2.1.}$$
$$-1.6 = y \qquad \text{Simplify.}$$

Verify that -1.6 is the solution. The solution set is $\{-1.6\}$.

[**YOU TRY 4**] Solve.

a) $-\dfrac{4}{9}z + 3 = -7$ b) $6.7 = -0.4t - 5.3$

5 Solve Equations with Variables on One Side of the Equal Sign by Combining Like Terms

So far, we have learned how to solve equations with only one variable term. How do we solve an equation with *more than one* variable term? We can generalize the steps for solving linear equations with the following procedure.

W Hint
In your own words, summarize the procedure for solving a linear equation.

Procedure How to Solve a Linear Equation

Step 1: **Clear parentheses** and **combine like terms** on each side of the equation.

Step 2: **Get the variable on one side of the equal sign and the constant on the other side of the equal sign** (isolate the variable) using the addition or subtraction property of equality.

Step 3: **Solve for the variable** using the multiplication or division property of equality.

Step 4: **Check the solution** in the original equation.

Now, let's solve some equations that require us to combine like terms.

EXAMPLE 5   Solve $4x + 9 - 6x + 2 = 17$.

Solution

Step 1: Because there are two x-terms on the left side of the equal sign, begin by combining like terms.
$$4x + 9 - 6x + 2 = 17$$
$$-2x + 11 = 17 \qquad \text{Combine like terms.}$$

Step 2: Isolate the variable.

$$-2x + 11 - 11 = 17 - 11 \qquad \text{Subtract 11 from each side.}$$
$$-2x = 6 \qquad \text{Combine like terms.}$$

Step 3: Solve for x using the division property of equality.

$$\frac{-2x}{-2} = \frac{6}{-2} \qquad \text{Divide each side by } -2.$$
$$x = -3 \qquad \text{Simplify.}$$

Step 4: Check:
$$4x + 9 - 6x + 2 = 17$$
$$4(-3) + 9 - 6(-3) + 2 = 17$$
$$-12 + 9 + 18 + 2 = 17$$
$$17 = 17 \quad \checkmark$$

The solution set is $\{-3\}$.

[YOU TRY 5] Solve $15 - 7u - 6 + 2u = -1$.

EXAMPLE 6 Solve $2(1 - 3h) - 5(2h + 3) = -21$.

Solution

Step 1: Clear the parentheses and combine like terms.

$$2(1 - 3h) - 5(2h + 3) = -21$$
$$2 - 6h - 10h - 15 = -21 \qquad \text{Distribute.}$$
$$-16h - 13 = -21 \qquad \text{Combine like terms.}$$

Step 2: Isolate the variable.

$$-16h - 13 + 13 = -21 + 13 \qquad \text{Add 13 to each side.}$$
$$-16h = -8 \qquad \text{Combine like terms.}$$

Step 3: Solve for h using the division property of equality.

$$\frac{-16h}{-16} = \frac{-8}{-16} \qquad \text{Divide each side by } -16.$$
$$h = \frac{1}{2} \qquad \text{Simplify.}$$

Step 4: The check is left to the student. The solution set is $\left\{\dfrac{1}{2}\right\}$.

[YOU TRY 6] Solve $-3(4y - 3) + 4(y + 1) = 15$.

EXAMPLE 7

Solve $\dfrac{1}{2}(3b + 8) + \dfrac{3}{4} = -\dfrac{1}{2}$.

Solution

Step 1: Clear the parentheses and combine like terms.

$$\frac{1}{2}(3b + 8) + \frac{3}{4} = -\frac{1}{2}$$

<table>
<tr><td>$\dfrac{3}{2}b + 4 + \dfrac{3}{4} = -\dfrac{1}{2}$</td><td>Distribute.</td></tr>
<tr><td>$\dfrac{3}{2}b + \dfrac{16}{4} + \dfrac{3}{4} = -\dfrac{1}{2}$</td><td>Get a common denominator for the like terms.</td></tr>
<tr><td>$\dfrac{3}{2}b + \dfrac{19}{4} = -\dfrac{1}{2}$</td><td>Combine like terms.</td></tr>
</table>

W Hint

If necessary, review operations with fractions in Section 1.1.

Step 2: Isolate the variable.

<table>
<tr><td>$\dfrac{3}{2}b + \dfrac{19}{4} - \dfrac{19}{4} = -\dfrac{1}{2} - \dfrac{19}{4}$</td><td>Subtract $\dfrac{19}{4}$ from each side.</td></tr>
<tr><td>$\dfrac{3}{2}b = -\dfrac{2}{4} - \dfrac{19}{4}$</td><td>Get a common denominator.</td></tr>
<tr><td>$\dfrac{3}{2}b = -\dfrac{21}{4}$</td><td>Simplify.</td></tr>
</table>

Step 3: Solve for b using the multiplication property of equality.

<table>
<tr><td>$\dfrac{2}{3} \cdot \dfrac{3}{2}b = \dfrac{2}{3} \cdot \left(-\dfrac{21}{4}\right)$</td><td>Multiply both sides by the reciprocal of $\dfrac{3}{2}$.</td></tr>
<tr><td>$b = \dfrac{\overset{1}{2}}{\underset{1}{3}} \cdot \left(-\dfrac{\overset{-7}{21}}{\underset{2}{4}}\right)$</td><td>Perform the multiplication.</td></tr>
<tr><td>$b = -\dfrac{7}{2}$</td><td>Simplify.</td></tr>
</table>

Step 4: The check is left to the student. The solution set is $\left\{-\dfrac{7}{2}\right\}$.

[YOU TRY 7]

Solve $\dfrac{1}{3}(2m - 1) + \dfrac{5}{9} = \dfrac{4}{3}$.

In the next section, we will learn how to solve equations that contain variables on both sides of the equal sign. Also, we will learn another way to solve an equation containing several fractions like the equation in Example 7.

ANSWERS TO [YOU TRY] EXERCISES

1) a) {14}   b) {−8}   2) a) {−5.3}   b) {25}   c) {56}   d) {−36}   3) {3}

4) a) $\left\{\dfrac{45}{2}\right\}$ b) {−30}   5) {2}   6) $\left\{-\dfrac{1}{4}\right\}$ 7) $\left\{\dfrac{5}{3}\right\}$

Get Ready

Perform the indicated operation.

1) $2.6 + 4.9$

2) $3.8 - 9.6$

3) $11.7 \div 3$

4) $5.4 \div 0.2$

5) $-9 \times (-0.6)$

6) 1.83×4

7) $\frac{5}{8} \cdot 24$

8) $-\frac{9}{4} \cdot 10$

9) $\frac{1}{3} - \frac{3}{4}$

10) $\frac{7}{10} + \frac{2}{5}$

Objective 1: Define a Linear Equation in One Variable

Identify each as an expression or an equation.

11) $9c + 4 - 2c$

12) $5n - 3 = 6$

13) $y + 10(8y + 1) = -13$

14) $4 + 2(5p - 3)$

15) Can we solve $-6x + 10x$? Why or why not?

16) Can we solve $-6x + 10x = 28$? Why or why not?

17) Which of the following are linear equations in one variable?

   a) $2k + 9 - 7k + 1$

   b) $-8 = \frac{3}{2}n - 1$

   c) $4(3t - 1) + 9t = 12$

   d) $w^2 + 13w + 36 = 0$

18) Explain how to check the solution of an equation.

Determine whether the given value is a solution to the equation.

19) $a - 4 = -9; \quad a = 5$

20) $-5c = -10, \quad c = 2$

21) $-12y = 8; \quad y = -\frac{2}{3}$

22) $5 = a + 14; \quad a = -9$

23) $1.3 = 2p - 1.7; \quad p = 1.5$

24) $20m + 3 = 16; \quad m = \frac{4}{5}$

Objective 2: Use the Addition and Subtraction Properties of Equality

Solve each equation, and check the solution.

25) $n - 5 = 12$

26) $z + 8 = -2$

27) $b + 10 = 4$

28) $x - 3 = 9$

29) $-16 = k - 12$

30) $23 = r + 14$

31) $a - 2.9 = -3.6$

32) $w + 4.7 = 9.1$

33) $-\frac{2}{5} = -\frac{5}{4} + c$

34) $-\frac{8}{9} = d + 2$

35) Write an equation that can be solved with the subtraction property of equality and that has a solution set of $\{-7\}$.

36) Write an equation that can be solved with the addition property of equality and that has a solution set of $\{10\}$.

Objective 3: Use the Multiplication and Division Properties of Equality

Solve each equation, and check the solution.

37) $2n = 8$

38) $9k = 72$

39) $-5z = 35$

40) $-8b = -24$

41) $-48 = -4r$

42) $-54 = 6m$

43) $-7 = -0.5d$

44) $-3.9 = 1.3p$

45) $-x = 1$

46) $-h = -3$

47) $\frac{a}{4} = 12$

48) $\frac{w}{5} = 4$

49) $\frac{w}{6} = -\frac{3}{4}$

50) $-\frac{a}{21} = -\frac{2}{3}$

51) $\frac{1}{5}q = -9$

52) $\frac{1}{8}y = 3$

53) $\frac{5}{12} = \frac{1}{4}c$

54) $-\frac{5}{9} = -\frac{1}{6}r$

55) $-\frac{5}{3}d = -30$

56) $-\frac{4}{7}w = -36$

Objective 4: Solve Equations of the Form $ax + b = c$

Solve each equation, and check the solution.

57) $5z + 8 = 43$

58) $2y - 5 = 3$

59) $8d - 15 = -15$

60) $10m + 7 = 7$

61) $-11 = 5t - 9$

62) $7 = 4k + 13$

63) $10 = 3 - 7y$

64) $-6 = 9 - 3p$

65) $\frac{1}{2}d + 7 = 12$

66) $\frac{1}{3}x + 4 = 11$

67) $\frac{4}{5}b - 9 = -13$

68) $-\frac{12}{7}r + 5 = 3$

69) $-1 = \frac{10}{11}c + 5$

70) $2 = -\frac{9}{4}a - 10$

71) $2 - \dfrac{5}{6}t = -2$ 72) $5 + \dfrac{3}{4}h = -1$

73) $\dfrac{3}{4} = \dfrac{1}{2} - \dfrac{1}{6}z$ 74) $1 = \dfrac{3}{5} - \dfrac{2}{3}k$

75) $0.2p + 9.3 = 5.7$ 76) $0.5x - 2.6 = 4.9$

(24) 77) $14.74 = -20.6 - 5.7u$ 78) $10.5 - 9.2m = -36.42$

Objective 5: Solve Equations with Variables on One Side of the Equal Sign by Combining Like Terms

79) Explain, in your own words, the steps for solving a linear equation.

80) What is the first step for solving $8n + 3 + 2n - 9 = 13$? Do not solve the equation.

Solve each equation.

Fill It In

Fill in the blanks with either the missing mathematical step or the reason for the given step.

81) $3x + 7 + 5x + 4 = 27$
     $8x + 11 = 27$ _____

     _____   Subtraction property of equality

     $8x = 16$ _____

     _____   Division property of equality

     _____   Simplify.

     The solution set is _____.

82) $5 - 2(3k + 1) + 2k = 23$

    _____   Distribute.
    _____   Combine like terms.

    $-4k + 3 - 3 = 23 - 3$ _____

    $-4k = 20$ _____

    $\dfrac{-4k}{-4} = \dfrac{20}{-4}$ _____

    _____   Simplify.

    The solution set is _____.

For Exercises 83 and 84, fill in the blank with *always*, *sometimes*, or *never* to make the statement true.

83) An equation containing fractions will _____ have a solution that is a fraction.

84) An equation that does *not* contain fractions will _____ have a solution that is a fraction.

Solve each equation, and check the solution.

(24) 85) $6a - 10 + 4a + 9 = 39$

86) $7m + 11 + 2m - 5 = 33$

87) $30 = 5c + 14 - 11c + 1$

88) $-42 = 4x - 17 + 5x + 8$

89) $\dfrac{1}{4}n + 2 + \dfrac{1}{2}n - \dfrac{3}{2} = \dfrac{11}{4}$

90) $\dfrac{1}{6} + \dfrac{1}{2}w - \dfrac{4}{3} + \dfrac{1}{3}w = \dfrac{1}{2}$

91) $4.2d - 1.7 - 2.2d + 4.3 = -1.4$

92) $5.9h + 2.8 - 3.7h - 3.9 = 1.1$

(24) 93) $7(b - 5) + 5(b + 4) = 45$

94) $4(z - 2) + 3(z + 8) = -12$

95) $-23 = 4(3x - 7) - (8x - 5)$

96) $38 = 9(2a + 3) - (10a - 7)$

97) $8 = 5(4n + 3) - 3(2n - 7) - 20$

98) $-4 = 3(2z - 5) - 2(5z - 1) + 9$

99) $2(7u - 3) - (u + 9) - 3(2u + 1) = 24$

100) $6(4h + 7) + 2(h - 5) - (h - 11) = 18$

101) $\dfrac{1}{3}(3w + 4) - \dfrac{2}{3} = -\dfrac{1}{3}$ 102) $\dfrac{3}{4}(2r - 5) + \dfrac{1}{2} = \dfrac{5}{4}$

103) $\dfrac{4}{3}(t + 1) - \dfrac{1}{6}(4t - 3) = 2$

104) $\dfrac{1}{4}(3x - 2) - \dfrac{1}{2}(x - 1) = -\dfrac{1}{7}$

R Rethink

R1) Why is it that we generally add numbers to or subtract numbers from both sides of an equation before we multiply or divide? (Hint: Compare to the order of operations.)

R2) What is the difference between an expression and an equation?

R3) Which exercises in this section do you need to practice more?

R4) Were there any problems that you could not do? If so, write them down or circle them and ask your instructor how to do them.

3.2 Solving Linear Equations Part II

P Prepare

What are your objectives for Section 3.2?

O Organize

How can you accomplish each objective?

1 Solve Equations Containing Variables on Both Sides of the Equal Sign	• Follow the procedure for solving equations that we learned in Section 3.1. • Complete the given examples on your own. • Complete You Trys 1 and 2.
2 Solve Equations Containing Fractions or Decimals	• Write the procedures for **Eliminating Fractions from an Equation** and **Eliminating Decimals from an Equation** in your own words, and add them to the procedure developed for solving linear equations in one variable. • Complete the given examples on your own. • Complete You Trys 3 and 4.
3 Solve Equations with No Solution or an Infinite Number of Solutions	• Complete and understand the given examples on your own. • Learn the **Outcomes When Solving Linear Equations,** and know how to interpret the results when solving an equation. • Complete You Try 5.
4 Use the Five Steps for Solving Applied Problems	• Summarize the **Five Steps for Solving Applied Problems** in your own words. • Complete the given examples on your own. • Complete You Trys 6 and 7.

W Work

Read the explanations, follow the examples, take notes, and complete the You Trys.

In Section 3.1, we learned how to solve equations containing variables on only one side of the equal sign. In this section, we will discuss how to solve equations containing variables on *both* sides of the equal sign.

1 Solve Equations Containing Variables on Both Sides of the Equal Sign

To solve an equation such as $3y - 11 + 7y = 6y + 9$, our goal is the same as it was when we solved equations in the previous section: Get the variables on one side of the equal sign and the constants on the other side. Let's use the steps we learned in Section 3.1.

EXAMPLE 1

Solve $3y - 11 + 7y = 6y + 9$.

Solution

Step 1: Combine like terms on the left side of the equal sign.

$$3y - 11 + 7y = 6y + 9$$
$$10y - 11 = 6y + 9 \qquad \text{Combine like terms.}$$

Step 2: Isolate the variable using the addition and subtraction properties of equality. Combine like terms so that there is a single variable term on one side of the equation and a constant on the other side.

$$10y - 6y - 11 = 6y - 6y + 9 \qquad \text{Subtract } 6y \text{ from each side.}$$
$$4y - 11 = 9 \qquad \text{Combine like terms.}$$
$$4y - 11 + 11 = 9 + 11 \qquad \text{Add 11 to each side.}$$
$$4y = 20 \qquad \text{Combine like terms.}$$

Step 3: Solve for y using the division property of equality.

$$\frac{4y}{4} = \frac{20}{4} \qquad \text{Divide each side by 4.}$$
$$y = 5 \qquad \text{Simplify.}$$

Step 4: Check:

$$3y - 11 + 7y = 6y + 9$$
$$3(5) - 11 + 7(5) = 6(5) + 9$$
$$15 - 11 + 35 = 30 + 9$$
$$39 = 39 \quad \checkmark$$

The solution set is $\{5\}$.

[YOU TRY 1]

Solve $-3k + 4 = 8k - 15 - 6k - 11$.

EXAMPLE 2

Solve $9t + 4 - (7t - 2) = t + 6(t + 1)$.

Solution

Step 1: Clear the parentheses and combine like terms.

$$9t + 4 - (7t - 2) = t + 6(t + 1)$$
$$9t + 4 - 7t + 2 = t + 6t + 6 \qquad \text{Distribute.}$$
$$2t + 6 = 7t + 6 \qquad \text{Combine like terms.}$$

 **Hint**

Are you working out the problems on your paper as you are reading the examples?

Step 2: Isolate the variable.

$$2t - 7t + 6 = 7t - 7t + 6 \qquad \text{Subtract } 7t \text{ from each side.}$$
$$-5t + 6 = 6 \qquad \text{Combine like terms.}$$
$$-5t + 6 - 6 = 6 - 6 \qquad \text{Subtract 6 from each side.}$$
$$-5t = 0 \qquad \text{Combine like terms.}$$

Step 3: Solve for t using the division property of equality.

$$\frac{-5t}{-5} = \frac{0}{-5} \qquad \text{Divide each side by } -5.$$
$$t = 0 \qquad \text{Simplify.}$$

Step 4: Check:

$$9t + 4 - (7t - 2) = t + 6(t + 1)$$
$$9(0) + 4 - [7(0) - 2] = 0 + 6[(0) + 1]$$
$$0 + 4 - (0 - 2) = 0 + 6(1)$$
$$4 - (-2) = 0 + 6$$
$$6 = 6 \quad \checkmark$$

The solution set is {0}.

[YOU TRY 2]

Solve $5 + 3(a + 4) = 7a - (9 - 10a) + 4$.

When equations contain fractions or decimals, they may appear difficult to solve. Let's learn how to eliminate the fractions or decimals so that it will be easier to solve such equations.

2 Solve Equations Containing Fractions or Decimals

To solve $\frac{1}{2}(3b + 8) + \frac{3}{4} = -\frac{1}{2}$ in Section 3.1, we began by using the distributive property to clear the parentheses, and we worked with the fractions throughout the solving process. But, there is another way we can solve equations containing several fractions. Before applying the steps for solving a linear equation, we can eliminate the fractions from the equation.

> **Procedure** Eliminating Fractions from an Equation
>
> To eliminate the fractions, determine the least common denominator for all the fractions in the equation. Then multiply both sides of the equation by the least common denominator (LCD).

Let's solve the equation we solved in Section 3.1 using this new approach.

EXAMPLE 3

Solve $\frac{1}{2}(3b + 8) + \frac{3}{4} = -\frac{1}{2}$.

Solution

The least common denominator of all the fractions in the equation is 4. Multiply both sides of the equation by 4 to eliminate the fractions.

$$4\left[\frac{1}{2}(3b + 8) + \frac{3}{4}\right] = 4\left(-\frac{1}{2}\right)$$

Step 1: Distribute the 4, clear the parentheses, and combine like terms.

$$4 \cdot \frac{1}{2}(3b + 8) + 4 \cdot \frac{3}{4} = -2 \qquad \text{Distribute.}$$
$$2(3b + 8) + 3 = -2 \qquad \text{Multiply.}$$
$$6b + 16 + 3 = -2 \qquad \text{Distribute.}$$
$$6b + 19 = -2 \qquad \text{Combine like terms.}$$

Step 2: Isolate the variable.

$$6b + 19 - 19 = -2 - 19 \qquad \text{Subtract 19 from each side.}$$
$$6b = -21 \qquad \text{Combine like terms.}$$

Step 3: Solve for b using the division property of equality.

$$\frac{6b}{6} = \frac{-21}{6} \qquad \text{Divide each side by 6.}$$
$$b = -\frac{7}{2} \qquad \text{Simplify.}$$

Step 4: The check is left to the student. The solution set is $\left\{-\frac{7}{2}\right\}$ This is the same as the result we obtained in Section 3.1, Example 7.

[YOU TRY 3] Solve $\frac{1}{3}(2m - 1) + \frac{5}{9} = \frac{4}{3}$.

Just as we can eliminate the fractions from an equation to make it easier to solve, we can eliminate decimals from an equation before applying the four-step equation-solving process.

Procedure Eliminating Decimals from an Equation

To eliminate the decimals from an equation, multiply both sides of the equation by the smallest power of 10 that will eliminate all decimals from the problem.

EXAMPLE 4

Solve $0.05a + 0.2(a + 3) = 0.1$

Solution

We want to eliminate the decimals. The number containing a decimal place farthest to the right is 0.05. The 5 is in the *hundredths* place. Therefore, multiply both sides of the equation by 100 to eliminate all decimals in the equation.

$$100[0.05a + 0.2(a + 3)] = 100(0.1)$$

Step 1: Distribute the 100, clear the parentheses, and combine like terms.

$$100 \cdot (0.05a) + 100[0.2(a + 3)] = 10 \qquad \text{Distribute.}$$
$$5a + 20(a + 3) = 10 \qquad \text{Multiply.}$$
$$5a + 20a + 60 = 10 \qquad \text{Distribute.}$$
$$25a + 60 = 10 \qquad \text{Combine like terms.}$$

Step 2: Isolate the variable.

$$25a + 60 - 60 = 10 - 60 \qquad \text{Subtract 60 from each side.}$$
$$25a = -50 \qquad \text{Combine like terms.}$$

Step 3: Solve for a using the division property of equality.

$$\frac{25a}{25} = \frac{-50}{25} \qquad \text{Divide each side by 25.}$$
$$a = -2 \qquad \text{Simplify.}$$

Step 4: The check is left to the student. The solution set is $\{-2\}$.

> **W Hint**
>
> In your own words, summarize the procedure for solving a linear equation and include the steps for eliminating fractions and decimals.

[**YOU TRY 4**]   Solve $0.08k - 0.2(k + 5) = -0.1$

3 Solve Equations with No Solution or an Infinite Number of Solutions

Does every equation have a solution? Consider the next example.

EXAMPLE 5

Solve $11w - 9 = 5w + 2(3w + 1)$.

Solution

$$11w - 9 = 5w + 2(3w + 1)$$
$$11w - 9 = 5w + 6w + 2 \qquad \text{Distribute.}$$
$$11w - 9 = 11w + 2 \qquad \text{Combine like terms.}$$
$$11w - 11w - 9 = 11w - 11w + 2 \qquad \text{Subtract } 11w \text{ from each side.}$$
$$-9 = 2 \qquad \text{False}$$

Notice that the variable has "dropped out." Is $-9 = 2$ a true statement? No! This means that there is no value for w that will make the statement true. This means that the equation has *no solution*. We say that the solution set is the **empty set,** or **null set,** and it is denoted by $\varnothing$.

We have seen that a linear equation may have one solution or no solution. There is a third possibility—a linear equation may have an infinite number of solutions.

EXAMPLE 6

Solve $10r - 3r + 15 = 7r + 15$.

Solution

$$10r - 3r + 15 = 7r + 15$$
$$7r + 15 = 7r + 15 \qquad \text{Combine like terms.}$$
$$7r - 7r + 15 = 7r - 7r + 5 \qquad \text{Subtract } 7r \text{ from each side.}$$
$$15 = 15 \qquad \text{True}$$

W Hint

Check to see if
$r = -2$, $r = 0$, and $r = 6$ are
solutions to the equation.

Here, the variable has "dropped out," and we are left with an equation, $15 = 15$, that is true. This means that any real number we substitute for r will make the original equation true. Therefore, this equation has an *infinite number of solutions*. The solution set is {**all real numbers**}.

[YOU TRY 5] Solve.

a) $9 - 4(3c + 1) = 15 - 12c - 10$ b) $12z + 7 - 10z = 2z + 5$

W Hint

Summarize these possible
outcomes in your notes.

Summary Outcomes When Solving Linear Equations

There are three possible outcomes when solving a linear equation. The equation may have

1) **one solution.** Solution set: {a real number}. An equation that is true for some values and not for others is called a **conditional equation.**

or

2) **no solution.** In this case, the variable will drop out, and there will be a false statement such as $-9 = 2$. Solution set: $\varnothing$. An equation that has no solution is called a **contradiction.**

or

3) **an infinite number of solutions.** In this case, the variable will drop out, and there will be a true statement such as $15 = 15$. Solution set: {all real numbers}. An equation that has all real numbers as its solution set is called an **identity.**

4   Use the Five Steps for Solving Applied Problems

Mathematical equations can be used to describe many situations in the real world. To do this, we must learn how to translate information presented in English into an algebraic equation. We will begin slowly and work our way up to more challenging problems. Yes, it may be difficult at first, but with patience and persistence, you can do it!

Although no single method will work for solving all applied problems, the following approach is suggested to help in the problem-solving process.

W Hint

Rewrite the steps for solving applied problems in your own words.

Procedure Five Steps for Solving Applied Problems

Step 1: **Read** the problem carefully, more than once if necessary, until you understand it. Draw a picture, if applicable. Identify what you are being asked to find.

Step 2: **Choose a variable** to represent an unknown quantity. If there are any other unknowns, define them in terms of the variable.

Step 3: **Translate** the problem from English into an equation using the chosen variable. Here are some suggestions for doing so:

- Restate the problem in your own words.
- Read and think of the problem in "small parts."
- Make a chart to separate these "small parts" of the problem to help you translate into mathematical terms.
- Write an equation in English, then translate it into an algebraic equation.

Step 4: **Solve** the equation.

Step 5: **Check** the answer in the original problem, and **interpret** the solution as it relates to the problem. Be sure your answer makes sense in the context of the problem.

EXAMPLE 7

Write the following statement as an equation, and find the number.
Nine more than twice a number is fifteen. Find the number.

Solution

Step 1: **Read** the problem carefully. We must find an unknown number.

Step 2: **Choose a variable** to represent the unknown.

$$\text{Let } x = \text{the number.}$$

Step 3: **Translate** the information that appears in English into an algebraic equation by rereading the problem slowly and "in parts."

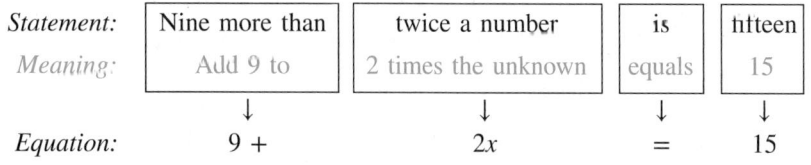

Statement:	Nine more than	twice a number	is	fifteen
Meaning:	Add 9 to	2 times the unknown	equals	15
	↓	↓	↓	↓
Equation:	9 +	2x	=	15

The equation is $9 + 2x = 15$.

Step 4: **Solve** the equation.

$$9 + 2x = 15$$
$$9 - 9 + 2x = 15 - 9 \quad \text{Subtract 9 from each side.}$$
$$2x = 6 \quad \text{Combine like terms.}$$
$$x = 3 \quad \text{Divide each side by 2.}$$

Step 5: **Check** the answer. Does the answer make sense? Nine more than twice three is $9 + 2(3) = 15$. The answer is correct. The number is 3.

[YOU TRY 6]

Write the following statement as an equation, and find the number. *Three more than twice a number is twenty-nine.*

Sometimes, dealing with subtraction in an application can be confusing. So let's look at an arithmetic problem first.

EXAMPLE 8

What is two less than seven?

Solution

To solve this problem, do we subtract $7 - 2$ or $2 - 7$? "Two less than seven" is written as $7 - 2$, and $7 - 2 = 5$. Five is two less than seven. To get the correct answer, the 2 is *subtracted from* the 7.

Keep this in mind as you read the next problem.

EXAMPLE 9

Write the following statement as an equation, and find the number.
Five less than three times a number is the same as the number increased by seven. Find the number.

Solution

Step 1: **Read** the problem carefully. We must find an unknown number.

Step 2: **Choose a variable** to represent the unknown.

$$\text{Let } x = \text{the number.}$$

Step 3: **Translate** the information that appears in English into an algebraic equation by rereading the problem slowly and "in parts."

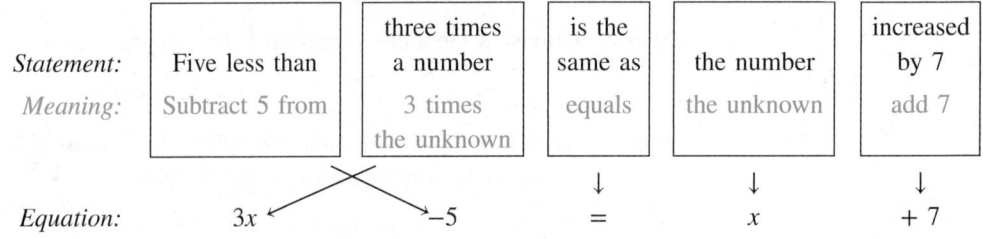

The equation is $3x - 5 = x + 7$.

Step 4: **Solve** the equation.

$$3x - 5 = x + 7$$
$$3x - x - 5 = x - x + 7 \qquad \text{Subtract } x \text{ from each side.}$$
$$2x - 5 = 7 \qquad \text{Combine like terms.}$$
$$2x - 5 + 5 = 7 + 5 \qquad \text{Add 5 to each side.}$$
$$2x = 12 \qquad \text{Combine like terms.}$$
$$x = 6 \qquad \text{Divide each side by 2.}$$

Step 5: **Check** the answer. Does it make sense? Five less than three times 6 is $3(6) - 5 = 13$. The number increased by seven is $6 + 7 = 13$. The answer is correct. The number is 6.

[**YOU TRY 7**] Write the following statement as an equation, and find the number.
Three less than five times a number is the same as the number increased by thirteen.

Using Technology

We can use a graphing calculator to solve a linear equation in one variable. First, enter the left side of the equation in Y_1 and the right side of the equation in Y_2. Then graph the equations. The x-coordinate of the point of intersection is the solution to the equation.

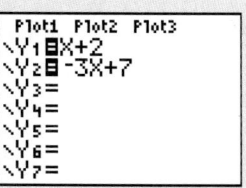

We will solve $x + 2 = -3x + 7$ algebraically and by using a graphing calculator, and then compare the results. First, use algebra to solve $x + 2 = -3x + 7$.

You should get $\frac{5}{4}$.

Next, use a graphing calculator to solve $x + 2 = -3x + 7$.

1) Enter $x + 2$ in Y_1 by pressing $\boxed{Y =}$ and entering $x + 2$ to the right of $\backslash Y_1 =$. Then press $\boxed{\text{ENTER}}$.

2) Enter $-3x + 7$ in Y_2 by pressing the $\boxed{Y =}$ key and entering $-3x + 7$ to the right of $\backslash Y_2 =$. Press $\boxed{\text{ENTER}}$.

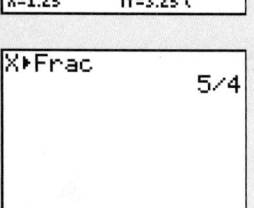

3) Press $\boxed{\text{ZOOM}}$ and select 6:ZStandard to graph the equations.

4) To find the intersection point, press 2nd $\boxed{\text{TRACE}}$ and select 5:intersect. Press $\boxed{\text{ENTER}}$ three times. The x-coordinate of the intersection point is shown on the left side of the screen and is stored in the variable x.

5) Return to the home screen by pressing 2nd $\boxed{\text{MODE}}$. Press X,T,Θ,n $\boxed{\text{ENTER}}$ to display the solution. Since the result in this case is a decimal value, we can convert it to a fraction by pressing X,T,Θ,n $\boxed{\text{MATH}}$, selecting Frac, then pressing $\boxed{\text{ENTER}}$.

The calculator then gives us the solution set $\left\{ \frac{5}{4} \right\}$.

Solve each equation algebraically, then verify your answer using a graphing calculator.

1) $x + 6 = -2x - 3$ 2) $2x + 3 = -x - 4$ 3) $\frac{5}{6}x + \frac{1}{2} = \frac{1}{6}x - \frac{3}{4}$

4) $0.3x - 1 = -0.2x - 5$ 5) $3x - 7 = -x + 5$ 6) $6x - 7 = 5$

ANSWERS TO $\boxed{\text{YOU TRY}}$ **EXERCISES**

1) $\{6\}$ 2) $\left\{ \frac{11}{7} \right\}$ 3) $\left\{ \frac{5}{3} \right\}$ 4) $\{-7.5\}$ 5) a) {all real numbers} b) $\varnothing$

6) $2x + 3 = 29$; 13   7) $5x - 3 = x + 13$; 4

ANSWERS TO TECHNOLOGY EXERCISES

1) $\{-3\}$ 2) $\left\{ -\frac{7}{3} \right\}$ 3) $\left\{ -\frac{15}{8} \right\}$ 4) $\{-8\}$ 5) $\{3\}$ 6) $\{2\}$

Objective 1: Solve Equations Containing Variables on Both Sides of the Equal Sign

 1) Explain, in your own words, the steps for solving a linear equation that has variables on both sides of the equal sign.

 2) What is the first step for solving $3 + 2(7x + 4) = 5 - 8x$? Do not solve the equation.

Solve each equation, and check the solution.

3) $5k - 6 = 7k - 8$

4) $3v + 14 = 9v - 22$

5) $-15w + 4 = 24 - 7w$

6) $-7x + 13 = 3 - 13x$

7) $18 - h + 5h - 11 = 9h + 19 - 3h$

8) $4m - 1 - 6m + 7 = 11m + 3 - 10m$

9) $1 + 5(4n - 7) = 4(7n - 3) - 30$

10) $10 + 2(z - 9) = 3(z + 1) - 6$

11) $2(1 - 8c) = 5 - 3(6c + 1) + 4c$

12) $13u + 6 - 5(2u - 3) = 1 + 4(u + 5)$

13) $9 - (8p - 5) + 4p = 6(2p + 1)$

14) $2(6d + 5) = 16 - (7d - 4) + 11d$

15) $-3(4r + 9) + 2(3r + 8) = r - (9r - 5)$

16) $2(3t - 4) - 6(t + 1) = -t + 4(t + 10)$

Objective 2: Solve Equations Containing Fractions or Decimals

 17) If an equation contains fractions, what is the first step you can perform to make it easier to solve?

 18) If an equation contains decimals, what is the first step you can perform to make it easier to solve?

 19) How can you eliminate the fractions from the equation $\frac{3}{8}x - \frac{1}{2} = \frac{1}{8}x + \frac{3}{4}$?

 20) How can you eliminate the decimals from the equation $0.02n + 0.1(n - 3) = 0.06$?

Solve each equation by first clearing the fractions or decimals.

21) $\frac{3}{8}x - \frac{1}{2} = \frac{1}{8}x + \frac{3}{4}$

22) $\frac{1}{2}c + \frac{7}{4} = \frac{5}{4}c - \frac{1}{2}$

23) $\frac{1}{3} - \frac{1}{2}m = \frac{1}{6}m + \frac{7}{9}$

24) $\frac{1}{15}p - \frac{1}{2} = \frac{1}{5}p - \frac{3}{10}$

25) $\frac{1}{3} + \frac{1}{9}(k + 5) - \frac{k}{4} = 2$

26) $\frac{5}{8}(2w + 3) + \frac{5}{4}w = \frac{3}{4}(4w + 1)$

27) $\frac{3}{4}(y + 7) + \frac{1}{2}(3y - 5) = \frac{9}{4}(2y - 1)$

28) $\frac{2}{3}(5z - 2) - \frac{4}{9}(3z - 2) = \frac{1}{3}(2z + 1)$

29) $\frac{1}{2}(4r + 1) - r = \frac{2}{5}(2r - 3) + \frac{3}{2}$

30) $\frac{2}{3}(3h - 5) + 1 = \frac{3}{2}(h - 2) + \frac{1}{6}h$

31) $0.06d + 0.13 = 0.31$

32) $0.09x - 0.14 = 0.4$

33) $0.04n - 0.05(n + 2) = 0.1$

34) $0.07t + 0.02(3t + 8) = -0.1$

35) $0.2(c - 4) + 1 = 0.15(c + 2)$

36) $0.12(5q - 1) - q = 0.15(7 - 2q)$

37) $0.35a - a = 0.03(5a + 4)$

38) $0.3(x - 2) + 1 = 0.25(x + 9)$

39) $0.07k + 0.15(200) = 0.09(k + 200)$

40) $0.2p + 0.08(120) = 0.16(p + 120)$

Objective 3: Solve Equations with No Solution or an Infinite Number of Solutions

For Exercises 41 and 42, fill in the blank with *always, sometimes,* or *never* to make the statement true.

41) An equation will _____ have one solution.

42) When the variable is eliminated from an equation, the solution set is _____ ∅.

43) How do you know that an equation has no solution?

44) How do you know that the solution set of an equation is {all real numbers}?

Determine whether each of the following equations has a solution set of {all real numbers} or has no solution, ∅.

45) $9(c + 6) - 2c = 4c + 1 + 3c$

46) $-21n + 22 = 3(4 - 7n) + 10$

(24) 47) $5t + 2(t + 3) - 4t = 4(t + 1) - (t - 2)$

48) $8z + 11 + 5z - 9 = 16 - 6(3 - 2z) + z$

(24) 49) $\dfrac{5}{6}k - \dfrac{2}{3} = \dfrac{1}{6}(5k - 4) + \dfrac{1}{2}$

50) $0.4y + 0.3(20 - y) = 0.1y + 6$

Mixed Exercises: Sections 3.1 and 3.2
The following set of exercises contains equations from Sections 3.1 and 3.2. Solve each equation.

51) $\dfrac{n}{5} = 20$ 52) $z + 18 = -5$

53) $-19 = 6 - p$ 54) $-a = 34$

55) $-5.4 = -0.9m$ 56) $\dfrac{15}{7}h = 25$

57) $51 = 4y - 13$

58) $3c + 8 = 5c + 11$

59) $9 - (7k - 2) + 2k = 4(k + 3) + 5$

60) $0.3t + 0.18(5000 - t) = 0.21(5000)$

61) $-\dfrac{5}{4}r + 17 = 7$

62) $-2.3 = 2.4z + 1.3$

63) $8(3t + 4) = 10t - 3 + 7(2t + 5)$

64) $-6 - (a + 9) + 7 = 3a + 2(4a - 1)$

65) $\dfrac{5}{3}w + \dfrac{2}{5} = w - \dfrac{7}{3}$

66) $2d + 7 = -4d + 3(2d - 5)$

67) $7(2q + 3) - 3(q + 5) = 6$

68) $-11 = \dfrac{4}{5}k - 17$

69) $0.16h + 0.4(2000) = 0.22(2000 + h)$

70) $\dfrac{4}{9} + \dfrac{2}{3}(c - 1) + \dfrac{5}{9}c = \dfrac{2}{9}(5c + 3)$

71) $-9r + 4r - 11 + 2 = 3r + 7 - 8r + 9$

72) $2u - 4.6 = -4.6$

73) $\dfrac{1}{2}(2r + 9) - \dfrac{1}{3}(r + 12) = 1$

74) $t + 18 = 3(5 - t) + 4t + 3$

Objective 4: Use the Five Steps for Solving Applied Problems

75) What are the five steps for solving applied problems?

76) If you are solving an applied problem in which you have to find the length of a side of a rectangle, would a solution of -18 be reasonable? Explain your answer.

Write each statement as an equation, and find the number.

77) Twelve more than a number is five.

78) Fifteen more than a number is nineteen.

79) Nine less than a number is twelve.

80) Fourteen less than a number is three.

81) The product of seven and a number is sixty-three.

82) The product of eleven and a number is fifty-five.

83) Five more than twice a number is seventeen.

84) Eighteen more than twice a number is eight.

(24) 85) Three times a number decreased by eight is forty.

86) Five less than four times a number is forty-three.

87) Three-fourths of a number is thirty-three.

88) Two-thirds of a number is twenty-six.

89) Nine less than half a number is three.

90) Two less than one-fourth of a number is three.

(24) 91) Three less than twice a number is the same as the number increased by eight.

92) Twelve less than five times a number is the same as the number increased by sixteen.

93) If twenty-four is subtracted from a number, the result is the same as the number divided by nine.

94) If forty-five is subtracted from a number, the result is the same as the number divided by four.

95) If two-thirds of a number is added to the number, the result is twenty-five.

96) If three-eighths of a number is added to twice the number, the result is thirty-eight.

Write a statement, in English, that translates to the given equation.

97) $5x - 9 = 16$

98) $4 = \dfrac{3}{2}x + 7$

99) $x + 8 = 2x + 1$

100) $\dfrac{x}{6} = \dfrac{1}{3}x - 4$

R **Rethink**

R1) How do you know which number to use to clear the decimals or fractions from an equation?

R2) Do you find it more difficult to solve equations containing decimals or fractions? Why do you think this is true?

R3) Which exercises were easy for you, and which were difficult? In the exercises you identified as difficult, what did you find difficult about them?

3.3 Applications of Linear Equations

P **Prepare**

O **Organize**

What are your objectives for Section 3.3?	How can you accomplish each objective?
1 Solve Problems Involving General Quantities	• Use the **Five Steps for Solving Applied Problems** to complete the given examples on your own. • Complete You Trys 1 and 2.
2 Solve Problems Involving Lengths	• Use the **Five Steps for Solving Applied Problems** to complete the given example on your own. • Complete You Try 3.
3 Solve Consecutive Integer Problems	• Understand how to represent *consecutive integers* and *consecutive odd/even integers* using a variable and expressions. • Use the **Five Steps for Solving Applied Problems** to complete the given examples on your own. • Complete You Trys 4 and 5.

 **W** **Work**

Read the explanations, follow the examples, take notes, and complete the You Trys.

In the previous section, we learned the Five Steps for Solving Applied Problems and used this procedure to solve problems involving unknown numbers. Now we will apply this problem-solving technique to other types of applications.

1 Solve Problems Involving General Quantities

EXAMPLE 1

Write an equation, and solve.

Swimmers Michael Phelps and Katie Ledecky both competed in the 2012 Olympics in London and in the 2016 Olympics in Rio de Janiero, where they won a total of 18 medals. Phelps won six more medals than Ledecky. How many Olympic medals did each athlete win? (espn.go.com)

Solution

Step 1: **Read** the problem carefully, and identify what we are being asked to find.

We must find the number of medals each Olympian won.

Step 2: **Choose a variable** to represent an unknown, and define the other unknown in terms of this variable.

In the statement "Phelps won six more medals than Ledecky," the number of medals that Michael Phelps won is expressed *in terms of* the number of medals won by Katie Ledecky. Therefore, let

$$x = \text{the number of medals Ledecky won}$$

Define the other unknown (the number of medals that Michael Phelps won) in terms of x. The statement "Phelps won six more medals than Ledecky" means

$$\text{number of Ledecky's medals} + 6 = \text{number of Phelps' medals}$$
$$x + 6 = \text{number of Phelps' medals}$$

Step 3: **Translate** the information that appears in English into an algebraic equation. One approach is to restate the problem in your own words.

Since these two athletes won a total of 18 medals, we can think of the situation in this problem as:

The number of medals Ledecky won *plus* the number of medals Phelps won *was* 18.

Let's write this as an equation.

**Hint**

Notice that we use only one variable and then use an expression ($x + 6$) to represent the other quantity.

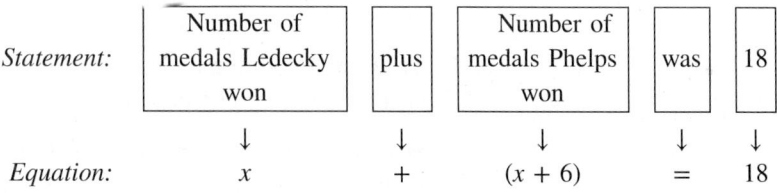

The equation is $x + (x + 6) = 18$.

Step 4: **Solve** the equation.

$$x + (x + 6) = 18$$
$$2x + 6 = 18$$
$$2x + 6 - 6 = 18 - 6 \qquad \text{Subtract 6 from each side.}$$
$$2x = 12 \qquad \text{Combine like terms.}$$
$$\frac{2x}{2} = \frac{12}{2} \qquad \text{Divide each side by 2.}$$
$$x = 6 \qquad \text{Simplify.}$$

Step 5: **Check** the answer, and **interpret** the solution as it relates to the problem.

Since x represents the number of medals that Katie Ledecky won, she won 6 medals.

The expression $x + 6$ represents the number of medals Michael Phelps won, so he won $x + 6 = 6 + 6 = 12$ medals.

The answer makes sense because the total number of medals they won was $6 + 12 = 18$.

[YOU TRY 1]

Write an equation, and solve.

An employee at a cellular phone store is doing inventory. The store has 23 more conventional cell phones in stock than smart phones. If the store has a total of 73 phones, how many of each type of phone is in stock?

EXAMPLE 2

Write an equation, and solve.

Nick has half as many songs on his playlist as Mariah. Together they have a total of 4887 songs. How many songs does each of them have?

Solution

Step 1: **Read** the problem carefully, and identify what we are being asked to find.

We must find the number of songs on Nick's playlist and the number on Mariah's playlist.

Step 2: **Choose a variable** to represent an unknown, and define the other unknown in terms of this variable.

In the sentence "Nick has half as many songs on his playlist as Mariah," the number of songs Nick has is expressed *in terms of* the number of songs Mariah has. Therefore, let

x = the number of songs on Mariah's playlist

Define the other unknown in terms of x.

$\frac{1}{2}x$ = the number of songs on Nick's playlist

Step 3: **Translate** the information that appears in English into an algebraic equation. One approach is to restate the problem in your own words.

Since Mariah and Nick have a total of 4887 songs, we can think of the situation in this problem as:

The number of Mariah's songs *plus* the number of Nick's songs *equals* 4887.

Let's write this as an equation.

	Number of Mariah's songs	plus	Number of Nick's songs	equals	4887
Statement:	↓	↓	↓	↓	↓
Equation:	x	$+$	$\frac{1}{2}x$	$=$	4887

The equation is $x + \frac{1}{2}x = 4887$.

Step 4: **Solve** the equation.

$$x + \frac{1}{2}x = 4887$$

$$\frac{3}{2}x = 4887 \qquad \text{Combine like terms.}$$

$$\frac{2}{3} \cdot \frac{3}{2}x = \frac{2}{3} \cdot 4887 \qquad \text{Multiply by the reciprocal of } \frac{3}{2}.$$

$$x = 3258 \qquad \text{Multiply.}$$

Step 5: **Check** the answer, and **interpret** the solution as it relates to the problem.

Mariah has 3258 songs on her playlist.

The expression $\frac{1}{2}x$ represents the number of songs on Nick's playlist, so there are $\frac{1}{2}(3258) = 1629$ songs on Nick's playlist.

The answer makes sense because the total number of songs on their playlists is $3258 + 1629 = 4887$ songs.

> [**YOU TRY 2**] Write an equation, and solve.
> Terrance and Janay are in college. Terrance has earned twice as many credits as Janay. How many credits does each student have if together they have earned 51 semester hours?

---

2 Solve Problems Involving Lengths

EXAMPLE 3 Write an equation, and solve.
A plumber has a section of PVC pipe that is 12 ft long. He needs to cut it into two pieces so that one piece is 2 ft shorter than the other. How long will each piece be?

Solution

Step 1: **Read** the problem carefully, and identify what we are being asked to find.

We must find the length of each of two pieces of pipe.

A picture will be very helpful in this problem.

Step 2: **Choose a variable** to represent an unknown, and define the other unknown in terms of this variable.

One piece of pipe must be 2 ft shorter than the other piece. Therefore, let

$$x = \text{the length of one piece}$$

Define the other unknown in terms of x.

$$x - 2 = \text{the length of the second piece}$$

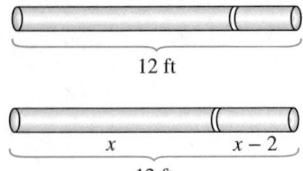

12 ft

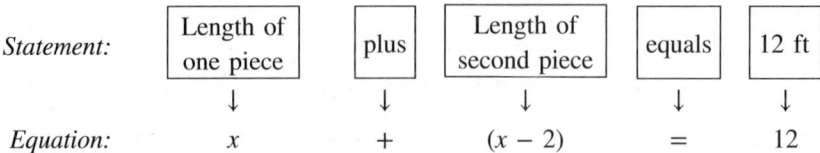

x   *x − 2*

12 ft

Step 3: **Translate** the information that appears in English into an algebraic equation. Let's label the picture with the expressions representing the unknowns and then restate the problem in our own words.

From the picture we can see that the

length of one piece plus the length of the second piece equals 12 ft

Let's write this as an equation.

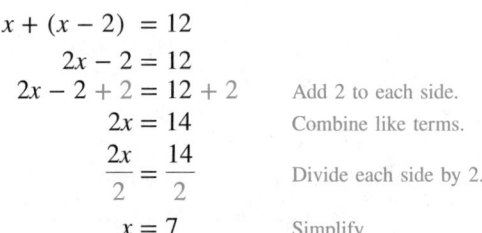

The equation is $x + (x - 2) = 12$.

Step 4: **Solve** the equation.

$$x + (x - 2) = 12$$
$$2x - 2 = 12$$
$$2x - 2 + 2 = 12 + 2 \qquad \text{Add 2 to each side.}$$
$$2x = 14 \qquad \text{Combine like terms.}$$
$$\frac{2x}{2} = \frac{14}{2} \qquad \text{Divide each side by 2.}$$
$$x = 7 \qquad \text{Simplify.}$$

W Hint

How does this type of problem differ from the problem in the previous objective?

Step 5: **Check** the answer, and **interpret** the solution as it relates to the problem.

One piece of pipe is 7 ft long.

The expression $x - 2$ represents the length of the other piece of pipe, so the length of the other piece is $x - 2 = 7 - 2 = 5$ ft.

The answer makes sense because the length of the original pipe was 7 ft + 5 ft = 12 ft.

[YOU TRY 3]

Write an equation, and solve.

An electrician has a 20-ft wire. He needs to cut the wire so that one piece is 4 ft shorter than the other. What will be the length of each piece?

3 Solve Consecutive Integer Problems

Consecutive means one after the other, in order. In this section, we will look at consecutive integers, consecutive even integers, and consecutive odd integers.

Consecutive integers differ by 1. Look at the consecutive integers 5, 6, 7, and 8. If $x = 5$, then $x + 1 = 6$, $x + 2 = 7$, and $x + 3 = 8$. Therefore, to define the unknowns for consecutive integers, let

$$x = \text{first integer}$$
$$x + 1 = \text{second integer}$$
$$x + 2 = \text{third integer}$$
$$x + 3 = \text{fourth integer}$$

and so on.

EXAMPLE 4

The sum of three consecutive integers is 87. Find the integers.

Solution

Step 1: **Read** the problem carefully, and identify what we are being asked to find.

We must find three consecutive integers with a sum of 87.

Step 2: **Choose a variable** to represent an unknown, and define the other unknowns in terms of this variable.

There are three unknowns. We will let x represent the first consecutive integer and then define the other unknowns in terms of x.

$$x = \text{the first integer}$$

Define the other unknowns in terms of x.

$$x + 1 = \text{the second integer} \qquad x + 2 = \text{the third integer}$$

Step 3: **Translate** the information that appears in English into an algebraic equation. What does the original statement mean?

"The sum of three consecutive integers is 87" means that when the three numbers are *added*, the sum is 87.

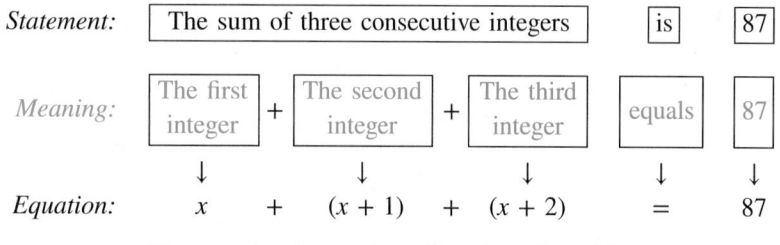

The equation is $x + (x + 1) + (x + 2) = 87$.

Step 4: **Solve** the equation.

$$x + (x + 1) + (x + 2) = 87$$
$$3x + 3 = 87$$
$$3x + 3 - 3 = 87 - 3 \qquad \text{Subtract 3 from each side.}$$
$$3x = 84 \qquad \text{Combine like terms.}$$
$$\frac{3x}{3} = \frac{84}{3} \qquad \text{Divide each side by 3.}$$
$$x = 28 \qquad \text{Simplify.}$$

Step 5: **Check** the answer, and **interpret** the solution as it relates to the problem.

The first integer is 28. The second integer is 29 since $x + 1 = 28 + 1 = 29$, and the third integer is 30 since $x + 2 = 28 + 2 = 30$.

The answer makes sense because their sum is $28 + 29 + 30 = 87$.

[**YOU TRY 4**] The sum of three consecutive integers is 162. Find the integers.

Next, let's look at **consecutive even integers,** which are even numbers that differ by 2, such as -10, -8, -6, and -4. If x is the first even integer, we have

-10	-8	-6	-4
x	$x + 2$	$x + 4$	$x + 6$

> **W Hint**
> Get in the habit of writing an equation in English before writing a math equation.

W Hint

Both consecutive even and consecutive odd integers are two units apart.

Therefore, to define the unknowns for consecutive even integers, let

$$x = \text{the first even integer}$$
$$x + 2 = \text{the second even integer}$$
$$x + 4 = \text{the third even integer}$$
$$x + 6 = \text{the fourth even integer}$$

and so on.

We also use the same reasoning to define the unknowns for **consecutive odd integers** such as 9, 11, 13, and 15.

EXAMPLE 5

The sum of two consecutive odd integers is 19 more than five times the larger integer. Find the integers.

Solution

Step 1: **Read** the problem carefully, and identify what we are being asked to find.

We must find two consecutive odd integers.

Step 2: **Choose a variable** to represent an unknown, and define the other unknown in terms of this variable.

There are two unknowns. We will let x represent the first consecutive odd integer and then define the other unknown in terms of x.

$$x = \text{the first odd integer}$$
$$x + 2 = \text{the second odd integer}$$

Step 3: **Translate** the information that appears in English into an algebraic equation. Read the problem slowly and carefully, breaking it into small parts.

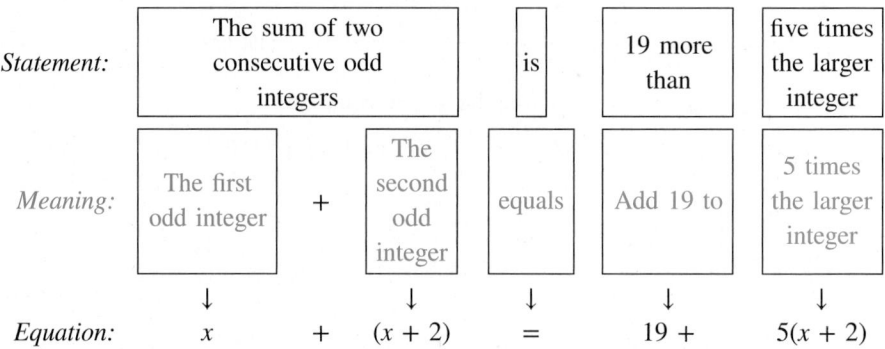

The equation is $x + (x + 2) = 19 + 5(x + 2)$.

Step 4: **Solve** the equation.

W Hint

Remember to use the procedure for solving linear equations containing variables on both sides of the equal sign.

$$x + (x + 2) = 19 + 5(x + 2)$$

$2x + 2 = 19 + 5x + 10$	Combine like terms; distribute.
$2x + 2 = 5x + 29$	Combine like terms.
$2x + 2 - 2 = 5x + 29 - 2$	Subtract 2 from each side.
$2x = 5x + 27$	Combine like terms.
$2x - 5x = 5x - 5x + 27$	Subtract $5x$ from each side.
$-3x = 27$	Combine like terms.
$\dfrac{-3x}{-3} = \dfrac{27}{-3}$	Divide each side by -3.
$x = -9$	Simplify.

Step 5: **Check** the answer, and **interpret** the solution as it relates to the problem.

The first odd integer is -9. The second integer is -7 since $x + 2 = -9 + 2 = -7$.

Check these numbers in the original statement of the problem. The sum of -9 and -7 is -16. Then, 19 more than five times the larger integer is $19 + 5(-7) = 19 + (-35) = -16$. The numbers are -9 and -7.

$$\left[\text{ YOU TRY 5 }\right]$$ The sum of two consecutive even integers is 16 less than three times the larger number. Find the integers.

ANSWERS TO $\left[\text{ YOU TRY }\right]$ EXERCISES

1) smart phones: 25; conventional phones: 48   2) Janay: 17 hr; Terrance: 34 hr
3) 8 ft and 12 ft   4) 53, 54, 55   5) 12 and 14

E Evaluate ## 3.3 Exercises   Do the exercises, and check your work.

Objective 1: Solve Problems Involving General Quantities

1) During the month of June, a car dealership sold 14 more compact cars than SUVs. Write an expression for the number of compact cars sold if c SUVs were sold.

2) During a Little League game, the Tigers scored 3 more runs than the Eagles. Write an expression for the number of runs the Tigers scored if the Eagles scored r runs.

3) A restaurant had 37 fewer customers on a Wednesday night than on a Thursday night. If there were c customers on Thursday, write an expression for the number of customers on Wednesday.

4) After a storm rolled through Omaha, the temperature dropped 15 degrees. If the temperature before the storm was t degrees, write an expression for the temperature after the storm.

5) Due to the increased use of email to send documents, the shipping expenses of a small business in 2019 were half of what they were in 2000. Write an expression for the cost of shipping in 2019 if the cost in 2000 was s dollars.

6) A coffee shop serves three times as many cups of regular coffee as decaffeinated coffee. If the shop serves d cups of decaffeinated coffee, write an expression for the number of cups of regular coffee it sells.

7) An electrician cuts a 14-ft wire into two pieces. If one is x feet long, how long is the other piece?

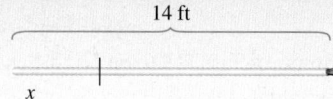
14 ft

x

8) Ralph worked for a total of 8.5 hr one day, some at his office and some at home. If he worked h hours in his office, write an expression for the number of hours he worked at home.

9) If you are asked to find the number of children in a class, why would 26.5 not be a reasonable answer?

10) If you are asked to find the length of a piece of wire, why would -7 not be a reasonable answer?

11) If you are asked to find consecutive odd integers, why would -10 not be a reasonable answer?

12) If you are asked to find the number of workers at an ice cream shop, why would $5\frac{1}{4}$ not be a reasonable answer?

Solve using the five-step method. See Examples 1 and 2.

13) The wettest April (greatest rainfall amount) for Albuquerque, NM, was recorded in 1905. The amount was 1.2 in. more than the amount recorded for the second-wettest April, in 2004. If the total rainfall for these two months was 7.2 in., how much rain fell in April of each year? (www.srh.noaa.gov)

14) Bo-Lin applied to three more colleges than his sister Liling. Together they applied to 13 schools. To how many colleges did each apply?

15) In their 2018 Stanley Cup–clinching game, the Capitals had two more shots on goal than the Golden Knights. Together, they had 64 shots. How many shots did each team have? (www.nhl.com)

16) In 2018, an Apple MacBook weighed about 13.9 lb less than the Apple Macintosh Portable did in 1989. Find the weight of each computer if they weighed a total of 18.1 lb. (http://oldcomputers.net; www.apple.com)

17) A 12-oz cup of regular coffee at Starbucks has 13 times the amount of caffeine found in the same-sized serving of decaffeinated coffee. Together they contain 280 mg of caffeine. How much caffeine is in each type of coffee? (www.starbucks.com)

18) A farmer plants soybeans and corn on his 540 acres of land. He plants twice as many acres with soybeans as with corn. How many acres are planted with each crop?

Arthur Tilley/Getty Images

19) In the sophomore class at Dixon High School, the number of students taking French is two-thirds of the number taking Spanish. How many students are studying each language if the total number of students in French and Spanish is 310?

20) A serving of salsa contains one-sixth of the number of calories of the same-sized serving of guacamole. Find the number of calories in each snack if they contain a total of 175 calories.

Objective 2: Solve Problems Involving Lengths
Solve using the five-step method. See Example 3.

21) A plumber has a 36-in. pipe. He must cut it into two pieces so that one piece is 14 in. longer than the other. How long is each piece?

22) A 40-in. board is to be cut into two pieces so that one piece is 8 in. shorter than the other. Find the length of each piece.

23) Trisha has a 28.5-in. piece of wire to make a necklace and a bracelet. She has to cut the wire so that the piece for the necklace will be twice as long as the piece for the bracelet. Find the length of each piece.

24) Ethan has a 20-ft piece of rope that he will cut into two pieces. One piece will be one-fourth the length of the other piece. Find the length of each piece of rope.

25) Derek orders a 6-ft sub sandwich for himself and two friends. Cory wants his piece to be 2 ft longer than Tamara's piece, and Tamara wants half as much as Derek. Find the length of each person's sub.

26) A 24-ft pipe must be cut into three pieces. The longest piece will be twice as long as the shortest piece, and the medium-sized piece will be 4 ft longer than the shortest piece. Find the length of each piece of pipe.

Objective 3: Solve Consecutive Integer Problems
Solve using the five-step method. See Examples 4 and 5.

27) The sum of three consecutive integers is 126. Find the integers.

28) The sum of two consecutive integers is 171. Find the integers.

29) Find two consecutive even integers such that twice the smaller is 16 more than the larger.

30) Find two consecutive odd integers such that the smaller one is 12 more than one-third the larger.

31) Find three consecutive odd integers such that their sum is five more than four times the largest integer.

32) Find three consecutive even integers such that their sum is 12 less than twice the smallest.

33) Two consecutive page numbers in a book add up to 215. Find the page numbers.

34) The addresses on the west side of Hampton Street are consecutive even numbers. Two consecutive house numbers add up to 7446. Find the addresses of these two houses.

Mixed Exercises: Objectives 1–3
Solve using the five-step method.

35) In a fishing derby, Jimmy caught six more trout than his sister Kelly. How many fish did each person catch if they caught a total of 20 fish?

36) Five times the sum of two consecutive integers is two more than three times the larger integer. Find the integers.

Big Cheese Photo/PunchStock

37) A 16-ft steel beam is to be cut into two pieces so that one piece is 1 ft longer than twice the other. Find the length of each piece.

38) A plumber has a 9-ft piece of copper pipe that has to be cut into three pieces. The longest piece will be 4 ft longer than the shortest piece. The medium-sized piece will be three times the length of the shortest. Find the length of each piece of pipe.

39) One year, the attendance at the Lollapalooza Festival was about 60,000 more than three times the attendance at Bonnaroo that year. The total number of people attending those festivals was about 380,000. How many people went to each event? (www.forbes.com)

Ingram Publishing/ SuperStock

40) A cookie recipe uses twice as much flour as sugar. If the total amount of these ingredients is $2\frac{1}{4}$ cups, how much sugar and how much flour are in these cookies?

41) The sum of three consecutive page numbers in a book is 174. What are the page numbers?

42) At a ribbon-cutting ceremony, the mayor cuts a 12-ft ribbon into two pieces so that the length of one piece is 2 ft shorter than the other. Find the length of each piece.

43) Charlie has 218 fewer Facebook friends than Ileana. The number of Deepa's Facebook friends is 28 more than half of Ileana's. Together, Charlie, Deepa, and Ileana have 2175 friends on Facebook. How many Facebook friends does each of them have?

44) Find three consecutive odd integers such that three times the middle number is 23 more than the sum of the other two.

45) A builder is installing hardwood floors. He has to cut a 72-in. piece into three separate pieces so that the smallest piece is one-third the length of the longest piece, and the third piece is 12 in. shorter than the longest. How long is each piece?

46) In 2015, there were 292 more new cases of tuberculosis in the United States than in 2016. In 2017, the number of new cases was 163 less than in 2016. If the total number of new TB cases in those three years was 27,897, how many people had TB each year? (www.cdc.gov)

47) Three of the top-selling albums, in total units, for 2017 were Ed Sheeran's *Divide*, Taylor Swift's *Reputation*, and Drake's *More Life*. Ed Sheeran sold 428,000 more albums than Taylor Swift, while Drake sold 109,000 fewer albums than Swift. Find the number of albums sold by each artist if together they sold 7,327,000 albums. (www.billboard.com)

48) Workers cutting down a large tree have a rope that is 33 ft long. They need to cut it into two pieces so that one piece is half the length of the other piece. How long is each piece of rope?

49) One-sixth of the smallest of three consecutive even integers is three less than one-tenth the sum of the other even integers. Find the integers.

50) Caedon's mom is a math teacher, and when he asks her on which pages he can find the magazine article on LeBron James, she says, "The article is on three consecutive pages so that 62 less than four times the last page number is the same as the sum of all the page numbers." On what page does the LeBron James article begin?

R Rethink

R1) When solving an application problem, how do you determine which unknown to let the variable *x* represent?

R2) Where have you recently encountered a problem similar to those you just solved? Write an application problem and have a classmate solve it.

3.4 Applications Involving Percents

O Organize

What are your objectives for Section 3.4?	How can you accomplish each objective?
1 Solve Basic Percent Problems	• Compare the difference between using arithmetic and using an equation to find a percentage. • Complete the given example on your own. • Complete You Try 1.
2 Solve Applications Involving Percents	• Adapt the **Five Steps for Solving Applied Problems** to help solve problems that contain percents. • Complete the given examples on your own. • Complete You Trys 2 and 3.
3 Solve Applications Involving Percent Change	• Change a percent into its decimal form when solving an equation. • Follow the procedure for solving applied problems. • Complete the given examples on your own. • Complete You Trys 4 and 5.
4 Solve Applications Involving Simple Interest	• Know the formula for calculating *simple interest*. • Be able to identify the *principal, interest earned, annual interest rate,* and the *time* to be substituted into the formula. One of those values is usually not given. • Don't forget to convert all rates to their decimal form when using them in an equation. • Follow the procedure for solving applied problems when needed. • Complete the given examples on your own. • Complete You Trys 6–8.
5 Solve Mixture Problems	• Follow the procedure for solving an applied problem. • Use the decimal form of the percent in the equation. • Complete the given example on your own. • Complete You Try 9.

 **W Work**   **Read the explanations, follow the examples, take notes, and complete the You Trys.**

Problems involving percents are everywhere—in stores, in banks, in laboratories, and in many more places. In this section, we will learn how to solve some algebraic problems involving percents. Before doing this, let's look at a problem involving only arithmetic so that you can see the relationship between an arithmetic problem and an algebraic problem.

1 Solve Basic Percent Problems

EXAMPLE 1

a) Use multiplication to find 20% of 120.

b) Use an equation to find 20% of 120.

Solution

a) The *of* in 20% *of* 120 means multiply. Change 20% to a decimal, then multiply.

$$20\% \text{ of } 120 = 0.20 \cdot 120 = 24$$

W Hint

How are these two approaches different?

b) Let's think of finding 20% of 120 as the question, "What is 20% of 120?" Let $x =$ the unknown quantity.

$$x = \text{the unknown quantity, } 20\% \text{ of } 120$$

Write the question in English, then write an equation.

English: What   is   20%   of   120?
          ↓   ↓   ↓   ↓   ↓
Equation: x =   0.20   ·   120

Change the percent to a decimal.

The equation is $x = 0.20 \cdot 120$. Solve.

$$x = 24 \quad \text{Multiply.}$$

Therefore, 24 is 20% of 120.

[YOU TRY 1]

a) Use multiplication to find 15% of 300.

b) Use an equation to find 15% of 300.

Using an equation to solve Example 1 may be longer than just using multiplication, but sometimes we use an equation like this to solve an applied problem. Also, using an equation is a more efficient way to find other quantities in a percent problem.

2 Solve Applications Involving Percents

EXAMPLE 2

The Montez family is driving to Disneyland for their family vacation, and they have already driven 220 mi. Mr. Montez has calculated that they have completed approximately 75% of their trip. Find the total distance from their home to Disneyland. (Round the answer to the nearest mile.)

Solution

Step 1: **Read** the problem carefully, and identify what you are being asked to find.

We must find the distance from the Montez family's home to Disneyland.

Step 2: **Choose a variable** to represent the unknown.

$$x = \text{the distance from home to Disneyland}$$

Step 3: **Translate** the information that appears in English into an equation. Since the family has already driven 220 mi, we can think of this problem as

220 miles is 75% of the distance from home to Disneyland.

Statement:	220 miles	is	75%	of	the distance from home to Disneyland.

$$\downarrow \quad\quad \downarrow \quad \downarrow \quad \downarrow \quad\quad\quad \downarrow$$

Equation:	220	=	0.75	·		x

The equation is $220 = 0.75 \cdot x$.

Step 4: **Solve** the equation.

$$220 = 0.75x$$
$$\frac{220}{0.75} = \frac{0.75}{0.75}x \quad \text{Divide both sides by 0.75.}$$
$$293.\overline{3} = x \quad \text{Perform the division.}$$

Step 5: **Check** the answer, and **interpret** the solution as it relates to the problem.

Rounding to the nearest mile, the total distance between their home and Disneyland is 293 mi. To check the answer, find 75% of 293: $0.75 \cdot 293 = 219.75$. This is close to 220 mi, but not exact, because we rounded the distance to the nearest mile.

[YOU TRY 2] At noon on election day, approximately 32% of registered voters in a particular district had already voted. If 501 people voted before noon, how many registered voters live in that district? (Round to the nearest person.)

Sometimes, we are asked to find a percentage in an applied problem.

EXAMPLE 3 Lara bought a sweater on sale for $36.00 that normally sells for $60.00. What percent of the original price did Lara save?

Solution

Step 1: **Read** the problem carefully, and identify what we are being asked to find.

We must find the percent of the original price that was saved.

Step 2: **Choose a variable** to represent the unknown.

$$x = \text{the percent savings}$$

Step 3: **Translate** the information that appears in English into an equation. Since we must find the *percent* saved, we must first find the *amount* of money Lara saved.

Amount of money saved = Original price − Sale price
Amount of money saved = $60.00   − $36.00 = $24.00

Now, we can think of the problem this way:

The amount Lara saved is what percent of the original price?

Statement:	Amount Lara saved	is	what percent	of	the original price

$$\downarrow \quad\quad \downarrow \quad\quad \downarrow \quad\quad \downarrow \quad\quad \downarrow$$

Equation:	24	=	x	·	60

The equation is $24 = x \cdot 60$ or $24 = 60x$.

Radius Images/Alamy Stock Photo

Step 4: **Solve** the equation.

$$24 = 60x$$

$$\frac{24}{60} = \frac{60}{60}x \quad \text{Divide both sides by 60.}$$

$$0.4 = x \quad \text{Perform the division.}$$

Step 5: **Check** the answer, and **interpret** the solution as it relates to the problem.

Change $x = 0.4$ to a percent: 40%. Lara saved 40% on the sweater.

Check: The amount Lara saved was 40% of $60.00 or $0.40 \cdot 60 = \$24.00$. The sale price of the sweater was $60.00 − $24.00 = $36.00.

[**YOU TRY 3**] Shane bought a video game on sale for $48.75; it normally costs $65.00. What percent did he save on the original price?

3 Solve Applications Involving Percent Change

Next, we will look at a common arithmetic problem involving percents. It is important to understand the *procedure* for solving this arithmetic example so that we can apply the same procedure for solving an algebra problem.

EXAMPLE 4   A digital scale that normally sells for $32.00 is marked down 30%. What is the sale price?

Solution

Concentrate on the *procedure* used to obtain the answer. This is the same procedure we will use to solve algebra problems with percent increase and percent decrease.

Sale price = Original price − Amount of discount

How much is the discount? It is 30% *of* $32.00.

Change the percent to a decimal. The amount of the discount is calculated by multiplying:

Amount of discount = (Rate of discount)(Original price)

Amount of discount =   (0.30) · ($32.00) = $9.60

Sale price = Original price − Amount of discount

$$= \quad \$32.00 \quad − \quad (0.30)(\$32.00)$$
$$= \quad \$32.00 \quad − \quad \$9.60$$
$$= \quad \$22.40$$

The sale price is $22.40.

> **W Hint**
>
> Remember that to convert a percent to a decimal, you move the decimal two places to the left.

[**YOU TRY 4**]   A dress shirt that normally sells for $39.00 is marked down 25%. What is the sale price?

Next, let's solve an algebra problem involving a markdown or percent decrease.

EXAMPLE 5 The sale price of a snowboard is $256.00 after a 20% discount. What was the original price of the snowboard?

Solution

Step 1: **Read** the problem carefully, and identify what we are being asked to find.

We must find the original price of the snowboard.

Step 2: **Choose a variable** to represent the unknown.

x = the original price of the snowboard

Step 3: **Translate** the information that appears in English into an algebraic equation. One way to figure out how to write an algebraic equation is to relate this problem to the arithmetic problem in Example 4. To find the sale price of the digital scale in Example 4 we found that

Sale price = Original price − Amount of discount

where we found the amount of the discount by multiplying the rate of the discount by the original price. We will write an algebraic equation using the same procedure.

> **W Hint**
> Remember that a *sale price* will always be less than the *original price*.

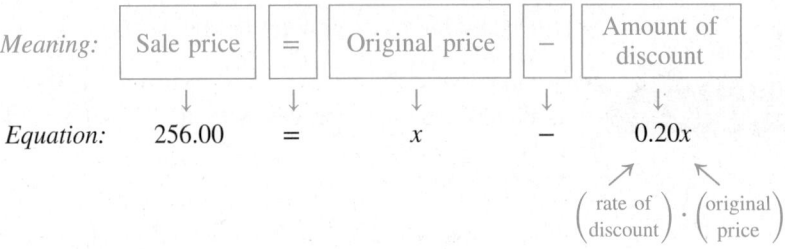

The equation is $256.00 = x - 0.20x$.

Step 4: **Solve** the equation.

$$256.00 = x - 0.20x$$
$$256.00 = 0.80x \qquad \text{Combine like terms.}$$
$$\frac{256.00}{0.80} = \frac{0.80x}{0.80} \qquad \text{Divide each side by 0.80.}$$
$$x = 320.00 \qquad \text{Simplify.}$$

Step 5: **Check** the answer, and **interpret** the solution as it relates to the problem.

The original price of the snowboard was $320.00.

The answer makes sense because the amount of the discount is
$(0.20)(\$320.00) = \$64.00$, which makes the sale price $\$320.00 - \$64.00 = \$256.00$.

[YOU TRY 5] A video game is on sale for $35.00 after a 30% discount. What was the original price of the video game?

4 Solve Applications Involving Simple Interest

When customers invest their money in bank accounts, their accounts earn interest. There are different ways to calculate the amount of interest earned from an investment, and in this section we will discuss *simple interest*. **Simple interest** calculations are based on the initial amount of money deposited in an account. This is known as the **principal.**

The formula used to calculate simple interest is $I = PRT$, where

I = interest (simple) earned
P = principal (initial amount invested)
R = annual interest rate (expressed as a decimal)
T = amount of time the money is invested (in years)

We will begin with two arithmetic problems. The procedures used will help you understand more clearly how we arrive at the algebraic equation in Example 8.

EXAMPLE 6

If $600 is invested for 1 year in an account earning 4% simple interest, how much interest will be earned?

Solution

We are given that $P = \$600$, $R = 0.04$, $T = 1$. We need to find I.

$$I = PRT$$
$$I = (600)(0.04)(1)$$
$$I = 24$$

The interest earned will be $24.

[YOU TRY 6]

If $1400 is invested for 1 year in an account earning 3% simple interest, how much interest will be earned?

EXAMPLE 7

Gavin invests $1000 in an account earning 6% interest and $7000 in an account earning 3% interest. After 1 year, how much interest will he have earned?

Solution

Gavin will earn interest from two accounts. Therefore,

Total interest earned = Interest from 6% account + Interest from 3% account

$$\text{Total interest earned} = \underset{P \quad\quad R \quad\quad T}{(1000)(0.06)(1)} + \underset{P \quad\quad R \quad\quad T}{(7000)(0.03)(1)}$$
$$= \quad\quad 60 \quad\quad + \quad\quad 210$$
$$= \$270$$

Gavin will earn a total of $270 in interest from the two accounts.

[YOU TRY 7]

Taryn invests $2500 in an account earning 4% interest and $6000 in an account earning 5.5% interest. After 1 year, how much interest will she have earned?

In the next example, we will use the same procedure for solving an algebraic problem that we used for solving the arithmetic problems in Examples 6 and 7.

EXAMPLE 8

Samira had $8000 to invest. She invested some of it in a savings account that paid 4% simple interest and the rest in a certificate of deposit that paid 6% simple interest. In 1 year, she earned a total of $360 in interest. How much did Samira invest in each account?

Solution

Step 1: **Read** the problem carefully, and identify what we are being asked to find.

We must find the amounts Samira invested in the 4% account and in the 6% account.

Step 2: **Choose a variable** to represent an unknown, and define the other unknown in terms of this variable.

Let $x =$ amount Samira invested in the 4% account.

How do we write an expression, in terms of x, for the amount invested in the 6% account?

Total invested   Amount invested in 4% account
    ↓   ↓
   8000   − x = Amount invested in the 6% account

We define the unknowns as:

$x =$ amount Samira invested in the 4% account
$8000 - x =$ amount Samira invested in the 6% account

Step 3: **Translate** the information that appears in English into an algebraic equation. Use the "English equation" we used in Example 7. Remember, since $T = 1$, we can compute the interest using $I = PR$.

Total interest earned = Interest from 4% account + Interest from 6% account
                              P R   P   R
    360   = $(x)(0.04)$ + $(8000 - x)(0.06)$

The equation is $360 = 0.04x + 0.06(8000 - x)$.

We can also get the equation by organizing the information in a table:

Amount Invested, in Dollars P	Interest Rate R	Interest Earned After 1 Year I
x	0.04	$0.04x$
$8000 - x$	0.06	$0.06(8000 - x)$

Total interest earned = Interest from 4% account + Interest from 6% account
    360   = $0.04x$ + $0.06(8000 - x)$

The equation is $360 = 0.04x + 0.06(8000 - x)$.

Either way of organizing the information will lead us to the correct equation.

Step 4: **Solve** the equation. Begin by multiplying both sides of the equation by 100 to eliminate the decimals.

$$360 = 0.04x + 0.06(8000 - x)$$
$$100(360) = 100[0.04x + 0.06(8000 - x)]$$
$$36,000 = 4x + 6(8000 - x) \qquad \text{Multiply by 100.}$$
$$36,000 = 4x + 48,000 - 6x \qquad \text{Distribute.}$$
$$36,000 = -2x + 48,000 \qquad \text{Combine like terms.}$$
$$-12,000 = -2x \qquad \text{Subtract 48,000.}$$
$$6000 = x \qquad \text{Divide by } -2.$$

Step 5: **Check** the answer, and **interpret** the solution as it relates to the problem.

Samira invested $6000 at 4% interest. The amount invested at 6% is $8000 − *x* or $8000 − $6000 = $2000.

Check:

Total interest earned = Interest from 4% account + Interest from 6% account

360	=	6000(0.04)	+	2000(0.06)
	=	240	+	120
	= 360			

[**YOU TRY 8**] Jeff inherited $10,000 from his grandfather. He invested part of it at 3% simple interest and the rest at 5% simple interest. Jeff earned a total of $440 in interest after 1 year. How much did he deposit in each account?

5 Solve Mixture Problems

Percents can also be used to solve mixture problems.

EXAMPLE 9 A chemist needs to make 24 liters (L) of an 8% acid solution. She will make it from some 6% acid solution and some 12% acid solution that is in the storeroom. How much of the 6% solution and the 12% solution should she use?

Solution

Step 1: **Read** the problem carefully, and identify what we are being asked to find.

We must find the amount of 6% acid solution and the amount of 12% acid solution she should use.

Step 2: **Choose a variable** to represent an unknown, and define the other unknown in terms of this variable. Let

x = the number of liters of 6% acid solution needed

Define the other unknown (the amount of 12% acid solution needed) in terms of *x*. Since she wants to make a total of 24 L of acid solution,

$24 - x$ = the number of liters of 12% acid solution needed

Step 3: **Translate** the information that appears in English into an algebraic equation.

Let's begin by arranging the information in a table. *To obtain the expression in the last column, multiply the percent of acid in the solution by the number of liters of solution to get the number of liters of acid in the solution.*

	Percent of Acid in Solution (as a decimal)	Liters of Solution	Liters of Acid in Solution
Mix these {	0.06	x	$0.06x$
	0.12	$24 - x$	$0.12(24 - x)$
to make →	0.08	24	$0.08(24)$

Now, write an equation in English. Since we make the 8% solution by mixing the 6% and 12% solutions,

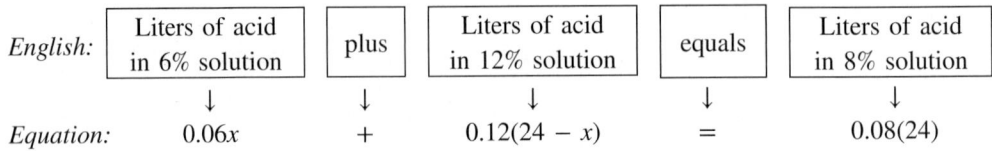

English: Liters of acid in 6% solution | plus | Liters of acid in 12% solution | equals | Liters of acid in 8% solution

↓   ↓   ↓   ↓   ↓

Equation: $0.06x$ + $0.12(24 - x)$ = $0.08(24)$

The equation is $0.06x + 0.12(24 - x) = 0.08(24)$.

Step 4: **Solve** the equation.

$$0.06x + 0.12(24 - x) = 0.08(24)$$
$$100[0.06x + 0.12(24 - x)] = 100[0.08(24)] \quad \text{Multiply by 100 to eliminate decimals.}$$
$$6x + 12(24 - x) = 8(24)$$
$$6x + 288 - 12x = 192 \quad \text{Distribute.}$$
$$-6x + 288 = 192 \quad \text{Combine like terms.}$$
$$-6x = -96 \quad \text{Subtract 288 from each side.}$$
$$x = 16 \quad \text{Divide by } -6.$$

Step 5: **Check** the answer, and **interpret** the solution as it relates to the problem.

The chemist needs 16 L of the 6% solution.

Find the other unknown, the amount of 12% solution needed.

$$24 - x = 24 - 16 = 8 \text{ L of 12\% solution.}$$

Check:

Acid in 6% solution + Acid in 12% solution = Acid in 8% solution
$$0.06(16) \quad + \quad 0.12(8) \quad = \quad 0.08(24)$$
$$0.96 \quad + \quad 0.96 \quad = \quad 1.92$$
$$1.92 = 1.92$$

[YOU TRY 9]

Write an equation and solve.

How many milliliters (mL) of a 10% alcohol solution and how many milliliters of a 20% alcohol solution must be mixed to obtain 30 mL of a 16% alcohol solution?

ANSWERS TO [YOU TRY] EXERCISES

1) a) 45 b) 45   2) 1566 registered voters   3) 25%   4) $29.25   5) $50.00   6) $42

7) $430   8) $3000 at 3% and $7000 at 5%

9) 12 mL of the 10% solution and 18 mL of the 20% solution

Get Ready

Write each percent as a decimal.

1) 40%   2) 83%

3) 2.7%   4) 61.9%

5) 5%   6) 6%

7) 200%   8) 110%

Objective 1: Solve Basic Percent Problems

Write an equation and solve.

9) What is 14% of 52?

10) What is 31% of 65?

11) 92 is 23% of what number?

12) 30 is 12% of what number?

13) 6.5% of what number is 9.75?

14) 8.4% of what number is 15.12?

15) 75 is what percent of 125?

16) 90 is what percent of 225?

17) What percent of 32 is 48?

18) What percent of 76 is 95?

Objective 2: Solve Applications Involving Percents

19) A baby stroller with a regular price of $69.00 is on sale at 20% off.

    a) Find the amount of the discount.

    b) Find the sale price.

20) The Johnson family has decided to give 5% of their annual income to charity. If their annual income is $140,000,

    a) how much do they donate to charity?

    b) how much of their income remains?

Solve using the Five Steps for Solving Applied Problems.

21) Korina buys a new phone for $148.00 and also has to pay 6.5% sales tax. Find the total cost of the phone.

22) Antoine buys a jacket for $86.00 plus 5.5% sales tax. Find the total cost of the jacket.

23) In March, Bruno paid $4.40 for a calendar, and this was 25% of the original price. What was the original price of the calendar?

24) After the holidays, Pilar paid $1.95 for a roll of wrapping paper. This was 30% of the original price. What was the original price of the wrapping paper?

25) Corey saved $105.00 on a surfboard with a regular price of $525.00. What percent of the original price did he save?

John Lund/Drew Kelly/Blend Images

26) A USB thumb drive that regularly sells for $23.00 was marked down by $3.45. What was the percent markdown off the original price?

27) Sameet buys a computer for $460.00 plus $22.08 in sales tax. Find the sales tax rate.

28) Alhaji buys a bike for $170.00 plus $10.54 in sales tax. Find the sales tax rate.

Objective 3: Solve Applications Involving Percent Change

Find the sale price of each item.

29) A TV with a regular price of $299.00 is on sale at 20% off.

30) The regular price of a pair of ballet pointe shoes is $104.00, and it is marked down 15%.

31) A three-ton floor jack is 35% off its regular price of $160.00.

32) A ticket to a comedy club has been discounted 30% from its regular price of $45.00.

Solve using the Five Steps for Solving Applied Problems.

33) Katrina paid $25.50 for a box fan that was marked down 15%. Find the original price of the box fan.

34) Manuela bought a computer printer for $96.00. This was 20% off the regular price. What was the regular price of the printer?

35) In 2019, there were about 1224 acres of farmland in Crane County. This is 32% less than the number of acres of farmland in 2010. Calculate the number of acres of farmland in Crane County in 2010.

MaxyM/Shutterstock

36) This year, a company has 136 fewer employees than last year for a drop of 15% in the number of employees. How many people were employed at this company last year?

37) Jet Fi's salary this year is 12% higher than it was three years ago. If he earns $43,120 this year, what did he earn three years ago?

38) Two hundred four teams participated in the 2012 Summer Olympics in London. This was 155% more than the number of teams that competed in the 1980 Summer Olympics in Moscow. How many teams participated in the Moscow Olympic Games? (www.olympic.org)

Objective 4: Solve Applications Involving Simple Interest
Solve.

39) Jenna invests $800 in an account for one year earning 4% simple interest. How much interest was earned from this account?

40) Last year, Mr. Jaworski deposited $11,000 in an account earning 7% simple interest for one year. How much interest was earned?

41) Rachel has a total of $5500 to invest for one year. She deposits $4000 into an account earning 6.5% annual simple interest and the rest into an account earning 8% annual simple interest. How much interest did Rachel earn?

42) Maurice plans to invest a total of $9000 for one year. In the account earning 5.2% simple interest he will deposit $6000, and in an account earning 7% simple interest he will deposit the rest. How much interest will Maurice earn?

Solve using the Five Steps for Solving Applied Problems.

43) Javier receives a $12,000 signing bonus when he accepts a new job. He will invest some of it in an account earning 4% simple interest and the rest at 3% simple interest. If he will earn $410 in interest after one year, how much will Javier invest in each account?

44) Angelica invested part of her $15,000 inheritance in an account earning 5% simple interest and the rest in an account earning 4% simple interest. How much did she invest in each account if she earned $680 in total interest after one year?

45) Last year, Taz invested a total of $7500 in two accounts earning simple interest. He invested some of it at 9.5% and the rest at 6.5%. How much did he invest in each account if he earned a total of $577.50 in interest last year?

46) Midori has $3000 to invest. She deposits a portion of it into an account earning 4% simple interest and the rest at 6.5% simple interest. After one year, she has earned $170 in interest. How much did Midori deposit into each account?

47) Arpana's money earned $204 in interest after one year. She invested some of her money in an account earning 6% simple interest and $450 more than that amount in an account earning 5% simple interest. Find the amount Arpana invested in each account.

PhotosIndia.com/Glow Images

48) Jamaal invested some money in an account earning 3% simple interest and $1500 more than that amount in an account earning 4% simple interest. After one year, he had earned a total of $235 in interest. How much did Jamaal deposit in each account?

Objective 5: Solve Mixture Problems

Solve.

49) How many ounces of alcohol are in 50 oz of a 6% alcohol solution?

50) How many milliliters of acid are in 50 mL of a 5% acid solution?

51) Seventy-five milliliters of a 10% acid solution are mixed with 30 mL of a 2.5% acid solution. How much acid is in the mixture?

52) Fifty ounces of a 9% alcohol solution are mixed with 60 oz of a 7% alcohol solution. How much alcohol is in the mixture?

Solve using the Five Steps for Solving Applied Problems.

53) How many ounces of a 2% acid solution and how many ounces of an 8% acid solution must be mixed to make 36 oz of a 6% acid solution?

54) How many milliliters of an 8% hydrogen peroxide solution and how many milliliters of a 2% hydrogen peroxide solution should be mixed to obtain 300 mL of a 4% hydrogen peroxide solution?

55) How many liters of a 25% antifreeze solution must be mixed with 4 L of a 60% antifreeze solution to make a mixture that is 45% antifreeze?

56) How many milliliters of a 17% alcohol solution must be added to 40 mL of a 3% alcohol solution to make a 12% alcohol solution?

57) How many milliliters of pure alcohol and how many milliliters of a 4% alcohol solution must be combined to make 480 mL of an 8% alcohol solution?

58) How much pure acid must be added to 6 gal of a 4% acid solution to make a 20% acid solution?

59) The Nut Hut sells a mix consisting of almonds and cashews. How many pounds of almonds, which sells for $6.00 per pound, should be mixed with 2 lb of cashews, which sell for $9.00 per pound, to get a mix worth $7.20 per pound?

60) Creative Coffees blends its coffees for customers. How much of the Aromatic coffee, which sells for $6.00 per pound, and how much of the Hazelnut, which sells for $9.00 per pound, should be mixed to make 3 lb of the Smooth blend to be sold for $8.75 per pound?

Ingram Publishing/SuperStock

Mixed Exercises: Objectives 1–5

Solve using the Five Steps for Solving Applied Problems.

61) Huang's bonus last year was $6720, and this was 12% of his base salary. What is his base salary?

62) Diego inherited $20,000 and put some of it into an account earning 4% simple interest and the rest into an account earning 7% simple interest. He earned a total of $1130 in interest after one year. How much did he deposit into each account?

63) Tessa earns $284 in interest from 1-year investments. She invested some money in an account earning 6% simple interest, and she deposited $1500 more than that amount into an account paying 5% simple interest. How much did Tessa invest in each account?

64) Gil marks up the prices of fishing poles by 55%. Determine what Gil paid his supplier for his best-selling fishing pole if Gil charges his customers $124.

65) How many milliliters of a 4% acid solution and how many milliliters of a 10% acid solution must be mixed to obtain 54 mL of a 6% acid solution?

66) A store owner plans to make 10 lb of a candy mix worth $1.92/lb. How many pounds of gummi bears worth $2.40/lb and how many pounds of jelly beans worth $1.60/lb must be combined to make the candy mix?

67) In her gift shop, Kyung-Ah sells all stuffed animals for 60% more than what she paid her supplier. If one of these toys sells for $14.00 in her shop, what did it cost Kyung-Ah?

68) By the end of August, 287 students had already taken their senior class pictures. If there are 350 students in the class, what percentage had already had their pictures taken?

69) How many ounces of pure orange juice and how many ounces of a citrus fruit drink containing 5% fruit juice should be mixed to get 76 oz of a fruit drink that is 25% fruit juice?

Evgeny Karandaev/iStockphoto/
Getty Images

70) Melina has $7500 to invest. She will invest some of it in a long-term IRA paying 4% simple interest and the rest in a short-term CD earning 2.5% simple interest. After one year, Melina's investments have earned $225 in interest. How much did Melina invest in each account?

R Rethink

R1) Which types of percent problems were the easiest and which were the most difficult? What made some easier or harder than others?

R2) What was the last use of a percent that you saw outside the classroom? How was it used?

3.5 Geometry Applications and Solving Formulas

P Prepare

O Organize

What are your objectives for Section 3.5?	How can you accomplish each objective?
1 Substitute Values into a Formula, and Find the Unknown Variable	• Understand and be able to identify a *formula*. • Complete the given example on your own, and create a procedure. • Complete You Try 1.
2 Solve Problems Using Formulas from Geometry	• Review the geometry formulas in Section 1.3. • Use the **Five Steps for Solving Applied Problems** to complete the given examples on your own. • Complete You Trys 2 and 3.
3 Solve Problems Involving Angle Measures	• Review the special relationships between angles in Section 1.3. • Use the **Five Steps for Solving Applied Problems** to complete the given examples on your own. • Complete You Trys 4–6.
4 Solve a Formula for a Specific Variable	• Understand the similarities between solving for a variable in an equation that contains numbers and one that contains only other variables. • Complete the given examples on your own. • Complete You Trys 7 and 8.

Read the explanations, follow the examples, take notes, and complete the You Trys.

A **formula** is a rule containing variables and mathematical symbols to state relationships between certain quantities.

Some examples of formulas we have used already are

$$P = 2l + 2w \qquad A = \frac{1}{2}bh \qquad C = 2\pi r$$

In this section we will solve problems using *formulas,* and then we will learn how to solve a formula for a specific variable.

1 Substitute Values into a Formula, and Find the Unknown Variable

EXAMPLE 1

The formula for the area of a triangle is $A = \frac{1}{2}bh$. If $A = 30$ when $b = 8$, find h.

Solution

The only unknown variable is h since we are given the values of A and b. Substitute $A = 30$ and $b = 8$ into the formula, and solve for h.

$$A = \frac{1}{2}bh$$

$$30 = \frac{1}{2}(8)h \qquad \text{Substitute the given values.}$$

Since h is the only remaining variable in the equation, we can solve for it.

$$30 = 4h \qquad \text{Multiply.}$$
$$\frac{30}{4} = \frac{4h}{4} \qquad \text{Divide by 4.}$$
$$\frac{15}{2} = h \qquad \text{Simplify.}$$

[YOU TRY 1]

The area of a trapezoid is $A = \frac{1}{2}h(b_1 + b_2)$. If $A = 21$ when $b_1 = 10$ and $b_2 = 4$, find h.

2 Solve Problems Using Formulas from Geometry

Next we will solve applied problems using concepts and formulas from geometry. Unlike in Example 1, you will not be given a formula. You will need to know the geometry formulas that we reviewed in Section 1.3.

EXAMPLE 2

A soccer field is in the shape of a rectangle and has an area of 9000 yd². Its length is 120 yd. What is the width of the field?

Solution

Step 1: **Read** the problem carefully, and identify what we are being asked to find.

We must find the length of the soccer field.

A picture will be very helpful in this problem.

Area = 9000 yd²

120 yd

Step 2: **Choose a variable** to represent the unknown.

w = the width of the soccer field

Label the picture with the length, 120 yd, and the width, w.

Step 3: **Translate** the information that appears in English into an algebraic equation. We will use a known geometry formula. How do we know which formula to use? List the information we are given and what we want to find:

The field is in the shape of a rectangle; its area = 9000 yd² and its length = 120 yd. We must find the width. Which formula involves the area, length, and width of a rectangle?

$$A = lw$$

Substitute the known values into the formula for the area of a rectangle, and solve for w.

$$A = lw$$
$$9000 = 120w \qquad \text{Substitute the known values.}$$

Step 4: **Solve** the equation.

$$9000 = 120w$$
$$\frac{9000}{120} = \frac{120w}{120} \qquad \text{Divide by 120.}$$
$$75 = w \qquad \text{Simplify.}$$

> **W Hint**
> On a sheet of paper, make a list of all the geometric formulas that are used in this section.

> **W Hint**
> Remember to include the correct units in your answer!

Step 5: **Check** the answer, and **interpret** the solution as it relates to the problem.

If w = 75 yd, then $l \cdot w$ = 120 yd · 75 yd = 9000 yd². Therefore, the width of the soccer field is 75 yd.

[YOU TRY 2] The area of a rectangular room is 270 ft². Find the length of the room if the width is 15 ft.

EXAMPLE 3

Stewart wants to put a rectangular safety fence around his backyard pool. He calculates that he will need 120 ft of fencing and that the length will be 4 ft longer than the width. Find the dimensions of the safety fence.

Solution

Step 1: **Read** the problem carefully, and identify what we are being asked to find.

We must find the length and width of the safety fence.

Draw a picture.

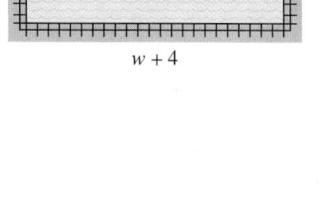

Perimeter = 120 ft

w

$w + 4$

Step 2: **Choose a variable** to represent an unknown, and define the other unknown in terms of this variable.

The length is 4 ft longer than the width. Therefore, let

$$w = \text{the width of the safety fence}$$

Define the other unknown in terms of w.

$$w + 4 = \text{the length of the safety fence}$$

Label the picture with the expressions for the width and length.

Step 3: **Translate** the information that appears in English into an algebraic equation.

Use a known geometry formula. What does the 120 ft of fencing represent? *Since the fencing will go around the pool, the 120 ft represents the perimeter of the rectangular safety fence.* We need to use a formula that involves the length, width, and perimeter of a rectangle. The formula we will use is

$$P = 2l + 2w$$

Substitute the known values and expressions into the formula.

$$P = 2l + 2w$$
$$120 = 2(w + 4) + 2w \qquad \text{Substitute.}$$

Step 4: **Solve** the equation.

$$120 = 2(w + 4) + 2w$$
$$120 = 2w + 8 + 2w \qquad \text{Distribute.}$$
$$120 = 4w + 8 \qquad \text{Combine like terms.}$$
$$120 - 8 = 4w + 8 - 8 \qquad \text{Subtract 8 from each side.}$$
$$112 = 4w \qquad \text{Combine like terms.}$$
$$\frac{112}{4} = \frac{4w}{4} \qquad \text{Divide each side by 4.}$$
$$28 = w \qquad \text{Simplify.}$$

Step 5: **Check** the answer, and **interpret** the solution as it relates to the problem.

The width of the safety fence is 28 ft. The length is $w + 4 = 28 + 4 = 32$ ft.

The answer makes sense because the perimeter of the fence is
$2(32 \text{ ft}) + 2(28 \text{ ft}) = 64 \text{ ft} + 56 \text{ ft} = 120 \text{ ft}$.

[YOU TRY 3] Marina wants to make a rectangular dog run in her backyard. It will take 46 feet of fencing to enclose it, and the length will be 1 foot less than three times the width. Find the dimensions of the dog run.

3 Solve Problems Involving Angle Measures

Recall from Section 1.3 that the sum of the angle measures in a triangle is 180°. We will use this fact in our next example.

EXAMPLE 4 Find the missing angle measures.

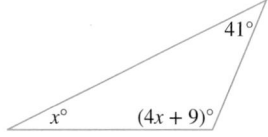

Solution

Step 1: **Read** the problem carefully, and identify what we are being asked to find.

Find the missing angle measures.

Step 2: The unknowns are already defined. We must find *x,* the measure of one angle, and then $4x + 9$, the measure of the other angle.

Step 3: **Translate** the information into an algebraic equation. Since the sum of the angles in a triangle is 180°, we can write

	Measure of one angle	plus	Measure of second angle	plus	Measure of third angle	is	180°
English:	↓	↓	↓	↓	↓	↓	↓
Equation:	x	$+$	41	$+$	$4x + 9$	$=$	180

The equation is $x + 41 + (4x + 9) = 180$.

Step 4: **Solve** the equation.

$$x + 41 + (4x + 9) = 180$$
$$5x + 50 = 180 \qquad \text{Combine like terms.}$$
$$5x + 50 - 50 = 180 - 50 \qquad \text{Subtract 50 from each side.}$$
$$5x = 130 \qquad \text{Combine like terms.}$$
$$\frac{5x}{5} = \frac{130}{5} \qquad \text{Divide each side by 5.}$$
$$x = 26 \qquad \text{Simplify.}$$

Step 5: **Check** the answer, and **interpret** the solution as it relates to the problem.

One angle, *x*, has a measure of 26°. The other unknown angle measure is $4x + 9 = 4(26) + 9 = 113°$.

The answer makes sense because the sum of the angle measures is $26° + 41° + 113° = 180°$.

[YOU TRY 4] Find the missing angle measures.

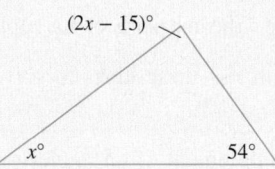

Let's look at another type of problem involving angle measures.

EXAMPLE 5 Find the measure of each indicated angle.

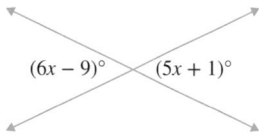

Solution

The indicated angles are *vertical angles,* and **vertical angles** have the same measure. (See Section 1.3.) Since their measures are the same, set $6x - 9 = 5x + 1$ and solve for *x*.

$$6x - 9 = 5x + 1$$
$$6x - 9 + 9 = 5x + 1 + 9 \qquad \text{Add 9 to each side.}$$
$$6x - 5x + 10 \qquad \text{Combine like terms.}$$
$$6x - 5x = 5x - 5x + 10 \qquad \text{Subtract } 5x \text{ from each side.}$$
$$x = 10 \qquad \text{Combine like terms.}$$

Be careful! Although $x = 10$, the angle measure is *not* 10. To find the angle measures, substitute $x = 10$ into the expressions for the angles.

The measure of the angle on the left is $6x - 9 = 6(10) - 9 = 51°$. The other angle measure is also 51° since these are vertical angles. We can verify this by substituting 10 into the expression for the other angle, $5x + 1$: $5x + 1 = 5(10) + 1 = 51°$.

[YOU TRY 5] Find the measure of each indicated angle.

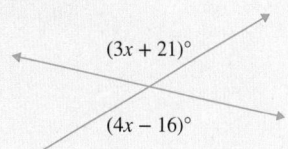

In Section 1.3, we learned that two angles are **complementary** if the sum of their angles is 90°, and two angles are **supplementary** if the sum of their angles is 180°.

For example, if the measure of $\angle A$ is 71°, then

    a)   the measure of its complement is $90° - 71° = 19°$.

    b)   the measure of its supplement is $180° - 71° = 109°$.

Now let's say the measure of an angle is x. Using the same reasoning as above,

    a)   the measure of its complement is $90 - x$.

    b)   the measure of its supplement is $180 - x$.

We will use these ideas to solve the problem in Example 6.

EXAMPLE 6

The supplement of an angle is 34° more than twice the complement of the angle. Find the measure of the angle.

Solution

Step 1: **Read** the problem carefully, and identify what we are being asked to find.

               We must find the measure of the angle.

Step 2: **Choose a variable** to represent an unknown, and define the other unknowns in terms of this variable.

This problem has three unknowns: the measures of the angle, its complement, and its supplement. Choose a variable to represent the *original angle,* then define the other unknowns in terms of this variable.

$$x = \text{the measure of the angle}$$

W Hint

On a sheet of paper, make a list of all the cases where angle measurements are used in this section.

Define the other unknowns in terms of x.

$$90 - x = \text{the measure of the complement}$$
$$180 - x = \text{the measure of the supplement}$$

Step 3: **Translate** the information that appears in English into an algebraic equation.

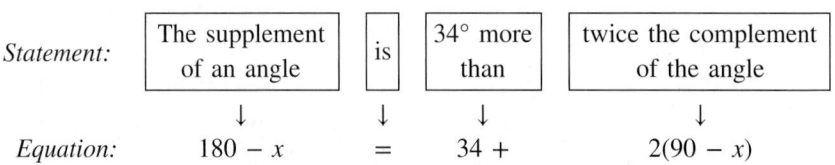

Statement:	The supplement of an angle	is	34° more than	twice the complement of the angle
	↓	↓	↓	↓
Equation:	$180 - x$	$=$	$34 +$	$2(90 - x)$

The equation is $180 - x = 34 + 2(90 - x)$.

Step 4: **Solve** the equation.

$$180 - x = 34 + 2(90 - x)$$
$$180 - x = 34 + 180 - 2x \qquad \text{Distribute.}$$
$$180 - x = 214 - 2x \qquad \text{Combine like terms.}$$
$$180 - 180 - x = 214 - 180 - 2x \qquad \text{Subtract 180 from each side.}$$
$$-x = 34 - 2x \qquad \text{Combine like terms.}$$
$$-x + 2x = 34 - 2x + 2x \qquad \text{Add } 2x \text{ to each side.}$$
$$x = 34 \qquad \text{Simplify.}$$

Step 5: **Check** the answer, and **interpret** the solution as it relates to the problem.

The measure of the angle is 34°.

To check the answer, we first need to find its complement and supplement. The complement is 90° − 34° = 56°, and its supplement is 180° − 34° = 146°. Now we can check these values in the original statement: The supplement is 146°. Thirty-four degrees more than twice the complement is 34° + 2(56°) = 34° + 112° = 146°.

[YOU TRY 6]   Twice the complement of an angle is 18° less than the supplement of the angle. Find the measure of the angle.

---

4 Solve a Formula for a Specific Variable

The formula $P = 2l + 2w$ allows us to find the perimeter of a rectangle when we know its length (l) and width (w). But what if we were solving problems where we repeatedly needed to find the value of w? Then, we could rewrite $P = 2l + 2w$ so that it is solved for w:

$$w = \frac{P - 2l}{2}$$

Doing this means that we have *solved the formula $P = 2l + 2w$ for the specific variable w.*

Solving a formula for a specific variable may seem confusing at first because the formula contains more than one letter. Keep in mind that we will solve for a specific variable the same way we have been solving equations up to this point.

We'll start by solving $3x + 4 = 19$ step-by-step for x and then applying the same procedure to solving $ax + b = c$ for x.

EXAMPLE 7   Solve $3x + 4 = 19$ and $ax + b = c$ for x.

Solution

Look at these equations carefully, and notice that they have the same form. Read the parts of the solution in numerical order.

Part 1 Solve $3x + 4 = 19$.

Don't quickly run through the solution of this equation. **The emphasis here is on the steps used to solve the equation and why we use those steps!**

$$3\boxed{x} + 4 = 19$$

We are solving for x. We'll put a box around it. What is the first step? "Get rid of" what is being added to the $3x$; that is, "get rid of" the 4 on the left. Subtract 4 from each side.

$$3\boxed{x} + 4 - 4 = 19 - 4$$

Combine like terms.

$$3\boxed{x} = 15$$

Part 3 We need to solve $3\boxed{x} = 15$ for x. We need to eliminate the 3 on the left. Since x is being multiplied by 3, we will **divide** each side by 3.

$$\frac{3\boxed{x}}{3} = \frac{15}{3}$$

Simplify.

$$x = 5$$

The solution set is $\{5\}$.

W Hint

Why can the solution be written two different ways?

Part 2 Solve $ax + b = c$ for x.

Since we are solving for x, we'll put a box around it.

$$a\boxed{x} + b = c$$

The goal is to get the x on a side by itself. What do we do first? As in Part 1, "get rid of" what is being added to the ax term; that is, "get rid of" the b on the left. Since b is being added to ax, we will subtract it from each side. (We are performing the same steps as in Part 1!)

$$a\boxed{x} + b - b = c - b$$

Combine like terms.

$$a\boxed{x} = c - b$$

We cannot combine the terms on the right, so it remains $c - b$.

Part 4 Now, we have to solve $a\boxed{x} = c - b$ for x. We need to eliminate the a on the left. Since x is being multiplied by a, we will **divide** each side by a.

$$\frac{a\boxed{x}}{a} = \frac{c - b}{a}$$

These are the same steps used in Part 3!

Simplify.

$$\frac{a\boxed{x}}{a} = \frac{c - b}{a}$$

$$x = \frac{c - b}{a} \quad \text{or} \quad \frac{c}{a} - \frac{b}{a}$$

Note

To obtain the result $x = \dfrac{c}{a} - \dfrac{b}{a}$, we distributed the a in the denominator to each term in the numerator. Either form of the answer is correct.

When you are solving a formula for a specific variable, think about the steps you use to solve an equation in one variable.

[YOU TRY 7] Solve $rt - n = k$ for t.

EXAMPLE 8

$U = \frac{1}{2}LI^2$ is a formula used in physics. Solve this equation for L.

Solution

$$U = \frac{1}{2}\boxed{L}I^2 \qquad \text{Solve for } L. \text{ Put it in a box.}$$

$$2U = 2 \cdot \frac{1}{2}\boxed{L}I^2 \qquad \text{Multiply by 2 to eliminate the fraction.}$$

$$\frac{2U}{I^2} = \frac{\boxed{L}I^2}{I^2} \qquad \text{Divide each side by } I^2.$$

$$\frac{2U}{I^2} = L \qquad \text{Simplify.}$$

EXAMPLE 9

$A = \frac{1}{2}h(b_1 + b_2)$ is the formula for the area of a trapezoid. Solve it for b_1.

Solution

There are two ways to solve this for b_1.

Method 1: We will put b_1 in a box to remind us that this is what we must solve for. In Method 1, we will start by eliminating the fraction.

$$2A = 2 \cdot \frac{1}{2}h(\boxed{b_1} + b_2) \qquad \text{Multiply each side by 2.}$$

$$2A = h(\boxed{b_1} + b_2) \qquad \text{Simplify.}$$

$$\frac{2A}{h} = \frac{h(\boxed{b_1} + b_2)}{h} \qquad \text{Divide each side by } h.$$

$$\frac{2A}{h} = \boxed{b_1} + b_2$$

$$\frac{2A}{h} - b_2 = \boxed{b_1} + b_2 - b_2 \qquad \text{Subtract } b_2 \text{ from each side.}$$

$$\frac{2A}{h} - b_2 = b_1 \qquad \text{Simplify}$$

Method 2: Another way to solve $A = \frac{1}{2}h(b_1 + b_2)$ for b_1 is to begin by distributing $\frac{1}{2}h$ on the right.

$$A = \frac{1}{2}h\boxed{b_1} + \frac{1}{2}hb_2 \qquad \text{Distribute.}$$

$$2A = 2\left(\frac{1}{2}h\boxed{b_1} + \frac{1}{2}hb_2\right) \qquad \text{Multiply by 2 to eliminate the fractions.}$$

$$2A = h\boxed{b_1} + hb_2 \qquad \text{Distribute.}$$

$$2A - hb_2 = h\boxed{b_1} + hb_2 - hb_2 \qquad \text{Subtract } hb_2 \text{ from each side.}$$

$$2A - hb_2 = h\boxed{b_1} \qquad \text{Simplify.}$$

$$\frac{2A - hb_2}{h} = \frac{h\boxed{b_1}}{h} \qquad \text{Divide by } h.$$

$$\frac{2A - hb_2}{h} = b_1 \qquad \text{Simplify.}$$

Therefore, b_1 can be written as $b_1 = \dfrac{2A}{h} - \dfrac{hb_2}{h}$ or $b_1 = \dfrac{2A}{h} - b_2$. These two forms are equivalent.

$$\left[\text{YOU TRY 8} \right]$$ Solve for the indicated variable.

a) $t = \dfrac{qr}{s}$ for q b) $R = t(k - c)$ for c

ANSWERS TO $\boxed{\text{YOU TRY}}$ **EXERCISES**

1) 3   2) 18 ft   3) 6 ft × 17 ft   4) 47°, 79°   5) 132°, 132°   6) 18°

7) $t = \dfrac{k + n}{r}$ 8) a) $q = \dfrac{st}{r}$ b) $c = \dfrac{kt - R}{t}$ or $c = k - \dfrac{R}{t}$

E Evaluate **3.5** Exercises   Do the exercises, and check your work.

Get Ready

1) A rectangular box has length l, width w, and height h. What is the formula for its volume, V?

2) A circle has radius r.

  a) What is the formula for its area, A?

  b) What is the formula for its circumference, C?

3) How do you find the perimeter of a triangle?

4) What is the relationship between vertical angles?

Objective 1: Substitute Values into a Formula, and Find the Unknown Variable

5) If you are using the formula $A = \dfrac{1}{2}bh$, is it reasonable to get an answer of $h = -6$? Explain your answer.

6) If you are finding the area of a rectangle and the lengths of the sides are given in inches, the area of the rectangle would be expressed in which unit?

7) If you are asked to find the volume of a sphere and the radius is given in centimeters, the volume would be expressed in which unit?

8) If you are asked to find the perimeter of a football field and the length and width are given in yards, the perimeter of the field would be expressed in which unit?

Determine whether each statement is true or false.

9) To find the area of a triangle, the units of the base and height must be the same before they are substituted into the formula.

10) For any formula involving the irrational number π, the value of at least one of the variables must be an irrational number.

Substitute the given values into the formula and solve for the remaining variable.

11) $A = lw$; If $A = 44$ when $l = 16$, find w.

12) $A = \dfrac{1}{2}bh$; If $A = 21$ when $h = 14$, find b.

13) $I = PRT$; If $I = 240$ when $R = 0.04$ and $T = 2$, find P.

14) $I = PRT$; If $I = 600$ when $P = 2500$ and $T = 4$, find R.

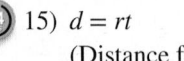 15) $d = rt$
   (Distance formula: $distance = rate \cdot time$);
   If $d = 150$ when $r = 60$, find t.

16) $d = rt$
   (Distance formula: $distance = rate \cdot time$);
   If $r = 36$ and $t = 0.75$, find d.

17) $C = 2\pi r$; If $r = 4.6$, find C.

18) $C = 2\pi r$; If $C = 15\pi$, find r.

19) $P = 2l + 2w$; If $P = 11$ when $w = \dfrac{3}{2}$, find l.

20) $P = s_1 + s_2 + s_3$ (Perimeter of a triangle);
   If $P = 11.6$ when $s_2 = 2.7$ and $s_3 = 3.8$, find s_1.

21) $V = lwh$; If $V = 52$ when $l = 6.5$ and $h = 2$, find w.

22) $V = \dfrac{1}{3}Ah$ (Volume of a pyramid); If $V = 16$ when $A = 24$, find h.

23) $V = \dfrac{1}{3}\pi r^2 h$; If $V = 48\pi$ when $r = 4$, find h.

24) $V = \dfrac{1}{3}\pi r^2 h$; If $V = 50\pi$ when $r = 5$, find h.

25) $S = 2\pi r^2 + 2\pi rh$ (Surface area of a right circular cylinder); If $S = 154\pi$ when $r = 7$, find h.

26) $S = 2\pi r^2 + 2\pi rh$; If $S = 132\pi$ when $r = 6$, find h.

27) $A = \dfrac{1}{2}h(b_1 + b_2)$; If $A = 136$ when $b_1 = 7$ and $h = 16$, find b_2.

28) $A = \dfrac{1}{2}h(b_1 + b_2)$; If $A = 1.5$ when $b_1 = 3$ and $b_2 = 1$, find h.

Objective 2: Solve Problems Using Formulas from Geometry

Use a known formula to solve. See Example 2.

29) The area of a tennis court is 2808 ft^2. Find the length of the court if it is 36 ft wide.

30) A rectangular tabletop has an area of 13.5 ft^2. What is the width of the table if it is 4.5 ft long?

31) A rectangular flower box holds 1232 in^3 of soil. Find the height of the box if it is 22 in. long and 7 in. wide.

32) A rectangular storage box is 2.5 ft wide, 4 ft long, and 1.5 ft high. What is the storage capacity of the box?

33) The center circle on a soccer field has a radius of 10 yd. What is the area of the center circle? Use 3.14 for π.

34) The face of the clock on Big Ben in London has a radius of 11.5 ft. What is the area of this circular clock face? Use 3.14 for π. (www.bigben.freeservers.com)

Spaces Images/Blend Images

35) Abbas drove 134 miles on the highway in 2 hr. What was his average speed?

36) If Reza drove 108 miles at 72 mph, without stopping, for how long did she drive?

37) A stainless steel garbage can is in the shape of a right circular cylinder. If its radius is 6 in. and its volume is 864π in^3, what is the height of the can?

38) A coffee can in the shape of a right circular cylinder has a volume of 50π in^3. Find the height of the can if its diameter is 5 in.

39) A flag is in the shape of a triangle and has an area of 6 ft^2. Find the length of the base if its height is 4 ft.

40) A championship banner hanging from the rafters of a stadium is in the shape of a triangle and has an area of 20 ft^2. How long is the banner if its base is 5 ft?

Use a known formula to solve. See Example 3.

41) Vivian is making a rectangular wooden picture frame that will have a width that is 10 in. shorter than its length. If she will use 92 in. of wood, what are the dimensions of the frame?

42) A construction crew is making repairs next to a school, so they have to enclose the rectangular area with a fence. They determine that they will need 176 ft of fencing for the work area, which is 22 ft longer than it is wide. Find the dimensions of the fenced area.

43) The "lane" on a basketball court is a rectangle that has a perimeter of 62 ft. Find the dimensions of the "lane" given that its length is 5 ft less than twice the width.

Lawrence M. Sawyer/Getty Images

44) A rectangular whiteboard in a classroom is twice as long as it is high. Its perimeter is 24 ft. What are the dimensions of the whiteboard?

45) One base of a trapezoid is 2 in. longer than three times the other base. Find the lengths of the bases if the trapezoid is 5 in. high and has an area of 25 in^2.

46) A caution flag on the side of a road is shaped like a trapezoid. One base of the trapezoid is 1 ft shorter than the other base. Find the lengths of the bases if the trapezoid is 4 ft high and has an area of 10 ft^2.

47) A triangular sign in a store window has a perimeter of 5.5 ft. Two of the sides of the triangle are the same length while the third side is 1 ft longer than those sides. Find the lengths of the sides of the sign.

48) A triangle has a perimeter of 31 in. The longest side is 1 in. less than twice the shortest side, and the third side is 4 in. longer than the shortest side. Find the lengths of the sides.

Objective 3: Solve Problems Involving Angle Measures

Find the missing angle measures.

49)

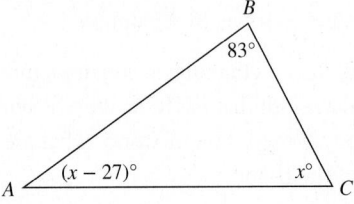

50)

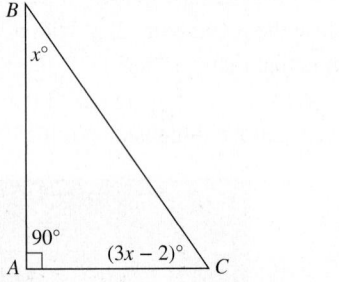

51)

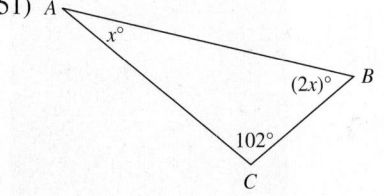

52)

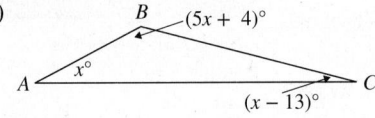

53)

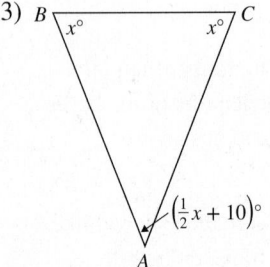

54)

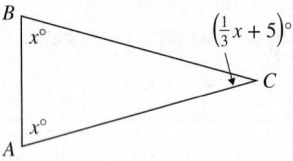

Find the measure of each indicated angle.

55)

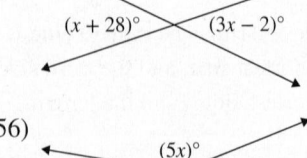

56)

57)

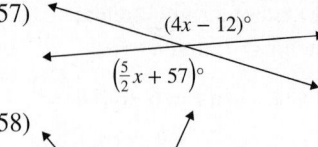

58)

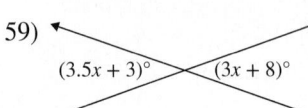

59)

60)

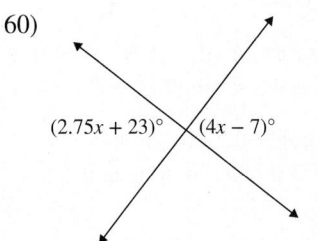

61)

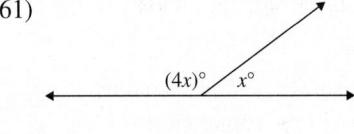

62)

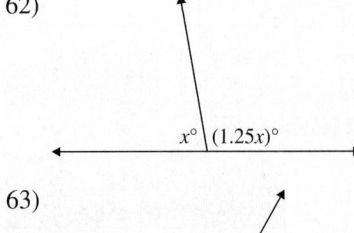

63)

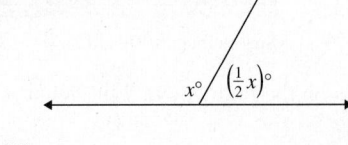

64)

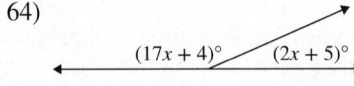

65)

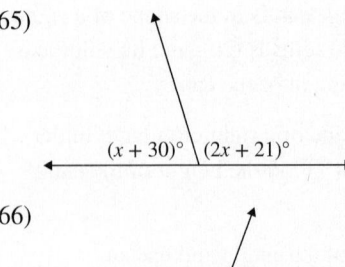

66)

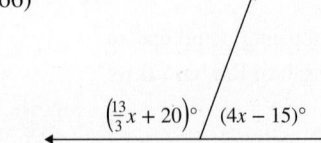

67) If x = the measure of an angle, write an expression for its supplement.

68) If x = the measure of an angle, write an expression for its complement.

Write an equation and solve.

 69) The supplement of an angle is 63° more than twice the measure of its complement. Find the measure of the angle.

70) Twice the complement of an angle is 49° less than its supplement. Find the measure of the angle.

71) Six times an angle is 12° less than its supplement. Find the measure of the angle.

72) An angle is 1° less than 12 times its complement. Find the measure of the angle.

73) Four times the complement of an angle is 40° less than twice the angle's supplement. Find the angle, its complement, and its supplement.

74) Twice the supplement of an angle is 30° more than eight times its complement. Find the angle, its complement, and its supplement.

75) The sum of an angle and half its supplement is seven times its complement. Find the measure of the angle.

76) The sum of an angle and three times its complement is 62° more than its supplement. Find the measure of the angle.

77) The sum of four times an angle and twice its complement is 270°. Find the angle.

78) The sum of twice an angle and half its supplement is 192°. Find the angle.

Objective 4: Solve a Formula for a Specific Variable

79) Solve for x.
   a) $x + 16 = 37$
   b) $x + h = y$
   c) $x + r = c$

80) Solve for t.
   a) $t - 8 = 17$
   b) $t - p = z$
   c) $t - k = n$

81) Solve for c.
   a) $8c = 56$
   b) $ac = d$
   c) $mc = v$

82) Solve for k.
   a) $9k = 54$
   b) $nk = t$
   c) $wk = h$

83) Solve for a.
   a) $\dfrac{a}{4} = 11$
   b) $\dfrac{a}{y} = r$
   c) $\dfrac{a}{w} = d$

84) Solve for d.
   a) $\dfrac{d}{6} = 3$
   b) $\dfrac{d}{t} = q$
   c) $\dfrac{d}{x} = a$

 85) Solve for d.
   a) $8d - 7 = 17$
   b) $kd - a = z$

86) Solve for w.
   a) $5w + 18 = 3$
   b) $pw + r = \pi$

87) Solve for h.
   a) $9h + 23 = 17$
   b) $qh + v = n$

88) Solve for b.
   a) $12b - 5 = 17$
   b) $mb - c = a$

Solve each formula for the indicated variable.

89) $F = ma$ for m (Physics)

90) $C = 2\pi r$ for r

91) $n = \dfrac{c}{v}$ for c (Physics)

92) $f = \dfrac{R}{2}$ for R (Physics)

93) $E = \sigma T^4$ for σ (Meteorology)

94) $p = \rho g y$ for ρ (Geology)

95) $V = \dfrac{1}{3}\pi r^2 h$ for h

96) $d = rt$ for r

97) $R = \dfrac{E}{I}$ for E (Electricity)

98) $A = \dfrac{1}{2}bh$ for b

99) $I = PRT$ for R

100) $I = PRT$ for P

101) $P = 2l + 2w$ for l

102) $A = P + PRT$ for T (Finance)

103) $H = \dfrac{D^2 N}{2.5}$ for N (Auto mechanics)

104) $V = \dfrac{AH}{3}$ for A (Geometry)

105) $A = \dfrac{1}{2}h(b_1 + b_2)$ for b_2

106) $A = \pi(R^2 - r^2)$ for r^2 (Geometry)

107) The perimeter, P, of a rectangle is $P = 2l + 2w$, where l = length and w = width.

    a) Solve $P = 2l + 2w$ for w.

    b) Find the width of the rectangle with perimeter 28 cm and length 11 cm.

108) The area, A, of a triangle is $A = \dfrac{1}{2}bh$, where b = length of the base and h = height.

    a) Solve $A = \dfrac{1}{2}bh$ for h.

    b) Find the height of the triangle that has an area of 39 cm^2 and a base of length 13 cm.

109) The formula $C = \dfrac{5}{9}(F - 32)$ can be used to convert from degrees Fahrenheit, F, to degrees Celsius, C.

    a) Solve this formula for F.

    b) The average high temperature in Paris, France, in May is 20°C. Use the result in part a) to find the equivalent temperature in degrees Fahrenheit. (www.bbc.co.uk)

110) The average low temperature in Buenos Aires, Argentina, in June is 5°C. Use the result in Exercise 109 a) to find the equivalent temperature in degrees Fahrenheit. (www.bbc.co.uk)

R Rethink

R1) What kinds of landscape features are often rectangular in shape?

R2) What kinds of jobs use angle measurements?

R3) What kinds of sports require that a player project a ball using some type of strategic angle measurement?

3.6 Applications of Linear Equations to Proportions, Money Problems, and $d = rt$

P Prepare

What are your objectives for Section 3.6?	How can you accomplish each objective?
1 Use Ratios	• Write the definition of *ratio* in your own words. • Complete the given examples on your own. • Complete You Trys 1 and 2.
2 Solve Proportions and Applications	• Write the definition of *proportion* in your own words. • Understand what *cross products* are and how to use them. • Write the definition of *similar triangles* in your own words. • Complete the given examples on your own. • Complete You Trys 3–6.
3 Solve Problems Involving Money	• Be able to write the amount of money in *dollars* and *cents*. Be consistent with using one form when setting up your equation. • Know that the value of a quantity of coins is very different from the number of coins. • Follow the procedure for solving an applied problem. • Complete the given examples on your own. • Complete You Trys 7 and 8.
4 Solve Problems Involving Distance, Rate, and Time	• Learn the formula for calculating distance. • Follow the procedure for solving an applied problem. • Make a chart to keep track of your *distance, rate, and time*. • Complete the given examples on your own. • Complete You Trys 9 and 10.

W Work

Read the explanations, follow the examples, take notes, and complete the You Trys.

1 Use Ratios

We hear about *ratios* and use them in many ways in everyday life. For example, if a survey on cell phone use revealed that 80 teenagers prefer texting their friends while 25 prefer calling their friends, we could write the ratio of teens who prefer texting to teens who prefer calling as

$$\frac{\text{Number who prefer texting}}{\text{Number who prefer calling}} = \frac{80}{25} = \frac{16}{5}$$

SECTION 3.6 Applications of Linear Equations to Proportions, Money Problems, and $d = rt$ 187

Here is a formal definition of a ratio:

Definition

A **ratio** is a quotient of two quantities. The ratio of the number x to the number y, where $y \neq 0$, can be written as $\dfrac{x}{y}$, x to y, or $x : y$.

EXAMPLE 1

Write the ratio of 4 feet to 2 yards.

Solution

Write each quantity with the same units. Let's change yards to feet. Since there are 3 feet in 1 yard,

$$2 \text{ yards} = 2 \cdot 3 \text{ feet} = 6 \text{ feet}$$

Then the ratio of 4 feet to 2 yards is

$$\frac{4 \text{ feet}}{2 \text{ yd}} = \frac{4 \text{ feet}}{6 \text{ feet}} = \frac{4}{6} = \frac{2}{3}$$

[**YOU TRY 1**]   Write the ratio of 3 feet to 24 inches.

We can use ratios to help us figure out which item in a store gives us the most value for our money. To do this, we will determine the *unit price* of each item. The **unit price** is the ratio of the price of the item to the amount of the item.

EXAMPLE 2

A store sells Haagen-Dazs vanilla ice cream in three different sizes. The sizes and prices are listed here. Which size is the best buy?

Size	Price
4 oz	$1.00
14 oz	$3.49
28 oz	$7.39

Solution

For each carton of ice cream, we must find the unit price, or how much the ice cream costs per ounce. We will find the unit price by dividing.

$$\text{Unit price} = \frac{\text{Price of ice cream}}{\text{Number of ounces in the container}} = \text{Cost per ounce}$$

Size	Unit Price
4 oz	$\dfrac{\$1.00}{4 \text{ oz}} = \$0.250$ per oz
14 oz	$\dfrac{\$3.49}{14 \text{ oz}} = \$0.249$ per oz
28 oz	$\dfrac{\$7.39}{28 \text{ oz}} = \$0.264$ per oz

We round the answers to the thousandths place because, as you can see, there is not much difference in the unit price. Since the 14-oz carton of ice cream has the smallest unit price, it is the best buy.

A store sells Gatorade fruit punch in three different sizes. A 20-oz bottle costs $1.00, a 32-oz bottle sells for $1.89, and the price of a 128-oz bottle is $5.49. Which size is the best buy, and what is its unit price?

2 Solve Proportions and Applications

We have learned that a ratio is a way to compare two quantities. If two ratios are equivalent, like $\frac{4}{6}$ and $\frac{2}{3}$, we can set them equal to make a *proportion*.

Definition

A **proportion** is a statement that two ratios are equal.

How can we be certain that a proportion is true? We can find the **cross products.** If the cross products are equal, then the proportion is true. If the cross products are not equal, then the proportion is false.

Property

Cross Products   If $\frac{a}{b} = \frac{c}{d}$, then $ad = bc$ provided that $b \neq 0$ and $d \neq 0$.

We will see later in the book that finding the cross products is the same as multiplying both sides of the equation by a common denominator of the fractions.

EXAMPLE 3

Determine whether each proportion is true or false.

a) $\frac{5}{7} = \frac{15}{21}$ b) $\frac{2}{9} = \frac{7}{36}$

Solution

W Hint
Notice that cross products are used only across an equal sign!

a)   Find the cross products.

$$\frac{5}{7} \diagup\!\!\!\!\diagdown \frac{15}{21}$$ Multiply. Multiply.

$5 \cdot 21 = 7 \cdot 15$ Set the cross products equal.
$105 = 105$ True

The cross products are equal, so the proportion is true.

b)   Find the cross products.

$$\frac{2}{9} \diagup\!\!\!\!\diagdown \frac{7}{36}$$ Multiply. Multiply.

$2 \cdot 36 = 9 \cdot 7$ Set the cross products equal.
$72 = 63$ False

The cross products are not equal, so the proportion is false.

[YOU TRY 3]

Determine whether each proportion is true or false.

a) $\frac{4}{9} = \frac{24}{56}$ b) $\frac{3}{8} = \frac{12}{32}$

We can use cross products to solve equations.

EXAMPLE 4

Solve each proportion.

a) $\dfrac{16}{24} = \dfrac{x}{3}$ b) $\dfrac{k+2}{2} = \dfrac{k-4}{5}$

Solution

Find the cross products.

a) $\dfrac{16}{24} \underset{\text{Multiply.}}{\overset{\text{Multiply.}}{\bcancel{=}}} \dfrac{x}{3}$

b) $\dfrac{k+2}{2} \underset{\text{Multiply.}}{\overset{\text{Multiply.}}{\bcancel{=}}} \dfrac{k-4}{5}$

W Hint

Be sure to use parentheses when the numerator or denominator contains more than one term.

a)
$16 \cdot 3 = 24 \cdot x$	Set the cross products equal.
$48 = 24x$	Multiply.
$2 = x$	Divide by 24.

The solution set is $\{2\}$.

b)
$5(k+2) = 2(k-4)$	Set the cross products equal.
$5k + 10 = 2k - 8$	Distribute.
$3k + 10 = -8$	Subtract $2k$.
$3k = -18$	Subtract 10.
$k = -6$	Divide by 3.

The solution set is $\{-6\}$.

[YOU TRY 4]

Solve each proportion.

a) $\dfrac{2}{3} = \dfrac{w}{27}$ b) $\dfrac{b-6}{12} = \dfrac{b+2}{20}$

Proportions are often used to solve real-world problems. When we solve problems by setting up a proportion, we must be sure that the numerators contain the same quantities and the denominators contain the same quantities.

EXAMPLE 5

Write an equation and solve.

Cailen is an artist, and she wants to make turquoise paint by mixing the green and blue paints that she already has. To make turquoise, she will have to mix 4 parts of green with 3 parts of blue. If she uses 6 oz of green paint, how much blue paint should she use?

Solution

Step 1: **Read** the problem carefully, and identify what we are being asked to find.

We must find the amount of blue paint needed.

Step 2: **Choose a variable** to represent the unknown.

x = the number of ounces of blue paint

W Hint

Again, use the same five-step method for solving applied problems.

Step 3: **Translate** the information that appears in English into an algebraic equation. Write a proportion. We will write our ratios in the form of

$\dfrac{\text{Amount of green paint}}{\text{Amount of blue paint}}$ so that the numerators contain the same quantities and the denominators contain the same quantities.

Amount of green paint $\rightarrow$ $\dfrac{4}{3} = \dfrac{6}{x}$ $\leftarrow$ Amount of green paint
Amount of blue paint $\rightarrow$ $\phantom{\dfrac{4}{3} = \dfrac{6}{x}}$ $\leftarrow$ Amount of blue paint

The equation is $\dfrac{4}{3} = \dfrac{6}{x}$.

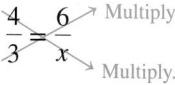

Step 4: **Solve** the equation.

$$\frac{4}{3} = \frac{6}{x}$$ Multiply.

Multiply.

$4x = 6 \cdot 3$ Set the cross products equal.

$4x = 18$ Multiply.

$x = 4.5$ Divide by 4.

Step 5: **Check** the answer, and **interpret** the solution as it relates to the problem.

Cailen should mix 4.5 oz of blue paint with the 6 oz of green paint to make the turquoise paint she needs. The check is left to the student.

Ken Cavanagh/McGraw-Hill Education

[**YOU TRY 5**]   Write an equation and solve.

If 3 lb of coffee costs \$21.60, how much would 5 lb of the same coffee cost?

Another application of proportions is for solving similar triangles.

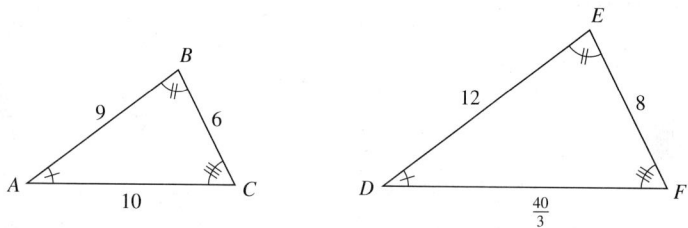

$$m\angle A = m\angle D, \quad m\angle B = m\angle E, \quad \text{and} \quad m\angle C = m\angle F$$

We say that $\triangle ABC$ and $\triangle DEF$ are *similar triangles*. Two triangles are **similar** if they have the same shape, the corresponding angles have the same measure, and the corresponding sides are proportional.

The ratio of each of the corresponding sides is $\frac{3}{4}$:

$$\frac{9}{12} = \frac{3}{4}; \quad \frac{6}{8} = \frac{3}{4}; \quad \frac{10}{\frac{40}{3}} = 10 \cdot \frac{3}{40} = \frac{3}{4}$$

We can use a proportion to find the length of an unknown side in two similar triangles.

EXAMPLE 6   Given the following similar triangles, find x.

Solution

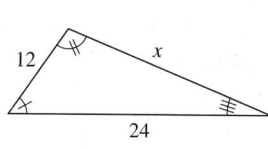

$$\frac{12}{18} = \frac{x}{30}$$ Set the ratios of two corresponding sides equal to each other. (Set up a proportion.)

$12 \cdot 30 = 18 \cdot x$ Solve the proportion.

$360 = 18x$ Multiply.

$20 = x$ Divide by 18.

> **W Hint**
> Note that there is more than one way to set up the proportion to solve for x.

[YOU TRY 6] Given the following similar triangles, find x.

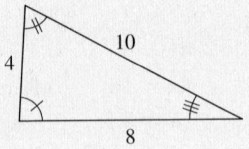
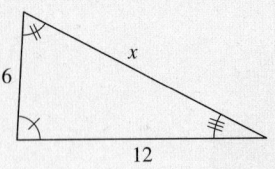

3 Solve Problems Involving Money

Many application problems involve thinking about the number of coins or bills and their values. Let's look at how arithmetic and algebra problems involving these ideas are related.

EXAMPLE 7 Determine the amount of money you have in cents *and* in dollars if you have

a) 8 nickels   b) 7 quarters   c) 8 nickels and 7 quarters

Solution

You may be able to do these problems "in your head," but it is very important that we understand the *procedure* that is used to do this arithmetic problem so that we can apply the same procedure to algebra. So, read this carefully!

Parts a) and b):   Let's begin with part a), finding the value of 8 nickels.

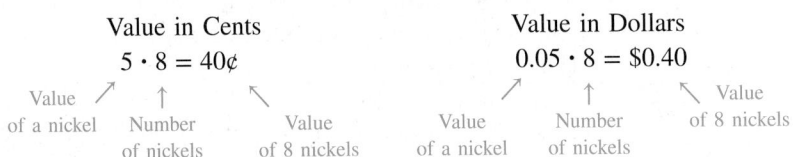

Here's how we find the value of 7 quarters:

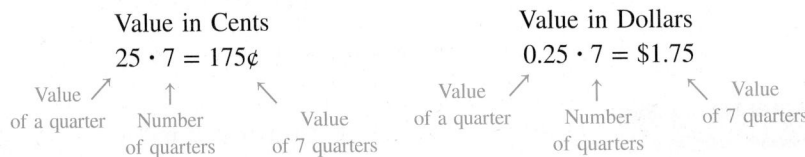

W Hint
On a sheet of paper, make a list of the different types of tables used in this section. This will make your homework easier!

A table can help us organize the information, so let's put both part a) and part b) in a table so that we can see a pattern.

Value of the Coins (in cents)

	Value of the Coin	Number of Coins	Total Value of the Coins
Nickels	5	8	$5 \cdot 8 = 40$
Quarters	25	7	$25 \cdot 7 = 175$

Value of the Coins (in dollars)

	Value of the Coin	Number of Coins	Total Value of the Coins
Nickels	0.05	8	$0.05 \cdot 8 = 0.40$
Quarters	0.25	7	$0.25 \cdot 7 = 1.75$

Notice that each time we want to find the total value of the coins, we find it by multiplying.

$$\text{Value of the coin} \cdot \text{Number of coins} = \text{Total value of the coins}$$

c) Now let's write an equation in English to find the total value of the 8 nickels and 7 quarters.

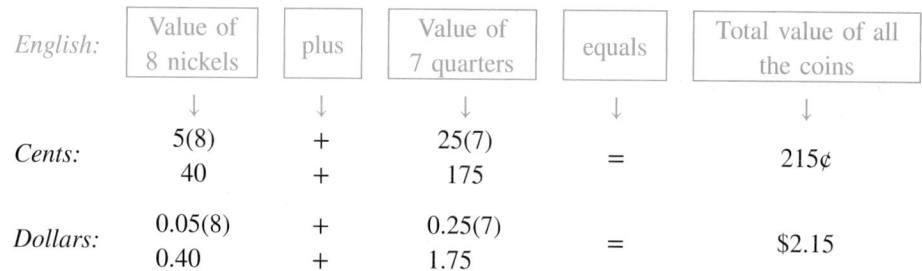

English:	Value of 8 nickels	plus	Value of 7 quarters	equals	Total value of all the coins
	↓	↓	↓	↓	↓
Cents:	5(8) 40	+ +	25(7) 175	=	215¢
Dollars:	0.05(8) 0.40	+ +	0.25(7) 1.75	=	$2.15

We will use the same procedure that we just used to solve these arithmetic problems to write algebraic expressions to represent the value of a collection of coins.

EXAMPLE 8

Write expressions for the amount of money you have in cents *and* in dollars if you have

a) n nickels   b) q quarters   c) n nickels and q quarters

Solution

Parts a) and b)   Let's use tables just like we did in Example 7. We will put parts a) and b) in the same table.

Value of the Coins (in cents)

	Value of the Coin	Number of Coins	Total Value of the Coins
Nickels	5	n	$5 \cdot n = 5n$
Quarters	25	q	$25 \cdot q = 25q$

Value of the Coins (in dollars)

	Value of the Coin	Number of Coins	Total Value of the Coins
Nickels	0.05	n	$0.05 \cdot n = 0.05n$
Quarters	0.25	q	$0.25 \cdot q = 0.25q$

If you have n nickels, then the expression for the amount of money in cents is $5n$. The amount of money in dollars is $0.05n$. If you have q quarters, then the expression for the amount of money in cents is $25q$. The amount of money in dollars is $0.25q$.

c)   Write an equation in English to find the total value of n nickels and q quarters. It is based on the same idea that we used in Example 7.

English:	Value of n nickels	plus	Value of q quarters	equals	Total value of all the coins
	↓	↓	↓	↓	↓
Equation in cents:	$5n$	+	$25q$	=	$5n + 25q$
Equation in dollars:	$0.05n$	+	$0.25q$	=	$0.05n + 0.25q$

The expression in cents is $5n + 25q$. The expression in dollars is $0.05n + 0.25q$.

[YOU TRY 7]

Determine the amount of money you have in cents *and* in dollars if you have

a) 11 dimes   b) 20 pennies   c) 8 dimes and 46 pennies

d) d dimes   e) p pennies   f) d dimes and p pennies

Next, we'll apply this idea of the value of different denominations of money to an application problem.

EXAMPLE 9

Jamaal has only dimes and quarters in his piggy bank. When he counts the change, he finds that he has $18.60 and that there are twice as many quarters as dimes. How many dimes and quarters are in his bank?

Solution

Step 1: **Read** the problem carefully, and identify what we are being asked to find.

We must find the number of dimes and quarters in the bank.

Step 2: **Choose a variable** to represent an unknown, and define the other unknown in terms of this variable.

In the statement "there are twice as many quarters as dimes," the number of quarters is expressed *in terms of* the number of dimes. Therefore, let

$$d = \text{the number of dimes}$$

Define the other unknown (the number of quarters) in terms of d:

$$2d = \text{the number of quarters}$$

Stockdisc/Getty Images

Step 3: **Translate** the information that appears in English into an algebraic equation.

Let's begin by making a table to write an expression for the value of the dimes and the value of the quarters. We will write the expression in terms of dollars because the total value of the coins, $18.60, is given in dollars.

	Value of the Coin	Number of Coins	Total Value of the Coins
Dimes	0.10	d	$0.10d$
Quarters	0.25	$2d$	$0.25 \cdot (2d)$

W Hint

Remember, writing an equation in English will help you to write it using algebra.

Write an equation in English and substitute the expressions we found in the table and the total value of the coins to get an algebraic equation.

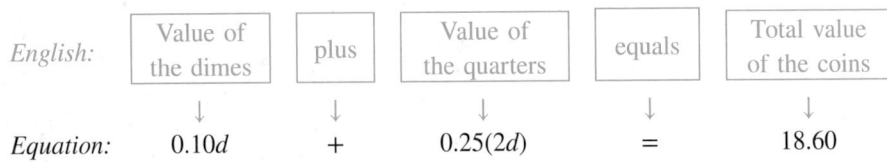

English:	Value of the dimes	plus	Value of the quarters	equals	Total value of the coins
	↓	↓	↓	↓	↓
Equation:	$0.10d$	$+$	$0.25(2d)$	$=$	18.60

Step 4: **Solve** the equation.

$$0.10d + 0.25(2d) = 18.60$$
$$100[0.10d + 0.25(2d)] = 100(18.60) \qquad \text{Multiply by 100 to eliminate the decimals.}$$
$$10d + 25(2d) = 1860 \qquad \text{Distribute.}$$
$$10d + 50d = 1860 \qquad \text{Multiply.}$$
$$60d = 1860 \qquad \text{Combine like terms.}$$
$$\frac{60d}{60} = \frac{1860}{60} \qquad \text{Divide each side by 60.}$$
$$d = 31 \qquad \text{Simplify.}$$

Step 5: **Check** the answer, and **interpret** the solution as it relates to the problem.

There were 31 dimes and 2(31) = 62 quarters in the bank.

Check: The value of the dimes is $0.10(31) = $3.10, and the value of the quarters is $0.25(62) = $15.50. Their total is $3.10 + $15.50 = $18.60.

[**YOU TRY 8**] A collection of coins consists of pennies and nickels. There are five fewer nickels than there are pennies. If the coins are worth a total of $4.97, how many of each type of coin is in the collection?

4 Solve Problems Involving Distance, Rate, and Time

If you drive at 50 mph for 4 hr, how far will you drive? One way to get the answer is to use the formula

$$\text{Distance} = \text{Rate} \cdot \text{Time}$$
$$\text{or}$$
$$d = rt$$

$$d = (50 \text{ mph}) \cdot (4 \text{ hr})$$
$$\text{Distance traveled} = 200 \text{ mi}$$

Notice that the rate is in miles per *hour* and the time is in *hours*. The units must be consistent in this way. If the time in this problem had been expressed in minutes, it would have been necessary to convert minutes to hours. We will use the formula $d = rt$ to solve two algebraic applications.

EXAMPLE 10 Two planes leave St. Louis, one flying east and the other flying west. The westbound plane travels 100 mph faster than the eastbound plane, and after 1.5 hours they are 750 miles apart. Find the speed of each plane.

Solution

Step 1: **Read** the problem carefully, and identify what we are being asked to find.

We must find the speed of the eastbound and westbound planes.

We will draw a picture to help us see what is happening in this problem.

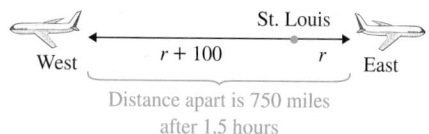

St. Louis

West $r + 100$ r East

Distance apart is 750 miles
after 1.5 hours

Step 2: **Choose a variable** to represent an unknown, and define the other unknown in terms of this variable.

The westbound plane is traveling 100 mph faster than the eastbound plane, so let

$$r = \text{the rate of the eastbound plane}$$
$$r + 100 = \text{the rate of the westbound plane}$$

Label the picture.

Step 3: **Translate** the information that appears in English into an algebraic equation.

Let's make a table using the equation $d = rt$. Fill in the time, 1.5 hr, and the rates first, then multiply those together to fill in the values for the distance.

	d	r	t
Eastbound	$1.5r$	r	1.5
Westbound	$1.5(r + 100)$	$r + 100$	1.5

We will write an equation in English to help us write an algebraic equation. The picture shows that

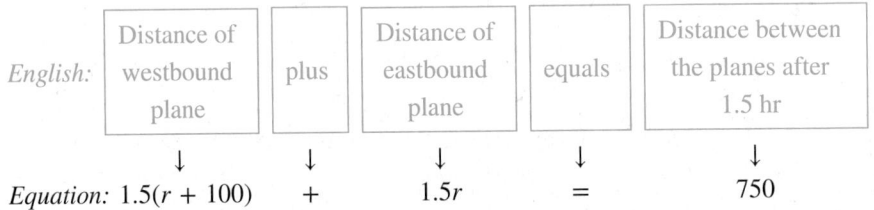

English:	Distance of westbound plane	plus	Distance of eastbound plane	equals	Distance between the planes after 1.5 hr
	↓	↓	↓	↓	↓
Equation:	$1.5(r + 100)$	$+$	$1.5r$	$=$	750

The expressions for the distances in the equation come from the table.

The equation is $1.5(r + 100) + 1.5r = 750$.

Step 4: **Solve** the equation.

$$1.5(r + 100) + 1.5r = 750$$
$$10[1.5(r + 100) + 1.5r] = 10(750) \qquad \text{Multiply by 10 to eliminate the decimals.}$$
$$15(r + 100) + 15r = 7500 \qquad \text{Distribute.}$$
$$15r + 1500 + 15r = 7500 \qquad \text{Distribute.}$$
$$30r + 1500 = 7500 \qquad \text{Combine like terms.}$$
$$30r = 6000 \qquad \text{Subtract 1500.}$$
$$\frac{30r}{30} = \frac{6000}{30} \qquad \text{Divide each side by 30.}$$
$$x = 200 \qquad \text{Simplify.}$$

Step 5: **Check** the answer, and **interpret** the solution as it relates to the problem.

The speed of the eastbound plane is 200 mph, and the speed of the westbound plane is $200 + 100 = 300$ mph.

Check to see that $1.5(200) + 1.5(300) = 300 + 450 = 750$ miles.

[YOU TRY 9]

Write an equation and solve.

Two drivers leave Albany, Oregon, on Interstate 5. Dhaval heads south traveling 4 mph faster than Pradeep, who is driving north. After $\frac{1}{2}$ hr, they are 62 miles apart. How fast is each man driving?

EXAMPLE 11

Alex and Jenny are taking a cross-country road trip on their motorcycles. Jenny leaves a rest area first traveling at 60 mph. Alex leaves 30 min later, traveling on the same highway, at 70 mph. How long will it take Alex to catch Jenny?

Solution

Step 1: **Read** the problem carefully, and identify what we are being asked to find.

We must determine how long it takes Alex to catch Jenny.

We will draw a picture to help us see what is happening in this problem.

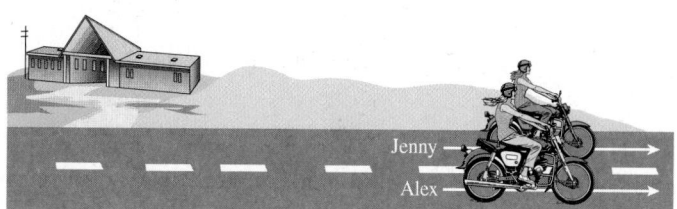

> **W Hint**
>
> Remember that it is always useful to sketch a picture to help you visualize these problems!

Since both girls leave the same rest area and travel on the same highway, when Alex catches Jenny they have driven the *same* distance.

Step 2: **Choose a variable** to represent an unknown, and define the other unknown in terms of this variable.

Alex's time is in terms of Jenny's time, so let

t = the number of hours Jenny has been riding when Alex catches her

Alex leaves 30 minutes ($\frac{1}{2}$ hour) after Jenny, so Alex travels $\frac{1}{2}$ hour *less than* Jenny.

$t - \frac{1}{2}$ = the number of hours it takes Alex to catch Jenny

Step 3: **Translate** the information that appears in English into an algebraic equation.

Let's make a table using the equation $d = rt$. Fill in the time and the rates first, then multiply those together to fill in the value for the distance.

	d	r	t
Jenny	$60t$	60	t
Alex	$70\left(t - \frac{1}{2}\right)$	70	$t - \frac{1}{2}$

We will write an equation in English to help us write an algebraic equation. The picture shows that

English:	Jenny's distance	is the same as	Alex's distance
	↓	↓	↓
Equation:	$60t$	$=$	$70\left(t - \frac{1}{2}\right)$

The equation is $60t = 70\left(t - \frac{1}{2}\right)$.

Step 4: **Solve** the equation.

$$60t = 70\left(t - \frac{1}{2}\right)$$

$$60t = 70t - 35 \qquad \text{Distribute.}$$

$$-10t = -35 \qquad \text{Subtract } 70t.$$

$$\frac{-10t}{-10} = \frac{-35}{-10} \qquad \text{Divide each side by } -10.$$

$$t = 3.5 \qquad \text{Simplify.}$$

Step 5: **Check** the answer, and **interpret** the solution as it relates to the problem.

Remember, Jenny's time is t. Alex's time is $t - \frac{1}{2} = 3\frac{1}{2} - \frac{1}{2} = 3$ hr.

It will take Alex 3 hr to catch Jenny.

Check to see that Jenny travels 60 mph · (3.5 hr) = 210 miles, and Alex travels 70 mph · (3 hr) = 210 miles. The girls travel the same distance.

[YOU TRY 10] Brad leaves home driving 40 mph. Angelina leaves the house 30 minutes later driving the same route at 50 mph. How long will it take Angelina to catch Brad?

ANSWERS TO [YOU TRY] EXERCISES

1) $\frac{3}{2}$ 2) 128-oz bottle; $0.043/oz   3) a) false b) true   4) a) {18} b) {18}   5) $36.00

6) 15   7) a) 110¢, $1.10 b) 20¢, $0.20 c) 126¢, $1.26 d) 10d cents, 0.10d dollars e) 1p cents, 0.01p dollars f) 10d + 1p cents, 0.10d + 0.01p dollars   8) 87 pennies, 82 nickels   9) Dhaval: 64 mph, Pradeep: 60 mph   10) 2 hr

---

E Evaluate **3.6** Exercises Do the exercises, and check your work.

Objective 1: Use Ratios

1) Write three ratios that are equivalent to $\frac{3}{4}$.

2) Is 0.65 equivalent to the ratio 13 to 20? Explain.

3) Is a percent a type of ratio? Explain.

4) Write 57% as a ratio.

Write as a ratio in lowest terms.

5) 16 girls to 12 boys

6) 9 managers to 90 employees

7) 4 coaches to 50 team members

8) 30 blue marbles to 18 red marbles

9) 20 feet to 80 feet

10) 7 minutes to 4 minutes

11) 2 feet to 36 inches

12) 30 minutes to 3 hours

13) 18 hours to 2 days

14) 20 inches to 3 yards

A store sells the same product in different sizes. Determine which size is the best buy based on the unit price of each item.

15) Batteries

Number	Price
8	$ 6.29
16	$12.99

16) Cat litter

Size	Price
30 lb	$ 8.48
50 lb	$12.98

17) Mayonnaise

Size	Price
8 oz	$2.69
15 oz	$3.59
48 oz	$8.49

18) Applesauce

Size	Price
16 oz	$1.69
24 oz	$2.29
48 oz	$3.39

19) Cereal

Size	Price
11 oz	$4.49
16 oz	$5.15
24 oz	$6.29

20) Shampoo

Size	Price
14 oz	$3.19
25 oz	$5.29
32 oz	$6.99

Objective 2: Solve Proportions and Applications

21) What is the difference between a ratio and a proportion?

22) In the proportion $\dfrac{a}{b} = \dfrac{c}{d}$, can $b = 0$? Explain.

Determine whether each proportion is true or false.

 23) $\dfrac{4}{7} = \dfrac{20}{35}$

24) $\dfrac{54}{64} = \dfrac{7}{8}$

25) $\dfrac{72}{54} = \dfrac{8}{7}$

26) $\dfrac{120}{140} = \dfrac{30}{35}$

27) $\dfrac{8}{10} = \dfrac{2}{\frac{5}{2}}$

28) $\dfrac{3}{4} = \dfrac{\frac{1}{2}}{\frac{2}{3}}$

Solve each proportion.

29) $\dfrac{8}{36} = \dfrac{c}{9}$

30) $\dfrac{n}{3} = \dfrac{20}{15}$

 31) $\dfrac{w}{15} = \dfrac{32}{12}$

32) $\dfrac{8}{14} = \dfrac{d}{21}$

33) $\dfrac{40}{24} = \dfrac{30}{a}$

34) $\dfrac{10}{x} = \dfrac{12}{54}$

35) $\dfrac{2}{k} = \dfrac{9}{12}$

36) $\dfrac{15}{27} = \dfrac{m}{6}$

37) $\dfrac{3z + 10}{14} = \dfrac{2}{7}$

38) $\dfrac{8t - 9}{20} = \dfrac{3}{4}$

39) $\dfrac{r + 7}{9} = \dfrac{r - 5}{3}$

40) $\dfrac{b + 6}{5} = \dfrac{b + 10}{15}$

41) $\dfrac{3h + 15}{16} = \dfrac{2h + 5}{4}$

42) $\dfrac{a + 7}{8} = \dfrac{4a - 11}{6}$

43) $\dfrac{4m - 1}{6} = \dfrac{6m}{10}$

44) $\dfrac{9w + 8}{10} = \dfrac{5 - 3w}{12}$

Set up a proportion and solve.

45) If 4 containers of yogurt cost $2.36, find the cost of 6 containers of yogurt.

46) Find the cost of 3 scarves if 2 scarves cost $29.00.

47) A marinade for chicken uses 2 parts of lime juice for every 3 parts of orange juice. If the marinade uses $\dfrac{1}{3}$ cup of lime juice, how much orange juice should be used?

48) The ratio of salt to baking soda in a cookie recipe is 0.75 to 1. If a recipe calls for $1\dfrac{1}{2}$ teaspoons of salt, how much baking soda is in the cookie dough?

49) A 12-oz serving of Mountain Dew contains 55 mg of caffeine. How much caffeine is in an 18-oz serving of Mountain Dew? (www.energyfiend.com)

50) An 8-oz serving of Red Bull energy drink contains about 80 mg of caffeine. Approximately how much caffeine is in 12 oz of Red Bull? (www.energyfiend.com)

51) Approximately 9 out of 10 smokers began smoking before the age of 21. In a group of 400 smokers, about how many of them started before they reached their 21st birthday? (www.lungusa.org)

52) Ridgemont High School administrators estimate that 2 out of 3 members of its student body attended the homecoming football game. If there are 1941 students in the school, how many went to the game?

53) At the end of a week, Ernest put 20 lb of yard waste and some kitchen scraps on the compost pile. If the ratio of yard waste to kitchen scraps was 5 to 2, how many pounds of kitchen scraps did he put on the pile?

54) On a map of the United States, 1 inch represents 120 miles. If two cities are 3.5 inches apart on the map, what is the actual distance between the two cities?

55) On August 28, 2018, the exchange rate was such that $20.00 (American) was worth 17.10 Euros. How many Euros could you get for $50.00? (www.xe.com)

56) On August 28, 2018, the exchange rate was such that 100 British pounds were worth $128.70 (American). How many dollars could you get for 280 British pounds? (www.xe.com)

Given the following similar triangles, find *x*.

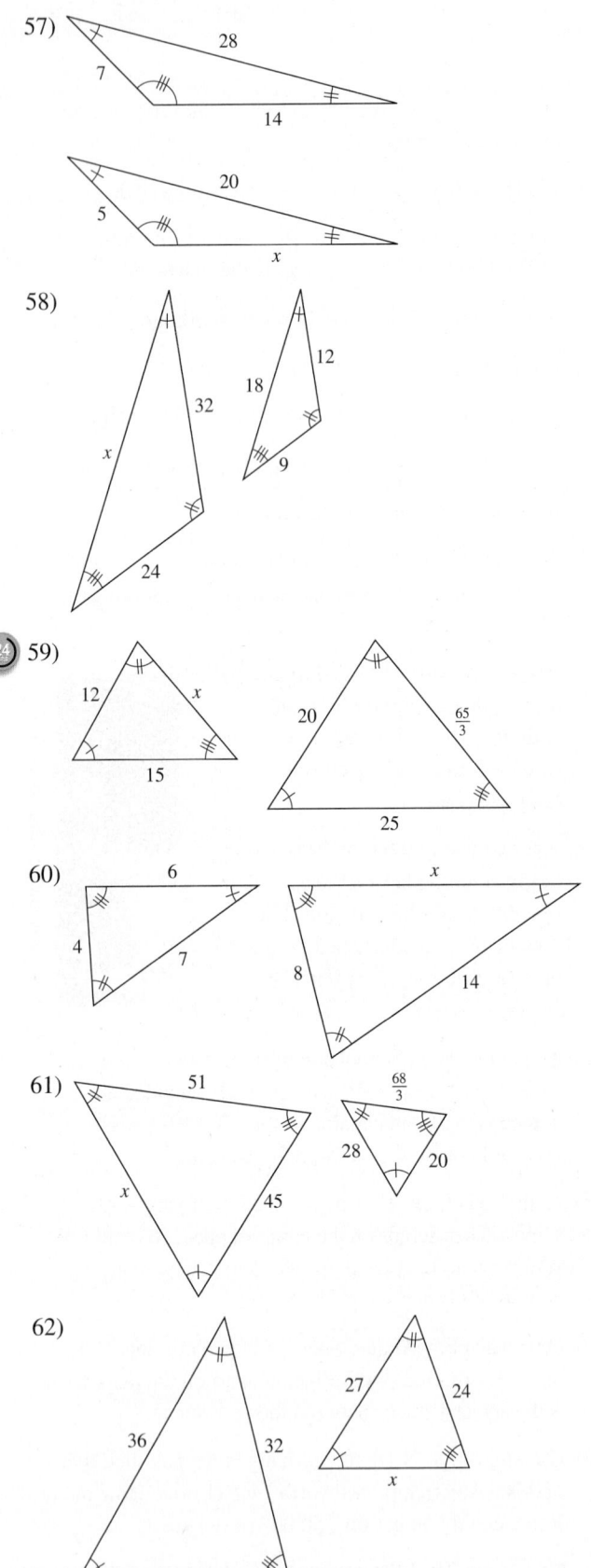

57)

58)

59)

60)

61)

62)

For Exercises 63–68, determine the amount of money
a) in dollars and b) in cents given the following quantities.

63) 8 dimes

64) 32 nickels

65) 217 pennies

66) 12 quarters

67) 9 quarters and 7 dimes

68) 89 pennies and
    14 nickels

For Exercises 69–74, write an expression which represents
the amount of money in a) dollars and b) cents given the
following quantities.

69) *q* quarters

70) *p* pennies

71) *d* dimes

72) *n* nickels

73) *p* pennies and *q* quarters

74) *n* nickels and *d* dimes

Solve using the Five Steps for Solving Applied Problems.
See Example 9.

75) Dustin and Dart combine their coins to find they have all
    nickels and quarters. They have 8 more quarters than
    nickels, and the coins are worth a total of $4.70. How
    many nickels and quarters do they have?

76) Danika saves all of her pennies and nickels in a jar.
    One day she counted them and found that there were
    131 coins worth $3.43. How many pennies and how
    many nickels were in the jar?

77) Kyung Soo has been saving her babysitting money. She
    has $69.00 consisting of $5 bills and $1 bills.
    If she has a total of 25 bills, how many $5 bills and
    how many $1 bills does she have?

78) A bank employee is servicing the ATM after a busy
    Friday night. She finds the machine contains only
    $20 bills and $10 bills and that there are twice as many
    $20 bills remaining as there are $10 bills. If there is a
    total of $550.00 left in the machine, how many of the
    bills are twenties, and how many are tens?

79) A movie theater charges $9.00 for adults and $7.00 for
    children. The total revenue for a particular movie is
    $475.00. Determine the number of each type of ticket
    sold if the number of children's tickets sold was half
    the number of adult tickets sold.

80) At the post office, Ronald buys 12 more 50¢ stamps
    than 35¢ stamps. If he spends $12.80 on the stamps,
    how many of each type did he buy?

Objective 4: Solve Problems Involving Distance, Rate, and Time

Solve using the Five Steps for Solving Applied Problems. See Examples 10 and 11.

81) Two cars leave Indianapolis, one driving east and the other driving west. The eastbound car travels 8 mph slower than the westbound car, and after 3 hr they are 414 mi apart. Find the speed of each car.

82) Two planes leave San Francisco, one flying north and the other flying south. The southbound plane travels 50 mph faster than the northbound plane, and after 2 hours they are 900 miles apart. Find the speed of each plane.

83) Maureen and Yvette leave the gym to go to work traveling the same route, but Maureen leaves 10 min after Yvette. If Yvette drives 60 mph and Maureen drives 72 mph, how long will it take Maureen to catch Yvette?

84) Vinay and Sadiva leave the same location traveling the same route, but Sadiva leaves 20 minutes after Vinay. If Vinay drives 30 mph and Sadiva drives 36 mph, how long will it take Sadiva to catch Vinay?

85) A passenger train and a freight train leave cities 400 mi apart and travel toward each other. The passenger train is traveling 20 mph faster than the freight train. Find the speed of each train if they pass each other after 5 hr.

86) A freight train passes the Old Towne train station at 11:00 A.M. going 30 mph. Ten minutes later a passenger train, headed in the same direction on an adjacent track, passes the same station at 45 mph. At what time will the passenger train catch the freight train?

87) A truck and a car leave the same intersection traveling in the same direction. The truck is traveling at 35 mph, and the car is traveling at 45 mph. In how many minutes will they be 6 mi apart?

88) At noon, a truck and a car leave the same intersection traveling in the same direction. The truck is traveling at 30 mph, and the car is traveling at 42 mph. At what time will they be 9 mi apart?

89) Ajay is traveling north on a road while Rohan is traveling south on the same road. They pass by each other at 3 P.M., Ajay driving 30 mph and Rohan driving 40 mph. At what time will they be 105 miles apart?

90) When Lucas and Max pass each other on their bikes going in opposite directions, Lucas is riding at 22 mph, and Max is pedaling at 18 mph. If they continue at those speeds, after how long will they be 100 mi apart?

91) At noon, a cargo van crosses an intersection at 30 mph. At 12:30 P.M., a car crosses the same intersection traveling in the opposite direction. At 1 P.M., the van and car are 54 miles apart. How fast is the car traveling?

92) A freight train passes the Naperville train station at 9:00 A.M. going 30 mph. Ten minutes later a passenger train, headed in the same direction on an adjacent track, passes the same station at 45 mph. At what time will the passenger train catch the freight train?

Mixed Exercises: Objectives 2–4

Solve using the five-step method.

93) At the end of her shift, a cashier has a total of $6.30 in dimes and quarters. There are 7 more dimes than quarters. How many of each of these coins does she have?

Blend Images/Alamy Stock Photo

94) Paloma leaves Mateo's house traveling 30 mph. Mateo leaves 15 minutes later, trying to catch up to Paloma, going 40 mph. If they drive along the same route, how long will it take Mateo to catch Paloma?

95) Sherri is riding her bike at 10 mph when Bill passes her going in the opposite direction at 14 mph. How long will it take before the distance between them is 6 miles?

96) If the exchange rate between the American dollar and the Japanese yen is such that $4.00 = 442 yen, how many yen could be exchanged for $70.00?

97) Approximately 94 out of every 1000 Americans have diabetes. The population of the United States is about 325,000,000. How many Americans have diabetes? (http://diabetes.niddk.nih.gov)

98) At Ralph's grocery store, green peppers cost $0.88 each and red peppers cost $0.95 each. Chung-Hee buys twice as many green peppers as red peppers and spends $5.42. How many green peppers and how many red peppers did he buy?

99) A jet flying at an altitude of 35,000 ft passes over a small plane flying at 10,000 ft headed in the same direction. The jet is flying twice as fast as the small plane, and 30 minutes later they are 100 mi apart. Find the speed of each plane.

100) At the end of his shift, Bruno had $340 worth of tips, all in $10 and $20 bills. If he had two more $20 bills than $10 bills, how many of each bill did Bruno have?

R Rethink

R1) Which objective is the most difficult for you?

R2) In which of your future courses, other than a math course, do you think you will need to solve applications problems similar to those in this section?

R3) Why is it useful to sometimes organize the given application data in a table?

3.7 Linear Inequalities in One Variable

P Prepare

O Organize

What are your objectives for Section 3.7?	How can you accomplish each objective?
1 Use Graphs and Set and Interval Notations	• Be able to recognize a *linear inequality*. • Understand how to write an answer in *set notation* and in *interval notation*. • Be able to determine when to use a bracket and when to use a parenthesis in *interval notation*. • Complete the given example on your own. • Complete You Try 1.
2 Solve Inequalities Using the Addition and Subtraction Properties of Inequality	• Learn the **Addition and Subtraction Properties of Inequality,** and know when they should be used to solve inequalities. • Complete the given example on your own. • Complete You Try 2.
3 Solve Inequalities Using the Multiplication Property of Inequality	• Learn the **Multiplication Property of Inequality,** and understand that we reverse the inequality symbol when we *multiply* or *divide* by a negative number. • Complete the given example on your own. • Complete You Try 3.
4 Solve Inequalities Using a Combination of the Properties	• Know all the properties of inequalities and when to use each property. • Complete the given example on your own. • Complete You Try 4.

(continued)

What are your objectives for Section 3.7?	How can you accomplish each objective?
5 Solve Compound Inequalities Containing Three Parts	• Be able to identify *compound inequalities*. • Understand the process for solving *three-part inequalities*. • Know how the solutions of *three-part inequalities* are graphed on a number line and written in *interval notation*. • Complete the given examples on your own. • Complete You Trys 5–7.
6 Solve Applications Involving Linear Inequalities	• Learn the phrases used with the different inequality symbols. • Follow the procedure for solving applied problems. • Complete the given example on your own. • Complete You Try 8.

W Work   **Read the explanations, follow the examples, take notes, and complete the You Trys.**

Recall the inequality symbols

$<$ "is less than" $\leq$ "is less than or equal to"

$>$ "is greater than" $\geq$ "is greater than or equal to"

We will use the symbols to form *linear inequalities in one variable*.

While an equation states that two expressions are equal, an *inequality* states that two expressions are not necessarily equal. Here is a comparison of an equation and an inequality:

Equation	Inequality
$3x - 8 = 13$	$3x - 8 \leq 13$

Definition

A **linear inequality in one variable** can be written in the form $ax + b < c$, $ax + b \leq c$, $ax + b > c$, or $ax + b \geq c$ where a, b, and c are real numbers and $a \neq 0$.

The solution to a linear inequality is a set of numbers that can be represented in one of three ways:

1) On a graph

2) In *set notation*

3) In *interval notation*

In this section, we will learn how to solve linear inequalities in one variable and how to represent the solution in each of those three ways.

1 Use Graphs and Set and Interval Notations

EXAMPLE 1

Graph each inequality and express the solution in set notation and interval notation.

a) $x \leq -1$ b) $t > 4$

Solution

a) $x \leq -1$:

Graphing $x \leq -1$ means that we are finding the solution set of $x \leq -1$. What value(s) of x will make the inequality true? The largest solution is -1. Then, any number *less than* -1 will make $x \leq -1$ true. We represent this **on the number line** as follows:

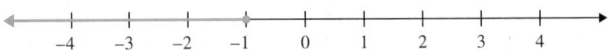

The graph illustrates that the solution is the set of all numbers less than and including -1.

    Notice that the dot on -1 is shaded. This tells us that -1 is included in the solution set. The shading to the left of -1 indicates that *any* real number (not just integers) in this region is a solution.

    We can express the solution set in **set notation** this way: $\{x \mid x \leq -1\}$.

In **interval notation** we write $(-\infty, -1]$

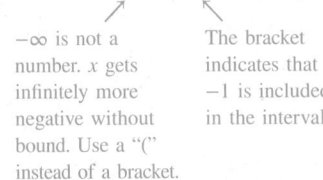

> **Note**
> The variable does not appear anywhere in interval notation.

b) $t > 4$:

We will plot 4 as an *open circle* on the number line because the symbol is ">" and *not* "$\geq$." The inequality $t > 4$ means that we must find the set of all numbers, t, *greater than* (but *not* equal to) 4. Shade to the right of 4.

 The graph illustrates that the solution is the set of all numbers greater than 4 but not including 4.

> **W Hint**
> Remember that we will never include ∞ or $-\infty$ in the solution set. Therefore, ∞ and $-\infty$ will always get a parenthesis when writing the solution set in interval notation.

    We can express the solution set in *set notation* this way: $\{t \mid t > 4\}$

In *interval notation* we write

$$(4, \infty)$$

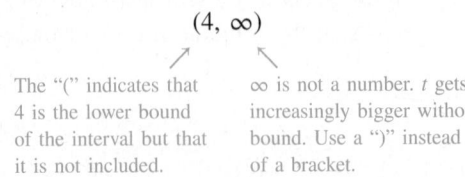

Note

Hints for using interval notation:
1) The variable never appears in interval notation.
2) A number *included* in the solution set gets a bracket: $x \le -1 \to (-\infty, -1]$
3) A number *not included* in the solution set gets a parenthesis:
   $t > 4 \to (4, \infty)$
4) The symbols $-\infty$ and ∞ *always* get parentheses.
5) The smaller number is always placed to the left. The larger number is placed to the right.
6) Even if we are not asked to graph the solution set, the graph may be helpful in writing the interval notation correctly.

[YOU TRY 1]

Graph each inequality, and express the solution in interval notation.

a) $k \ge -7$ b) $c < 5$

Another way to graph an inequality on a number line is to use the bracket [ ] and parentheses ( ) notation that we use for interval notation. A bracket indicates that the endpoint *is* included in the solution, and a parenthesis indicates that the endpoint is *not* included in the solution. Therefore, we can also graph the inequalities in Example 1 as follows:

$x \le -1$

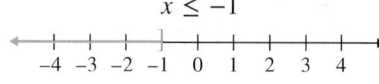

$t > 4$

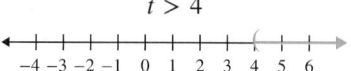

W Hint

Be sure that you draw the brackets and parentheses carefully on your number lines so that you *and your professor* can tell the difference between them!

Note

Because both methods (open/closed circles and brackets/parentheses) are commonly used on number lines, we must choose one to use in this book. We will use open and closed circles, but be aware of the equivalent number lines using brackets and parentheses.

2 Solve Inequalities Using the Addition and Subtraction Properties of Inequality

The addition and subtraction properties of equality help us to solve equations. Similar properties hold for inequalities as well.

W Hint

Notice that whenever you *add* or *subtract* numbers to both sides of an inequality, the direction of the inequality is unchanged.

Property Addition and Subtraction Properties of Inequality

Let a, b, and c be real numbers. Then,

1) $a < b$ and $a + c < b + c$ are equivalent

and

2) $a < b$ and $a - c < b - c$ are equivalent.

Adding the same number to both sides of an inequality or subtracting the same number from both sides of an inequality will not change the solution. These properties hold for any of the inequality symbols.

EXAMPLE 2

Solve $y - 8 \geq -5$. Graph the solution set, and write the answer in interval and set notations.

Solution

$$y - 8 \geq -5$$
$$y - 8 + 8 \geq -5 + 8 \qquad \text{Add 8 to each side.}$$
$$y \geq 3$$

The solution set in interval notation is $[3, \infty)$. In set notation we write $\{y | y \geq 3\}$.

[YOU TRY 2]

Solve $k - 10 \geq -4$. Graph the solution set, and write the answer in interval and set notations.

---

3 Solve Inequalities Using the Multiplication Property of Inequality

While the addition and subtraction properties for solving equations and inequalities work the same way, this is not true for multiplication and division. Let's see why.

Begin with an inequality we know is true: $2 < 5$. Multiply both sides by a *positive* number, say 3.

$$2 < 5 \qquad \text{True}$$
$$3(2) < 3(5) \qquad \text{Multiply by 3.}$$
$$6 < 15 \qquad \text{True}$$

Begin again with $2 < 5$. Multiply both sides by a *negative* number, say -3.

$$2 < 5 \qquad \text{True}$$
$$-3(2) < -3(5) \qquad \text{Multiply by } -3.$$
$$-6 < -15 \qquad \text{False}$$

To make $-6 < -15$ into a *true* statement, we must *reverse the direction of the inequality symbol.*

$$-6 > -15 \qquad \text{True}$$

If you begin with a true inequality and *divide* by a positive number or by a negative number, the results will be the same as above since division can be defined in terms of multiplication. This leads us to the multiplication property of inequality.

Property Multiplication Property of Inequality

Let a, b, and c be real numbers.

1) If c is a *positive* number, then $a < b$ and $ac < bc$ are equivalent inequalities and have the same solutions.

2) If c is a *negative* number, then $a < b$ and $ac > bc$ are equivalent inequalities and have the same solutions.

It is also true that if $c > 0$ and $a < b$, then $\dfrac{a}{c} < \dfrac{b}{c}$. If $c < 0$ and $a < b$, then $\dfrac{a}{c} > \dfrac{b}{c}$.

For the most part, the procedures used to solve linear inequalities are the same as those for solving linear equations **except** *when you multiply or divide an inequality by a negative number, you must reverse the direction of the inequality symbol.*

EXAMPLE 3

Solve each inequality. Graph the solution set, and write the answer in interval and set notations.

a) $-5w \le 20$ b) $5w \le -20$

Solution

a) $-5w \le 20$

First, divide each side by -5. *Since we are dividing by a negative number, we must remember to reverse the direction of the inequality symbol.*

$$-5w \le 20$$
$$\frac{-5w}{-5} \ge \frac{20}{-5} \qquad \text{Divide by } -5, \text{ so reverse the inequality symbol.}$$
$$w \ge -4$$

W Hint

When solving an inequality, remember that we must reverse the direction of the inequality symbol whenever we multiply or divide by a negative number!

Interval notation: $[-4, \infty)$

Set notation: $\{w | w \ge -4\}$

b) $5w \le -20$

First, divide by 5. Since we are dividing by a *positive* number, the inequality symbol remains the same.

$$5w \le -20$$
$$\frac{5w}{5} \le \frac{-20}{5} \qquad \text{Divide by 5. Do } not \text{ reverse the inequality symbol.}$$
$$w \le -4$$

Interval notation: $(-\infty, -4]$

Set notation: $\{w | w \le -4\}$

$\left[\text{YOU TRY 3}\right]$

Solve $-\dfrac{1}{4}m < 3$. Graph the solution set, and write the answer in interval and set notations.

4 Solve Inequalities Using a Combination of the Properties

Often it is necessary to combine the properties to solve an inequality.

EXAMPLE 4

Solve $4(5 - 2d) + 11 < 2(d + 3)$. Graph the solution set, and write the answer in interval and set notations.

Solution

$$4(5 - 2d) + 11 < 2(d + 3)$$
$$20 - 8d + 11 < 2d + 6 \qquad \text{Distribute.}$$
$$31 - 8d < 2d + 6 \qquad \text{Combine like terms.}$$

$$31 - 8d - 2d < 2d - 2d + 6 \qquad \text{Subtract } 2d \text{ from each side.}$$
$$31 - 10d < 6$$
$$31 - 31 - 10d < 6 - 31 \qquad \text{Subtract 31 from each side.}$$
$$-10d < -25$$
$$\frac{-10d}{-10} > \frac{-25}{-10} \qquad \begin{array}{l}\text{Divide both sides by } -10.\\ \text{Reverse the inequality symbol.}\end{array}$$
$$d > \frac{5}{2} \qquad \text{Simplify.}$$

W Hint

In this example, notice *when* the direction of the inequality symbol changed.

To graph the inequality, think of $\frac{5}{2}$ as $2\frac{1}{2}$.

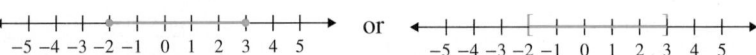

Interval notation: $\left(\dfrac{5}{2}, \infty\right)$. Set notation: $\left\{d \,\middle|\, d > \dfrac{5}{2}\right\}$.

[YOU TRY 4] Solve $4(p + 2) + 1 > 2(3p + 10)$. Graph the solution set and write the answer in interval and set notations.

---

5 Solve Compound Inequalities Containing Three Parts

A **compound inequality** contains more than one inequality symbol. Some types of compound inequalities are

$$-5 < b + 4 < 1, \qquad t \leq \frac{1}{2} \ \text{ or } \ t \geq 3, \qquad \text{and} \qquad 2z + 9 < 5 \text{ and } z - 1 > 6$$

W Hint

When the variable term is between two numbers, both inequality signs are pointing in the same direction.

In this section, we will learn how to solve the first type of compound inequality, also called a **three-part inequality**. In Section 3.8 we will discuss the last two.

Consider the inequality $-2 \leq x \leq 3$. We can think of this in two ways:

1) x is *between* -2 and 3, and -2 and 3 are included in the interval.

or

2) We can break up $-2 \leq x \leq 3$ into the two inequalities $-2 \leq x$ *and* $x \leq 3$.

Either way we think about $-2 \leq x \leq 3$, the meaning is the same. On a number line, the inequality would be represented as

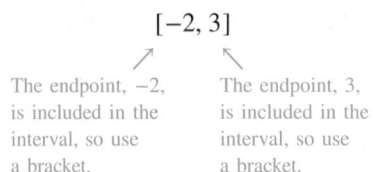

Notice that the **lower bound** of the interval on the number line is -2 (including -2), and the **upper bound** is 3 (including 3). Therefore, we can write the interval notation as

$$[-2, 3]$$

The endpoint, -2, is included in the interval, so use a bracket.   The endpoint, 3, is included in the interval, so use a bracket.

The set notation to represent $-2 \leq x \leq 3$ is $\{x | -2 \leq x \leq 3\}$.

Next, we will solve the inequality $-5 < b + 4 < 1$. To solve a three-part inequality you must remember that *whatever operation you perform on one part of the inequality must be performed on all three parts.* All properties of inequalities apply.

EXAMPLE 5

Solve $-5 < b + 4 < 1$. Graph the solution set, and write the answer in interval notation.

Solution

$$-5 < b + 4 < 1$$
$$-5 - 4 < b + 4 - 4 < 1 - 4 \qquad \text{Subtract 4 from each part of the inequality.}$$
$$-9 < b < -3$$

The graph of the solution set is ← |—|—|—◇—|—|—|—|—|—|—◇—|—|—| → Every real number
$\quad\quad\quad\quad\quad\quad\quad\quad\quad\quad\quad$ $-12\ -11\ -10\ -9\ -8\ -7\ -6\ -5\ -4\ -3\ -2\ -1\ \ 0\ \ 1$

in the shaded region makes the original inequality true. In interval notation, we write $(-9, -3)$.

Note

Use parentheses in the interval notation because -9 and -3 are not included in the solution set.

YOU TRY 5

Solve $-2 \le 7k - 9 \le 19$. Graph the solution set, and write the answer in interval notation.

We can eliminate fractions in an inequality by multiplying by the LCD of all of the fractions.

EXAMPLE 6

Solve $-\dfrac{7}{3} < \dfrac{1}{2}y - \dfrac{1}{3} \le \dfrac{1}{2}$. Graph the solution set, and write the answer in interval notation.

Solution

The LCD of the fractions is 6. Multiply by 6 to eliminate the fractions.

$$-\dfrac{7}{3} < \dfrac{1}{2}y - \dfrac{1}{3} \le \dfrac{1}{2}$$

$$6\left(-\dfrac{7}{3}\right) < 6\left(\dfrac{1}{2}y - \dfrac{1}{3}\right) \le 6\left(\dfrac{1}{2}\right) \qquad \text{Multiply all parts of the inequality by 6.}$$

$$-14 < 3y - 2 \le 3$$

$$-14 + 2 < 3y - 2 + 2 \le 3 + 2 \qquad \text{Add 2 to each part.}$$

$$-12 < 3y \le 5 \qquad \text{Combine like terms.}$$

$$-\dfrac{12}{3} < \dfrac{3y}{3} \le \dfrac{5}{3} \qquad \text{Divide each part by 3.}$$

$$-4 < y \le \dfrac{5}{3} \qquad \text{Simplify.}$$

W Hint

Remember, we can eliminate fractions in inequalities just like we do in equations!

$\quad\quad\quad\quad\quad\quad\quad\quad\quad\quad\quad\quad$ $\frac{5}{3}$
← |—|—◇—|—|—|—|—|—|—●—|—|—| → $\quad$ Interval notation: $\left(-4, \dfrac{5}{3}\right]$
$-6\ -5\ -4\ -3\ -2\ -1\ \ 0\ \ 1\ \ 2\ \ 3\ \ 4\ \ 5\ \ 6$

YOU TRY 6

Solve $-\dfrac{3}{4} < \dfrac{1}{3}z - \dfrac{3}{4} \le \dfrac{5}{4}$. Graph the solution set, and write the answer in interval notation.

Remember, if we multiply or divide an inequality by a negative number, we reverse the direction of the inequality symbol. When solving a compound inequality like these, reverse *both* symbols.

EXAMPLE 7

Solve $11 < -3x + 2 < 17$. Graph the solution set, and write the answer in interval notation.

Solution

$$11 < -3x + 2 < 17$$
$$11 - 2 < -3x + 2 - 2 < 17 - 2 \qquad \text{Subtract 2 from each part.}$$
$$9 < -3x < 15$$
$$\frac{9}{-3} > \frac{-3x}{-3} > \frac{15}{-3} \qquad \text{When we divide by a negative number, reverse the direction of the inequality symbol.}$$
$$-3 > x > -5 \qquad \text{Simplify.}$$

W Hint

In this example, notice *when* the direction of inequality changed.

Think carefully about what $-3 > x > -5$ means. It means "x is less than -3 *and* x is greater than -5." This is especially important to understand when writing the correct interval notation.

The graph of the solution set is

$$\overset{\longleftarrow\,+\,+\,+\,\diamond\,+\,+\,\diamond\,+\,+\,+\,+\,+\,+\,\longrightarrow}{\scriptstyle -7\ -6\ -5\ -4\ -3\ -2\ -1\ \ 0\ \ 1\ \ 2\ \ 3}$$

Even though we got $-3 > x > -5$ as our result, -5 is actually the lower bound of the solution set and -3 is the upper bound. The inequality $-3 > x > -5$ can also be written as $-5 < x < -3$.

The solution in interval notation is $(-5, -3)$.

↑ ↑
Lower bound on the left   Upper bound on the right

[YOU TRY 7]

Solve $4 < -2x - 4 < 10$. Graph the solution set, and write the answer in interval notation.

6 Solve Applications Involving Linear Inequalities

W Hint

Make a note of these four phrases. This may help you remember which inequality symbol to use when doing your homework.

Certain phrases in applied problems indicate the use of inequality symbols:

at least:	$\geq$	no less than:	$\geq$
at most:	$\leq$	no more than:	$\leq$

There are others. Next, we will look at an example of a problem involving the use of an inequality symbol. We will use the same steps that were used to solve applications involving equations.

EXAMPLE 8

Joe Amici wants to have his son's birthday party at Kiddie Fun Factory. The cost of a party is $175 for the first 10 children plus $3.50 for each additional child. If Joe can spend at most $200, find the greatest number of children who can attend the party.

Solution

Step 1: **Read** the problem carefully. We must find the greatest number of children who can attend the party.

Pressmaster/Shutterstock

Step 2: **Choose a variable** to represent the unknown quantity. We know that the first 10 children will cost $175, but we do not know how many *additional* guests Joe can afford to invite.

x = number of children **over** the first 10 who attend the party

Step 3: **Translate** from English to an algebraic inequality.

English:	Cost of first 10 children	+	Cost of additional children	is at most	$200
	↓		↓	↓	↓
Inequality:	175	+	3.50x	≤	200

The inequality is $175 + 3.50x \leq 200$.

Step 4: **Solve** the inequality.

$$175 + 3.50x \leq 200$$
$$3.50x \leq 25 \qquad \text{Subtract 175.}$$
$$x \leq 7.142 \ldots \qquad \text{Divide by 3.50.}$$

Step 5: **Check** the answer, and **interpret** the solution as it relates to the problem.

The result was $x \leq 7.142 \ldots$, where x represents the number of additional children who can attend the party. Since it is not possible to have 7.142 … people and $x \leq 7.142 \ldots$, in order to stay within budget, Joe can afford to pay for at most 7 additional guests *over* the initial 10.

Therefore, the greatest number of people who can attend the party is

The first 10	+	additional	=	total
↓		↓		↓
10	+	7	=	17

At most, 17 children can attend the birthday party. Does the answer make sense?

$$\text{Total Cost of Party} = \$175 + \$3.50(7)$$
$$= \$175 + \$24.50$$
$$= \$199.50$$

We can see that one more guest (at a cost of $3.50) would put Joe over budget.

[YOU TRY 8]

For $4.00 per month, Van can send or receive 200 text messages. Each additional message costs $0.05. If Van can spend at most $9.00 per month on text messages, find the greatest number he can send or receive each month.

ANSWERS TO [YOU TRY] EXERCISES

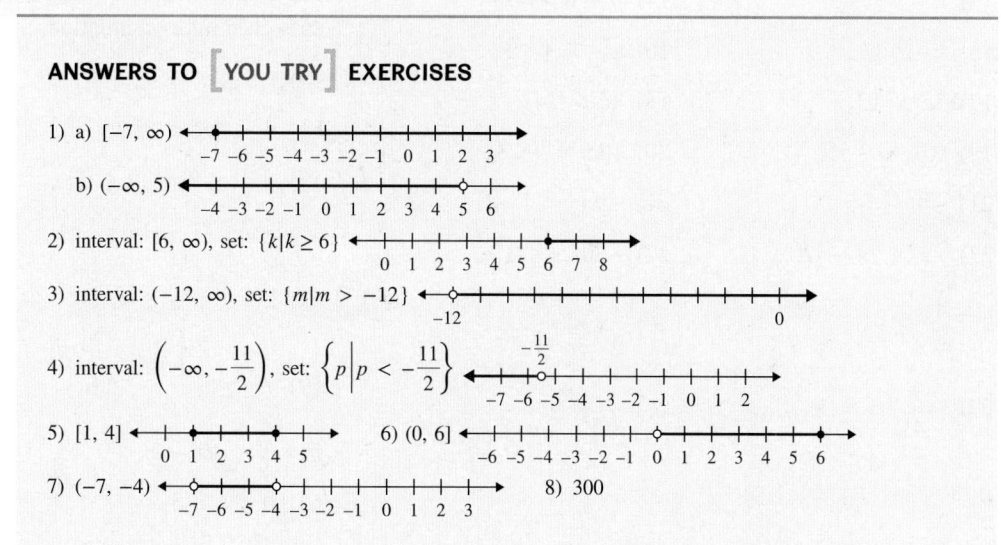

1) a) $[-7, \infty)$
   b) $(-\infty, 5)$
2) interval: $[6, \infty)$, set: $\{k | k \geq 6\}$
3) interval: $(-12, \infty)$, set: $\{m | m > -12\}$
4) interval: $\left(-\infty, -\frac{11}{2}\right)$, set: $\left\{p \middle| p < -\frac{11}{2}\right\}$
5) $[1, 4]$ 6) $(0, 6]$
7) $(-7, -4)$ 8) 300

Objective 1: Use Graphs and Set and Interval Notations

1) When do you use parentheses when writing a solution set in interval notation?

2) When do you use brackets when writing a solution set in interval notation?

Write each set of numbers in interval notation.

3)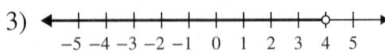
   $-5\ -4\ -3\ -2\ -1\ \ 0\ \ 1\ \ 2\ \ 3\ \ 4\ \ 5$

4) $-4\ -3\ -2\ -1\ \ 0\ \ 1\ \ 2\ \ 3\ \ 4$

5) $-4\ -3\ -2\ -1\ \ 0\ \ 1\ \ 2\ \ 3\ \ 4$

6) $-4\ -3\ -2\ -1\ \ 0\ \ 1\ \ 2\ \ 3\ \ 4$

Graph the inequality. Express the inequality in
a) set notation and b) interval notation.

7) $x \geq 3$

8) $t \geq -4$

9) $c < -1$

10) $r < \dfrac{5}{2}$

11) $w > -\dfrac{11}{3}$

12) $p \leq 2$

Mixed Exercises: Objectives 2 and 3

Solve each inequality. Graph the solution set and write the answer in a) set notation and b) interval notation. See Examples 2 and 3.

13) $r - 9 \leq -5$

14) $p + 6 \geq 4$

15) $y + 5 \geq 1$

16) $n - 8 \leq -3$

17) $3c > 12$

18) $8v > 24$

19) $15k < -55$

20) $16m < -28$

21) $-4b \leq 32$

22) $-9a \geq 27$

23) $-14w > -42$

24) $-30t < -18$

25) $\dfrac{1}{5}z \geq -3$

26) $\dfrac{1}{3}x < -2$

27) $-\dfrac{9}{4}y < -18$

28) $-\dfrac{2}{5}p \geq 4$

Objective 4: Solve Inequalities Using a Combination of the Properties

Solve each inequality. Graph the solution set and write the answer in interval notation. See Example 4.

29) $8z + 19 > 11$

30) $5x - 2 \leq 18$

31) $12 - 7t \geq 15$

32) $-1 - 4p < 5$

33) $-23 - w < -20$

34) $16 - h \geq 9$

35) $6(7y + 4) - 10 > 2(10y + 13)$

36) $7a + 4(5 - a) \leq 4 - 5a$

37) $9c + 17 > 14c - 3$

38) $-11n + 6 \leq 16 - n$

39) $\dfrac{8}{3}(2k + 1) > \dfrac{1}{6}k + \dfrac{8}{3}$

40) $\dfrac{1}{2}(c - 3) + \dfrac{3}{4}c \geq \dfrac{1}{2}(2c + 3) + \dfrac{3}{8}$

41) $0.04x + 0.12(10 - x) \geq 0.08(10)$

42) $0.09m + 0.05(8) \leq 0.07(m + 8)$

Objective 5: Solve Compound Inequalities Containing Three Parts

Graph the inequality. Express the inequality in
a) set notation and b) interval notation.

43) $1 \leq n \leq 4$

44) $-3 \leq g \leq 2$

45) $-2 < a < 1$

46) $-4 < d < 0$

47) $\dfrac{1}{2} < z \leq 3$

48) $-2 \leq y < 3$

Solve each inequality. Graph the solution set and write the answer in interval notation. See Examples 5–7.

49) $-8 \leq a - 5 \leq -4$

50) $1 \leq t + 3 \leq 7$

51) $9 < 6n < 18$

52) $-10 < 2x < 7$

53) $-19 \leq 7p + 9 \leq 2$

54) $-5 \leq 3k - 11 \leq 4$

55) $-6 \leq 4c - 13 < -1$

56) $-11 < 6m + 1 \leq -3$

57) $2 < \dfrac{3}{4}u + 8 < 11$

58) $2 \leq \dfrac{5}{2}y - 3 \leq 7$

59) $-\dfrac{1}{2} \leq \dfrac{5d + 2}{6} \leq 0$

60) $2 < \dfrac{2b + 7}{3} < 5$

61) $3 < 19 - 2j \leq 9$

62) $-13 \leq 14 - 9h < 5$

63) $0 \leq 4 - 3w \leq 7$

64) $-6 < -5 - z < 0$

Mixed Exercises: Objectives 2–5

Solve each inequality. Write the answer in interval notation.

65) $k + 11 > 4$

66) $5 < x + 9 < 12$

67) $-12p \geq -16$

68) $2w + 7 \geq 13$

69) $5(2b - 3) - 7b > 5b + 9$

70) $8 - m < 14$

71) $-12 < \dfrac{8}{5}t + 12 \leq 6$

72) $0.29 \geq 0.04a + 0.05$

73) $\dfrac{5}{4}(k + 4) + \dfrac{1}{4} \geq \dfrac{5}{6}(k + 3) - 1$

74) $-3 \leq 6c - 1 \leq 5$

75) $4 < 4 - 7y \leq 18$

76) $9z \leq -18$

Objective 6: Solve Applications Involving Linear Inequalities

Write an inequality for each problem and solve. See Example 8.

 77) Carson's Parking Garage charges $4.00 for the first 3 hr plus $1.50 for each additional half-hour. Ted has only $11.50 for parking. For how long can Ted park his car in this garage?

78) Oscar makes a large purchase at Home Depot and plans to rent one of its trucks to take his supplies home. The most he wants to spend on the truck rental is $50.00. If Home Depot charges $19.00 for the first 75 min and $5.00 for each additional 15 min, for how long can Oscar keep the truck and remain within his budget? (www.homedepot.com)

79) A taxi charges $2.00 plus $0.25 for every $\dfrac{1}{5}$ of a mile. How many miles can you go if you have $12.00?

John Foxx/Getty Images

80) A taxi charges $2.50 plus $0.20 for every $\dfrac{1}{4}$ of a mile. How many miles can you go if you have $12.50?

81) Melinda's first two test grades in Psychology were 87 and 94. What does she need to make on the third test to maintain an average of at least 90?

82) Russell's first three test scores in Geography were 86, 72, and 81. What does he need to make on the fourth test to maintain an average of at least 80?

R Rethink

R1) Which exercises in this section do you find most challenging?

R2) When must you switch the direction of the inequality?

R3) If $a < b < c$, then why must it be true that $c > b > a$?

R4) What kind of paid services do you use that place a limit on the number of times you use the service? If there is a cost for going over the limit, how much is it?

3.8 Compound Inequalities in One Variable

P Prepare

O Organize

What are your objectives for Section 3.8?	How can you accomplish each objective?
1 Find the Intersection and Union of Two Sets	<ul><li>Understand the meaning of *union* and *intersection*.</li><li>Know the symbols used to indicate *union* and *intersection*.</li><li>Complete the given example on your own.</li><li>Complete You Try 1.</li></ul>
2 Solve Compound Inequalities Containing the Word *And*	<ul><li>Learn the procedure for **Solving a Compound Inequality Containing *and*.**</li><li>Complete the given examples on your own.</li><li>Complete You Trys 2 and 3.</li></ul>
3 Solve Compound Inequalities Containing the Word *Or*	<ul><li>Follow the procedure for **Solving a Compound Inequality Containing *or*.**</li><li>Complete the given example on your own.</li><li>Complete You Try 4.</li></ul>
4 Solve Special Compound Inequalities	<ul><li>Review the meaning of *union* and *intersection*.</li><li>Know the procedure for **Solving a Compound Inequality.**</li><li>Complete the given example on your own.</li><li>Complete You Try 5.</li></ul>
5 Application of Intersection and Union	<ul><li>Understand the meaning of *and* and *or* in this type of application.</li><li>Complete the given example on your own.</li><li>Complete You Try 6.</li></ul>

W Work

Read the explanations, follow the examples, take notes, and complete the You Trys.

In Section 3.7, we learned how to solve a compound inequality like $-8 \leq 3x + 4 \leq 13$. In this section, we will discuss how to solve compound inequalities like these:

$$t \leq \frac{1}{2} \quad \text{or} \quad t \geq 3 \qquad \text{and} \qquad 2z + 9 < 5 \text{ and } z - 1 > 6$$

But first, we must talk about set notation and operations.

1 Find the Intersection and Union of Two Sets

EXAMPLE 1   Let $A = \{1, 2, 3, 4, 5, 6\}$ and $B = \{3, 5, 7, 9, 11\}$.

The **intersection** of sets A and B is the set of numbers that are elements of A **and** of B. The *intersection* of A and B is denoted by $A \cap B$.

$A \cap B = \{3, 5\}$ because 3 and 5 are found in both A and B.

The **union** of sets A and B is the set of numbers that are elements of A **or** of B. The *union* of A and B is denoted by $A \cup B$. The set $A \cup B$ consists of the elements in A *or* in B *or* in *both*.

$$A \cup B = \{1, 2, 3, 4, 5, 6, 7, 9, 11\}$$

Note

Although the elements 3 and 5 appear in both set A and in set B, we do not write them twice in the set $A \cup B$.

[**YOU TRY 1**] Let $A = \{2, 4, 6, 8, 10\}$ and $B = \{1, 2, 5, 6, 9, 10\}$. Find $A \cap B$ and $A \cup B$.

Note

The word "*and*" indicates *intersection*, while the word "*or*" indicates *union*. This same principle holds when solving compound inequalities involving "*and*" or "*or*."

2 Solve Compound Inequalities Containing the Word *And*

EXAMPLE 2

Solve the compound inequality $c + 5 \geq 3$ and $8c \leq 32$. Graph the solution set, and write the answer in interval notation.

Solution

Step 1: Identify the inequality as "*and*" or "*or*" and understand what that means. These two inequalities are connected by "*and*." That means the solution set will consist of the values of c that make *both* inequalities true. The solution set will be the *intersection* of the solution sets of $c + 5 \geq 3$ and $8c \leq 32$.

Step 2: Solve each inequality separately.

$$c + 5 \geq 3 \qquad \text{and} \qquad 8c \leq 32$$
$$c \geq -2 \qquad \text{and} \qquad c \leq 4$$

Step 3: Graph the solution set to each inequality on its own number line even if the problem does not require you to graph the solution set. This will help you visualize the solution set of the compound inequality.

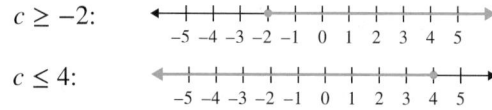

Step 4: Look at the number lines and think about where the solution set for the compound inequality would be graphed.

Since this is an "*and*" inequality, the solution set of $c + 5 \geq 3$ and $8c \leq 32$ consists of the numbers that are solutions to *both* inequalities. We can visualize it this way: if we take the number line above representing $c \geq -2$ and place it on top of the number line representing $c \leq 4$, what shaded areas would overlap (intersect)?

$$c \geq -2 \text{ and } c \leq 4: \qquad \xleftarrow{\;\;} \begin{array}{c} | \; | \; | \; | \; | \; | \; | \; | \; | \; | \; | \\ -5 \; -4 \; -3 \; -2 \; -1 \;\; 0 \;\; 1 \;\; 2 \;\; 3 \;\; 4 \;\; 5 \end{array} \xrightarrow{\;\;}$$

They intersect between -2 and 4, *including* those endpoints.

Step 5: Write the answer in interval notation.

The final number line illustrates that the solution to $c + 5 \geq 3$ and $8c \leq 32$ is $[-2, 4]$. The graph of the solution set is the final number line above.

Here are the steps to follow when solving a compound inequality.

Procedure Steps for Solving a Compound Inequality

Step 1: Identify the inequality as "*and*" or "*or*" and understand what that means.

Step 2: Solve each inequality separately.

Step 3: Graph the solution set to each inequality on its own number line even if the problem does not explicitly tell you to graph it. This will help you to visualize the solution to the compound inequality.

Step 4: Use the separate number lines to graph the solution set of the compound inequality.

    a) If it is an "*and*" inequality, the solution set consists of the regions on the separate number lines that would *overlap* (intersect) if one number line was placed on top of the other.

    b) If it is an "*or*" inequality, the solution set consists of the *total* (union) of what would be shaded if you took the separate number lines and put one on top of the other.

Step 5: Use the graph of the solution set to write the answer in interval notation.

[**YOU TRY 2**] Solve the compound inequality $y - 2 \leq 1$ and $7y > -28$. Graph the solution set, and write the answer in interval notation.

EXAMPLE 3 Solve the compound inequality $7y + 2 > 37$ and $5 - \dfrac{1}{3}y < 6$. Write the solution set in interval notation.

Solution

Step 1: This is an "*and*" inequality. The solution set will be the *intersection* of the solution sets of the separate inequalities $7y + 2 > 37$ and $5 - \dfrac{1}{3}y < 6$.

Step 2: We must solve each inequality separately.

$$7y + 2 > 37 \qquad \text{and} \qquad 5 - \frac{1}{3}y < 6$$

Subtract 2. $\quad 7y > 35 \qquad$ and $\qquad -\dfrac{1}{3}y < 1 \qquad$ Subtract 5, then multiply both sides by -3.

Divide by 7. $\quad\;\; y > 5 \qquad$ and $\qquad\;\; y > -3 \qquad$ Reverse the direction of the inequality symbol.

W Hint

Remember that the solution set to an "and" compound inequality is found by looking at the *intersection* of the solution sets of two different inequalities.

Step 3: Graph the solution sets separately so that it is easier to find their intersection.

$y > 5$:

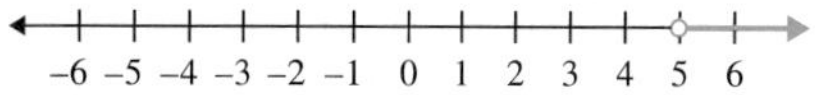

$y > -3$:

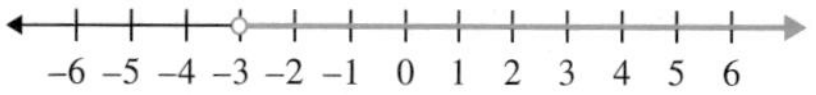

Step 4: If we were to put the number lines above on top of each other, where would they intersect?

$y > 5$ and $y > -3$:

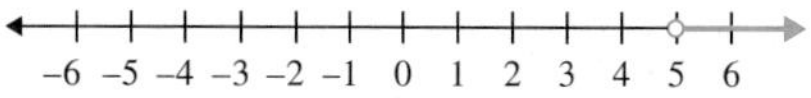

Step 5: The solution, shown in the shaded region in Step 4, is $(5, \infty)$.

[YOU TRY 3]

Solve each compound inequality and write the answer in interval notation.

a) $4x - 3 > 1$ and $x + 6 < 13$ b) $-\frac{4}{5}m > -8$ and $2m + 5 \leq 12$

3 Solve Compound Inequalities Containing the Word *Or*

Recall that the word "*or*" indicates the union of two sets.

EXAMPLE 4

Solve the compound inequality $6p + 5 \leq -1$ or $p - 3 \geq 1$. Write the answer in interval notation.

Solution

Step 1: These two inequalities are joined by "*or*." Therefore, the solution set will consist of the values of p that are in the solution set of $6p + 5 \leq -1$ *or* in the solution set of $p - 3 \geq 1$ *or* in *both* solution sets.

Step 2: Solve each inequality separately.

$$6p + 5 \leq -1 \quad \text{or} \quad p - 3 \geq 1$$
$$6p \leq -6$$
$$p \leq -1 \quad \text{or} \quad p \geq 4$$

W Hint

Remember that the solution set to an "or" compound inequality is found by looking at the *union* of the solution sets of two different inequalities.

Step 3: Graph the solution sets separately so that it is easier to find the *union* of the sets.

$p \leq -1$:

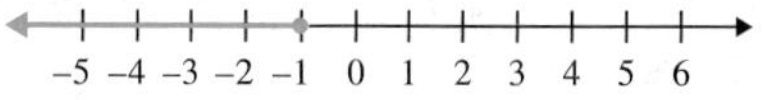

$p \geq 4$:

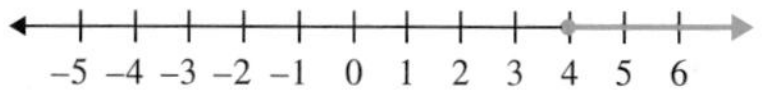

Step 4: The solution set of the compound inequality $6p + 5 \leq -1$ or $p - 3 \geq 1$ consists of the numbers which are solutions to the first inequality *or* the second inequality *or* both. We can visualize it this way: if we put the number lines on top of each other, the solution set of the compound inequality is the **total** (union) of what is shaded.

$p \leq -1$ or $p \geq 4$:

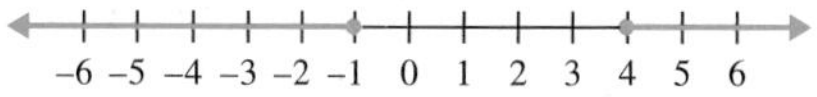

Step 5: The solution, shown in Step 4, is written as $(-\infty, -1] \cup [4, \infty)$.

↑ Use the *union* symbol for "or."

Solve $t + 8 \geq 14$ or $\dfrac{3}{2}t < 6$ and write the solution in interval notation.

4 Solve Special Compound Inequalities

EXAMPLE 5

Solve each compound inequality, and write the answer in interval notation.

a) $k - 5 < -2$ or $4k + 9 > 6$ b) $\dfrac{1}{2}w \geq 3$ and $1 - w \geq 0$

Solution

a) $k - 5 < -2$ or $4k + 9 > 6$

Step 1: The solution to this "*or*" inequality is the *union* of the solution sets of $k - 5 < -2$ and $4k + 9 > 6$.

Step 2: Solve each inequality separately.

$$k - 5 < -2 \qquad \text{or} \qquad 4k + 9 > 6$$
$$4k > -3$$
$$k < 3 \qquad \text{or} \qquad k > -\frac{3}{4}$$

Hint

Do you see why it is helpful to graph each inequality separately?

Step 3: $k < 3$:

$k > -\dfrac{3}{4}$:

Step 4: $k < 3$ or $k > -\dfrac{3}{4}$:

If the number lines in Step 3 were placed on top of each other, the *total* (union) of what would be shaded is the entire number line. This represents all real numbers.

Step 5: The solution set of the compound inequality is $(-\infty, \infty)$.

b) $\dfrac{1}{2}w \geq 3$ and $1 - w \geq 0$

Step 1: The solution to this "*and*" inequality is the *intersection* of the solution sets of $\dfrac{1}{2}w \geq 3$ and $1 - w \geq 0$.

Step 2: Solve each inequality separately.

$$\frac{1}{2}w \geq 3 \qquad \text{and} \qquad 1 - w \geq 0$$
$$1 \geq w \qquad \text{Add } w.$$
$$\text{Multiply by 2.} \qquad w \geq 6 \qquad \text{and} \qquad w \leq 1 \qquad \text{Rewrite } 1 \geq w \text{ as } w \leq 1.$$

Step 3: $w \geq 6$:

$w \leq 1$:

Step 4: $w \geq 6$ and $w \leq 1$:

If the number lines in Step 3 were placed on top of each other, the shaded regions would *not* intersect. Therefore, the solution set is the empty set, $\varnothing$.

Step 5: The solution set of $\dfrac{1}{2}w \geq 3$ and $1 - w \geq 0$ is $\varnothing$.

[YOU TRY 5]

Solve the compound inequalities, and write the solution in interval notation.

a) $-3w \leq w - 6$ and $5w < 4$ b) $9z - 8 \leq -8$ or $z + 7 \geq 2$

5 Application of Intersection and Union

EXAMPLE 6

The following table of selected NBA teams contains the number of times they have appeared in the play-offs as well as the number of NBA championships they have won through the 2017–2018 season.

Team	Play-Off Appearances	Championships
Boston Celtics	55	17
Chicago Bulls	35	6
Cleveland Cavaliers	22	1
Golden State Warriors	34	6
Los Angeles Lakers	49	11
New York Knicks	42	2

(www.basketball-reference.com)

List the elements of the set that satisfy the given information.

a) The set of teams with more than 20 play-off appearances and more than 5 championships

b) The set of teams with fewer than 30 play-off appearances or more than 8 championships

Solution

a) Because the two conditions in this statement are connected by *and*, we must find the team or teams that satisfy *both* conditions. The set of teams is

{Boston Celtics, Chicago Bulls, Los Angeles Lakers, Golden State Warriors}

b) Because the two conditions in this statement are connected by *or*, we must find the team or teams that satisfy either the first condition, *or* the second condition, *or* both. The set of teams is

{Boston Celtics, Cleveland Cavaliers, Los Angeles Lakers}

[YOU TRY 6]

Use the table in Example 6, and list the elements of the set that satisfy the given information.

a) The set of teams with fewer than 40 play-off appearances and at least 5 championships

b) The set of teams with more than 40 play-off appearances or one championship

---

ANSWERS TO [YOU TRY] EXERCISES

1) $A \cap B = \{2, 6, 10\}$, $A \cup B = \{1, 2, 4, 5, 6, 8, 9, 10\}$ 2) ![number line with open circle at -3 and filled circle at 3] $-5\ -4\ -3\ -2\ -1\ 0\ 1\ 2\ 3\ 4\ 5$

$(-4, 3]$ 3) a) $(1, 7)$ b) $\left(-\infty, \dfrac{7}{2}\right]$ 4) $(-\infty, 4) \cup [6, \infty)$ 5) a) $\varnothing$ b) $(-\infty, \infty)$

6) a) {Chicago Bulls, Golden State Warriors}

b) {Boston Celtics, Cleveland Cavaliers, Los Angeles Lakers, New York Knicks}

Objective 1: Find the Intersection and Union of Two Sets

1) Given sets A and B, explain how to find $A \cap B$.

2) Given sets X and Y, explain how to find $X \cup Y$.

Given sets $A = \{2, 4, 6, 8, 10\}$, $B = \{1, 3, 5\}$, $X = \{8, 10, 12, 14\}$, and $Y = \{5, 6, 7, 8, 9\}$ find

3) $A \cap X$

4) $X \cap Y$

5) $A \cup Y$

6) $B \cup Y$

7) $X \cap B$

8) $B \cap A$

9) $A \cup B$

10) $X \cup Y$

Each number line represents the solution set of an inequality. Graph the *intersection* of the solution sets and write the intersection in interval notation.

11) $x \geq -3$:

$x \leq 2$:

12) $n \leq 4$:

$n \geq 0$:

13) $t < 3$:

$t > -1$:

14) $y > -4$:

$y < -2$:

15) $c > 1$:

$c \geq 3$:

16) $p < 2$:

$p < -1$:

17) $z \leq 0$:

$z \geq 2$:

18) $g \geq -1$:

$g < -\dfrac{5}{2}$:

Mixed Exercises: Objectives 2 and 4

Solve each compound inequality. Graph the solution set, and write the answer in interval notation.

19) $a \leq 5$ and $a \geq 2$

20) $k > -3$ and $k < 4$

21) $b - 7 > -9$ and $8b < 24$

22) $3x \leq 1$ and $x + 11 \geq 4$

23) $5w + 9 \leq 29$ and $\dfrac{1}{3}w - 8 > -9$

24) $4y - 11 > -7$ and $\dfrac{3}{2}y + 5 \leq 14$

25) $2m + 15 \geq 19$ and $m + 6 < 5$

26) $d - 1 > 8$ and $3d - 12 < 4$

27) $r - 10 > -10$ and $3r - 1 > 8$

28) $2t - 3 \leq 6$ and $5t + 12 \leq 17$

29) $9 - n \leq 13$ and $n - 8 \leq -7$

30) $c + 5 \geq 6$ and $10 - 3c \geq -5$

Objective 1: Find the Intersection and Union of Two Sets

Each number line represents the solution set of an inequality. Graph the *union* of the solution sets and write the union in interval notation.

31) $p < -1$:

$p > 5$:

32) $z < 2$:

$z > 6$:

33) $a \leq \dfrac{5}{3}$:

$a > 4$:

34) $v \leq -3$:

$v \geq \dfrac{11}{4}$:

35) $y > 1$:

$y > 3$:

36) $x \leq -6$:

$x \leq -2$:

37) $c < \dfrac{7}{2}$:

$c \geq -2$:

38) $q \leq 3$:

$q > -2.7$:

Mixed Exercises: Objectives 3 and 4

Solve each compound inequality. Graph the solution set, and write the answer in interval notation.

39) $z < -1$ or $z > 3$

40) $x \le -4$ or $x \ge 0$

 41) $6m \le 21$ or $m - 5 > 1$

42) $a + 9 > 7$ or $8a \le -44$

43) $3t + 4 > -11$ or $t + 19 > 17$

44) $5y + 8 \le 13$ or $2y \le -6$

45) $-2v - 5 \le 1$ or $\dfrac{7}{3}v < -14$

46) $k - 11 < -4$ or $-\dfrac{2}{9}k \le -2$

47) $c + 3 \ge 6$ or $\dfrac{4}{5}c \le 10$

48) $\dfrac{8}{3}g \ge -12$ or $2g + 1 \le 7$

49) $7 - 6n \ge 19$ or $n + 14 < 11$

50) $d - 4 > -7$ or $-6d \le 2$

Mixed Exercises: Objectives 2–4

The following exercises contain *and* and *or* inequalities. Solve each inequality, and write the answer in interval notation.

51) $4n + 7 \le 9$ and $n + 6 \ge 1$

52) $8t - 5 \ge 11$ or $-\dfrac{2}{5}t \ge 6$

53) $\dfrac{4}{3}x + 5 < 2$ or $x + 3 \ge 8$

54) $p + 10 < -3$ and $8p - 7 > 11$

55) $\dfrac{8}{3}w < -16$ or $w + 9 > -4$

56) $5y - 2 > 8$ and $\dfrac{3}{4}y + 2 < 11$

57) $7 - r > 7$ and $0.3r < 6$

58) $2c - 9 \le -3$ or $10c + 1 \ge 7$

59) $3 - 2k > 11$ and $\dfrac{1}{2}k + 5 \ge 1$

60) $6 - 5a \ge 1$ or $0.8a > 8$

Objective 5: Application of Intersection and Union

The following table lists the net worth (in billions of dollars) of some of the wealthiest women in the world for the years 2014 and 2018. (www.forbes.com)

Name	Net Worth in 2014	Net Worth in 2018
Laurene Powell Jobs	14.0	18.8
Susanne Klatten	17.4	25.0
Jacqueline Mars	20.0	23.6
Sandra Ortega Mera	6.1	7.0
Alice Walton	34.3	46.0

List the elements of the set that satisfy the given information.

61) The set of women with a net worth more than $15 billion in 2014 and in 2018.

62) The set of women with a net worth more than $12 billion in 2014 and less than $20 billion in 2018.

63) The set of women with a net worth less than $10 billion in 2014 or more than $20 billion in 2018.

64) The set of women with a net worth more than $18 billion in 2014 or more than $28 billion in 2018.

R Rethink

R1) How does the *intersection* of two sets differ from the *union* of two sets?

R2) What is the mathematical symbol for the *union*, and what is the mathematical symbol for the *intersection* of two sets?

R3) How does the "and" compound inequality differ from the "or" compound inequality?

R4) Which exercises in this section do you find most challenging?

Group Activity – Percents

Your favorite store is having a sale. You run across a shirt that originally sold for $47, but is on a "40% OFF" rack. A pair of jeans that originally sold for $60 is advertised for "30% OFF." Your eye is attracted to a pair of designer shoes on sale for $35. You decide to purchase the shirt, jeans, and shoes.

1) What is the sale price of the shirt? What is the sale price of the jeans?

2) What is the subtotal of all your items?

3) You discover a coupon for 25% off a purchase of $50.00 or more and give it to the cashier. What is the pre-tax cost of your purchase?

4) If the sales tax is 5%, what is the total amount you are charged?

5) Your credit card has an annual percentage rate (APR) of 24% and a minimum payment of $10.00. If you make the minimum payment every month, use the following table to determine how long it will take to pay off your purchase.

Month	Balance	Payment	Remaining Balance	Interest	New Balance
1		$10.00			
2					
3					
4					
5					
6					
7					
8					
9					
10					

6) How much do you end up paying for your purchase if you use your credit card?

If you are like most college students, you have a lot of demands on your time. In addition to course obligations both inside and out of class, you may have a job, a family to care for, plus an active social life. Therefore, it's important that you make the most of your time.

To discover your personal time style, rate how well each of the statements below describes you. Use this rating scale:

1 = Doesn't describe me at all
2 = Describes me only slightly
3 = Describes me fairly well
4 = Describes me very well

	1	2	3	4
1. I often wake up later than I should.				
2. I am usually late for classes and appointments.				
3. I am always in a rush getting places.				
4. I put off big tasks and assignments until the last minute.				
5. My friends often comment on my lateness.				
6. I am easily interrupted, putting aside what I'm doing for something new.				
7. When I look at a clock, I'm often surprised at how late it is.				
8. I often forget appointments and have to reschedule them.				
9. When faced with a big task, I feel overwhelmed and turn my mind away from it until later.				
10. At the end of the day, I have no idea where the time went.				

Rate yourself by adding up the points you assigned yourself. Use this scale to assess your time style:

10–15 = Very efficient time user
16–20 = Efficient time user
21–30 = Time use needs work
31–40 = Victim of time

If you scored a 20 or higher, you are likely making your schedule even more hectic than it needs to be. If you make a point of paying attention to the clock and sticking to a schedule, you will likely feel less stressed and find it easier to accomplish your daily tasks.

Chapter 3: Summary

Definition/Procedure	Example

3.1 Solving Linear Equations Part I

The Addition and Subtraction Properties of Equality

1) If $a = b$, then $a + c = b + c$.
2) If $a = b$, then $a - c = b - c$.

Solve $3 + b = 20$

$$3 - 3 + b = 20 - 3 \quad \text{Subtract 3 from each side.}$$
$$b = 17$$

The solution set is $\{17\}$.

The Multiplication and Division Properties of Equality

1) If $a = b$, then $ac = bc$.
2) If $a = b$, then $\dfrac{a}{c} = \dfrac{b}{c}$ $(c \neq 0)$.

Solve $\dfrac{3}{5}m = 6$

$$\frac{5}{3} \cdot \frac{3}{5}m = \frac{5}{3} \cdot 6 \quad \text{Multiply each side by } \frac{5}{3}.$$
$$m = 10$$

The solution set is $\{10\}$.

How to Solve a Linear Equation

Step 1: **Clear parentheses** and **combine like terms** on each side of the equation.

Step 2: **Get the variable on one side of the equal sign and the constant on the other side of the equal sign** (isolate the variable) using the addition or subtraction property of equality.

Step 3: **Solve for the variable** using the multiplication or division property of equality.

Step 4: **Check the solution** in the original equation.

Solve $2(c + 2) + 11 = 5c + 9$.

$2c + 4 + 11 = 5c + 9$	Distribute.
$2c + 15 = 5c + 9$	Combine like terms.
$2c - 5c + 15 = 5c - 5c + 9$	Get variable terms on one side.
$-3c + 15 = 9$	
$-3c = -6$	Get constants on one side.
$\dfrac{-3c}{-3} = \dfrac{-6}{-3}$	Division property of equality.
$c = 2$	

The solution set is $\{2\}$.

3.2 Solving Linear Equations Part II

Solve Equations Containing Fractions or Decimals

To eliminate the fractions, determine the least common denominator (LCD) for all of the fractions in the equation. Then, multiply both sides of the equation by the LCD.

To eliminate the decimals from an equation, multiply both sides of the equation by the smallest power of 10 that will eliminate all decimals from the problem.

Solve $\dfrac{3}{4}y - 3 = \dfrac{1}{4}y - \dfrac{2}{3}$

$12\left(\dfrac{3}{4}y - 3\right) = 12\left(\dfrac{1}{4}y - \dfrac{2}{3}\right)$	Multiply each side of the equation by 12.
$9y - 36 = 3y - 8$	Distribute.
$9y - 3y - 36 = 3y - 3y - 8$	Get the y-terms on one side.
$6y - 36 = -8$	
$6y - 36 + 36 = -8 + 36$	Get the constants on the other side.
$6y = 28$	
$\dfrac{6y}{6} = \dfrac{28}{6}$	Divide each side by 6.
$y = \dfrac{28}{6} = \dfrac{14}{3}$	Simplify.

The solution set is $\left\{\dfrac{14}{3}\right\}$.

Definition/Procedure	Example
Steps for Solving Applied Problems *Step 1:* **Read** and reread the problem. Draw a picture, if applicable. *Step 2:* **Choose a variable** to represent an unknown. Define other unknown quantities in terms of the variable. *Step 3:* **Translate** from English into math. *Step 4:* **Solve** the equation. *Step 5:* **Check** the answer in the original problem, and **interpret** the solution as it relates to the problem.	Nine less than twice a number is the same as the number plus thirteen. Find the number. *Step 1:* **Read** the problem carefully, then read it again. *Step 2:* **Choose a variable** to represent the unknown. $$x = \text{the number}$$ *Step 3:* "Nine less than twice a number is the same as the number plus thirteen" means $2x - 9 = x + 13$. *Step 4:* **Solve** the equation. $$2x - 9 = x + 13$$ $$2x - 9 + 9 = x + 13 + 9$$ $$2x = x + 22$$ $$x = 22$$ *Step 5:* The number is 22. The **check** is left to the student.

3.3 Applications of Linear Equations

The **Five Steps for Solving Applied Problems** can be used to solve problems involving general quantities, lengths, and consecutive integers.	The sum of three consecutive even integers is 72. Find the integers. *Step 1:* **Read** the problem carefully, then read it again. *Step 2:* **Choose** a variable, and define the unknowns. $$x = \text{the first even integer}$$ $$x + 2 = \text{the second even integer}$$ $$x + 4 = \text{the third even integer}$$ *Step 3:* "The sum of three consecutive even integers is 72" means First even $+$ Second even $+$ Third even $=$ 72 x $+$ $(x + 2)$ $+$ $(x + 4)$ $=$ 72 Equation: $x + (x + 2) + (x + 4) = 72$ *Step 4:* **Solve** $x + (x + 2) + (x + 4) = 72$ $$3x + 6 = 72$$ $$3x + 6 - 6 = 72 - 6$$ $$3x = 66$$ $$\frac{3x}{3} = \frac{66}{3}$$ $$x = 22$$ *Step 5:* Find the values of all the unknowns. $$x = 22, \quad x + 2 = 24, \quad x + 4 = 26$$ The numbers are 22, 24, and 26. **Check** to verify the sum is 72: $22 + 24 + 26 = 72$.

Definition/Procedure	Example

3.4 Applications Involving Percents

We can use an equation to solve many different types of applications involving percents.

Use the Five-Step Process to solve the problem.

A Russell Wilson wall decal is on sale for $17 after a 15% discount. Find the original price.

Solution

Step 1: **Read** the problem carefully, then read it again.

Step 2: **Choose** a variable to represent the unknown.
$$x = \text{the original price of the decal}$$

Step 3: **Translate** the information that appears in English into an equation.

Original price of the decal	−	Amount of the discount	=	Sale price of the decal
x	−	$0.15x$	=	17

Equation: $x - 0.15x = 17$

Step 4: **Solve** the equation.
$$x - 0.15x = 17$$
$$0.85x = 17$$
$$\frac{0.85x}{0.85} = \frac{17}{0.85}$$
$$x = 20$$

Step 5: The original price of the Russell Wilson wall decal was $20.

3.5 Geometry Applications and Solving Formulas

Formulas from geometry can be used to solve applications.

A rectangular bulletin board has an area of 180 in². It is 12 in. wide. Find its length.

Use $A = lw$.   Formula for the area of a rectangle

$A = 180 \text{ in}^2$, $w = 12$ in.   Find l.

$$A = lw$$
$$180 = l(12) \quad \text{Substitute values into } A = lw.$$
$$\frac{180}{12} = \frac{l(12)}{12}$$
$$15 = l$$

The length is 15 in.

To solve a formula for a specific variable, think about the steps involved in solving a linear equation in one variable.

Solve $C = kr - w$ for k.
$$C + w = \boxed{k}r - w + w \qquad \text{Add } w \text{ to each side.}$$
$$C + w = \boxed{k}r$$
$$\frac{C + w}{r} = \frac{\boxed{k}r}{r} \qquad \text{Divide each side by } r.$$
$$\frac{C + w}{r} = k$$

Definition/Procedure	Example

3.6 Applications of Linear Equations to Proportions, Money Problems, and $d = rt$

A **ratio** is a quotient of two quantities. The ratio of x to y, where $y \neq 0$, can be written as $\frac{x}{y}$, x to y, or $x : y$.

The ratio of 12 in. to 16 in. is $\frac{12 \text{ in.}}{16 \text{ in.}} = \frac{3}{4}$.   Write in lowest terms.

A **proportion** is a statement that two ratios are equal.

To solve a proportion, set the cross products equal to each other and solve the equation.

Solve $\dfrac{n-3}{4} = \dfrac{n+1}{5}$.

$$\frac{n-3}{4} \times \frac{n+1}{5}$$ Multiply.
   Multiply.

$5(n - 3) = 4(n + 1)$ Set the cross products equal.
$5n - 15 = 4n + 4$ Distribute.
$n - 15 = 4$ Subtract $4n$.
$n = 19$ Add 15.

The solution set is $\{19\}$.

We can use the Five-Step Method to solve problems involving proportions as well as other applications.

If Geri can watch 4 movies in 3 weeks, how long will it take her to watch 7 movies?

Step 1: **Read** the problem carefully twice.
Step 2: **Choose** a variable to represent the unknown.
$x =$ number of weeks to watch 7 movies

Step 3: Set up a proportion. $\dfrac{4 \text{ movies}}{3 \text{ weeks}} = \dfrac{7 \text{ movies}}{x \text{ weeks}}$

Equation: $\dfrac{4}{3} = \dfrac{7}{x}$

Step 4: **Solve** $\dfrac{4}{3} \times \dfrac{7}{x}$.   Multiply.
   Multiply.

$4x = 3(7)$ Set cross products equal.

$\dfrac{4x}{4} = \dfrac{21}{4}$ Divide by 4

$x = \dfrac{21}{4} = 5\dfrac{1}{4}$

Step 5: It will take Geri $5\dfrac{1}{4}$ weeks to watch 7 movies. The check is left to the student.

3.7 Linear Inequalities in One Variable

A **linear inequality in one variable** can be written in the form $ax + b < c$, $ax + b \leq c$, $ax + b > c$, or $ax + b \geq c$.

We solve linear inequalities in very much the same way we solve linear equations *except that when we multiply or divide by a negative number, we must reverse the direction of the inequality symbol.*

We can graph the solution set, write the solution in set notation, or write the solution in interval notation.

Solve $x - 9 \leq -7$. Graph the solution set, and write the answer in both set notation and interval notation.

$$x - 9 \leq -7$$
$$x - 9 + 9 < -7 + 9 \qquad \text{Add 9 to each side.}$$
$$x \leq 2$$

$\{x | x \leq 2\}$ Set notation
$(-\infty, 2]$ Interval notation

Definition/Procedure	Example
A **three-part inequality** states that one number is between two other numbers. To **solve a three-part inequality,** remember that each operation you perform on one part of the inequality must be performed on all three parts.	Solve $1 \leq 3k + 7 < 19$. $$1 \leq 3k + 7 < 19$$ $$1 - 7 \leq 3k + 7 - 7 < 19 - 7 \quad \text{Subtract 7.}$$ $$-6 \leq 3k < 12$$ $$\frac{-6}{3} \leq \frac{3k}{3} < \frac{12}{3} \quad \text{Divide by 3.}$$ $$-2 \leq k < 4$$ The graph of the solution set is −5 −4 −3 −2 −1 0 1 2 3 4 5 In interval notation, the solution is $[-2, 4)$.

3.8 Compound Inequalities in One Variable

The solution set of a compound inequality joined by **"and"** will be the **intersection** of the solution sets of the individual inequalities.	Solve the compound inequality $5x - 2 \geq -17$ and $x + 8 \leq 9$. $$5x - 2 \geq -17 \quad \text{and} \quad x + 8 \leq 9$$ $$5x \geq -15$$ $$x \geq -3 \quad \text{and} \quad x \leq 1$$ −4 −3 −2 −1 0 1 2 3 4 Solution in interval notation: $[-3, 1]$
The solution set of a compound inequality joined by **"or"** will be the **union** of the solution sets of the individual inequalities.	Solve the compound inequality $x - 3 < -1$ or $7x > 42$. $$x - 3 < -1 \quad \text{or} \quad 7x > 42$$ $$x < 2 \quad \text{or} \quad x > 6$$ 0 1 2 3 4 5 6 7 8 9 Solution in interval notation: $(-\infty, 2) \cup (6, \infty)$

Chapter 3: Review Exercises

(3.1 and 3.2) Determine whether the given value is a solution to the equation.

1) $\frac{3}{2}k - 5 = 1; \quad k = -4$

2) $5 - 2(3p + 1) = 9p - 2; \quad p = \frac{1}{3}$

3) How do you know that an equation has no solution?

4) What can you do to make it easier to solve an equation with fractions?

Solve each equation.

5) $h + 14 = -5$

6) $w - 9 = 16$

7) $-7g = 56$

8) $-0.78 = -0.6t$

9) $4 = \frac{c}{9}$

10) $-\frac{10}{3}y = 16$

11) $23 = 4m - 7$

12) $\frac{1}{6}v - 7 = -3$

13) $4c + 9 + 2(c - 12) = 15$

14) $\frac{5}{9}x + \frac{1}{6} = -\frac{3}{2}$

15) $2z + 11 = 8z + 15$

16) $8 - 5(2y - 3) = 14 - 9y$

17) $k + 3(2k - 5) = 4(k - 2) - 7$

18) $10 - 7b = 4 - 5(2b + 9) + 3b$

19) $0.18a + 0.1(20 - a) = 0.14(20)$

20) $16 = -\frac{12}{5}d$

21) $3(r + 4) - r = 2(r + 6)$

22) $\frac{1}{2}(n - 5) - 1 = \frac{2}{3}(n - 6)$

Write each statement as an equation, and find the number.

23) Nine less than twice a number is twenty-five.

24) One more than two-thirds of a number is the same as the number decreased by three.

(3.3) Solve using the five-step method.

25) Kendrick received 24 fewer emails on Friday than he did on Thursday. If he received a total of 126 emails on those two days, how many did he get on each day?

26) The number of Michael Jackson solo albums sold the week after his death was 42.2 times the number sold the previous week. If a total of 432,000 albums were sold during those two weeks, how many albums were sold the week after his death? (http://abcnews.go.com)

27) A plumber has a 36-in. pipe that he has to cut into two pieces so that one piece is 8 in. longer than the other. Find the length of each piece.

Corbis

28) The sum of three consecutive integers is 249. Find the integers.

(3.4) Solve using the five-step method.

29) Today's typical hip implant weighs about 50% less than it did 20 years ago. If an implant weighs about 3 lb today, how much did it weigh 20 years ago?

30) In Fall 2016, 31,861 students were enrolled at the University of Colorado–Boulder. This is approximately 4.2% less than the number of students enrolled in Fall 2017. How many students attended the university in Fall 2017? Round the answer to the nearest whole number. (www.colorado.edu)

31) Jose had $6000 to invest. He put some of it into a savings account earning 2% simple interest and the rest into an account earning 4% simple interest. If he earned $210 of interest in 1 year, how much did he invest in each account?

32) How many milliliters of a 10% hydrogen peroxide solution and how many milliliters of a 2% hydrogen peroxide solution should be mixed to obtain 500 mL of a 4% hydrogen peroxide solution?

(3.5) Substitute the given values into the formula, and solve for the remaining variable.

33) $P = 2l + 2w$; If $P = 32$ when $l = 9$, find w.

34) $V = \frac{1}{3}\pi r^2 h$; If $V = 60\pi$ when $r = 6$, find h.

Use a known formula to solve.

35) The base of a triangle measures 12 in. If the area of the triangle is 42 in², find the height.

36) The Statue of Liberty holds a tablet in her left hand that is inscribed with the date, in Roman numerals, that the Declaration of Independence was signed. The length of this rectangular tablet is 120 in. more than the width, and the perimeter of the tablet is 892 in. What are the dimensions of the tablet? (www.nps.gov)

37) Find the missing angle measures.

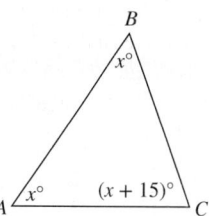

Find the measure of each indicated angle.

38)
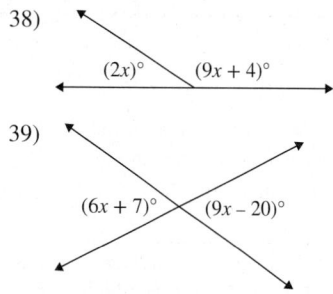

39)

Solve using the five-step method.

40) The sum of the supplement of an angle and twice the angle is 10° more than four times the measure of its complement. Find the measure of the angle.

Solve for the indicated variable.

41) $p - n = z$ for p

42) $r = ct + a$ for t

43) $A = \frac{1}{2}bh$ for b

44) $M = \frac{1}{4}k(d + D)$ for D

(3.6)

45) Can 15% be written as a ratio? Explain.

46) What is the difference between a ratio and a proportion?

47) Write the ratio of 12 girls to 15 boys in lowest terms.

48) A store sells olive oil in three different sizes. Which size is the best buy, and what is its unit price?

Size	Price
17 oz	$ 8.69
25 oz	$11.79
101 oz	$46.99

Solve each proportion.

49) $\frac{x}{15} = \frac{8}{10}$

50) $\frac{2c + 3}{6} = \frac{c - 4}{2}$

Set up a proportion and solve.

51) The 2017 Youth Risk Behavior Survey found that about 3 out of 10 high school students drank some amount of alcohol in the 30 days preceding the survey. If a high school has 2400 students, how many would be expected to have used alcohol within a 30-day period? (www.cdc.gov)

52) Given these two similar triangles, find x.

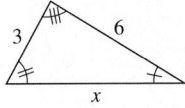 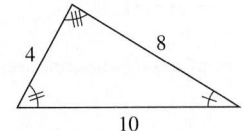

Solve using the five-step method.

53) Peter leaves Mitchell's house traveling 30 mph. Mitchell leaves 15 min later, trying to catch up to Peter, going 40 mph. If they drive along the same route, how long will it take Mitchell to catch Peter?

54) A collection of coins contains 91 coins, all dimes and quarters. If the value of the coins is $14.05, determine the number of each type of coin in the collection.

(3.7) Solve each inequality. Graph the solution set, and write the answer in interval notation.

55) $z + 6 \geq 14$

56) $-10y + 7 > 32$

57) $w + 8 > 5$

58) $-6k \leq 15$

59) $5x - 2 \leq 18$

60) $0.03c + 0.09(6 - c) > 0.06(6)$

61) $-15 < 4p - 7 \leq 5$

62) $-1 \leq \dfrac{5 - 3x}{2} \leq 0$

63) $3(3c + 8) - 7 > 2(7c + 1) - 5$

64) $-19 \leq 7p + 9 \leq 2$

65) $-3 < \dfrac{3}{4}a - 6 \leq 0$

66) $3 < -4t + 1 < 9$

67) **Write an inequality and solve.** Gia's scores on her first three History tests were 94, 88, and 91. What does she need to make on her fourth test to have an average of at least 90?

(3.8)

68) $A = \{10, 20, 30, 40, 50\}$ $B = \{20, 25, 30, 35\}$

   a) Find $A \cup B$.   b) Find $A \cap B$.

Solve each compound inequality. Graph the solution set, and write the answer in interval notation.

69) $a + 6 \leq 9$ and $7a - 2 \geq 5$

70) $3r - 1 > 5$ or $-2r \geq 8$

71) $8 - y < 9$ or $\dfrac{1}{10}y > \dfrac{3}{5}$

72) $x + 12 \leq 9$ and $0.2x \geq 3$

The following table lists the number of hybrid vehicles sold in the United States by certain manufacturers in April and May of 2018. (www.hybridcars.com)

Manufacturer	Number Sold in April	Number Sold in May
Chevrolet	242	291
Ford	3,847	5,570
Honda	631	2,598
Lexus	1,823	2,569
Toyota	12,443	14,402

List the elements of the set that satisfy the given information.

73) The set of manufacturers who sold more than 5000 hybrid vehicles in April and May

74) The set of manufacturers who sold more than 3000 hybrid vehicles in April or fewer than 1500 in May

Mixed Exercises: Solving Equations, Inequalities, and Applications

Solve each equation or inequality. Graph the solution set of an inequality, and also write it in interval notation.

75) $-8k + 13 = -7$

76) $-7 - 4(3w - 2) = 1 - 9w$

77) $29 = -\dfrac{4}{7}m + 5$

78) $\dfrac{c}{20} = \dfrac{18}{12}$

79) $-8m > 4$

80) $-3 \leq 5t + 2 < 7$

81) $10p + 11 = 5(2p + 3) - 1$

82) $0.14a + 0.06(36 - a) = 0.12(36)$

83) $3n - 17 > 5 + 2(5n + 3)$

84) $6 < 2v + 9 < 13$

85) $\dfrac{2x + 9}{5} = \dfrac{x + 1}{2}$

86) $14 = 8 - h$

87) $-1 \leq 9 - \dfrac{5}{3}y \leq 4$

88) $10 - z \leq 8$

89) $\dfrac{5}{6} - \dfrac{3}{4}(r + 2) = \dfrac{1}{2}r + \dfrac{7}{12}$

90) $\dfrac{1}{4}d + \dfrac{9}{4} = 1 + \dfrac{1}{4}(d + 5)$

Solve using the five-step procedure.

91) The sum of an angle and three times its complement is 33° more than its supplement. Find the measure of the angle.

92) A library offers free tutoring after school for children in grades 1–5. The number of students who attended on Friday was half the number who attended on Thursday. How many students came for tutoring each day if the total number of students served on both days was 42?

93) The sum of two consecutive odd integers is 21 less than three times the larger integer. Find the numbers.

94) The perimeter of a triangle is 35 cm. One side is 3 cm longer than the shortest side, and the longest side is twice as long as the shortest. How long is each side of the triangle?

95) A 2008 poll revealed that 9 out of 50 residents of Quebec, Canada, wanted to secede from the rest of the country. If 2000 people were surveyed, how many said they would like to see Quebec separate from the rest of Canada? (www.cbc.ca)

96) In 2017, there were 751,000 emotional support animals on U.S. commercial flights. This is approximately 56.1% more than in 2016. How many emotional support animals flew on U.S. commercial flights in 2016? Round the answer to the nearest thousand. (abcnews. go.com)

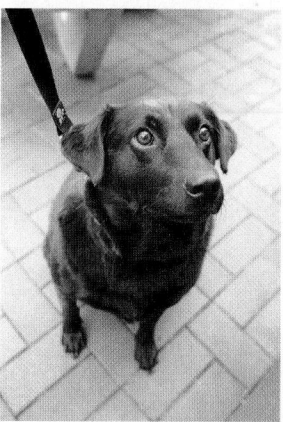

Holly Hildreth/McGraw-Hill Education

Chapter 3:   Test

Solve each equation.

1) $\frac{8}{3}n - 11 = 5$

2) $3c - 2 = 8c + 13$

3) $0.06x + 0.14(x - 5) = 0.10(23)$

4) $\frac{1}{2} - \frac{1}{6}(x - 5) = \frac{1}{3}(x + 1) + \frac{2}{3}$

5) $7(3k + 4) = 11k + 8 + 10k - 20$

6) $\frac{9 - w}{4} = \frac{3w + 1}{2}$

7) What is the difference between a ratio and a proportion?

Solve using the five-step method.

8) Ramon's tuition bill for Spring semester of 2020 was $174 more than his bill for the Fall 2019 semester. He paid a total of $2784 for both semesters. Find the amount of Ramon's tuition bill each semester.

9) The sum of three consecutive even integers is 114. Find the numbers.

10) Ray buys 14 gallons of gas and pays $40.60. His wife, Debra, goes to the same gas station later that day and buys 11 gallons of the same gasoline. How much did she spend?

11) The tray table on the back of an airplane seat is in the shape of a rectangle. It is 5 in. longer than it is wide and has a perimeter of 50 in. Find the dimensions.

12) Given the following similar triangles, find x.

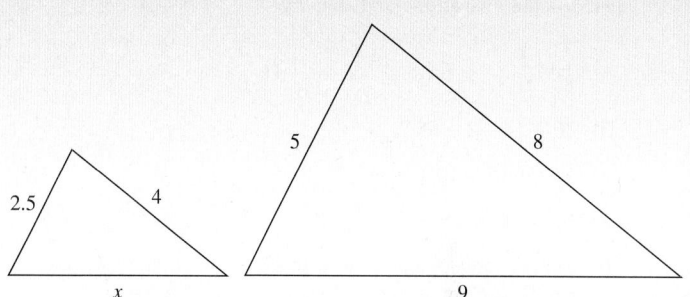

13) Motor oil is available in three types: regular, synthetic, and synthetic blend, which is a mixture of regular oil with synthetic. Bob decides to make 5 qt of his own synthetic blend. How many quarts of regular oil costing $1.20 per quart and how many quarts of synthetic oil costing $3.40 per quart should he mix so that the blend is worth $1.86 per quart?

14) In 2018, Wisconsin had 9 drive-in theaters. This is 82% less than the number of drive-ins in 1967. Determine the number of drive-in theaters in Wisconsin in 1967. (www.drive-inthruwisconsin.com)

15) Two cars leave the same location at the same time, one traveling east and the other going west. The westbound car is driving 6 mph faster than the eastbound car. After 2.5 hr they are 345 miles apart. What is the speed of each car?

16) How much pure acid must be added to 6 gal of a 4% acid solution to make 20% acid solution?

Solve for the indicated variable.

17) $B = \dfrac{an}{4}$ for a

18) $S = 2\pi r^2 + 2\pi rh$ for h

19) Find the measure of each indicated angle.

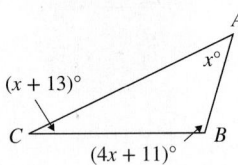

Solve. Graph the solution set, and write the answer in interval notation.

20) $6m + 19 \le 7$

21) $1 - 2(3x - 5) < 2x + 5$

22) $-5 \le \dfrac{3}{2}a - 5 < 4$

23) $3 < 3 - 2c < 9$

24) *Write an inequality and solve:* Rawlings Builders will rent a forklift for \$46.00 per day plus \$9.00 per hour. If they have at most \$100.00 allotted for a one-day rental, for how long can they keep the forklift and remain within budget?

25) Given sets $A = \{1, 2, 3, 6, 12\}$ and $B = \{1, 2, 9, 12\}$, find each of the following.

   a) $A \cup B$ b) $A \cap B$

Solve each compound inequality. Write the answer in interval notation.

26) $3n + 5 > 12$ or $\dfrac{1}{4}n < -2$

27) $y - 8 \le -5$ and $2y \ge 0$

28) $6 - p < 10$ or $p - 7 < 2$

Chapter 3:   Cumulative Review for Chapters 1–3

Perform the operations and simplify.

1) $\dfrac{3}{8} - \dfrac{5}{6}$ 2) $\dfrac{5}{8} \cdot 12$

3) $26 - 14 \div 2 + 5 \cdot 7$

4) $-82 + 15 + 10(1 - 3)$

5) $-39 - |7 - 15|$

6) Find the area of a triangle with a base of length 9 cm and height of 6 cm.

Given the set of numbers $\left\{\dfrac{3}{4}, -5, \sqrt{11}, 2.5, 0, 9, 0.\overline{4}\right\}$**, identify**

7) the integers

8) the rational numbers

9) the whole numbers

10) Which property is illustrated by $6(5 + 2) = 6 \cdot 5 + 6 \cdot 2$?

11) Does the commutative property apply to the subtraction of real numbers? Explain.

12) Combine like terms.

   $11y^2 - 14y + 6y^2 + y - 5y$

Simplify. The answer should not contain any negative exponents.

13) $\dfrac{35r^{16}}{28r^4}$ 14) $(-2m^5)^3 (3m^9)^2$

15) $(-12z^{10})\left(\dfrac{3}{8}z^{-16}\right)$ 16) $\left(\dfrac{10c^{12}d^2}{5c^9d^{-3}}\right)^{-2}$

17) Write 0.00000895 in scientific notation.

Solve.

18) $8t - 17 = 10t + 6$

19) $\dfrac{3}{2}n + 14 = 20$

20) $3(7w - 5) - w = -7 + 4(5w - 2)$

21) $\dfrac{x + 3}{10} = \dfrac{2x - 1}{4}$

22) $-\dfrac{1}{2}c + \dfrac{1}{5}(2c - 3) = \dfrac{3}{10}(2c + 1) - \dfrac{3}{4}c$

Solve using the five-step method.

23) Stu and Phil were racing from Las Vegas back to Napa Valley. Stu can travel 140 miles by train in the time it takes Phil to travel 120 miles by car. What are the speeds of the train and the car if the train is traveling 10 mph faster than the car?

Digital Vision/Getty Images

Solve. Write the answer in interval notation.

24) $7k + 4 \ge 9k + 16$

25) $8x \le -24$ or $4x - 5 \ge 6$

Linear Equations in Two Variables and Functions

Get Ready

We have already learned how to evaluate expressions and solve linear equations in one variable. Brush up on these skills to prepare for learning about linear equations in *two* variables.

1) **Write $\frac{12}{28}$ in lowest terms:** $\frac{12}{28} = \frac{12 \div 4}{28 \div 4} = \frac{3}{7}$. $\frac{3}{7}$ is in lowest terms because the only common factor of 3 and 7 is 1.

2) The **reciprocal** of a number, $\frac{a}{b}$, is $\frac{b}{a}$ provided that $a \neq 0$ and $b \neq 0$.

   Example: The reciprocal of $\frac{5}{8}$ is $\frac{8}{5}$.

3) To **evaluate an algebraic expression,** such as $3k - 1$ for $k = 6$, we substitute the value for the variable and simplify: $3(6) - 1 = 18 - 1 = 17$.

4) To **solve an equation** means to find the value of the variable that makes the equation true.

   Example: Solve $2x + 11 = 5$.

   $$2x + 11 - 11 = 5 - 11$$
   $$2x = -6$$
   $$\frac{2x}{2} = \frac{-6}{2}$$
   $$x = -3 \qquad \text{The solution set is } \{-3\}.$$

In this chapter, use the following *Basic Skills Worksheets* to prepare students for future chapters: WS2 Powers (to prepare for operations with polynomials in Chapter 6) and WS3 Prefactoring (so that students can practice the mental arithmetic skills needed to factor polynomials in Chapter 7).

Get Ready Exercises

Write each fraction in lowest terms.

1) $\dfrac{-10}{6}$ 2) $\dfrac{5}{20}$ 3) $\dfrac{-18}{-24}$ 4) $\dfrac{54}{9}$

Find the reciprocal of each number.

5) $\dfrac{2}{3}$ 6) $-\dfrac{11}{7}$ 7) -4 8) $\dfrac{1}{2}$

9) Evaluate $-5x + 4$ for $x = 2$.

10) Evaluate $t^2 - 7t + 3$ for a) $t = 4$ and b) $t = -1$.

Solve each equation.

11) $4x = 28$

12) $11 + 5y = -19$

13) $2 = \dfrac{3}{2}x - 7$

14) $0 = -5x$

15) $0 = -\dfrac{1}{4}x + 9$

16) $-x + 8 = 13$

17) $5 - y = 12$

18) $0 = \dfrac{10}{3}x - 8$

19) $7y = 0$

20) $1 - \dfrac{4}{5}y = 11$

Answers

1) $-\dfrac{5}{3}$ 2) $\dfrac{1}{4}$ 3) $\dfrac{3}{4}$ 4) 6 5) $\dfrac{3}{2}$ 6) $-\dfrac{7}{11}$ 7) $-\dfrac{1}{4}$ 8) 2 9) -6 10) a) -9 b) 11 11) $\{7\}$ 12) $\{-6\}$

13) $\{6\}$ 14) $\{0\}$ 15) $\{36\}$ 16) $\{-5\}$ 17) $\{-7\}$ 18) $\left\{\dfrac{12}{5}\right\}$ 19) $\{0\}$ 20) $\left\{-\dfrac{25}{2}\right\}$

Study Strategies Taking Notes in Class

Taking good notes in class is an essential skill for college success. Let's learn how to use the P.O.W.E.R. framework to take good notes.

- *Prepare means to explicitly state a goal.* **I will take good notes.**

- *Organize means to **organize** the physical and mental tools you need to achieve your goal.*
- Complete the emPOWERme survey that appears before the Chapter Summary to increase your awareness of your current note-taking style.
- Buy one notebook for each class. Bring your book, math notebook, and pencils or pens to class.
- Find a seat that allows you to see and hear the instructor clearly.
- Turn off your cell phone and put it away to minimize distractions.

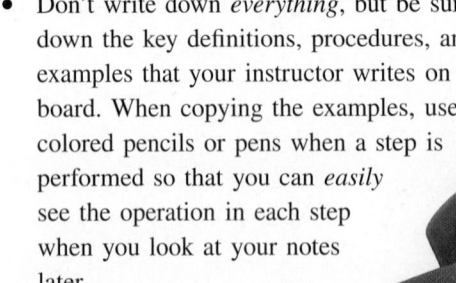

- *Work means to **do the work** that needs to be done to achieve the goal.*
- Be an active listener. Think about what your instructor is saying, and ask questions if you do not understand something.
- Take notes in a notebook that is only for the class you are in. At the top of the paper, write down the date as well as the section number for the day's lesson. Write neatly!
- Don't write down *everything*, but be sure to write down the key definitions, procedures, and examples that your instructor writes on the board. When copying the examples, use colored pencils or pens when a step is performed so that you can *easily* see the operation in each step when you look at your notes later.
- Include comments *in your own words* that may help you to better understand a procedure or example.

Image Source/Getty Images

- *Evaluate* means you should **evaluate** what you have done. Did you achieve your goal?
- Look over your notes as class is ending. Do you understand everything you wrote? Is there anything that is incomplete or that you do not understand? If so, ask another student or your instructor about it before you leave class.

- **Rethink** means to **rethink** and reflect upon your goal.
- Sometime after class, especially before doing your homework, read over your notes. Are they easy to understand? Could one of your classmates understand what you have written? If not, what could you do differently to take better notes?

4.1 Introduction to Linear Equations in Two Variables

P Prepare

O Organize

What are your objectives for Section 4.1?	How can you accomplish each objective?
1 Define a Linear Equation in Two Variables	• Write the definition of *linear equation in two variables* in your own words, and write an example. • Understand the *Cartesian coordinate system* or *rectangular coordinate system*.
2 Decide Whether an Ordered Pair Is a Solution of a Given Equation	• Know how to substitute values into an equation. • Complete the given example on your own. • Complete You Try 1.
3 Complete Ordered Pairs for a Given Equation	• Be able to substitute either the x-value or the y-value, and solve for the missing variable. • Complete the given examples on your own. • Complete You Trys 2 and 3.
4 Plot Ordered Pairs	• Draw the *Cartesian coordinate system* in your notes with the quadrants, x-axis, y-axis, and origin labeled. • Follow the examples to create a procedure for plotting points. • Complete You Trys 4 and 5.
5 Solve Applied Problems Involving Ordered Pairs	• Follow the example to know what questions to ask when solving. • Complete the given example on your own.

Read the explanations, follow the examples, take notes, and complete the You Trys.

Graphs are everywhere—online, in newspapers, and in books. The accompanying graph shows how many billions of dollars consumers spent shopping online during each quarter of 2017.

We can get different types of information from this graph. For example, in the first quarter of 2017, shoppers spent about $98 billion online, while in the last quarter they spent about $142 billion. The graph also illustrates that, as time goes on, people spend more money shopping online.

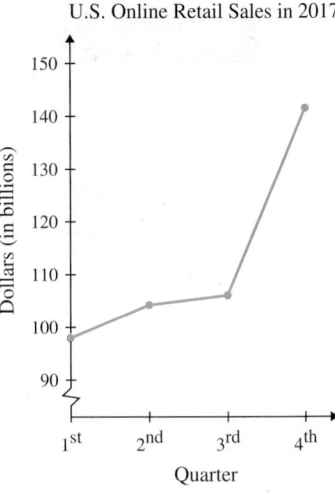

U.S. Online Retail Sales in 2017

Source: www.census.gov

1 Define a Linear Equation in Two Variables

Later in this section, we will see that graphs like the preceding one are based on the *Cartesian coordinate system*, also known as the *rectangular coordinate system*, which gives us a way to graphically represent the relationship between two quantities. We will also learn about different ways to represent relationships between two quantities, like year and online spending, when we learn about *linear equations in two variables*. Let's begin with a definition.

Definition

A **linear equation in two variables** can be written in the form $Ax + By = C$ where A, B, and C are real numbers and where both A and B do not equal zero.

Some examples of linear equations in two variables are

$$5x - 2y = 11 \qquad y = \frac{3}{4}x + 1 \qquad -3a + b = 2 \qquad y = x \qquad x = -3$$

(We can write $x = -3$ as $x + 0y = -3$; therefore it is a linear equation in two variables.)

2 Decide Whether an Ordered Pair Is a Solution of a Given Equation

A solution to a linear equation in two variables is written as an *ordered pair* so that when the values are substituted for the appropriate variables, we obtain a true statement.

EXAMPLE 1

Determine whether each ordered pair is a solution of $5x - 2y = 11$.

a) $(1, -3)$ b) $\left(\dfrac{3}{5}, 4\right)$

Solution

a)   Solutions to the equation $5x - 2y = 11$ are written in the form (x, y), where (x, y) is called an **ordered pair.** Therefore, the ordered pair $(1, -3)$ means that $x = 1$ and $y = -3$.

$$(1, -3)$$

x-coordinate y-coordinate

To determine whether $(1, -3)$ is a solution of $5x - 2y = 11$, we substitute 1 for x and -3 for y. Remember to put these values in parentheses.

$$5x - 2y = 11$$
$$5(1) - 2(-3) = 11 \qquad \text{Substitute 1 for } x \text{ and } -3 \text{ for } y.$$
$$5 + 6 = 11 \qquad \text{Multiply.}$$
$$11 = 11 \qquad \text{True}$$

Since substituting $x = 1$ and $y = -3$ into the equation gives the true statement $11 = 11$, $(1, -3)$ *is a solution* of $5x - 2y = 11$. We say that $(1, -3)$ *satisfies* $5x - 2y = 11$.

b) The ordered pair $\left(\dfrac{3}{5}, 4\right)$ tells us that $x = \dfrac{3}{5}$ and $y = 4$.

$$5x - 2y = 11$$
$$5\left(\dfrac{3}{5}\right) - 2(4) = 11 \qquad \text{Substitute } \dfrac{3}{5} \text{ for } x \text{ and 4 for } y.$$
$$3 - 8 = 11 \qquad \text{Multiply.}$$
$$-5 = 11 \qquad \text{False}$$

Since substituting $\left(\dfrac{3}{5}, 4\right)$ into the equation gives the false statement $-5 = 11$, the ordered pair is *not* a solution to the equation.

[**YOU TRY 1**]

Determine whether each ordered pair is a solution of the equation $y = -\dfrac{3}{4}x + 5$.

a) $(12, -4)$ b) $(0, 7)$ c) $(-8, 11)$

If the variables in the equation are not x and y, then the variables in the ordered pairs are written in alphabetical order. For example, solutions to $-3a + b = 2$ are ordered pairs of the form (a, b).

3 Complete Ordered Pairs for a Given Equation

Often, we are given the value of one variable in an equation and we can find the value of the other variable that makes the equation true.

EXAMPLE 2

Complete the ordered pair $(-3, \quad)$ for $y = 2x + 10$.

Solution

To complete the ordered pair $(-3, \quad)$, we must find the value of y from $y = 2x + 10$ when $x = -3$.

$$y = 2x + 10$$
$$y = 2(-3) + 10 \qquad \text{Substitute } -3 \text{ for } x.$$
$$y = -6 + 10$$
$$y = 4$$

When $x = -3$, $y = 4$. The ordered pair is $(-3, 4)$.

[**YOU TRY 2**]

Complete the ordered pair $(5, \quad)$ for $y = 3x - 7$.

If we want to complete more than one ordered pair for a particular equation, we can organize the information in a **table of values.**

EXAMPLE 3 Complete the table of values for each equation, and write the information as ordered pairs.

a) $-x + 3y = 8$

x	y
1	
	-4
	$\frac{2}{3}$

b) $y = 2$

x	y
7	
-5	
0	

Solution

a) $-x + 3y = 8$

x	y
1	
	-4
	$\frac{2}{3}$

The first ordered pair is (1,   ), and we must find y.

$$-x + 3y = 8$$
$$-(1) + 3y = 8 \qquad \text{Substitute 1 for } x.$$
$$-1 + 3y = 8$$
$$3y = 9 \qquad \text{Add 1 to each side.}$$
$$y = 3 \qquad \text{Divide by 3.}$$

The ordered pair is (1, 3).

The second ordered pair is (, -4), and we must find x.

The ordered pair is (-20, -4).

$$-x + 3y = 8$$
$$-x + 3(-4) = 8 \qquad \text{Substitute } -4 \text{ for } y.$$
$$-x + (-12) = 8 \qquad \text{Multiply.}$$
$$-x = 20 \qquad \text{Add 12 to each side.}$$
$$x = -20 \qquad \text{Divide by } -1.$$

The third ordered pair is $\left(\ ,\frac{2}{3}\right)$, and we must find x.

The ordered pair is $\left(-6, \frac{2}{3}\right)$.

$$-x + 3y = 8$$
$$-x + 3\left(\frac{2}{3}\right) = 8 \qquad \text{Substitute } \frac{2}{3} \text{ for } y.$$
$$-x + 2 = 8 \qquad \text{Multiply.}$$
$$-x = 6 \qquad \text{Subtract 2 from each side.}$$
$$x = -6 \qquad \text{Divide by } -1.$$

As you complete each ordered pair, fill in the table of values. The completed table will look like this:

x	y
1	3
-20	-4
-6	$\frac{2}{3}$

b) $y = 2$

x	y
7	
-5	
0	

The first ordered pair is (7,   ), and we must find y. The equation $y = 2$ means that *no matter the value of x, y always equals* 2. Therefore, when $x = 7$, $y = 2$.

The ordered pair is (7, 2).

Since $y = 2$ for every value of x, we can complete the table of values as follows:

The ordered pairs are (7, 2), (-5, 2), and (0, 2).

x	y
7	2
-5	2
0	2

W Hint

For the equation $y = 2$, you will see that the y-values never change and are always equal to 2.

[YOU TRY 3] Complete the table of values for each equation, and write the information as ordered pairs.

a) $x - 2y = 9$

x	y
5	
12	
	−7
	$\dfrac{5}{2}$

b) $x = -3$

x	y
	1
	3
	−8

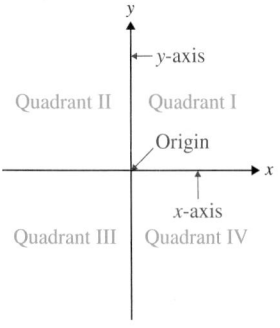

Quadrant II | Quadrant I

Origin

x-axis

Quadrant III | Quadrant IV

4 Plot Ordered Pairs

When we completed the table of values for the last two equations, we were finding solutions to each linear equation in two variables.

How can we represent the solutions graphically? We will use the **Cartesian coordinate system,** also known as the **rectangular coordinate system,** to graph the ordered pairs, (x, y).

In the Cartesian coordinate system, we have a horizontal number line, called the **x-axis,** and a vertical number line, called the **y-axis.**

The x-axis and y-axis in the Cartesian coordinate system determine a flat surface called a **plane.** The axes divide this plane into four **quadrants,** as shown in the figure. The point at which the x-axis and y-axis intersect is called the **origin.** The arrow at one end of the x-axis and one end of the y-axis indicates the positive direction on each axis.

Ordered pairs can be represented by **points** in the plane. Therefore, to graph the ordered pair $(4, 2)$ we *plot the point* $(4, 2)$. We will do this in Example 4.

W Hint
You move counterclockwise from Quadrant I to move through the quadrants in order.

EXAMPLE 4

Plot the point $(4, 2)$.

Solution

Since $x = 4$, we say that the *x-coordinate* of the point is 4. Likewise, the *y-coordinate* is 2.

The *origin* has coordinates $(0, 0)$. The **coordinates** of a point tell us how far from the origin, in the x-direction and y-direction, the point is located. So, the coordinates of the point $(4, 2)$ tell us that to locate the point we do the following:

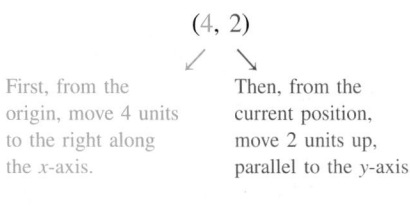

First, from the origin, move 4 units to the right along the x-axis.

Then, from the current position, move 2 units up, parallel to the y-axis.

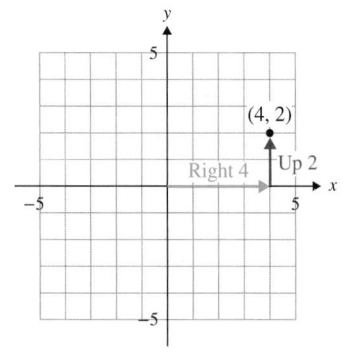

W Hint
Always start at the origin to first move horizontally and then move vertically.

EXAMPLE 5

Plot the points.

a) (−2, 5)   b) (1, −4)   c) $\left(\dfrac{5}{2}, 3\right)$

d) (−5, −2)   e) (0, 1)   f) (−4, 0)

Solution

The points are plotted on the graph below.

a)   (−2, 5)   b)   (1, −4)
                    ↙   ↘   ↙   ↘
               First   Then   First   Then
          From the origin,   From the   From the origin,   From the
          move left 2 units   current position,   move right 1 unit   current position,
          on the x-axis.   move 5 units   on the x-axis.   move 4 units
                              up, parallel to   down, parallel
                              the y-axis.   to the y-axis.

> **Ⓦ Hint**
>
> Negative x-values make
> you move left, and
> negative y-values make
> you move down.

c) $\left(\dfrac{5}{2}, 3\right)$

Think of $\dfrac{5}{2}$ as $2\dfrac{1}{2}$. From the origin, move right $2\dfrac{1}{2}$

units, then up 3 units.

d) (−5, −2) From the origin, move left 5 units, then
              down 2 units.

e) (0, 1)   The x-coordinate of 0 means that we
              don't move in the x-direction
              (horizontally). From the origin, move
              up 1 on the y-axis.

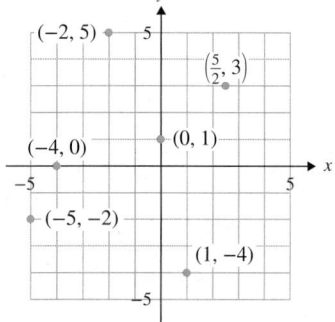

f) (−4, 0)   From the origin, move left 4 units. Since the y-coordinate is zero, we do not
              move in the y-direction (vertically).

[YOU TRY 4]   Plot the points.

a) (3, 1)   b) (−2, 4)   c) (0, −5)   d) (2, 0)   e) (−4, −3)   f) $\left(1, \dfrac{7}{2}\right)$

📌 **Note**

The coordinate system should always be labeled to indicate how many units
each mark represents.

We can graph sets of ordered pairs for a linear equation in two variables.

EXAMPLE 6

Complete the table of values for $2x - y = 5$, then plot the points.

x	y
0	
1	
	3

Solution

The first ordered pair is (0,), and we must find y.

$$2x - y = 5$$
$$2(0) - y = 5 \quad \text{Substitute 0 for } x.$$
$$0 - y = 5$$
$$-y = 5$$
$$y = -5 \quad \text{Divide by } -1.$$

The ordered pair is (0, −5).

The second ordered pair is (1,), and we must find y.

$$2x - y = 5$$
$$2(1) - y = 5 \quad \text{Substitute 1 for } x.$$
$$2 - y = 5$$
$$-y = 3 \quad \text{Subtract 2 from each side.}$$
$$y = -3 \quad \text{Divide by } -1.$$

The ordered pair is (1, −3).

The third ordered pair is (, 3), and we must find x.

$$2x - y = 5$$
$$2x - (3) = 5 \quad \text{Substitute 3 for } y.$$
$$2x = 8 \quad \text{Add 3 to each side.}$$
$$x = 4 \quad \text{Divide by 2.}$$

The ordered pair is (4, 3).

Each of the points (0, −5), (1, −3), and (4, 3) satisfies the equation $2x - y = 5$.

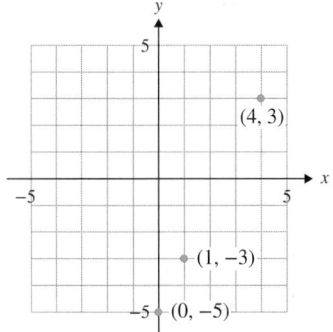

[YOU TRY 5]

Complete the table of values for $3x + y = 1$, then plot the points.

x	y
0	
−1	
	−5

---

5 Solve Applied Problems Involving Ordered Pairs

Next, we will look at an application of ordered pairs.

EXAMPLE 7

The length of an 18-year-old female's hair is measured to be 250 millimeters (mm) (almost 10 in.). The length of her hair after x days can be approximated by

$$y = 0.30x + 250$$

where y is the length of her hair in millimeters.

a) Find the length of her hair (i) 10 days, (ii) 60 days, and (iii) 90 days after the initial measurement, and write the results as ordered pairs.

b) Graph the ordered pairs.

c) How long would it take for her hair to reach a length of 274 mm (almost 11 in.)?

Solution

a) The problem states that in the equation $y = 0.30x + 250$,

x = number of days after the hair was measured
y = length of the hair (in millimeters)

x	y
10	
60	
90	

We must determine the length of her hair after 10 days, 60 days, and 90 days. We can organize the information in a table of values.

i) $x = 10$: $y = 0.30x + 250$
             $y = 0.30(10) + 250$ Substitute 10 for x.
             $y = 3 + 250$ Multiply.
             $y = 253$

After 10 days, her hair is 253 mm long. We can write this as the ordered pair (10, 253).

ii) $x = 60$: $y = 0.30x + 250$
              $y = 0.30(60) + 250$ Substitute 60 for x.
              $y = 18 + 250$ Multiply.
              $y = 268$

After 60 days, her hair is 268 mm long. We can write this as the ordered pair (60, 268).

iii) $x = 90$: $y = 0.30x + 250$
               $y = 0.30(90) + 250$ Substitute 90 for x.
               $y = 27 + 250$ Multiply.
               $y = 277$

After 90 days, her hair is 277 mm long. We can write this as the ordered pair (90, 277).

We can complete the table of values:

x	y
10	253
60	268
90	277

The ordered pairs are (10, 253), (60, 268), and (90, 277).

b) Graph the ordered pairs.

The x-axis represents the number of days after the hair was measured. Since it does not make sense to talk about a negative number of days, we will not continue the x-axis in the negative direction.

The y-axis represents the length of the female's hair. Likewise, a negative number does not make sense in this situation, so we will not continue the y-axis in the negative direction.

The scales on the x-axis and y-axis are different. This is because the size of the numbers they represent are quite different.

Here are the ordered pairs we must graph: (10, 253), (60, 268), and (90, 277).

The *x*-values are 10, 60, and 90, so we will let each mark in the *x*-direction represent 10 units.

The *y*-values are 253, 268, and 277. While the numbers are rather large, they do not actually differ by much. We will begin labeling the *y*-axis at 250, but each mark in the *y*-direction will represent 3 units. Because there is a large jump in values from 0 to 250 on the *y*-axis, we indicate this with "⸏" on the axis between the 0 and 250.

Notice also that we have labeled both axes. The ordered pairs are plotted on the following graph.

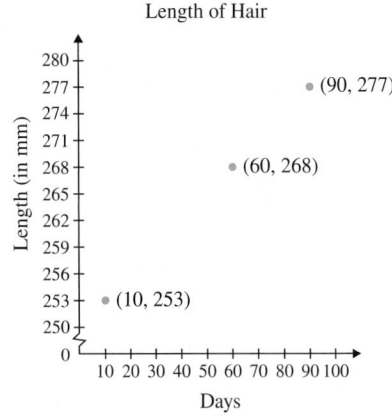

c) We must determine how many days it would take for the hair to grow to a length of 274 mm.

The length, 274 mm, is the *y*-value. We must find the value of *x* that corresponds to $y = 274$ since *x* represents the number of days.

The equation relating *x* and *y* is $y = 0.30x + 250$. We will substitute 274 for *y* and solve for *x*.

$$y = 0.30x + 250$$
$$274 = 0.30x + 250$$
$$24 = 0.30x$$
$$80 - x$$

It will take 80 days for her hair to grow to a length of 274 mm.

ANSWERS TO [YOU TRY] **EXERCISES**

1) a) yes b) no c) yes   2) (5, 8)   3) a) $(5, -2)$, $\left(12, \frac{3}{2}\right)$, $(-5, -7)$, $\left(14, \frac{5}{2}\right)$
b) $(-3, 1)$, $(-3, 3)$, $(-3, -8)$
4)   5) $(0, 1)$, $(-1, 4)$, $(2, -5)$

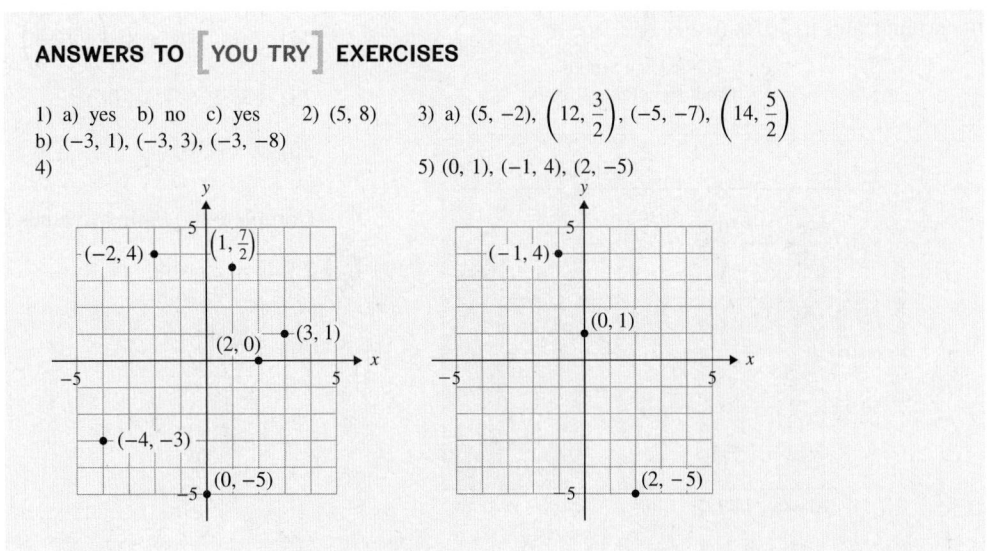

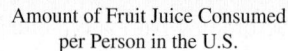

Mixed Exercises: Objectives 1 and 2

The graph shows the number of gallons of fruit juice consumed per person for the years 2010–2015.

(U.S. Dept of Agriculture)

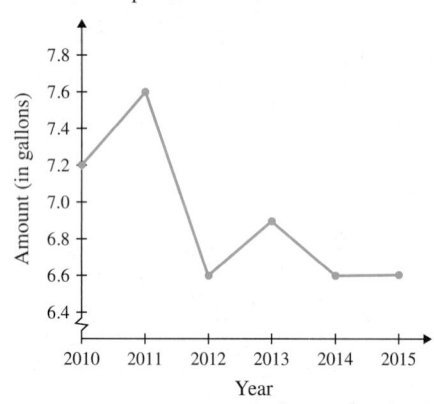

Amount of Fruit Juice Consumed
per Person in the U.S.

1) How many gallons of fruit juice were consumed per person in 2010?

2) During which year was the consumption level about 6.9 gal per person?

3) During which years was consumption the same, and how much juice was consumed each of these years?

4) During which year did people drink the most fruit juice?

5) What was the general consumption trend from 2013 to 2014?

6) Compare the consumption level in 2011 with that in 2012.

The bar graph shows the public high school graduation rate in certain states in 2016. (www.edweek.org)

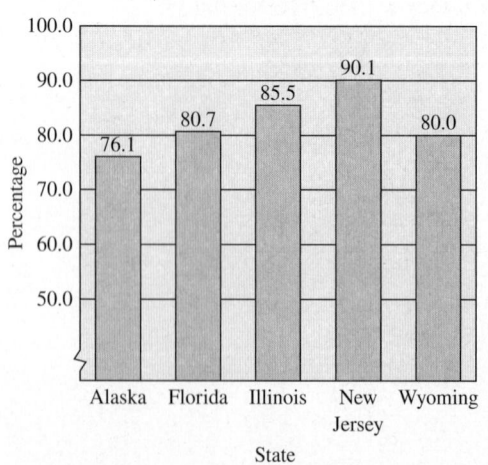

Public High School
Graduation Rate, 2016

7) Which state had the highest graduation rate, and what percentage of its public high school students graduated?

8) Which states graduated between 75% and 85% of their students?

9) How does the graduation rate of Florida compare with that of New Jersey?

10) Which state had a graduation rate of about 85.5%?

11) Explain the difference between a linear equation in one variable and a linear equation in two variables. Give an example of each.

12) True or False: $3x + 6y^2 = -1$ is a linear equation in two variables.

Determine whether each ordered pair is a solution of the given equation.

 13) $2x + 5y = 1$; $(-2, 1)$ 14) $2x + 7y = -4$; $(2, -5)$

15) $-3x - 2y = -15$; $(7, -3)$ 16) $y = 5x - 6$; $(3, 9)$

17) $y = -\dfrac{3}{2}x - 7$; $(8, 5)$ 18) $5y = \dfrac{2}{3}x + 1$; $(6, 1)$

19) $y = -7$; $(9, -7)$ 20) $x = 8$; $(-10, 8)$

Objective 3: Complete Ordered Pairs for a Given Equation

Complete the ordered pair for each equation.

21) $y = 3x - 7$; $(4,\ \ )$ 22) $y = -2x + 3$; $(6,\ \ )$

23) $2x - 15y = 13$; $\left(\ \ , -\dfrac{4}{3}\right)$

24) $-x + 10y = 8$; $\left(\ \ , \dfrac{2}{5}\right)$

25) $x = 5$; $(\ \ , -200)$

26) $y = -10$; $(12,\ \ )$

Complete the table of values for each equation.

27) $y = 2x - 4$

x	y
0	
1	
-1	
-2	

28) $y = -5x + 1$

x	y
0	
1	
2	
-1	

29) $y = 4x$

x	y
0	
$\frac{1}{2}$	
	12
	−20

30) $y = 9x - 8$

x	y
0	
$-\frac{1}{3}$	
	−17
	1

31) $5x + 4y = -8$

x	y
0	
	0
1	
$-\frac{12}{5}$	

32) $2x - y = 12$

x	y
	0
0	
	−2
$\frac{5}{2}$	

33) $y = -2$

x	y
0	
−3	
8	
17	

34) $x = 20$

x	y
	0
	3
	−4
	−9

35) Explain, in words, how to complete the table of values for $x = -13$.

x	y
	0
	2
	−1

36) Explain, in words, how to complete the ordered pair (, −3) for $y = -x - 2$.

Objective 4: Plot Ordered Pairs

Name each point with an ordered pair, and identify the quadrant in which each point lies.

37)

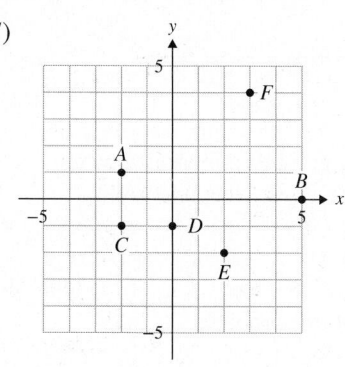

38)

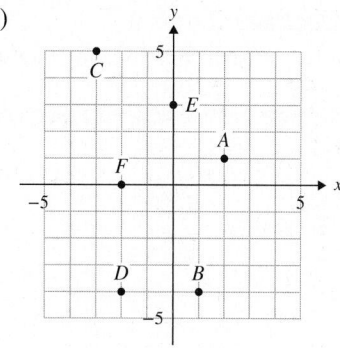

Graph each ordered pair, and explain how you plotted the points.

39) (2, 4) 40) (4, 1)

41) (−3, −5) 42) (−2, 1)

Graph each ordered pair.

43) (−6, 1) 44) (−2, −3)

45) (0, −1) 46) (4, −5)

47) (0, 4) 48) (−5, 0)

49) (−2, 0) 50) (0, −1)

51) $\left(-2, \frac{3}{2}\right)$ 52) $\left(\frac{4}{3}, 3\right)$

53) $\left(3, -\frac{1}{4}\right)$ 54) $\left(-2, -\frac{9}{4}\right)$

55) $\left(0, -\frac{11}{5}\right)$ 56) $\left(-\frac{9}{2}, -\frac{2}{3}\right)$

Fill in the blank with *always, sometimes,* or *never* to make the statement true.

57) When writing an ordered pair for a linear equation of the form $Ax + By = C$, the y-coordinate is _____ placed before the x-coordinate as in (y, x).

58) A linear equation will _____ contain a squared variable term.

59) A linear equation will _____ contain two different variable terms.

60) A linear equation will _____ contain one variable term.

Mixed Exercises: Objectives 3 and 4

Complete the table of values for each equation, and plot the points.

61) $y = -4x + 3$

x	y
0	
	0
2	
	7

62) $y = -3x + 4$

x	y
0	
	0
	-2
-1	

63) $y = x$

x	y
0	
-1	
	3
	-5

64) $y = -2x$

x	y
	0
	4
2	
-3	

65) $3x + 4y = 12$

x	y
0	
	0
1	
	6

66) $2x - 3y = 6$

x	y
0	
	0
	2
2	

67) $y + 1 = 0$

x	y
0	
1	
-3	
-1	

68) $x = 3$

x	y
	0
	-2
	3
	1

69) $y = \frac{1}{4}x + 2$

x	y
0	
-2	
4	
-1	

70) $y = -\frac{5}{2}x + 3$

x	y
0	
4	
2	
1	

71) For $y = \frac{2}{3}x - 7$,

   a) find y when $x = 3$, $x = 6$, and $x = -3$. Write the results as ordered pairs.

   b) find y when $x = 1$, $x = 5$, and $x = -2$. Write the results as ordered pairs.

   c) why is it easier to find the y-values in part a) than in part b)?

72) Which ordered pair is a solution to every linear equation of the form $y = mx$, where m is a real number?

Fill in the blank with *positive*, *negative*, or *zero*.

73) The x-coordinate of every point in quadrant III is _____.

74) The y-coordinate of every point in quadrant I is _____.

75) The y-coordinate of every point in quadrant II is _____.

76) The x-coordinate of every point in quadrant II is _____.

77) The x-coordinate of every point in quadrant I is _____.

78) The y-coordinate of every point in quadrant IV is _____.

79) The y-coordinate of every point on the x-axis is _____.

80) The x-coordinate of every point on the y-axis is _____.

Objective 5: Solve Applied Problems Involving Ordered Pairs

81) The graph shows the number of people who visited Las Vegas from 2011 to 2017. (www.lvcva.com)

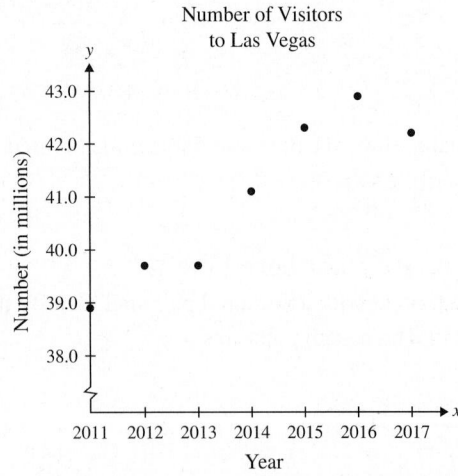

   a) If a point on the graph is represented by the ordered pair (x, y), then what do x and y represent?

   b) What does the ordered pair (2015, 42.3) represent in the context of this problem?

   c) Approximately how many people went to Las Vegas in 2017?

d) In which years were there approximately the same number of visitors, and how many visitors were there?

e) Approximately how many more people visited Las Vegas in 2016 than in 2011?

f) Represent the following with an ordered pair: During which year did Las Vegas have the most visitors, and how many visitors were there?

82) The graph shows the average amount of time people spent commuting to work in the Los Angeles metropolitan area from 2011 to 2016.
(American Community Survey, U.S. Census Bureau)

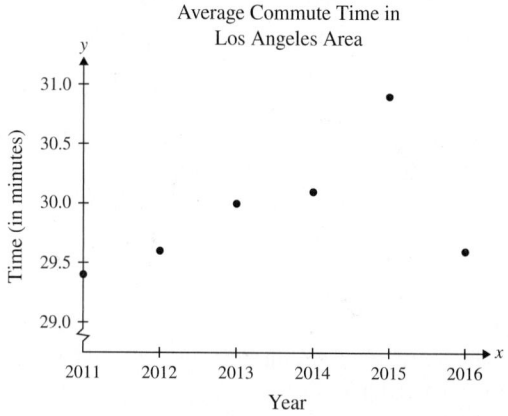

Average Commute Time in Los Angeles Area

a) If a point on the graph is represented by the ordered pair (x, y), then what do x and y represent?

b) What does the ordered pair (2016, 29.6) represent in the context of this problem?

c) Which year during this time period had the shortest commute? What was the approximate commute time?

d) When was the average commute 30.9 min?

e) Write an ordered pair to represent when the average commute time was 30.0 min.

83) The percentage of deadly crashes involving alcohol in Minnesota is given in the table. (Minnesota Dept of Public Safety)

Year	Percentage
2011	37
2012	33
2013	30
2014	31
2015	33
2016	31

a) Write the information as ordered pairs (x, y), where x represents the year and y represents the percentage of accidents involving alcohol.

b) Label a coordinate system, choose an appropriate scale, and graph the ordered pairs.

c) Explain the meaning of the ordered pair (2016, 31) in the context of the problem.

84) The average annual salary of a social worker is given in the table. (www.bls.gov)

Year	Salary
2014	$49,150
2015	$49,670
2016	$50,710
2017	$60,900

a) Write the information as ordered pairs (x, y) where x represents the year and y represents the average annual salary.

b) Label a coordinate system, choose an appropriate scale, and graph the ordered pairs.

c) Explain the meaning of the ordered pair (2016, $50,710) in the context of the problem.

85) The amount of sales tax paid by consumers in Seattle in 2018 is given by $y = 0.101x$, where x is the price of an item in dollars and y is the amount of tax to be paid.

Tetra Images/Getty Images

a) Complete the table of values, and write the information as ordered pairs.

x	y
100.00	
140.00	
210.70	
250.00	

b) Label a coordinate system, choose an appropriate scale, and graph the ordered pairs.

c) Explain the meaning of the ordered pair (140.00, 14.14) in the context of the problem.

d) How much tax would a customer pay if the cost of an item was $210.70?

e) Look at the graph. Is there a pattern indicated by the points?

f) If a customer paid $20.20 in sales tax, what was the cost of the item purchased?

86) Kyle is driving from Atlanta to Oklahoma City. His distance from Atlanta, y (in miles), is given by $y = 66x$, where x represents the number of hours driven.

a) Complete the table of values, and write the information as ordered pairs.

x	y
1	
1.5	
2	
4.5	

b) Label a coordinate system, choose an appropriate scale, and graph the ordered pairs.

c) Explain the meaning of the ordered pair (4.5, 297) in the context of the problem.

d) Look at the graph. Is there a pattern indicated by the points?

e) What does the 66 in $y = 66x$ represent?

f) How many hours of driving time will it take for Kyle to get to Oklahoma City if the distance between Atlanta and Oklahoma City is about 860 miles?

R Rethink

R1) If the x-coordinate of an ordered pair is 0, on which axis does the point lie?

R2) If the y-coordinate of an ordered pair is 0, on which axis does the point lie?

R3) Why do we need two number lines when plotting ordered pairs?

R4) Which objective is the most difficult for you? Which exercises do you need to work again to fully master?

4.2 Graphing by Plotting Points and Finding Intercepts

P Prepare

O Organize

What are your objectives for Section 4.2?	How can you accomplish each objective?
1 Graph a Linear Equation by Plotting Points	• Learn the properties for **Solutions of Linear Equations in Two Variables** and **The Graph of a Linear Equation in Two Variables.** • Create a procedure after following the examples for graphing linear equations by plotting points. • Complete the given example on your own. • Complete You Try 1.
2 Graph a Linear Equation in Two Variables by Finding the Intercepts	• Write the definitions of *x*- and *y*-intercept in your own words. • Learn the procedure for **Finding Intercepts,** and write it in your notes. • Complete the given example on your own. • Complete You Try 2.
3 Graph a Linear Equation of the Form $Ax + By = 0$	• Write the property **The Graph of $Ax + By = 0$** in your own words, and learn it. • Complete the given example on your own. • Complete You Try 3.
4 Graph Linear Equations of the Forms $x = a$ and $y = b$	• Write the properties **The Graph of $x = a$** and **The Graph of $y = b$** in your own words, and learn them. • Complete the given examples on your own. • Complete You Trys 4 and 5.
5 Model Data with a Linear Equation	• Do not let "bigger numbers" scare you. Follow the example to learn how to scale the Cartesian coordinate system. • Follow the example, and then complete the example in your notes without looking at the solution.

W Work

Read the explanations, follow the examples, take notes, and complete the You Trys.

1 Graph a Linear Equation by Plotting Points

In Example 3 of Section 4.1, we found that the ordered pairs (1, 3), (−20, −4), and $\left(-6, \frac{2}{3}\right)$ are three solutions to the equation $-x + 3y = 8$. But how many solutions does the equation have? *It has an infinite number of solutions.* Every linear equation in two variables has an infinite number of solutions because we can choose any real number for one of the variables and we will get another real number for the other variable.

Property Solutions of Linear Equations in Two Variables

Every linear equation in two variables has an infinite number of solutions, and the solutions are ordered pairs.

How can we represent all of the solutions to a linear equation in two variables? We can represent them with a graph, and that graph is a *line*.

Property The Graph of a Linear Equation in Two Variables

The graph of a linear equation in two variables, $Ax + By = C$, is a straight **line.** Each point on the line is a solution to the equation.

EXAMPLE 1

Graph $-x + 2y = 4$.

Solution

We will find three ordered pairs that satisfy the equation. Let's complete a table of values for $x = 0$, $x = 2$, and $x = -4$.

$x = 0$:

$$-x + 2y = 4$$
$$-(0) + 2y = 4$$
$$2y = 4$$
$$y = 2$$

$x = 2$:

$$-x + 2y = 4$$
$$-(2) + 2y = 4$$
$$-2 + 2y = 4$$
$$2y = 6$$
$$y = 3$$

$x = -4$:

$$-x + 2y = 4$$
$$-(-4) + 2y = 4$$
$$4 + 2y = 4$$
$$2y = 0$$
$$y = 0$$

We get the table of values

x	y
0	2
2	3
-4	0

Plot the points (0, 2), (2, 3), and (-4, 0), and draw the line through them.

W Hint

If possible, use graph paper when graphing lines. Don't forget to label the x-axis and y-axis and to draw arrowheads on both ends of your graphed line!

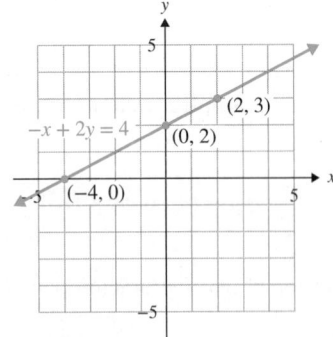

The line represents all solutions to the equation $-x + 2y = 4$. Every point on the line is a solution to the equation. The arrows on the ends of the line indicate that the line extends indefinitely in each direction. Although it is true that we need to find only two points to graph a line, it is best to plot at least three as a check.

[YOU TRY 1]

Graph each line.

a) $3x + 2y = 6$ b) $y = 4x - 8$

2 Graph a Linear Equation in Two Variables by Finding the Intercepts

In Example 1, the line crosses the *x*-axis at −4 and crosses the *y*-axis at 2. These points are called **intercepts.**

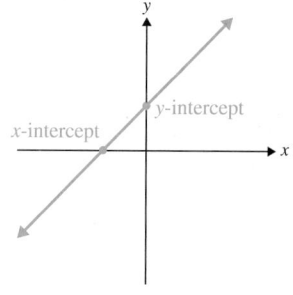

Definitions

The **x-intercept** of the graph of an equation is the point where the graph intersects the *x*-axis.

The **y-intercept** of the graph of an equation is the point where the graph intersects the *y*-axis.

What is the *y*-coordinate of any point on the *x*-axis? *It is zero.* Likewise, the *x*-coordinate of any point on the *y*-axis is zero. (See the graph at the right for some points on the axes.)

We use these facts to find the intercepts of the graph of an equation.

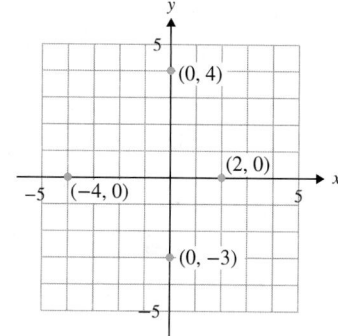

Procedure Finding Intercepts

To find the *x-intercept* of the graph of an equation, let $y = 0$ and solve for x.

To find the *y-intercept* of the graph of an equation, let $x = 0$ and solve for y.

Finding intercepts is very helpful for graphing linear equations in two variables.

EXAMPLE 2

Graph $y = -\dfrac{1}{3}x + 1$ by finding the intercepts and one other point.

Solution

We will begin by finding the intercepts.

x-intercept: Let $y = 0$, and solve for x.

$$0 = -\frac{1}{3}x + 1$$

$$-1 = -\frac{1}{3}x$$

$$3 = x \qquad \text{Multiply both sides by −3 to solve for } x.$$

The *x*-intercept is (3, 0).

y-intercept: Let $x = 0$, and solve for y.

$$y = -\frac{1}{3}(0) + 1$$

$$y = 0 + 1$$

$$y = 1$$

The *y*-intercept is (0, 1).

We must find another point. Let's look closely at the equation $y = -\frac{1}{3}x + 1$. The coefficient of x is $-\frac{1}{3}$. If we choose a value for x that is a multiple of 3 (the denominator of the fraction), then $-\frac{1}{3}x$ will not be a fraction.

Let $x = -3$. $y = -\frac{1}{3}x + 1$

$$y = -\frac{1}{3}(-3) + 1$$

$$y = 1 + 1$$

$$y = 2$$

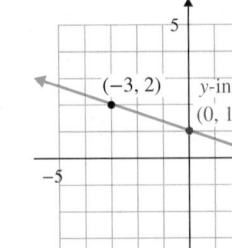

The third point is $(-3, 2)$.

Plot the points, and draw the line through them.

$\left[\text{YOU TRY 2}\right]$ Graph $y = \frac{4}{3}x - 2$ by finding the intercepts and one other point.

---

3 Graph a Linear Equation of the Form $Ax + By = 0$

Sometimes the x- and y-intercepts are the same point.

EXAMPLE 3

Graph $-2x + y = 0$.

Solution

If we begin by finding the x-intercept, let $y = 0$ and solve for x.

$$-2x + y = 0$$
$$-2x + (0) = 0$$
$$-2x = 0$$
$$x = 0$$

The x-intercept is $(0, 0)$. But this is the same as the y-intercept since we find the y-intercept by substituting 0 for x and solving for y. Therefore, *the x- and y-intercepts are the same point.*

Instead of the intercepts giving us two points on the graph of $-2x + y = 0$, we have only one. We will find two other points on the line.

$x = 2$: $-2x + y = 0$ $x = -2$: $-2x + y = 0$
        $-2(2) + y = 0$ $-2(-2) + y = 0$
          $-4 + y = 0$ $4 + y = 0$
                $y = 4$ $y = -4$

Hint

Graphs from equations of this form will always pass through the origin.

The ordered pairs (2, 4) and (−2, −4) are also solutions to the equation. Plot all three points on the graph, and draw the line through them.

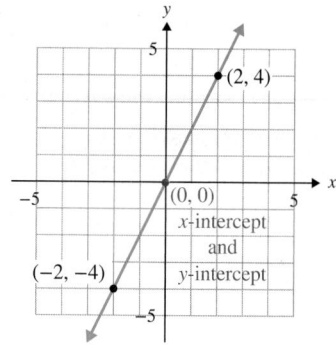

Property The Graph of $Ax + By = 0$

If A and B are nonzero real numbers, then the graph of $Ax + By = 0$ is a line passing through the origin, (0, 0).

[YOU TRY 3]

Graph $x - y = 0$.

4 Graph Linear Equations of the Forms $x = a$ and $y = b$

An equation like $x = -2$ is a linear equation in two variables since it can be written in the form $x + 0y = -2$. The same is true for $y = 3$. It can be written as $0x + y = 3$. Let's see how we can graph these equations.

EXAMPLE 4

Graph $x = -2$.

Solution

The equation $x = -2$ means that *no matter the value of y, x always equals* −2. We can make a table of values where we choose any value for y, but x is always −2.

Plot the points, and draw a line through them. The graph of $x = -2$ is a **vertical line.**

x	y
−2	0
−2	1
−2	−2

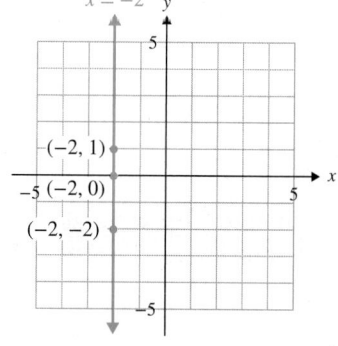

We can generalize the result as follows:

Hint

If x is constant, then the graphed line is vertical, perpendicular to the x-axis.

Property The Graph of $x = a$

If a is a constant, then the graph of $x = a$ is a **vertical line** going through the point $(a, 0)$.

[YOU TRY 4] Graph $x = 2$.

EXAMPLE 5 Graph $y = 3$.

Solution

The equation $y = 3$ means that *no matter the value of x, y always equals* 3. Make a table of values where we choose any value for x, but y is always 3.

Plot the points, and draw a line through them. The graph of $y = 3$ is a **horizontal line.**

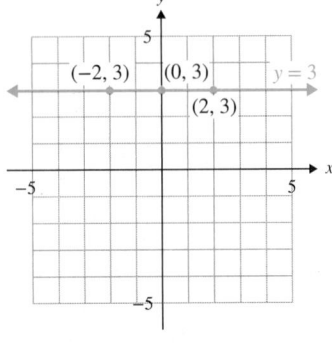

x	y
0	3
2	3
−2	3

W Hint

If y is constant, then the graphed line is horizontal, perpendicular to the y-axis.

We can generalize the result as follows:

Property The Graph of $y = b$

If b is a constant, then the graph of $y = b$ is a **horizontal line** going through the point $(0, b)$.

[YOU TRY 5] Graph $y = -4$.

5 Model Data with a Linear Equation

Linear equations are often used to model (or describe mathematically) real-world data. We can use these equations to learn what has happened in the past or predict what will happen in the future.

EXAMPLE 6 The annual cost of college tuition and fees at a private, four-year institution can be modeled by

$$y = 1172.50x + 24{,}909$$

where x is the number of years after 2015 and y is the cost of tuition and fees, in dollars.

a) Find the y-intercept of the graph of this equation and explain its meaning.

b) Find the approximate cost of tuition and fees in 2016 and 2019. Write the information as ordered pairs.

c) Graph $y = 1172.50x + 24{,}909$.

d) Use the graph to approximate the cost of tuition and fees in 2020. Is this the same result as when you use the equation to estimate the cost?

Hero/Corbis/Glow Images

Solution

a) To find the *y*-intercept, let $x = 0$.

$$y = 1172.50(0) + 24{,}909$$
$$y = 24{,}909$$

The *y*-intercept is (0, 24,909). What does this represent?

The problem states that *x* is the number of years *after* 2015. Therefore, $x = 0$ represents zero years after 2015, which is the year 2015.

The *y*-intercept (0, 24,909) tells us that in 2015, the cost of tuition and fees at this private four-year institution was $24,909.

b) The approximate cost of tuition and fees in

2016: First, realize that $x \neq 2016$. *x* is the number of years *after* 2015. Since 2016 is 1 year after 2015, $x = 1$. Let $x = 1$ in $y = 1172.50x + 24{,}909$ and find *y*.

$$y = 1172.50(1) + 24{,}909$$
$$y = 1172.50 + 24{,}909$$
$$y = 26{,}081.50$$

In 2016, the approximate cost of college tuition and fees at this school was $26,081.50. We can write this information as the ordered pair (1, 26,081.50).

2019: Begin by finding *x*. 2019 is 4 years after 2015, so $x = 4$.

$$y = 1172.50(4) + 24{,}909$$
$$y = 4690 + 24{,}909$$
$$y = 29{,}599$$

In 2019, the approximate cost of college tuition and fees at this private four-year school was $29,599.

The ordered pair (4, 29,599) can be written from this information.

c) We will plot the points (0, 24,909), (1, 26,081.50), and (4, 29,599). Label the axes, and choose an appropriate scale for each.

The *x*-coordinates of the ordered pairs range from 0 to 4, so we will let each mark in the *x*-direction represent 1 unit. (Sometimes, we let each tick mark represent more than one unit.) The *y*-coordinates of the ordered pairs range from 24,909 to 29,599. We will let each mark in the *y*-direction represent 2000 units.

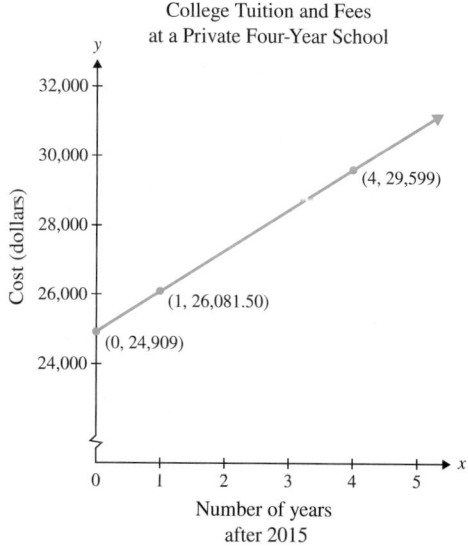

College Tuition and Fees at a Private Four-Year School

d) Using the graph to estimate the cost of tuition and fees in 2020, we locate $x = 5$ on the *x*-axis since 2020 is 5 years after 2015. When $x = 5$, we move straight up the graph to $y \approx 30{,}700$. Our approximation from the graph is $30,700.

If we use the equation and let $x = 5$, we get

$$y = 1172.50x + 24{,}909$$
$$y = 1172.50(5) + 24{,}909$$
$$y = 30{,}771.50$$

From the equation, we find that the cost of college tuition and fees at this private four-year school was about 30,771.50. The numbers are not exactly the same, but they are close.

A graphing calculator can be used to graph an equation and to verify information that we find using algebra. We will graph the equation $y = -\frac{1}{2}x + 2$ and then find the intercepts both algebraically and using the calculator.

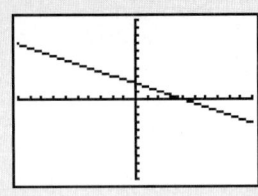

First, enter the equation into the calculator. Press ZOOM and select 6:Zstandard to graph the equation.

1) Find, algebraically, the y-intercept of the graph of $y = -\frac{1}{2}x + 2$. Is it consistent with the graph of the equation?

2) Find, algebraically, the x-intercept of the graph of $y = -\frac{1}{2}x + 2$. Is it consistent with the graph of the equation?

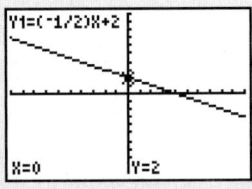

Now let's verify the intercepts using the graphing calculator. To find the y-intercept, press TRACE after displaying the graph. The cursor is automatically placed at the center x-value on the screen, which is at the point $(0, 2)$ as shown in the top graph. To find the x-intercept, press TRACE, type 4, and press ENTER. The calculator displays $(4, 0)$ as shown in the second graph. This is consistent with the intercepts found in 1 and 2, using algebra.

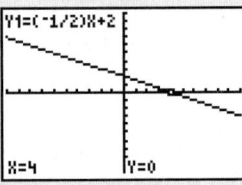

Use algebra to find the x- and y-intercepts of the graph of each equation. Then, use the graphing calculator to verify your results.

1) $y = 2x - 4$ 2) $y = x + 3$ 3) $y = -x + 5$

4) $2x - 5y = 10$ 5) $3x + 4y = 24$ 6) $3x - 7y = 21$

ANSWERS TO [YOU TRY] EXERCISES

1) a)

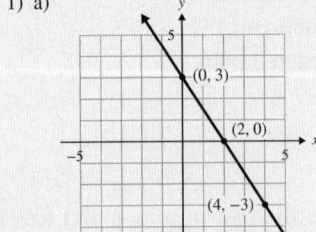

b)

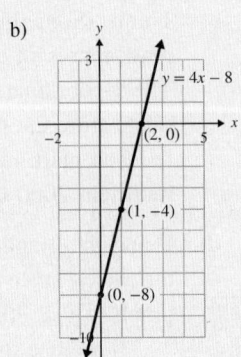

2)

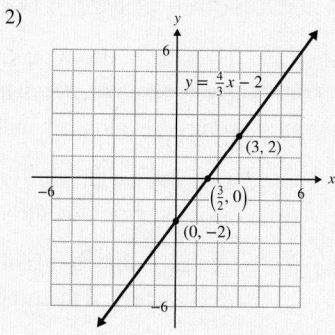

$y = \frac{4}{3}x - 2$
(3, 2)
$\left(\frac{3}{2}, 0\right)$
(0, −2)

3)

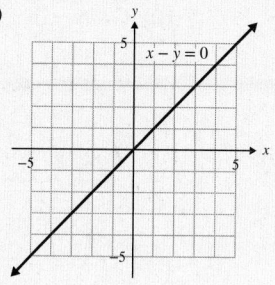

$x - y = 0$

4)

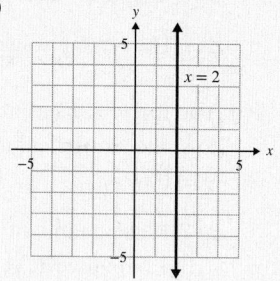

$x = 2$

5)

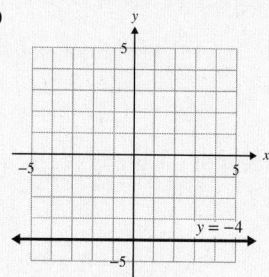

$y = -4$

ANSWERS TO TECHNOLOGY EXERCISES

1) (2, 0), (0, −4)   2) (−3, 0), (0, 3)   3) (5, 0), (0, 5)

4) (5, 0), (0, −2)   5) (8, 0), (0, 6)   6) (7, 0), (0, −3)

 **Evaluate** **4.2** Exercises   Do the exercises, and check your work.

Objective 1: Graph a Linear Equation by Plotting Points

1) The graph of a linear equation in two variables is a _____.

2) Every linear equation in two variables has how many solutions?

Complete the table of values, and graph each equation.

 3) $y = -2x + 4$

x	y
0	
−1	
2	
3	

4) $y = 3x - 2$

x	y
0	
1	
2	
−1	

5) $y = \frac{3}{2}x + 7$

x	y
0	
2	
−2	
−4	

6) $y = -\frac{5}{3}x + 3$

x	y
0	
−3	
3	
6	

7) $2x = 3 - y$

x	y
	0
0	
$\frac{1}{2}$	
	5

8) $-x + 5y = 10$

x	y
0	
	0
	4
−3	

9) $x = -\dfrac{4}{9}$

x	y
5	
0	
-1	
-2	

10) $y + 5 = 0$

x	y
	0
	-3
	-1
	2

Mixed Exercises: Objectives 1–4

11) What is the y-intercept of the graph of an equation? How do you find it?

12) What is the x-intercept of the graph of an equation? How do you find it?

Graph each equation by finding the intercepts and at least one other point.

13) $y = x - 1$

14) $y = -x + 3$

15) $3x - 4y = 12$

16) $2x - 7y = 14$

17) $x = -\dfrac{4}{3}y - 2$

18) $x = \dfrac{5}{4}y - 5$

19) $2x - y = 8$

20) $3x + y = -6$

21) $y = -x$

22) $y = 3x$

23) $4x - 3y = 0$

24) $6y - 5x = 0$

25) $x = 5$

26) $y = -4$

27) $y = 0$

28) $x = 0$

29) $x - \dfrac{4}{3} = 0$

30) $y + 1 = 0$

31) $4x - y = 9$

32) $x + 3y = -5$

For Exercises 33–36, fill in the blank with *always, sometimes,* or *never* to make the statement true.

33) The x-intercept of the graph of an equation will _____ have a y-coordinate of 0.

34) The y-intercept of the graph of an equation will _____ have a y-coordinate of 0.

35) The graph of a linear equation will _____ pass through all four quadrants of the Cartesian coordinate system.

36) If a linear equation contains only one variable, its graph will _____ pass through the origin.

37) Which ordered pair is a solution to every linear equation of the form $Ax + By = 0$?

38) True or False: The graph of $Ax + By = 0$ will always pass through the origin.

Objective 5: Model Data with a Linear Equation

39) The cost of downloading popular songs from iTunes is given by $y = 1.29x$, where x represents the number of songs downloaded and y represents the cost, in dollars.

a) Make a table of values using $x = 0, 4, 7,$ and 12, and write the information as ordered pairs.

b) Explain the meaning of each ordered pair in the context of the problem.

c) Graph the equation. Use an appropriate scale.

d) How many songs could you download for $11.61?

40) The force, y, measured in newtons (N), required to stretch a particular spring x meters is given by $y = 100x$.

a) Make a table of values using $x = 0, 0.5, 1.0,$ and 1.5, and write the information as ordered pairs.

b) Explain the meaning of each ordered pair in the context of the problem.

c) Graph the equation. Use an appropriate scale.

d) If the spring was pulled with a force of 80 N, how far did it stretch?

41) The number of doctorate degrees awarded in electrical engineering in the United States from 2000 to 2015 can be modeled by $y = 45x + 1327$, where x represents the number of years after 2000, and y represents the number of doctorate degrees awarded. The actual data are graphed here. (www.nsf.gov)

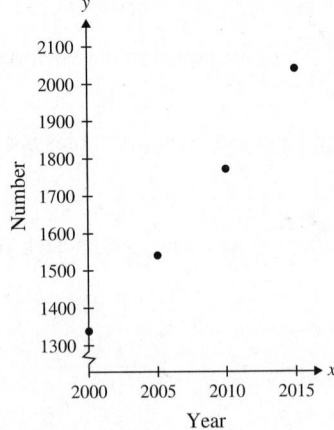

Number of Electrical Engineering Doctorates Awarded in the U.S.

a) From the graph, estimate the number of electrical engineering doctorates awarded in 2005 and 2015.

b) Determine the number of degrees awarded during the same years using the equation. Are the numbers close?

c) Graph the line that models the data given on the original graph.

d) What is the *y*-intercept of the graph of this equation, and what does it represent? How close is it to the actual point plotted on the given graph?

e) If the trend continues, how many electrical engineering doctorates will be awarded in 2022? Use the equation.

42) The number of females participating in high school sports in the United States from 2014 to 2018 can be modeled by $y = 41x + 3258$, where *x* represents the number of years after 2014, and *y* represents the number of female participants, in thousands. The actual data are graphed here. (www.nfhs.org)

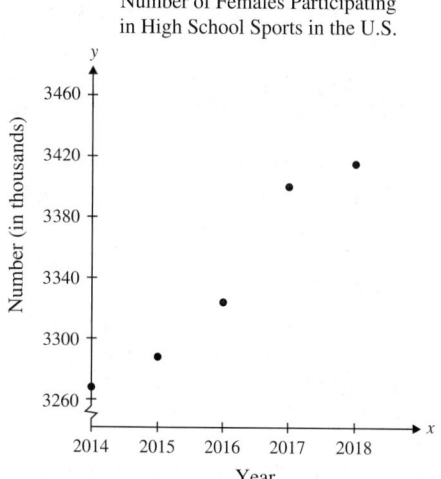

Number of Females Participating
in High School Sports in the U.S.

a) From the graph, estimate the number of high school female athletes in 2017 and 2018.

b) Using the equation, determine the number of high school female athletes in 2017 and 2018, and compare the result to part a). Are the numbers close?

c) Use the equation to graph the line that models the data given on the original graph.

d) What is the *y*-intercept of the graph of this equation, and what does it represent? How close is it to the actual point plotted on the given graph?

e) If the trend continues, how many high school females can be expected to participate in sports in 2025? Use the equation.

R Rethink

R1) In which other courses, besides math, have you had to use an equation of a line?

R2) When using the intercepts to graph a line, why should you always find a third point?

R3) How can you tell whether a line is going to be drawn vertically?

R4) How can you tell whether a line is going to be drawn horizontally?

R5) Explain how graph paper has helped you to draw your graphs neatly.

4.3 The Slope of a Line

What are your objectives for Section 4.3?	How can you accomplish each objective?
1 Understand the Concept of Slope	• Write the property for **Slope of a Line** in your own words. • Complete the given example on your own. • Complete You Try 1.
2 Find the Slope of a Line Given Two Points on the Line	• Learn the formula for finding **The Slope of a Line,** and write it in your notes. • Understand and learn the property that explains **Positive and Negative Slopes.** • Complete the given example on your own. • Complete You Try 2.
3 Use Slope to Solve Applied Problems	• Follow the example to know what questions to ask when solving applied problems. • Complete the given example on your own.
4 Find the Slope of Horizontal and Vertical Lines	• Write the property for **Slopes of Horizontal and Vertical Lines** in your own words. • Complete the given example on your own. • Complete You Try 3.
5 Use Slope and One Point on a Line to Graph the Line	• Follow Example 5 to create a procedure for using the slope and one point to graph the line. • Complete You Try 4.

W Work   Read the explanations, follow the examples, take notes, and complete the You Trys.

1   Understand the Concept of Slope

In Section 4.2, we learned to graph lines by plotting points. You may have noticed that some lines are steeper than others. Their "slants" are different, too.

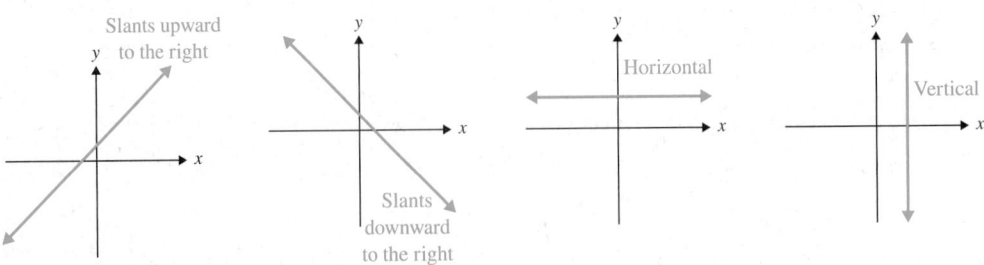

We can describe the steepness of a line with its *slope*.

Property Slope of a Line

The **slope** of a line measures its steepness. It is the ratio of the vertical change in y to the horizontal change in x. Slope is denoted by m.

We can also think of slope as a rate of change. *Slope* is the rate of change between two points. More specifically, it describes the rate of change in y to the change in x.

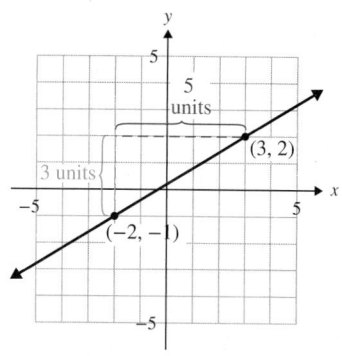

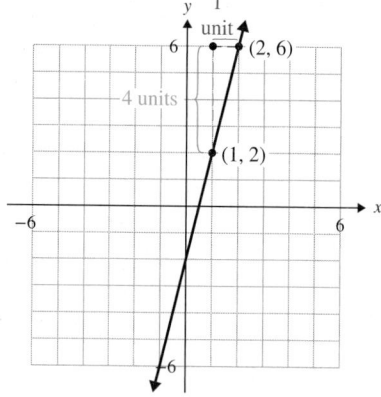

> **W Hint**
>
> Use two clearly defined points on the line to find the slope.

$$\text{Slope} = \frac{3}{5} \quad \leftarrow \text{Vertical change} \atop \leftarrow \text{Horizontal change}$$

$$\text{Slope} = 4 \text{ or } \frac{4}{1} \quad \leftarrow \text{Vertical change} \atop \leftarrow \text{Horizontal change}$$

> **W Hint**
>
> As the magnitude of the slope gets larger, the line gets steeper.

For example, in the graph on the left, the line changes 3 units vertically for every 5 units it changes horizontally. Its slope is $\frac{3}{5}$. The line on the right changes 4 units vertically for every 1 unit of horizontal change. It has a slope of $\frac{4}{1}$ or 4.

Notice that the line with slope 4 is steeper than the line that has a slope of $\frac{3}{5}$.

EXAMPLE 1

A sign along a highway through the Rocky Mountains is shown on the left. What does it mean?

Solution

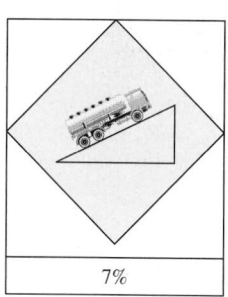

Percent means "out of 100." Therefore, we can write 7% as $\frac{7}{100}$. We can interpret $\frac{7}{100}$ as the ratio of the vertical change in the road to horizontal change in the road.

$$\text{The slope of the road is } \frac{7}{100}. \quad \leftarrow \text{Vertical change} \atop \leftarrow \text{Horizontal change}$$

The highway rises 7 ft for every 100 horizontal feet.

[YOU TRY 1]

The slope of a conveyer belt is $\frac{5}{12}$, where the dimensions of the ramp are in inches. What does this mean?

2 Find the Slope of a Line Given Two Points on the Line

Here is line *L*. The points (x_1, y_1) and (x_2, y_2) are two points on line *L*. *We will find the ratio of the vertical change in y to the horizontal change in x between the points (x_1, y_1) and (x_2, y_2).*

To get from (x_1, y_1) to (x_2, y_2), we move *vertically* to point *P* then *horizontally* to (x_2, y_2). The *x*-coordinate of point *P* is x_1, and the *y*-coordinate of *P* is y_2.

When we moved *vertically* from (x_1, y_1) to point $P(x_1, y_2)$, how far did we go? We moved a vertical distance $y_2 - y_1$.

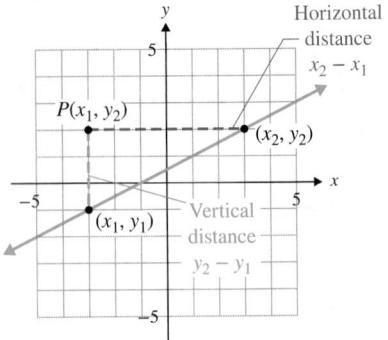

> **Note**
> The vertical change is $y_2 - y_1$ and is called the **rise**.

Then we moved *horizontally* from point $P(x_1, y_2)$ to (x_2, y_2). How far did we go? We moved a horizontal distance $x_2 - x_1$.

> **Note**
> The horizontal change is $x_2 - x_1$ and is called the **run**.

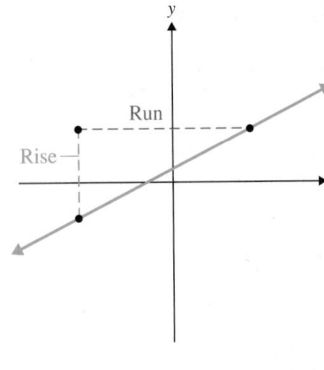

We said that the slope of a line is the ratio of the vertical change (rise) to the horizontal change (run). Therefore,

> **Formula** The Slope of a Line
>
> The **slope**, *m*, of a line containing the points (x_1, y_1) and (x_2, y_2) is given by
>
> $$m = \frac{\text{Vertical change}}{\text{Horizontal change}} = \frac{y_2 - y_1}{x_2 - x_1}$$

W Hint

Notice that the slope of a line is a ratio. It is the ratio of $(y_2 - y_1)$ to $(x_2 - x_1)$.

We can also think of slope as:

$$\frac{\text{Rise}}{\text{Run}} \quad \text{or} \quad \frac{\text{Change in } y}{\text{Change in } x}.$$

Let's look at some different ways to determine the slope of a line.

EXAMPLE 2

Determine the slope of each line.

a)

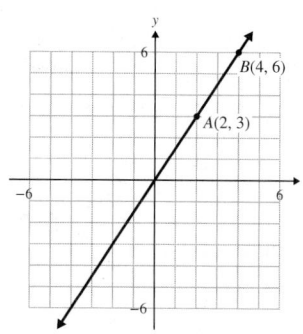

b)

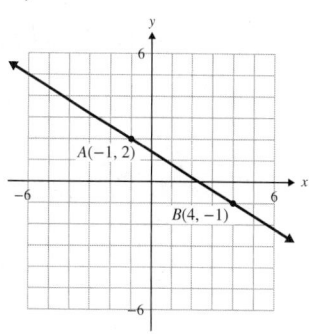

Solution

a) We will find the slope in two ways.

i) First, we will find the vertical change and the horizontal change by counting these changes as we go from A to B.

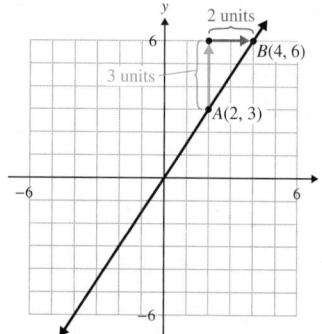

Vertical change (change in y) from A to B: 3 units

Horizontal change (change in x) from A to B: 2 units

$$\text{Slope} = \frac{\text{Change in } y}{\text{Change in } x} = \frac{3}{2} \quad \text{or} \quad m = \frac{3}{2}$$

W Hint

If a line slopes *upward* from left to right, the slope is positive. If the line slopes *downward* from left to right, the slope is negative.

ii) We can also find the slope using the formula.
Let $(x_1, y_1) = (2, 3)$ and $(x_2, y_2) = (4, 6)$.

$$m = \frac{y_2 - y_1}{x_2 - x_1} = \frac{6 - 3}{4 - 2} = \frac{3}{2}$$

You can see that we get the same result either way we find the slope.

b) i) First, find the slope by counting the vertical change and horizontal change as we go from A to B.

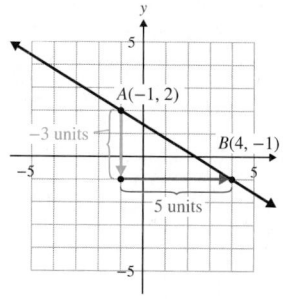

Vertical change (change in y) from A to B: -3 units

Horizontal change (change in x) from A to B: 5 units

$$\text{Slope} = \frac{\text{Change in } y}{\text{Change in } x} = \frac{-3}{5} = -\frac{3}{5}$$

$$\text{or } m = -\frac{3}{5}$$

ii) We can also find the slope using the formula.

Let $(x_1, y_1) = (-1, 2)$ and $(x_2, y_2) = (4, -1)$.

$$m = \frac{y_2 - y_1}{x_2 - x_1} = \frac{-1 - 2}{4 - (-1)} = \frac{-3}{5} = -\frac{3}{5}$$

Again, we obtain the same result using either method for finding the slope.

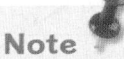

Note

The slope of $-\dfrac{3}{5}$ can be thought of as $\dfrac{-3}{5}$, $\dfrac{3}{-5}$, or $-\dfrac{3}{5}$.

$\left[\text{YOU TRY 2}\right]$ Determine the slope of each line by

a)   counting the vertical change and horizontal change.   b)   using the formula.

a)

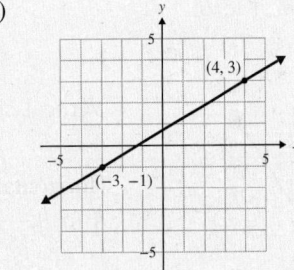

b)

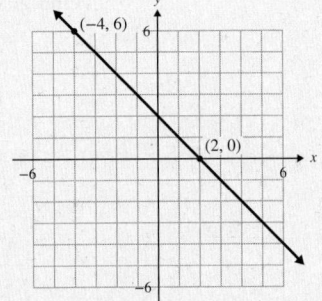

Notice that in Example 2a), the line has a positive slope and slants upward from left to right. As the value of x increases, the value of y increases as well. The line in Example 2b) has a negative slope and slants downward from left to right. Notice, in this case, that as the line goes from left to right, the value of x increases while the value of y decreases. We can summarize these results with the following general statements.

Property Positive and Negative Slopes

A line with a **positive slope** slants upward from left to right. As the value of x increases, the value of y increases as well.

A line with a **negative slope** slants downward from left to right. As the value of x increases, the value of y decreases.

3 Use Slope to Solve Applied Problems

EXAMPLE 3

The graph models the number of members of a certain health club from 2015 to 2019.

a) How many members did the club have in 2015? in 2019?

b) What does the sign of the slope of the line segment mean in the context of the problem?

c) Find the slope of the line segment, and explain what it means in the context of the problem.

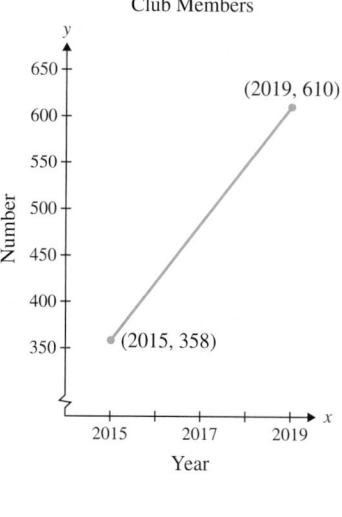

Number of Health Club Members

Solution

a) We can determine the number of members by reading the graph. In 2015, there were 358 members, and in 2019 there were 610 members.

b) The positive slope tells us that from 2015 to 2019 the number of members was increasing.

c) Let $(x_1, y_1) = (2015, 358)$ and $(x_2, y_2) = (2019, 610)$.

$$\text{Slope} = \frac{y_2 - y_1}{x_2 - x_1} = \frac{610 - 358}{2019 - 2015} = \frac{252}{4} = 63$$

The slope of the line is 63. Therefore, the number of members of the health club increased by 63 per year between 2015 and 2019.

4 Find the Slope of Horizontal and Vertical Lines

EXAMPLE 4

Find the slope of the line containing each pair of points.
a) $(-4, 1)$ and $(2, 1)$ b) $(2, 4)$ and $(2, -3)$

Solution

a) Let $(x_1, y_1) = (-4, 1)$ and $(x_2, y_2) = (2, 1)$.

$$m = \frac{y_2 - y_1}{x_2 - x_1} = \frac{1 - 1}{2 - (-4)} = \frac{0}{6} = 0$$

W Hint

Horizontal lines are going neither up nor down; therefore, they have a slope of zero.

If we plot the points, we see that they lie on a horizontal line. Each point on the line has a y-coordinate of 1, so $y_2 - y_1$ *always* equals zero. **The slope of every horizontal line is zero.**

b) Let $(x_1, y_1) = (2, 4)$ and $(x_2, y_2) = (2, -3)$.

$$m = \frac{y_2 - y_1}{x_2 - x_1} = \frac{-3 - 4}{2 - 2} = \frac{-7}{0} \quad \text{undefined}$$

We say that the slope is undefined. Plotting these points gives us a vertical line. Each point on the line has an x-coordinate of 2, so $x_2 - x_1$ *always* equals zero.

The slope of every vertical line is undefined.

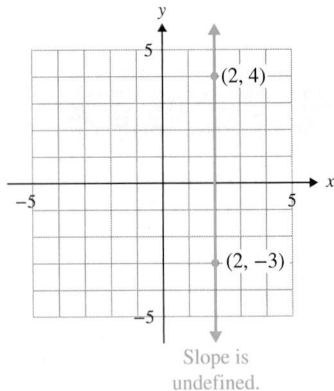

Slope is undefined.

[YOU TRY 3]

Find the slope of the line containing each pair of points.

a) (4, 9) and (−3, 9)   b) (−7, 2) and (−7, 0)

Property Slopes of Horizontal and Vertical Lines

The slope of a horizontal line, $y = b$, is **zero**. The slope of a vertical line, $x = a$, is **undefined.** We can also say that a vertical line has **no slope.** (*a* and *b* are constants.)

5 Use Slope and One Point on a Line to Graph the Line

We have seen how we can find the slope of a line given two points on the line. Now, we will see how we can use the slope and *one* point on the line to graph the line.

EXAMPLE 5

Graph the line containing the point

a) (−1, −2) with a slope of $\frac{3}{2}$.

b) (0, 1) with a slope of −3.

Solution

a) Plot the point.

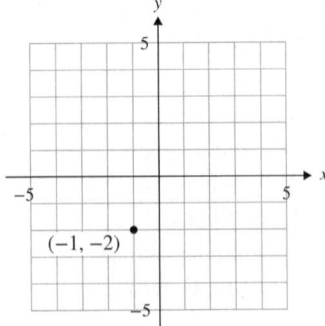

Use the slope to find another point on the line.

$$m = \frac{3}{2} = \frac{\text{Change in } y}{\text{Change in } x}$$

To get from the point (−1, −2) to another point on the line, move up 3 units in the y-direction and right 2 units in the x-direction.

Plot this second point, and draw a line through the two points.

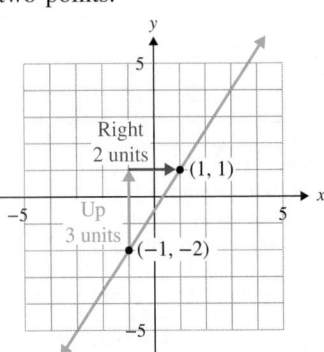

b) Plot the point (0, 1).

What does the slope, $m = -3$, mean?

$$m = -3 = \frac{-3}{1} = \frac{\text{Change in } y}{\text{Change in } x}$$

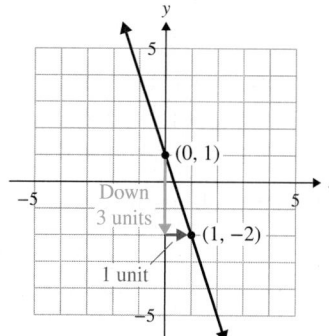

To get from (0, 1) to another point on the line, we will move *down* 3 units in the *y*-direction and *right* 1 unit in the *x*-direction. We end up at (1, −2).

Plot this point, and draw a line through (0, 1) and (1, −2).

In part b), we could have written $m = -3$ as $m = \frac{3}{-1}$. This would have given us a different point on the same line.

YOU TRY 4

Graph the line containing the point

a)   (−2, 1) with a slope of $-\frac{3}{2}$.   b)   (0, −3) with a slope of 2.

c)   (3, 2) with an undefined slope.

Using Technology

When we look at the graph of a linear equation, we should be able to estimate its slope. Use the equation $y = x$ as a guideline.

Step 1: Graph the equation $y = x$.

We can make the graph a thick line (so we can tell it apart from the others) by moving the arrow all the way to the left and pressing ENTER:

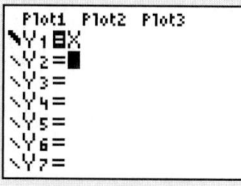 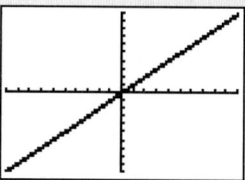

Step 2: Keeping this equation, graph the equation $y = 2x$:

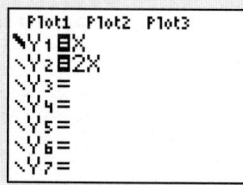 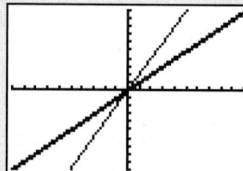

  a. Is the new graph steeper or flatter than the graph of $y = x$?
  b. Make a guess as to whether $y = 3x$ will be steeper or flatter than $y = x$. Test your guess by graphing $y = 3x$.

Step 3: Clear the equation $y = 2x$, and graph the equation $y = 0.5x$:

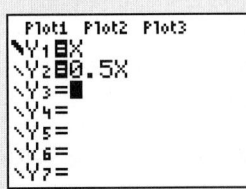 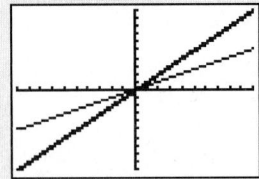

    a. Is the new graph steeper or flatter than the graph of $y = x$?

    b. Make a guess as to whether $y = 0.65x$ will be steeper or flatter than $y = x$. Test your guess by graphing $y = 0.65x$.

Step 4: Test similar situations, except with negative slopes: $y = -x$

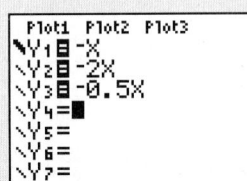 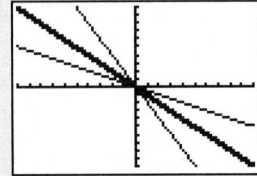

Did you notice that we have the same relationship, except in the opposite direction? That is, $y = 2x$ is steeper than $y = x$ in the positive direction, and $y = -2x$ is steeper than $y = -x$, but in the negative direction. And $y = 0.5x$ is flatter than $y = x$ in the positive direction, and $y = -0.5x$ is flatter than $y = -x$, but in the negative direction.

ANSWERS TO $\boxed{\text{YOU TRY}}$ **EXERCISES**

1) The belt rises 5 in. for every 12 horizontal inches.　　2) a) $m = \dfrac{4}{7}$　b) $m = -1$

3) a) $m = 0$　b) undefined

4) a)　　　　　　　　　　　b)　　　　　　　　　　　c)

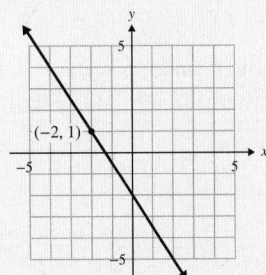 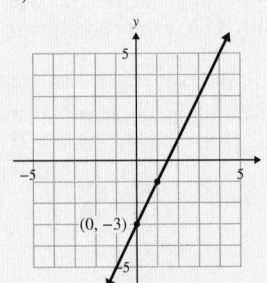 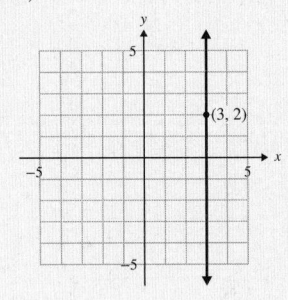

E Evaluate **4.3** Exercises　　　Do the exercises, and check your work.

Get Ready
Write each fraction in lowest terms.

1) $\dfrac{30}{54}$

2) $\dfrac{32}{12}$

3) $\dfrac{7}{-21}$

4) $\dfrac{-5}{40}$

5) $\dfrac{-84}{-12}$

6) $\dfrac{0}{3}$

Objective 1: Understand the Concept of Slope

 7) Explain the meaning of slope.

8) Describe the slant of a line with a negative slope.

9) Describe the slant of a line with a positive slope.

10) The slope of a horizontal line is _____.

11) The slope of a vertical line is _____.

12) If a line contains the points (x_1, y_1) and (x_2, y_2), write the formula for the slope of the line.

Mixed Exercises: Objectives 2 and 4

Determine the slope of each line by

    a)   counting the vertical change and the horizontal change as you move from one point to the other on the line;

       and

    b)   using the slope formula. (See Example 2.)

13) 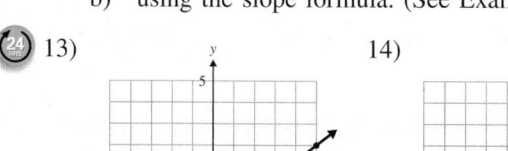   14)

15) 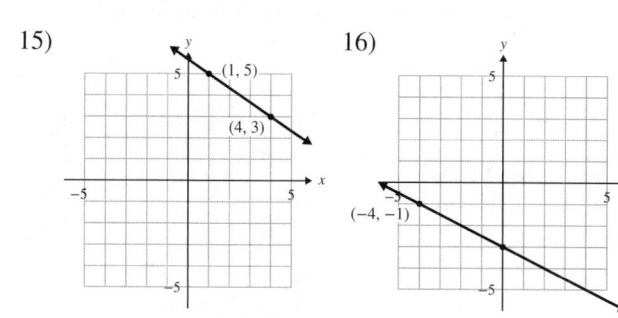   16)

17) 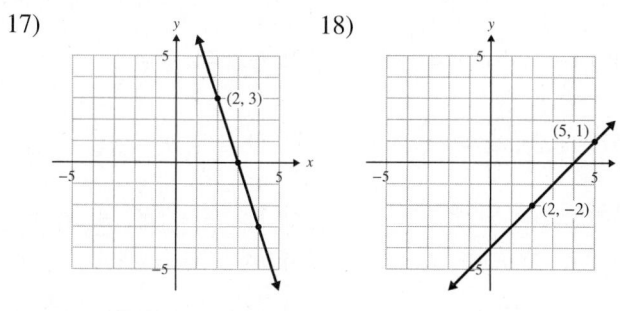   18)

19) 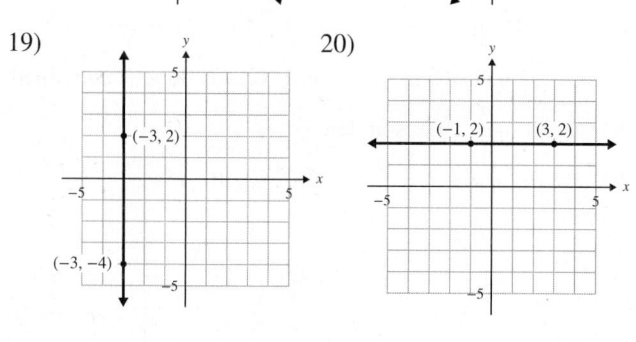   20)

21) Graph a line with a positive slope and a negative y-intercept.

22) Graph a line with a negative slope and a positive x-intercept.

Use the slope formula to find the slope of the line containing each pair of points.

23) $(2, 1)$ and $(0, -3)$ 24) $(0, 3)$ and $(9, 6)$

25) $(2, -6)$ and $(-1, 6)$ 26) $(-3, 9)$ and $(2, 4)$

27) $(-4, 3)$ and $(1, -8)$ 28) $(2, 0)$ and $(-5, 4)$

29) $(-2, -2)$ and $(-2, 7)$ 30) $(0, -6)$ and $(-9, -6)$

31) $(3, 5)$ and $(-1, 5)$ 32) $(1, 3)$ and $(1, -1)$

33) $\left(\dfrac{2}{3}, \dfrac{5}{2}\right)$ and $\left(-\dfrac{1}{2}, 2\right)$ 34) $\left(-\dfrac{1}{5}, \dfrac{3}{4}\right)$ and $\left(\dfrac{1}{3}, -\dfrac{3}{5}\right)$

35) $(3.5, -1.4)$ and $(7.5, 1.6)$

36) $(-1.7, 10.2)$ and $(0.8, -0.8)$

Objective 3: Use Slope to Solve Applied Problems

37) The longest run at Ski Dubai, an indoor ski resort in the Middle East, has a vertical drop of about 60 m with a horizontal distance of about 395 m. What is the slope of this ski run? (www.skidxb.com)

38) The federal government requires that all wheelchair ramps in new buildings have a maximum slope of $\dfrac{1}{12}$. Does the following ramp meet this requirement? Give a reason for your answer. (www.access-board.gov)

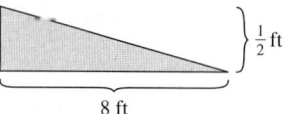

Use the following information for Exercises 39 and 40.

To minimize accidents, the Park District Risk Management Agency recommends that playground slides and sledding hills have a maximum slope of about 0.577. (Illinois Parks and Recreation)

39) Does this slide meet the agency's recommendations?

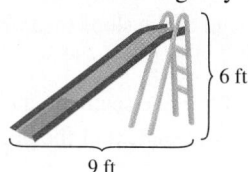

40) Does this sledding hill meet the agency's recommendations?

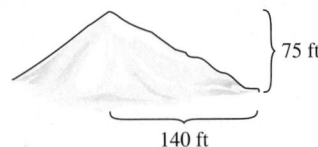

75 ft

140 ft

41) In Granby, Colorado, the first 50 ft of a driveway cannot have a slope of more than 5%. If the first 50 ft of a driveway rises 0.75 ft for every 20 ft of horizontal distance, does this driveway meet the building code? (http://co.grand.co.us)

Use the following information for Exercises 42–44.

The steepness (slope) of a roof on a house in a certain town cannot exceed $\frac{7}{12}$, also known as a 7:12 *pitch*.

The first number refers to the rise of the roof. The second number refers to how far over you must go (the run) to attain that rise.

42) Find the slope of a roof with a 12:20 pitch.

43) Find the slope of a roof with a 12:26 pitch.

44) Does the slope in Exercise 42 meet the town's building code? Give a reason for your answer.

45) The graph shows the approximate number of people in a metropolitan area injured in motor vehicle accidents from 2015 to 2019.

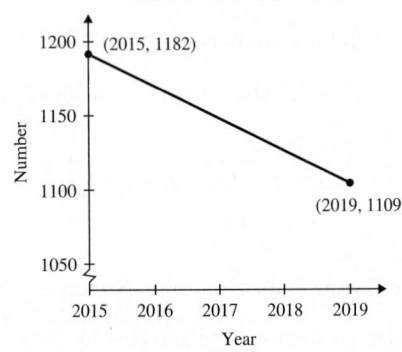

Number of People Injured in Motor Vehicle Accidents

a) Approximately how many people were injured in 2015? in 2019?

b) Without computing the slope, determine whether it is positive or negative.

c) What does the sign of the slope mean in the context of the problem?

d) Find the slope of the line segment, and explain what it means in the context of the problem.

46) The graph shows the approximate online and mail order revenue, in billions of dollars, from the sale of drugs, healthcare aids, and beauty aids in the United States from 2012 to 2016. (www.census.gov)

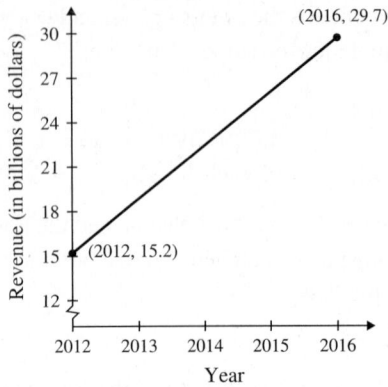

Online and Mail Order Revenue from Drugs, Health Aids, and Beauty Aids in the U.S.

a) What was the approximate revenue in 2012? in 2016?

b) Without computing the slope, determine whether it is positive or negative.

c) What does the sign of the slope mean in the context of the problem?

d) Find the slope of the line segment, and explain what it means in the context of the problem.

Objective 5: Use Slope and One Point on a Line to Graph the Line

Graph the line containing the given point and with the given slope.

47) $(2, 1)$; $m = \frac{3}{4}$

48) $(1, 2)$; $m = \frac{1}{3}$

49) $(-2, -3)$; $m = \frac{1}{4}$

50) $(-5, 0)$; $m = \frac{2}{5}$

51) $(1, 2)$; $m = -\frac{3}{4}$

52) $(1, -3)$; $m = -\frac{2}{5}$

53) $(-1, -3)$; $m = 3$

54) $(0, -2)$; $m = -2$

55) $(6, 2)$; $m = -4$

56) $(4, 3)$; $m = -5$

57) $(3, -4)$; $m = -1$

58) $(-1, -2)$; $m = 0$

59) $(-2, 3)$; $m = 0$

60) $(2, 0)$; slope is undefined.

61) $(-1, -4)$; slope is undefined.

62) $(0, 0)$; $m = 1$

63) $(0, 0)$; $m = -1$

R1) After completing the exercises, what steps could you take to graph a line if you knew only two points of the line?

R2) How would two lines with the same slope look on a graph?

R3) Are there any problems you could not do? If so, write them down or circle them and ask your instructor for help.

4.4 The Slope-Intercept Form of a Line

P Prepare

O Organize

What are your objectives for Section 4.4?	How can you accomplish each objective?
1 Define the Slope-Intercept Form of a Line	• Understand how the slope-intercept form of a line is derived. • Write the definition of the *slope-intercept form of a line* in your notes, and draw an example.
2 Graph a Line Expressed in Slope-Intercept Form	• Use Objective 5 from Section 4.3 to help graph the line. • Complete the given example on your own. • Complete You Try 1.
3 Rewrite an Equation in Slope-Intercept Form, and Graph the Line	• Solve for y to write equations in slope-intercept form. • Complete the given example on your own. • Complete You Try 2. • Review the different methods for graphing a line given its equation.
4 Use Slope to Determine Whether Two Lines Are Parallel or Perpendicular	• Understand the properties of **Parallel Lines** and **Perpendicular Lines.** • Complete the given examples on your own. • Develop steps to determine whether lines are parallel or perpendicular, and compare the steps. • Complete You Try 3.

W Work

Read the explanations, follow the examples, take notes, and complete the You Trys.

In Section 4.1, we learned that a linear equation in two variables can be written in the form $Ax + By = C$ (this is called **standard form**), where A, B, and C are real numbers and where both A and B do not equal zero. Equations of lines can take other forms, too, and we will look at one of those forms in this section.

1 Define the Slope-Intercept Form of a Line

We know that if (x_1, y_1) and (x_2, y_2) are points on a line, then the slope of the line is

$$m = \frac{y_2 - y_1}{x_2 - x_1}$$

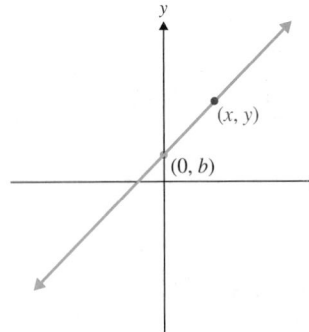

Recall that to find the y-intercept of a line, we let $x = 0$ and solve for y. Let one of the points on a line be the y-intercept $(0, b)$, where b is a number. Let another point on the line be (x, y). See the graph on the left.

Substitute the points $(0, b)$ and (x, y) into the slope formula:

Subtract y-coordinates.
↓
$$m = \frac{y_2 - y_1}{x_2 - x_1} = \frac{y - b}{x - 0} = \frac{y - b}{x}$$
↑
Subtract x-coordinates.

Solve $m = \dfrac{y - b}{x}$ for y.

$$mx = \frac{y - b}{x} \cdot x \qquad \text{Multiply by } x \text{ to eliminate the fraction.}$$

$$mx = y - b$$

$$mx + b = y - b + b \qquad \text{Add } b \text{ to each side to solve for } y.$$

$$mx + b = y$$

OR

$$y = mx + b \qquad \text{Slope-intercept form}$$

> **W Hint**
>
> Notice that in $y = mx + b$, the letter b represents the y-coordinate of the y-intercept.

Definition

The **slope-intercept form of a line** is $y = mx + b$, where m is the slope and $(0, b)$ is the y-intercept.

When an equation is in the form $y = mx + b$, we can quickly recognize the y-intercept and slope to graph the line.

2 Graph a Line Expressed in Slope-Intercept Form

EXAMPLE 1

Graph each equation.

a) $y = 4x - 3$

b) $y = \dfrac{1}{2}x$

Solution

Notice that each equation is in slope-intercept form, $y = mx + b$, where m is the slope and $(0, b)$ is the y-intercept.

a)   Graph $y = 4x - 3$.

Slope = 4, y-intercept is $(0, -3)$.

Plot the y-intercept first, then use the slope to locate another point on the line. Since the slope is 4, think of it as $\dfrac{4}{1}$.   ← Change in y
← Change in x

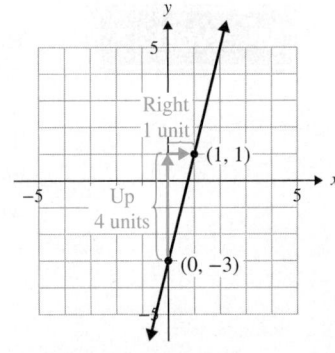

b) The equation $y = \dfrac{1}{2}x$ is the same as $y = \dfrac{1}{2}x + 0$.

Identify the slope and y-intercept.

$$\text{Slope} = \dfrac{1}{2}, \quad y\text{-intercept is } (0, 0).$$

Plot the y-intercept, then use the slope to locate another point on the line.

$\dfrac{1}{2}$ is equivalent to $\dfrac{-1}{-2}$, so we can use $\dfrac{-1}{-2}$ as the slope to locate yet another point on the line.

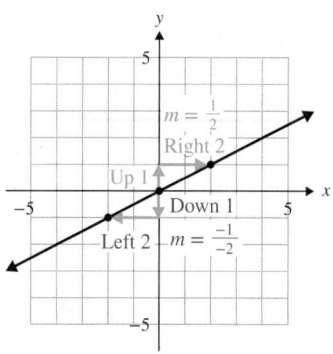

[**YOU TRY 1**] Graph each line using the slope and y-intercept.

a) $y = \dfrac{1}{4}x + 1$ b) $y = x - 3$ c) $y = -2x$

3 Rewrite an Equation in Slope-Intercept Form, and Graph the Line

Lines are not always written in slope-intercept form. They may be written in *standard form* (like $7x + 4y = 12$) or in another form such as $2x = 2y + 10$. We can put equations like these into slope-intercept form by solving the equation for y.

EXAMPLE 2 Write $7x + 4y = 12$ in slope-intercept form, and graph.

Solution

The slope-intercept form of a line is $y = mx + b$.
We must solve the equation for y.

$$7x + 4y = 12$$
$$4y = -7x + 12 \qquad \text{Add } -7x \text{ to each side.}$$
$$y = -\dfrac{7}{4}x + 3 \qquad \text{Divide each side by 4.}$$
$$\text{Slope} = -\dfrac{7}{4} \text{ or } \dfrac{-7}{4}; \quad y\text{-intercept is } (0, 3).$$

Plot the y-intercept, then use the slope to locate another point on the line.

$\left(\text{We could also have thought of the slope as } \dfrac{7}{-4}.\right)$

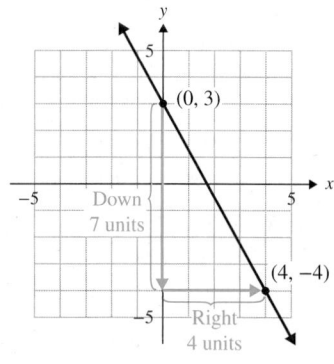

[**YOU TRY 2**] Put each equation into slope-intercept form, and graph.

a) $10x - 5y = -5$ b) $2x = -3 - 3y$

Summary Different Methods for Graphing a Line Given Its Equation

We have learned that we can use different methods for graphing lines. Given the equation of a line we can:

1) Make a table of values, plot the points, and draw the line through the points.

2) Find the *x*-intercept by letting $y = 0$ and solving for *x*, and find the *y*-intercept by letting $x = 0$ and solving for *y*. Plot the points, then draw the line through the points.

3) Put the equation into slope-intercept form, $y = mx + b$, identify the slope and *y*-intercept, then graph the line.

4 Use Slope to Determine Whether Two Lines Are Parallel or Perpendicular

Recall that two lines in a plane are **parallel** if they do not intersect. If we are given the equations of two lines, how can we determine whether they are parallel?

Here are the equations of two lines:

$$2x - 3y = -3 \qquad y = \frac{2}{3}x - 5$$

We will graph each line. To graph the first line, we write it in slope-intercept form.

$$-3y = -2x - 3 \qquad \text{Add } -2x \text{ to each side.}$$

$$y = \frac{-2}{-3}x - \frac{3}{-3} \qquad \text{Divide by } -3.$$

$$y = \frac{2}{3}x + 1 \qquad \text{Simplify.}$$

The slope-intercept form of the first line is $y = \frac{2}{3}x + 1$, and the second line is already in slope-intercept form, $y = \frac{2}{3}x - 5$. Now, graph each line.

These lines are parallel. Their slopes are the same, but they have different *y*-intercepts. (If the *y*-intercepts were the same, they would be the same line.) This is how we determine whether two (nonvertical) lines are parallel. They have the same slope, but different *y*-intercepts.

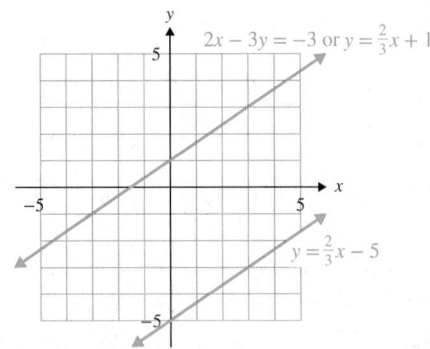

Property Parallel Lines

Parallel lines have the same slope. (If two lines are vertical, they are parallel. However, their slopes are undefined.)

EXAMPLE 3

Determine whether each pair of lines is parallel.

a) $2x + 8y = 12$
    $x + 4y = -20$

b) $y = -5x + 2$
    $5x - y = 7$

Solution

a) To determine whether the lines are parallel, we must find the slope of each line. If the slopes are the same, but the y-intercepts are different, the lines are parallel.

Write each equation in slope-intercept form.

$$2x + 8y = 12 \qquad\qquad x + 4y = -20$$
$$8y = -2x + 12 \qquad\qquad 4y = -x - 20$$
$$y = -\frac{2}{8}x + \frac{12}{8} \qquad\qquad y = -\frac{x}{4} - \frac{20}{4}$$
$$y = -\frac{1}{4}x + \frac{3}{2} \qquad\qquad y = -\frac{1}{4}x - 5$$

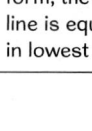

Hint

If the equation of a line is given in $Ax + By = C$ form, the slope of the line is equivalent to $-A/B$ in lowest terms.

Each line has a slope of $-\frac{1}{4}$. Their y-intercepts are different. Therefore, $2x + 8y = 12$ and $x + 4y = -20$ are parallel lines.

b) Again, we must find the slope of each line. $y = -5x + 2$ is already in slope-intercept form. Its slope is -5.

Write $5x - y = 7$ in slope-intercept form.

$$-y = -5x + 7 \qquad \text{Add } -5x \text{ to each side.}$$
$$y = 5x - 7 \qquad \text{Divide each side by } -1.$$

The slope of $y = -5x + 2$ is -5. The slope of $5x - y = 7$ is 5. The slopes are different; therefore, the lines are not parallel.

The slopes of two lines can tell us about another relationship between the lines. The slopes can tell us whether two lines are *perpendicular*.

Recall that two lines are **perpendicular** if they intersect at 90° angles.

The graphs of two perpendicular lines and their equations are on the left. We will see how their slopes are related.

Find the slope of each line by writing them in slope-intercept form.

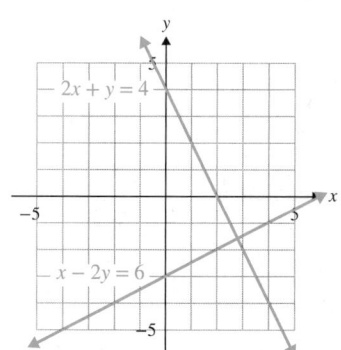

Line 1: $2x + y = 4$

$$y = -2x + 4$$

$$m = -2$$

Line 2: $x - 2y = 6$
$$-2y = -x + 6$$
$$y = \frac{-x}{-2} + \frac{6}{-2}$$
$$y = \frac{1}{2}x - 3$$
$$m = \frac{1}{2}$$

How are the slopes related? They are **negative reciprocals.** That is, if the slope of one line is a, then the slope of a line perpendicular to it is $-\frac{1}{a}$. This is how we determine whether two lines are perpendicular (where neither one is vertical).

Property Perpendicular Lines

Perpendicular lines have slopes that are negative reciprocals of each other.

EXAMPLE 4

Determine whether each pair of lines is perpendicular.

a) $15x - 12y = -4$ b) $2x - 9y = -9$
     $4x - 5y = 10$ $9x + 2y = 8$

Solution

a) To determine whether the lines are perpendicular, we must find the slope of each line. If the slopes are negative reciprocals, then the lines are perpendicular.

Write each equation in slope-intercept form.

$$
\begin{aligned}
15x - 12y &= -4 \\
-12y &= -15x - 4 \\
y &= \frac{-15}{-12}x - \frac{4}{-12} \\
y &= \frac{5}{4}x + \frac{1}{3} \\
m &= \frac{5}{4}
\end{aligned}
\qquad
\begin{aligned}
4x - 5y &= 10 \\
-5y &= -4x + 10 \\
y &= \frac{-4}{-5}x + \frac{10}{-5} \\
y &= \frac{4}{5}x - 2 \\
m &= \frac{4}{5}
\end{aligned}
$$

The slopes are reciprocals, but they are not *negative* reciprocals. Therefore, the lines are *not* perpendicular.

b) Begin by writing each equation in slope-intercept form so that we can find their slopes.

W Hint

Notice that if two lines are perpendicular, the product of their slopes will always equal −1.

$$
\begin{aligned}
2x - 9y &= -9 \\
-9y &= -2x - 9 \\
y &= \frac{-2}{-9}x - \frac{9}{-9} \\
y &= \frac{2}{9}x + 1 \\
m &= \frac{2}{9}
\end{aligned}
\qquad
\begin{aligned}
9x + 2y &= 8 \\
2y &= -9x + 8 \\
y &= -\frac{9}{2}x + \frac{8}{2} \\
y &= -\frac{9}{2}x + 4 \\
m &= -\frac{9}{2}
\end{aligned}
$$

The slopes are negative reciprocals; therefore, the lines are perpendicular.

YOU TRY 3

Determine whether each pair of lines is parallel, perpendicular, or neither.

a) $5x - y = -2$ b) $y = \frac{8}{3}x + 9$ c) $x + 2y = 8$ d) $x = 7$

     $3x + 15y = -20$ $-32x + 12y = 15$ $2x = 4y + 3$ $y = -4$

ANSWERS TO [YOU TRY] EXERCISES

1) a)

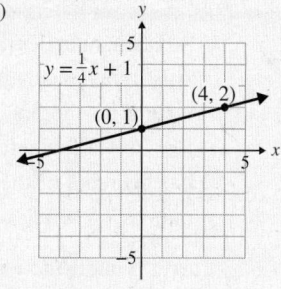

b)

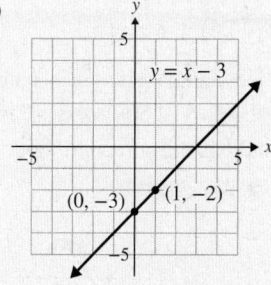

c)

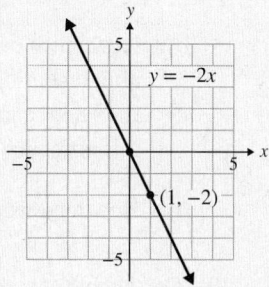

2) a)

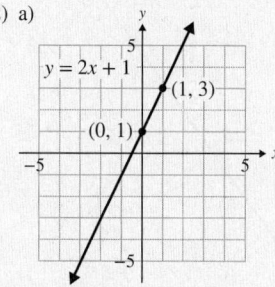

b)

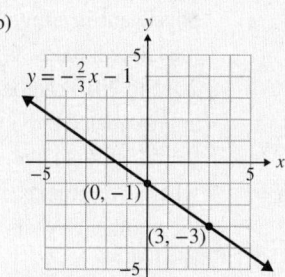

3) a) perpendicular
b) parallel
c) neither
d) perpendicular

E Evaluate **4.4** Exercises   Do the exercises, and check your work.

Mixed Exercises: Objectives 1 and 2

1) The slope-intercept form of a line is $y = mx + b$. What is the slope? What is the y-intercept?

2) How do you put an equation that is in standard form, $Ax + By = C$, into slope-intercept form?

Each of the following equations is in slope-intercept form. Identify the slope and the y-intercept, then graph each line using this information.

 3) $y = \dfrac{2}{5}x - 6$

4) $y = \dfrac{7}{5}x - 1$

5) $y = -\dfrac{3}{2}x + 3$ 6) $y = -\dfrac{1}{3}x + 2$

7) $y = \dfrac{3}{4}x + 2$ 8) $y = \dfrac{2}{3}x + 5$

9) $y = -2x - 3$ 10) $y = 3x - 1$

11) $y = 5x$ 12) $y = -2x + 5$

13) $y = -\dfrac{3}{2}x - \dfrac{7}{2}$

14) $y = \dfrac{3}{5}x + \dfrac{3}{4}$

15) $y = 6$

16) $y = -4$

Objective 3: Rewrite an Equation in Slope-Intercept Form, and Graph the Line

Put each equation into slope-intercept form, if possible, and graph.

 17) $x + 3y = -6$ 18) $x + 2y = -8$

19) $4x + 3y = 21$ 20) $2x - 5y = 5$

21) $2 = x + 3$

22) $x + 12 = 4$

23) $2x = 18 - 3y$

24) $98 = 49y - 28x$

25) $y + 2 = -3$

26) $y + 3 = 3$

27) Kolya has a part-time job, and his gross pay can be described by the equation $P = 12.50h$, where P is his gross pay, in dollars, and h is the number of hours worked.

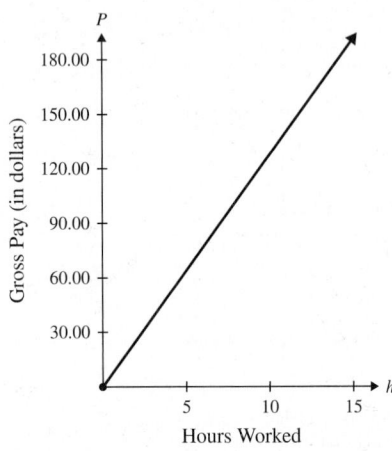

Kolya's Gross Pay

a) What is the P-intercept? What does it mean in the context of the problem?

b) What is the slope? What does it mean in the context of the problem?

c) Use the graph to find Kolya's gross pay when he works 12 hr. Confirm your answer using the equation.

28) The number of people y, in millions, leaving on cruises from the United States from 2012 to 2016 can be approximated by $y = 0.41x + 9.92$, where x is the number of years after 2012. (www.statista.com)

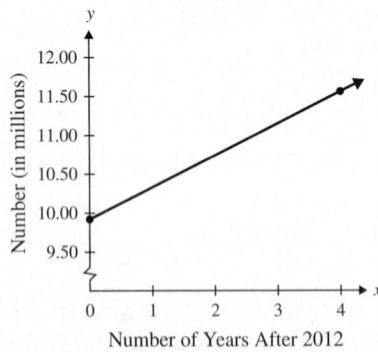

Number of People Leaving on
Cruises from the United States

a) What is the y-intercept? What does it mean in the context of the problem?

b) What is the slope? What does it mean in the context of the problem?

c) Use the graph to determine how many people left on cruises from the United States in 2015. Confirm your answer using the equation.

29) A Tasmanian devil is a marsupial that lives in Australia. Once a joey leaves its mother's pouch, its weight for the first 8 weeks can be approximated by $y = 2x + 18$, where x represents the number of weeks it has been out of the pouch and y represents its weight, in ounces. (Wikipedia and Animal Planet)

a) What is the y-intercept, and what does it represent?

b) How much does a Tasmanian devil weigh 3 weeks after emerging from the pouch?

c) Explain the meaning of the slope in the context of this problem.

d) How long would it take for a joey to weigh 32 oz?

30) The number of active physicians in a region in the United States, y, from 2015 to 2019 can be approximated by $y = 51.8x + 2424.4$, where x represents the number of years after 2015.

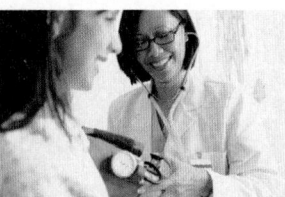

Fuse/Getty Images

a) What is the y-intercept and what does it represent?

b) How many doctors were practicing in 2019?

c) Explain the meaning of the slope in the context of this problem.

d) If the current trend continues, how many practicing doctors would this region have in 2028?

31) On a certain day in 2018, the exchange rate between the American dollar and the Indian rupee was given by $r = 72.00d$, where d represents the number of dollars and r represents the number of rupees.

a) What is the r-intercept and what does it represent?

b) What is the slope? What does it mean in the context of the problem?

c) If Madhura is going to India to visit her family, how many rupees could she get for $70.00?

d) How many dollars could be exchanged for 3240 rupees?

32) The value of a car, v, in dollars, t years after it is purchased is given by $v = -1800t + 20{,}000$.

    a)   What is the v-intercept, and what does it represent?

    b)   What is the slope? What does it mean in the context of the problem?

    c)   What is the car worth after 3 years?

    d)   When will the car be worth $11,000?

Write the slope-intercept form for the equation of a line with the given slope and y-intercept.

33) $m = -4$; y-int: $(0, 7)$

34) $m = -7$; y-int: $(0, 4)$

35) $m = \dfrac{9}{5}$; y-int: $(0, -3)$

36) $m = \dfrac{7}{4}$; y-int: $(0, -2)$

37) $m = -\dfrac{5}{2}$; y-int: $(0, -1)$

38) $m = \dfrac{1}{4}$; y-int: $(0, 7)$

39) $m = 1$; y-int: $(0, 2)$

40) $m = -1$; y-int: $(0, 0)$

41) $m = 0$; y-int: $(0, 0)$

42) $m = 0$; y-int: $(0, -8)$

Objective 4: Use Slope to Determine Whether Two Lines Are Parallel or Perpendicular

43) How do you know whether two lines are perpendicular?

44) How do you know whether two lines are parallel?

Determine whether each statement is true or false.

45) If two nonvertical lines are parallel, then the lines have different y-intercepts.

46) If two lines are perpendicular, then both lines must cross the x-axis.

47) If two lines are perpendicular, then one of the lines must have a positive slope and the other must have a negative slope.

48) If two lines are parallel and neither intersects the x-axis, then both lines have a slope of 0.

Determine whether each pair of lines is parallel, perpendicular, or neither.

49) $y = -x - 5$
    $y = x + 8$

50) $y = \dfrac{3}{4}x + 2$
    $y = \dfrac{3}{4}x - 1$

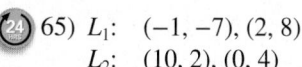 51) $y = \dfrac{2}{9}x + 4$
    $4x - 18y = 9$

52) $y = \dfrac{4}{5}x + 2$
    $5x + 4y = 12$

53) $3x - y = 4$
    $2x - 5y = -9$

54) $-4x + 3y = -5$
    $4x - 6y = -3$

55) $-x + y = -21$
    $y = 2x + 5$

56) $x + 3y = 7$
    $y = 3x$

57) $x + 7y = 4$
    $y - 7x = 4$

58) $5y - 3x = 1$
    $3x - 5y = -8$

59) $y = -\dfrac{1}{2}x$
    $x + 2y = 4$

60) $-4x + 6y = 5$
    $2x - 3y = -12$

61) $x = -1$
    $y = 6$

62) $y = 12$
    $y = 4$

63) $x = -4.3$
    $x = 0$

64) $x = 7$
    $y = 0$

Lines L_1 and L_2 contain the given points. Determine whether lines L_1 and L_2 are parallel, perpendicular, or neither.

65) L_1: $(-1, -7), (2, 8)$
    L_2: $(10, 2), (0, 4)$

66) L_1: $(0, -3), (-4, -11)$
    L_2: $(-2, 0), (3, 10)$

67) L_1: $(1, 10), (3, 8)$
    L_2: $(2, 4), (-5, -17)$

68) L_1: $(-1, 4), (2, -8)$
    L_2: $(8, 5), (0, 3)$

69) L_1: $(-3, 6), (4, -1)$
    L_2: $(-6, -5), (-10, -1)$

70) L_1: $(5, -5), (7, 11)$
    L_2: $(-3, 0), (6, 3)$

71) L_1: $(-6, 2), (-6, 1)$
    L_2: $(4, 0), (4, -5)$

72) L_1: $(8, 1), (7, 1)$
    L_2: $(12, -1), (-2, -1)$

73) L_1: $(7, 2), (7, 5)$
    L_2: $(-2, 0), (1, 0)$

74) L_1: $(-6, 4), (-6, -1)$
    L_2: $(-1, 10), (-3, 10)$

R1) Which equation from this chapter is the easiest for you to remember?

R2) Which equation from this chapter is the most difficult for you to remember?

R3) Using the standard form $Ax + By = C$, can you determine the slope and y-intercept of a line by inspection? Hint: Solve the equation $Ax + By = C$ for y.

R4) Which exercises in this section do you find most challenging?

4.5 Writing an Equation of a Line

P Prepare   **O** Organize

What are your objectives for Section 4.5?	How can you accomplish each objective?
1 Rewrite an Equation in Standard Form	• Complete the given example on your own, and create a procedure. • Complete You Try 1.
2 Write an Equation of a Line Given Its Slope and y-intercept	• Learn the procedure for **Writing an Equation of a Line Given Its Slope and y-intercept,** and write it in your notes. • Complete the given example on your own. • Complete You Try 2.
3 Use the Point-Slope Formula to Write an Equation of a Line Given Its Slope and a Point on the Line	• Learn the formula for **Point-Slope Form of a Line,** and write it in your notes. • Complete the given example on your own, and notice the different scenarios you might see. • Complete You Try 3.
4 Use the Point-Slope Formula to Write an Equation of a Line Given Two Points on the Line	• Learn the steps for writing an equation of a line given two points on the line. • Complete the given example on your own. • Complete You Try 4.
5 Write Equations of Horizontal and Vertical Lines	• Learn the formulas for **Equations of Horizontal and Vertical Lines,** and write them in your notes. • Complete the given example on your own. • Complete You Try 5. • Write the summary for *Writing Equations of Lines* in your own words.
6 Write an Equation of a Line That Is Parallel or Perpendicular to a Given Line	• Follow the examples, and create a step-by-step procedure for finding and writing the equations of parallel and perpendicular lines. • Complete You Trys 6 and 7.
7 Write a Linear Equation to Model Real-World Data	• Follow the example, and then complete the example in your notes without looking at the solution.

Read the explanations, follow the examples, take notes, and complete the You Trys.

So far in this chapter, we have been graphing lines given their equations. Now we will write an equation of a line when we are given information about it.

1 Rewrite an Equation in Standard Form

In Section 4.4, we practiced writing equations of lines in slope-intercept form. Here we will discuss how to write a line in **standard form,** $Ax + By = C$, **with the additional conditions that A, B, and C are integers and A is positive.**

EXAMPLE 1

Rewrite each linear equation in standard form.

a) $3x + 8 = -2y$

b) $y = -\dfrac{3}{4}x + \dfrac{1}{6}$

Solution

a) In standard form, the x- and y-terms are on the same side of the equation.

$$3x + 8 = -2y$$
$$3x = -2y - 8 \qquad \text{Subtract 8 from each side.}$$
$$3x + 2y = -8 \qquad \text{Add 2y to each side; the equation is now in standard form.}$$

W Hint

Review how to eliminate fractions from an equation as demonstrated in Section 3.2.

b) Since an equation $Ax + By = C$ is considered to be in standard form when A, B, and C are integers, the first step in writing $y = -\dfrac{3}{4}x + \dfrac{1}{6}$ in standard form is to eliminate the fractions.

$$y = -\dfrac{3}{4}x + \dfrac{1}{6}$$
$$12 \cdot y = 12\left(-\dfrac{3}{4}x + \dfrac{1}{6}\right) \qquad \text{Multiply both sides of the equation by 12.}$$
$$12y = -9x + 2$$
$$9x + 12y = 2 \qquad \text{Add 9x to each side.}$$

The standard form is $9x + 12y = 2$.

[YOU TRY 1]

Rewrite each equation in standard form.

a) $5x = 3 + 11y$

b) $y = \dfrac{1}{3}x - 7$

In the rest of this section, we will learn how to write equations of lines given information about their graphs.

2 Write an Equation of a Line Given Its Slope and y-intercept

> **Procedure** Write an Equation of a Line Given Its Slope and y-intercept
>
> If we are given the slope and y-intercept of a line, use $y = mx + b$ and substitute those values into the equation.

EXAMPLE 2

Find an equation of the line with slope $= -6$ and y-intercept $(0, 5)$.

Solution

Since we are told the slope and y-intercept, use $y = mx + b$.

$$m = -6 \quad \text{and} \quad b = 5$$

Substitute these values into $y = mx + b$ to get $y = -6x + 5$.

[YOU TRY 2]

Find an equation of the line with slope $= \dfrac{5}{8}$ and y-intercept $(0, -9)$.

W Hint

Remember, a y-intercept will always have an x-coordinate of 0.

3 Use the Point-Slope Formula to Write an Equation of a Line Given Its Slope and a Point on the Line

When we are given the slope of a line and a point on that line, we can use another method to find its equation. This method comes from the formula for the slope of a line.

Let (x_1, y_1) be a given point on a line, and let (x, y) be any other point on the same line, as shown in the figure. The slope of that line is

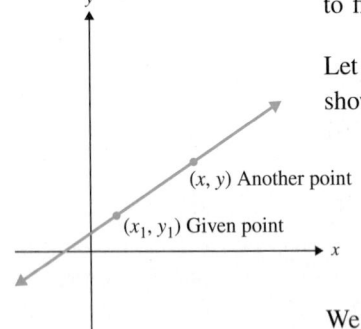

$$m = \frac{y - y_1}{x - x_1} \qquad \text{Definition of slope}$$

$$m(x - x_1) = y - y_1 \qquad \text{Multiply each side by } x - x_1.$$

$$y - y_1 = m(x - x_1) \qquad \text{Rewrite the equation.}$$

We have found the *point-slope form* of the equation of a line.

W Hint

The point-slope formula will help us write an equation of a line. We will not express our final answer in this form. We will write our answer in either slope-intercept form or in standard form.

> **Formula** Point-Slope Form of a Line
>
> The **point-slope form of a line** is
>
> $$y - y_1 = m(x - x_1)$$
>
> where (x_1, y_1) is a point on the line and m is its slope.

EXAMPLE 3

A line has slope -4 and contains the point $(1, 5)$. Find the standard form for the equation of the line.

Solution

Although we are told to find the *standard form* for the equation of the line, we do not try to immediately "jump" to standard form. First, ask yourself, *"What information am I given?"*

We are given the *slope* and a *point on the line*. Therefore, we will begin by using the point-slope formula. Our *last* step will be to put the equation in standard form.

Use $y - y_1 = m(x - x_1)$. Substitute -4 for m. Substitute $(1, 5)$ for (x_1, y_1).

$$y - y_1 = m(x - x_1)$$
$$y - 5 = -4(x - 1) \qquad \text{Substitute 1 for } x_1 \text{ and 5 for } y_1; \text{ let } m = -4.$$
$$y - 5 = -4x + 4 \qquad \text{Distribute.}$$

To write the answer in standard form, we must get the x- and y-terms on the same side of the equation so that the coefficient of x is positive.

$$4x + y - 5 = 4 \qquad \text{Add } 4x \text{ to each side.}$$
$$4x + y = 9 \qquad \text{Add 5 to each side.}$$

The standard form of the equation is $4x + y = 9$.

W Hint

Notice that when you write the equation of a line using the point-slope formula, you will always substitute the coordinates of the point for x_1 and y_1.

YOU TRY 3

a) A line has slope -8 and contains the point $(-4, 5)$. Find the standard form for the equation of the line.

b) Find an equation of the line containing the point $(5, 3)$ with slope $= 2$. Express the answer in slope-intercept form.

---

4 Use the Point-Slope Formula to Write an Equation of a Line Given Two Points on the Line

We are now ready to discuss how to write an equation of a line when we are given two points on a line.

To write an equation of a line given two points on the line,

a) use the points to find the slope of the line

then

b) use the slope and *either one* of the points in the point-slope formula.

EXAMPLE 4

Write an equation of the line containing the points $(4, 9)$ and $(2, 6)$. Express the answer in slope-intercept form.

Solution

We are given two points on the line, so first, we will find the slope.

$$m = \frac{6 - 9}{2 - 4} = \frac{-3}{-2} = \frac{3}{2}$$

We will use the slope and *either one* of the points in the point-slope formula. (Each point will give the same result.) We will use $(4, 9)$.

Substitute $\dfrac{3}{2}$ for m. Substitute $(4, 9)$ for (x_1, y_1).

$$y - y_1 = m(x - x_1)$$

$$y - 9 = \frac{3}{2}(x - 4) \qquad \text{Substitute 4 for } x_1 \text{ and 9 for } y_1; \text{ let } m = \frac{3}{2}.$$

$$y - 9 = \frac{3}{2}x - 6 \qquad \text{Distribute.}$$

We must write our answer in slope-intercept form, $y = mx + b$, so solve the equation for y.

$$y = \frac{3}{2}x + 3 \qquad \text{Add 9 to each side to solve for } y.$$

The equation is $y = \dfrac{3}{2}x + 3$.

[**YOU TRY 4**] Find the slope-intercept form for the equation of the line containing the points $(4, 2)$ and $(1, -5)$.

5 Write Equations of Horizontal and Vertical Lines

Earlier we learned that the slope of a horizontal line is zero and that it has equation $y = b$, where b is a constant. The slope of a vertical line is undefined, and its equation is $x = a$, where a is a constant.

Formula Equations of Horizontal and Vertical Lines

Equation of a Horizontal Line: The equation of a horizontal line containing the point (a, b) is $y = b$.
Equation of a Vertical Line: The equation of a vertical line containing the point (a, b) is $x = a$.

EXAMPLE 5 Write an equation of the horizontal line containing the point $(7, -1)$.

Solution

The equation of a horizontal line has the form $y = b$, where b is the y-coordinate of the point. The equation of the line is $y = -1$.

[**YOU TRY 5**] Write an equation of the horizontal line containing the point $(3, -8)$.

Summary Writing Equations of Lines

If you are given

1) **the slope and y-intercept of the line,** use $y = mx + b$ and substitute those values into the equation.

2) **the slope of the line and a point on the line,** use the point-slope formula:

$$y - y_1 = m(x - x_1)$$

Substitute the slope for m and the point you are given for (x_1, y_1). Write your answer in slope-intercept or standard form.

3) **two points on the line,** find the slope of the line and then use the slope and *either one* of the points in the point-slope formula. Write your answer in slope-intercept or standard form.

The equation of a **horizontal line** containing the point (a, b) is **$y = b$.**

The equation of a **vertical line** containing the point (a, b) is **$x = a$.**

6 Write an Equation of a Line That Is Parallel or Perpendicular to a Given Line

In Section 4.4, we learned that parallel lines have the same slope, and perpendicular lines have slopes that are negative reciprocals of each other. We will use this information to write the equation of a line that is parallel or perpendicular to a given line.

EXAMPLE 6

A line contains the point $(2, -2)$ and is parallel to the line $y = \frac{1}{2}x + 1$. Write the equation of the line in slope-intercept form.

Solution

Let's look at the graph on the left to help us understand what is happening in this example. We must find the equation of the line in red. It is the line containing the point $(2, -2)$ that is parallel to $y = \frac{1}{2}x + 1$.

The line $y = \frac{1}{2}x + 1$ has $m = \frac{1}{2}$. Therefore, the red line will have $m = \frac{1}{2}$ as well.

We know the slope, $\frac{1}{2}$, and a point on the line, $(2, -2)$, so we use the point-slope formula to find its equation.

Substitute $\frac{1}{2}$ for m. Substitute $(2, -2)$ for (x_1, y_1).

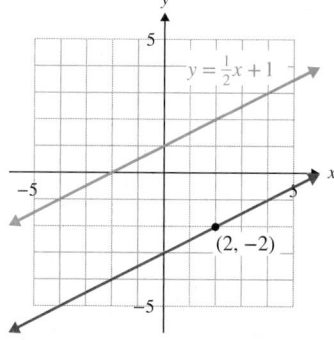

$$y - y_1 = m(x - x_1)$$

$$y - (-2) = \frac{1}{2}(x - 2) \qquad \text{Substitute 2 for } x_1 \text{ and } -2 \text{ for } y_1; \text{ let } m = \frac{1}{2}.$$

$$y + 2 = \frac{1}{2}x - 1 \qquad \text{Distribute.}$$

$$y = \frac{1}{2}x - 3 \qquad \text{Subtract 2 from each side.}$$

The equation is $y = \frac{1}{2}x - 3$.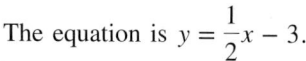

A line contains the point $(-6, 2)$ and is parallel to the line $y = -\dfrac{3}{2}x + \dfrac{1}{4}$. Write the equation of the line in slope-intercept form.

EXAMPLE 7

Find the standard form for the equation of the line that contains the point $(-4, 3)$ and that is perpendicular to $3x - 4y = 8$.

Solution

Begin by finding the slope of $3x - 4y = 8$ by putting it into slope-intercept form.

$$3x - 4y = 8$$
$$-4y = -3x + 8 \qquad \text{Add } -3x \text{ to each side.}$$
$$y = \frac{-3}{-4}x + \frac{8}{-4} \qquad \text{Divide by } -4.$$
$$y = \frac{3}{4}x - 2 \qquad \text{Simplify.}$$
$$m = \frac{3}{4}$$

W Hint

Are you writing out the example as you are reading it?

Then, determine the slope of the line containing $(-4, 3)$ by finding the *negative reciprocal* of the slope of the given line. $\qquad m_{\text{perpendicular}} = -\dfrac{4}{3}$

The line we want has $m = -\dfrac{4}{3}$ and contains the point $(-4, 3)$. Use the point-slope formula to find an equation of the line.

Substitute $-\dfrac{4}{3}$ for m.

Substitute $(-4, 3)$ for (x_1, y_1).

$$y - y_1 = m(x - x_1)$$
$$y - 3 = -\frac{4}{3}(x - (-4)) \qquad \text{Substitute } -4 \text{ for } x_1 \text{ and } 3 \text{ for } y_1; \text{ let } m = -\frac{4}{3}.$$
$$y - 3 = -\frac{4}{3}(x + 4)$$
$$y - 3 = -\frac{4}{3}x - \frac{16}{3} \qquad \text{Distribute.}$$

Since we are asked to write the equation in standard form, eliminate the fractions by multiplying each side by 3.

$$3(y - 3) = 3\left(-\frac{4}{3}x - \frac{16}{3}\right)$$
$$3y - 9 = -4x - 16 \qquad \text{Distribute.}$$
$$3y = -4x - 7 \qquad \text{Add 9 to each side.}$$
$$4x + 3y = -7 \qquad \text{Add } 4x \text{ to each side.}$$

The equation is $4x + 3y = -7$.

[YOU TRY 7]

Find the equation of the line perpendicular to $5x - y = -6$ containing the point $(10, 0)$. Write the equation in standard form.

---

7 Write a Linear Equation to Model Real-World Data

Equations of lines are often used to describe real-world situations. We will look at an example in which we must find the equation of a line when we are given some data.

EXAMPLE 8

Since 2012, the number of charging points for electric vehicles in the United States has increased by about 6925 per year. There were approximately 40,000 charging points in 2016. (www.forbes.com)

Anton Gvozdikov/Shutterstock

a) Write a linear equation to model these data. Let x represent the number of years after 2012, and let y represent the number of electric vehicle charging points.

b) Find the number of charging points in 2012 and in 2017.

Solution

a) The statement "the number of charging points … has increased by about 6925 per year" tells us the rate of change of the number with respect to time. Therefore, this is the *slope*. It will be *positive* because the number is *increasing*.

$$m = 6925$$

The statement "There were approximately 40,000 charging points in 2016" gives us a point on the line.

Let x = the number of years after 2012. Then, the year 2016 corresponds to $x = 4$.

Let y = the number of electric vehicle charging points. Then, 40,000 charging points corresponds to $y = 40,000$.

A point on the line is (4, 40,000).

Use the *slope* and the *point* in the *point-slope formula* to write an equation of the line that models the data in the problem.

Substitute 6925 for m. Substitute (4, 40,000) for (x_1, y_1).

$$y - y_1 = m(x - x_1)$$

$y - 40,000 = 6925(x - 4)$	Substitute 4 for x_1 and 40,000 for y_1.
$y - 40,000 = 6925x - 27,700$	Distribute.
$y = 6925x + 12,300$	Add 40,000 to each side.

The equation is $y = 6925x + 12{,}300$.

b) **To determine the number of charging points in 2012,** let $x = 0$ in the equation because x = the number of years *after* 2012.

$y = 6925(0) + 12,300$	Substitute 0 for x.
$y = 12,300$	

In 2012, there were approximately 12,300 electric vehicle charging points in the United States. Notice that the equation we found in part a) is in slope-intercept form, $y = mx + b$, and our result is b. That is because when we find the y-intercept, we let $x = 0$.

To determine the number of charging points in 2017, let $x = 5$ because 2017 is 5 years *after* 2012.

$y = 6925(5) + 12,300$	Substitute 5 for x.
$y = 34,625 + 12,300$	Multiply.
$y = 46,925$	

In 2017, there were approximately 46,925 electric vehicle charging points in the United States.

We can use a graphing calculator to explore what we have learned about perpendicular lines.

1. Graph the line $= -2x + 4$. What is its slope?

2. Find the slope of the line perpendicular to the graph of $y = -2x + 4$.

3. Find the equation of the line perpendicular to $y = -2x + 4$ that passes through the point (6, 0). Express the equation in slope-intercept form.

4. Graph both the original equation and the equation of the perpendicular line:

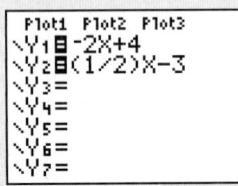
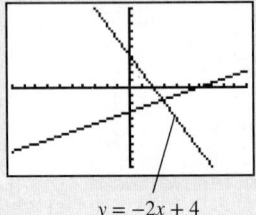

$y = -2x + 4$

5. Do the lines above appear to be perpendicular?

6. Press ZOOM and choose 5:Zsquare.

$y = \frac{1}{2}x - 3$

7. Do the graphs look perpendicular now? Because the viewing window on a graphing calculator is a rectangle, *squaring* the window will give a more accurate picture of the graphs of the equations.

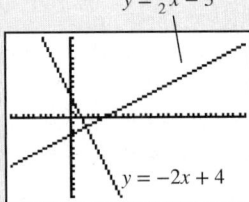

$y = -2x + 4$

ANSWERS TO YOU TRY **EXERCISES**

1) a) $5x - 11y = 3$ b) $x - 3y = 21$ 2) $y = \frac{5}{8}x - 9$ 3) a) $8x + y = -27$ b) $y = 2x - 7$

4) $y = \frac{7}{3}x - \frac{22}{3}$ 5) $y = -8$ 6) $y = -\frac{3}{2}x - 7$ 7) $x + 5y = 10$

ANSWERS TO TECHNOLOGY EXERCISES

1) -2 2) $\frac{1}{2}$ 3) $y = \frac{1}{2}x - 3$ 5) No, because they do not meet at 90° angles.

7) Yes, because they meet at 90° angles.

E Evaluate   **4.5**   Evaluate   Exercises   Do the exercises, and check your work.

Objective 1: Rewrite an Equation in Standard Form
Rewrite each equation in standard form.

1) $y = -2x - 4$
2) $y = 3x + 5$
3) $x = y + 1$
4) $x = -4y - 9$
5) $y = \frac{4}{5}x + 1$
6) $y = \frac{2}{3}x - 6$
7) $y = -\frac{1}{3}x - \frac{5}{4}$
8) $y = -\frac{1}{4}x + \frac{2}{5}$

Objective 2: Write an Equation of a Line Given Its Slope and y-intercept

9) Explain how to find an equation of a line when you are given the slope and y-intercept of the line.

Find an equation of the line with the given slope and y-intercept. Express your answer in the indicated form.

10) $m = -3$, y-int: (0, 3); slope-intercept form

11) $m = -7$, y-int: $(0, 2)$; slope-intercept form

12) $m = 1$, y-int: $(0, -4)$; standard form

13) $m = -4$, y-int: $(0, 6)$; standard form

14) $m = -\dfrac{2}{5}$, y-int: $(0, -4)$; standard form

15) $m = \dfrac{2}{7}$, y-int: $(0, -3)$; standard form

16) $m = 1$, y-int: $(0, 0)$; slope-intercept form

17) $m = -1$, y-int: $(0, 0)$; slope-intercept form

18) $m = \dfrac{5}{9}$, y-int: $\left(0, -\dfrac{1}{3}\right)$; standard form

Objective 3: Use the Point-Slope Formula to Write an Equation of a Line Given Its Slope and a Point on the Line

19) a) If (x_1, y_1) is a point on a line with slope m, then the point-slope formula is _____.

   b) Explain how to find an equation of a line when you are given the slope and a point on the line.

Find an equation of the line containing the given point with the given slope. Express your answer in the indicated form.

20) $(2, 3)$, $m = 4$; slope-intercept form

21) $(5, 7)$, $m = 1$; slope-intercept form

22) $(-2, 5)$, $m = -3$; slope-intercept form

23) $(4, -1)$, $m = -5$; slope-intercept form

24) $(-1, -2)$, $m = 2$; standard form

25) $(-2, -1)$, $m = 4$; standard form

26) $(9, 3)$, $m = -\dfrac{1}{3}$; standard form

27) $(-5, 8)$, $m = \dfrac{2}{5}$; standard form

28) $(-2, -3)$, $m = \dfrac{1}{8}$; slope-intercept form

29) $(5, 1)$, $m = -\dfrac{5}{4}$; slope-intercept form

30) $(4, 0)$, $m = -\dfrac{3}{16}$; standard form

31) $(-3, 0)$, $m = \dfrac{5}{6}$; standard form

32) $\left(\dfrac{1}{4}, -1\right)$, $m = 3$; slope-intercept form

Objective 4: Use the Point-Slope Formula to Write an Equation of a Line Given Two Points on the Line

33) Explain how to find an equation of a line when you are given two points on the line.

Find an equation of the line containing the two given points. Express your answer in the indicated form.

34) $(-2, 1)$ and $(8, 11)$; slope-intercept form

35) $(-1, 7)$ and $(3, -5)$; slope-intercept form

36) $(6, 8)$ and $(-1, -4)$; slope-intercept form

37) $(4, 5)$ and $(7, 11)$; slope-intercept form

38) $(2, -1)$ and $(5, 1)$; standard form

39) $(-2, 4)$ and $(1, 3)$; slope-intercept form

40) $(-1, 10)$ and $(3, -2)$; standard form

41) $(-5, 1)$ and $(4, -2)$; standard form

42) $(4.2, 1.3)$ and $(-3.4, -17.7)$; slope-intercept form

43) $(-3, -11)$ and $(3, -1)$; standard form

44) $(-6, 0)$ and $(3, -1)$; standard form

45) $(-2.3, 8.3)$ and $(5.1, -13.9)$; slope-intercept form

46) $(-7, -4)$ and $(14, 2)$; standard form

Write the slope-intercept form of the equation of each line, if possible.

47)

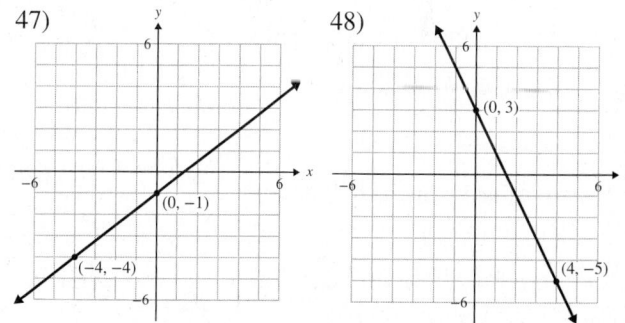

48)

49)

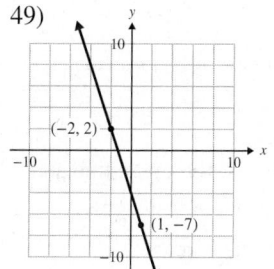

50)

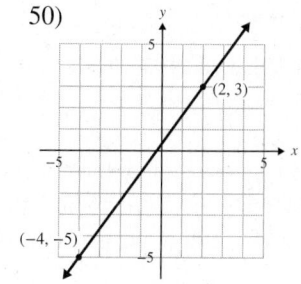

51)

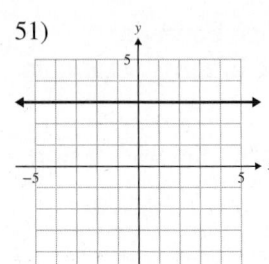

52)

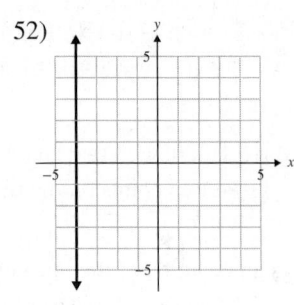

73) $y = 4x + 2$; $(-1, -4)$; standard form

74) $y = \dfrac{2}{3}x - 6$; $(6, 6)$; standard form

75) $x + 2y = 22$; $(-4, 7)$; standard form

76) $3x + 5y = -6$; $(-5, 8)$; standard form

77) $15x - 3y = 1$; $(-2, -12)$; slope-intercept form

78) $x + 6y = 12$; $(-6, 8)$; slope-intercept form

Mixed Exercises: Objectives 2–5

Write the slope-intercept form of the equation of the line, if possible, given the following information.

53) contains $(-4, 7)$ and $(2, -1)$

54) $m = 2$ and contains $(-3, -2)$

55) $m = 1$ and contains $(3, 5)$

56) $m = \dfrac{7}{5}$ and y-intercept $(0, -4)$

57) y-intercept $(0, 6)$ and $m = 7$

58) contains $(-3, -3)$ and $(1, -7)$

59) vertical line containing $(3, 5)$

60) vertical line containing $\left(\dfrac{1}{2}, 6\right)$

61) horizontal line containing $(2, 3)$

62) horizontal line containing $(5, -4)$

63) $m = -4$ and y-intercept $(0, -4)$

64) $m = -\dfrac{2}{3}$ and contains $(3, -1)$

65) $m = -3$ and contains $(10, -10)$

66) contains $(0, 3)$ and $(5, 0)$

67) contains $(-4, -4)$ and $(2, -1)$

68) $m = -1$ and y-intercept $(0, 0)$

Objective 6: Write an Equation of a Line That Is Parallel or Perpendicular to a Given Line

69) What can you say about the equations of two parallel lines?

70) What can you say about the equations of two perpendicular lines?

Write an equation of the line *parallel* to the given line and containing the given point. Write the answer in slope-intercept form or in standard form, as indicated.

71) $y = 4x + 9$; $(0, 2)$; slope-intercept form

72) $y = 8x + 3$; $(0, -3)$; slope-intercept form

Write an equation of the line *perpendicular* to the given line and containing the given point. Write the answer in slope-intercept form or in standard form, as indicated.

79) $y = -\dfrac{2}{3}x + 4$; $(4, 2)$; slope-intercept form

80) $y = -\dfrac{5}{3}x + 10$; $(10, 5)$; slope-intercept form

81) $y = -5x + 1$; $(10, 0)$; standard form

82) $y = \dfrac{1}{4}x - 9$; $(-1, 7)$; standard form

83) $y = x$; $(4, -9)$; slope-intercept form

84) $x + y = 9$; $(4, 4)$; slope-intercept form

85) $x + 3y = 18$; $(4, 2)$; standard form

86) $12x - 15y = 10$; $(16, -25)$; standard form

Write the slope-intercept form (if possible) of the equation of the line meeting the given conditions.

87) parallel to $3x + y = 8$ containing $(-4, 0)$

88) perpendicular to $x - 5y = -4$ containing $(3, 5)$

89) perpendicular to $y = x - 2$ containing $(2, 9)$

90) parallel to $y = 4x - 1$ containing $(-3, -8)$

91) parallel to $y = 1$ containing $(-3, 4)$

92) parallel to $x = -3$ containing $(-7, -5)$

93) perpendicular to $x = 0$ containing $(9, 2)$

94) perpendicular to $y = 4$ containing $(-4, -5)$

95) perpendicular to $21x - 6y = 2$ containing $(4, -1)$

96) parallel to $-3x + 4y = 8$ containing $(9, 4)$

97) parallel to $y = 0$ containing $\left(4, -\dfrac{3}{2}\right)$

98) perpendicular to $y = \dfrac{7}{3}$ containing $(-7, 9)$

Objective 7: Write a Linear Equation to Model Real-World Data

99) The graph shows the number of employees in health care and social assistance in the United States from

2008 to 2013. x represents the number of years after 2008 so that x = 0 represents 2008, x = 1 represents 2009, and so on. Let y represent the number of these employees, in millions. (www.census.gov)

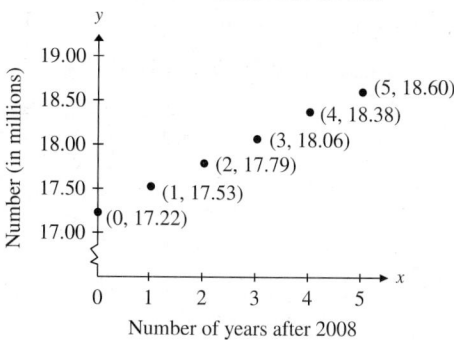

Number of Employees in Health Care and Social Assistance in the U.S.

a) Write a linear equation to model these data. Use the data points for 2008 and 2013.

b) Explain the meaning of the slope in the context of this problem.

c) If this current trend continues, find the number of employees in health care and social assistance in 2020.

100) The graph shows a dieter's weight over a 12-week period. Let y represent his weight x weeks after beginning his diet.

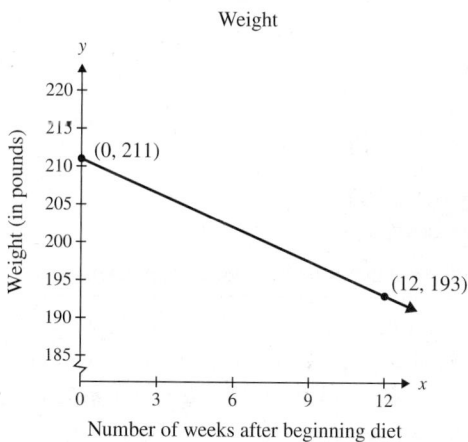

Weight

a) Write a linear equation to model these data. Use the data points for week 0 and week 12.

b) What is the meaning of the slope in the context of this problem?

c) If he keeps losing weight at the same rate, what will he weigh 13 weeks after he started his diet?

101) In 2017, a grocery store chain had an advertising budget of $500,000 per year. Every year since then its budget has been cut by $15,000 per year. Let y represent the advertising budget, in dollars, x years after 2017.

a) Write a linear equation to model these data.

b) Explain the meaning of the slope in the context of the problem.

c) What will the advertising budget be in 2020?

d) If the current trend continues, in what year will the advertising budget be $365,000?

102) A temperature of −10°C is equivalent to 14°F, while 15°C is the same as 59°F. Let F represent the temperature on the Fahrenheit scale and C represent the temperature on the Celsius scale.

a) Write a linear equation to convert from degrees Celsius to degrees Fahrenheit. That is, write an equation for F in terms of C.

b) Explain the meaning of the slope in the context of the problem.

c) Convert 24°C to degrees Fahrenheit.

d) Change 95°F to degrees Celsius.

103) A kitten weighs an average of 100 g at birth and should gain about 8 g per day for the first few weeks of life. Let y represent the weight of a kitten, in grams, x days after birth. (http://veterinarymedicine.dvm360.com)

a) Write a linear equation to model these data.

b) Explain the meaning of the slope in the context of the problem.

c) How much would an average kitten weigh 5 days after birth? 2 weeks after birth?

Siede Preis/Getty Images

d) How long would it take for a kitten to reach a weight of 284 g?

104) In 2009, the Pampered Pups Salon charged $50.00 for basic grooming for a large dog. The price changed at the same rate every year, and in 2019 the price was $65.00. Let y represent the cost for basic grooming x years after 2009.

a) Write a linear equation to model these data.

b) Explain the meaning of the slope in the context of the problem.

c) Find the cost to get a large dog groomed in 2012.

d) In what year was the cost $62.00?

105) If a woman wears a size 6 on the U.S. shoe size scale, her European size is 38. A U.S. women's size 8.5 corresponds to a European size 42. Let A represent the U.S. women's shoe size, and let E represent that size on the European scale.

a) Write a linear equation that models the European shoe size in terms of the U.S. shoe size.

b) If a woman's U.S. shoe size is 7.5, what is her European shoe size? (Round to the nearest unit.)

106) If a man's foot is 11.5 in. long, his U.S. shoe size is 12.5. A man wears a size 8 if his foot is 10 in. long. Let L represent the length of a man's foot, and let S represent his shoe size.

a) Write a linear equation that describes the relationship between shoe size in terms of the length of a man's foot.

b) If a man's foot is 10.5 in. long, what is his shoe size?

R Rethink

R1) Which concepts from the previous sections did you have to learn before completing this exercise set?

R2) What is the least amount of information you need about a line in order to write its equation or graph it?

R3) Why do equations of vertical or horizontal lines involve only one variable term?

4.6 Introduction to Functions

P Prepare

O Organize

What are your objectives for Section 4.6?	How can you accomplish each objective?
1 Define and Identify Relations, Functions, Domain, and Range	• Learn the definitions of *relation*, *domain*, *range*, and *function*, and write them in your own words. • Be able to determine when a *relation* is also a *function*. • Understand **The Vertical Line Test.** • Complete the given examples on your own. • Complete You Trys 1 and 2.
2 Given an Equation, Determine Whether y Is a Function of x and Find the Domain	• Know the property for determining whether y *is a function of* x. • Know the procedure for **Finding the Domain of a Relation,** and write it in your own words. • Complete the given examples on your own. • Complete You Trys 3 and 4.

(continued)

What are your objectives for Section 4.6?	How can you accomplish each objective?
3 Use Function Notation	• Write the definition of *function notation* in your own words. • Be able to evaluate functions for real numbers, variables, and expressions. • Be able to find function values for functions represented by a set of ordered pairs, a correspondence, or a graph. • Complete the given example on your own. • Complete You Trys 5–8.
4 Define and Graph a Linear Function	• Learn the definition of *linear function*. Write your own example in your notes. • Be able to graph a linear function using the slope and *y*-intercept. • Complete the given example on your own. • Complete You Try 9.
5 Solve Problems Using Linear Functions	• Know how to label the graph for a linear function. • Be able to choose a meaningful variable for the independent variable and the function. • Understand the meaning of function values. • Complete the given example on your own.

 **Work**

Read the explanations, follow the examples, take notes, and complete the You Trys.

If you are driving on a highway at a constant speed of 60 mph, the distance you travel depends on the amount of time spent driving.

Driving Time	Distance Traveled
1 hr	60 mi
2 hr	120 mi
2.5 hr	150 mi
3 hr	180 mi

Steve Allen/Getty Images

We can express these relationships with the ordered pairs

$$(1, 60) \quad (2, 120) \quad (2.5, 150) \quad (3, 180)$$

where the first coordinate represents the driving time (in hours), and the second coordinate represents the distance traveled (in miles).

We can also describe this relationship with the equation

$$y = 60x$$

where y is the distance traveled, in miles, and x is the number of hours spent driving.

The distance traveled *depends on* the amount of time spent driving. Therefore, the distance traveled is the **dependent variable,** and the driving time is the **independent variable.** In terms of x and y, since the value of y *depends on* the value of x, y is the *dependent variable* and x is the *independent variable*.

1 Define and Identify Relations, Functions, Domain, and Range

If we form a set of ordered pairs from the ones listed earlier, we get the *relation* {(1, 60), (2, 120), (2.5, 150), (3, 180)}.

Definition

A **relation** is any set of ordered pairs.

Definition

The **domain** of a relation is the set of all values of the independent variable (the first coordinates in the set of ordered pairs). The **range** of a relation is the set of all values of the dependent variable (the second coordinates in the set of ordered pairs).

The domain of the given relation is {1, 2, 2.5, 3}. The range of the relation is {60, 120, 150, 180}.

The relation {(1, 60), (2, 120), (2.5, 150), (3, 180)} is also a *function* because every first coordinate corresponds to *exactly one* second coordinate. A function is a very important concept in mathematics.

 Hint

Remember that only special types of relations can be classified as functions.

Definition

A **function** is a special type of relation. If each element of the domain corresponds to *exactly one* element of the range, then the relation is a function.

Relations and functions can be represented in another way—as a *correspondence* or a *mapping* from one set, the domain, to another, the range. In this representation, the domain is the set of all values in the first set, and the range is the set of all values in the second set. Our previously stated definition of a function still holds.

EXAMPLE 1

Identify the domain and range of each relation, and determine whether each relation is a function.

a) {(2, 0), (3, 1), (6, 2), (6, −2)}

b) $\left\{ (-2, -6), (0, -5), \left(1, -\dfrac{9}{2}\right), (4, -3), \left(5, -\dfrac{5}{2}\right) \right\}$

c)

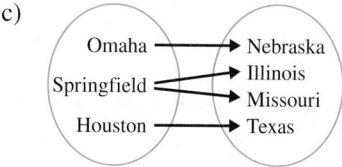

Solution

a) The *domain* is the set of first coordinates, {2, 3, 6}. (We write the 6 in the set only once even though it appears in two ordered pairs.) The *range* is the set of second coordinates, {0, 1, 2, −2}.

To determine whether or not this relation is a function ask yourself, *"Does every first coordinate correspond to exactly one second coordinate?" No.* In the ordered pairs (6, 2) and (6, −2), the same first coordinate, 6, corresponds to two different second coordinates, 2 and −2. Therefore, this relation is *not* a function.

b) The *domain* is {−2, 0, 1, 4, 5}. The *range* is $\left\{ -6, -5, -\dfrac{9}{2}, -3, -\dfrac{5}{2} \right\}$.

Ask yourself, "Does every first coordinate correspond to *exactly one* second coordinate?" *Yes.* This relation *is* a function.

c) The *domain* is {Omaha, Springfield, Houston}. The *range* is {Nebraska, Illinois, Missouri, Texas}.

One of the elements in the domain, Springfield, corresponds to *two* elements in the range, Illinois and Missouri. Therefore, this relation is *not* a function.

YOU TRY 1

Identify the domain and range of each relation, and determine whether each relation is a function.

a) {(−1, −3), (1, 1), (2, 3), (4, 7)}   b) {(−12, −6), (−12, 6), (−1, $\sqrt{3}$), (0, 0)}

c)

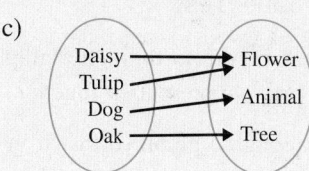

We stated earlier that a relation is a function if each element of the domain corresponds to *exactly one* element of the range.

If the ordered pairs of a relation are such that the first coordinates represent *x*-values and the second coordinates represent *y*-values (the ordered pairs are in the form (*x*, *y*)), then we can think of the definition of a function in this way:

Definition

A relation is a **function** if each *x*-value corresponds to exactly one *y*-value.

What does a function look like when it is graphed? Following are the graphs of the ordered pairs in the relations of Example 1a) and 1b).

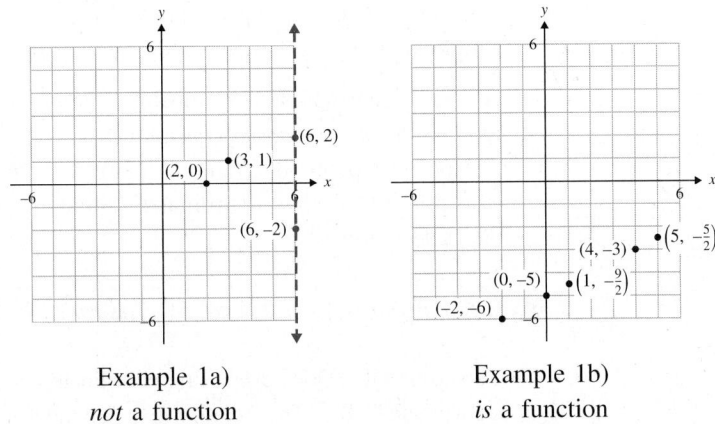

Example 1a)
not a function

Example 1b)
is a function

The relation in Example 1a) is *not* a function since the *x*-value of 6 corresponds to *two different y*-values, 2 and −2. Notice that we can draw a vertical line that intersects the graph in more than one point—the line through (6, 2) and (6, −2).

The relation in Example 1b), however, *is* a function—each *x*-value corresponds to only one *y*-value. Anywhere we draw a vertical line through the points on the graph of this relation, the line intersects the graph in *exactly one point*.

This leads us to the **vertical line test** for a function.

Procedure The Vertical Line Test

If there is no vertical line that can be drawn through a graph so that it intersects the graph more than once, then the graph represents a function.

If a vertical line *can* be drawn through a graph so that it intersects the graph more than once, then the graph does *not* represent a function.

EXAMPLE 2

Use the vertical line test to determine whether each graph, in blue, represents a function. Identify the domain and range.

a)

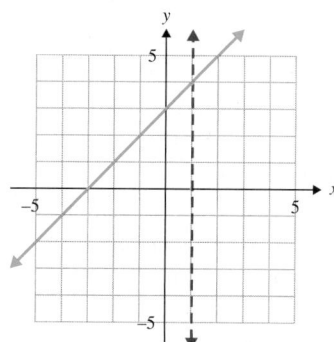

b)

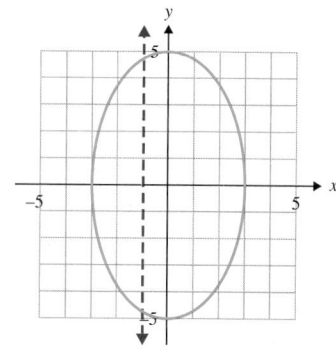

Solution

a)   Anywhere a vertical line is drawn through the graph, the line will intersect the graph only once. *This graph represents a function.*

The arrows on the graph indicate that the graph continues without bound.

The domain of this function is the set of *x*-values on the graph. Since the graph continues indefinitely in the *x*-direction, the domain is the set of all real numbers. *The domain is* $(-\infty, \infty)$.

The range of this function is the set of *y*-values on the graph. Since the graph continues indefinitely in the *y*-direction, the range is the set of all real numbers. *The range is* $(-\infty, \infty)$.

b)   This graph fails the vertical line test because we can draw a vertical line through the graph that intersects it more than once. *This graph does* not *represent a function.*

The set of *x*-values on the graph includes all real numbers from −3 to 3. *The domain is* $[-3, 3]$.

The set of *y*-values on the graph includes all real numbers from −5 to 5. *The range is* $[-5, 5]$.

W Hint

Notice that the graph of a line, except for a vertical line, will always pass the vertical line test. Therefore, every nonvertical line represents a function.

[YOU TRY 2]

Use the vertical line test to determine whether each relation is also a function. Then, identify the domain and range.

a)

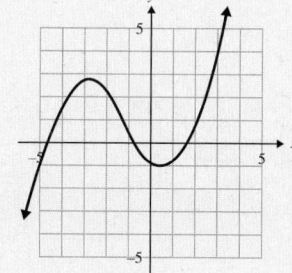

b)

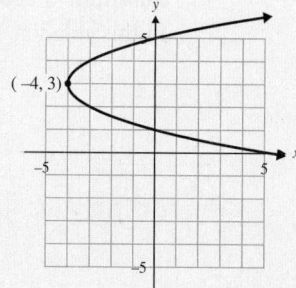

2 Given an Equation, Determine Whether y Is a Function of x and Find the Domain

We can also represent relations and functions with equations. The example given at the beginning of the section illustrates this.

The equation $y = 60x$ describes the distance traveled (y, in miles) after x hours of driving at 60 mph. If $x = 2$, $y = 60(2) = 120$. If $x = 3$, $y = 60(3) = 180$, and so on. For *every* value of x that could be substituted into $y = 60x$ there is *exactly one* corresponding value of y. Therefore, $y = 60x$ is a function.

Furthermore, we can say that *y is a function of x*. In the function described by $y = 60x$, the value of y *depends* on the value of x. That is, x is the independent variable and y is the dependent variable.

Property y Is a Function of x

If a function describes the relationship between x and y so that x is the independent variable and y is the dependent variable, then we say that *y is a function of x*.

EXAMPLE 3 Determine whether each relation describes y as a function of x.

a) $y = x + 2$ b) $y^2 = x$

Solution

a) To begin, substitute a couple of values for x and solve for y to get an idea of what is happening in this relation.

$x = 0$	$x = 3$	$x = -4$
$y = x + 2$	$y = x + 2$	$y = x + 2$
$y = 0 + 2$	$y = 3 + 2$	$y = -4 + 2$
$y = 2$	$y = 5$	$y = -2$

The ordered pairs $(0, 2)$, $(3, 5)$, and $(-4, -2)$ satisfy $y = x + 2$. Each of the values substituted for x has *one* corresponding y-value. In this case, when *any* value is substituted for x there will be *exactly* one corresponding value of y. Therefore, $y = x + 2$ *is* a function.

b) Substitute a couple of values for x, and solve for y to get an idea of what is happening in this relation.

$x = 0$	$x = 4$	$x = 9$
$y^2 = x$	$y^2 = x$	$y^2 = x$
$y^2 = 0$	$y^2 = 4$	$y^2 = 9$
$y = 0$	$y = \pm 2$	$y = \pm 3$

The ordered pairs $(0, 0)$, $(4, 2)$, $(4, -2)$, $(9, 3)$, and $(9, -3)$ satisfy $y^2 = x$. Since $2^2 = 4$ and $(-2)^2 = 4$, $x = 4$ corresponds to two different y-values, 2 and -2. Likewise, $x = 9$ corresponds to the two different y-values of 3 and -3 since $3^2 = 9$ and $(-3)^2 = 9$. Finding one such example is enough to determine that $y^2 = x$ is *not* a function.

[YOU TRY 3]
Determine whether each relation describes y as a function of x.

a) $y = 3x - 5$ b) $y^2 = x + 1$

Next, we will discuss how to determine the domain of a relation written as an equation.

Sometimes, it is helpful to ask yourself, "Is there any number that *cannot* be substituted for x?"

EXAMPLE 4

Determine the domain of each relation, and determine whether each relation describes y as a function of x.

a) $y = \dfrac{1}{x}$ b) $y = \dfrac{7}{x - 3}$ c) $y = -2x + 6$

Solution

a) To determine the domain of $y = \dfrac{1}{x}$ ask yourself, "Is there any number that *cannot* be substituted for x?" Yes. *x cannot equal zero because a fraction is undefined if its denominator equals zero.*

The domain contains all real numbers *except* 0. We can write the domain in interval notation as $(-\infty, 0) \cup (0, \infty)$.

$y = \dfrac{1}{x}$ *is a function* since each value of x in the domain will have only one corresponding value of y.

W Hint

Remember that a fraction with a denominator of 0 is undefined. Therefore, we must exclude any x-values from the domain of the function that make the denominators equal to 0.

b) Ask yourself, "Is there any number that *cannot* be substituted for x in $y = \dfrac{7}{x - 3}$?"

Look at the denominator. When will it equal 0? Set the denominator equal to 0, and solve for x.

$$x - 3 = 0 \quad \text{Set the denominator} = 0.$$
$$x = 3 \quad \text{Solve.}$$

When $x = 3$, the denominator of $y = \dfrac{7}{x - 3}$ equals zero. The domain contains all real numbers *except* 3. Write the domain in interval notation as $(-\infty, 3) \cup (3, \infty)$.

$y = \dfrac{7}{x - 3}$ *is a function.* For every value that can be substituted for x there is only one corresponding value of y.

c) *Is there any number that cannot be substituted for x in $y = -2x + 6$? No.* Any real number can be substituted for x, and $y = -2x + 6$ will be defined.

The domain consists of all real numbers, which can be written as $(-\infty, \infty)$.

Every value substituted for x will have exactly one corresponding y-value. $y = -2x + 6$ *is a function.*

Procedure Finding the Domain of a Relation

The domain of a relation that is written as an equation, where y is in terms of x, is the set of all real numbers that can be substituted for the independent variable, x. When determining the domain of a relation, it can be helpful to keep these tips in mind.

1) Ask yourself, "Is there any number that *cannot* be substituted for x?"

2) If x is in the denominator of a fraction, determine what value of x will make the denominator equal 0 by setting the expression equal to zero. Solve for x. This x-value is *not* in the domain.

The domain consists of all real numbers that can be substituted for x.

[YOU TRY 4] Determine the domain of each relation, and determine whether each relation describes y as a function of x.

a) $y = x - 9$ b) $y = -x^2 + 6$ c) $y = \dfrac{4}{x + 1}$

---

3 Use Function Notation

We can use *function notation* to name functions. If a relation is a function, then $f(x)$ can be used in place of y. In this case, $f(x)$ *is the same as* y.

For example, $y = x + 3$ is a function. We can also write $y = x + 3$ as $f(x) = x + 3$. *They mean the same thing.*

W Hint
Remember that only when a relation is classified as a function, do we replace y with $f(x)$.

Definition

$y = f(x)$ is called **function notation,** and it is read as "y equals f of x." $y = f(x)$ means that y is a function of x (y depends on x).

EXAMPLE 5

a) Evaluate $y = x + 3$ for $x = 2$.   b) If $f(x) = x + 3$, find $f(2)$.

Solution

a) To evaluate $y = x + 3$ for $x = 2$ means to substitute 2 for x and find the corresponding value of y.

$$y = x + 3$$
$$y = 2 + 3 \quad \text{Substitute 2 for } x.$$
$$y = 5$$

W Hint
Notice that $f(2)$ indicates that we replace the x-value with 2.

When $x = 2$, $y = 5$. We can also say that the ordered pair $(2, 5)$ satisfies $y = x + 3$.

---

b) To find $f(2)$ (read as "f of 2") means to find the value of the function when $x = 2$.

$$f(x) = x + 3$$
$$f(2) = 2 + 3 \qquad \text{Substitute 2 for } x.$$
$$f(2) = 5$$

We can also say that the ordered pair (2, 5) satisfies $f(x) = x + 3$ where the ordered pair represents $(x, f(x))$.

Note

Example 5 illustrates that evaluating $y = x + 3$ for $x = 2$ and finding $f(2)$ when $f(x) = x + 3$ is *exactly* the same thing. Remember, $f(x)$ is another name for y.

[YOU TRY 5]   a)   Evaluate $y = -2x + 4$ for $x = 1$.   b)   If $f(x) = -2x + 4$, find $f(1)$.

Different letters can be used to name functions. $g(x)$ is read as "g of x," $h(x)$ is read as "h of x," and so on. Also, the function notation does *not* indicate multiplication; $f(x)$ does *not* mean f times x.

BE CAREFUL $f(x)$ does *not* mean f times x.

Sometimes, we call evaluating a function for a certain value *finding a function value*.

EXAMPLE 6   Let $f(x) = 6x - 5$ and $g(x) = x^2 - 8x + 3$. Find the following function values.

a) $f(3)$ b) $f(0)$ c) $g(-1)$

Solution

a)   "Find $f(3)$" means to find the value of the function when $x = 3$. Substitute 3 for x.

$$f(x) = 6x - 5$$
$$f(3) = 6(3) - 5 = 18 - 5 = 13$$
$$f(3) = 13$$

We can also say that the ordered pair (3, 13) satisfies $f(x) = 6x - 5$.

b)   To find $f(0)$, substitute 0 for x in the function $f(x)$.

$$f(x) = 6x - 5$$
$$f(0) = 6(0) - 5 = 0 - 5 = -5$$
$$f(0) = -5$$

The ordered pair (0, −5) satisfies $f(x) = 6x - 5$.

c) To find $g(-1)$, substitute -1 for every x in the function $g(x)$.

$$g(x) = x^2 - 8x + 3$$
$$g(-1) = (-1)^2 - 8(-1) + 3 = 1 + 8 + 3 = 12$$
$$g(-1) = 12$$

The ordered pair $(-1, 12)$ satisfies $g(x) = x^2 - 8x + 3$.

[YOU TRY 6] Let $f(x) = -4x + 1$ and $h(x) = 2x^2 + 3x - 7$. Find the following function values.

a) $f(5)$ b) $f(-2)$ c) $h(0)$ d) $h(3)$

We can also find function values for functions represented by a set of ordered pairs, a correspondence, or a graph.

EXAMPLE 7

Find $f(4)$ for each function.

a) $f = \{(-2, -11), (0, -5), (3, 4), (4, 7)\}$

b) 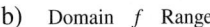 Domain f Range
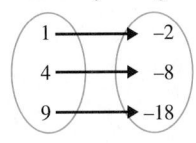

c)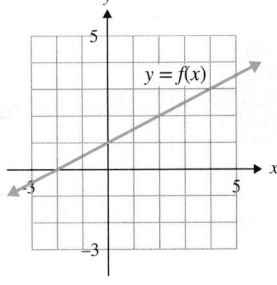

Solution

a) Since this function is expressed as a set of ordered pairs, finding $f(4)$ means finding the y-coordinate of the ordered pair with x-coordinate 4. The ordered pair with x-coordinate 4 is $(4, 7)$, so $f(4) = 7$.

b) In this function, the element 4 in the domain corresponds to the element -8 in the range. Therefore, $f(4) = -8$.

c) To find $f(4)$ from the graph of this function means to find the y-coordinate of the point on the line that has an x-coordinate of 4. Find 4 on the x-axis. Then, go straight up to the graph and move to the left to read the y-coordinate of the point on the graph where the x-coordinate is 4. That y-coordinate is 3. So, $f(4) = 3$.

W Hint

Remember that $f(4)$ indicates that 4 is in the domain of the function. $f(4)$ is equal to the range value that corresponds to 4.

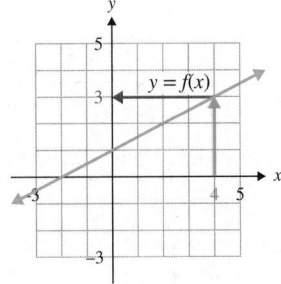

[YOU TRY 7] Find $f(2)$ for each function.

a) $f = \{(-5, 8), (-1, 2), (2, -3), (6, -9)\}$

b)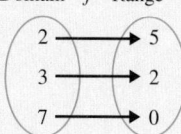

c)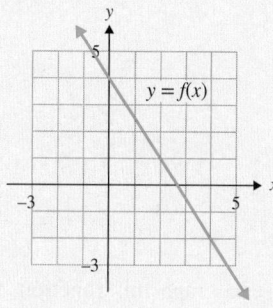

Functions can be evaluated for variables or expressions.

EXAMPLE 8 | Let $h(x) = 5x + 3$. Find each of the following and simplify.

a) $h(c)$ b) $h(t - 4)$

Solution

a) Finding $h(c)$ (read as *h of c*) means to substitute c for x in the function h, and simplify the expression as much as possible.

$$h(x) = 5x + 3$$
$$h(c) = 5c + 3 \qquad \text{Substitute } c \text{ for } x.$$

W Hint

Notice that $h(c)$ indicates that we replace the x-value with c.

b) Finding $h(t - 4)$ (read as *h of t minus 4*) means to substitute $t - 4$ for x in function h, and simplify the expression as much as possible. *Since $t - 4$ contains two terms, we must put it in parentheses.*

$$h(x) = 5x + 3$$
$$h(t - 4) = 5(t - 4) + 3 \qquad \text{Substitute } t - 4 \text{ for } x.$$
$$h(t - 4) = 5t - 20 + 3 \qquad \text{Distribute.}$$
$$h(t - 4) = 5t - 17 \qquad \text{Combine like terms.}$$

W Hint

Notice that $h(t - 4)$ indicates that we replace the x-value with $t - 4$.

[YOU TRY 8] Let $f(x) = 2x - 7$. Find each of the following and simplify.

a) $f(k)$ b) $f(p + 3)$

---

4 Define and Graph a Linear Function

Earlier in this chapter, we learned that a linear equation can have the form $y = mx + b$. Similarly, a *linear function* has the form $f(x) = mx + b$.

W Hint

Remember that a vertical line has undefined slope. Its equation cannot be written in $y = mx + b$ form. The graph of a vertical line does not represent a function.

Definition

A **linear function** has the form $f(x) = mx + b$, where m and b are real numbers, m is the *slope* of the line, and $(0, b)$ is the *y-intercept*.

EXAMPLE 9

Graph $f(x) = -\dfrac{1}{3}x - 1$ using the slope and

y-intercept.

Solution

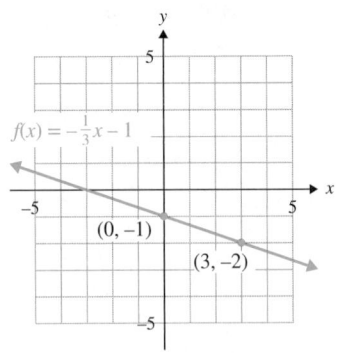

$$f(x) = -\dfrac{1}{3}x - 1$$

$$\uparrow \qquad \uparrow$$

$$m = -\dfrac{1}{3} \qquad y\text{-int: } (0, -1)$$

To graph this function, first plot the y-intercept, $(0, -1)$, then use the slope to locate another point on the line.

[**YOU TRY 9**]

Graph $f(x) = \dfrac{3}{4}x - 2$ using the slope and y-intercept.

5 Solve Problems Using Linear Functions

The independent variable of a function does not have to be x. When using functions to model real-life problems, we often choose a more "meaningful" letter to represent a quantity. For example, if the independent variable represents time, we may use the letter t instead of x. The same is true for naming the function.

No matter what letter is chosen for the independent variable, *the horizontal axis is used to represent the values of the independent variable, and the vertical axis represents the function values.*

EXAMPLE 10

In 2017, Instagram posts received approximately 2.9 million likes per minute. The function $L(t) = 2.9t$ tells us how many likes, $L(t)$, in *millions*, there are on Instagram after t minutes. (www.brandwatch.com)

a) How many Instagram likes are there in 10 min?

b) How many Instagram likes are there in 1 hr?

c) How long would it take for 696 million likes to be registered on Instagram?

d) What is the smallest value t could equal?

e) Graph the function.

Solution

a) To determine the number of likes in 10 min, let $t = 10$ and find $L(10)$.

$$L(t) = 2.9t$$
$$L(10) = 2.9(10) \qquad \text{Substitute 10 for } t.$$
$$L(10) = 29 \qquad\qquad \text{Multiply.}$$

$L(t)$ is in millions, so **the number of likes in 10 min is 29 million.**

b) To determine the number of likes in 1 hr, do we let $t = 1$ and find $L(1)$? *No.* Because t is in *minutes*, we must change 1 hr to minutes before substituting for t.

$$1 \text{ hr} = 60 \text{ min}$$

Let $t = 60$, and find $L(60)$.

$$L(t) = 2.9t$$
$$L(60) = 2.9(60) \qquad \text{Substitute 60 for } t.$$
$$L(60) = 174 \qquad \text{Multiply.}$$

$L(t)$ is in millions, so **there are 174 million likes on Instagram in 1 hr.**

c) Remember that t represents time, in minutes. Because we are asked *how long* it would take to get 696 million likes, we will solve for t when $L(t) = 696$.

$$L(t) = 2.9t$$
$$696 = 2.9t \qquad \text{Substitute 696 for } L(t).$$
$$240 = t \qquad \text{Divide both sides by 2.9.}$$

It will take 240 min, or 4 hr, to get 696 million likes on Instagram.

d) Because t represents *minutes*, the smallest value that makes sense for t is 0. (This means that in 0 min, there are 0 likes.)

e) t is the independent variable and $L(t)$ represents the function values. So, we can write the information obtained in parts a), b), and c) as ordered pairs of the form $(t, L(t))$: $(10, 29)$, $(60, 174)$, and $(240, 696)$. We learned in part d) that when $t = 0$, there are 0 likes. Therefore, the ordered pair $(0, 0)$ is also on the graph of the function.

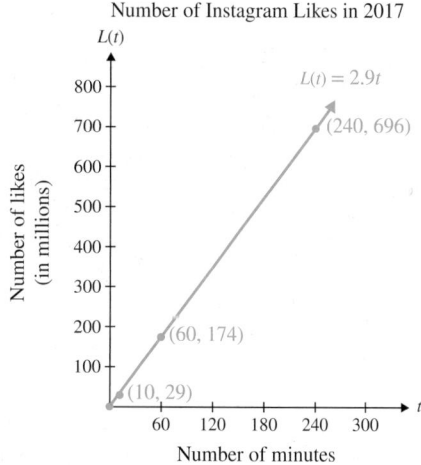

Number of Instagram Likes in 2017

Using Technology

A graphing calculator can be used to represent a function as a graph and also as a table of values. Consider the function $f(x) = 2x - 5$. To graph the function, press $\boxed{Y =}$, then type $2x - 5$ to the right of $\setminus Y1 =$. Press $\boxed{\text{ZOOM}}$ and select 6:ZStandard to graph the equation as shown on the left below. We can select a point on the graph. For example, press $\boxed{\text{TRACE}}$, type 4, and press $\boxed{\text{ENTER}}$. The point $(4, 3)$ is displayed on the screen as shown below on the right.

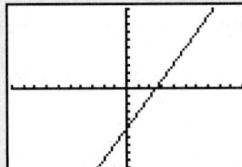 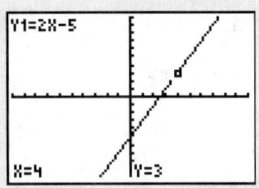

The function can also be represented as a table on a graphing calculator. To set up the table, press 2nd WINDOW, move the cursor after TblStart = , and enter a number such as 0 to set the starting x-value for the table. Enter 1 after ΔTbl = to set the increment between x-values as shown on the left below. Then press 2nd GRAPH to display the table as shown on the right below.

 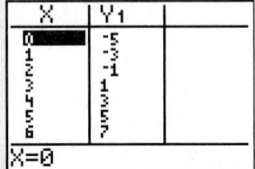

The point (4, 3) is represented in the table above as well as on the graph.

Given the function, find the function value on a graph and a table using a graphing calculator.

1) $f(x) = 3x - 4; f(2)$ 2) $f(x) = 4x - 1; f(1)$ 3) $f(x) = -3x + 7; f(1)$
4) $f(x) = 2x + 5; f(-1)$ 5) $f(x) = 2x - 7; f(-1)$ 6) $f(x) = -x + 5; f(4)$

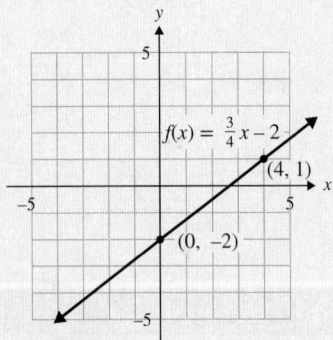

Objective 1: Define and Identify Relations, Functions, Domain, and Range

1) a)   What is a relation?

   b)   What is a function?

   c)   Give an example of a relation that is also a function.

2) Give an example of a relation that is *not* a function.

Identify the domain and range of each relation, and determine whether each relation is a function.

3) $\{(5, 13), (-2, 6), (1, 4), (-8, -3)\}$

4) $\{(0, -3), (1, -4), (1, -2), (16, -5), (16, -1)\}$

5) $\{(9, -1), (25, -3), (1, 1), (9, 5), (25, 7)\}$

6) $\left\{ (-4, -2), \left(-3, -\dfrac{1}{2}\right), \left(-1, -\dfrac{1}{2}\right), (0, -2) \right\}$

7)
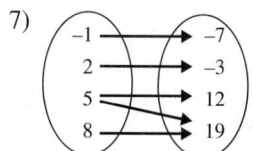

8)
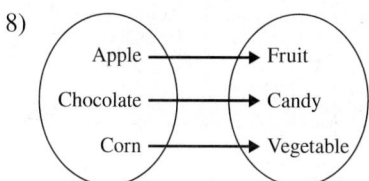

9)
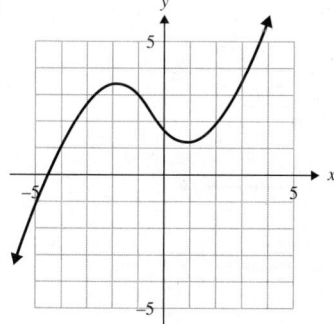

10)
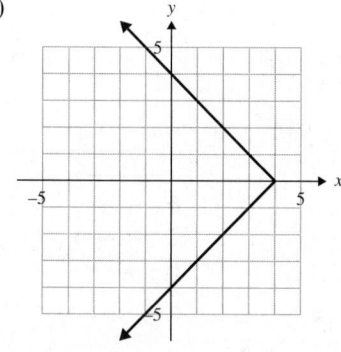

11)
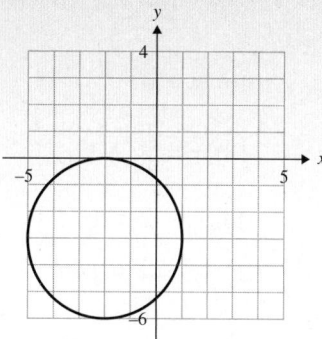

12)
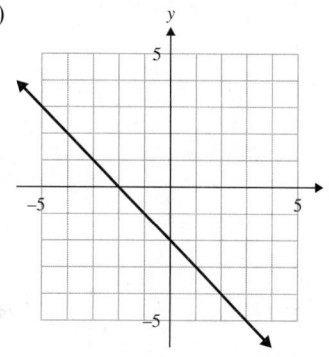

13)
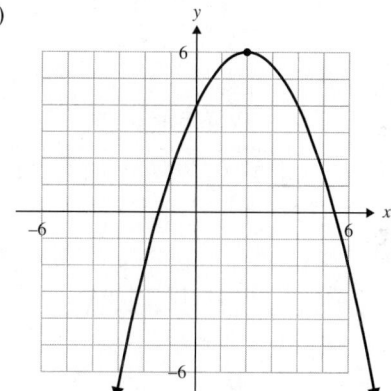

14)
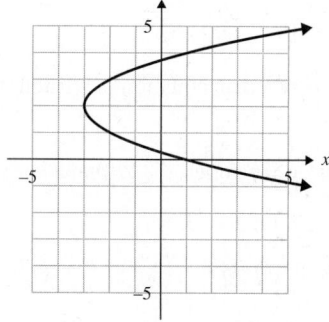

Objective 2: Given an Equation, Determine Whether y Is a Function of x and Find the Domain

Determine whether each relation describes y as a function of x.

15) $y = x - 9$

16) $y = x + 4$

17) $y = 2x + 7$

18) $y = \dfrac{2}{3}x + 1$

19) $x = y^4$

20) $x = y^2 - 3$

21) $y^2 = x - 4$

22) $y^2 = x + 9$

Determine the domain of each relation, and determine whether each relation describes y as a function of x.

23) $y = x - 5$

24) $y = 2x + 1$

25) $y = x^3 + 2$

26) $y = -x^3 + 4$

27) $x = |y|$

28) $x = y^4$

29) $y = -\dfrac{8}{x}$

30) $y = \dfrac{5}{x}$

 31) $y = \dfrac{9}{x + 4}$

32) $y = \dfrac{2}{x - 7}$

33) $y = \dfrac{3}{x - 5}$

34) $y = \dfrac{1}{x + 10}$

35) $y = \dfrac{6}{5x - 3}$

36) $y = -\dfrac{4}{9x + 8}$

37) $y = \dfrac{15}{3x + 4}$

38) $y = \dfrac{5}{6x - 1}$

39) $y = -\dfrac{5}{9 - 3x}$

40) $y = \dfrac{1}{-6 + 4x}$

41) $y = \dfrac{x}{12}$

42) $y = \dfrac{x + 8}{7}$

Objective 3: Use Function Notation

43) Explain what it means when an equation is written in the form $y = f(x)$.

44) Does $y = f(x)$ mean "$y = f$ times x"? Explain.

45) a) Evaluate $y = 5x - 8$ for $x = 3$.

   b) If $f(x) = 5x - 8$, find $f(3)$.

46) a) Evaluate $y = -3x - 2$ for $x = -4$.

   b) If $f(x) = -3x - 2$, find $f(-4)$.

Let $f(x) = -4x + 7$ and $g(x) = x^2 + 9x - 2$. Find the following function values.

47) $f(5)$

48) $f(2)$

49) $f(0)$

50) $f\left(-\dfrac{3}{2}\right)$

51) $g(4)$

52) $g(1)$

53) $g(-1)$

54) $g(0)$

55) $g\left(-\dfrac{1}{2}\right)$

56) $g\left(\dfrac{1}{3}\right)$

57) $f(6) - g(6)$

58) $f(-4) - g(-4)$

For each function f in Exercises 59–64, find $f(-1)$ and $f(4)$.

59) $f = \left\{(-8, -1), \left(-1, \dfrac{5}{2}\right), (4, 5), (10, 8)\right\}$

60) $f = \{(-3, 16), (-1, 10), (0, 7), (1, 4), (4, -5)\}$

61)

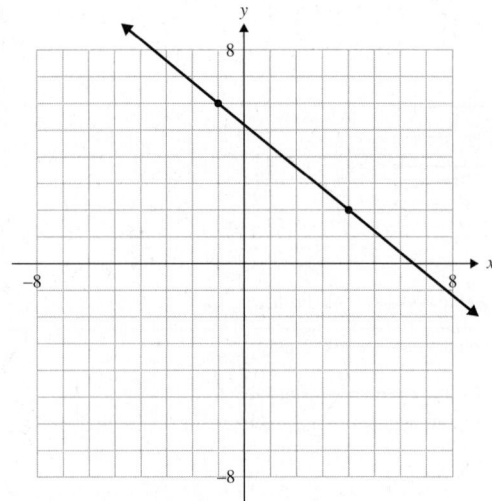

62)

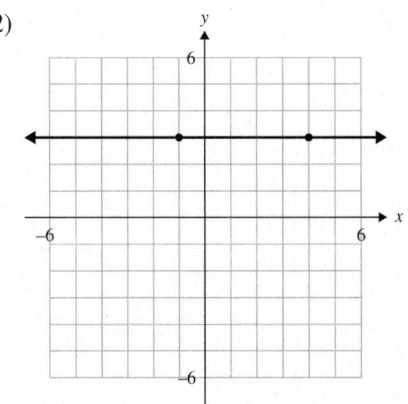

63) Domain f Range

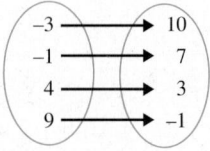

64) Domain f Range

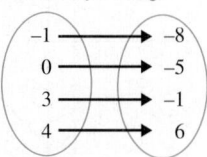

65) $f(x) = -3x - 2$. Find x so that $f(x) = 10$.

66) $f(x) = 5x + 4$. Find x so that $f(x) = 9$.

67) $g(x) = \dfrac{2}{3}x + 1$. Find x so that $g(x) = 5$.

68) $h(x) = -\dfrac{1}{2}x - 6$. Find x so that $h(x) = -2$.

Fill It In

Fill in the blanks with either the missing mathematical step or the reason for the given step.

69) Let $f(x) = -9x + 2$. Find $f(n - 3)$.

_____   Substitute $n - 3$ for x.

$= -9n + 27 + 2$ _____

_____   Simplify.

70) Let $f(x) = 4x - 5$. Find $f(k + 6)$.

$f(k + 6) = 4(k + 6) - 5$ _____

_____   Distribute.

_____   Simplify.

71) $f(x) = -7x + 2$ and $g(x) = x^2 - 5x + 12$. Find each of the following, and simplify.

    a) $f(c)$ b) $f(t)$

    c) $f(a + 4)$ d) $f(z - 9)$

    e) $g(k)$ f) $g(m)$

    g) $f(x + h)$ h) $f(x + h) - f(x)$

72) $f(x) = 5x + 6$ and $g(x) = x^2 - 3x - 11$. Find each of the following, and simplify.

    a) $f(n)$ b) $f(p)$

    c) $f(w + 8)$ d) $f(r - 7)$

    e) $g(b)$ f) $g(s)$

    g) $f(x + h)$ h) $f(x + h) - f(x)$

Objective 4: Define and Graph a Linear Function

Graph each function by making a table of values and plotting points.

73) $f(x) = x - 4$ 74) $f(x) = x + 2$

75) $f(x) = \dfrac{2}{3}x + 2$ 76) $g(x) = -\dfrac{3}{5}x + 2$

77) $h(x) = -3$ 78) $g(x) = 1$

Graph each function by finding the x- and y-intercepts and one other point.

79) $g(x) = 3x + 3$ 80) $k(x) = -2x + 6$

81) $f(x) = -\dfrac{1}{2}x + 2$ 82) $f(x) = \dfrac{1}{3}x + 1$

83) $h(x) = x$ 84) $f(x) = -x$

Graph each function using the slope and y-intercept.

85) $f(x) = -4x - 1$ 86) $f(x) = -x + 5$

87) $g(x) = -\dfrac{1}{4}x - 2$ 88) $h(x) = \dfrac{3}{5}x - 2$

89) $g(x) = 2x + \dfrac{1}{2}$ 90) $h(x) = 3x + 1$

Graph each function.

91) $s(t) = -\dfrac{1}{3}t - 2$ 92) $k(d) = d - 1$

93) $A(r) = -3r$ 94) $N(t) = 3.5t + 1$

Objective 5: Solve Problems Using Linear Functions

95) A truck on the highway travels at a constant speed of 54 mph. The distance, D (in miles), that the truck travels after t hr can be defined by the function

$$D(t) = 54t$$

    a) How far will the truck travel after 2 hr?

    b) How far will the truck travel after 4 hr?

    c) How long does it take the truck to travel 135 mi?

    d) Graph the function.

96) The velocity of an object, v (in feet per second), of an object during free fall t sec after being dropped can be defined by the function

$$v(t) = 32t$$

    a) Find the velocity of an object 1 sec after being dropped.

    b) Find the velocity of an object 3 sec after being dropped.

    c) When will the object be traveling at 256 ft/sec?

    d) Graph the function.

97) If gasoline costs \$3.50 per gallon, then the cost, C (in dollars), of filling a gas tank with g gal of gas is defined by

$$C(g) = 3.50g$$

a) Find $C(8)$, and explain what this means in the context of the problem.

b) Find $C(15)$, and explain what this means in the context of the problem.

Gary He/McGraw-Hill Education

c) Find g so that $C(g) = 42$, and explain what this means in the context of the problem.

98) Jenelle earns \$7.50 per hour at her part-time job. Her total earnings, E (in dollars), for working t hr can be defined by the function

$$E(t) = 7.50t$$

a) Find $E(10)$, and explain what this means in the context of the problem.

b) Find $E(15)$, and explain what this means in the context of the problem.

c) Find t so that $E(t) = 210$, and explain what this means in the context of the problem.

99) A $16 \times$ DVD recorder can transfer 21.13 MB (megabytes) of data per second onto a recordable DVD. The function $D(t) = 21.13t$ describes how much data, D (in megabytes), is recorded on a DVD in t sec. (www.osta.org)

a) How much data is recorded after 12 sec?

b) How much data is recorded after 1 min?

c) How long would it take to record 422.6 MB of data?

d) Graph the function.

100) The average hourly wage of an embalmer in Missouri in 2017 was \$24.75. Seth's earnings, E (in dollars), for working t hr in a week can be defined by the function $E(t) = 24.75t$. (www.bls.gov)

a) How much does Seth earn if he works 30 hr?

b) How much does Seth earn if he works 27 hr?

c) How many hours would Seth have to work to make \$866.25?

d) If Seth can work at most 40 hr per week, what is the domain of this function?

e) Graph the function.

 101) Law enforcement agencies use a computerized system called AFIS (Automated Fingerprint Identification System) to identify fingerprints found at crime scenes. One AFIS system can compare 30,000 fingerprints per second. The function

$$F(s) = 30s$$

describes how many fingerprints, $F(s)$ in thousands, are compared after s sec.

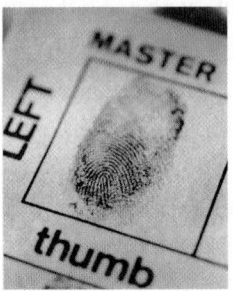
Steve Cole/Getty Images

a) How many fingerprints can be compared in 2 sec?

b) How long would it take AFIS to search through 105,000 fingerprints?

102) Refer to the function in Exercise 101 to answer the following questions.

a) How many fingerprints can be compared in 3 sec?

b) How long would it take AFIS to search through 45,000 fingerprints?

103) Refer to the function in Example 10 in Section 4.6 to determine the following.

a) Find $L(15)$, and explain what this means in the context of the problem.

b) Find $L(180)$, and explain what this means in the context of the problem.

c) Find t so that $L(t) = 1392$, and explain what this means in the context of the problem.

104) Refer to the function in Exercise 99 to determine the following.

a) Find $D(10)$, and explain what this means in the context of the problem.

b) Find $D(120)$, and explain what this means in the context of the problem.

c) Find t so that $D(t) = 633.9$, and explain what this means in the context of the problem.

105) The graph shows the amount, A, of ibuprofen in Sasha's bloodstream t hr after she takes two tablets for a headache.

Amount of Ibuprofen in
Sasha's Bloodstream

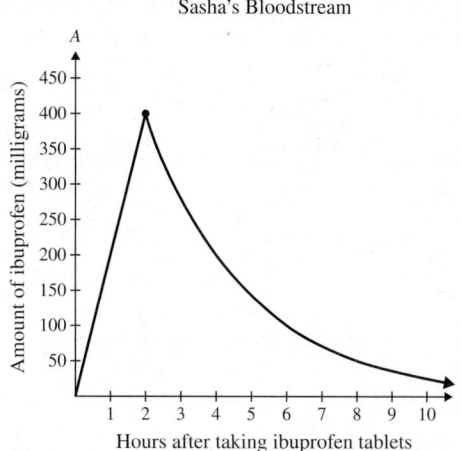

Hours after taking ibuprofen tablets

a) How long after taking the tablets will the amount of ibuprofen in her bloodstream be the greatest? How much ibuprofen is in her bloodstream at this time?

b) When will there be 100 mg of ibuprofen in Sasha's bloodstream?

c) How much of the drug is in her bloodstream after 4 hr?

d) Call this function A. Find $A(8)$, and explain what it means in the context of the problem.

106) The graph shows the number of gallons (in millions), G, of water entering a water treatment plant t hr after midnight on a certain day.

Amount of Water Entering a
Water Treatment Plant

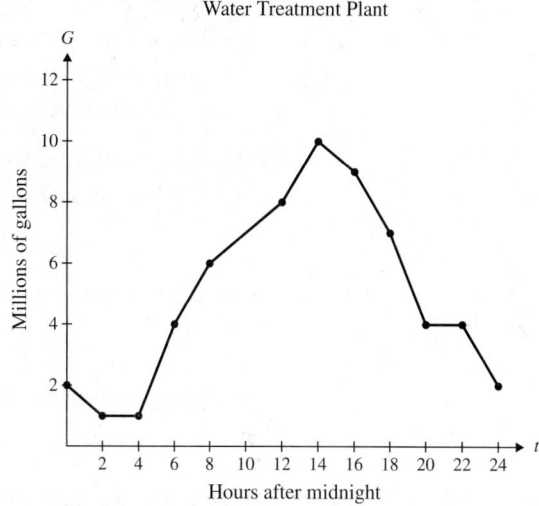

Hours after midnight

a) Identify the domain and range of this function.

b) How many gallons of water enter the facility at noon? At 10 P.M.?

c) At what time did the most water enter the treatment plant? How much water entered the treatment plant at this time?

d) At what time did the least amount of water enter the treatment plant?

e) Call this function G, find $G(18)$, and explain what it means in the context of the problem.

R Rethink

R1) What is the difference between a relation and a function?

R2) Are all functions considered to be relations?

R3) How do you determine if a graph represents a function?

R4) Is (a, b) an ordered pair of $f(b) = a$ or $f(a) = b$?

R5) If $h(3) = 7$, is 3 in the domain or range of the function?

R6) Which objective is the most difficult for you?

Group Activity – Wind Chill Temperature (*www.weather.gov*)

As you walk across your college campus in October, you comment on the refreshing autumn weather. Then the wind kicks in, and you begin to curse the coming of winter. You notice that the greater the wind, the colder you feel. The temperature with the wind that you feel on your skin is known as the wind chill temperature.

Mathematically, the wind chill temperature T_{WC} is a function of the wind speed V ($V > 3$ mph) and the air temperature T ($T \leq 50°F$):

$$T_{WC} = 35.74 + 0.6215T - (35.75 \cdot V^{0.16}) + (0.4275 \cdot V^{0.16})T$$

We can group the terms as follows:

$$T_{WC} = [(0.4275 \cdot V^{0.16})T + 0.6215T] + [35.74 - (35.75 \cdot V^{0.16})]$$

Factoring T from the first group yields:

$$T_{WC} = [(0.4275 \cdot V^{0.16}) + 0.6215]T + [35.74 - (35.75 \cdot V^{0.16})]$$

For a specific value of V, this becomes a linear equation.

1) Find the equation when $V = 15$ mph.

   a) Using the equation, fill in the following table and graph the data. Round to the nearest tenth.

T (air temperature)	T_{WC} (when V = 15 mph)
−40°F	
−30°F	
−20°F	
−10°F	
0°F	
10°F	
20°F	
30°F	
40°F	
50°F	

■ Frostbite occurs in 5 min

■ Frostbite occurs in 10 min

■ Frostbite occurs in 30 min

   b) What does the slope represent?

2) Some data for wind speeds of 5 mph and 50 mph are given below.

a) Complete the table using your knowledge of a linear equation.

T (air temperature)	T_{WC} (when V = 5 mph)	T_{WC} (when V = 50 mph)
−40°F		
−30°F		
−20°F		
−10°F		
0°F		
10°F	1°F	
20°F		
30°F	25°F	12°F
40°F		26°F
50°F		

b) Verify the slope using the formula for V = 5 mph and V = 50 mph. What could be a plausible reason for any discrepancy?

3) Look up a wind chill chart. How do your results compare?

4) What was the lowest wind chill registered in your city?

"What does that say?" "What calculation was I doing there?" "Why do I have 10 pages of notes for a 40-minute class?"

For too many students, these are the questions that arise when they review their class notes. It is essential to develop your note-taking abilities so that your notes are clear and organized and so that they reflect the most important material that was covered in class. To get a sense of your current note-taking abilities, take a set of notes from a recent class and evaluate it based on the following criteria.

Statement	Not Even Slightly	Slightly	Moderately	Pretty Well	Very Well
1. I can read my notes (i.e., they are legible).					
2. Someone else can read my notes.					
3. My notes are complete; I missed nothing important.					
4. My notes represent the key points that were covered in class.					
5. My notes reflect what the instructor emphasizes.					
6. The instructor's key points are clear and understandable.					
7. The notes contain only important points, with no extraneous material.					
8. I understand not only the notes but the class content they reflect.					
9. Using only the notes, I will be able to reconstruct the essential content of the class at the end of the term.					

- What do your answers tell you about the effectiveness of your note-taking skills?

- What might you do differently the next time you take notes?

Chapter 4: Summary

Definition/Procedure	Example

4.1 Introduction to Linear Equations in Two Variables

A **linear equation in two variables** can be written in the form $Ax + By = C$, where A, B, and C are real numbers and where both A and B do not equal zero.

To determine whether an ordered pair is a solution of an equation, substitute the values for the variables.

Is $(4, -1)$ a solution of $3x - 5y = 17$?

$$3x - 5y = 17$$
$$3(4) - 5(-1) = 17 \quad \text{Substitute 4 for } x \text{ and } -1 \text{ for } y.$$
$$12 + 5 = 17$$
$$17 = 17 \ \checkmark$$

Yes, $(4, -1)$ is a solution.

4.2 Graphing by Plotting Points and Finding Intercepts

The graph of a linear equation in two variables, $Ax + By = C$, is a straight line. Each point on the line is a solution to the equation.

We can graph the line by plotting the points and drawing the line through them.

Graph $y = \dfrac{1}{3}x + 2$ by plotting points.

Make a table of values. Plot the points, and draw a line through them.

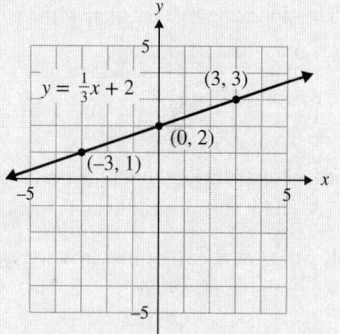

x	y
0	2
3	3
-3	1

The **x-intercept** of an equation is the point where the graph intersects the x-axis. To find the *x-intercept* of the graph of an equation, let $y = 0$ and solve for x.

The **y-intercept** of an equation is the point where the graph intersects the y-axis. To find the *y-intercept* of the graph of an equation, let $x = 0$ and solve for y.

x- and y-intercepts should be written as ordered pairs.

Graph $2x + 5y = -10$ by finding the intercepts and another point on the line.

x-intercept: Let $y = 0$, and solve for x.

$$2x + 5(0) = -10$$
$$2x = 10$$
$$x = -5$$

The x-intercept is $(-5, 0)$.

y-intercept: Let $x = 0$, and solve for y.

$$2(0) + 5y = -10$$
$$5y = -10$$
$$y = -2$$

The y-intercept is $(0, -2)$. Another point on the line is $(5, -4)$. Plot the points, and draw the line through them.

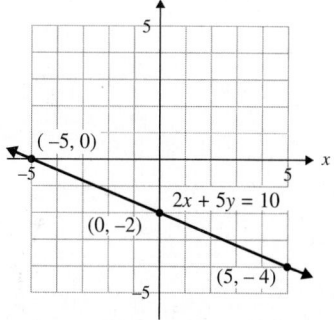

Definition/Procedure	Example
If a is a constant, then the graph of $x = a$ is a **vertical line** going through the point $(a, 0)$. If b is a constant, then the graph of $y = b$ is a **horizontal line** going through the point $(0, b)$.	Graph $x = -2$. Graph $y = 4$. 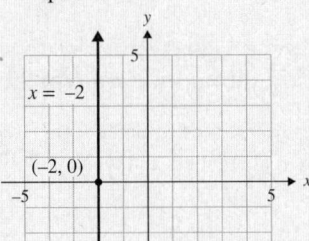 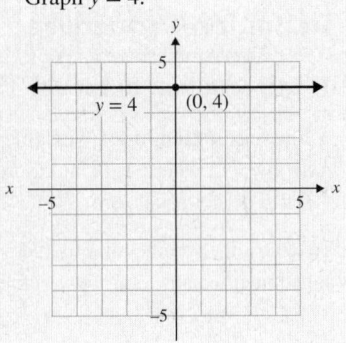

4.3 The Slope of a Line

The **slope** of a line is the ratio of the vertical change in y to the horizontal change in x. Slope is denoted by m. The slope of a line containing the points (x_1, y_1) and (x_2, y_2) is $$m = \frac{y_2 - y_1}{x_2 - x_1}.$$ The slope of a horizontal line is zero. The slope of a vertical line is undefined.	Find the slope of the line containing the points $(4, -3)$ and $(-1, 5)$. $$m = \frac{y_2 - y_1}{x_2 - x_1}$$ $$= \frac{5 - (-3)}{-1 - 4} = \frac{8}{-5} = -\frac{8}{5}$$ The slope of the line is $-\dfrac{8}{5}$.
If we know the slope of a line and a point on the line, we can graph the line.	Graph the line containing the point $(-2, 3)$ with a slope of $-\dfrac{5}{6}$. Start with the point $(-2, 3)$, and use the slope to plot another point on the line. $$m = \frac{-5}{6} = \frac{\text{Change in } y}{\text{Change in } x}$$ 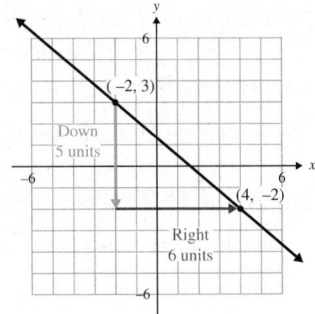

Definition/Procedure	Example

4.4 The Slope-Intercept Form of a Line

The **slope-intercept form of a line** is $y = mx + b$, where m is the slope and $(0, b)$ is the y-intercept.

If a line is written in slope-intercept form, we can use the y-intercept and the slope to graph the line.

Write the equation in slope-intercept form and graph it.

$$8x - 3y = 6$$
$$-3y = -8x + 6$$
$$y = \frac{-8}{-3}x + \frac{6}{-3}$$
$$y = \frac{8}{3}x - 2 \qquad \text{Slope-intercept form}$$

$m = \dfrac{8}{3}$, y-intercept $(0, -2)$

Plot $(0, -2)$, then use the slope to locate another point on the line. We will think of the slope as

$$m = \frac{8}{3} = \frac{\text{Change in } y}{\text{Change in } x}.$$

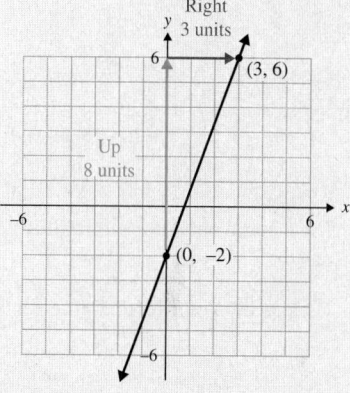

Parallel lines have the same slope.

Perpendicular lines have slopes that are negative reciprocals of each other.

Determine whether the lines $2x + y = 18$ and $x - 2y = 7$ are parallel, perpendicular, or neither.

Put each line into slope-intercept form to find their slopes.

$$
\begin{array}{ll}
2x + y = 18 & x - 2y = 7 \\
y = -2x + 18 & -2y = -x + 7 \\
 & y = \frac{1}{2}x - \frac{7}{2} \\
m = -2 & m = \frac{1}{2}
\end{array}
$$

The lines are *perpendicular* since their slopes are negative reciprocals of each other.

4.5 Writing an Equation of a Line

To write the equation of a line given its slope and y-intercept, use $y = mx + b$ and substitute those values into the equation.

Find an equation of the line with slope $= 5$ and y-intercept $(0, -3)$.

Use $y = mx + b$.
$y = 5x - 3 \qquad$ Substitute 5 for m and -3 for b.

If (x_1, y_1) is a point on a line and m is the slope of the line, then the equation of the line is given by $y - y_1 = m(x - x_1)$.
This is the **point-slope formula**.

If we are given the slope of the line and a point on the line, we can use the point-slope formula to find an equation of the line.

Find an equation of the line containing the point $(7, -2)$ with slope $= 3$. Express the answer in standard form.

Use $y - y_1 = m(x - x_1)$.

Substitute 3 for m. Substitute $(7, -2)$ for (x_1, y_1).

$$
\begin{array}{ll}
y - (-2) = 3(x - 7) & \text{Substitute the values.} \\
y + 2 = 3x - 21 & \text{Distribute.} \\
-3x + y = -23 & \\
3x - y = 23 & \text{Standard form}
\end{array}
$$

Definition/Procedure	Example

To write an equation of a line given two points on the line,

a) use the points to find the slope of the line

then

b) use the slope and *either one* of the points in the point-slope formula.

Find an equation of the line containing the points $(4, 1)$ and $(-4, 5)$. Express the answer in slope-intercept form.

$$m = \frac{5-1}{-4-4} = \frac{4}{-8} = -\frac{1}{2}$$

We will use $m = -\frac{1}{2}$ and the point $(4, 1)$ in the point-slope formula.

$$y - y_1 = m(x - x_1)$$

Substitute $-\frac{1}{2}$ for m. Substitute $(4, 1)$ for (x_1, y_1).

$$y - 1 = -\frac{1}{2}(x - 4) \qquad \text{Substitute.}$$

$$y - 1 = -\frac{1}{2}x + 2 \qquad \text{Distribute.}$$

$$y = -\frac{1}{2}x + 3 \qquad \text{Slope-intercept form}$$

The equation of a **horizontal line** containing the point (a, b) is $y = b$.

The equation of a **vertical line** containing the point (a, b) is $x = a$.

The equation of a horizontal line containing the point $(3, -2)$ is $y = -2$.

The equation of a vertical line containing the point $(6, 4)$ is $x = 6$.

To write an equation of the line parallel or perpendicular to a given line, we must first find the slope of the given line.

Write an equation of the line parallel to $4x - 5y = 20$ containing the point $(4, -3)$. Express the answer in slope-intercept form.

Find the slope of $4x - 5y = 20$.
$$-5y = -4x + 20$$
$$y = \frac{4}{5}x - 4 \qquad m = \frac{4}{5}$$

The slope of the parallel line is also $\frac{4}{5}$. Since this line contains $(4, -3)$, use the point-slope formula to write its equation.

$$y - y_1 = m(x - x_1)$$
$$y - (-3) = \frac{4}{5}(x - 4) \qquad \text{Substitute values.}$$
$$y + 3 = \frac{4}{5}x - \frac{16}{5} \qquad \text{Distribute.}$$
$$y = \frac{4}{5}x - \frac{31}{5} \qquad \text{Slope-intercept form}$$

Definition/Procedure	Example

4.6 Introduction to Functions

A **relation** is any set of ordered pairs. A relation can also be represented as a correspondence or mapping from one set to another.	Relations: a) $\{(-4, -12), (-1, -3), (3, 9), (5, 15)\}$ b) 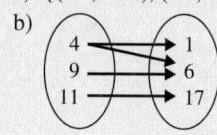
The **domain** of a relation is the set of values of the independent variable (the first coordinates in the set of ordered pairs). The **range** of a relation is the set of all values of the dependent variable (the second coordinates in the set of ordered pairs).	In a) above, the domain is $\{-4, -1, 3, 5\}$, and the range is $\{-12, -3, 9, 15\}$. In b) above, the domain is $\{4, 9, 11\}$, and the range is $\{1, 6, 17\}$.
A **function** is a relation in which each element of the domain corresponds to *exactly one* element of the range.	The relation above in a) *is* a function. The relation above in b) *is not* a function.
The Vertical Line Test	This graph represents a function. Anywhere a vertical line is drawn, it will intersect the graph only once. 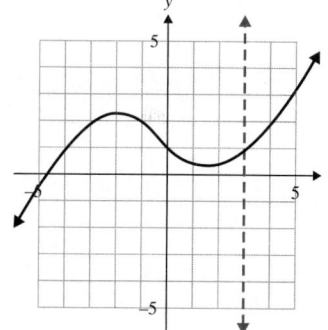 This is *not* the graph of a function. A vertical line can be drawn so that it intersects the graph more than once. 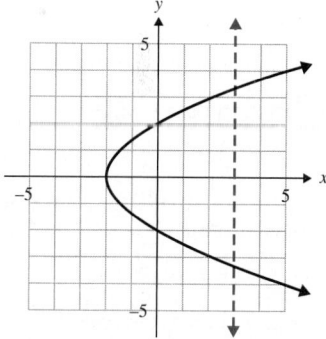
The **domain** of a relation that is written as an equation, where y is in terms of x, is the set of all real numbers that can be substituted for the independent variable, x. When determining the domain of a relation, it can be helpful to keep these tips in mind. 1) Ask yourself, "Is there any number that *cannot* be substituted for x?" 2) If x is in the denominator of a fraction, determine what value of x will make the denominator equal 0 by setting the denominator equal to zero. Solve for x. This x-value is *not* in the domain.	Determine the domain of $f(x) = \dfrac{9}{x + 8}$. $x + 8 = 0$ Set the denominator $= 0$. $x = -8$ Solve. When $x = -8$, the denominator of $f(x) = \dfrac{9}{x + 8}$ equals zero. The domain contains all real numbers *except* -8. The domain of the function is $(-\infty, -8) \cup (-8, \infty)$.

Definition/Procedure	Example
Function Notation If a function describes the relationship between x and y so that x is the independent variable and y is the dependent variable, then y is a function of x. $y = f(x)$ is called **function notation** and it is read as "y equals f of x." Finding a function value means evaluating the function for the given value of the variable.	If $f(x) = 9x - 4$, find $f(2)$. Substitute 2 for x and evaluate. $$f(2) = 9(2) - 4 = 18 - 4 = 14$$ $$f(2) = 14$$
A **linear function** has the form $$f(x) = mx + b$$ where m and b are real numbers, m is the *slope* of the line, and $(0, b)$ is the *y-intercept*.	Graph $f(x) = -3x + 4$ using the slope and y-intercept. The slope is -3 and the y-intercept is $(0, 4)$. Plot the y-intercept, and use the slope to locate another point on the line. 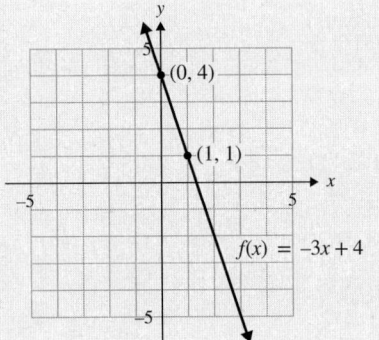

Chapter 4:   Review Exercises

(4.1) Determine whether each ordered pair is a solution of the given equation.

1) $5x - y = 13$; $(2, -3)$

2) $2x + 3y = 8$; $(-1, 5)$

3) $y = -\dfrac{4}{3}x + \dfrac{7}{3}$; $(4, -3)$

4) $x = 6$; $(6, 2)$

Complete the ordered pair for each equation.

5) $y = -2x + 4$; $(-5,\ )$

6) $y = \dfrac{5}{2}x - 3$; $(6,\ )$

7) $y = -9$; $(7,\ )$

8) $8x - 7y = -10$; $(\ , 4)$

Complete the table of values for each equation.

9) $y = x - 14$

x	y
0	
6	
−3	
−8	

10) $3x - 2y = 9$

x	y
	0
0	
2	
	−1

Plot the ordered pairs on the same coordinate system.

11) a) $(4, 0)$ b) $(-2, 3)$

    c) $(5, 1)$ d) $(-1, -4)$

12) a) $(0, -3)$ b) $(-4, 4)$

    c) $(1, \frac{3}{2})$ d) $(-\frac{1}{3}, -2)$

13) The cost of renting a pick-up for one day is given by $y = 0.5x + 45.00$, where x represents the number of miles driven and y represents the cost, in dollars.

a) Complete the table of values, and write the information as ordered pairs.

x	y
10	
18	
29	
36	

b) Label a coordinate system, choose an appropriate scale, and graph the ordered pairs.

c) Explain the meaning of the ordered pair (58, 74) in the context of the problem.

14) Fill in the blank with positive, negative, or zero.

a) The x-coordinate of every point in quadrant III is _____.

b) The y-coordinate of every point in quadrant II is _____.

(4.2) Complete the table of values and graph each equation.

15) $y = -2x + 4$ 16) $2x + 3y = 6$

x	y
0	
1	
2	
3	

x	y
0	
3	
-2	
-3	

Graph each equation by finding the intercepts and at least one other point.

17) $x - 2y = 2$ 18) $3x - y = -3$

19) $y = -\dfrac{1}{2}x + 1$ 20) $2x + y = 0$

21) $y = 4$ 22) $x = -1$

(4.3) Determine the slope of each line.

23)

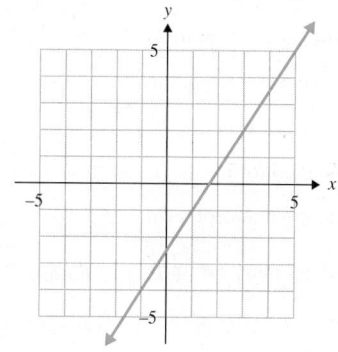

24)

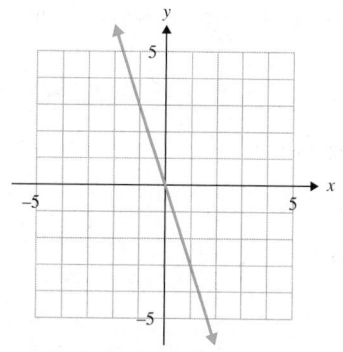

Use the slope formula to find the slope of the line containing each pair of points.

25) (5, 8) and (1, −12)

26) (−3, 4) and (1, −1)

27) (−7, −2) and (2, 4)

28) (7, 3) and (15, 1)

29) $\left(-\dfrac{1}{4}, 1\right)$ and $\left(\dfrac{3}{4}, -6\right)$

30) (3.7, 2.3) and (5.8, 6.5)

31) (−2, 5) and (4, 5)

32) (−9, 3) and (−9, 2)

33) Christine collects old record albums. The graph shows the value of an original, autographed copy of one of her albums from 1980.

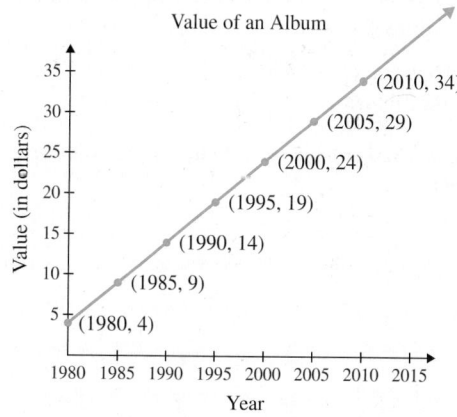

a) How much did she pay for the album in 1980?

b) Is the slope of the line positive or negative? What does the sign of the slope mean in the context of the problem?

c) Find the slope. What does it mean in the context of the problem?

Graph the line containing the given point and with the given slope.

34) $(3, -4)$; $m = 2$

35) $(-2, 2)$; $m = -3$

36) $(1, 3)$; $m = -\dfrac{1}{2}$

37) $(-4, 1)$; slope undefined

38) $(-2, -3)$; $m = 0$

(4.4) Identify the slope and y-intercept, then graph the line.

39) $y = -x + 5$

40) $y = 4x - 2$

41) $y = \dfrac{2}{5}x - 6$

42) $y = -\dfrac{1}{2}x + 5$

43) $x + 3y = -6$

44) $18 = 6y - 15x$

45) $x + y = 0$

46) $y + 6 = 1$

47) Sirena received a gift card to her favorite coffee shop for her birthday. Each time she goes there she buys the same drink, and with tax it costs $2.95. The amount of money remaining on the gift card is given by the linear equation $y = -2.95x + 50.00$, where y represents the amount of money remaining on the card after Sirena buys x drinks.

   a) What is the y-intercept? What does it mean in the context of the problem?

   b) What is the slope, and what does it mean in the context of the problem?

   c) How much money remains on the card after Sirena buys 5 drinks?

   d) How many drinks has Sirena purchased if the balance on the gift card is $8.70?

Determine whether each pair of lines is parallel, perpendicular, or neither.

48) $\quad y = \dfrac{3}{5}x - 8$

   $\quad 5x + 3y = 3$

49) $\quad x - 4y = 20$

   $\quad -x + 4y = 6$

50) $\quad 5x + y = 4$

   $\quad 2x + 10y = 1$

51) $x = 7$

   $y = -3$

52) Write the point-slope formula for the equation of a line with slope m and which contains the point (x_1, y_1).

(4.5) Write the slope-intercept form of the equation of the line, if possible, given the following information.

53) $m = 6$ and contains $(-1, 4)$

54) $m = -5$ and y-intercept $(0, -3)$

55) $m = -\dfrac{3}{4}$ and y-intercept $(0, 7)$

56) contains $(-4, 2)$ and $(-2, 5)$

57) contains $(4, 1)$ and $(6, -3)$

58) $m = \dfrac{2}{3}$ and contains $(5, -2)$

59) horizontal line containing $(3, 7)$

60) vertical line containing $(-5, 1)$

Write the standard form of the equation of the line given the following information.

61) contains $(4, 5)$ and $(-1, -10)$

62) $m = -\dfrac{1}{2}$ and contains $(3, 0)$

63) $m = \dfrac{5}{2}$ and contains $\left(1, -\dfrac{3}{2}\right)$

64) contains $(-4, 1)$ and $(4, 3)$

65) $m = -4$ and y-intercept $(0, 0)$

66) $m = -\dfrac{3}{7}$ and y-intercept $(0, 1)$

67) contains $(6, 1)$ and $(2, 5)$

68) $m = \dfrac{3}{4}$ and contains $\left(-2, \dfrac{7}{2}\right)$

69) Mr. Romanski works as an advertising consultant, and his salary has been growing linearly. In 2012, he earned $62,000, and in 2019, he earned $86,500. Let y represent Mr. Romanski's salary, in dollars, x years after 2012.

   a) Write a linear equation to model these data.

   b) Explain the meaning of the slope in the context of the problem.

   c) How much did he earn in 2015?

   d) If the trend continues, in what year could he expect to earn $100,500?

Write an equation of the line parallel to the given line and containing the given point. Write the answer in slope-intercept form or in standard form, as indicated.

70) $y = 2x + 10$; $(2, -5)$; slope-intercept form

71) $y = -8x + 8$; $(-1, 14)$; slope-intercept form

72) $3x + y = 5$; $(-3, 5)$; standard form

73) $x - 2y = 6$; $(4, 11)$; standard form

74) $3x + 4y = 1$; $(-1, 2)$; slope-intercept form

75) $x + 5y = 10$; $(15, 7)$; slope-intercept form

Write an equation of the line *perpendicular* to the given line and containing the given point. Write the answer in slope-intercept form or in standard form, as indicated.

76) $y = -\dfrac{1}{5}x + 7$; (1, 7); slope-intercept form

77) $y = -x + 9$; (3, −9); slope-intercept form

78) $4x - 3y = 6$; (8, −5); slope-intercept form

79) $2x + 3y = -3$; (−4, −4); slope-intercept form

80) $x + 8y = 8$; (−2, −7); standard form

81) Write an equation of the line parallel to $y = 5$ containing (8, 4).

82) Write an equation of the line perpendicular to $x = -2$ containing (4, −3).

(4.6) Identify the domain and range of each relation, and determine whether each relation is a function.

83) $\{(-3, 1), (5, 3), (5, -3), (12, 4)\}$

84)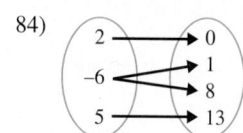

85)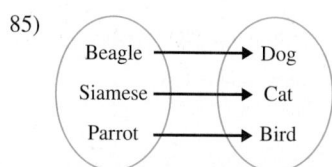

86)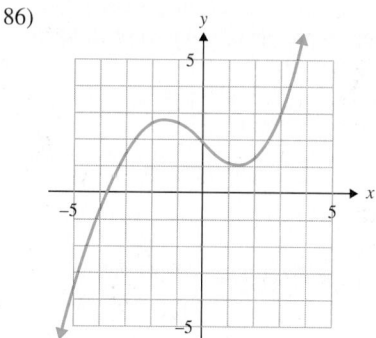

87)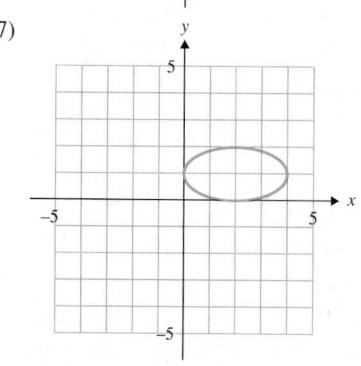

Determine the domain of each relation, and determine whether each relation describes *y* as a function of *x*.

88) $y = 4x - 7$

89) $y = \dfrac{8}{x + 3}$

90) $y = \dfrac{15}{x}$

91) $y^2 = x$

92) $y = x^2 - 6$

93) $y = \dfrac{5}{7x - 2}$

For each function, *f*, find *f*(3) and *f*(−2).

94) $f = \{(-7, -2), (-2, -5), (1, -10), (3, -14)\}$

95)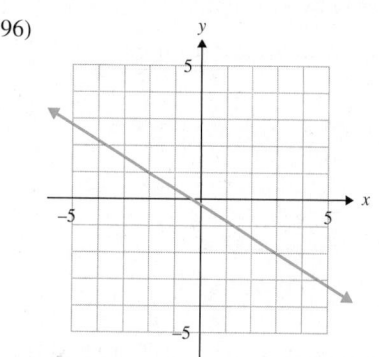

96)

97) Let $f(x) = 5x - 12$, $g(x) = x^2 + 6x + 5$. Find each of the following and simplify.

    a) $f(4)$ b) $f(-3)$

    c) $g(3)$ d) $g(0)$

    e) $f(a)$ f) $g(t)$

    g) $f(k + 8)$ h) $f(c - 2)$

    i) $f(x + h)$ j) $f(x + h) - f(x)$

98) $h(x) = -3x + 7$. Find x so that $h(x) = 19$.

99) $f(x) = \dfrac{3}{2}x + 5$. Find x so that $f(x) = \dfrac{11}{2}$.

100) Graph $f(x) = -2x + 6$ by making a table of values and plotting points.

101) Graph each function using the slope and y-intercept.

    a) $f(x) = \dfrac{2}{3}x - 1$ b) $f(x) = -3x + 2$

102) Graph $g(x) = \dfrac{3}{2}x + 3$ by finding the x- and y-intercepts and one other point.

Graph each function.

103) $h(c) = -\dfrac{5}{2}c + 4$

104) $D(t) = 3t$

105) Alexandra and Justin are planning their wedding. The function $C(x) = 102x + 3000$ represents the cost of the room rental and dinner when x is the number of people attending the reception, *including* the bride and groom.

    a) Find the cost of this part of the wedding if Alex and Justin invite 51 people.

    b) How many people can they invite if they have budgeted $10,000 for this part of their wedding day?

106) A jet travels at a constant speed of 420 mph. The distance D (in miles) that the jet travels after t hr can be defined by the function

$$D(t) = 420t$$

    a) Find $D(2)$, and explain what this means in the context of the problem.

    b) Find t so that $D(t) = 2100$, and explain what this means in the context of the problem.

Chapter 4: Test

1) Is $(-3, -2)$ a solution of $2x - 7y = 8$?

2) Complete the table of values and graph $y = \dfrac{3}{2}x - 2$.

x	y
0	
-2	
4	
	1

3) Fill in the blanks with *positive* or *negative*. In quadrant IV, the x-coordinate of every point is _____ and the y-coordinate is _____.

4) For $3x - 4y = 6$,

    a) find the x-intercept.

    b) find the y-intercept.

    c) find one other point on the line.

    d) graph the line.

5) Graph $y = -3$.

6) Graph $x + y = 0$.

Find the slope of the line containing the given points.

7) $(3, -1)$ and $(-5, 9)$

8) $(8, 6)$ and $(11, 6)$

9) Graph the line containing the point $(-1, 4)$ with slope $-\dfrac{3}{2}$.

10) Graph the line containing the point $(2, 3)$ with an undefined slope.

11) Put $3x - 2y = 10$ into slope-intercept form. Then, graph the line.

12) Write the slope-intercept form for the equation of the line with slope 7 and y-intercept $(0, -10)$.

13) Write the standard form for the equation of a line with slope $-\dfrac{1}{3}$ containing the point $(-3, 5)$.

14) Write the slope-intercept form for the equation of the line containing the points $(-5, -11)$ and $(1, 4)$.

15) Determine whether $4x + 18y = 9$ and $9x - 2y = -6$ are parallel, perpendicular, or neither.

16) Find the slope-intercept form of the equation of the line

    a) perpendicular to $y = 2x - 9$ containing $(-6, 10)$.

    b) parallel to $3x - 4y = -4$ containing $(11, 8)$.

Use the graph for Exercises 17–22.

The graph shows the number of children attending a neighborhood school from 2014–2019. Let y represent the number of children attending the school x years after 2014.

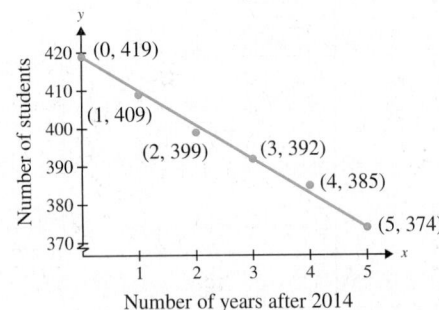

School Population

17) According to the graph, how many children attended this school in 2016?

18) Write a linear equation (in slope-intercept form) to model these data. Use the data points for 2014 and 2019.

19) Use the equation in Exercise 18 to determine the number of students attending the school in 2016. How does your answer in Exercise 17 compare to the number predicted by the equation?

20) Explain the meaning of the slope in the context of the problem.

21) What is the y-intercept? What does it mean in the context of the problem?

22) If the current trend continues, how many children can be expected to attend this school in 2022?

Identify the domain and range of each relation, and determine whether each relation is a function.

23) $\{(-2, -5), (1, -1), (3, 1), (8, 4)\}$

24)

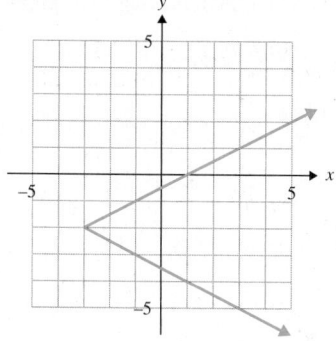

For each function, (a) determine the domain. (b) Is y a function of x?

25) $y = \dfrac{7}{3}x - 5$

26) $y = \dfrac{8}{2x - 5}$

27) For each function, f, find $f(2)$.

a) $f = \{(-3, -8), (0, -5), (2, -3), (7, 2)\}$

b)

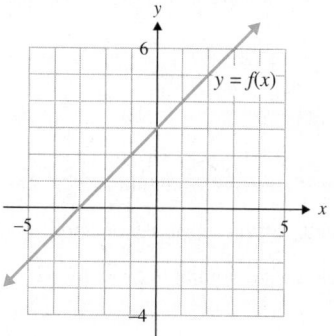

28) Let $f(x) = -4x + 2$ and $g(x) = x^2 - 3x + 7$. Find each of the following, and simplify.

a) $f(6)$ b) $g(2)$

c) $g(t)$ d) $f(h - 7)$

Graph the function.

29) $h(x) = -\dfrac{3}{4}x + 5$

30) A USB 3.1 Gen 2 device can transfer data at a rate of 10 GB/sec (gigabytes/second). Let $f(t) = 10t$ represent the number of gigabytes of data that can be transferred in t sec. (www.usb.org)

a) How many gigabytes of a file can be transferred in 3 sec?

b) Find t so that $f(t) = 132$, and explain what this means in the context of the problem.

Chapter 4: Cumulative Review for Chapters 1–4

1) Find all factors of 90.

2) Write $\dfrac{336}{792}$ in lowest terms.

3) A rectangular picture frame measures 7 in. by 12.5 in. Find the perimeter of the frame.

Evaluate.

4) -3^4

5) $\dfrac{24}{35} \cdot \dfrac{49}{60}$

6) Simplify completely. The answer should contain only positive exponents.

a) $2k^4 \cdot 9k^{10}$ b) $\dfrac{7a^{-3}b^8}{42a^6b^{-2}}$ c) $\left(-\dfrac{5x^9y^2}{3xy^{-4}}\right)^{-3}$

7) Write 0.00000673 in scientific notation.

8) The complement of $29°$ is _____.

9) Evaluate $2x + 11y$ when $x = 4$ and $y = -2$.

10) Which property of real numbers does $16 + 2 = 2 + 16$ illustrate?

11) Combine like terms and simplify: $33b + 18 - 9b + 3 - 10b$

Solve.

12) $-\dfrac{2}{5}y + 9 = 15$

13) $\dfrac{3}{2}(7c - 5) - 1 = \dfrac{2}{3}(2c + 1)$

14) $7 + 2(p - 6) = 8(p + 3) - 6p + 1$

15) Solve. Write the solution in interval notation.
$3x + 14 \le 7x + 4$

16) The Chase family put their house on the market for $306,000. This is 10% less than what they paid for it 3 years ago. What did they pay for the house?

17) Find the missing angle measures.

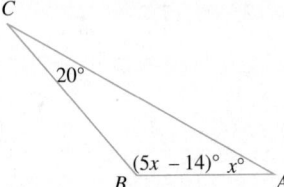

18) Determine whether $(5, -10)$ is a solution to $3x + 2y = -5$.

19) Fill in the blank with *positive, negative,* or *zero.*

   a) The *y*-coordinate of every point in quadrant III is _____.

   b) The *x*-coordinate of every point on the *y*-axis is _____.

20) Find the slope of the line containing the points $(4, -11)$ and $(10, 5)$.

21) Graph $3x + y = 4$.

22) Write an equation of the line with slope $\dfrac{3}{8}$ containing the point $(16, 5)$. Express the answer in standard form.

23) Write an equation of the line perpendicular to $4x + 3y = -6$ containing the point $(-8, -6)$. Express the answer in slope-intercept form.

24) Let $f(x) = 8x + 3$. Find each of the following, and simplify.

   a) $f(-5)$ b) $f(a)$

   c) $f(t + 2)$

25) Graph each function.

   a) $f(x) = 2$ b) $h(x) = -\dfrac{1}{4}x + 2$

Solving Systems of Linear Equations

Get Ready

Let's prepare to learn how to solve systems of linear equations.

1) The **opposite,** or **additive inverse,** of a number is the *negative* of the number.

   Examples: The opposite of 2 is −2.
   The opposite of −3 is −(−3) = 3.

2) To **evaluate an algebraic expression,** we substitute the value (whether it is a number or an expression) for the variable and simplify:

   Example: Evaluate $2x + 5$ for $x = 3$: $2(3) + 5 = 6 + 5 = 11$

   Example: Evaluate $2x + 5$ for $x = a − 1$:
   $$2(a − 1) + 5 = 2a − 2 + 5 = 2a + 3$$

3) We **solve a formula for a specific variable** using the same steps for solving an equation.

   Example: Solve $−x + 3y = 8$ for x.

   $$−x + 3y − 3y = 8 − 3y$$

   $$−x = 8 − 3y$$

   $$\frac{−x}{−1} = \frac{8 − 3y}{−1} = \frac{8}{−1} − \frac{3y}{−1}$$

   $$x = −8 + 3y \quad \text{or} \quad x = 3y − 8$$

In this chapter, use the following *Basic Skills Worksheets* to prepare students for future chapters: WS2 Powers (to prepare for operations with polynomials in Chapter 6) and WS3 Prefactoring (so that students can practice the mental arithmetic skills needed to factor polynomials in Chapter 7).

Find the opposite of each number.

1) 9   2) -1 3) -4 4) 7

5) Evaluate $-3x - 7$ for   a) $x = 4$ b) $x = -9$ c) $x = y + 2$

6) Evaluate $8y + 1$ for   a) $y = -6$ b) $y = \dfrac{3}{2}$ c) $y = 5x - 4$

Solve each equation for x. Simplify the answer.

7) $x + y = 12$ 8) $4x - 6y = -1$ 9) $5y - x = 10$

Solve each equation for y. Simplify the answer.

10) $x + y = 12$ 11) $4x - 6y = -1$ 12) $5y - x = 10$

Answers

1) -9 2) 1   3) 4   4) -7 5) a) -19 b) 20   c) $-3y - 13$ 6) a) -47 b) 13   c) $40x - 31$ 7) $x = 12 - y$ 8) $x = \dfrac{3}{2}y - \dfrac{1}{4}$ 9) $x = 5y - 10$ 10) $y = 12 - x$ 11) $y = \dfrac{2}{3}x + \dfrac{1}{6}$ 12) $y = \dfrac{1}{5}x + 2$

Study Strategies   Taking Math Tests

Taking a math test can be stressful, and it is very different from doing your homework or taking a quiz. You can use the P.O.W.E.R. framework to perform your best on math tests.

 **Prepare**

- **I will get a good grade on my math test.** (Your goal should include the grade you want.)

 **Organize**

- Complete the emPOWERme survey that appears before the Chapter Summary to learn how you currently prepare for and take a math test.
- Attend a review session or organize a study group.
- Find a quiet, **distraction-free** place to study.
- Begin studying a few days before the test by reading your notes and working out some problems in every section.
- Make a list of questions about problems or topics you do not understand. Ask about them in class, in your instructor's office, or at a tutoring center.
- Do the review that your instructor has given the class or the Chapter Review in the book. Ask questions if you have any.
- When you think you are ready for the test, do the Chapter Test that appears at the end of the chapter in the book *without looking back at your notes or the book*. If you cannot do the problems without looking back, then you are not ready for the test.
- Get a good night's sleep before the test.
- The day of the test, warm up for it by *doing* (not just reading) a few problems before walking into the classroom.
- Arrive early for the test so that you do not feel rushed and stressed. Be sure you have everything you need such as pencils, an eraser, and a calculator (if that is allowed).

 **Work**

- It's time for the test. Stay calm, focused, and confident. If you have prepared well, believe that you will do well!
- When you get the test, read all of the directions *carefully*.
- Look over the whole test to get an idea of how many questions it has.

- Do the easiest problems first. Circle the ones you do not know how to do and come back to them later.
- Write neatly and in an organized way, and show all of your work.
- Double-check your work if you have extra time. Sometimes, it is better to *redo a problem* and see if you get the same answer than to just read over the work you have already done.
- Keep track of time so that you can pace yourself.

E Evaluate
- Your instructor returns the tests. Did you get the grade you wanted?

R Rethink
- If you got the grade you wanted, then congratulations! Ask yourself and think about what you did that helped you do well on the test.
- If you did not achieve your goal, ask yourself, *"Why not?"* (Be honest!) What could you do differently next time to improve the outcome?
- Be sure you understand the mistakes you made in the problems you did wrong. Know (and write down) how to do the problems correctly.

5.1 Solving Systems by Graphing

P Prepare

O Organize

What are your objectives for Section 5.1?	How can you accomplish each objective?
1 Determine Whether an Ordered Pair Is a Solution of a System	• Write the definition of a *solution of a system* in your own words, and write an example. • Know how to substitute values into an equation. • Complete the given example on your own. • Complete You Try 1.
2 Solve a Linear System by Graphing	• Understand that a system with one solution is considered a *consistent system* with *independent equations*. • Complete the given example on your own. • Complete You Try 2.
3 Solve a Linear System by Graphing: Special Cases	• Know that the graphs of some equations in a system might not intersect. • Understand that a system with no solution is considered an *inconsistent system* with *independent equations*. • Understand that a system with an infinite number of solutions is considered a *consistent system* with *dependent equations*. • Complete the given examples on your own. • Complete You Try 3. • Summarize how to solve systems by graphing by writing the procedure for **Solving a System by Graphing** in your own words and drawing examples for the three different cases.
4 Determine the Number of Solutions of a System Without Graphing	• Follow the explanation to use slopes and *y*-intercepts to determine the number of solutions to a system of equations. • Complete the given example on your own. • Complete You Try 4.

Read the explanations, follow the examples, take notes, and complete the You Trys.

What is a system of linear equations? A **system of linear equations** consists of two or more linear equations with the same variables. In Sections 5.1–5.3, we will learn how to solve systems of two equations in two variables. Some examples of such systems are

$$2x + 5y = 5 \qquad\qquad y = \frac{1}{3}x - 8 \qquad\qquad -3x + y = 1$$
$$x + 4y = -1 \qquad\qquad 5x - 6y = 10 \qquad\qquad x = -2$$

In the third system, we see that $x = -2$ is written with only one variable. However, we can think of it as an equation in two variables by writing it as $x + 0y = -2$.

It is also possible to solve systems with more than two equations. In Sections 5.5 and 5.6, we will learn two methods for solving systems of three linear equations.

1 Determine Whether an Ordered Pair Is a Solution of a System

We will begin our work with systems of equations by determining whether an ordered pair is a solution of the system.

Definition

A **solution of a system** of two equations in two variables is an ordered pair that is a solution of each equation in the system.

EXAMPLE 1

Determine whether $(2, 3)$ is a solution of each system of equations.

a) $y = x + 1$ b) $4x - 5y = -7$
    $x + 2y = 8$ $3x + y = 4$

Solution

a) If $(2, 3)$ is a solution of $\begin{array}{l} y = x + 1 \\ x + 2y = 8 \end{array}$ then when we substitute 2 for x and 3 for y, the ordered pair will make each equation true.

$$y = x + 1 \qquad\qquad\qquad x + 2y = 8$$
$$3 \stackrel{?}{=} 2 + 1 \quad \text{Substitute.} \qquad 2 + 2(3) \stackrel{?}{=} 8 \quad \text{Substitute.}$$
$$\qquad\qquad\qquad\qquad\qquad 2 + 6 \stackrel{?}{=} 8$$
$$3 = 3 \quad \text{True} \qquad\qquad\qquad 8 = 8 \quad \text{True}$$

Since $(2, 3)$ is a solution of each equation, it is a solution of the system.

b) We will substitute 2 for x and 3 for y to see whether $(2, 3)$ satisfies (is a solution of) each equation.

W Hint

When given the ordered pair $(2, 3)$, write down $x = 2$ and $y = 3$ on scratch paper. This may help you remember which value is x, and which is y.

$$4x - 5y = -7 \qquad\qquad\qquad 3x + y = 4$$
$$4(2) - 5(3) \stackrel{?}{=} -7 \quad \text{Substitute.} \qquad 3(2) + 3 \stackrel{?}{=} 4 \quad \text{Substitute.}$$
$$8 - 15 \stackrel{?}{=} -7 \qquad\qquad\qquad 6 + 3 \stackrel{?}{=} 4$$
$$-7 = -7 \quad \text{True} \qquad\qquad\qquad 9 = 4 \quad \text{False}$$

Although $(2, 3)$ is a solution of the first equation, it does *not* satisfy $3x + y = 4$. Therefore, $(2, 3)$ is *not* a solution of the system.

YOU TRY 1 Determine whether $(-4, 3)$ is a solution of each system of equations.

a) $3x + 5y = 3$

   $-2x - y = -5$

b) $y = \dfrac{1}{2}x + 5$

   $-x + 3y = 13$

Let's begin solving systems of equations by graphing.

2 Solve a Linear System by Graphing

To **solve a system of equations in two variables** means to find the ordered pair (or pairs) that satisfies each equation in the system.

Recall from Chapter 4 that the graph of a linear equation is a line. This line represents all solutions of the equation.

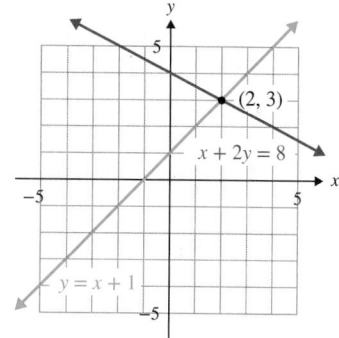

> **W Hint**
>
> Notice that when you are solving a system of two linear equations, you are trying to find the intersection of the two lines.

If two lines intersect at one point, that point of intersection is a solution of each equation.

For example, the graph shows the lines representing the two equations in Example 1a). The solution to that system is their point of intersection, $(2, 3)$.

Definition

When solving a system of equations by graphing, the point of intersection is the solution of the system. If a system has at least one solution, we say that the system is **consistent.** The equations are **independent** if the system has one solution.

EXAMPLE 2 Solve the system by graphing.

$$y = \frac{1}{3}x - 2$$

$$2x + 3y = 3$$

Solution

Graph each line on the same axes. The first equation is in slope-intercept form, and we see that $m = \dfrac{1}{3}$ and $b = -2$. Its graph is in blue.

Let's graph $2x + 3y = 3$ by plotting points.

x	y
0	1
−3	3
3	−1

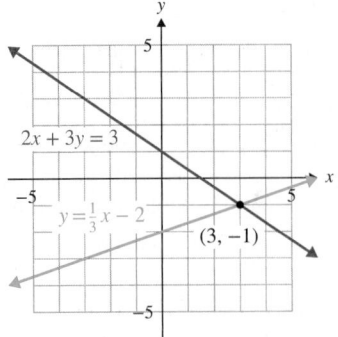

The point of intersection is $(3, -1)$, so the solution is $(3, -1)$.

The solution of the system is $(3, -1)$. The system is consistent.

W Hint

When drawing lines, be sure to extend them beyond the boundaries of your grid! Using a straightedge to graph the lines is a good idea.

Note

It is important that you use a straightedge to graph the lines. If the graph is not precise, it will be difficult to correctly locate the point of intersection. Furthermore, if the solution of a system contains numbers that are not integers, it may be impossible to accurately read the point of intersection. This is one reason why solving a system by graphing is not always the best way to find the solution. But it can be a useful method, and it is one that is used to solve problems not only in mathematics, but also in areas such as business, economics, and chemistry.

[YOU TRY 2] Solve the system by graphing. $\quad 3x + 2y = 2$
$$y = \frac{1}{2}x - 3$$

3 Solve a Linear System by Graphing: Special Cases

Do two lines *always* intersect? No! Then if we are trying to solve a system of two linear equations by graphing and the graphs do not intersect, what does this tell us about the solution to the system?

EXAMPLE 3 Solve the system by graphing. $\quad -2x - y = 1$
$$2x + y = 3$$

Solution

Graph each line on the same axes.

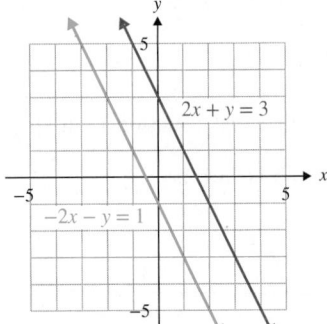

The lines are parallel; they will never intersect. Therefore, there is *no solution* to the system. We write the solution set as ∅.

W Hint

When the graphs of the lines in a system are parallel, there is no solution.

Definition

When solving a system of equations by graphing, if the lines are parallel, then the system has **no solution.** We write this as ∅. Furthermore, a system that has no solution is **inconsistent,** and the equations are **independent.**

What if the graphs of the equations in a system are the same line?

EXAMPLE 4

Solve the system by graphing. $\qquad y = \dfrac{2}{3}x + 2$

$$12y - 8x = 24$$

Solution

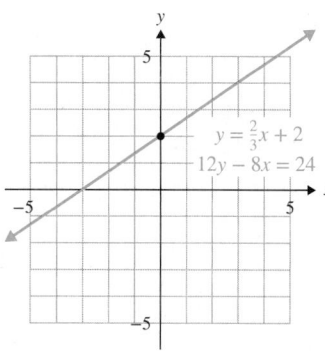

If we write the second equation in slope-intercept form, we see that it is the same as the first equation. This means that the graph of each equation is the same line. Therefore, each point on the line satisfies each equation. The system has an *infinite number of solutions* of

the form $y = \dfrac{2}{3}x + 2$.

The solution set is $\left\{ (x, y) \,\middle|\, y = \dfrac{2}{3}x + 2 \right\}$.

We read this as "the set of all ordered pairs (x, y) such that $y = \dfrac{2}{3}x + 2$."

We could have used either equation to write the solution set in Example 4. However, we will use either the equation that is written in slope-intercept form or the equation written in standard form with integer coefficients that have no common factor other than 1.

W Hint

When writing the solution set for a system with an *infinite number of solutions,* always include one of the equations as part of the solution set.

Definition

When solving a system of equations by graphing, if the graph of each equation is the same line, then the system has an **infinite number of solutions.** The system is **consistent,** and the equations are **dependent.**

We will summarize what we have learned so far about solving a system of linear equations by graphing.

Procedure Solving a System by Graphing

To solve a system by graphing, graph each line on the same axes.

1) If the lines intersect at a **single point,** then the point of intersection is the solution of the system. The system is *consistent,* and the equations are *independent.* (See Figure 5.1a)

2) If the lines are **parallel,** then the system has *no solution.* We write the solution set as ∅. The system is *inconsistent.* The equations are *independent.* (See Figure 5.1b)

3) If the graphs are the **same line,** then the system has an *infinite number of solutions.* We say that the system is *consistent,* and the equations are *dependent.* (See Figure 5.1c)

Figure 5.1

Hint

Use an acronym to help you remember the three cases. For example CI for one, II for none, CD for infinite.

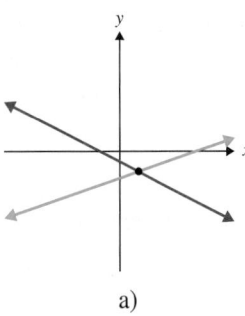

a)

One solution—the point
of intersection
Consistent system
Independent equations

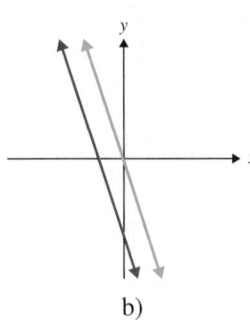

b)

No solution
Inconsistent system
Independent equations

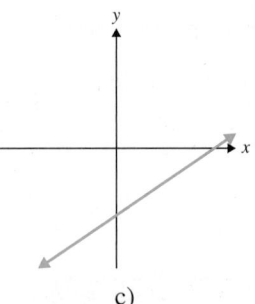

c)

Infinite number
of solutions
Consistent system
Dependent equations

YOU TRY 3

Solve each system by graphing.

a) $-x = y + 4$
    $x + y = 1$

b) $2x - 6y = 9$
    $-4x + 12y = -18$

---

4 Determine the Number of Solutions of a System Without Graphing

The graphs of lines can lead us to the solution of a system. We can also determine the number of solutions a system has without graphing.

We saw in Example 4 that if a system has lines with the same slope and the same y-intercept (they are the same line), then the system has an *infinite number of solutions*.

Example 3 shows that if a system contains lines with the same slope and different y-intercepts, then the lines are parallel and the system has *no solution*.

Finally, we learned in Example 2 that if the lines in a system have different slopes, then they will intersect and the system has *one solution*.

EXAMPLE 5

Without graphing, determine whether each system has no solution, one solution, or an infinite number of solutions.

a) $y = \dfrac{3}{4}x + 7$
    $5x + 8y = -8$

b) $4x - 8y = 10$
    $-6x + 12y = -15$

c) $9x + 6y = -13$
    $3x + 2y = 4$

Solution

a) The first equation is already in slope-intercept form, so write the second equation in slope-intercept form.

$$5x + 8y = -8$$
$$8y = -5x - 8$$
$$y = -\frac{5}{8}x - 1$$

The slopes, $\dfrac{3}{4}$ and $-\dfrac{5}{8}$, are different; therefore, this system has *one solution*.

---

b) Write each equation in slope-intercept form.

$$4x - 8y = 10$$
$$-8y = -4x + 10$$
$$y = \frac{-4}{-8}x + \frac{10}{-8}$$
$$y = \frac{1}{2}x - \frac{5}{4}$$

$$-6x + 12y = -15$$
$$12y = 6x - 15$$
$$y = \frac{6}{12}x - \frac{15}{12}$$
$$y = \frac{1}{2}x - \frac{5}{4}$$

The equations are the same: they have the same slope and y-intercept. Therefore, this system has an *infinite number of solutions.*

c) Write each equation in slope-intercept form.

$$9x + 6y = -13$$
$$6y = -9x - 13$$
$$y = \frac{-9}{6}x - \frac{13}{6}$$
$$y = -\frac{3}{2}x - \frac{13}{6}$$

$$3x + 2y = 4$$
$$2y = -3x + 4$$
$$y = \frac{-3}{2}x + \frac{4}{2}$$
$$y = -\frac{3}{2}x + 2$$

The equations have the same slope but different y-intercepts. If we graphed them, the lines would be parallel. Therefore, this system has *no solution.*

[YOU TRY 4]

Without graphing, determine whether each system has no solution, one solution, or an infinite number of solutions.

a) $-2x = 4y - 8$

   $x + 2y = -6$

b) $y = -\frac{5}{6}x + 1$

   $10x + 12y = 12$

c) $-5x + 3y = 12$

   $3x - y = 2$

Using Technology

In this section, we have learned that the solution of a system of equations is the point at which their graphs intersect. We can solve a system by graphing using a graphing calculator. On the calculator, we will solve the following system by graphing:

$$x + y = 5$$
$$y = 2x - 3$$

Begin by entering each equation using the $\boxed{Y=}$ key.
Before entering the first equation, we must solve for y.

$$x + y = 5$$
$$y = -x + 5$$

Enter $-x + 5$ in Y1 and $2x - 3$ in Y2, press $\boxed{\text{ZOOM}}$, and select 6: ZStandard to graph the equations.

Since the lines intersect, the system has a solution. How can we find that solution? Once you see from the graph that the lines intersect, press $\boxed{\text{2nd}}$ $\boxed{\text{TRACE}}$. Select 5: intersect and then press $\boxed{\text{ENTER}}$ three times. The screen will move the cursor to the point of intersection and display the solution to the system of equations on the bottom of the screen.

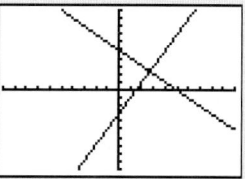

To obtain the exact solution to the system of equations, first return to the home screen by pressing 2nd MODE. To display the x-coordinate of the solution, press X,T,Θ,n MATH ENTER ENTER, and to display the y-coordinate of the solution, press ALPHA 1 MATH ENTER ENTER. The solution to the system is $\left(\dfrac{8}{3}, \dfrac{7}{3}\right)$.

Use a graphing calculator to solve each system.

1) $y = x + 4$
   $y = -x + 2$

2) $y = -3x + 7$
   $y = x - 5$

3) $y = -4x - 2$
   $y = x + 5$

4) $5x + y = -1$
   $4x - y = 2$

5) $5x + 2y = 7$
   $2x + 4y = 3$

6) $3x + 2y = -2$
   $-x - 3y = -5$

ANSWERS TO [YOU TRY] EXERCISES

1) a) no b) yes

2) $(2, -2)$

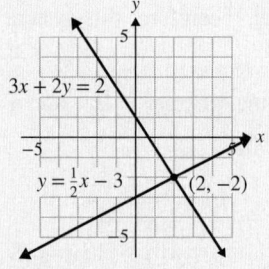

3) a) ∅

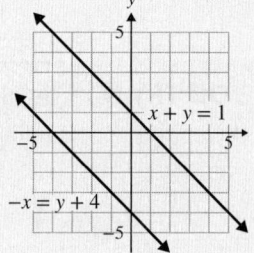

b) infinite number of solutions of the form $\{(x, y) \mid 2x - 6y = 9\}$

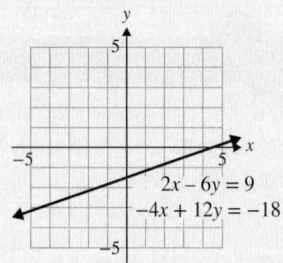

4) a) no solution   b) infinite number of solutions   c) one solution

ANSWERS TO TECHNOLOGY EXERCISES

1) $(-1, 3)$

2) $(3, -2)$

3) $\left(-\dfrac{7}{5}, \dfrac{18}{5}\right)$

4) $\left(\dfrac{1}{9}, -\dfrac{14}{9}\right)$

5) $\left(\dfrac{11}{8}, \dfrac{1}{16}\right)$

6) $\left(-\dfrac{16}{7}, \dfrac{17}{7}\right)$

Objective 1: Determine Whether an Ordered Pair Is a Solution of a System

Determine whether the ordered pair is a solution of the system of equations.

1) $x + 2y = -6$
    $-x - 3y = 13$
    $(8, -7)$

2) $y - x = 4$
    $x + 3y = 8$
    $(-1, 3)$

 3) $5x + y = 21$
    $2x - 3y = 11$
    $(4, 1)$

4) $7x + 2y = 14$
    $-5x + 6y = -12$
    $(2, 0)$

5) $5y - 4x = -5$
    $6x + 2y = -21$
    $\left(-\dfrac{5}{2}, -3\right)$

6) $x = 9y - 7$
    $18y = 7x + 4$
    $\left(-1, \dfrac{2}{3}\right)$

7) $y = -x + 11$
    $x = 5y - 2$
    $(0, 9)$

8) $x = -y$
    $y = \dfrac{5}{8}x - 13$
    $(8, -8)$

Mixed Exercises: Objectives 2 and 3

9) If you are solving a system of equations by graphing, how do you know whether the system has no solution?

10) If you are solving a system of equations by graphing, how do you know whether the system has an infinite number of solutions?

Solve each system of equations by graphing. If the system is inconsistent or the equations are dependent, identify this.

 11) $y = -\dfrac{2}{3}x + 3$
    $y = x - 2$

12) $y = \dfrac{1}{2}x + 2$
    $y = 2x - 1$

13) $y = x + 1$
    $y = -\dfrac{1}{2}x + 4$

14) $y = -2x + 3$
    $y = x - 3$

15) $x + y = -1$
    $x - 2y = 14$

16) $2x - 3y = 6$
    $x + y = -7$

17) $x - 2y = 7$
    $-3x + y = -1$

18) $-x + 2y = 4$
    $3x + 4y = -12$

19) $\dfrac{3}{4}x - y = 0$
    $3x - 4y = 20$

20) $y = -x$
    $4x + 4y = 2$

21) $y = \dfrac{1}{3}x - 2$
    $4x - 12y = 24$

22) $5x + 5y = 5$
    $x + y = 1$

23) $x = 8 - 4y$
    $3x + 2y = 4$

24) $x - y = 0$
    $7x - 3y = 12$

 25) $y = -3x + 1$
    $12x + 4y = 4$

26) $2x - y = 1$
    $-2x + y = -3$

27) $x + y = 0$
    $y = \dfrac{1}{2}x + 3$

28) $x = -2$
    $y = -\dfrac{5}{2}x - 1$

29) $-3x + y = -4$
    $y = -1$

30) $5x + 2y = 6$
    $-15x - 6y = -18$

31) $y = \dfrac{3}{5}x - 6$
    $-3x + 5y = 10$

32) $y - x = -2$
    $2x + y = -5$

Write a system of equations so that the given ordered pair is a solution of the system.

33) $(2, 5)$

34) $(3, 1)$

35) $(-4, -3)$

36) $(6, -1)$

37) $\left(-\dfrac{1}{3}, 4\right)$

38) $\left(0, \dfrac{3}{2}\right)$

For Exercises 39–42, determine which ordered pair could be a solution to the system of equations that is graphed. Explain why you chose that ordered pair.

39)

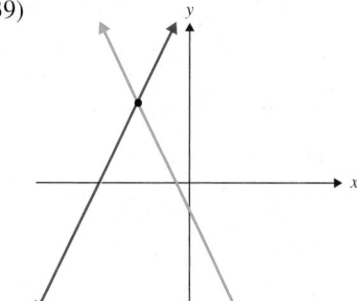

A. $(2, -6)$ B. $(3, 4)$
C. $(-3, 4)$ D. $(-2, -3)$

40)

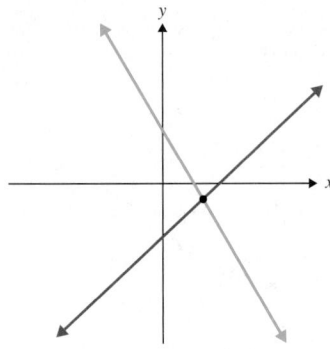

A. $\left(\dfrac{7}{2}, -\dfrac{1}{2}\right)$ B. $(-4, -1)$

C. $\left(\dfrac{9}{4}, \dfrac{3}{4}\right)$ D. $\left(-\dfrac{10}{3}, -\dfrac{2}{3}\right)$

41)

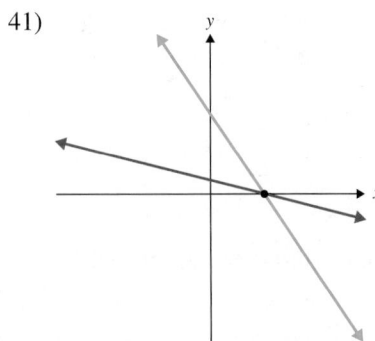

A. $(0, 3.8)$ B. $(4.1, 0)$
C. $(-2.1, 0)$ D. $(0, 5)$

42)

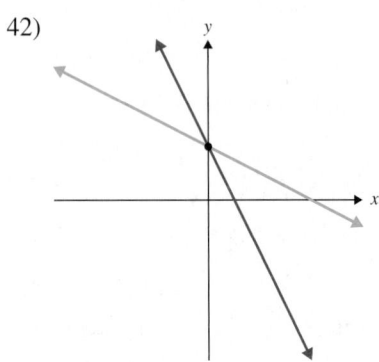

A. $(4, 0)$ B. $\left(\dfrac{1}{3}, 0\right)$

C. $(0, -3)$ D. $(0, 2)$

Objective 4: Determine the Number of Solutions of a System Without Graphing

43) How do you determine, *without graphing*, that a system of equations has exactly one solution?

44) How do you determine, *without graphing*, that a system of equations has no solution?

Without graphing, determine whether each system has no solution, one solution, or an infinite number of solutions.

45) $y = 5x - 4$
$y = -3x + 7$

46) $y = \dfrac{2}{3}x + 9$
$y = \dfrac{2}{3}x + 1$

47) $y = -\dfrac{3}{8}x + 1$
$6x + 16y = -9$

48) $y = -\dfrac{1}{4}x + 3$
$2x + 8y = 24$

49) $-15x + 9y = 27$
$10x - 6y = -18$

50) $7x - y = 6$
$x + y = 13$

51) $3x + 12y = 9$
$x - 4y = 3$

52) $6x - 4y = -10$
$-21x + 14y = 35$

53) $x = 5$
$x = -1$

54) $y = x$
$y = 2$

55) The graph shows the percentage of electric cars and petroleum-electric hybrid cars registered in a city from 2015 to 2019.

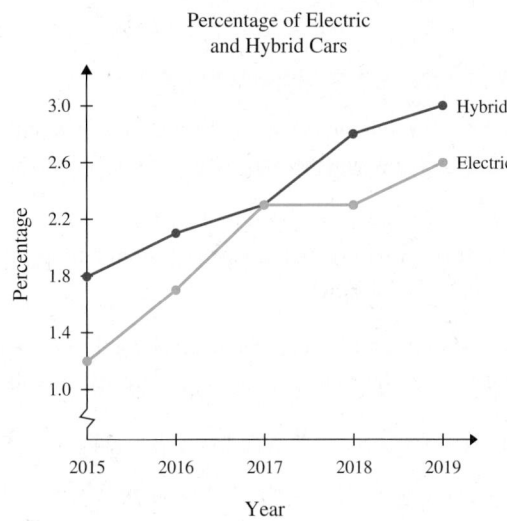

a) Was the percentage of electric cars ever more than the percentage of hybrid cars?

b) Write the point of intersection of the graphs as an ordered pair and explain its meaning.

c) During which years did the percentage of electric cars remain the same?

d) During which years did the percentage of hybrid cars increase the most? How can this be related to the slope of the line segment?

56) The graph shows the percentage of white males and females, 25 years and older, who completed four or more years of college in 2012–2017. (www.census.gov)

Percentage Completing Four or More
Years of College

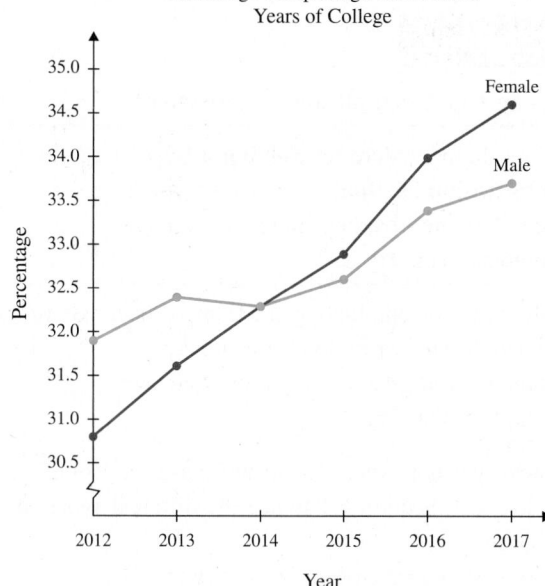

a) In which year was the percentage for males about 1.1% more than the percentage for females? Approximately what was the percentage for each?

b) Write the data point for females in 2013 as an ordered pair, and explain its meaning.

c) When was the percentage of female completers greater than the percentage of male completers?

d) Write the point of intersection of the graphs as an ordered pair, and explain its meaning.

Solve by graphing. Given the functions $f(x)$ and $g(x)$, determine the value of x for which $f(x) = g(x)$.

57) $f(x) = x - 2$, $g(x) = -\dfrac{2}{3}x + 3$

58) $f(x) = x + 3$, $g(x) = -2x - 3$

59) $f(x) = 2x$, $g(x) = -2x - 4$

60) $f(x) = -\dfrac{5}{4}x + 4$, $g(x) = -1$

Solve each system using a graphing calculator.

61) $y = -2x + 2$
    $y = x - 7$

62) $y = x + 1$
    $y = 3x + 3$

63) $x - y = 3$
    $x + 4y = 8$

64) $2x + 3y = 3$
    $y - x = -4$

65) $4x + 5y = -17$
    $3x - 7y = 4.45$

66) $-5x + 6y = 22.8$
    $3x - 2y = -5.2$

R Rethink

R1) What are you really looking for when you solve a system of linear equations?

R2) What happens when you graph a system of linear equations that has an infinite number of solutions?

R3) What is the solution to a system of linear equations when the two lines are parallel?

R4) Which objective is the most difficult for you, and why?

5.2 Solving Systems by the Substitution Method

P Prepare

O Organize

What are your objectives for Section 5.2?	How can you accomplish each objective?
1 Solve a Linear System by Substitution	• Write the procedure for **Solving a System by the Substitution Method** in your own words. • Complete the given examples on your own. • Complete You Try 1.
2 Solve a System Containing Fractions or Decimals	• Add steps for eliminating fractions or decimals to the procedure you wrote for Objective 1. • Complete the given example on your own. • Complete You Try 2.
3 Solve a System by Substitution: Special Cases	• Know that not every system will have one solution, and review Section 5.1 to see the different types of systems. • Complete the given examples on your own. • Complete You Try 3.

 **Work**

Read the explanations, follow the examples, take notes, and complete the You Trys.

In Section 5.1, we learned to solve a system of equations by graphing. This method, however, is not always the *best* way to solve a system. If your graphs are not precise, you may read the solution incorrectly. And, if a solution consists of numbers that are not integers, like $\left(\frac{2}{3}, -\frac{1}{4}\right)$, it may not be possible to accurately identify the point of intersection of the graphs.

1 Solve a Linear System by Substitution

Another way to solve a system of equations is to use the *substitution method.* When we use the **substitution method,** we solve one of the equations for one of the variables in terms of the other. Then we substitute that expression into the other equation. We can do this because solving a system means finding the ordered pair, or pairs, that satisfy *both* equations. *The substitution method is especially good when one of the variables has a coefficient of 1 or −1.*

EXAMPLE 1   Solve the system using substitution.

$$2x + 3y = -1$$
$$y = 2x - 3$$

Solution

The second equation, $y = 2x - 3$, is already solved for y; it tells us that y *equals* $2x - 3$. Therefore, we can substitute $2x - 3$ for y in the first equation, then solve for x.

$2x + 3y = -1$	First equation
$2x + 3(2x - 3) = -1$	Substitute.
$2x + 6x - 9 = -1$	Distribute.
$8x - 9 = -1$	Combine like terms.
$8x = 8$	Add 9 to each side.
$x = 1$	

W Hint

Write out all of the steps in the example as you are reading it!

We have found that $x = 1$, but we still need to find y. Substitute $x = 1$ into *either* equation, and solve for y. In this case, we will substitute $x = 1$ into the second equation since it is already solved for y.

$y = 2x - 3$	Second equation
$y = 2(1) - 3$	Substitute.
$y = 2 - 3$	
$y = -1$	

W Hint

Try graphing both of these equations. You will see that $(1, -1)$ is the point of intersection.

Check $x = 1$, $y = -1$ in *both* equations.

$2x + 3y = -1$		$y = 2x - 3$	
$2(1) + 3(-1) \stackrel{?}{=} -1$	Substitute.	$-1 \stackrel{?}{=} 2(1) - 3$	Substitute.
$2 - 3 \stackrel{?}{=} -1$		$-1 \stackrel{?}{=} 2 - 3$	
$-1 = -1$	True	$-1 = -1$	True

We write the solution of the system as an ordered pair, $(1, -1)$.

Let's summarize the steps we use to solve a system by the substitution method.

Procedure Solving a System by the Substitution Method

Step 1: Solve one of the equations for one of the variables. If possible, solve for a variable that has a coefficient of 1 or −1.

Step 2: Substitute the expression found in *Step 1* into the *other* equation. The equation you obtain should contain only one variable.

Step 3: Solve the equation you obtained in *Step 2*.

Step 4: Substitute the value found in *Step 3* into either of the equations to obtain the value of the other variable.

Step 5: Check the values in each of the original equations, and write the solution as an ordered pair.

EXAMPLE 2

Solve the system by the substitution method.

$$x - 2y = 7 \qquad (1)$$
$$2x + 3y = -21 \qquad (2)$$

Solution

We will follow the steps listed in the Procedure box.

Step 1: For which variable should we solve? The x in the first equation is the only variable with a coefficient of 1 or -1. Therefore, we will solve the first equation for x.

$$x - 2y = 7 \qquad \text{First equation (1)}$$
$$x = 2y + 7 \qquad \text{Add } 2y.$$

W Hint

Notice that, in the first equation in Example 2, the coefficient of x is 1 so that it is easiest to begin by solving for this variable.

Step 2: Substitute $2y + 7$ for the x in equation (2).

$$2x + 3y = -21 \qquad \text{Second equation (2)}$$
$$2(2y + 7) + 3y = -21 \qquad \text{Substitute.}$$

Step 3: Solve this new equation for y.

$$2(2y + 7) + 3y = -21$$
$$4y + 14 + 3y = -21 \qquad \text{Distribute.}$$
$$7y + 14 = -21 \qquad \text{Combine like terms.}$$
$$7y = -35 \qquad \text{Subtract 14 from each side.}$$
$$y = -5$$

Step 4: To determine the value of x, we can substitute -5 for y in either equation. We will use equation (1).

$$x - 2y = 7 \qquad \text{Equation (1)}$$
$$x - 2(-5) = 7 \qquad \text{Substitute.}$$
$$x + 10 = 7$$
$$x = -3$$

Step 5: The check is left to the reader. The solution of the system is $(-3, -5)$.

[YOU TRY 1]

Solve the system by the substitution method.

$$-3x + 4y = -2$$
$$6x - y = -3$$

If no variable in the system has a coefficient of 1 or -1, solve for any variable.

2 Solve a System Containing Fractions or Decimals

If a system contains an equation with fractions, first multiply the equation by the least common denominator to eliminate the fractions. Likewise, if an equation in the system contains decimals, begin by multiplying the equation by the lowest power of 10 that will eliminate the decimals.

EXAMPLE 3

Solve the system by the substitution method.

$$\frac{3}{10}x - \frac{1}{5}y = 1 \quad \text{(1)}$$

$$-\frac{1}{12}x + \frac{1}{3}y = \frac{5}{6} \quad \text{(2)}$$

Solution

Before applying the steps for solving the system, eliminate the fractions in each equation.

$$\frac{3}{10}x - \frac{1}{5}y = 1 \qquad\qquad -\frac{1}{12}x + \frac{1}{3}y = \frac{5}{6}$$

$$10\left(\frac{3}{10}x - \frac{1}{5}y\right) = 10 \cdot 1 \quad \substack{\text{Multiply by}\\\text{the LCD: 10.}} \qquad 12\left(-\frac{1}{12}x + \frac{1}{3}y\right) = 12 \cdot \frac{5}{6} \quad \substack{\text{Multiply by}\\\text{the LCD: 12.}}$$

$$3x - 2y = 10 \quad \text{(3)} \quad \text{Distribute.} \qquad\qquad -x + 4y = 10 \quad \text{(4)} \quad \text{Distribute.}$$

From the original equations, we obtain an equivalent system of equations.

$$3x - 2y = 10 \quad \text{(3)}$$

$$-x + 4y = 10 \quad \text{(4)}$$

W Hint

Review the *eliminating fractions from an equation* procedure that you learned in Section 3.2.

Now, we will work with equations (3) and (4).

Apply the steps for solving the system:

Step 1: The x in equation (4) has a coefficient of -1. Solve this equation for x.

$$-x + 4y = 10 \qquad \text{Equation (4)}$$

$$-x = 10 - 4y \qquad \text{Subtract } 4y.$$

$$x = -10 + 4y \qquad \text{Divide by } -1.$$

Step 2: Substitute $-10 + 4y$ for x in equation (3).

$$3x - 2y = 10 \qquad \text{Equation (3)}$$

$$3(-10 + 4y) - 2y = 10 \qquad \text{Substitute.}$$

Step 3: Solve the equation above for y.

$$3(-10 + 4y) - 2y = 10$$

$$-30 + 12y - 2y = 10 \qquad \text{Distribute.}$$

$$-30 + 10y = 10 \qquad \text{Combine like terms.}$$

$$10y = 40 \qquad \text{Add 30 to each side.}$$

$$y = 4 \qquad \text{Divide by 10.}$$

Step 4: Find x by substituting 4 for y in either equation (3) or (4). Let's use equation (4) since it has smaller coefficients.

$$-x + 4y = 10 \qquad \text{Equation (4)}$$

$$-x + 4(4) = 10 \qquad \text{Substitute.}$$

$$-x + 16 = 10 \qquad \text{Multiply.}$$

$$-x = -6 \qquad \text{Subtract 16 from each side.}$$

$$x = 6 \qquad \text{Divide by } -1.$$

Step 5: Check $x = 6$ and $y = 4$ in the original equations. The solution of the system is $(6, 4)$.

Solve each system by the substitution method.

a) $-\dfrac{1}{6}x + \dfrac{1}{3}y = \dfrac{2}{3}$

   $\dfrac{3}{2}x - \dfrac{5}{2}y = -7$

b) $0.1x + 0.03y = -0.05$

   $0.1x - 0.1y = 0.6$

3 Solve a System by Substitution: Special Cases

We saw in Section 5.1 that a system may have no solution or an infinite number of solutions. If we are solving a system by graphing, we know that a system has no solution if the lines are parallel, and a system has an infinite number of solutions if the graphs are the same line.

When we solve a system by *substitution*, how do we know whether the system is inconsistent or dependent? Read Examples 4 and 5 to find out.

EXAMPLE 4

Solve the system by substitution.
$$3x + y = 5 \qquad (1)$$
$$12x + 4y = -7 \qquad (2)$$

Solution

Step 1: $y = -3x + 5$ Solve equation (1) for y.

Step 2: $12x + 4y = -7$ Substitute $-3x + 5$ for y in equation (2).

$12x + 4(-3x + 5) = -7$

Step 3: $12x + 4(-3x + 5) = -7$ Solve the resulting equation for x.

$12x - 12x + 20 = -7$ Distribute.

$20 = -7$ False

> **W Hint**
>
> Remember that a system of linear equations has *no solution* when the two lines are parallel.

Because the variables drop out and we get a false statement, there is no solution to the system. The system is inconsistent, so the solution set is $\varnothing$.

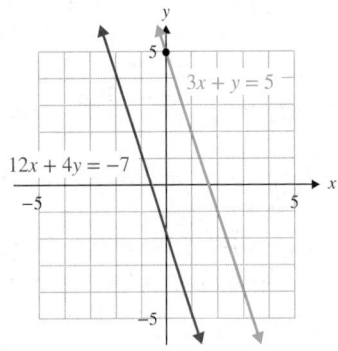

The graph of the equations in the system supports our work. The lines are parallel; therefore, the system has no solution.

EXAMPLE 5

Solve the system by substitution.

$$2x - 6y = 10 \quad (1)$$
$$x = 3y + 5 \quad (2)$$

Solution

Step 1: Equation (2) is already solved for x.

Step 2:
$$2x - 6y = 10 \quad \text{Substitute } 3y + 5 \text{ for } x \text{ in equation (1).}$$
$$2(3y + 5) - 6y = 10$$

Step 3:
$$2(3y + 5) - 6y = 10 \quad \text{Solve the resulting equation for y.}$$
$$6y + 10 - 6y = 10 \quad \text{Distribute.}$$
$$10 = 10 \quad \text{True}$$

Because the variables drop out and we get a true statement, the system has an infinite number of solutions. The equations are dependent, and the solution set is $\{(x, y)|x = 3y + 5\}$.

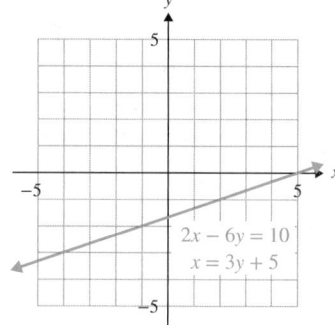

$$2x - 6y = 10$$
$$x = 3y + 5$$

W Hint

Remember that a system of linear equations has *an infinite number of solutions* when the two equations represent the same line.

The graph shows that the equations in the system are the same line; therefore, the system has an infinite number of solutions.

Note

When you are solving a system of equations and the variables drop out:
1) If you get a *false statement*, like $20 = -7$, then the system has *no solution* and is *inconsistent*.
2) If you get a *true statement*, like $10 = 10$, then the system has an *infinite number of solutions*. The equations are *dependent*.

[YOU TRY 3]

Solve each system by substitution.

a) $-20x + 5y = 3$
    $4x - y = -1$

b) $x - 4y = -7$
    $-2x + 8y = 14$

ANSWERS TO [YOU TRY] EXERCISES

1) $\left(-\dfrac{2}{3}, -1\right)$ 2) a) $(-8, -2)$ b) $(1, -5)$ 3) a) $\varnothing$ b) $\{(x, y)|x - 4y = -7\}$

Mixed Exercises: Objectives 1 and 3

1) If you were asked to solve this system by substitution, why would it be easiest to begin by solving for y in the first equation?

$$7x + y = 1$$
$$-2x + 5y = 9$$

2) When is the best time to use substitution to solve a system?

3) When solving a system of linear equations, how do you know whether the system has no solution?

4) When solving a system of linear equations, how do you know whether the system has an infinite number of solutions?

Solve each system by substitution.

5) $y = 4x - 3$
$5x + y = 15$

6) $y = 3x + 10$
$-5x + 2y = 14$

7) $x = 7y + 11$
$4x - 5y = -2$

8) $x = 9 - y$
$-3x + 4y = 8$

9) $x + 2y = -3$
$4x + 5y = -6$

10) $x + 4y = 1$
$5x + 3y = 5$

11) $2y - 7x = -14$
$4x - y = 7$

12) $-2x - y = 3$
$3x + 2y = -3$

13) $9y - 18x = 5$
$2x - y = 3$

14) $2x + 30y = 9$
$x = 6 - 15y$

15) $x - 2y = 10$
$3x - 6y = 30$

16) $6x + y = -6$
$-12x - 2y = 12$

17) $10x + y = -5$
$-5x + 2y = 10$

18) $2y - x = 4$
$x + 6y = 8$

19) $x = -\dfrac{3}{5}y + 7$
$x + 4y = 24$

20) $y = \dfrac{3}{2}x - 5$
$2x - y = 5$

21) $4y = 2x + 4$
$2y - x = 2$

22) $3x + y = -12$
$6x = 10 - 2y$

23) $2x + 3y = 6$
$5x + 2y = -7$

24) $2x - 5y = -4$
$8x - 9y = 6$

25) $6x - 7y = -4$
$9x - 2y = 11$

26) $4x + 6y = -13$
$7x - 4y = -1$

27) $18x + 6y = -66$
$12x + 4y = -19$

28) $6y - 15x = -12$
$5x - 2y = 4$

Objective 2: Solve a System Containing Fractions or Decimals

29) If an equation in a system contains fractions, what should you do first to make the system easier to solve?

30) If an equation in a system contains decimals, what should you do first to make the system easier to solve?

Solve each system by substitution.

31) $\dfrac{1}{4}x - \dfrac{1}{2}y = 1$
$\dfrac{2}{3}x + \dfrac{1}{6}y = \dfrac{25}{6}$

32) $\dfrac{2}{9}x + \dfrac{2}{9}y = 2$
$\dfrac{7}{4}x - \dfrac{1}{8}y = \dfrac{3}{4}$

33) $\dfrac{1}{6}x + \dfrac{4}{3}y = \dfrac{13}{3}$
$\dfrac{2}{5}x + \dfrac{3}{2}y = \dfrac{18}{5}$

34) $\dfrac{1}{10}x + \dfrac{1}{2}y = \dfrac{1}{5}$
$-\dfrac{1}{3}x + \dfrac{1}{2}y = \dfrac{3}{2}$

35) $\dfrac{x}{10} - \dfrac{y}{2} = \dfrac{13}{10}$
$\dfrac{x}{3} + \dfrac{5}{4}y = -\dfrac{3}{2}$

36) $-\dfrac{x}{3} + \dfrac{y}{2} = \dfrac{5}{3}$
$\dfrac{x}{5} - \dfrac{4}{5}y = -1$

37) $y - \dfrac{5}{2}x = -2$
$\dfrac{3}{4}x - \dfrac{3}{10}y = \dfrac{3}{5}$

38) $-\dfrac{2}{15}x - \dfrac{1}{3}y = \dfrac{2}{3}$
$\dfrac{2}{3}x + \dfrac{5}{3}y = \dfrac{1}{2}$

39) $\dfrac{3}{4}x + \dfrac{1}{2}y = 6$
$x = 3y + 8$

40) $\dfrac{5}{3}x - \dfrac{4}{3}y = -\dfrac{4}{3}$
$y = 2x + 4$

41) $0.2x - 0.1y = 0.1$
$0.01x + 0.04y = 0.23$

42) $0.01x - 0.09y = -0.5$
$0.02x + 0.05y = 0.38$

43) $0.6x - 0.1y = 1$
$-0.4x + 0.5y = -1.1$

44) $0.8x + 0.7y = -1.7$
$0.6x - 0.1y = 0.6$

45) $0.02x + 0.01y = -0.44$
$-0.1x - 0.2y = 4$

46) $0.3x + 0.1y = 3$
$0.01x - 0.05y = -0.06$

47) $2.8x + 0.7y = 0.1$
$0.04x + 0.01y = -0.06$

48) $0.1x - 0.3y = -1.2$
$1.5y - 0.5x = 6$

Mixed Exercises: Objectives 1–3

Solve by substitution. Begin by combining like terms.

49) $8 + 2(3x - 5) - 7x + 6y = 16$
$9(y - 2) + 5x - 13y = -4$

50) $3 + 4(2y - 9) + 5x - 2y = 8$
$3(x + 3) - 4(2y + 1) - 2x = 4$

51) $10(x + 3) - 7(y + 4) = 2(4x - 3y) + 3$
$10 - 3(2x - 1) + 5y = 3y - 7x - 9$

52) $7x + 3(y - 2) = 7y + 6x - 1$
$18 + 2(x - y) = 4(x + 2) - 5y$

(24) 53) $-(y + 3) = 5(2x + 1) - 7x$
$x + 12 - 8(y + 2) = 6(2 - y)$

54) $9y - 4(2y + 3) = -2(4x + 1)$
$16 - 5(2x + 3) = 2(4 - y)$

55) Jamari wants to rent a cargo trailer to move his son into an apartment when he returns to college. A+ Rental charges $0.60 per mile while Rock Bottom Rental charges $70 plus $0.25 per mile. Let x = the number of miles driven, and let y = the cost of the rental. The cost of renting a cargo trailer from each company can be expressed with the following equations:

A+ Rental: $y = 0.60x$
Rock Bottom Rental: $y = 0.25x + 70$

a) How much would it cost Jamari to rent a cargo trailer from each company if he will drive a total of 160 mi?

b) How much would it cost Jamari to rent a trailer from each company if he planned to drive 300 mi?

c) Solve the system of equations using the substitution method, and explain the meaning of the solution.

d) Graph the system of equations, and explain when it is cheaper to rent a cargo trailer from A+ and when it is cheaper to rent it from Rock Bottom Rental. When is the cost the same?

56) To rent a pressure washer, Walsh Rentals charges $16.00 per hour while Discount Company charges $24.00 plus $12.00 per hour. Let x = the number of hours, and let y = the cost of the rental. The cost of renting a pressure washer from each company can be expressed with the following equations:

Walsh Rentals: $y = 16.00x$
Discount Company: $y = 12.00x + 24$

a) How much would it cost to rent a pressure washer from each company if it would be used for 4 hr?

b) How much would it cost to rent a pressure washer from each company if it would be rented for 9 hr?

c) Solve the system of equations using the substitution method, and explain the meaning of the solution.

d) Graph the system of equations, and explain when it is cheaper to rent a pressure washer from Walsh and when it is cheaper to rent it from Discount. When is the cost the same?

For Exercises 57 to 60, write a system of equations that has the given ordered pair as its only solution.

57) $(-3, 8)$

58) $(4, -1)$

59) $\left(\dfrac{1}{4}, \dfrac{5}{2} \right)$

60) $\left(-\dfrac{2}{3}, -\dfrac{1}{2} \right)$

R Rethink

R1) So far, this chapter has demonstrated two methods for finding the intersection of two lines. Which method is easier for you?

R2) Suppose you are using the substitution method and the variables drop out. If your final statement is true,

what does this mean? What does it mean if your final statement is false?

R3) Which exercises were easy and which were difficult? What made the hard problems difficult for you?

5.3 Solving Systems by the Elimination Method

P Prepare | O Organize

What are your objectives for Section 5.3?	How can you accomplish each objective?
1 Solve a Linear System Using the Elimination Method	• Follow Examples 1–3 in order to fully understand how the elimination method works. • Write the procedure for **Solving a System of Two Linear Equations by the Elimination Method** in your own words. • Complete the given examples on your own. • Complete You Trys 1–3.
2 Solve a Linear System Using the Elimination Method: Special Cases	• Understand that not all systems will have one solution, and review Section 5.1 to see the different types of systems. • Complete the given examples on your own. • Complete You Trys 4 and 5.
3 Use the Elimination Method Twice to Solve a Linear System	• Complete the given example on your own, and recognize when using elimination twice will be an easier way to solve a system. • Complete You Try 6.

Read the explanations, follow the examples, take notes, and complete the You Trys.

1 Solve a Linear System Using the Elimination Method

The next technique we will learn for solving a system of equations is the **elimination method.** (This is also called the **addition method.**) It is based on the addition property of equality that says that we can add the *same* quantity to each side of an equation and preserve the equality.

$$\text{If } a = b, \text{ then } a + c = b + c.$$

We can extend this idea by saying that we can add *equal* quantities to each side of an equation and still preserve the equality.

$$\text{If } a = b \text{ and } c = d, \text{ then } a + c = b + d.$$

The objective of the elimination method is to add the equations (or multiples of one or both of the equations) so that one variable is eliminated. Then, we can solve for the remaining variable.

EXAMPLE 1

Solve the system using the elimination method.

$$x + y = 11 \quad \text{(1)}$$
$$x - y = -5 \quad \text{(2)}$$

Solution

The left side of each equation is equal to the right side of each equation. Therefore, if we add the left sides together and add the right sides together, we can set them equal. We will add these equations vertically. The *y*-terms are eliminated, enabling us to solve for *x*.

$$\begin{array}{ll} x + y = 11 & \text{(1)} \\ + x - y = -5 & \text{(2)} \\ \hline 2x + 0y = 6 & \text{Add equations (1) and (2).} \\ 2x = 6 & \text{Simplify.} \\ x = 3 & \text{Divide by 2.} \end{array}$$

> **W Hint**
>
> Are you writing out the steps in the example as you are reading it?

Now we substitute $x = 3$ into either equation to find the value of *y*. Here, we will use equation (1).

$$\begin{array}{ll} x + y = 11 & \text{Equation (1)} \\ 3 + y = 11 & \text{Substitute 3 for } x. \\ y = 8 & \text{Subtract 3.} \end{array}$$

Check $x = 3$ and $y = 8$ in *both* equations.

$$\begin{array}{ll} x + y = 11 & \\ 3 + 8 \overset{?}{=} 11 & \text{Substitute.} \\ 11 = 11 & \text{True} \end{array} \qquad \begin{array}{ll} x - y = -5 & \\ 3 - 8 \overset{?}{=} -5 & \text{Substitute.} \\ -5 = -5 & \text{True} \end{array}$$

The solution of the system is (3, 8).

[YOU TRY 1]

Solve the system using the elimination method.

$$3x + y = 10$$
$$x - y = 6$$

In Example 1, simply adding the equations eliminated a variable. But what can we do if we *cannot* eliminate a variable just by adding the equations together?

EXAMPLE 2

Solve the system using the elimination method.

$$2x + 5y = 5 \quad \text{(1)}$$
$$x + 4y = 7 \quad \text{(2)}$$

Solution

Just adding these equations will *not* eliminate a variable. The multiplication property of equality tells us that multiplying both sides of an equation by the same quantity results in an equivalent equation. If we multiply equation (2) by -2, the coefficient of *x* will be -2.

$$\begin{array}{ll} -2(x + 4y) = -2(7) & \text{Multiply equation (2) by } -2. \\ -2x - 8y = -14 & \text{New, equivalent equation} \end{array}$$

> **W Hint**
>
> Notice that we multiply the second equation by -2 so that the coefficients of *x* are opposites. This way, when you add the two equations together, the *x*'s are eliminated.

Original System		**Rewrite the System**
$2x + 5y = 5$		$2x + 5y = 5$
$x + 4y = 7$	$\longrightarrow$	$-2x - 8y = -14$

Add the equations in the rewritten system. The x is eliminated.

$$2x + 5y = 5$$
$$+ \underline{-2x - 8y = -14}$$
$$0x - 3y = -9 \qquad \text{Add the equations.}$$
$$-3y = -9 \qquad \text{Simplify.}$$
$$y = 3 \qquad \text{Solve for } y.$$

Substitute $y = 3$ into (1) or (2) to find x. We will use equation (2).

$$x + 4y = 7 \qquad \text{Equation (2)}$$
$$x + 4(3) = 7 \qquad \text{Substitute 3 for } y.$$
$$x + 12 = 7$$
$$x = -5$$

The solution is $(-5, 3)$. Verify that $(-5, 3)$ satisfies equations (1) and (2).

$\left[\; \textbf{YOU TRY 2} \;\right]$ Solve the system using the elimination method.

$$8x - y = -5$$
$$-6x + 2y = 15$$

Next we summarize the steps for solving a system using the elimination method.

Procedure Solving a System of Two Linear Equations by the Elimination Method

Step 1: Write each equation in the form $Ax + By = C$.

Step 2: Determine which variable to eliminate. If necessary, multiply one or both of the equations by a number so that the coefficients of the variable to be eliminated are negatives of one another.

Step 3: Add the equations, and solve for the remaining variable.

Step 4: Substitute the value found in *Step 3* into either of the original equations to find the value of the other variable.

Step 5: Check the solution in each of the original equations.

EXAMPLE 3   Solve the system using the elimination method.

$$2x = 9y + 4 \qquad (1)$$
$$3x - 7 = 12y \qquad (2)$$

Solution

Step 1: **Write each equation in the form $Ax + By = C$.**

$2x = 9y + 4 \qquad (1)$	$3x - 7 = 12y \qquad (2)$
$2x - 9y = 4 \qquad \text{Subtract } 9y.$	$3x - 12y = 7 \qquad \text{Subtract } 12y \text{ and add 7.}$

When we rewrite the equations in the form $Ax + By = C$, we get

$$2x - 9y = 4 \quad \text{(3)}$$
$$3x - 12y = 7 \quad \text{(4)}$$

Step 2: **Determine which variable to eliminate from equations (3) and (4).** Often, it is easier to eliminate the variable with the smaller coefficients. Therefore, *we will eliminate x.*

The least common multiple of 2 and 3 (the *x*-coefficients) is 6. Before we add the equations, one *x*-coefficient should be 6, and the other should be −6. Multiply equation (3) by 3 and equation (4) by −2.

Rewrite the System

$$3(2x - 9y) = 3(4) \quad \text{3 times (3)}$$
$$-2(3x - 12y) = -2(7) \quad \text{−2 times (4)}$$

$\longrightarrow$

$$6x - 27y = 12$$
$$-6x + 24y = -14$$

W Hint

Once again, notice that the coefficients of *x* are made to be opposite in sign so that they can be eliminated when the equations are added.

Step 3: **Add the resulting equations to eliminate *x*. Solve for *y*.**

$$\begin{array}{r} 6x - 27y = 12 \\ + \underline{-6x + 24y = -14} \\ -3y = -2 \\ y = \dfrac{2}{3} \end{array}$$

Step 4: **Substitute $y = \dfrac{2}{3}$ into equation (1) and solve for *x*.**

$$2x = 9y + 4 \qquad \text{Equation (1)}$$
$$2x = 9\left(\dfrac{2}{3}\right) + 4 \qquad \text{Substitute.}$$
$$2x = 6 + 4 \qquad \text{Multiply.}$$
$$2x = 10 \qquad \text{Add.}$$
$$x = 5$$

Step 5: **Check** to verify that $\left(5, \dfrac{2}{3}\right)$ satisfies each of the original equations. The solution is $\left(5, \dfrac{2}{3}\right)$.

[YOU TRY 3] Solve the system using the elimination method.

$$5x = 2y - 14$$
$$4x + 3y = 21$$

---

2 Solve a Linear System Using the Elimination Method: Special Cases

We have seen in Sections 5.1 and 5.2 that some systems have no solution, and some have an infinite number of solutions. How does the elimination method illustrate these results?

EXAMPLE 4

Solve the system using the elimination method.

$$4y = 10x + 3 \qquad (1)$$
$$6y - 15x = -8 \qquad (2)$$

Solution

Step 1: **Write each equation in the form $Ax + By = C$.**

$$4y = 10x + 3 \qquad\qquad -10x + 4y = 3 \qquad (3)$$
$$6y - 15x = -8 \qquad\longrightarrow\qquad -15x + 6y = -8 \qquad (4)$$

Step 2: **Determine which variable to eliminate from equations (3) and (4).** Eliminate y. The least common multiple of 4 and 6, the y-coefficients, is 12. One y-coefficient must be 12, and the other must be -12.

Rewrite the System

$$-3(-10x + 4y) = -3(3) \qquad\qquad 30x - 12y = -9$$
$$2(-15x + 6y) = 2(-8) \qquad\longrightarrow\qquad -30x + 12y = -16$$

Step 3: **Add the equations.**

$$30x - 12y = -9$$
$$+ \underline{-30x + 12y = -16}$$
$$0 = -25 \qquad \text{False}$$

The variables drop out, and we get a false statement. Therefore, the system is inconsistent, and the solution set is $\varnothing$.

W Hint

Notice that, just as in Section 5.2, if the variables drop out and you end up with a false statement, there is *no solution*.

[YOU TRY 4]

Solve the system using the elimination method.

$$24x + 6y = -7$$
$$4y + 3 = -16x$$

EXAMPLE 5

Solve the system using the elimination method.

$$12x - 18y = 9 \qquad (1)$$
$$y = \frac{2}{3}x - \frac{1}{2} \qquad (2)$$

Solution

Step 1: **Write equation (2) in the form $Ax + By = C$.**

$$y = \frac{2}{3}x - \frac{1}{2} \qquad\qquad \text{Equation (2)}$$

$$6y = 6\left(\frac{2}{3}x - \frac{1}{2}\right) \qquad\qquad \text{Multiply by 6 to eliminate fractions.}$$

$$6y = 4x - 3$$
$$-4x + 6y = -3 \qquad (3) \qquad \text{Rewrite as } Ax + By = C.$$

We can rewrite $y = \frac{2}{3}x - \frac{1}{2}$ as $-4x + 6y = -3$, equation (3).

Step 2: **Determine which variable to eliminate from equations (1) and (3).**

$$12x - 18y = 9 \quad \text{(1)}$$
$$-4x + 6y = -3 \quad \text{(3)}$$

Eliminate x. Multiply equation (3) by 3.

$$12x - 18y = 9 \quad \text{(1)}$$
$$-12x + 18y = -9 \quad \text{3 times (3)}$$

Step 3: **Add the equations.**

$$
\begin{array}{r}
12x - 18y = 9 \\
+ \ -12x + 18y = -9 \\
\hline
0 = 0 \quad \text{True}
\end{array}
$$

The variables drop out, and we get a true statement. The equations are dependent, so there are an infinite number of solutions. The solution set is $\left\{ (x, y) \mid y = \dfrac{2}{3}x - \dfrac{1}{2} \right\}$.

[YOU TRY 5] Solve the system using the elimination method.

$$-6x + 8y = 4$$
$$3x - 4y = -2$$

---

3 Use the Elimination Method Twice to Solve a Linear System

Sometimes, applying the elimination method *twice* is the best strategy.

EXAMPLE 6 Solve using the elimination method.

$$5x - 6y = 2 \quad \text{(1)}$$
$$9x + 4y = -3 \quad \text{(2)}$$

Solution

Each equation is written in the form $Ax + By = C$, so we begin with **Step 2.**

Step 2: We will eliminate y from equations (1) and (2).

Rewrite the System

$$2(5x - 6y) = 2(2)$$
$$3(9x + 4y) = 3(-3)$$
$$\longrightarrow$$
$$10x - 12y = 4$$
$$27x + 12y = -9$$

Step 3: Add the resulting equations to eliminate y. Solve for x.

$$
\begin{array}{r}
10x - 12y = \ \ 4 \\
+ \ 27x + 12y = -9 \\
\hline
37x = -5
\end{array}
$$

$$x = -\dfrac{5}{37} \quad \text{Solve for } x.$$

Normally, we would substitute $x = -\dfrac{5}{37}$ into equation (1) or equation (2) and solve for y.

This time, however, working with a number like $-\dfrac{5}{37}$ would be difficult, so *we will use the elimination method a second time.*

Go back to the original equations, (1) and (2), and use the elimination method again but eliminate the other variable, x. Then, solve for y.

Eliminate x from the original system;

$$5x - 6y = 2 \qquad (1)$$
$$9x + 4y = -3 \qquad (2)$$

Rewrite the System

$$-9(5x - 6y) = -9(2) \qquad \longrightarrow \qquad -45x + 54y = -18$$
$$5(9x + 4y) = 5(-3) \qquad\qquad\qquad 45x + 20y = -15$$

Add the equations.

$$
\begin{array}{r}
-45x + 54y = -18 \\
+\quad 45x + 20y = -15 \\
\hline
74y = -33
\end{array}
$$

$$y = -\dfrac{33}{74} \qquad \text{Solve for } y.$$

Check to verify that the solution is $\left(-\dfrac{5}{37}, -\dfrac{33}{74}\right)$.

[YOU TRY 6] Solve using the elimination method.

$$-9x + 2y = -3$$
$$2x - 5y = 4$$

ANSWERS TO [YOU TRY] EXERCISES

1) $(4, -2)$ 2) $\left(\dfrac{1}{2}, 9\right)$ 3) $(0, 7)$ 4) $\varnothing$

5) infinite number of solutions of the form $\{(x, y) \mid 3x - 4y = -2\}$ 6) $\left(\dfrac{7}{41}, -\dfrac{30}{41}\right)$

E Evaluate **5.3** Exercises   Do the exercises, and check your work.

Mixed Exercises: Objectives 1 and 2

1) What is the first step you would use to solve this system by elimination if you wanted to eliminate y?

$$5x + y = 2$$
$$3x - y = 6$$

2) What is the first step you would use to solve this system by elimination if you wanted to eliminate x?

$$4x - 3y = 14$$
$$8x - 11y = 18$$

For Exercises 3 and 4, fill in the blank with *always, sometimes,* or *never* to make the statement true.

3) A system of equations with two different lines written in the form $Ax + By = 0$ will _____ have $(0, 0)$ as the solution.

4) If both variables are eliminated while solving a system of equations using elimination, the system will _____ have an infinite number of solutions.

Solve each system using the elimination method.

5) $x - y = -3$
$2x + y = 18$

6) $x + 3y = 1$
$-x + 5y = -9$

7) $-x + 2y = 2$
$x - 7y = 8$

8) $4x - y = -15$
$3x + y = -6$

⟨24⟩ 9) $x + 4y = 1$
$3x - 4y = -29$

10) $5x - 4y = -10$
$-5x + 7y = 25$

11) $-8x + 5y = -16$
$4x - 7y = 8$

12) $7x + 6y = 3$
$3x + 2y = -1$

13) $4x + 15y = 13$
$3x + 5y = 16$

14) $12x + 7y = 7$
$-3x + 8y = 8$

15) $9x - 7y = -14$
$4x + 3y = 6$

16) $5x - 2y = -6$
$4x + 5y = -18$

17) $-9x + 2y = -4$
$6x - 3y = 11$

18) $12x - 2y = 3$
$8x - 5y = -9$

⟨24⟩ 19) $9x - y = 2$
$18x - 2y = 4$

20) $-4x + 7y = 13$
$12x - 21y = -5$

21) $x = 12 - 4y$
$2x - 7 = 9y$

22) $5x + 3y = -11$
$y = 6x + 4$

23) $4y = 9 - 3x$
$5x - 16 = -6y$

24) $8x = 6y - 1$
$10y - 6 = -4x$

25) $2x - 9 = 8y$
$20y - 5x = 6$

26) $3x + 2y = 6$
$4y = 12 - 6x$

27) $6x - 11y = -1$
$-7x + 13y = 2$

28) $10x - 4y = 7$
$12x - 3y = -15$

29) $9x + 6y = -2$
$-6x - 4y = 11$

30) $4x - 9y = -3$
$36y - 16x = 12$

31) What is the first step in solving this system by the elimination method? DO NOT SOLVE.

$$\frac{x}{4} + \frac{y}{2} = -1$$

$$\frac{3}{8}x + \frac{5}{3}y = -\frac{7}{12}$$

32) What is the first step in solving this system by the elimination method? DO NOT SOLVE.

$$0.1x + 2y = -0.8$$
$$0.03x + 0.10y = 0.26$$

Solve each system by elimination.

33) $\frac{4}{5}x - \frac{1}{2}y = -\frac{3}{2}$
$2x - \frac{1}{4}y = \frac{1}{4}$

34) $\frac{1}{3}x - \frac{4}{5}y = \frac{13}{15}$
$\frac{1}{6}x - \frac{3}{4}y = -\frac{1}{2}$

35) $\frac{5}{4}x - \frac{1}{2}y = \frac{7}{8}$
$\frac{2}{5}x - \frac{1}{10}y = -\frac{1}{2}$

36) $\frac{1}{2}x - \frac{11}{8}y = -1$
$-\frac{2}{5}x + \frac{3}{10}y = \frac{4}{5}$

⟨24⟩ 37) $\frac{x}{4} + \frac{y}{2} = -1$
$\frac{3}{8}x + \frac{5}{3}y = -\frac{7}{12}$

38) $\frac{x}{12} - \frac{y}{6} = \frac{2}{3}$
$\frac{x}{4} + \frac{y}{3} = 2$

39) $\frac{x}{12} - \frac{y}{8} = \frac{7}{8}$
$y = \frac{2}{3}x - 7$

40) $\frac{5}{3}x + \frac{1}{3}y = \frac{2}{3}$
$\frac{3}{4}x + \frac{3}{20}y = -\frac{5}{4}$

41) $-\frac{1}{2}x + \frac{5}{4}y = \frac{3}{4}$
$\frac{2}{5}x - \frac{1}{2}y = -\frac{1}{10}$

42) $y = 2 - 4x$
$\frac{1}{3}x - \frac{3}{8}y = \frac{5}{8}$

43) $0.08x + 0.07y = -0.84$
$0.32x - 0.06y = -2$

44) $0.06x + 0.05y = 0.58$
$0.18x - 0.13y = 1.18$

45) $0.1x + 2y = -0.8$
$0.03x + 0.10y = 0.26$

46) $0.6x - 0.1y = 0.5$
$0.1x - 0.03y = -0.01$

⟨24⟩ 47) $-0.4x + 0.2y = 0.1$
$0.6x - 0.3y = 1.5$

48) $x - 0.5y = 0.2$
$-0.3x + 0.15y = -0.06$

49) $0.04x + 0.03y = 0.16$
$0.6x + 0.2y = 1.15$

50) $-0.5x + 0.8y = 0.3$
$0.03x + 0.1y = -0.24$

51) $17x - 16(y + 1) = 4(x - y)$
$19 - 10(x + 2) = -4(x + 6) - y + 2$

52) $28 - 4(y + 1) = 3(x - y) + 4$
$-5(x + 4) - y + 3 = 28 - 5(2x + 5)$

53) $5 - 3y = 6(3x + 4) - 8(x + 2)$
$6x - 2(5y + 2) = -7(2y - 1) - 4$

54) $5(y + 3) = 6(x + 1) + 6x$
$7 - 3(2 - 3x) - y = 2(3y + 8) - 5$

55) $6(x - 3) + x - 4y = 1 + 2(x - 9)$
$4(2y - 3) + 10x = 5(x + 1) - 4$

56) $8y + 2(4x + 5) - 5x = 7y - 11$
$11y - 3(x + 2) = 16 + 2(3y - 4) - x$

Objective 3: Use the Elimination Method Twice to Solve a Linear System

Solve each system using the elimination method twice.

 57) $4x + 5y = -6$
$3x + 8y = 15$

58) $8x - 4y = -21$
$-5x + 6y = 12$

59) $4x + 9y = 7$
$6x + 11y = -14$

60) $10x + 3y = 18$
$9x - 4y = 5$

Find k so that the given ordered pair is a solution of the given system.

61) $x + ky = 17$; $(5, 4)$
$2x - 3y = -2$

62) $kx + y = -13$; $(-1, -8)$
$9x - 2y = 7$

63) $3x + 4y = -9$; $(-7, 3)$
$kx - 5y = 41$

64) $4x + 3y = -7$; $(2, -5)$
$3x + ky = 16$

65) Given the following system of equations,
$$x - y = 5$$
$$x - y = c$$

find c so that the system has

a) an infinite number of solutions.

b) no solution.

66) Given the following system of equations,
$$2x + y = 9$$
$$2x + y = c$$

find c so that the system has

a) an infinite number of solutions.

b) no solution.

67) Given the following system of equations,
$$9x + 12y = -15$$
$$ax + 4y = -5$$

find a so that the system has

a) an infinite number of solutions.

b) exactly one solution.

68) Given the following system of equations,
$$-2x + 7y = 3$$
$$4x + by = -6$$

find b so that the system has

a) an infinite number of solutions.

b) exactly one solution.

Extension

Let a, b, and c represent nonzero constants. Solve each system for x and y.

69) $-5x + 4by = 6$
$5x + 3by = 8$

70) $ax - 6y = 4$
$-ax + 9y = 2$

71) $3ax + by = 4$
$ax - by = -5$

72) $2ax + by = c$
$ax + 3by = 4c$

R Rethink

R1) In this chapter, you have learned three methods for finding the intersection of two lines. Which of the three methods do you prefer and why?

R2) Suppose you want to eliminate the y-values in a system of linear equations. What do you

need to do before you add the two equations together?

R3) Are there any problems you could not do? If so, write them down or circle them and ask your instructor for help.

Putting It All Together

What are your objectives?	How can you accomplish each objective?
1 Review the concepts of Sections 5.1–5.3	• Understand the summary for **Choosing Between Substitution and the Elimination Method to Solve a System,** and write it in your own words. • Complete the You Try.

W Work   **Read the explanations, follow the examples, take notes, and complete the You Try.**

1 Review the Concepts of Sections 5.1–5.3

We have learned three methods for solving systems of linear equations:

1) Graphing   2) Substitution   3) Elimination

How do we know which method is best for a particular system? We will answer this question by summarizing what we have learned so far.

First, solving a system by graphing is the least desirable of the methods. The point of intersection can be difficult to read, especially if one of the numbers is a fraction. But, the graphing method is important in certain situations and is one you should know.

Summary Choosing Between Substitution and the Elimination Method to Solve a System

1) If at least one equation is solved for a variable and contains no fractions, **use substitution.**

$$-5x + 2y = -8$$
$$x = 4y + 16$$

2) If a variable has a coefficient of 1 or -1, you can solve for that variable and **use substitution.**

$$4x + y = 10$$
$$-3x - 8y = 7$$

Or, leave each equation in the form $Ax + By = C$ and **use elimination.** Either approach is good and is a matter of personal preference.

3) If no variable has a coefficient of 1 or -1, **use elimination.**

$$4x - 5y = -3$$
$$6x + 8y = 11$$

Remember, if an equation contains fractions or decimals, begin by eliminating them. Then, decide which method to use following the guidelines listed here.

Putting It All Together Exercises

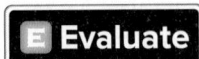 Do the exercises, and check your work.

Objective 1: Review the Concepts of Sections 5.1–5.3

Decide which method to use to solve each system, substitution or elimination, and explain why this method was chosen. Then solve the system.

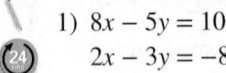

1) $8x - 5y = 10$
   $2x - 3y = -8$

2) $x = 2y - 7$
   $8x - 3y = 9$

3) $12x - 5y = 18$
   $8x + y = -1$

4) $11x + 10y = -4$
   $9x - 5y = 2$

5) $y - 4x = -11$
   $x = y + 8$

6) $4x - 5y = 4$
   $y = \dfrac{3}{4}x - \dfrac{1}{2}$

Solve each system using either the substitution or elimination method.

7) $4x + 5y = 24$
   $x - 3y = 6$

8) $6y - 5x = 22$
   $-9x - 8y = 2$

9) $6x + 15y = -1$
   $9x = 10y - 8$

10) $x + 2y = 9$
    $7x - y = 3$

11) $10x + 4y = 7$
    $15x + 6y = -2$

12) $y = -6x + 5$
    $12x + 2y = 10$

13) $10x + 9y = 4$
    $x = -\dfrac{1}{2}$

14) $6x - 4y = 11$
    $\dfrac{3}{2}x + \dfrac{1}{4}y = \dfrac{7}{8}$

15) $7y - 2x = 13$
    $3x - 2y = 6$

16) $y = 6$
    $12x + y = 8$

17) $\dfrac{2}{5}x + \dfrac{4}{5}y = -2$
    $\dfrac{1}{6}x + \dfrac{1}{6}y = \dfrac{1}{3}$

18) $5x + 4y = 14$
    $y = -\dfrac{8}{5}x + 7$

19) $-0.3x + 0.1y = 0.4$
    $0.01x + 0.05y = 0.2$

20) $0.01x - 0.06y = 0.03$
    $0.4x + 0.3y = -1.5$

21) $-6x + 2y = -10$
    $21x - 7y = 35$

22) $\dfrac{5}{3}x + \dfrac{4}{3}y = \dfrac{2}{3}$
    $10x + 8y = -5$

23) $2 = 5y - 8x$
    $y = \dfrac{3}{2}x - \dfrac{1}{2}$

24) $\dfrac{5}{6}x - \dfrac{3}{4}y = \dfrac{2}{3}$
    $\dfrac{1}{3}x + 2y = \dfrac{10}{3}$

25) $2x - 3y = -8$
    $7x + 10y = 4$

26) $6x = 9 - 13y$
    $4x + 3y = -2$

27) $6(2x - 3) = y + 4(x - 3)$
    $5(3x + 4) + 4y = 11 - 3y + 27x$

28) $3 - 5(x - 4) = 2(1 - 4y) + 2$
    $2(x + 10) + y + 1 = 3x + 5(y + 6) - 17$

29) $2y - 2(3x + 4) = -5(y - 2) - 17$
    $4(2x + 3) = 10 + 5(y + 1)$

30) $x - y + 23 = 2y + 3(2x + 7)$
    $9y - 8 + 4(x + 2) = 2(4x - 1) - 3x + 10y$

31) $y = -4x$
    $10x + 2y = -5$

32) $x = \dfrac{2}{3}y$
    $9x - 5y = -6$

For Exercises 33 to 36, write a system of two equations in x and y so that the given ordered pair is the only solution of the system.

33) $(-5, 2)$

34) $(0, -7)$

35) $\left(\dfrac{3}{4}, \dfrac{1}{3}\right)$

36) $\left(-\dfrac{1}{2}, -\dfrac{4}{5}\right)$

Solve each system by graphing.

 37) $y = \dfrac{1}{2}x + 1$

     $x + y = 4$

38) $x + y = -3$

    $y = 3x + 1$

39) $x + y = 0$

    $x - 2y = -12$

40) $2y - x = 6$

    $y = 2x$

41) $2x - 3y = 3$

    $y = \dfrac{2}{3}x + 1$

42) $y = -\dfrac{5}{2}x - 3$

    $10x + 4y = -12$

Solve each system using a graphing calculator.

43) $8x - 6y = -7$

    $4x - 16y = 3$

44) $4x + 3y = -9$

    $2x + y = 2$

R Rethink

R1) What are the advantages and disadvantages of each method for solving a system of equations?

R2) Which characteristics of the equations in a system will help you determine whether you will solve a system by substitution or by elimination?

R3) Which method of solving systems do you prefer? Why?

R4) How will choosing the most efficient way to solve a system help you on a test?

5.4 Applications of Systems of Two Equations

P Prepare

O Organize

What are your objectives for Section 5.4?	How can you accomplish each objective?
1 Solve Problems Involving General Quantities	• Understand the procedure for **Solving an Applied Problem Using a System of Equations,** and write it in your own words. • Complete the given example on your own. • Complete You Try 1.
2 Solve Geometry Problems	• Review geometry formulas, if necessary, given in Section 1.3. • Complete the given example on your own. • Complete You Try 2.
3 Solve Problems Involving Cost	• Complete the given example on your own. • Complete You Try 3.
4 Solve Mixture Problems	• Complete the given example on your own. • Complete You Try 4.
5 Solve Distance, Rate, and Time Problems	• Complete the given example on your own. • Complete You Try 5.

 **Read the explanations, follow the examples, take notes, and complete the You Trys.**

In Section 3.2, we introduced the five-step method for solving applied problems. Here, we modify the method for problems with *two* unknowns and *two* equations.

> ## Procedure Solving an Applied Problem Using a System of Equations
>
> **Step 1: Read** the problem carefully, more than once if necessary. Draw a picture, if applicable. Identify what you are being asked to find.
>
> **Step 2: Choose variables** to represent the unknown quantities. Label any pictures with the variables.
>
> **Step 3: Write a system of equations using two variables.** It may be helpful to begin by writing the equations in words.
>
> **Step 4: Solve** the system.
>
> **Step 5: Check** the answer in the original problem, and **interpret** the solution as it relates to the problem. Be sure your answer makes sense in the context of the problem.

1 Solve Problems Involving General Quantities

EXAMPLE 1

Write a system of equations, and solve.

As of March 1, 2018, Pink Floyd's album, *The Dark Side of the Moon*, spent more weeks on the Billboard 200 chart for top-selling albums than any other album in history. It was on the chart 427 more weeks than the second-place album, *Legend*, by Bob Marley and the Wailers. If they were on the charts for a total of 1447 weeks, how many weeks did each album spend on the Billboard 200 chart? (www.billboard.com)

Solution

Step 1: Read the problem carefully, and identify what we are being asked to find.

We must find the number of weeks each album was on the chart.

Step 2: Choose variables to represent the unknown quantities.

x = the number of weeks *The Dark Side of the Moon* was on the Billboard 200 chart

y = the number of weeks *Legend* was on the Billboard 200 chart

Step 3: Write a system of equations using two variables. First, let's think of the equations in English. Then we will translate them into algebraic equations.

To get one equation, use the information that says these two albums were on the Billboard 200 chart for a total of 1447 weeks. Write an equation in words, then translate it into an algebraic equation.

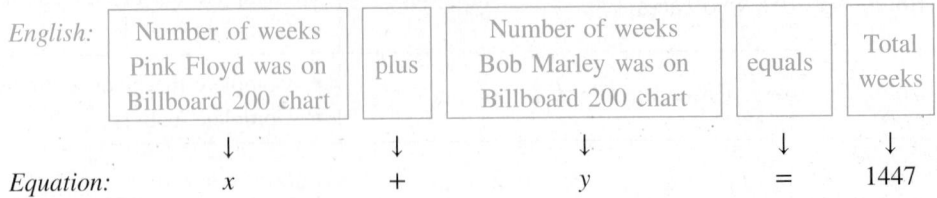

English:	Number of weeks Pink Floyd was on Billboard 200 chart	plus	Number of weeks Bob Marley was on Billboard 200 chart	equals	Total weeks
	↓	↓	↓	↓	↓
Equation:	x	$+$	y	$=$	1447

The first equation is $x + y = 1447$.

To get the second equation, use the information that says the Pink Floyd album was on the chart 427 weeks more than the Bob Marley album.

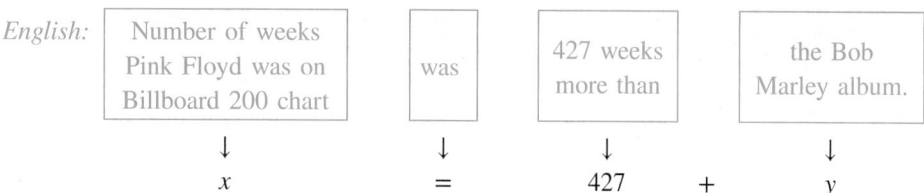

English:

Number of weeks Pink Floyd was on Billboard 200 chart	was	427 weeks more than	the Bob Marley album.
↓	↓	↓	↓
x	$=$	427 $+$	y

The second equation is $x = 427 + y$.

The system of equations is $x + y = 1447$
$$x = 427 + y.$$

Step 4: **Solve** the system.

$$x + y = 1447$$
$$(427 + y) + y = 1447 \qquad \text{Substitute.}$$
$$427 + 2y = 1447 \qquad \text{Combine like terms.}$$
$$2y = 1020 \qquad \text{Subtract 427 from each side.}$$
$$\frac{2y}{2} = \frac{1020}{2} \qquad \text{Divide each side by 2.}$$
$$y = 510 \qquad \text{Simplify.}$$

Find x by substituting $y = 510$ into $x = 427 + y$.

$$x = 427 + 510 = 937$$

The solution to the system is $(937, 510)$.

Step 5: **Check** the answer, and **interpret** the solution as it relates to the problem.

The Dark Side of the Moon was on the Billboard 200 for 937 weeks, and *Legend* was on the chart for 510 weeks.

They were on the chart for a total of $937 + 510 = 1447$ weeks, and the Pink Floyd album was on there $937 - 510 = 427$ weeks longer than the other album.

W Hint

Try using the elimination method to solve this system of linear equations. Next, decide which method is easier for you!

[**YOU TRY 1**] Write a system of equations, and solve.

In 2017, Casper, WY, had about 3100 more residents than Carson City, NV. Find the population of each city if together they had approximately 112,500 residents. (www.census.gov)

Next we will see how we can use two variables and a system of equations to solve geometry problems.

2 Solve Geometry Problems

EXAMPLE 2

Write a system of equations, and solve.

A builder installed a rectangular window in a new house and needs 182 in. of trim to go around it on the inside of the house. Find the dimensions of the window if the width is 23 in. less than the length.

Solution

Step 1: **Read** the problem carefully, and identify what we are being asked to find. Draw a picture.

We must find the length and width of the window.

Step 2: **Choose variables** to represent the unknown quantities.

$$w = \text{the width of the window}$$
$$l = \text{the length of the window}$$

Label the picture with the variables.

Step 3: **Write a system of equations using two variables.**

To get one equation, we know that the width is 23 in. less than the length. We can write the equation $w = l - 23$.

If it takes 182 in. of trim to go around the window, this is the *perimeter* of the rectangular window. Use the equation for the perimeter of a rectangle.

$$2l + 2w = 182$$

The system of equations is
$$w = l - 23$$
$$2l + 2w = 182.$$

Step 4: **Solve** the system.

$$2l + 2w = 182$$
$$2l + 2(l - 23) = 182 \qquad \text{Substitute.}$$
$$2l + 2l - 46 = 182 \qquad \text{Distribute.}$$
$$4l - 46 = 182 \qquad \text{Combine like terms.}$$
$$4l = 228 \qquad \text{Add 46 to each side.}$$
$$\frac{4l}{4} = \frac{228}{4} \qquad \text{Divide each side by 4.}$$
$$l = 57 \qquad \text{Simplify.}$$

Find w by substituting $l = 57$ into $w = l - 23$.

$$w = 57 - 23 = 34$$

The solution to the system is (57, 34). (The ordered pair is written as (l, w), in alphabetical order.)

Step 5: **Check** the answer, and **interpret** the solution as it relates to the problem.

The length of the window is 57 in., and the width is 34 in. The check is left to the student.

[YOU TRY 2]

Write a system of equations and solve.

The top of a rectangular desk is twice as long as it is wide. If the perimeter of the desk is 162 in., find its dimensions.

3 Solve Problems Involving Cost

EXAMPLE 3

Write a system of equations, and solve.

    Ari buys two mezzanine tickets to a Broadway play and four tickets to the top of the Empire State Building for $504. Lloyd spends $723 on four mezzanine tickets and three tickets to the top of the Empire State Building. Find the cost of a ticket to each attraction. (www.esbnyc.com)

Solution

Step 1: **Read** the problem carefully, and identify what we are being asked to find.

    We must find the cost of a ticket to a Broadway play and to the top of the Empire State Building.

Step 2: **Choose variables** to represent the unknown quantities.

$$x = \text{the cost of a ticket to a Broadway play}$$
$$y = \text{the cost of a ticket to the Empire State Building}$$

Step 3: **Write a system of equations using two variables.** First, let's think of the equations in English. Then we will translate them into algebraic equations.

    First, use the information about Ari's purchase.

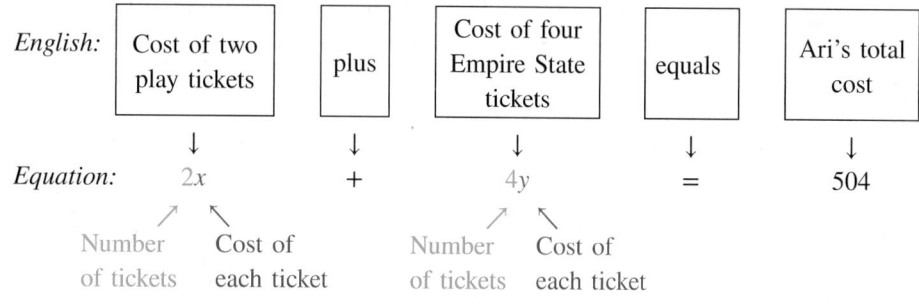

    One equation is $2x + 4y = 504$.

    Next, use the information about Lloyd's purchase.

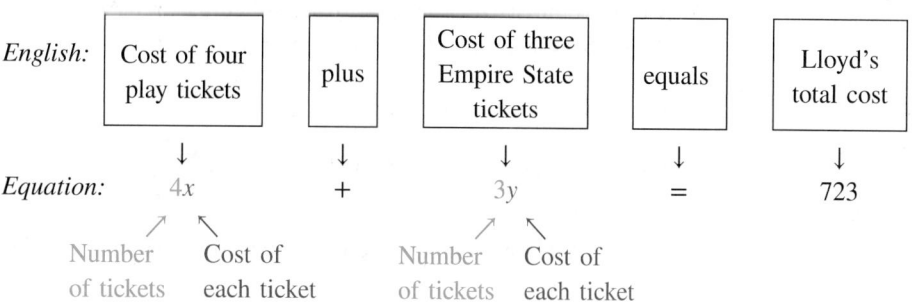

    The other equation is $4x + 3y = 723$.

    The system of equations is $2x + 4y = 504$
$$4x + 3y = 723.$$

Step 4: **Solve** the system.

Use the elimination method. Multiply the first equation by -2 to eliminate x.

$$-4x - 8y = -1008$$
$$+\ \underline{4x + 3y = \quad 723}$$
$$-5y = \quad -285 \qquad \text{Add the equations.}$$
$$y = 57$$

Find x. We will substitute $y = 57$ into $2x + 4y = 504$.

$$2x + 4(57) = 504 \qquad \text{Substitute.}$$
$$2x + 228 = 504 \qquad \text{Multiply.}$$
$$2x = 276 \qquad \text{Subtract 228 from each side.}$$
$$x = 138$$

The solution to the system is (138, 57).

Step 5: **Check** the answer, and **interpret** the solution as it relates to the problem.

A Broadway play ticket costs $138.00, and a ticket to the top of the Empire State Building costs $57.00.

The check is left to the student.

$\left[\ \textbf{YOU TRY 3}\ \right]$ Write a system of equations, and solve.

Torie bought three scarves and a belt for $105 while Liz bought two scarves and two belts for $98. Find the cost of a scarf and a belt.

Now we will learn how to solve mixture problems using two variables and a system of two equations.

4 Solve Mixture Problems

EXAMPLE 4

A pharmacist needs to make 200 mL of a 10% hydrogen peroxide solution. She will make it from some 8% hydrogen peroxide solution and some 16% hydrogen peroxide solution that are in the storeroom. How much of the 8% solution and 16% solution should she use?

Solution

Step 1: **Read** the problem carefully, and identify what we are being asked to find. Draw a picture.

We must find the amount of 8% solution and 16% solution she should use.

W Hint

Remember that a hydrogen peroxide *solution* contains both pure hydrogen peroxide and pure water.

Amount of 8%   Amount of 16%   200 mL of 10%
solution, x solution, y solution

Step 2: **Choose variables** to represent the unknown quantities.

x = amount of 8% solution needed

y = amount of 16% solution needed

Step 3: **Write a system of equations using two variables.**

Let's begin by arranging the information in a table. To obtain the expression in the last column, multiply the percent of hydrogen peroxide in the solution by the amount of solution to get the amount of pure hydrogen peroxide in the solution.

	Percent of Hydrogen Peroxide in Solution (as a decimal)	Amount of Solution	Amount of Pure Hydrogen Peroxide in Solution
Mix these {	0.08	x	$0.08x$
	0.16	y	$0.16y$
to make →	0.10	200	$0.10(200)$

To get one equation, use the information in the second column. It tells us that

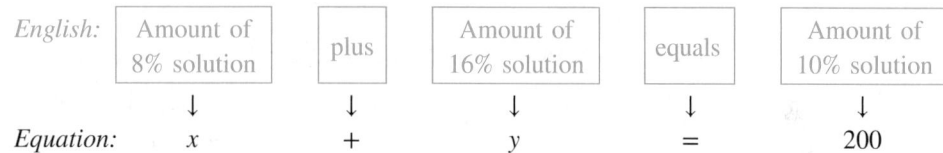

English: Amount of 8% solution | plus | Amount of 16% solution | equals | Amount of 10% solution

Equation: $\quad x \quad + \quad y \quad = \quad 200$

The equation is $x + y = 200$.

To get the second equation, use the information in the third column. It tells us that

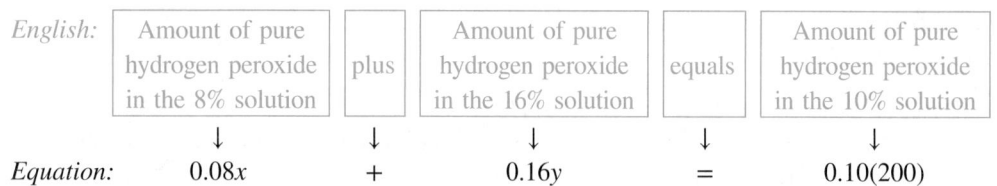

English: Amount of pure hydrogen peroxide in the 8% solution | plus | Amount of pure hydrogen peroxide in the 16% solution | equals | Amount of pure hydrogen peroxide in the 10% solution

Equation: $\quad 0.08x \quad + \quad 0.16y \quad = \quad 0.10(200)$

The equation is $0.08x + 0.16y = 0.10(200)$.

The system of equations is
$$x + y = 200$$
$$0.08x + 0.16y = 0.10(200).$$

Step 4: **Solve** the system.

Multiply the second equation by 100 to eliminate the decimals. Our system becomes

$$x + y = 200$$
$$8x + 16y = 2000$$

Use the elimination method. Multiply the first equation by -8 to eliminate x.

$$-8x - 8y = -1600$$
$$+ \quad 8x + 16y = \quad 2000$$
$$\overline{\qquad\qquad\quad 8y = 400}$$
$$y = 50$$

Find x. Substitute $y = 50$ into $x + y = 200$.

$$x + 50 = 200$$
$$x = 150$$

The solution to the system is (150, 50).

W Hint

Remember that x and y must have a sum of 200, and neither should be negative.

Step 5: **Check** the answer, and **interpret** the solution as it relates to the problem.

The pharmacist needs 150 mL of the 8% solution and 50 mL of the 16% solution. Check the answers in the original problem to verify that they are correct.

[**YOU TRY 4**] Write an equation, and solve.

How many milliliters of a 5% acid solution and how many milliliters of a 17% acid solution must be mixed to obtain 60 mL of a 13% acid solution?

5 Solve Distance, Rate, and Time Problems

EXAMPLE 5   Write an equation, and solve.

Two cars leave Kearney, Nebraska, one driving east and the other heading west. The eastbound car travels 4 mph faster than the westbound car, and after 2.5 hours they are 330 miles apart. Find the speed of each car.

Solution

Step 1: **Read** the problem carefully, and identify what we are being asked to find.

We must find the speed of the eastbound and westbound cars. We will draw a picture to help us see what is happening in this problem. After 2.5 hr, the positions of the cars look like this:

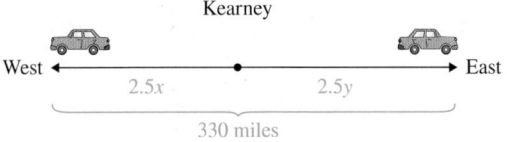

W Hint

Sketch a picture to visualize distance, rate, and time problems.

Step 2: **Choose variables** to represent the unknown quantities.

$$x = \text{the speed of the westbound car}$$
$$y = \text{the speed of the eastbound car}$$

Step 3: **Write a system of equations using two variables.**

Let's make a table using the equation $d = rt$. Fill in the time, 2.5 hr, and the rates first, then multiply those together to fill in the values for the distance.

	d	r	t
Westbound	2.5x	x	2.5
Eastbound	2.5y	y	2.5

Label the picture with the expressions for the distances.

To get one equation, look at the picture and think about the distance between the cars after 2.5 hr.

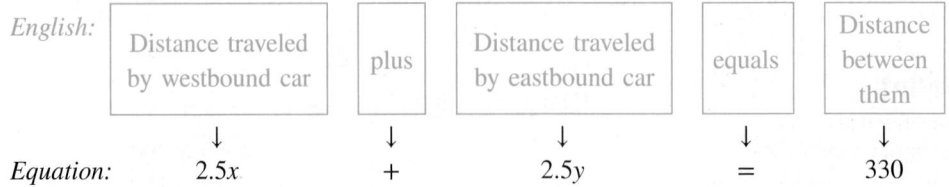

The equation is $2.5x + 2.5y = 330$.

To get the second equation, use the information that says the eastbound car goes 4 mph faster than the westbound car.

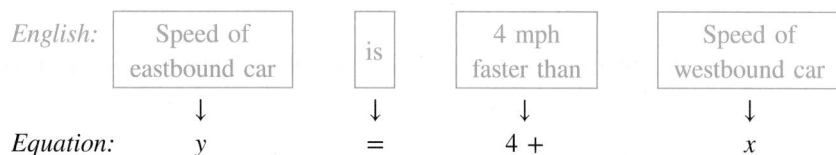

English: | Speed of eastbound car | is | 4 mph faster than | Speed of westbound car |

Equation: $\quad y \qquad\qquad = \qquad 4 + \qquad\qquad x$

The equation is $y = 4 + x$.

The system of equations is $2.5x + 2.5y = 330$
$$y = 4 + x.$$

Step 4: **Solve** the system.

Use substitution.

$$2.5x + 2.5y = 330$$
$$2.5x + 2.5(4 + x) = 330 \qquad \text{Substitute } 4 + x \text{ for } y.$$
$$2.5x + 10 + 2.5x = 330 \qquad \text{Distribute.}$$
$$5x + 10 = 330 \qquad \text{Combine like terms.}$$
$$5x = 320 \qquad \text{Subtract 10 from each side.}$$
$$x = 64$$

Find y by substituting $x = 64$ into $y = 4 + x$.
$$y = 4 + 64 = 68$$

The solution to the system is $(64, 68)$.

Step 5: **Check** the answer, and **interpret** the solution as it relates to the problem.

The speed of the westbound car is 64 mph, and the speed of the eastbound car is 68 mph.

Check.

Distance of westbound car   Distance of eastbound car

$$2.5(64) \quad + \quad 2.5(68) = 160 + 170 = 330 \text{ mi}$$

> **W Hint**
> Remember that both x and y should be positive values when working with distance, rate, and time problems.

[YOU TRY 5]

Write an equation, and solve.

Two planes leave the same airport, one headed north and the other headed south. The northbound plane goes 100 mph slower than the southbound plane. Find each of their speeds if they are 1240 miles apart after 2 hours.

ANSWERS TO [YOU TRY] EXERCISES

1) Carson City: 54,700; Casper: 57,800   2) width: 27 in.; length: 54 in.
3) scarf: $28; belt: $21   4) 20 mL of 5% solution; 40 mL of 17% solution
5) northbound plane: 260 mph; southbound plane: 360 mph

Objective 1: Solve Problems Involving General Quantities

Write a system of equations, and solve.

1) The sum of two numbers is 87, and one number is eleven more than the other. Find the numbers.

2) One number is half another number. The sum of the two numbers is 141. Find the numbers.

3) Through 2016, *Star Wars: The Force Awakens* earned more money on its opening weekend than any other movie. It earned $39.2 million more than the second place movie, *Jurassic World,* did on its opening weekend. Together, they earned $456.8 million. How much did each movie earn on opening weekend? (www.boxofficemojo.com)

4) In the 1976–1977 season, Kareem Abdul-Jabbar led all players in blocked shots. He blocked 50 more shots than Bill Walton, who finished in second place. How many shots did each man block if they rejected a total of 472 shots? (www.nba.com)

5) Through 2018, Jay-Z had been nominated for 14 fewer BET Awards than Beyoncé. Determine how many nominations each performer received if they got a total of 84 nominations. (www.bet.com)

6) From 1965 to 2000, twice as many people immigrated to the United States from the Philippines as from Vietnam. The total number of immigrants from these two countries was 2,100,000. How many people came to the United States from each country? (www.ellisisland.org)

7) According to U.S. Census projections, approximately half as many people in the United States will speak Vietnamese at home as will speak Chinese in 2020. If a total of about 4,680,000 people speak these languages at home, how many will speak Vietnamese and how many will speak Chinese? (www.census.gov)

8) During one week, a hardware store sold 27 fewer "regular" incandescent lightbulbs than energy-efficient compact fluorescent light (CFL) bulbs. How many of each type of bulb was sold if the store sold a total of 79 of these two types of lightbulbs?

Mark Steinmetz/McGraw-Hill Education

9) On April 12, 1961, Yuri Gagarin of the Soviet Union became the first person in space when he piloted the Vostok 1 mission. The next month, Alan B. Shepard became the first American in space in the Freedom 7 space capsule. The two of them spent a total of about 123 min in space, with Gagarin logging 93 more min than Shepard. How long did each man spend in space? (www-pao.ksc.nasa.gov, www.enchantedlearning.com)

10) Mr. Monet has 85 students in his Art History lecture. For their assignment on impressionists, one-fourth as many students chose to recreate an impressionist painting as chose to write a paper. How many students will be painting, and how many will be writing papers?

Objective 2: Solve Geometry Problems

Write a system of equations, and solve.

11) Find the dimensions of a rectangular door that has a perimeter of 220 in. if the width is 50 in. less than the height of the door.

12) The length of a rectangle is 3.5 in. more than its width. If the perimeter is 23 in., what are the dimensions of the rectangle?

13) An iPod Touch is rectangular in shape and has a perimeter of 343.6 mm. Find its length and width given that it is 48.2 mm longer than it is wide.

14) Eliza needs 332 in. of a decorative border to sew around a rectangular quilt she just made. Its width is 26 in. less than its length. Find the dimensions of the quilt.

15) A rectangular horse corral is bordered on one side by a barn as pictured here. The length of the corral is 1.5 times the width. If 119 ft of fencing was used to make the corral, what are its dimensions?

16) The length of a rectangular mirror is twice its width. Find the dimensions of the mirror if its perimeter is 246 cm.

17) Find the measures of angles x and y if the measure of angle x is three-fifths the measure of angle y and if the angles are related according to the figure.

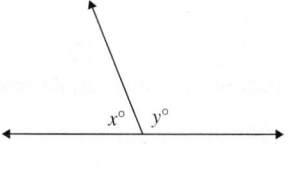

18) Find the measures of angles x and y if the measure of angle y is two-thirds the measure of angle x and if the angles are related according to the figure.

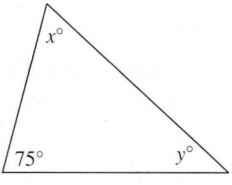

Objective 3: Solve Problems Involving Cost
Write a system of equations, and solve.

19) Kenny and Kyle are huge Colorado Avalanche fans. Kenny buys a T-shirt and two souvenir hockey pucks for $36.00, and Kyle spends $64.00 on two T-shirts and three pucks. Find the price of a T-shirt and the price of a puck.

20) Hannah and Elijah go to a vintage art and music store. Hannah buys two albums and three books for $12.50, and Elijah buys three albums and one book for $13.50. Find the price of an album and the price of a book if each of the albums had the same price and each of the books had the same price.

21) Ned and Catelyn and their friends are big fans of *Game of Thrones*. Ned buys four Hodor mugs and two Daenerys figurines for $79.94, while Catelyn buys three Hodor mugs and three Daenerys figurines for $74.94. Find the cost of each item. (store.hbo.com)

22) Manny and Hiroki buy tickets in advance to some Los Angeles Dodgers games. Manny buys three left-field pavilion seats and six club seats for $735. Hiroki spends $805 on eight left-field pavilion seats and five club seats. Find the cost of each type of ticket. (www.dodgers.com)

23) Carol orders five White Castle hamburgers and a medium order of french fries for $6.25, and Momar orders four hamburgers and two medium fries for $7.58. Find the cost of a hamburger and the cost of a medium order of french fries at White Castle. (White Castle menu)

Burke/Triolo/Brand X Pictures

24) Phuong buys New Jersey lottery tickets every Friday. One day she spent $17.00 on four Gold Rush tickets and three Crossword tickets. The next Friday, she bought three Gold Rush tickets and six Crossword tickets for $24.00. How much did she pay for each type of lottery ticket?

25) Lakeisha is selling wrapping paper products for a school fund-raiser. Her mom buys four rolls of wrapping paper and three packages of gift bags for $52.00. Her grandmother spends $29.00 on three rolls of wrapping paper and one package of gift bags. Find the cost of a roll of wrapping paper and a package of gift bags.

26) Alberto is selling popcorn to raise money for his Cub Scout den. His dad spends $86.00 on two tins of popcorn and three tins of caramel corn. His neighbor buys two tins of popcorn and one tin of caramel corn for $48.00. How much does each type of treat cost?

Objective 4: Solve Mixture Problems
Write a system of equations, and solve.

27) How many ounces of a 9% alcohol solution and how many ounces of a 17% alcohol solution must be mixed to obtain 12 oz of a 15% alcohol solution?

28) How many milliliters of a 15% acid solution and how many milliliters of a 3% acid solution must be mixed to get 45 mL of a 7% alcohol solution?

29) How many liters of pure acid and how many liters of a 25% acid solution should be mixed to get 10 L of a 40% acid solution?

30) How many ounces of pure cranberry juice and how many ounces of a citrus fruit drink containing 10% fruit juice should be mixed to get 120 oz of a fruit drink that is 25% fruit juice?

31) How many ounces of Asian Treasure tea that sells for $7.50/oz should be mixed with Pearadise tea that sells for $5.00/oz so that a 60-oz mixture is obtained that will sell for $6.00/oz?

32) How many pounds of peanuts that sell for $1.80 per pound should be mixed with cashews that sell for $4.50 per pound so that a 10-pound mixture is obtained that will sell for $2.61 per pound?

33) During a late-night visit to Taco Bell, Giovanni orders three Crunchy Tacos and a Beef Chalupa Supreme. His order contains 1490 mg of sodium. Jurgis orders two Crunchy Tacos and two Beef Chalupa Supremes, and

his order contains 1740 mg of sodium. How much sodium is in each item? (www.tacobell.com)

Burke/Triolo/Brand X Pictures

34) Five White Castle hamburgers and one medium order of french fries contain 1300 calories. Four hamburgers and two orders of fries contain 1760 calories. Determine how many calories are in a White Castle hamburger and in a medium order of french fries. (www.whitecastle.com)

35) Mahmud invested $6000 in two accounts, some of it at 2% simple interest, the rest in an account earning 4% simple interest. How much did he invest in each account if he earned $190 in interest after 1 year?

36) Marijke inherited $15,000 and puts some of it into an account earning 5% simple interest and the rest in an account earning 4% simple interest. She earns a total of $660 in interest after 1 year. How much did she deposit into each account?

37) Oscar purchased 16 stamps. He bought some $0.49 stamps and some $0.34 stamps and spent $7.24. How many of each type of stamp did he buy?

38) Kelly saves all of her dimes and nickels in a jar on her desk. When she counts her money, she finds that she has 133 coins worth a total of $10.45. How many dimes and how many nickels does she have?

Write a system of equations, and solve.

(24) 39) Sheldon and Amy leave the same location but head in opposite directions on their bikes. Sheldon rides 1 mph faster than Amy, and after 3 hr they are 51 mi apart. How fast was each of them riding?

40) A passenger train and a freight train leave cities 400 mi apart and travel toward each other. The passenger train is traveling 16 mph faster than the freight train. Find the speed of each train if they pass each other after 5 hr.

41) A small plane leaves an airport and heads south, while a jet takes off at the same time heading north. The speed of the small plane is 160 mph less than the speed of the jet. If they are 1280 mi apart after 2 hr, find the speeds of both planes.

Getty Images/Digital Vision

42) Tyreese and Justine start jogging toward each other from opposite ends of a trail 6.5 mi apart. They meet after 30 min. Find their speeds if Tyreese jogs 3 mph faster than Justine.

43) Lori and Rick leave from opposite ends of a bike trail 9 mi apart and travel toward each other. Lori is traveling 2 mph slower than Rick. Find each of their speeds if they meet after 30 min.

44) Stanley and Phyllis leave the office and travel in opposite directions. Stanley drives 6 mph slower than Phyllis, and after 1 hr they are 76 mi apart. How fast was each person driving?

Other types of distance, rate, and time problems involve a boat traveling upstream and downstream, and a plane traveling with and against the wind. To solve problems like these, we will still use a table to help us organize our information, but we must understand what is happening in such problems.

Let's think about the case of a boat traveling upstream and downstream.

Let x = the speed of the boat in still water and let y = the speed of the current.

When the boat is going *downstream* (*with* the current), the boat is being pushed along by the current so that

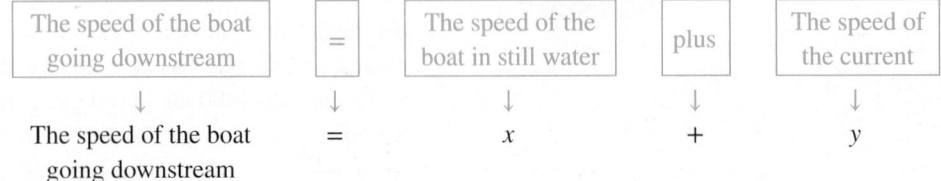

The speed of the boat going downstream	=	The speed of the boat in still water	plus	The speed of the current
↓	↓	↓	↓	↓
The speed of the boat going downstream	=	x	+	y

When the boat is going *upstream* (*against* the current), the boat is being slowed down by the current so that

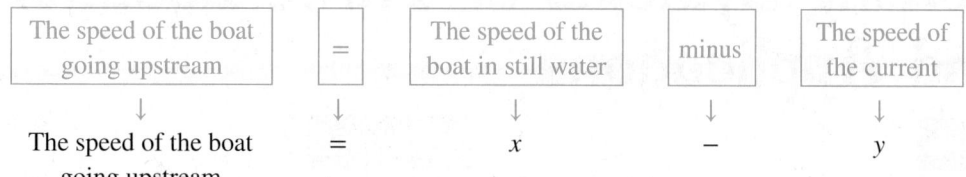

The speed of the boat going upstream	=	The speed of the boat in still water	minus	The speed of the current
↓	↓	↓	↓	↓
The speed of the boat going upstream	=	x	−	y

Use this idea to solve Exercises 45–50.

45) It takes 2 hr for a boat to travel 14 mi downstream. The boat can travel 10 mi upstream in the same amount of time. Find the speed of the boat in still water and the speed of the current. (Hint: Use the information in the following table, and write a system of equations.)

	d	r	t
Downstream	14	$x + y$	2
Upstream	10	$x - y$	2

46) A boat can travel 15 mi downstream in 0.75 hr. It takes the same amount of time for the boat to travel 9 mi upstream. Find the speed of the boat in still water and the speed of the current. (Hint: Use the information in the following table, and write a system of equations.)

	d	r	t
Downstream	15	$x + y$	0.75
Upstream	9	$x - y$	0.75

47) It takes 5 hr for a boat to travel 80 mi downstream. The boat can travel the same distance back upstream in 8 hr. Find the speed of the boat in still water and the speed of the current.

48) A boat can travel 12 mi downstream in 1.5 hr. It takes 3 hr for the boat to travel back to the same spot going upstream. Find the speed of the boat in still water and the speed of the current.

49) A jet can travel 1000 mi against the wind in 2.5 hr. Going with the wind, the jet could travel 1250 mi in the same amount of time. Find the speed of the jet in still air and speed of the wind.

50) It takes 2 hr for a small plane to travel 390 mi with the wind. Going against the wind, the plane can travel 330 mi in the same amount of time. Find the speed of the plane in still air and the speed of the wind.

R Rethink

R1) If you have two unknowns in an application problem, how many equations do you need to solve the problem?

R2) In which of your future courses do you think you will need to solve a system of linear equations?

R3) In Chapter 3, we solved applications similar to many of these using one variable and one equation.

Do you prefer to solve applications using one variable and one equation or two variables and two equations? Why?

R4) Choose six problems and redo them without looking back at the book or your notes for help. Were you able to do them on your own?

5.5 Solving Systems of Three Equations and Applications

P Prepare

What are your objectives for Section 5.5?

O Organize

How can you accomplish each objective?

What are your objectives for Section 5.5?	How can you accomplish each objective?
1 Understand Systems of Three Equations in Three Variables	• Know how to recognize a *system of three equations in three variables*. • Know what a *three-dimensional coordinate system* looks like and what an *ordered triple* is. • Be aware of the different possible solutions of a system of three equations.
2 Solve Systems of Linear Equations in Three Variables	• Learn the procedure for **Solving a System of Linear Equations in Three Variables.** • Complete the given example on your own. • Complete You Try 1.
3 Solve Special Systems in Three Variables	• Be aware that when solving a system in three variables, variables may "drop out," which may indicate *no solution* or *infinite solutions*. • Be able to determine the difference between *no solution* and *infinite solutions* and know what those solutions mean. • Complete the given example on your own. • Complete You Try 2.
4 Solve a System with Missing Terms	• Notice that the procedure for **Solving a System of Linear Equations in Three Variables** is modified when equations are missing terms. • Complete the given example on your own. • Complete You Try 3.
5 Solve Problems Involving Three Variables	• Extend the procedure for **Solving an Applied Problem Using a System of Equations in Two Variables** to include a third variable and third equation. • Complete the given example on your own. • Complete You Try 4.

W Work

Read the explanations, follow the examples, take notes, and complete the You Trys.

In this section, we will learn how to solve a system of linear equations in *three* variables.

1 Understand Systems of Three Equations in Three Variables

Definition

A **linear equation in three variables** is an equation of the form $Ax + By + Cz = D$ where A, B, and C are not all zero and where A, B, C, and D are real numbers. Solutions of this type of an equation are **ordered triples** of the form (x, y, z).

An example of a linear equation in three variables is

$$2x - y + 3z = 12.$$

This equation has infinitely many solutions. Here are a few:

$(5, 1, 1)$	since	$2(5) - (1) + 3(1) = 12$
$(3, 0, 2)$	since	$2(3) - (0) + 3(2) = 12$
$(6, -3, -1)$	since	$2(6) - (-3) + 3(-1) = 12$

Ordered triples, like $(1, 2, 3)$ and $(3, 0, 2)$, are graphed on a three-dimensional coordinate system, as shown to the right. Notice that the ordered triples are *points*.

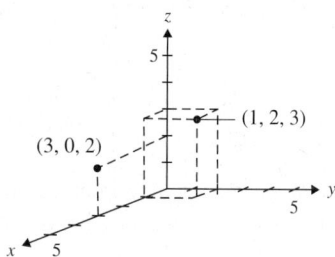

The graph of a linear equation in three variables is a *plane*.

A **solution of a system of linear equations in three variables** is an *ordered triple* that satisfies each equation in the system. Like systems of linear equations in two variables, systems of linear equations in *three* variables can have *one* solution, *no* solution, or *infinitely many* solutions.

Here is an example of a system of linear equations in three variables:

$$x + 4y + 2z = 10$$
$$3x - y + z = 6$$
$$2x + 3y - z = -4$$

In Section 5.1, we solved systems of linear equations in *two* variables by graphing. Since the graph of an equation like $x + 4y + 2z = 10$ is a *plane*, however, solving a system in three variables by graphing would not be practical. But let's look at the graphs of systems of linear equations in three variables that have one solution, no solution, or an infinite number of solutions.

One solution:

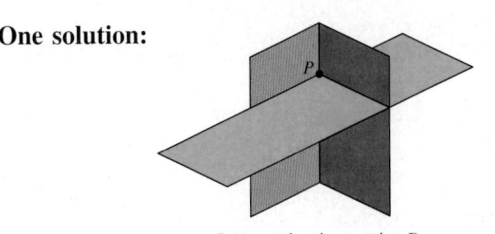

Intersection is at point *P*.

All three planes intersect at one point; this point is the solution of the system.

W Hint
Write down a description of each of these three cases using your own words.

No solution:

None of the planes may intersect or *two* of the planes may intersect, but if there is no solution to the system, *all three planes* do not have a common point of intersection.

Infinite number of solutions:

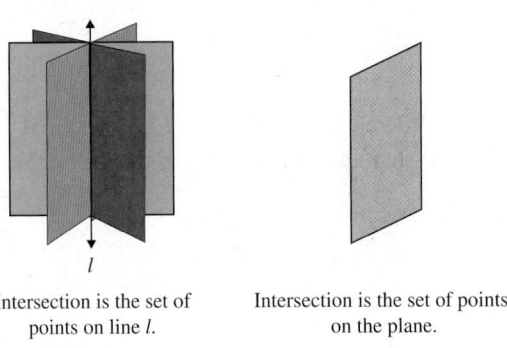

Intersection is the set of points on line *l*.

Intersection is the set of points on the plane.

The three planes may intersect so that they have a line or a plane in common. The solution to the system is the infinite set of points on the line or the plane, respectively.

2 Solve Systems of Linear Equations in Three Variables

First we will learn how to solve a system in which each equation has three variables.

Procedure Solving a System of Linear Equations in Three Variables

Step 1: **Label** the equations ①, ②, and ③.

Step 2: **Choose a variable to eliminate. Eliminate** this variable from *two* sets of *two* equations using the elimination method. You will obtain two equations containing the same two variables. Label one of these new equations [A] and the other [B].

Step 3: Use the elimination method to **eliminate a variable from equations [A] and [B].** You have now found the value of one variable.

Step 4: **Find the value of another variable** by substituting the value found in *Step 3* into equation [A] or [B] and solving for the second variable.

Step 5: **Find the value of the third variable** by substituting the values of the two variables found in *Steps 3* and *4* into equation ①, ②, or ③.

Step 6: **Check** the solution in each of the original equations, and **write the solution as an ordered triple.**

EXAMPLE 1

Solve ① $x + 2y - 2z = 3$
② $2x + y + 3z = 1$
③ $x - 2y + z = -10$

Solution

Steps 1 and 2: We have already **labeled** the equations. We'll **choose** to eliminate the variable y from *two* sets of *two* equations:

a) Add equations ① and ③ to eliminate y. Label the resulting equation Ⓐ.

$$
\begin{array}{ll}
① & x + 2y - 2z = 3 \\
③ + & x - 2y + z = -10 \\
\hline
Ⓐ & 2x \quad\ - z = -7
\end{array}
$$

> **W Hint**
> This example has many steps. It will be very helpful to write it out as you are reading it!

b) Multiply equation ② by 2, and add it to equation ③ to eliminate y. Label the resulting equation Ⓑ.

$$
\begin{array}{ll}
2 \times ② & 4x + 2y + 6z = 2 \\
③ + & x - 2y + z = -10 \\
\hline
Ⓑ & 5x \quad\quad + 7z = -8
\end{array}
$$

Note
Equations Ⓐ and Ⓑ contain only *two* variables and they are the same variables, x and z.

Step 3: Use the elimination method to **eliminate a variable from equations Ⓐ and Ⓑ**. We will eliminate z from Ⓐ and Ⓑ. Multiply Ⓐ by 7, and add it to Ⓑ.

$$
\begin{array}{ll}
7 \times Ⓐ & 14x - 7z = -49 \\
Ⓑ + & 5x + 7z = -8 \\
\hline
& 19x \quad\quad = -57 \\
& \boxed{x = -3} \quad \text{Divide by 19.}
\end{array}
$$

Step 4: **Find the value of another variable** by substituting $x = -3$ into equation Ⓐ or Ⓑ. We will use Ⓐ since it has smaller coefficients.

$$
\begin{array}{ll}
Ⓐ & 2x - z = -7 \\
& 2(-3) - z = -7 & \text{Substitute } -3 \text{ for } x. \\
& -6 - z = -7 & \text{Multiply.} \\
& \boxed{z = 1} & \text{Add 6, and divide by } -1.
\end{array}
$$

Step 5: **Find the value of the third variable** by substituting $x = -3$ and $z = 1$ into equation ①, ②, or ③. We will use equation ①.

$$
\begin{array}{ll}
① & x + 2y - 2z = 3 \\
& -3 + 2y - 2(1) = 3 & \text{Substitute } -3 \text{ for } x \text{ and 1 for } z. \\
& -3 + 2y - 2 = 3 & \text{Multiply.} \\
& 2y - 5 = 3 & \text{Combine like terms.} \\
& \boxed{y = 4} & \text{Add 5, and divide by 2.}
\end{array}
$$

Step 6: **Check** the solution, $(-3, 4, 1)$, in each of the original equations, and **write the solution.**

① $\quad\quad\; x + 2y - 2z = 3$
$-3 + 2(4) - 2(1) \overset{?}{=} 3$
$\quad\quad -3 + 8 - 2 \overset{?}{=} 3$
$\quad\quad\quad\quad\quad 3 = 3$
$\quad\quad\quad\quad\quad$ True

② $\quad\quad\; 2x + y + 3z = 1$
$2(-3) + 4 + 3(1) \overset{?}{=} 1$
$\quad\quad -6 + 4 + 3 \overset{?}{=} 1$
$\quad\quad\quad\quad\quad 1 = 1$
$\quad\quad\quad\quad\quad$ True

③ $\quad\quad\; x - 2y + z = -10$
$-3 - 2(4) + 1 \overset{?}{=} -10$
$\quad\quad -3 - 8 + 1 \overset{?}{=} -10$
$\quad\quad\quad\quad -10 = -10$
$\quad\quad\quad\quad\quad$ True

The solution is $(-3, 4, 1)$.

$\left[\text{ YOU TRY 1 }\right]$ Solve $\quad x + 2y + 3z = -11$
$\quad\quad\quad\quad\quad 3x - y + z = 0$
$\quad\quad\quad\quad -2x + 3y - z = 4$

3 Solve Special Systems in Three Variables

Some systems in three variables have no solution, and some have an infinite number of solutions.

EXAMPLE 2

Solve ① $-3x + 2y - z = 5$
② $\quad x + 4y + z = -4$
③ $\;9x - 6y + 3z = -2$

Solution

Steps 1 and 2: We have already *labeled* the equations. The *variable we choose to eliminate is z, the easiest.*

a) Add equations ① and ② to eliminate z. Label the resulting equation $\boxed{A}$.

$$\begin{array}{r} ①\quad -3x + 2y - z = 5 \\ ②\; + \quad\quad x + 4y + z = -4 \\ \hline \boxed{A}\quad -2x + 6y \quad\quad = 1 \end{array}$$

W Hint

Label the equations when you are solving systems of three equations. It will help you organize your work!

b) Multiply equation ① by 3, and add it to equation ③ to eliminate z. Label the resulting equation $\boxed{B}$.

$$① \; -3x + 2y - z = 5 \longrightarrow 3 \times ① \quad\quad -9x + 6y - 3z = 15$$
$$③ \; + \quad\quad 9x - 6y + 3z = -2$$
$$\boxed{B} \quad\quad\quad\quad\quad\quad\quad\quad 0 = 13 \quad \text{False}$$

Since the variables are eliminated and we get the false statement $0 = 13$, equations ① and ③ have no ordered triple that satisfies each equation.

The system is inconsistent, so there is no solution. The solution set is ∅.

Note

If the variables are eliminated and you get a false statement, there is *no solution* to the system. The system is inconsistent, so the solution set is ∅.

EXAMPLE 3

Solve ① $-4x - 2y + 8z = -12$
② $\quad 2x + y - 4z = 6$
③ $\quad 6x + 3y - 12z = 18$

Solution

Steps 1 and 2: We label the equations and choose a variable, y, to eliminate.

a) Multiply equation ② by 2 and add it to equation ①. Label the resulting equation $\boxed{A}$.

$$
\begin{array}{rl}
2 \times ② & 4x + 2y - 8z = 12 \\
① & \underline{+\,{-4x} - 2y + 8z = -12} \\
\boxed{A} & \phantom{+{-4x} - 2y + 8z = }0 = 0 \qquad \text{True}
\end{array}
$$

W Hint

Notice that just like in Section 5.2, Objective 3, if the variables drop out and you end up with a true statement, there are *an infinite number of solutions.*

The variables were eliminated, and we obtained the true statement $0 = 0$. This is because equation ① is a multiple of equation ②.

Notice, also, that equation ③ is a multiple of equation ②.

The equations in this system are dependent. There are an infinite number of solutions, and we write the solution set as $\{(x, y, z) \mid 2x + y - 4z = 6\}$. The equations all have the same graph.

[YOU TRY 2] Solve each system of equations.

a) $\quad 8x + 20y - 4z = -16$ b) $\quad x + 4y - 3z = 2$
$\quad -6x - 15y + 3z = 12$ $2x - 5y + 2z = -8$
$\quad\quad 2x + 5y - z = -4$ $-3x - 12y + 9z = 7$

---

4 Solve a System with Missing Terms

EXAMPLE 4

Solve ① $5x - 2y = 6$
② $\quad y + 2z = 1$
③ $\quad 3x - 4z = -8$

Solution

First, notice that while this *is* a system of three equations in three variables, none of the equations contains three variables. Furthermore, each equation is "missing" a different variable.

Note

We will use many of the *ideas* outlined in the steps for solving a system of three equations, but we will use *substitution* rather than the elimination method.

Step 1: Label the equations ①, ②, ③.

Step 2: The goal of *Step 2* is to obtain two equations that contain the same two variables. We will modify this step from the way it was outlined on p. 374.

In order to obtain *two* equations with the same *two* variables, we will use *substitution*.

Since y in equation ② is the only variable in the system with a coefficient of 1, we will solve equation ② for y.

$$② \quad y + 2z = 1$$
$$y = 1 - 2z \qquad \text{Subtract } 2z.$$

Substitute $y = 1 - 2z$ into equation ① to obtain an equation containing the variables x and z. Simplify. Label the resulting equation $\boxed{A}$.

$$① \qquad 5x - 2y = 6$$
$$5x - 2(1 - 2z) = 6 \qquad \text{Substitute } 1 - 2z \text{ for } y.$$
$$5x - 2 + 4z = 6 \qquad \text{Distribute.}$$
$$\boxed{A} \qquad 5x + 4z = 8 \qquad \text{Add 2.}$$

Step 3: The goal of *Step 3* is to solve for one of the variables. Equations $\boxed{A}$ and ③ contain only x and z.

We will eliminate z from $\boxed{A}$ and ③. Add the two equations to eliminate z, then solve for x.

$$\boxed{A} \quad\;\; 5x + 4z = 8$$
$$③ + \underline{\;3x - 4z = -8\;}$$
$$8x \qquad\quad = 0$$
$$\boxed{x = 0} \qquad \text{Divide by 8.}$$

Step 4: Find the value of another variable by substituting $x = 0$ into either $\boxed{A}$, ①, or ③.

$$\boxed{A} \quad\;\; 5x + 4z = 8$$
$$5(0) + 4z = 8 \qquad \text{Substitute 0 for } x.$$
$$4z = 8$$
$$\boxed{z = 2} \qquad \text{Divide by 4.}$$

Step 5: Find the value of the third variable by substituting $x = 0$ into ① or $z = 2$ into ②.

$$① \quad\;\; 5x - 2y = 6$$
$$5(0) - 2y = 6 \qquad \text{Substitute 0 for } x.$$
$$-2y = 6$$
$$\boxed{y = -3} \qquad \text{Divide by } -2.$$

Step 6: Check the solution $(0, -3, 2)$ in each of the original equations. The check is left to the student. The solution is $(0, -3, 2)$.

$\left[\text{ YOU TRY 3 }\right]$ Solve $x + 2y = 8$
$2y + 3z = 1$
$3x - z = -3$

---

5 Solve Problems Involving Three Variables

To solve applications involving a system of three equations in three variables, we will extend the method used for two equations in two variables.

EXAMPLE 5

Write a system of equations, and solve.

Ingrid bought 21 flowers for $99 to make a bouquet. The lilies cost $5 each, roses cost $6 each, and the tulips cost $2 each. If she bought twice as many tulips as lilies, how many of each type of flower did she buy?

Solution

Step 1: **Read** the problem carefully. We must determine the number of each type of flower Ingrid bought.

Step 2: **Choose variables** to represent the unknown quantities.

$$L = \text{the number of lilies Ingrid bought}$$
$$R = \text{the number of roses Ingrid bought}$$
$$T = \text{the number of tulips Ingrid bought}$$

Step 3: **Write a system of equations using the variables.**

To write one equation, we will use the information that says Ingrid bought 21 flowers:

$$L + R + T = 21 \qquad \text{Equation (1)}$$

To write a second equation, we will use the information that says Ingrid paid $99:

$$5L + 6R + 2T = 99 \qquad \text{Equation (2)}$$

Juice Images/Glow Images

To write the third equation, we will use the information that says Ingrid bought twice as many tulips as lilies:

$$T = 2L \qquad \text{Equation (3)}$$

The system is
  ① $L + R + T = 21$
  ② $5L + 6R + 2T = 99$
  ③ $T = 2L$

Step 4: **Solve** the system. Because equation ③ tells us that $T = 2L$, we can substitute $2L$ for T in equations ① and ② and combine like terms to get equations $\boxed{A}$ and $\boxed{B}$:

① $L + R + T = 21$ ② $5L + 6R + 2T = 99$
    $L + R + (2L) = 21$ $5L + 6R + 2(2L) = 99$
    $\boxed{A}$ $3L + R = 21$ $\boxed{B}$ $9L + 6R = 99$

Use elimination with equations $\boxed{A}$ and $\boxed{B}$ to find the values of R and L. Multiply equation $\boxed{A}$ by -6 and add it to $\boxed{B}$:

$$-6 \cdot \boxed{A}: \qquad -18L - 6R = -126$$
$$\boxed{B}: \qquad \underline{+\ 9L + 6R = 99}$$
$$-9L = -27 \qquad \text{Add the equations.}$$
$$L = 3 \qquad \text{Divide by } -9.$$

To solve for T, substitute $L = 3$ into equation ③, $T = 2L$: $T = 2(3) = 6$.

To solve for R, substitute $L = 3$ into equation $\boxed{A}$, $3L + R = 21$, and solve for R:

$$3(3) + R = 21 \qquad \text{Substitute 3 for } L.$$
$$9 + R = 21 \qquad \text{Multiply.}$$
$$R = 12 \qquad \text{Subtract 9.}$$

We have found that $L = 3$, $R = 12$, and $T = 6$. The solution of the system is $(3, 12, 6)$.

Step 5: **Check** and **interpret** the solution.

Ingrid bought 3 lilies, 12 roses, and 6 tulips. The check is left to the student.

 **YOU TRY 4**

Write a system of equations, and solve.

    Amelia, Bella, and Carmen are sisters. Bella is 5 yr older than Carmen, and Amelia's age is 5 yr less than twice Carmen's age. The sum of their ages is 48. How old is each girl?

ANSWERS TO [**YOU TRY**] **EXERCISES**

1) $(2, 1, -5)$ 2) a) $\{(x, y, z) \mid 2x + 5y - z = -4\}$ b) $\varnothing$ 3) $(-2, 5, -3)$
4) Amelia: 19; Bella: 17; Carmen: 12

E Evaluate **5.5** Exercises   Do the exercises, and check your work.

Objective 1: Understand Systems of Three Equations in Three Variables

Determine whether the ordered triple is a solution of the system.

 1) $3x + y + 2z = 2$
      $-2x - y + z = 5$
       $x + 2y - z = -11$
      $(1, -5, 2)$

2) $4x + 3y - 7z = -6$
    $x - 2y + 5z = -3$
   $-x + y + 2z = 7$
    $(-2, 3, 1)$

3) $-x + y - 2z = 2$
    $3x - y + 5z = 4$
    $2x + 3y - z = 7$
    $(0, 6, 2)$

4) $6x - y + 4z = 4$
    $-2x + y - z = 5$
    $2x - 3y + z = 2$
    $\left(-\dfrac{1}{2}, -3, 1\right)$

5) Write a system of equations in x, y, and z so that the ordered triple $(4, -1, 2)$ is a solution of the system.

6) Find the value of c so that $(6, 0, 5)$ is a solution of the system $2x - 5y - 3z = -3$
                  $-x + y + 2z = 4$
           $-2x + 3y + cz = 8$

Objective 2: Solve Systems of Linear Equations in Three Variables

Solve each system. See Example 1.

 7) $x + 3y + z = 3$
    $4x - 2y + 3z = 7$
   $-2x + y - z = -1$

8) $x - y + 2z = -7$
  $-3x - 2y + z = -10$
   $5x + 4y + 3z = 4$

9) $5x + 3y - z = -2$
  $-2x + 3y + 2z = 3$
    $x + 6y + z = -1$

10) $-2x - 2y + 3z = 2$
    $3x + 3y - 5z = -3$
     $-x + y - z = 9$

11) $3a + 5b - 3c = -4$
     $a - 3b + c = 6$
   $-4a + 6b + 2c = -6$

12) $a - 4b + 2c = -7$
   $3a - 8b + c = 7$
  $6a - 12b + 3c = 12$

Objective 3: Solve Special Systems in Three Variables

Solve each system. Identify any systems that are inconsistent or that have dependent equations. See Examples 2 and 3.

 13) $a - 5b + c = -4$
    $3a + 2b - 4c = -3$
   $6a + 4b - 8c = 9$

14) $-a + 2b - 12c = 8$
$-6a + 2b - 8c = -3$
$3a - b + 4c = 4$

15) $-15x - 3y + 9z = 3$
$5x + y - 3z = -1$
$10x + 2y - 6z = -2$

16) $-4x + 10y - 16z = -6$
$-6x + 15y - 24z = -9$
$2x - 5y + 8z = 3$

17) $-3a + 12b - 9c = -3$
$5a - 20b + 15c = 5$
$-a + 4b - 3c = -1$

18) $3x - 12y + 6z = 4$
$-x + 4y - 2z = 7$
$5x + 3y + z = -2$

Objective 4: Solve a System with Missing Terms

Solve each system. See Example 4.

19) $5x - 2y + z = -5$
$x - y - 2z = 7$
$4y + 3z = 5$

20) $-x + z = 9$
$-2x + 4y - z = 4$
$7x + 2y + 3z = -1$

21) $a + 15b = 5$
$4a + 10b + c = -6$
$-2a - 5b - 2c = -3$

22) $2x - 6y - 3z = 4$
$-3y + 2z = -6$
$-x + 3y + z = -1$

23) $x + 2y + 3z = 4$
$-3x + y = -7$
$4y + 3z = -10$

24) $-3a + 5b + c = -4$
$a + 5b = 3$
$4a - 3c = -11$

24) 25) $-5x + z = -3$
$4x - y = -1$
$3y - 7z = 1$

26) $a + b = 1$
$a - 5c = 2$
$b + 2c = -4$

27) $4a + 2b = -11$
$-8a - 3c = -7$
$b + 2c = 1$

28) $3x + 4y = -6$
$-x + 3z = 1$
$2y + 3z = -1$

Mixed Exercises: Objectives 2–4

Solve each system. Identify any systems that are inconsistent or that have dependent equations.

29) $6x + 3y - 3z = -1$
$10x + 5y - 5z = 4$
$x - 3y + 4z = 6$

30) $2x + 3y - z = 0$
$x - 4y - 2z = -5$
$-4x + 5y + 3z = -4$

31) $7x + 8y - z = 16$
$-\dfrac{1}{2}x - 2y + \dfrac{3}{2}z = 1$
$\dfrac{4}{3}x + 4y - 3z = -\dfrac{2}{3}$

32) $3a + b - 2c = -3$
$9a + 3b - 6c = -9$
$-6a - 2b + 4c = 6$

33) $2a - 3b = -4$
$3b - c = 8$
$-5a + 4c = -4$

34) $5x + y - 2z = -2$
$-\dfrac{1}{2}x - \dfrac{3}{4}y + 2z = \dfrac{5}{4}$
$x - 6z = 3$

35) $-4x + 6y + 3z = 3$
$-\dfrac{2}{3}x + y + \dfrac{1}{2}z = \dfrac{1}{2}$
$12x - 18y - 9z = -9$

36) $x - \dfrac{5}{2}y + \dfrac{1}{2}z = \dfrac{5}{4}$
$x + 3y - z = 4$
$-6x + 15y - 3z = -1$

37) $a + b + 9c = -3$
$-5a - 2b + 3c = 10$
$4a + 3b + 6c = -15$

38) $2x + 3y = 2$
$-3x + 4z = 0$
$y - 5z = -17$

39) $x + 5z = 10$
$4y + z = -2$
$3x - 2y = 2$

40) $a + 3b - 8c = 2$
$-2a - 5b + 4c = -1$
$4a + b + 16c = -4$

41) $2x - y + 4z = -1$
$x + 3y + z = -5$
$-3x + 2y = 7$

42) $-2a + 3b = 3$
$a + 5c = -1$
$b - 2c = -5$

43) Given the following two equations, write a third equation to obtain a system of three equations in x, y, and z so that the system has no solution.

$$x + 3y - 2z = -9$$
$$2x - 5y + z = 1$$

44) Given the following two equations, write a third equation to obtain a system of three equations in x, y, and z so that the system has an infinite number of solutions.

$$9x - 12y + 3z = 21$$
$$-3x + 4y - z = -7$$

Objective 5: Solve Problems Involving Three Variables

Write a system of equations, and solve.

(24) 45) Moe buys two hot dogs, two orders of fries, and a large soda for $9.00. Larry buys two hot dogs, one order of fries, and two large sodas for $9.50, and Curly spends $11.00 on three hot dogs, two orders of fries, and a large soda. Find the price of a hot dog, an order of fries, and a large soda.

Dan Kosmayer/Shutterstock

46) A movie theater charges $9.00 for an adult's ticket, $7.00 for a ticket for seniors 60 and over, and $6.00 for a child's ticket. For a particular movie, the theater sold a total of 290 tickets, which brought in $2400. The number of seniors' tickets sold was twice the number of children's tickets sold. Determine the number of adults', seniors', and children's tickets sold.

47) A Chocolate Chip Peanut Crunch Clif Bar contains 4 fewer grams of protein than a Chocolate Peanut Butter Balance Bar Plus. A Chocolate Peanut Butter Protein Plus PowerBar contains 6 more grams of protein than the Balance Bar Plus. All three bars contain a total of 44 g of protein. How many grams

of protein are in each type of bar? (www.clifbar.com, www.balance.com, www.powerbar.com)

48) A 1-tablespoon serving size of Hellman's Real Mayonnaise has 55 more calories than the same serving size of Hellman's Light Mayonnaise. Miracle Whip and Hellman's Light have the same number of calories in a 1-tablespoon serving size. If the three spreads have a total of 160 calories in one serving, determine the number of calories in one serving of each. (product labels)

49) At the beginning of the 2018 season, the Golden State Warriors, Los Angeles Lakers, and New York Knicks were the most valuable franchises in the NBA. Their total value was $10.0 billion. The Knicks were worth $0.5 billion more than the Warriors, and the Warriors were worth $0.2 billion less than the Lakers. Find the value of each team. (www.forbes.com)

50) A family spends $108 on tickets to go up to the crown at the Statue of Liberty. They buy half as many adult tickets as children's tickets and one more adult ticket than senior tickets. An adult ticket costs $21.50, a senior ticket costs $17.00, and a child's ticket costs $12.00. How many of each type of ticket did they buy? (www.nps.gov)

Luca Amedei/Shutterstock

51) The price of tickets to a Chicago Cubs game at Wrigley Field varies depending on their dates. At the beginning of the 2018 season, Bill, Corrinne, and Jason bought tickets for themselves in the bleachers for several games. Bill spent $294 on four bronze dates, four silver dates, and three gold dates. Corrinne's total was $230 for four bronze dates, three silver dates, and two gold dates. Jason spent $173 on three bronze dates, three silver dates, and one gold date. How much did it cost to sit in the bleachers at Wrigley Field on a bronze date, silver date, and gold date in 2018? (www.mlb.com)

52) The three top-earning *Hunger Games* movies in the United States during their opening weekends were *The Hunger Games*, *The Hunger Games: Catching Fire*, and *The Hunger Games: Mockingjay — Part 1*. *Catching Fire* earned $36.2 million more than *Mockingjay — Part 1*, and *Mockingjay — Part 1* earned $30.6 million less than

The Hunger Games. During their opening weekends, the three movies earned a total of $432.5 million. How much did each movie earn during its opening weekend?
(www.boxofficemojo.com)

53) The measure of the largest angle of a triangle is twice the measure of the middle angle. The smallest angle measures 28° less than the middle angle. Find the measures of the angles of the triangle. (Hint: Recall that the sum of the measures of the angles of a triangle is 180°.)

54) The measure of the smallest angle of a triangle is one-third the measure of the largest angle. The middle angle measures 30° less than the largest angle. Find the measures of the angles of the triangle. (Hint: Recall that the sum of the measures of the angles of a triangle is 180°.)

55) The smallest angle of a triangle measures 44° less than the largest angle. The sum of the two smaller angles is 20° more than the measure of the largest angle. Find the measures of the angles of the triangle.

56) The sum of the measures of the two smaller angles of a triangle is 40° less than the measure of the largest angle. The measure of the largest angle is twice the measure of the middle angle. Find the measures of the angles of the triangle.

57) The perimeter of a triangle is 29 cm. The longest side is 5 cm longer than the shortest side, and the sum of

the two smaller sides is 5 cm more than the longest side. Find the lengths of the sides of the triangle.

58) The smallest side of a triangle is half the length of the longest side. The sum of the two smaller sides is 2 in. more than the longest side. Find the lengths of the sides if the perimeter is 58 in.

Extension

Extend the concepts of this section to solve each system. Write the solution in the form (a, b, c, d).

59)
$$a - 2b - c + d = 0$$
$$-a + 2b + 3c + d = 6$$
$$2a + b + c - d = 8$$
$$a - b + 2c + 3d = 7$$

60)
$$-a + 4b + 3c - d = 4$$
$$2a + b - 3c + d = -6$$
$$a + b + c + d = 0$$
$$a - b + 2c - d = -1$$

61)
$$3a + 4b + c - d = -7$$
$$-3a - 2b - c + d = 1$$
$$a + 2b + 3c - 2d = 5$$
$$2a + b + c - d = 2$$

62)
$$3a - 4b + c + d = 12$$
$$-3a + 2b - c + 3d = -4$$
$$a - 2b + 2c - d = 2$$
$$-a + 4b + c + d = 8$$

R **Rethink**

R1) When solving a system of linear equations in three variables, how do you know which variable to eliminate first?

R2) How does this section relate to the Putting It All Together section?

R3) When solving a system of linear equations with three variables, what is the minimum number of equations needed to solve the system?

R4) Which objective is the most difficult for you?

R5) When solving these application problems, were any of your answers negative? If not, can application problems have negative answers?

5.6 Solving Systems of Linear Equations Using Matrices

What are your objectives for Section 5.6?	How can you accomplish each objective?
1 Learn the Vocabulary Associated with Gaussian Elimination	• Know the definitions of a *matrix* and an *augmented matrix* and the arrangement of numbers in the matrix from a given equation. • Understand the goal of the **Gaussian elimination method** and the final form of the matrix—*row echelon form.*
2 Solve a System of Two Equations Using Gaussian Elimination	• Learn the *matrix row operations.* • Label the row operation you will be doing before you begin each step. • Recognize the final matrix when there are 1's on the diagonal. • Learn how to write a system of equations from the final matrix. • Complete the given example on your own.
3 Solve a System of Any Number of Equations Using Gaussian Elimination	• Learn the procedure for **How to Solve a System of Equations Using Gaussian Elimination.** • Label the row operation you will be doing before you begin each step. • Recognize the final matrix when there are 1's on the diagonal. • Learn how to write a system of equations from the final matrix. • Complete the given example on your own. • Complete You Trys 1 and 2.

Read the explanations, follow the examples, take notes, and complete the You Trys.

We have learned how to solve systems of linear equations by graphing, substitution, and the elimination method. In this section, we will learn how to use *row operations* and *Gaussian elimination* to solve systems of linear equations. We begin by defining some terms.

1 Learn the Vocabulary Associated with Gaussian Elimination

A **matrix** is a rectangular array of numbers. (The plural of *matrix* is *matrices*.) Each number in the matrix is an **element** of the matrix. An example of a matrix is

$$\begin{array}{ccc} \text{Column 1} & \text{Column 2} & \text{Column 3} \\ \downarrow & \downarrow & \downarrow \end{array}$$

$$\begin{array}{c} \text{Row 1} \rightarrow \\ \text{Row 2} \rightarrow \end{array} \begin{bmatrix} 3 & -1 & 4 \\ 0 & 2 & -5 \end{bmatrix}$$

We can represent a system of equations in an *augmented matrix*. An **augmented matrix** has a vertical line to distinguish between different parts of the equation. For example, we can represent the system below with the augmented matrix shown here:

$$\begin{array}{ll} 5x + 4y = 1 & \text{Equation (1)} \\ x - 3y = 6 & \text{Equation (2)} \end{array} \qquad \left[\begin{array}{cc|c} 5 & 4 & 1 \\ 1 & -3 & 6 \end{array}\right] \begin{array}{l} \text{Row 1} \\ \text{Row 2} \end{array}$$

Notice that the vertical line separates the system's coefficients from its constants on the other side of the = sign. The system needs to be in standard form, so the first column in the matrix represents the *x*-coefficients. The second column represents the *y*-coefficients, and the column on the right represents the constants.

Gaussian elimination is the process of using row operations on an augmented matrix to solve the corresponding system of linear equations. It is a variation of the elimination method and can be very efficient. Computers often use augmented matrices and row operations to solve systems.

The goal of Gaussian elimination is to obtain a matrix of the form $\left[\begin{array}{cc|c} 1 & a & b \\ 0 & 1 & c \end{array}\right]$ or $\left[\begin{array}{ccc|c} 1 & a & b & d \\ 0 & 1 & c & e \\ 0 & 0 & 1 & f \end{array}\right]$ when solving a system of two or three equations, respectively. Notice the 1's along the **diagonal** of the matrix and zeros below the diagonal.

We say a matrix is in **row echelon form** when it has 1's along the diagonal and 0's below the diagonal. We get matrices in row echelon form by performing row operations. When we rewrite a matrix that is in row echelon form back into a system, its solution is easy to find.

2 Solve a System of Two Equations Using Gaussian Elimination

The row operations we can perform on augmented matrices are similar to the operations we use to solve a system of equations using the elimination method.

Definition Matrix Row Operations

Performing the following row operations on a matrix produces an equivalent matrix.

1) Interchanging two rows
2) Multiplying every element in a row by a nonzero real number
3) Replacing a row by the sum of it and the multiple of another row

Let's use these operations to solve a system using Gaussian elimination. Notice the similarities between this method and the elimination method.

EXAMPLE 1

Solve using Gaussian elimination. $\quad x + 5y = -1$
$$2x - y = 9$$

Solution

Begin by writing the system as an augmented matrix. $\quad \begin{bmatrix} 1 & 5 & | & -1 \\ 2 & -1 & | & 9 \end{bmatrix}$

We will use the 1 in Row 1 to make the element below it a zero. If we multiply the 1 by -2 (to get -2) and add it to the 2, we get zero. We must do this operation to the entire row. Denote this as $-2R_1 + R_2 \rightarrow R_2$. (Read as, "$-2$ times Row 1 plus Row 2 makes the new Row 2.") We get a new Row 2.

W Hint

Write out the example as you are reading it!

Use this $\rightarrow \begin{bmatrix} \boxed{1} & 5 & | & -1 \\ \boxed{2} & -1 & | & 9 \end{bmatrix} \xrightarrow{-2R_1 + R_2 \rightarrow R_2} \begin{bmatrix} 1 & 5 & | & -1 \\ -2(1) + 2 & -2(5) + (-1) & | & -2(-1) + 9 \end{bmatrix}$

to make ↗
this 0. $= \begin{bmatrix} 1 & 5 & | & -1 \\ 0 & -11 & | & 11 \end{bmatrix}$ $\quad$ Multiply each element of Row 1 by -2, and add it to the corresponding element of Row 2.

Note

We are not *making* a new Row 1, so it stays the same.

We have obtained the first 1 on the diagonal with a 0 below it. Next we need a 1 on the diagonal in Row 2.

This column is
in the correct form.
↓

$\begin{bmatrix} 1 & 5 & | & -1 \\ 0 & \boxed{-11} & | & 11 \end{bmatrix} \xrightarrow{-\frac{1}{11}R_2 \rightarrow R_2} \begin{bmatrix} 1 & 5 & | & -1 \\ 0 & 1 & | & -1 \end{bmatrix}$ $\quad$ Multiply each element of Row 2 by $-\dfrac{1}{11}$ to get a 1 on the diagonal.

↑
Make this 1.

We have obtained the final matrix because there are 1's on the diagonal and a 0 below. The matrix is in row echelon form. From this matrix, write a system of equations. The last row gives us the value of y.

W Hint

Notice on this matrix that 1's on the diagonal with a zero below means that you have eliminated the x in the linear equation represented by Row 2.

$\begin{bmatrix} 1 & 5 & | & -1 \\ 0 & 1 & | & -1 \end{bmatrix}$ $\quad \begin{array}{l} 1x + 5y = -1 \\ 0x + 1y = -1 \end{array}$ or $\quad \begin{array}{ll} x + 5y = -1 & \text{Equation (1)} \\ y = -1 & \text{Equation (2)} \end{array}$

$x + 5(-1) = -1$ $\quad$ Substitute -1 for y in equation (1).
$x - 5 = -1$ $\quad$ Multiply.
$x = 4$ $\quad$ Add 5.

The solution is $(4, -1)$. Check by substituting $(4, -1)$ into both equations of the original system.

3 Solve a System of Any Number of Equations Using Gaussian Elimination

Here are the steps for using Gaussian elimination to solve a system of any number of equations. **Our goal is to obtain a matrix with 1's along the diagonal and 0's below— row echelon form.**

Procedure How to Solve a System of Equations Using Gaussian Elimination

Step 1: Write the system as an *augmented matrix.*

Step 2: Use row operations to make the *first entry in column 1* be a 1.

Step 3: Use row operations to make *all entries below the 1 in column 1* be 0's.

Step 4: Use row operations to make the *second entry in column 2* be a 1.

Step 5: Use row operations to make *all entries below the 1 in column 2* be 0's.

Step 6: Continue this procedure until the matrix is in *row echelon form*—1's along the diagonal and 0's below.

Step 7: Write the matrix in *Step 6* as a *system of equations.*

Step 8: *Solve the system* from *Step 7.* The last equation in the system will give you the value of one of the variables; find the values of the other variables by using substitution.

Step 9: *Check the solution* in each equation of the original system.

[**YOU TRY 1**] Solve the system using Gaussian elimination.
$$x - y = -1$$
$$-3x + 5y = 9$$

Next we will solve a system of three equations using Gaussian elimination.

EXAMPLE 2

Solve using Gaussian elimination.

$$2x + y - z = -3$$
$$x + 2y - 3z = 1$$
$$-x - y + 2z = 2$$

Solution

Step 1: Write the system as an *augmented matrix.*

$$\left[\begin{array}{ccc|c} 2 & 1 & -1 & -3 \\ 1 & 2 & -3 & 1 \\ -1 & -1 & 2 & 2 \end{array}\right]$$

W Hint

Are you writing out this example as you are reading it?

Step 2: To make the *first entry in column 1* be a 1, we *could* multiply Row 1 by $\frac{1}{2}$, but this would make the rest of the entries in the first row fractions. Instead, recall that we can interchange two rows. If we interchange Row 1 and Row 2, the first entry in column 1 will be 1.

$$\begin{array}{c} R_1 \leftrightarrow R_2 \\ \text{Interchange} \\ \text{Row 1 and Row 2.} \end{array} \left[\begin{array}{ccc|c} ① & 2 & -3 & 1 \\ 2 & 1 & -1 & -3 \\ -1 & -1 & 2 & 2 \end{array}\right]$$

Step 3: We want to make *all the entries below the 1 in column 1* be 0's. To obtain a 0 in place of the 2 in column 1, multiply the 1 by -2 (to get -2) and add it to the 2. Perform that same operation on the entire row to obtain the new Row 2.

$$\begin{array}{c} \text{Use this} \rightarrow \\ \text{to make} \rightarrow \\ \text{this zero.} \end{array} \left[\begin{array}{ccc|c} ① & 2 & -3 & 1 \\ \boxed{2} & 1 & -1 & -3 \\ -1 & -1 & 2 & 2 \end{array}\right] \begin{array}{c} -2R_1 + R_2 \rightarrow R_2 \\ -2 \text{ times Row 1} + \\ \text{Row 2 = new Row 2} \end{array} \left[\begin{array}{ccc|c} 1 & 2 & -3 & 1 \\ 0 & -3 & 5 & -5 \\ -1 & -1 & 2 & 2 \end{array}\right]$$

To obtain a 0 in place of the -1 in column 1, add the 1 and the -1. Perform that same operation on the entire row to obtain a new Row 3.

Use this →
to make
this zero. →
$$\left[\begin{array}{ccc|c} ① & 2 & -3 & 1 \\ 0 & -3 & 5 & -5 \\ \boxed{-1} & -1 & 2 & 2 \end{array}\right]$$
$R_1 + R_3 \rightarrow R_3$
Row 1 + Row 3 = new Row 3
$$\left[\begin{array}{ccc|c} 1 & 2 & -3 & 1 \\ 0 & -3 & 5 & -5 \\ 0 & 1 & -1 & 3 \end{array}\right]$$

Step 4: Next, we want the *second entry in column 2* to be a 1. We *could* multiply Row 2 by $-\dfrac{1}{3}$ to get the 1, but the other entries would be fractions. Instead, interchanging Row 2 and Row 3 will give us a 1 on the diagonal and keep 0's in column 1. (Sometimes, though, fractions are unavoidable.)

$R_2 \leftrightarrow R_3$
Interchange Rows 2 and 3.
$$\left[\begin{array}{ccc|c} 1 & 2 & -3 & 1 \\ 0 & ① & -1 & 3 \\ 0 & -3 & 5 & -5 \end{array}\right]$$

Step 5: We want to make *all the entries below the 1 in column 2* be 0's. To obtain a 0 in place of -3 in column 2, multiply the 1 above it by 3 (to get 3) and add it to -3. Perform that same operation on the entire row to obtain a new Row 3.

Use this →
to make →
this zero.
$$\left[\begin{array}{ccc|c} 1 & 2 & -3 & 1 \\ 0 & ① & -1 & 3 \\ 0 & \boxed{-3} & 5 & -5 \end{array}\right]$$
$3R_2 + R_3 \rightarrow R_3$
3 times Row 2 + Row 3 = new Row 3
$$\left[\begin{array}{ccc|c} 1 & 2 & -3 & 1 \\ 0 & 1 & -1 & 3 \\ 0 & 0 & 2 & 4 \end{array}\right]$$

We have completed *Step 5* because there is only one entry below the 1 in column 2.

Step 6: *Continue this procedure.* The last entry in column 3 needs to be a 1. (This is the last 1 we need along the diagonal.) Multiply Row 3 by $\dfrac{1}{2}$ to obtain the last 1.

$\dfrac{1}{2}R_3 \rightarrow R_3$
Multiply Row 3 by $\dfrac{1}{2}$.
$$\left[\begin{array}{ccc|c} 1 & 2 & -3 & 1 \\ 0 & 1 & -1 & 3 \\ 0 & 0 & 1 & 2 \end{array}\right]$$

W Hint

Notice on this matrix that 1's on the diagonal with zeros below means that you have eliminated both x and y in the linear equation that is represented by Row 3.

We are done performing row operations because there are 1's on the diagonal and zeros below.

Step 7: Write the matrix in *Step 6* as a *system of equations.*

$$\begin{array}{ll} 1x + 2y - 3z = 1 & x + 2y - 3z = 1 \\ 0x + 1y - 1z = 3 \quad \text{or} & \quad\quad y - z = 3 \\ 0x + 0y + 1z = 2 & \quad\quad\quad\quad z = 2 \end{array}$$

Step 8: *Solve the system* in *Step 7.* The last row tells us that $z = 2$. Substitute $z = 2$ into the equation above it ($y - z = 3$) to get the value of y: $y - 2 = 3$, so $y = 5$.

Substitute $y = 5$ and $z = 2$ into $x + 2y - 3z = 1$ to solve for x.

$$\begin{array}{ll} x + 2y - 3z = 1 & \\ x + 2(5) - 3(2) = 1 & \text{Substitute values.} \\ x + 10 - 6 = 1 & \text{Multiply.} \\ x + 4 = 1 & \text{Subtract.} \\ x = -3 & \end{array}$$

The solution of the system is $(-3, 5, 2)$.

Step 9: *Check the solution* in each equation of the original system. The check is left to the student.

This procedure may seem long and complicated at first, but as you practice and become more comfortable with the steps, you will see that it is actually quite efficient.

[YOU TRY 2] Solve the system using Gaussian elimination.

$$x + 3y - 2z = 10$$
$$3x + 2y + z = 9$$
$$-x + 4y - z = -1$$

If we are performing Gaussian elimination and obtain a matrix that produces a false equation as shown, then the system has *no solution*. The system is *inconsistent*.

$$\begin{bmatrix} 1 & -6 & | & 9 \\ 0 & 0 & | & 8 \end{bmatrix} \quad 0x - 0y = 8 \quad \text{False}$$

If, however, we obtain a matrix that produces a row of zeros as shown, then the system has an *infinite number of solutions*. The system is *consistent* with *dependent* equations. We write its solution as we did in previous sections.

$$\begin{bmatrix} 1 & 5 & | & -1 \\ 0 & 0 & | & 0 \end{bmatrix} \quad 0x - 0y = 0 \quad \text{True}$$

Using Technology

In this section, we have learned how to solve a system of three equations using Gaussian elimination. The row operations used to convert an augmented matrix to row echelon form can be performed on a graphing calculator.

Follow the nine-step method given in the text to solve the system using Gaussian elimination:

$$x + 2y - 3z = 1$$
$$y - z = 3$$
$$-2y + 4z = -4$$

Step 1: Write the system as an augmented matrix:

$$\begin{bmatrix} 1 & 2 & -3 & | & 1 \\ 0 & 1 & -1 & | & 3 \\ 0 & -2 & 4 & | & -4 \end{bmatrix}$$

Store the matrix in matrix [A] using a graphing calculator. Press `2nd` `x⁻¹` to select [A]. Press the right arrow key two times and press `ENTER` to select EDIT. Press `3` `ENTER` then `4` `ENTER` to enter the number of rows and number of columns in the augmented matrix. Enter the coefficients one row at a time as follows: `1` `ENTER` `2` `ENTER` `(−)` `3` `ENTER` `1` `ENTER` `0` `ENTER` `1` `ENTER` `(−)` `1` `ENTER` `3` `ENTER` `0` `ENTER` `(−)` `2` `ENTER` `4` `ENTER` `(−)` `4` `ENTER`. Press `2nd` `MODE` to return to the home screen. Press `2nd` `x⁻¹` `ENTER` `ENTER` to display matrix [A].

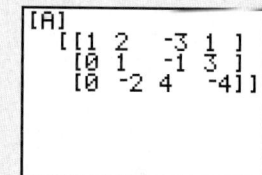

Notice that we can omit *Steps 2–4* because we already have two 1's on the diagonal and 0's below the first 1.

Step 5: Get the element in Row 3, column 2, to be 0. Multiply Row 2 by the opposite of the number in Row 3, column 2, and add to Row 3. The graphing calculator row operation used to multiply a row by a nonzero number and add to another row is ***row−(nonzero number, matrix name, first row, second row)**.

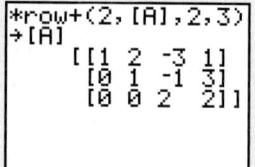

In this case, we have *row−(2, [A], 2, 3). To enter this row operation on your calculator, press `2nd` `x⁻¹`, then press the right arrow to access the MATH menu. Scroll down to option F and press `ENTER` to display *row+(then enter `2` `,` `2nd` `x⁻¹` `ENTER` `,` `2` `,` `3` `)` as shown. Store the result back in matrix [A] by pressing `STO>` `2nd` `x⁻¹` `ENTER` `ENTER`.

Step 6: To make the last number on the diagonal be 1, multiply Row 3 by $\frac{1}{2}$. The graphing calculator row operation used to multiply a row by a nonzero number is ***row(nonzero number, matrix name, row)**. In this case, we have *row(1/2, [A], 3). On your calculator, press 2nd x^{-1}, then press the right arrow to access the MATH menu. Scroll down to option E and press ENTER to display *row(then enter 1 ÷ 2 , 2nd x^{-1} ENTER , 3) as shown.

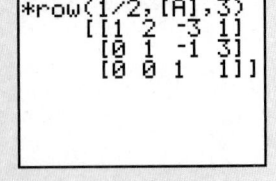

Step 7: Write the matrix from *Step 6* as:

$$\begin{bmatrix} 1 & 2 & -3 & | & 1 \\ 0 & 1 & -1 & | & 3 \\ 0 & 0 & 1 & | & 1 \end{bmatrix} \qquad \begin{aligned} 1x + 2y - 3z &= 1 \\ 0x + 1y - 1z &= 3 \\ 0x + 0y + 1z &= 1 \end{aligned} \quad \text{or} \quad \begin{aligned} x + 2y - 3z &= 1 \\ y - z &= 3 \\ z &= 1 \end{aligned}$$

Step 8: Solve the system using substitution to obtain the solution $x = -4$, $y = 4$, $z = 1$ or $(-4, 4, 1)$.

Step 9: Check the solution.

Using Row Echelon Form

The row echelon form shown above is not unique. Another row echelon form can be obtained *in one step* using a graphing calculator. Given the original augmented matrix stored in [A], press 2nd x^{-1}, then press the right arrow, scroll down to option A, and press ENTER to display ref(which stands for row echelon form, and press ENTER. Press 2nd x^{-1} ENTER) ENTER to show the matrix in row echelon form.

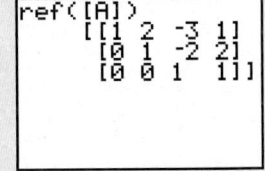

Using Reduced Row Echelon Form

The **reduced row echelon form** of an augmented matrix contains 1's on the diagonal and 0's *above* and *below* the 1's. We can find this using row operations as shown in the 9-step process, or directly in one step. Given the original augmented matrix stored in [A], press 2nd x^{-1}, then press the right arrow, scroll down to option B, and press ENTER to display ref(which stands for reduced row-echelon form, and press ENTER. Press 2nd x^{-1} ENTER) ENTER to show the matrix in reduced row echelon form.

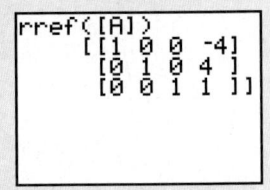

Write a system of equations from the matrix that is in reduced row echelon form.

$$\begin{bmatrix} 1 & 0 & 0 & | & -4 \\ 0 & 1 & 0 & | & 4 \\ 0 & 0 & 1 & | & 1 \end{bmatrix} \qquad \begin{aligned} 1x + 0y + 0z &= -4 \\ 0x + 1y + 0z &= 4 \\ 0x + 0y + 1z &= 1 \end{aligned} \quad \text{or} \quad \begin{aligned} x &= -4 \\ y &= 4 \\ z &= 1 \end{aligned}$$

Use a graphing calculator to solve each system using Gaussian elimination.

1) $\begin{aligned} x + 2y &= 1 \\ 3x - y &= 17 \end{aligned}$
    2) $\begin{aligned} x - 5y &= -3 \\ 2x - 7y &= 3 \end{aligned}$
    3) $\begin{aligned} -5x + 2y &= -4 \\ 3x - y &= 8 \end{aligned}$

4) $\begin{aligned} 3x - 5y - 3z &= 6 \\ -x + 3y + 2z &= 1 \\ -2x + 7y + 5z &= 6 \end{aligned}$
    5) $\begin{aligned} 3x + 2y + z &= 9 \\ -5x - 2y - z &= -7 \\ 4x + y + z &= 3 \end{aligned}$
    6) $\begin{aligned} 2x - y + 2z &= -4 \\ -x + y - 2z &= 7 \\ -3x + y - z &= -1 \end{aligned}$

E Evaluate 5.6 Exercises   Do the exercises, and check your work.

Objective 1: Learn the Vocabulary Associated with Gaussian Elimination

Write each system in an augmented matrix.

 1) $x - 7y = 15$
$4x + 3y = -1$

2) $x + 6y = 4$
$-5x + y = -3$

3) $x + 6y - z = -2$
$3x + y + 4z = 7$
$-x - 2y + 3z = 8$

4) $x + 2y - 7z = 3$
$3x - 5y = -1$
$-x + 2z = -4$

Write a system of linear equations in x and y represented by each augmented matrix.

5) $\begin{bmatrix} 3 & 10 & | & -4 \\ 1 & -2 & | & 5 \end{bmatrix}$

6) $\begin{bmatrix} 1 & -1 & | & 6 \\ -4 & 7 & | & 2 \end{bmatrix}$

7) $\begin{bmatrix} 1 & -6 & | & 8 \\ 0 & 1 & | & -2 \end{bmatrix}$

8) $\begin{bmatrix} 1 & 2 & | & 11 \\ 0 & 1 & | & 3 \end{bmatrix}$

Write a system of linear equations in x, y, and z represented by each augmented matrix.

9) $\begin{bmatrix} 1 & -3 & 2 & | & 7 \\ 4 & -1 & 3 & | & 0 \\ -2 & 2 & -3 & | & -9 \end{bmatrix}$

10) $\begin{bmatrix} 1 & 4 & -3 & | & -5 \\ -1 & 2 & 5 & | & 8 \\ 6 & -2 & -1 & | & 3 \end{bmatrix}$

11) $\begin{bmatrix} 1 & 5 & 2 & | & 14 \\ 0 & 1 & -8 & | & 2 \\ 0 & 0 & 1 & | & -3 \end{bmatrix}$

12) $\begin{bmatrix} 1 & 4 & -7 & | & -11 \\ 0 & 1 & 3 & | & -1 \\ 0 & 0 & 1 & | & 6 \end{bmatrix}$

Objective 2: Solve a System of Two Equations Using Gaussian Elimination

Solve each system using Gaussian elimination. Identify any inconsistent systems or dependent equations.

 13) $x + 4y = -1$
$3x + 5y = 4$

14) $x - 3y = 1$
$-3x + 7y = 3$

15) $x + 4y = -6$
$2x + 5y = 0$

16) $x - 3y = 9$
$-6x + 5y = 11$

17) $4x - 3y = 6$
$x + y = -2$

18) $-4x + 5y = -3$
$x - 8y = -6$

Objective 3: Solve a System of Any Number of Equations Using Gaussian Elimination

 19) $x + y - z = -5$
$4x + 5y - 2z = 0$
$8x - 3y + 2z = -4$

20) $x - 2y + 2z = 3$
$2x - 3y + z = 13$
$-4x - 5y - 6z = 8$

21) $x - 3y + 2z = -1$
$3x - 8y + 4z = 6$
$-2x - 3y - 6z = 1$

22) $x - 2y + z = -2$
$2x - 3y + z = 3$
$3x - 6y + 2z = 1$

23) $-4x - 3y + z = 5$
$x + y - z = -7$
$6x + 4y + z = 12$

24) $6x - 9y - 2z = 7$
$-3x + 4y + z = -4$
$x - y - z = 1$

25) $x - 3y + z = -4$
$4x + 5y - z = 0$
$2x - 6y + 2z = 1$

26) $x - y + 3z = 1$
$5x - 5y + 15z = 5$
$-4x + 4y - 12z = -4$

Extension

Extend the concepts of this section to solve these systems using Gaussian elimination.

27) $a + b + 3c + d = -1$
$-a + c - d = 7$
$2a + 3b + 9c - 2d = 7$
$a - 2b + c + 3d = -11$

28) $a - 2b - c + 3d = 15$
$2a - 3b + c + 4d = 22$
$-a + 4b + 6c + 7d = -3$
$3a + 2b - c - d = -7$

29) $w - 3x + 2y - z = -2$
$-3w + 8x - 5y + z = 2$
$2w - x + y + 3z = 7$
$w - 2x + y + 2z = 3$

30) $w + x - 4y + 2z = -21$
$3w + 2x + y - z = 6$
$-2w - x - 2y + 6z = -30$
$-w + 3x + 4y + z = 1$

R1) How is Gaussian elimination similar to the elimination method for solving a system of equations?

R2) If you had to solve a system of linear equations in two variables, would you prefer to use the *elimination method* shown in Section 5.3 or *Gaussian elimination*? Why?

R3) How would you explain to a friend how to solve a system of two linear equations using Gaussian elimination?

R4) When solving a system of linear equations using *Gaussian elimination,* does it matter in which row you represent each equation? In other words, does the top equation have to be placed in Row 1?

Group Activity – Incorporating Systems of Equations into Topics from Chapters 1–5

1) Find the values of a and b so the perimeter of the parallelogram is 50 units and the perimeter of the trapezoid is 68 units.

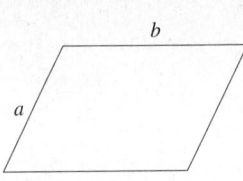
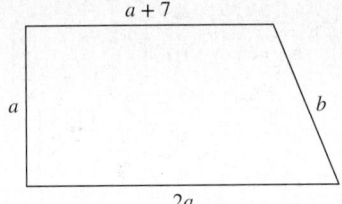

2) Let $\dfrac{x}{y}$ represent a fraction.

If the numerator is increased by 5 and the denominator is decreased by 3, the resulting fraction is $\dfrac{1}{4}$.

However, if the numerator is decreased by $\dfrac{22}{5}$ and the denominator is tripled, the resulting fraction is $-\dfrac{2}{5}$. Find the fraction $\dfrac{x}{y}$.

3) Does the line passing through the points $(-2, 4)$ and $(6, 8)$ intersect a line with a slope of 5 and y-intercept of 3? If so, find the coordinates of that point.

em POWER me What Is My Test-Taking Style?

Successfully taking math, or *any* tests, is about more than actually *taking* the test. Taking a test includes *preparing* for the test, and preparation is, probably, more important than most students realize. Take this survey to learn about your studying and test-taking habits. How do *you* take math tests?

☐ I know whether it is a quiz, test, exam, or something else.

☐ I know exactly which sections, chapters, or material will be on the test.

☐ I know how much time we will have to take the test.

☐ I know, beforehand, whether we can use a calculator, notes, or none of these on the test.

☐ I attend review sessions or study groups.

☐ I do the review my instructor has given to the class or the Chapter Review in the book, and I make a list of questions about it and get them answered.

☐ I wait until the night before the test before I start studying.

☐ I begin studying a couple or a few days before the test.

☐ I study in a quiet place that is free from distractions.

☐ I read my notes and rework problems on a separate piece of paper.

☐ I read my notes and highlight or note important definitions, procedures, and formulas.

☐ A couple of days before the test, I make a list of the types of problems I do not understand. I ask about them when we review for the test in class, or I go to my instructor's office or to the tutoring center for help.

☐ The night before the test, I study by reading my notes and book and by *doing* problems.

☐ When I think I am ready for the test, I work out a couple of problems from each section *without looking back at my notes or book* to see just how ready I am.

☐ I get a good night's sleep before the test.

☐ I warm up for the test by working out a few problems before I walk into the classroom to take the test.

☐ I arrive early for the test so that I do not feel rushed.

☐ I have everything I need to take the test: pencils, eraser, scratch paper, and a calculator.

☐ I read the directions, carefully, before starting the test.

☐ I look over all of the questions on the test before I begin.

☐ I do the easiest questions first.

☐ I write neatly and show *all* of my work.

☐ I ask the instructor a question if something on the test is unclear.

☐ I keep track of time during the test so that I can pace myself.

☐ Other: _____

Think about the items that you have, and have *not*, checked in this survey. What can you learn from it?

Chapter 5: Summary

Definition/Procedure	Example

5.1 Solving Systems by Graphing

A **system of linear equations** consists of two or more linear equations with the same variables. A **solution of a system** of two equations in two variables is an ordered pair that is a solution of each equation in the system.

Determine whether $(4, 2)$ is a solution of the system

$$x + 2y = 8$$
$$-3x + 4y = -4$$

$x + 2y = 8$	$-3x + 4y = -4$
$4 + 2(2) \overset{?}{=} 8$ Substitute.	$-3(4) + 4(2) \overset{?}{=} -4$ Substitute.
$4 + 4 \overset{?}{=} 8$	$-12 + 8 \overset{?}{=} -4$
$8 = 8$ True	$-4 = -4$ True

Since $(4, 2)$ is a solution of each equation in the system, **yes,** it is a solution of the system.

To **solve a system by graphing,** graph each line on the same axes.

a) If the lines intersect at a **single point,** then this point is the solution of the system. The system is **consistent.**

b) If the lines are **parallel,** then the system has **no solution.** We write the solution set as $\emptyset$. The system is **inconsistent.**

c) If the graphs are the **same line,** then the system has an **infinite number of solutions.** The equations are **dependent.**

Solve by graphing.
$$y = -\frac{1}{2}x + 2$$
$$5x + 3y = -1$$

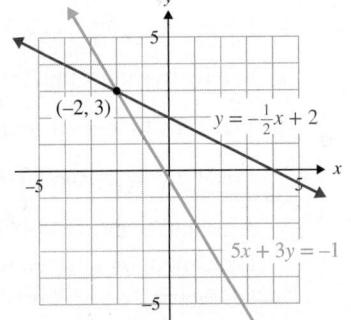

The solution of the system is $(-2, 3)$.

The system is **consistent.**

5.2 Solving Systems by the Substitution Method

Steps for Solving a System by the Substitution Method

Step 1: Solve one of the equations for one of the variables. If possible, solve for a variable that has a coefficient of 1 or -1.

Step 2: Substitute the expression in *Step 1* into the *other* equation. The equation you obtain should contain only one variable.

Step 3: Solve the equation in *Step 2*.

Step 4: Substitute the value found in *Step 3* into either of the equations to obtain the value of the other variable.

Step 5: Check the values in the original equations.

Solve by the substitution method.
$$7x - 3y = 8$$
$$x + 2y = -11$$

Step 1: Solve for x in the second equation since its coefficient is 1.
$$x = -2y - 11$$

Step 2: Substitute $-2y - 11$ for the x in the first equation.
$$7(-2y - 11) - 3y = 8$$

Step 3: Solve the equation above for y.

$7(-2y - 11) - 3y = 8$	
$-14y - 77 - 3y = 8$	Distribute.
$-17y - 77 = 8$	Combine like terms.
$-17y = 85$	Add 77 to each side.
$y = -5$	Divide by -17.

Step 4: Substitute $y = -5$ into the equation in *Step 1* to find x.

$x = -2(-5) - 11$	Substitute -5 for y.
$x = 10 - 11$	Multiply.
$x = -1$	

Step 5: The solution is $(-1, -5)$. Verify this by substituting $(-1, -5)$ into each of the original equations.

Definition/Procedure	Example
If the variables drop out and a false equation is obtained, the system has **no solution. The system is inconsistent, and the solution set is** $\varnothing$.	Solve by the substitution method. $\quad 2x - 8y = 9$ $\qquad\qquad\qquad\qquad\qquad\qquad\qquad x = 4y + 2$ **Step 1:** The second equation is solved for x. **Step 2:** Substitute $4y + 2$ for x in the first equation. $$2(4y + 2) - 8y = 9$$ **Step 3:** Solve the equation above for y. $$\begin{aligned} 2(4y + 2) - 8y &= 9 \\ 8y + 4 - 8y &= 9 \qquad \text{Distribute.} \\ 4 &= 9 \qquad \text{False} \end{aligned}$$ The system has no solution. The solution set is $\varnothing$.
If the variables drop out and a true equation is obtained, the system has an **infinite number of solutions. The equations are dependent.**	Solve by the substitution method. $\qquad\qquad y = x - 3$ $\qquad\qquad\qquad\qquad\qquad\qquad\qquad 3x - 3y = 9$ **Step 1:** The first equation is solved for y. **Step 2:** Substitute $x - 3$ for y in the second equation. $$3x - 3(x - 3) = 9$$ **Step 3:** Solve the equation above for x. $$\begin{aligned} 3x - 3(x - 3) &= 9 \\ 3x - 3x + 9 &= 9 \qquad \text{Distribute.} \\ 9 &= 9 \qquad \text{True} \end{aligned}$$ The system has an infinite number of solutions of the form $\{(x, y) \mid y = x - 3\}$.

5.3 Solving Systems by the Elimination Method

Steps for Solving a System of Two Linear Equations by the Elimination Method	Solve using the elimination method. $\quad 4x + 5y = -7$ $\qquad\qquad\qquad\qquad\qquad\qquad\qquad\qquad -5x - 6y = 8$
Step 1: Write each equation in the form $Ax + By = C$. **Step 2:** Determine which variable to eliminate. If necessary, multiply one or both of the equations by a number so that the coefficients of the variable to be eliminated are negatives of one another. **Step 3:** Add the equations, and solve for the remaining variable. **Step 4:** Substitute the value found in *Step 3* into either of the original equations to find the value of the other variable. **Step 5:** Check the solution in each of the original equations.	Eliminate x. Multiply the first equation by 5, and multiply the second equation by 4 to rewrite the system with equivalent equations. **Rewrite the system** $$\begin{aligned} 5(4x + 5y) = 5(-7) \quad &\rightarrow \quad 20x + 25y = -35 \\ 4(-5x - 6y) = 4(8) \quad &\rightarrow \quad -20x - 24y = 32 \end{aligned}$$ Add the equations: $\qquad\begin{aligned} 20x + 25y &= -35 \\ + \; \underline{-20x - 24y} &= \underline{32} \\ y &= -3 \end{aligned}$ Substitute $y = -3$ into either of the original equations, and solve for x. $$\begin{aligned} 4x + 5y &= -7 \\ 4x + 5(-3) &= -7 \\ 4x - 15 &= -7 \\ 4x &= 8 \\ x &= 2 \end{aligned}$$ The solution is $(2, -3)$. Verify this by substituting $(2, -3)$ into each of the original equations.

Definition/Procedure	Example

5.4 Applications of Systems of Two Equations

Use the **Five Steps for Solving Applied Problems** outlined in the section to solve an applied problem.

Step 1: **Read** the problem carefully. Draw a picture, if applicable. Identify what you are being asked to find.

Step 2: **Choose variables** to represent the unknown quantities. If applicable, label the picture with the variables.

Step 3: **Write a system of equations using two variables.** It may be helpful to begin by writing an equation in words.

Step 4: **Solve the system.**

Step 5: **Check** the answer in the original problem, and **interpret** the solution as it relates to the problem.

Amana spent $40.20 at a second-hand movie and music store when she purchased some DVDs and CDs. Each DVD cost $6.30, and each CD cost $2.50. How many DVDs and CDs did she buy if she purchased 10 items all together?

Step 1: Read the problem carefully.
Step 2: Choose variables.

$$x = \text{number of DVDs she bought}$$
$$y = \text{number of CDs she bought}$$

Step 3: One equation involves the *cost* of the items:

$$\text{Cost DVDs} + \text{Cost CDs} = \text{Total cost}$$
$$6.30x \quad + \quad 2.50y \quad = \quad 40.20$$

The second equation involves the number of items:

$$\frac{\text{Number of}}{\text{DVDs}} + \frac{\text{Number of}}{\text{CDs}} = \frac{\text{Total number}}{\text{of items}}$$
$$x \quad + \quad y \quad = \quad 10$$

The system is $6.30x + 2.50y = 40.20$.
$$x + y = 10$$

Step 4: Multiply by 10 to eliminate the decimals in the first equation, and then solve the system using substitution.

$$10(6.30x + 2.50y) = 10(40.20) \quad \text{Eliminate decimals.}$$
$$63x + 25y = 402$$

Solve the system $\begin{matrix} 63x + 25y = 402 \\ x + y = 10 \end{matrix}$ to determine that the solution is (4, 6).

Step 5: Amana bought 4 DVDs and 6 CDs. Verify the solution.

5.5 Solving Systems of Three Equations and Applications

A **linear equation in three variables** is an equation of the form $Ax + By + Cz = D$, where A, B, and C are not all zero and where A, B, C, and D are real numbers. Solutions of this type of an equation are **ordered triples** of the form (x, y, z).

$5x + 3y + 9z = -2$

One solution of this equation is $(-1, -2, 1)$ because substituting the values for x, y, and z satisfies the equation.

$$5x + 3y + 9z = -2$$
$$5(-1) + 3(-2) + 9(1) \stackrel{?}{=} -2$$
$$-5 - 6 + 9 \stackrel{?}{=} -2$$
$$-2 = -2 \quad \text{True}$$

Definition/Procedure	Example

Solving a System of Three Linear Equations in Three Variables

Step 1: **Label** the equations ①, ②, and ③.

Step 2: **Choose a variable to eliminate. Eliminate** this variable from *two* sets of *two* equations using the elimination method. You will obtain two equations containing the same two variables. Label one of these new equations $\boxed{A}$ and the other $\boxed{B}$.

Step 3: Use the elimination method to **eliminate a variable from equations** $\boxed{A}$ **and** $\boxed{B}$. You have now found the value of one variable.

Step 4: **Find the value of another variable** by substituting the value found in *Step 3* into equation $\boxed{A}$ or $\boxed{B}$ and solving for the second variable.

Step 5: **Find the value of the third variable** by substituting the values of the two variables found in *Steps 3* and *4* into equation ①, ②, or ③.

Step 6: **Check** the solution in each of the original equations, and **write the solution as an ordered triple.**

Solve ① $x + 2y + 3z = 5$
② $4x - 2y - z = -1$
③ $-3x + y + 4z = -12$

Step 1: Label the equations ①, ②, and ③.

Step 2: We will eliminate y from *two* sets of *two* equations.
  a) Add equations ① *and* ② to eliminate y. Label the resulting equation $\boxed{A}$.

$$\begin{array}{ll} ① & x + 2y + 3z = 5 \\ ②\ + & 4x - 2y - z = -1 \\ \hline \boxed{A} & 5x + \qquad 2z = 4 \end{array}$$

  b) To eliminate y again, multiply equation ③ by 2 and add it to equation ②. Label the resulting equation $\boxed{B}$.

$$\begin{array}{ll} 2 \times ③ & -6x + 2y + 8z = -24 \\ ②\ + & 4x - 2y - z = -1 \\ \hline \boxed{B} & -2x + \qquad 7z = -25 \end{array}$$

Step 3: Eliminate x from $\boxed{A}$ and $\boxed{B}$. Multiply $\boxed{A}$ by 2 and $\boxed{B}$ by 5. Add the resulting equations.

$$\begin{array}{ll} 2 \times \boxed{A} & 10x + 4z = 8 \\ 5 \times \boxed{B}\ + & -10x + 35z = -125 \\ \hline & 39z = -117 \\ & \boxed{z = -3} \end{array}$$

Step 4: Substitute $z = -3$ into either $\boxed{A}$ or $\boxed{B}$.

$$\begin{array}{lll} \boxed{A} & 5x + 2z = 4 & \\ & 5x + 2(-3) = 4 & \text{Substitute } -3 \text{ for } z. \\ & 5x - 6 = 4 & \text{Multiply.} \\ & 5x = 10 & \text{Add 6.} \\ & \boxed{x = 2} & \text{Divide by 5.} \end{array}$$

Step 5: Substitute $x = 2$ and $z = -3$ into ① to solve for y.

$$\begin{array}{lll} \boxed{1} & x + 2y + 3z = 5 & \\ & 2 + 2y + 3(-3) = 5 & \text{Substitute.} \\ & 2 + 2y - 9 = 5 & \text{Multiply.} \\ & 2y - 7 = 5 & \text{Combine like terms.} \\ & 2y = 12 & \text{Add 7.} \\ & \boxed{y = 6} & \text{Divide by 2.} \end{array}$$

Step 6: The solution is $(2, 6, -3)$. The check is left to the student.

Definition/Procedure	Example

5.6 Solving Systems of Linear Equations Using Matrices

An **augmented matrix** contains a vertical line to separate different parts of the matrix.	An example of an augmented matrix is $\begin{bmatrix} 1 & 4 & \mid & -9 \\ 2 & -3 & \mid & 8 \end{bmatrix}$.

Matrix Row Operations

Performing the following row operations on a matrix produces an equivalent matrix.

1) Interchanging two rows
2) Multiplying every element in a row by a nonzero real number
3) Replacing a row by the sum of it and the multiple of another row

Gaussian elimination is the process of performing row operations on a matrix to put it into *row echelon* form.

A matrix is in **row echelon form** when it has 1's along the diagonal and 0's below.

$$\begin{bmatrix} 1 & a & \mid & b \\ 0 & 1 & \mid & c \end{bmatrix} \qquad \begin{bmatrix} 1 & a & b & \mid & d \\ 0 & 1 & c & \mid & e \\ 0 & 0 & 1 & \mid & f \end{bmatrix}$$

Solve using Gaussian elimination.

$$x - y = 5$$
$$2x + 7y = 1$$

Write the system in an augmented matrix. Then, perform row operations to get it into row echelon form.

$$\begin{bmatrix} 1 & -1 & \mid & 5 \\ 2 & 7 & \mid & 1 \end{bmatrix} \xrightarrow{-2R_1 + R_2 \to R_2} \begin{bmatrix} 1 & -1 & \mid & 5 \\ 0 & 9 & \mid & -9 \end{bmatrix} \xrightarrow{\frac{1}{9}R_2 \to R_2} \begin{bmatrix} 1 & -1 & \mid & 5 \\ 0 & 1 & \mid & -1 \end{bmatrix}$$

The matrix is in row echelon form since it has 1's on the diagonal and a 0 below.

Write a system of equations from the matrix that is in row echelon form.

$$\begin{bmatrix} 1 & -1 & \mid & 5 \\ 0 & 1 & \mid & -1 \end{bmatrix} \quad \begin{matrix} 1x - 1y = 5 \\ 0x + 1y = -1 \end{matrix} \quad \text{or} \quad \begin{matrix} x - y = 5 \\ y = -1 \end{matrix}$$

Solving the system, we obtain the solution $(4, -1)$.

Chapter 5: Review Exercises

(5.1) Determine whether the ordered pair is a solution of the system of equations.

1) $x - 5y = 13$
   $2x + 7y = 20$
   $(-4, -5)$

2) $8x + 3y = 16$
   $10x - 6y = 7$
   $\left(\dfrac{3}{2}, \dfrac{4}{3}\right)$

3) If you are solving a system of equations by graphing, how do you know whether the system has no solution?

Solve each system by graphing.

4) $y = \dfrac{1}{2}x + 1$
   $x + y = 4$

5) $x - 3y = 9$
   $-x + 3y = 6$

6) $6x - 3y = 12$
   $-2x + y = -4$

7) $-x + 2y = 1$
   $2x + 3y = -9$

Without graphing, determine whether each system has no solution, one solution, or an infinite number of solutions.

8) $8x + 9y = -2$
   $x - 4y = 1$

9) $y = -\dfrac{5}{2}x + 3$
   $5x + 2y = 6$

10) The graph shows the hotel occupancy rates in Iceland and Spain for June through September of 2016. (ec.europa.eu)

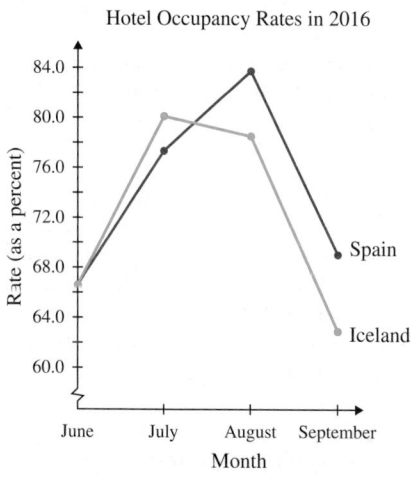

Hotel Occupancy Rates in 2016

a) Which country had a higher occupancy rate in September, and approximately what was it?

b) Write the point of intersection of the graphs as an ordered pair, and explain its meaning.

c) When was Iceland's occupancy rate greater than that of Spain, and by how much?

d) Which line segment has the steepest positive slope? What does this mean in the context of the problem?

(5.2) Solve each system by the substitution method.

11) $9x - 2y = 8$
    $y = 2x + 1$

12) $y = -6x + 5$
    $12x + 2y = 10$

13) $-x + 8y = 19$
    $4x - 3y = 11$

14) $-12x + 7y = 9$
    $8x - y = -6$

(5.3) Solve each system using the elimination method.

15) $x - 7y = 3$
    $-x + 5y = -1$

16) $5x + 4y = -23$
    $3x - 8y = -19$

17) $-10x + 4y = -8$
    $5x - 2y = 4$

18) $7x - 4y = 13$
    $6x - 5y = 8$

Solve each system using the elimination method twice.

19) $2x + 9y = -6$
    $5x + y = 3$

20) $7x - 4y = 10$
    $6x + 3y = 8$

(5.2–5.3)

21) When is the best time to use substitution to solve a system?

22) If an equation in a system contains fractions, what should you do first to make the system easier to solve?

Solve each system by either the substitution or elimination method.

23) $6x + y = -8$
    $9x + 7y = -1$

24) $4y - 5x = -23$
    $2x + 3y = -23$

25) $15x - 6y = -7$
    $-10x + 4y = 9$

26) $-x + 5y = 4$
    $8x - 40y = -32$

27) $\dfrac{1}{3}x - \dfrac{2}{9}y = -\dfrac{2}{3}$
    $\dfrac{5}{12}x + \dfrac{1}{3}y = 1$

28) $0.02x - 0.01y = 0.13$
    $-0.1x + 0.4y = 1.8$

29) $6(2x - 3) = y + 4(x - 3)$
    $5(3x + 4) + 4y = 11 - 3y + 27x$

30) $x - 3y = 36$
    $y = \dfrac{5}{3}x$

31) $y = -\dfrac{9}{7}x + \dfrac{6}{7}$
    $18x + 14y = 12$

32) $\dfrac{3}{4}x - \dfrac{5}{4}y = \dfrac{7}{8}$
    $4 - 2(x + 5) - y = 3(1 - 2y) + x$

(5.4) Write a system of equations, and solve.

33) One day in the school cafeteria, the number of children who bought white milk was twice the number who bought chocolate milk. How many cartons of each type of milk were sold if the cafeteria sold a total of 141 cartons of milk?

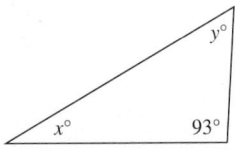
Ken Cavanagh/McGraw-Hill Education

34) How many ounces of a 7% acid solution and how many ounces of a 23% acid solution must be mixed to obtain 20 oz of a 17% acid solution?

35) Edwin and Camille leave from opposite ends of a jogging trail 7 miles apart and travel toward each other. Edwin jogs 2 mph faster than Camille, and they meet after half an hour. How fast does each of them jog?

36) At a movie theater concession stand, three candy bars and two small sodas cost $14.00. Four candy bars and three small sodas cost $19.50. Find the cost of a candy bar and the cost of a small soda.

37) The width of a rectangle is 5 cm less than the length. Find the dimensions of the rectangle if the perimeter is 38 cm.

38) Two planes leave the same airport and travel in opposite directions. The northbound plane flies 40 mph slower than the southbound plane. After 1.5 hours they are 1320 miles apart. Find the speed of each plane.

39) Shawanna saves her quarters and dimes in a piggy bank. When she opens it, she has 63 coins worth a total of $11.55. How many of each type of coin does she have?

40) Find the measure of angles x and y if the measure of angle x is half the measure of angle y.

(Triangle with angles $y°$, $x°$, and $93°$)

41) At a ski shop, two packs of hand warmers and one pair of socks cost $27.50. Five packs of hand warmers and three pairs of socks cost $78.00. Find the cost of a pack of hand warmers and a pair of socks.

42) A 7 P.M. spinning class has 9 more members than a 10 A.M. spinning class. The two classes have a total of 71 students. How many are in each class?

(5.5) Determine whether the ordered triple is a solution of the system.

43) $x - 6y + 4z = 13$
$5x + y + 7z = 8$
$2x + 3y - z = -5$
$(-3, -2, 1)$

44) $-4x + y + 2z = 1$
$x - 3y - 4z = 3$
$-x + 2y + z = -7$
$(0, -5, 3)$

Solve each system using one of the methods in Section 5.5. Identify any inconsistent systems or dependent equations.

45) $2x - 5y - 2z = 3$
$x + 2y + z = 5$
$-3x - y + 2z = 0$

46) $x - 2y + 2z = 6$
$x + 4y - z = 0$
$5x + 3y + z = -3$

47) $5a - b + 2c = -6$
$-2a - 3b + 4c = -2$
$a + 6b - 2c = 10$

48) $2x + 3y - 15z = 5$
$-3x - y + 5z = 3$
$-x + 6y - 10z = 12$

49) $4x - 9y + 8z = 2$
$x + 3y = 5$
$6y + 10z = -1$

50) $-a + 5b - 2c = -3$
$3a + 2c = -3$
$2a + 10b = -2$

51) $x + 3y - z = 0$
$11x - 4y + 3z = 8$
$5x + 15y - 5z = 1$

52) $4x + 2y + z = 0$
$8x + 4y + 2z = 0$
$16x + 8y + 4z = 0$

53) $12a - 8b + 4c = 8$
$3a - 2b + c = 2$
$-6a + 4b - 2c = -4$

54) $3x - 12y - 6z = -8$
$x + y - z = 5$
$-4x + 16y + 8z = 10$

55) $5y + 2z = 6$
$-x + 2y = -1$
$4x - z = 1$

56) $2a - b = 4$
$3b + c = 8$
$-3a + 2c = -5$

57) $8x + z = 7$
$3y + 2z = -4$
$4x - y = 5$

58) $6y - z = -2$
$x + 3y = 1$
$-3x + 2z = 8$

Write a system of equations, and solve.

59) One serving (12 fl oz) of Propel has 10 mg more sodium than one serving of Powerade. One serving of Gatorade has the same amount of sodium as the same serving size of Propel. Together the three drinks have 470 mg of sodium. How much sodium is in one serving of each drink? (Product labels)

60) In 2017, the institutions with the greatest number of active cardholders were Citigroup Inc., JPMorgan Chase & Co., and American Express Co. Together, they held approximately 43% of the market. Citigroup's market share was 6% more than that of American Express, and JPMorgan's share was 2% less than Citigroup's. Approximately what percent share did each company hold in 2017? (www.valuepenguin.com)

61) One Friday, Serena, Blair, and Chuck were busy texting their friends. Together, they sent a total of 140 text messages. Blair sent 15 more texts then Serena, while Chuck sent half as many as Serena. How many texts did each person send that day?

62) Arpana bought 28 items for a breakfast meeting. The cost was $29.20 before tax. The donuts cost $1.00 each, bagels cost $1.30 each, and the bananas cost $0.70 each. If she bought twice as many donuts as bananas, how many of each breakfast item did she buy?

Nancy R. Cohen/Getty Images

63) A family of six people goes to an ice cream store every Sunday after dinner. One week, they order two ice cream cones, three shakes, and one sundae for $13.50. The next week they get three cones, one shake, and two sundaes for $13.00. The week after that they spend $11.50 on one shake, one sundae, and four ice cream cones. Find the price of an ice cream cone, a shake, and a sundae.

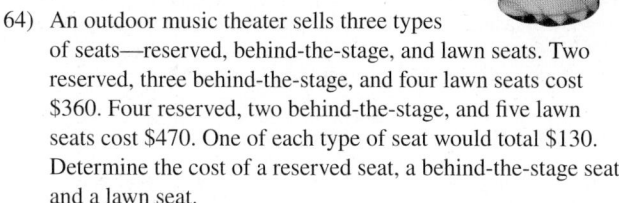

Brand X Pictures/Punchstock

64) An outdoor music theater sells three types of seats—reserved, behind-the-stage, and lawn seats. Two reserved, three behind-the-stage, and four lawn seats cost $360. Four reserved, two behind-the-stage, and five lawn seats cost $470. One of each type of seat would total $130. Determine the cost of a reserved seat, a behind-the-stage seat, and a lawn seat.

65) The measure of the smallest angle of a triangle is one-third the measure of the middle angle. The measure of the largest angle is 70° more than the measure of the smallest angle. Find the measures of the angles of the triangle.

66) The perimeter of a triangle is 40 in. The longest side is twice the length of the shortest side, and the sum of the two smaller sides is four inches longer than the longest side. Find the lengths of the sides of the triangles.

(5.6) Solve each system using Gaussian elimination.

67) $x - y = -11$
$2x + 9y = 0$

68) $x - 8y = -13$
$4x + 9y = -11$

69) $5x + 3y = 5$
$-x + 8y = -1$

70) $3x + 5y = 5$
$-4x - 9y = 5$

71) $x - 3y - 3z = -7$
$2x - 5y - 3z = 2$
$-3x + 5y + 4z = -1$

72) $x - 3y + 5z = 3$
$2x - 5y + 6z = -3$
$3x + 2y + 2z = 3$

Chapter 5: Test

1) Determine whether $\left(-\dfrac{2}{3}, 4\right)$ is a solution of the system $9x + 5y = 14$
$-6x - y = 0$.

Solve each system by graphing.

2) $y = -x + 2$
$3x - 4y = 20$

3) $3y - 6x = 6$
$2x - y = 1$

4) The graph shows the unemployment rate in the civilian labor force in Indiana and Massachusetts from 2014 to 2017. (www.bls.gov)

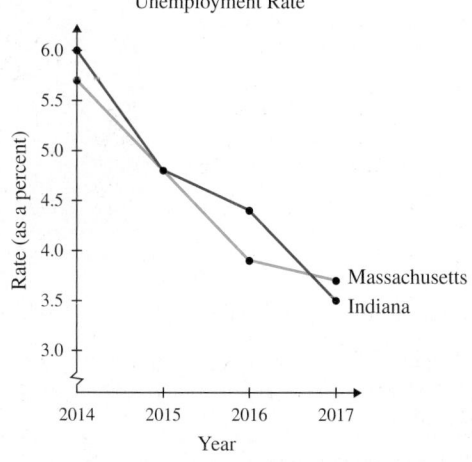

a) During what year and in which state was the smallest percentage of people unemployed? Approximately what percent of the state's population was unemployed at that time?

b) Write the point of intersection of the graphs as an ordered pair, and explain its meaning.

c) When was Massachusetts' unemployment rate less than that of Indiana?

d) Which line segment has the most negative slope? How can this be explained in the context of the problem?

Solve each system by the substitution method.

5) $3x - 10y = -10$
$x + 8y = -9$

6) $y = \dfrac{1}{2}x - 3$
$4x - 8y = 24$

Solve each system by the elimination method.

7) $2x + 5y = 11$
$7x - 5y = 16$

8) $3x + 4y = 24$
$7x - 3y = -18$

9) $-6x + 9y = 14$
$4x - 6y = 5$

Solve each system using any method.

10) $11x - y = -14$
$-9x + 7y = -38$

11) $\dfrac{5}{8}x + \dfrac{1}{4}y = \dfrac{1}{4}$
$\dfrac{1}{3}x + \dfrac{1}{2}y = -\dfrac{4}{3}$

12) $7 - 4(2x + 3) = x + 7 - y$
$5(x - y) + 20 = 8(2 - x) + x - 12$

13) $-x + 4y + 3z = 6$
$3x - 2y + 6z = -18$
$x + y + 2z = -1$

14) Write a system of equations in two variables that has $(5, -1)$ as its only solution.

Write a system of equations, and solve.

15) The area of Yellowstone National Park is about 1.1 million fewer acres than the area of Death Valley National Park. If they cover a total of 5.5 million acres, how big is each park? (www.nps.gov)

16) The Mahmood and Kuchar families take their kids to an amusement park. The Mahmoods buy one adult ticket and two children's tickets for $85.00. The Kuchars spend $150.00 on two adult and three children's tickets. How much did they pay for each type of ticket?

imac/Alamy Stock Photo

17) The width of a rectangle is half its length. Find the dimensions of the rectangle if the perimeter is 114 cm.

18) How many milliliters of a 12% alcohol solution and how many milliliters of a 30% alcohol solution must be mixed to obtain 72 mL of a 20% alcohol solution?

19) Rory and Lorelai leave Stars Hollow, Connecticut, and travel in opposite directions. Rory drives 4 mph faster than Lorelai, and after 1.5 hr they are 120 miles apart. How fast was each driving?

20) The measure of the smallest angle of a triangle is 9° less than the measure of the middle angle. The largest angle is 30° more than the sum of the two smaller angles. Find the measures of the angles of the triangle.

Solve using Gaussian elimination.

21) $x + 5y = -4$
$3x + 2y = 14$

22) $-3x + 5y + 8z = 0$
$x - 3y + 4z = 8$
$2x - 4y - 3z = 3$

Chapter 5: Cumulative Review for Chapters 1–5

Perform the operations and simplify.

1) $\dfrac{7}{15} + \dfrac{9}{10}$

2) $4\dfrac{1}{5} \div \dfrac{9}{20}$

3) $3(5 - 7)^3 + 18 \div 6 - 8$

4) Find the area of the triangle.

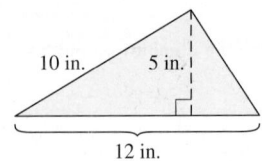

10 in.　5 in.

12 in.

5) Simplify $-3(4x^2 + 5x - 1)$.

6) Find the measure of each indicated angle.

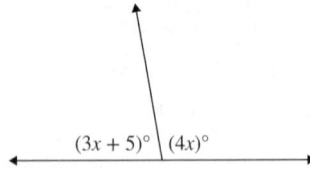

$(3x + 5)°$　$(4x)°$

Solve each equation.

7) $y + 9 = 34$

8) $-8 = -\dfrac{w}{6}$

9) $16 = -6m + 3$

10) $0.04(3p - 2) - 0.02p = 0.1(p + 3)$

11) $11 - 3(2k - 1) = 2(6 - k)$

12) Solve. Write the answer in interval notation.

$$-5 < 4v - 9 < 15$$

13) Write an equation and solve.

In 2018, the average gas mileage for new vehicles in the United States was 15% more than for new cars in 2011. If the average gas mileage for new cars in 2018 was 25.1 mpg, what was the average gas mileage of a new car in 2011? (https://consumerfed.org)

Thinkstock/Jupiterimages

14) The area, A, of a trapezoid is $A = \dfrac{1}{2}h(b_1 + b_2)$, where

h = height of the trapezoid, b_1 = length of one base of the trapezoid, and b_2 = length of the second base of the trapezoid.

a) Solve the equation for h.

b) Find the height of the trapezoid that has an area of 39 cm² and bases of length 8 cm and 5 cm.

15) Graph $2x - 3y = 9$.

16) Find the *x*- and *y*-intercepts of the graph of $x - 8y = 16$.

17) Write the slope-intercept form of the equation of the line containing $(3, 2)$ and $(-9, -1)$.

18) Determine whether the lines are parallel, perpendicular, or neither.
$$10x + 18y = 9$$
$$9x - 5y = 17$$

Solve each system of equations.

19) $9x - 3y = 6$
$3x - 2y = -8$

20) $3(2x - 1) - (y + 10) = 2(2x - 3) - 2y$
$3x + 13 = 4x - 5(y - 3)$

21) $x + 2y = 4$
$-3x - 6y = 6$

22) $-\dfrac{1}{4}x - \dfrac{3}{4}y = \dfrac{1}{6}$
$\dfrac{1}{2}x + \dfrac{3}{2}y = -\dfrac{1}{3}$

23) $y = 4x + 1$
$2x - y = 3$

Write a system of equations, and solve.

24) Dhaval used twice as many 6-foot boards as 4-foot boards when he made a treehouse for his children. If he used a total of 48 boards, how many of each size did he use?

Sandra Ivany/BrandX/ Getty Images

25) Pilar loves to watch animal videos on You Tube. Her favorite puppy video has twice as many views as her favorite pig video. Her favorite kitty video has 1.6 million more views than her favorite puppy video. All three videos have a total of 7.1 million views. How many views has each video had?

Polynomials

Get Ready

Here is a quick review of some topics we need to know for Chapter 6.

1) **Like terms** contain the same variables with the same exponents. We can **add and subtract,** or **combine,** only those terms that are *like* terms.

Example: Combine like terms to simplify $7x^2 + 10x - 4 + 2x^2 - 5x + 1$.

$$7x^2 + 10x - 4 + 2x^2 - 5x + 1 = 7x^2 + 2x^2 + 10x - 5x - 4 + 1$$

Rewrite like terms together.

$$= 9x^2 + 5x - 3$$ Simplify.

2) The **rules of exponents** help us simplify some expressions.

Examples: Simplify using the rules of exponents. Assume all variables represent nonzero real numbers. The answer should contain only positive exponents.

a) $y^4 \cdot y^3 = y^{4+3} = y^7$ b) $(a^2)^6 = a^{2 \cdot 6} = a^{12}$ c) $(2p)^5 = 2^5 \cdot p^5 = 32p^5$

d) $\left(\dfrac{n}{5}\right)^3 = \dfrac{n^3}{5^3} = \dfrac{n^3}{125}$ e) $8^0 = 1$ f) $\dfrac{x^9}{x^7} = x^{9-7} = x^2$

g) $w^{-6} = \dfrac{1}{w^6}$ h) $\dfrac{c^{-7}}{d^{-1}} = \dfrac{d}{c^7}$ i) $\left(\dfrac{m}{3}\right)^{-4} = \left(\dfrac{3}{m}\right)^4 = \dfrac{3^4}{m^4} = \dfrac{81}{m^4}$

In this chapter, use the following *Basic Skills Worksheets* to prepare students for this, and future, chapters: WS2 Powers (to prepare for operations with polynomials) and WS3 Prefactoring (so that students can practice the mental arithmetic skills needed to factor polynomials in Chapter 7).

Get Ready Exercises

1) Identify the pairs of like terms from each group of terms.

   a) $9n, -4n^2, n, 8n^3, 5n^2, n^3$

   b) $-2hk^2, 6h^2k, 3h^2k^2, h^2k, -h^2k^2, 6hk^2$

Combine like terms to simplify each expression.

2) $8z - 11 + 2z + 4$

3) $2x + 5y + 3 + 9x - y - 10$

4) $c^2 - 9c + 7 + 3c^2 - c - 7$

5) $-3t^3 + 4t^2 - 5 + 4t^3 - t^2 - 6 + 2t^2 + 9$

6) $a^2b^2 - 6a^2b + 8ab + 5a^2b^2 + a^2b - 3ab^2$

Simplify using the rules of exponents. Assume all variables represent nonzero real numbers. The answer should contain only positive exponents.

7) $\dfrac{m^8}{m^3}$

8) $(w^7)^4$

9) $h^5 \cdot h^6$

10) z^{-9}

11) $\left(\dfrac{2}{c}\right)^6$

12) $(-4)^0$

13) $\dfrac{b^{-5}}{a^{-2}}$

14) $(3y)^3$

15) $\left(\dfrac{k}{7}\right)^{-2}$

16) $-6t^4 \cdot 8t^{-1}$

17) $\dfrac{x^4}{x^{10}}$

18) $(9n^8)^2$

 **Study Strategies** Doing Math Homework

Whether homework counts as part of your grade or not, doing homework on a regular basis is essential to learning math and to being successful in a math course. And you should not just do your homework to get it done; you should do it in a way that helps you truly learn the material. We will use the steps of P.O.W.E.R. to learn how to *effectively and efficiently* do math homework.

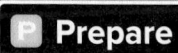

- **I will do my homework effectively and efficiently.**

- Complete the emPOWERme survey that appears before the Chapter Summary to learn about some of your homework habits.
- Realize that you should be doing the homework to *truly learn the material* and not just to get it done.
- Determine what you will need to do your homework: your class notes, book, pencils, pens, paper, a calculator, and computer, and have those materials with you.
- Think about *how* you learn best: Is it in a completely quiet room? Is it in a room with ambient noise, like a coffee shop or a tutoring center? Is it while listening to music? Choose that environment for doing your homework.
- Study in a place with a table that has enough space for you to open your book and notes *in addition to* having enough space to write.
- Set aside enough time to do your homework, and do it soon after it is assigned so that you do not forget what you learned in class. At the very least, do the 24-hr problems noted with this icon ⏰ in the exercise set within 24 hours of leaving class if you must do the rest of the homework at a later time.

- It's time to do your homework. Sit at a table and spread out everything you need.
- Turn off your cell phone and put it away to minimize distractions.
- Read over your notes and examples in the book to remind you of what you learned in class. *Then*, start the homework your instructor has assigned, write neatly, and show all of your work.

- Do your homework in a designated math notebook.
- If you cannot use a calculator on your quizzes and exams, do not use one while doing your homework. The same is true for homework websites. Do not use them just to complete the homework. Remember, the purpose of doing homework is for learning the math!
- Circle any problems that you could not do, and ask about them in class.

- After going over the homework in class, how did you do? Do you have a true understanding of the concepts in the assignment?
- Make corrections and notes on your paper for any problems that you did wrong. Be sure you understand your mistakes and the correct way to do the problems.

- Keep all of your corrected homework in the same place so that you can refer to it later when studying for a quiz or exam.
- Even after making corrections, do you still have questions about anything in the assignment? If so, see your instructor during office hours or go to the tutoring center.
- If you feel like you had a very good understanding of how to do the exercises in the homework, think about what you did well to learn the material. Would you use the same approach next time? What might you do differently?
- If you had difficulty completing the homework, ask yourself why. Were your notes difficult to understand? Did you wait too long before you started to do your homework? Think about adjustments that need to be made.

Dean Drobot/Shutterstock

6.1 Review of the Rules of Exponents

What is your objective for Section 6.1?	**How can you accomplish the objective?**
1 Review the Rules of Exponents	• Understand and apply the **Rules of Exponents** by writing a summary in your notes using your own examples. • Complete the given example on your own. • Complete You Try 1.

W Work   **Read the explanations, follow the examples, take notes, and complete the You Try.**

In Chapter 2, we learned the rules of exponents. We will review them here to prepare us for the topics in the rest of this chapter—adding, subtracting, multiplying, and dividing polynomials.

Rules of Exponents

For real numbers a and b and integers m and n, the following rules apply:

Summary The Rules of Exponents

Rule	Example
Product Rule: $a^m \cdot a^n = a^{m+n}$	$y^6 \cdot y^9 = y^{6+9} = y^{15}$
Basic Power Rule: $(a^m)^n = a^{mn}$	$(k^4)^7 = k^{28}$
Power Rule for a Product: $(ab)^n = a^n b^n$	$(9t)^2 = 9^2 t^2 = 81t^2$
Power Rule for a Quotient:	
$\left(\dfrac{a}{b}\right)^n = \dfrac{a^n}{b^n}$, where $b \neq 0$.	$\left(\dfrac{2}{r}\right)^5 = \dfrac{2^5}{r^5} = \dfrac{32}{r^5}$
Zero Exponent: If $a \neq 0$, then $a^0 = 1$.	$(7)^0 = 1$
Negative Exponent:	
For $a \neq 0$, $a^{-n} = \left(\dfrac{1}{a}\right)^n = \dfrac{1}{a^n}$.	$\left(\dfrac{3}{4}\right)^{-3} = \left(\dfrac{4}{3}\right)^3 = \dfrac{4^3}{3^3} = \dfrac{64}{27}$
If $a \neq 0$ and $b \neq 0$, then $\dfrac{a^{-m}}{b^{-n}} = \dfrac{b^n}{a^m}$;	$\dfrac{x^{-6}}{y^{-3}} = \dfrac{y^3}{x^6}$
also, $\left(\dfrac{a}{b}\right)^{-m} = \left(\dfrac{b}{a}\right)^m$.	$\left(\dfrac{8c}{d}\right)^{-2} = \left(\dfrac{d}{8c}\right)^2 = \dfrac{d^2}{64c^2}$
Quotient Rule: If $a \neq 0$, then $\dfrac{a^m}{a^n} = a^{m-n}$.	$\dfrac{2^9}{2^4} = 2^{9-4} = 2^5 = 32$

W Hint

How does this summary compare to the summary you made in your notes for Chapter 2?

When we use the rules of exponents to simplify an expression, we must remember to use the order of operations.

EXAMPLE 1

Simplify. Assume all variables represent nonzero real numbers. The answer should contain only positive exponents.

a) $(7k^{10})(-2k)$

b) $\dfrac{(-4)^5 \cdot (-4)^2}{(-4)^4}$

c) $\dfrac{10x^5 y^{-3}}{2x^2 y^5}$

d) $\left(\dfrac{c^2}{2d^4}\right)^{-5}$

e) $4(3p^5 q)^2$

Solution

a) $(7k^{10})(-2k) = -14k^{10+1}$ Multiply coefficients and add the exponents.

                 $= -14k^{11}$ Simplify.

b) $\dfrac{(-4)^5 \cdot (-4)^2}{(-4)^4} = \dfrac{(-4)^{5+2}}{(-4)^4} = \dfrac{(-4)^7}{(-4)^4}$ Product rule—the bases in the numerator are the same, so add the exponents.

                      $= (-4)^{7-4}$ Quotient rule

                      $= (-4)^3$

                      $= -64$ Evaluate.

W Hint

If you need more detailed explanations, go back and review Chapter 2.

c) $\dfrac{10x^5y^{-3}}{2x^2y^5} = 5x^{5-2}y^{-3-5}$ Divide coefficients and subtract the exponents.

$= 5x^3y^{-8}$ Simplify.

$= \dfrac{5x^3}{y^8}$ Write the answer with only positive exponents.

d) $\left(\dfrac{c^2}{2d^4}\right)^{-5} = \left(\dfrac{2d^4}{c^2}\right)^5$ Take the reciprocal of the base and make the exponent positive.

$= \dfrac{2^5d^{20}}{c^{10}}$ Power rule

$= \dfrac{32d^{20}}{c^{10}}$ Simplify.

e)   In this expression, a quantity is raised to a power and that quantity is being multiplied by 4. Remember what the order of operations says: Perform exponents before multiplication.

$4(3p^5q)^2 = 4(9p^{10}q^2)$ Apply the power rule *before* multiplying factors.

$= 36p^{10}q^2$ Multiply.

[YOU TRY 1] Simplify. Assume all variables represent nonzero real numbers. The answer should contain only positive exponents.

a) $(-6u^2)(-4u^3)$

b) $\dfrac{8^3 \cdot 8^4}{8^5}$

c) $\dfrac{8n^9}{12n^5}$

d) $(3y^{-9})^2(2y^7)$

e) $\left(\dfrac{3a^3b^{-4}}{2ab^6}\right)^{-4}$

ANSWERS TO [YOU TRY] EXERCISE

1) a) $24u^5$ b) 64 c) $\dfrac{2n^4}{3}$ d) $\dfrac{18}{y^{11}}$ e) $\dfrac{16b^{40}}{81a^8}$

E Evaluate **6.1** Exercises   Do the exercises, and check your work.

Objective 1: Review the Rules of Exponents

State which exponent rule must be used to simplify each exercise. Then simplify.

 1) $\dfrac{k^{10}}{k^4}$

2) $p^5 \cdot p^2$

3) $(2h)^4$

4) $\left(\dfrac{5}{w}\right)^3$

Evaluate using the rules of exponents.

5) $2^2 \cdot 2^4$

6) $(-3)^2 \cdot (-3)$

7) $\dfrac{(-4)^8}{(-4)^5}$

8) $\dfrac{2^{10}}{2^6}$

9) 6^{-1}

10) $(12)^{-2}$

11) $\left(\dfrac{1}{9}\right)^{-2}$

12) $\left(-\dfrac{1}{5}\right)^{-3}$

(24) 13) $\left(\dfrac{3}{2}\right)^{-4}$

14) $\left(\dfrac{7}{9}\right)^{-2}$

15) $6^0 + \left(-\dfrac{1}{2}\right)^{-5}$

16) $\left(\dfrac{1}{4}\right)^{-2} + \left(\dfrac{1}{4}\right)^0$

17) $\dfrac{8^5}{8^7}$

18) $\dfrac{2^7}{2^{12}}$

Simplify. Assume all variables represent nonzero real numbers. The answer should not contain negative exponents.

19) $t^5 \cdot t^8$

20) $n^{10} \cdot n^6$

21) $(-8c^4)(2c^5)$

22) $(3w^9)(-7w)$

23) $(z^6)^4$

24) $(y^3)^2$

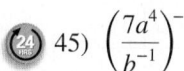 25) $(5p^{10})^3$

26) $(-6m^4)^2$

27) $\left(-\dfrac{2}{3}a^7b\right)^3$

28) $\left(\dfrac{7}{10}r^2s^5\right)^2$

29) $\dfrac{f^{11}}{f^7}$

30) $\dfrac{u^9}{u^8}$

31) $\dfrac{35v^9}{5v^8}$

32) $\dfrac{36k^8}{12k^5}$

33) $\dfrac{9d^{10}}{54d^6}$

34) $\dfrac{7m^4}{56m^2}$

35) $\dfrac{x^3}{x^9}$

36) $\dfrac{v^2}{v^5}$

37) $\dfrac{m^2}{m^3}$

38) $\dfrac{t^3}{t^3}$

39) $\dfrac{45k^{-2}}{30k^2}$

40) $\dfrac{22n^{-9}}{55n^{-3}}$

41) $5(2m^4n^7)^2$

42) $2(-3a^8b)^3$

43) $(6y^2)(2y^3)^2$

44) $(-c^4)(5c^9)^3$

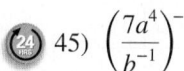 45) $\left(\dfrac{7a^4}{b^{-1}}\right)^{-2}$

46) $\left(\dfrac{3t^{-3}}{2u}\right)^{-4}$

47) $\dfrac{a^{-12}b^7}{a^{-9}b^2}$

48) $\dfrac{mn^{-4}}{m^9n^7}$

49) $\dfrac{(x^2y^{-3})^4}{x^5y^8}$

50) $\dfrac{10r^{-6}t}{(4r^{-5}t^4)^3}$

51) $\dfrac{12a^6bc^{-9}}{(3a^2b^{-7}c^4)^2}$

52) $\dfrac{(-7k^2m^{-3}n^{-1})^2}{14km^{-2}n^2}$

53) $(xy^{-3})^{-5}$

54) $-(s^{-6}t^2)^{-4}$

55) $\left(\dfrac{a^2b}{4c^2}\right)^{-3}$

56) $\left(\dfrac{2s^3}{rt^4}\right)^{-5}$

57) $\left(\dfrac{7h^{-1}k^9}{21h^{-5}k^5}\right)^{-2}$

58) $\left(\dfrac{24m^8n^{-3}}{16mn}\right)^{-3}$

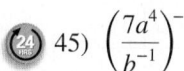 59) $\left(\dfrac{15cd^{-4}}{5c^3d^{-10}}\right)^{-3}$

60) $\left(\dfrac{10x^{-5}y}{20x^5y^{-3}}\right)^{-2}$

61) $\dfrac{(2u^{-5}v^2w^4)^{-5}}{(u^6v^{-7}w^{-10})^2}$

62) $\dfrac{(a^{-10}b^{-5}c^2)^4}{6(a^9bc^{-4})^{-2}}$

Write expressions for the area and perimeter for each rectangle.

63)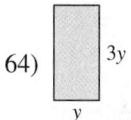
5x, 2x

64) 3y, y

65) $\frac{3}{4}p$, $\frac{1}{4}p$

66)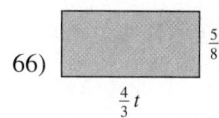
$\frac{4}{3}t$, $\frac{5}{8}t$

Simplify. Assume that the variables represent nonzero integers.

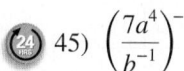 67) $k^{4a} \cdot k^{2a}$

68) $r^{9y} \cdot r^y$

69) $(g^{2x})^4$

70) $(t^{5c})^3$

71) $\dfrac{x^{7b}}{x^{4b}}$

72) $\dfrac{m^{10u}}{m^{3u}}$

73) $(2r^{6m})^{-3}$

74) $(5a^{-2x})^{-2}$

R Rethink

R1) How well did you remember the rules from Chapter 2?

R2) Why do you think you needed to review the rules of exponents before learning about polynomials, the main topic of the chapter?

6.2 Addition and Subtraction of Polynomials

P Prepare

What are your objectives for Section 6.2?

O Organize

How can you accomplish each objective?

1 Learn the Vocabulary Associated with Polynomials	• Write the definition of *polynomial* in your own words. • Understand the defining features of a polynomial, such as descending powers of *x*, constant, degree of a term, and degree of a polynomial. • Be able to compare and contrast a *monomial*, a *binomial*, and a *trinomial*. • Complete the given example on your own. • Complete You Try 1.
2 Evaluate Polynomials	• Complete the given example on your own. • Complete You Try 2.
3 Add Polynomials	• Learn the procedure for **Adding Polynomials,** and write an example in your notes. • Complete the given examples on your own. • Complete You Try 3.
4 Subtract Polynomials	• Learn the procedure for **Subtracting Polynomials,** and write an example in your notes. • Complete the given examples on your own. • Complete You Try 4.
5 Add and Subtract Polynomials in More Than One Variable	• Know how to use the previous procedures, and line up the like terms. • Complete the given example on your own. • Complete You Try 5.
6 Define and Evaluate a Polynomial Function	• Learn the definition of *polynomial function*. • Use parentheses when evaluating a polynomial function. • Complete the given example on your own. • Complete You Try 6.

W Work

Read the explanations, follow the examples, take notes, and complete the You Trys.

1 Learn the Vocabulary Associated with Polynomials

In Section 1.7, we defined an *algebraic expression* as a collection of numbers, variables, and grouping symbols connected by operation symbols such as $+$, $-$, $\times$, and $\div$.

An example of an algebraic expression is

$$5x^3 + \frac{7}{4}x^2 - x + 9$$

The *terms* of this algebraic expression are $5x^3, \frac{7}{4}x^2, -x$, and 9. A *term* is a number or a variable or a product or quotient of numbers and variables.

The expression $5x^3 + \frac{7}{4}x^2 - x + 9$ is also a *polynomial*.

Definition

A **polynomial in x** is the sum of a finite number of terms of the form ax^n, where n is a whole number and a is a real number. (The exponents must be whole numbers.)

Let's look more closely at the polynomial $5x^3 + \frac{7}{4}x^2 - x + 9$.

1) The polynomial is written **in descending powers of x** since the powers of x decrease from left to right. Generally, we write polynomials in descending powers of the variable.

2) Recall that the term without a variable is called a **constant.** The constant is 9. The **degree of a term** equals the exponent on its variable. (If a term has more than one variable, the degree equals the *sum* of the exponents on the variables.) We will list each term, its coefficient, and its degree.

Term	Coefficient	Degree
$5x^3$	5	3
$\frac{7}{4}x^2$	$\frac{7}{4}$	2
$-x$	-1	1
9	9	0

$(9 = 9x^0)$

3) The **degree of the polynomial** equals the highest degree of any nonzero term. The degree of $5x^3 + \frac{7}{4}x^2 - x + 9$ is 3. Or, we say that this is a **third-degree polynomial.**

EXAMPLE 1

Decide whether each expression *is* or *is not* a polynomial. If it is a polynomial, identify each term and the degree of each term. Then, find the degree of the polynomial.

a) $-8p^4 + 5.7p^3 - 9p^2 - 13$ b) $4c^2 - \frac{2}{5}c + 3 + \frac{6}{c^2}$

c) $a^3b^3 + 4a^3b^2 - ab + 1$ d) $7n^6$

Solution

a) The expression $-8p^4 + 5.7p^3 - 9p^2 - 13$ *is* a polynomial in p. Its terms have whole-number exponents and real coefficients. The term with the highest degree is $-8p^4$, so the degree of the polynomial is 4.

Term	Degree
$-8p^4$	4
$5.7p^3$	3
$-9p^2$	2
-13	0

b) The expression $4c^2 - \frac{2}{5}c + 3 + \frac{6}{c^2}$ is *not* a polynomial because one of its terms has a variable in the denominator.

$\left(\frac{6}{c^2} = 6c^{-2}; \text{ the exponent } -2 \text{ is not a whole number.} \right)$

c) The expression $a^3b^3 + 4a^3b^2 - ab + 1$ is a polynomial because the variables have whole-number exponents and the coefficients are real numbers. Since this is a polynomial in two variables, we find the degree of each term by adding the exponents. The first term, a^3b^3, has the highest degree, 6, so the polynomial has degree 6.

Term	Degree
a^3b^3	6
$4a^3b^2$	5
$-ab$	2
1	0

Add the exponents to get the degree.

d) The expression $7n^6$ *is* a polynomial even though it has only one term. The degree of the term is 6, and that is the degree of the polynomial as well.

$\left[\text{YOU TRY 1}\right]$ Decide whether each expression *is* or *is not* a polynomial. If it is a polynomial, identify each term and the degree of each term. Then, find the degree of the polynomial.

a) $d^4 + 7d^3 + \dfrac{3}{d}$

b) $k^3 - k^2 - 3.8k + 10$

c) $5x^2y^2 + \dfrac{1}{2}xy - 6x - 1$

d) $2r + 3r^{1/2} - 7$

The polynomial in Example 1d) is $7n^6$ and has one term. We call $7n^6$ a *monomial*. A **monomial** is a polynomial that consists of one term (*mono* means one). Some other examples of monomials are

$$y^2, \qquad -4t^5, \qquad x, \qquad m^2n^2, \qquad \text{and} \qquad -3$$

A **binomial** is a polynomial that consists of exactly two terms (*bi* means two). Some examples are

$$w + 2, \qquad 4z^2 - 11, \qquad a^4 - b^4, \qquad \text{and} \qquad -8c^3d^2 + 3cd$$

A **trinomial** is a polynomial that consists of exactly three terms (*tri* means three). Here are some examples:

$$x^2 - 3x - 40, \qquad 2q^4 - 18q^2 + 10q, \qquad \text{and} \qquad 6a^4 + 29a^2b + 28b^2$$

Expressions have different values depending on the value of the variable(s). The same is true for polynomials.

2 Evaluate Polynomials

EXAMPLE 2

Evaluate the trinomial $n^2 - 7n + 4$ when

a) $n = 3$

b) $n = -2$

Solution

a) Substitute 3 for n in $n^2 - 7n + 4$. Remember to put 3 in parentheses.

$$n^2 - 7n + 4 = (3)^2 - 7(3) + 4 \qquad \text{Substitute.}$$
$$= 9 - 21 + 4$$
$$= -8 \qquad \text{Add.}$$

W Hint

Remember that *evaluate* means *find the value of the expression* for a given value of the variable. You are *not* solving for a variable.

b) Substitute -2 for n in $n^2 - 7n + 4$. Put -2 in parentheses.

$$n^2 - 7n + 4 = (-2)^2 - 7(-2) + 4 \qquad \text{Substitute.}$$
$$= 4 + 14 + 4$$
$$= 22 \qquad \text{Add.}$$

[YOU TRY 2]

Evaluate $t^2 - 9t - 6$ when

a) $t = 5$ b) $t = -4$

Recall in Section 1.7 we said that **like terms** contain the same variables with the same exponents. We add or subtract like terms by adding or subtracting the coefficients and leaving the variable(s) and exponent(s) the same. We use the same idea for adding and subtracting polynomials.

3 Add Polynomials

Procedure Adding Polynomials

To add polynomials, add like terms.

We can set up an addition problem vertically or horizontally.

EXAMPLE 3

Add $(m^3 - 9m^2 + 5m - 4) + (2m^3 + 3m^2 - 1)$.

Solution

We will add these vertically. Line up like terms in columns and add.

$$\begin{array}{r} m^3 - 9m^2 + 5m - 4 \\ + \; 2m^3 + 3m^2 \qquad\;\; - 1 \\ \hline 3m^3 - 6m^2 + 5m - 5 \end{array}$$

[YOU TRY 3]

Add $(6b^3 - 11b^2 + 3b + 3) + (-2b^3 - 6b^2 + b - 8)$.

EXAMPLE 4

Add $10k^2 + 2k - 1$ and $5k^2 + 7k + 9$.

Solution

Let's add these horizontally. Put the polynomials in parentheses since each contains more than one term. Use the associative and commutative properties to rewrite like terms together.

$$(10k^2 + 2k - 1) + (5k^2 + 7k + 9) = (10k^2 + 5k^2) + (2k + 7k) + (-1 + 9)$$
$$= 15k^2 + 9k + 8 \qquad \text{Combine like terms.}$$

W Hint

Remember that the exponents remain unchanged when we add like terms.

4 Subtract Polynomials

To subtract two polynomials such as $(8x + 3) - (5x - 4)$, we will be using the distributive property to clear the parentheses in the second polynomial.

EXAMPLE 5

Subtract $(8x + 3) - (5x - 4)$.

Solution

$$
\begin{aligned}
(8x + 3) - (5x - 4) &= (8x + 3) - 1(5x - 4) \\
&= (8x + 3) + (-1)(5x - 4) \qquad \text{Change } -1 \text{ to } + (-1). \\
&= (8x + 3) + (-5x + 4) \qquad \text{Distribute.} \\
&= 3x + 7 \qquad\qquad\qquad\quad \text{Combine like terms.}
\end{aligned}
$$

In Example 5, notice that we changed the sign of each term in the second polynomial and then added it to the first.

Procedure Subtracting Polynomials

To subtract two polynomials, change the sign of each term in the second polynomial. Then, add the polynomials.

Let's see how we use this rule to subtract polynomials.

EXAMPLE 6

Subtract $(-6w^3 - w^2 + 10w + 1) - (2w^3 - 4w^2 + 9w - 5)$ vertically.

Solution

To subtract vertically, line up like terms in columns.

$$
\begin{array}{r}
-6w^3 - w^2 + 10w + 1 \\
- \underline{(2w^3 - 4w^2 + 9w - 5)}
\end{array}
\qquad
\begin{array}{l}
\text{Change the signs in the} \\
\text{second polynomial, and} \\
\text{add the polynomials.}
\end{array}
\qquad
\begin{array}{r}
-6w^3 - w^2 + 10w + 1 \\
+ \underline{-2w^3 + 4w^2 - 9w + 5} \\
-8w^3 + 3w^2 + w + 6
\end{array}
$$

[YOU TRY 4] Subtract $(-7h^2 - 8h + 1) - (-3h^2 + h - 4)$.

5 Add and Subtract Polynomials in More Than One Variable

To add and subtract polynomials in more than one variable, remember that like terms contain the same variables with the same exponents.

EXAMPLE 7

Perform the indicated operation.

a) $(a^2b^2 + 2a^2b - 13ab - 4) + (9a^2b^2 - 5a^2b - ab + 17)$

b) $(6tu - t + 2u + 5) - (4tu + 8t - 2)$

Solution

a) $(a^2b^2 + 2a^2b - 13ab - 4) + (9a^2b^2 - 5a^2b - ab + 17)$
$= 10a^2b^2 - 3a^2b - 14ab + 13$ Combine like terms.

b) $(6tu - t + 2u + 5) - (4tu + 8t - 2) = (6tu - t + 2u + 5) - 4tu - 8t + 2$
$= 2tu - 9t + 2u + 7$ Combine like terms.

[YOU TRY 5]

Perform the indicated operation.

a) $(-12x^2y^2 + xy - 6y + 1) - (-4x^2y^2 - 10xy + 3y + 6)$

b) $(3.6m^3n^2 + 8.1mn - 10n) + (8.5m^3n^2 - 11.2mn + 4.3)$

6 Define and Evaluate a Polynomial Function

In Chapter 4 we learned about linear functions of the form $f(x) = mx + b$. A linear function is a special type of polynomial function, which we will study now.

Definition

A **polynomial function of degree n** is given by
$f(x) = a_nx^n + a_{n-1}x^{n-1} + \cdots + a_1x + a_0$, where $a_n, a_{n-1}, \cdots, a_1$, and a_0 are real numbers, $a_n \neq 0$, and n is a whole number.

Look at the polynomial $2x^2 - 5x + 7$. If we substitute 3 for x, the *only* value of the expression is 10:

$$2(3)^2 - 5(3) + 7 = 2(9) - 15 + 7 = 18 - 15 + 7 = 10$$

It is true that polynomials have different values depending on what value is substituted for the variable. It is also true that for any value we substitute for x in a polynomial like $2x^2 - 5x + 7$ there will be *only one value* of the expression. Since each value substituted for the variable produces *only one value* of the expression, we can use function notation to represent a polynomial like $2x^2 - 5x + 7$.

$f(x) = 2x^2 - 5x + 7$ is a *polynomial function* since $2x^2 - 5x + 7$ is a polynomial. Therefore, finding $f(3)$ when $f(x) = 2x^2 - 5x + 7$ is the same as evaluating $2x^2 - 5x + 7$ when $x = 3$.

EXAMPLE 6

If $f(x) = x^3 - 4x^2 + 3x + 1$, find $f(-2)$.

Solution

$$f(x) = x^3 - 4x^2 + 3x + 1$$
$$f(-2) = (-2)^3 - 4(-2)^2 + 3(-2) + 1 \quad \text{Substitute } -2 \text{ for } x.$$
$$f(-2) = -8 - 4(4) - 6 + 1$$
$$f(-2) = -8 - 16 - 6 + 1$$
$$f(-2) = -29$$

$\left[\text{YOU TRY 6} \right]$ If $g(t) = 2t^4 + t^3 - 7t^2 + 12$, find $g(-1)$.

ANSWERS TO $\boxed{\text{YOU TRY}}$ **EXERCISES**

1) a) not a polynomial   b) polynomial of degree 3   c) polynomial of degree 4

Term	Degree
k^3	3
$-k^2$	2
$-3.8k$	1
10	0

Term	Degree
$5x^2y^2$	4
$\frac{1}{2}xy$	2
$-6x$	1
-1	0

d) not a polynomial   2) a) -26 b) 46   3) $4b^3 - 17b^2 + 4b - 5$ 4) $-4h^2 - 9h + 5$
5) a) $-8x^2y^2 + 11xy - 9y - 5$ b) $12.1m^3n^2 - 3.1mn - 10n + 4.3$ 6) 6

$\boxed{\text{E Evaluate}}$ **6.2** Exercises   Do the exercises, and check your work.

Objective 1: Learn the Vocabulary Associated with Polynomials

Is the given expression a polynomial? Why or why not?

1) $-2p^2 - 5p + 6$

2) $8r^3 + 7r^2 - t + \frac{4}{5}$

3) $c^3 + 5c^2 + 4c^{-1} - 8$

4) $9a^5$

5) $f^{3/4} + 6f^{2/3} + 1$

6) $7y - 1 + \frac{3}{y}$

Determine whether each is a monomial, a binomial, or a trinomial.

7) $4x - 1$

8) $-5q^2$

9) $m^2n^2 - mn + 13$

10) $11c^2 + 3c$

11) 8

12) $k^5 + 2k^3 + 8k$

For Exercises 13 and 14, answer *always, sometimes,* or *never.*

13) If a, b, and c are any integer values except zero, then the expression $ax^2 + bx + c$ can *always, sometimes,* or *never* be classified as a trinomial.

14) If a, b, and c are all different integer values, then the expression $ax^2 + bx + c$ can *always, sometimes,* or *never* be classified as a binomial.

15) How do you determine the degree of a polynomial in one variable?

16) Write a third-degree polynomial in one variable.

17) How do you determine the degree of a term in a polynomial in more than one variable?

18) Write a fourth-degree monomial in x and y.

For each polynomial, identify each term in the polynomial, the coefficient and degree of each term, and the degree of the polynomial.

(24) 19) $3y^4 + 7y^3 - 2y + 8$ 20) $6a^2 + 2a - 11$

21) $-4x^2y^3 - x^2y^2 + \frac{2}{3}xy + 5y$

22) $3c^2d^2 + 0.7c^2d + cd - 1$

Objective 2: Evaluate Polynomials

Evaluate each polynomial when a) $r = 3$ and b) $r = -1$.

23) $2r^2 - 7r + 4$ 24) $2r^3 + 5r - 6$

Evaluate each polynomial when $x = 5$ and $y = -2$.

25) $9x + 4y$ 26) $-2x + 3y + 16$

(24) 27) $x^2y^2 - 5xy + 2y$ 28) $-2xy^2 + 7xy + 12y - 6$

29) $\frac{1}{2}xy - 4x - y$ 30) $x^2 - y^2$

For Exercises 31 and 32, answer *always, sometimes,* or *never.*

31) If x is any real number, a polynomial of the form $x^4 + x^2 + 1$ will *always, sometimes,* or *never* have a negative value.

32) If x is any real number, a polynomial of the form $x^3 + x^2 + x + 1$ will *always, sometimes,* or *never* have a negative value.

33) Bob will make a new gravel road from the highway to his house. The cost of building the road, y (in dollars), includes the cost of the gravel and is given by $y = 60x + 380$, where x is the number of hours he rents the equipment needed to complete the job.

a) Evaluate the binomial when $x = 5$, and explain what it means in the context of the problem.

b) If he keeps the equipment for 9 hours, how much will it cost to build the road?

c) If it cost \$860.00 to build the road, for how long did Bob rent the equipment?

34) An object is thrown upward so that its height, y (in feet), x seconds after being thrown is given by
$$y = -16x^2 + 48x + 64$$

a) Evaluate the polynomial when $x = 2$, and explain what it means in the context of the problem.

b) What is the height of the object 3 seconds after it is thrown?

c) Evaluate the polynomial when $x = 4$, and explain what it means in the context of the problem.

Objective 3: Add Polynomials
Add like terms.

35) $5c^2 + 9c - 16c^2 + c - 3c$

36) $-4y^3 + 3y^5 + 17y^5 + 6y^3 - 5y^5$

37) $6.7t^2 - 9.1t^6 - 2.5t^2 + 4.8t^6$

38) $\frac{5}{4}w^3 + \frac{3}{8}w^4 - \frac{2}{3}w^4 - \frac{5}{6}w^3$

39) $7a^2b^2 + 4ab^2 - 16ab^2 - a^2b^2 + 5ab^2$

40) $x^5y^2 - 14xy + 6xy + 5x^5y^2 + 8xy$

Add the polynomials.

41) $\quad 5n - 8$
$+ \underline{4n + 3}$

42) $\quad 9d + 14$
$+ \underline{2d + \ 5}$

43) $\quad 9r^2 + 16r + 2$
$+ \underline{3r^2 - 10r + 9}$

44) $\quad m^2 - 3m - 8$
$+ \underline{2m^2 + 7m + 1}$

45) $\quad b^2 - 8b - 14$
$+ \underline{3b^2 + 8b + 11}$

46) $\quad 8g^2 + \ g + 5$
$+ \underline{5g^2 - 6g - 5}$

47) $\quad \frac{5}{6}w^4 - \frac{2}{3}w^2 \qquad + \frac{1}{2}$
$+ \ \underline{-\frac{4}{9}w^4 + \frac{1}{6}w^2 - \frac{3}{8}w - 2}$

48) $\quad -1.7p^3 - 2p^2 + 3.8p - \ 6$
$+ \ \underline{\quad 6.2p^3 \qquad\quad - 1.2p + 14}$

49) $(6m^2 - 5m + 10) + (-4m^2 + 8m + 9)$

50) $(3t^4 - 2t^2 + 11) + (t^4 + t^2 - 7)$

51) $\left(-2c^4 - \frac{7}{10}c^3 + \frac{3}{4}c - \frac{2}{9}\right)$
$\quad + \left(12c^4 + \frac{1}{2}c^3 - c + 3\right)$

52) $\left(\frac{7}{4}y^3 - \frac{3}{8}\right) + \left(\frac{5}{6}y^3 + \frac{7}{6}y^2 - \frac{9}{16}\right)$

53) $(2.7d^3 + 5.6d^2 - 7d + 3.1)$
$\quad + (-1.5d^3 + 2.1d^2 - 4.3d - 2.5)$

54) $(0.2t^4 - 3.2t + 4.1)$
$\quad + (-2.7t^4 + 0.8t^3 - 6.4t + 3.9)$

Objective 4: Subtract Polynomials
Subtract the polynomials.

55) $\quad 15w + 7$
$- \underline{\ 3w + 11}$

56) $\quad 12a - 8$
$- \underline{\ 2a + 9}$

57) $\quad\ y - 6$
$- \underline{\ 2y - 8}$

58) $\quad 6p + \ 1$
$- \underline{9p - 17}$

59) $\quad 3b^2 - 8b + 12$
$- \underline{\ 5b^2 + 2b - \ 7}$

60) $\quad -7d^2 + 15d + 6$
$- \underline{\ 8d^2 + \ 3d - 9}$

61) $\quad\ f^4 - 6f^3 + 5f^2 - 8f + 13$
$- \underline{\ -3f^4 + 8f^3 - \ f^2 \qquad + \ 4}$

62) $\quad 11x^4 + x^3 - 9x^2 + 2x - 4$
$- \underline{\ -3x^4 + x^3 \qquad\quad - \ x + 1}$

63) $\quad 10.7r^2 + 1.2r + \ \ 9$
$- \underline{\ 4.9r^2 - 5.3r - 2.8}$

64) $\quad -\frac{11}{10}m^3 + \frac{1}{2}m - \frac{5}{8}$
$\quad - \ \underline{\ \frac{2}{5}m^3 + \frac{1}{7}m - \frac{5}{6}}$

65) $(j^2 + 16j) - (-6j^2 + 7j + 5)$

66) $(-3p^2 + p + 4) - (4p^2 + p + 1)$

67) $(17s^5 - 12s^2) - (9s^5 + 4s^4 - 8s^2 - 1)$

68) $(-5d^4 - 8d^2 + d + 3) - (-3d^4 + 17d^3 - 6d^2 - 20)$

69) $\left(-\frac{3}{8}r^2 + \frac{2}{9}r + \frac{1}{3}\right) - \left(-\frac{7}{16}r^2 - \frac{5}{9}r + \frac{7}{6}\right)$

70) $(3.8t^5 + 7.5t - 9.6) - (-1.5t^5 + 2.9t^2 - 1.1t + 3.4)$

71) Explain, in your own words, how to subtract two polynomials.

72) Do you prefer adding and subtracting polynomials vertically or horizontally? Why?

73) Will the sum of two trinomials always be a trinomial? Why or why not? Give an example.

74) Write a third-degree polynomial in x that does not contain a second-degree term.

Mixed Exercises: Objectives 3 and 4

Perform the indicated operations.

75) $(8a^4 - 9a^2 + 17) - (15a^4 + 3a^2 + 3)$

76) $(-11n^2 - 8n + 21) + (4n^2 + 15n - 3) + (7n^2 - 10)$

77) $(w^3 + 5w^2 + 3) - (6w^3 - 2w^2 + w + 12) + (9w^3 + 7)$

78) $(3r + 2) - (r^2 + 5r - 1) - (-9r^3 - r + 6)$

79) $\left(y^3 - \dfrac{3}{4}y^2 - 5y + \dfrac{3}{7}\right)$
$\quad + \left(\dfrac{1}{3}y^3 - y^2 + 8y - \dfrac{1}{2}\right)$

80) $\left(\dfrac{3}{5}c^4 + c^3 - \dfrac{3}{2}c^2 + 1\right)$
$\quad + \left(c^4 - 6c^3 - \dfrac{1}{4}c^2 + 6c - 1\right)$

81) $(3m^3 - 5m^2 + m + 12) - [(7m^3 + 4m^2 - m + 11)$
$\quad + (-5m^3 - 2m^2 + 6m + 8)]$

82) $(j^2 - 13j - 9) - [(-7j^2 + 10j - 2) + (4j^2 - 11j - 6)]$

Perform the indicated operations.

83) Find the sum of $p^2 - 7$ and $8p^2 + 2p - 1$.

84) Add $12n - 15$ to $5n + 4$.

85) Subtract $z^6 - 8z^2 + 13$ from $6z^6 + z^2 + 9$.

86) Subtract $-7x^2 + 8x + 2$ from $2x^2 + x$.

87) Subtract the sum of $6p^2 + 1$ and $3p^2 - 8p + 4$ from $2p^2 + p + 5$.

88) Subtract $17g^3 + 2g - 10$ from the sum of $5g^3 + g^2 + g$ and $3g^3 - 2g - 7$.

For Exercises 89 and 90, answer *always, sometimes,* or *never.*

89) Finding the difference of two second-degree trinomials will *always, sometimes,* or *never* result in a second-degree polynomial.

90) Finding the sum of two second-degree trinomials will *always, sometimes,* or *never* result in a polynomial having a degree greater than 2.

Objective 5: Add and Subtract Polynomials in More Than One Variable

Each of the polynomials is a polynomial in two variables. Perform the indicated operations.

91) $(5w + 17z) - (w + 3z)$

92) $(-4g - 7h) + (9g + h)$

93) $(ac + 8a + 6c) + (-6ac + 4a - c)$

94) $(11rt - 6r + 2) - (10rt - 7r + 12t + 2)$

95) $(-6u^2v^2 + 11uv + 14)$
$\quad - (-10u^2v^2 - 20uv + 18)$

96) $(-7j^2k^2 + 9j^2k - 23jk^2 + 13)$
$\quad + (10j^2k^2 + 5j^2k - 17)$

97) $(12x^3y^2 - 5x^2y^2 + 9x^2y - 17) + (5x^3y^2 + x^2y - 1)$
$\quad - (6x^2y^2 + 10x^2y + 2)$

98) $(r^3s^2 + 2r^2s^2 + 10) - (7r^3s^2 + 18r^2s^2 - 9)$
$\quad + (11r^3s^2 - 3r^2s^2 - 4)$

Find the polynomial that represents the perimeter of each rectangle.

99)
$2x + 7$

$x - 4$

100)
$a^2 + 3a - 4$

$a^2 - 5a + 1$

101)
$5p^2 - 2p + 3$

$p - 6$

102)
$\frac{2}{3}m + 4$

$\frac{2}{3}m + 4$

Objective 6: Define and Evaluate a Polynomial Function

103) If $f(x) = 5x^2 + 7x - 8$, find
a) $f(-3)$ b) $f(1)$

104) If $h(a) = -a^2 - 3a + 10$, find
a) $h(5)$ b) $h(-4)$

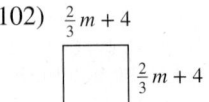

105) If $P(t) = t^3 - 3t^2 + 2t + 5$, find
a) $P(4)$ b) $P(0)$

106) If $G(c) = 3c^4 + c^2 - 9c - 4$, find
a) $G(0)$ b) $G(-1)$

107) If $f(x) = \dfrac{1}{3}x + 5$, find x so that $f(x) = 7$.

108) If $H(z) = -4z + 9$, find z so that $H(z) = 11$.

109) If $r(k) = \dfrac{2}{5}k - 3$, find k so that $r(k) = 13$.

110) If $Q(a) = 6a - 1$, find a so that $Q(a) = -9$.

111) According to Twitter statistics in 2018, the function $T(s) = 5800s$ approximates the number of tweets, $T(s)$, sent in s seconds. (www.disruptiveadvertising.com)
a) How many tweets are sent in 5 seconds?
b) How many tweets are sent in one hour?
c) How long would it take for 522,000 tweets to be sent?

112) The cost of using a car from a car sharing company can be described by the function $C(m) = 2.50m$, where m is the number of miles the car is driven and $C(m)$ is the cost, in dollars.

x9626/Shutterstock

a) How much does it cost to drive the car 8 miles?
b) How far did a customer drive a car if the cost was $107.50?

R Rethink

R1) What does it mean to evaluate a polynomial expression?

R2) What are the differences between a binomial, a trinomial, and a polynomial?

R3) Compare and contrast what you knew about adding like terms to adding and subtracting polynomials.

R4) Were there any problems you could not do? If so, write them down or circle them, then ask your instructor for help.

6.3 Multiplication of Polynomials

P Prepare

O Organize

What are your objectives for Section 6.3?	How can you accomplish each objective?
1 Multiply a Monomial and a Polynomial	• Know which property to use to complete this objective. • Complete the given example on your own. • Complete You Try 1.
2 Multiply Two Polynomials	• Follow the explanation to fully understand the procedure for **Multiplying Polynomials,** and write an example in your notes. • Complete the given examples on your own. • Complete You Try 2.
3 Multiply Two Binomials Using FOIL	• Know what FOIL means, and be able to use it to multiply two binomials. • Complete the given example on your own. • Complete You Try 3.
4 Find the Product of More Than Two Polynomials	• Multiply using both methods to decide which you like better. • Complete the given example on your own. • Complete You Try 4.
5 Find the Product of Binomials of the Form $(a + b)(a - b)$	• Follow the explanation to fully understand the procedure for **The Product of the Sum and Difference of Two Terms,** and write an example in your notes. • Complete the given example on your own. • Complete You Try 5.
6 Square a Binomial	• Follow the explanation to fully understand the procedure for **The Square of a Binomial,** and write an example in your notes. • Complete the given example on your own. • Complete You Try 6.
7 Multiply Other Binomials	• Understand what the word *expand* means. • Multiply using methods already learned in this chapter. • Complete the given example on your own. • Complete You Try 7.

We have already learned that when multiplying two monomials, we multiply the coefficients and add the exponents of the same bases:

$$4c^5 \cdot 3c^6 = 12c^{11} \qquad -3x^2y^4 \cdot 7xy^3 = -21x^3y^7$$

In this section, we will discuss how to multiply other types of polynomials.

1 Multiply a Monomial and a Polynomial

To multiply a monomial and a polynomial, we use the distributive property.

EXAMPLE 1

Multiply $2k^2(6k^2 + 5k - 3)$.

Solution

$$2k^2(6k^2 + 5k - 3) = (2k^2)(6k^2) + (2k^2)(5k) + (2k^2)(-3) \qquad \text{Distribute.}$$
$$= 12k^4 + 10k^3 - 6k^2 \qquad \text{Multiply.}$$

[YOU TRY 1] Multiply $5z^4(4z^3 - 7z^2 - z + 8)$.

2 Multiply Two Polynomials

To multiply two polynomials, we use the distributive property repeatedly. For example, to multiply $(2x - 3)(x^2 + 7x + 4)$, we multiply each term in the second polynomial by $(2x - 3)$.

$$(2x - 3)(x^2 + 7x + 4) = (2x - 3)(x^2) + (2x - 3)(7x) + (2x - 3)(4) \qquad \text{Distribute.}$$

Next, we distribute again.

W Hint

Count how many times the distributive property is used.

$$(2x - 3)(x^2) + (2x - 3)(7x) + (2x - 3)(4)$$
$$= (2x)(x^2) - (3)(x^2) + (2x)(7x) - (3)(7x) + (2x)(4) - (3)(4)$$
$$= 2x^3 - 3x^2 + 14x^2 - 21x + 8x - 12 \qquad \text{Multiply.}$$
$$= 2x^3 + 11x^2 - 13x - 12 \qquad \text{Combine like terms.}$$

This process of repeated distribution leads us to the following rule.

Procedure Multiplying Polynomials

To multiply two polynomials, multiply each term in the second polynomial by each term in the first polynomial. Then combine like terms. The answer should be written in descending powers.

Let's use this procedure to multiply the polynomials in Example 2.

EXAMPLE 2

Multiply $(n^2 + 5)(2n^3 + n - 9)$.

Solution

Multiply each term in the second polynomial by each term in the first.

$(n^2 + 5)(2n^3 + n - 9)$
$= (n^2)(2n^3) + (n^2)(n) + (n^2)(-9) + (5)(2n^3) + (5)(n) + (5)(-9)$ Distribute.
$= 2n^5 + n^3 - 9n^2 + 10n^3 + 5n - 45$ Multiply.
$= 2n^5 + 11n^3 - 9n^2 + 5n - 45$ Combine like terms.

Polynomials can be multiplied vertically as well.

EXAMPLE 3

Multiply $(a^3 - 4a^2 + 5a - 1)(3a + 7)$ vertically.

Solution

Set up the multiplication problem like you would for whole numbers:

$$
\begin{array}{r}
a^3 - 4a^2 + 5a - 1 \\
\times \qquad\qquad 3a + 7 \\
\hline
7a^3 - 28a^2 + 35a - 7 \\
3a^4 - 12a^3 + 15a^2 - 3a \qquad\quad \\
\hline
3a^4 - 5a^3 - 13a^2 + 32a - 7
\end{array}
$$

Multiply each term in $a^3 - 4a^2 + 5a - 1$ by 7.
Multiply each term in $a^3 - 4a^2 + 5a - 1$ by $3a$.
Line up like terms in the same column. Add.

[YOU TRY 2]

Multiply.

a) $(9x + 5)(2x^2 - x - 3)$ b) $\left(t^2 - \dfrac{2}{3}t - 4\right)(4t^2 + 6t - 5)$

3   Multiply Two Binomials Using FOIL

Multiplying two binomials is one of the most common types of polynomial multiplication used in algebra. A method called **FOIL** is one that is often used to multiply two binomials, and it comes from using the distributive property.

Let's use the distributive property to multiply $(x + 6)(x + 4)$.

$(x + 6)(x + 4) = (x + 6)(x) + (x + 6)(4)$ Distribute.
$= x(x) + 6(x) + x(4) + 6(4)$ Distribute.
$= x^2 + 6x + 4x + 24$ Multiply.
$= x^2 + 10x + 24$ Combine like terms.

To be sure that each term in the first binomial has been multiplied by each term in the second binomial, we can use FOIL. **FOIL** stands for **F**irst **O**uter **I**nner **L**ast. Let's see how we can apply FOIL to the example above:

$$
(x + 6)(x + 4) = (x + 6)(x + 4) = \overset{F}{x \cdot x} + \overset{O}{x \cdot 4} + \overset{I}{6 \cdot x} + \overset{L}{6 \cdot 4}
$$
$= x^2 + 4x + 6x + 24$ Multiply.
$= x^2 + 10x + 24$ Combine like terms.

First   Last   Inner   Outer

You can see that we get the same result.

EXAMPLE 4

Use FOIL to multiply the binomials.

a) $(p + 5)(p - 2)$ b) $(4r - 3)(r - 1)$

c) $(a + 4b)(a - 3b)$ d) $(2x + 9)(3y + 5)$

Solution

a) $(p + 5)(p - 2) = (p + 5)(p - 2) = p(p) + p(-2) + 5(p) + 5(-2)$ Use FOIL.

$\qquad\qquad\qquad\qquad\qquad = p^2 - 2p + 5p - 10$ Multiply.

$\qquad\qquad\qquad\qquad\qquad = p^2 + 3p - 10$ Combine like terms.

Notice that the middle terms are like terms, so we can combine them.

W Hint

Work out the example on your paper as you are reading it.

b) $(4r - 3)(r - 1) = (4r - 3)(r - 1) = 4r(r) + 4r(-1) - 3(r) - 3(-1)$ Use FOIL.

$\qquad\qquad\qquad\qquad\qquad = 4r^2 - 4r - 3r + 3$ Multiply.

$\qquad\qquad\qquad\qquad\qquad = 4r^2 - 7r + 3$ Combine like terms.

The middle terms are like terms, so we can combine them.

c) $(a + 4b)(a - 3b) = a(a) + a(-3b) + 4b(a) + 4b(-3b)$ Use FOIL.

$\qquad\qquad\qquad\quad = a^2 - 3ab + 4ab - 12b^2$ Multiply.

$\qquad\qquad\qquad\quad = a^2 + ab - 12b^2$ Combine like terms.

As in parts a) and b), we combined the middle terms.

d) $(2x + 9)(3y + 5) = 2x(3y) + 2x(5) + 9(3y) + 9(5)$ Use FOIL.

$\qquad\qquad\qquad\quad = 6xy + 10x + 27y + 45$ Multiply.

In this case, the middle terms were not like terms, so we could not combine them.

[YOU TRY 3]

Use FOIL to multiply the binomials.

a) $(n + 8)(n + 5)$ b) $(3k + 7)(k - 4)$

c) $(x - 2y)(x - 6y)$ d) $(5c - 8)(2d + 1)$

With practice, you should get to the point where you can find the product of two binomials "in your head." Remember that, as in the case of parts a)–c) in Example 4, it is often possible to combine the middle terms.

4 Find the Product of More Than Two Polynomials

EXAMPLE 5

Multiply $5d^2(4d - 3)(2d - 1)$.

Solution

We can approach this problem a couple of ways.

Method 1

Begin by multiplying the binomials, *then* multiply by the monomial.

$5d^2(4d - 3)(2d - 1) = 5d^2(8d^2 - 4d - 6d + 3)$ Use FOIL to multiply the binomials.

$\qquad\qquad\qquad\quad = 5d^2(8d^2 - 10d + 3)$ Combine like terms.

$\qquad\qquad\qquad\quad = 40d^4 - 50d^3 + 15d^2$ Distribute.

Begin by multiplying $5d^2$ by $(4d - 3)$, then multiply *that* product by $(2d - 1)$.

$$5d^2(4d - 3)(2d - 1) = (20d^3 - 15d^2)(2d - 1) \quad \text{Multiply } 5d^2 \text{ by } (4d - 3).$$
$$= 40d^4 - 20d^3 - 30d^3 + 15d^2 \quad \text{Use FOIL to multiply.}$$
$$= 40d^4 - 50d^3 + 15d^2 \quad \text{Combine like terms.}$$

The result is the same. These may be multiplied by whichever method you prefer.

[YOU TRY 4] Multiply $-6x^3(x + 5)(3x - 4)$.

There are special types of binomial products that come up often in algebra. We will look at these next.

5 Find the Product of Binomials of the Form $(a + b)(a - b)$

Let's find the product $(y + 6)(y - 6)$. Using FOIL, we get

$$(y + 6)(y - 6) = y^2 - 6y + 6y - 36$$
$$= y^2 - 36$$

Notice that the middle terms, the y-terms, drop out. In the result, $y^2 - 36$, the first term (y^2) is the square of y and the last term (36) is the square of 6. The resulting polynomial is a *difference of squares*. This pattern always holds when multiplying two binomials of the form $(a + b)(a - b)$.

Procedure The Product of the Sum and Difference of Two Terms

$$(a + b)(a - b) = a^2 - b^2$$

EXAMPLE 6 Multiply.

a) $(z + 9)(z - 9)$ b) $(2 + c)(2 - c)$

c) $(5n^2 - 8)(5n^2 + 8)$ d) $\left(\dfrac{3}{4}t + u\right)\left(\dfrac{3}{4}t - u\right)$

Solution

a) The product $(z + 9)(z - 9)$ is in the form $(a + b)(a - b)$, where $a = z$ and $b = 9$. Use the rule that says $(a + b)(a - b) = a^2 - b^2$.

$$(z + 9)(z - 9) = z^2 - 9^2$$
$$= z^2 - 81$$

b) $(2 + c)(2 - c) = 2^2 - c^2 \quad a = 2 \text{ and } b = c$
$$= 4 - c^2$$

Be very careful on a problem like this. The answer is $4 - c^2$, NOT $c^2 - 4$; subtraction is not commutative.

c) $(5n^2 - 8)(5n^2 + 8) = (5n^2 + 8)(5n^2 - 8)$ Commutative property

$$= (5n^2)^2 - 8^2$$ $a = 5n^2$ and $b = 8$; put $5n^2$ in parentheses.

$$= 25n^4 - 64$$

d) $\left(\dfrac{3}{4}t + u\right)\left(\dfrac{3}{4}t - u\right) = \left(\dfrac{3}{4}t\right)^2 - u^2$ $a = \dfrac{3}{4}t$ and $b = u$; put $\dfrac{3}{4}t$ in parentheses.

$$= \dfrac{9}{16}t^2 - u^2$$

Multiply.

a) $(k + 7)(k - 7)$ b) $(3c^2 - 4)(3c^2 + 4)$

c) $(8 - p)(8 + p)$ d) $\left(\dfrac{5}{2}m + n\right)\left(\dfrac{5}{2}m - n\right)$

---

6   Square a Binomial

Another type of special binomial product is a **binomial square** such as $(x + 5)^2$. $(x + 5)^2$ means $(x + 5)(x + 5)$. Therefore, we can use FOIL to multiply.

$$(x + 5)^2 = (x + 5)(x + 5) = x^2 + 5x + 5x + 25$$ Use FOIL.

$$= x^2 + 10x + 25$$ Note that $10x = 2(x)(5)$.

Notice that the outer and inner products are the same. When we add those terms, we see that the middle term of the result is *twice* the product of the terms in each binomial.

In the expansion of $(x + 5)^2$, $10x$ is $2(x)(5)$.

The *first* term in the result is the square of the *first* term in the binomial, and the *last* term in the result is the square of the *last* term in the binomial. We can express these relationships with the following formulas:

> **W Hint**
>
> Do you see that when multiplying these binomials using FOIL, the middle terms are always the same?

Procedure   The Square of a Binomial

$$(a + b)^2 = a^2 + 2ab + b^2$$
$$(a - b)^2 = a^2 - 2ab + b^2$$

We can think of the formulas in words as:

To square a binomial, you square the first term, square the second term, then multiply 2 times the first term times the second term and add.

Finding the products $(a + b)^2 = a^2 + 2ab + b^2$ and $(a - b)^2 = a^2 - 2ab + b^2$ is also called *expanding* the binomial squares $(a + b)^2$ and $(a - b)^2$.

$$(a + b)^2 \neq a^2 + b^2 \quad \text{and} \quad (a - b)^2 \neq a^2 - b^2.$$

EXAMPLE 7

Expand.

a) $(q + 4)^2$
b) $(u - 10)^2$
c) $(4x + 7y)^2$
d) $[(3x - y) + 1]^2$

Solution

a) $(q + 4)^2 = q^2 \quad + \quad 2(q)(4) \quad + \quad 4^2$ $\qquad a = q, b = 4$

          ↑   ↑   ↑

    Square the   Two times   Square the
    first term   first term   second term
             times second
                term

$$= q^2 + 8q + 16$$

Notice, $(q + 4)^2 \neq q^2 + 16$. Do not "distribute" the power of 2 to each term in the binomial!

b) $(u - 10)^2 = u^2 \quad - \quad 2(u)(10) \quad + \quad (10)^2$ $\qquad a = u, b = 10$

          ↑   ↑   ↑

    Square the   Two times   Square the
    first term   first term   second term
             times second
                term

W Hint

Using your own words, write down the pattern you see in these examples.

$$= u^2 - 20u + 100$$

c) $(4x + 7y)^2 = (4x)^2 + 2(4x)(7y) + (7y)^2$ $\qquad a = 4x, b = 7y$
$$= 16x^2 + 56xy + 49y^2$$

d) Although this expansion looks complicated, we use the same method. Here, $a = 3x - y$ and $b = 1$.

$[(3x - y) + 1]^2 = (3x - y)^2 + 2(3x - y)(1) + (1)^2$ $\quad (a + b)^2 = a^2 + 2ab + b^2$
$= (3x)^2 - 2(3x)(y) + (y)^2 + 2(3x - y)(1) + (1)^2$ $\quad$ Expand $(3x - y)^2$.
$= 9x^2 - 6xy + y^2 + 6x - 2y + 1$ $\quad$ Simplify.

[YOU TRY 6]

Expand.

a) $(t + 7)^2$
b) $(b - 12)^2$
c) $(2a + 5b)^2$
d) $[(3c - d) + 4]^2$

7   Multiply Other Binomials

To find other products of binomials, we use techniques we have already discussed.

EXAMPLE 8

Expand.

a) $(a + 2)^3$
b) $[(3c + d) + 2n][(3c + d) - 2n]$

Solution

a) Just as $x^2 \cdot x = x^3$, $(a + 2)^2 \cdot (a + 2) = (a + 2)^3$. So we can think of $(a + 2)^3$ as $(a + 2)^2(a + 2)$.

$(a + 2)^3 = (a + 2)^2(a + 2)$
$= (a^2 + 4a + 4)(a + 2)$ $\qquad$ Square the binomial.
$= a^3 + 2a^2 + 4a^2 + 8a + 4a + 8$ $\qquad$ Multiply.
$= a^3 + 6a^2 + 12a + 8$ $\qquad$ Combine like terms.

b) Notice that $[(3c + d) + 2n][(3c + d) - 2n]$ has the form $(a + b)(a - b)$, where $a = 3c + d$ and $b = 2n$. So we can use $(a + b)(a - b) = a^2 - b^2$.

$$[(3c + d) + 2n][(3c + d) - 2n] = (3c + d)^2 - (2n)^2$$
$$= (3c)^2 + 2(3c)(d) + (d)^2 - (2n)^2 \quad \text{Expand } (3c + d)^2.$$
$$= 9c^2 + 6cd + d^2 - 4n^2 \quad \text{Simplify.}$$

[YOU TRY 7]   Expand.

a) $(n - 3)^3$ b) $[(x - 2y) + 3z][(x - 2y) - 3z]$

ANSWERS TO [YOU TRY] EXERCISES

1) $20z^7 - 35z^6 - 5z^5 + 40z^4$ 2) a) $18x^3 + x^2 - 32x - 15$ b) $4t^4 + \dfrac{10}{3}t^3 - 25t^2 - \dfrac{62}{3}t + 20$

3) a) $n^2 + 13n + 40$ b) $3k^2 - 5k - 28$ c) $x^2 - 8xy + 12y^2$ d) $10cd + 5c - 16d - 8$

4) $-18x^5 - 66x^4 + 120x^3$ 5) a) $k^2 - 49$ b) $9c^4 - 16$ c) $64 - p^2$ d) $\dfrac{25}{4}m^2 - n^2$

6) a) $t^2 + 14t + 49$ b) $b^2 - 24b + 144$ c) $4a^2 + 20ab + 25b^2$ d) $9c^2 - 6cd + d^2 + 24c - 8d + 16$

7) a) $n^3 - 9n^2 + 27n - 27$ b) $x^2 - 4xy + 4y^2 - 9z^2$

[E Evaluate] **6.3** Exercises   Do the exercises, and check your work.

Objective 1: Multiply a Monomial and a Polynomial

1) Explain how to multiply a monomial and a binomial.

2) Explain how to multiply two binomials.

Multiply.

3) $(3m^5)(8m^3)$

4) $(2k^6)(7k^3)$

5) $(-8c)(4c^5)$

6) $\left(-\dfrac{2}{9}z^3\right)\left(\dfrac{3}{4}z^9\right)$

Multiply.

7) $5a(2a - 7)$

8) $3y(10y - 1)$

9) $6v^3(v^2 - 4v - 2)$

10) $8f^5(f^2 - 3f - 6)$

11) $-9b^2(4b^3 - 2b^2 - 6b - 9)$

12) $-4h^7(5h^6 + 4h^3 + 11h - 3)$

13) $3a^2b(ab^2 + 6ab - 13b + 7)$

14) $4x^6y^2(-5x^2y + 11xy^2 - xy + 2y - 1)$

15) $-\dfrac{3}{5}k^4(15k^2 + 20k - 3)$

16) $\dfrac{3}{4}t^5(12t^3 - 20t^2 + 9)$

Objective 2: Multiply Two Polynomials

For Exercises 17 and 18, answer *always, sometimes,* or *never.*

17) The product of two first-degree binomials will *always, sometimes,* or *never* be a second-degree polynomial.

18) Multiplying a second-degree binomial by a third-degree trinomial will *always, sometimes,* or *never* result in a third-degree polynomial.

Multiply.

19) $(c + 4)(6c + 7)$

20) $(d + 8)(5d + 9)$

21) $(f - 5)(3f^2 + 2f - 4)$

22) $(k - 2)(9k^2 - 4k - 12)$

23) $(4x^3 - x^2 + 6x + 2)(2x - 5)$

24) $(3m^3 + 3m^2 - 4m - 9)(4m - 7)$

25) $\left(\dfrac{1}{3}y^2 + 4\right)(12y^2 + 7y - 9)$

26) $\left(\dfrac{3}{5}q^2 - 1\right)(10q^2 - 7q + 20)$

27) $(4h^2 - h + 2)(-6h^3 + 5h^2 - 9h)$

28) $(n^4 + 8n^2 - 5)(n^2 - 3n - 4)$

Multiply both horizontally and vertically. Which method do you prefer and why?

29) $(3y - 2)(5y^2 - 4y + 3)$

30) $(2p^2 + p - 4)(5p + 3)$

Objective 3: Multiply Two Binomials Using FOIL

31) What do the letters in the word FOIL represent?

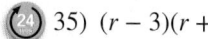 32) Can FOIL be used to expand $(x + 8)^2$? Explain your answer.

Use FOIL to multiply.

33) $(w + 5)(w + 7)$

34) $(u + 5)(u + 3)$

35) $(r - 3)(r + 9)$

36) $(w - 12)(w - 4)$

37) $(y - 7)(y - 1)$

38) $(g + 4)(g - 8)$

39) $(3p + 7)(p - 2)$

40) $(5u + 1)(u + 7)$

41) $(7n^2 + 4)(3n + 1)$

42) $(4y^3 - 3)(7y + 6)$

43) $(5 - 4w)(3 - w)$

44) $(2 - 3r)(4 - 5r)$

45) $(4a - 5b)(3a + 4b)$

46) $(3c + 2d)(c - 5d)$

47) $\left(v + \dfrac{1}{3}\right)\left(v + \dfrac{3}{4}\right)$

48) $(0.5p - 0.3q)(0.7p - 0.4q)$

49) $\left(\dfrac{1}{2}a + 5b^2\right)\left(\dfrac{2}{3}a - b^2\right)$

50) $\left(\dfrac{3}{4}x - y^2\right)\left(\dfrac{1}{3}x + 4y^2\right)$

Write an expression for a) the perimeter of each figure and b) the area of each figure.

51)

![rectangle with width $y + 5$ and height $y - 3$]

$y - 3$

$y + 5$

52)

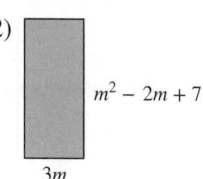

$m^2 - 2m + 7$

$3m$

Objective 4: Find the Product of More Than Two Polynomials

53) To find the product $2(n + 6)(n - 1)$, Raman begins by multiplying $2(n + 6)$ and then he multiplies that result by $(n - 1)$. Peggy begins by multiplying $(n + 6)(n - 1)$ and multiplies that result by 2. Who is right?

54) Find the product $(3a + 2)(a - 4)(a - 2)$

   a) by first multiplying $(3a + 2)(a - 4)$ and then multiplying that result by $(a - 2)$.

   b) by first multiplying $(a - 4)(a - 2)$ and then multiplying that result by $(3a + 2)$.

   c) What do you notice about the results?

Multiply. See Example 5.

55) $3(y + 4)(5y - 2)$

56) $4(7 - 3z)(2z - 1)$

57) $12g^2(2g + 5)(-g + 1)$

58) $-7r^2(r - 9)(r - 2)$

59) $(c + 3)(c + 4)(c - 1)$

60) $(x - 5)(x - 2)(x + 3)$

61) $10n\left(\dfrac{1}{2}n^2 + 3\right)(n^2 + 5)$

62) $12k\left(\dfrac{1}{4}k^2 - \dfrac{2}{3}\right)(k^2 + 1)$

63) $(r + t)(r - 2t)(2r - t)$

64) $(x + y)(x - 2y)(x + 3y)$

Objective 5: Find the Product of Binomials of the Form $(a + b)(a - b)$

Find the following special products. See Example 6.

65) $(3m + 2)(3m - 2)$ 66) $(5y - 4)(5y + 4)$

67) $(7a - 8)(7a + 8)$ 68) $(4x - 11)(4x + 11)$

69) $(2p + 7q)(2p - 7q)$ 70) $(6a - b)(6a + b)$

71) $\left(n + \dfrac{1}{2}\right)\left(n - \dfrac{1}{2}\right)$ 72) $\left(b - \dfrac{1}{5}\right)\left(b + \dfrac{1}{5}\right)$

73) $\left(\dfrac{2}{3} - k\right)\left(\dfrac{2}{3} + k\right)$

74) $\left(\dfrac{4}{3} + z\right)\left(\dfrac{4}{3} - z\right)$

75) $(0.3x - 0.4y)(0.3x + 0.4y)$

76) $(1.2a + 0.8b)(1.2a - 0.8b)$

77) $(5x^2 + 4)(5x^2 - 4)$

78) $(9k^2 + 3l^2)(9k^2 - 3l^2)$

Objective 6: Square a Binomial
Expand.

79) $(y + 8)^2$ 80) $(b + 6)^2$

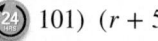 81) $(t - 11)^2$ 82) $(g - 5)^2$

83) $(4w + 1)^2$ 84) $(7n + 2)^2$

85) $(2d - 5)^2$ 86) $(3p - 5)^2$

87) $(6a - 5b)^2$ 88) $(7x + 6y)^2$

89) $(c^2 - 9)^2$ 90) $\left(\dfrac{3}{8}x + 2\right)^2$

91) $[(3m + n) + 2]^2$ 92) $[(2c - d) + 7]^2$

93) $[(x - 4) - y]^2$ 94) $[(3r + 2) - t]^2$

95) Does $4(t + 3)^2 = (4t + 12)^2$? Why or why not?

96) Explain, in words, how to find the product $3(z - 4)^2$, then find the product.

Find the product.

97) $6(x + 1)^2$ 98) $2(k + 5)^2$

99) $2a(a + 3)^2$ 100) $-3(m - 1)^2$

Objective 7: Multiply Other Binomials
Expand.

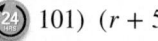 101) $(r + 5)^3$ 102) $(w + 4)^3$

103) $(s - 2)^3$ 104) $(q - 1)^3$

105) $(y + 2)^4$ 106) $(b + 3)^4$

107) $[(v - 5w) + 4][(v - 5w) - 4]$

108) $[(4p + 3q) + 1][(4p + 3q) - 1]$

109) $[(2a + b) + c][(2a + b) - c]$

110) $[(x - 3y) - 2z][(x - 3y) + 2z]$

Mixed Exercises: Objectives 1–7

111) Does $(x + 5)^2 = x^2 + 25$? Why or why not?

112) Does $(y - 3)^3 = y^3 - 27$? Why or why not?

Find each product or sum or difference.

113) $(c - 12)(c + 7)$ 114) $(3y^2 - 8z)(3y^2 + 8z)$

115) $(2k - 9)(5k^2 + 4k - 1)$ 116) $(m^3 + 12)^2$

117) $\left(\dfrac{1}{6} - h\right)\left(\dfrac{1}{6} + h\right)$

118) $3(7p^3 + 4p^2 + 2) - (5p^3 - 18p - 11)$

119) $(3c + 1)^3$

120) $(4w - 5)(2w - 3)$

121) $\left(\dfrac{3}{8}p^7\right)\left(\dfrac{3}{4}p^4\right)$

122) $xy(2x - y)(x - 3y)(x - 2y)$

123) $4(5y^2 - 8y - 3) - 3(2y^2 + y - 10)$

124) $-5z(z - 3)^2$

125) Express the volume of the cube as a polynomial.

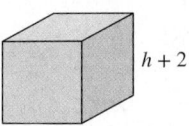

$h + 2$

126) Express the area of the square as a polynomial.

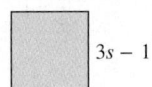

$3s - 1$

127) Express the area of the shaded region as a polynomial.

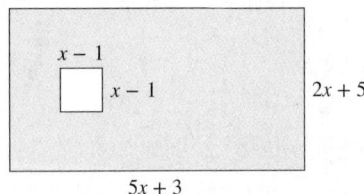

$x - 1$ $x - 1$ $2x + 5$

$5x + 3$

128) Express the area of the triangle as a polynomial.

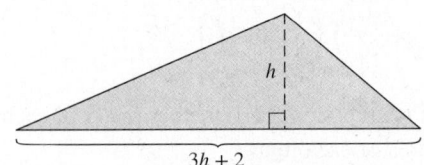

h

$3h + 2$

R1) Write a paragraph that explains why you can use FOIL to multiply two binomials.

R2) How many special products did you encounter in this section? Have you mastered their formulas?

R3) What is the most common mistake you make when multiplying polynomials?

R4) Which exercises in this section do you find most challenging? Redo some of those problems to get more practice.

6.4 Division of Polynomials

P **Prepare**

O **Organize**

What are your objectives for Section 6.4?	How can you accomplish each objective?
1 Divide a Polynomial by a Monomial	• Learn the procedure for **Dividing a Polynomial by a Monomial.** • Review the quotient rules for exponents. • Complete the given examples on your own. • Complete You Trys 1 and 2.
2 Divide a Polynomial by a Polynomial	• Review long division of whole numbers by writing out Example 3 in your notes. • Learn the long division process for polynomials. • Notice the similarities between whole number division and polynomial long division. • Complete the given examples on your own. • Complete You Trys 3–6.

W **Work**

Read the explanations, follow the examples, take notes, and complete the You Trys.

The last operation with polynomials we need to discuss is *division* of polynomials. We will consider this in two parts:

1) Dividing a polynomial by a monomial

$$\text{Examples:} \quad \frac{12a^2 - a + 15}{3}, \quad \frac{-48m^3 + 30m^2 - 8m}{8m^2}$$

and

2) Dividing a polynomial by a polynomial

$$\text{Examples:} \quad \frac{n^2 + 14n + 48}{n + 6}, \quad \frac{27z^3 - 1}{3z - 1}$$

1 Divide a Polynomial by a Monomial

The procedure for dividing a polynomial by a monomial is based on the procedure for adding or subtracting fractions.

To add $\dfrac{4}{15} + \dfrac{7}{15}$, we do the following:

$$\frac{4}{15} + \frac{7}{15} = \frac{4 + 7}{15} \qquad \text{Add numerators, keep the common denominator.}$$
$$= \frac{11}{15}$$

Reversing the process above we can write $\dfrac{11}{15} = \dfrac{4 + 7}{15} = \dfrac{4}{15} + \dfrac{7}{15}$. We can generalize this result and say that $\dfrac{a + b}{c} = \dfrac{a}{c} + \dfrac{b}{c}$ $(c \neq 0)$.

> **W Hint**
> Write this procedure in your notes, in your own words.

> **Procedure** Dividing a Polynomial by a Monomial
>
> To divide a polynomial by a monomial, divide *each term* in the polynomial by the monomial and simplify.

EXAMPLE 1

Divide.

a) $\dfrac{40a^2 - 25a + 10}{5}$

b) $\dfrac{9x^3 + 30x^2 + 3x}{3x}$

Solution

a) First, note that the polynomial is being divided by a *monomial*. That means we divide each term in the numerator by the monomial 5.

$$\frac{40a^2 - 25a + 10}{5} = \frac{40a^2}{5} - \frac{25a}{5} + \frac{10}{5}$$
$$= 8a^2 - 5a + 2$$

> **W Hint**
> Notice that when dividing a trinomial by a monomial, the result is a trinomial.

Let's label the components of our division problem the same way as when we divide with integers.

$$\text{Dividend} \rightarrow \frac{40a^2 - 25a + 10}{5} = 8a^2 - 5a + 2 \leftarrow \text{Quotient}$$
$$\text{Divisor} \rightarrow$$

We can check our answer by multiplying the quotient by the divisor. The answer should be the dividend.

$$Check: \quad 5(8a^2 - 5a + 2) = 40a^2 - 25a + 10 \quad \checkmark$$

The quotient $8a^2 - 5a + 2$ is correct.

b) $\dfrac{9x^3 + 30x^2 + 3x}{3x} = \dfrac{9x^3}{3x} + \dfrac{30x^2}{3x} + \dfrac{3x}{3x}$ Divide each term in numerator by $3x$.

$$= 3x^2 + 10x + 1 \qquad \text{Apply the quotient rule for exponents.}$$

Students will often incorrectly "cancel out" $\dfrac{3x}{3x}$ and get nothing. But $\dfrac{3x}{3x} = 1$ since a quantity divided by itself equals one.

$Check:$ $3x(3x^2 + 10x + 1) = 9x^3 + 30x^2 + 3x$ ✓   The quotient is correct.

Note

In Example 1b), x cannot equal zero because then the denominator of $\dfrac{9x^3 + 30x^2 + 3x}{3x}$ would equal zero. Remember, a fraction is undefined when its denominator equals zero!

[YOU TRY 1]

Divide $\dfrac{24t^5 - 6t^4 - 54t^3}{6t^2}$.

EXAMPLE 2

Divide $(15m + 45m^3 - 4 + 18m^2) \div (9m^2)$.

Solution

Although this example is written differently, it is the same as the previous examples. Notice, however, the terms in the numerator are not written in descending powers. Rewrite them in descending powers before dividing.

$$\frac{15m + 45m^3 - 4 + 18m^2}{9m^2} = \frac{45m^3 + 18m^2 + 15m - 4}{9m^2}$$

$$= \frac{45m^3}{9m^2} + \frac{18m^2}{9m^2} + \frac{15m}{9m^2} - \frac{4}{9m^2}$$

$$= 5m + 2 + \frac{5}{3m} - \frac{4}{9m^2} \qquad \text{Apply the quotient rule, and simplify.}$$

The quotient is not a polynomial since m and m^2 appear in denominators. The quotient of polynomials is not necessarily a polynomial.

[YOU TRY 2]

Divide $(6u^2 + 40 - 24u^3 - 8u) \div (8u^2)$.

2 Divide a Polynomial by a Polynomial

When dividing a polynomial by a polynomial containing two or more terms, we use $long$ $division$ of $polynomials$. This method is similar to long division of whole numbers. We will look at a long division problem here so that we can compare the procedure with polynomial long division.

EXAMPLE 3

Divide 4593 by 8.

Solution

$$\begin{array}{r} 5 \\ 8)\overline{4593} \\ -40\downarrow \\ \hline 59 \end{array}$$

1) How many times does 8 divide into 45 evenly?   5
2) Multiply $5 \times 8 = 40$.
3) Subtract $45 - 40 = 5$.
4) Bring down the 9.

Start the process again.

$$\begin{array}{r} 57 \\ 8)\overline{4593} \\ -40 \\ \hline 59 \\ -56 \\ \hline 33 \end{array}$$

1) How many times does 8 divide into 59 evenly?   7
2) Multiply $7 \times 8 = 56$.
3) Subtract $59 - 56 = 3$.
4) Bring down the 3.

Do the procedure again.

$$\begin{array}{r} 574 \\ 8)\overline{4593} \\ -40 \\ \hline 59 \\ -56 \\ \hline 33 \\ -32 \\ \hline 1 \end{array}$$

1) How many times does 8 divide into 33 evenly?   4
2) Multiply $4 \times 8 = 32$.
3) Subtract $33 - 32 = 1$.
4) There are no more numbers to bring down, so the remainder is 1.

W Hint

Notice that when you have a remainder, your final answer is a mixed number.

Write the result.

$$4593 \div 8 = 574\frac{1}{8} \quad \begin{array}{l} \leftarrow \text{Remainder} \\ \leftarrow \text{Divisor} \end{array}$$

Check: $(8 \times 574) + 1 = 4592 + 1 = 4593$ ✓

[YOU TRY 3]

Divide 3827 by 6.

To divide two polynomials, we use a long division process similar to that of Example 3.

EXAMPLE 4

Divide $\dfrac{3x^2 + 19x + 20}{x + 5}$.

Solution

First, notice that we are dividing by a binomial. That tells us to use long division of polynomials.

We will work with the x in $x + 5$ like we worked with the 8 in Example 3.

W Hint

Write down the procedure for dividing a polynomial by a binomial in your notes, in your own words.

$$\begin{array}{r} 3x \\ x + 5)\overline{3x^2 + 19x + 20} \\ -(3x^2 + 15x) \quad \downarrow \\ \hline 4x + 20 \end{array}$$

1) By what do we multiply x to get $3x^2$? $3x$
   Line up terms in the quotient according to exponents, so write $3x$ above $19x$.
2) Multiply $3x$ by $(x + 5)$: $3x(x + 5) = 3x^2 + 15x$.
3) Subtract $(3x^2 + 19x) - (3x^2 + 15x) = 4x$.
4) Bring down the $+20$.

Start the process again. Remember, work with the x in $x + 5$ like we worked with the 8 in Example 3.

Notice that *like terms* are always lined up in the same columns.

 Hint

Notice that *like terms* are always lined up in the same columns.

$$
\begin{array}{r}
3x + 4 \\
x + 5{\overline{\smash{\big)}\,3x^2 + 19x + 20}} \\
\underline{-(3x^2 + 15x)} \\
4x + 20 \\
\underline{-(4x + 20)} \\
0
\end{array}
$$

1) By what do we multiply x to get $4x$?   4
   Write $+4$ above $+20$.
2) Multiply 4 by $(x + 5)$: $4(x + 5) = 4x + 20$.
3) Subtract $(4x + 20) - (4x + 20) = 0$.
4) There are no more terms. The remainder is 0.

Write the result.

$$\frac{3x^2 + 19x + 20}{x + 5} = 3x + 4$$

Check: $(x + 5)(3x + 4) = 3x^2 + 4x + 15x + 20 = 3x^2 + 19x + 20$ ✓

[YOU TRY 4]

Divide.

a) $\dfrac{x^2 + 11x + 24}{x + 8}$ b) $\dfrac{2x^2 + 23x + 45}{x + 9}$

Next, we will look at a division problem with a remainder.

EXAMPLE 5

Divide $\dfrac{-11c + 16c^3 + 19 - 38c^2}{2c - 5}$.

Solution

When we write our long division problem, the polynomial in the numerator must be rewritten so that the exponents are in descending order. Then, perform the long division.

$$
\begin{array}{r}
8c^2 \\
2c - 5{\overline{\smash{\big)}\,16c^3 - 38c^2 - 11c + 19}} \\
\underline{-(16c^3 - 40c^2)} \downarrow \\
2c^2 - 11c
\end{array}
$$

1) By what do we multiply $2c$ to get $16c^3$? $8c^2$
2) Multiply $8c^2(2c - 5) = 16c^3 - 40c^2$.
3) Subtract.
   $(16c^3 - 38c^2) - (16c^3 - 40c^2)$
   $= 16c^3 - 38c^2 - 16c^3 + 40c^2$
   $= 2c^2$
4) Bring down the $-11c$.

Notice how the parentheses and the distributive property are used in this example.

 Hint

Notice how the parentheses and the distributive property are used in this example.

Repeat the process.

$$
\begin{array}{r}
8c^2 + c \\
2c - 5{\overline{\smash{\big)}\,16c^3 - 38c^2 - 11c + 19}} \\
\underline{-(16c^3 - 40c^2)} \\
2c^2 - 11c \\
\underline{-(2c^2 - 5c)} \\
-6c + 19
\end{array}
$$

1) By what do we multiply $2c$ to get $2c^2$? c
2) Multiply $c(2c - 5) = 2c^2 - 5c$.
3) Subtract.
   $(2c^2 - 11c) - (2c^2 - 5c)$
   $= 2c^2 - 11c - 2c^2 + 5c$
   $= -6c$
4) Bring down the $+19$.

$$8c^2 + c - 3$$
$$2c - 5 \overline{)\,16c^3 - 38c^2 - 11c + 19}$$
$$\underline{-(16c^3 - 40c^2)}$$
$$2c^2 - 11c$$
$$\underline{-(2c^2 - 5c)}$$
$$-6c + 19$$
$$\underline{-(-6c + 15)}$$
$$4$$

1) By what do we multiply $\underline{2c}$ to get $-6c$? -3
2) Multiply $-3(2c - 5) = -6c + 15$.
3) Subtract.
$$(-6c + 19) - (-6c + 15)$$
$$= -6c + 19 + 6c - 15 = 4$$

Hint
Are you writing out the example as you are reading it? Do you *understand* each step as it is being performed?

We are done with the long division process. How do we know that? Because the degree of 4 (degree zero) is less than the degree of $2c - 5$ (degree one) we cannot divide anymore. *The remainder is* 4.

The answer is $\dfrac{16c^3 - 38c^2 - 11c + 19}{2c - 5} = 8c^2 + c - 3 + \dfrac{4}{2c - 5}$

Check: $(2c - 5)(8c^2 + c - 3) + 4 = 16c^3 + 2c^2 - 6c - 40c^2 - 5c + 15 + 4$
$$= 16c^2 - 38c^2 - 11c + 19 \quad \checkmark$$

[YOU TRY 5]   Divide $-23t^2 - 2 + 20t^3 - 11t$ by $5t + 3$.

As we saw in Example 5, we must write our polynomials so that the exponents are in descending order. We have to watch out for something else as well—missing terms. **If a polynomial is missing one or more terms, we put them into the polynomial with coefficients of zero.**

EXAMPLE 6

Divide.

a) $x^3 + 125$ by $x + 5$

b) $t^4 + 3t^3 + 6t^2 + 11t + 5$ by $t^2 + 2$

Solution

a)   The degree of the polynomial $x^3 + 125$ is three, but it is missing the x^2-term and the x-term. We will insert these terms into the polynomial by giving them coefficients of zero.

$$x^3 + 125 = x^3 + 0x^2 + 0x + 125$$

Hint
In your notes, summarize the procedures for dividing by a monomial and dividing by a polynomial with two or more terms. Include an example of each.

Divide.
$$x^2 - 5x + 25$$
$$x + 5 \overline{)\,x^3 + 0x^2 + 0x + 125}$$
$$\underline{-(x^3 + 5x^2)}$$
$$-5x^2 + 0x$$
$$\underline{-(-5x^2 - 25x)}$$
$$25x + 125$$
$$\underline{-(25x + 125)}$$
$$0$$

Therefore, $(x^3 + 125) \div (x + 5) = x^2 - 5x + 25$

Check: $(x + 5)(x^2 - 5x + 25) = x^3 - 5x^2 + 25x + 5x^2 - 25x + 125$
$$= x^3 + 125 \quad \checkmark$$

b) In this case, the divisor, $t^2 + 2$, is missing a t-term. Rewrite it as $t^2 + 0t + 2$ and divide.

$$
\begin{array}{r}
t^2 + 3t + 4 \\
t^2 + 0t + 2 \overline{)\, t^4 + 3t^3 + 6t^2 + 11t + 5} \\
\underline{-(t^4 + 0t^3 + 2t^2)} \\
3t^3 + 4t^2 + 11t \\
\underline{-(3t^3 + 0t^2 + 6t)} \\
4t^2 + 5t + 5 \\
\underline{-(4t^2 + 0t + 8)} \\
5t - 3 \quad \leftarrow \text{Remainder}
\end{array}
$$

We are done with the long division process because the degree of $5t - 3$ (degree 1) is less than the degree of the divisor, $t^2 + 0t + 2$ (degree 2).

Write the answer as $t^2 + 3t + 4 + \dfrac{5t - 3}{t^2 + 2}$. The check is left to the student.

YOU TRY 6

Divide.

a) $\dfrac{4m^3 + 17m^2 - 38}{m + 3}$ b) $\dfrac{p^4 + 6p^3 + 3p^2 + 10p + 1}{p^2 + 1}$

ANSWERS TO YOU TRY EXERCISES

1) $4t^3 - t^2 - 9t$ 2) $-3u + \dfrac{3}{4} - \dfrac{1}{u} + \dfrac{5}{u^2}$ 3) $637\dfrac{5}{6}$ 4) a) $x + 3$ b) $2x + 5$

5) $4t^2 - 7t + 2 - \dfrac{8}{5t + 3}$ 6) a) $4m^2 + 5m - 15 + \dfrac{7}{m + 3}$ b) $p^2 + 6p + 2 + \dfrac{4p - 1}{p^2 + 2}$

E Evaluate **6.4** Exercises   Do the exercises, and check your work.

Label the dividend, divisor, and quotient of each division problem.

1) $\dfrac{12c^3 + 20c^2 - 4c}{4c} = 3c^2 + 5c - 1$

2) $2p + 3 \overline{)\, 10p^3 + p^2 - 25p - 6} \quad 5p^2 - 7p - 2$

3) Explain, in your own words, how to divide a polynomial by a monomial.

4) When do you use long division to divide polynomials?

Objective 1: Divide a Polynomial by a Monomial

Divide.

5) $\dfrac{28k^4 + 8k^3 - 40k^2}{4k^2}$

6) $\dfrac{4a^5 - 10a^4 + 6a^3}{2a^3}$

7) $\dfrac{18u^7 + 18u^5 + 45u^4 - 72u^2}{9u^2}$

8) $\dfrac{-15m^6 + 10m^5 + 20m^4 - 35m^3}{5m^3}$

9) $(35d^5 - 7d^2) \div (-7d^2)$

10) $(-32q^6 - 8q^3 + 4q^2) \div (-4q^2)$

11) $\dfrac{9w^5 + 42w^4 - 6w^3 + 3w^2}{6w^3}$

12) $\dfrac{-54j^5 + 30j^3 - 9j^2 + 15}{9j}$

13) $(10v^7 - 36v^5 - 22v^4 - 5v^2 + 1) \div (4v^4)$

14) $(60z^5 + 3z^4 - 10z) \div (5z^2)$

Divide.

15) $\dfrac{90a^4b^3 + 60a^3b^3 - 40a^3b^2 + 100a^2b^2}{10ab^2}$

16) $\dfrac{24x^6y^6 - 54x^5y^4 - x^3y^3 + 12x^3y^2}{6x^2y}$

17) $(9t^5u^4 - 63t^4u^4 - 108t^3u^4 + t^3u^2) \div (-9tu^2)$

18) $(-45c^8d^6 - 15c^6d^5 + 60c^3d^5 + 30c^3d^3) \div (-15c^3d^2)$

19) Irene divides $16t^3 - 36t^2 + 4t$ by $4t$ and gets a quotient of $4t^2 - 9t$. Is this correct? Why or why not?

20) Kinh divides $\dfrac{15x^2 + 12x}{12x}$ and gets a quotient of $15x^2$. What was his mistake? What is the correct answer?

Objective 2: Divide a Polynomial by a Polynomial
Divide.

21) $17)\overline{8976}$

22) $14)\overline{5194}$

23) $6)\overline{949}$

24) $4)\overline{857}$

25) $9)\overline{3937}$

26) $8)\overline{4189}$

Divide.

27) $\dfrac{g^2 + 9g + 20}{g + 5}$

28) $\dfrac{n^2 + 13n + 40}{n + 8}$

29) $\dfrac{p^2 + 8p + 12}{p + 2}$

30) $\dfrac{v^2 + 13v + 12}{v + 1}$

31) $\dfrac{k^2 + 4k - 45}{k + 9}$

32) $\dfrac{m^2 - 6m - 27}{m + 3}$

33) $\dfrac{h^2 + 5h - 24}{h - 3}$

34) $\dfrac{u^2 - 11u + 30}{u - 5}$

35) $\dfrac{4a^3 - 24a^2 + 29a + 15}{2a - 5}$

36) $\dfrac{28b^3 - 26b^2 + 41b - 15}{7b - 3}$

37) $(p + 45p^2 - 1 + 18p^3) \div (6p + 1)$

38) $(17z^2 - 10 - 12z^3 + 32z) \div (4z + 5)$

39) $(6t^2 - 7t + 4) \div (t - 5)$

40) $(7d^2 + 57d - 4) \div (d + 9)$

41) $\dfrac{61z + 12z^3 - 37 + 44z^2}{3z + 5}$

42) $\dfrac{23k^3 + 22k - 8 + 6k^4 + 44k^2}{6k - 1}$

43) $\dfrac{w^3 + 64}{w + 4}$

44) $\dfrac{a^3 - 27}{a - 3}$

45) $(16r^3 + 58r^2 - 9) \div (8r - 3)$

46) $(50c^3 + 7c + 4) \div (5c + 2)$

47) $\dfrac{15t^4 - 40t^3 - 33t^2 + 10t + 2}{5t^2 - 1}$

48) $\dfrac{18v^4 - 15v^3 - 18v^2 + 13v - 10}{3v^2 - 4}$

Mixed Exercises: Objectives 1 and 2
Fill in the blank with *always*, *sometimes*, or *never* to make the statement true.

49) The quotient of two polynomials is _____ a polynomial.

50) A trinomial divided by a monomial is _____ a binomial.

Divide.

51) $\dfrac{6x^4y^4 + 30x^4y^3 - x^2y^2 + 3xy}{6x^2y^2}$

52) $\dfrac{12v^2 - 23v + 14}{3v - 2}$

53) $\dfrac{-8g^4 + 49g^2 + 36 - 25g - 2g^3}{4g - 9}$

54) $(12c^2 + 6c - 30c^3 + 48c^4) \div (-6c)$

55) $\dfrac{6t^2 - 43t - 20}{t - 8}$

56) $\dfrac{-14u^3v^3 + 7u^2v^3 + 21uv + 56}{7u^2v}$

57) $(8n^3 - 125) \div (2n - 5)$

58) $(12a^4 - 19a^3 + 22a^2 - 9a - 20) \div (3a - 4)$

59) $(13x^2 - 7x^3 + 6 + 5x^4 - 14x) \div (x^2 + 2)$

60) $(18m^4 - 66m^3 + 39m^2 + 11m - 7) \div (6m^2 - 1)$

61) $\dfrac{-12a^3 + 9a^2 - 21a}{-3a}$

62) $\dfrac{64r^3 + 27}{4r + 3}$

63) $\dfrac{10h^4 - 6h^3 - 49h^2 + 27h + 19}{2h^2 - 9}$

64) $\dfrac{16w^2 - 3 - 7w + 15w^4 - 5w^3}{5w^2 + 7}$

65) $\dfrac{6d^4 + 19d^3 - 8d^2 - 61d - 40}{2d^2 + 7d + 5}$

66) $\dfrac{8x^4 + 2x^3 - 13x^2 - 53x + 14}{2x^2 + 5x + 7}$

67) $\dfrac{9c^4 - 82c^3 - 41c^2 + 9c + 16}{c^2 - 10c + 4}$

68) $\dfrac{15n^4 - 16n^3 - 31n^2 + 50n - 22}{5n^2 - 7n + 2}$

69) $\dfrac{k^4 - 81}{k^2 + 9}$ 70) $\dfrac{b^4 - 16}{b^2 - 4}$

71) $\dfrac{49a^4 - 15a^2 - 14a^3 + 5a^6}{-7a^3}$

72) $\dfrac{9q^2 + 26q^4 + 8 - 6q - 4q^3}{2q^2}$

73) $\dfrac{2v^2 - 7 + v + 12v^4 - 15v^3}{3v^2 - 1}$

74) $\dfrac{2r^3 - 8r + 14r^4 - 17 - 29r^2}{2r^2 - 5}$

75) $\left(x^2 + \dfrac{13}{2}x + 3 \right) \div (2x + 1)$

76) $\left(k^2 + \dfrac{11}{3}k + 2 \right) \div (3k + 2)$

77) $\left(2w^2 + \dfrac{10}{3}w - 8 \right) \div (3w - 4)$

78) $\left(3y^2 - \dfrac{41}{4}y + 9 \right) \div (4y - 3)$

For each rectangle, find a polynomial that represents the missing side.

79)

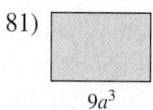

$y - 6$

Find the length if the area is given by $4y^2 - 23y - 6$.

80)

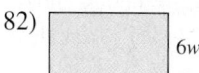

$3x + 2$

Find the width if the area is given by $6x^2 + x - 2$.

81)

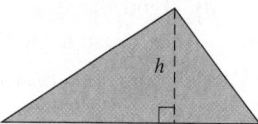

$9a^3$

Find the width if the area is given by $18a^5 - 45a^4 + 9a^3$.

82)

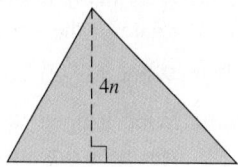

$6w$

Find the length if the area is given by $9w^3 + 6w^2 - 24w$.

83) Find the base of the triangle if the area is given by $6h^3 + 3h^2 + h$.

h

84) Find the base of the triangle if the area is given by $6n^3 - 2n^2 + 10n$.

$4n$

R Rethink

R1) Explain how long division of whole numbers and long division of polynomials are similar.

R2) What is the most common mistake you make when performing polynomial long division?

R3) When performing polynomial long division, why do you line up like terms in the same column?

R4) Which exercises in this section do you find most challenging?

Group Activity — Polynomials and Polynomial Functions

1) Pascal's triangle is an array of numbers used to expand binomials of the form $(a + b)^n$. Let's begin by noticing some patterns of the triangle.

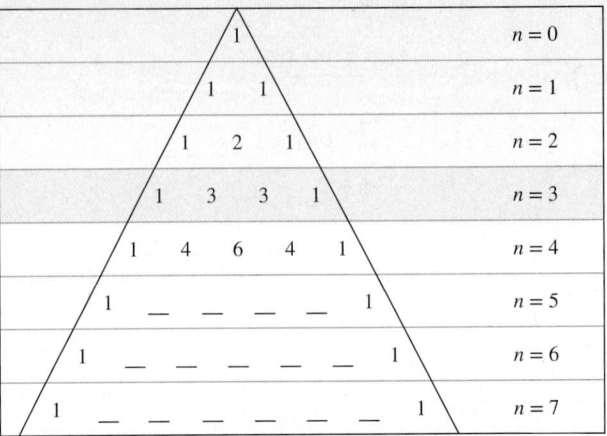

a) Fill in the last three rows of Pascal's triangle.

b) How does the number of terms in each row compare to the value of n?

c) If n is even, how many terms are there? If n is odd, how many terms are there?

d) Find the sum of all the numbers in each row of the triangle. Find a simple expression for the sum of all the numbers in row n.

e) Draw a vertical line down the middle of the triangle. Do you notice any symmetry?

2) Let $n = 3$ to obtain $(a + b)^3$. Verify the following:
$(a + b)^3 = a^3 + 3a^2b + 3ab^2 + b^3$.

a) How many terms are obtained when you multiply $(a + b)^3$? How does this relate to the value of n?

b) What do you notice about the coefficients of each term?

c) What happens to the exponents of a and b in the expansion? [Note: a^3 and b^3 can be written as a^3b^0 and a^0b^3, respectively.]

d) Add the exponents of a and b for each term. How does this relate to the value of n?

3) Complete the binomial expansion for $(a + b)^9$ by filling in the missing terms.

$(a + b)^9 = a^9 + 9a^8b +$ _____ $+ 84a^6b^3 + 126a^5b^4 +$ _____ $+$ _____ $+$
$36a^2b^7 +$ _____ $+$ _____

4) Expand the following using Pascal's triangle.

a) $(a + b)^4$ b) $(x + 2)^4$ c) $(x - 2)^4$ d) $(3x + 2y)^4$ e) $(x^2 + 2y)^4$

5) Simplify each expression without expanding any term.

a) $(6x + 1)^3 + 3(6x + 1)^2(4 - 6x) + 3(6x + 1)(4 - 6x)^2 + (4 - 6x)^3$

b) $(5x^2 - 3)^6 + 6(5x^2 - 3)^5(3 - 4x^2) + 15(5x^2 - 3)^4(3 - 4x^2)^2 +$
$20(5x^2 - 3)^3(3 - 4x^2)^3 + 15(5x^2 - 3)^2(3 - 4x^2)^4 +$
$6(5x^2 - 3)(3 - 4x^2)^5 + (3 - 4x^2)^6$

6) Find the number of terms in each expression after expanding. [Hint: Use your knowledge of the laws of exponents and your answer to 2a).]

a) $[(a + b)^3]^5$

b) $(a + b)^3(a + b)^5$

c) $(x + y)\left[\dfrac{(x + y)^6}{(x + y)^{-5}}\right]^4$

d) $(a + b)^4 - (a + b)^2$

Are you doing your homework in an environment that maximizes your chances for success? To find out, circle the number that best applies for each question using the following scale:

1) I keep my cell phone on my desk when I do my homework, in case anyone calls me.

Highly accurate   4   3   2   1   Not true at all

2) I send text messages while doing my homework.

Highly accurate   4   3   2   1   Not true at all

3) I do my homework in a place where I am distracted by many people.

Highly accurate   4   3   2   1   Not true at all

4) The TV is on when I do my homework.

Highly accurate   4   3   2   1   Not true at all

5) The desk I use to do my homework is a total mess.

Highly accurate   4   3   2   1   Not true at all

6) I often do my homework in bed.

Highly accurate   4   3   2   1   Not true at all

7) I only have one pen or pencil handy with me when I start my homework.

Highly accurate   4   3   2   1   Not true at all

8) I try to do my homework on the bus so that I can get it over with before I get home.

Highly accurate   4   3   2   1   Not true at all

9) I do my homework on any piece of paper that I can find and not always in a notebook.

Highly accurate   4   3   2   1   Not true at all

10) If I haven't finished my homework, I do it right before class starts.

Highly accurate   4   3   2   1   Not true at all

Scoring: Total the numbers you have circled.

If the score is below 15, you do your homework in an environment that allows you to focus and perform your best. If your score is 16–25, you could make some changes to the place where you do your homework that would improve your performance.

If your score is above 25, the environment in which you do your homework is probably negatively affecting your performance. Consider doing your homework in the library or some other quiet place outside your home.

Chapter 6:   Summary

Definition/Procedure	Example

6.1 Review of the Rules of Exponents

For real numbers a and b and integers m and n, the following rules apply:

Product rule: $a^m \cdot a^n = a^{m+n}$

$p^3 \cdot p^5 = p^{3+5} = p^8$

Power rules:

a) $(a^m)^n = a^{mn}$

a) $(c^4)^3 = c^{4 \cdot 3} = c^{12}$

b) $(ab)^n = a^n b^n$

b) $(2g)^5 = 2^5 g^5 = 32g^5$

c) $\left(\dfrac{a}{b}\right)^n = \dfrac{a^n}{b^n} \ (b \neq 0)$

c) $\left(\dfrac{t}{4}\right)^3 = \dfrac{t^3}{4^3} = \dfrac{t^3}{64}$

Zero exponent: $a^0 = 1$ if $a \neq 0$

$9^0 = 1$

Negative exponent:

a) $a^{-n} = \left(\dfrac{1}{a}\right)^n = \dfrac{1}{a^n} \ (a \neq 0)$

a) $6^{-2} = \left(\dfrac{1}{6}\right)^2 = \dfrac{1}{6^2} = \dfrac{1}{36}$

b) $\dfrac{a^{-m}}{b^{-n}} = \dfrac{b^n}{a^m} \ (a \neq 0, b \neq 0)$

b) $\dfrac{a^{-5}}{b^{-3}} = \dfrac{b^3}{a^5}$

Quotient rule: a) $\dfrac{a^m}{a^n} = a^{m-n} \ (a \neq 0)$

$\dfrac{k^{14}}{k^4} = k^{14-4} = k^{10}$

6.2 Addition and Subtraction of Polynomials

A **polynomial in x** is the sum of a finite number of terms of the form ax^n where n is a whole number and a is a real number.

The **degree of a term** equals the exponent on its variable. If a term has more than one variable, the degree equals the *sum* of the exponents on the variables.

The **degree of the polynomial** equals the highest degree of any nonzero term.

Identify each term in the polynomial, the coefficient and degree of each term, and the degree of the polynomial.
$3m^4n^2 - m^3n^2 + 2m^2n^3 + mn - 5n$

Term	Coeff.	Degree
$3m^4n^2$	3	6
$-m^3n^2$	-1	5
$2m^2n^3$	2	5
mn	1	2
$-5n$	-5	1

The degree of the polynomial is 6.

To **add polynomials,** add like terms. Polynomials may be added horizontally or vertically.

Add the polynomials.
$(4q^2 + 2q - 12) + (-5q^2 + 3q + 8)$
$= [4q^2 + (-5q^2)] + (2q + 3q) + (-12 + 8) = -q^2 + 5q - 4$

To **subtract two polynomials,** change the sign of each term in the second polynomial. Then add the polynomials.

Subtract the polynomials.
$(4t^3 - 7t^2 + 4t + 4) - (12t^3 - 8t^2 + 3t + 9)$
$= (4t^3 - 7t^2 + 4t + 4) + (-12t^3 + 8t^2 - 3t - 9)$
$= -8t^3 + t^2 + t - 5$

$f(x) = 3x^2 + 8x - 4$ is an example of a **polynomial function** since $3x^2 + 8x - 4$ is a polynomial and since each real number that is substituted for x produces only one value for the expression. Finding $f(4)$ is the same as evaluating $3x^2 + 8x - 4$ when $x = 4$.

If $f(x) = 3x^2 + 8x - 4$, find $f(4)$.

$f(4) = 3(4)^2 + 8(4) - 4$ Substitute 4 for x.
$= 3(16) + 32 - 4$
$= 48 + 32 - 4$
$= 76$

Definition/Procedure	Example

6.3 Multiplication of Polynomials

Definition/Procedure	Example
When multiplying a **monomial** and a **polynomial,** use the distributive property.	Multiply. $5y^3(-2y^2 + 8y - 3)$ $= (5y^3)(-2y^2) + (5y^3)(8y) + (5y^3)(-3)$ $= -10y^5 + 40y^4 - 15y^3$
To **multiply two polynomials,** multiply each term in the second polynomial by each term in the first polynomial. Then combine like terms.	Multiply. $(5p + 2)(p^2 - 3p + 6)$ $= (5p)(p^2) + (5p)(-3p) + (5p)(6)$ $\qquad + (2)(p^2) + (2)(-3p) + (2)(6)$ $= 5p^3 - 15p^2 + 30p + 2p^2 - 6p + 12$ $= 5p^3 - 13p^2 + 24p + 12$
Multiplying Two Binomials We can use FOIL to multiply two binomials. **FOIL** stands for First, **O**uter, **I**nner, **L**ast. Then add like terms.	Use FOIL to multiply $(4a - 5)(a + 3)$. $(4a - 5)(a + 3) = 4a^2 + 12a - 5a - 15$ $\qquad\qquad\qquad = 4a^2 + 7a - 15$
Special Products 1) $(a + b)(a - b) = a^2 - b^2$ 2) $(a + b)^2 = a^2 + 2ab + b^2$ 3) $(a - b)^2 = a^2 - 2ab + b^2$	1) Multiply. $(c + 9)(c - 9) = c^2 - 9^2 = c^2 - 81$ 2) Expand. $(x + 4)^2 = x^2 + 2(x)(4) + 4^2$ $\qquad\qquad\qquad = x^2 + 8x + 16$ 3) Expand. $(6v - 7)^2 = (6v)^2 - 2(6v)(7) + 7^2$ $\qquad\qquad\qquad\quad = 36v^2 - 84v + 49$

6.4 Division of Polynomials

Definition/Procedure	Example
To **divide a polynomial by a monomial,** divide *each term* in the polynomial by the monomial and simplify.	Divide $\dfrac{22s^4 + 6s^3 - 7s^2 + 3s - 8}{4s^2}$. $= \dfrac{22s^4}{4s^2} + \dfrac{6s^3}{4s^2} - \dfrac{7s^2}{4s^2} + \dfrac{3s}{4s^2} - \dfrac{8}{4s^2}$ $= \dfrac{11s^2}{2} + \dfrac{3s}{2} - \dfrac{7}{4} + \dfrac{3}{4s} - \dfrac{2}{s^2}$
To **divide a polynomial by another polynomial** containing two or more terms, use *long division*.	Divide $\dfrac{10w^3 + 2w^2 + 13w + 18}{5w + 6}$. $\begin{array}{r} 2w^2 - 2w + 5 \\ 5w + 6 \overline{)\ 10w^3 + 2w^2 + 13w + 18} \\ \underline{-(10w^3 + 12w^2)}\ \downarrow \\ -10w^2 + 13w \\ \underline{-(-10w^2 - 12w)}\ \downarrow \\ 25w + 18 \\ \underline{-(25w + 30)} \\ -12 \rightarrow \text{Remainder} \end{array}$ $\dfrac{10w^3 + 2w^2 + 13w + 18}{5w + 6} = 2w^2 - 2w + 5 - \dfrac{12}{5w + 6}$

Chapter 6: Review Exercises

(6.1) Evaluate using the rules of exponents.

1) $\dfrac{3^{10}}{3^6}$

2) 8^{-2}

3) $\left(\dfrac{5}{4}\right)^{-3}$

4) $-4^0 + 7^0$

Simplify. Assume all variables represent nonzero real numbers. Write the answers with positive exponents.

5) $(z^6)^3$

6) $(-5c^4)^2$

7) $(-9t)(6t^6)$

8) $\dfrac{6m^{10}}{24m^6}$

9) $\dfrac{k^3}{k^{11}}$

10) $\dfrac{d^{-6}}{d^3}$

11) $(-2a^2b)^3(5a^{-12}b)$

12) $\dfrac{x^5y^{-3}}{x^8y^{-4}}$

13) $\left(\dfrac{3pq^{-10}}{2p^{-2}q^5}\right)^{-2}$

14) $(7c^{-8}d^2)(3c^{-2}d)^2$

15) $\left(\dfrac{40}{21}x^{10}\right)(3x^{-12})\left(\dfrac{49}{20}x^2\right)$

16) $\left(\dfrac{4r^{-3}t}{s^2}\right)^{-3}\left(\dfrac{3t^{-5}s}{r^2}\right)^{-2}\left(\dfrac{2rs^2}{t^3}\right)^4$

Simplify. Assume that the variables represent nonzero integers. Write the final answer so that the exponents have positive coefficients.

17) $(y^{2p})^3$

18) $\dfrac{w^{-12a}}{w^{-3a}}$

19) True or False: $-x^2 = (-x)^2$ for every real number value of x. Explain your answer.

20) True or False: $(5y)^{-3} = \dfrac{-125}{y^3}$ if $y \neq 0$. Explain your answer.

(6.2) For Exercises 21 and 22, identify each term in the polynomial, the coefficient and degree of each term, and the degree of the polynomial.

21) $7s^3 - 9s^2 + s + 6$

22) $a^2b^3 + 7ab^2 - 2ab + 9b$

23) Evaluate $2r^2 - 8r - 11$ for $r = -3$.

24) Evaluate $p^3q^2 + 4pq^2 - pq - 2q + 9$ for $p = -1$ and $q = 4$.

25) If $h(x) = 5x^2 - 3x - 6$, find

   a) $h(-2)$

   b) $h(0)$

26) $f(t) = \dfrac{2}{5}t + 4$. Find t so that $f(t) = \dfrac{4}{5}$.

Add or subtract as indicated.

27) $(6c^2 + 2c - 8) - (8c^2 + c - 13)$

28) $(-2m^2 - m + 11) + (6m^2 - 12m + 1)$

29) $\begin{array}{r} 6.7j^3 - 1.4j^2 + \quad j - 5.3 \\ + \ 3.1j^3 + 5.7j^2 + 2.4j + 4.8 \\ \hline \end{array}$

30) $\begin{array}{r} -4.2p^3 + 12.5p^2 - 7.2p + 6.1 \\ - \ \ 1.3p^3 - \ \ 3.3p^2 + 2.5p + 4.3 \\ \hline \end{array}$

31) $\left(\dfrac{3}{5}k^2 + \dfrac{1}{2}k + 4\right) - \left(\dfrac{1}{10}k^2 + \dfrac{3}{2}k - 2\right)$

32) $\left(\dfrac{2}{7}u^2 - \dfrac{5}{8}u + \dfrac{4}{3}\right) + \left(\dfrac{3}{7}u^2 + \dfrac{3}{8}u - \dfrac{11}{12}\right)$

33) Subtract $4x^2y^2 - 7x^2y + xy + 5$ from $x^2y^2 + 2x^2y - 4xy + 11$.

34) Find the sum of $7xy + 2x - 3y - 11$ and $-3xy + 5y + 1$ and subtract it from $-5xy - 9x + y + 4$.

35) Dylan is paid an annual salary plus 12% commission for the construction equipment he sells. His annual earnings, $E(x)$, can be described by the function $E(x) = 0.12x + 38{,}000$, where x is the amount, in dollars, that he sells.

   a) What was the dollar value of the equipment Dylan sold last year if he earned $56,000?

   b) If Dylan's sales total is $190,000 this year, what will he earn?

36) Write a fifth-degree polynomial in x that does not contain a third-degree term.

Find the polynomial that represents the perimeter of each rectangle.

37) $d^2 + 6d + 2$; $d^2 - 3d + 1$

38) $7m - 1$; $3m + 5$

(6.3) Multiply.

39) $3r(8r - 13)$

40) $-5m^2(7m^2 - 4m + 8)$

41) $(4w + 3)(-8w^3 - 2w + 1)$

42) $\left(2t^2 - \dfrac{1}{3}\right)(-9t^2 + 7t - 12)$

43) $(y - 3)(y - 9)$

44) $(3p + 4)(3p + 1)$

45) $(5d^2 + 2)(6d + 5)$

46) $-(a - 13)(a + 10)$

47) $6pq^2(7p^3q^2 + 11p^2q^2 - pq + 4)$

48) $9x^3y^4(-6x^2y + 2xy^2 + 8x - 1)$

49) $(2x - 9y)(2x + y)$

50) $(7r + 3s)(r - s)$

51) $(x^2 + 5x - 12)(10x^4 - 3x^2 + 6)$

52) $(3m^2 - 4m + 2)(m^2 + m - 5)$

53) $-3(5u - 11)(u + 4)$

54) $4f^2(2f - 7)(f - 6)$

55) $(z + 3)(z + 1)(z + 4)$

56) $(p + 2)(p + 5)(p + 4)$

57) $\left(\frac{2}{7}d + 3\right)\left(\frac{1}{2}d - 8\right)$

58) $\left(\frac{3}{10}t - 6\right)\left(\frac{2}{9}t - 5\right)$

Expand.

59) $(c + 4)^2$

60) $(x - 12)^2$

61) $(4p - 3)^2$

62) $(9 - 2y)^2$

63) $(x - 3)^3$

64) $(p + 4)^3$

65) $(2a + 5b)^3$

66) $(4x - 3y)^3$

67) $[(m - 3) + n]^2$

68) $[(5r + t) + 7]^2$

Find the special products.

69) $(p - 13)(p + 13)$

70) $\left(\frac{1}{4}n - 5\right)\left(\frac{1}{4}n + 5\right)$

71) $\left(\frac{9}{2} + \frac{5}{6}x\right)\left(\frac{9}{2} - \frac{5}{6}x\right)$

72) $\left(\frac{6}{11} - r^2\right)\left(\frac{6}{11} + r^2\right)$

73) $\left(3a^2 - \frac{1}{2}b\right)\left(3a^2 + \frac{1}{2}b\right)$

74) $-4(2d - 7)^2$

75) $3u(u + 4)^2$

76) $[(2p + 5) + q][(2p + 5) - q]$

77) Write an expression for the a) area and b) perimeter of the rectangle.

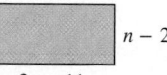

$n - 2$

$2n + 11$

78) Express the volume of the cube as a polynomial.

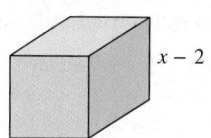

$x - 2$

79) When $3y^2 - 7y - 4$ is multiplied by a monomial, the result is $24y^5 - 56y^4 - 32y^3$. Find the monomial.

80) Is the product of two binomials always a binomial? Explain.

(6.4) Divide.

81) $\dfrac{12t^6 - 30t^5 - 15t^4}{3t^4}$

82) $\dfrac{42p^4 + 12p^3 - 18p^2 + 6p}{-6p}$

83) $\dfrac{w^2 + 9w + 20}{w + 4}$

84) $\dfrac{a^2 - 2a - 24}{a - 6}$

85) $\dfrac{8r^3 + 22r^2 - r - 15}{2r + 5}$

86) $\dfrac{-36h^3 + 99h^2 + 4h + 1}{12h - 1}$

87) $\dfrac{14t^4 + 28t^3 - 21t^2 + 20t}{14t^3}$

88) $\dfrac{48w^4 - 30w^3 + 24w^2 + 3w}{6w^2}$

89) $(14v + 8v^2 - 3) \div (4v + 9)$

90) $(-8 + 12r^2 - 19r) \div (3r - 1)$

91) $\dfrac{6v^4 - 14v^3 + 25v^2 - 21v + 24}{2v^2 + 3}$

92) $\dfrac{8t^4 + 20t^3 - 30t^2 - 65t + 13}{4t^2 - 13}$

93) $\dfrac{c^3 - 8}{c - 2}$

94) $\dfrac{g^3 + 64}{g + 4}$

95) $\dfrac{-4 + 13k + 18k^3}{3k + 2}$

96) $\dfrac{10 + 12m^3 - 34m^2}{6m + 1}$

97) $(20x^4y^4 - 48x^2y^4 - 12xy^2 + 15x) \div (-12xy^2)$

98) $(3u^4 - 31u^3 - 13u^2 + 76u - 30) \div (u^2 - 11u + 5)$

99) $(10n^4 + n^3 + 28n^2 + 27n - 13) \div (5n^2 + 3n - 2)$

100) $(6c^4 + 13c^3 - 21c^2 - 9c + 10) \div (2c^2 + 5c - 4)$

Determine whether the statement is true or false.

101) The quotient of two polynomials is always a polynomial.

102) A third-degree polynomial divided by a first-degree polynomial will always be a second-degree polynomial.

103) Find the base of the triangle if the area is given by $12a^2 + 3a$ sq units.

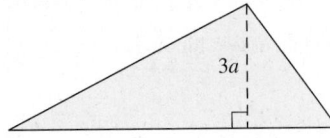

104) Find the length of the rectangle if the area is given by $28x^3 - 51x^2 + 34x - 8$ sq units.

$7x - 4$

Mixed Exercises

Perform the operations and simplify. Assume that all variables represent nonzero real numbers. The answer should not contain negative exponents.

105) $\begin{array}{r} 18c^3 + 7c^2 - 11c + 2 \\ + \quad 2c^2 - 19c^2 \qquad - 1 \\ \hline \end{array}$

106) $\dfrac{15a - 11 + 14a^2}{7a - 3}$

107) $(12 - 7w)(12 + 7w)$

108) $(5p - 9)(2p^2 - 4p - 7)$

109) $5(-2r^7 t^9)^3$

110) $5(7k^2 + k - 9) - 2(-3k^2 + 10k - 4)$

111) $(39a^6 b^6 + 21a^4 b^5 - 5a^3 b^4 + a^2 b) \div (3a^3 b^3)$

112) $\dfrac{(6x^{-4}y^5)^{-2}}{(3xy^{-2})^{-4}}$

113) $(h - 5)^3$

114) $\left(\dfrac{1}{8}m - \dfrac{2}{3}n\right)^2$

115) $\dfrac{-23c + 41 + 2c^3}{c + 4}$

116) $-7d^3(5d^2 + 12d - 8)$

117) $\left(\dfrac{5}{y^4}\right)^{-3}$

118) $(27q^3 + 8) \div (3q + 2)$

119) $(6p^4 + 11p^3 - 20p^2 - 17p + 20) \div (3p^2 + p - 4)$

120) $\left(\dfrac{3b^{-2}c}{a^5}\right)^{-3}\left(\dfrac{4a^{-2}b}{c^4}\right)\left(\dfrac{2ab^3}{c^2}\right)$

Chapter 6: Test

1) Evaluate each expression.

   a) -3^4

   b) 2^{-5}

   c) $-6^0 - 9^0$

   d) $\left(\dfrac{3}{10}\right)^{-3}$

   e) $\dfrac{2^{11}}{2^{17}}$

Simplify. Assume all variables represent nonzero real numbers. Write the answers with positive exponents.

2) $(-3p^4)(10p^8)$

3) $\dfrac{a^5 b}{a^9 b^7}$

4) $(2y^{-4})^6\left(\dfrac{1}{2}y^5\right)^3$

5) $\left(\dfrac{36xy^8}{54x^3 y^{-1}}\right)^{-2}$

6) $t^{10k} \cdot t^{3k}$

7) Given the polynomial $6n^3 + 6n^2 - n - 7$,

   a) what is the coefficient of n?

   b) what is the degree of the polynomial?

8) What is the degree of the polynomial $6a^4 b^5 + 11a^4 b^3 - 2a^3 b + 5ab^2 - 3$?

9) Evaluate $-2r^2 + 7s$ when $r = -4$ and $s = 5$.

10) Write a fourth-degree polynomial in x that does not contain a first-degree term.

11) Let $f(x) = 6x - 11$ and $g(x) = x^3 - 5x^2 - 8x + 4$.

   a) Find x so that $f(x) = 0$.

   b) Find $f\left(\dfrac{2}{3}\right)$.

   c) Find $g(-3)$.

Perform the indicated operation(s).

12) $4h^3(6h^2 - 3h + 1)$

13) $(7a^3 b^2 + 9a^2 b^2 - 4ab + 8)$
    $+ (5a^3 b^2 - 12a^2 b^2 + ab + 1)$

14) Subtract $6y^2 - 5y + 13$ from $15y^2 - 8y + 6$.

15) $6(-n^3 + 4n - 2) - 3(2n^3 + 5n^2 + 8n - 1)$
    $+ 4n(n^2 - 7n + 3)$

16) $(u - 5)(u - 9)$

17) $(4g + 3)(2g + 1)$

18) $\left(v + \dfrac{2}{5}\right)\left(v - \dfrac{2}{5}\right)$

19) $(3x - 7y)(2x + y)$

20) $(4t^2 + 1)(9t - 7)$

21) $(5 - 6n)(2n^2 + 3n - 8)$

22) $2y(y + 6)^2$

Expand.

23) $(3m - 4)^2$

24) $\left(\dfrac{4}{3}x + y\right)^2$

25) $[(5a - b) - 3]^2$

26) $(t - 2)^3$

Divide.

27) $\dfrac{w^2 + 9w + 18}{w + 6}$

28) $\dfrac{24m^6 - 40m^5 + 8m^4 - 6m^3}{8m^4}$

29) $(22p - 50 + 18p^3 - 45p^2) \div (3p - 7)$

30) $\dfrac{y^3 - 27}{y - 3}$

31) $(2r^4 + 3r^3 + 6r^2 + 15r - 20) \div (r^2 + 5)$

32) Fill in the blank with *always*, *sometimes*, or *never* to make the statement true.

a) The sum of two binomials is _____ a binomial.

b) A polynomial divided by a monomial is _____ a polynomial.

c) The product of two monomials is _____ a binomial.

33) Write an expression for a) the area and b) the perimeter of the rectangle.

$3d + 1$

$d - 5$

34) An online restaurant reservation company charges restaurants a fee for diners to use the company's website to make a reservation. Restaurants pay the company $75.00 per month plus $0.65 per reservation. The restaurant's monthly fee, F (in dollars), can be described by the function $F(x) = 0.65x + 75.00$, where x is the number of reservations made using the online system.

a) How much must a restaurant pay the company if 72 reservations were made using this online system in June?

b) In March, a restaurant's fee for using this service was $109.45. How many reservations were made on this website?

Monty Rakusen/Getty Images

Chapter 6:   Cumulative Review for Chapters 1–6

1) Given the set of numbers

$$\left\{\dfrac{3}{8}, -15, 2.1, \sqrt{17}, 41, 0.\overline{52}, 0, 9.32087326...\right\}$$

list the

a) whole numbers

b) integers

c) rational numbers

2) Evaluate $-3^4 + 2 \cdot 9 \div (-3)$.

3) Divide $3\dfrac{1}{8}$ by $1\dfrac{7}{24}$.

Solve.

4) $-\dfrac{18}{7}m - 9 = 21$

5) $5(u + 3) + 2u = 1 + 7(u - 2)$

6) $\dfrac{5}{6} - \dfrac{1}{2}(2p + 1) = \dfrac{1}{3}(p + 3)$

7) Solve $5y - 16 \geq 8y - 1$. Write the answer in interval notation.

8) *Write an equation in one variable and solve.* How many milliliters of a 12% alcohol solution and how many milliliters of a 4% alcohol solution must be mixed to obtain 60 ml of a 10% alcohol solution?

9) Find the x- and y-intercepts of $3x - 8y = 24$, and sketch a graph of the equation.

10) Graph $y = -4$.

11) Write an equation of the line containing the points $(-4, 7)$ and $(2, -11)$. Express the answer in standard form.

12) Write an equation of the line perpendicular to $4x - y = 1$ containing the point $(8, 1)$. Express the answer in slope-intercept form.

13) Solve this system using the elimination method.
$$3x - 4y = -17$$
$$x + 2y = -4$$

14) *Write a system of two equations in two variables, and solve.* The length of a rectangle is 1 cm less than three times its width. The perimeter of the rectangle is 94 cm. What are the dimensions of the figure?

15) Let $f(x) = -4x + 7$ and $g(x) = x^2 - 6x - 2$.

    a) Find $f(5)$.

    b) Find $g(-3)$.

    c) Find $f(a - 9)$.

    d) Find x so that $f(x) = 17$.

Simplify. The answers should not contain negative exponents.

16) $-8(2a^7)^2$

17) $c^{10} \cdot c^7$

18) $\left(\dfrac{4p^{-12}}{p^{-5}} \right)^3$

Perform the indicated operations.

19) $(6q^2 + 7q - 1) - 4(2q^2 - 5q + 8) + 3(-9q - 4)$

20) $(n - 7)(n + 8)$

21) $(3a - 11)(3a + 11)$

22) $\dfrac{12a^4b^4 - 18a^3b + 60ab + 6b}{12a^3b^2}$

23) $(5p^3 - 14p^2 - 10p + 5) \div (p - 3)$

24) $5c(c - 4)^2$

25) $\dfrac{8z^3 + 1}{2z + 1}$

Design Icon Credits: Notes box icon (push pin): ©Shutterstock/ksevgi; Using Technology icon (graphing calculator): ©mbbirdy/Getty Images.

Factoring Polynomials

Get Ready

Which basic skills do we need so that we can learn how to factor polynomials?

1) *Quick* recall of the **basic multiplication facts** is *essential* for factoring polynomials.

   Examples: $5 \cdot 3 = 15$, $-4 \cdot 9 = -36$, $-12 \cdot (-7) = 84$

2) Being able to *quickly* recognize how to **write a number as the product of two of its factors** is also an important skill for factoring polynomials. (When we use the word *factors*, we mean natural numbers. If we want to include negative numbers, we must say so.)

   Example: Find all positive and negative factor pairs of 8.

   $$1 \cdot 8 = 8 \qquad\qquad -1 \cdot (-8) = 8$$
   $$2 \cdot 4 = 8 \qquad\qquad -2 \cdot (-4) = 8$$

   The factor pairs of 8 are $1 \cdot 8$, $2 \cdot 4$, $-1 \cdot (-8)$, and $-2 \cdot (-4)$.

In this chapter, use the following *Basic Skills Worksheets* to prepare students for this, and future, chapters: WS4 Divisibility Rules (to help students recognize factors of natural numbers) and WS3 Prefactoring (so that students can practice the mental arithmetic skills needed to factor polynomials in this chapter).

Get Ready Exercises

Multiply.

1) $8 \cdot 9$ 2) $-6 \cdot 2$ 3) $7 \cdot (-5)$ 4) $-4 \cdot (-8)$ 5) $-12 \cdot (-6)$

6) $3 \cdot 11$ 7) $-2 \cdot 10$ 8) $-9 \cdot (-6)$ 9) $11 \cdot 11$ 10) $7 \cdot 7$

Find all positive and negative factor pairs of each number.

11) 15 12) 14 13) 24 14) 40

15) -56 16) -81 17) -45 18) -42

Answers

1) 72 2) -12 3) -35 4) 32 5) 72 6) 33 7) -20 8) 54 9) 121 10) 49 11) $1 \cdot 15$, $-1 \cdot (-15)$, $3 \cdot 5$, $-3 \cdot (-5)$
12) $1 \cdot 14$, $-1 \cdot (-14)$, $2 \cdot 7$, $-2 \cdot (-7)$ 13) $1 \cdot 24$, $-1 \cdot (-24)$, $2 \cdot 12$, $-2 \cdot (-12)$, $3 \cdot 8$, $-3 \cdot (-8)$, $4 \cdot 6$, $-4 \cdot (-6)$
14) $1 \cdot 40$, $-1 \cdot (-40)$, $2 \cdot 20$, $-2 \cdot (-20)$, $4 \cdot 10$, $-4 \cdot (-10)$, $5 \cdot 8$, $-5 \cdot (-8)$, $1 \cdot 56$, $-1 \cdot (-56)$, $2 \cdot 28$, $-2 \cdot (-28)$,
$-4 \cdot 14$, $4 \cdot (-14)$, $-7 \cdot 8$, $7 \cdot (-8)$ 16) $-1 \cdot 81$, $1 \cdot (-81)$, $-3 \cdot 27$, $3 \cdot (-27)$, $-9 \cdot 9$ 17) $-1 \cdot 45$, $1 \cdot (-45)$, $-3 \cdot 15$,
$3 \cdot (-15)$, $-5 \cdot 9$, $5 \cdot (-9)$ 18) $-1 \cdot 42$, $1 \cdot (-42)$, $-2 \cdot 21$, $2 \cdot (-21)$, $-3 \cdot 14$, $3 \cdot (-14)$, $-6 \cdot 7$, $6 \cdot (-7)$

A **mindset** is a person's set of beliefs. People who think that their beliefs and qualities *cannot* be changed have a **fixed mindset.** For example, a student with a fixed mindset may tell himself, "I'm not a math person. No matter what I do, I'll never be good at math." People with a **growth mindset** believe that their beliefs, abilities, and personalities *can* change by using strategies and working hard. Such a person would say, "I haven't been good at math in the past, but if I get help from my instructor and try new strategies, I can do it." People with a fixed mindset tend to believe that events are out of their control and they can do nothing about them, while those with a growth mindset feel they have control over outcomes.

Carol Dweck is a psychology professor and expert in mindsets. Her research has shown that people's attitudes and strongly held beliefs greatly influence their ability to achieve their goals, and that the "*view you adopt for yourself* profoundly affects the way you lead your life."*

Her research has also revealed that, with a growth mindset, a person's abilities and personalities can change. Among other things, this involves coming up with a plan to succeed; your mindset can hold you back from achieving your goals, or it can put you on a path to success.

So how can *you* adopt (or maintain) a growth mindset? Let's use the P.O.W.E.R. framework to see how to develop a growth mindset to help you achieve your goals in school *and* in life.

- **I will develop a growth mindset to help me to succeed in my math class, in other courses, and in life.**

- Complete the emPOWERme survey that appears before the Chapter Summary to learn whether you have more of a fixed or a growth mindset.
- Write down some examples of when you have had a fixed mindset and when you have had a growth mindset.
- Understand that your brain can actually grow new connections as you learn new things. *Truly believe* that core traits and beliefs *can* change with hard work and a plan. Tell yourself that you can succeed!
- Know that most successful people do not become that way "just because they are smart." The fact is, most successful people achieve their goals because they know their strengths and weaknesses, establish clear goals, work hard and smart, and ask for help from others when they need it.

- Write a few sentences describing how you feel about learning math. What did you write about that indicates a fixed mindset, and what points to a growth mindset? How can your fixed mindset items hold you back from learning?
- Make a list of your strengths and weaknesses. (For example, *I come to class every day* is a strength. *I don't always do my homework* is a weakness.) Be honest!
- Write down a long-term goal for this course. Also write down short-term goals that could help you achieve your long-term goal. Tell yourself that you *can* achieve those goals!
- Identify a role model who has achieved a goal similar to what you want to achieve. Find out how that person came to be successful, and make a list of what *you* can do to achieve your goal.

Image Source/Getty Images

*Dweck, Carol S. (2016). *Mindset: The New Psychology of Success*. New York: Random House.

- Talk with your instructor and come up with a plan for succeeding in the course. Include strategies for using your strengths and for improving your weaknesses. Accept constructive criticism as an opportunity to learn.
- Work *hard and smart* to achieve your goals. Try something new if what you were doing was not working.
- If you do poorly on a test, talk with your instructor and devise new strategies for success. Don't give up!

- Take the emPOWERme survey again. Answer honestly! Did you have more answers that corresponded to a growth mindset than to a fixed mindset?
- Are you achieving your short-term and long-term goals in the course?
- Do you have more confidence in your ability to succeed at difficult tasks than you did before? Do you feel a stronger sense of control over the outcomes?

- Do you have more of a growth mindset than you did before? Why or why not?
- List some strategies that helped you achieve your goals as well as strategies that did not work well.
- How can you apply what you have learned about using a growth mindset to succeed in a math class to achieving goals outside the classroom?

7.1 The Greatest Common Factor and Factoring by Grouping

P Prepare	**O Organize**
What are your objectives for Section 7.1?	**How can you accomplish each objective?**
1 Find the GCF of a Group of Monomials	• Write the definition of a GCF in your own words. • Notice the similarities between factoring a number and finding the GCF, and finding the GCF of monomials. • Complete the given examples on your own. • Complete You Trys 1 and 2.
2 Factor Out the Greatest Common Monomial Factor	• Understand that factoring a polynomial is the opposite of multiplying a polynomial. • In your notes, write the steps for factoring out the GCF in your own words. • Work the given examples on your own by applying the procedure for factoring out the GCF. • Complete You Trys 3 and 4.
3 Factor Out the Greatest Common Binomial Factor	• Follow the same procedure you used for the previous examples. • Complete the examples on your own. • Complete You Try 5.
4 Factor by Grouping	• Complete the examples on your own, and notice the new steps added in each example. • Complete You Trys 6–8. • Write the **Procedure for Factoring by Grouping** in your own words.

 **Work**   **Read the explanations, follow the examples, take notes, and complete the You Trys.**

In Section 1.1, we discussed writing a number as the product of factors:

$$18 = 3 \cdot 6$$

Product Factor Factor

To **factor** an integer is to write it as the product of two or more integers. Therefore, 18 can also be factored in other ways:

$$18 = 1 \cdot 18 \qquad 18 = 2 \cdot 9 \qquad 18 = -1 \cdot (-18)$$
$$18 = -2 \cdot (-9) \qquad 18 = -3 \cdot (-6) \qquad 18 = 2 \cdot 3 \cdot 3$$

The last **factorization**, $2 \cdot 3 \cdot 3$ or $2 \cdot 3^2$, is called the **prime factorization** of 18 since all of the factors are prime numbers. (See Section 1.1.) The factors of 18 are 1, 2, 3, 6, 9, 18, −1, −2, −3, −6, −9, and −18.

We can also write the factors as ±1, ±2, ±3, ±6, ±9, and ±18. (Read ±1 as "plus or minus 1.")

In this chapter, we will learn how to factor polynomials, a skill that is used in many ways throughout algebra.

1   Find the GCF of a Group of Monomials

 **Hint**
Write the definition of GCF in your own words.

Definition
The **greatest common factor (GCF)** of a group of two or more integers is the *largest* common factor of the numbers in the group.

For example, if we want to find the GCF of 18 and 24, we can list their positive factors.

18: 1, 2, 3, 6, 9, 18
24: 1, 2, 3, 4, 6, 8, 12, 24

The greatest common factor of 18 and 24 is 6. We can also use prime factors.

We begin our study of factoring polynomials by discussing how to find the greatest common factor of a group of monomials.

EXAMPLE 1   Find the greatest common factor of x^4 and x^6.

Solution

We can write x^4 as $1 \cdot x^4$, and we can write x^6 as $x^4 \cdot x^2$. The largest power of x that is a factor of both x^4 and x^6 is x^4. Therefore, the GCF is x^4.

Notice that the power of 4 in the GCF is the smallest of the powers when comparing x^4 and x^6. This will always be true.

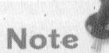
Note
The exponent on the variable in the GCF will be the *smallest* exponent appearing on the variable in the group of terms.

EXAMPLE 2

Find the greatest common factor for each group of terms.

a) $24n^5$, $8n^9$, $16n^3$ b) $-15x^{10}y$, $25x^6y^8$

c) $49a^4b^5$, $21a^3$, $35a^2b^4$

Solution

a) The GCF of the coefficients, 24, 8, and 16, is 8. The smallest exponent on n is 3, so n^3 is part of the GCF.

$$\text{The GCF is } 8n^3.$$

Hint

Complete the example on your own and notice the two-step process.

b) The GCF of the coefficients, -15 and 25, is 5. The smallest exponent on x is 6, so x^6 is part of the GCF. The smallest exponent on y is 1, so y is part of the GCF.

$$\text{The GCF is } 5x^6y.$$

c) The GCF of the coefficients is 7. The smallest exponent on a is 2, so a^2 is part of the GCF. There is no b in the term $21a^3$, so there will be no b in the GCF.

$$\text{The GCF is } 7a^2.$$

[YOU TRY 2] Find the greatest common factor for each group of terms.

a) $18w^6$, $45w^{10}$, $27w^5$ b) $-14hk^3$, $18h^4k^2$ c) $54c^5d^5$, $66c^8d^3$, $24c^2$

2 Factor Out the Greatest Common Monomial Factor

Earlier we said that to **factor an integer** is to write it as the product of two or more integers. To **factor a polynomial** is to write it as a product of two or more polynomials. Throughout this chapter, we will study different factoring techniques. We will begin by discussing how to factor out the greatest common factor.

EXAMPLE 3

Factor out the greatest common factor from $3y + 15$.

Solution

We will use the distributive property to factor out the greatest common factor from $3y + 15$. First, identify the GCF of $3y$ and 15: The GCF is 3.

Then, rewrite each term as a product of two factors with one factor being 3.

$$3y + 15 = (3)(y) + (3)(5)$$
$$= 3(y + 5) \qquad \text{Distributive property}$$

When we factor $3y + 15$, we get $3(y + 5)$. We can check our result by multiplying.

$$3(y + 5) = 3 \cdot y + 3 \cdot 5 = 3y + 15$$

Procedure Steps for Factoring Out the Greatest Common Factor

1) Identify the GCF of all of the terms of the polynomial.
2) Rewrite each term as the product of the GCF and another factor.
3) Use the distributive property to factor out the GCF from the terms of the polynomial.
4) Check the answer by multiplying the factors. The result should be the original polynomial.

EXAMPLE 4

Factor out the greatest common factor.

a) $28p^5 + 12p^4 + 4p^3$ b) $w^8 - 7w^6$

c) $6a^5b^3 + 30a^5b^2 - 54a^4b^2 - 6a^3b$

Solution

a) Identify the GCF of all of the terms: GCF $= 4p^3$.

$$28p^5 + 12p^4 + 4p^3 = (4p^3)(7p^2) + (4p^3)(3p) + (4p^3)(1)$$ Rewrite each term using the GCF as one of the factors.

$$= 4p^3(7p^2 + 3p + 1)$$ Distributive property

Check by multiplying: $4p^3(7p^2 + 3p + 1) = 28p^5 + 12p^4 + 4p^3$ ✓

b) The GCF of the two terms is w^6.

$$w^8 - 7w^6 = (w^6)(w^2) - (w^6)(7)$$ Rewrite each term using the GCF as one of the factors.

$$= w^6(w^2 - 7)$$ Distributive property

Check: $w^6(w^2 - 7) = w^8 - 7w^6$ ✓

W **Hint**

Complete each You Try by
referring to the example
before it.

c) The GCF of all of the terms is $6a^3b$.

$$6a^5b^3 + 30a^5b^2 - 54a^4b^2 - 6a^3b$$

$$= (6a^3b)(a^2b^2) + (6a^3b)(5a^2b) - (6a^3b)(9ab) - (6a^3b)(1)$$ Rewrite using the GCF.

$$= 6a^3b(a^2b^2 + 5a^2b - 9ab - 1)$$ Distributive property

Check: $6a^3b(a^2b^2 + 5a^2b - 9ab - 1) = 6a^5b^3 + 30a^5b^2 - 54a^4b^2 - 6a^3b$ ✓

[**YOU TRY 3**]

Factor out the greatest common factor.

a) $2w + 16$ b) $3u^6 + 36u^5 + 15u^4$ c) $z^5 - 9z^2$

d) $45r^4t^3 + 36r^4t^2 + 18r^3t^2 - 9r^2t$

Sometimes we need to take out a negative factor.

EXAMPLE 5

Factor out $-2k$ from $-6k^4 + 10k^3 - 8k^2 + 2k$.

Solution

$-6k^4 + 10k^3 - 8k^2 + 2k$
$= (-2k)(3k^3) + (-2k)(-5k^2) + (-2k)(4k) + (-2k)(-1)$ Rewrite using $-2k$ as one of the factors.

$= -2k[3k^3 + (-5k^2) + 4k + (-1)]$ Distributive property
$= -2k(3k^3 - 5k^2 + 4k - 1)$ Rewrite $+(-5k^2)$ as $-5k^2$ and $+(-1)$ as -1.

Check: $-2k(3k^3 - 5k^2 + 4k - 1) = -6k^4 + 10k^3 - 8k^2 + 2k$ ✓

BE CAREFUL

When taking out a negative factor, be very careful with the signs!

[YOU TRY 4]

Factor out $-y^2$ from $-y^4 + 10y^3 - 8y^2$.

3 Factor Out the Greatest Common Binomial Factor

Until now, all of the GCFs have been monomials. Sometimes, however, the greatest common factor is a *binomial*.

EXAMPLE 6

Factor out the greatest common factor.

a) $a(b + 5) + 8(b + 5)$ b) $x(y + 3) - (y + 3)$

Solution

a)   In the polynomial $\underbrace{a(b + 5)}_{\text{term}} + \underbrace{8(b + 5)}_{\text{term}}$, $a(b + 5)$ is a term and $8(b + 5)$ is a term.

What do these terms have in common? $b + 5$

The GCF of $a(b + 5)$ and $8(b + 5)$ is $b + 5$. Use the distributive property to factor out $b + 5$.

$a(b + 5) + 8(b + 5) = (b + 5)(a + 8)$ Distributive property

Check: $(b + 5)(a + 8) = (b + 5)a + (b + 5)8$ Distribute.
$ = a(b + 5) + 8(b + 5)$ Commutative property

b)   Begin by writing $x(y + 3) - (y + 3)$ as $x(y + 3) - 1(y + 3)$. The GCF is $y + 3$.

$x(y + 3) - 1(y + 3) = (y + 3)(x - 1)$ Distributive property

The check is left to the student.

W Hint

In your notes, use different color pens to write out the examples as they are written in the book. This will help you see the steps being performed.

BE CAREFUL It is important to write -1 in front of $(y + 3)$. Otherwise, the following mistake is often made:

$$x(y + 3) - (y + 3) = (y + 3)x \quad \text{This is incorrect!}$$

The correct factor is $x - 1$, *not* x.

YOU TRY 5

Factor out the GCF.

a) $c(d - 8) + 2(d - 8)$ b) $k(k^2 + 15) - 7(k^2 + 15)$

c) $u(v + 2) - (v + 2)$

Taking out a binomial factor leads us to our next method of factoring—factoring by grouping.

4 Factor by Grouping

When we are asked to factor a polynomial containing four terms, we often try to **factor by grouping.**

EXAMPLE 7

Factor by grouping.

a) $rt + 7r + 2t + 14$ b) $3xz - 4yz + 18x - 24y$ c) $n^3 + 8n^2 - 5n - 40$

Solution

a) Begin by grouping terms together so that each group has a common factor.

$$\underbrace{rt + 7r} + \underbrace{2t + 14}$$

$$\downarrow \qquad\qquad \downarrow$$

Factor out r to get $r(t + 7)$. $= r(t + 7) + 2(t + 7)$ Factor out 2 to get $2(t + 7)$.
$$= (t + 7)(r + 2)$$ Factor out $(t + 7)$.

Check by multiplying: $(t + 7)(r + 2) = rt + 7r + 2t + 14$ ✓

W Hint

Complete each example, and notice the new steps compared with those in the example before it.

b) Group terms together so that each group has a common factor.

$$\underbrace{3xz - 4yz} + \underbrace{18x - 24y}$$

$$\downarrow \qquad\qquad \downarrow$$

Factor out z to get $z(3x - 4y)$. $= z(3x - 4y) + 6(3x - 4y)$ Factor out 6 to get $6(3x - 4y)$.
$$= (3x - 4y)(z + 6)$$ Factor out $(3x - 4y)$.

Check by multiplying: $(3x - 4y)(z + 6) = 3xz - 4yz + 18x - 24$ ✓

c) Group terms together so that each group has a common factor.

$$\underbrace{n^3 + 8n^2} \underbrace{-5n - 40}$$

$$\downarrow \qquad\qquad \downarrow$$

Factor out n^2 to get $n^2(n + 8)$. $= n^2(n + 8) - 5(n + 8)$ Factor out -5 to get $-5(n + 8)$.
$$= (n + 8)(n^2 - 5)$$ Factor out $(n + 8)$.

We must factor out −5, *not* 5, from the second group so that the binomial factors for both groups are the same! (If we had factored out 5, then the factorization of the second group would have been $5(-n - 8)$.)

Check: $(n + 8)(n^2 - 5) = n^3 + 8n^2 - 5n - 40$ ✓

[YOU TRY 6]

Factor by grouping.

a) $xy + 4x + 10y + 40$ b) $5pr - 8qr + 10p - 16q$

c) $w^3 + 9w^2 - 6w - 54$

Sometimes we have to rearrange the terms before we can factor.

EXAMPLE 8

Factor $12c^2 - 2d + 3c - 8cd$ completely.

Solution

Group terms together so that each group has a common factor.

$$12c^2 - 2d + 3c - 8cd$$

Factor out 2 to get $2(6c^2 - d)$. $= 2(6c^2 - d) + c(3 - 8d)$ Factor out c to get $c(3 - 8d)$.

These groups do not have common factors! Let's rearrange the terms in the original polynomial and group the terms differently.

$$12c^2 + 3c \quad - \quad 8cd - 2d$$

Factor out $3c$ to get $3c(4c + 1)$. $= 3c(4c + 1) - 2d(4c + 1)$ Factor out $-2d$ to get $-2d(4c + 1)$.
$= (4c + 1)(3c - 2d)$ Factor out $(4c + 1)$.

$12c^2 - 2d + 3c - 8cd$ factors to $(4c + 1)(3c - 2d)$. The check is left to the student.

Note

Often, there is more than one way to rearrange the terms so that the polynomial can be factored by grouping.

[YOU TRY 7]

Factor $3k^2 - 48m + 8k - 18km$ completely.

Often, we have to combine the two factoring techniques we have learned here. That is, we begin by factoring out the GCF, and then we factor by grouping. Let's summarize how to factor a polynomial by grouping and then look at another example.

Procedure Steps for Factoring by Grouping

1) Before trying to factor by grouping, look at each term in the polynomial and ask yourself, *"Can I factor out a GCF first?"* If so, factor out the GCF from all of the terms.
2) Make two groups of two terms so that each group has a common factor.
3) Take out the common factor in each group of terms.
4) Factor out the common binomial factor using the distributive property.
5) Check the answer by multiplying the factors.

EXAMPLE 9

Factor $7h^4 + 7h^3 - 42h^2 - 42h$ completely.

Solution

Notice that this polynomial has four terms. This is a clue for us to try factoring by grouping. *However,* look at the polynomial carefully and ask yourself, *"Can I factor out a GCF?" Yes! Therefore, the first step in factoring this polynomial is to factor out 7h.*

$$7h^4 + 7h^3 - 42h^2 - 42h = 7h(h^3 + h^2 - 6h - 6) \qquad \text{Factor out the GCF, } 7h.$$

The polynomial in parentheses has four terms. Try to factor it by grouping.

$$7h(\underbrace{h^3 + h^2}\ \underbrace{- 6h - 6})$$

$$= 7h[h^2(h + 1) - 6(h + 1)] \qquad \text{Take out the common factor in each group.}$$

$$= 7h(h + 1)(h^2 - 6) \qquad \text{Factor out } (h + 1) \text{ using the distributive property.}$$

Check by multiplying: $7h(h + 1)(h^2 - 6) = 7h(h^3 + h^2 - 6h - 6)$
$$= 7h^4 + 7h^3 - 42h^2 - 42h \quad \checkmark$$

YOU TRY 8

Factor $12t^3 + 12t^2 - 3t^2u - 3tu$ completely.

Remember, seeing a polynomial with four terms is a clue to try factoring by grouping. Not all polynomials will factor this way, however. We will learn other techniques later, and some polynomials must be factored using methods learned in later courses.

ANSWERS TO YOU TRY EXERCISES

1) y^5 2) a) $9w^5$ b) $2hk^2$ c) $6c^2$ 3) a) $2(w + 8)$ b) $3u^4(u^2 + 12u + 5)$ c) $z^2(z^3 - 9)$
d) $9r^2t(5r^2t^2 + 4r^2t + 2rt - 1)$ 4) $-y^2(y^2 - 10y + 8)$ 5) a) $(d - 8)(c + 2)$
b) $(k^2 + 15)(k - 7)$ c) $(v + 2)(u - 1)$ 6) a) $(y + 4)(x + 10)$ b) $(5p - 8q)(r + 2)$
c) $(w + 9)(w^2 - 6)$ 7) $(3k + 8)(k - 6m)$ 8) $3t(t + 1)(4t - u)$

Objective 1: Find the GCF of a Group of Monomials

Find the greatest common factor of each group of terms.

1) $28, 21c$

2) $9t, 36$

3) $18p^3, 12p^2$

4) $32z^5, 56z^3$

5) $12n^6, 28n^{10}, 36n^7$

6) $63b^4, 45b^7, 27b$

7) $35a^3b^2, 15a^2b$

8) $10x^5y^4, 2x^4y^4$

9) $21r^3s^6, 63r^3s^2, -42r^4s^5$

10) $-60p^2q^2, 36pq^5, 96p^3q^3$

11) $a^2b^2, 3ab^2, 6a^2b$

12) $n^3m^4, -n^3m^4, -n^4$

13) $c(k-9), 5(k-9)$

14) $a^2(h+8), b^2(h+8)$

15) Explain how to find the GCF of a group of terms.

16) What does it mean to factor a polynomial?

Objective 2: Factor Out the Greatest Common Monomial Factor

Determine whether each expression is written in factored form.

17) $5p(p+9)$

18) $8h^2 - 24h$

19) $18w^2 + 30w$

20) $-3z^2(2z+7)$

21) $a^2b^2(-4ab)$

22) $c^3d - (2c+d)$

Factor out the greatest common factor. Be sure to check your answer.

23) $2w + 10$

24) $3y + 18$

25) $18z^2 - 9$

26) $14h - 12h^2$

27) $100m^3 - 30m$

28) $t^5 - t^4$

29) $r^9 + r^2$

30) $\frac{1}{2}a^2 + \frac{3}{2}a$

31) $\frac{1}{5}y^2 + \frac{4}{5}y$

32) $9a^3 + 2b^2$

33) $s^7 - 4t^3$

34) $14u^7 + 63u^6 - 42u^5$

35) $10n^5 - 5n^4 + 40n^3$

36) $3d^8 - 33d^7 - 24d^6 + 3d^5$

37) $40p^6 + 40p^5 - 8p^4 + 8p^3$

38) $44m^3n^3 - 24mn^4$

39) $63a^3b^3 - 36a^3b^2 + 9a^2b$

40) $8p^4q^3 + 8p^3q^3 - 72p^2q^2$

41) Factor out -6 from $-30n - 42$.

42) Factor out $-c$ from $-9c^3 + 2c^2 - c$.

43) Factor out $-4w^3$ from $-12w^5 - 16w^3$.

44) Factor out $-m$ from $-6m^3 - 3m^2 + m$.

45) Factor out -1 from $-k + 3$.

46) Factor out -1 from $-p - 10$.

Objective 3: Factor Out the Greatest Common Binomial Factor

Factor out the common binomial factor.

47) $u(t-5) + 6(t-5)$

48) $c(b+9) + 2(b+9)$

49) $y(6x+1) - z(6x+1)$

50) $s(4r-3) - t(4r-3)$

51) $p(q+12) + (q+12)$

52) $8x(y-2) + (y-2)$

53) $5h^2(9k+8) - (9k+8)$

54) $3a(4b+1) - (4b+1)$

Objective 4: Factor by Grouping

Factor by grouping.

55) $ab + 2a + 7b + 14$

56) $cd + 8c - 5d - 40$

57) $3rt + 4r - 27t - 36$

58) $5pq + 15p - 6q - 18$

59) $8b^2 + 20bc + 2bc^2 + 5c^3$

60) $4a^3 - 12ab + a^2b - 3b^2$

61) $fg - 7f + 4g - 28$

62) $xy - 8y - 7x + 56$

63) $st - 10s - 6t + 60$

64) $cd + 3c - 11d - 33$

65) $5tu + 6t - 5u - 6$

66) $qr + 3q - r - 3$

67) $36g^4 + 3gh - 96g^3h - 8h^2$

68) $40j^3 + 72jk - 55j^2k - 99k^2$

69) Explain, in your own words, how to factor by grouping.

70) What should be the first step in factoring $6ab + 24a + 18b + 54$?

Factor completely. You may need to begin by factoring out the GCF first or by rearranging terms.

Fill It In

Fill in the blanks with either the missing mathematical step or the reason for the given step.

71) $4xy + 12x + 20y + 60$

$4xy + 12x + 20y + 60$

$= $ _____   Factor out the GCF.

$= 4[x(y + 3) + 5(y + 3)]$ _____

_____

$= $ _____   Take out the binomial factor.

72) $2m^2n - 4m^2 - 18mn + 36m$

$2m^2n - 4m^2 - 18mn + 36m$

$= $ _____   Factor out the GCF.

$= 2m[m(n - 2) - 9(n - 2)]$ _____

_____

$= $ _____   Take out the binomial factor.

73) $3cd + 6c + 21d + 42$

74) $5xy + 15x - 5y - 15$

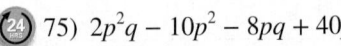 75) $2p^2q - 10p^2 - 8pq + 40p$

76) $3uv^2 - 24uv + 3v^2 - 24v$

77) $10st + 5s - 12t - 6$

78) $8pq + 12p + 10q + 15$

79) $3a^3 - 21a^2b - 2ab + 14b^2$

80) $8c^3 + 32c^2d + cd + 4d^2$

81) $8u^2v^2 + 16u^2v + 10uv^2 + 20uv$

82) $10x^2y^2 - 5x^2y - 60xy^2 + 30xy$

Mixed Exercises: Objectives 1–4

Factor completely.

83) $3mn + 21m + 10n + 70$

84) $4yz + 7z - 20y - 35$

85) $16b - 24$

86) $2yz^3 + 14yz^2 + 3z^3 + 21z^2$

87) $cd + 6c - 4d - 24$

88) $5x^3 - 30x^2y^2 + xy - 6y^3$

89) $6a^4b + 12a^4 - 8a^3b - 16a^3$

90) $6x^2 + 48x^3$ 91) $7cd + 12 + 28c + 3d$

92) $2uv + 12u - 7v - 42$ 93) $dg - d + g - 1$

94) $2ab - 10a - 12b + 60$

95) $x^4y^2 + 12x^3y^3$

96) $8u^2 - 16uv^2 + 3uv - 6v^3$

97) $4mn + 8m + 12n + 24$

98) $5c^2 - 20$

99) Factor out -2 from $-6p^2 - 20p + 2$.

100) Factor out $-5g$ from $-5g^3 + 50g^2 - 25g$.

R Rethink

R1) Do you understand how to factor by grouping?

R2) Write down a problem you got wrong or that was difficult for you to do. Think about where you made a mistake or why it was hard, then redo the problem without looking at your previous work.

R3) How would you explain the process of factoring these expressions to a friend?

7.2 Factoring Trinomials of the Form $x^2 + bx + c$

What are your objectives for Section 7.2?	How can you accomplish each objective?
1 Practice Arithmetic Skills Needed for Factoring Trinomials	• Follow the approach in the examples to come up with a solution. • In your notes, make a chart that summarizes the approach used in the example. • Complete the given example on your own. • Complete You Try 1.
2 Factor a Trinomial of the Form $x^2 + bx + c$	• Notice that the process of factoring is the opposite of multiplying. • Write the procedure for **Factoring a Polynomial of the Form $x^2 + bx + c$** in your notes. How does it compare to the chart you made for Objective 1? • Complete the given example on your own. • Complete You Try 2.
3 More on Factoring a Trinomial of the Form $x^2 + bx + c$	• Add the step of *"Can I factor out a GCF?"* as the first step in the procedure for Objective 2. • Complete the given example on your own. • Complete You Try 3.
4 Factor a Trinomial Containing Two Variables	• Use the same procedure as before. • Complete the given example on your own. • Complete You Try 4.

W Work

Read the explanations, follow the examples, take notes, and complete the You Trys.

One of the factoring problems encountered most often in algebra is the factoring of trinomials. In this section, we will discuss how to factor a trinomial of the form $x^2 + bx + c$; notice that the coefficient of the squared term is 1.

Let's begin with arithmetic skills we need to be able to factor a trinomial of the form $x^2 + bx + c$.

1 Practice Arithmetic Skills Needed for Factoring Trinomials

EXAMPLE 1 Find two integers whose

a) product is 15 and sum is 8.

b) product is 24 and sum is −10.

c) product is −28 and sum is 3.

Solution

a) If the product of two numbers is *positive* 15 and the sum of the numbers is *positive* 8, *then the two numbers will be positive*. (The product of two positive numbers is positive, and their sum is positive as well.)

First, list the pairs of *positive* integers whose product is 15—the *factors* of 15. Then, find the *sum* of those factors.

Factors of 15	Sum of the Factors
$1 \cdot 15$	$1 + 15 = 16$
$3 \cdot 5$	$3 + 5 = 8$

The product of 3 and 5 is 15, and their sum is 8.

W Hint

Follow the approach used in the solution of each example.

b) If the product of two numbers is *positive* 24 and the sum of those numbers is *negative* 10, *then the two numbers will be negative*. (The product of two negative numbers is positive, while the sum of two negative numbers is negative.)

First, list the pairs of negative numbers that are factors of 24. Then, find the sum of those factors. You can stop making your list when you find the pair that works.

Factors of 24	Sum of the Factors
$-1 \cdot (-24)$	$-1 + (-24) = -25$
$-2 \cdot (-12)$	$-2 + (-12) = -14$
$-3 \cdot (-8)$	$-3 + (-8) = -11$
$-4 \cdot (-6)$	$-4 + (-6) = -10$

The product of -4 and -6 is 24, and their sum is -10.

c) If two numbers have a product of *negative* 28 and their sum is *positive* 3, *one number must be positive and one number must be negative*. (The product of a positive number and a negative number is negative, while the sum of the numbers can be either positive *or* negative.)

First, list pairs of factors of -28. Then, find the sum of those factors.

Factors of -28	Sum of the Factors
$-1 \cdot 28$	$-1 + 28 = 27$
$1 \cdot (-28)$	$1 + (-28) = -27$
$-4 \cdot 7$	$-4 + 7 = 3$

The product of -4 and 7 is -28, and their sum is 3.

[YOU TRY 1] Find two integers whose

a) product is 21 and sum is 10.

b) product is -18 and sum is -3.

c) product is 20 and sum is -12.

Note

You should try to get to the point where you can come up with the correct numbers *in your head* without making a list.

2 Factor a Trinomial of the Form $x^2 + bx + c$

The process of factoring is the opposite of multiplying. Let's see how this will help us understand how to factor a trinomial of the form $x^2 + bx + c$.

Multiply $(x + 4)(x + 5)$ using FOIL.

Hint

Does this look similar to the procedure in Example 1?

$$(x + 4)(x + 5) = x^2 + 5x + 4x + 4 \cdot 5 \qquad \text{Multiply using FOIL.}$$
$$= x^2 + (5 + 4)x + 20 \qquad \text{Use the distributive property,}$$
$$= x^2 + 9x + 20 \qquad \text{and multiply } 4 \cdot 5.$$

$$(x + 4)(x + 5) = x^2 + 9x + 20 \quad \longleftarrow \quad \text{The } product \text{ of 4 and 5 is 20.}$$

The *sum* of 4 and 5 is 9.

So, if we were asked to *factor* $x^2 + 9x + 20$, we need to think of two integers whose *product* is 20 and whose *sum* is 9. Those numbers are 4 and 5. The *factored form* of $x^2 + 9x + 20$ is $(x + 4)(x + 5)$.

Hint

Compare the chart you made for Objective 1 to the procedure.

> **Procedure** Factoring a Polynomial of the Form $x^2 + bx + c$
>
> To factor a polynomial of the form $x^2 + bx + c$, find two integers m and n whose product is c and whose sum is b. Then, $x^2 + bx + c = (x + m)(x + n)$.
>
> 1) If b and c are positive, then both m and n must be positive.
> 2) If c is positive and b is negative, then both m and n must be negative.
> 3) If c is negative, then one integer, m, must be positive and the other integer, n, must be negative.
>
> You can check the answer by multiplying the binomials. The result should be the original polynomial.

EXAMPLE 2

Factor, if possible.

a) $x^2 + 7x + 12$ b) $p^2 - 9p + 14$ c) $w^2 + w - 30$

d) $a^2 - 3a - 54$ e) $c^2 - 6c + 9$ f) $y^2 + 11y + 35$

Solution

a) To factor $x^2 + 7x + 12$, we must find two integers whose *product* is 12 and whose *sum* is 7. Both integers will be positive.

Factors of 12	Sum of the Factors
$1 \cdot 12$	$1 + 12 = 13$
$2 \cdot 6$	$2 + 6 = 8$
$3 \cdot 4$	$3 + 4 = 7$

The numbers are 3 and 4. Therefore, $x^2 + 7x + 12 = (x + 3)(x + 4)$.

Check: $(x + 3)(x + 4) = x^2 + 4x + 3x + 12 = x^2 + 7x + 12$ ✓

b) To factor $p^2 - 9p + 14$, find two integers whose *product* is 14 and whose *sum* is -9. Since 14 is positive and the coefficient of p is a negative number, -9, both integers will be negative.

Factors of 14	Sum of the Factors
$-1 \cdot (-14)$	$-1 + (-14) = -15$
$-2 \cdot (-7)$	$-2 + (-7) = -9$

The numbers are -2 and -7. So, $p^2 - 9p + 14 = (p - 2)(p - 7)$.

Check: $(p - 2)(p - 7) = p^2 - 7p - 2p + 14 = p^2 - 9p + 14$ ✓

c) $w^2 + w - 30$

The coefficient of w is 1, so we can think of this trinomial as $w^2 + 1w - 30$.

Find two integers whose *product* is -30 and whose *sum* is 1. Since the last term in the trinomial is negative, one of the integers must be positive and the other must be negative.

Try to find these integers mentally. Two numbers with a product of *positive* 30 are 5 and 6. We need a product of -30, so either the 5 is negative or the 6 is negative.

Factors of -30	Sum of the Factors
$-5 \cdot 6$	$-5 + 6 = 1$

The numbers are -5 and 6. Therefore, $w^2 + w - 30 = (w - 5)(w + 6)$.

Check: $(w - 5)(w + 6) = w^2 + 6w - 5w - 30 = w^2 + w - 30$ ✓

d) To factor $a^2 - 3a - 54$, find two integers whose *product* is -54 and whose *sum* is -3. Since the last term in the trinomial is negative, one of the integers must be positive and the other must be negative.

Find the integers mentally. First, think about two integers whose product is *positive* 54: 1 and 54, 2 and 27, 3 and 18, 6 and 9. One number must be positive and the other negative, however, to get our product of -54, and they must add up to -3.

Factors of -54	Sum of the Factors
$-6 \cdot 9$	$-6 + 9 = 3$
$6 \cdot (-9)$	$6 + (-9) = -3$

The numbers are 6 and -9: $a^2 - 3a - 54 = (a + 6)(a - 9)$.

The check is left to the student.

e) To factor $c^2 - 6c + 9$, notice that the *product*, 9, is positive and the *sum*, -6, is negative. So both integers must be negative. The numbers that multiply to 9 and add to -6 are the same number, -3 and -3: $(-3) \cdot (-3) = 9$ and $-3 + (-3) = -6$.

$$\text{So } c^2 - 6c + 9 = (c - 3)(c - 3) \text{ or } (c - 3)^2.$$

Either form of the factorization is correct.

f) To factor $y^2 + 11y + 35$, find two integers whose *product* is 35 and whose *sum* is 11. We are looking for two positive numbers.

Factors of 35	Sum of the Factors
$1 \cdot 35$	$1 + 35 = 36$
$5 \cdot 7$	$5 + 7 = 12$

There are no such factors! Therefore, $y^2 + 11y + 35$ does not factor using the methods we have learned here. We say that it is **prime.**

Note

We say that trinomials like $y^2 + 11y + 35$ are **prime** if they cannot be factored using the method presented here.

In later mathematics courses, you may learn how to factor such polynomials using other methods so that they are not considered prime.

[**YOU TRY 2**] Factor, if possible.

a) $m^2 + 11m + 28$ b) $c^2 - 16c + 48$ c) $a^2 - 5a - 21$

d) $r^2 - 4r - 45$ e) $r^2 + 5r - 24$ f) $h^2 + 12h + 36$

---

3 More on Factoring a Trinomial of the Form $x^2 + bx + c$

Sometimes it is necessary to factor out the GCF before applying this method for factoring trinomials.

Note

From this point on, the *first* step in factoring *any* polynomial should be to ask yourself, "Can I factor out a greatest common factor?"

Since some polynomials can be factored more than once, after performing one factorization, ask yourself, "Can I factor again?" If so, factor again. If not, you know that the polynomial has been completely factored.

EXAMPLE 3

Factor $4n^3 - 12n^2 - 40n$ completely.

Solution

Ask yourself, *"Can I factor out a GCF?"* Yes. The GCF is $4n$.

$$4n^3 - 12n^2 - 40n = 4n(n^2 - 3n - 10)$$

W Hint

Add *"Can I factor out a GCF?"* as the first step in this section's procedure.

Look at the trinomial and ask yourself, *"Can I factor again?"* Yes. The integers whose product is -10 and whose sum is -3 are -5 and 2. Therefore,

$$4n^3 - 12n^2 - 40n = 4n(n^2 - 3n - 10)$$
$$= 4n(n - 5)(n + 2)$$

Ask yourself, *"Can I factor again?"* No. Therefore, $4n^3 - 12n^2 - 40n = 4n(n - 5)(n + 2)$.

Check: $4n(n - 5)(n + 2) = 4n(n^2 + 2n - 5n - 10)$
$$= 4n(n^2 - 3n - 10)$$
$$= 4n^3 - 12n^2 - 40n \checkmark$$

[YOU TRY 3]

Factor completely.

a) $7p^4 + 42p^3 + 56p^2$ b) $3a^2b - 33ab + 90b$

4 Factor a Trinomial Containing Two Variables

If a trinomial contains two variables and we cannot take out a GCF, the trinomial may still be factored according to the method outlined in this section.

EXAMPLE 4

Factor $x^2 + 12xy + 32y^2$ completely.

Solution

Ask yourself, *"Can I factor out a GCF?"* No. Notice that the first term is x^2. Let's rewrite the trinomial as

$$x^2 + 12yx + 32y^2$$

so that we can think of $12y$ as the coefficient of x. Find two expressions whose product is $32y^2$ and whose sum is $12y$. They are $4y$ and $8y$ since $4y \cdot 8y = 32y^2$ and $4y + 8y = 12y$.

$$x^2 + 12xy + 32y^2 = (x + 4y)(x + 8y)$$

We cannot factor $(x + 4y)(x + 8y)$ any more, so this is the complete factorization. The check is left to the student.

[YOU TRY 4]

Factor completely.

a) $m^2 + 10mn + 16n^2$ b) $5a^3 + 40a^2b - 45ab^2$

ANSWERS TO [YOU TRY] EXERCISES

1) a) 3, 7 b) −6, 3 c) −2, −10   2) a) $(m + 4)(m + 7)$ b) $(c - 12)(c - 4)$ c) prime
d) $(r - 9)(r + 5)$ e) $(r + 8)(r - 3)$ f) $(h + 6)(h + 6)$ or $(h + 6)^2$ 3) a) $7p^2(p + 4)(p + 2)$
b) $3b(a - 5)(a - 6)$ 4) a) $(m + 2n)(m + 8n)$ b) $5a(a - b)(a + 9b)$

Objective 1: Practice Arithmetic Skills Needed for Factoring Trinomials

1) Find two integers whose

	PRODUCT IS	and whose SUM IS	ANSWER
a)	10	7	
b)	−56	−1	
c)	−5	4	
d)	36	−13	

2) Find two integers whose

	PRODUCT IS	and whose SUM IS	ANSWER
a)	42	−13	
b)	−14	13	
c)	54	15	
d)	−21	−4	

Objective 2: Factor a Trinomial of the Form $x^2 + bx + c$

3) If $x^2 + bx + c$ factors to $(x + m)(x + n)$ and if c is positive and b is negative, what do you know about the signs of m and n?

4) If $x^2 + bx + c$ factors to $(x + m)(x + n)$ and if b and c are positive, what do you know about the signs of m and n?

5) When asked to factor a polynomial, what is the first question you should ask yourself?

6) What does it mean to say that a polynomial is prime?

7) After factoring a polynomial, what should you ask yourself to be sure that the polynomial is completely factored?

8) How do you check the factorization of a polynomial?

Complete the factorization.

9) $n^2 + 7n + 10 = (n + 5)(\quad)$

10) $p^2 + 11p + 28 = (p + 4)(\quad)$

11) $c^2 - 16c + 60 = (c - 6)(\quad)$

12) $t^2 - 12t + 27 = (t - 9)(\quad)$

13) $x^2 + x - 12 = (x - 3)(\quad)$

14) $r^2 - 8r - 9 = (r + 1)(\quad)$

Factor completely, if possible. Check your answer.

15) $g^2 + 8g + 12$

16) $p^2 + 9p + 14$

17) $y^2 + 10y + 16$

18) $a^2 + 11a + 30$

19) $w^2 - 17w + 72$

20) $d^2 - 14d + 33$

21) $b^2 - 3b - 4$

22) $t^2 + 2t - 48$

23) $z^2 + 6z - 11$

24) $x^2 - 7x - 15$

25) $c^2 - 13c + 36$

26) $h^2 - 13h + 12$

27) $m^2 + 4m - 60$

28) $v^2 - 4v - 45$

29) $r^2 - 4r - 96$

30) $a^2 - 21a + 110$

31) $q^2 + 12q + 42$

32) $d^2 - 15d + 32$

33) $x^2 + 16x + 64$

34) $c^2 - 10c + 25$

35) $n^2 - 2n + 1$

36) $w^2 + 20w + 100$

37) $24 + 14d + d^2$

38) $10 + 7k + k^2$

39) $-56 + 12a + a^2$

40) $63 + 21w + w^2$

Objective 3: More on Factoring a Trinomial of the Form $x^2 + bx + c$

Factor completely, if possible. Check your answer.

41) $2k^2 - 22k + 48$

42) $6v^2 + 54v + 120$

43) $50h + 35h^2 + 5h^3$

44) $3d^3 - 33d^2 - 36d$

45) $r^4 + r^3 - 132r^2$

46) $2n^4 - 40n^3 + 200n^2$

47) $7q^3 - 49q^2 - 42q$

48) $8b^4 + 24b^3 + 16b^2$

49) $3z^4 + 24z^3 + 48z^2$

50) $-36w + 6w^2 + 2w^3$

51) $xy^3 - 2xy^2 - 63xy$

52) $2c^3d - 14c^2d - 24cd$

Factor completely by first taking out −1 and then factoring the trinomial, if possible. Check your answer.

53) $-m^2 - 12m - 35$

54) $-x^2 - 15x - 36$

55) $-c^2 - 3c + 28$

56) $-t^2 + 2t + 48$

57) $-z^2 + 13z - 30$

58) $-n^2 + 16n - 55$

59) $-p^2 + p + 56$

60) $-w^2 - 2w + 3$

Objective 4: Factor a Trinomial Containing Two Variables

Factor completely. Check your answer.

61) $x^2 + 7xy + 12y^2$

62) $a^2 + 11ab + 18b^2$

63) $c^2 - 7cd - 8d^2$

64) $p^2 + 6pq - 72q^2$

65) $u^2 - 14uv + 45v^2$ 66) $h^2 - 8hk + 7k^2$

67) $m^2 + 4mn - 21n^2$ 68) $a^2 - 6ab - 40b^2$

69) $a^2 + 24ab + 144b^2$ 70) $g^2 + 6gh + 5h^2$

Determine whether each polynomial is factored completely. If it is not, explain why and factor it completely.

71) $3x^2 + 21x + 30 = (3x + 6)(x + 5)$

72) $6a^2 + 24a - 72 = 6(a + 6)(a - 2)$

73) $n^4 - 3n^3 - 108n^2 = n^2(n - 12)(n + 9)$

74) $9y^3 - 45y^2 + 54y = (y - 2)(9y^2 - 27y)$

Mixed Exercises: Objectives 2–4

Factor completely. Begin by asking yourself, *"Can I factor out a GCF?"*

75) $2x^2 + 16x + 30$ 76) $3c^2 + 21c + 18$

77) $n^2 - 6n + 8$ 78) $a^2 + a - 6$

79) $m^2 + 7mn - 44n^2$ 80) $a^2 + 10ab + 24b^2$

81) $h^2 - 10h + 32$ 82) $z^2 + 9z + 36$

83) $4q^3 - 28q^2 + 48q$ 84) $3w^3 - 9w^2 - 120w$

85) $-k^2 - 18k - 81$ 86) $-y^2 + 8y - 16$

87) $4h^5 + 32h^4 + 28h^3$ 88) $3r^4 - 6r^3 - 45r^2$

89) $k^2 + 21k + 108$ 90) $j^2 - 14j - 15$

91) $p^3q - 17p^2q^2 + 70pq^3$ 92) $u^3v^2 - 2u^2v^3 - 15uv^4$

93) $a^2 + 9ab + 24b^2$ 94) $m^2 - 8mn - 35n^2$

95) $x^2 - 13xy + 12y^2$ 96) $p^2 - 3pq - 40q^2$

97) $5v^5 + 55v^4 - 45v^3$ 98) $6t^4 + 42t^3 + 48t^2$

99) $6x^3y^2 - 48x^2y^2 - 54xy^2$

100) $2c^2d^4 - 18c^2d^3 + 28c^2d^2$

101) $36 - 13z + z^2$ 102) $121 + 22w + w^2$

103) $a^2b^2 + 13ab + 42$ 104) $h^2k^2 + 8hk - 20$

105) $(x + y)z^2 + 7(x + y)z - 30(x + y)$

106) $(m + n)k^2 + 17(m + n)k + 66(m + n)$

107) $(a - b)c^2 - 11(a - b)c + 28(a - b)$

108) $(r - t)u^2 - 4(r - t)u - 45(r - t)$

109) $(p + q)r^2 + 24r(p + q) + 144(p + q)$

110) $(a + b)d^2 - 8(a + b)d + 16(a + b)$

R Rethink

R1) Do you have quick recall of the multiplication facts from 1 to 12, or do you need more practice? If so, practice them online or by using flash cards.

R2) Were you able to do most of the factoring in your head?

R3) Could you complete similar exercises without looking at your notes?

7.3 Factoring Trinomials of the Form $ax^2 + bx + c$ ($a \neq 1$)

P Prepare

O Organize

What are your objectives for Section 7.3?	How can you accomplish each objective?
1 Factor $ax^2 + bx + c$ ($a \neq 1$) by Grouping	• Write your own procedure for **Factoring** $ax^2 + bx + c$ ($a \neq 1$) **by Grouping** in your notes by following Objective 1's introduction. • Complete the given examples on your own. • Complete You Trys 1 and 2.
2 Factor $ax^2 + bx + c$ ($a \neq 1$) by Trial and Error	• Summarize in your notes how you would factor by trial and error. • Complete the given examples on your own. • Complete You Trys 3 and 4.
3 Factor Using Substitution	• Understand the method of using *substitution*. • Complete the given example on your own. • Complete You Try 5.

W Work

Read the explanations, follow the examples, take notes, and complete the You Trys.

In the previous section, we learned that we could factor $2x^2 + 10x + 8$ by first taking out the GCF of 2 and then factoring the trinomial.

$$2x^2 + 10x + 8 = 2(x^2 + 5x + 4) = 2(x + 4)(x + 1)$$

In this section, we will learn how to factor a trinomial like $2x^2 + 11x + 15$ where we *cannot* factor out the leading coefficient of 2.

1 Factor $ax^2 + bx + c$ ($a \neq 1$) by Grouping

Sum is 11.

To factor $2x^2 + 11x + 15$, first find the product of 2 and 15. Then, find two integers

Product: $2 \cdot 15 = 30$

whose *product* is 30 and whose *sum* is 11. The numbers are 6 and 5. Rewrite the middle term, $11x$, as $6x + 5x$, then factor by grouping.

$$
\begin{aligned}
2x^2 + 11x + 15 &= \underline{2x^2 + 6x} + \underline{5x + 15} \\
&= 2x(x + 3) + 5(x + 3) \quad \text{\small Take out the common factor from each group.} \\
&= (x + 3)(2x + 5) \quad \text{\small Factor out } (x + 3).
\end{aligned}
$$

Check: $(x + 3)(2x + 5) = 2x^2 + 5x + 6x + 15 = 2x^2 + 11x + 15$ ✓

EXAMPLE 1

Factor completely.

a) $8k^2 + 14k + 3$ b) $6c^2 - 17c + 12$ c) $7x^2 - 34xy - 5y^2$

Solution

a) Since we cannot factor out a GCF (the GCF = 1), we begin with a new method.

Sum is 14.
$\downarrow$
$8k^2 + 14k + 3$
Product: 8 · 3 = 24

Think of two integers whose *product* is 24 and whose *sum* is 14. The integers are 2 and 12.
Rewrite the middle term, $14k$, as $2k + 12k$. Factor by grouping.

$$8k^2 + 14k + 3 = \underbrace{8k^2 + 2k} + \underbrace{12k + 3}$$
$$= 2k(4k + 1) + 3(4k + 1) \qquad \text{Take out the common factor from each group.}$$
$$= (4k + 1)(2k + 3) \qquad \text{Factor out } (4k + 1).$$

Check by multiplying: $(4k + 1)(2k + 3) = 8k^2 + 14k + 3$ ✓

W Hint

Write the procedure, in your own words, for factoring these polynomials by grouping.

b)
Sum is −17.
$\downarrow$
$6c^2 - 17c + 12$
Product: 6 · 12 = 72

Think of two integers whose *product* is 72 and whose *sum* is −17. (Both numbers will be negative.) The integers are −9 and −8.
Rewrite the middle term, $-17c$, as $-9c - 8c$. Factor by grouping.

$$6c^2 - 17c + 12 = \underbrace{6c^2 - 9c} \underbrace{- 8c + 12}$$
$$= 3c(2c - 3) - 4(2c - 3) \qquad \text{Take out the common factor from each group.}$$
$$= (2c - 3)(3c - 4) \qquad \text{Factor out } (2c - 3).$$

Check: $(2c - 3)(3c - 4) = 6c^2 - 17c + 12$ ✓

c)
Sum is −34.
$\downarrow$
$7x^2 - 34xy - 5y^2$
Product: 7 · (−5) = −35

The integers whose *product* is −35 and whose *sum* is −34 are −35 and 1.
Rewrite the middle term, $-34xy$, as $-35xy + xy$. Factor by grouping.

$$7x^2 - 34xy - 5y^2 = \underbrace{7x^2 - 35xy} + \underbrace{xy - 5y^2}$$
$$= 7x(x - 5y) + y(x - 5y) \qquad \text{Take out the common factor from each group.}$$
$$= (x - 5y)(7x + y) \qquad \text{Factor out } (x - 5y).$$

Check: $(x - 5y)(7x + y) = 7x^2 - 34xy - 5y^2$ ✓

$\left[\text{YOU TRY 1}\right]$ Factor completely.

a) $4p^2 + 16p + 15$ b) $10y^2 - 13y + 4$ c) $5a^2 - 29ab - 6b^2$

EXAMPLE 2 Factor $12n^2 + 64n - 48$ completely.

Solution

It is tempting to jump right in and multiply $12 \cdot (-48) = -576$ and try to think of two integers with a product of −576 and a sum of 64. However, first ask yourself, *"Can I factor out a GCF?"* Yes! We can factor out 4.

$$12n^2 + 64n - 48 = 4(3n^2 + 16n - 12) \qquad \text{Factor out 4.}$$
Product: 3 · (−12) = −36

Now factor $3n^2 + 16n - 12$ by finding two integers whose *product* is -36 and whose *sum* is 16. The numbers are 18 and -2.

$$= 4(\underbrace{3n^2 + 18n}_{} \underbrace{- 2n - 12}_{})$$

$$= 4[3n(n + 6) - 2(n + 6)] \qquad \text{Take out the common factor from each group.}$$

$$= 4(n + 6)(3n - 2) \qquad \text{Factor out } (n + 6).$$

Check by multiplying: $4(n + 6)(3n - 2) = 4(3n^2 + 16n - 12)$

$$= 12n^2 + 64n - 48 \quad \checkmark$$

[YOU TRY 2]

Factor completely.

a) $24h^2 - 54h - 15$ b) $20d^3 + 38d^2 + 12d$

---

2 Factor $ax^2 + bx + c$ $(a \neq 1)$ by Trial and Error

At the beginning of this section, we factored $2x^2 + 11x + 15$ by grouping. Now we will factor it by trial and error, which is just reversing the process of FOIL.

EXAMPLE 3

Factor $2x^2 + 11x + 15$ completely.

Solution

Can we factor out a GCF? No. So try to factor $2x^2 + 11x + 15$ as the product of two binomials. Notice that all terms are positive, so all factors will be positive.

Begin with the squared term, $2x^2$. Which two expressions with integer coefficients can we multiply to get $2x^2$? $2x$ and x. Put these in the binomials.

$$2x^2 + 11x + 15 = (2x \quad)(x \quad) \qquad 2x \cdot x = 2x^2$$

Next, look at the last term, 15. What are the pairs of positive integers that multiply to 15? They are 15 and 1 as well as 5 and 3.

Try these numbers as the last terms of the binomials. The middle term, $11x$, comes from finding the sum of the products of the outer terms and inner terms.

First Try

$2x^2 + 11x + 15 \overset{?}{=} (2x + 15)(x + 1)$ Incorrect!

$15x$

These must both be $11x$. $+ \underline{2x}$
                                    $17x$

Try again. Switch the 15 and the 1.

$2x^2 + 11x + 15 \overset{?}{=} (2x + 1)(x + 15)$ Incorrect!

$1x$

These must both be $11x$. $+ \underline{30x}$
                                    $31x$

Try using 5 and 3. $2x^2 + 11x + 15 \overset{?}{=} (2x + 5)(x + 3)$ Correct!

$5x$

These must both be $11x$. $+ \underline{6x}$
                                    $11x$

Therefore, $2x^2 + 11x + 15 = (2x + 5)(x + 3)$. Check by multiplying.

EXAMPLE 4

Factor $3t^2 - 29t + 18$ completely.

Solution

Can we factor out a GCF? No. To get a product of $3t^2$, we will use $3t$ and t.

$$3t^2 - 29t + 18 = (3t \quad)(t \quad) \qquad 3t \cdot t = 3t^2$$

Since the last term is positive and the middle term is negative, we want pairs of negative integers that multiply to 18. The pairs are -1 and -18, -2 and -9, and -3 and -6. Try these numbers as the last terms of the binomials. The middle term, $-29t$, comes from finding the sum of the products of the outer terms and inner terms.

$$3t^2 - 29t + 18 \overset{?}{=} (3t - 1)(t - 18) \qquad \text{Incorrect!}$$

These must both be $-29t$.

$$-t$$
$$+ (-54t)$$
$$-55t$$

Try again. Switch the -1 and the -18: $\quad 3t^2 - 29t + 18 \overset{?}{=} (3t - 18)(t - 1)$

Without multiplying, we know that this choice is incorrect. How? In the factor $(3t - 18)$, a 3 can be factored out to get $3(t - 6)$. But, we concluded that we could not factor out a GCF from the original polynomial, $3t^2 - 29t + 18$. Therefore, it will not be possible to take out a common factor from one of the binomial factors.

> **Note**
>
> If you cannot factor out a GCF from the original polynomial, then you cannot take out a factor from one of the binomial factors either.

Try using -2 and -9. $\quad 3t^2 - 29t + 18 = (3t - 2)(t - 9) \qquad \text{Correct!}$

These must both be $-29t$.

$$-2t$$
$$+ (-27t)$$
$$-29t$$

So, $3t^2 - 29t + 18 = (3t - 2)(t - 9)$. Check by multiplying.

[YOU TRY 3]

Factor completely.

a) $2k^2 + 17k + 8$ b) $6z^2 - 23z + 20$

EXAMPLE 5

Factor completely.

a) $16a^2 + 62a - 8$ b) $-2c^2 + 3c + 20$

Solution

a) Ask yourself, *"Can I take out a common factor?"* Yes!

$$16a^2 + 62a - 8 = 2(8a^2 + 31a - 4)$$

Now, try to factor $8a^2 + 31a - 4$. To get a product of $8a^2$, we can try either $8a$ *and* a or $4a$ *and* $2a$. Let's start by trying $8a$ *and* a.

$$8a^2 + 31a - 4 = (8a \quad)(a \quad)$$

List pairs of integers that multiply to −4: 4 and −1, −4 and 1, 2 and −2.

   Try 4 and −1. Do not put 4 in the same binomial as $8a$ since then it would be possible to factor out 2. But, 2 does not factor out of $8a^2 + 31a - 4$. Put the 4 in the same binomial as a.

$$8a^2 + 31a - 4 \overset{?}{=} (8a - 1)(a + 4)$$

$$\begin{array}{c} -a \\ + \dfrac{32a}{31a} \quad \text{Correct} \end{array}$$

Don't forget that the very first step was to factor out a 2. Therefore,

$$16a^2 + 62a - 8 = 2(8a^2 + 31a - 4) = 2(8a - 1)(a + 4)$$

Check by multiplying.

b) Since the coefficient of the squared term is negative, begin by factoring out −1. (There is no other common factor except 1.)

$$-2c^2 + 3c + 20 = -1(2c^2 - 3c - 20)$$

Try to factor $2c^2 - 3c - 20$. To get a product of $2c^2$, we will use $2c$ and c in the binomials.

$$2c^2 - 3c - 20 = (2c\quad)(c\quad)$$

We need pairs of integers whose product is −20. They are 1 and −20, −1 and 20, 2 and −10, −2 and 10, 4 and −5, −4 and 5.

Do *not* start with 1 and −20 or −1 and 20 because the middle term, −3c, is not very large. Using 1 and −20 or −1 and 20 would likely result in a larger middle term.

Think about 2 and −10 *and* −2 and 10. *These will not work because if we put any of these numbers in the factor containing 2c, then it will be possible to factor out 2.*

Try 4 and −5. Do not put 4 in the same binomial as $2c$ since then it would be possible to factor out 2.

$$2c^2 - 3c - 20 \overset{?}{=} (2c - 5)(c + 4)$$

$$\begin{array}{c} -5c \\ + \dfrac{8c}{3c} \quad \text{This must equal } -3c. \quad \text{Incorrect} \end{array}$$

Only the sign of the sum is incorrect. *Change the signs in the binomials to get the correct sum.*

$$2c^2 - 3c - 20 \overset{?}{=} (2c + 5)(c - 4)$$

$$\begin{array}{c} 5c \\ + \dfrac{(-8c)}{-3c} \quad \text{Correct} \end{array}$$

Remember that we factored out −1 to begin the problem.

$$-2c^2 + 3c + 20 = -1(2c^2 - 3c - 20) = -(2c + 5)(c - 4)$$

Check by multiplying.

W Hint

Be sure to read the reasoning in this example, and apply it to the You Try!

[**YOU TRY 4**]   Factor completely.

a) $10y^2 - 58y + 40$ b) $-4n^2 - 5n + 6$

We have seen two methods for factoring $ax^2 + bx + c \ (a \neq 1)$: factoring by grouping and factoring by trial and error. In either case, remember to begin by taking out a common factor from all terms whenever possible.

3 Factor Using Substitution

Some polynomials can be factored using a method called **substitution.** We will use this method in Example 6.

EXAMPLE 6

Factor $3(x + 4)^2 + 11(x + 4) + 10$ completely.

Solution

Notice that the binomial $x + 4$ appears as a *squared quantity* and a *linear quantity*. We will use another letter to represent this quantity. We can use any letter except x. Let's use the letter u.

$$\text{Let } u = x + 4. \text{ Then, } u^2 = (x + 4)^2.$$

Substitute the u and u^2 into the original polynomial, and factor.

$$3(x + 4)^2 + 11(x + 4) + 10 =$$
$$3u^2 \qquad + 11u \qquad + 10 = (3u + 5)(u + 2)$$

Hint

Try using different color pencils or pens, as done here, when writing this problem in your notes. That can help you to see the substitution.

$$\qquad\qquad = [3(x + 4) + 5][(x + 4) + 2] \qquad \text{Since the original polynomial was in terms of } x, \text{ substitute } x + 4 \text{ for } u.$$

$$\qquad\qquad = (3x + 12 + 5)(x + 4 + 2) \qquad \text{Distribute.}$$

$$\qquad\qquad = (3x + 17)(x + 6) \qquad \text{Combine like terms.}$$

BE CAREFUL The final factorization is **not** the one containing the substitution variable, u. We must go back and replace u with the expression it represented so that the factorization is in terms of the original variable.

[YOU TRY 5]

Factor $2(x - 3)^2 + 3(x - 3) - 35$ completely.

Using Technology

We found some ways to narrow down the possibilities when factoring $ax^2 + bx + c \ (a \neq 1)$ using the trial and error method.

We can also use a graphing calculator to help with the process. Consider the trinomial $2x^2 - 9x - 35$. Enter the trinomial into Y_1 and press $\boxed{\text{ZOOM}}$; then enter 6 to display the graph in the standard viewing window.

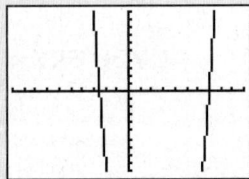

Look on the graph for the x-intercept (if any) that appears to be an integer. It appears that 7 is an x-intercept.

To check whether 7 is an x-intercept, press TRACE then 7 and press ENTER. As shown on the graph, when $x = 7$, $y = 0$, so 7 is an x-intercept.

When an x-intercept is an integer, then x minus that x-intercept is a factor of the trinomial. In this case, $x - 7$ is a factor of $2x^2 - 9x - 35$. We can then complete the factoring as $(2x + 5)(x - 7)$, since we must multiply -7 by 5 to obtain -35.

Find an x-intercept using a graphing calculator and factor the trinomial.

1) $3x^2 + 11x - 4$ 2) $2x^2 + x - 15$ 3) $5x^2 + 6x - 8$

4) $2x^2 - 5x + 3$ 5) $4x^2 - 3x - 10$ 6) $14x^2 - x - 4$

ANSWERS TO YOU TRY EXERCISES

1) a) $(2p + 5)(2p + 3)$ b) $(5y - 4)(2y - 1)$ c) $(5a + b)(a - 6b)$
2) a) $3(2h - 5)(4h + 1)$ b) $2d(5d + 2)(2d + 3)$ 3) a) $(2k + 1)(k + 8)$ b) $(3z - 4)(2z - 5)$
4) a) $2(5y - 4)(y - 5)$ b) $-(4n - 3)(n + 2)$ 5) $(2x - 13)(x + 2)$

ANSWERS TO TECHNOLOGY EXERCISES

1) $(3x - 1)(x + 4)$ 2) $(2x - 5)(x + 3)$ 3) $(x + 2)(5x - 4)$ 4) $(x - 1)(2x - 3)$
5) $(x - 2)(4x + 5)$ 6) $(2x + 1)(7x - 4)$

E Evaluate   **7.3**   Exercises   Do the exercises, and check your work.

Objective 1: Factor $ax^2 + bx + c$ ($a \neq 1$) by Grouping

1) Find two integers whose

	PRODUCT IS	and whose SUM IS	ANSWER
a)	-50	5	
b)	27	-28	
c)	12	8	
d)	-72	-6	

2) Find two integers whose

	PRODUCT IS	and whose SUM IS	ANSWER
a)	18	19	
b)	-132	1	
c)	-30	-13	
d)	63	-16	

Factor by grouping.

 3) $3c^2 + 12c + 8c + 32$

4) $5y^2 + 15y + 2y + 6$

5) $6k^2 - 6k - 7k + 7$

6) $4r^2 - 4r + 9r - 9$

7) $6x^2 - 27xy + 8xy - 36y^2$

8) $14p^2 - 8pq - 7pq + 4q^2$

9) When asked to factor a polynomial, what is the first question you should ask yourself?

10) After factoring a polynomial, what should you ask yourself to be sure that the polynomial is factored completely?

11) Find the polynomial that factors to $(4k + 9)(k + 2)$.

12) Find the polynomial that factors to $(6m - 5)(2m - 3)$.

Complete the factorization.

13) $5t^2 + 13t + 6 = (5t + 3)($ $)$

14) $4z^2 + 29z + 30 = (4z + 5)($ $)$

(24) 15) $6a^2 - 11a - 10 = (2a - 5)($ $)$

16) $15c^2 - 23c + 4 = (3c - 4)($ $)$

17) $12x^2 - 25xy + 7y^2 = (4x - 7y)($ $)$

18) $12r^2 - 52rt - 9t^2 = (6r + t) ($ $)$

Factor by grouping. See Example 1.

19) $2h^2 + 13h + 15$ 20) $3z^2 + 13z + 14$

21) $7y^2 - 11y + 4$ 22) $5a^2 - 21a + 18$

(24) 23) $5b^2 + 9b - 18$ 24) $11m^2 - 18m - 8$

25) $6p^2 + p - 2$ 26) $8c^2 - 22c + 5$

27) $4t^2 + 16t + 15$ 28) $10k^2 + 23k + 12$

29) $9x^2 - 13xy + 4y^2$ 30) $6a^2 + ab - 5b^2$

Objective 2: Factor $ax^2 + bx + c$ $(a \neq 1)$ by Trial and Error

31) How do we know that $(2x - 4)$ cannot be a factor of $2x^2 + 13x - 24$?

32) How do we know that $(3p + 2)$ cannot be a factor of $6p^2 - 25p + 14$?

Factor by trial and error. See Examples 3 and 4.

33) $2r^2 + 9r + 10$ 34) $3q^2 + 10q + 8$

(24) 35) $3u^2 - 23u + 30$ 36) $7m^2 - 15m + 8$

37) $7a^2 + 31a - 20$ 38) $5x^2 - 11x - 36$

39) $6y^2 + 23y + 10$ 40) $8u^2 + 18u + 7$

41) $9w^2 + 20w - 21$ 42) $10h^2 - 59h - 6$

43) $8c^2 - 42c + 27$ 44) $15v^2 - 16v + 4$

45) $4k^2 + 40k + 99$ 46) $4n^2 - 41n + 10$

47) $20b^2 - 32b - 45$ 48) $14g^2 + 27g - 20$

49) $2r^2 + 13rt - 24t^2$ 50) $3c^2 - 17cd - 6d^2$

51) $6a^2 - 25ab + 4b^2$ 52) $6x^2 + 31xy + 18y^2$

53) Factor $4z^2 + 5z - 6$ using each method. Do you get the same answer? Which method do you prefer? Why?

54) Factor $10a^2 + 27a + 18$ using each method. Do you get the same answer? Which method do you prefer? Why?

Use the trinomial $ax^2 + bx + c$ $(a > 0)$ to answer *always, sometimes,* or *never* to Exercises 55 and 56.

55) If the product of a and c is negative, both factors will have a minus sign between terms.

56) If the product of a and c is positive, both factors will have a minus sign between terms.

Factor completely by first taking out a negative common factor. See Example 5.

57) $-n^2 - 8n + 48$ 58) $-c^2 - 16c - 63$

(24) 59) $-10z^2 + 19z - 6$ 60) $-7a^2 + 4a + 3$

61) $-3z^3 - 15z^2 + 198z$

62) $-20m^3 - 120m^2 - 135m$

Objective 3: Factor Using Substitution

Use substitution to factor each polynomial.

(24) 63) $(n + 5)^2 + 6(n + 5) - 27$

64) $(p - 6)^2 + 11(p - 6) + 28$

65) $(k - 3)^2 - 9(k - 3) + 8$

66) $(t + 4)^2 - 10(t + 4) - 24$

67) $2(w + 1)^2 - 13(w + 1) + 15$

68) $3(c - 9)^2 + 14(c - 9) + 16$

69) $6(2y - 1)^2 - 5(2y - 1) - 4$

70) $10(3a + 2)^2 - 19(3a + 2) + 6$

Mixed Exercises: Objectives 1–3

Factor completely.

71) $3p^2 - 16p - 12$ 72) $2t^2 - 19t + 24$

73) $9k^2 - 42k + 49$ 74) $25p^2 + 20p + 4$

75) $2d^2 + 2d - 40$ 76) $6c^2 + 42c + 72$

77) $7x^2 - 17xy + 6y^2$ 78) $5a^2 + 23ab + 12b^2$

79) $4q^3 - 28q^2 + 48q$ 80) $12x^3 + 15x^2 - 18x$

81) $6y^2 - 10y + 3$ 82) $9z^2 + 14z + 8$

83) $-10z^2 + 19z - 6$

84) $4(2h + 1)^2 - 3(2h + 1) - 22$

85) $12p^2(q - 1)^2 - 49p(q - 1)^2 + 49(q - 1)^2$

86) $-h^2 - 3h + 54$

87) $3(b + 5)^2 + 4(b + 5) - 20$

88) $(x + y)t^2 - 4(x + y)t - 21(x + y)$

89) $30r^4t^2 + 23r^3t^2 + 3r^2t^2$

90) $8m^2n^3 + 4m^2n^2 - 60m^2n$

R1) How have your arithmetic skills helped you complete these exercises?

R2) Which method of factoring did you prefer while completing the exercises? Why?

R3) How much more practice will you need to master the objectives of this section?

7.4 Factoring Special Trinomials and Binomials

P Prepare

O Organize

What are your objectives for Section 7.4?	How can you accomplish each objective?
1 Factor a Perfect Square Trinomial	• Learn the formula for **Factoring a Perfect Square Trinomial.** • Review all of the *perfect squares.* • Always begin by asking yourself, *"Can I factor out a GCF?"* and, after factoring, always ask yourself, *"Can I factor again?"* • Complete the given examples on your own. • Complete You Try 1.
2 Factor the Difference of Two Squares	• Learn the formula for **Factoring the Difference of Two Squares.** • Always begin by asking yourself, *"Can I factor out a GCF?"* and, after factoring, always ask yourself, *"Can I factor again?"* • Recognize the difference between the *sum of two squares* and the *difference of two squares.* • Complete the given examples on your own. • Complete You Trys 2 and 3.
3 Factor the Sum and Difference of Two Cubes	• Learn the formula for **Factoring the Sum and Difference of Two Cubes.** • Review all the *perfect cubes.* • Always begin by asking yourself, *"Can I factor out a GCF?"* and, after factoring, always ask yourself, *"Can I factor again?"* • Complete the given examples on your own. • Complete You Trys 4 and 5.

W Work

Read the explanations, follow the examples, take notes, and complete the You Trys.

1 Factor a Perfect Square Trinomial

Recall that we can square a binomial using the formulas

$$(a + b)^2 = a^2 + 2ab + b^2$$
$$(a - b)^2 = a^2 - 2ab + b^2$$

For example, $(x + 5)^2 = x^2 + 2x(5) + 5^2 = x^2 + 10x + 25$.

Since factoring a polynomial means writing the polynomial as a product of its factors, $x^2 + 10x + 25$ factors to $(x + 5)^2$.

The expression $x^2 + 10x + 25$ is a *perfect square trinomial*. A **perfect square trinomial** is a trinomial that results from squaring a binomial.

We can use the factoring methods presented in Sections 7.2 and 7.3 to factor a perfect square trinomial, or we can learn to recognize the special pattern that appears in these trinomials. Above we stated that $x^2 + 10x + 25$ factors to $(x + 5)^2$. How are the terms of the trinomial and binomial related?

Compare $x^2 + 10x + 25$ to $(x + 5)^2$.

x^2 is the square of x, the first term in the binomial.

25 is the square of 5, the last term in the binomial.

We get the term $10x$ by doing the following:

$$10x = 2 \quad \cdot \quad x \quad \cdot \quad 5$$

Two   First term   Last term in binomial
times   in binomial

This follows directly from how we found $(x + 5)^2$ using the formula.

Formula Factoring a Perfect Square Trinomial

$$a^2 + 2ab + b^2 = (a + b)^2$$
$$a^2 - 2ab + b^2 = (a - b)^2$$

Note

In order for a trinomial to be a perfect square, two of its terms must be perfect squares.

EXAMPLE 1

Factor $t^2 + 12t + 36$ completely.

Solution

We cannot take out a common factor, so let's see if this follows the pattern of a perfect square trinomial.

$$t^2 + 12t + 36$$

What do you square to get t^2? t $(t)^2$ $(6)^2$ What do you square to get 36?   6

Does the middle term equal $2 \cdot t \cdot 6$? *Yes.*

$$2 \cdot t \cdot 6 = 12t$$

Therefore, $t^2 + 12t + 36 = (t + 6)^2$. Check by multiplying.

EXAMPLE 2

Factor completely.

a) $n^2 - 14n + 49$ b) $4p^3 + 24p^2 + 36p$

c) $9k^2 + 30k + 25$ d) $4c^2 + 20c + 9$

Solution

a) We cannot take out a common factor. However, since the middle term is negative and the first and last terms are positive, the sign in the binomial will be a minus ($-$) sign. Does this fit the pattern of a perfect square trinomial?

$$n^2 - 14n + 49$$

What do you square to get n^2? n $(n)^2$ $(7)^2$ What do you square to get 49?   7

Does the middle term equal $2 \cdot n \cdot 7$? *Yes:* $2 \cdot n \cdot 7 = 14n$

Since there is a minus sign in front of $14n$, $n^2 - 14n + 49$ fits the pattern of $a^2 - 2ab + b^2 = (a - b)^2$ with $a = n$ and $b = 7$.

Therefore, $n^2 - 14n + 49 = (n - 7)^2$. Check by multiplying.

W Hint

The coefficients of the first and last terms in the trinomials are both perfect squares.

W Hint

Don't forget to ask yourself, "Can I factor out a GCF?"

b) From $4p^3 + 24p^2 + 36p$ we *can* begin by taking out the GCF of $4p$.

$$4p^3 + 24p^2 + 36p = 4p(p^2 + 6p + 9)$$

What do you square to get p^2? p $(p)^2$ $(3)^2$ What do you square to get 9?   3

Does the middle term equal $2 \cdot p \cdot 3$? *Yes:* $2 \cdot p \cdot 3 = 6p$.

$$4p^3 + 24p^2 + 36p = 4p(p^2 + 6p + 9) = 4p(p + 3)^2$$

Check by multiplying.

c) We cannot take out a common factor. Since the first and last terms of $9k^2 + 30k + 25$ are perfect squares, let's see if this is a perfect square trinomial.

$$9k^2 + 30k + 25$$

What do you square to get $9k^2$? $3k$ $(3k)^2$ $(5)^2$ What do you square to get 25?   5

Does the middle term equal $2 \cdot 3k \cdot 5$? *Yes:* $2 \cdot 3k \cdot 5 = 30k$.

Therefore, $9k^2 + 30k + 25 = (3k + 5)^2$. Check by multiplying.

d) We cannot take out a common factor. The first and last terms of $4c^2 + 20c + 9$ are perfect squares. Is this a perfect square trinomial?

$$4c^2 + 20c + 9$$

What do you square to get $4c^2$? $2c$ $(2c)^2$ $(3)^2$ What do you square to get 9?   3

Does the middle term equal $2 \cdot 2c \cdot 3$? *No:* $2 \cdot 2c \cdot 3 = 12c$

This is *not* a perfect square trinomial. Applying a method from Section 7.3, we find that the trinomial does factor, however.

$4c^2 + 20c + 9 = (2c + 9)(2c + 1)$. Check by multiplying.

Factor completely.

a) $w^2 + 8w + 16$ b) $a^2 - 20a + 100$ c) $4d^2 - 36d + 81$

2   Factor the Difference of Two Squares

Another common type of factoring problem is a **difference of two squares.** Some examples of this type of binomial are

$$y^2 - 9, \qquad 25m^2 - 16n^2, \qquad 64 - t^2, \qquad \text{and} \qquad h^4 - 16.$$

Notice that in each binomial, the terms are being *subtracted,* and each term is a perfect square.

In Section 6.3, we saw that

$$(a + b)(a - b) = a^2 - b^2.$$

If we reverse the procedure, we get the factorization of the difference of two squares.

Formula   Factoring the Difference of Two Squares

$$a^2 - b^2 = (a + b)(a - b)$$

Don't forget that we can check all factorizations by multiplying.

EXAMPLE 3

Factor completely.

a) $y^2 - 9$ b) $25m^2 - 16n^2$ c) $w^2 - \dfrac{9}{64}$ d) $c^2 + 36$

Solution

a)   First, notice that $y^2 - 9$ is the difference of two terms *and* those terms are perfect squares. We can use the formula $a^2 - b^2 = (a + b)(a - b)$.

Identify a and b.

$$y^2 - 9$$
$$\downarrow \qquad \downarrow$$

What do you square $(y)^2 \quad (3)^2$ What do you square
to get y^2? y to get 9?   3

Then, $a = y$ and $b = 3$. Therefore, $y^2 - 9 = (y + 3)(y - 3)$.

W Hint

Notice in this example that the coefficients of the first and last terms of the binomials are both perfect squares!

b)   Look carefully at $25m^2 - 16n^2$. Each term *is* a perfect square, and they are being subtracted.

Identify a and b.

$$25m^2 - 16n^2$$
$$\downarrow \qquad \quad \downarrow$$

What do you square $(5m)^2 \quad (4n)^2$ What do you square
to get $25m^2$? $5m$ to get $16n^2$? $4n$

Then, $a = 5m$ and $b = 4n$. So, $25m^2 - 16n^2 = (5m + 4n)(5m - 4n)$.

c) Each term in $w^2 - \dfrac{9}{64}$ is a perfect square, and they are being subtracted.

$$w^2 - \frac{9}{64}$$

$$\downarrow \qquad \downarrow$$

What do you square to get w^2? w $\qquad (w)^2 \quad \left(\dfrac{3}{8}\right)^2 \qquad$ What do you square to get $\dfrac{9}{64}$? $\dfrac{3}{8}$

So, $a = w$ and $b = \dfrac{3}{8}$. Therefore, $w^2 - \dfrac{9}{64} = \left(w + \dfrac{3}{8}\right)\left(w - \dfrac{3}{8}\right)$.

d) Each term in $c^2 + 36$ is a perfect square, but the expression is the *sum* of two squares. This polynomial does not factor.

$$c^2 + 36 \ne (c + 6)(c - 6) \text{ since } (c + 6)(c - 6) = c^2 - 36.$$
$$c^2 + 36 \ne (c + 6)(c + 6) \text{ since } (c + 6)(c + 6) = c^2 + 12c + 36.$$

So, $c^2 + 36$ is prime.

Note

If the sum of two squares does not contain a common factor, then it cannot be factored.

YOU TRY 2

Factor completely.

a) $r^2 - 25$
b) $49p^2 - 121q^2$
c) $x^2 - \dfrac{25}{144}$
d) $h^2 + 1$

Remember that sometimes we can factor out a GCF first. And, after factoring once, ask yourself, *"Can I factor again?"*

EXAMPLE 4

Factor completely.

a) $128t - 2t^3$
b) $5x^2 + 45$
c) $h^4 - 16$

Solution

a) Ask yourself, *"Can I take out a common factor?"* Yes. Factor out $2t$.

$$128t - 2t^3 = 2t(64 - t^2)$$

Now ask yourself, *"Can I factor again?"* Yes. $64 - t^2$ is the difference of two squares. Identify a and b.

$$64 - t^2$$
$$\downarrow \quad \downarrow$$
$$(8)^2 \ (t)^2$$

So, $a = 8$ and $b = t$. $64 - t^2 = (8 + t)(8 - t)$.

Therefore, $128t - 2t^3 = 2t(8 + t)(8 - t)$.

 **BE CAREFUL** $(8 + t)(8 - t)$ is *not* the same as $(t + 8)(t - 8)$ because subtraction is not commutative. While $8 + t = t + 8$, $8 - t$ *does not equal* $t - 8$. You must write the terms in the correct order.

Another way to see that they are not equivalent is to multiply $(t + 8)(t - 8)$. $(t + 8)(t - 8) = t^2 - 64$. This is not the same as $64 - t^2$.

b) Ask yourself, *"Can I take out a common factor?"* Yes. Factor out 5.

$$5(x^2 + 9)$$

"Can I factor again?" No; $x^2 + 9$ is the *sum* of two squares.
Therefore, $5x^2 + 45 = 5(x^2 + 9)$.

c) The terms in $h^4 - 16$ have no common factors, but they are perfect squares. Identify a and b.

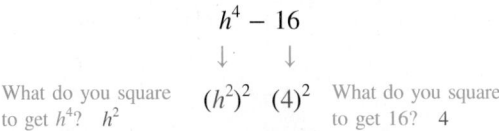

So, $a = h^2$ and $b = 4$. Therefore, $h^4 - 16 = (h^2 + 4)(h^2 - 4)$.
Can we factor again?

$h^2 + 4$ is the *sum* of two squares. It will not factor.

$h^2 - 4$ is the difference of two squares, so it *will* factor.

$h^2 - 4$
$\downarrow \quad \downarrow \qquad h^2 - 4 = (h + 2)(h - 2)$
$(h)^2 \quad (2)^2$
$a = h \text{ and } b = 2$

Therefore, $h^4 - 16 = (h^2 + 4)(h^2 - 4) = (h^2 + 4)(h + 2)(h - 2)$.

W Hint

Remember: After you factor once, ask yourself, *"Can I factor again?"*

[YOU TRY 3]

Factor completely.

a) $27k - 12k^3$ b) $y^4 - 1$ c) $2n^2 + 72$

---

3 Factor the Sum and Difference of Two Cubes

We can understand where we get the formulas for factoring the sum and difference of two cubes by looking at two products.

$$(a + b)(a^2 - ab + b^2) = a(a^2 - ab + b^2) + b(a^2 - ab + b^2) \qquad \text{Distributive property}$$
$$= a^3 - a^2b + ab^2 + a^2b - ab^2 + b^3 \qquad \text{Distribute.}$$
$$= a^3 + b^3 \qquad \text{Combine like terms.}$$

So, $(a + b)(a^2 - ab + b^2) = a^3 + b^3$, the sum of two cubes.

Now, let's multiply $(a - b)(a^2 + ab + b^2)$.

$$\begin{aligned}(a - b)(a^2 + ab + b^2) &= a(a^2 + ab + b^2) - b(a^2 + ab + b^2) & \text{Distributive property}\\ &= a^3 + a^2b + ab^2 - a^2b - ab^2 - b^3 & \text{Distribute.}\\ &= a^3 - b^3 & \text{Combine like terms.}\end{aligned}$$

So, $(a - b)(a^2 + ab + b^2) = a^3 - b^3$, the difference of two cubes.

The formulas for factoring the sum and difference of two cubes, then, are as follows:

W **Hint**

Write down any patterns you see in the formulas. This may help you remember them.

Formula Factoring the Sum and Difference of Two Cubes

$$a^3 + b^3 = (a + b)(a^2 - ab + b^2)$$
$$a^3 - b^3 = (a - b)(a^2 + ab + b^2)$$

Note

Notice that each factorization is the product of a binomial and a trinomial. To factor the sum and difference of two cubes

Step 1: Identify a and b.

Step 2: Place them in the binomial factor, and write the trinomial based on a and b.

Step 3: Simplify.

EXAMPLE 5

Factor completely.

a) $n^3 + 8$ b) $c^3 - 64$ c) $125r^3 + 27s^3$

Solution

a) Use Steps 1–3 to factor.

Step 1: Identify a and b.

$$n^3 + 8$$

What do you cube to get n^3? n $(n)^3$ $(2)^3$ What do you cube to get 8? 2

W **Hint**

Write out the example as you are reading it!

So, $a = n$ and $b = 2$.

Step 2: Remember, $a^3 + b^3 = (a + b)(a^2 - ab + b^2)$.
Write the binomial factor, then write the trinomial.

Same sign   Square a.   Product of a and b Square b.

$$n^3 + 8 = (n + 2)[(n)^2 - (n)(2) + (2)^2]$$

Opposite sign

Step 3: Simplify: $n^3 + 8 = (n + 2)(n^2 - 2n + 4)$

b) **Step 1:** Identify a and b.

$$c^3 - 64$$

What do you cube
to get c^3? c

$(c)^3$ $(4)^3$ What do you cube
to get 64?   4

So, $a = c$ and $b = 4$.

Step 2: Write the binomial factor, then write the trinomial. Remember,
$a^3 - b^3 = (a - b)(a^2 + ab + b^2)$.

Square a.   Square b.
Product
Same sign   of a and b

$$c^3 - 64 = (c - 4)[(c)^2 + (c)(4) + (4)^2]$$

Opposite
sign

Step 3: Simplify: $c^3 - 64 = (c - 4)(c^2 + 4c + 16)$

c) $125r^3 + 27s^3$

Step 1: Identify a and b.

$$125r^3 + 27s^3$$

What do you cube
to get $125r^3$? $5r$

$(5r)^3$ $(3s)^3$ What do you cube
to get $27s^3$? $3s$

So, $a = 5r$ and $b = 3s$.

Step 2: Write the binomial factor, then write the trinomial. Remember,
$a^3 + b^3 = (a + b)(a^2 - ab + b^2)$.

Square a.   Square b.
Product
Same sign   of a and b

$$125r^3 + 27s^3 = (5r + 3s)[(5r)^2 - (5r)(3s) + (3s)^2]$$

Opposite
sign

Step 3: Simplify: $125r^3 + 27s^3 = (5r + 3s)(25r^2 - 15rs + 9s^2)$

[**YOU TRY 4**]   Factor completely.

a) $r^3 + 1$ b) $p^3 - 1000$ c) $64x^3 - 125y^3$

Just as in the other factoring problems we've studied so far, the first step in factoring *any* polynomial should be to ask ourselves, *"Can I factor out a GCF?"*

EXAMPLE 6   Factor $3d^3 - 81$ completely.

Solution

"Can I factor out a GCF?" Yes. The GCF is 3.

$$3d^3 - 81 = 3(d^3 - 27)$$

Factor $d^3 - 27$. Use $a^3 - b^3 = (a - b)(a^2 + ab + b^2)$.

$$d^3 - 27 = (d - 3)[(d)^2 + (d)(3) + (3)^2]$$

$$\downarrow \qquad \downarrow$$

$$(d)^3 - (3)^3 = (d - 3)(d^2 + 3d + 9)$$

$$3d^3 - 81 = 3(d^3 - 27)$$
$$= 3(d - 3)(d^2 + 3d + 9)$$

[YOU TRY 5]

Factor completely.

a) $4t^3 + 4$ b) $72a^3 - 9b^6$

As always, the first thing you should do when factoring is ask yourself, *"Can I factor out a GCF?"* and the last thing you should do is ask yourself, *"Can I factor again?"* Now we will summarize the factoring methods discussed in this section.

Summary Special Factoring Rules

Perfect square trinomials: $a^2 + 2ab + b^2 = (a + b)^2$
$a^2 - 2ab + b^2 = (a - b)^2$

Difference of two squares: $a^2 - b^2 = (a + b)(a - b)$

Sum of two cubes: $a^3 + b^3 = (a + b)(a^2 - ab + b^2)$

Difference of two cubes: $a^3 - b^3 = (a - b)(a^2 + ab + b^2)$

ANSWERS TO [YOU TRY] EXERCISES

1) a) $(w + 4)^2$ b) $(a - 10)^2$ c) $(2d - 9)^2$ 2) a) $(r + 5)(r - 5)$ b) $(7p + 11q)(7p - 11q)$
c) $\left(x - \dfrac{5}{12}\right)\left(x + \dfrac{5}{12}\right)$ d) prime   3) a) $3k(3 + 2k)(3 - 2k)$ b) $(y^2 + 1)(y + 1)(y - 1)$
c) $2(n^2 + 36)$ 4) a) $(r + 1)(r^2 - r + 1)$ b) $(p - 10)(p^2 + 10p + 100)$
c) $(4x - 5y)(16x^2 + 20xy + 25y^2)$ 5) a) $4(t + 1)(t^2 - t + 1)$ b) $9(2a - b^2)(4a^2 + 2ab^2 + b^4)$

E Evaluate **7.4** Exercises   Do the exercises, and check your work.

Objective 1: Factor a Perfect Square Trinomial

1) Find the following.

a) 6^2 b) 10^2

c) 4^2 d) 11^2

e) 3^2 f) 8^2

g) 12^2 h) $\left(\dfrac{1}{2}\right)^2$

i) $\left(\dfrac{3}{5}\right)^2$

2) What is a perfect square trinomial?

3) Fill in the blank with a term that has a positive coefficient.

a) $(\underline{\quad})^2 = n^4$ b) $(\underline{\quad})^2 = 25t^2$

c) $(\underline{\quad})^2 = 49k^2$ d) $(\underline{\quad})^2 = 16p^4$

e) $(\underline{\quad})^2 = \dfrac{1}{9}$ f) $(\underline{\quad})^2 = \dfrac{25}{4}$

4) If x^n is a perfect square, then n is divisible by what number?

5) What perfect square trinomial factors to $(z + 9)^2$?

6) What perfect square trinomial factors to $(2b - 7)^2$?

7) Why isn't $9c^2 - 12c + 16$ a perfect square trinomial?

8) Why isn't $k^2 + 6k + 8$ a perfect square trinomial?

Factor completely.

9) $t^2 + 16t + 64$

10) $x^2 + 12x + 36$

11) $g^2 - 18g + 81$

12) $q^2 - 22q + 121$

13) $4y^2 + 12y + 9$

14) $49r^2 + 14r + 1$

15) $9k^2 - 24k + 16$

16) $16b^2 - 24b + 9$

17) $25n^2 + 10n + 4$

18) $9c^2 - 12c + 16$

19) $a^2 + \dfrac{2}{3}a + \dfrac{1}{9}$

20) $m^2 + m + \dfrac{1}{4}$

21) $v^2 - 3v + \dfrac{9}{4}$

22) $h^2 - \dfrac{4}{5}h + \dfrac{4}{25}$

23) $x^2 + 6xy + 9y^2$

24) $36t^2 - 60tu + 25u^2$

25) $9a^2 - 12ab + 4b^2$

26) $81k^2 + 18km + m^2$

27) $4f^2 + 24f + 36$

28) $9j^2 - 18j + 9$

29) $2p^4 - 24p^3 + 72p^2$

30) $5r^3 + 40r^2 + 80r$

31) $-18d^2 - 60d - 50$

32) $-28z^2 + 28z - 7$

33) $12c^3 + 3c^2 + 27c$

34) $100n^4 - 8n^3 + 64n^2$

Objective 2: Factor the Difference of Two Squares

35) What binomial factors to

a) $(x + 4)(x - 4)$?

b) $(4 + x)(4 - x)$?

36) What binomial factors to

a) $(y - 9)(y + 9)$?

b) $(9 - y)(9 + y)$?

Factor completely.

37) $x^2 - 9$

38) $q^2 - 49$

39) $n^2 - 121$

40) $d^2 - 81$

41) $m^2 + 64$

42) $q^2 + 9$

43) $y^2 - \dfrac{1}{25}$

44) $t^2 - \dfrac{1}{100}$

45) $c^2 - \dfrac{9}{16}$

46) $m^2 - \dfrac{4}{25}$

47) $36 - h^2$

48) $4 - b^2$

49) $169 - a^2$

50) $121 - w^2$

51) $\dfrac{49}{64} - j^2$

52) $\dfrac{144}{49} - r^2$

53) $100m^2 - 49$

54) $36x^2 - 25$

55) $16p^2 - 81$

56) $9a^2 - 1$

57) $4t^2 + 25$

58) $64z^2 + 9$

59) $\dfrac{1}{4}k^2 - \dfrac{4}{9}$

60) $\dfrac{1}{36}d^2 - \dfrac{4}{49}$

61) $b^4 - 64$

62) $u^4 - 49$

63) $144m^2 - n^4$

64) $64p^2 - 25q^4$

65) $r^4 - 1$

66) $k^4 - 81$

67) $16h^4 - g^4$

68) $b^4 - a^4$

69) $4a^2 - 100$

70) $3p^2 - 48$

71) $2m^2 - 128$

72) $6j^2 - 6$

73) $45r^4 - 5r^2$

74) $32n^5 - 200n^3$

Objective 3: Factor the Sum and Difference of Two Cubes

75) Find the following.

a) 4^3

b) 1^3

c) 10^3

d) 3^3

e) 5^3

f) 2^3

76) If x^n is a perfect cube, then n is divisible by what number?

77) Fill in the blank.

a) $(\underline{\quad})^3 = y^3$

b) $(\underline{\quad})^3 = 8c^3$

c) $(\underline{\quad})^3 = 125r^3$

d) $(\underline{\quad})^3 = x^6$

78) If x^n is a perfect square *and* a perfect cube, then n is divisible by what number?

Complete the factorization.

79) $x^3 + 27 = (x + 3)(\qquad\qquad)$

80) $t^3 - 125 = (t - 5)(\qquad\qquad)$

Factor completely.

81) $d^3 + 1$

82) $n^3 + 125$

83) $p^3 - 27$

84) $g^3 - 8$

85) $k^3 + 64$

86) $z^3 - 1000$

87) $t^3 + \dfrac{1}{8}$

88) $x^3 - \dfrac{1}{27}$

89) $27m^3 - 125$

90) $64c^3 + 1$

91) $125y^3 - 8$

92) $27a^3 + 64$

93) $1000c^3 - d^3$

94) $125v^3 + w^3$

95) $8j^3 + 27k^3$

96) $125m^3 - 27n^3$

97) $64x^3 + 125y^3$

98) $27a^3 - 1000b^3$

99) $6c^3 + 48$

100) $9k^3 - 9$

101) $7v^3 - 7000w^6$

102) $216a^6 + 64b^3$

103) $p^6 - 1$

104) $h^6 - 64$

Extend the concepts of this section to factor completely.

105) $(x + 5)^2 - (x - 2)^2$

106) $(r - 6)^2 - (r + 1)^2$

107) $(2p + 3)^2 - (p + 4)^2$

108) $(3d - 2)^2 - (d - 5)^2$

109) $(t + 5)^3 + 8$

110) $(c - 2)^3 + 27$

111) $(k - 9)^3 - 1$

112) $(y + 3)^3 - 125$

113) $a^2 + 14a + 49 - b^2$

114) $m^2 + 8m + 16 - n^2$

115) $4x^2 - 4xy + y^2 - 25$

116) $9c^2 - 12cd + 4d^2 - 1$

R Rethink

R1) How do you know if a trinomial is a *perfect square?*

R2) When does the product of two binomials result in the *difference of two squares?*

R3) Write down four examples of binomials that are *differences of two squares.*

R4) Write down four examples of binomials that are *differences of two cubes.*

R5) How do you know when to factor a binomial as a difference of squares and when to factor a binomial as a difference of cubes?

Putting It All Together

P Prepare

O Organize

What is your objective for Putting It All Together?	How can you accomplish the objective?
1 Learn Strategies for Factoring a Given Polynomial	• Review the summary on **How To Factor a Polynomial,** and write it in your notes. • Learn to recognize the characteristics of a polynomial so that you can determine which method to use to factor it. • Be sure to check your answer. • Complete the given examples on your own. • Complete You Try 1.

W Work

Read the explanations, follow the examples, take notes, and complete the You Try.

1 Learn Strategies for Factoring a Given Polynomial

In this chapter, we have discussed several different types of factoring problems:

1) Factoring out a GCF (Section 7.1)

2) Factoring by grouping (Section 7.1)

3) Factoring a trinomial of the form $x^2 + bx + c$ (Section 7.2)

4) Factoring a trinomial of the form $ax^2 + bx + c$ (Section 7.3)

5) Factoring a perfect square trinomial (Section 7.4)

6) Factoring the difference of two squares (Section 7.4)

7) Factoring the sum and difference of two cubes (Section 7.4)

We have practiced the factoring methods separately in each section, but how do we know which factoring method to use given many different types of polynomials together? We will discuss some strategies in this section. First, recall the steps for factoring *any* polynomial:

Summary To Factor a Polynomial

1) *Always* begin by asking yourself, *"Can I factor out a GCF?"* If so, factor it out.

2) Look at the expression to decide if it will factor further. Apply the appropriate method to factor. If there are

   a) *two terms,* see if it is a difference of two squares or the sum or difference of two cubes as in Section 7.4.

   b) *three terms,* see if it can be factored using the methods of Section 7.2 or 7.3 *or* determine if it is a perfect square trinomial (Section 7.4).

   c) *four terms,* see if it can be factored by grouping as in Section 7.1.

3) After factoring, *always* look carefully at the result and ask yourself, *"Can I factor it again?"* If so, factor again.

Next, we will discuss how to decide which factoring method should be used to factor a particular polynomial.

EXAMPLE 1

Factor completely.

a) $12a^2 - 27b^2$ b) $y^2 - y - 30$ c) $mn^2 - 4m + 5n^2 - 20$

d) $p^2 - 16p + 64$ e) $8x^2 + 26x + 20$ f) $27k^3 + 8$

g) $t^2 + 36$

Solution

a) *"Can I factor out a GCF?"* is the first thing you should ask yourself. *Yes.* Factor out 3.

$$12a^2 - 27b^2 = 3(4a^2 - 9b^2)$$

 **Hint**

Write out the examples in your notes, and include information about *why* a given procedure was used.

Ask yourself, *"Can I factor again?"* Examine $4a^2 - 9b^2$. It has two terms that are being subtracted, and each term is a perfect square. $4a^2 - 9b^2$ is the difference of squares.

$$4a^2 - 9b^2 = (2a + 3b)(2a - 3b)$$
$$\quad\downarrow\qquad\downarrow$$
$$(2a)^2\ (3b)^2$$

$$12a^2 - 27b^2 = 3(4a^2 - 9b^2)$$
$$= 3(2a + 3b)(2a - 3b)$$

"Can I factor again?" No. It is completely factored.

b) *"Can I factor out a GCF?"* No. To factor $y^2 - y - 30$, think of two numbers whose *product* is -30 and *sum* is -1. The numbers are -6 and 5.

$$y^2 - y - 30 = (y - 6)(y + 5)$$

"Can I factor again?" No. It is completely factored.

c) Look at $mn^2 - 4m + 5n^2 - 20$. *"Can I factor out a GCF?"* No. Notice that this polynomial has *four terms*. When a polynomial has *four terms*, think about *factoring by grouping*.

$$\underline{mn^2 - 4m} + \underline{5n^2 - 20}$$
$$\qquad\downarrow\qquad\qquad\downarrow$$
$$= m(n^2 - 4) + 5(n^2 - 4) \qquad \text{Take out the common factor from each pair of terms.}$$
$$= (n^2 - 4)(m + 5) \qquad \text{Factor out } (n^2 - 4) \text{ using the distributive property.}$$

Examine $(n^2 - 4)(m + 5)$ and ask yourself, *"Can I factor again?"* Yes! $(n^2 - 4)$ is the difference of two squares. Factor again.

$$(n^2 - 4)(m + 5) = (n + 2)(n - 2)(m + 5)$$

"Can I factor again?" No. It is completely factored.

$$mn^2 - 4m + 5n^2 - 20 = (n + 2)(n - 2)(m + 5)$$

 **Note**

Seeing four terms is a clue to try factoring by grouping.

d) We cannot take out a GCF from $p^2 - 16p + 64$. It is a trinomial, and notice that the first and last terms are perfect squares. *Is this a perfect square trinomial?*

$$p^2 - 16p + 64$$
$$\downarrow\qquad\qquad\downarrow$$
$$(p)^2\qquad\quad (8)^2$$

Does the middle term equal $2 \cdot p \cdot (8)$? Yes: $2 \cdot p \cdot (8) = 16p$.
Use $a^2 - 2ab + b^2 = (a - b)^2$ with $a = p$ and $b = 8$.
Then, $p^2 - 16p + 64 = (p - 8)^2$.

"Can I factor again?" No. It is completely factored.

e) It is tempting to jump right in and try to factor $8x^2 + 26x + 20$ as the product of two binomials, but ask yourself, *"Can I take out a GCF?"* Yes! Factor out 2.

$$8x^2 + 26x + 20 = 2(4x^2 + 13x + 10)$$

"Can I factor again?" Yes.

$$2(4x^2 + 13x + 10) = 2(4x + 5)(x + 2)$$

"Can I factor again?" No. So, $8x^2 + 26x + 20 = 2(4x + 5)(x + 2)$.

f) We cannot take out a GCF from $27k^3 + 8$. Notice that $27k^3 + 8$ has two terms, so think about squares and cubes. Neither term is a perfect square *and* the positive terms are being added, so this *cannot* be the difference of squares.

Is each term a perfect cube? *Yes!* $27k^3 + 8$ is the sum of two cubes. We will factor $27k^3 + 8$ using $a^3 + b^3 = (a + b)(a^2 - ab + b^2)$ with $a = 3k$ and $b = 2$.

$$27k^3 + 8 = (3k + 2)[(3k)^2 - (3k)(2) + (2)^2]$$
$$\qquad\qquad \downarrow \qquad \downarrow$$
$$\qquad\quad (3k)^3 \ \ (2)^3$$
$$= (3k + 2)(9k^2 - 6k + 4)$$

"Can I factor again?" No. It is completely factored.

g) Look at $t^2 + 36$ and ask yourself, *"Can I factor out a GCF?"* No. The binomial $t^2 + 36$ is the *sum* of two squares, so it does not factor. This polynomial is prime.

[**YOU TRY 1**] Factor completely.

a) $3p^2 + p - 10$ b) $2n^3 - n^2 + 12n - 6$ c) $4k^4 + 36k^3 + 32k^2$

d) $48 - 3y^4$ e) $8r^3 - 125$

ANSWERS TO [**YOU TRY**] **EXERCISES**

1) a) $(3p - 5)(p + 2)$ b) $(n^2 + 6)(2n - 1)$ c) $4k^2(k + 8)(k + 1)$ d) $3(4 + y^2)(2 + y)(2 - y)$
e) $(2r - 5)(4r^2 + 10r + 25)$

Putting It All Together Exercises [E **Evaluate**] Do the exercises, and check your work.

Objective 1: Learn Strategies for Factoring a Given Polynomial

1) What is the first thing you should ask yourself when you have to factor a polynomial?

2) After you factor a polynomial, what should you ask yourself?

Factor completely.

3) $m^2 + 16m + 60$ 4) $h^2 - 36$

5) $uv + 6u + 9v + 54$ 6) $2y^2 + 5y - 18$

 7) $3k^2 - 14k + 8$ 8) $n^2 - 14n + 49$

9) $16d^6 + 8d^5 + 72d^4$ 10) $b^2 - 3bc - 4c^2$

11) $60w^3 + 70w^2 - 50w$ 12) $7c^3 - 7$

13) $t^3 + 1000$ 14) $pq - 6p + 4q - 24$

15) $49 - p^2$ 16) $h^2 - 15h + 56$

17) $4x^2 + 4xy + y^2$ 18) $27c - 18$

19) $3z^4 - 21z^3 - 24z^2$ 20) $9a^2 + 6a - 8$

21) $4b^2 + 1$

22) $5abc - 15ac + 10bc - 30c$

23) $40x^3 - 135$ 24) $81z^2 + 36z + 4$

25) $c^2 - \dfrac{1}{4}$ 26) $v^2 + 3v + 4$

27) $45s^2t + 4 - 36s^2 - 5t$ 28) $12c^5d - 75cd^3$

29) $k^2 + 9km + 18m^2$ 30) $64r^3 + 8$

31) $z^2 - 3z - 88$ 32) $40f^4g^4 + 8f^3g^3 + 16fg^2$

33) $80y^2 - 40y + 5$ 34) $4t^2 - t - 5$

35) $20c^2 + 26cd + 6d^2$ 36) $x^2 - \dfrac{9}{49}$

37) $n^4 - 16m^4$ 38) $k^2 - 21k + 108$

39) $2a^2 - 10a - 72$ 40) $x^2y - 4y + 7x^2 - 28$

41) $r^2 - r + \dfrac{1}{4}$

42) $v^3 - 125$

(24) 43) $28gh + 16g - 63h - 36$

44) $-24x^3 + 30x^2 - 9x$

45) $8b^2 - 14b - 15$

46) $50u^2 + 60u + 18$

47) $55a^6b^3 + 35a^5b^3 - 10a^4b - 20a^2b$

48) $64 - u^2$

49) $2d^2 - 9d + 3$

50) $2v^4w + 14v^3w^2 + 12v^2w^3$

51) $9p^2 - 24pq + 16q^2$

52) $c^4 - 16$

53) $30y^2 + 37y - 7$

54) $g^2 + 49$

(24) 55) $80a^3 - 270b^3$

56) $26n^6 - 39n^4 + 13n^3$

57) $rt - r - t + 1$

58) $h^2 + 10h + 25$

59) $4g^2 - 4$

60) $25a^2 - 55ab + 24b^2$

(24) 61) $3c^2 - 24c + 48$

62) $9t^4 + 64u^2$

63) $144k^2 - 121$

64) $125p^3 - 64q^3$

65) $-48g^2 - 80g - 12$

66) $5d^2 + 60d + 55$

67) $q^3 + 1$

68) $9x^2 + 12x + 4$

69) $81u^4 - v^4$

70) $45v^2 + 9vw^2 + 30vw + 6w^3$

71) $11f^2 + 36f + 9$

72) $4y^3 - 4y^2 - 80y$

73) $2j^{11} - j^3$

74) $d^2 - \dfrac{169}{100}$

75) $w^2 - 2w - 48$

76) $16a^2 - 40a + 25$

(24) 77) $k^2 + 100$

78) $24y^3 + 375$

79) $m^2 + 4m + 4$

80) $r^2 - 15r + 54$

81) $100c^4 - 36c^2$

82) $9t^2 - 64$

83) $x^3 - \dfrac{8}{125}$

84) $h^3 + \dfrac{27}{64}$

85) $(2z + 1)y^2 + 6(2z + 1)y - 55(2z + 1)$

86) $(a + b)c^2 - 5(a + b)c - 24(a + b)$

87) $(r - 4)^2 + 11(r - 4) + 28$

88) $(n + 3)^2 - 2(n + 3) - 35$

89) $(3p - 4)^2 - 5(3p - 4) - 36$

90) $(5w - 2)^2 - 8(5w - 2) + 12$

91) $(4k + 1)^2 - (3k + 2)^2$

92) $(5z + 3)^2 - (3z - 1)^2$

93) $(x + y)^2 - (2x - y)^2$

94) $(3s - t)^2 - (2s + t)^2$

95) $n^2 + 12n + 36 - p^2$

96) $h^2 - 10h + 25 - k^2$

97) $x^2 - 2xy + y^2 - z^2$

98) $a^2 + 2ab + b^2 - c^2$

R Rethink

R1) Are you able to write down the factoring methods without using your notes?

R2) Explain how *you* determine which method to use to factor a polynomial.

R3) Would you be able to do these problems without looking back in the sections?

R4) Which types of problems did you understand well, and which were difficult for you? What do you think made them understandable or difficult?

7.5 Solving Quadratic Equations by Factoring

What are your objectives for Section 7.5?	How can you accomplish each objective?
1 Solve a Quadratic Equation of the Form $ab = 0$	• Learn the definition of a *quadratic equation*. • Learn the *zero product rule*. • Complete the given example on your own. • Complete You Try 1.
2 Solve Quadratic Equations by Factoring	• Know how to write a quadratic equation in *standard form*. • Learn the procedure for **Solving a Quadratic Equation by Factoring**. • Complete the given examples on your own. • Complete You Trys 2 and 3.
3 Solve Higher Degree Equations by Factoring	• Be able to recognize a higher degree equation. • Follow the same procedure for **Solving a Quadratic Equation by Factoring**. • Complete the given example on your own. • Complete You Try 4.

W Work

Read the explanations, follow the examples, take notes, and complete the You Trys.

In Section 3.1 we began our study of linear equations in one variable. A *linear equation in one variable* is an equation that can be written in the form $ax + b = 0$, where a and b are real numbers and $a \neq 0$.

In this section, we will learn how to solve *quadratic equations*.

Definition

A **quadratic equation** can be written in the form $ax^2 + bx + c = 0$, where a, b, and c are real numbers and $a \neq 0$.

When a quadratic equation is written in the form $ax^2 + bx + c = 0$, we say that it is in **standard form.** But quadratic equations can be written in other forms too.

Some examples of quadratic equations are

$$x^2 + 12x + 27 = 0, \quad 2p(p - 5) = 0, \quad \text{and} \quad (c + 1)(c - 8) = 3.$$

Quadratic equations are also called *second-degree equations* because the highest power on the variable is 2.

There are many different ways to solve quadratic equations. In this section, we will learn how to solve them by factoring; other methods will be discussed later in this book.

Solving a quadratic equation by factoring is based on the *zero product rule:* if the product of two quantities is zero, then one or both of the quantities is zero.

For example, if $5y = 0$, then $y = 0$. If $p \cdot 4 = 0$, then $p = 0$. If $ab = 0$, then either $a = 0$, $b = 0$, or *both a and b* equal zero.

Definition

Zero product rule: If $ab = 0$, then $a = 0$ or $b = 0$.

We will use this idea to solve quadratic equations by factoring.

1 Solve a Quadratic Equation of the Form $ab = 0$

EXAMPLE 1

Solve each equation.

a) $x(x + 8) = 0$ b) $(4y - 3)(y + 9) = 0$

Solution

a) The zero product rule says that at least one of the factors on the left must equal zero in order for the *product* to equal zero.

$$x(x + 8) = 0$$

$$x = 0 \quad \text{or} \quad x + 8 = 0 \qquad \text{Set each factor equal to 0.}$$
$$x = -8 \qquad \text{Solve.}$$

Check the solutions in the original equation:

If $x = 0$,	If $x = -8$,
$0(0 + 8) \overset{?}{=} 0$	$-8(-8 + 8) \overset{?}{=} 0$
$0(8) = 0$ ✓	$-8(0) = 0$ ✓

The solution set is $\{-8, 0\}$.

> **W Hint**
>
> When x is isolated as a factor, one of the solutions will always be equal to zero.

Note

It is important to remember that the factor x gives us the solution 0.

b) At least one of the factors on the left must equal zero for the *product* to equal zero.

$$(4y - 3)(y + 9) = 0$$

$$4y - 3 = 0 \quad \text{or} \quad y + 9 = 0 \qquad \text{Set each factor equal to 0.}$$
$$4y = 3 \qquad\qquad\quad y = -9 \qquad \text{Solve each equation.}$$
$$y = \frac{3}{4}$$

Check in the original equation:

If $y = \dfrac{3}{4}$,

$$\left[4\left(\dfrac{3}{4}\right) - 3\right]\left[\dfrac{3}{4} + 9\right] \stackrel{?}{=} 0$$

$$(3 - 3)\left(\dfrac{39}{4}\right) \stackrel{?}{=} 0$$

$$0\left(\dfrac{39}{4}\right) = 0 \quad \checkmark$$

If $y = -9$,

$$[4(-9) - 3][-9 + 9] \stackrel{?}{=} 0$$

$$-39(0) = 0 \quad \checkmark$$

The solution set is $\left\{-9, \dfrac{3}{4}\right\}$.

[YOU TRY 1]   Solve each equation.

a) $c(c - 9) = 0$ b) $(5t + 2)(t - 7) = 0$

---

2 Solve Quadratic Equations by Factoring

If the equation is in standard form, $ax^2 + bx + c = 0$, begin by factoring the expression.

EXAMPLE 2   Solve $m^2 - 6m - 40 = 0$.

Solution

$$m^2 - 6m - 40 = 0$$
$$(m - 10)(m + 4) = 0 \qquad \text{Factor.}$$

$$m - 10 = 0 \quad \text{or} \quad m + 4 = 0 \qquad \text{Set each factor equal to zero.}$$
$$m = 10 \quad \text{or} \qquad m = -4 \qquad \text{Solve.}$$

Check in the original equation:

If $m = 10$,
$$(10)^2 - 6(10) - 40 \stackrel{?}{=} 0$$
$$100 - 60 - 40 = 0 \quad \checkmark$$

If $m = -4$,
$$(-4)^2 - 6(-4) - 40 \stackrel{?}{=} 0$$
$$16 + 24 - 40 = 0 \quad \checkmark$$

The solution set is $\{-4, 10\}$.

Here are the steps to use to solve a quadratic equation by factoring:

Procedure Solving a Quadratic Equation by Factoring

1)   Write the equation in the form $ax^2 + bx + c = 0$ (standard form) so that all terms are on one side of the equal sign and zero is on the other side.

2)   Factor the expression.

3)   Set each factor equal to zero, and solve for the variable. (Use the zero product rule.)

4)   Check the answer(s).

[YOU TRY 2]   Solve $h^2 + 9h + 18 = 0$.

EXAMPLE 3

Solve each equation by factoring.

a) $2r^2 + 3r = 20$ b) $6d^2 = -42d$ c) $k^2 = -12(k + 3)$

d) $2(x^2 + 5) + 5x = 6x(x - 1) + 16$ e) $(z - 8)(z - 4) = 5$

Solution

a) Begin by writing $2r^2 + 3r = 20$ in standard form, $ar^2 + br + c = 0$.

$$2r^2 + 3r - 20 = 0 \quad \text{Standard form}$$
$$(2r - 5)(r + 4) = 0 \quad \text{Factor.}$$

$$2r - 5 = 0 \quad \text{or} \quad r + 4 = 0 \quad \text{Set each factor equal to zero.}$$
$$2r = 5$$
$$r = \frac{5}{2} \quad \text{or} \quad r = -4 \quad \text{Solve.}$$

Check in the original equation:

If $r = \frac{5}{2}$,

$$2\left(\frac{5}{2}\right)^2 + 3\left(\frac{5}{2}\right) \overset{?}{=} 20$$

$$2\left(\frac{25}{4}\right) + \frac{15}{2} \overset{?}{=} 20$$

$$\frac{25}{2} + \frac{15}{2} \overset{?}{=} 20$$

$$\frac{40}{2} = 20 \quad \checkmark$$

If $r = -4$,
$$2(-4)^2 + 3(-4) \overset{?}{=} 20$$
$$2(16) - 12 \overset{?}{=} 20$$
$$32 - 12 = 20 \quad \checkmark$$

The solution set is $\left\{-4, \frac{5}{2}\right\}$.

b) Write $6d^2 = -42d$ in standard form.

$$6d^2 + 42d = 0 \quad \text{Standard form}$$
$$6d(d + 7) = 0 \quad \text{Factor.}$$

$$6d = 0 \quad \text{or} \quad d + 7 = 0 \quad \text{Set each factor equal to zero.}$$
$$d = 0 \quad \text{or} \quad d = -7 \quad \text{Solve.}$$

Check. The solution set is $\{-7, 0\}$.

Because both terms in $6d^2 = -42d$ are divisible by 6, we could have started part b) by dividing by 6:

$$\frac{6d^2}{6} = \frac{-42d}{6} \quad \text{Divide by 6.}$$
$$d^2 = -7d$$
$$d^2 + 7d = 0 \quad \text{Write in standard form.}$$
$$d(d + 7) = 0 \quad \text{Factor.}$$

$$d = 0 \quad \text{or} \quad d + 7 = 0 \quad \text{Set each factor equal to zero.}$$
$$d = -7 \quad \text{Solve.}$$

The solution set is $\{-7, 0\}$. We get the same result.

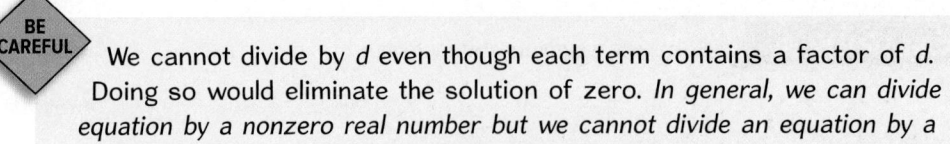

BE CAREFUL We cannot divide by *d* even though each term contains a factor of *d*. Doing so would eliminate the solution of zero. *In general, we can divide an equation by a nonzero real number but we cannot divide an equation by a variable because we may eliminate a solution, and we may be dividing by zero.*

c) To solve $k^2 = -12(k + 3)$, begin by writing the equation in standard form.

$$k^2 = -12k - 36 \qquad \text{Distribute.}$$
$$k^2 + 12k + 36 = 0 \qquad \text{Write in standard form.}$$
$$(k + 6)^2 = 0 \qquad \text{Factor.}$$

Because $(k + 6)^2 = 0$ means $(k + 6)(k + 6) = 0$, setting each factor equal to zero will result in the same value for *k*.

$$k + 6 = 0 \qquad \text{Set } k + 6 = 0.$$
$$k = -6 \qquad \text{Solve.}$$

Check. The solution set is $\{-6\}$.

d) We will have to perform several steps to write the equation in standard form.

$$2(x^2 + 5) + 5x = 6x(x - 1) + 16$$
$$2x^2 + 10 + 5x = 6x^2 - 6x + 16 \qquad \text{Distribute.}$$

Move the terms on the left side of the equation to the right side so that the coefficient of x^2 is positive.

$$0 = 4x^2 - 11x + 6 \qquad \text{Write in standard form.}$$
$$0 = (4x - 3)(x - 2) \qquad \text{Factor.}$$

$$4x - 3 = 0 \quad \text{or} \quad x - 2 = 0 \qquad \text{Set each factor equal to zero.}$$
$$4x = 3$$
$$x = \frac{3}{4} \quad \text{or} \quad x = 2 \qquad \text{Solve.}$$

The check is left to the student. The solution set is $\left\{ \dfrac{3}{4}, 2 \right\}$.

e) It is tempting to solve $(z - 8)(z - 4) = 5$ like this:

$$(z - 8)(z - 4) = 5$$

$$z - 8 = 5 \quad \text{or} \quad z - 4 = 5 \qquad \text{This is incorrect!}$$

One side of the equation must equal zero in order to set each factor equal to zero. Begin by multiplying on the left.

$$(z - 8)(z - 4) = 5$$
$$z^2 - 12z + 32 = 5 \qquad \text{Multiply using FOIL.}$$
$$z^2 - 12z + 27 = 0 \qquad \text{Standard form}$$
$$(z - 9)(z - 3) = 0 \qquad \text{Factor.}$$

$$z - 9 = 0 \quad \text{or} \quad z - 3 = 0 \qquad \text{Set each factor equal to zero.}$$
$$z = 9 \quad \text{or} \quad z = 3 \qquad \text{Solve.}$$

The check is left to the student. The solution set is $\{3, 9\}$.

| YOU TRY 3 |

Solve.

a) $w^2 + 4w - 5 = 0$ b) $29b = 5(b^2 + 4)$ c) $(a + 6)(a + 4) = 3$

d) $3t^2 = -24t$ e) $(2y + 1)^2 + 5 = y^2 + 2(y + 7)$

3 Solve Higher Degree Equations by Factoring

Sometimes, equations that are not quadratics can be solved by factoring as well.

EXAMPLE 4

Solve each equation.

a) $(2x - 1)(x^2 - 9x - 22) = 0$ b) $4w^3 - 100w = 0$

Solution

a) This is *not* a quadratic equation because if we multiplied the factors on the left we would get $2x^3 - 19x^2 - 35x + 22 = 0$. This is a *cubic* equation because the degree of the polynomial on the left is 3.

The original equation is the product of two factors, so we can use the zero product rule.

W Hint

The degree of the polynomial is the same as the greatest number of solutions the equation *might* have.

$$(2x - 1)(x^2 - 9x - 22) = 0$$
$$(2x - 1)(x - 11)(x + 2) = 0 \qquad \text{Factor.}$$

$2x - 1 = 0$ or $x - 11 = 0$ or $x + 2 = 0$ Set each factor equal to zero.
$2x = 1$

$x = \dfrac{1}{2}$ or $x = 11$ or $x = -2$ Solve.

The check is left to the student. The solution set is $\left\{ -2, \dfrac{1}{2}, 11 \right\}$.

b) The GCF of the terms in the equation is $4w$. Remember, however, that *we can divide an equation by a constant but we cannot divide an equation by a variable.* Dividing by a variable may eliminate a solution and may mean we are dividing by zero. So let's begin by dividing each term by 4.

$$\frac{4w^3}{4} - \frac{100w}{4} = \frac{0}{4} \qquad \text{Divide by 4.}$$
$$w^3 - 25w = 0 \qquad \text{Simplify.}$$
$$w(w^2 - 25) = 0 \qquad \text{Factor out } w.$$
$$w(w + 5)(w - 5) = 0 \qquad \text{Factor } w^2 - 25.$$

$w = 0$ or $w + 5 = 0$ or $w - 5 = 0$ Set each factor equal to zero.
$w = -5$ $w = 5$ Solve.

Check. The solution set is $\{0, -5, 5\}$.

| YOU TRY 4 |

Solve.

a) $(c + 10)(2c^2 + 5c - 7) = 0$ b) $r^4 = 25r^2$

In this section, it was possible to solve all of the equations by factoring. In the *Using Technology* box we show the relationship between solving a quadratic equation by factoring and solving it using a graphing calculator. In Chapter 11 we will learn other methods for solving quadratic equations.

Using Technology

In this section, we learned how to solve a quadratic equation by factoring. We can use a graphing calculator to solve a quadratic equation as well. Let's see how the two are related by using the equation $x^2 - x - 6 = 0$.

$$x^2 - x - 6 = 0$$
$$(x + 2)(x - 3) = 0$$
$$x + 2 = 0 \quad \text{or} \quad x - 3 = 0$$
$$x = -2 \qquad x = 3$$

The solution set is $\{-2, 3\}$.

Next, solve $x^2 - x - 6 = 0$ using a graphing calculator. Recall from Chapter 4 that to find the x-intercepts of the graph of an equation we let $y = 0$ and solve the equation for x. If we let $y = x^2 - x - 6$, then solving $x^2 - x - 6 = 0$ is the same as finding the x-intercepts of the graph of $y = x^2 - x - 6$. X-intercepts are also called zeros of the equation since they are the values of x that make $y = 0$.

Use $\boxed{Y=}$ to enter $y = x^2 - x - 6$ into the calculator, press $\boxed{ZOOM}$, and then enter 6 to display the graph using the standard viewing window as shown at right. We obtain a graph called a parabola, and we can see that it has two x-intercepts. If the scale for each tick mark on the graph is 1, then it appears that the x-intercepts are -2 and 3. To verify press $\boxed{TRACE}$, type -2, and press $\boxed{ENTER}$. Since $x = -2$ and $y = 0$, $x = -2$ is an x-intercept.

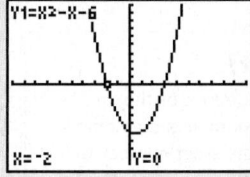

While still in "Trace mode," type 3 and press $\boxed{ENTER}$. Since $x = 3$ and $y = 0$, $x = 3$ is an x-intercept.

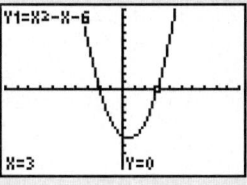

Sometimes an x-intercept is not an integer.

Solve $2x^2 + x - 15 = 0$ using a graphing calculator. Enter $2x^2 + x - 15$ into the calculator, and press $\boxed{GRAPH}$. The x-intercept on the right side of the graph is between two tick marks, so it is not an integer. To find the x-intercept, press $\boxed{2nd}$ $\boxed{TRACE}$ and select 2:zero. Move the cursor to the left of one of the intercepts and press $\boxed{ENTER}$, then move the cursor again, so that it is to the right of the same intercept and press $\boxed{ENTER}$. Press $\boxed{ENTER}$ one more time, and the calculator will reveal the intercept and, therefore one solution to the equation.

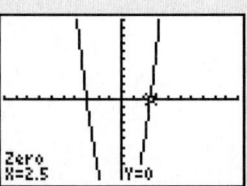

Press 2nd MODE to return to the home screen. Press X,T,θ,n MATH ENTER ENTER to display the x-intercept in fraction form: $x = \dfrac{5}{2}$. Since the other x-intercept appears to be -3, press TRACE -3 ENTER to reveal that $x = -3$ and $y = 0$.

```
X▸Frac
                    5/2
```

Solve using a graphing calculator.

1) $x^2 - 3x - 4 = 0$
2) $2x^2 + 5x - 3 = 0$
3) $x^2 - 5x - 14 = 0$
4) $3x^2 - 17x + 10 = 0$
5) $x^2 - 7x + 12 = 0$
6) $2x^2 - 9x + 10 = 0$

ANSWERS TO [YOU TRY] EXERCISES

1) a) $\{0, 9\}$ b) $\left\{-\dfrac{2}{5}, 7\right\}$ 2) $\{-6, -3\}$ 3) a) $\{-5, 1\}$ b) $\left\{\dfrac{4}{5}, 5\right\}$ c) $\{-7, -3\}$

d) $\{-8, 0\}$ e) $\left\{-2, \dfrac{4}{3}\right\}$ 4) a) $\left\{-10, -\dfrac{7}{2}, 1\right\}$ b) $\{0, 5, -5\}$

ANSWERS TO TECHNOLOGY EXERCISES

1) $\{-1, 4\}$ 2) $\left\{\dfrac{1}{2}, -3\right\}$ 3) $\{7, -2\}$ 4) $\left\{\dfrac{2}{3}, 5\right\}$ 5) $\{4, 3\}$ 6) $\left\{2, \dfrac{5}{2}\right\}$

E Evaluate **7.5** Exercises   Do the exercises, and check your work.

Objective 1: Solve a Quadratic Equation of the Form $ab = 0$

 1) Explain the zero product rule.

2) When Ivan solves $c(c + 7) = 0$, he gets a solution set of $\{-7\}$. Is this correct? Why or why not?

Solve each equation.

3) $(m + 9)(m - 8) = 0$

4) $(a + 10)(a + 4) = 0$

5) $(q - 4)(q - 7) = 0$

6) $(x - 5)(x + 2) = 0$

7) $(4z + 3)(z - 9) = 0$

8) $(2n + 1)(n - 13) = 0$

9) $-5r(r - 8) = 0$

10) $11s(s + 15) = 0$

11) $(6x - 5)^2 = 0$

12) $(d + 7)^2 = 0$

13) $(4h + 7)(h + 3) = 0$

14) $(8p - 5)(3p - 11) = 0$

15) $\left(y + \dfrac{3}{2}\right)\left(y - \dfrac{1}{4}\right) = 0$

16) $\left(t - \dfrac{9}{8}\right)\left(t + \dfrac{5}{6}\right) = 0$

17) $q(q - 2.5) = 0$

18) $w(w + 0.8) = 0$

Objective 2: Solve Quadratic Equations by Factoring

19) Can we solve $(y + 6)(y - 11) = 8$ by setting each factor equal to 8 like this: $y + 6 = 8$ or $y - 11 = 8$? Why or why not?

20) Explain two ways you could begin to solve $5n^2 - 10n - 40 = 0$.

Solve each equation.

21) $v^2 + 15v + 56 = 0$

22) $y^2 + 2y - 35 = 0$

23) $k^2 + 12k - 45 = 0$

24) $z^2 - 12z + 11 = 0$

25) $3y^2 - y - 10 = 0$

26) $4f^2 - 15f + 14 = 0$

27) $14w^2 + 8w = 0$

28) $10a^2 + 20a = 0$

29) $d^2 - 15d = -54$

30) $j^2 + 11j = -28$

31) $t^2 - 49 = 0$

32) $k^2 - 100 = 0$

33) $36 = 25n^2$

34) $16 = 169p^2$

35) $\frac{1}{6}m^2 = 10 - \frac{7}{6}m$

36) $\frac{1}{10}g^2 + 2 = \frac{6}{5}g$

37) $0.01x^2 + 0.64 = 0.16x$

38) $0.01w^2 + 0.81 = 0.18w$

39) $m^2 = 60 - 7m$

40) $g^2 + 20 = 12g$

41) $55w = -20w^2 - 30$

42) $4v = 14v^2 - 48$

43) $p^2 = 11p$

44) $d^2 = d$

45) $45k + 27 = 18k^2$

46) $104r + 36 = 12r^2$

47) $b(b - 4) = 96$

48) $54 = w(15 - w)$

49) $-63 = 4j(j - 8)$

50) $g(3g + 11) = 70$

51) $(y - 7)(y - 4) = 18$

52) $(t + 3)(t - 5) = 9$

53) $(6a + 1)(a + 1) = 6$

54) $(10z + 7)(z + 1) = 4$

55) $10x(x + 1) - 6x = 9(x^2 + 5)$

56) $5r(3r + 7) = 2(4r^2 - 21)$

57) $3(h^2 - 4) = 5h(h - 1) - 9h$

58) $5(5 + u^2) + 10 = 3u(2u + 1) - u$

59) $\frac{1}{2}(m + 1)^2 = -\frac{3}{4}m(m + 5) - \frac{5}{2}$

60) $\frac{1}{8}(2y - 3)^2 + \frac{1}{8}y = \frac{1}{8}(y - 5)^2 - \frac{3}{4}$

61) $\frac{1}{5}t(t - 5) + \frac{14}{15} = \frac{1}{3} - \frac{1}{15}t(t + 3)$

62) $\frac{1}{2}c(2 - c) - \frac{3}{2} = \frac{2}{5}c(c + 1) - \frac{7}{5}$

63) $3t(t - 5) + 14 = 5 - t(t + 3)$

64) $(2y - 3)^2 + y = (y - 5)^2 - 6$

65) $33 = -m(14 + m)$

66) $-84 = s(s + 19)$

67) $(3w + 2)^2 - (w - 5)^2 = 0$

68) $(2j - 7)^2 - (j + 3)^2 = 0$

69) $(q + 3)^2 - (2q - 5)^2 = 0$

70) $(6n + 5)^2 - (3n + 4)^2 = 0$

Objective 3: Solve Higher Degree Equations by Factoring

The following equations are not quadratic but can be solved by factoring and applying the zero product rule. Solve each equation.

71) $8y(y + 4)(2y - 1) = 0$

72) $-13b(12b + 7)(b - 11) = 0$

73) $(9p - 2)(p^2 - 10p - 11) = 0$

74) $(4f + 5)(f^2 - 3f - 18) = 0$

75) $(2r - 5)(r^2 - 6r + 9) = 0$

76) $(3x - 1)(x^2 - 16x + 64) = 0$

77) $m^3 = 64m$

78) $r^3 = 81r$

79) $5w^2 + 36w = w^3$

80) $14a^2 - 49a = a^3$

81) $2g^3 = 120g - 14g^2$

82) $36z - 24z^2 = -3z^3$

83) $45h = 20h^3$

84) $64d^3 = 100d$

85) $2s^2(3s + 2) + 3s(3s + 2) - 35(3s + 2) = 0$

86) $10n^2(n - 8) + n(n - 8) - 2(n - 8) = 0$

87) $10a^2(4a + 3) + 2(4a + 3) = 9a(4a + 3)$

88) $12d^2(7d - 3) = 5d(7d - 3) + 2(7d - 3)$

89) $t^3 + 6t^2 - 4t - 24 = 0$

90) $k^3 - 8k^2 - 9k + 72 = 0$

Mixed Exercises: Objectives 1–3

Solve each equation.

91) $p^2 + 5p - 84 = 0$

92) $\frac{1}{25}n^2 + \frac{4}{5}n + 4 = 0$

93) $18w^3 = 98w$

94) $6a^2 = 54a$

95) $(x - 6)(x - 5) = 12$

96) $9z^2 + 99 = 108z$

97) $\frac{1}{6}(2y + 3)^2 + \frac{1}{2}y = \frac{1}{6}y(y - 10) + \frac{1}{6}$

98) $3c^2(2c + 3) - 14c(2c + 3) - 5(2c + 3) = 0$

99) $38r + 19r^2 + 2r^3 = 24$

100) $0.6 = 2.3k - k^2$

Find the indicated values for the following polynomial functions.

101) $f(x) = x^2 + 10x + 21$. Find x so that $f(x) = 0$.

102) $h(t) = t^2 - 6t - 16$. Find t so that $h(t) = 0$.

103) $g(a) = 2a^2 - 13a + 24$. Find a so that $g(a) = 4$.

104) $Q(x) = 4x^2 - 4x + 9$. Find x so that $Q(x) = 8$.

105) $H(b) = b^2 + 3$. Find b so that $H(b) - 19$.

106) $f(z) = z^3 + 3z^2 - 54z + 5$. Find z so that $f(z) = 5$.

107) $h(k) = 5k^3 - 25k^2 + 20k$. Find k so that $h(k) = 0$.

108) $g(x) = 9x^2 - 10$. Find x so that $g(x) = -6$.

R1) Why do you factor a quadratic equation before
    solving it?

R2) Why do quadratic equations in this section have
    two solutions?

R3) Which types of equations in this section have more
    than two solutions?

R4) Can you just look at an equation and determine
    how many solutions it will have?

R5) Which exercises in this section do you find most
    challenging?

7.6 Applications of Quadratic Equations

P Prepare

O Organize

What are your objectives for Section 7.6?	How can you accomplish each objective?
1 Solve Problems Involving Geometry	• Review the **Steps for Solving Applied Problems.** • Use the procedure for **Solving a Quadratic Equation by Factoring.** • Complete the given example on your own. • Complete You Try 1.
2 Solve Problems Involving Consecutive Integers	• Review the discussion of consecutive integer problems in Section 3.3. • Be aware that negative solutions are allowed. • Complete the given example on your own. • Complete You Try 2.
3 Solve Problems Using the Pythagorean Theorem	• Learn the *Pythagorean theorem.* • Be able to recognize a right triangle. • Complete the given examples on your own. • Complete You Trys 3 and 4.
4 Solve an Applied Problem Using a Given Quadratic Equation	• Understand the meanings of the variables in the equation. • Draw a diagram, when appropriate. • Complete the given example on your own. • Complete You Try 5.

W Work

Read the explanations, follow the examples, take notes, and complete the You Trys.

In Chapters 3 and 5 we explored applications of linear equations. In this section we will
look at applications involving quadratic equations. Let's begin by restating the five steps for
solving applied problems.

Procedure Steps for Solving Applied Problems

Step 1: **Read** the problem carefully, more than once if necessary, until you understand it. Draw a picture, if applicable. Identify what you are being asked to find.

Step 2: **Choose a variable** to represent an unknown quantity. If there are any other unknowns, define them in terms of the variable.

Step 3: **Translate** the problem from English into an equation using the chosen variable.

Step 4: **Solve** the equation.

Step 5: **Check** the answer in the original problem, and **interpret** the solution as it relates to the problem. Be sure your answer makes sense in the context of the problem.

1 Solve Problems Involving Geometry

EXAMPLE 1

Solve. A rectangular vegetable garden is 7 ft longer than it is wide. What are the dimensions of the garden if it covers 60 ft²?

Solution

Step 1: **Read** the problem carefully. Draw a picture.

Step 2: **Choose a variable** to represent the unknown.

$$\text{Let} \quad w = \text{the width}$$
$$w + 7 = \text{the length}$$

W Hint

Review basic geometry formulas, if necessary.

Step 3: **Translate** the information that appears in English into an algebraic equation. We must find the length and width of the garden. We are told that the *area* is 60 ft², so let's use the formula for the area of a rectangle. Then, substitute the expressions above for the length and width and 60 for the area.

$$(length)(width) = Area$$
$$(w + 7)(w) = 60 \qquad \text{length} = w + 7, \text{ width} = w, \text{ area} = 60$$

Step 4: **Solve** the equation.

$$w^2 + 7w = 60 \qquad \text{Distribute.}$$
$$w^2 + 7w - 60 = 0 \qquad \text{Write the equation in standard form.}$$
$$(w + 12)(w - 5) = 0 \qquad \text{Factor.}$$

$$w + 12 = 0 \qquad \text{or} \qquad w - 5 = 0 \qquad \text{Set each factor equal to zero.}$$
$$w = -12 \qquad \text{or} \qquad w = 5 \qquad \text{Solve.}$$

Step 5: **Check** the answer, and **interpret** the solution as it relates to the problem. Since w represents the width of the garden, it cannot be a negative number. So, $w = -12$ cannot be the solution. Therefore, the width is 5 ft, which will make the height $5 + 7 = 12$ ft. The area, then, is $(12 \text{ ft}) \cdot (5 \text{ ft}) = 60 \text{ ft}^2$.

[YOU TRY 1]

Solve. The area of the surface of a desk is 8 ft². Find the dimensions of the rectangular desktop if the width is 2 ft less than the length.

2 Solve Problems Involving Consecutive Integers

In Chapter 3 we solved problems involving consecutive integers. Some applications involving consecutive integers lead to quadratic equations.

EXAMPLE 2

Solve. Twice the sum of three consecutive odd integers is 9 less than the product of the smaller two. Find the integers.

Solution

Step 1: **Read** the problem carefully, and identify what we are being asked to find.

We must find three consecutive odd integers.

Step 2: **Choose a variable** to represent an unknown, and define the other unknowns in terms of this variable.

$$x = \text{the first odd integer}$$
$$x + 2 = \text{the second odd integer}$$
$$x + 4 = \text{the third odd integer}$$

Step 3: **Translate** the information that appears in English into an algebraic equation. Read the problem slowly and carefully, breaking it into small parts.

Statement:
| Twice the sum of three consecutive odd integers | is | 9 less than | the product of the smaller two. |

Equation: $2[x + (x + 2) + (x + 4)] = x(x + 2) - 9$

Step 4: **Solve** the equation.

$$2[x + (x + 2) + (x + 4)] = x(x + 2) - 9$$

$\qquad\qquad\quad 2(3x + 6) = x^2 + 2x - 9$ Combine like terms; distribute.

$\qquad\qquad\qquad\; 6x + 12 = x^2 + 2x - 9$ Distribute.

$\qquad\qquad\qquad\qquad\quad 0 = x^2 - 4x - 21$ Write in standard form.

$\qquad\qquad\qquad\qquad\quad 0 = (x + 3)(x - 7)$ Factor.

$\quad x + 3 = 0 \qquad \text{or} \qquad x - 7 = 0$ Set each factor equal to zero.

$\qquad\quad x = -3 \qquad\qquad\qquad x = 7$ Solve.

W Hint

In this application problem, one of the solutions is negative. Also note that this particular problem has two sets of solutions.

Step 5: **Check** the answer, and **interpret** the solution as it relates to the problem.

We get two sets of solutions. If $x = -3$, then the other odd integers are -1 and 1. If $x = 7$, the other odd integers are 9 and 11.

Check these numbers in the original statement of the problem.

$$2[-3 + (-1) + 1] = (-3)(-1) - 9 \qquad\qquad 2[7 + 9 + 11] = (7)(9) - 9$$
$$2(-3) = 3 - 9 \qquad\qquad\qquad\qquad 2(27) = 63 - 9$$
$$-6 = -6 \qquad\qquad\qquad\qquad\qquad 54 = 54$$

[YOU TRY 2]

Solve. Find three consecutive even integers such that the product of the two smaller numbers is the same as twice the sum of the integers.

3 Solve Problems Using the Pythagorean Theorem

A **right triangle** is a triangle that contains a 90° **(right)** angle. We can label a right triangle as follows.

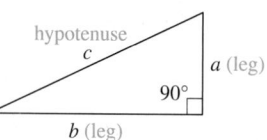

The side opposite the 90° angle is the longest side of the triangle and is called the **hypotenuse.** The other two sides are called the **legs.** The Pythagorean theorem states a relationship between the lengths of the sides of a right triangle. This is a very important relationship in mathematics and is one which is used in many different ways.

Definition Pythagorean Theorem

Given a right triangle with legs of length a and b and hypotenuse of length c,

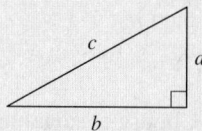

the Pythagorean theorem states that $a^2 + b^2 = c^2$ [or $(\text{leg})^2 + (\text{leg})^2 = (\text{hypotenuse})^2$].

The Pythagorean theorem is true *only* for right triangles.

EXAMPLE 3   Find the length of the missing side.

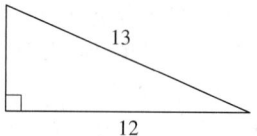

Solution

Since this is a right triangle, we can use the Pythagorean theorem to find the length of the side. Let a represent its length, and label the triangle.

The length of the hypotenuse is 13, so $c = 13$. a and 12 are legs. Let $b = 12$.

$$a^2 + b^2 = c^2 \qquad \text{Pythagorean theorem}$$
$$a^2 + (12)^2 = (13)^2 \qquad \text{Substitute values.}$$
$$a^2 + 144 = 169$$
$$a^2 - 25 = 0 \qquad \text{Write the equation in standard form.}$$
$$(a + 5)(a - 5) = 0 \qquad \text{Factor.}$$
$$a + 5 = 0 \quad \text{or} \quad a - 5 = 0 \qquad \text{Set each factor equal to 0.}$$
$$a = -5 \quad \text{or} \quad a = 5 \qquad \text{Solve.}$$

> **W Hint**
>
> Notice that the negative solution does not make sense because it represents the side length of a triangle.

$a = -5$ does not make sense as an answer because the length of a side of a triangle cannot be negative. Therefore, $a = 5$.

Check: $5^2 + (12)^2 \overset{?}{=} (13)^2$
$$25 + 144 = 169 \quad \checkmark$$

YOU TRY 3   Find the length of the missing side.

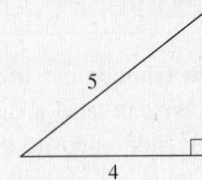

EXAMPLE 4

Solve. An animal holding pen situated between two buildings at a right angle with each other will have walls as two of its sides and a fence on the longest side. The side with the fence is 20 ft longer than the shortest side, while the third side is 10 ft longer than the shortest side. Find the length of the fence.

Solution

Step 1: **Read** the problem carefully, and identify what we are being asked to find. Draw a picture.

We must find the length of the fence.

Step 2: **Choose a variable** to represent an unknown, and define the other unknowns in terms of this variable. Draw and label the picture.

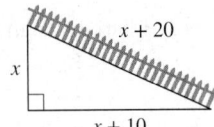

x = length of the shortest side (a leg)

$x + 10$ = length of the side along other building (a leg)

$x + 20$ = length of the fence (hypotenuse)

> **W Hint**
>
> Sketching a picture and labeling it is very useful in visualizing the application problem. This may make it easier to set up the equation.

Step 3: **Translate** the information that appears in English into an algebraic equation. We will use the Pythagorean theorem.

$$a^2 + b^2 = c^2 \qquad \text{Pythagorean theorem}$$
$$x^2 + (x + 10)^2 = (x + 20)^2 \qquad \text{Substitute.}$$

Step 4: **Solve** the equation.

$$x^2 + (x + 10)^2 = (x + 20)^2$$
$$x^2 + x^2 + 20x + 100 = x^2 + 40x + 400 \qquad \text{Multiply using FOIL.}$$
$$2x^2 + 20x + 100 = x^2 + 40x + 400$$
$$x^2 - 20x - 300 = 0 \qquad \text{Write in standard form.}$$
$$(x - 30)(x + 10) = 0 \qquad \text{Factor.}$$

$$x - 30 = 0 \quad \text{or} \quad x + 10 = 0 \qquad \text{Set each factor equal to 0.}$$
$$x = 30 \quad \text{or} \qquad x = -10 \qquad \text{Solve.}$$

Step 5: **Check** the answer, and **interpret** the solution as it relates to the problem.

The length of the shortest side, x, cannot be a negative number, so x cannot equal -10. Therefore, the length of the shortest side must be 30 ft.

The length of the side along the other building is $x + 10$, so $30 + 10 = 40$ ft.

The length of the fence is $x + 20$, so $30 + 20 = 50$ ft.

Do these lengths satisfy the Pythagorean theorem? Yes.

$$a^2 + b^2 = c^2$$
$$(30)^2 + (40)^2 \stackrel{?}{=} (50)^2$$
$$900 + 1600 = 2500 \quad \checkmark$$

Therefore, the length of the fence is 50 ft.

[YOU TRY 4]

Solve. A wire is attached to the top of a pole. The wire is 4 ft longer than the pole, and the distance from the wire on the ground to the bottom of the pole is 4 ft less than the height of the pole. Find the length of the wire and the height of the pole.

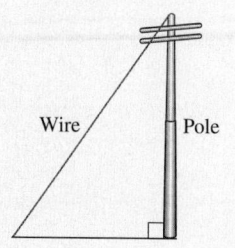

Next we will see how to use quadratic equations that model real-life situations.

4 Solve an Applied Problem Using a Given Quadratic Equation

EXAMPLE 5

A Little League baseball player throws a ball upward. The height h of the ball (in feet) t sec after the ball is released is given by the quadratic equation

$$h = -16t^2 + 30t + 4$$

a) What is the initial height of the ball?

b) How long does it take the ball to reach a height of 18 ft?

c) How long does it take for the ball to hit the ground?

Solution

 **Hint**

Try to sketch a picture for this problem.

a) We are asked to find the height at which the ball is released. Since t represents the number of seconds after the ball is thrown, $t = 0$ at the time of release.

Let $t = 0$, and solve for h.

$$h = -16(0)^2 + 30(0) + 4 \qquad \text{Substitute 0 for } t.$$
$$= 0 + 0 + 4$$
$$= 4$$

The initial height of the ball is 4 ft.

b) We must find the *time* it takes for the ball to reach a height of 18 ft.

Find t when $h = 18$.

$$h = -16t^2 + 30t + 4$$
$$18 = -16t^2 + 30t + 4 \qquad \text{Substitute 18 for } h.$$
$$0 = -16t^2 + 30t - 14 \qquad \text{Write in standard form.}$$
$$0 = 8t^2 - 15t + 7 \qquad \text{Divide by } -2.$$
$$0 = (8t - 7)(t - 1) \qquad \text{Factor.}$$

$$8t - 7 = 0 \quad \text{or} \quad t - 1 = 0 \qquad \text{Set each factor equal to 0.}$$
$$8t = 7$$
$$t = \frac{7}{8} \quad \text{or} \qquad t = 1 \qquad \text{Solve.}$$

How can two answers be possible? After $\frac{7}{8}$ sec the ball is 18 ft above the ground *on its way up,* and after 1 sec, the ball is 18 ft above the ground *on its way down.*

The ball reaches a height of 18 ft after $\frac{7}{8}$ sec *and* after 1 sec.

c) We must determine the amount of time it takes for the ball to hit the ground. When the ball hits the ground, how high off the ground is it? *It is 0 ft high.*

Find t when $h = 0$.

$$h = -16t^2 + 30t + 4$$
$$0 = -16t^2 + 30t + 4 \qquad \text{Substitute 0 for } h.$$
$$0 = 8t^2 - 15t - 2 \qquad \text{Divide by } -2.$$
$$0 = (8t + 1)(t - 2) \qquad \text{Factor.}$$

$$8t + 1 = 0 \quad \text{or} \quad t - 2 = 0 \qquad \text{Set each factor equal to 0.}$$
$$8t = -1$$
$$t = -\frac{1}{8} \quad \text{or} \qquad t = 2 \qquad \text{Solve.}$$

Since t represents time, t cannot equal $-\frac{1}{8}$. We reject that as a solution. Therefore, $t = 2$. The ball will hit the ground after 2 sec.

> **Note**
>
> In Example 5, the equation can also be written using function notation, $h(t) = -16t^2 + 30t + 4$, because the expression $-16t^2 + 30t + 4$ is a polynomial. Furthermore, $h(t) = -16t^2 + 30t + 4$ is a *quadratic function,* and we say that the height, h, is a function of the time, t. We will study quadratic functions in more detail in Chapter 12.

[YOU TRY 5]

An object is thrown upward from a building. The height h of the object (in feet) t sec after the object is released is given by the quadratic equation

$$h = -16t^2 + 36t + 36$$

a) What is the initial height of the object?

b) How long does it take the object to reach a height of 44 ft?

c) How long does it take for the object to hit the ground?

ANSWERS TO [YOU TRY] EXERCISES

1) width = 2 ft; length = 4 ft   2) 6, 8, 10 or −2, 0, 2   3) 3
4) length of wire = 20 ft; height of pole = 16 ft
5) a) 36 ft   b) 0.25 sec and 2 sec   c) 3 sec

Objective 1: Solve Problems Involving Geometry

Find the length and width of each rectangle.

1) Area = 36 in^2

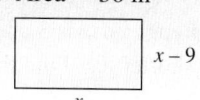

2) Area = 40 cm^2

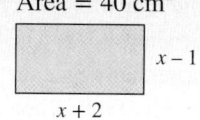

Find the base and height of each triangle.

3) Area = 12 cm^2

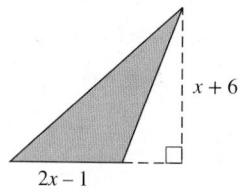

4) Area = 42 in^2

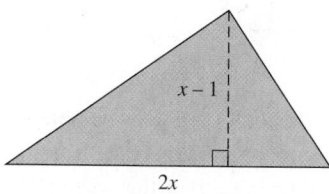

Find the base and height of each parallelogram.

5) Area = 18 in^2

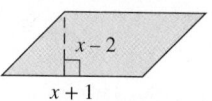

6) Area = 50 cm^2

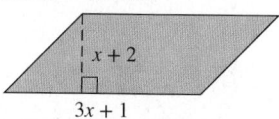

7) The volume of the box is 240 in^3. Find its length and width.

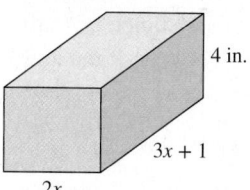

8) The volume of the box is 120 in^3. Find its width and height.

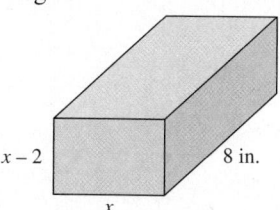

Write an equation, and solve.

9) A rectangular rug is 4 ft longer than it is wide. If its area is 45 ft^2, what are its length and width?

10) The surface of a rectangular bulletin board has an area of 300 in^2. Find its dimensions if it is 5 in. longer than it is wide.

11) Judy makes stained glass windows. She needs to cut a rectangular piece of glass with an area of 54 in^2 so that its width is 3 in. less than its length. Find the dimensions of the glass she must cut.

Apis|Abramis/Alamy Stock Photo

12) A rectangular painting is twice as long as it is wide. Find its dimensions if it has an area of 12.5 ft^2.

13) The volume of a rectangular storage box is 1440 in^3. It is 20 in. long, and it is half as tall as it is wide. Find the width and height of the box.

14) A rectangular aquarium is 15 in. high, and its length is 8 in. more than its width. Find the length and width if the volume of the aquarium is 3600 in^3.

Ken Cavanagh/McGraw-Hill Education

15) The height of a triangle is 3 cm more than its base. Find the height and base if its area is 35 cm^2.

16) The area of a triangle is 16 cm^2. Find the height and base if its height is half the length of the base.

Objective 2: Solve Problems Involving Consecutive Integers

Write an equation, and solve.

17) The product of two consecutive integers is 19 more than their sum. Find the integers.

18) The product of two consecutive odd integers is 1 less than three times their sum. Find the integers.

19) Find three consecutive even integers such that the sum of the smaller two is one-fourth the product of the second and third integers.

20) Find three consecutive integers such that the square of the smallest is 29 less than the product of the larger two.

21) Find three consecutive integers such that the square of the largest is 22 more than the product of the smaller two.

22) Find three consecutive odd integers such that the product of the smaller two is 15 more than four times the sum of the three integers.

Objective 3: Solve Problems Using the Pythagorean Theorem

23) In your own words, explain the Pythagorean theorem.

24) Can the Pythagorean theorem be used to find a in this triangle? Why or why not?

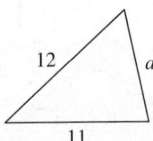

Use the Pythagorean theorem to find the length of the missing side.

25)

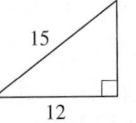

26)

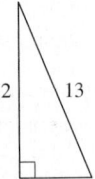

27)

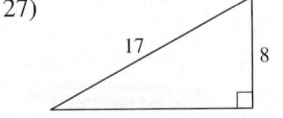

28)

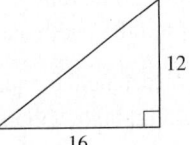

29)

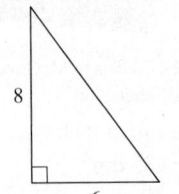

30)

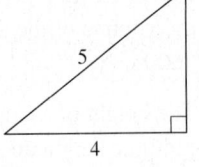

Find the lengths of the sides of each right triangle.

31)

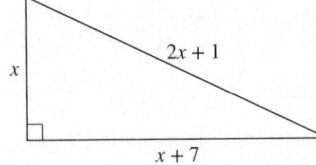

32)

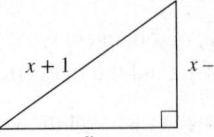

33)

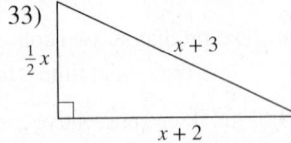

34)

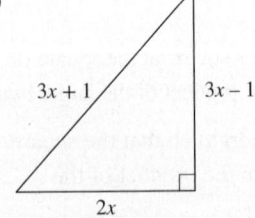

Write an equation, and solve.

35) The hypotenuse of a right triangle is 2 in. longer than the longer leg. The shorter leg measures 2 in. less than the longer leg. Find the measure of the longer leg of the triangle.

36) The longer leg of a right triangle is 7 cm more than the shorter leg. The length of the hypotenuse is 3 cm more than twice the length of the shorter leg. Find the length of the hypotenuse.

37) A 13-ft ladder is leaning against a wall. The distance from the top of the ladder to the bottom of the wall is 7 ft more than the distance from the bottom of the ladder to the wall. Find the distance from the bottom of the ladder to the wall.

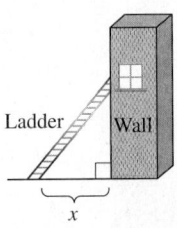

38) A wire is attached to the top of a pole. The pole is 2 ft shorter than the wire, and the distance from the wire on the ground to the bottom of the pole is 9 ft less than the length of the wire. Find the length of the wire and the height of the pole.

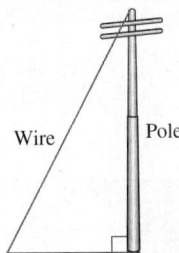

39) From a bike shop, Rana pedals due north while Yasmeen rides due west. When Yasmeen is 4 mi from the shop, the distance between her and Rana is two miles more than Rana's distance from the bike shop. Find the distance between Rana and Yasmeen.

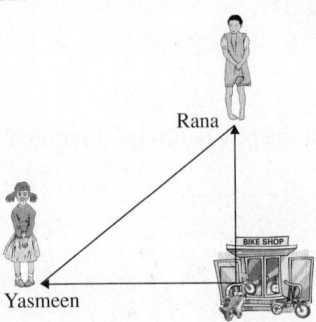

40) Henry and Allison leave home to go to work. Henry drives due east while his wife drives due south. At 8:30 A.M., Allison is 3 mi farther from home than Henry, and the distance between them is 6 mi more than Henry's distance from home. Find Henry's distance from his house.

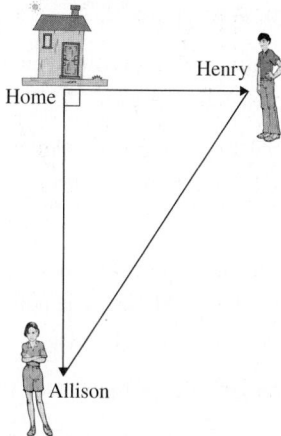

Objective 4: Solve Applied Problems Using a Given Quadratic Equation

Solve.

(24) 41) A rock is dropped from a cliff and into the ocean. The height h (in feet) of the rock after t sec is given by $h = -16t^2 + 144$.

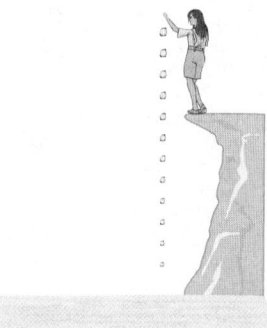

a) What is the initial height of the rock?

b) When is the rock 80 ft above the water?

c) How long does it take the rock to hit the water?

42) An object is launched from a platform with an initial velocity of 32 ft/sec. The height h (in feet) of the object after t sec is given by $h = -16t^2 + 32t + 20$.

a) What is the initial height of the object?

b) When is the object 32 ft above the ground?

c) How long does it take for the object to hit the ground?

Organizers of fireworks shows use quadratic and linear equations to help them design their programs. *Shells* contain the chemicals that produce the bursts we see in the sky. At a fireworks show the shells are shot from *mortars* and when the chemicals inside the shells ignite they explode, producing the brilliant bursts we see in the night sky.

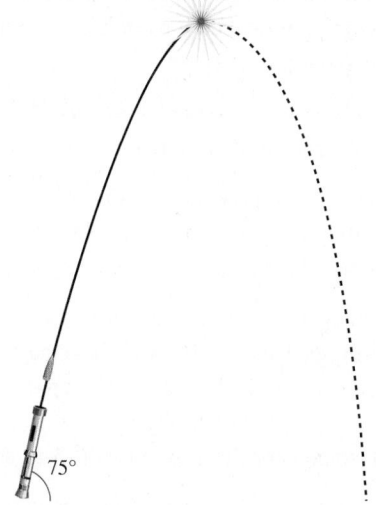

43) At a fireworks show, a 3-in. shell is shot from a mortar at an angle of 75°. The height, y (in feet), of the shell t sec after being shot from the mortar is given by the quadratic equation

$$y = -16t^2 + 144t$$

and the horizontal distance of the shell from the mortar, x (in feet), is given by the linear equation

$$x = 39t$$

(http://library.thinkquest.org/15384/physics/physics.html)

a) How high is the shell after 3 sec?

b) What is the shell's horizontal distance from the mortar after 3 sec?

c) The maximum height is reached when the shell explodes. How high is the shell when it bursts after 4.5 sec?

d) What is the shell's horizontal distance from its launching point when it explodes? (Round to the nearest foot.)

44) When a 10-in. shell is shot from a mortar at an angle of 75°, the height, y (in feet), of the shell t sec after being shot from the mortar is given by

$$y = -16t^2 + 264t$$

and the horizontal distance of the shell from the mortar, x (in feet), is given by

$$x = 71t$$

a) How high is the shell after 3 sec?

b) Find the shell's horizontal distance from the mortar after 3 sec.

c) The shell explodes after 8.25 sec. What is its height when it bursts?

d) What is the shell's horizontal distance from its launching point when it explodes? (Round to the nearest foot.)

e) Compare your answers to 43a) and 44a). What is the difference in their heights after 3 sec?

f) Compare your answers to 43c) and 44c). What is the difference in the shells' heights when they burst?

g) Assuming that the technicians timed the firings of the 3-in. shell and the 10-in. shell so that they exploded at the same time, how far apart would their respective mortars need to be so that the 10-in. shell would burst directly above the 3-in. shell?

45) The senior class at Richmont High School is selling t-shirts to raise money for its prom. The equation $R(p) = -25p^2 + 600p$ describes the revenue, R, in dollars, as a function of the price, p, in dollars, of a t-shirt. That is, the revenue is a function of price.

a) Determine the revenue if the group sells each shirt for $10.

b) Determine the revenue if the group sells each shirt for $15.

c) If the senior class hopes to have a revenue of $3600, how much should it charge for each t-shirt?

R **Rethink**

R1) What is an application problem?

R2) Do you think a calculator can solve quadratic equations for you? How will a calculator help you solve the problems in this application section?

R3) Why do you sometimes eliminate negative solutions when you are solving application problems involving quadratic equations?

46) A famous comedian will appear at a comedy club for one performance. The equation $R(p) = -5p^2 + 300p$ describes the relationship between the price of a ticket, p, in dollars, and the revenue, R, in dollars, from ticket sales. That is, the revenue is a function of price.

a) Determine the club's revenue from ticket sales if the price of a ticket is $40.

b) Determine the club's revenue from ticket sales if the price of a ticket is $25.

c) If the club is expecting its revenue from ticket sales to be $4500, how much should it charge for each ticket?

47) An object is launched upward from the ground with an initial velocity of 200 ft/sec. The height h (in feet) of the object after t sec is given by $h(t) = -16t^2 + 200t$.

a) Find the height of the object after 1 sec.

b) Find the height of the object after 4 sec.

c) When is the object 400 ft above the ground?

d) How long does it take for the object to hit the ground?

48) The equation $R(p) = -7p^2 + 700p$ describes the revenue from ticket sales, R, in dollars, as a function of the price, p, in dollars, of a ticket to a fundraising dinner. That is, the revenue is a function of price.

a) Determine the revenue if the ticket price is $40.

b) Determine the revenue if the group sells each ticket for $70.

c) If the goal of the organizers is to have ticket revenue of $17,500, how much should it charge for each ticket?

R4) What types of college courses do you think often use quadratic equations to solve application problems?

R5) Which exercises in this section do you find most challenging?

Group Activity – The Geometry of Special Binomials

The formulas for $a^2 - b^2$ and $a^3 - b^3$ can be derived from geometry. In Section 1.3, you learned the formulas for the area of a rectangle and the volume of a rectangular solid. In Section 7.1, you learned how to factor out a common binomial factor. Using these concepts, you can derive the formulas for factoring special binomials.

Difference of Two Squares

The area of the shaded region below can be found by subtracting the area of the smaller square from the area of the larger square: $a^2 - b^2$.

The area of the shaded region can also be found by breaking up the shaded region into rectangles and summing the areas of the rectangles.

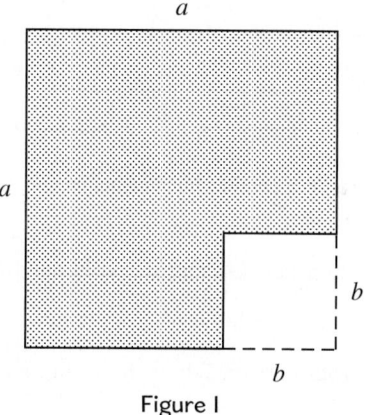

Figure I

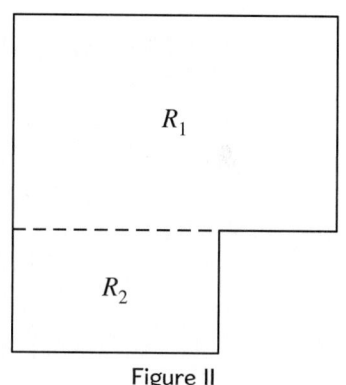

Figure II

1) This problem derives the formula for factoring a difference of two squares.

   a) Find the area of the rectangles R_1 and R_2 in terms of a and b. Leave in factored form.

   b) Rewrite the equation below using the areas from Figures I and II:

   $$\text{area of the shaded region} = \text{area of } R_1 + \text{area of } R_2$$

   c) In the equation from part b), factor the right-hand side to obtain the formula for factoring the expression $a^2 - b^2$.

   d) In the equation from part c), multiply and simplify the right-hand side to verify the formula.

2) Draw a geometric shape that corresponds to each of the following expressions, then factor.

   a) $x^2 - 5^2$

   b) $x^2 - 16$

   c) $(3x)^2 - 4$

   d) $25x^2 - 1$

Difference of Two Cubes

The volume of the solid below can be found by subtracting the volume of the smaller cube from the volume of the larger cube: $a^3 - b^3$

The volume can also be found by breaking up the solid into rectangular solids, and summing the individual volumes.

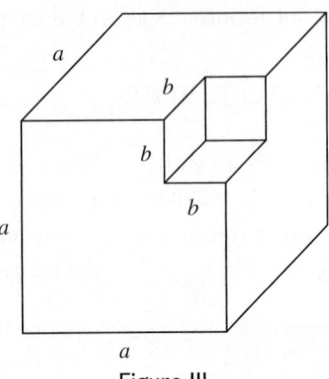

Figure III

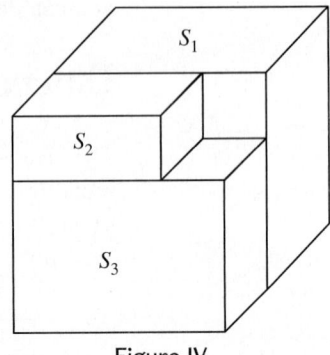

Figure IV

3) This problem derives the formula for factoring a difference of two cubes.

  a) Find the volume of the rectangular solids S_1, S_2, and S_3 in terms of a and b. Leave in factored form.

  b) Rewrite the equation below using the volumes from Figures III and IV:

  volume of the solid = volume of S_1 + volume of S_2 + volume of S_3

  c) In the equation from part b), factor the right-hand side to obtain the formula for factoring the expression $a^3 - b^3$.

  d) In the equation from part c), multiply and simplify the right-hand side to verify the formula.

4) Draw a geometric shape that corresponds to each of the following expressions, then factor.

  a) $x^3 - 2^3$

  b) $8x^3 - 27$

Sum of Two Cubes

5) The formula for factoring a sum of two cubes can also be obtained by breaking a solid (Figure V) into four rectangular solids to obtain the following equations:

$$a^3 + b^3 = a(a - b)^2 + ab(a - b) + ab(a - b) + b^2(a + b)$$

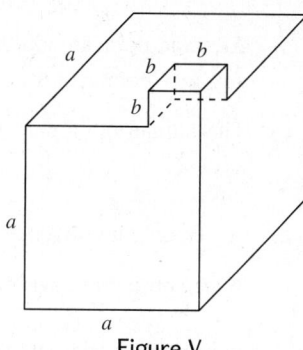

Figure V

a) Simplify $a(a - b)^2 + ab(a - b) + ab(a - b)$, and factor completely.

b) In the equation, replace $a(a - b)^2 + ab(a - b) + ab(a - b)$ with the factored form.

c) In the equation in part b), factor out $(a + b)$ to obtain the formula for factoring the expression $a^3 + b^3$.

For each numbered choice, circle the statement that most closely reflects what you believe. Be honest in your choices!

1) a) At some point in adulthood, the human brain stops growing and does not make any new connections.

   b) The human brain can make new connections and "grow" as we learn, even as adults.

2) a) A person's intelligence can change as they grow and learn.

   b) A person is born with a certain amount of intelligence, and it cannot change.

3) a) A person's basic personality cannot change.

   b) A person's personality can change.

4) a) Hard work, goal-setting, and strategic planning are the most important elements to becoming successful.

   b) Being smart and lucky are the most important elements to becoming successful.

5) a) People cannot change their strengths and weaknesses.

   b) It is possible for people to improve their strengths and weaknesses.

6) a) I can clearly identify my strengths and weaknesses.

   b) I do not know what my strengths and weaknesses are.

7) a) I prefer to do tasks that are easy for me, over and over, even if they become boring.

   b) I prefer to take on new and difficult tasks because I like to be challenged.

8) a) When someone criticizes me, I listen to what they have to say and try to learn from them.

   b) When someone criticizes me, I ignore them because they don't know what they are talking about.

9) a) When I encounter obstacles in my life, I give up because the situation is out of my hands and there is nothing I can do about it.

   b) When I encounter obstacles in my life, I figure out how to overcome them to achieve my goals.

10) a) I view failure as just a setback and as something I can learn from. Then, I try again.

    b) When I fail at something, I give up.

11) a)   If I get a bad grade on a test, I don't try harder in the future because, I think, *"Why bother? It's out of my control, and I can't do it."*

   b)   If I get a bad grade on a test, I talk with my instructor to figure out what I did wrong and to come up with strategies for improving my learning and test-taking for next time.

12) a)   When I don't understand what the instructor is doing in class, I tell myself that I am in class to learn what I don't know *yet* and that I *can* learn with the right strategies and attitude.

   b)   I feel stupid when I don't understand what the instructor is teaching in class.

13) a)   I have no long-term plan for my life.

   b)   I have a long-term plan for my life and a plan for achieving it.

14) a)   I have a role model whom I admire and to whom I can look to help me achieve a similar goal.

   b)   I have a role model, but I know I can never achieve anything close to what that person has achieved.

In the *odd*-numbered statements, #1 – #13, part a) indicates a *fixed mindset* while b) indicates a *growth mindset*. In the *even*-numbered statements, #2 – #14, part a) indicates a *growth mindset* while part b) indicates a *fixed mindset*. Next to each of #1 – #14, write an *F* or a *G* to note whether you circled the fixed mindset or growth mindset statement, and count the number of each.

Do your answers show a tendency toward a fixed or a growth mindset? (Most people have some of both.) What, exactly, are these mindsets, and what can they tell you about yourself?

A **mindset** is a person's set of beliefs. People who think that their beliefs and qualities cannot be changed have a **fixed mindset.** On the other hand, people with a **growth mindset** believe that their beliefs, abilities, and personalities can change by using strategies and working hard. In fact, research shows that the brain itself actually changes as you learn new things, no matter your age! So, adopting a growth mindset is important for learning and succeeding in school and in life. (For the original version, see Dweck, Carol S. [2016]. *Mindset: The New Psychology of Success.* New York: Random House.)

Chapter 7: Summary

Definition/Procedure	Example

7.1 The Greatest Common Factor and Factoring by Grouping

To **factor a polynomial** is to write it as a product of two or more polynomials.

To factor out a greatest common factor (GCF):

1) Identify the GCF of all of the terms of the polynomial.

2) Rewrite each term as the product of the GCF and another factor.

3) Use the distributive property to factor out the GCF from the terms of the polynomial.

4) Check the answer by multiplying the factors.

Factor out the greatest common factor.

$$6k^6 - 27k^5 + 15k^4$$

The GCF is $3k^4$.

$$6k^6 - 27k^5 + 15k^4 = (3k^4)(2k^2) - (3k^4)(9k) + (3k^4)(5)$$
$$= 3k^4(2k^2 - 9k + 5)$$

Check: $3k^4(2k^2 - 9k + 5) = 6k^6 - 27k^5 + 15k^4$ ✓

The first step in factoring any polynomial is to ask yourself, *"Can I factor out a GCF?"*

The last step in factoring any polynomial is to ask yourself, *"Can I factor again?"*

Try to **factor by grouping** when you are asked to factor a polynomial containing four terms.

1) Make two groups of two terms so that each group has a common factor.

2) Take out the common factor from each group of terms.

3) Factor out the common factor using the distributive property.

4) Check the answer by multiplying the factors.

Factor $10xy + 5y - 8x - 4$ completely.

Since the four terms have a GCF of 1, we will not factor out a GCF. Begin by grouping two terms together so that each group has a common factor.

$$\underbrace{10xy + 5y}\ \underbrace{- 8x - 4}$$

$= 5y(2x + 1) - 4(2x + 1)$ Take out the common factor.

$= (2x + 1)(5y - 4)$ Factor out $(2x + 1)$.

Check: $(2x + 1)(5y - 4) = 10xy + 5y - 8x - 4$ ✓

7.2 Factoring Trinomials of the Form $x^2 + bx + c$

Factoring $x^2 + bx + c$

If $x^2 + bx + c = (x + m)(x + n)$, then

1) if b and c are positive, then both m and n must be positive.

2) if c is positive and b is negative, then both m and n must be negative.

3) if c is negative, then one integer, m, must be positive and the other integer, n, must be negative.

Factor completely.

a) $t^2 + 9t + 20$

Think of two numbers whose *product* is 20 and whose *sum* is 9: **4 and 5.** Then,

$$t^2 + 9t + 20 = (t + 4)(t + 5)$$

b) $3s^3 - 33s^2 + 54s$

Begin by factoring out the GCF of $3s$.

$$3s^3 - 33s^2 + 54s = 3s(s^2 - 11s + 18) = 3s(s - 2)(s - 9)$$

7.3 Factoring Trinomials of the Form $ax^2 + bx + c$ (a ≠ 1)

Factoring $ax^2 + bx + c$ by **grouping.**

Factor $5t^2 + 18t - 8$ completely.

$$\text{Sum is 18.}$$
$$\downarrow$$
$$5t^2 + 18t - 8$$
$$\text{Product: } 5 \cdot (-8) = -40$$

Think of two integers whose *product* is -40 and whose *sum* is 18. **20 and -2**

Definition/Procedure	Example
	Factor by grouping. $5t^2 + 18t - 8 = \underbrace{5t^2 + 20t}_{} \underbrace{- 2t - 8}_{}$ Write $18t$ as $20t - 2t$. $= 5t(t + 4) - 2(t + 4)$ $= (t + 4)(5t - 2)$
Factoring $ax^2 + bx + c$ by trial and error. When approaching a problem in this way, we must keep in mind that we are reversing the FOIL process.	Factor completely. $4x^2 - 16x + 15$ $4x^2 - 16x + 15 = (2x - 3)(2x - 5)$ $-6x$ $+ -10x$ $-16x$ $4x^2 - 16x + 15 = (2x - 3)(2x - 5)$

7.4 Factoring Special Trinomials and Binomials

A **perfect square trinomial** is a trinomial that results from squaring a binomial.
 	
Factoring a Perfect Square Trinomial
 	
$a^2 + 2ab + b^2 = (a + b)^2$
$a^2 - 2ab + b^2 = (a - b)^2$
	Factor completely.
	a) $g^2 + 22g + 121 = (g + 11)^2$
	    $a = g$ $b = 11$
	
	b) $16d^2 - 8d + 1 = (4d - 1)^2$
	    $a = 4d$ $b = 1$

Factoring the Difference of Two Squares $a^2 - b^2 = (a + b)(a - b)$	Factor completely. $w^2 - 64 = (w + 8)(w - 8)$ $\downarrow$ $\downarrow$ $(w)^2$ $(8)^2$ $a = w, b = 8$
Factoring the Sum and Difference of Two Cubes $a^3 + b^3 = (a + b)(a^2 - ab + b^2)$ $a^3 - b^3 = (a - b)(a^2 + ab + b^2)$	Factor completely. $w^3 + 27 = (w + 3)[(w)^2 - (w)(3) + (3)^2]$ $\downarrow$ $\downarrow$ $(w)^3$ $(3)^3$ $a = w,$ $b = 3$ $w^3 + 27 = (w + 3)(w^2 - 3w + 9)$

7.5 Solving Quadratic Equations by Factoring

A **quadratic equation** can be written in the form $ax^2 + bx + c = 0$, where a, b, and c are real numbers and $a \neq 0$.	Some examples of quadratic equations are $5x^2 - 20 = 0$, $y^2 = 4y + 21$, $4(p - 2)^2 = 8 - 7p$
To solve a quadratic equation by factoring, use the **zero product rule:** If $ab = 0$, then $a = 0$ or $b = 0$.	Solve $(y + 7)(y - 4) = 0$ $y + 7 = 0$ or $y - 4 = 0$ Set each factor equal to zero. $y = -7$ or $y = 4$ Solve. The solution set is $\{-7, 4\}$.
Steps for Solving a Quadratic Equation by Factoring 1) Write the equation in the form $ax^2 + bx + c = 0$. 2) Factor the expression. 3) Set each factor equal to zero, and solve for the variable. 4) Check the answer(s) in the original equation.	Solve $2p^2 - 3p - 11 = -9$. $2p^2 - 3p - 2 = 0$ Standard form $(2p + 1)(p - 2) = 0$ Factor. $2p + 1 = 0$ or $p - 2 = 0$ $2p = -1$ $p = -\dfrac{1}{2}$ or $p = 2$ The solution set is $\left\{-\dfrac{1}{2}, 2\right\}$. Check the answers.

Definition/Procedure	Example

7.6 Applications of Quadratic Equations

Pythagorean Theorem

Given a right triangle with legs of length a and b and hypotenuse of length c,

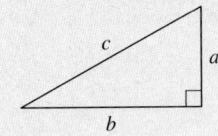

the Pythagorean theorem states that

$$a^2 + b^2 = c^2$$

Find the length of side a.

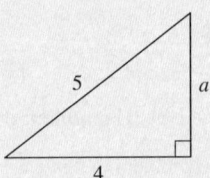

Let $b = 4$ and $c = 5$ in $a^2 + b^2 = c^2$.

$$a^2 + (4)^2 = (5)^2$$
$$a^2 + 16 = 25$$
$$a^2 - 9 = 0$$
$$(a + 3)(a - 3) = 0$$

$$a + 3 = 0 \quad \text{or} \quad a - 3 = 0$$
$$a = -3 \quad \text{or} \quad a = 3$$

Reject -3 as a solution because the length of a side cannot be negative.

Therefore, $a = 3$.

Chapter 7: Review Exercises

(7.1) Find the greatest common factor of each group of terms.

1) $40, 56$

2) $36y, 12y^2, 54y^2$

3) $15h^4, 45h^5, 20h^3$

4) $4c^4d^3, 20c^4d^2, 28c^2d$

Factor out the greatest common factor.

5) $63t + 45$

6) $21w^5 - 56w$

7) $2p^6 - 20p^5 + 2p^4$

8) $18a^3b^3 - 3a^2b^3 - 24ab^3$

9) $n(m + 8) - 5(m + 8)$

10) $x(9y - 4) + w(9y - 4)$

11) Factor out $-5r$ from $-15r^3 - 40r^2 + 5r$.

12) Factor out -1 from $-z^2 + 9z - 4$.

Factor by grouping.

13) $ab + 2a + 9b + 18$

14) $cd - 3c + 8d - 24$

15) $4xy - 28y - 3x + 21$

16) $hk^2 + 6h - k^2 - 6$

(7.2) Factor completely.

17) $q^2 + 10q + 24$

18) $t^2 - 12t + 27$

19) $z^2 - 6z - 72$

20) $h^2 + 6h - 7$

21) $m^2 - 13mn + 30n^2$

22) $a^2 + 11ab + 30b^2$

23) $4v^2 - 24v - 64$

24) $7c^2 - 7c - 84$

25) $-9w^4 - 9w^3 + 18w^2$

26) $-5x^3y + 25x^2y^2 - 20xy^3$

27) $(a + 1)b^2 - 11b(a + 1) + 18(a + 1)$

28) $(m - 3)n^2 + 12n(m - 3) + 35(m - 3)$

(7.3) Factor completely.

29) $3r^2 - 23r + 14$

30) $5k^2 + 11k + 6$

31) $4p^2 - 8p - 5$

32) $8d^2 + 29d - 12$

33) $12c^2 + 38c + 20$

34) $21n^2 - 54n + 24$

35) $10x^2 + 39xy - 27y^2$

36) $6a^2 - 19ab - 20b^2$

37) $(3c - 5)^2 + 10(3c - 5) + 24$

38) $2(k + 1)^2 - 15(k + 1) + 28$

(7.4) Factor completely.

39) $w^2 - 49$

40) $\dfrac{25}{9} - h^2$

41) $64t^2 - 25u^2$

42) $y^4 - 81$

43) $4b^2 + 9$

44) $12c^2 - 48d^2$

45) $r^2 + 12r + 36$

46) $9z^2 - 24z + 16$

47) $20k^2 - 60k + 45$

48) $25a^2 + 20ab + 4b^2$

49) $h^3 + 8$

50) $q^3 - 1$

51) $27p^3 - 64q^3$

52) $16c^3 + 250d^3$

(7.1–7.4) Mixed Exercises

For Exercises 53 and 54, answer *always*, *sometimes*, or *never*.

53) A binomial of the form $ax^2 + bx$ will *always*, *sometimes*, or *never* factor into a difference of two squares.

54) A binomial that is the product of two binomials is *always*, *sometimes*, or *never* a difference of two squares.

Factor completely, if possible.

55) $7r^2 + 8r - 12$

56) $-3y^2 - 60y - 300$

57) $\dfrac{9}{25} - x^2$

58) $81v^6 + 36v^5 - 9v^4$

59) $st - 5s - 8t + 40$

60) $n^2 - 11n + 30$

61) $w^5 - w^2$

62) $gh + 8g - 11h - 88$

63) $a^2 + 3a - 14$

64) $49k^2 - 144$

65) $(a - b)^2 - (a + b)^2$

66) $1000a^3 + 27b^3$

67) $6(y - 2)^2 - 13(y - 2) - 8$

68) $5a^2 + 22ab + 8b^2$

(7.5) Solve each equation.

69) $c(2c - 1) = 0$

70) $(4z + 7)^2 = 0$

71) $3x^2 + x = 2$

72) $f^2 - 1 = 0$

73) $n^2 = 12n + 45$

74) $10j^2 - 8 + 11j = 0$

75) $\dfrac{2}{3}m^2 - \dfrac{2}{3}m + \dfrac{1}{6} = 0$

76) $\dfrac{1}{2}k^2 - \dfrac{4}{3}k + \dfrac{8}{9} = 0$

77) $36 = 49d^2$

78) $-13w = w^2$

79) $8b + 64 = 2b^2$

80) $18 = a(9 - a)$

81) $y(5y - 9) = -4$

82) $(z + 2)^2 = -z(3z + 4) + 9$

83) $6a^3 - 3a^2 - 18a = 0$

84) $48 = 6r^2 + 12r$

85) $(h + 8)(h - 3) = -10$

86) $(2b - 1)(b - 6) = 13$

87) $c(5c - 1) + 8 = 4(20 + c^2)$

88) $15t^3 + 40t = 70t^2$

89) $n^3 + 5n^2 - 9n = 45$

90) $p^2(6p - 1) - 10p(6p - 1) + 21(6p - 1) = 0$

(7.6)

91) Find the base and height if the area of the triangle is 15 in^2.

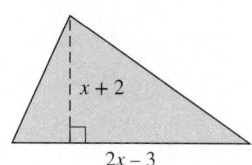

92) Find the length and width of the rectangle if its area is 28 cm^2.

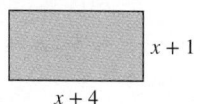

93) Find the height and length of the box if its volume is 96 in^3.

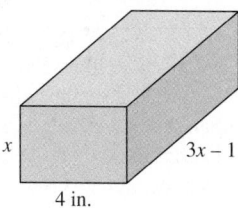

94) Find the length and width of the box if its volume is 360 in^3.

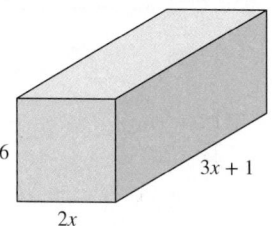

Use the Pythagorean theorem to find the length of the missing side.

95)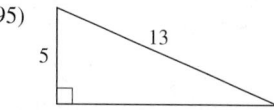

96)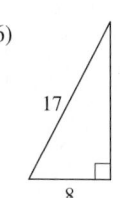

Write an equation, and solve.

97) A rectangular countertop has an area of 15 ft^2. If the width is 3.5 ft shorter than the length, what are the dimensions of the countertop?

98) Kelsey cuts a piece of fabric into a triangle to make a bandana for her dog. The base of the triangle is twice its height. Find the base and height if there is 144 in^2 of fabric.

Alley Cat Productions/ Brand X Pictures/Getty Images

99) The sum of three consecutive integers is one-third the square of the middle number. Find the integers.

100) Find two consecutive even integers such that their product is 6 more than 3 times their sum.

101) Seth builds a bike ramp in the shape of a right triangle. One leg is one inch shorter than the "ramp" while the other leg, the height of the ramp, is 8 in. shorter than the ramp. What is the height of the ramp?

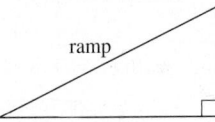

ramp

102) A car heads east from an intersection while a motorcycle travels south. After 20 min, the car is 2 mi farther from the intersection than the motorcycle. The distance between the two vehicles is 4 mi more than the motorcycle's distance from the intersection. What is the distance between the car and the motorcycle?

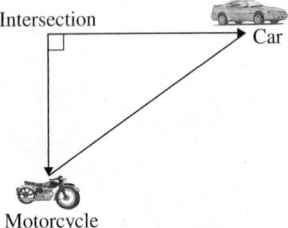

Intersection

Car

Motorcycle

103) An object is launched with an initial velocity of 96 ft/sec. The height h (in feet) of the object after t sec is given by $h = -16t^2 + 96t$.

   a) From what height is the object launched?

   b) When does the object reach a height of 128 ft?

   c) How high is the object after 3 sec?

   d) When does the object hit the ground?

Chapter 7:   Test

1) What is the first thing you should do when you are asked to factor a polynomial?

Factor completely.

2) $n^2 - 11n + 30$

3) $16 - b^2$

4) $5a^2 - 13a - 6$

5) $56p^6q^6 - 77p^4q^4 + 7p^2q^3$

6) $y^3 - 8z^3$

7) $2d^3 + 14d^2 - 36d$

8) $r^2 + 25$

9) $9h^2 + 24h + 16$

10) $24xy - 36x + 22y - 33$

11) $s^2 - 3st - 28t^2$

12) $16s^4 - 81t^4$

13) $4(3p + 2)^2 + 17(3p + 2) - 15$

14) $12b^2 - 44b + 35$

15) $m^{12} + m^9$

16) $w^3 + \dfrac{64}{27}$

17) $(a - b)^2 - (x + y)^2$

Solve each equation.

18) $b^2 + 7b + 12 = 0$

19) $25k = k^3$

20) $144m^2 = 25$

21) $(c - 5)(c + 2) = 18$

22) $4q(q - 5) + 14 = 11(2 + q)$

23) $24y^2 + 80 = 88y$

24) $\dfrac{1}{12}t^2 + \dfrac{3}{4} = \dfrac{1}{2}t$

25) $2x^3 + 7x^2 - 2x - 7 = 0$

Write an equation, and solve.

26) Find the width and height of the storage locker pictured below if its volume is 120 ft^3.

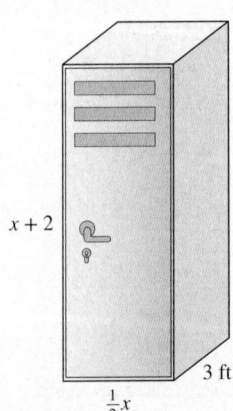

$x + 2$

3 ft

$\dfrac{1}{2}x$

27) Find three consecutive odd integers such that the sum of the three numbers is 60 less than the square of the largest integer.

28) Cory and Isaac leave an intersection with Cory jogging north and Isaac jogging west. When Isaac is 1 mi farther from the intersection than Cory, the distance between them is 2 mi more than Cory's distance from the intersection. How far is Cory from the intersection?

29) The length of a rectangular dog run is 4 ft more than twice its width. Find the dimensions of the run if it covers 96 ft^2.

30) An object is thrown upward with an initial velocity of 68 ft/sec. The height h (in feet) of the object t sec after it is thrown is given by

$$h = -16t^2 + 68t + 60$$

a) How long does it take for the object to reach a height of 120 ft?

b) What is the initial height of the object?

c) What is the height of the object after 2 sec?

d) How long does it take the object to hit the ground?

Chapter 7:   Cumulative Review for Chapters 1–7

Perform the indicated operation(s) and simplify.

1) $\dfrac{3}{8} - \dfrac{5}{6} + \dfrac{7}{12}$

2) $-\dfrac{15}{32} \cdot \dfrac{12}{25}$

Simplify. The answer should not contain any negative exponents.

3) $\dfrac{54t^5u^2}{36tu^8}$

4) $(8k^6)(-3k^4)$

5) Write 4.813×10^5 without exponents.

6) Solve $\dfrac{1}{3}(n - 2) + \dfrac{1}{4} = \dfrac{5}{12} + \dfrac{1}{6}n$

7) Solve for R.

$$A = P + PRT$$

8) *Write an equation in one variable, and solve.*
In 2018, the number of McDonald's restaurants in California was five less than fifty-two times the number of McDonald's in Rhode Island. Together, there were 1320 restaurants in the two states. Find the number of McDonald's in each state. (www.usatoday.com)

9) Solve. Write the answer in interval notation.

$$n + 9 \leq 5 \text{ or } 5n - 12 \geq 26$$

10) Graph $y = -\dfrac{3}{5}x + 7$.

11) Write the equation of the line perpendicular to $3x + y = 4$ containing the point $(-6, -1)$. Express the answer in slope-intercept form.

12) Use any method to solve this system of equations.

$$6(x + 2) + y = x - y - 2$$
$$5(2x - y + 1) = 2(x - y) - 5$$

Multiply and simplify.

13) $(6y + 5)(2y - 3)$

14) $(4p - 7)(2p^2 - 9p + 8)$

15) $(c + 8)^2$

16) Add $(4a^2b^2 - 17a^2b + 12ab - 11)$
$$+ (-a^2b^2 + 10a^2b - 5ab^2 + 7ab + 3).$$

Divide.

17) $\dfrac{12x^4 - 30x^3 - 14x^2 + 27x + 20}{2x - 5}$

18) $\dfrac{12r^3 + 4r^2 - 10r + 3}{4r^2}$

Factor completely, if possible.

19) $bc + 8b - 7c - 56$

20) $54q^2 - 144q + 42$

21) $y^2 + 1$

22) $t^4 - 81$

23) $x^3 - 125$

Solve.

24) $z^2 + 3z = 40$

25) $-12j(1 - 2j) = 16(5 + j)$

Rational Expressions

Get Ready

We will practice some operations with fractions to prepare for operations with rational expressions. Remember that answers are written in lowest terms.

1) To **multiply fractions,** we can take out the common factors before we multiply.

   Example: Multiply $\dfrac{8}{15} \cdot \dfrac{9}{16}$. $\dfrac{\overset{1}{\cancel{8}}}{\underset{5}{\cancel{15}}} \cdot \dfrac{\overset{3}{\cancel{9}}}{\underset{2}{\cancel{16}}} = \dfrac{1}{5} \cdot \dfrac{3}{2} = \dfrac{3}{10}$ Divide 8 and 16 by 8. Divide 15 and 9 by 3. Then, multiply.

2) To **divide fractions,** multiply the first fraction by the reciprocal of the second.

   Example: Divide $\dfrac{7}{4} \div \dfrac{8}{3}$. $\dfrac{7}{4} \div \dfrac{8}{3} = \dfrac{7}{4} \cdot \dfrac{3}{8} = \dfrac{21}{32}$

3) To **add and subtract fractions,** the fractions must first have a common denominator. Then, add or subtract the numerators and keep the same denominator.

   Examples: Add or subtract.

   a) $\dfrac{1}{6} + \dfrac{4}{9}$ The **least common denominator, LCD,** is 18.
   Write each fraction with a denominator of 18, then add.

   $\dfrac{1}{6} \cdot \dfrac{3}{3} = \dfrac{3}{18}, \dfrac{4}{9} \cdot \dfrac{2}{2} = \dfrac{8}{18}$ $\dfrac{1}{6} + \dfrac{4}{9} = \dfrac{3}{18} + \dfrac{8}{18} = \dfrac{11}{18}$

   b) $\dfrac{5}{21} - \dfrac{2}{3}$ The **LCD is 21.** Write $\dfrac{2}{3}$ with a denominator of 21.

   $\dfrac{2}{3} \cdot \dfrac{7}{7} = \dfrac{14}{21}$ $\dfrac{5}{21} - \dfrac{2}{3} = \dfrac{5}{21} - \dfrac{14}{21} = -\dfrac{9}{21} = -\dfrac{3}{7}$

   Divide 9 and 21 by 3 to write the answer in lowest terms.

In this chapter, use the following *Basic Skills Worksheets* to prepare students for this, and future, chapters: **WS8 Fractions (to give students more practice on adding and subtracting fractions), and WS2 Powers (to help students recognize powers of integers for Chapter 10).**

Get Ready Exercises

Perform the indicated operation. Write the answer in lowest terms.

1) $\dfrac{3}{4} \cdot \dfrac{5}{8}$ 2) $\dfrac{3}{10} \div \dfrac{2}{5}$ 3) $\dfrac{7}{18} + \dfrac{1}{6}$ 4) $\dfrac{7}{8} - \dfrac{5}{12}$

5) $\dfrac{2}{5} - \dfrac{8}{9}$ 6) $-\dfrac{55}{144} \cdot \dfrac{12}{77}$ 7) $-\dfrac{35}{16} \div \left(-\dfrac{5}{12}\right)$ 8) $\dfrac{3}{8} + \dfrac{35}{56}$

9) $\dfrac{9}{10} + \dfrac{3}{4}$ 10) $\dfrac{1}{25} - \dfrac{3}{10}$ 11) $\dfrac{12}{21} \div 16$ 12) $10 \cdot \dfrac{4}{5}$

13) $\dfrac{7}{12} - \dfrac{4}{9}$ 14) $\dfrac{1}{5} + \dfrac{2}{3}$ 15) $\dfrac{1}{8} \cdot 20$ 16) $\dfrac{7}{12} \div \dfrac{1}{18}$

Answers

1) $\dfrac{15}{32}$ 2) $\dfrac{3}{4}$ 3) $\dfrac{5}{9}$ 4) $\dfrac{11}{24}$ 5) $-\dfrac{22}{45}$ 6) $-\dfrac{5}{84}$ 7) $\dfrac{21}{4}$ 8) 1 9) $\dfrac{33}{20}$ 10) $-\dfrac{13}{50}$ 11) $\dfrac{1}{28}$ 12) 8 13) $\dfrac{5}{36}$ 14) $\dfrac{13}{15}$ 15) $\dfrac{5}{2}$ 16) $\dfrac{21}{2}$

 **Study Strategies** Know Your School

Understanding how your college or university works and where to find resources are critical components of being successful in college. Where do you go if you have a question about financial aid? Where can you find out about clubs that are of interest to you? Where do you go if you have questions about which classes you still need to take in order to graduate? Let's use the P.O.W.E.R. framework to help you get to know your school.

P Prepare

- **I will learn about the different offices, resources, and services at my school as well as how my school works.**

 **O Organize**

- Complete the emPOWERme survey that appears before the Chapter Summary to learn how well you know your school. Think about what you know and what you still need to learn.
- Know your school's website. Identify important places on the website such as where you can register for classes, pay your tuition online, and how to make an appointment to see an adviser.
- Identify places you need to know at your school: important offices where you can get questions answered, services offered at your school, organizations or clubs that you can join, etc.
- Identify important deadlines you should know such as registering for classes, paying tuition, and submitting financial aid applications.
- Identify academic services that you may want to use such as the library, a tutoring center, and a testing center.

CHAPTER 8   **Rational Expressions**   523

- Identify support services that may be helpful to you such as the campus health center, a veterans' support office, and an office to help students with disabilities.
- Be sure you have your syllabus for each class.

Work

- It's time to learn about some of the offices, resources, and support services at your school and to learn about some of the procedures. Some important, common offices and resources are listed here, but different schools have different names for these services. Some apply to all students, and some may not apply to you. For the items that apply to you, fill in the blanks with the information for *your* school.
- *My school's website is* _____. Among other things, know how you can register for classes and pay your tuition online.
- *The office where I can go to register for classes is called the* _____ *office. Its location is* _____. Know the deadlines for registering for classes.
- When you are choosing your classes, know where they are located. If the campus is large, be sure you have enough time to get from one class to the next.
- *The office where I can go to ask questions about my tuition bill or to pay my bill in person is called the* _____ *office. Its location is* _____. Know when tuition is due.
- *The office where I can go to ask questions about financial aid, get financial aid forms, or turn in forms is called the* _____ *office. Its location is* _____. Know when to get the financial aid process started and when the forms are due.
- *The office where I can talk with someone about choosing a major or get help choosing classes is called the* _____ *office. Its location is* _____.
- *The office where I can go to talk with someone about personal problems that are interfering with my success in college is called the* _____ *office. Its location is* _____.
- Does your school have an emergency alert system? If so, know how to sign up for it. You may have options of receiving an email, text, or automated phone call.
- Most institutions have a Tutoring Center or a Math Lab that is available to students free of charge. *The place I can go to get free help with my math class is called the* _____. *Its location is* _____, *and its hours are* _____. What are the procedures for seeing a tutor?
- *My math instructor's office is located in* _____, *and his/her office hours are* _____.
- Most schools have a testing center where students can go if they are allowed extra time on their tests or if they need to take a make-up test. *At my school, this center is called the* _____. *Its hours are* _____. What are the procedures for taking a test in their center?
- *The office where I can learn about the resources available to me as a student with a disability is called the* _____ *office. Its location is* _____.
- Most campuses have an office that helps veterans. *On my campus, that office is called the* _____. *Its location is* _____.
- Many campuses have a child care center for students with children. *I can get information about the child care center at* _____.
- *The place I can go to find out about campus clubs, events, and organizations is called the* _____.
- Most institutions have a place where students can get help finding a job or preparing for an interview. *At my school, this office is called the* _____ *office. Its location is* _____.

- Do you feel like you have a good understanding of how your school works and where to go for help? Did you learn the names and locations of the offices listed in the Work section?

- If you feel like you learned about the places and services that will be beneficial to you, be sure to use them when you need them!
- If you did not learn everything you think you need to know, ask one of your instructors or go to an information office at your school. Search the institution's website.
- Are there other services or resources that you feel like you need but did not learn about? Talk to one of your instructors and ask if there is such a place at your school.

Caia Image/Glow Images

8.1 Rational Expressions and Functions

Prepare

Organize

What are your objectives for Section 8.1?	How can you accomplish each objective?
1 Evaluate a Rational Expression	• Know the definition and be able to recognize a *rational expression*. • Complete the given example on your own. • Complete You Try 1.
2 Find the Values of the Variable That Make a Rational Expression Undefined or Equal to Zero	• Write down the steps you need to take to find the values of the variable that will make a rational expression equal to zero or undefined. • Complete the given example on your own. • Complete You Try 2.
3 Write a Rational Expression in Lowest Terms	• Learn the **Fundamental Property of Rational Expressions.** • Write the procedure for **Writing a Rational Expression in Lowest Terms** in your own words. • Complete the given example on your own. • Complete You Try 3.

What are your objectives for Section 8.1?	How can you accomplish each objective?
4 Simplify a Rational Expression of the Form $\dfrac{a-b}{b-a}$	• Follow the explanation so you can understand the note that outlines the principle for this objective. • Complete the given example on your own. • Complete You Try 4.
5 Write Equivalent Forms of a Rational Expression	• Understand that there can be many different ways to write a rational expression. • Complete the given example on your own. • Complete You Try 5.
6 Determine the Domain of a Rational Function	• Understand the definition of the *domain of a rational function*, and write it in your own words. • Review interval notation. • Complete the given example on your own. • Complete You Try 6.

 **Work**   **Read the explanations, follow the examples, take notes, and complete the You Trys.**

1   Evaluate a Rational Expression

In Section 1.4, we defined a **rational number** as the quotient of two integers where the denominator does not equal zero. Some examples of rational numbers are

$$\frac{7}{8}, \quad -\frac{2}{5}, \quad 18 \left(\text{since } 18 = \frac{18}{1} \right)$$

We can define a *rational expression* in a similar way. A **rational expression** is a quotient of two polynomials provided that the denominator does not equal zero. We state the definition formally next.

> **W Hint**
> Why can't $Q = 0$?

> ## Definition
> A **rational expression** is an expression of the form $\dfrac{P}{Q}$, where P and Q are polynomials and where $Q \neq 0$.

Some examples of rational expressions are

$$\frac{4k^3}{7}, \quad \frac{2n-1}{n+6}, \quad \frac{5}{t^2-3t-28}, \quad -\frac{9x+2y}{x^2+y^2}$$

We can *evaluate* rational expressions for given values of the variables as long as the values do not make any denominators equal zero.

EXAMPLE 1

Evaluate $\dfrac{x^2 - 9}{x + 1}$ (if possible) for each value of x.

a) $x = 7$ b) $x = 3$ c) $x = -1$

Solution

a) $\dfrac{x^2 - 9}{x + 1} = \dfrac{(7)^2 - 9}{(7) + 1}$ Substitute 7 for x.

$= \dfrac{49 - 9}{7 + 1} = \dfrac{40}{8} = 5$

b) $\dfrac{x^2 - 9}{x + 1} = \dfrac{(3)^2 - 9}{(3) + 1}$ Substitute 3 for x.

$= \dfrac{9 - 9}{3 + 1} = \dfrac{0}{4} = 0$

c) $\dfrac{x^2 - 9}{x + 1} = \dfrac{(-1)^2 - 9}{(-1) + 1}$ Substitute -1 for x.

$= \dfrac{1 - 9}{0} = \dfrac{-8}{0}$

Undefined

Remember, a fraction is **undefined** when its denominator equals zero. Therefore, we say that $\dfrac{x^2 - 9}{x + 1}$ is *undefined* when $x = -1$ since this value of x makes the denominator equal zero. So, x *cannot equal* -1 in this expression.

[YOU TRY 1]

Evaluate $\dfrac{k^2 - 1}{k + 11}$ (if possible) for each value of k.

a) $k = 9$ b) $k = -2$ c) $k = -11$ d) $k = 1$

---

2 Find the Values of the Variable That Make a Rational Expression Undefined or Equal to Zero

Parts b) and c) in Example 1 remind us about two important aspects of fractions and rational expressions.

W Hint

Remember that zero divided by any nonzero number equals zero.

Note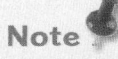

1) A fraction (rational expression) *equals zero* when its numerator equals zero.
2) A fraction (rational expression) is *undefined* when its denominator equals zero.

EXAMPLE 2

For each rational expression, for what values of the variable

   i)   does the expression equal zero?

   ii) is the expression undefined?

a) $\dfrac{m+8}{m-3}$ b) $\dfrac{5c}{6}$ c) $\dfrac{4}{z^2 - 5z - 36}$

Solution

a) i) $\dfrac{m+8}{m-3} = 0$ when its *numerator* equals zero. Set the numerator equal to zero, and solve for m.

$$m + 8 = 0$$
$$m = -8$$ Therefore, $\dfrac{m+8}{m-3} = 0$ when $m = -8$.

   ii) $\dfrac{m+8}{m-3}$ is *undefined* when its *denominator* equals zero. Set the denominator equal to zero, and solve for m.

$$m - 3 = 0$$
$$m = 3$$

$\dfrac{m+8}{m-3}$ is *undefined* when $m = 3$. This means that any real number *except* 3 can be substituted for m in this expression.

b) i) To determine the values of c that make $\dfrac{5c}{6} = 0$, set $5c = 0$ and solve.

$$5c = 0$$
$$c = 0$$ Divide by 5.

So, $\dfrac{5c}{6} = 0$ when $c = 0$.

   ii) $\dfrac{5c}{6}$ is *undefined* when the denominator equals zero. However, the denominator is 6, and $6 \neq 0$. Therefore, there *is no value of c* that makes $\dfrac{5c}{6}$ undefined. *We say that* $\dfrac{5c}{6}$ *is defined for all real numbers.*

c) i) $\dfrac{4}{z^2 - 5z - 36} = 0$ when the numerator equals zero. The numerator is 4, and $4 \neq 0$. Therefore, $\dfrac{4}{z^2 - 5z - 36}$ will *never* equal zero.

   ii) $\dfrac{4}{z^2 - 5z - 36}$ is *undefined* when $z^2 - 5z - 36 = 0$. Solve for z.

$$z^2 - 5z - 36 = 0$$
$$(z + 4)(z - 9) = 0 \qquad \text{Factor.}$$
$$z + 4 = 0 \quad \text{or} \quad z - 9 = 0 \qquad \text{Set each factor equal to 0.}$$
$$z = -4 \quad \text{or} \quad z = 9 \qquad \text{Solve.}$$

$\dfrac{4}{z^2 - 5z - 36}$ is undefined when $z = -4$ or $z = 9$. All real numbers *except* -4 and 9 can be substituted for z in this expression.

> **W Hint**
>
> In your notes and in your own words, explain how to determine when a rational expression equals zero and when it is undefined.

[YOU TRY 2] For each rational expression, for what values of the variable

     i)   does the expression equal zero?

     ii) is the expression undefined?

a) $\dfrac{v-6}{v+11}$ b) $\dfrac{9w}{w^2-12w+20}$ c) $\dfrac{x^2-25}{8}$ d) $\dfrac{1}{5q+4}$

All of the operations that can be performed with fractions can also be done with rational expressions. We begin our study of these operations with rational expressions by learning how to write a rational expression in lowest terms.

3 Write a Rational Expression in Lowest Terms

One way to think about writing a fraction such as $\dfrac{8}{12}$ in lowest terms is

$$\frac{8}{12}=\frac{2\cdot 4}{3\cdot 4}=\frac{2}{3}\cdot\frac{4}{4}=\frac{2}{3}\cdot 1=\frac{2}{3}$$

Since $\dfrac{4}{4}=1$, we can also think of reducing $\dfrac{8}{12}$ as $\dfrac{8}{12}=\dfrac{2\cdot\cancel{4}}{3\cdot\cancel{4}}=\dfrac{2}{3}$.

To write $\dfrac{8}{12}$ in lowest terms, we can *factor* the numerator and denominator, then *divide* the numerator and denominator by the common factor, 4. This is the approach we use to write a rational expression in lowest terms.

> **Definition** Fundamental Property of Rational Expressions
>
> If P, Q, and C are polynomials such that $Q\neq 0$ and $C\neq 0$, then
>
> $$\frac{PC}{QC}=\frac{P}{Q}$$

This property mirrors the example above since

$$\frac{PC}{QC}=\frac{P}{Q}\cdot\frac{C}{C}=\frac{P}{Q}\cdot 1=\frac{P}{Q}$$

Or, we can also think of the reducing procedure as dividing the numerator and denominator by the common factor, C.

$$\frac{P\cancel{C}}{Q\cancel{C}}=\frac{P}{Q}$$

W Hint

Is this procedure any different from writing a fraction in lowest terms?

Procedure Writing a Rational Expression in Lowest Terms

1) Completely **factor** the numerator and denominator.

2) **Divide** the numerator and denominator by the greatest common factor.

EXAMPLE 3

Write each rational expression in lowest terms.

a) $\dfrac{21r^{10}}{3r^4}$ b) $\dfrac{8a + 40}{3a + 15}$ c) $\dfrac{5n^2 - 20}{n^2 + 5n + 6}$

Solution

a) We can simplify $\dfrac{21r^{10}}{3r^4}$ using the quotient rule presented in Chapter 5.

$$\frac{21r^{10}}{3r^4} = 7r^6$$ Divide 21 by 3, and use the quotient rule: $\dfrac{r^{10}}{r^4} = r^{10-4} = r^6$.

b) $\dfrac{8a + 40}{3a + 15} = \dfrac{8\cancel{(a+5)}}{3\cancel{(a+5)}}$ Factor.

$\qquad\qquad = \dfrac{8}{3}$ Divide out the common factor, $a + 5$.

c) $\dfrac{5n^2 - 20}{n^2 + 5n + 6} = \dfrac{5(n^2 - 4)}{(n + 2)(n + 3)}$ Factor.

$\qquad\qquad = \dfrac{5\cancel{(n+2)}(n - 2)}{\cancel{(n+2)}(n + 3)}$ Factor completely.

$\qquad\qquad = \dfrac{5(n - 2)}{n + 3}$ Divide out the common factor, $n + 2$.

BE CAREFUL

Notice that we divide by *factors* not *terms*.

$$\frac{\cancel{x + 5}}{2\cancel{(x + 5)}} = \frac{1}{2}$$ $\qquad\qquad\qquad$ $\dfrac{x}{x + 5} \neq \dfrac{1}{5}$

Divide by the *factor* $x + 5$. $\qquad$ We cannot divide by x because the x in the denominator is a *term* in a sum.

[**YOU TRY 3**]

Write each rational expression in lowest terms.

a) $\dfrac{12b^8}{18b^3}$ b) $\dfrac{2h + 8}{7h + 28}$ c) $\dfrac{y^2 - 9y + 14}{9y^4 - 9y^3 - 18y^2}$

4 Simplify a Rational Expression of the Form $\dfrac{a-b}{b-a}$

Do you think that $\dfrac{x-4}{4-x}$ is in lowest terms? Let's look at it more closely to understand the answer.

$$\frac{x-4}{4-x} = \frac{x-4}{-1(-4+x)} \qquad \text{Factor } -1 \text{ out of the denominator.}$$

$$= \frac{1\cancel{(x-4)}}{-1\cancel{(x-4)}} \qquad \text{Rewrite } -4 + x \text{ as } x - 4.$$

$$= -1 \qquad \text{Divide out the common factor, } x - 4.$$

Therefore, $\dfrac{x-4}{4-x} = -1$.

We can generalize this result as follows:

> **Note**
>
> 1) $b - a = -1(a - b)$ and 2) $\dfrac{a-b}{b-a} = -1$
>
> The terms in the numerator and denominator in 2) differ only in sign. The rational expression simplifies to -1.

EXAMPLE 4

Write each rational expression in lowest terms.

a) $\dfrac{7-t}{t-7}$ b) $\dfrac{4h^2 - 25}{15 - 6h}$ c) $\dfrac{x+6}{x-6}$

Solution

a) $\dfrac{7-t}{t-7} = -1$ since $\dfrac{7-t}{t-7} = \dfrac{-1\cancel{(t-7)}}{\cancel{(t-7)}} = -1$

b) $\dfrac{4h^2 - 25}{15 - 6h} = \dfrac{(2h+5)(2h-5)}{3(5-2h)}$ Factor.

$$= \frac{(2h+5)\overset{-1}{\cancel{(2h-5)}}}{3\cancel{(5-2h)}} \qquad \frac{2h-5}{5-2h} = -1$$

$$= \frac{-1(2h+5)}{3}$$

$$= \frac{-2h - 5}{3} \quad \text{or} \quad -\frac{2h+5}{3} \qquad \text{Simplify.}$$

c) The expression $\dfrac{x+6}{x-6}$ is already in lowest terms. Notice that it is *not* in the form $\dfrac{a-b}{b-a}$.

YOU TRY 4

Write each rational expression in lowest terms.

a) $\dfrac{10-m}{m-10}$ b) $\dfrac{100 - 4k^2}{k^2 - 8k + 15}$ c) $\dfrac{c-3}{c+3}$

5 Write Equivalent Forms of a Rational Expression

The answer to Example 4b) can be written in several different ways. Recall that a fraction like $-\dfrac{1}{2}$ can also be written as $\dfrac{-1}{2}$ or $\dfrac{1}{-2}$. Likewise, when we are simplifying the expression $\dfrac{-1(2h + 5)}{3}$ we can distribute the -1 in the numerator to get $\dfrac{-2h - 5}{3}$.

Or, we can apply the -1 to the denominator to get $\dfrac{2h + 5}{-1 \cdot 3} = \dfrac{2h + 5}{-3}$. Or, we can write $\dfrac{-1(2h + 5)}{3}$ as $-1 \cdot \dfrac{2h + 5}{3} = -\dfrac{2h + 5}{3}$. We can write the answer to Example 4b) in any of these ways. You should be able to recognize equivalent forms of rational expressions because there isn't always just one way to write the correct answer.

EXAMPLE 5

Write $-\dfrac{5x - 8}{3 + x}$ in three different ways.

Solution

The negative sign in front of a fraction can be applied to the numerator or to the denominator. This can result in expressions that look quite different but that are, actually, equivalent.

i) Apply the negative sign to the denominator.

$$-\frac{5x - 8}{3 + x} = \frac{5x - 8}{-1(3 + x)}$$

$$= \frac{5x - 8}{-3 - x} \qquad \text{Distribute.}$$

ii) Apply the negative sign to the numerator.

$$-\frac{5x - 8}{3 + x} = \frac{-1(5x - 8)}{3 + x} = \frac{-5x + 8}{3 + x}$$

iii) Apply the negative sign to the numerator, distribute the -1, and rewrite addition as subtraction.

$$-\frac{5x - 8}{3 + x} = \frac{-1(5x - 8)}{3 + x}$$

$$= \frac{-5x + 8}{3 + x} \qquad \text{Distribute.}$$

$$= \frac{8 - 5x}{3 + x} \qquad \text{Rewrite } -5x + 8 \text{ as } 8 - 5x.$$

Therefore, $\dfrac{5x - 8}{-3 - x}$, $\dfrac{-5x + 8}{3 + x}$, and $\dfrac{8 - 5x}{3 + x}$ are *all* equivalent forms of $-\dfrac{5x - 8}{3 + x}$.

Keep this idea of equivalent forms of rational expressions in mind when checking your answers against the answers in the back of the book. Sometimes students believe their answer is wrong because it "looks different" when, in fact, it is an *equivalent form* of the given answer!

Write $\dfrac{-(1-t)}{5t-8}$ in three different ways.

6 Determine the Domain of a Rational Function

We can combine what we have learned about rational expressions with what we learned about functions in Section 4.6. $f(x) = \dfrac{x+3}{x-8}$ is an example of a **rational function** because $\dfrac{x+3}{x-8}$ is a rational expression and each value that can be substituted for x will produce *only one* value for the expression.

Recall from Chapter 4 that the domain of a function $f(x)$ is the set of all real numbers that can be substituted for x. Since a rational expression is undefined when its denominator equals zero, we define the domain of a rational function as follows.

Definition

The **domain of a rational function** consists of all real numbers *except* the values of the variable that make the denominator equal zero.

Therefore, to determine the domain of a rational function we set the denominator equal to zero and solve for the variable. Any value that makes the denominator equal to zero is *not* in the domain of the function.

To determine the domain of a rational function, sometimes it is helpful to ask yourself, "Is there any number that *cannot* be substituted for the variable?"

EXAMPLE 6

Determine the domain of each rational function.

a) $f(x) = \dfrac{x+3}{x-8}$ b) $g(c) = \dfrac{6}{c^2+3c-4}$ c) $h(n) = \dfrac{4n^2-9}{7}$

Solution

a) To determine the domain of $f(x) = \dfrac{x+3}{x-8}$, ask yourself, "Is there any number that *cannot* be substituted for x?" Yes. $f(x)$ is *undefined* when the denominator equals zero. Set the denominator equal to zero, and solve for x.

$$x - 8 = 0 \quad \text{Set the denominator} = 0.$$
$$x = 8 \quad \text{Solve.}$$

When $x = 8$, the denominator of $f(x) = \dfrac{x+3}{x-8}$ equals zero. The domain contains all real numbers *except* 8. Write the domain in interval notation as $(-\infty, 8) \cup (8, \infty)$.

b) To determine the domain of $g(c) = \dfrac{6}{c^2 + 3c - 4}$, ask yourself, "Is there any number that *cannot* be substituted for c?" Yes. $g(c)$ is *undefined* when its *denominator* equals zero. Set the denominator equal to zero and solve for c.

$$c^2 + 3c - 4 = 0 \qquad \text{Set the denominator} = 0.$$
$$(c + 4)(c - 1) = 0 \qquad \text{Factor.}$$
$$c + 4 = 0 \quad \text{or} \quad c - 1 = 0 \qquad \text{Set each factor equal to 0.}$$
$$c = -4 \quad \text{or} \quad c = 1 \qquad \text{Solve.}$$

When $c = -4$ or $c = 1$, the denominator of $g(c) = \dfrac{6}{c^2 + 3c - 4}$ equals zero. The domain contains all real numbers *except* -4 and 1. Write the domain in interval notation as $(-\infty, -4) \cup (-4, 1) \cup (1, \infty)$.

c) Ask yourself, "Is there any number that *cannot* be substituted for n?" No! Looking at the denominator, we see that the number 7 will never equal zero. Therefore, there is *no value* of n that makes $h(n) = \dfrac{4n^2 - 9}{7}$ undefined. Any real number may be substituted for n and the function will be defined.

The domain of the function is the set of all real numbers. We can write the domain in interval notation as $(-\infty, \infty)$.

[YOU TRY 6]   Determine the domain of each rational function.

a) $h(t) = \dfrac{9}{t + 5}$

b) $f(x) = \dfrac{2x - 3}{x^2 - 8x + 12}$

c) $g(a) = \dfrac{a + 4}{10}$

ANSWERS TO [YOU TRY] EXERCISES

1) a) 4 b) $\dfrac{1}{3}$ c) undefined d) 0   2) a) i) 6 ii) -11 b) i) 0 ii) 2, 10

c) i) $-5, 5$ ii) defined for all real numbers d) i) never equals zero ii) $-\dfrac{4}{5}$

3) a) $\dfrac{2b^5}{3}$ b) $\dfrac{2}{7}$ c) $\dfrac{y - 7}{9y^2(y + 1)}$ 4) a) -1 b) $-\dfrac{4(5 + k)}{k - 3}$ c) $\dfrac{c - 3}{c + 3}$

5) Some possibilities are $\dfrac{t - 1}{5t - 8}, -\dfrac{1 - t}{5t - 8}, \dfrac{1 - t}{8 - 5t}$.   6) a) $(-\infty, -5) \cup (-5, \infty)$

b) $(-\infty, 2) \cup (2, 6) \cup (6, \infty)$ c) $(-\infty, \infty)$

E Evaluate **8.1** Exercises   Do the exercises, and check your work.

Objective 1: Evaluate a Rational Expression

1) When is a fraction or a rational expression undefined?

2) When does a fraction or a rational expression equal 0?

Evaluate (if possible) for a) $x = 3$ and b) $x = -2$.

3) $\dfrac{2x - 1}{5x + 2}$

4) $\dfrac{3(x^2 + 1)}{x^2 + 2x + 1}$

Evaluate (if possible) for a) $z = 1$ and b) $z = -3$.

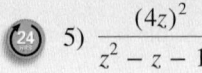 5) $\dfrac{(4z)^2}{z^2 - z - 12}$

6) $\dfrac{3(z^2 - 9)}{z^2 + 8}$

7) $\dfrac{15 + 5z}{16 - z^2}$

8) $\dfrac{4z - 3}{z^2 + 6z - 7}$

Objective 2: Find the Values of the Variable That Make a Rational Expression Undefined or Equal to Zero

9) How do you determine the values of the variable for which a rational expression is undefined?

10) If $x^2 + 9$ is the numerator of a rational expression, can that expression equal zero? Give a reason.

Determine the value(s) of the variable for which

   a)   the expression equals zero.

   b)   the expression is undefined.

11) $\dfrac{m + 4}{3m}$

12) $\dfrac{-y}{y + 3}$

13) $\dfrac{2w - 7}{4w + 1}$

14) $\dfrac{3x + 13}{2x + 13}$

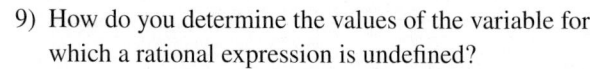 15) $\dfrac{11v - v^2}{5v - 9}$

16) $-\dfrac{r + 5}{r^2 - 100}$

17) $\dfrac{8}{p}$

18) $\dfrac{22}{m - 1}$

19) $-\dfrac{7k}{k^2 + 9k + 20}$

20) $\dfrac{4}{3f^2 - 13f + 10}$

21) $\dfrac{g^2 + 9g + 18}{9g}$

22) $\dfrac{6m - 11}{10}$

23) $\dfrac{4y}{y^2 + 9}$

24) $\dfrac{q^2 + 49}{7}$

Objective 3: Write a Rational Expression in Lowest Terms
Write each rational expression in lowest terms.

25) $\dfrac{7x(x - 11)}{3(x - 11)}$

26) $\dfrac{24(g + 3)}{6(g + 3)(g - 5)}$

27) $\dfrac{24g^2}{56g^4}$

28) $\dfrac{99d^7}{9d^3}$

29) $\dfrac{4d - 20}{5d - 25}$

30) $\dfrac{12c - 3}{8c - 2}$

31) $\dfrac{-39u^2 - 26}{30u^2 + 20}$

32) $\dfrac{-15v^2 + 12}{40v^2 - 32}$

33) $\dfrac{g^2 - g - 56}{g + 7}$

34) $\dfrac{3k^2 - 36k + 96}{k - 8}$

35) $\dfrac{r + 9}{r^2 + 7r - 18}$

36) $\dfrac{t - 5}{t^2 - 25}$

37) $\dfrac{q^2 - 25}{2q^2 - 7q - 15}$

38) $\dfrac{6p^2 + 11p - 10}{9p^2 - 4}$

39) $\dfrac{w^3 + 125}{5w^2 - 25w + 125}$

40) $\dfrac{4m^3 - 4}{m^2 + m + 1}$

41) $\dfrac{4u^2 - 20u + 4uv - 20v}{13u + 13v}$

42) $\dfrac{ab + 3a - 6b - 18}{b^2 - 9}$

43) $\dfrac{x^2 - y^2}{x^3 - y^3}$

44) $\dfrac{a^3 + b^3}{a^2 - b^2}$

Objective 4: Simplify a Rational Expression of the Form $\dfrac{a - b}{b - a}$

45) Any rational expression of the form $\dfrac{a - b}{b - a}$ $(a \neq b)$ reduces to what?

46) Does $\dfrac{h + 4}{h - 4} = -1$?

Write each rational expression in lowest terms.

47) $\dfrac{8 - q}{q - 8}$

48) $\dfrac{m - 15}{15 - m}$

49) $\dfrac{m^2 - 121}{11 - m}$

50) $\dfrac{k - 9}{162 - 2k^2}$

51) $\dfrac{36 - 42x}{7x^2 + 8x - 12}$

52) $\dfrac{a^2 - 6a - 27}{9 - a}$

53) $\dfrac{16 - 4b^2}{b - 2}$

54) $\dfrac{45 - 9v}{v^2 - 25}$

55) $\dfrac{y^3 - 3y^2 + 2y - 6}{21 - 7y}$

56) $\dfrac{3t^3 - t^2 + 12t - 4}{1 - 9t^2}$

57) $\dfrac{8t^3 - 27}{9 - 4t^2}$

58) $\dfrac{x^3 - y^3}{y^2 - x^2}$

Mixed Exercises: Objectives 1–4
Write each rational expression in lowest terms.

59) $\dfrac{4a^2 - 9}{4a^2 - 12a + 9}$

60) $\dfrac{36n^3}{42n^9}$

61) $\dfrac{b^2 + 6b - 72}{4b^2 + 52b + 48}$

62) $\dfrac{k^2 - 16k + 64}{k^3 - 8k^2 + 9k - 72}$

63) $\dfrac{28h^4 - 56h^3 + 7h}{7h}$

64) $\dfrac{5p^2 - 13p + 6}{32 - 8p^2}$

65) $\dfrac{9c^2 - 27c + 81}{c^3 + 27}$

66) $\dfrac{z^2 + 5z - 36}{64 - z^3}$

67) $\dfrac{-5v - 10}{v^3 - v^2 - 4v + 4}$

68) $\dfrac{38x^2 + 38}{-12x^2 - 12}$

Objective 5: Write Equivalent Forms of a Rational Expression

Find three equivalent forms of each rational expression.

(24) 69) $-\dfrac{u+7}{u-2}$

70) $-\dfrac{8y-1}{2y+5}$

71) $-\dfrac{9-5t}{2t-3}$

72) $\dfrac{w-6}{-4w+7}$

73) $\dfrac{-12m}{m^2-3}$

74) $\dfrac{-9x-11}{18-x}$

Write each expression in lowest terms

    a)   using long division.

    b)   using the methods of this section.

75) $\dfrac{4y^2-11y+6}{y-2}$

76) $\dfrac{2x^2+x-28}{x+4}$

77) $\dfrac{8a^3+125}{2a+5}$

78) $\dfrac{27t^3-8}{3t-2}$

Recall that the area of a rectangle is $A = lw$, where w = width and l = length. Solving for the width, we get $w = \dfrac{A}{l}$ and solving for the length gives us $l = \dfrac{A}{w}$.

Find the missing side in each rectangle.

79) Area $= 3x^2 + 8x + 4$

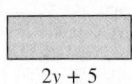
$x+2$

Find the length.

80) Area $= 2y^2 - 3y - 20$

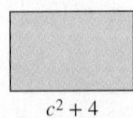
$2y+5$

Find the width.

81) Area $= 2c^3 + 4c^2 + 8c + 16$

c^2+4

Find the width.

82) Area $= 3n^3 - 12n^2 - n + 4$

$n-4$

Find the length.

Recall that the area of a triangle is $A = \dfrac{1}{2}bh$, where b = length of the base and h = height. Solving for the height, we get $h = \dfrac{2A}{b}$. Find the height of each triangle.

83) Area $= 3k^2 + 13k + 4$

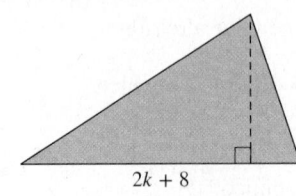
$2k+8$

84) Area $= 6p^2 + 52p + 32$

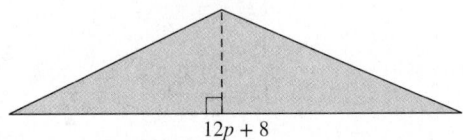
$12p+8$

Objective 6: Determine the Domain of a Rational Function

For each rational function,

    a)   find $f(-2)$, if possible.

    b)   find x so that $f(x) = 0$.

    c)   determine the domain of the function.

(24) 85) $f(x) = \dfrac{x+8}{x+6}$

86) $f(x) = \dfrac{x}{3x-1}$

87) $f(x) = \dfrac{5x-3}{x+2}$

88) $f(x) = \dfrac{9}{x-1}$

89) $f(x) = \dfrac{6}{x^2+6x+5}$

90) $f(x) = \dfrac{2x-1}{x^2+x-12}$

Determine the domain of each rational function.

91) $f(p) = \dfrac{1}{p-7}$

92) $h(z) = \dfrac{z+8}{z+3}$

93) $f(a) = \dfrac{6a}{7-2a}$

94) $k(r) = \dfrac{r}{5r+2}$

95) $g(t) = \dfrac{3t - 4}{t^2 - 9t + 8}$

96) $r(c) = \dfrac{c + 9}{c^2 - c - 42}$

97) $h(w) = \dfrac{w + 7}{w^2 - 81}$

98) $k(t) = \dfrac{t}{t^2 - 14t + 33}$

99) $A(c) = \dfrac{8}{c^2 + 6}$

100) $C(n) = \dfrac{3n + 1}{2}$

101) Write your own example of a rational function, $f(x)$, that has a domain of $(-\infty, -8) \cup (-8, \infty)$.

102) Write your own example of a rational function, $g(x)$, that has a domain of $(-\infty, -5) \cup (-5, 6) \cup (6, \infty)$.

R Rethink

R1) Explain how good factoring skills helped you move through these exercises.

R2) Look at Exercises 75–78. Why can you simplify these rational expressions using either method? Which do you prefer, and why?

R3) Think about how writing a fraction in lowest terms helps you multiply and divide fractions. Do you think the same will follow for rational expressions?

8.2 Multiplying and Dividing Rational Expressions

P Prepare

O Organize

What are your objectives for Section 8.2?	How can you accomplish each objective?
1 Multiply Rational Expressions	• Notice the steps you take to multiply fractions. • Understand how to **Multiply Rational Expressions** and compare to the steps taken in Example 1. • Complete the given examples on your own. • Complete You Trys 1 and 2.
2 Divide Rational Expressions	• Understand the principle behind the procedure for **Dividing Rational Expressions.** • Complete the given examples on your own. • Complete You Trys 3 and 4.

W Work

Read the explanations, follow the examples, take notes, and complete the You Trys.

1 Multiply Rational Expressions

Before we multiply rational expressions, let's review how to multiply fractions.

EXAMPLE 1

Multiply $\dfrac{9}{16} \cdot \dfrac{8}{15}$.

Solution

We could multiply numerators, multiply denominators, then simplify by dividing out common factors. *Or* we can factor the numerators and denominators, then divide out the common factors before multiplying.

$$\frac{9}{16} \cdot \frac{8}{15} = \frac{\overset{1}{\cancel{3}} \cdot 3}{2 \cdot \underset{1}{\cancel{8}}} \cdot \frac{\overset{1}{\cancel{8}}}{\underset{1}{\cancel{3}} \cdot 5} \qquad \text{Factor, and divide out common factors.}$$

$$= \frac{3}{2 \cdot 5} \qquad \text{Multiply.}$$

$$= \frac{3}{10} \qquad \text{Simplify.}$$

YOU TRY 1

Multiply $\dfrac{8}{35} \cdot \dfrac{5}{12}$.

Multiplying rational expressions works the same way.

> **Procedure** Multiplying Rational Expressions
>
> If $\dfrac{P}{Q}$ and $\dfrac{R}{T}$ are rational expressions, then $\dfrac{P}{Q} \cdot \dfrac{R}{T} = \dfrac{PR}{QT}$.

 **Hint**

Is this the same procedure you used to multiply fractions?

> **Note**
>
> We can also follow these steps to multiply rational expressions:
>
> 1) Factor the numerators and denominators.
> 2) Divide out common factors.
> 3) Multiply.
>
> *All products should be written in lowest terms.*

EXAMPLE 2

Multiply.

a) $\dfrac{18}{y^7} \cdot \dfrac{y^4}{9}$ 　　 b) $\dfrac{9c + 45}{6c^{10}} \cdot \dfrac{c^6}{c^2 - 25}$

c) $\dfrac{2n^2 - 11n - 6}{n^2 - 2n - 24} \cdot \dfrac{n^2 + 8n + 16}{2n^2 + n}$

Solution

a) $\dfrac{18}{y^7} \cdot \dfrac{y^4}{9} = \dfrac{\overset{2}{\cancel{18}}}{\cancel{y^4} \cdot y^3} \cdot \dfrac{\cancel{y^4}}{\cancel{9}}$ Factor; divide out common factors.

$\qquad\qquad = \dfrac{2}{y^3}$ Multiply.

b) $\dfrac{9c + 45}{6c^{10}} \cdot \dfrac{c^6}{c^2 - 25} = \dfrac{\overset{3}{\cancel{9}}\cancel{(c+5)}}{\underset{2}{\cancel{6c^6}} \cdot c^4} \cdot \dfrac{\cancel{c^6}}{\cancel{(c+5)}(c-5)}$ Factor; divide out common factors.

$\qquad\qquad\qquad = \dfrac{3}{2c^4(c-5)}$ Multiply.

c) $\dfrac{2n^2 - 11n - 6}{n^2 - 2n - 24} \cdot \dfrac{n^2 + 8n + 16}{2n^2 + n} = \dfrac{\cancel{(2n+1)}\cancel{(n-6)}}{\cancel{(n+4)}\cancel{(n-6)}} \cdot \dfrac{\overset{(n+4)}{\cancel{(n+4)^2}}}{n\cancel{(2n+1)}}$ Factor; divide out common factors.

$\qquad\qquad\qquad\qquad = \dfrac{n + 4}{n}$ Multiply.

[YOU TRY 2] Multiply.

a) $\dfrac{n^7}{20} \cdot \dfrac{10}{n^2}$ b) $\dfrac{d^2}{d^2 - 4} \cdot \dfrac{4d - 8}{12d^5}$ c) $\dfrac{h^2 + 10h + 25}{3h^2 - 4h} \cdot \dfrac{3h^2 + 5h - 12}{h^2 + 8h + 15}$

---

2 Divide Rational Expressions

When we divide rational numbers, we multiply by a reciprocal. For example,

$\dfrac{7}{4} \div \dfrac{3}{8} = \dfrac{7}{\underset{1}{\cancel{4}}} \cdot \dfrac{\overset{2}{\cancel{8}}}{3} = \dfrac{14}{3}$. We divide rational expressions the same way.

    To divide rational expressions, we multiply the first rational expression by the reciprocal of the second rational expression.

Procedure Dividing Rational Expressions

If $\dfrac{P}{Q}$ and $\dfrac{R}{T}$ are rational expressions with Q, R, and T not equal to zero, then

$$\dfrac{P}{Q} \div \dfrac{R}{T} = \dfrac{P}{Q} \cdot \dfrac{T}{R} = \dfrac{PT}{QR}$$

Multiply the first rational expression by the reciprocal of the second rational expression.

EXAMPLE 3 Divide.

a) $\dfrac{15a^7}{b^3} \div \dfrac{3a^4}{b^9}$ b) $\dfrac{t^2 - 16t + 63}{t^2} \div (t - 7)^2$

c) $\dfrac{x^2 - 9}{x^2 + 3x - 10} \div \dfrac{24 - 8x}{x^2 + 9x + 20}$

Solution

a) $\dfrac{15a^7}{b^3} \div \dfrac{3a^4}{b^9} = \dfrac{\overset{5a^3}{\cancel{15a^7}}}{\cancel{b^3}} \cdot \dfrac{\overset{b^6}{\cancel{b^9}}}{\cancel{3a^4}}$ Multiply by the reciprocal; divide out common factors.

$= 5a^3 b^6$ Multiply.

Notice that we used the *quotient rule* for exponents to reduce:

$$\dfrac{a^7}{a^4} = a^3, \quad \dfrac{b^9}{b^3} = b^6$$

W Hint

Are you writing out all of the steps as you read the example?

b) $\dfrac{t^2 - 16t + 63}{t^2} \div (t - 7)^2 = \dfrac{(t - 9)\cancel{(t - 7)}^{\,1}}{t^2} \cdot \dfrac{1}{\underset{(t-7)}{\cancel{(t - 7)^2}}}$ Since $(t - 7)^2$ can be written as $\dfrac{(t - 7)^2}{1}$, its reciprocal is $\dfrac{1}{(t - 7)^2}$.

$= \dfrac{t - 9}{t^2(t - 7)}$ Divide out common factors, and multiply.

c) $\dfrac{x^2 - 9}{x^2 + 3x - 10} \div \dfrac{24 - 8x}{x^2 + 9x + 20} = \dfrac{x^2 - 9}{x^2 + 3x - 10} \cdot \dfrac{x^2 + 9x + 20}{24 - 8x}$ Multiply by the reciprocal.

$= \dfrac{(x + 3)\overset{-1}{\cancel{(x - 3)}}}{\cancel{(x + 5)}(x - 2)} \cdot \dfrac{(x + 4)\cancel{(x + 5)}}{8\underset{1}{\cancel{(3 - x)}}}$ Factor; $\dfrac{x - 3}{3 - x} = -1$.

$= -\dfrac{(x + 3)(x + 4)}{8(x - 2)}$ Divide out common factors, and multiply.

[YOU TRY 3] Divide.

a) $\dfrac{k^3}{12h^7} \div \dfrac{k^4}{16h^2}$ b) $\dfrac{w^2 + 4w - 45}{w} \div (w + 9)^2$

c) $\dfrac{2m^2 - m - 15}{1 - m^2} \div \dfrac{m^2 - 10m + 21}{12m - 12}$

Remember that a fraction, itself, represents division. That is, $\dfrac{30}{5} = 30 \div 5 = 6$.

We can write division problems involving fractions and rational expressions in a similar way.

540   CHAPTER 8   **Rational Expressions**

EXAMPLE 4

Divide.

a) $\dfrac{\dfrac{8}{35}}{\dfrac{16}{45}}$ b) $\dfrac{\dfrac{3w+2}{5}}{\dfrac{9w^2-4}{10}}$

Solution

> **W Hint**
>
> Don't be intimidated! You've done problems similar to these before.

a) $\dfrac{\dfrac{8}{35}}{\dfrac{16}{45}}$ means $\dfrac{8}{35} \div \dfrac{16}{45}$. Then,

$$\frac{8}{35} \div \frac{16}{45} = \frac{8}{35} \cdot \frac{45}{16} \qquad \text{Multiply by the reciprocal.}$$

$$= \frac{\overset{1}{8}}{\underset{7}{35}} \cdot \frac{\overset{9}{45}}{\underset{2}{16}} \qquad \text{Divide 8 and 16 by 8. Divide 45 and 35 by 5.}$$

$$= \frac{9}{14} \qquad \text{Multiply.}$$

b) $\dfrac{\dfrac{3w+2}{5}}{\dfrac{9w^2-4}{10}}$ means $\dfrac{3w+2}{5} \div \dfrac{9w^2-4}{10}$. Then,

$$\frac{3w+2}{5} \div \frac{9w^2-4}{10} = \frac{3w+2}{5} \cdot \frac{10}{9w^2-4} \qquad \text{Multiply by the reciprocal.}$$

$$= \frac{\overset{1}{\cancel{3w+2}}}{\underset{1}{\cancel{5}}} \cdot \frac{\overset{2}{\cancel{10}}}{\underset{1}{\cancel{(3w+2)}}(3w-2)} \qquad \text{Factor; divide out common factors.}$$

$$= \frac{2}{3w-2} \qquad \text{Multiply.}$$

[YOU TRY 4] Divide.

a) $\dfrac{\dfrac{4}{45}}{\dfrac{20}{27}}$ b) $\dfrac{\dfrac{25u^2-9}{24}}{\dfrac{5u+3}{16}}$

ANSWERS TO [YOU TRY] EXERCISES

1) $\dfrac{2}{21}$ 2) a) $\dfrac{n^5}{2}$ b) $\dfrac{1}{3d^3(d+2)}$ c) $\dfrac{h+5}{h}$

3) a) $\dfrac{4}{3h^5 k}$ b) $\dfrac{w-5}{w(w+9)}$ c) $-\dfrac{12(2m+5)}{(m+1)(m-7)}$ 4) a) $\dfrac{3}{25}$ b) $\dfrac{2(5u-3)}{3}$

Objective 1: Multiply Rational Expressions

Multiply.

1) $\dfrac{5}{6} \cdot \dfrac{7}{9}$

2) $\dfrac{4}{11} \cdot \dfrac{2}{3}$

(24) 3) $\dfrac{6}{15} \cdot \dfrac{25}{42}$

4) $\dfrac{12}{21} \cdot \dfrac{7}{4}$

5) $\dfrac{16b^5}{3} \cdot \dfrac{4}{36b}$

6) $\dfrac{26}{25r^3} \cdot \dfrac{15r^6}{2}$

7) $\dfrac{21s^4}{15t^2} \cdot \dfrac{5t^4}{42s^{10}}$

8) $\dfrac{15u^4}{14v^2} \cdot \dfrac{7v^7}{20u^8}$

9) $\dfrac{9c^4}{42c} \cdot \dfrac{35}{3c^3}$

10) $-\dfrac{10}{8x^7} \cdot \dfrac{24x^9}{9x}$

11) $\dfrac{5t^2}{(3t-2)^2} \cdot \dfrac{3t-2}{10t^3}$

12) $\dfrac{11(z+5)^5}{6(z-4)} \cdot \dfrac{3}{22(z+5)}$

13) $\dfrac{4u-5}{9u^3} \cdot \dfrac{6u^5}{(4u-5)^3}$

14) $\dfrac{5k+6}{2k^3} \cdot \dfrac{12k^5}{(5k+6)^4}$

(24) 15) $\dfrac{6}{n+5} \cdot \dfrac{n^2+8n+15}{n+3}$

16) $\dfrac{9p^2-1}{12} \cdot \dfrac{9}{9p+3}$

17) $\dfrac{18y-12}{4y^2} \cdot \dfrac{y^2-4y-5}{3y^2+y-2}$

18) $\dfrac{12v-3}{8v+12} \cdot \dfrac{2v^2-5v-12}{3v-12}$

19) $(c-6) \cdot \dfrac{5}{c^2-6c}$

20) $(r^2+r-2) \cdot \dfrac{18r^2}{3r^2+6r}$

21) $\dfrac{7x}{11-x} \cdot (x^2-121)$

22) $\dfrac{4b}{2b^2-3b-5} \cdot (b+1)^2$

23) $\dfrac{r^3+27}{4t+20} \cdot \dfrac{rt+5r-2t-10}{r^2-9}$

24) $\dfrac{5-2m}{mn+7m+4n+28} \cdot \dfrac{n+7}{8m^3-125}$

Objective 2: Divide Rational Expressions

Divide.

(24) 25) $\dfrac{20}{9} \div \dfrac{10}{27}$

26) $\dfrac{4}{5} \div \dfrac{12}{7}$

27) $42 \div \dfrac{9}{2}$

28) $\dfrac{18}{7} \div 6$

(24) 29) $\dfrac{12}{5m^5} \div \dfrac{21}{8m^{12}}$

30) $\dfrac{12k^3}{35} \div \dfrac{42k^6}{25}$

31) $-\dfrac{50g}{7h^3} \div \dfrac{15g^4}{14h}$

32) $-\dfrac{c^{12}}{b} \div \dfrac{c^2}{6b}$

33) $\dfrac{2(k-2)}{21k^6} \div \dfrac{(k-2)^2}{28}$

34) $\dfrac{18}{(x+4)^3} \div \dfrac{36(x-7)}{x+4}$

35) $\dfrac{16q^5}{p+7} \div \dfrac{2q^4}{(p+7)^2}$

36) $\dfrac{(2a-5)^2}{32a^5} \div \dfrac{2a-5}{8a^3}$

37) $\dfrac{q+8}{q} \div \dfrac{q^2+q-56}{5}$

38) $\dfrac{4y^2-25}{10} \div \dfrac{18y-45}{18}$

39) $\dfrac{z^2+18z+80}{2z+1} \div (z+8)^2$

40) $\dfrac{6w^2-30w}{7} \div (w-5)^2$

41) $\dfrac{36a-12}{16} \div (9a^2-1)$

42) $\dfrac{h^2-21h+108}{4h} \div (144-h^2)$

(24) 43) $\dfrac{8d^2-8d+8}{25-4d^2} \div \dfrac{d^3+1}{2d^2-3d-5}$

44) $\dfrac{x^3+y^3}{6x^3-6x^2y+6xy^2} \div \dfrac{x^2-y^2}{9}$

45) $\dfrac{7n^2-14n}{8n} \div \dfrac{n^2+4n-12}{4n+24}$

46) $\dfrac{4j+24}{9} \div \dfrac{j^2-36}{9j-54}$

47) $\dfrac{4c-9}{2c^2-8c} \div \dfrac{12c-27}{c^2-3c-4}$

48) $\dfrac{p+13}{p+3} \div \dfrac{p^3+13p^2}{p^2-5p-24}$

49) $\dfrac{t^2+7t-18}{5t+5} \div \dfrac{8-t^3}{t+1}$

50) $\dfrac{27-z^3}{z^2-11z+24} \div \dfrac{z^3+3z^2+9z}{4}$

51) Explain how to multiply rational expressions.

52) Explain how to divide rational expressions.

53) Find the missing polynomial in the denominator of
$$\dfrac{9h+45}{h^4} \cdot \dfrac{h^3}{} = \dfrac{9}{h(h-2)}.$$

54) Find the missing monomial in the numerator of
$$\dfrac{}{m^2-81} \cdot \dfrac{3m-27}{2m^2} = \dfrac{15m^3}{m+9}.$$

55) Find the missing binomial in the numerator of
$$\frac{4z^2 - 49}{z^2 - 3z - 40} \div \frac{}{z + 5} = \frac{2z + 7}{8 - z}.$$

56) Find the missing polynomial in the denominator of
$$\frac{12x^4}{50x^2 + 40x} \div \frac{x^3 + 2x^2 + x + 2}{} = \frac{6x^3}{5(x^2 + 1)}.$$

Divide.

57) $\dfrac{\dfrac{25}{42}}{\dfrac{8}{21}}$

58) $\dfrac{\dfrac{9}{35}}{\dfrac{4}{15}}$

59) $\dfrac{\dfrac{5}{24}}{\dfrac{15}{4}}$

60) $\dfrac{\dfrac{4}{3}}{\dfrac{2}{9}}$

61) $\dfrac{\dfrac{3d + 7}{24}}{\dfrac{3d + 7}{6}}$

62) $\dfrac{\dfrac{8s - 7}{4}}{\dfrac{8s - 7}{16}}$

63) $\dfrac{\dfrac{16r + 24}{r^3}}{\dfrac{12r + 18}{r}}$

64) $\dfrac{\dfrac{44m - 33}{3m^2}}{\dfrac{8m - 6}{m}}$

65) $\dfrac{\dfrac{a^2 - 25}{3a^{11}}}{\dfrac{4a + 20}{a^3}}$

66) $\dfrac{\dfrac{4z - 8}{z^8}}{\dfrac{z^2 - 4}{z^6}}$

67) $\dfrac{\dfrac{16x^2 - 25}{x^7}}{\dfrac{36x - 45}{6x^3}}$

68) $-\dfrac{\dfrac{16a^2}{3a^2 + 2a}}{\dfrac{12}{9a^2 - 4}}$

Mixed Exercises: Objectives 1 and 2
Perform the operations and simplify.

69) $\dfrac{c^2 + c - 30}{9c + 9} \cdot \dfrac{c^2 + 2c + 1}{c^2 - 25}$

70) $\dfrac{d^2 + 3d - 54}{d - 12} \cdot \dfrac{d^2 - 10d - 24}{7d + 63}$

71) $\dfrac{3x + 2}{9x^2 - 4} \div \dfrac{4x}{15x^2 - 7x - 2}$

72) $\dfrac{b^2 - 10b + 25}{8b - 40} \div \dfrac{2b^2 - 5b - 25}{2b + 5}$

73) $\dfrac{3k^2 - 12k}{12k^2 - 30k - 72} \cdot (2k + 3)^2$

74) $\dfrac{4a^3}{a^2 + a - 72} \cdot (a^2 - a - 56)$

75) $\dfrac{30}{4y^2 - 4x^2} \div \dfrac{10x^2 + 10xy + 10y^2}{x^3 - y^3}$

76) $\dfrac{4n^2 - 1}{10n^3} \div \dfrac{2n^2 - 7n - 4}{6n^5}$

77) $\dfrac{4h^3}{h^2 - 64} \cdot \dfrac{8h - h^2}{12}$

78) $\dfrac{c^2 - 36}{c + 6} \div \dfrac{30 - 5c}{c - 9}$

79) $\dfrac{54x^8}{22x^3y^2} \div \dfrac{36xy^5}{11x^2y}$

80) $\dfrac{28cd^9}{2c^3d} \cdot \dfrac{5d^2}{84c^{10}d^2}$

81) $\dfrac{a^2 - 4a}{6a + 54} \cdot \dfrac{a^2 + 13a + 36}{16 - a^2}$

82) $\dfrac{64 - u^2}{40 - 5u} \div \dfrac{u^2 + 10u + 16}{2u + 3}$

83) $\dfrac{r^3 + 8}{r + 2} \cdot \dfrac{7}{3r^2 - 6r + 12}$

84) $\dfrac{2t^2 - 6t + 18}{5t - 5} \cdot \dfrac{t^2 - 9}{t^3 + 27}$

85) $\dfrac{2a^2}{a^2 + a - 20} \cdot \dfrac{a^3 + 5a^2 + 4a + 20}{2a^2 + 8}$

86) $\dfrac{18x^4}{x^3 + 3x^2 - 9x - 27} \cdot \dfrac{6x^2 + 19x + 3}{18x^2 + 3x}$

87) $\dfrac{3m^2 + 8m + 4}{4} \div (12m + 8)$

88) $\dfrac{w^2 - 17w + 72}{3w} \div (w - 8)$

89) $\dfrac{4j^2 - 21j + 5}{j^3} \div \left(\dfrac{3j + 2}{j^3 - j^2} \cdot \dfrac{j^2 - 6j + 5}{j} \right)$

90) $\dfrac{t^3 - 8}{t - 2} \div \left(\dfrac{3t + 11}{5t + 15} \cdot \dfrac{t^2 + 2t + 4}{3t^2 + 11t} \right)$

91) If the area of a rectangle is $\dfrac{3x}{2y^2}$ and the width is $\dfrac{y}{8x^4}$, what is the length of the rectangle?

92) If the area of a triangle is $\dfrac{2n}{n^2 - 4n + 3}$ and the height is $\dfrac{n + 3}{n - 1}$, what is the length of the base of the triangle?

R1) Why do you think you learned to multiply and divide rational expressions before learning to add and subtract rational expressions? (Hint: Compare this with adding and subtracting fractions.)

R2) Which exercises could you do easily and which gave you more trouble? Why?

8.3 Finding the Least Common Denominator

P Prepare

O Organize

What are your objectives for Section 8.3?	How can you accomplish each objective?
1 Find the Least Common Denominator for a Group of Rational Expressions	• Notice that the skills you learned involving fractions apply again! • Write the procedure for **Finding the LCD** in your notes. • Complete the given examples on your own. • Complete You Trys 1–3.
2 Rewrite Rational Expressions Using Their LCD	• Compare the solution for Example 4 to the procedure for **Writing Rational Expressions as Equivalent Expressions with the LCD.** • Write the procedure in your own words. • Complete the given examples on your own. • Complete You Trys 4 and 5.

W Work Read the explanations, follow the examples, take notes, and complete the You Trys.

1 Find the Least Common Denominator for a Group of Rational Expressions

Recall that to add or subtract fractions, they must have a common denominator. Similarly, rational expressions must have common denominators in order to be added or subtracted. In this section, we will discuss how to find the least common denominator (LCD) of rational expressions.

We begin by looking at the fractions $\frac{3}{8}$ and $\frac{5}{12}$. By inspection, we can see that the LCD = 24. But, *why* is that true? Let's write each of the denominators, 8 and 12, as the product of their prime factors:

$$8 = 2 \cdot 2 \cdot 2 = 2^3$$
$$12 = 2 \cdot 2 \cdot 3 = 2^2 \cdot 3$$

The LCD will contain each factor the *greatest* number of times it appears in any single factorization.

> *The LCD will contain* 2^3 because 2 appears *three* times in the factorization of 8.
>
> *The LCD will contain* 3 because 3 appears *one* time in the factorization of 12.
>
> The LCD, then, is the product of the factors we have identified.

$$\text{LCD of } \frac{3}{8} \text{ and } \frac{5}{12} = 2^3 \cdot 3 = 8 \cdot 3 = 24$$

This is the same result that we obtained just by inspecting the two denominators.

We use the same procedure to find the least common denominator of rational expressions.

Procedure Finding the Least Common Denominator (LCD)

Step 1: Factor the denominators.

Step 2: The LCD will contain each unique factor the *greatest* number of times it appears in any single factorization.

Step 3: The LCD is the *product* of the factors identified in Step 2.

EXAMPLE 1

Find the LCD of each pair of rational expressions.

a) $\dfrac{17}{24}, \dfrac{5}{36}$ b) $\dfrac{1}{12n}, \dfrac{10}{21n}$ c) $\dfrac{8}{49c^3}, \dfrac{13}{14c^2}$

Solution

a) Follow the steps for finding the least common denominator.

Step 1: Factor the denominators.

$$24 = 2 \cdot 2 \cdot 2 \cdot 3 = 2^3 \cdot 3$$
$$36 = 2 \cdot 2 \cdot 3 \cdot 3 = 2^2 \cdot 3^2$$

Step 2: The LCD will contain each unique factor the *greatest* number of times it appears in any factorization. *The LCD will contain* 2^3 *and* 3^2.

Step 3: The LCD is the *product* of the factors in Step 2.

$$\text{LCD} = 2^3 \cdot 3^2 = 8 \cdot 9 = 72$$

b) *Step 1:* Factoring the denominators of $\dfrac{1}{12n}$ and $\dfrac{10}{21n}$ gives us

$$12n = 2 \cdot 2 \cdot 3 \cdot n = 2^2 \cdot 3 \cdot n$$
$$21n = 3 \cdot 7 \cdot n$$

Step 2: The LCD will contain each unique factor the *greatest* number of times it appears in any factorization. *It will contain* 2^2, 3, 7, *and* n.

Step 3: The LCD is the *product* of the factors in Step 2.

$$\text{LCD} = 2^2 \cdot 3 \cdot 7 \cdot n = 84n$$

c) **Step 1:** Factoring the denominators of $\dfrac{8}{49c^3}$ and $\dfrac{13}{14c^2}$ gives us

$$49c^3 = 7 \cdot 7 \cdot c^3 = 7^2 \cdot c^3$$
$$14c^2 = 2 \cdot 7 \cdot c^2$$

Step 2: The LCD will contain each unique factor the *greatest* number of times it appears in any factorization. *It will contain* 2, 7^2, *and* c^3.

Step 3: The LCD is the *product* of the factors in Step 2.

$$\text{LCD} = 2 \cdot 7^2 \cdot c^3 = 98c^3$$

[YOU TRY 1] Find the LCD of each pair of rational expressions.

a) $\dfrac{14}{15}, \dfrac{11}{18}$ b) $\dfrac{3}{14d}, \dfrac{7}{10d}$ c) $\dfrac{20}{27h^2}, \dfrac{1}{6h^4}$

EXAMPLE 2   Find the LCD of each group of rational expressions.

a) $\dfrac{4}{k}, \dfrac{6}{k+3}$ b) $\dfrac{7}{a-6}, \dfrac{2a}{a^2+3a-54}$ c) $\dfrac{3p}{p^2+4p+4}, \dfrac{1}{5p^2+10p}$

Solution

a) The denominators of $\dfrac{4}{k}$ and $\dfrac{6}{k+3}$ are already in simplest form. It is important to recognize that k *and* $k+3$ *are different factors.*

The LCD will be the product of k and $k+3$:   **LCD** $= k(k+3)$

Usually, we leave the LCD in this form; we do not distribute.

b) **Step 1:** Factor the denominators of $\dfrac{7}{a-6}$ and $\dfrac{2a}{a^2+3a-54}$.

$a-6$ cannot be factored. $a^2+3a-54 = (a-6)(a+9)$

Step 2: The LCD will contain each unique factor the *greatest* number of times it appears in any factorization. *It will contain* $a-6$ *and* $a+9$.

Step 3: The LCD is the *product* of the factors identified in Step 2.

LCD $= (a-6)(a+9)$

c) **Step 1:** Factor the denominators of $\dfrac{3p}{p^2+4p+4}$ and $\dfrac{1}{5p^2+10p}$.

$p^2+4p+4 = (p+2)^2$ and $5p^2+10p = 5p(p+2)$

Step 2: The unique factors are 5, p, and $p+2$, with $p+2$ *appearing at most twice.* *The factors we will use in the LCD are* 5, p, *and* $(p+2)^2$.

Step 3: LCD $= 5p(p+2)^2$

W Hint

Just because the problem looks complicated does *not* mean you can't do it. Break it down one step at a time, and it will be easier to manage.

[YOU TRY 2] Find the LCD of each group of rational expressions.

a) $\dfrac{6}{w}, \dfrac{9w}{w+1}$ b) $\dfrac{1}{r+8}, \dfrac{5r}{r^2+r-56}$ c) $\dfrac{4b}{b^2-18b+81}, \dfrac{3b+2}{8b^2-72b}$

What is the least common denominator of $\dfrac{9}{x-7}$ and $\dfrac{5}{7-x}$? Is it $(x-7)(7-x)$? Read Example 3 to find out.

EXAMPLE 3

Find the LCD of $\dfrac{9}{x-7}$ and $\dfrac{5}{7-x}$.

Solution

Recall from Section 8.1 that $a - b = -1(b - a)$. So, $7 - x = -(x - 7)$, and we can rewrite $\dfrac{5}{7-x}$ as $\dfrac{5}{-(x-7)} = -\dfrac{5}{x-7}$. Therefore, we can now think of our task as finding the LCD of $\dfrac{9}{x-7}$ and $-\dfrac{5}{x-7}$. The least common denominator of $\dfrac{9}{x-7}$ and $\dfrac{5}{7-x}$ is $x - 7$.

[YOU TRY 3]

Find the LCD of $\dfrac{2}{k-5}$ and $\dfrac{13}{5-k}$.

2 Rewrite Rational Expressions Using Their LCD

As we know from our previous work with fractions, after *determining* the least common denominator, we must *rewrite* those fractions as equivalent fractions with the LCD so that they can be added or subtracted.

EXAMPLE 4

Identify the LCD of $\dfrac{1}{6}$ and $\dfrac{8}{9}$, and rewrite each as an equivalent fraction with the LCD as its denominator.

Solution

The LCD of $\dfrac{1}{6}$ and $\dfrac{8}{9}$ is 18. We must rewrite each fraction with a denominator of 18.

$\dfrac{1}{6}$: By what number should we multiply 6 to get 18?   3

$$\dfrac{1}{6} \cdot \dfrac{3}{3} = \dfrac{3}{18} \qquad \text{Multiply the numerator } and \text{ denominator by 3 to obtain an equivalent fraction.}$$

$\dfrac{8}{9}$: By what number should we multiply 9 to get 18?   2

$$\dfrac{8}{9} \cdot \dfrac{2}{2} = \dfrac{16}{18} \qquad \text{Multiply the numerator } and \text{ denominator by 2 to obtain an equivalent fraction.}$$

Therefore, $\dfrac{1}{6} = \dfrac{3}{18}$ and $\dfrac{8}{9} = \dfrac{16}{18}$.

[YOU TRY 4]

Identify the LCD of $\dfrac{7}{12}$ and $\dfrac{2}{9}$, and rewrite each as an equivalent fraction with the LCD as its denominator.

The procedure for rewriting rational expressions as equivalent expressions with the least common denominator is very similar to the process used in Example 4.

Procedure Writing Rational Expressions as Equivalent Expressions with the Least Common Denominator

Step 1: Identify and write down the LCD.

Step 2: Look at each rational expression (with its denominator in factored form) and compare its denominator with the LCD. Ask yourself, "What factors are missing?"

Step 3: Multiply the numerator and denominator by the "missing" factors to obtain an equivalent rational expression with the desired LCD.

Note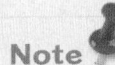

Use the distributive property to multiply the terms in the numerator, but leave the denominator as the product of factors. (We will see why this is done in Section 8.4.)

EXAMPLE 5

Identify the LCD of each pair of rational expressions, and rewrite each as an equivalent expression with the LCD as its denominator.

a) $\dfrac{5}{12z}, \dfrac{4}{9z^3}$

b) $\dfrac{m}{m-4}, \dfrac{3}{m+8}$

c) $\dfrac{2}{3x^2-6x}, \dfrac{9x}{x^2-7x+10}$

d) $\dfrac{r}{r-6}, \dfrac{2}{6-r}$

Solution

a) *Step 1:* Identify and write down the LCD of $\dfrac{5}{12z}$ and $\dfrac{4}{9z^3}$: **LCD $= 36z^3$**

Step 2: Compare the denominators of $\dfrac{5}{12z}$ and $\dfrac{4}{9z^3}$ to the LCD and ask yourself, "What's missing?"

$\dfrac{5}{12z}$: $12z$ *is "missing" the factors 3 and z^2.*

$\dfrac{4}{9z^3}$: $9z^3$ *is "missing" the factor 4.*

Step 3: Multiply the numerator and denominator by $3z^2$.

Multiply the numerator and denominator by 4.

$$\dfrac{5}{12z} \cdot \dfrac{3z^2}{3z^2} = \dfrac{15z^2}{36z^3} \qquad \dfrac{4}{9z^3} \cdot \dfrac{4}{4} = \dfrac{16}{36z^3}$$

$$\dfrac{5}{12z} = \dfrac{15z^2}{36z^3} \quad \text{and} \quad \dfrac{4}{9z^3} = \dfrac{16}{36z^3}$$

b) *Step 1:* Identify and write down the LCD of $\dfrac{m}{m-4}$ and $\dfrac{3}{m+8}$:

LCD $= (m-4)(m+8)$

Step 2: Compare the denominators of $\dfrac{m}{m-4}$ and $\dfrac{3}{m+8}$ to the LCD and ask yourself, "What's missing?"

$\dfrac{m}{m-4}$: $m-4$ *is "missing"*
the factor $m+8$.

$\dfrac{3}{m+8}$: $m+8$ *is "missing"*
the factor $m-4$.

Step 3: Multiply the numerator and denominator by $m+8$.

$$\dfrac{m}{m-4} \cdot \dfrac{m+8}{m+8} = \dfrac{m(m+8)}{(m-4)(m+8)}$$
$$= \dfrac{m^2+8m}{(m-4)(m+8)}$$

Multiply the numerator and denominator by $m-4$.

$$\dfrac{3}{m+8} \cdot \dfrac{m-4}{m-4} = \dfrac{3(m-4)}{(m+8)(m-4)}$$
$$= \dfrac{3m-12}{(m-4)(m+8)}$$

Notice that we multiplied the factors in the numerator but left the denominator in factored form.

$$\dfrac{m}{m-4} = \dfrac{m^2+8m}{(m-4)(m+8)} \quad \text{and} \quad \dfrac{3}{m+8} = \dfrac{3m-12}{(m-4)(m+8)}$$

c) **Step 1:** Identify and write down the LCD of $\dfrac{2}{3x^2-6x}$ and $\dfrac{9x}{x^2-7x+10}$.

First, we must factor the denominators.

$$\dfrac{2}{3x^2-6x} = \dfrac{2}{3x(x-2)}, \quad \dfrac{9x}{x^2-7x+10} = \dfrac{9x}{(x-2)(x-5)}$$

We will work with the factored forms of the expressions.

$$\textbf{LCD} = \mathbf{3x(x-2)(x-5)}$$

Step 2: Compare the denominators of $\dfrac{2}{3x(x-2)}$ and $\dfrac{9x}{(x-2)(x-5)}$ to the LCD and ask yourself, "What's missing?"

$\dfrac{2}{3x(x-2)}$: $3x(x-2)$ *is*
"missing" the
factor $x-5$.

$\dfrac{9x}{(x-2)(x-5)}$: $(x-2)(x-5)$ *is*
"missing" $3x$.

Step 3: Multiply the numerator and denominator by $x-5$.

$$\dfrac{2}{3x(x-2)} \cdot \dfrac{x-5}{x-5} = \dfrac{2(x-5)}{3x(x-2)(x-5)}$$
$$= \dfrac{2x-10}{3x(x-2)(x-5)}$$

Multiply the numerator and denominator by $3x$.

$$\dfrac{9x}{(x-2)(x-5)} \cdot \dfrac{3x}{3x} = \dfrac{27x^2}{3x(x-2)(x-5)}$$

$$\dfrac{2}{3x^2-6x} = \dfrac{2x-10}{3x(x-2)(x-5)} \quad \text{and} \quad \dfrac{9x}{x^2-7x+10} = \dfrac{27x^2}{3x(x-2)(x-5)}$$

d) To find the LCD of $\dfrac{r}{r-6}$ and $\dfrac{2}{6-r}$, recall that $6-r$ can be rewritten as $-(r-6)$. So,

$$\frac{2}{6-r} = \frac{2}{-(r-6)} = -\frac{2}{r-6}$$

Therefore, the LCD of $\dfrac{r}{r-6}$ and $-\dfrac{2}{r-6}$ is $r-6$.

The expression $\dfrac{r}{r-6}$ already has the LCD, while $\dfrac{2}{6-r} = -\dfrac{2}{r-6}$.

[YOU TRY 5] Identify the least common denominator of each pair of rational expressions, and rewrite each as an equivalent expression with the LCD as its denominator.

a) $\dfrac{3}{10a^6}, \dfrac{7}{8a^5}$ b) $\dfrac{6}{n+10}, \dfrac{n}{2n-3}$ c) $\dfrac{v-9}{v^2+15v+44}, \dfrac{8}{5v^2+55v}$

d) $\dfrac{k}{7-k}, \dfrac{4}{k-7}$

ANSWERS TO [YOU TRY] EXERCISES

1) a) 90 b) 70d c) $54h^4$ 2) a) $w(w+1)$ b) $(r+8)(r-7)$ c) $8b(b-9)^2$

3) $k-5$ 4) LCD = 36; $\dfrac{7}{12} = \dfrac{21}{36}, \dfrac{2}{9} = \dfrac{8}{36}$ 5) a) LCD = $40a^6$; $\dfrac{3}{10a^6} = \dfrac{12}{40a^6}, \dfrac{7}{8a^5} = \dfrac{35a}{40a^6}$

b) LCD = $(2n-3)(n+10)$; $\dfrac{6}{n+10} = \dfrac{12n-18}{(2n-3)(n+10)}, \dfrac{n}{2n-3} = \dfrac{n^2+10n}{(2n-3)(n+10)}$

c) LCD = $5v(v+4)(v+11)$; $\dfrac{v-9}{v^2+15v+44} = \dfrac{5v^2-45v}{5v(v+4)(v+11)}, \dfrac{8}{5v^2+55v} = \dfrac{8v+32}{5v(v+4)(v+11)}$

d) LCD = $k-7$; $\dfrac{k}{7-k} = -\dfrac{k}{k-7}, \dfrac{4}{k-7} = \dfrac{4}{k-7}$

E Evaluate **8.3** Exercises   Do the exercises, and check your work.

Objective 1: Find the Least Common Denominator for a Group of Rational Expressions

Find the LCD of each group of fractions.

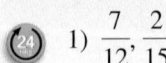

1) $\dfrac{7}{12}, \dfrac{2}{15}$

2) $\dfrac{3}{8}, \dfrac{9}{7}$

3) $\dfrac{27}{40}, \dfrac{11}{10}, \dfrac{5}{12}$

4) $\dfrac{19}{8}, \dfrac{1}{12}, \dfrac{3}{32}$

5) $\dfrac{3}{n^7}, \dfrac{5}{n^{11}}$

6) $\dfrac{4}{c^2}, \dfrac{8}{c^3}$

7) $\dfrac{13}{14r^4}, \dfrac{3}{4r^7}$

8) $\dfrac{11}{6p^4}, \dfrac{3}{10p^9}$

9) $-\dfrac{5}{6z^5}, \dfrac{7}{36z^2}$

10) $\dfrac{5}{24w^5}, -\dfrac{1}{4w^{10}}$

11) $\dfrac{7}{10m}, \dfrac{9}{22m^4}$

12) $-\dfrac{3}{2k^2}, \dfrac{5}{14k^5}$

13) $\dfrac{4}{24x^3y^2}, \dfrac{11}{6x^3y}$

14) $\dfrac{3}{10a^4b^2}, \dfrac{8}{15ab^4}$

15) $\dfrac{4}{11}, \dfrac{8}{z-3}$

16) $\dfrac{3}{n+8}, \dfrac{1}{5}$

17) $\dfrac{10}{w}, \dfrac{6}{2w+1}$

18) $\dfrac{1}{y}, -\dfrac{6}{6y+1}$

19) What is the first step for finding the LCD of $\dfrac{9}{8t-10}$ and $\dfrac{3t}{20t-25}$?

20) Is $(h-9)(9-h)$ the LCD of $\dfrac{2h}{h-9}$ and $\dfrac{4}{9-h}$? Explain your answer.

Find the LCD of each group of fractions.

21) $\dfrac{8}{5c-5}, \dfrac{9}{2c-2}$

22) $\dfrac{5}{7k+14}, \dfrac{11}{4k+8}$

23) $\dfrac{2}{9p^4-6p^3}, \dfrac{3}{3p^6-2p^5}$

24) $\dfrac{21}{6a^2-8a}, \dfrac{13}{18a^3-24a^2}$

25) $\dfrac{4m}{m-7}, \dfrac{2}{m-3}$

26) $\dfrac{5}{r+9}, \dfrac{7}{r-1}$

27) $\dfrac{11}{z^2+11z+24}, \dfrac{7z}{z^2+5z-24}$

28) $\dfrac{7x}{x^2-12x+35}, \dfrac{x}{x^2-x-20}$

(24) 29) $\dfrac{n-1}{n+4}, \dfrac{2n+7}{n^2+8n+16}$

30) $\dfrac{k-2}{k+9}, \dfrac{5k+1}{k^2+18k+81}$

31) $\dfrac{14t}{t^2-3t-18}, -\dfrac{6}{t^2-36}, \dfrac{t}{t^2+9t+18}$

32) $\dfrac{6w}{w^2-10w+16}, \dfrac{3}{w^2-7w-8}, \dfrac{4w}{w^2-w-2}$

33) $\dfrac{6}{a-8}, \dfrac{7}{8-a}$

34) $\dfrac{6}{b-3}, \dfrac{5}{3-b}$

35) $\dfrac{12}{y-x}, \dfrac{5y}{x-y}$

36) $\dfrac{u}{v-u}, \dfrac{8}{u-v}$

Objective 2: Rewrite Rational Expressions Using Their LCD

37) Explain, in your own words, how to rewrite $\dfrac{4}{x+9}$ as an equivalent rational expression with a denominator of $(x+9)(x-3)$.

38) Explain, in your own words, how to rewrite $\dfrac{7}{5-m}$ as an equivalent rational expression with a denominator of $m-5$.

Rewrite each rational expression with the indicated denominator.

39) $\dfrac{7}{12} = \dfrac{}{48}$

40) $\dfrac{5}{7} = \dfrac{}{42}$

41) $\dfrac{8}{z} = \dfrac{}{9z}$

42) $\dfrac{-6}{b} = \dfrac{}{4b}$

(24) 43) $\dfrac{3}{8k} = \dfrac{}{56k^4}$

44) $\dfrac{5}{3p^4} = \dfrac{}{9p^6}$

45) $\dfrac{6}{5t^5u^2} = \dfrac{}{10t^7u^5}$

46) $\dfrac{13}{6cd^2} = \dfrac{}{24c^3d^3}$

47) $\dfrac{7}{3r+4} = \dfrac{}{r(3r+4)}$

48) $\dfrac{8}{m-8} = \dfrac{}{m(m-8)}$

49) $\dfrac{v}{4(v-3)} = \dfrac{}{16v^5(v-3)}$

50) $\dfrac{a}{5(2a+7)} = \dfrac{}{15a(2a+7)}$

51) $\dfrac{9x}{x+6} = \dfrac{}{(x+6)(x-5)}$

52) $\dfrac{5b}{b+3} = \dfrac{}{(b+3)(b+7)}$

53) $\dfrac{z-3}{2z-5} = \dfrac{}{(2z-5)(z+8)}$

54) $\dfrac{w+2}{4w-1} = \dfrac{}{(4w-1)(w-4)}$

(24) 55) $\dfrac{5}{3-p} = \dfrac{}{p-3}$

56) $\dfrac{10}{10-n} = \dfrac{}{n-10}$

57) $-\dfrac{8c}{6c-7} = \dfrac{}{7-6c}$

58) $-\dfrac{g}{3g-2} = \dfrac{}{2-3g}$

Identify the least common denominator of each pair of rational expressions, and rewrite each as an equivalent rational expression with the LCD as its denominator.

59) $\dfrac{8}{15}, \dfrac{1}{6}$

60) $\dfrac{3}{8}, \dfrac{5}{12}$

61) $\dfrac{4}{u}, \dfrac{8}{u^3}$

62) $\dfrac{9}{d^5}, \dfrac{7}{d^2}$

63) $\dfrac{9}{8n^6}, \dfrac{2}{3n^2}$

64) $\dfrac{5}{8a}, \dfrac{7}{10a^5}$

65) $\dfrac{6}{4a^3b^5}, \dfrac{6}{a^4b}$

66) $\dfrac{3}{x^3y}, \dfrac{6}{5xy^5}$

67) $\dfrac{r}{5}, \dfrac{2}{r-4}$

68) $\dfrac{t}{5t-1}, \dfrac{8}{7}$

69) $\dfrac{m}{m+7}, \dfrac{3}{m}$

70) $\dfrac{5}{c}, \dfrac{4}{c+2}$

71) $\dfrac{a}{30a-15}, \dfrac{1}{12a-6}$

72) $\dfrac{7}{24x-16}, \dfrac{x}{18x-12}$

73) $\dfrac{9}{k-9}, \dfrac{5k}{k+3}$

74) $\dfrac{6}{h+1}, \dfrac{11h}{h+7}$

75) $\dfrac{3}{a+2}, \dfrac{2a}{3a+4}$

76) $\dfrac{b}{6b-5}, \dfrac{8}{b-9}$

77) $\dfrac{9y}{y^2 - y - 42}, \dfrac{3}{2y^2 + 12y}$

78) $\dfrac{12q}{q^2 - 6q - 16}, \dfrac{4}{2q^2 - 16q}$

79) $\dfrac{c}{c^2 + 9c + 18}, \dfrac{11}{c^2 + 12c + 36}$

80) $\dfrac{z}{z^2 - 8z + 16}, \dfrac{9z}{z^2 + 4z - 32}$

81) $\dfrac{11}{g - 3}, \dfrac{4}{3 - g}$

82) $\dfrac{6}{n - 9}, \dfrac{1}{9 - n}$

83) $\dfrac{4}{3x - 4}, \dfrac{7x}{16 - 9x^2}$

84) $\dfrac{12}{5k - 2}, \dfrac{4k}{4 - 25k^2}$

85) $-\dfrac{9}{h^3 + 8}, \dfrac{2h}{5h^2 - 10h + 20}$

86) $\dfrac{5x}{x^3 - y^3}, \dfrac{7}{8x - 8y}$

87) $\dfrac{2}{z^2 + 3z}, \dfrac{6}{3z^2 + 9z}, \dfrac{8}{z^2 + 6z + 9}$

88) $\dfrac{4}{w^2 - 4w}, \dfrac{6}{7w^2 - 28w}, \dfrac{11}{w^2 - 8w + 16}$

89) $\dfrac{t}{t^2 - 13t + 30}, \dfrac{6}{t - 10}, \dfrac{7}{t^2 - 9}$

90) $-\dfrac{2}{a + 2}, \dfrac{a}{a^2 - 4}, \dfrac{15}{a^2 - 3a + 2}$

R **Rethink**

R1) Where could you use some help to master the objectives presented in this section?

R2) When you got stuck on a "harder" problem, what did you do to help move forward on the problem?

8.4 Adding and Subtracting Rational Expressions

P **Prepare**

O **Organize**

What are your objectives for Section 8.4?	How can you accomplish each objective?
1 Add and Subtract Rational Expressions with a Common Denominator	• Understand the procedure for **Adding and Subtracting Rational Expressions.** • Complete the given examples on your own. • Complete You Trys 1 and 2.
2 Add and Subtract Rational Expressions with Different Denominators	• Write the **Steps for Adding and Subtracting Rational Expressions with Different Denominators** in your own words. • Complete the given examples on your own. • Complete You Trys 3 and 4.
3 Add and Subtract Rational Expressions with Special Denominators	• Keep in mind that $a - b = -(b - a)$. • Complete the given example on your own. • Complete You Try 5.

We know that in order to add or subtract fractions, they must have a common denominator. The same is true for rational expressions.

1 Add and Subtract Rational Expressions with a Common Denominator

Let's first look at fractions and rational expressions with common denominators.

EXAMPLE 1

Add or subtract.

a) $\dfrac{8}{11} - \dfrac{5}{11}$

b) $\dfrac{2x}{4x-9} + \dfrac{5x+3}{4x-9}$

Solution

a) Because the fractions have the same denominator, subtract the terms in the numerator and keep the common denominator.

$$\frac{8}{11} - \frac{5}{11} = \frac{8-5}{11} = \frac{3}{11} \qquad \text{Subtract terms in the numerator.}$$

b) Because the rational expressions have the same denominator, add the terms in the numerator and keep the common denominator.

$$\frac{2x}{4x-9} + \frac{5x+3}{4x-9} = \frac{2x + (5x+3)}{4x-9} \qquad \text{Add terms in the numerator.}$$

$$= \frac{7x+3}{4x-9} \qquad \text{Combine like terms.}$$

We can generalize the procedure for adding and subtracting rational expressions that have a common denominator as follows.

Procedure Adding and Subtracting Rational Expressions

If $\dfrac{P}{Q}$ and $\dfrac{R}{Q}$ are rational expressions with $Q \neq 0$, then

1) $\dfrac{P}{Q} + \dfrac{R}{Q} = \dfrac{P+R}{Q}$ and   2) $\dfrac{P}{Q} - \dfrac{R}{Q} = \dfrac{P-R}{Q}$

YOU TRY 1

Add or subtract.

a) $\dfrac{11}{12} - \dfrac{7}{12}$

b) $\dfrac{6h}{5h-2} + \dfrac{3h+8}{5h-2}$

All answers to a sum or difference of rational expressions should be in lowest terms. Sometimes it is necessary to simplify our result to lowest terms by factoring the numerator and dividing the numerator and denominator by the greatest common factor.

EXAMPLE 2

Subtract $\dfrac{c^2 - 3}{c(c + 4)} - \dfrac{5 - 2c}{c(c + 4)}$.

Solution

$$\dfrac{c^2 - 3}{c(c + 4)} - \dfrac{5 - 2c}{c(c + 4)} = \dfrac{(c^2 - 3) - (5 - 2c)}{c(c + 4)} \qquad \text{Subtract terms in the numerator.}$$

$$= \dfrac{c^2 - 3 - 5 + 2c}{c(c + 4)} \qquad \text{Distribute.}$$

$$= \dfrac{c^2 + 2c - 8}{c(c + 4)} \qquad \text{Combine like terms.}$$

$$= \dfrac{(c + 4)(c - 2)}{c(c + 4)} \qquad \text{Factor the numerator.}$$

$$= \dfrac{c - 2}{c} \qquad \text{Divide out the common factors.}$$

[YOU TRY 2]

Add or subtract.

a) $\dfrac{19}{32w} + \dfrac{9}{32w}$ b) $\dfrac{m^2 - 5}{m(m + 6)} - \dfrac{3m + 49}{m(m + 6)}$

BE CAREFUL

After combining like terms in the numerator, ask yourself, *"Can I factor the numerator?"* If so, factor it. Sometimes, the expression can be simplified by dividing the numerator and denominator by the greatest common factor.

2 Add and Subtract Rational Expressions with Different Denominators

If we are asked to add or subtract rational expressions with different denominators, we must begin by rewriting each expression with the least common denominator. Then, add or subtract. Write the answer in lowest terms.

Using the procedure studied in Section 8.3, here are the steps to follow to add or subtract rational expressions with different denominators.

Hint

Write this procedure in your own words.

Procedure Steps for Adding and Subtracting Rational Expressions with Different Denominators

1) Factor the denominators.

2) Write down the LCD.

3) Rewrite each rational expression as an equivalent rational expression with the LCD.

4) Add or subtract the numerators and keep the common denominator in factored form.

5) After combining like terms in the numerator, ask yourself, *"Can I factor it?"* If so, factor.

6) Divide out common factors, if possible. The final answer should be written in lowest terms.

EXAMPLE 3

Add or subtract.

a) $\dfrac{t+6}{4} + \dfrac{t-8}{12}$

b) $\dfrac{3}{10a} - \dfrac{7}{8a^2}$

c) $\dfrac{7n-30}{n^2-36} + \dfrac{n}{n+6}$

Solution

a) The LCD is **12**. $\dfrac{t-8}{12}$ already has the LCD.

Rewrite $\dfrac{t+6}{4}$ with the LCD: $\dfrac{t+6}{4} \cdot \dfrac{3}{3} = \dfrac{3(t+6)}{12}$

$\dfrac{t+6}{4} + \dfrac{t-8}{12} = \dfrac{3(t+6)}{12} + \dfrac{t-8}{12}$ Write each expression with the LCD.

$= \dfrac{3(t+6) + (t-8)}{12}$ Add the numerators.

$= \dfrac{3t + 18 + t - 8}{12}$ Distribute.

$= \dfrac{4t + 10}{12}$ Combine like terms.

Ask yourself, *"Can I factor the numerator?"* Yes.

$= \dfrac{\overset{1}{\cancel{2}}(2t+5)}{\underset{6}{\cancel{12}}}$ Factor.

$= \dfrac{2t+5}{6}$ Divide out the common factor of 2.

Hint

Are you writing out the example as you read it?

W Hint

Use different colored pens to take notes. This will help you see the steps being performed, just like in these examples.

b) The LCD of $\dfrac{3}{10a}$ and $\dfrac{7}{8a^2}$ is $\mathbf{40a^2}$. Rewrite each expression with the LCD.

$$\dfrac{3}{10a}\cdot\dfrac{4a}{4a}=\dfrac{12a}{40a^2}\quad\text{and}\quad\dfrac{7}{8a^2}\cdot\dfrac{5}{5}=\dfrac{35}{40a^2}$$

$$\dfrac{3}{10a}-\dfrac{7}{8a^2}=\dfrac{12a}{40a^2}-\dfrac{35}{40a^2}\qquad\text{Write each expression with the LCD.}$$

$$=\dfrac{12a-35}{40a^2}\qquad\text{Subtract the numerators.}$$

"Can I factor the numerator?" No. The expression is in simplest form since the numerator and denominator have no common factors.

c) First, factor the denominator of $\dfrac{7n-30}{n^2-36}$ to get $\dfrac{7n-30}{(n+6)(n-6)}$.

The LCD of $\dfrac{7n-30}{(n+6)(n-6)}$ and $\dfrac{n}{n+6}$ is $\mathbf{(n+6)(n-6)}$.

Rewrite $\dfrac{n}{n+6}$ with the LCD: $\dfrac{n}{n+6}\cdot\dfrac{n-6}{n-6}=\dfrac{n(n-6)}{(n+6)(n-6)}$

$$\dfrac{7n-30}{n^2-36}+\dfrac{n}{n+6}=\dfrac{7n-30}{(n+6)(n-6)}+\dfrac{n}{n+6}\qquad\text{Factor the denominator.}$$

$$=\dfrac{7n-30}{(n+6)(n-6)}+\dfrac{n(n-6)}{(n+6)(n-6)}\qquad\text{Write each expression with the LCD.}$$

$$=\dfrac{7n-30+n(n-6)}{(n+6)(n-6)}\qquad\text{Add the numerators.}$$

$$=\dfrac{7n-30+n^2-6n}{(n+6)(n-6)}\qquad\text{Distribute.}$$

$$=\dfrac{n^2+n-30}{(n+6)(n-6)}\qquad\text{Combine like terms.}$$

Ask yourself, *"Can I factor the numerator?"* Yes.

$$=\dfrac{(n+6)(n-5)}{(n+6)(n-6)}\qquad\text{Factor.}$$

$$=\dfrac{n-5}{n-6}\qquad\text{Write in lowest terms.}$$

[YOU TRY 3] Add or subtract.

a) $\dfrac{t+4}{5}+\dfrac{2t-7}{15}$ b) $\dfrac{7}{12v}-\dfrac{9}{16v^2}$ c) $\dfrac{15h-8}{h^2-64}+\dfrac{h}{h+8}$

EXAMPLE 4

Subtract $\dfrac{6r}{r^2 + 10r + 16} - \dfrac{3r + 4}{r^2 + 3r - 40}$.

Solution

Factor the denominators, then write down the LCD.

$$\frac{6r}{r^2 + 10r + 16} = \frac{6r}{(r+8)(r+2)}, \qquad \frac{3r+4}{r^2 + 3r - 40} = \frac{3r+4}{(r+8)(r-5)}$$

LCD$= (r + 8)(r + 2)(r - 5)$

Rewrite each expression with the LCD.

$$\frac{6r}{(r+8)(r+2)} \cdot \frac{r-5}{r-5} = \frac{6r(r-5)}{(r+8)(r+2)(r-5)}$$

$$\frac{3r+4}{(r+8)(r-5)} \cdot \frac{r+2}{r+2} = \frac{(3r+4)(r+2)}{(r+8)(r+2)(r-5)}$$

$$\frac{6r}{r^2 + 10r + 16} - \frac{3r+4}{r^2 + 3r - 40}$$

$$= \frac{6r}{(r+8)(r+2)} - \frac{3r+4}{(r+8)(r-5)} \qquad \text{Factor the denominators.}$$

$$= \frac{6r(r-5)}{(r+8)(r+2)(r-5)} - \frac{(3r+4)(r+2)}{(r+8)(r+2)(r-5)} \qquad \text{Write each expression with the LCD.}$$

$$= \frac{6r(r-5) - (3r+4)(r+2)}{(r+8)(r+2)(r-5)} \qquad \text{Subtract the numerators.}$$

$$= \frac{6r^2 - 30r - (3r^2 + 10r + 8)}{(r+8)(r+2)(r-5)} \qquad \text{Distribute. You } must \text{ use parentheses.}$$

$$= \frac{6r^2 - 30r - 3r^2 - 10r - 8}{(r+8)(r+2)(r-5)} \qquad \text{Distribute.}$$

$$= \frac{3r^2 - 40r - 8}{(r+8)(r+2)(r-5)} \qquad \text{Combine like terms.}$$

Ask yourself, *"Can I factor the numerator?"* No. The expression is in simplest form since the numerator and denominator have no common factors.

BE CAREFUL

In Example 4, when you move from

$$\frac{6r(r-5) - (3r+4)(r+2)}{(r+8)(r+2)(r-5)} \quad \text{to} \quad \frac{6r^2 - 30r - (3r^2 + 10r + 8)}{(r+8)(r+2)(r-5)}$$

you *must* use parentheses because the entire quantity $3r^2 + 10r + 8$ is being subtracted from $6r^2 - 30r$.

[YOU TRY 4]

Subtract $\dfrac{4z}{z^2 + 10z + 21} - \dfrac{3z + 5}{z^2 - z - 12}$.

3 Add and Subtract Rational Expressions with Special Denominators

EXAMPLE 5

Add or subtract.

a) $\dfrac{z}{z-9} - \dfrac{8}{9-z}$

b) $\dfrac{4}{7-w} + \dfrac{10}{w^2-49}$

Solution

a) Recall that $a - b = -(b - a)$. The least common denominator of $\dfrac{z}{z-9}$ and $\dfrac{8}{9-z}$ is $z - 9$ or $9 - z$. We will use **LCD $= z - 9$.**

Rewrite $\dfrac{8}{9-z}$ with the LCD:

W Hint

What did you learn in Sections 8.1 and 7.4?

$$\dfrac{8}{9-z} = \dfrac{8}{-(z-9)} = -\dfrac{8}{z-9}$$

$$\dfrac{z}{z-9} - \dfrac{8}{9-z} = \dfrac{z}{z-9} - \left(-\dfrac{8}{z-9}\right) \qquad \text{Write each expression with the LCD.}$$

$$= \dfrac{z}{z-9} + \dfrac{8}{z-9} \qquad \text{Distribute.}$$

$$= \dfrac{z+8}{z-9} \qquad \text{Add the numerators.}$$

b) Factor the denominator of $\dfrac{10}{w^2-49}$: $\quad \dfrac{10}{w^2-49} = \dfrac{10}{(w+7)(w-7)}$

Rewrite $\dfrac{4}{7-w}$ with a denominator of $w - 7$:

$$\dfrac{4}{7-w} = \dfrac{4}{-(w-7)} = -\dfrac{4}{w-7}$$

Now we must find the LCD of $\dfrac{10}{(w+7)(w-7)}$ and $-\dfrac{4}{w-7}$.

$$\textbf{LCD} = (w+7)(w-7)$$

Rewrite $-\dfrac{4}{w-7}$ with the LCD.

$$-\dfrac{4}{w-7} \cdot \dfrac{w+7}{w+7} = -\dfrac{4(w+7)}{(w+7)(w-7)} = \dfrac{-4(w+7)}{(w+7)(w-7)}$$

$$\dfrac{4}{7-w} + \dfrac{10}{w^2-49} = -\dfrac{4}{w-7} + \dfrac{10}{(w+7)(w-7)}$$

$$= \dfrac{-4(w+7)}{(w+7)(w-7)} + \dfrac{10}{(w+7)(w-7)} \qquad \text{Write each expression with the LCD.}$$

$$= \dfrac{-4(w+7)+10}{(w+7)(w-7)} \qquad \text{Add the numerators.}$$

$$= \dfrac{-4w-28+10}{(w+7)(w-7)} \qquad \text{Distribute.}$$

$$= \dfrac{-4w-18}{(w+7)(w-7)} \qquad \text{Combine like terms.}$$

Ask yourself, *"Can I factor the numerator?"* Yes.

$$= \frac{-2(2w+9)}{(w+7)(w-7)} \quad \text{Factor.}$$

Although the numerator factors, the numerator and denominator do not contain any common factors. The result, $\dfrac{-2(2w+9)}{(w+7)(w-7)}$, is in simplest form.

[YOU TRY 5] Add or subtract.

a) $\dfrac{n}{n-12} - \dfrac{7}{12-n}$ b) $\dfrac{15}{4-t} + \dfrac{20}{t^2-16}$

W Hint

Did you check your answers? Were you able to figure out where you might have made any mistakes?

ANSWERS TO [YOU TRY] EXERCISES

1) a) $\dfrac{1}{3}$ b) $\dfrac{9h+8}{5h-2}$ 2) a) $\dfrac{7}{8w}$ b) $\dfrac{m-9}{m}$ 3) a) $\dfrac{t+1}{3}$ b) $\dfrac{28v-27}{48v^2}$

c) $\dfrac{h-1}{h-8}$ 4) $\dfrac{z^2-42z-35}{(z+7)(z+3)(z-4)}$ 5) a) $\dfrac{n+7}{n-12}$ b) $\dfrac{-5(3t+8)}{(t+4)(t-4)}$

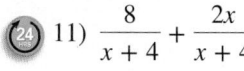 **8.4** Exercises   Do the exercises, and check your work.

Objective 1: Add and Subtract Rational Expressions with a Common Denominator

Add or subtract.

1) $\dfrac{5}{16} + \dfrac{9}{16}$

2) $\dfrac{5}{7} - \dfrac{3}{7}$

3) $\dfrac{11}{14} - \dfrac{3}{14}$

4) $\dfrac{1}{10} + \dfrac{9}{10}$

5) $\dfrac{5}{p} - \dfrac{23}{p}$

6) $\dfrac{6}{a} + \dfrac{3}{a}$

7) $\dfrac{7}{3c} + \dfrac{8}{3c}$

8) $\dfrac{10}{3k^2} - \dfrac{2}{3k^2}$

9) $\dfrac{6}{z-1} + \dfrac{z}{z-1}$

10) $\dfrac{4n}{n+9} - \dfrac{6}{n+9}$

11) $\dfrac{8}{x+4} + \dfrac{2x}{x+4}$

12) $\dfrac{5m}{m+7} + \dfrac{35}{m+7}$

13) $\dfrac{25t+17}{t(4t+3)} - \dfrac{5t+2}{t(4t+3)}$

14) $\dfrac{9w-20}{w(2w-5)} - \dfrac{20-7w}{w(2w-5)}$

15) $\dfrac{d^2+15}{(d+5)(d+2)} + \dfrac{8d-3}{(d+5)(d+2)}$

16) $\dfrac{2r+15}{(r-5)(r+4)} + \dfrac{r^2-10r}{(r-5)(r+4)}$

Objective 2: Add and Subtract Rational Expressions with Different Denominators

17) For $\dfrac{4}{9b^2}$ and $\dfrac{5}{6b^4}$,

a) find the LCD.

b) explain, in your own words, how to rewrite each expression with the LCD.

c) rewrite each expression with the LCD.

18) For $\dfrac{8}{x-3}$ and $\dfrac{2}{x}$,

a) find the LCD.

b) explain, in your own words, how to rewrite each expression with the LCD.

c) rewrite each expression with the LCD.

Add or subtract.

19) $\dfrac{3}{8} + \dfrac{2}{5}$

20) $\dfrac{5}{12} - \dfrac{1}{8}$

21) $\dfrac{4t}{3} + \dfrac{3}{2}$

22) $\dfrac{14x}{15} - \dfrac{5x}{6}$

23) $\dfrac{10}{3h^3} + \dfrac{2}{5h}$

24) $\dfrac{5}{8u} - \dfrac{2}{3u^2}$

25) $\dfrac{3}{2f^2} - \dfrac{7}{f}$

26) $\dfrac{8}{5a} + \dfrac{2}{5a^2}$

27) $\dfrac{13}{y+3} + \dfrac{3}{y}$

28) $\dfrac{3}{k} + \dfrac{11}{k+9}$

29) $\dfrac{15}{d-8} - \dfrac{4}{d}$

30) $\dfrac{14}{r-5} - \dfrac{3}{r}$

31) $\dfrac{9}{c-4} + \dfrac{6}{c+8}$

32) $\dfrac{2}{z+5} + \dfrac{1}{z+2}$

33) $\dfrac{m}{3m+5} - \dfrac{2}{m-10}$

34) $\dfrac{x}{x+4} - \dfrac{3}{2x+1}$

35) $\dfrac{8u+2}{u^2-1} + \dfrac{3u}{u+1}$

36) $\dfrac{t}{t+7} + \dfrac{9t-35}{t^2-49}$

37) $\dfrac{7g}{g^2+10g+16} + \dfrac{3}{g^2-64}$

38) $\dfrac{b}{b^2-25} + \dfrac{8}{b^2-3b-40}$

39) $\dfrac{5a}{a^2-6a-27} - \dfrac{2a+1}{a^2+2a-3}$

40) $\dfrac{3c}{c^2+4c-12} - \dfrac{2c-5}{c^2+2c-24}$

41) $\dfrac{2x}{x^2+x-20} - \dfrac{4}{x^2+2x-15}$

42) $\dfrac{3m}{m^2+10m+24} - \dfrac{2}{m^2+3m-4}$

43) $\dfrac{4b+1}{3b-12} + \dfrac{5b}{b^2-b-12}$

44) $\dfrac{k+12}{2k-18} + \dfrac{3k}{k^2-12k+27}$

45) $\dfrac{n-4}{8n^3+27} - \dfrac{6}{4n^2-9}$

46) $\dfrac{2}{9w^2-16} - \dfrac{w+5}{27w^3-64}$

Objective 3: Add and Subtract Rational Expressions with Special Denominators

47) Is $(x-6)(6-x)$ the LCD for $\dfrac{9}{x-6} + \dfrac{4}{6-x}$?

Why or why not?

48) When Lamar adds $\dfrac{u}{7-2u} + \dfrac{5}{2u-7}$, he gets $\dfrac{u-5}{7-2u}$, but when he checks his answer in the back of the book, it says that the answer is $\dfrac{5-u}{2u-7}$. Which is the correct answer?

Add or subtract.

49) $\dfrac{16}{q-4} + \dfrac{10}{4-q}$

50) $\dfrac{8}{z-9} + \dfrac{4}{9-z}$

51) $\dfrac{11}{f-7} - \dfrac{15}{7-f}$

52) $\dfrac{5}{a-b} - \dfrac{4}{b-a}$

53) $\dfrac{7}{x-4} + \dfrac{x-1}{4-x}$

54) $\dfrac{10}{m-5} + \dfrac{m+21}{5-m}$

55) $\dfrac{8}{3-a} + \dfrac{a+5}{a-3}$

56) $\dfrac{9}{6-n} + \dfrac{n+3}{n-6}$

57) $\dfrac{3}{2u-3v} - \dfrac{6u}{3v-2u}$

58) $\dfrac{3c}{11b-5c} - \dfrac{9}{5c-11b}$

59) $\dfrac{8}{x^2-9} + \dfrac{2}{3-x}$

60) $\dfrac{4}{8-y} + \dfrac{12}{y^2-64}$

61) $\dfrac{a}{4a^2-9} - \dfrac{4}{3-2a}$

62) $\dfrac{3b}{9b^2-25} - \dfrac{3}{5-3b}$

63) $\dfrac{5}{1-c} - \dfrac{c^2}{c^3-1}$

64) $\dfrac{7}{2-k} - \dfrac{6k}{k^3-8}$

Mixed Exercises: Objectives 2 and 3
Perform the indicated operations.

65) $\dfrac{5}{a^2-2a} + \dfrac{8}{a} - \dfrac{10a}{a-2}$

66) $\dfrac{3}{j^2+6j} + \dfrac{2j}{j+6} - \dfrac{2}{3j}$

67) $\dfrac{c}{c^2-8c+16} - \dfrac{5}{c^2-c-12}$

68) $\dfrac{n}{n^2+11n+30} - \dfrac{6}{n^2+10n+25}$

69) $\dfrac{9}{4a+4b} + \dfrac{8}{a-b} - \dfrac{6a}{a^2-b^2}$

70) $\dfrac{1}{x+y} + \dfrac{x}{x^2-y^2} - \dfrac{10}{5x-5y}$

71) $\dfrac{2v+1}{6v^2-29v-5} - \dfrac{v-2}{3v^2-13v-10}$

72) $\dfrac{n+2}{4n^2+11n-3} - \dfrac{n-3}{2n^2+7n+3}$

73) $\dfrac{g-5}{5g^2-30g} + \dfrac{g}{2g^2-17g+30} - \dfrac{6}{2g^2-5g}$

74) $\dfrac{y+6}{y^2-4y} + \dfrac{y}{2y^2-13y+20} - \dfrac{1}{2y^2-5y}$

75) $\dfrac{7}{8k^3+1} + \dfrac{k^2}{4k^2-2k+1}$

76) $\dfrac{4}{3w+2} + \dfrac{w}{27w^3+8}$

77) $\dfrac{3b-1}{b^2+8b} + \dfrac{b}{3b^2+25b+8} + \dfrac{2}{3b^2+b}$

78) $\dfrac{2k+7}{k^2-4k} + \dfrac{9k}{2k^2-15k+28} + \dfrac{15}{2k^2-7k}$

79) $\dfrac{a}{a^2-b^2} + \dfrac{b}{a^3-b^3}$

80) $\dfrac{r}{r^3+t^3} + \dfrac{t}{r^2-t^2}$

For each rectangle, find a rational expression in simplest form to represent its a) area and b) perimeter.

81)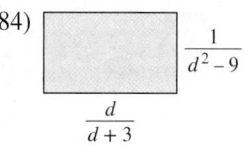

$\dfrac{k-4}{4}$

$\dfrac{8}{k+1}$

82)

$\dfrac{4}{r-3}$

$\dfrac{r+1}{6}$

83)

$\dfrac{6}{h^2+9h+20}$

$\dfrac{h}{h+5}$

84)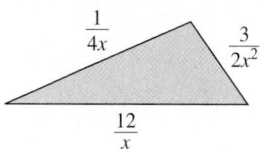

$\dfrac{1}{d^2-9}$

$\dfrac{d}{d+3}$

85) Find a rational expression in simplest form to represent the perimeter of the triangle.

$\dfrac{1}{4x}$ $\dfrac{3}{2x^2}$

$\dfrac{12}{x}$

86) The total resistance of a set of resistors in parallel in an electrical circuit can be found using the formula $\dfrac{1}{R_T} = \dfrac{1}{R_1} + \dfrac{1}{R_2}$, where R_1 = the resistance in resistor 1, R_2 = the resistance in resistor 2, and R_T = the total resistance in the circuit. (Resistance is measured in ohms.) For the given circuit,

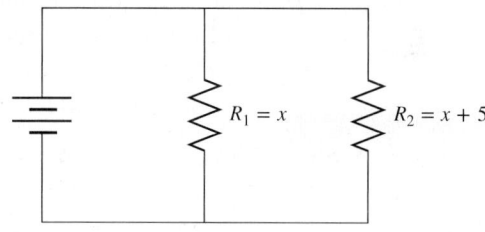

$R_1 = x$ $R_2 = x + 5$

a) find the sum $\dfrac{1}{x} + \dfrac{1}{x+5}$.

b) find an expression for the total resistance, R_T, in the circuit.

c) if $R_1 = 10$ ohms, what is the total resistance in the circuit?

R Rethink

R1) For which problems could you use more practice?

R2) How often did you encounter a solution that could be simplified? What were some clues that alerted you that you may need to simplify?

Putting It All Together

P Prepare	**O Organize**
What is your objective?	**How can you accomplish the objective?**
1 Review the Concepts Presented in Sections 8.1–8.4	• Be sure that you can apply the techniques you have learned in the previous sections. • Understand when a common denominator is needed and when it is not. • Complete the given examples on your own. • Complete You Try 1.

W Work   **Read the explanations, follow the examples, take notes, and complete the You Trys.**

In Section 8.1, we defined a rational expression, evaluated expressions, and discussed how to write a rational expression in lowest terms. We also learned that a rational expression *equals zero* when its *numerator equals zero*, and a rational expression is *undefined* when its *denominator equals zero*.

EXAMPLE 1   Determine the values of c for which $\dfrac{c+8}{c^2-25}$

a)   equals zero.   b)   is undefined.

Solution

a) $\dfrac{c+8}{c^2-25}$ equals zero when its *numerator* equals zero.

Let $c + 8 = 0$, and solve for c.

$$c + 8 = 0$$
$$c = -8$$

$\dfrac{c+8}{c^2-25}$ equals zero when $c = -8$.

> **W Hint**
> Be certain to follow all of the examples! The exercises will be a mixture of problems.

b) $\dfrac{c+8}{c^2-25}$ is undefined when its *denominator* equals zero.

Let $c^2 - 25 = 0$, and solve for c.

$$c^2 - 25 = 0$$
$$(c + 5)(c - 5) = 0 \qquad \text{Factor.}$$
$$c + 5 = 0 \quad \text{or} \quad c - 5 = 0 \qquad \text{Set each factor equal to zero.}$$
$$c = -5 \quad \text{or} \qquad c = 5 \qquad \text{Solve.}$$

$\dfrac{c+8}{c^2-25}$ is undefined when $c = 5$ or $c = -5$. So, $c \neq 5$ and $c \neq -5$ in the expression.

In Sections 8.2–8.4, we learned how to multiply, divide, add, and subtract rational expressions. Now we will practice these operations together so that we will learn to recognize the techniques needed to perform these operations.

562   CHAPTER 8   Rational Expressions

EXAMPLE 2

Divide $\dfrac{t^2 - 3t - 28}{16t^2 - 81} \div \dfrac{t^2 - 7t}{54 - 24t}$.

Solution

Do we need a common denominator to divide? *No.* A common denominator is needed to add or subtract but not to multiply or divide.

To divide, multiply the first rational expression by the reciprocal of the second expression, then factor, divide out common factors, and multiply.

$$\dfrac{t^2 - 3t - 28}{16t^2 - 81} \div \dfrac{t^2 - 7t}{54 - 24t} = \dfrac{t^2 - 3t - 28}{16t^2 - 81} \cdot \dfrac{54 - 24t}{t^2 - 7t} \qquad \text{Multiply by the reciprocal.}$$

$$= \dfrac{(t + 4)\cancel{(t - 7)}}{(4t + 9)\cancel{(4t - 9)}} \cdot \dfrac{6\overset{-1}{\cancel{(9 - 4t)}}}{t\cancel{(t - 7)}} \qquad \text{Factor, and divide out common factors.}$$

$$= -\dfrac{6(t + 4)}{t(4t + 9)} \qquad \text{Multiply.}$$

Recall that $\dfrac{9 - 4t}{4t - 9} = -1.$

EXAMPLE 3

Add $\dfrac{x}{x + 2} + \dfrac{4}{3x - 1}$.

Solution

To add or subtract rational expressions, we need a common denominator.

$$\mathbf{LCD = (x + 2)(3x - 1)}$$

Rewrite each expression with the LCD.

$$\dfrac{x}{x + 2} \cdot \dfrac{3x - 1}{3x - 1} = \dfrac{x(3x - 1)}{(x + 2)(3x - 1)}, \qquad \dfrac{4}{3x - 1} \cdot \dfrac{x + 2}{x + 2} = \dfrac{4(x + 2)}{(x + 2)(3x - 1)}$$

$$\dfrac{x}{x + 2} + \dfrac{4}{3x - 1} = \dfrac{x(3x - 1)}{(x + 2)(3x - 1)} + \dfrac{4(x + 2)}{(x + 2)(3x - 1)} \qquad \begin{array}{l}\text{Write each expression}\\\text{with the LCD.}\end{array}$$

$$= \dfrac{x(3x - 1) + 4(x + 2)}{(x + 2)(3x - 1)} \qquad \text{Add the numerators.}$$

$$= \dfrac{3x^2 - x + 4x + 8}{(x + 2)(3x - 1)} \qquad \text{Distribute.}$$

$$= \dfrac{3x^2 + 3x + 8}{(x + 2)(3x - 1)} \qquad \text{Combine like terms.}$$

Although this numerator will not factor, remember that sometimes it *is* possible to factor the numerator and simplify the result.

Putting It All Together Exercises

E Evaluate   Do the exercises, and check your work.

Objective 1: Review the Concepts Presented in Sections 8.1–8.4

Evaluate, if possible, for a) $x = -3$ and b) $x = 2$.

1) $\dfrac{x + 3}{3x + 4}$

2) $\dfrac{x}{x - 2}$

3) $\dfrac{5x - 3}{x^2 + 10x + 21}$

4) $-\dfrac{x^2}{x^2 - 12}$

Determine the values of the variable for which

  a) the expression is undefined.

  b) the expression equals zero.

5) $-\dfrac{5w}{w^2 - 36}$

6) $\dfrac{m - 4}{2m^2 + 11m + 15}$

7) $\dfrac{3 - 5b}{b^2 + 2b - 8}$

8) $\dfrac{5k - 8}{64 - k^2}$

9) $\dfrac{12}{5r}$

10) $\dfrac{t - 15}{t^2 + 4}$

Write each rational expression in lowest terms.

11) $\dfrac{12w^{16}}{3w^5}$

12) $\dfrac{42n^3}{18n^8}$

13) $\dfrac{m^2 + 6m - 27}{2m^2 + 2m - 24}$

14) $\dfrac{2j + 20}{2j^2 + 10j - 100}$

15) $\dfrac{12 - 15n}{5n^2 + 6n - 8}$

16) $\dfrac{-x - y}{xy + y^2 + 5x + 5y}$

17) $\dfrac{125t^3 + 8}{5t + 2}$

18) $\dfrac{4a^3 - 4b^3}{2a^2 + 4ab - 6b^2}$

Perform the operations, and simplify.

19) $\dfrac{4c^2 + 4c - 24}{c + 3} \div \dfrac{3c - 6}{8}$

20) $\dfrac{6}{f + 11} - \dfrac{2}{f}$

21) $\dfrac{4j}{j^2 - 81} + \dfrac{3}{j^2 - 3j - 54}$

22) $\dfrac{27a^4}{8b} \cdot \dfrac{40b^2}{81a^2}$

23) $\dfrac{12y^7}{4z^6} \cdot \dfrac{8z^4}{72y^6}$

24) $\dfrac{3}{q^2 - q - 20} + \dfrac{8q}{q^2 + 11q + 28}$

25) $\dfrac{x}{2x^2 - 7x - 4} - \dfrac{x + 3}{4x^2 + 4x + 1}$

26) $\dfrac{n - 4}{4n - 44} \cdot \dfrac{121 - n^2}{n + 11}$

27) $\dfrac{16 - m^2}{m + 4} \div \dfrac{8m - 32}{m + 7}$

28) $\dfrac{16}{r - 7} + \dfrac{4}{7 - r}$

29) $\dfrac{3xy - 24x - 5y + 40}{y^2 - 64} \div \dfrac{27x^3 - 125}{9x}$

30) $\dfrac{\dfrac{10d}{d + 11}}{\dfrac{5d^7}{3d + 33}}$

31) $\dfrac{9}{d+3} + \dfrac{8}{d^2}$

32) $\dfrac{3a^2 - 6a + 12}{5a - 10} \cdot \dfrac{a^2 - 4}{a^3 + 8}$

33) $\dfrac{\dfrac{9k^2 - 1}{14k}}{\dfrac{3k - 1}{21k^4}}$

34) $\dfrac{13}{4z} - \dfrac{1}{3z}$

35) $\dfrac{2w}{25 - w^2} + \dfrac{w - 3}{w^2 - 12w + 35}$

36) $\dfrac{12a^4}{10a - 30} \div \dfrac{4a}{a^3 - 3a^2 + 5a - 15}$

37) $\dfrac{10}{x - 8} + \dfrac{4}{x + 3}$

38) $\dfrac{1}{4y} + \dfrac{8}{6y^4}$

39) $\dfrac{2h^2 + 11h + 5}{8} \div (2h + 1)^2$

40) $\dfrac{b^2 - 15b + 36}{b^2 - 8b - 48} \cdot (b + 4)^2$

41) $\dfrac{3m}{7m - 4n} - \dfrac{20n}{4n - 7m}$

42) $\dfrac{10d}{8c - 10d} + \dfrac{8c}{10d - 8c}$

43) $\dfrac{2p + 3}{p^2 + 7p} - \dfrac{4p}{p^2 - p - 56} + \dfrac{5}{p^2 - 8p}$

44) $\dfrac{6u + 1}{3u^2 - 2u} - \dfrac{u}{3u^2 + u - 2} + \dfrac{10}{u^2 + u}$

45) $\dfrac{6t + 6}{3t^2 - 24t} \cdot (t^2 - 7t - 8)$

46) $\dfrac{3r^2 + r - 14}{5r^3 - 10r^2} \div (9r^2 - 49)$

47) $\dfrac{b^2 - 10a^2}{125a^3 + 8b^3} + \dfrac{3}{20a + 8b}$

48) $\dfrac{t - 8}{27t^3 - 1} - \dfrac{2}{1 - 9t^2}$

49) $\dfrac{\dfrac{3c^3}{8c + 40}}{\dfrac{9c}{c + 5}}$

50) $\dfrac{\dfrac{6v - 30}{4}}{\dfrac{v - 5}{3}}$

51) $\dfrac{f - 8}{f - 4} - \dfrac{4}{4 - f}$

52) $\dfrac{12p}{4p^2 + 11p + 6} - \dfrac{5}{p^2 - 4p - 12}$

53) $\dfrac{5x^3}{9y^8} \div \dfrac{20x^4}{3y}$

54) $\dfrac{4r}{t^5} \div \dfrac{36r^5}{t^2}$

55) $\left(\dfrac{3m}{3m - 1} - \dfrac{4}{m + 4} \right) \cdot \dfrac{9m^2 - 1}{21m^2 + 28}$

56) $\left(\dfrac{2c}{c + 8} + \dfrac{4}{c - 2} \right) \div \dfrac{6}{4c + 32}$

57) $\dfrac{3}{k^2 + 3k} - \dfrac{4}{3k} + \dfrac{1}{k + 3}$

58) $\dfrac{3}{w^2 - w} + \dfrac{4}{5w} - \dfrac{3}{w - 1}$

59) Find a rational expression in simplest form to represent the a) area and b) perimeter of the rectangle.

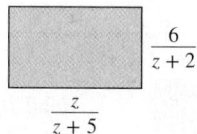

60) Find a rational expression in simplest form to represent the perimeter of the triangle.

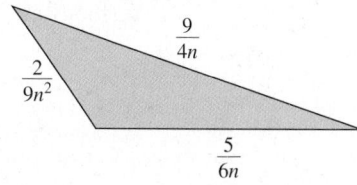

R **Rethink**

R1) When you need to factor, does it feel comfortable?

R2) Discuss the importance of mastering each concept in Sections 8.1–8.4, in order.

8.5 Simplifying Complex Fractions

|

What are your objectives for Section 8.5?	How can you accomplish each objective?
1 Simplify a Complex Fraction with One Term in the Numerator and One Term in the Denominator	• Learn the definition of a *complex fraction*, and know how to identify a complex fraction. • Follow Example 1 to help develop a procedure you will use to simplify. • Write the procedure to **Simplify a Complex Fraction with One Term in the Numerator and One Term in the Denominator** in your own words. • Complete You Try 1.
2 Simplify a Complex Fraction with More Than One Term in the Numerator and/or Denominator by Rewriting It as a Division Problem	• Recognize that there are two ways to simplify these types of expressions. • Write the procedure to **Simplify a Complex Fraction Using Method 1** in your own words. • Complete the given example on your own. • Complete You Try 2.
3 Simplify a Complex Fraction with More Than One Term in the Numerator and/or Denominator by Multiplying by the LCD	• Write the procedure to **Simplify a Complex Fraction Using Method 2** in your own words. • Realize that when given a choice between using Method 1 and Method 2, you want to choose the method that will make the expression "easier" to simplify. • Complete the given examples on your own. • Complete You Trys 3 and 4.
4 Simplify Rational Expressions Containing Negative Exponents	• Use the rules for making negative exponents positive. • Be able to determine which method for **Simplifying Complex Fractions** should be used. • Complete the given example on your own. • Complete You Try 5.

W Work   **Read the explanations, follow the examples, take notes, and complete the You Trys.**

In algebra, we sometimes encounter fractions that contain fractions in their numerators, denominators, or both. These are called *complex fractions*. Some examples of complex fractions are

$$\frac{\frac{3}{7}}{\frac{9}{2}}, \quad \frac{\frac{1}{8}+\frac{5}{6}}{2-\frac{2}{3}}, \quad \frac{\frac{4}{xy^2}}{\frac{1}{x}-\frac{1}{y}}, \quad \frac{\frac{5a-15}{4}}{\frac{a-3}{a}}$$

Definition

A **complex fraction** is a rational expression that contains one or more fractions in its numerator, its denominator, or both.

A complex fraction is not considered to be an expression in simplest form. In this section, we will learn how to simplify two different types of complex fractions:

1) complex fractions with *one term* in the numerator and *one term* in the denominator

2) complex fractions that have *more than one term* in their numerators, their denominators, or both

1 Simplify a Complex Fraction with One Term in the Numerator and One Term in the Denominator

We studied these expressions in Section 8.2 when we learned how to divide fractions. We will look at another example.

EXAMPLE 1

Simplify $\dfrac{\dfrac{5a-15}{4}}{\dfrac{a-3}{a}}$

⟵ This is the numerator.
⟵ This is the main fraction bar.
⟵ This is the denominator.

Solution

This complex fraction contains one term in the numerator and one term in the denominator. To simplify, rewrite as a division problem, multiply by the reciprocal, and simplify.

$$\frac{\dfrac{5a-15}{4}}{\dfrac{a-3}{a}} = \frac{5a-15}{4} \div \frac{a-3}{a} \qquad \text{Rewrite as a division problem.}$$

$$= \frac{5a-15}{4} \cdot \frac{a}{a-3} = \frac{5(a-3)}{4} \cdot \frac{a}{a-3} = \frac{5(a-3)}{4} \cdot \frac{a}{a-3} = \frac{5a}{4}$$

[YOU TRY 1]

Simplify $\dfrac{\dfrac{9}{z^2-64}}{\dfrac{3z+3}{z+8}}$.

Let's summarize how to simplify this first type of complex fraction.

Procedure Simplify a Complex Fraction Containing One Term in the Numerator and One Term in the Denominator

To simplify a complex fraction containing one term in the numerator and one term in the denominator:

1) Rewrite the complex fraction as a division problem.

2) Perform the division by multiplying the first fraction by the reciprocal of the second (that is, multiply the numerator of the complex fraction by the reciprocal of the denominator).

2 Simplify a Complex Fraction with More Than One Term in the Numerator and/or Denominator by Rewriting It as a Division Problem

When a complex fraction has more than one term in the numerator and/or the denominator, we can use one of two methods to simplify.

Procedure Simplify a Complex Fraction Using Method 1

1) Combine the terms in the numerator and combine the terms in the denominator so that each contains only one fraction.

2) Rewrite as a division problem.

3) Perform the division by multiplying the first fraction by the reciprocal of the second.

EXAMPLE 2

Simplify.

a) $\dfrac{\dfrac{1}{4} + \dfrac{2}{3}}{2 - \dfrac{1}{2}}$ b) $\dfrac{\dfrac{5}{a^2 b}}{\dfrac{a}{b} + \dfrac{1}{a}}$

Solution

a) The numerator, $\dfrac{1}{4} + \dfrac{2}{3}$, contains two terms; the denominator, $2 - \dfrac{1}{2}$, contains two terms. We will add the terms in the numerator and subtract the terms in the denominator so that the numerator and denominator will each contain one fraction.

$$\frac{\dfrac{1}{4} + \dfrac{2}{3}}{2 - \dfrac{1}{2}} = \frac{\dfrac{3}{12} + \dfrac{8}{12}}{\dfrac{4}{2} - \dfrac{1}{2}} = \frac{\dfrac{11}{12}}{\dfrac{3}{2}}$$

Add the fractions in the numerator.

Subtract the fractions in the denominator.

W Hint

Compare the solutions to parts a) and b). Did it help you understand Method 1 better?

Rewrite as a division problem, multiply by the reciprocal, and simplify.

$$\frac{11}{12} \div \frac{3}{2} = \frac{11}{\underset{6}{\cancel{12}}} \cdot \frac{\overset{1}{\cancel{2}}}{3} = \frac{11}{18}$$

b) The numerator, $\dfrac{5}{a^2b}$, contains one term; the denominator, $\dfrac{a}{b} + \dfrac{1}{a}$, contains two terms. We will add the terms in the denominator so that it, like the numerator, will contain only one term. The LCD of the expressions in the denominator is ab.

$$\dfrac{\dfrac{5}{a^2b}}{\dfrac{a}{b} + \dfrac{1}{a}} = \dfrac{\dfrac{5}{a^2b}}{\dfrac{a}{b} \cdot \dfrac{a}{a} + \dfrac{1}{a} \cdot \dfrac{b}{b}} = \dfrac{\dfrac{5}{a^2b}}{\dfrac{a^2}{ab} + \dfrac{b}{ab}} = \dfrac{\dfrac{5}{a^2b}}{\dfrac{a^2 + b}{ab}}$$

Rewrite as a division problem, multiply by the reciprocal, and simplify.

$$\dfrac{5}{a^2b} \div \dfrac{a^2 + b}{ab} = \dfrac{5}{a^2b} \cdot \dfrac{ab}{a^2 + b} = \dfrac{5}{a(a^2 + b)}$$

[YOU TRY 2]

Simplify.

a) $\dfrac{\dfrac{5}{8} - \dfrac{1}{2}}{1 + \dfrac{1}{4}}$ 　　　 b) $\dfrac{\dfrac{6}{r^2t^2}}{\dfrac{2}{r} - \dfrac{r}{t}}$

---

3 Simplify a Complex Fraction with More Than One Term in the Numerator and/or Denominator by Multiplying by the LCD

Another method we can use to simplify complex fractions involves multiplying the numerator and denominator of the complex fraction by the LCD of *all* of the fractions in the expression.

Procedure Simplify a Complex Fraction Using Method 2

1) Identify and write down the LCD of *all* of the fractions in the complex fraction.
2) Multiply the numerator and denominator of the complex fraction by the LCD.
3) Simplify.

We will simplify the complex fractions we simplified in Example 2 using Method 2.

EXAMPLE 3

Simplify using Method 2.

a) $\dfrac{\dfrac{1}{4} + \dfrac{2}{3}}{2 - \dfrac{1}{2}}$ 　　　 b) $\dfrac{\dfrac{5}{a^2b}}{\dfrac{a}{b} + \dfrac{1}{a}}$

Solution

a) List all of the fractions in the complex fraction: $\dfrac{1}{4}, \dfrac{2}{3}, \dfrac{1}{2}$. Write down their LCD:

LCD = 12.

Multiply the numerator and denominator of the complex fraction by the LCD, 12, then simplify.

$$\dfrac{12\left(\dfrac{1}{4}+\dfrac{2}{3}\right)}{12\left(2-\dfrac{1}{2}\right)}$$

We are multiplying the expression by $\dfrac{12}{12}$, which equals 1.

$$=\dfrac{12\cdot\dfrac{1}{4}+12\cdot\dfrac{2}{3}}{12\cdot2-12\cdot\dfrac{1}{2}}$$ Distribute.

$$=\dfrac{3+8}{24-6}=\dfrac{11}{18}$$ Simplify.

This is the same result we obtained in Example 2 using Method 1.

> **Note**
>
> In the denominator, we multiplied the 2 by 12 even though 2 is not a fraction. Remember, *all* terms, not just the fractions, must be multiplied by the LCD.

b) List all of the fractions in the complex fraction: $\dfrac{5}{a^2b}, \dfrac{a}{b}, \dfrac{1}{a}$. Write down the LCD:

LCD = a^2b.

Multiply the numerator and denominator of the complex fraction by the LCD, a^2b, then simplify.

$$\dfrac{a^2b\left(\dfrac{5}{a^2b}\right)}{a^2b\left(\dfrac{a}{b}+\dfrac{1}{a}\right)}$$

We are multiplying the expression by $\dfrac{a^2b}{a^2b}$, which equals 1.

$$=\dfrac{a^2b\cdot\dfrac{5}{a^2b}}{a^2b\cdot\dfrac{a}{b}+a^2b\cdot\dfrac{1}{a}}$$ Distribute.

$$=\dfrac{5}{a^3+ab}=\dfrac{5}{a(a^2+b)}$$ Simplify.

If the numerator and denominator can be factored, factor them. Sometimes, you can divide by a common factor to simplify.

Notice that the result is the same as what we obtained in Example 2 using Method 1.

Hint

Write out the example as you are reading it. Use different colored pens so that you can see the steps being done more easily.

Simplify using Method 2.

a) $\dfrac{\dfrac{5}{8} - \dfrac{1}{2}}{1 + \dfrac{1}{4}}$
b) $\dfrac{\dfrac{6}{r^2 t^2}}{\dfrac{2}{r} - \dfrac{r}{t}}$

W Hint

Be sure to do this You Try using Method 2 even though it is the same as You Try 2!

You should be familiar with both methods for simplifying complex fractions containing two terms in the numerator or denominator. After a lot of practice, you will be able to decide which method is better for a particular problem.

EXAMPLE 4

Determine which method to use to simplify each complex fraction, then simplify.

a) $\dfrac{\dfrac{4}{x} + \dfrac{1}{x+3}}{\dfrac{2}{x+3} - \dfrac{1}{x}}$
b) $\dfrac{\dfrac{n^2 - 1}{7n + 28}}{\dfrac{6n - 6}{n^2 - 16}}$

Solution

a) This complex fraction contains two terms in the numerator and two terms in the denominator. Let's use Method 2: Multiply the numerator and denominator by the LCD of all of the fractions.

LCD = $x(x + 3)$.

W Hint

Try doing this using Method 1. How "messy" and complicated does it become?

Multiply the numerator and denominator of the complex fraction by the LCD, $x(x + 3)$, then simplify.

$$\dfrac{x(x+3)\left(\dfrac{4}{x} + \dfrac{1}{x+3}\right)}{x(x+3)\left(\dfrac{2}{x+3} - \dfrac{1}{x}\right)} = \dfrac{x(x+3) \cdot \dfrac{4}{x} + x(x+3) \cdot \dfrac{1}{x+3}}{x(x+3) \cdot \dfrac{2}{x+3} - x(x+3) \cdot \dfrac{1}{x}}$$

Multiply the numerator and denominator by $x(x + 3)$ and distribute.

$$= \dfrac{4(x + 3) + x}{2x - (x + 3)}$$

Multiply.

$$= \dfrac{4x + 12 + x}{2x - x - 3} = \dfrac{5x + 12}{x - 3}$$

Distribute and combine like terms.

b) The complex fraction $\dfrac{\dfrac{n^2 - 1}{7n + 28}}{\dfrac{6n - 6}{n^2 - 16}}$ contains one term in the numerator, $\dfrac{n^2 - 1}{7n + 28}$, and

one term in the denominator, $\dfrac{6n - 6}{n^2 - 16}$. To simplify, rewrite it as a division problem,

multiply by the reciprocal, and simplify.

$$\dfrac{\dfrac{n^2 - 1}{7n + 28}}{\dfrac{6n - 6}{n^2 - 16}} = \dfrac{n^2 - 1}{7n + 28} \div \dfrac{6n - 6}{n^2 - 16} \qquad \text{Rewrite as a division problem.}$$

$$= \dfrac{n^2 - 1}{7n + 28} \cdot \dfrac{n^2 - 16}{6n - 6} \qquad \text{Multiply by the reciprocal.}$$

$$= \dfrac{(n + 1)\cancel{(n - 1)}}{7\cancel{(n + 4)}} \cdot \dfrac{\cancel{(n + 4)}(n - 4)}{6\cancel{(n - 1)}} \qquad \text{Factor and divide out common factors.}$$

$$= \dfrac{(n + 1)(n - 4)}{42} \qquad \text{Multiply.}$$

[**YOU TRY 4**] Determine which method to use to simplify each complex fraction, then simplify.

a) $\dfrac{\dfrac{8}{k} - \dfrac{1}{k + 5}}{\dfrac{3}{k + 5} + \dfrac{5}{k}}$ b) $\dfrac{\dfrac{c^2 - 9}{8c - 56}}{\dfrac{2c + 6}{c^2 - 49}}$

---

4 Simplify Rational Expressions Containing Negative Exponents

If a rational expression contains a negative exponent, rewrite it with positive exponents and simplify.

EXAMPLE 5

Simplify $\dfrac{x^{-2} - 3y^{-1}}{1 + x^{-1}}$.

Solution

First, rewrite the expression with positive exponents: $\dfrac{x^{-2} - 3y^{-1}}{1 + x^{-1}} = \dfrac{\dfrac{1}{x^2} - \dfrac{3}{y}}{1 + \dfrac{1}{x}}$

> **W Hint**
> Look at Chapter 2 or Section 6.1 if you need to review the rules of exponents.

Next we have to simplify. Identify the LCD of all the fractions: x^2y

Multiply the numerator and denominator of the complex fraction by x^2y, and simplify.

$$\frac{x^2y\left(\dfrac{1}{x^2} - \dfrac{3}{y}\right)}{x^2y\left(1 + \dfrac{1}{x}\right)} = \frac{x^2y \cdot \dfrac{1}{x^2} - x^2y \cdot \dfrac{3}{y}}{x^2y \cdot 1 + x^2y \cdot \dfrac{1}{x}}$$ Multiply and distribute.

$$= \frac{y - 3x^2}{x^2y + xy}$$ Multiply.

$$= \frac{y - 3x^2}{xy(x + 1)}$$ Factor.

[YOU TRY 5]

Simplify $\dfrac{4a^{-2} + b^{-1}}{a^{-3} + b}$.

ANSWERS TO [YOU TRY] EXERCISES

1) $\dfrac{3}{(z - 8)(z + 1)}$ 2) a) $\dfrac{1}{10}$ b) $\dfrac{6}{rt(2t - r^2)}$ 3) a) $\dfrac{1}{10}$ b) $\dfrac{6}{rt(2t - r^2)}$

4) a) $\dfrac{7k + 40}{8k + 25}$ b) $\dfrac{(c - 3)(c + 7)}{16}$ 5) $\dfrac{a(4b + a^2)}{b(1 + a^3b)}$

E Evaluate **8.5** Exercises   Do the exercises, and check your work.

1) Explain, in your own words, two ways to simplify $\dfrac{\dfrac{2}{9}}{\dfrac{5}{18}}$.

Then, simplify it both ways. Which method do you prefer and why?

2) Explain, in your own words, two ways to simplify $\dfrac{\dfrac{3}{2} - \dfrac{1}{5}}{\dfrac{1}{10} + \dfrac{3}{5}}$. Then, simplify it both ways. Which method do you prefer and why?

Objective 1: Simplify a Complex Fraction with One Term in the Numerator and One Term in the Denominator

Simplify completely.

3) $\dfrac{\dfrac{7}{10}}{\dfrac{5}{4}}$

4) $\dfrac{\dfrac{3}{8}}{\dfrac{4}{3}}$

 5) $\dfrac{\dfrac{a^2}{b}}{\dfrac{a}{b^3}}$

6) $\dfrac{\dfrac{u^5}{v^2}}{\dfrac{u^2}{v}}$

7) $\dfrac{\dfrac{s^3}{t^3}}{\dfrac{s^4}{t}}$

8) $\dfrac{\dfrac{x^4}{y}}{\dfrac{x^2}{y^3}}$

9) $\dfrac{\dfrac{14m^5n^4}{9}}{\dfrac{35mn^6}{3}}$

10) $\dfrac{\dfrac{11b^4c^2}{4}}{\dfrac{55bc}{8}}$

11) $\dfrac{\dfrac{t - 6}{5}}{\dfrac{t - 6}{t}}$

12) $\dfrac{\dfrac{m - 3}{m}}{\dfrac{m - 3}{16}}$

13) $\dfrac{\dfrac{8}{y^2 - 64}}{\dfrac{6}{y + 8}}$

14) $\dfrac{\dfrac{g^2 - 36}{15}}{\dfrac{g - 6}{45}}$

15) $\dfrac{\dfrac{25w - 35}{w^5}}{\dfrac{30w - 42}{w}}$

16) $\dfrac{\dfrac{d^3}{16d - 24}}{\dfrac{d}{40d - 60}}$

17) $\dfrac{\dfrac{2x}{x+7}}{\dfrac{2}{x^2+4x-21}}$

18) $\dfrac{\dfrac{c^2-7c-8}{6c}}{\dfrac{c-8}{c}}$

19) $\dfrac{\dfrac{a^3-125b^3}{a^3+5a^2b+25ab^2}}{\dfrac{25b^2-a^2}{a}}$

20) $\dfrac{\dfrac{x^2-xy+y^2}{12}}{\dfrac{x^3+y^3}{3}}$

37) $\dfrac{\dfrac{r}{s^2}+\dfrac{1}{rs}}{\dfrac{s}{r}+\dfrac{1}{r^2}}$

38) $\dfrac{\dfrac{n}{m^3}+\dfrac{m}{n}}{\dfrac{3}{n}-\dfrac{m}{n^4}}$

39) $\dfrac{1+\dfrac{4}{t-3}}{\dfrac{t}{t-3}+\dfrac{2}{t^2-9}}$

40) $\dfrac{1-\dfrac{4}{t+5}}{\dfrac{4}{t^2-25}+\dfrac{t}{t-5}}$

Mixed Exercises: Objectives 2 and 3

Simplify using Method 1 then by using Method 2. Think about which method you prefer and why.

21) $\dfrac{\dfrac{1}{4}+\dfrac{3}{2}}{\dfrac{2}{3}+\dfrac{1}{2}}$

22) $\dfrac{\dfrac{7}{9}-\dfrac{1}{3}}{2+\dfrac{1}{9}}$

23) $\dfrac{\dfrac{7}{c}+\dfrac{2}{d}}{1-\dfrac{5}{c}}$

24) $\dfrac{\dfrac{r}{s}-2}{\dfrac{1}{s}+\dfrac{3}{r}}$

25) $\dfrac{\dfrac{5}{z-2}-\dfrac{1}{z+1}}{\dfrac{1}{z-2}+\dfrac{4}{z+1}}$

26) $\dfrac{\dfrac{6}{w+4}+\dfrac{4}{w-1}}{\dfrac{5}{w-1}+\dfrac{3}{w+4}}$

27) In Exercises 21–26, which types of complex fractions did you prefer to simplify using Method 1? Why?

28) In Exercises 21–26, which types of complex fractions did you prefer to simplify using Method 2? Why?

Simplify using either Method 1 or Method 2.

29) $\dfrac{9+\dfrac{5}{y}}{\dfrac{9y+5}{8}}$

30) $\dfrac{4-\dfrac{12}{m}}{\dfrac{4m-12}{9}}$

31) $\dfrac{x-\dfrac{7}{x}}{x-\dfrac{11}{x}}$

32) $\dfrac{\dfrac{4}{c}-c}{3+\dfrac{8}{c}}$

33) $\dfrac{\dfrac{4}{3}+\dfrac{2}{5}}{\dfrac{1}{6}-\dfrac{2}{3}}$

34) $\dfrac{\dfrac{1}{4}-\dfrac{5}{6}}{\dfrac{3}{8}+\dfrac{1}{3}}$

35) $\dfrac{\dfrac{4}{x}-\dfrac{4}{y}}{\dfrac{3}{x^2}-\dfrac{3}{y^2}}$

36) $\dfrac{\dfrac{2}{a}-\dfrac{2}{b}}{\dfrac{1}{a^2}-\dfrac{1}{b^2}}$

Mixed Exercises: Objectives 1–3

Simplify completely.

41) $\dfrac{b+\dfrac{1}{b}}{b-\dfrac{3}{b}}$

42) $\dfrac{\dfrac{z+6}{4}}{\dfrac{z+6}{z}}$

43) $\dfrac{\dfrac{m}{n^2}}{\dfrac{m^4}{n}}$

44) $\dfrac{\dfrac{z^2+1}{5}}{z+\dfrac{1}{z}}$

45) $\dfrac{\dfrac{h^2-1}{4h-12}}{\dfrac{7h+7}{h^2-9}}$

46) $\dfrac{\dfrac{r^2+13r+40}{r^2-6r}}{\dfrac{r^2+2r-48}{3r}}$

47) $\dfrac{\dfrac{6}{x+3}-\dfrac{4}{x-1}}{\dfrac{2}{x-1}+\dfrac{1}{x+2}}$

48) $\dfrac{\dfrac{c^2}{d}+\dfrac{2}{c^2d}}{\dfrac{d}{c}-\dfrac{c}{d}}$

49) $\dfrac{\dfrac{r^2-6}{20}}{r-\dfrac{6}{r}}$

50) $\dfrac{\dfrac{1}{6}}{\dfrac{7}{8}}$

51) $\dfrac{\dfrac{a-4}{12}}{\dfrac{a-4}{a}}$

52) $\dfrac{\dfrac{8}{w}-w}{1+\dfrac{6}{w}}$

53) $\dfrac{\dfrac{5}{6}}{\dfrac{9}{15}}$

54) $\dfrac{\dfrac{5}{h+2}+\dfrac{7}{2h-3}}{\dfrac{1}{h-3}+\dfrac{3}{2h-3}}$

55) $\dfrac{\dfrac{5}{2n+1}+1}{\dfrac{1}{n+3}+\dfrac{2}{2n+1}}$

56) $\dfrac{\dfrac{y^4}{z^3}}{\dfrac{y^6}{z^4}}$

57) $\dfrac{\dfrac{c^2}{16c^3+54}}{\dfrac{2c^2+3c}{16c^2-24c+36}}$

58) $\dfrac{\dfrac{8}{k^2-1}}{\dfrac{4}{k-1}-\dfrac{1}{k}}$

Objective 4: Simplify Rational Expressions Containing Negative Exponents

Simplify.

 59) $\dfrac{w^{-1} - v^{-1}}{2w^{-2} + v^{-1}}$

60) $\dfrac{4p^{-2} + q^{-1}}{p^{-1} + q^{-1}}$

61) $\dfrac{8x^{-2}}{x^{-1} - y^{-2}}$

62) $\dfrac{3d^{-1}}{2c^{-2} - d^{-1}}$

63) $\dfrac{a^{-3} + b^{-2}}{2b^{-2} - 7}$

64) $\dfrac{r^{-2} - t^{-2}}{5 + 7t^{-3}}$

65) $\dfrac{4m^{-1} - n^{-1}}{n^{-1} + m}$

66) $\dfrac{h^{-3} + 9}{k^{-2} - h}$

For Exercises 67 and 68, let $f(x) = \dfrac{1}{x}$.

67) Complete the table of values. As x gets larger, the value of $f(x)$ gets closer to what number?

x	f(x)
1	
2	
3	
10	
100	
1000	

68) Complete the table of values. As x gets smaller, what happens to the value of $f(x)$?

x	f(x)
1	
$\frac{1}{2}$	
$\frac{1}{3}$	
$\frac{1}{10}$	
$\frac{1}{100}$	
$\frac{1}{1000}$	

R Rethink

R1) Can all complex fractions be simplified using either Method 1 or Method 2? Explain.

R2) Can all complex fractions be simplified?

R3) Which method for simplifying complex fractions do you prefer? Why?

R4) Do you need to factor at all when you use either of these methods?

8.6 Solving Rational Equations

What are your objectives for Section 8.6?	How can you accomplish each objective?
1 Differentiate Between Rational Expressions and Rational Equations	• Know the difference between an *expression* and an *equation*. • Complete the given example on your own. • Complete You Try 1.
2 Solve Rational Equations	• Write the procedure for **How to Solve a Rational Equation** in your own words. • Use your procedure to follow the examples, and complete them on your own. • Complete You Trys 2–5.
3 Solve a Proportion	• Be able to recognize a proportion and write the definition of a *proportion* (it should include the word *ratio*). • Complete the given example on your own. • Complete You Try 6.
4 Solve an Equation for a Specific Variable	• Follow the explanation to develop your own procedure to **Solve an Equation for a Specific Variable.** • Complete the given examples on your own. • Complete You Trys 7 and 8.

W Work

Read the explanations, follow the examples, take notes, and complete the You Trys.

A **rational equation** is an equation that contains a rational expression. Some examples of rational equations are

$$\frac{1}{2}a + \frac{7}{10} = \frac{3}{5}a - 4, \qquad \frac{8}{p+7} - \frac{p}{p-10} = 2, \qquad \frac{3n}{n^2 + 10n + 16} + \frac{5}{n+8} = \frac{1}{n+2}$$

1 Differentiate Between Rational Expressions and Rational Equations

In Chapter 2, we solved rational equations like the first one above, and we learned how to add and subtract rational expressions in Section 8.4. Let's summarize the difference between the two because this is often a point of confusion for students.

W Hint

Remember: An equation contains an = sign, but an expression does not.

Summary Expressions Versus Equations

1) *The sum or difference of rational expressions does* not *contain an = sign.* To add or subtract, rewrite each expression with the LCD, and *keep the denominator* while performing the operations.

2) *An equation contains an = sign.* To solve an equation containing rational expressions, *multiply* the equation by the LCD of all fractions to *eliminate* the denominators, then solve.

EXAMPLE 1

Determine whether each is an equation or is a sum or difference of expressions. Then, solve the equation or find the sum or difference.

a) $\dfrac{c-5}{6} + \dfrac{c}{8} = \dfrac{3}{2}$ b) $\dfrac{c-5}{6} + \dfrac{c}{8}$

Solution

a)   This is an *equation* because it contains an = sign. We will *solve* for c using the method we learned in Chapter 3: Eliminate the denominators by multiplying by the LCD of all of the expressions.   **LCD = 24**

$$24\left(\dfrac{c-5}{6} + \dfrac{c}{8}\right) = 24 \cdot \dfrac{3}{2}$$ Multiply by the LCD of 24 to eliminate the denominators.

$$4(c-5) + 3c = 36$$ Distribute and eliminate denominators.

$$4c - 20 + 3c = 36$$ Distribute.

$$7c - 20 = 36$$ Combine like terms.

$$7c = 56$$

$$c = 8$$

Check to verify that the solution set is {8}.

b) $\dfrac{c-5}{6} + \dfrac{c}{8}$ is *not* an equation to be solved because it does *not* contain an = sign.

It is a sum of rational expressions. Rewrite each expression with the LCD, then add, *keeping the denominators* while performing the operations.

LCD = 24

$$\dfrac{(c-5)}{6} \cdot \dfrac{4}{4} = \dfrac{4(c-5)}{24} \qquad \dfrac{c}{8} \cdot \dfrac{3}{3} = \dfrac{3c}{24}$$

$$\dfrac{c-5}{6} + \dfrac{c}{8} = \dfrac{4(c-5)}{24} + \dfrac{3c}{24}$$ Rewrite each expression with a denominator of 24.

$$= \dfrac{4(c-5) + 3c}{24}$$ Add the numerators.

$$= \dfrac{4c - 20 + 3c}{24}$$ Distribute.

$$= \dfrac{7c - 20}{24}$$ Combine like terms.

[YOU TRY 1]   Determine whether each is an equation or is a sum or difference of expressions. Then solve the equation or find the sum or difference.

a) $\dfrac{m+1}{6} - \dfrac{m}{2}$ b) $\dfrac{m+1}{6} - \dfrac{m}{2} = \dfrac{5}{6}$

2 Solve Rational Equations

Let's list the steps we use to solve a rational equation. Then we will look at more examples.

Procedure How to Solve a Rational Equation

1) If possible, factor all denominators.
2) Write down the LCD of all of the expressions.
3) Multiply both sides of the equation by the LCD to *eliminate* the denominators.
4) Solve the equation.
5) Check the solution(s) in the original equation. If a proposed solution makes a denominator equal 0, then it is rejected as a solution.

EXAMPLE 2

Solve $\dfrac{t}{16} + \dfrac{2}{t} = \dfrac{3}{4}$.

Solution

Since this is an *equation,* we will eliminate the denominators by multiplying the equation by the LCD of all of the expressions. **LCD = 16t**

$$16t\left(\frac{t}{16} + \frac{2}{t}\right) = 16t\left(\frac{3}{4}\right) \qquad \text{Multiply both sides of the equation by the LCD, } 16t.$$

$$\cancel{16t}\left(\frac{t}{\cancel{16}}\right) + 16\cancel{t}\left(\frac{2}{\cancel{t}}\right) = \overset{4}{\cancel{16t}}\left(\frac{3}{\cancel{4}}\right) \qquad \text{Distribute and divide out common factors.}$$

$$t^2 + 32 = 12t$$
$$t^2 - 12t + 32 = 0 \qquad \text{Subtract } 12t.$$
$$(t-8)(t-4) = 0 \qquad \text{Factor.}$$

$$t - 8 = 0 \quad \text{or} \quad t - 4 = 0$$
$$t = 8 \quad \text{or} \qquad t = 4$$

Check: $t = 8$ $t = 4$

$$\frac{t}{16} + \frac{2}{t} \overset{?}{=} \frac{3}{4} \qquad\qquad \frac{t}{16} + \frac{2}{t} \overset{?}{=} \frac{3}{4}$$

$$\frac{8}{16} + \frac{2}{8} \overset{?}{=} \frac{3}{4} \qquad\qquad \frac{4}{16} + \frac{2}{4} \overset{?}{=} \frac{3}{4}$$

$$\frac{2}{4} + \frac{1}{4} = \frac{3}{4} \checkmark \qquad\qquad \frac{1}{4} + \frac{2}{4} = \frac{3}{4} \checkmark$$

> **W Hint**
> Remember when a rational expression is undefined.

The solution set is {4, 8}.

[YOU TRY 2]

Solve $\dfrac{d}{3} + \dfrac{4}{d} = \dfrac{13}{3}$.

It is *very* important to check the proposed solution. Sometimes, what appears to be a solution actually is not.

EXAMPLE 3

Solve $2 - \dfrac{9}{k+9} = \dfrac{k}{k+9}$.

Solution

Since this is an *equation,* we will eliminate the denominators by multiplying the equation by the LCD of all of the expressions.   **LCD = $k + 9$**

$$(k+9)\left(2 - \frac{9}{k+9}\right) = (k+9)\left(\frac{k}{k+9}\right)$$ Multiply both sides of the equation by the LCD, $k + 9$.

$$(k+9)2 - (k+9)\cdot\frac{9}{k+9} = (k+9)\left(\frac{k}{k+9}\right)$$ Distribute and divide out common factors.

$$2k + 18 - 9 = k$$ Multiply.
$$2k + 9 = k$$
$$9 = -k$$ Subtract $2k$.
$$-9 = k$$ Divide by -1.

Check: $2 - \dfrac{9}{(-9)+9} \overset{?}{=} \dfrac{-9}{(-9)+9}$ Substitute -9 for k in the original equation.

$$2 - \frac{9}{0} = \frac{-9}{0}$$

Since $k = -9$ makes the denominator equal zero, -9 cannot be a solution to the equation. Therefore, this equation has no solution. The solution set is $\varnothing$.

BE CAREFUL   *Always* check what *appears* to be the solution or solutions to an equation containing rational expressions. If one of these values makes a denominator zero, then it *cannot* be a solution to the equation.

[YOU TRY 3]   Solve $\dfrac{3m}{m-4} - 1 = \dfrac{12}{m-4}$.

EXAMPLE 4

Solve $\dfrac{1}{4} - \dfrac{1}{a+2} = \dfrac{a+18}{4a^2 - 16}$.

Solution

This is an *equation.* Eliminate the denominators by multiplying by the LCD. Begin by factoring the denominator of $\dfrac{a+18}{4a^2 - 16}$.

$$\frac{1}{4} - \frac{1}{a+2} = \frac{a+18}{4(a+2)(a-2)}$$ Factor the denominator.

$$\textbf{LCD} = \textbf{4(a + 2)(a - 2)}$$

$$4(a+2)(a-2)\left(\frac{1}{4} - \frac{1}{a+2}\right) = 4(a+2)(a-2)\left(\frac{a+18}{4(a+2)(a-2)}\right)$$ Multiply both sides of the equation by the LCD.

Distribute and divide out common factors.

$$4(a+2)(a-2)\left(\frac{1}{4}\right) - 4(a+2)(a-2)\left(\frac{1}{a+2}\right) = 4(a+2)(a-2)\left(\frac{a+18}{4(a+2)(a-2)}\right)$$

$$(a + 2)(a - 2) - 4(a - 2) = a + 18 \qquad \text{Multiply.}$$
$$a^2 - 4 - 4a + 8 = a + 18 \qquad \text{Distribute.}$$
$$a^2 - 4a + 4 = a + 18 \qquad \text{Combine like terms.}$$
$$a^2 - 5a - 14 = 0 \qquad \text{Subtract } a, \text{ and subtract } 18.$$
$$(a - 7)(a + 2) = 0 \qquad \text{Factor.}$$
$$a - 7 = 0 \quad \text{or} \quad a + 2 = 0 \qquad \text{Set each factor equal to zero.}$$
$$a = 7 \quad \text{or} \quad a = -2 \qquad \text{Solve.}$$

Look at the factored form of the equation. If $a = 7$, no denominator will equal zero. If $a = -2$, *however, two of the denominators will equal zero. Therefore, we must reject* $a = -2$ *as a solution.* Check only $a = 7$.

$$\text{Check:} \quad \frac{1}{4} - \frac{1}{7 + 2} \overset{?}{=} \frac{7 + 18}{4(7)^2 - 16} \qquad \text{Substitute } a = 7 \text{ into the original equation.}$$

$$\frac{1}{4} - \frac{1}{9} \overset{?}{=} \frac{25}{180} \qquad \text{Simplify.}$$

$$\frac{9}{36} - \frac{4}{36} \overset{?}{=} \frac{5}{36} \qquad \text{Get a common denominator, and simplify } \frac{25}{180}.$$

$$\frac{5}{36} = \frac{5}{36} \quad \checkmark \qquad \text{Subtract.}$$

The solution set is $\{7\}$.

The previous problem is a good example of why it is necessary to check all "solutions" to equations containing rational expressions.

[**YOU TRY 4**] Solve $\dfrac{1}{3} - \dfrac{1}{z + 2} = \dfrac{z + 14}{3z^2 - 12}$.

EXAMPLE 5 Solve $\dfrac{11}{6h^2 + 48h + 90} = \dfrac{h}{3h + 15} + \dfrac{1}{2h + 6}$.

Solution

Since this is an *equation,* we will eliminate the denominators by multiplying by the LCD. Begin by factoring all denominators, then identify the LCD.

$$\frac{11}{6(h + 5)(h + 3)} = \frac{h}{3(h + 5)} + \frac{1}{2(h + 3)} \qquad \textbf{LCD} = \textbf{6(h + 5)(h + 3)}$$

$$6(h + 5)(h + 3)\left(\frac{11}{6(h + 5)(h + 3)}\right) = 6(h + 5)(h + 3)\left(\frac{h}{3(h + 5)} + \frac{1}{2(h + 3)}\right) \qquad \text{Multiply by the LCD.}$$

$$\cancel{6(h+5)(h+3)}\left(\frac{11}{\cancel{6(h+5)(h+3)}}\right) = \overset{2}{\cancel{6}}(\cancel{h+5})(h+3)\left(\frac{h}{3\cancel{(h+5)}}\right) + \overset{3}{\cancel{6}}(h+5)\cancel{(h+3)}\left(\frac{1}{2\cancel{(h+3)}}\right) \qquad \text{Distribute.}$$

$$11 = 2h(h + 3) + 3(h + 5) \qquad \text{Multiply.}$$
$$11 = 2h^2 + 6h + 3h + 15 \qquad \text{Distribute.}$$
$$11 = 2h^2 + 9h + 15 \qquad \text{Combine like terms.}$$
$$0 = 2h^2 + 9h + 4 \qquad \text{Subtract 11.}$$
$$0 = (2h + 1)(h + 4) \qquad \text{Factor.}$$

$$2h + 1 = 0 \quad \text{or} \quad h + 4 = 0$$

$$h = -\frac{1}{2} \quad \text{or} \quad h = -4 \qquad \text{Solve.}$$

W Hint

Did you check all of your answers by hand for You Trys 2–5?

You can see from the factored form of the equation that neither $h = -\dfrac{1}{2}$ nor $h = -4$ will make a denominator zero. Check the values in the original equation to verify that the solution set is $\left\{ -4, -\dfrac{1}{2} \right\}$.

[**YOU TRY 5**]

Solve $\dfrac{5}{6n^2 + 18n + 12} = \dfrac{n}{2n + 2} + \dfrac{1}{3n + 6}$.

3 Solve a Proportion

EXAMPLE 6

Solve $\dfrac{18}{r + 7} = \dfrac{6}{r - 1}$.

Solution

This rational equation is also a *proportion*. A **proportion** is a statement that two ratios are equal. We can solve this proportion as we have solved the other equations in this section, by multiplying both sides of the equation by the LCD. Or, recall from Section 3.6 that *we can solve a proportion by setting the cross products equal to each other.*

$$\dfrac{18}{r + 7} = \dfrac{6}{r - 1}$$

Multiply.   Multiply.

$$18(r - 1) = 6(r + 7) \qquad \text{Set the cross products equal to each other.}$$
$$18r - 18 = 6r + 42 \qquad \text{Distribute.}$$
$$12r = 60$$
$$r = 5 \qquad \text{Solve.}$$

The proposed solution, $r = 5$, does *not* make a denominator equal zero. Check to verify that the solution set is $\{5\}$.

[**YOU TRY 6**]

Solve $\dfrac{7}{d + 3} = \dfrac{14}{3d + 5}$.

4 Solve an Equation for a Specific Variable

In Chapter 3, we learned how to solve an equation for a specific variable. For example, to solve $2l + 2w = P$ for w, we do the following:

$$2l + 2\boxed{w} = P \qquad \text{Put a box around } w, \text{ the variable for which we are solving.}$$
$$2\boxed{w} = P - 2l \qquad \text{Subtract } 2l.$$
$$w = \dfrac{P - 2l}{2} \qquad \text{Divide by 2.}$$

Next we discuss how to solve for a specific variable in a rational expression.

EXAMPLE 7

Solve $z = \dfrac{n}{d - D}$ for d.

Solution

Note that the equation contains a lowercase d and an uppercase D. These represent different quantities, so be sure to write them correctly. Put d in a box.

Since d is in the denominator of the rational expression, multiply both sides of the equation by $d - D$ to eliminate the denominator.

$$z = \frac{n}{\boxed{d} - D}$$
Put d in a box.

$$(\boxed{d} - D)z = (\boxed{d} - D)\left(\frac{n}{\boxed{d} - D}\right)$$
Multiply both sides by $d - D$ to eliminate the denominator.

$$\boxed{d}z - Dz = n$$
Distribute.

$$\boxed{d}z = n + Dz$$
Add Dz.

$$d = \frac{n + Dz}{z}$$
Divide by z.

[YOU TRY 7]

Solve $v = \dfrac{k}{m + M}$ for m.

EXAMPLE 8

Solve $\dfrac{1}{x} + \dfrac{1}{y} = \dfrac{1}{z}$ for y.

Solution

Put the y in a box. The LCD of all of the fractions is xyz. Multiply both sides of the equation by xyz.

$$\frac{1}{x} + \frac{1}{\boxed{y}} = \frac{1}{z}$$
Put y in a box.

$$x\boxed{y}z\left(\frac{1}{x} + \frac{1}{\boxed{y}}\right) = x\boxed{y}z\left(\frac{1}{z}\right)$$
Multiply both sides by xyz to eliminate the denominator.

$$x\boxed{y}z \cdot \frac{1}{x} + x\boxed{y}z \cdot \frac{1}{\boxed{y}} = x\boxed{y}z \cdot \left(\frac{1}{z}\right)$$
Distribute.

$$\boxed{y}z + xz = x\boxed{y}$$
Divide out common factors.

Since we are solving for y and there are terms containing y on each side of the equation, we must get yz and xy on one side of the equation and xz on the other side.

$$xz = x\boxed{y} - \boxed{y}z$$
Subtract yz from each side.

To isolate y, we will *factor* y out of each term on the right-hand side of the equation.

$$xz = \boxed{y}(x - z)$$
Factor out y.

$$\frac{xz}{x - z} = y$$
Divide by $x - z$.

[YOU TRY 8]

Solve $\dfrac{1}{x} + \dfrac{1}{y} = \dfrac{1}{z}$ for z.

We can use a graphing calculator to solve a rational equation in one variable. First, enter the left side of the equation in Y_1 and the right side of the equation in Y_2. Then enter $Y_1 - Y_2$ in Y_3. Then graph the equation in Y_3. The zeros or x-intercepts of the graph are the solutions to the equation.

We will solve $\dfrac{2}{x+5} - \dfrac{3}{x-2} = \dfrac{4x}{x^2+3x-10}$ using a graphing calculator.

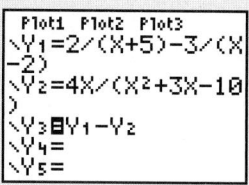

1) Enter $\dfrac{2}{x+5} - \dfrac{3}{x-2}$ by entering $2/(x+5) - 3/(x-2)$ in Y_1.

2) Enter $\dfrac{4x}{x^2+3x-10}$ by entering $4x/(x^2+3x-10)$ in Y_2.

3) Enter $Y_1 - Y_2$ in Y_3 as follows: press <kbd>VARS</kbd>, select Y-VARS using the right arrow key, and press <kbd>ENTER</kbd> <kbd>ENTER</kbd> to select Y_1. Then press <kbd>−</kbd>. Press <kbd>VARS</kbd>, select Y-VARS using the right arrow key, press <kbd>ENTER</kbd> <kbd>2</kbd> to select Y_2. Then press <kbd>ENTER</kbd>.

4) Move the cursor onto the = sign just right of $/Y_1$ and press <kbd>ENTER</kbd> to deselect Y_1. Repeat to deselect Y_2. Press <kbd>GRAPH</kbd> to graph $Y_1 - Y_2$.

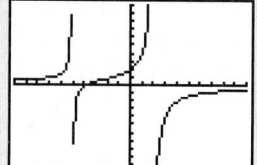

5) Press <kbd>2ⁿᵈ</kbd> <kbd>TRACE</kbd> 2:zero, move the cursor to the left of the zero, and press <kbd>ENTER</kbd>; move the cursor to the right of the zero, and press <kbd>ENTER</kbd>; and move the cursor close to the zero, and press <kbd>ENTER</kbd> to display the zero.

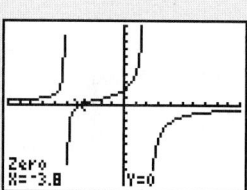

6) Press <kbd>X,T,Θ,n</kbd> <kbd>MATH</kbd> <kbd>ENTER</kbd> <kbd>ENTER</kbd> to display the zero $x = -\dfrac{19}{5}$.

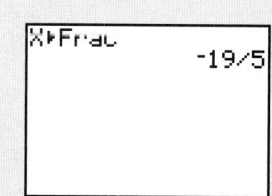

If there is more than one zero, repeat Steps 5 and 6 above for each zero.

Solve each equation using a graphing calculator.

1) $\dfrac{2x}{x-3} + \dfrac{1}{x+5} = \dfrac{4-2x}{x^2+2x-15}$

2) $\dfrac{4}{x-3} + \dfrac{5}{x+3} = \dfrac{15}{x^2-9}$

3) $\dfrac{2}{x+2} + \dfrac{4}{x-5} = \dfrac{16}{x^2-3x-10}$

4) $\dfrac{1}{x-7} + \dfrac{3}{x+4} = \dfrac{7}{x^2-3x-28}$

5) $\dfrac{6}{x+1} = \dfrac{5x+3}{x^2-x-2} - \dfrac{x}{x-2}$

6) $\dfrac{4}{x+3} - \dfrac{x}{x+2} = \dfrac{3x}{x^2+5x+6}$

E Evaluate **8.6** Exercises   Do the exercises, and check your work.

Objective 1: Differentiate Between Rational Expressions and Rational Equations

1) When solving an equation containing rational expressions, do you keep the LCD throughout the problem or do you eliminate the denominators?

2) When adding or subtracting two rational expressions, do you keep the LCD throughout the problem or do you eliminate the denominators?

Determine whether each is an equation or a sum or difference of expressions. Then solve the equation or find the sum or difference.

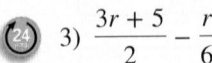

3) $\dfrac{3r + 5}{2} - \dfrac{r}{6}$

4) $\dfrac{m}{12} + \dfrac{m - 8}{3}$

5) $\dfrac{3h}{2} + \dfrac{4}{3} = \dfrac{2h + 3}{3}$

6) $\dfrac{7f - 24}{12} = f + \dfrac{1}{2}$

7) $\dfrac{3}{a^2} + \dfrac{1}{a + 11}$

8) $\dfrac{z}{z - 5} - \dfrac{4}{z}$

9) $\dfrac{8}{b - 11} - 5 = \dfrac{3}{b - 11}$

10) $1 + \dfrac{2}{c + 5} = \dfrac{11}{c + 5}$

Mixed Exercises: Objectives 2 and 3

Values that make the denominators equal zero cannot be solutions of an equation. Find *all* of the values that make the denominators zero and which, therefore, cannot be solutions of each equation. Do NOT solve the equation.

11) $\dfrac{k + 3}{k - 2} + 1 = \dfrac{7}{k}$

12) $\dfrac{t}{t + 12} - \dfrac{5}{t} = 3$

13) $\dfrac{8}{p + 3} - \dfrac{6}{p} = \dfrac{p}{p^2 - 9}$

14) $\dfrac{7}{d^2 - 64} + \dfrac{6}{d} = \dfrac{8}{d + 8}$

15) $\dfrac{9h}{h^2 - 5h - 36} + \dfrac{1}{h + 4} = \dfrac{h + 7}{3h - 27}$

16) $\dfrac{v + 8}{v^2 - 8v + 12} - \dfrac{5}{3v - 4} = \dfrac{2v}{v - 6}$

Solve each equation.

17) $\dfrac{a}{3} + \dfrac{7}{12} = \dfrac{1}{4}$

18) $\dfrac{y}{2} - \dfrac{4}{3} = \dfrac{1}{6}$

19) $\dfrac{1}{4}j - j = -4$

20) $\dfrac{1}{3}h + h = -4$

21) $\dfrac{8m - 5}{24} = \dfrac{m}{6} - \dfrac{7}{8}$

22) $\dfrac{13u - 1}{20} = \dfrac{3u}{5} - 1$

23) $\dfrac{8}{3x + 1} = \dfrac{2}{x + 3}$

24) $\dfrac{4}{5t + 2} = \dfrac{2}{2t - 1}$

25) $\dfrac{r + 1}{2} = \dfrac{4r + 1}{5}$

26) $\dfrac{w}{3} = \dfrac{6w - 4}{9}$

27) $\dfrac{13}{3} = \dfrac{x}{3} + \dfrac{4}{x}$

28) $\dfrac{v}{18} = \dfrac{1}{v} + \dfrac{1}{6}$

54) $\dfrac{5}{m^2 - 36} = \dfrac{4}{m^2 + 6m}$

29) $\dfrac{23}{z} + 8 = -\dfrac{25}{z}$

30) $\dfrac{18}{a} - 2 = \dfrac{10}{a}$

55) $\dfrac{b + 3}{3b - 18} - \dfrac{b + 2}{b - 6} = \dfrac{b}{3}$

31) $\dfrac{5q}{q + 1} - 2 = \dfrac{5}{q + 1}$

32) $\dfrac{n}{n + 3} + 5 = \dfrac{12}{n + 3}$

56) $\dfrac{3y - 2}{y + 2} = \dfrac{y}{4} + \dfrac{1}{4y + 8}$

33) $\dfrac{2}{s + 6} + 4 = \dfrac{2}{s + 6}$

34) $\dfrac{u}{u - 5} + 3 = \dfrac{5}{u - 5}$

57) $\dfrac{4}{n + 1} = \dfrac{10}{n^2 - 1} - \dfrac{5}{n - 1}$

35) $\dfrac{8}{r} - 1 = \dfrac{6}{r}$

36) $\dfrac{11}{g} + 3 = -\dfrac{10}{g}$

58) $\dfrac{2}{c - 6} - \dfrac{24}{c^2 - 36} = -\dfrac{3}{c + 6}$

37) $z + \dfrac{12}{z} = -8$

38) $y - \dfrac{28}{y} = 3$

59) $-\dfrac{a}{5} = \dfrac{3}{a + 8}$

39) $\dfrac{15}{b} = 8 - b$

40) $n = 13 - \dfrac{12}{n}$

60) $\dfrac{u}{7} = \dfrac{2}{9 - u}$

41) $\dfrac{8}{c + 2} - \dfrac{12}{c - 4} = \dfrac{2}{c + 2}$

61) $\dfrac{8}{p + 2} + \dfrac{p}{p + 1} = \dfrac{5p + 2}{p^2 + 3p + 2}$

42) $\dfrac{2}{m - 1} + \dfrac{1}{m + 4} = \dfrac{4}{m + 4}$

62) $\dfrac{6}{x - 1} + \dfrac{x}{x + 3} = \dfrac{2x + 28}{x^2 + 2x - 3}$

43) $\dfrac{9}{c - 8} - \dfrac{15}{c} = 1$

63) $\dfrac{-14}{3a^2 + 15a - 18} = \dfrac{a}{a - 1} + \dfrac{2}{3a + 18}$

44) $\dfrac{6}{r + 5} - \dfrac{2}{r} = -1$

64) $\dfrac{3}{2n^2 + 10n + 8} = \dfrac{n}{2n + 2} + \dfrac{1}{n + 1}$

45) $\dfrac{3}{p - 4} + \dfrac{8}{p + 4} = \dfrac{13}{p^2 - 16}$

65) $\dfrac{3}{f + 4} = \dfrac{f}{f + 6} - \dfrac{2}{f^2 + 10f + 24}$

46) $\dfrac{5}{w - 7} - \dfrac{8}{w + 7} = \dfrac{52}{w^2 - 49}$

66) $\dfrac{11}{c + 9} = \dfrac{c}{c - 4} - \dfrac{36 - 8c}{c^2 + 5c - 36}$

47) $\dfrac{9}{k + 5} - \dfrac{4}{k + 1} = \dfrac{10}{k^2 + 6k + 5}$

67) $\dfrac{b}{b^2 + b - 6} + \dfrac{3}{b^2 + 9b + 18} = \dfrac{8}{b^2 + 4b - 12}$

48) $\dfrac{3}{a + 2} + \dfrac{10}{a^2 - 6a - 16} = \dfrac{5}{a - 8}$

68) $\dfrac{h}{h^2 + 2h - 8} + \dfrac{4}{h^2 + 8h - 20} = \dfrac{4}{h^2 + 14h + 40}$

49) $\dfrac{12}{g^2 - 9} + \dfrac{2}{g + 3} = \dfrac{7}{g - 3}$

69) $\dfrac{r}{r^2 + 8r + 15} - \dfrac{2}{r^2 + r - 6} = \dfrac{2}{r^2 + 3r - 10}$

50) $\dfrac{9}{t + 4} + \dfrac{8}{t^2 - 16} = \dfrac{1}{t - 4}$

70) $\dfrac{5}{t^2 + 5t - 6} - \dfrac{t}{t^2 + 10t + 24} = \dfrac{1}{t^2 + 3t - 4}$

51) $\dfrac{x^2}{2} = \dfrac{x^2 - 6x}{3}$

71) $\dfrac{k}{k^2 - 6k - 16} - \dfrac{12}{5k^2 - 65k + 200} = \dfrac{28}{5k^2 - 15k - 50}$

52) $\dfrac{k^2}{3} = \dfrac{k^2 + 3k}{4}$

72) $\dfrac{q}{q^2 + 4q - 32} + \dfrac{2}{q^2 - 14q + 40} = \dfrac{6}{q^2 - 2q - 80}$

53) $\dfrac{3}{t^2} = \dfrac{6}{t^2 + 8t}$

73) $\dfrac{3b}{b + 7} - 6 = \dfrac{3}{b + 7}$

74) $\dfrac{c}{c - 5} - 5 = \dfrac{20}{c - 5}$

75) $\dfrac{5}{p-3} - \dfrac{7}{p^2 - 7p + 12} = \dfrac{8}{p-4}$

76) $\dfrac{8}{x^2 + 2x - 15} = \dfrac{6}{x-3} + \dfrac{4}{x+5}$

77) The formula $P = \dfrac{1}{f}$, where f is the focal length of a lens in meters and P is the power of the lens in diopters, is used to determine an eyeglass prescription. If the power of a lens is 2.5 diopters, what is the focal length of the lens?

78) Magnetic stripe patterns on the ocean floor help geologists study the rates at which oceanic plates are spreading. Geologists use the formula

$$Rate\ of\ spreading\ for\ stripes = \dfrac{width\ of\ stripe}{time\ duration}$$

a) If a magnetic stripe on a plate on the ocean floor is 75 miles wide and formed over 2 million years, find the rate at which the plate is spreading.

b) If it continues to spread at this rate, how long will it have taken to become 90 miles wide?

79) The average profit a king salmon fisherman receives is given by $A(x) = 6 - \dfrac{1500}{x}$, where x is the number of pounds of king salmon. How many pounds of salmon must he catch so that the profit he earns per pound (the average profit) is $4.00?

80) The average profit earned from the sale of x purses is given by $A(x) = 36 - \dfrac{8400}{x}$. How many purses must be sold so that the profit earned per purse (the average profit) is $20.00?

Objective 4: Solve an Equation for a Specific Variable

Solve for the indicated variable.

81) $W = \dfrac{CA}{m}$ for m

82) $V = \dfrac{nRT}{P}$ for P

83) $a = \dfrac{rt}{2b}$ for b

84) $y = \dfrac{kx}{z}$ for z

85) $B = \dfrac{t+u}{3x}$ for x

86) $Q = \dfrac{n-k}{5r}$ for r

87) $d = \dfrac{t}{z-n}$ for n

88) $z = \dfrac{a}{b+c}$ for b

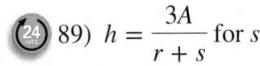 89) $h = \dfrac{3A}{r+s}$ for s

90) $A = \dfrac{4r}{q-t}$ for t

91) $r = \dfrac{kx}{y-az}$ for y

92) $w = \dfrac{na}{kc+b}$ for c

93) $\dfrac{1}{t} = \dfrac{1}{r} - \dfrac{1}{s}$ for r

94) $\dfrac{1}{R_1} + \dfrac{1}{R_2} = \dfrac{1}{R_3}$ for R_2

95) $\dfrac{5}{x} = \dfrac{1}{y} - \dfrac{4}{z}$ for z

96) $\dfrac{2}{A} + \dfrac{1}{C} = \dfrac{3}{B}$ for C

R Rethink

R1) In your own words, explain the difference between a rational expression and a rational equation.

R2) Do you have a good understanding of how to solve rational equations?

R3) Select a problem that you thought was difficult, and write the reasoning for each step you took to solve it.

8.7 Applications of Rational Equations

P Prepare

O Organize

What are your objectives for Section 8.7?	How can you accomplish each objective?
1 Solve Problems Involving Proportions	• Use the **Steps for Solving Applied Problems.** • Understand how to write a *proportion.* • Complete the given examples on your own. • Complete You Try 1.
2 Solve Problems Involving Distance, Rate, and Time	• Use the **Steps for Solving Applied Problems.** • Review how to use the *distance formula, $d = rt$.* • Make a chart to organize your information. • Complete the given example on your own. • Complete You Try 2.
3 Solve Problems Involving Work	• Use the **Steps for Solving Applied Problems.** • Learn the procedure for **Solving Work Problems.** • Complete the given example on your own. • Complete You Try 3.

W Work

Read the explanations, follow the examples, take notes, and complete the You Trys.

We have studied applications of linear and quadratic equations. Now we turn our attention to applications involving equations with rational expressions. We will continue to use the Steps for Solving Applied Problems outlined in Section 3.2.

1 Solve Problems Involving Proportions

We first solved applications involving proportions in Section 3.6. Recall that a **proportion** is a statement that two ratios are equal.

Let's begin this section by solving another problem using a proportion.

EXAMPLE 1

Mariusz Szczawinski/Alamy
Stock Photo

Write an equation, and solve.

One morning at a coffee shop, the ratio of the number of customers who ordered regular coffee to the number who ordered decaffeinated coffee was 4 to 1. If the number of people who ordered regular coffee was 126 more than the number who ordered decaf, how many people ordered each type of coffee?

Solution

Step 1: **Read** the problem carefully, and identify what we are being asked to find.

We must find the number of customers who ordered regular coffee and the number who ordered decaffeinated coffee.

Step 2: **Choose a variable** to represent the unknown, and define the other unknown in terms of this variable.

$$x = \text{number of people who ordered decaffeinated coffee}$$
$$x + 126 = \text{number of people who ordered regular coffee}$$

Step 3: **Translate** the information that appears in English into an algebraic equation. Write a proportion. We will write our ratios so that the numerators contain the same quantities and the denominators contain the same quantities.

$$\text{number who ordered regular coffee} \rightarrow \frac{4}{1} = \frac{x + 126}{x} \leftarrow \text{number who ordered regular coffee}$$
$$\text{number who ordered decaf coffee} \rightarrow \qquad\qquad \leftarrow \text{number who ordered decaf coffee}$$

The equation is $\dfrac{4}{1} = \dfrac{x + 126}{x}$.

Step 4: Solve the equation.

$$\frac{4}{1} = \frac{x + 126}{x}$$

$$4x = 1(x + 126) \qquad \text{Set the cross products equal.}$$
$$4x = x + 126$$
$$3x = 126 \qquad\qquad \text{Subtract } x.$$
$$x = 42 \qquad\qquad\quad \text{Divide by 3.}$$

Step 5: **Check** the answer, and **interpret** the solution as it relates to the problem.

Therefore, 42 customers ordered decaffeinated coffee and $42 + 126 = 168$ people ordered regular coffee. The check is left to the student.

[YOU TRY 1] *Write an equation, and solve.*

During one week at a bookstore, the ratio of the number of romance novels sold to the number of travel books sold was 5 to 3. Determine the number of each type of book sold if customers bought 106 more romance novels than travel books.

---

2 Solve Problems Involving Distance, Rate, and Time

In Chapter 3 we solved problems involving distance (d), rate (r), and time (t).

The basic formula is $d = rt$. We can solve this formula for r and then for t to obtain

$$r = \frac{d}{t} \qquad \text{and} \qquad t = \frac{d}{r}.$$

In this section, we will encounter problems involving boats going with and against a current, and planes going with and against the wind. Both scenarios use the same idea.

Say a boat's speed is 18 mph in still water. If that same boat had a 4 mph current pushing *against* it, how fast would it be traveling? (The current will cause the boat to slow down.)

$$\text{Speed } against \text{ the current} = 18 \text{ mph} \quad - \quad 4 \text{ mph}$$
$$= 14 \text{ mph}$$

$$\frac{\text{Speed } against}{\text{the current}} = \frac{\text{Speed in}}{\text{still water}} - \frac{\text{Speed of}}{\text{the current}}$$

If the speed of the boat in still water is 18 mph and a 4 mph current is *pushing* the boat, how fast would the boat be traveling *with* the current? (The current will cause the boat to travel faster.)

$$\text{Speed } \textit{with} \text{ the current} = 18 \text{ mph} + 4 \text{ mph}$$
$$= 22 \text{ mph}$$

$$\begin{array}{c} \text{Speed } \textit{with} \\ \text{the current} \end{array} = \begin{array}{c} \text{Speed in} \\ \text{still water} \end{array} + \begin{array}{c} \text{Speed of} \\ \text{the current} \end{array}$$

A boat traveling *against* the current is said to be traveling *upstream*. A boat traveling *with* the current is said to be traveling *downstream*. We will use these ideas in Example 2.

EXAMPLE 2

Write an equation, and solve.

A boat can travel 8 mi downstream in the same amount of time it can travel 6 mi upstream. If the speed of the current is 2 mph, what is the speed of the boat in still water?

Solution

Step 1: **Read** the problem carefully, and identify what we are being asked to find.

First, we must understand that "8 mi downstream" means 8 mi *with the current,* and "6 mi upstream" means 6 mi *against the current.*

We must find the speed of the boat in still water.

Step 2: **Choose a variable** to represent the unknown, and define the other unknowns in terms of this variable.

$$x = \text{the speed of the boat in still water}$$
$$x + 2 = \text{the speed of the boat } \textit{with} \text{ the current (downstream)}$$
$$x - 2 = \text{the speed of the boat } \textit{against} \text{ the current (upstream)}$$

Step 3: **Translate** from English to an algebraic equation. Use a table to organize the information.

First, fill in the distances and the rates (or speeds).

	d	r	t
Downstream	8	$x + 2$	
Upstream	6	$x - 2$	

Next we must write expressions for the time it takes the boat to go downstream and upstream. We know that $d = rt$, so if we solve for t we get $t = \dfrac{d}{r}$.

Substitute the information from the table to get the expressions for the time.

$$\text{Downstream: } t = \frac{d}{r} = \frac{8}{x + 2} \qquad \text{Upstream: } t = \frac{d}{r} = \frac{6}{x - 2}$$

Put these expressions into the table.

	d	r	t
Downstream	8	$x + 2$	$\dfrac{8}{x + 2}$
Upstream	6	$x - 2$	$\dfrac{6}{x - 2}$

The problem states that it takes the boat the *same amount of time* to travel 8 mi downstream as it does to go 6 mi upstream. We can write an equation in English:

$$\frac{\text{Time for boat to go}}{\text{8 mi downstream}} = \frac{\text{Time for boat to go}}{\text{6 mi upstream}}$$

Looking at the table, we can write the algebraic equation using the expressions for time. The equation is $\dfrac{8}{x+2} = \dfrac{6}{x-2}$.

Step 4: **Solve** the equation.

$$\frac{8}{x+2} \diagdown\!\!\!\diagup \frac{6}{x-2}$$ Multiply.
Multiply.

$$8(x-2) = 6(x+2)$$ Set the cross products equal.
$$8x - 16 = 6x + 12$$ Distribute.
$$2x = 28$$
$$x = 14$$ Solve.

Step 5: **Check** the answer, and **interpret** the solution as it relates to the problem.

The speed of the boat in still water is 14 mph.

Check: The speed of the boat going downstream is $14 + 2 = 16$ mph, so the time to travel downstream is

$$t = \frac{d}{r} = \frac{8}{16} = \frac{1}{2}\ \text{hr}$$

The speed of the boat going upstream is $14 - 2 = 12$ mph, so the time to travel upstream is

$$t = \frac{d}{r} = \frac{6}{12} = \frac{1}{2}\ \text{hr}$$

So, time upstream = time downstream. ✓

[YOU TRY 2]

Write an equation, and solve.

It takes a boat the same amount of time to travel 10 mi upstream as it does to travel 15 mi downstream. Find the speed of the boat in still water if the speed of the current is 4 mph.

---

3 Solve Problems Involving Work

W Hint

Be sure you understand these concepts before you do Example 3.

Suppose it takes Brian 5 hr to paint his bedroom. What is the *rate* at which he does the job?

$$\text{rate} = \frac{1\ \text{room}}{5\ \text{hr}} = \frac{1}{5}\ \text{room/hr}$$

Brian works at a rate of $\dfrac{1}{5}$ of a room per hour.

In general, we can say that if it takes t units of time to do a job, then the *rate* at which the job is done is $\dfrac{1}{t}$ job per unit of time.

This idea of *rate* is what we use to determine how long it can take for two or more people or things to do a job.

Let's assume, again, that Brian can paint his room in 5 hr. At this rate, how much of the job can he do in 2 hr?

$$\begin{array}{ccc} \text{Fractional part of} & = & \text{Rate of} \cdot \text{Amount of} \\ \text{the job done} & & \text{work} \quad \text{time worked} \end{array}$$

$$= \frac{1}{5} \cdot 2$$

$$= \frac{2}{5}$$

He can paint $\frac{2}{5}$ of the room in 2 hr.

Procedure Solving Work Problems

The basic equation used to solve work problems is:

$$\begin{array}{ccc} \text{Fractional part of a job} & + & \text{Fractional part of a job} & = 1 \text{ (whole job)} \\ \text{done by one person or thing} & & \text{done by another person or thing} & \end{array}$$

EXAMPLE 3

Write an equation, and solve.

If Brian can paint his bedroom in 5 hr but his brother, Doug, could paint the room on his own in 4 hr, how long would it take for the two of them to paint the room together?

Solution

Step 1: **Read** the problem carefully, and identify what we are being asked to find.

We must determine how long it would take Brain and Doug to paint the room together.

Step 2: **Choose a variable** to represent the unknown.

t — the number of hours to paint the room together

Step 3: **Translate** the information that appears in English into an algebraic equation.

Let's write down their rates:

$$\text{Brian's rate} = \frac{1}{5} \text{ room/hr (since the job takes him 5 hr)}$$

$$\text{Doug's rate} = \frac{1}{4} \text{ room/hr (since the job takes him 4 hr)}$$

It takes them t hr to paint the room together. Recall that

$$\begin{array}{ccc} \text{Fractional part} & = & \text{Rate of} \cdot \text{Amount of} \\ \text{of job done} & & \text{work} \quad \text{time worked} \end{array}$$

$$\text{Brian's fractional part} = \frac{1}{5} \cdot t = \frac{1}{5}t$$

$$\text{Doug's fractional part} = \frac{1}{4} \cdot t = \frac{1}{4}t$$

The equation we can write comes from

$$\underset{\substack{\text{Fractional part of the}\\\text{job done by Brian}}}{} \quad + \quad \underset{\substack{\text{Fractional part of the}\\\text{job done by Doug}}}{} = 1 \text{ (whole job)}$$

$$\frac{1}{5}t \quad + \quad \frac{1}{4}t \quad = \quad 1$$

Step 4: **Solve** the equation.

$$20\left(\frac{1}{5}t + \frac{1}{4}t\right) = 20(1) \qquad \text{Multiply by the LCD of 20 to eliminate the fractions.}$$

$$20\left(\frac{1}{5}t\right) + 20\left(\frac{1}{4}t\right) = 20(1) \qquad \text{Distribute.}$$

$$4t + 5t = 20 \qquad \text{Multiply.}$$

$$9t = 20 \qquad \text{Combine like terms.}$$

$$t = \frac{20}{9} \qquad \text{Divide by 9.}$$

Step 5: **Check** the answer, and **interpret** the solution as it relates to the problem.

Brian and Doug could paint the room together in $\frac{20}{9}$ hr or $2\frac{2}{9}$ hr.

$$\textit{Check:} \quad \underset{\substack{\text{Fractional part of the}\\\text{job done by Brian}}}{} \quad + \quad \underset{\substack{\text{Fractional part of the}\\\text{job done by Doug}}}{} \quad = \quad 1 \text{ whole job}$$

$$\frac{1}{5}\cdot\left(\frac{20}{9}\right) \quad + \quad \frac{1}{4}\cdot\left(\frac{20}{9}\right) \quad \overset{?}{=} \quad 1$$

$$\frac{4}{9} \quad + \quad \frac{5}{9} \quad = \quad 1 \ \checkmark$$

[YOU TRY 3]

Write an equation, and solve.

Krutesh can mow a lawn in 2 hr while it takes Stefan 3 hr to mow the same lawn. How long would it take for them to mow the lawn if they worked together?

ANSWERS TO [YOU TRY] EXERCISES

1) 159 travel books, 265 romance novels   2) 20 mph   3) $\frac{6}{5}$ hr or $1\frac{1}{5}$ hr

---

E Evaluate **8.7** Exercises   Do the exercises, and check your work.

Objective 1: Solve Problems Involving Proportions

Write an equation for each, and solve. See Examples 1 and 2.

1) Hector buys batteries for his keyboard and mouse. He knows that 3 packs of batteries cost $4.26. How much should Hector expect to pay for 5 packs of batteries?

2) If 8 oz of granola costs $1.98, find the cost of 20 oz of granola.

 3) A 12-oz serving of Coca-Cola Classic contains 34 mg of caffeine. How much caffeine is in a 20-oz serving of Coke? Round to the nearest mg.

(www.coca-colaproductfacts.com)

4) A doughnut factory can make 1980 doughnuts in 1.5 hr. How long would it take to make 6600 doughnuts?

5) A nurse sets an intravenous fluid drip rate at 1000 mL every 8 hr. How much fluid would the patient receive in 3 hr?

6) A medication is to be given to a patient at a rate of 2.8 mg for every 40 lb of body weight. How much medication should be given to a patient who weighs 190 lb?

7) At a motocross race, the ratio of male spectators to female spectators was 10 to 3. If there were 370 male spectators, how many females were in the crowd?

8) The ratio of students in a history lecture who took notes in pen to those who took notes in pencil was 8 to 3. If 72 students took notes in pen, how many took notes in pencil?

9) In a gluten-free flour mixture, the ratio of potato-starch flour to tapioca flour is 2 to 1. If a mixture contains 3 more cups of potato-starch flour than tapioca flour, how much of each type of flour is in the mixture?

10) Rosa Cruz won an election over her opponent by a ratio of 6 to 5. If her opponent received 372 fewer votes than she did, how many votes did each candidate receive?

11) The ancient Greeks believed that the rectangle most pleasing to the eye, the golden rectangle, had sides in which the ratio of its length to its width was  Royalty-Free/Corbis approximately 8 to 5. They erected many buildings using this golden ratio, including the Parthenon. The marble floor of a museum foyer is to be designed as a golden rectangle. If its width is to be 18 ft less than its length, find the length and width of the foyer.

12) To obtain a particular color, a painter mixed two colors in a ratio of 7 parts blue to 3 parts yellow. If he used 8 fewer gallons of yellow than blue, how many gallons of blue paint did he use?

13) Ms. Hiramoto has invested her money so that the ratio of the amount in bonds to the amount in stocks is 3 to 2. If she has $4000 more invested in bonds than in stocks, how much does she have invested in each?

14) At a wildlife refuge, the ratio of deer to rabbits is 4 to 9. Determine the number of deer and rabbits at the refuge if there are 40 more rabbits than deer.

15) In a small town, the ratio of households with pets to those without pets is 5 to 4. If 271 more households have pets than do not, how many households have pets?

16) An industrial cleaning solution calls for 5 parts water to 2 parts concentrated cleaner. If a worker uses 15 more quarts of water than concentrated cleaner to make a solution,

   a) how much concentrated cleaner did she use?

   b) how much water did she use?

   c) how much solution did she make?

Objective 2: Solve Problems Involving Distance, Rate, and Time

17) If the speed of a boat in still water is 10 mph,

   a) what is its speed going *against* a 3 mph current?

   b) what is its speed *with* a 3 mph current?

18) If an airplane travels at a constant rate of 300 mph,

   a) what is its speed going *into* a 25 mph wind?

   b) what is its speed going *with* a 25 mph wind?

19) If an airplane travels at a constant rate of x mph,

   a) what is its speed going *with* a 30 mph wind?

   b) what is its speed going *against* a 30 mph wind?

20) If the speed of a boat in still water is 13 mph,

   a) what is its speed going *against* a current with a rate of x mph?

   b) what is its speed going *with* a current with a rate of x mph?

Write an equation for each, and solve. See Example 3.

21) A current flows at 5 mph. A boat can travel 20 mi downstream in the same amount of time it can go 12 mi upstream. What is the speed of the boat in still water?

22) With a current flowing at 4 mph, a boat can travel 32 mi with the current in the same amount of time it can go 24 mi against the current. Find the speed of the boat in still water.

23) A boat travels at 16 mph in still water. It takes the same amount of time for the boat to travel 15 mi downstream as to go 9 mi upstream. Find the speed of the current.

24) A boat can travel 12 mi downstream in the time it can go 6 mi upstream. If the speed of the boat in still water is 9 mph, what is the speed of the current?  Ariel Skelley/Blend Images LLC

25) An airplane flying at constant speed can fly 350 mi with the wind in the same amount of time it can fly 300 mi against the wind. What is the speed of the plane if the wind blows at 20 mph?

26) When the wind is blowing at 25 mph, a plane flying at a constant speed can travel 500 mi with the wind in the same amount of time it can fly 400 mi against the wind. Find the speed of the plane.

27) In still water the speed of a boat is 10 mph. Against the current it can travel 4 mi in the same amount of time it can travel 6 mi with the current. What is the speed of the current?

28) Flying at a constant speed, a plane can travel 800 mi with the wind in the same amount of time it can fly 650 mi against the wind. If the wind blows at 30 mph, what is the speed of the plane?

Objective 3: Solve Problems Involving Work

29) Toby can finish a computer programming job in 4 hr. What is the rate at which he does the job?

30) It takes Crystal 3 hr to paint her backyard fence. What is the rate at which she works?

31) Eloise can fertilize her lawn in t hr. What is the rate at which she does this job?

32) It takes Manu twice as long to clean a pool as it takes Anders. If it takes Anders t hr to clean the pool, at what rate does Manu do the job?

Write an equation for each, and solve. See Example 3.

33) It takes Arlene 2 hr to trim the bushes at a city park while the same job takes Andre 3 hr. How long would it take for them to do the job together?

34) A hot water faucet can fill a sink in 8 min while it takes the cold water faucet only 6 min. How long would it take to fill the sink if both faucets were on?

35) Jermaine and Sue must put together notebooks for each person attending a conference. Working alone it would take Jermaine 5 hr while it would take Sue 8 hr. How long would it take for them to assemble the notebooks together?

36) The Williams family has two printers on which they can print out their vacation pictures. The larger printer can print all of the pictures in 3 hr, while it would take 5 hr on the smaller printer. How long would it take to print the vacation pictures using both printers?

37) A faucet can fill a tub in 12 min. The leaky drain can empty the tub in 30 min. If the faucet is on and the drain is leaking, how long would it take to fill the tub?

38) It takes Deepak 50 min to shovel snow from his sidewalk and driveway. When he works with his brother, Kamal, it takes only 30 min. How long would it take Kamal to do the shoveling himself?

(24) 39) Fatima and Antonio must cut out shapes for an art project at a day-care center. Fatima can do the job twice as fast as Antonio. Together, it takes 2 hr to cut out all of the shapes. How long would it take Fatima to cut out the shapes herself?

40) It takes Burt three times as long as Phong to set up a new alarm system. Together they can set it up in 90 min. How long would it take Phong to set up the alarm system by himself?

41) Working together it takes 2 hr for a new worker and an experienced worker to paint a billboard. If the new employee worked alone, it would take him 6 hr. How long would it take the experienced worker to paint the billboard by himself?

42) Audrey can address party invitations in 40 min, while it would take her mom 1 hr. How long would it take for them to address the invitations together?

43) Homer uses the moving walkway to get to his gate at the airport. He can travel 126 ft when he is walking on the moving walkway in the same amount of time it would take for him to walk only 66 ft on the floor next to it. If the walkway is moving at 2 ft/sec, how fast does Homer walk?

44) Another walkway at the airport moves at $2\frac{1}{2}$ ft/sec. If Bart can travel 140 ft when he is walking on the moving walkway in the same amount of time he can walk 80 ft on the floor next to it, how fast does Bart walk?

R Rethink

R1) Which of the application problems are hardest for you to understand?

R2) Why do the units for time have to be the same within a work problem?

R3) Write an example of a work problem that you encountered this past week.

8.8 Variation

P Prepare

O Organize

What are your objectives for Section 8.8?	How can you accomplish each objective?
1 Solve Direct Variation Problems	• Learn the definition of *direct variation*. • Learn the procedure for **Solving a Variation Problem.** • Complete the given examples on your own. • Complete You Trys 1–3.
2 Solve Inverse Variation Problems	• Learn the definition of *inverse variation*. • Follow the procedure for **Solving a Variation Problem.** • Complete the given examples on your own. • Complete You Trys 4 and 5.
3 Solve Joint Variation Problems	• Learn the definition of *joint variation*. • Follow the procedure for **Solving a Variation Problem.** • Complete the given example on your own. • Complete You Try 6.
4 Solve Combined Variation Problems	• Follow the procedure for **Solving a Variation Problem.** • Complete the given example on your own. • Complete You Try 7.

W Work   Read the explanations, follow the examples, take notes, and complete the You Trys.

1 Solve Direct Variation Problems

In Section 4.6 we discussed the following situation:

If you are driving on a highway at a constant speed of 60 mph, the distance you travel depends on the amount of time spent driving.

Let y = the distance traveled, in miles, and let x = the number of hours spent driving. An equation relating x and y is $y = 60x$, and y is a function of x.

We can make a table of values relating x and y. We can say that the distance traveled, y, is *directly proportional to* the time spent traveling, x. Or y *varies directly as* x.

x	y
1	60
1.5	90
2	120
3	180

Definition

Direct Variation: *y* varies directly as *x* (or ***y* is directly proportional to *x***) means

$$y = kx$$

where k is a nonzero real number. **k is called the constant of variation.**

If two quantities vary directly and $k > 0$, then as one quantity increases the other increases as well. And, as one quantity decreases, the other decreases.

In our example of driving distance, $y = 60x$, 60 is the *constant of variation*. Given information about how variables are related, we can write an equation and solve a variation problem.

EXAMPLE 1

Suppose y varies directly as x. If $y = 18$ when $x = 3$,

a) find the constant of variation, k.

b) write a variation equation relating x and y using the value of k found in a).

c) find y when $x = 11$.

Solution

a) To find the constant of variation, write a *general* variation equation relating x and y. *y varies directly as x means* $y = kx$.

We are told that $y = 18$ when $x = 3$. Substitute these values into the equation, and solve for k.

$$y = kx$$
$$18 = k(3) \qquad \text{Substitute 3 for } x \text{ and 18 for } y.$$
$$6 = k \qquad \text{Divide by 3.}$$

b) The *specific* variation equation is the equation obtained when we substitute 6 for k in $y = kx$: Therefore, $y = 6x$.

c) To find y when $x = 11$, substitute 11 for x in $y = 6x$ and evaluate.

$$y = 6x$$
$$= 6(11) \qquad \text{Substitute 11 for } x.$$
$$= 66 \qquad \text{Multiply.}$$

W Hint

Learn this procedure, and write it in your notes.

Procedure Solving a Variation Problem

Step 1: Write the *general* variation equation.

Step 2: Find k by substituting the known values into the equation and solving for k.

Step 3: Write the *specific* variation equation by substituting the value of k into the *general* variation equation.

Step 4: Use the specific variation equation to solve the problem.

YOU TRY 1

Suppose y varies directly as x. If $y = 40$ when $x = 5$,

a) find the constant of variation.

b) write the specific variation equation relating x and y.

c) find y when $x = 3$.

EXAMPLE 2

Suppose p varies directly as the square of z. If $p = 12$ when $z = 2$, find p when $z = 5$.

Solution

Step 1: Write the *general* variation equation.

p varies directly as the *square* of z means $p = kz^2$.

Step 2: Find k using the known values: $p = 12$ when $z = 2$.

$$p = kz^2$$
$$12 = k(2)^2 \qquad \text{Substitute 2 for } z \text{ and 12 for } p.$$
$$12 = k(4)$$
$$3 = k$$

W Hint

Write out these examples, using the *steps*, as you are reading them.

Step 3: Substitute $k = 3$ into $p = kz^2$ to get the *specific* variation equation, $p = 3z^2$.

Step 4: We are asked to find p when $z = 5$. Substitute $z = 5$ into $p = 3z^2$ to get p.

$$p = 3z^2$$
$$= 3(5)^2 \qquad \text{Substitute 5 for } z.$$
$$= 3(25)$$
$$= 75$$

YOU TRY 2

Suppose w varies directly as the cube of n. If $w = 135$ when $n = 3$, find w when $n = 2$.

EXAMPLE 3

A theater's nightly revenue varies directly as the number of tickets sold. If the revenue from the sale of 80 tickets is \$3360, find the revenue from the sale of 95 tickets.

Solution

Let n = the number of tickets sold, and let R = revenue.

We will follow the four steps for solving a variation problem.

Step 1: Write the *general* variation equation: $R = kn$.

Step 2: Find k using the known values: $R = 3360$ when $n = 80$.

$$R = kn$$
$$3360 = k(80) \qquad \text{Substitute 80 for } n \text{ and 3360 for } R.$$
$$42 = k \qquad \text{Divide by 80.}$$

Step 3: Substitute $k = 42$ into $R = 42n$ to get the *specific* variation equation, $R = 42n$.

Step 4: We must find the revenue from the sale of 95 tickets. Substitute $n = 95$ into $R = 42n$ to find R.

$$R = 42n$$
$$R = 42(95)$$
$$R = 3990$$

The revenue is \$3990.

YOU TRY 3

The cost to carpet a room varies directly as the area of the room. If it costs \$525.00 to carpet a room of area 210 ft^2, how much would it cost to carpet a room of area 288 ft^2?

2 Solve Inverse Variation Problems

If two quantities vary *inversely* (are *inversely* proportional) then as one value increases, the other decreases. Likewise, as one value decreases, the other increases.

> ### Definition
>
> **Inverse Variation: *y* varies inversely as *x* (or *y* is inversely proportional to *x*) means**
>
> $$y = \frac{k}{x}$$
>
> where *k* is a nonzero real number. ***k* is the constant of variation.**

A good example of inverse variation is the relationship between the time, *t*, it takes to travel a given distance, *d*, as a function of the rate (or speed), *r*. We can define this relationship as $t = \dfrac{d}{r}$. As the rate, *r*, increases, the time, *t*, that it takes to travel *d* mi decreases. Likewise, as *r* decreases, the time, *t*, that it takes to travel *d* mi increases. Therefore, *t* varies *inversely* as *r*.

EXAMPLE 4

Suppose *q* varies inversely as *h*. If *q* = 4 when *h* = 15, find *q* when *h* = 10.

Solution

Step 1: Write the *general* variation equation, $q = \dfrac{k}{h}$.

Step 2: Find *k* using the known values: *q* = 4 when *h* = 15.

$$q = \frac{k}{h}$$

$$4 = \frac{k}{15} \qquad \text{Substitute 15 for } h \text{ and 4 for } q.$$

$$60 = k \qquad \text{Multiply by 15.}$$

Step 3: Substitute *k* = 60 into $q = \dfrac{k}{h}$ to get the *specific* variation equation, $q = \dfrac{60}{h}$.

Step 4: Substitute 10 for *h* in $q = \dfrac{60}{h}$ to find *q*.

$$q = \frac{60}{10} = 6$$

[YOU TRY 4]

Suppose *m* varies inversely as the square of *v*. If *m* = 1.5 when *v* = 4, find *m* when *v* = 2.

EXAMPLE 5

The intensity of light (in lumens) varies inversely as the square of the distance from the source. If the intensity of the light is 40 lumens 5 ft from the source, what is the intensity of the light 4 ft from the source?

Solution

Let d = distance from the source (in feet), and let I = intensity of the light (in lumens).

Step 1: Write the *general* variation equation, $I = \dfrac{k}{d^2}$.

Step 2: Find k using the known values: $I = 40$ when $d = 5$.

$$I = \frac{k}{d^2}$$

$$40 = \frac{k}{(5)^2} \qquad \text{Substitute 5 for } d \text{ and 40 for } I.$$

$$40 = \frac{k}{25}$$

$$1000 = k \qquad \text{Multiply by 25.}$$

Comstock/JupiterImages

Step 3: Substitute $k = 1000$ into $I = \dfrac{k}{d^2}$ to get the *specific* variation equation, $I = \dfrac{1000}{d^2}$.

Step 4: Find the intensity, I, of the light 4 ft from the source. Substitute $d = 4$ into $I = \dfrac{1000}{d^2}$ to find I.

$$I = \frac{1000}{(4)^2} = \frac{1000}{16} = 62.5$$

The intensity of the light is 62.5 lumens.

$\left[\text{YOU TRY 5} \right]$ If the voltage in an electrical circuit is held constant (stays the same), then the current in the circuit varies inversely as the resistance. If the current is 40 amps when the resistance is 3 ohms, find the current when the resistance is 8 ohms.

3 Solve Joint Variation Problems

If a variable varies directly as the *product* of two or more other variables, the first variable *varies jointly* as the other variables.

Definition

Joint Variation: y varies jointly as x and z means $y = kxz$ where k is a nonzero real number.

EXAMPLE 6

For a given amount invested in a bank account (called the principal), the interest earned varies jointly as the interest rate (expressed as a decimal) and the time the principal is in the account. If Graham earns $80 in interest when he invests his money for 1 yr at 4%, how much interest would the same principal earn if he invested it at 5% for 2 yr?

Solution

Let r = interest rate (as a decimal)
t = the number of years the principal is invested
I = interest earned

Step 1: Write the *general* variation equation, $I = krt$.

Step 2: Find k using the known values: $I = 80$ when $t = 1$ and $r = 0.04$.

$$I = krt$$
$$80 = k(0.04)(1) \qquad \text{Substitute the values into } I = krt.$$
$$80 = 0.04k$$
$$2000 = k \qquad \text{Divide by 0.04.}$$

(The amount he invested, the principal, is $2000.)

Step 3: Substitute $k = 2000$ into $I = krt$ to get the *specific* variation equation,
$I = 2000rt$.

Step 4: Find the interest Graham would earn if he invested $2000 at 5% interest for 2 yr. Let $r = 0.05$ and $t = 2$. Solve for I.

$$I = 2000(0.05)(2) \qquad \text{Substitute 0.05 for } r \text{ and 2 for } t.$$
$$= 200 \qquad \text{Multiply.}$$

Graham would earn $200.

[YOU TRY 6]

The volume of a box of constant height varies jointly as its length and width. A box with a volume of 9 ft^3 has a length of 3 ft and a width of 2 ft. Find the volume of a box with the same height, if its length is 4 ft and its width is 3 ft.

4 Solve Combined Variation Problems

A combined variation problem involves both direct and inverse variation.

EXAMPLE 7

Suppose y varies directly as the square root of x and inversely as z. If $y = 12$ when $x = 36$ and $z = 5$, find y when $x = 81$ and $z = 15$.

Solution

Step 1: Write the *general* variation equation.

$$y = \frac{k\sqrt{x}}{z} \qquad \begin{array}{l} \leftarrow y \text{ varies directly as the square root of } x. \\ \leftarrow y \text{ varies inversely as } z. \end{array}$$

Step 2: Find k using the known values: $y = 12$ when $x = 36$ and $z = 5$.

$$12 = \frac{k\sqrt{36}}{5} \qquad \text{Substitute the values.}$$
$$60 = 6k \qquad \text{Multiply by 5; } \sqrt{36} = 6.$$
$$10 = k \qquad \text{Divide by 6.}$$

Step 3: Substitute $k = 10$ into $y = \dfrac{k\sqrt{x}}{z}$ to get the specific variation equation,
$$y = \frac{10\sqrt{x}}{z}.$$

Step 4: Find y when $x = 81$ and $z = 15$.
$$y = \frac{10\sqrt{81}}{15} \qquad \text{Substitute 81 for } x \text{ and 15 for } z.$$
$$y = \frac{10 \cdot 9}{15} = \frac{90}{15} = 6$$

[YOU TRY 7] Suppose a varies directly as b and inversely as the square of c. If $a = 28$ when $b = 12$ and $c = 3$, find a when $b = 36$ and $c = 4$.

ANSWERS TO [YOU TRY] EXERCISES

1) a) 8 b) $y = 8x$ c) 24   2) 40   3) $720.00   4) 6   5) 15 amps   6) 18 ft^3 7) 47.25

E Evaluate 8.8 Exercises   Do the exercises, and check your work.

Mixed Exercises: Objectives 1–4

1) If z varies directly as y, then as y increases, the value of z _____.

2) If a varies inversely as b, then as b increases, the value of a _____.

Decide whether each equation represents direct, inverse, joint, or combined variation.

3) $y = 6x$

4) $c = 4ab$

5) $f = \dfrac{15}{t}$

6) $z = 3\sqrt{x}$

7) $p = \dfrac{8q^2}{r}$

8) $w = \dfrac{11}{v^2}$

Write a general variation equation using k as the constant of variation.

9) M varies directly as n.

10) q varies directly as r.

11) h varies inversely as j.

12) R varies inversely as B.

13) T varies inversely as the square of c.

14) b varies directly as the cube of w.

15) s varies jointly as r and t.

16) C varies jointly as A and D.

17) Q varies directly as the square root of z and inversely as m.

18) r varies directly as d and inversely as the square of L.

19) Suppose z varies directly as x. If $z = 63$ when $x = 7$,
   a) find the constant of variation.
   b) write the specific variation equation relating z and x.
   c) find z when $x = 6$.

20) Suppose A varies directly as D. If $A = 12$ when $D = 3$,
   a) find the constant of variation.
   b) write the specific variation equation relating A and D.
   c) find A when $D = 11$.

 21) Suppose N varies inversely as y. If $N = 4$ when $y = 12$,
   a) find the constant of variation.
   b) write the specific variation equation relating N and y.
   c) find N when $y = 3$.

22) Suppose j varies inversely as m. If $j = 7$ when $m = 9$,
   a) find the constant of variation.
   b) write the specific variation equation relating j and m.
   c) find j when $m = 21$.

23) Suppose Q varies directly as the square of r and inversely as w. If $Q = 25$ when $r = 10$ and $w = 20$,

   a) find the constant of variation.

   b) write the specific variation equation relating Q, r, and w.

   c) find Q when $r = 6$ and $w = 4$.

24) Suppose y varies jointly as a and the square root of b. If $y = 42$ when $a = 3$ and $b = 49$,

   a) find the constant of variation.

   b) write the specific variation equation relating y, a, and b.

   c) find y when $a = 4$ and $b = 9$.

Solve.

25) If B varies directly as R, and $B = 35$ when $R = 5$, find B when $R = 8$.

26) If q varies directly as p, and $q = 10$ when $p = 4$, find q when $p = 10$.

27) If L varies inversely as the square of h, and $L = 8$ when $h = 3$, find L when $h = 2$.

28) If w varies inversely as d, and $w = 3$ when $d = 10$, find w when $d = 5$.

29) If y varies jointly as x and z, and $y = 60$ when $x = 4$ and $z = 3$, find y when $x = 7$ and $z = 2$.

30) If R varies directly as P and inversely as the square of Q, and $R = 5$ when $P = 10$ and $Q = 4$, find R when $P = 18$ and $Q = 3$.

Solve each problem by writing a variation equation.

31) Kosta is paid hourly at his job. His weekly earnings vary directly as the number of hours worked. If Kosta earned $437.50 when he worked 35 hr, how much would he earn if he worked 40 hr?

32) If distance is held constant, the time it takes to travel that distance is inversely proportional to the speed at which one travels. If it takes 14 hr to travel the given distance at 60 mph, how long would it take to travel the same distance at 70 mph?

33) The cost of manufacturing a certain brand of spiral notebook is inversely proportional to the number produced. When 16,000 notebooks are produced, the cost per notebook

PhotoDisc/Getty Images

is $0.60. What is the cost of each notebook when 12,000 are produced?

34) The surface area of a cube varies directly as the square of the length of one of its sides. A cube has a surface area of 54 in^2 when the length of each side is 3 in. What is the surface area of a cube with a side of length 6 in.?

35) The power in an electrical system varies jointly as the current and the square of the resistance. If the power is 100 watts when the current is 4 amps and the resistance is 5 ohms, what is the power when the current is 5 amps and the resistance is 6 ohms?

36) The force exerted on an object varies jointly as the mass and acceleration of the object. If a 20-newton force is exerted on an object of mass 10 kg and an acceleration of 2 m/sec^2, how much force is exerted on a 50 kg object with an acceleration of 8 m/sec^2?

37) The volume of a cylinder varies jointly as its height and the square of its radius. The volume of a cylindrical can is 108π cm^3 when its radius is 3 cm and it is 12 cm high. Find the volume of a cylindrical can with a radius of 4 cm and a height of 3 cm.

38) The kinetic energy of an object varies jointly as its mass and the square of its speed. When a roller coaster car with a mass of 1000 kg is traveling at 15 m/sec, its kinetic energy is 112,500 J (joules). What is the kinetic energy of the same car when it travels at 18 m/sec?

39) The frequency of a vibrating string varies inversely as its length. If a 5-ft-long piano string vibrates at 100 cycles/sec, what is the frequency of a piano string that is 2.5 ft long?

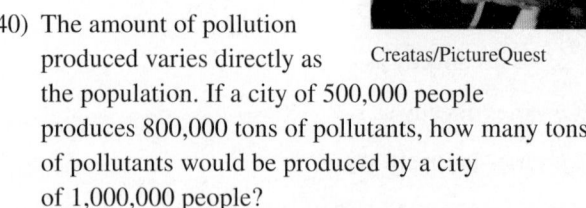

Creatas/PictureQuest

40) The amount of pollution produced varies directly as the population. If a city of 500,000 people produces 800,000 tons of pollutants, how many tons of pollutants would be produced by a city of 1,000,000 people?

41) The resistance of a wire varies directly as its length and inversely as its cross-sectional area. A wire of length 40 cm and cross-sectional area 0.05 cm^2 has a resistance of 2 ohms. Find the resistance of 60 cm of the same type of wire.

42) When a rectangular beam is positioned horizontally, the maximum weight that it can support varies jointly as its width and the square of its thickness and inversely as its length. A beam is $\frac{3}{4}$ ft wide, $\frac{1}{3}$ ft thick, and 8 ft long, and it can support 17.5 tons. How much weight can a similar beam support if it is 1 ft wide, $\frac{1}{2}$ ft thick and 12 ft long?

43) Hooke's law states that the force required to stretch a spring is proportional to the distance that the spring is stretched from its original length. A force of 200 lb is required to stretch a spring 5 in. from its natural length. How much force is needed to stretch the spring 8 in. beyond its natural length?

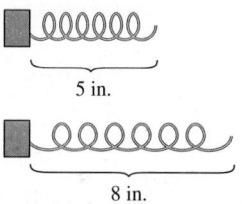

44) The weight of an object on Earth varies inversely as the square of its distance from the center of the Earth. If an object weighs 300 lb on the surface of the Earth (4000 mi from the center), what is the weight of the object if it is 800 mi above the Earth? (Round to the nearest pound.)

R Rethink

R1) Can you think of an example of a direct variation problem that you encountered this week?

R2) What is the purpose of the constant of variation, k?

R3) Which variation problem is hardest for you to solve? Why?

Group Activity – Graphs of Rational Equations

1) On your own paper, graph $y = \frac{1}{x}$ by filling in the table below and plotting points.

Row A	x	-1	-2	-5	-10	-100	-1000
Row B	$y = \frac{1}{x}$						
	(x, y)						

Row C	x	-0.5	-0.2	-0.1	-0.01	-0.001	
Row D	$y = \frac{1}{x}$						
	(x, y)						

Row E	x	0.5	0.2	0.1	0.01	0.001	
Row F	$y = \frac{1}{x}$						
	(x, y)						

Row G	x	1	2	5	10	100	1000
Row H	$y = \dfrac{1}{x}$						
	(x, y)						

2) Use the table to answer parts a)–d). As you answer the following, refer to your graph to see how this corresponds to each part of the graph.

   a) Look at the values of x in Row A. Notice that the absolute values of these numbers get very large. We say that "x approaches negative infinity." Symbolically, we write $x \to -\infty$. What happens to the values of y as $x \to -\infty$? Do they approach a certain number? What number?

   b) Look at the values of x in Row C. Notice that the absolute values of these numbers get very small. We say that "x approaches zero from the left." Symbolically, we write $x \to 0^-$. What happens to the values of y as $x \to 0^-$? Do they approach a certain number, or do they get very large? Express this symbolically.

   c) From Row E, the values of x get closer to zero, or $x \to 0^+$. What happens to the values of y?

   d) From Row G, the values of x get very large, or $x \to \infty$. What happens to the values of y?

3) What happens when $x = 0$? Does the graph have a y-intercept?

4) Are there any values of x that make y equal to zero? Does the graph have an x-intercept?

5) Use this graph to answer parts a)–d).

   a) As $x \to -\infty$, $y \to$ _____

   b) As $x \to \infty$, $y \to$ _____

   c) As $x \to 0^+$, $y \to$ _____

   d) As $x \to 0^-$, $y \to$ _____

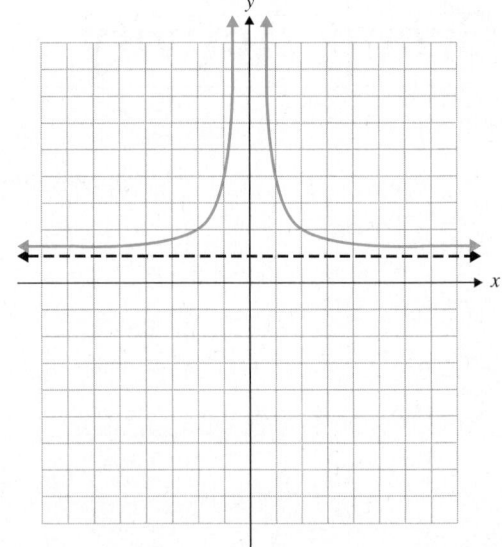

em POWER me My School

Every school—whether it's a high school, community college, college, or university—operates under its own set of rules and procedures. Understanding how your school works and where to go for help are essential parts of being successful in college. It's important to understand how your school works so that, for example, you know where and when to turn in your financial aid application and you know where to get help if you have questions about choosing the classes you need for graduation. Take this survey to learn how well you know your school. Check all boxes that apply.

- ❑ I know the address of my school's website.
- ❑ I can navigate the school's website to find most information that I need.
- ❑ I am aware of whether my school has a handbook containing useful information.
- ❑ I have signed up to receive emergency campus messages by email, text, or automated phone call.
- ❑ I am aware of important dates such as when to register for classes, when tuition is due, and when financial aid forms are due.
- ❑ I know where to register for classes on campus.
- ❑ On campus, I know where to ask questions about financial aid.
- ❑ I can locate the bookstore.
- ❑ I know the location of the library.
- ❑ I know the difference between an adviser and a counselor.
- ❑ I know the location of the campus health center.
- ❑ I know the location of student services offices that might be of interest to me. Some examples are veterans' support services, the office to help students with disabilities, and child care.
- ❑ I know the locations of all of my instructors' offices as well as their office hours.
- ❑ I know the location of the tutoring center/math lab, and I know their procedures for getting help when I need it.
- ❑ I can locate the Testing Center, and I know its rules and hours of operation.
- ❑ I am aware of clubs, organizations, and activities on campus, and I know where to go to become involved in those that interest me.
- ❑ I know the location of the office where I can go if I have questions about or want help finding a job.

Think about the items that you have, and have *not*, checked in this survey. Which apply to you and might contribute to your success in college? In the Study Strategies at the beginning of this chapter, you will learn how to get to know your school.

Chapter 8:   Summary

Definition/Procedure	Example

8.1 Rational Expressions and Functions

A **rational expression** is an expression of the form $\dfrac{P}{Q}$, where P and Q are polynomials and where $Q \neq 0$.

We can *evaluate* rational expressions.

Evaluate $\dfrac{5a - 8}{a + 3}$ for $a = 2$.

$$\frac{5(2) - 8}{2 + 3} = \frac{10 - 8}{5} = \frac{2}{5}$$

How to Determine When a Rational Expression Equals Zero and When It Is Undefined

1) To determine what values of the variable make the expression equal zero, set the numerator equal to zero and solve for the variable.

2) To determine what values of the variable make the expression undefined, set the denominator equal to zero and solve for the variable.

For what value(s) of x is $\dfrac{x - 7}{x + 9}$

a) equal to zero?   b) undefined?

a) $\dfrac{x - 7}{x + 9} = 0$ when $x - 7 = 0$.

$$x - 7 = 0$$
$$x = 7$$

When $x = 7$, the expression equals zero.

b) $\dfrac{x - 7}{x + 9}$ is undefined when its denominator equals zero.

Solve $x + 9 = 0$.

$$x + 9 = 0$$
$$x = -9$$

When $x = -9$, the expression is undefined.

Writing a Rational Expression in Lowest Terms

1) Completely *factor* the numerator and denominator.

2) *Divide* the numerator and denominator by the greatest common factor.

Simplify $\dfrac{3r^2 - 10r + 8}{2r^2 - 8}$.

$$\frac{3r^2 - 10r + 8}{2r^2 - 8} = \frac{(3r - 4)(r - 2)}{2(r + 2)(r - 2)} = \frac{3r - 4}{2(r + 2)}$$

Simplifying $\dfrac{a - b}{b - a}$.

A rational expression of the form $\dfrac{a - b}{b - a}$ will simplify to -1.

Simplify $\dfrac{5 - w}{w^2 - 25}$.

$$\frac{5 - w}{w^2 - 25} = \frac{\overset{-1}{\cancel{5 - w}}}{(w + 5)\cancel{(w - 5)}} = -\frac{1}{w + 5}$$

Rational Functions

$f(x) = \dfrac{x - 9}{x + 2}$ is a rational function because $\dfrac{x - 9}{x + 2}$ is a rational expression and because each value that can be substituted for x will produce only one value for the expression.

The **domain** of a rational function consists of all real numbers except the value(s) of the variable which make the denominator equal zero.

Determine the domain of $f(x) = \dfrac{x - 9}{x + 2}$.

$$x + 2 = 0 \quad \text{Set the denominator} = 0.$$
$$x = -2 \quad \text{Solve.}$$

When $x = -2$, the denominator of $f(x) = \dfrac{x - 9}{x + 2}$ equals zero. The domain contains all real numbers *except* -2. Write the domain in interval notation as $(-\infty, -2) \cup (-2, \infty)$.

Definition/Procedure	Example

8.2 Multiplying and Dividing Rational Expressions

Multiplying Rational Expressions

1) Factor numerators and denominators.

2) Divide out common factors.

3) Multiply.

All answers should be written in lowest terms.

Multiply $\dfrac{16v^4}{v^2 + 10v + 21} \cdot \dfrac{3v + 21}{4v}$.

$$\dfrac{16v^4}{v^2 + 10v + 21} \cdot \dfrac{3v + 21}{4v} = \dfrac{\overset{4}{\cancel{16v^3} \cdot \cancel{v}}}{(v + 3)\cancel{(v + 7)}} \cdot \dfrac{3\cancel{(v + 7)}}{\cancel{4v}}$$

$$= \dfrac{12v^3}{v + 3}$$

Dividing Rational Expressions

To **divide** rational expressions, multiply the first expression by the reciprocal of the second.

Divide $\dfrac{2x^2 + 5x}{x + 4} \div \dfrac{4x^2 - 25}{12x - 30}$.

$$\dfrac{2x^2 + 5x}{x + 4} \div \dfrac{4x^2 - 25}{12x - 30} = \dfrac{2x^2 + 5x}{x + 4} \cdot \dfrac{12x - 30}{4x^2 - 25}$$

$$= \dfrac{x\cancel{(2x + 5)}}{x + 4} \cdot \dfrac{6\cancel{(2x - 5)}}{\cancel{(2x + 5)}\cancel{(2x - 5)}} = \dfrac{6x}{x + 4}$$

8.3 Finding the Least Common Denominator

How to Find the Least Common Denominator (LCD)

1) Factor the denominators.

2) The LCD will contain each unique factor the greatest number of times it appears in any single factorization.

3) The LCD is the *product* of the factors identified in Step 2.

Find the LCD of $\dfrac{9b}{b^2 + 8b}$ and $\dfrac{6}{b^2 + 16a + 64}$.

1) $b^2 + 8b = b(b + 8)$, $\quad b^2 + 16a + 64 = (b + 8)^2$

2) The factors we will use in the LCD are b and $(b + 8)^2$.

3) $\text{LCD} = b(b + 8)^2$

8.4 Adding and Subtracting Rational Expressions

Adding and Subtracting Rational Expressions

1) Factor the denominators.

2) Write down the LCD.

3) Rewrite each rational expression as an equivalent rational expression with the LCD.

4) Add or subtract the numerators, and keep the common denominator in factored form.

5) After combining like terms in the numerator, ask yourself, *"Can I factor it?"* If so, factor.

6) Divide out common factors, if possible. The final answer should be written in lowest terms.

Add $\dfrac{y}{y + 7} + \dfrac{10y - 28}{y^2 - 49}$.

1) Factor the denominator of $\dfrac{10y - 28}{y^2 - 49}$.

$$\dfrac{10y - 28}{y^2 - 49} = \dfrac{10y - 28}{(y + 7)(y - 7)}$$

2) The LCD is $(y + 7)(y - 7)$.

3) Rewrite $\dfrac{y}{y + 7}$ with the LCD.

$$\dfrac{y}{y + 7} \cdot \dfrac{y - 7}{y - 7} = \dfrac{y(y - 7)}{(y + 7)(y - 7)}$$

4) $\dfrac{y}{y + 7} + \dfrac{10y - 28}{y^2 - 49} = \dfrac{y(y - 7)}{(y + 7)(y - 7)} + \dfrac{10y - 28}{(y + 7)(y - 7)}$

$$= \dfrac{y(y - 7) + 10y - 28}{(y + 7)(y - 7)}$$

$$= \dfrac{y^2 - 7y + 10y - 28}{(y + 7)(y - 7)}$$

$$= \dfrac{y^2 + 3y - 28}{(y + 7)(y - 7)}$$

5) $\qquad = \dfrac{\cancel{(y + 7)}(y - 4)}{\cancel{(y + 7)}(y - 7)}$ Factor.

6) $\qquad = \dfrac{y - 4}{y - 7}$ Divide out common factors.

Definition/Procedure	Example

8.5 Simplifying Complex Fractions

A **complex fraction** is a rational expression that contains one or more fractions in its numerator, its denominator, or both.

Some examples of complex fractions are

$$\frac{\frac{9}{16}}{\frac{3}{4}}, \quad \frac{\frac{b+3}{2}}{\frac{6b+18}{7}}, \quad \frac{\frac{1}{x}-\frac{1}{y}}{1-\frac{x}{y}}$$

To simplify a complex fraction containing one term in the numerator and one term in the denominator,

1) Rewrite the complex fraction as a division problem.

2) Perform the division by multiplying the first fraction by the reciprocal of the second.

Simplify $\dfrac{\frac{b+3}{2}}{\frac{6b+18}{7}}$.

$$\frac{\frac{b+3}{2}}{\frac{6b+18}{7}} = \frac{b+3}{2} \div \frac{6b+18}{7}$$

$$= \frac{b+3}{2} \cdot \frac{7}{6(b+3)} = \frac{b+3}{2} \cdot \frac{7}{6(b+3)} = \frac{7}{12}$$

To simplify complex fractions containing more than one term in the numerator and/or the denominator,

Method 1

1) Combine the terms in the numerator and combine the terms in the denominator so that each contains only one fraction.

2) Rewrite as a division problem.

3) Perform the division.

Method 1

Simplify $\dfrac{\frac{1}{x}-\frac{1}{y}}{1-\frac{x}{y}}$.

$$\frac{\frac{1}{x}-\frac{1}{y}}{1-\frac{x}{y}} = \frac{\frac{y}{xy}-\frac{x}{xy}}{\frac{y}{y}-\frac{x}{y}} = \frac{\frac{y-x}{xy}}{\frac{y-x}{y}} = \frac{y-x}{xy} \div \frac{y-x}{y}$$

$$= \frac{y-x}{xy} \cdot \frac{y}{y-x} = \frac{1}{x}$$

Method 2

1) Write down the LCD of *all* of the fractions in the complex fraction.

2) Multiply the numerator and denominator of the complex fraction by the LCD.

3) Simplify.

Method 2

Simplify $\dfrac{\frac{1}{x}-\frac{1}{y}}{1-\frac{x}{y}}$.

1) LCD = xy

2) Multiply the numerator and denominator by

the LCD: $\dfrac{xy\left(\frac{1}{x}-\frac{1}{y}\right)}{xy\left(1-\frac{x}{y}\right)}$

$$\frac{xy\left(\frac{1}{x}-\frac{1}{y}\right)}{xy\left(1-\frac{x}{y}\right)} = \frac{xy \cdot \frac{1}{x} - xy \cdot \frac{1}{y}}{xy \cdot 1 - xy \cdot \frac{x}{y}} \quad \text{Distribute.}$$

3) $$= \frac{y-x}{xy-x^2} \quad \text{Simplify.}$$

$$= \frac{y-x}{x(y-x)} = \frac{1}{x}$$

Definition/Procedure	Example

8.6 Solving Rational Equations

An **equation** contains an = sign, an **expression** does not.

How to Solve a Rational Equation

1) If possible, factor all denominators.

2) Write down the LCD of all of the expressions.

3) Multiply both sides of the equation by the LCD to *eliminate* the denominators.

4) Solve the equation.

5) Check the solution(s) in the original equation. If a proposed solution makes a denominator equal 0, then it is rejected as a solution.

Solve $\dfrac{n}{n+6} + 1 = \dfrac{18}{n+6}$.

This is an *equation* because it contains an = sign. We must eliminate the denominators. Identify the LCD of all of the expressions in the equation.

LCD = (n + 6)

Multiply both sides of the equation by $(n + 6)$.

$$(n+6)\left(\frac{n}{n+6} + 1\right) = (n+6)\left(\frac{18}{n+6}\right)$$

$$(n+6) \cdot \left(\frac{n}{n+6}\right) + (n+6) \cdot 1 = (n+6) \cdot \frac{18}{n+6}$$

$$n + n + 6 = 18$$
$$2n + 6 = 18$$
$$2n = 12$$
$$n = 6$$

The solution set is $\{6\}$.

The check is left to the student.

Solve an Equation for a Specific Variable

Solve $x = \dfrac{3b}{n+m}$ for n.

Since we are solving for n, put it in a box.

$$x = \frac{3b}{\boxed{n} + m}$$

$$(\boxed{n} + m)x = (\boxed{n} + m) \cdot \frac{3b}{\boxed{n} + m}$$

$$(\boxed{n} + m)x = 3b$$

$$\boxed{n}x + mx = 3b$$

$$\boxed{n}x = 3b - mx$$

$$n = \frac{3b - mx}{x}$$

8.7 Applications of Rational Equations

Use the **Five Steps for Solving Applied Problems** outlined in Section 3.2.

Write an equation and solve.

Jeff can wash and wax his car in 3 hours, but it takes his dad only 2 hours to wash and wax the car. How long would it take the two of them to wash and wax together?

Step 1: **Read** the problem carefully.

Step 2: **Choose a variable** to represent the unknown.
  t = number of hours to wash and wax the car together.

Definition/Procedure	Example
	Step 3: **Translate** from English into an algebraic equation.

$$\text{Jeff's rate} = \frac{1}{3}\text{ wash/hr} \qquad \text{Dad's rate} = \frac{1}{2}\text{ wash/hr}$$

$$\text{Fractional part} = \text{Rate} \quad \cdot \quad \text{Time}$$

$$\text{Jeff's part} = \frac{1}{3} \quad \cdot \quad t \quad = \frac{1}{3}t$$

$$\text{Dad's part} = \frac{1}{2} \quad \cdot \quad t \quad = \frac{1}{2}t$$

$$\underset{\text{job by Jeff}}{\text{Fractional}} + \underset{\text{job by his dad}}{\text{Fractional}} = \underset{\text{job}}{\text{1 whole}}$$

$$\frac{1}{3}t \quad + \quad \frac{1}{2}t \quad = \quad 1$$

Equation: $\frac{1}{3}t + \frac{1}{2}t = 1$

Step 4: **Solve** the equation.

$$6\left(\frac{1}{3}t + \frac{1}{2}t\right) = 6(1) \qquad \text{Multiply by 6, the LCD.}$$

$$6 \cdot \frac{1}{3}t + 6 \cdot \frac{1}{2}t = 6(1) \qquad \text{Distribute.}$$

$$2t + 3t = 6 \qquad \text{Multiply.}$$

$$5t = 6$$

$$t = \frac{6}{5}$$

Step 5: **Interpret** the solution as it relates to the problem.

Jeff and his dad could wash and wax the car together in $\frac{6}{5}$ hours or $1\frac{1}{5}$ hours.

The **check** is left to the student.

8.8 Variation

Definition/Procedure	Example
Direct Variation **y varies directly as x** (or **y is directly proportional to x**) means $$y = kx$$ where k is a nonzero real number. k is called the **constant of variation.**	The circumference, C, of a circle is given by $C = 2\pi r$. C varies directly as r, where $k = 2\pi$.
Inverse Variation **y varies inversely as x** (or **y is inversely proportional to x**) means $$y = \frac{k}{x}$$ where k is a nonzero real number.	The time, t (in hours), it takes to drive 600 mi is inversely proportional to the rate, r, at which you drive. $$t = \frac{600}{r}$$ where $k = 600$.

Definition/Procedure	Example
Joint Variation **y varies jointly as x and z** means $y = kxz$, where k is a nonzero real number.	For a given amount, called the principal, deposited in a bank account, the interest earned, I, varies jointly as the interest rate, r, and the time, t, the principal is in the account. $$I = 1000rt$$ $k = 1000$, the principal.
Combined Variation A **combined variation** problem involves both direct and inverse variation.	The resistance of a wire, R, varies directly as its length, L, and inversely as its cross-sectional area, A. $$R = \frac{0.002L}{A}$$ The constant of variation, k, is 0.002. This is the resistivity of the material from which the wire was made.
Solving a Variation Problem **Step 1:** Write the *general* variation equation. **Step 2:** Find k by substituting the known values into the equation and solving for k. **Step 3:** Write the *specific* variation equation by substituting the value of k into the *general* variation equation. **Step 4:** Use the specific variation equation to solve the problem.	The cost of manufacturing a certain soccer ball is inversely proportional to the number produced. When 15,000 are made, the cost per ball is \$4.00. What is the cost to manufacture each soccer ball when 25,000 are produced? Let n = number of soccer balls produced and let C = cost of producing each ball **Step 1:** Write the *general* variation equation: $C = \dfrac{k}{n}$ **Step 2:** Find k using $C = 4$ when $n = 15{,}000$. $$4 = \frac{k}{15{,}000}$$ $$60{,}000 = k$$ **Step 3:** Write the *specific* variation equation: $C = \dfrac{60{,}000}{n}$ **Step 4:** Find the cost, C, per ball when $n = 25{,}000$. $$C = \frac{60{,}000}{25{,}000} \quad \text{Substitute 25,000 for } n.$$ $$C = 2.4$$ The cost per ball is \$2.40.

Chapter 8: Review Exercises

(8.1) Evaluate, if possible, for a) $n = 5$ and b) $n = -2$.

1) $\dfrac{n^2 - 3n - 10}{3n + 2}$

2) $\dfrac{3n - 2}{n^2 - 4}$

Determine the value(s) of the variable for which
a) the expression equals zero.
b) the expression is undefined.

3) $\dfrac{2s}{4s + 11}$

4) $\dfrac{2c^2 - 3c - 9}{c^2 - 7c}$

5) $\dfrac{15}{4t^2 - 9}$

6) $\dfrac{15 - 5d}{d^2 + 25}$

For each rational function,
a) find $f(5)$.
b) find x so that $f(x) = 0$.
c) determine the domain of the function.

7) $f(x) = \dfrac{x + 9}{5x - 1}$

8) $f(x) = \dfrac{8}{x^2 - 100}$

Determine the domain of each rational function.

9) $h(a) = \dfrac{9a}{a^2 - 2a - 24}$

10) $k(t) = \dfrac{6t - 1}{t^2 + 7}$

Write each rational expression in lowest terms.

11) $\dfrac{77k^9}{7k^3}$

12) $\dfrac{18c - 66}{39c - 143}$

13) $\dfrac{r^2 - 14r + 48}{4r^2 - 24r}$

14) $\dfrac{y^2 + 8y - yz - 8z}{yz - 3y - z^2 + 3z}$

15) $\dfrac{11 - x}{x^2 - 121}$

16) $\dfrac{3t^2 + 5tu + 2u^2}{27t^3 + 8u^3}$

Find three equivalent forms of each rational expression.

17) $-\dfrac{4n + 1}{5 - 3n}$

18) $-\dfrac{u - 8}{u + 2}$

Find the missing side in each rectangle.

19) Area $= 2b^2 + 13b + 21$

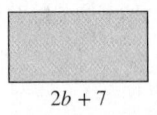

2b + 7

Find the width.

20) Area $= 3x^2 - 8x - 3$

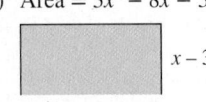

x − 3

Find the length.

(8.2) Perform the operations and simplify.

21) $\dfrac{64}{45} \cdot \dfrac{27}{56}$

22) $\dfrac{6}{25} \div \dfrac{9}{10}$

23) $\dfrac{t + 6}{4} \cdot \dfrac{2(t + 2)}{(t + 6)^2}$

24) $\dfrac{4m^3}{30n} \div \dfrac{20m^6}{3n^5}$

25) $\dfrac{3x^2 + 11x + 8}{15x + 40} \div \dfrac{9x + 9}{x - 3}$

26) $\dfrac{6w - 1}{6w^2 + 5w - 1} \cdot \dfrac{3w + 3}{12w}$

27) $\dfrac{r^2 - 16r + 63}{2r^3 - 18r^2} \div (r - 7)^2$

28) $(h^2 + 10h + 24) \cdot \dfrac{h}{h^2 + h - 12}$

29) $\dfrac{24k}{k^3 - k^2 + k - 1} \cdot \dfrac{1 - k^3}{8k^2 + 8k + 8}$

30) $\dfrac{25 - a^2}{9a^2 - 6a + 1} \div \dfrac{4a - 20}{3a - 1}$

31) $\dfrac{3p^5}{20q^2} \cdot \dfrac{4q^3}{21p^7}$

32) $\dfrac{c^3 - d^3}{c^3 + c^2d + cd^2} \div \dfrac{d^2 - c^2}{c}$

Divide.

33) $\dfrac{\dfrac{9}{8}}{\dfrac{15}{4}}$

34) $\dfrac{\dfrac{2r + 10}{r^2}}{\dfrac{r^2 - 25}{4r}}$

35) $\dfrac{\dfrac{3s + 8}{12}}{\dfrac{3s + 8}{4}}$

36) $\dfrac{\dfrac{16m - 8}{m^2}}{\dfrac{12m - 6}{m^4}}$

(8.3) Find the LCD of each group of fractions.

37) $\dfrac{9}{10}, \dfrac{7}{15}, \dfrac{6}{5}$

38) $\dfrac{2}{9x^2y}, \dfrac{13}{4xy^4}$

39) $\dfrac{3}{k^5}, \dfrac{11}{k^2}$

40) $\dfrac{3}{2m}, \dfrac{4}{m + 4}$

41) $\dfrac{1}{4x + 9}, \dfrac{3x}{x - 7}$

42) $\dfrac{8}{3d^2 - d}, \dfrac{11}{9d - 3}$

43) $\dfrac{w}{w - 5}, \dfrac{11}{5 - w}$

44) $\dfrac{6m}{m^2 - n^2}, \dfrac{n}{n - m}$

45) $\dfrac{3c - 11}{c^2 + 9c + 20}, \dfrac{8c}{c^2 - 2c - 35}$

46) $\dfrac{6}{x^2 + 7x}, \dfrac{1}{2x^2 + 14x}, \dfrac{13}{x^2 + 14x + 49}$

Rewrite each rational expression with the indicated denominator.

47) $\dfrac{3}{5y} = \dfrac{}{20y^3}$

48) $\dfrac{4k}{k - 9} = \dfrac{}{(k - 6)(k - 9)}$

49) $\dfrac{6}{2z + 5} = \dfrac{}{z(2z + 5)}$

50) $\dfrac{n}{9 - n} = \dfrac{}{n - 9}$

51) $\dfrac{t - 3}{3t + 1} = \dfrac{}{(3t + 1)(t + 4)}$

Identify the LCD of each group of fractions, and rewrite each as an equivalent fraction with the LCD as its denominator.

52) $\dfrac{4}{5a^3b}, \dfrac{3}{8ab^5}$

53) $\dfrac{8c}{c^2 + 5c - 24}, \dfrac{5}{c^2 - 6c + 9}$

54) $\dfrac{6}{p + 9}, \dfrac{3}{p}$

55) $\dfrac{7}{2q^2 - 12q}, \dfrac{3q}{36 - q^2}, \dfrac{q - 5}{2q^2 + 12q}$

56) $\dfrac{1}{g - 12}, \dfrac{6}{12 - g}$

(8.4) Add or subtract.

57) $\dfrac{5}{9c} + \dfrac{7}{9c}$

58) $\dfrac{5}{6z^2} + \dfrac{9}{12z}$

59) $\dfrac{9}{10u^2v} - \dfrac{1}{8u^3v}$

60) $\dfrac{3m}{m - 4} - \dfrac{1}{m - 4}$

61) $\dfrac{n}{3n - 5} - \dfrac{4}{n}$

62) $\dfrac{8}{t + 2} + \dfrac{8}{t}$

63) $\dfrac{9}{y + 2} - \dfrac{5}{y - 3}$

64) $\dfrac{7d - 3}{d^2 + 3d - 28} + \dfrac{3d}{5d + 35}$

65) $\dfrac{k - 3}{k^2 + 14k + 49} - \dfrac{2}{k^2 + 7k}$

66) $\dfrac{8p + 3}{2p + 2} - \dfrac{6}{p^2 - 3p - 4}$

67) $\dfrac{t + 9}{t - 18} - \dfrac{11}{18 - t}$

68) $\dfrac{x^2}{x^2 - y^2} + \dfrac{x}{y - x}$

69) $\dfrac{4w}{w^2 + 11w + 24} - \dfrac{3w - 1}{2w^2 - w - 21}$

70) $\dfrac{2a + 7}{a^2 - 6a + 9} + \dfrac{6}{a^2 + 2a - 15}$

71) $\dfrac{b}{9b^2 - 4} + \dfrac{b + 1}{6b^2 - 4b} - \dfrac{1}{6b + 4}$

72) $\dfrac{d + 4}{d^2 + 3d} + \dfrac{d}{5d^2 + 12d - 9} - \dfrac{8}{5d^2 - 3d}$

73) Find a rational expression in simplest form to represent the
a) area and b) perimeter of the rectangle.

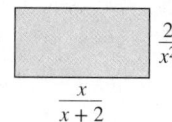

74) Find a rational expression in simplest form to represent the
perimeter of the triangle.

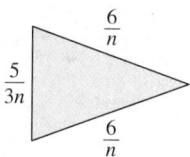

(8.5) Simplify completely.

75) $\dfrac{\dfrac{x}{y}}{\dfrac{x^3}{y^7}}$

76) $\dfrac{p + \dfrac{4}{p}}{\dfrac{9}{p} + p}$

77) $\dfrac{\dfrac{4}{5} - \dfrac{2}{3}}{\dfrac{1}{2} + \dfrac{1}{6}}$

78) $\dfrac{\dfrac{4q}{7q + 70}}{\dfrac{q^3}{8q + 80}}$

79) $\dfrac{1 - \dfrac{1}{y - 8}}{\dfrac{2}{y + 4} + 1}$

80) $\dfrac{\dfrac{10}{21}}{\dfrac{16}{9}}$

81) $\dfrac{1 + \dfrac{1}{r - t}}{\dfrac{1}{r^2 - t^2} + \dfrac{1}{r + t}}$

82) $\dfrac{\dfrac{z}{z + 2} + \dfrac{1}{z^2 - 4}}{1 - \dfrac{3}{z + 2}}$

83) $\dfrac{2x^{-2} + y^{-1}}{x^{-1} - y^{-2}}$

84) $\dfrac{12a^{-1}}{4a + b^{-2}}$

(8.6) Solve each equation.

85) $\dfrac{5a + 4}{15} = \dfrac{a}{5} + \dfrac{4}{5}$

86) $\dfrac{2}{y - 7} = \dfrac{8}{y + 5}$

87) $\dfrac{m}{7} = \dfrac{5}{m + 2}$

88) $-\dfrac{w}{20} = \dfrac{1}{5w} + \dfrac{1}{4}$

89) $\dfrac{1}{3n} = \dfrac{7}{12} + \dfrac{n}{6}$

90) $\dfrac{16}{9c - 27} + \dfrac{2c - 4}{c - 3} = \dfrac{c}{9}$

91) $\dfrac{r}{r + 5} + 4 = \dfrac{5}{r + 5}$

92) $\dfrac{3}{j + 9} + \dfrac{j}{j - 3} = \dfrac{2j^2 + 2}{j^2 + 6j - 27}$

93) $\dfrac{5}{t^2 + 10t + 24} + \dfrac{5}{t^2 + 3t - 18} = \dfrac{t}{t^2 + t - 12}$

94) $p - \dfrac{20}{p} = 8$

95) $\dfrac{3}{x + 1} = \dfrac{6x}{x^2 - 1} - \dfrac{4}{x - 1}$

96) $\dfrac{9}{4k^2 + 28k + 48} = \dfrac{k}{4k + 16} + \dfrac{9}{8k + 24}$

Solve for the indicated variable.

97) $R = \dfrac{s + T}{D}$ for D

98) $A = \dfrac{2p}{c}$ for c

99) $w = \dfrac{N}{c - ak}$ for k

100) $n = \dfrac{t}{a + b}$ for a

101) $\dfrac{1}{R_1} + \dfrac{1}{R_2} = \dfrac{1}{R_3}$ for R_1

102) $\dfrac{1}{r} = \dfrac{1}{s} + \dfrac{1}{t}$ for s

(8.7) Write an equation and solve.

103) A boat can travel 8 miles downstream in the same amount
of time it can travel 6 miles upstream. If the speed of the
boat in still water is 14 mph, what is the speed of the
current?

104) The ratio of saturated fat to total fat in a Starbucks
grande Caramel Frappuccino is 3 to 5. If there are 6 more
grams of total fat in the drink than there are grams of
saturated fat, how much total fat is in a Caramel
Frappuccino? (www.starbucks.com)

105) Kendall and Kylie must put together notebooks for each
person attending a conference. Working alone, it would take
Kendall 5 hours while it would take Kylie 8 hours. How
long would it take for them to assemble the notebooks
together?

106) When the wind is blowing at 40 mph, a plane flying at a
constant speed can travel 800 mi with the wind in the same
amount of time it can fly 600 mi against the wind. Find the
speed of the plane.

(8.8) Solve each problem by writing a variation equation.

107) Suppose c varies directly as m. If $c = 56$ when $m = 8$, find c when $m = 3$.

108) Suppose A varies jointly as t and r. If $A = 15$ when $t = \dfrac{1}{2}$ and $r = 5$, find A when $t = 3$ and $r = 4$.

109) Suppose z varies inversely as the cube of w. If $z = 16$ when $w = 2$, find z when $w = 4$.

110) Suppose p varies directly as n and inversely as the square of d. If $p = 42$ when $n = 7$ and $d = 2$, find p when $n = 12$ and $d = 3$.

111) The surface area of a cube varies directly as the square of the length of one of its sides. A cube has a surface area of 54 cm^2 when the length of each side is 3 cm. What is the surface area of a cube with a side of length 6 cm?

112) The frequency of a vibrating piano string varies inversely as its length. If a 4-ft-long string vibrates at 125 cycles/second, what is the frequency of a piano string that is 2 feet long?

C Squared Studios/Getty Images

Mixed Exercises: Sections 8.2–8.6

Perform the operations, and simplify.

113) $\dfrac{5n}{2n - 1} - \dfrac{2n + 3}{n + 2}$

114) $\dfrac{27w^3}{3w^2 + w - 4} \cdot \dfrac{2 - 2w}{15w}$

115) $\dfrac{2a^2 + 9a + 10}{4a - 7} \div (2a + 5)^2$

116) $\dfrac{5}{8b} + \dfrac{2}{9b^4}$

117) $\dfrac{1}{8 - r} + \dfrac{16}{r^2 - 64}$

118) $\dfrac{\dfrac{7}{x} + \dfrac{8}{y}}{1 - \dfrac{6}{y}}$

119) $\dfrac{9p^2 + 6pq + 4q^2}{27p^3 - 8q^3} \cdot \dfrac{9p^2 - 4q^2}{3p^2 + 5pq + 2q^2}$

120) $\dfrac{4}{2m^2 - 3m - 20} - \dfrac{m - 5}{m^3 - 64}$

Solve.

121) $\dfrac{a}{4} = \dfrac{5}{a} - 2$

122) $\dfrac{1}{2} - \dfrac{x + 5}{2x^2 - 2} = \dfrac{1}{x + 1}$

123) $\dfrac{h}{5} = \dfrac{h - 3}{h + 1} + \dfrac{12}{5h + 5}$

124) $\dfrac{5w}{6} - \dfrac{2}{3} = -\dfrac{1}{6}$

125) $\dfrac{8}{3g^2 - 7g - 6} - \dfrac{8}{3g + 2} = -\dfrac{4}{g - 3}$

126) $\dfrac{4k}{k + 16} = \dfrac{4}{k + 1}$

Chapter 8:   Test

1) Evaluate, if possible, for $k = -4$.

   $\dfrac{5k + 8}{k^2 + 16}$

Determine the values of the variable for which
a) the expression is undefined.
b) the expression equals zero.

2) $\dfrac{2c - 9}{c + 10}$

3) $\dfrac{n^2 + 1}{n^2 - 5n - 36}$

4) Determine the domain of $g(x) = \dfrac{x + 9}{2x + 3}$.

Write each rational expression in lowest terms.

5) $\dfrac{21t^8u^2}{63t^{12}u^5}$

6) $\dfrac{24h^2 - 6h}{4h^2 - 33h + 8}$

7) $\dfrac{z^2 - 9}{z^3 - 27}$

8) Write three equivalent forms of $\dfrac{8 - m}{4m - 5}$.

9) Identify the LCD of $\dfrac{2z}{z + 6}$ and $\dfrac{9}{z}$.

Perform the operations, and simplify.

10) $\dfrac{8}{15r} + \dfrac{2}{15r}$

11) $\dfrac{28a^9}{b^2} \div \dfrac{20a^{15}}{b^3}$

12) $\dfrac{5h}{12} - \dfrac{7h}{9}$

13) $\dfrac{6}{c + 2} + \dfrac{c}{3c + 5}$

14) $\dfrac{k^3 - 9k^2 + 2k - 18}{4k - 24} \cdot \dfrac{k^2 + 3k - 54}{81 - k^2}$

15) $\dfrac{8d^2 + 24d}{20} \div (d + 3)^2$

16) $\dfrac{2t-5}{t-7}+\dfrac{t+9}{7-t}$

17) $\dfrac{3}{2v^2-7v+6}-\dfrac{v+4}{v^2+7v-18}$

Simplify completely.

18) $\dfrac{1-\dfrac{1}{m+2}}{\dfrac{m}{m+2}-\dfrac{1}{m}}$

19) $\dfrac{\dfrac{5x+5y}{x^2y^2}}{\dfrac{20}{xy}}$

Solve each equation.

20) $\dfrac{3r+1}{2}+\dfrac{1}{10}=\dfrac{6r}{5}$

21) $\dfrac{1}{n+3}=\dfrac{n+4}{2}$

22) $\dfrac{28}{w^2-4}=\dfrac{7}{w-2}-\dfrac{5}{w+2}$

23) $\dfrac{3}{x+8}+\dfrac{x}{x-4}=\dfrac{7x+9}{x^2+4x-32}$

24) Solve for b.

$\dfrac{1}{a}+\dfrac{1}{b}=\dfrac{1}{c}$

25) Write an expression for the base of the triangle if the area is given by $12k^2+28k$.

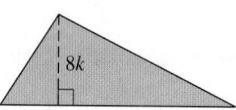

26) Find all values that cannot be solutions to the equation $\dfrac{3}{5}-\dfrac{x+2}{4x-1}=\dfrac{7}{x}$. Do not solve the equation.

Write an equation for each, and solve.

27) Every Sunday night, the equipment at a restaurant must be taken apart and cleaned. Ricardo can do this job twice as fast as Michael. When they work together, they can do the cleaning in 2 hr. How long would it take each man to do the job on his own?

28) A current flows at 4 mph. If a boat can travel 12 mi downstream in the same amount of time it can go 6 mi upstream, find the speed of the boat in still water.

29) Suppose n varies jointly as r and the square of s. If $n=72$ when $r=2$ and $s=3$, find n when $r=3$ and $s=5$.

30) If the temperature remains the same, the volume of a gas is inversely proportional to the pressure. If the volume of a gas is 6.25 L (liters) at a pressure of 2 atm (atmospheres), what is the volume of the gas at 1.25 atm?

Chapter 8: Cumulative Review for Chapters 1–8

1) Find the area of the triangle.

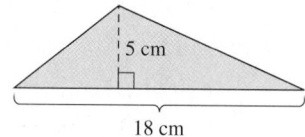

18 cm

2) Evaluate $72-30\div6+4(3^2-10)$.

3) Write an equation and solve.

The length of a rectangular garden is 4 ft longer than the width. Find the dimensions of the garden if its perimeter is 28 ft.

Solve each inequality. Write the answer in interval notation.

4) $19-8w>5$

5) $4\le\dfrac{3}{5}t+4\le13$

6) Find the x- and y-intercepts of $4x-3y=6$, and graph the equation.

7) Find the slope of the line containing the points $(4,1)$ and $(-2,9)$.

8) Solve the system.

$5x+4y=5$
$7x-6y=36$

Simplify. The answer should not contain negative exponents.

9) $(2p^3)^5$

10) $(5y^2)^{-3}$

Multiply and simplify.

11) $(2n-3)^2$

12) $(8a+b)(8a-b)$

Divide.

13) $\dfrac{45h^4-25h^3+15h^2-10}{15h^2}$

14) $\dfrac{5k^3+18k^2-11k-8}{k+4}$

Factor completely.

15) $4d^2+4d-15$

16) $3z^4-48$

17) $rt+8t-r-8$

18) Solve $x(x+16)=x-36$.

19) For what values of a is $\dfrac{7a + 2}{a^2 - 6a}$

    a) undefined?

    b) equal to zero?

20) Write $\dfrac{3c^2 + 21c - 54}{c^2 + 3c - 54}$ in lowest terms.

Perform the operations, and simplify.

21) $\dfrac{10n^2}{n^2 - 8n + 16} \cdot \dfrac{3n^2 - 14n + 8}{10n - 15n^2}$

22) $\dfrac{6}{y + 5} - \dfrac{3}{y}$

23) Simplify $\dfrac{\dfrac{2}{r - 8} + 1}{1 - \dfrac{3}{r - 8}}$.

24) Solve $\dfrac{1}{v - 1} + \dfrac{2}{5v - 3} = \dfrac{37}{5v^2 - 8v + 3}$.

25) Suppose h varies inversely as the square of p. If $h = 12$ when $p = 2$, find h when $p = 4$.

More Equations and Inequalities

Get Ready

Let's review the inequality symbols and the meaning of absolute value before we solve absolute value equations and inequalities.

1) We can use the following **inequality symbols** to compare quantities.

$<$ less than $\qquad\le$ less than or equal to
$>$ greater than $\qquad\ge$ greater than or equal to

Example: Determine whether each statement is true or false.

a) $-8 < 5$ True
b) $3 \ge 7$ False; $3 < 7$
c) $9 > 4$ True
d) $6 \le 6$ True

2) The **absolute value** of a number is *the distance of the number from zero.* The absolute value of a number is never negative. The absolute value of a number, x, is denoted by $|x|$.

Example: Evaluate.   a) $|2|$ b) $|-5|$ c) $|0|$

a) $|2| = 2$ because the distance from 2 to 0 is 2.

b) $|-5| = 5$ because the distance from -5 to 0 is 5.

c) $|0| = 0$ because the distance from 0 to 0 is 0.

In this chapter, use the following *Basic Skills Worksheet* to prepare students for Chapter 10: WS2 Powers (to help students recognize powers of integers). This will make it easier for students to simplify radicals.

Get Ready Exercises

Determine whether each statement is true or false.

1) $10 > 3$ 2) $-2 < -8$ 3) $-1 \le -4$ 4) $13 > 9$ 5) $7 \ge 7$ 6) $0 \le 0$

Evaluate.

7) $|6|$ 8) $|14|$ 9) $|-7|$ 10) $\left|-\dfrac{2}{3}\right|$

Determine whether each statement is true or false.

11) $|-9| \le 4$ 12) $|3| > -2$ 13) $8 > |8|$ 14) $15 \le |-15|$

15) $|-13| \ge 13$ 16) $|-4| < -4$ 17) $1 < \left|\dfrac{7}{6}\right|$ 18) $-0.5 \ge |0|$

Find three positive numbers and three negative numbers, x, that are solutions of each compound inequality.

19) $-3 \le x \le 3$ 20) $x \le -6$ or $x \ge 6$

Answers

1) True 2) False 3) False 4) True 5) True 6) True 7) 6 8) 14 9) 7 10) $\dfrac{2}{3}$ 11) False 12) True 13) False 14) True 15) True 16) False 17) True 18) False 19) Answers may vary. Some solutions are -3, -2, -1, 1, 2, and 3. 20) Answers may vary. Some solutions are -8, -7, -6, 6, 7, and 8.

Study Strategies   The Writing Process

Many students enter college thinking that most people are born either good at math or good at communicating, but that few people are "naturally" good at both. In reality, everyone needs to put in time and effort if they want to develop their skills in these areas. And, being good at writing *and* at math are important skills for college and beyond. Here are some strategies you can apply to becoming a better writer.

- Your goal will vary depending upon what kind of writing assignment you will do. Are you going to write a paper to persuade someone about a particular topic? Are you writing to share your opinion? Are you writing a research paper? Are you doing a creative writing assignment? **Think about, and write down, the purpose and goal for what you will write.**

- Complete the emPOWERme survey that appears before the Chapter Summary to learn about how clearly you write.
- Start the assignment many days, or weeks, ahead of the due date so that you have time to do research (if necessary), write an outline, write the paper, revise it, and reread it before turning it in.
- Find a quiet, distraction-free place to work.
- Gather the tools you will need such as paper, a pen or pencil, books, and a computer.
- Do a free-writing exercise: Write continuously on your topic, whatever comes to mind, for 5 or 10 minutes.
- If necessary, do research. This can include research on the Internet or at the library. Be sure you know how to cite sources correctly.
- Define the main point, or thesis, of your piece of writing. Everything in the writing should relate to this idea in some way.
- Looking at the ideas generated in the free-writing exercise and in your research, make an outline to ensure that your writing is in a logical order and will have a solid structure.

- Using your outline, write the first draft. Give yourself permission to be creative and make mistakes. Don't worry about spelling and punctuation at this point—just try to get your thinking down on paper.
- Once you have written your first draft, put it away overnight. Reread it the next day, when it is not as fresh in your mind, and look for ways to improve it. Keep in mind the goal you defined at the beginning of this process.

- Check your spelling and punctuation. Use spellcheck on the computer.
- If you used anyone else's ideas in your writing, make sure you have given them proper credit.
- *After* you have checked your spelling and punctuation, ask someone else to read what you have written. Ask for their honest feedback, and keep that in mind if you revise your work again.

- You have finished your writing. Depending upon your goal, your evaluate step may differ. Are you happy with the finished product? Was what you wrote easily understood by others? If you wrote a story, did you or your readers react to it the way you had hoped? If your writing was a class assignment, did you get the grade that you wanted?

- If you are happy with what you wrote, think about what you did in the process of writing that worked well. Also think about what you might be able to improve next time.
- If you are not happy with the outcome, think about what you did in the process that did not work well for you. Think about what you could do differently, too. Did you leave the assignment for the last minute? Should you have done more research? If you didn't write an outline before writing your piece, consider writing one in the future.

9.1 Absolute Value Equations

P Prepare

O Organize

What are your objectives for Section 9.1?	How can you accomplish each objective?
1 Understand the Meaning of an Absolute Value Equation	• Understand *absolute value* as it relates to a number line. • Learn the procedure for **Solving an Absolute Value Equation.** • Complete the given example on your own. • Complete You Try 1.
2 Solve an Equation of the Form $\|ax + b\| = k$	• Follow the procedure for **Solving an Absolute Value Equation for $k > 0$.** • Be sure to look at the value of k. • Complete the given examples on your own. • Complete You Trys 2 and 3.
3 Solve an Equation of the Form $\|ax + b\| = \|cx + d\|$	• Learn the procedure for **Solving an Absolute Value Equation** of this form. • Understand how to set up problems containing two absolute values. • Complete the given example on your own. • Complete You Try 4.

Read the explanations, follow the examples, take notes, and complete the You Trys.

In Section 1.4 we learned that the absolute value of a number describes its *distance from zero*.

$$|5| = 5 \quad \text{and} \quad |-5| = 5$$

5 units from zero   5 units from zero

$$-7 \; -6 \; -5 \; -4 \; -3 \; -2 \; -1 \quad 0 \quad 1 \quad 2 \quad 3 \quad 4 \quad 5 \quad 6 \quad 7$$

We use this idea of *distance from zero* to solve absolute value equations and inequalities.

1 Understand the Meaning of an Absolute Value Equation

EXAMPLE 1   Solve $|x| = 3$.

Solution

Since the equation contains an absolute value, **solve $|x| = 3$** means *"Find the number or numbers whose distance from zero is 3."*

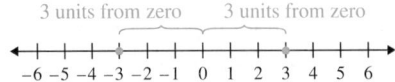

Those numbers are 3 and -3. Each of them is 3 units from zero. The solution set is $\{-3, 3\}$.

Check: $|3| = 3, |-3| = 3$ ✓

[YOU TRY 1]   Solve $|y| = 8$.

W Hint

Write the procedure for solving an absolute value equation in your own words.

Procedure Solving an Absolute Value Equation for $k > 0$

If P represents an expression and k is a positive real number, then to solve $|P| = k$ we rewrite the absolute value equation as the *compound equation*

$$P = k \quad \text{or} \quad P = -k$$

and solve for the variable. P can represent expressions like x, $3a + 2$, $\frac{1}{4}t - 9$, and so on.

---

2 Solve an Equation of the Form $|ax + b| = k$

EXAMPLE 2   Solve each equation.

a) $|m + 1| = 5$ b) $\left|\frac{3}{2}t + 7\right| + 5 = 6$

Solution

a)   Solving $|m + 1| = 5$ means, *"Find the number or numbers that can be substituted for m so that the quantity $m + 1$ is 5 units from 0."*

$m + 1$ will be 5 units from zero if $m + 1 = 5$ or if $m + 1 = -5$, since both 5 and -5 are 5 units from zero. Therefore, we can solve the equation this way:

$$|m + 1| = 5$$

$$m + 1 = 5 \quad \text{or} \quad m + 1 = -5 \qquad \text{Set the quantity inside the absolute value equal to 5 and } -5.$$

$$m = 4 \quad \text{or} \quad m = -6 \qquad \text{Solve.}$$

$$\text{Check: } m = 4: \quad |4 + 1| \overset{?}{=} 5 \qquad \bigg| \qquad m = -6: \quad |-6 + 1| \overset{?}{=} 5$$

$$|5| = 5 \ \checkmark \qquad \qquad \qquad |-5| = 5 \ \checkmark$$

The solution set is $\{-6, 4\}$.

b) Before we rewrite this equation as a compound equation, we must *isolate* the absolute value (get the absolute value on a side by itself).

$$\left|\frac{3}{2}t + 7\right| + 5 = 6$$

$$\left|\frac{3}{2}t + 7\right| = 1 \qquad \text{Subtract 5 to get the absolute value on a side by itself.}$$

W Hint

Isolate the absolute value on one side of the equation before "splitting" the equation.

$$\frac{3}{2}t + 7 = 1 \qquad \text{or} \qquad \frac{3}{2}t + 7 = -1 \qquad \text{Set the quantities inside the absolute value equal to 1 and } -1.$$

$$\frac{3}{2}t = -6 \qquad \qquad \frac{3}{2}t = -8 \qquad \text{Subtract 7.}$$

$$\frac{2}{3} \cdot \frac{3}{2}t = \frac{2}{3} \cdot (-6) \qquad \frac{2}{3} \cdot \frac{3}{2}t = \frac{2}{3} \cdot (-8) \qquad \text{Multiply by } \frac{2}{3}.$$

$$t = -4 \qquad \text{or} \qquad t = -\frac{16}{3} \qquad \text{Solve.}$$

The check is left to the student. The solution set is $\left\{-\frac{16}{3}, -4\right\}$.

[YOU TRY 2]

Solve each equation.

a) $|c - 4| = 3$ b) $\left|\frac{1}{4}n - 3\right| + 2 = 5$

Be sure to think about the value of k when solving $|ax + b| = k$.

EXAMPLE 3   Solve $|4y - 11| = -9$.

Solution

This equation says that the absolute value of the quantity $4y - 11$ equals *negative* 9. Can an absolute value be negative? No! This equation has *no solution*.

The solution set is $\varnothing$.

[YOU TRY 3]   Solve $|d + 3| = -5$.

---

3 Solve an Equation of the Form $|ax + b| = |cx + d|$

Another type of absolute value equation involves two absolute values.

Procedure Solve $|ax + b| = |cx + d|$

If P and Q are expressions, then to solve $|P| = |Q|$ we rewrite the absolute value equation as the *compound equation*

$$P = Q \quad \text{or} \quad P = -Q$$

and solve for the variable.

EXAMPLE 4

Solve $|2w - 3| = |w + 9|$.

Solution

This equation is true when the quantities inside the absolute values are the *same* or when they are *negatives* of each other.

$$|2w - 3| = |w + 9|$$

The quantities are the same   **or**   the quantities are negatives of each other.

$2w - 3 = w + 9$	$2w - 3 = -(w + 9)$
$w = 12$	$2w - 3 = -w - 9$
	$3w = -6$
	$w = -2$

W Hint

Notice that the absolute value equation is split into two separate equations, and consider why the distributive property is necessary.

Check: $w = 12$: $|2(12) - 3| \overset{?}{=} |12 + 9|$ | $w = -2$: $|2(-2) - 3| \overset{?}{=} |-2 + 9|$

$|24 - 3| \overset{?}{=} |21|$ $|-4 - 3| \overset{?}{=} |7|$

$|21| = 21$ ✓ $|-7| = 7$ ✓

The solution set is $\{-2, 12\}$.

 **BE CAREFUL**   In Example 4 and other examples like it, you *must* put parentheses around the expression with the negative as in $-(w + 9)$.

[YOU TRY 4]   Solve $|c + 7| = |3c - 1|$.

We can use a graphing calculator to solve an equation by entering one side of the equation as Y_1 and the other side as Y_2. Then graph the equations. Remember that absolute value equations like the ones found in this section can have 0, 1, or 2 solutions. *The x-coordinates of their points of intersection are the solutions to the equation.*

We will solve $|3x - 1| = 5$ algebraically and by using a graphing calculator, and then compare the results.

First, use algebra to solve $|3x - 1| = 5$. You should get $\left\{-\dfrac{4}{3}, 2\right\}$.

Next, use a graphing calculator to solve $|3x - 1| = 5$.

We will enter $|3x - 1|$ as Y_1 and 5 as Y_2. To enter $Y_1 = |3x - 1|$,

1) Press the $\boxed{Y=}$ key, so that the cursor is to the right of $\backslash Y_1 =$.

2) Press $\boxed{MATH}$ and then press the right arrow, to highlight **NUM.** Also highlighted is 1:abs (which stands for *absolute value*).

3) Press $\boxed{ENTER}$ and you are now back on the $\backslash Y_1 =$ screen. Enter $3x - 1$ with a closing parenthesis so that you have now entered $Y_1 = \text{abs}(3x - 1)$.

4) Press the down arrow to enter $\backslash Y_2 = 5$.

5) Press $\boxed{GRAPH}$.

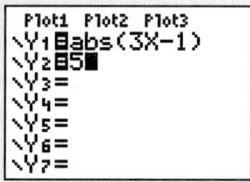 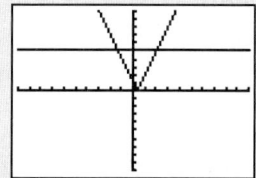

The graphs intersect at two points because there are two solutions to this equation. *Remember that the solutions to the equation are the x-coordinates of the points of intersection.*

To find these x-coordinates we will use the INTERSECT feature.

To find the left-hand intersection point, press $\boxed{2nd}$ $\boxed{TRACE}$ and select 5:intersect. Press $\boxed{ENTER}$. Move the cursor close to the point on the left and press $\boxed{ENTER}$ three times. You get the result in the screen below on the left.

To find the right-hand intersect point, press $\boxed{2nd}$ $\boxed{TRACE}$, select 5:intersect, and press $\boxed{ENTER}$. Move the cursor close to the point, and press $\boxed{ENTER}$ three times. You will see the screen that is below on the right.

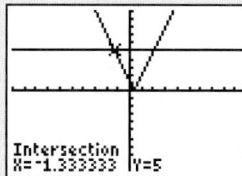 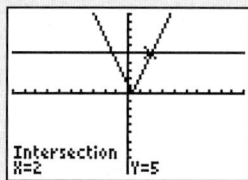

The screen on the left shows $x = -1.333333$. This is the calculator's approximation of $x = -1.\overline{3}$, the decimal equivalent of $x = -\dfrac{4}{3}$, one of the solutions found using algebra.

The screen on the right shows $x = 2$ as a solution, the same solution we obtained algebraically.

The calculator gives us a solution set of $\{-1.333333, 2\}$, while the solution set found using algebra is $\left\{-\dfrac{4}{3}, 2\right\}$.

Solve each equation algebraically, then verify your answer using a graphing calculator.

1) $|x - 1| = 2$ 2) $|x + 4| = 6$ 3) $|2x + 3| = 3$

4) $|4x - 5| = 1$ 5) $|3x + 7| - 6 = -8$ 6) $|6 - x| + 3 = 3$

ANSWERS TO $\boxed{\text{YOU TRY}}$ **EXERCISES**

1) $\{-8, 8\}$ 2) a) $\{1, 7\}$ b) $\{0, 24\}$ 3) $\varnothing$ 4) $\left\{-\dfrac{3}{2}, 4\right\}$

ANSWERS TO TECHNOLOGY EXERCISES

1) $\{-1, 3\}$ 2) $\{-10, 2\}$ 3) $\{-3, 0\}$ 4) $\{1, 1.5\}$ 5) $\varnothing$ 6) $\{6\}$

E Evaluate **9.1** **Exercises**   Do the exercises, and check your work.

Objective 1: Understand the Meaning of an Absolute Value Equation

1) In your own words, explain the *meaning* of the absolute value of a number.

2) Does $|x| = -8$ have a solution? Why or why not?

Objective 2: Solve an Equation of the Form $|ax + b| = k$

Solve.

3) $|q| = 6$ 4) $|z| = 7$

5) $|q - 5| = 3$ 6) $|a + 2| = 13$

7) $|4t - 5| = 7$ 8) $|9x - 8| = 10$

9) $1 = |12c + 5|$ 10) $11 = |4 - 5k|$

11) $\left|\dfrac{2}{3}b + 3\right| = 13$ 12) $\left|\dfrac{3}{4}h + 8\right| = 7$

13) $6 = \left|4 - \dfrac{3}{5}d\right|$ 14) $\dfrac{3}{4} = \left|\dfrac{3}{2}r + 5\right|$

15) $|m - 5| = -3$ 16) $|2k + 7| = -15$

17) $|z - 6| + 4 = 20$ 18) $|q + 3| - 1 = 14$

19) $13 = |2a + 5| + 8$ 20) $10 = |6t - 11| + 5$

21) $|w + 14| = 0$ 22) $|5h + 7| = -5$

23) $|8n + 11| = -1$ 24) $|4p - 3| = 0$

25) $|5b + 3| + 6 = 19$ 26) $1 = |7 - 8x| - 4$

27) $\left|\dfrac{5}{4}k + 2\right| + 9 = 7$ 28) $|3m - 1| + 5 = 2$

Objective 3: Solve an Equation of the Form $|ax + b| = |cx + d|$

Solve the following equations containing two absolute values.

29) $|s + 9| = |2s + 5|$

30) $|j - 8| = |4j - 7|$

31) $|3z + 2| = |6 - 5z|$

32) $|1 - 2a| = |10a + 3|$

33) $\left|\dfrac{3}{2}x - 1\right| = |x|$

34) $|y| = \left|\dfrac{4}{7}y + 12\right|$

35) $\left|\dfrac{1}{4}t - \dfrac{5}{2}\right| = \left|5 - \dfrac{1}{2}t\right|$

36) $\left|k + \dfrac{1}{6}\right| = \left|\dfrac{2}{3}k + \dfrac{1}{2}\right|$

37) Write an absolute value equation that means x is 9 *units from zero.*

38) Write an absolute value equation that means y is 6 *units from zero.*

39) Write an absolute value equation that has a solution set of $\left\{-\dfrac{1}{2}, \dfrac{1}{2}\right\}$.

40) Write an absolute value equation that has a solution set of $\{-1.4, 1.4\}$.

Mixed Exercises: Objectives 2 and 3
Solve.

41) $11 = |7 - v| + 4$

42) $|2q + 9| = 13$

43) $\left|\dfrac{3}{5}p + 3\right| - 7 = -5$

44) $2.8 = -1.4 + |3 - 0.2y|$

45) $|10h - 3| = 0$

46) $|6z + 1| = |4z + 15|$

47) $|1.8a - 3| = |4.2 - 1.2a|$

48) $-7 = |5w + 8|$

49) $15 + |2k + 1| = 6$

50) $\left|\dfrac{3}{4} - \dfrac{5}{6}t\right| + \dfrac{1}{2} = \dfrac{7}{6}$

51) $\left|9 - \dfrac{3}{2}n\right| = 1$

52) $|r - 3| + 8 = 8$

53) $\left|\dfrac{1}{3}g - 2\right| = \left|\dfrac{7}{9}g + \dfrac{1}{6}\right|$

54) $|0.6 + 7y| + 12 = 9$

55) $7.6 = |2.8d + 3.5| + 7.6$

56) $|x + 10| = |5 - 4x|$

R **Rethink**

R1) Which objectives could you use more help on?

R2) How would you explain to a classmate when an absolute value equation has 0, 1, or 2 solutions?

R3) Think about the equation $y = |x|$. What do you think its graph would look like? Do you think the graph is ever below the x-axis? Why or why not?

9.2 Absolute Value Inequalities

What are your objectives for Section 9.2?	How can you accomplish each objective?
1 Solve Absolute Value Inequalities Containing < or ≤	• Learn the procedure for **Solving Absolute Value Inequalities** of this form. • Know what the solution represents on the number line. • Complete the given examples on your own. • Complete You Trys 1 and 2.
2 Solve Absolute Value Inequalities Containing > or ≥	• Learn the procedure for **Solving Absolute Value Inequalities** of this form. • Know what the solution represents on the number line. • Complete the given examples on your own. • Complete You Trys 3 and 4.
3 Solve Special Cases of Absolute Value Inequalities	• Review the meaning of *absolute value*. • Carefully interpret the meaning of the inequality symbol when solving these absolute value inequalities. • Complete the given example on your own. • Complete You Try 5.
4 Solve an Applied Problem Using an Absolute Value Inequality	• Read the problem carefully. • Choose the correct inequality symbol to use in your problem. • Follow the procedure for **Solving Absolute Value Inequalities.** • Complete the given example on your own.

W Work

Read the explanations, follow the examples, take notes, and complete the You Trys.

Now we will learn how to solve **absolute value inequalities.** Some examples of absolute value inequalities are

$$|t| < 6, \qquad |n + 2| \le 5, \qquad |3k - 1| > 11, \qquad \left|5 - \frac{1}{2}y\right| \ge 3$$

1 Solve Absolute Value Inequalities Containing < or ≤

What does it mean to solve $|x| \le 3$? It means to find the set of all real numbers whose distance from zero is 3 *units or less.*

$$3 \text{ is } 3 \text{ units from } 0.$$
$$-3 \text{ is } 3 \text{ units from } 0.$$

Any number *between* 3 and −3 is less than 3 units from zero. For example, if $x = 1$, $|1| \le 3$. If $x = -2$, $|-2| \le 3$. We can represent the solution set on a number line as

We can write the solution set in interval notation as $[-3, 3]$.

W Hint

Notice that a *three-part inequality* is used to solve absolute value inequalities of this form.

Procedure Solve $|P| \leq k$ or $|P| < k$

Let P be an expression and let k be a positive real number. To solve $|P| \leq k$, solve the three-part inequality $-k \leq P \leq k$. ($<$ may be substituted for $\leq$.)

EXAMPLE 1

Solve $|t| < 6$. Graph the solution set, and write the answer in interval notation.

Solution

We must find the set of all real numbers whose distance from zero is less than 6. We can do this by solving the three-part inequality $-6 < t < 6$.

We can represent this on a number line as

We can write the solution set in interval notation as $(-6, 6)$. Any number between -6 and 6 will satisfy the inequality.

[YOU TRY 1]

Solve. Graph the solution set, and write the answer in interval notation.

$$|u| < 9$$

EXAMPLE 2

Solve each inequality. Graph the solution set, and write the answer in interval notation.

a) $|n + 2| \leq 5$ b) $|4 - 5p| < 16$

Solution

a) We must find the set of all real numbers, n, so that $n + 2$ is less than or equal to 5 units from zero. To solve $|n + 2| \leq 5$, we must solve the three-part inequality

$$-5 \leq n + 2 \leq 5$$
$$-7 \leq n \leq 3 \qquad \text{Subtract 2.}$$

The number line representation is

In interval notation, we write the solution as $[-7, 3]$. Any number between -7 and 3 (and including those endpoints) will satisfy the inequality.

b) Solve the three-part inequality.

$$-16 < 4 - 5p < 16$$
$$-20 < -5p < 12 \qquad \text{Subtract 4.}$$
$$4 > p > -\frac{12}{5} \qquad \text{Divide by } -5 \text{, and change the direction of the inequality symbols.}$$

This inequality means p *is less than* 4 *and greater than* $-\dfrac{12}{5}$. We can rewrite it as

$-\dfrac{12}{5} < p < 4$.

The number line representation of the solution set is

In interval notation, we write $\left(-\dfrac{12}{5}, 4\right)$.

[YOU TRY 2] Solve each inequality. Graph the solution set, and write the answer in interval notation.

a) $|6k + 5| \leq 13$ b) $|5 - 2w| < 9$

2 Solve Absolute Value Inequalities Containing $>$ or $\geq$

To solve $|x| \geq 4$ means to find the set of all real numbers whose distance from zero is 4 *units or more.*

$$4 \text{ is } 4 \text{ units from } 0.$$
$$-4 \text{ is } 4 \text{ units from } 0.$$

Any number greater than 4 *or* less than -4 is more than 4 units from zero.

For example, if $x = 6$, $|6| \geq 4$. If $x = -5$, then $|-5| = 5$ and $5 \geq 4$. We can represent the solution set to $|x| \geq 4$ as

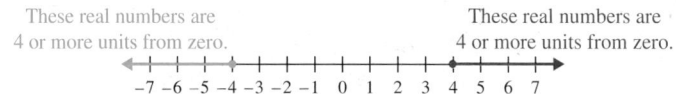

These real numbers are   These real numbers are
4 or more units from zero.   4 or more units from zero.

The solution set consists of two separate regions, so we can write a compound inequality using *or.*

$$x \leq -4 \quad \text{or} \quad x \geq 4$$

In interval notation, we write $(-\infty, -4] \cup [4, \infty)$.

W Hint

Notice that an "*or*" compound inequality is used to solve absolute value inequalities of this form.

Procedure   Solve $|P| \geq k$ or $|P| > k$

Let P be an expression and let k be a positive, real number. To solve $|P| \geq k$ ($>$ may be substituted for $\geq$), solve the compound inequality $P \geq k$ or $P \leq -k$.

EXAMPLE 3   Solve $|r| > 2$. Graph the solution set, and write the answer in interval notation.

Solution

We must find the set of all real numbers whose distance from zero is greater than 2. The solution is the compound inequality $r > 2$ or $r < -2$.

On the number line, we can represent the solution set as

In interval notation, we write $(-\infty, -2) \cup (2, \infty)$. Any number in the shaded region will satisfy the inequality. For example, to the right of 2, if $r = 3$, then $|3| > 2$. To the left of -2, if $r = -4$, then $|-4| > 2$.

[YOU TRY 3] Solve $|d| \geq 5$. Graph the solution set, and write the answer in interval notation.

EXAMPLE 4

Solve each inequality. Graph the solution set, and write the answer in interval notation.

a) $|3k - 1| > 11$ b) $|c + 6| + 10 \geq 12$

Solution

a) To solve $|3k - 1| > 11$ means to find the set of all real numbers, k, so that $3k - 1$ is more than 11 units from zero on the number line. We will solve the compound inequality

$$3k - 1 > 11 \qquad \text{or} \qquad 3k - 1 < -11.$$

↑   ↑

$3k - 1$ is more than 11 units away $3k - 1$ is more than 11 units away
from zero to the *right* of zero.   from zero to the *left* of zero.

$$
\begin{array}{llll}
3k - 1 > 11 & \text{or} & 3k - 1 < -11 & \\
3k > 12 & \text{or} & 3k < -10 & \text{Add 1.} \\
k > 4 & \text{or} & k < -\dfrac{10}{3} & \text{Divide by 3.}
\end{array}
$$

On the number line, we get

$$\overset{\longleftarrow \;+\;+\;+\;+\;+\;\circ\;+\;+\;+\;+\;+\;+\;+\;\diamond\;+\;+\;+\;\longrightarrow}{\quad -7\;-6\;-5\;-4\;-3\;-2\;-1\;\;0\;\;1\;\;2\;\;3\;\;4\;\;5\;\;6\;\;7}$$

From the number line, we can write the interval notation $\left(-\infty, -\dfrac{10}{3}\right) \cup (4, \infty)$.

Any number in the shaded region will satisfy the inequality.

Notice that we first isolate the absolute value on one side of the inequality before we write the "*or*" compound inequality.

b) Begin by getting the absolute value on a side by itself.

$$
\begin{array}{llll}
|c + 6| + 10 \geq 12 & & & \\
|c + 6| \geq 2 & & & \text{Subtract 10.} \\
c + 6 \geq 2 & \text{or} & c + 6 \leq -2 & \text{Rewrite as a compound inequality.} \\
c \geq -4 & \text{or} & c \leq -8 & \text{Subtract 6.}
\end{array}
$$

The graph of the solution set is

$$\overset{\longleftarrow \;+\;+\;\bullet\;+\;+\;+\;+\;\bullet\;+\;+\;+\;+\;+\;+\;\longrightarrow}{-10\;-9\;-8\;-7\;-6\;-5\;-4\;-3\;-2\;-1\;\;0\;\;1\;\;2}$$

The interval notation is $(-\infty, -8] \cup [-4, \infty)$.

[YOU TRY 4] Solve each inequality. Graph the solution set and write the answer in interval notation.

a) $|8q + 9| \geq 7$ b) $|k + 8| - 5 \geq 9$

Example 5 illustrates why it is important to understand what the absolute value inequality means before trying to solve it.

3 Solve Special Cases of Absolute Value Inequalities

EXAMPLE 5

Solve each inequality.

a) $|z + 3| < -6$ b) $|2s - 1| \geq 0$ c) $|4d + 7| + 9 \leq 9$

Solution

a) Look carefully at this inequality, $|z + 3| < -6$. It says that the absolute value of a quantity, $z + 3$, is *less than* a negative number. Since the absolute value of a quantity is always zero or positive, this inequality has *no solution*.

The solution set is $\varnothing$.

b) $|2s - 1| \geq 0$ says that the absolute value of a quantity, $2s - 1$, is greater than or equal to zero. An absolute value is *always* greater than or equal to zero, so *any* value of s will make the inequality true.

The solution set consists of all real numbers, which we can write in interval notation as $(-\infty, \infty)$.

c) Begin by isolating the absolute value. $|4d + 7| + 9 \leq 9$

$$|4d + 7| \leq 0 \qquad \text{Subtract 9.}$$

The absolute value of a quantity can *never be less than zero* but it *can equal zero*. To solve this, we must solve $4d + 7 = 0$.

$$4d + 7 = 0$$
$$4d = -7$$
$$d = -\frac{7}{4}$$

The solution set is $\left\{ -\frac{7}{4} \right\}$.

[YOU TRY 5] Solve each inequality.

a) $|p + 4| \geq 0$ 　　　　 b) $|5n - 7| < -2$ 　　　　 c) $|6y - 1| + 3 \leq 3$

---

4 Solve an Applied Problem Using an Absolute Value Inequality

EXAMPLE 6

On an assembly line, a machine is supposed to fill a can with 19 oz of soup. However, the possibility for error is ±0.25 oz. Let x represent the range of values for the amount of soup in the can. Write an absolute value inequality to represent the range for the number of ounces of soup in the can, then solve the inequality and explain the meaning of the answer.

Solution

If the *actual* amount of soup in the can is x and there is supposed to be 19 oz in the can, then the error in the amount of soup in the can is $|x - 19|$. If the possible error is ±0.25 oz, then we can write the inequality

$$|x - 19| \leq 0.25$$
$$-0.25 \leq x - 19 \leq 0.25 \qquad \text{Solve.}$$
$$18.75 \leq x \leq 19.25$$

The actual amount of soup in the can is between 18.75 and 19.25 oz.

Andrew Resek/McGraw-Hill Education

1) $(-9, 9)$

2) a) $\left[-3, \dfrac{4}{3}\right]$ b) $(-2, 7)$

3) $(-\infty, -5] \cup [5, \infty)$

4) a) $(-\infty, -2] \cup \left[-\dfrac{1}{4}, \infty\right)$

b) $(-\infty, -22] \cup [6, \infty)$

5) a) $(-\infty, \infty)$ b) $\varnothing$ c) $\left\{\dfrac{1}{6}\right\}$

E Evaluate **9.2** Exercises   Do the exercises, and check your work.

Mixed Exercises: Objectives 1 and 2

Graph each inequality on a number line, and represent the sets of numbers using interval notation.

1) $-1 \le p \le 5$

2) $7 < t < 11$

 3) $y < 2$ or $y > 9$

4) $a \le -8$ or $a \ge \dfrac{1}{2}$

5) $n \le -\dfrac{9}{2}$ or $n \ge \dfrac{3}{5}$

6) $-\dfrac{1}{4} \le q \le \dfrac{11}{4}$

Mixed Exercises: Objectives 1 and 3

Solve each inequality. Graph the solution set, and write the answer in interval notation.

7) $|m| \le 7$

8) $|c| < 1$

9) $|3k| < 12$

10) $\left|\dfrac{5}{4}z\right| \le 30$

11) $|w - 2| < 4$

12) $|k - 6| \le 2$

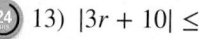 13) $|3r + 10| \le 4$

14) $|4a + 1| \le 12$

15) $|7 - 6p| \le 3$

16) $|17 - 9d| < 8$

17) $|5q + 11| < 0$

18) $|6t + 16| < 0$

19) $|8m - 15| \le -5$

20) $|2x + 7| \le -12$

21) $|2v + 5| + 3 < 14$

22) $|8c - 3| + 15 < 20$

23) $\left|\dfrac{3}{2}h + 6\right| - 2 \le 10$

24) $7 + \left|\dfrac{8}{3}u - 9\right| < 12$

Mixed Exercises: Objectives 2 and 3

Solve each inequality. Graph the solution set, and write the answer in interval notation.

25) $|t| \ge 7$

26) $|p| > 3$

27) $|d + 10| \ge 4$

28) $|q - 7| > 12$

29) $|4v - 3| \ge 9$

30) $|6a + 19| > 11$

31) $|17 - 6x| > 5$

32) $|1 - 4g| \ge 10$

33) $|8k + 5| \ge 0$

34) $|5b - 6| \ge 0$

35) $|z - 3| \ge -5$

36) $|3r + 10| > -11$

37) $|w + 6| - 4 \ge 2$

38) $|2m - 1| + 4 > 5$

39) $-3 + \left|\dfrac{5}{6}n + \dfrac{1}{2}\right| \ge 1$

40) $\left|\dfrac{3}{2}y - \dfrac{5}{4}\right| + 9 > 11$

41) Explain why $|3t - 7| < 0$ has no solution.

42) Explain why $|4l + 9| \le -10$ has no solution.

43) Explain why the solution to $|2x + 1| \ge -3$ is $(-\infty, \infty)$.

44) Explain why the solution to $|7y - 3| \ge 0$ is $(-\infty, \infty)$.

Mixed Exercises: Objectives 1–3

45) Write an absolute value inequality that means *y is at most 8 units from zero.*

46) Write an absolute value inequality that means *a is more than 5 units from zero.*

Exercises 47 to 68 contain absolute value equations, linear inequalities, and both types of absolute value inequalities. Solve each. Write the solution set for equations in set notation, and use interval notation for inequalities.

47) $|2v + 9| > 3$

48) $\left|\dfrac{5}{3}a + 2\right| = 8$

49) $3 = |4t + 5|$

50) $|4k + 9| \leq 5$

51) $9 \leq |7 - 8q|$

52) $|2p - 5| - 12 = 11$

53) $2(x - 8) + 10 < 4x$

54) $\dfrac{1}{2}n + 11 < 8$

55) $|6y + 5| \leq -9$

56) $8 \leq |5v + 2|$

57) $\left|\dfrac{4}{3}x + 1\right| = \left|\dfrac{5}{3}x + 8\right|$

58) $|7z - 8| \leq 0$

59) $|4 - 9t| + 2 = 1$

60) $|5b - 11| - 18 < -10$

61) $-\dfrac{3}{5} \geq \dfrac{5}{2}a - \dfrac{1}{2}$

62) $4 + 3(2r - 5) > 9 - 4r$

63) $|6k + 17| > -4$

64) $|5 - w| \geq 3$

65) $5 \geq |c + 8| - 2$

66) $0 \leq |4a + 1|$

67) $|5h - 8| > 7$

68) $\left|\dfrac{2}{3}y - 1\right| = \left|\dfrac{3}{2}y + 4\right|$

Objective 4: Solve an Applied Problem Using an Absolute Value Inequality

(24) 69) A gallon of milk should contain 128 oz. The possible error in this measurement, however, is ± 0.75 oz. Let a represent the range of values for the amount of milk in the container. Write an absolute value inequality to represent the range for the number of ounces of milk in the container, then solve the inequality and explain the meaning of the answer.

70) Dawn buys a 27-oz box of cereal. The possible error in this amount, however, is ± 0.5 oz. Let c represent the range of values for the amount of cereal in the box. Write an absolute value inequality to represent the range for the number of ounces of cereal in the box, then solve the inequality and explain the meaning of the answer.

Andrew Resek/McGraw-Hill Education

71) Emmanuel spent $38 on a birthday gift for his son. He plans on spending within $5 of that amount on his daughter's birthday gift. Let b represent the range of values for the amount he will spend on his daughter's gift. Write an absolute value inequality to represent the range for the amount of money Emmanuel will spend on his daughter's birthday gift, then solve the inequality and explain the meaning of the answer.

72) An employee at a home-improvement store is cutting a window shade for a customer. The customer wants the shade to be 32 in. wide. If the machine's possible error in cutting the shade is $\pm \dfrac{1}{16}$ in., write an absolute value inequality to represent the range for the width of the window shade, and solve the inequality. Explain the meaning of the answer. Let w represent the range of values for the width of the shade.

R Rethink

R1) Which objective is the most difficult for you?

R2) Which type of absolute value inequality requires the use of a three-part inequality to obtain the solution?

R3) Which type of absolute value inequality requires the use of a compound inequality to reach the solution?

9.3 Linear and Compound Linear Inequalities in Two Variables

What are your objectives for Section 9.3?	How can you accomplish each objective?
1 Graph a Linear Inequality in Two Variables	• Know the definition of a *linear inequality in two variables*. • Learn the two methods for graphing a linear inequality in two variables, and write them in your own words. • Understand what it means when a region is shaded and when a region is not shaded. • Complete the given examples on your own. • Complete You Trys 1 and 2.
2 Graph a Compound Linear Inequality in Two Variables	• Learn the procedure for **Graphing Compound Linear Inequalities in Two Variables.** • Understand the difference between the *intersection* of two solution sets and the *union* of two solution sets. • Choose a test point to check the shaded solution. • Complete the given examples on your own. • Complete You Trys 3 and 4.
3 Solve a Linear Programming Problem	• Be able to apply the given procedure to any problem. • Complete the given example on your own.

W Work

Read the explanations, follow the examples, take notes, and complete the You Trys.

In Chapter 3, we learned how to solve linear inequalities in *one variable* such as $2x - 3 \geq 5$.

We will begin this section by learning how to graph the solution set of linear inequalities in *two variables*. Then we will learn how to graph the solution set of *systems* of linear inequalities in two variables.

1 Graph a Linear Inequality in Two Variables

Definition

A **linear equality in two variables** is an inequality that can be written in the form $Ax + By \geq C$ or $Ax + By \leq C$ where A, B, and C are real numbers and where A and B are not both zero. (> and < may be substituted for $\geq$ and $\leq$.)

Here are some examples of linear inequalities in two variables.

$$x + y \geq 3, \qquad y < \frac{1}{4}x + 3, \qquad x \leq 2, \qquad y > -4$$

W Hint

Read very carefully. Do you *understand* what you are reading?

The solutions to linear inequalities in two variables, such as $x + y \geq 3$, are *ordered pairs* of the form (x, y) that make the inequality true. We graph a linear inequality in two variables on a rectangular coordinate system.

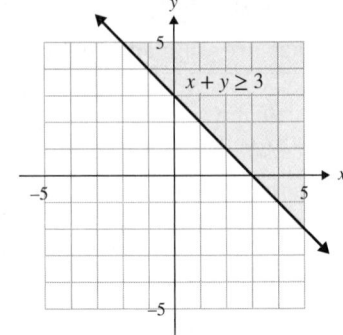

The points $(5, 2)$, $(1, 4)$, and $(3, 0)$ are some of the points that satisfy $x + y \geq 3$. There are infinitely many solutions. The points $(0, 0)$, $(-4, 1)$, and $(2, -3)$ are three of the points that do *not* satisfy $x + y \geq 3$. There are infinitely many points that are not solutions.

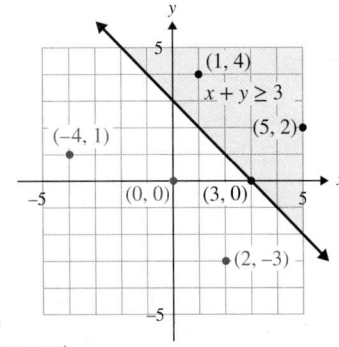

Points in the shaded region and on the line are in the solution set.

The points in the unshaded region are *not* in the solution set.

As we saw in the preceding graph, the line divides the *plane* into two regions or **half planes.** The line $x + y = 3$ is the **boundary line** between the two half planes. We will use this boundary line to graph a linear inequality in two variables. Notice that the boundary line is written as an equation; it uses an equal sign.

Procedure Graph a Linear Inequality in Two Variables Using a Test Point

1) **Graph the boundary line.** If the inequality contains $\geq$ or $\leq$, make this boundary line *solid.* If the inequality contains $>$ or $<$, make it *dotted.*

2) **Choose a test point not on the line, and shade the appropriate region.** Substitute the test point into the inequality. If $(0, 0)$ is not on the line, it is an easy point to test in the inequality.

    a) If it *makes the inequality true,* shade the region *containing* the test point. All points in the shaded region are part of the solution set.

    b) If the test point *does not satisfy the inequality,* shade the region on the *other* side of the line. All points in the shaded region are part of the solution set.

EXAMPLE 1 Graph $3x + 4y \leq -8$.

Solution

1) Graph the boundary line $3x + 4y = -8$ as a solid line.

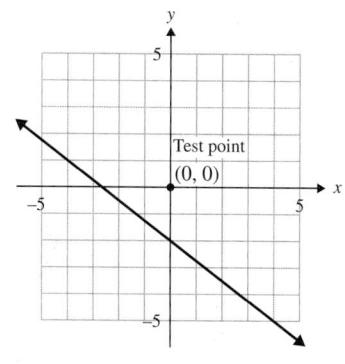

2) Choose a test point not on the line and substitute it into the inequality to determine whether it makes the inequality true.

> **W Hint**
>
> If the inequality includes the *equals* condition, the boundary line is drawn solid. Points on this line make the inequality true.

Test Point	Substitute into $3x + 4y \leq -8$
$(0, 0)$	$3(0) + 4(0) \leq -8$
	$0 \leq -8$ False

Since the test point $(0, 0)$ does *not* satisfy the inequality we will shade the region that does *not* contain the point $(0, 0)$.

All points on the line and in the shaded region satisfy the inequality $3x + 4y \leq -8$.

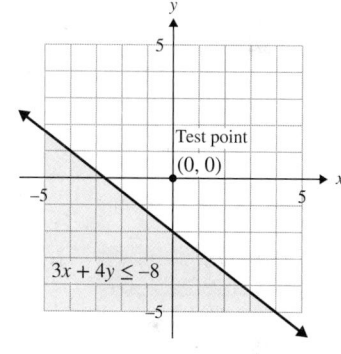

EXAMPLE 2 Graph $-x + 2y > -4$.

Solution

1) Since the inequality symbol is $>$, graph a *dotted* boundary line, $-x + 2y = -4$. (This means that the points *on* the line are not part of the solution set.)

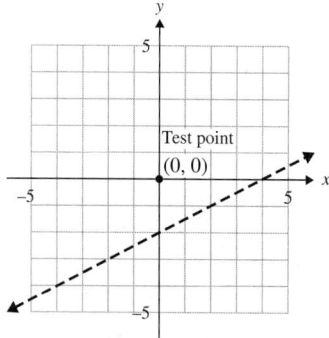

2) Choose a test point not on the line and substitute it into the inequality to determine whether it makes the inequality true.

> **W Hint**
>
> If the inequality *does not* include the equals condition, the boundary line is dotted. Points on this line make the inequality false.

Test Point	Substitute into $-x + 2y > -4$
$(0, 0)$	$-(0) + 2(0) > -4$
	$0 > -4$ True

Since the test point $(0, 0)$ satisfies the inequality, shade the region containing that point.

All points in the shaded region satisfy the inequality $-x + 2y > -4$.

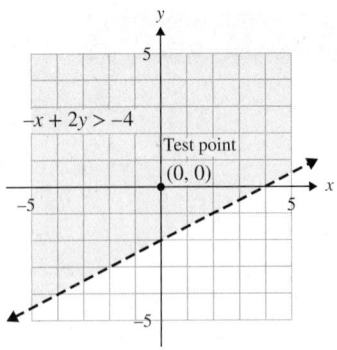

[YOU TRY 1] Graph each inequality.

a) $2x + y \leq 4$ b) $x + 4y > 12$

If we write the inequality in *slope-intercept form,* we can decide which region to shade without using test points.

> **Procedure** Graph a Linear Inequality in Two Variables Using the Slope-Intercept Method
>
> 1) Write the inequality in the form $y \geq mx + b$ ($y > mx + b$) or $y \leq mx + b$ ($y < mx + b$), and graph the boundary line $y = mx + b$.
> 2) If the inequality is in the form $y \geq mx + b$ or $y > mx + b$, shade *above* the line.
> 3) If the inequality is in the form $y \leq mx + b$ or $y < mx + b$, shade *below* the line.

EXAMPLE 3

Graph each inequality using the slope-intercept method.

a) $y < -\dfrac{1}{3}x + 5$ b) $2x - y \leq -2$

Solution

a) The inequality $y < -\dfrac{1}{3}x + 5$ is already in slope-intercept form.

Graph the boundary line $y = -\dfrac{1}{3}x + 5$ as a *dotted line.*

Since $y < -\dfrac{1}{3}x + 5$ has a *less than* symbol, shade *below* the line. All points in the shaded region satisfy $y < -\dfrac{1}{3}x + 5$.

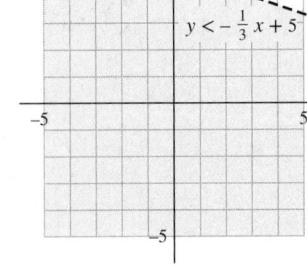

We can choose a point such as $(0, 0)$ in the shaded region as a check. Substituting this point into $y < -\dfrac{1}{3}x + 5$ gives us $0 < -\dfrac{1}{3}(0) + 5$, or $0 < 5$, which is true.

b) Solve $2x - y \leq -2$ for y.

$$2x - y \leq -2$$
$$-y \leq -2x - 2 \qquad \text{Subtract } 2x.$$
$$y \geq 2x + 2 \qquad \text{Divide by } -1, \text{ and change the}$$
$$\text{direction of the inequality symbol.}$$

Graph $y = 2x + 2$ as a *solid line*.

Since $y \geq 2x + 2$ has a *greater than or equal to* symbol, shade *above* the line.

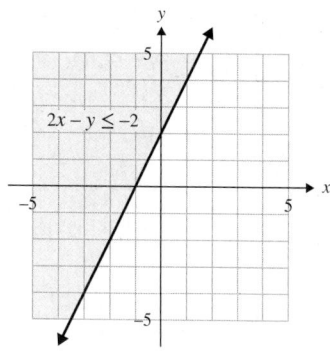

All points on the line and in the shaded region satisfy $2x - y \leq -2$.

<table>
<tr><td>

[YOU TRY 2]

</td><td>

Graph each inequality using the slope-intercept method.

a) $y \geq -\dfrac{3}{4}x - 6$ b) $5x - 2y > 4$

</td></tr>
</table>

2 Graph a Compound Linear Inequality in Two Variables

Linear inequalities in two variables are called *compound linear inequalities* if they are connected by the words *and* or *or*.

The solution set of a compound inequality containing *and* is the *intersection* of the solution sets of the inequalities. The solution set of a compound inequality containing *or* is the *union* of the solution sets of the inequalities.

Procedure Graphing Compound Linear Inequalities in Two Variables

1) Graph each inequality separately on the same axes. Shade lightly.
2) If the inequality contains *and*, the solution set is the *intersection* of the shaded regions. Heavily shade this region.
3) If the inequality contains *or*, the solution set is the *union* (total) of the shaded regions. Heavily shade this region.

EXAMPLE 4   Graph $x \le 2$ and $2x + 3y > 3$.

Solution

To graph $x \le 2$, graph the boundary line $x = 2$ as a solid line. The x-values are *less than* 2 to the *left* of 2, so shade the region to the left of the line $x = 2$.

Graph $2x + 3y > 3$. Use a dotted boundary line.

The region shaded blue in the third graph, along with the solid boundary line of the region, is the solution set of the compound inequality; it is the *intersection* of the shaded regions of the two graphs to the left.

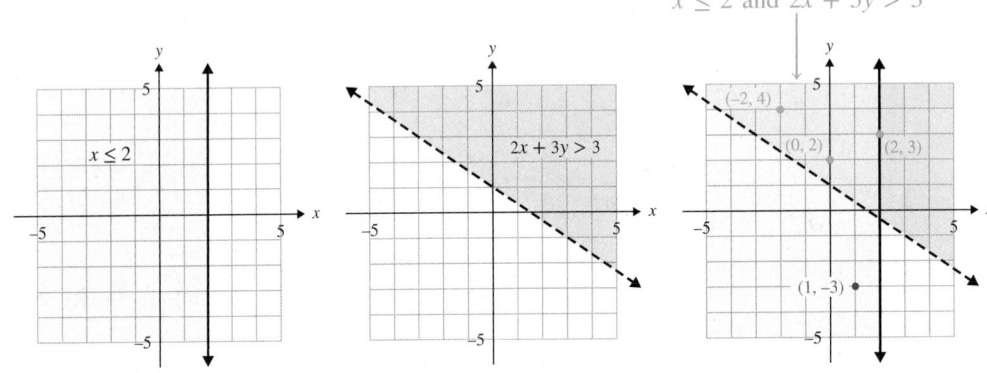

W Hint

Notice that the final solution is the set of ordered pairs that lie in the overlapping shaded region.

Any point in the solution set must satisfy *both* inequalities, and any point *not* in the solution set will not satisfy *both* inequalities. We check three test points next. (See the graph.)

Test Point	Substitute into $x \le 2$	Substitute into $2x + 3y > 3$	Solution?
$(-2, 4)$	$-2 \le 2$ True	$2(-2) + 3(4) > 3$ $8 > 3$ True	Yes
$(0, 2)$	$0 \le 2$ True	$2(0) + 3(2) > 3$ $6 > 3$ True	Yes
$(1, -3)$	$1 \le 2$ True	$2(1) + 3(-3) > 3$ $-7 > 3$ False	No

Although we show three separate graphs in Example 4, it is customary to graph everything on the same axes, shading lightly at first, then to heavily shade the region that is the graph of the compound inequality.

[YOU TRY 3]   Graph the compound inequality $y \le 3x - 1$ and $y + 2x \le 4$.

EXAMPLE 5

Graph $y \leq \frac{1}{2}x$ or $2x + y \geq 2$.

Solution

Graph each inequality separately. The solution set of the compound inequality will be the *union* (total) of the shaded regions.

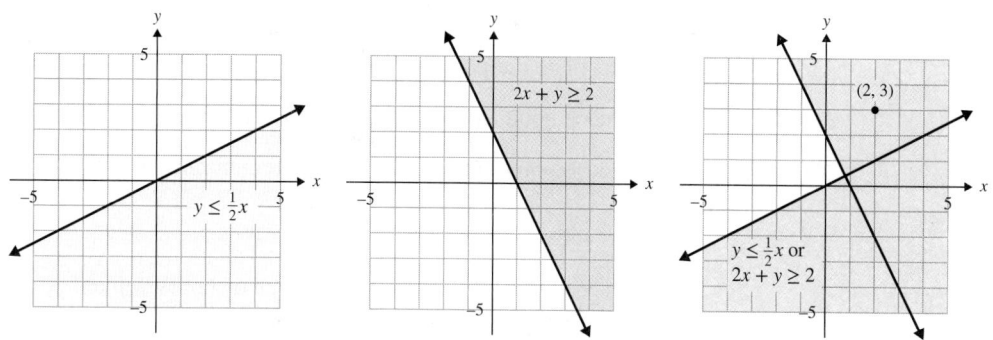

Any point in the shaded region of the *third graph* will be a solution to the compound inequality $y \leq \frac{1}{2}x$ or $2x + y \geq 2$. This means the point must satisfy $y \leq \frac{1}{2}x$ *or* $2x + y \geq 2$ *or* both. One point in the shaded region is $(2, 3)$.

Test Point	Substitute into $y \leq \frac{1}{2}x$	Substitute into $2x + y \geq 2$	Solution?
$(2, 3)$	$3 \leq \frac{1}{2}(2)$ $3 \leq 1$ False	$2(2) + 3 \geq 2$ $7 \geq 2$ True	Yes

Although $(2, 3)$ does not satisfy $y \leq \frac{1}{2}x$, it *does* satisfy $2x + y \geq 2$, so it *is* a solution of the compound inequality.

Choose a point in the region that is *not* shaded to verify that it does not satisfy either inequality.

[YOU TRY 4] Graph the compound inequality $x \geq -4$ or $x - 3y \leq -3$.

3 Solve a Linear Programming Problem

A practical application of linear inequalities in two variables is a process called **linear programming.** Companies use linear programming to determine the best way to use their machinery, employees, and other resources.

A linear programming problem may consist of several inequalities called **constraints.** Constraints describe the conditions that the variables must meet. The graph of the *intersection* of these inequalities is called the **feasible region**—the ordered pairs that are the possible solutions to the problem.

EXAMPLE 6

During a particular week, a company wants Harvey and Amy to work at most 40 hours between them.

Let x = the number of hours Harvey works
y = the number of hours Amy works

W Hint

Do you think these steps could be used for every linear programming problem?

a) Write the linear inequalities that describe the constraints on the number of hours available to work.

b) Graph the feasible region (solution set of the intersection of the inequalities), which describes the possible number of hours each person can work.

c) Find a point in the feasible region and discuss its meaning.

d) Find a point outside the feasible region and discuss its meaning.

Solution

a) Since x and y represent the number of hours worked, x and y cannot be negative. We can write $x \geq 0$ and $y \geq 0$.

Together they can work at most 40 hours. We can write $x + y \leq 40$.

The inequalities that describe the constraints on the number of hours available are $x \geq 0$ *and* $y \geq 0$ *and* $x + y \leq 40$. We want to find the *intersection* of these inequalities.

b) The graphs of $x \geq 0$ *and* $y \geq 0$ give us the set of points in the first quadrant since x and y are both positive here.

Graph $x + y \leq 40$. This will be the region *below and including* the line $x + y = 40$ in quadrant I.

The feasible region is shown here.

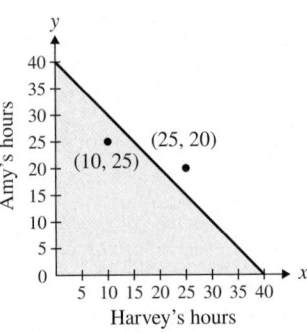

c) One point in the feasible region is (10, 25). It represents Harvey working 10 hours and Amy working 25 hours. It satisfies all three inequalities.

Test Point	Substitute into $x \geq 0$	Substitute into $y \geq 0$	Substitute into $x + y \leq 40$
(10, 25)	$10 \geq 0$ True	$25 \geq 0$ True	$10 + 25 \leq 40$ $35 \leq 40$ True

d) One point outside the feasible region is (25, 20). It represents Harvey working 25 hours and Amy working 20 hours. This is not possible since it does not satisfy the inequality $x + y \leq 40$.

Test Point	Substitute into $x \geq 0$	Substitute into $y \geq 0$	Substitute into $x + y \leq 40$
(25, 20)	$25 \geq 0$ True	$20 \geq 0$ True	$25 + 20 \leq 40$ $45 \leq 40$ False

To graph a linear inequality in two variables using a
graphing calculator, first solve the inequality for y. Then
graph the boundary line found by replacing the inequality
symbol with an = symbol. For example,
to graph the inequality $2x - y \leq 5$, solve it for
$y \geq 2x - 5$. Graph the boundary equation $y = 2x - 5$
using a solid line since the inequality symbol is $\leq$.
Press $\boxed{Y =}$, enter $2x - 5$ in Y_1, press $\boxed{ZOOM}$, and select 6:ZStandard to graph the
equation as shown.

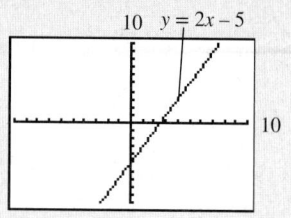

If the inequality symbol is $\leq$, shade below the boundary line. If the inequality
symbol is $\geq$, shade above it. To shade above the line, press $\boxed{Y =}$ and move the
cursor to the left of Y_1 using the left arrow key. Press $\boxed{ENTER}$ twice and then move
the cursor to the next line as shown below left. Press $\boxed{GRAPH}$ to graph the
inequality as shown below right.

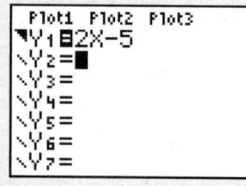
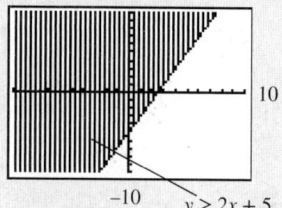

To shade below the line, press $\boxed{Y =}$ and move the cursor to the left of Y_1 using the
left arrow key. Keep pressing $\boxed{ENTER}$ until you see ◣ next to Y_1, then move the
cursor to the next line as shown below left. Press $\boxed{GRAPH}$ to graph the inequality
$y \leq 2x - 5$ as shown below right.

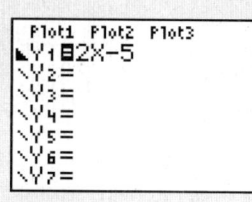
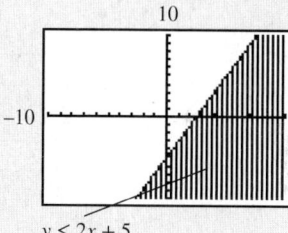

Graph the linear inequalities in two variables.

1) $y \leq 5x - 2$ 2) $y \geq x - 4$ 3) $x - 2y \leq 6$

4) $y - x \geq 5$ 5) $y \leq -4x + 1$ 6) $y \geq 3x - 6$

ANSWERS TO $\boxed{\text{YOU TRY}}$ **EXERCISES**

1) a)
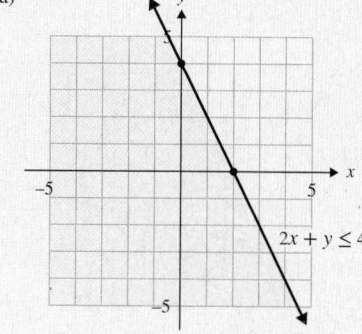

b)
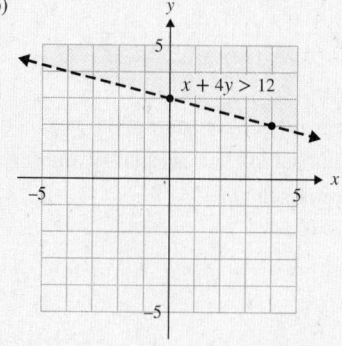

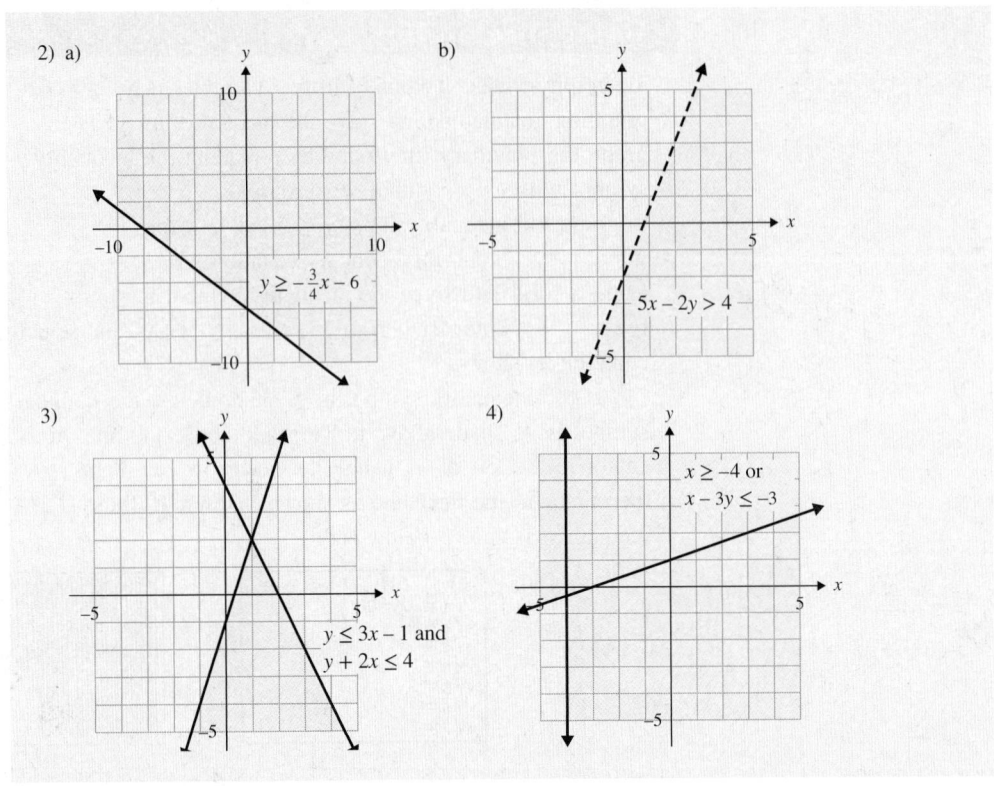

2) a) $y \geq -\frac{3}{4}x - 6$

b) $5x - 2y > 4$

3) $y \leq 3x - 1$ and $y + 2x \leq 4$

4) $x \geq -4$ or $x - 3y \leq -3$

ANSWERS TO TECHNOLOGY EXERCISES

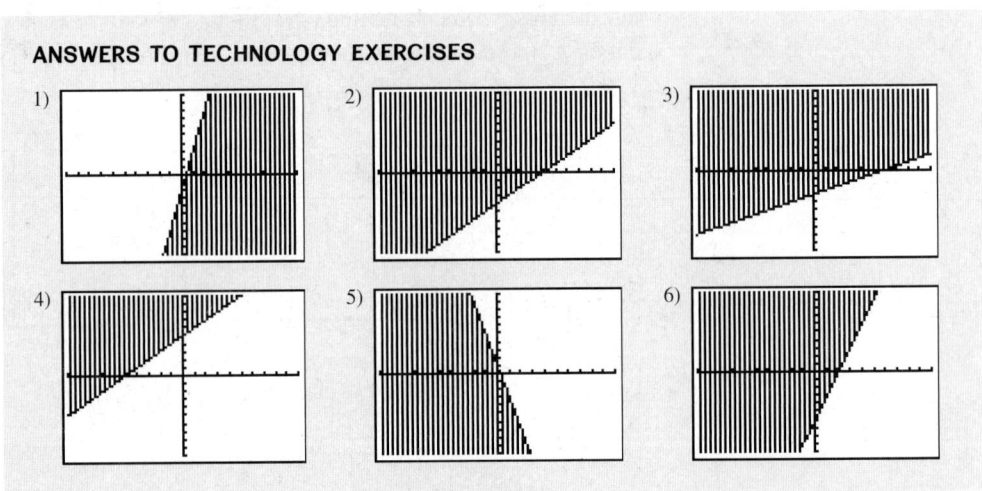

1) 2) 3) 4) 5) 6)

Do the exercises, and check your work.

Objective 1: Graph a Linear Inequality in Two Variables

The graphs of linear inequalities are given next. For each, find three points that satisfy the inequality and three that are not in the solution set.

1)

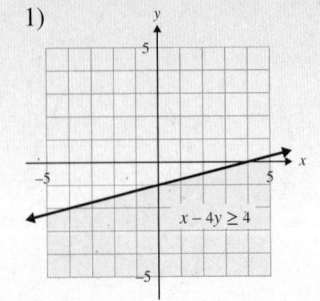

$x - 4y \geq 4$

2)

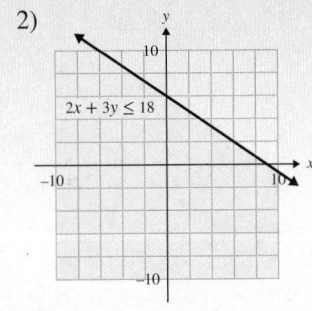

$2x + 3y \leq 18$

3)

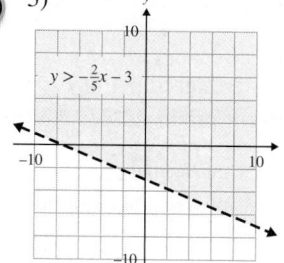

$y > -\frac{2}{5}x - 3$

4)

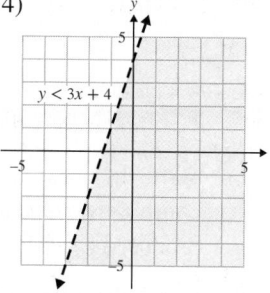

$y < 3x + 4$

5)

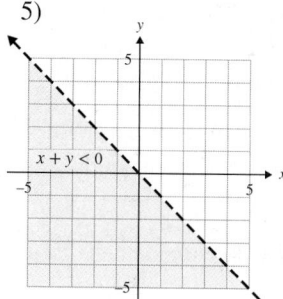

$x + y < 0$

6)

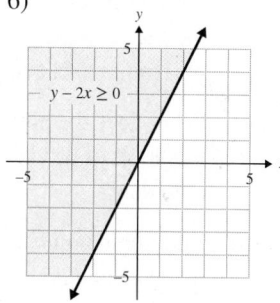

$y - 2x \geq 0$

7) Will the boundary line you draw to graph $3x - 4y < 5$ be solid or dotted?

8) Are points on solid boundary lines included in the inequality's solution set?

Graph the inequalities. Use a test point.

 9) $2x + y \geq 6$

10) $4x + y \leq 3$

11) $y < x + 2$

12) $y > \frac{1}{2}x - 1$

13) $2x - 7y \leq 14$

14) $4x + 3y < 15$

15) $y < x$

16) $y \geq 3x$

17) $y \geq -5$

18) $x < 1$

19) Should you shade the region above or below the boundary line for the inequality $y \leq 7x + 2$?

20) Should you shade the region above or below the boundary line for the inequality $y > 2x + 4$?

Use the slope-intercept method to graph each inequality.

21) $y \leq 4x - 3$

22) $y \geq \frac{5}{2}x - 8$

23) $y > \frac{2}{5}x - 4$

24) $y < \frac{1}{4}x + 1$

25) $6x + y > 3$

26) $2x + y > -5$

27) $9x - 3y \leq -21$

28) $3x + 5y < -20$

29) $x > 2y$

30) $x - y \leq 0$

31) To graph an inequality like $y \geq \frac{1}{3}x + 2$, would you rather use a test point or use the slope-intercept form?

32) To graph an inequality like $7x + 2y < 10$, would you rather use a test point or the slope-intercept form? Why?

Graph using either a test point or the slope-intercept method.

33) $y > -\frac{3}{4}x + 1$

34) $y \leq \frac{1}{3}x - 6$

35) $5x + 2y < -8$

36) $4x + y < 7$

37) $5x - 3y \geq -9$

38) $9x - 3y \leq 21$

39) $x > 2$

40) $y \leq 4$

41) $3x - 4y > 12$

42) $6x - y \leq 2$

Objective 2: Graph a Compound Linear Inequality in Two Variables

43) Is (3, 5) in the solution set of the compound inequality $x - y \geq -6$ and $2x + y < 7$? Why or why not?

44) Is (3, 5) in the solution set of the compound inequality $x - y \geq -6$ or $2x + y < 7$? Why or why not?

Graph each compound inequality.

45) $x \leq 4$ and $y \geq -\frac{3}{2}x + 3$

46) $y \leq \frac{1}{4}x + 2$ and $y \geq -1$

47) $y < x + 4$ and $y \geq -3$

48) $x < 3$ and $y > \frac{2}{3}x - 1$

49) $2x - 3y < -9$ and $x + 6y < 12$

50) $5x - 3y > 9$ and $2x + 3y \leq 12$

51) $y \leq -x - 1$ or $x \geq 6$

52) $y \leq 2$ or $y \leq \frac{4}{5}x + 2$

53) $y \leq 4$ or $4y - 3x \geq -8$

54) $x + 3y \geq 3$ or $x \geq -2$

55) $y > -\frac{2}{3}x + 1$ or $-2x + 5y \leq 0$

56) $y > x - 4$ or $3x + 2y \geq 12$

57) $x \geq 5$ and $y \leq -3$

58) $x \leq 6$ and $y \geq 1$

59) $y < 4$ or $x \geq -3$

60) $x \geq 2$ or $y \geq -6$

61) $2x + 5y < 15$ or $y \leq \frac{3}{4}x - 1$

62) $y - 2x \leq 1$ and $y \geq -\dfrac{1}{5}x - 2$

63) $y \geq \dfrac{2}{3}x - 4$ and $4x + y \leq 3$

64) $y < 5x + 2$ or $x + 4y < 12$

Objective 3: Solve a Linear Programming Problem

(24) 65) During the school year, Tazia earns money by babysitting and tutoring. She can work at most 15 hr per week.

Let x = number of hours Tazia babysits
    y = number of hours Tazia tutors

a) Write the linear inequalities that describe the constraints on the number of hours Tazia can work per week.

b) Graph the feasible region that describes how her hours can be distributed between babysitting and tutoring.

c) Find three points in the feasible region and discuss their meanings.

d) Find one point outside the feasible region and discuss its meaning.

66) A machine in a factory can be calibrated to fill either large or small bags of potato chips. The machine will run at most 12 hr per day.

Let x = number of hours the machine fills large bags
    y = number of hours the machine fills small bags

a) Write the linear inequalities that describe the constraints on the number of hours the machine fills the bags each day.

b) Graph the feasible region that describes how the hours can be distributed between filling the large and small bags of chips.

c) Find three points in the feasible region and discuss their meanings.

d) Find one point outside the feasible region and discuss its meaning.

67) A lawn mower company produces a push mower and a riding mower. Company analysts predict that, for next spring, the company will need to produce at least 150 push mowers and 100 riding mowers per day, but they can produce at most 250 push mowers and 200 riding mowers per day. To satisfy demand,

they will have to ship a total of at least 300 mowers per day.

Let p = number of push mowers produced per day
    r = number of riding mowers produced per day

a) Write the linear inequalities that describe the constraints on the number of mowers that can be produced per day.

b) Graph the feasible region that describes how production can be distributed between the riding mowers and the push mowers. Let p be the horizontal axis and r be the vertical axis.

c) What does the point (175, 110) represent? Will this level of production meet the needs of the company?

d) Find three points in the feasible region and discuss their meanings.

e) Find one point outside the feasible region and discuss its meaning.

68) A dog food company produces adult dog food and puppy food. The company estimates that for next month, it will need to produce at least 12,000 pounds of adult dog food and 8000 pounds of puppy food per day. The factory can produce at most 18,000 pounds of adult dog food and 14,000 pounds of puppy food per day. The company will need to ship a total of at least 25,000 pounds of dog food per day to its customers.

Let a = pounds of adult dog food produced per day
    p = pounds of puppy food produced per day

a) Write the linear inequalities that describe the constraints on the number of pounds of dog food that can be produced per day.

b) Graph the feasible region that describes how production can be distributed between the adult dog food and the puppy food. Let a be the horizontal axis and p be the vertical axis.

c) What does the point (17,000, 9000) represent? Will this level of production meet the needs of the company?

d) Find three points in the feasible region and discuss their meanings.

e) Find one point outside the feasible region and discuss its meaning.

R1) Why do you need to use a test point that is not on the boundary line when solving a linear inequality in two variables?

R2) How do you represent the solution set of a system of linear inequalities in two variables?

R3) When do you draw a boundary line solid?

R4) When do you draw a boundary line dashed?

R5) Which objective is the most difficult for you?

Group Activity – Signs of Inequalities

1) Fill in the following sign tables for multiplication and division.

×	Positive	Negative
Positive		
Negative		

÷	Positive	Negative
Positive		
Negative		

2) Fill in the blank (positive or negative).

a) If a is positive, a^2 is _____.

b) If a is negative, a^2 is _____ and a^3 is _____.

c) If a is positive and b is negative, then $a \cdot b$ is _____.

3) Given that $a > 0, b > 0, c < 0$, and $d < 0$, fill in the correct inequality symbol, $<$ or $>$.

a) $a \cdot b$ _____ 0

b) $a + b$ _____ 0

c) $\dfrac{a \cdot b}{c}$ _____ 0

d) $\dfrac{a \cdot b}{c^2}$ _____ 0

e) $a + d$ _____ $c + d$

f) $a \cdot c$ _____ $d \cdot c$

g) $c \cdot d$ _____ $c^2 \cdot d$

h) $c^3 \cdot d^3$ _____ $(c + d)^3$

4) Given that $a > b, c > d$, and $n < 0$, fill in the correct inequality symbol, $<$ or $>$.

a) $(a - b)(c - d)$ _____ 0

b) $(a - b)(d - c)$ _____ 0

c) $-a$ _____ $-b$

d) an _____ bn

e) $\dfrac{1}{d}$ _____ $\dfrac{1}{c}$

f) $\dfrac{b - a}{d - c}$ _____ $d - c$

5) Given that $a > b, a > 0, b > 0, n > 0$, and n is an integer, fill in the correct inequality symbol, $<$ or $>$.

a) a^n _____ b^n

b) a^n _____ $-b^n$

c) a^{-n} _____ b^{-n}

d) $-a^{-n}$ _____ b^{-n}

When an instructor asks students to write in a math class, the students often say, "But this is a math class! Why do we have to write?" Writing is a skill that is important no matter what career you pursue: nurses must write instructions for patients, mechanics write up bills with explanations of work being done, and even engineers, who use a lot of math, must write reports to explain the *meaning* of calculations or to submit bids on projects their companies hope to win. So, how well do *you* write? Try this to assess your writing skills.

Choose a math topic to explain, in writing, using words *and* math to solve the problem. Write explanations as you move from step to step. Write neatly, and use complete sentences and correct grammar and punctuation. Assume that the person reading your explanation has limited knowledge of the concept. Choose from the following topics and problems, or use a topic given to you by your instructor.

1) Solve a linear equation containing fractions: $\frac{1}{2}x - \frac{1}{3}(x + 1) = \frac{2}{3}x + \frac{1}{4}$.

2) Solve an absolute value equation: $\left|\frac{4}{3}c - 9\right| - 11 = -8$.

3) Solve an absolute value inequality: $|7p + 2| + 5 \geq 14$.

4) Write an equation of a line: Write an equation of a line containing the points $(-2, 9)$ and $(3, 7)$. Write the answer in standard form.

5) Solve a system of equations: $\begin{array}{l} 4x - 3y = 5 \\ -10x + 8y = -11 \end{array}$.

6) Simplify: $2(-3a^{-6}b^5)^4$.

7) Factor by grouping: $9n^3 - 18n^2 - n + 2$.

8) Solve a quadratic equation: $(k - 4)(k + 9) = 30$.

9) Divide: $\dfrac{6k^3 + 11k^2 + 17k + 60}{3k + 7}$.

10) Determine the values of w for which the expression is undefined: $\dfrac{4w + 1}{2w^2 - 3w - 20}$.

When you are finished, exchange papers with a classmate and critique what each of you has written. Be honest, but kind, when evaluating it. Did your classmate understand your explanation, or was something unclear? What did he or she like about what you wrote, and what could be improved? Did you find that *you* understood the topic better after writing about it?

Chapter 9: Summary

Definition/Procedure	Example

9.1 Absolute Value Equations

Absolute Value Equations

If P represents an expression and k is a positive, real number, then to solve $|P| = k$ we rewrite the absolute value equation as the *compound equation* $P = k$ or $P = -k$ and solve for the variable.

Solve $|4a + 10| = 18$.

$$|4a + 10| = 18$$

$4a + 10 = 18$ or $4a + 10 = -18$
$4a = 8$ $\qquad\qquad 4a = -28$
$a = 2$ or $\qquad a = -7$

Check the solutions in the original equation.
The solution set is $\{-7, 2\}$.

9.2 Absolute Value Inequalities

Inequalities Containing < or ≤

Let P be an expression and let k be a positive, real number. To solve $|P| \leq k$, solve the three-part inequality
$$-k \leq P \leq k$$
(< may be substituted for ≤.)

Solve $|x - 3| \leq 2$. Graph the solution set, and write the answer in interval notation.

$$-2 \leq x - 3 \leq 2$$
$$1 \leq x \leq 5 \quad \text{Add 3.}$$

$$-2 \; -1 \; 0 \; 1 \; 2 \; 3 \; 4 \; 5 \; 6 \; 7$$

In interval notation, we write $[1, 5]$.

Inequalities Containing > or ≥

Let P be an expression and let k be a positive, real number. To solve $|P| \geq k$, solve the compound inequality
$$P \geq k \text{ or } P \leq -k$$
(> may be substituted for ≥.)

Solve $|2n - 5| > 1$. Graph the solution set, and write the answer in interval notation.

$2n - 5 > 1$ or $2n - 5 < -1$ $\quad$ Solve.
$2n > 6$ or $\qquad 2n < 4$ $\qquad$ Add 5.
$n > 3$ or $\qquad n < 2$ $\qquad$ Divide by 2.

$$-3 \; -2 \; -1 \; 0 \; 1 \; 2 \; 3 \; 4 \; 5 \; 6$$

In interval notation, we write $(\infty, 2) \cup (3, \infty)$.

Definition/Procedure	Example

9.3 Linear and Compound Linear Inequalities in Two Variables

A **linear inequality in two variables** is an inequality that can be written in the form $Ax + By \geq C$ or $Ax + By \leq C$, where A, B, and C are real numbers and where A and B are not both zero. ($>$ and $<$ may be substituted for $\geq$ and $\leq$.)	Some examples of linear inequalities in two variables are $$x + 3y \leq 2, \quad y > -\frac{2}{3}x + 5, \quad y \geq -1, \quad x < 4$$

Graph a Linear Inequality in Two Variables Using a Test Point

1) *Graph the boundary line.*
   a) If the inequality contains $\geq$ or $\leq$, make the boundary line *solid*.
   b) If the inequality contains $>$ or $<$, make the boundary line *dotted*.
2) *Choose a test point not on the line, and shade the appropriate region.* Substitute the test point into the inequality. If $(0, 0)$ is not on the line, it is an easy point to test in the inequality.
   a) If it *makes the inequality true,* shade the region *containing* the test point. All points in the shaded region are part of the solution set.
   b) If the test point *does not satisfy the inequality,* shade the region on the *other* side of the line. All points in the shaded region are part of the solution set.

Graph $2x + y > -3$.

1) Graph the boundary line as a *dotted* line.
2) Choose a test point not on the line, and substitute it into the inequality to determine whether it makes the inequality true.

Test Point	Substitute into $2x + y > -3$
$(0, 0)$	$2(0) + (0) > -3$
	$0 > -3$ True

Since the test point satisfies the inequality, shade the region containing $(0, 0)$.

All points in the shaded region satisfy $2x + y > -3$.

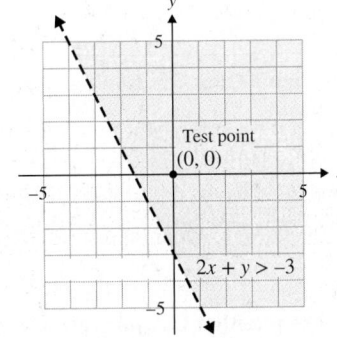

Graph a Linear Inequality in Two Variables Using the Slope-Intercept Method

1) Write the inequality in the form $y \geq mx + b$ ($y > mx + b$) or $y \leq mx + b$ ($y < mx + b$), and graph the boundary line $y = mx + b$.

2) If the inequality is in the form $y \geq mx + b$ or $y > mx + b$, shade *above* the line.

3) If the inequality is in the form $y \leq mx + b$ or $y < mx + b$, shade *below* the line.

Graph $-x + 3y \leq 6$ using the slope-intercept method.

Write the inequality in slope-intercept form by solving $-x + 3y \leq 6$ for y.

$$-x + 3y \leq 6$$
$$3y \leq x + 6$$
$$y \leq \frac{1}{3}x + 2$$

Graph $y = \frac{1}{3}x + 2$ as a *solid line.*

Since $y \leq \frac{1}{3}x + 2$ has a $\leq$ symbol, shade *below* the line.

All points on the line and in the shaded region satisfy $-x + 3y \leq 6$.

Definition/Procedure	Example
Graphing Compound Linear Inequalities in Two Variables 1) Graph each inequality separately on the same axes. Shade lightly. 2) If the inequality contains *and*, the solution set is the *intersection* of the shaded regions. Heavily shade this region. 3) If the inequality contains *or*, the solution set is the *union* (total) of the shaded regions. Heavily shade this region.	Graph the compound inequality $y \geq -4x + 3$ and $y \geq 1$. Since the inequality contains *and*, the solution set is the *intersection* of the shaded regions. Any point in the shaded area will satisfy *both* inequalities. 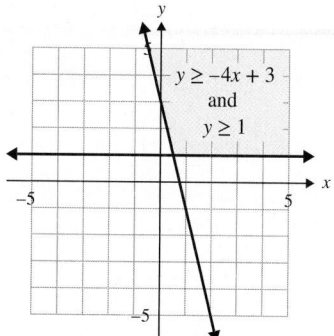

Chapter 9:   Review Exercises

(9.1) Solve.

1) $|m| = 9$

2) $\left|\frac{1}{2}c\right| = 5$

3) $|7t + 3| = 4$

4) $|4 - 3y| = 12$

5) $|8p + 11| - 7 = -3$

6) $|5k + 3| - 8 = 4$

7) $\left|4 - \frac{5}{3}x\right| = \frac{1}{3}$

8) $\left|\frac{2}{3}w + 6\right| = \frac{5}{2}$

9) $|7r - 6| = |8r + 2|$

10) $|3z - 4| = |5z - 6|$

11) $|2a - 5| = -10$

12) $|h + 6| - 12 = -20$

13) $|9d + 4| = 0$

14) $|6q - 7| = 0$

15) Write an absolute value equation which means *a is 4 units from zero.*

16) Write an absolute value equation which means *t is 7 units from zero.*

(9.2) Solve each inequality. Graph the solution set, and write the answer in interval notation.

17) $|c| \leq 3$

18) $|w + 1| < 11$

19) $|4t| > 8$

20) $|2v - 7| \geq 15$

21) $|12r + 5| \geq 7$

22) $|3k - 11| < 4$

23) $|4 - a| < 9$

24) $|2 - 5q| > 6$

25) $|4c + 9| - 8 \leq -2$

26) $|3m + 5| + 2 \geq 7$

27) $|5y + 12| - 15 \geq -8$

28) $3 + |z - 6| \leq 13$

29) $|k + 5| > -3$

30) $|4q - 9| < 0$

31) $|12s + 1| \leq 0$

32) Write an absolute value inequality that means *n is at least 4 units from zero.*

33) Write an absolute value inequality that means *x is less than 9 units from zero.*

34) A radar gun indicated that a pitcher threw a 93 mph fastball. The radar gun's possible error in measuring the speed of a pitch is ± 1 mph. Write an absolute value inequality to represent the range for the speed of the pitch, and solve the inequality. Explain the meaning of the answer. Let *s* represent the range of values for the speed of the pitch.

Richard Wear/Getty Images

(9.3) Graph each linear inequality in two variables.

35) $y \leq -2x + 7$

36) $y \geq -\frac{3}{2}x + 2$

37) $-3x + 4y > 12$

38) $5x - 2y \geq 8$

39) $y < x$

40) $x \geq 4$

Graph each compound inequality.

41) $y \geq \frac{3}{4}x - 4$ and $y \leq -5$

42) $y < \frac{5}{4}x - 5$ or $y < -3$

43) $4x - y < -1$ or $y > \frac{1}{2}x + 5$

44) $2x + 5y \leq 10$ and $y \geq \frac{1}{3}x + 4$

45) $4x + 2y \geq -6$ and $y \leq 2$

46) $2x + y \leq 3$ or $6x + y > 4$

Chapter 9:   Test

Solve.

1) $|4y - 9| = 11$

2) $7 = \left|6 - \dfrac{3}{8}d\right| - 5$

3) $|3k + 5| = |k - 11|$

4) $\left|\dfrac{1}{2}n - 1\right| = -8$

5) Write an absolute value equation that means *x is 8 units from zero.*

6) Explain why the solution to $|0.8a + 1.3| \geq 0$ is $(-\infty, \infty)$.

Solve each inequality. Graph the solution set and write the answer in interval notation.

7) $|c| > 4$

8) $|2z - 7| \leq 9$

9) $|4m + 9| - 8 \geq 5$

10) $\left|\dfrac{2}{3}w - 4\right| + 10 < 4$

11) A scale in a doctor's office has a possible error of ± 0.75 lb. If Thanh's weight is measured as 168 lb, write an absolute value inequality to represent the range for his weight, and solve the inequality. Let *w* represent the range of values for Thanh's weight. Explain the meaning of the answer.

12) Write an absolute value equation that has a solution set of $\{-17, 17\}$.

13) Write an absolute value inequality with a solution of $[-4, 4]$.

Graph each inequality.

14) $y \geq 3x + 1$

15) $2x - 5y > 10$

Graph each compound inequality.

16) $-2x + 3y \geq -12$ and $x \leq 3$

17) $y < -x$ or $2x - y > 1$

Chapter 9:   Cumulative Review for Chapters 1–9

Perform the operations and simplify.

1) $5 \times 6 - 36 \div 3^2$

2) $\dfrac{5}{12} - \dfrac{7}{8}$

Evaluate.

3) 3^4

4) 2^5

5) $\left(\dfrac{1}{8}\right)^2$

6) 4^{-3}

7) Write 0.00000914 in scientific notation.

8) Solve $8 - 3(2y - 5) = 4y + 1$.

9) Solve $3 - \dfrac{2}{7}n \geq 9$. Write the answer in interval notation.

10) Write an equation and solve.

How many ounces of a 9% alcohol solution must be added to 8 oz of a 3% alcohol solution to obtain a 5% alcohol solution?

11) Write the slope-intercept form of the line containing $(7, 2)$ with slope $\dfrac{1}{3}$.

12) Solve by graphing.

$$2x + y = -1$$
$$y = 3x - 6$$

Multiply and simplify.

13) $-4p^2(3p^2 - 7p - 1)$

14) $(2k + 5)(2k - 5)$

15) $(t + 8)^2$

16) Divide $\dfrac{6c^3 + 7c^2 - 38c + 24}{3c - 4}$.

Factor completely.

17) $9m^2 - 121$

18) $z^2 - 14z + 48$

Solve.

19) $a^2 + 6a + 9 = 0$

20) $2(x^2 - 4) = -(7x + 4)$

21) Subtract $\dfrac{1}{r^2 - 25} - \dfrac{r + 3}{2r + 10}$.

22) Multiply and simplify $\dfrac{w^2 - 3w - 54}{w^3 - 8w^2} \cdot \dfrac{w}{w + 6}$.

23) Solve $\left|\dfrac{1}{4}q - 7\right| - 8 = -5$.

24) Solve. Graph the solution set and write the answer in interval notation.

$$|9v + 4| > 14$$

25) Graph the compound inequality

$$3x + 4y > 16 \text{ or } y < \dfrac{1}{5}x + 1.$$

Design Icon Credits: Notes box icon (push pin): ©Shutterstock/ksevgi; Using Technology icon (graphing calculator): ©mbbirdy/Getty Images.

Radicals and Rational Exponents

C H A P T E R# 10

Get Ready

Let's review powers of integers and some operations with fractions to prepare for this chapter.

1) It is important to know some powers of integers *by heart* and to understand how to work with negative signs, too. (See Section 1.2 for a table of powers.)

$$7^2 = 49, \quad 4^3 = 64, \quad (-9)^2 = 81, \quad (-2)^5 = -32$$

2) To **multiply fractions,** we can take out the common factors before we multiply.

Example: Multiply $6 \cdot \dfrac{4}{3}$. $\dfrac{6}{1} \cdot \dfrac{4}{3} = \dfrac{2}{1} \cdot \dfrac{4}{1} = \dfrac{8}{1} = 8$ Write 6 as $\dfrac{6}{1}$. Divide 6 and 3 by 3. Then, multiply and simplify.

3) To **add and subtract fractions,** the fractions must first have a common denominator. Then, add or subtract the numerators and keep the same denominator.

Examples: Add or subtract.

a) $\dfrac{1}{2} + \dfrac{3}{5}$ The **least common denominator, LCD,** is 10.

Write each fraction with a denominator of 10, then add.

$\dfrac{1}{2} \cdot \dfrac{5}{5} = \dfrac{5}{10}$; $\dfrac{3}{5} \cdot \dfrac{2}{2} = \dfrac{6}{10}$ $\dfrac{1}{2} + \dfrac{3}{5} = \dfrac{5}{10} + \dfrac{6}{10} = \dfrac{11}{10}$

b) $\dfrac{3}{4} - \dfrac{11}{12}$ The **LCD is 12.** Write $\dfrac{3}{4}$ with a denominator of 12.

$\dfrac{3}{4} \cdot \dfrac{3}{3} = \dfrac{9}{12}$ $\dfrac{3}{4} - \dfrac{11}{12} = \dfrac{9}{12} - \dfrac{11}{12} = -\dfrac{2}{12} = -\dfrac{1}{6}$ Divide 2 and 12 by 2 to write the answer in lowest terms.

Use the following *Basic Skills Worksheets* along with this chapter: WS2 Powers (to practice quick recall of powers of integers) and, after Section 10.1, WS5 Roots (to practice quick recall of roots of integers).

10.1 Finding Roots
10.2 Rational Exponents
10.3 Simplifying Expressions Containing Square Roots
10.4 Simplifying Expressions Containing Higher Roots
10.5 Adding, Subtracting, and Multiplying Radicals
10.6 Dividing Radicals
Putting It All Together
10.7 Solving Radical Equations
10.8 Complex Numbers
Group Activity
emPOWERme: My Study Group Experience

Get Ready Exercises

Evaluate.

1) 6^2

2) 12^2

3) 5^3

4) 2^6

5) $(-10)^2$

6) $(-3)^3$

7) $(-2)^4$

8) $(-4)^3$

Perform the indicated operation. Write the answer in lowest terms.

9) $\dfrac{7}{8} \cdot \dfrac{2}{3}$

10) $\dfrac{4}{9} \cdot \dfrac{9}{10}$

11) $\dfrac{5}{6} - \dfrac{3}{8}$

12) $\dfrac{2}{15} + \dfrac{7}{10}$

13) $\dfrac{1}{2} + \dfrac{9}{16}$

14) $\dfrac{5}{21} - \dfrac{2}{3}$

15) $2 \cdot \dfrac{3}{4}$

16) $\dfrac{1}{6} \cdot 8$

17) $-\dfrac{3}{2} \cdot 7$

18) $-\dfrac{2}{5} \cdot \dfrac{1}{4}$

19) $\dfrac{1}{8} - \left(-\dfrac{3}{4}\right)$

20) $-\dfrac{7}{6} - \left(-\dfrac{2}{7}\right)$

Answers

1) 36 2) 144 3) 125 4) 64 5) 100 6) −27 7) 16 8) −64 9) $\dfrac{7}{12}$ 10) $\dfrac{2}{5}$ 11) $\dfrac{11}{24}$ 12) $\dfrac{5}{6}$ 13) $\dfrac{17}{16}$ 14) $-\dfrac{3}{7}$ 15) $\dfrac{3}{2}$ 16) $\dfrac{4}{3}$ 17) $-\dfrac{21}{2}$ 18) $-\dfrac{1}{10}$ 19) $\dfrac{7}{8}$ 20) $-\dfrac{37}{42}$

 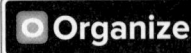

Study Strategies Working with a Study Group

There is a lot of truth to the old saying, "Two heads are better than one." Working with your fellow students allows you to share your knowledge and skills as you study for a test, collaborate on a project, or work toward a similar academic goal. However, simply gathering with a few classmates is not enough. To get the most out of your study group, apply these strategies using the P.O.W.E.R. framework.

P Prepare

- **I will form and work in a study group.**

O Organize

- Complete the emPOWERme survey that appears before the Chapter Summary to learn about your previous experience working in a group.
- Think about *why* you want to be part of a study group. Do you want to work with people who know more than you do so that you can get help from them? Do you want to be the person in the group who helps others because you learn best when you explain it to someone else?
- Based upon your answer to the question in the previous bullet, identify other students with whom you would like to work. Ask them, in class, if they are interested in forming a study group, or send them an email using your class' learning management system such as Blackboard or Moodle. Or, ask your instructor to make an announcement about studying together. Three to five people is usually a good size for a study group.
- Together, identify the group's goal. Are you going to meet to study for a specific test, or will you meet regularly to help each other with homework, quizzes, and tests throughout the term?
- Exchange phone numbers. Agree on a time and place to meet. Be sure the meeting place allows conversation. It can be a library, someone's house, a coffee shop, or a room on

campus. If you choose a room on campus, pick one that has a table that all of you can work at or with desks that can be rearranged the way you would like them. A whiteboard and chalkboard are great, too, so that you can work problems at the board and talk about them.

- Each group member should bring his or her textbook, notes, and other course materials. Bring a whiteboard marker or chalk if you will need them.
- Be sure that you have done the homework or studied for the test to the best of your ability *before* you meet with your study group. Make a list of specific questions about homework problems or topics and take them with you. Do not show up with a "teach me everything" attitude.
- If you are preparing for a test, think about assigning each member of the group a specific topic that he or she is responsible for explaining to the rest of the group.

- At the start of the study group, discuss your goals to ensure that everyone is on the same page and to ensure that everyone gets what they need out of the session. Determine how long you will meet so that you manage the time well and have enough time for everyone's questions.
- Be an active participant! Take turns asking the questions on your lists or asking for explanations of specific topics. Equally important, take turns *answering* questions and explaining topics to each other. You learn a lot when you have to explain something to others.
- If you are meeting in a room with a whiteboard or chalkboard, do problems at the board. This way, you can see everyone's work and discuss and ask questions about solving problems.
- Communicate with one another respectfully. Treat others the way you would like to be treated. Allow everyone to ask and answer questions.

- At the end of the study session, take a moment to discuss how well you accomplished your goals for the session. Did you get your questions answered? Do you understand concepts better *after* your meeting than you did before?
- Be honest with each other in discussing what worked well and what did not.

- If your study session went well, think about the things that made it effective so that you do those same things in the future. Did everyone come prepared with questions? Was everyone an active participant? Did you choose a room that was a great place to work?
- If you think that the study session was *not* helpful, think about why you feel that way. Was it disorganized? Were there too many people? Could you have chosen a better meeting place?
- Ask yourself whether a study group is right for you. While group study can be very helpful, some people prefer studying on their own.

Andersen Ross/Getty Images

10.1 Finding Roots

What are your objectives for Section 10.1?	How can you accomplish each objective?
1 Find Square Roots and Principal Square Roots	• Write the definitions of *square root, principal square root, negative square root, square root symbol* or *radical sign, radicand,* and *radical* in your notes with an example next to each. • Write the properties for **Radicands and Square Roots** in your notes, and learn them. • Complete the given examples on your own. • Complete You Trys 1 and 2.
2 Approximate the Square Root of a Whole Number	• Write down a procedure you would use to approximate the square root of a whole number. • Complete the given example on your own. • Complete You Try 3.
3 Find Higher Roots	• Know the definition of an *index*. • Understand the definitions of the *nth root*, and write examples in your notes that match the definitions. • Complete the given examples on your own. • Complete You Trys 4 and 5.
4 Evaluate $\sqrt[n]{a^n}$	• Write the procedure for **Evaluating** $\sqrt[n]{a^n}$ in your own words. • Complete the given example on your own. • Complete You Try 6.

W Work   **Read the explanations, follow the examples, take notes, and complete the You Trys.**

Recall that exponential notation represents repeated multiplication. For example,

$$3^2 \text{ means } 3 \cdot 3, \text{ so } 3^2 = 9.$$
$$2^4 \text{ means } 2 \cdot 2 \cdot 2 \cdot 2, \text{ so } 2^4 = 16.$$

In this chapter we will study the opposite, or inverse, procedure, finding **roots** of numbers.

1   Find Square Roots and Principal Square Roots

EXAMPLE 1   Find all square roots of 25.

Solution

To find a *square* root of 25 ask yourself, "What number do I *square* to get 25?" Or, "What number *multiplied by itself* equals 25?" One number is 5 since $5^2 = 25$. Another number is -5 since $(-5)^2 = 25$. So, 5 and -5 are square roots of 25.

[**YOU TRY 1**] Find all square roots of 64.

The $\sqrt{}$ symbol represents the *positive* square root, or the **principal square root,** of a nonnegative number. For example, $\sqrt{25} = 5$.

Notice that $\sqrt{25} = 5$, but $\sqrt{25} \neq -5$. The $\sqrt{}$ symbol represents *only* the principal square root (positive square root).

To find the **negative square root** of a nonnegative number we must put a negative sign in front of the $\sqrt{}$. For example, $-\sqrt{25} = -5$.

Next we will define some terms associated with the $\sqrt{}$ symbol.

The symbol $\sqrt{}$ is the **square root symbol** or the **radical sign.** The number under the radical sign is the **radicand.** The entire expression, $\sqrt{25}$, is called a **radical.**

$$\text{Radical sign} \rightarrow \underbrace{\sqrt{25}}_{\text{Radical}} \leftarrow \text{Radicand} = 25$$

EXAMPLE 2   Find each square root, if possible.

a) $\sqrt{100}$ b) $-\sqrt{16}$ c) $\sqrt{\dfrac{4}{25}}$ d) $-\sqrt{\dfrac{81}{49}}$ e) $\sqrt{-9}$

Solution

a) $\sqrt{100} = 10$ since $(10)^2 = 100$.

b) $-\sqrt{16}$ means $-1 \cdot \sqrt{16}$. Therefore, $-\sqrt{16} = -1 \cdot \sqrt{16} = -1 \cdot 4 = -4$.

c) $\sqrt{\dfrac{4}{25}} = \dfrac{2}{5}$ since $\left(\dfrac{2}{5}\right)^2 = \dfrac{4}{25}$.

d) $-\sqrt{\dfrac{81}{49}}$ means $-1 \cdot \sqrt{\dfrac{81}{49}}$. So, $-\sqrt{\dfrac{81}{49}} = -1 \cdot \sqrt{\dfrac{81}{49}} = -1 \cdot \dfrac{9}{7} = -\dfrac{9}{7}$.

e) To find $\sqrt{-9}$, ask yourself, "What number do I *square* to get -9?" *There is no such real number* since $3^2 = 9$ and $(-3)^2 = 9$. Therefore, $\sqrt{-9}$ is not a real number.

[W] **Hint**

In Example 2, we are finding the **principal square roots** of 100 and $\dfrac{4}{25}$ and the **negative square roots** of 16 and $\dfrac{81}{49}$.

[**YOU TRY 2**] Find each square root.

a) $\sqrt{9}$ b) $-\sqrt{144}$ c) $\sqrt{\dfrac{25}{36}}$ d) $-\sqrt{\dfrac{1}{64}}$ e) $\sqrt{-49}$

Let's review what we know about radicands and add a third fact.

> **Property** Radicands and Square Roots
>
> 1) If the radicand is a *perfect square,* the square root is a *rational* number.
>
>   *Example:* $\sqrt{16} = 4$ 16 is a perfect square. $\sqrt{\dfrac{100}{49}} = \dfrac{10}{7}$ $\dfrac{100}{49}$ is a perfect square.
>
> 2) If the radicand is a *negative number,* the square root is *not* a real number.
>
>   *Example:* $\sqrt{-25}$ is *not* a real number.
>
> 3) If the radicand is *positive and not a perfect square,* then the square root is an *irrational* number.
>
>   *Example:* $\sqrt{13}$ is irrational.   13 is not a perfect square.
>
> The square root of such a number is a real number that is a nonrepeating, nonterminating decimal. It is important to be able to approximate these square roots because sometimes it is necessary to estimate their places on a number line or on a Cartesian coordinate system when graphing.

For the purposes of graphing, approximating a radical to the nearest tenth is sufficient. A calculator with a $\sqrt{}$ key will give a better approximation of the radical.

2 Approximate the Square Root of a Whole Number

EXAMPLE 3

Approximate $\sqrt{13}$ to the nearest tenth, and plot it on a number line.

Solution

What is the largest perfect square that is *less than* 13?   **9**

What is the smallest perfect square that is *greater than* 13?   **16**

Since 13 is between 9 and 16 ($9 < 13 < 16$) it is true that $\sqrt{13}$ is between $\sqrt{9}$ and $\sqrt{16}$.

$$(\sqrt{9} < \sqrt{13} < \sqrt{16})$$
$$\sqrt{9} = 3$$
$$\sqrt{13} = ?$$
$$\sqrt{16} = 4$$

> **W Hint**
>
> Summarize the steps you would take to approximate the square root of a number that is not a perfect square.

$\sqrt{13}$ must be between 3 and 4. Numerically, 13 is closer to 16 than it is to 9. So, $\sqrt{13}$ will be closer to $\sqrt{16}$ than to $\sqrt{9}$. Check to see if 3.6 is a good approximation of $\sqrt{13}$. ($\approx$ means *approximately equal to.*)

$$\text{If } \sqrt{13} \approx 3.6, \text{ then } (3.6)^2 \approx 13.$$
$$(3.6)^2 = (3.6) \cdot (3.6) = 12.96$$

Is 3.7 a better approximation of $\sqrt{13}$?

$$\text{If } \sqrt{13} \approx 3.7, \text{ then } (3.7)^2 \approx 13$$
$$(3.7)^2 = (3.7) \cdot (3.7) = 13.69$$

3.6 is a better approximation of $\sqrt{13}$.

$$\sqrt{13} \approx 3.6$$

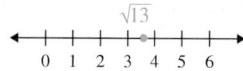

A calculator evaluates $\sqrt{13}$ as 3.6055513. Remember that this is only an approximation. We will discuss how to approximate radicals using a calculator later in this chapter.

[YOU TRY 3]

Approximate $\sqrt{29}$ to the nearest tenth, and plot it on a number line.

---

3 Find Higher Roots

We saw in Example 2a) that $\sqrt{100} = 10$ since $(10)^2 = 100$. We can also find higher roots of numbers like $\sqrt[3]{a}$ (read as "the cube root of a"), $\sqrt[4]{a}$ (read as "the fourth root of a"), $\sqrt[5]{a}$, etc. We will look at a few roots of numbers before learning important rules.

EXAMPLE 4

Find each root.

a) $\sqrt[3]{125}$ b) $\sqrt[5]{32}$

Solution

a) To find $\sqrt[3]{125}$ (the cube root of 125) ask yourself, "What number do I *cube* to get 125?" That number is 5.

$$\sqrt[3]{125} = 5 \text{ since } 5^3 = 125$$

Finding the cube root of a number is the *opposite*, or *inverse procedure*, of cubing a number.

W Hint

Could you write an equation that would help you find the root?

b) To find $\sqrt[5]{32}$ (the fifth root of 32) ask yourself, "What number do I raise to the *fifth power* to get 32?" That number is 2.

$$\sqrt[5]{32} = 2 \text{ since } 2^5 = 32$$

Finding the fifth root of a number and raising a number to the fifth power are *opposite*, or *inverse*, procedures.

[YOU TRY 4]

Find each root.

a) $\sqrt[3]{27}$ b) $\sqrt[3]{8}$

The symbol $\sqrt[n]{a}$ is read as "the *nth* root of a." If $\sqrt[n]{a} = b$, then $b^n = a$.

$$\text{Index} \rightarrow \underset{\text{Radical}}{\underbrace{\sqrt[n]{a}}} \leftarrow \text{Radicand} = a$$

We call n the **index** of the radical.

Note

When finding *square* roots we do not write $\sqrt[2]{a}$. The square root of *a* is written as $\sqrt{a}$, and the index is understood to be 2.

We know that a positive number, say 36, has a principal square root ($\sqrt{36}$, or 6) and a negative square root ($-\sqrt{36}$, or -6). This is true for all even roots of positive numbers: square roots, fourth roots, sixth roots, and so on. For example, 81 has a principal fourth root ($\sqrt[4]{81}$, or 3) and a negative fourth root ($-\sqrt[4]{81}$, or -3).

W Hint

Write examples in your notes to help you understand the different situations.

Definition   *n*th Root

For any *positive* number *a* and any *even* index *n*,
the **principal *n*th root** of *a* is $\sqrt[n]{a}$.
the **negative *n*th root** of *a* is $-\sqrt[n]{a}$.

$\overset{even}{\sqrt{positive}}$ = principal (positive) root
$-\overset{even}{\sqrt{positive}}$ = negative root

For any *negative* number *a* and any *even* index *n*,
there is **no** real *n*th root of *a*.

$\overset{even}{\sqrt{negative}}$ = no real root

For any number *a* and any *odd* index *n*,
there is **one** real *n*th root of *a*, $\sqrt[n]{a}$.

$\overset{odd}{\sqrt{any\ number}}$ = exactly one root

 **BE CAREFUL**

The definition means that $\sqrt[4]{81}$ *cannot* be -3 because $\sqrt[4]{81}$ is *defined* as the principal fourth root of 81, which must be positive. $\sqrt[4]{81} = 3$

EXAMPLE 5

Find each root, if possible.

a) $\sqrt[4]{16}$ b) $-\sqrt[4]{16}$ c) $\sqrt[4]{-16}$ d) $\sqrt[3]{64}$ e) $\sqrt[3]{-64}$

Solution

a) To find $\sqrt[4]{16}$ ask yourself, "What *positive* number do I raise to the *fourth power* to get 16?" Since $2^4 = 16$ and 2 is positive, $\sqrt[4]{16} = \mathbf{2}$.

b) In part a) we found that $\sqrt[4]{16} = 2$, so $-\sqrt[4]{16} = -(\sqrt[4]{16}) = \mathbf{-2}$.

c) To find $\sqrt[4]{-16}$ ask yourself, "What number do I raise to the *fourth power* to get -16?" There is no such real number since $2^4 = 16$ and $(-2)^4 = 16$. Therefore, $\sqrt[4]{-16}$ **has *no real root*.** (Recall from the definition that $\overset{even}{\sqrt{negative}}$ has no real root.)

d) To find $\sqrt[3]{64}$ ask yourself, "What number do I *cube* to get 64?" Since $4^3 = 64$ and since we know that $\overset{odd}{\sqrt{any\ number}}$ gives exactly one root, $\sqrt[3]{64} = \mathbf{4}$.

e) To find $\sqrt[3]{-64}$ ask yourself, "What number do I *cube* to get -64?" Since $(-4)^3 = -64$ and since we know that $\overset{odd}{\sqrt{any\ number}}$ gives exactly one root, $\sqrt[3]{-64} = \mathbf{-4}$.

[YOU TRY 5]

Find each root, if possible.

a) $\sqrt[6]{-64}$ b) $\sqrt[3]{-125}$ c) $-\sqrt[4]{81}$ d) $\sqrt[3]{1}$ e) $\sqrt[4]{81}$

4 Evaluate $\sqrt[n]{a^n}$

Earlier we said that the $\sqrt{}$ symbol represents only the *positive* square root of a number. For example, $\sqrt{9} = 3$. It is also true that $\sqrt{(-3)^2} = \sqrt{9} = 3$.

If a variable is in the radicand and we do not know whether the variable represents a positive number, then we must use the absolute value symbol to evaluate the radical. Then we know that the result will be a positive number. For example, $\sqrt{a^2} = |a|$.

What if the index is greater than 2? Let's look at how to find the following roots:

$$\sqrt[4]{(-2)^4} = \sqrt[4]{16} = 2 \qquad \sqrt[3]{(-4)^3} = \sqrt[3]{-64} = -4$$

When the index on the radical is any positive, even integer and we do not know whether the variable in the radicand represents a positive number, we must use the absolute value symbol to write the root. However, when the index is a positive, odd integer, we do not need to use the absolute value symbol.

Procedure Evaluating $\sqrt[n]{a^n}$

1) If n is a positive, *even* integer, then $\sqrt[n]{a^n} = |a|$.
2) If n is a positive, *odd* integer, then $\sqrt[n]{a^n} = a$.

EXAMPLE 6

Simplify.

a) $\sqrt{(-7)^2}$ b) $\sqrt{k^2}$ c) $\sqrt[3]{(-5)^3}$ d) $\sqrt[7]{n^7}$

e) $\sqrt[4]{(y-9)^4}$ f) $\sqrt[5]{(8p+1)^5}$

Solution

a) $\sqrt{(-7)^2} = |-7| = 7$ When the index is even, use the absolute value symbol to be certain that the result is not negative.

b) $\sqrt{k^2} = |k|$ When the index is even, use the absolute value symbol to be certain that the result is not negative.

W Hint

In your notes, summarize when you need the absolute value symbol and when you do not.

c) $\sqrt[3]{(-5)^3} = -5$ The index is odd, so the absolute value symbol is not necessary.

d) $\sqrt[7]{n^7} = n$ The index is odd, so the absolute value symbol is not necessary.

e) $\sqrt[4]{(y-9)^4} = |y-9|$ Even index: use the absolute value symbol to be certain that the result is not negative.

f) $\sqrt[5]{(8p+1)^5} = 8p+1$ Odd index: the absolute value symbol is not necessary.

[YOU TRY 6]

Simplify.

a) $\sqrt{(-12)^2}$ b) $\sqrt{w^2}$ c) $\sqrt[3]{(-3)^3}$ d) $\sqrt[5]{r^5}$

e) $\sqrt[6]{(t+4)^6}$ f) $\sqrt[7]{(4h-3)^7}$

We can evaluate square roots, cube roots, or even higher roots using a graphing calculator. A radical sometimes evaluates to an integer and sometimes must be approximated using a decimal.

To evaluate a square root:

For example, to evaluate $\sqrt{9}$ press 2ⁿᵈ x^2, enter the radicand 9, and then press) ENTER. The result is 3 as shown on the screen on the left below. When the radicand is a perfect square such as 9, 16, or 25, then the square root evaluates to a whole number. For example $\sqrt{16}$ evaluates to 4 and $\sqrt{25}$ evaluates to 5 as shown.

If the radicand of a square root is not a perfect square, then the result is a decimal approximation. For example, to evaluate $\sqrt{19}$ press 2ⁿᵈ x^2, enter the radicand 1 9, and then press) ENTER. The result is approximately 4.3589, rounded to four decimal places.

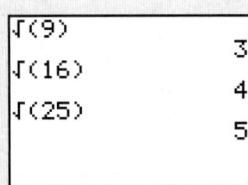
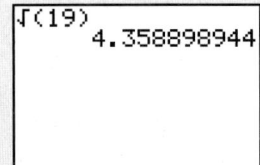

To evaluate a cube root:

For example, to evaluate $\sqrt[3]{27}$ press MATH 4, enter the radicand 2 7, and then press) ENTER. The result is 3 as shown.

If the radicand is a perfect cube such as 27, then the cube root evaluates to an integer. Since 28 is not a perfect cube, the cube root evaluates to approximately 3.0366.

To evaluate radicals with an index greater than 3:

For example, to evaluate $\sqrt[4]{16}$ enter the index 4, press MATH 5, enter the radicand 1 6, and press ENTER. The result is 2.

Since the fifth root of 18 evaluates to a decimal, the result is an approximation of 1.7826, rounded to four decimal places as shown.

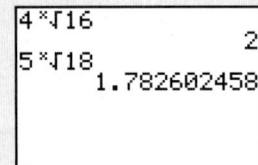

Evaluate each root using a graphing calculator. If necessary, approximate to the nearest tenth.

1) $\sqrt{25}$ 2) $\sqrt[3]{216}$ 3) $\sqrt{29}$ 4) $\sqrt{324}$ 5) $\sqrt[5]{1024}$ 6) $\sqrt[3]{343}$

ANSWERS TO [YOU TRY] EXERCISES

1) 8, −8   2) a) 3 b) −12 c) $\frac{5}{6}$ d) −$\frac{1}{8}$ e) not a real number

3) 5.4   [number line: $\sqrt{29}$ marked between 5 and 6, labels 0 1 2 3 4 5 6]   4) a) 3 b) 2 5) a) not a real number b) −5 c) −3 d) 1 e) 3

6) a) 12 b) $|w|$ c) −3 d) r e) $|t + 4|$ f) $4h − 3$

E Evaluate **10.1** Exercises   Do the exercises, and check your work.

Objective 1: Find Square Roots and Principal Square Roots

Decide whether each statement is true or false. If it is false, explain why.

1) $\sqrt{121} = 11$ and -11.

2) $\sqrt{81} = 9$

3) The square root of a negative number is a negative number.

4) The even root of a negative number is a negative number.

Find all square roots of each number.

5) 144

6) 2500

7) $\dfrac{36}{25}$

8) 0.01

Find each square root, if possible.

9) $\sqrt{49}$

10) $\sqrt{169}$

11) $\sqrt{-4}$

12) $\sqrt{-100}$

13) $\sqrt{\dfrac{81}{25}}$

14) $\sqrt{\dfrac{121}{4}}$

15) $-\sqrt{36}$

16) $-\sqrt{0.04}$

Objective 2: Approximate the Square Root of a Whole Number

Approximate each square root to the nearest tenth, and plot it on a number line.

17) $\sqrt{11}$

18) $\sqrt{3}$

19) $\sqrt{2}$

20) $\sqrt{5}$

21) $\sqrt{33}$

22) $\sqrt{39}$

23) $\sqrt{55}$

24) $\sqrt{72}$

Objective 3: Find Higher Roots

For Exercises 25 to 28, decide if each statement is true or false. If it is false, explain why.

25) The cube root of a negative number is a negative number.

26) The odd root of a negative number is not a real number.

27) Every nonnegative real number has two real, even roots.

28) $-\sqrt[4]{10,000} = -10$

29) Explain how to find $\sqrt[3]{64}$.

30) Explain how to find $\sqrt[4]{16}$.

31) Does $\sqrt[4]{-81} = -3$? Why or why not?

32) Does $\sqrt[3]{-8} = -2$? Why or why not?

Find each root, if possible.

33) $\sqrt[3]{125}$

34) $\sqrt[3]{27}$

35) $\sqrt[3]{-1}$

36) $\sqrt[3]{-8}$

37) $\sqrt[4]{81}$

38) $\sqrt[4]{16}$

39) $\sqrt[4]{-1}$

40) $\sqrt[4]{-81}$

41) $-\sqrt[4]{16}$

42) $-\sqrt[4]{1}$

43) $\sqrt[5]{-32}$

44) $-\sqrt[6]{64}$

45) $-\sqrt[3]{-1000}$

46) $-\sqrt[3]{-27}$

47) $\sqrt[6]{-64}$

48) $\sqrt[4]{-16}$

49) $\sqrt[3]{\dfrac{8}{125}}$

50) $\sqrt[4]{\dfrac{81}{16}}$

51) $\sqrt{60 - 11}$

52) $\sqrt{100 + 21}$

53) $\sqrt[3]{9 - 36}$

54) $\sqrt{1 - 9}$

55) $\sqrt{5^2 + 12^2}$

56) $\sqrt{3^2 + 4^2}$

Objective 4: Evaluate $\sqrt[n]{a^n}$

57) If n is a positive, even integer and we are not certain that $a \geq 0$, then we must use the absolute value symbol to evaluate $\sqrt[n]{a^n}$. That is, $\sqrt[n]{a^n} = |a|$. Why must we use the absolute value symbol?

58) If n is a positive, odd integer then $\sqrt[n]{a^n} = a$ for any value of a. Why don't we need to use the absolute value symbol?

Simplify.

59) $\sqrt{8^2}$

60) $\sqrt{5^2}$

61) $\sqrt{(-6)^2}$

62) $\sqrt{(-11)^2}$

63) $\sqrt{y^2}$

64) $\sqrt{d^2}$

65) $\sqrt[3]{5^3}$

66) $\sqrt[3]{(-4)^3}$

67) $\sqrt[3]{z^3}$

68) $\sqrt[7]{t^7}$

69) $\sqrt[4]{h^4}$

70) $\sqrt[6]{m^6}$

71) $\sqrt{(x+7)^2}$

72) $\sqrt{(a-9)^2}$

73) $\sqrt[3]{(2t-1)^3}$

74) $\sqrt[5]{(6r+7)^5}$

75) $\sqrt[4]{(3n+2)^4}$

76) $\sqrt[3]{(x-6)^3}$

77) $\sqrt[7]{(d-8)^7}$

78) $\sqrt[6]{(4y+3)^6}$

Solve each problem.

79) A stackable storage cube has a storage capacity of 8 ft³. The length of a side of the cube, s, is given by $s = \sqrt[3]{V}$, where V is the volume of the cube. How long is each side of the cube?

80) A die (singular of dice) has a surface area of 13.5 cm³. Since the die is in the shape of a cube, the length of a side of the die, s, is given by $s = \sqrt{\dfrac{A}{6}}$, where A is the surface area of the cube. How long is each side of the cube?

Mark Dierker/
McGraw-Hill
Education

81) If an object is dropped from the top of a building 160 ft tall, then the formula $t = \sqrt{\dfrac{160 - h}{16}}$

describes how many seconds, t, it takes for the object to reach a height of h ft above the ground. If a piece of ice falls off the top of this building, how long would it take to reach the ground? Give an exact answer and an approximation to two decimal places.

82) A circular flower garden has an area of 51 ft². The radius, r, of a circle in terms of its area is given by $r = \sqrt{\dfrac{A}{\pi}}$, where A is the area of the circle. What is the radius of the garden? Give an exact answer and an approximation to two decimal places.

83) The speed limit on a street in a residential neighborhood is 25 mph. A car involved in an accident on this road left skid marks 40 ft long. Accident investigators determine that the speed of the car can be described by the function $S(d) = \sqrt{22.5d}$, where $S(d)$ is the speed of the car, in miles per hour, at the time of impact and d is the length of the skid marks, in feet. Was the driver speeding at the time of the accident? Show all work to support your answer.

84) A car involved in an accident on a wet highway leaves skid marks 170 ft long. Accident investigators determine that the speed of the car, $S(d)$ in miles per hour, can be described by the function $S(d) = \sqrt{19.8d}$, where d is the length of the skid marks, in feet. The speed limit on the highway is 65 mph. Was the car speeding when the accident occurred? Show all work to support your answer.

R Rethink

R1) Would you be able to complete exercises similar to these on a quiz?

R2) Which section in the book could you review to help you better understand this section?

10.2 Rational Exponents

P Prepare

O Organize

What are your objectives for Section 10.2?	How can you accomplish each objective?
1 Evaluate Expressions of the Form $a^{1/n}$	• Write the definition of $a^{1/n}$ in your own words. • Complete the given example on your own. • Complete You Try 1.
2 Evaluate Expressions of the Form $a^{m/n}$	• Write the definition of $a^{m/n}$ in your own words. • Complete the given example on your own. • Complete You Try 2.
3 Evaluate Expressions of the Form $a^{-m/n}$	• Write the definition of $a^{-m/n}$ in your own words. • Complete the given example on your own. • Complete You Try 3.
4 Combine the Rules of Exponents	• Review the rules of exponents found in Section 6.1 to help you to follow the examples. • Complete the given examples on your own. • Complete You Trys 4 and 5.
5 Convert a Radical Expression to Exponential Form and Simplify	• Use the same rules of exponents to convert between radical and exponential forms. • Complete the given examples on your own. • Complete You Trys 6 and 7.

 **Work**

Read the explanations, follow the examples, take notes, and complete the You Trys.

1 Evaluate Expressions of the Form $a^{1/n}$

In this section, we will explain the relationship between radicals and rational exponents (fractional exponents). Sometimes, converting between these two forms makes it easier to simplify expressions.

Definition

If n is a positive integer greater than 1 and $\sqrt[n]{a}$ is a real number, then

$$a^{1/n} = \sqrt[n]{a}$$

(The denominator of the fractional exponent is the index of the radical.)

EXAMPLE 1

Write in radical form, and evaluate.

a) $8^{1/3}$ b) $49^{1/2}$ c) $81^{1/4}$

d) $-64^{1/6}$ e) $(-16)^{1/4}$ f) $(-125)^{1/3}$

Solution

a) The denominator of the fractional exponent is the index of the radical. Therefore, $8^{1/3} = \sqrt[3]{8} = 2$.

b) The denominator in the exponent of $49^{1/2}$ is 2, so the index on the radical is 2, meaning *square* root.

$$49^{1/2} = \sqrt{49} = 7$$

c) $81^{1/4} = \sqrt[4]{81} = 3$

d) $-64^{1/6} = -(64^{1/6}) = -\sqrt[6]{64} = -2$

e) $(-16)^{1/4} = \sqrt[4]{-16}$, which is not a real number. Remember, the even root of a negative number is not a real number.

f) $(-125)^{1/3} = \sqrt[3]{-125} = -5$ The odd root of a negative number is a negative number.

$\left[\text{YOU TRY 1}\right]$ Write in radical form, and evaluate.

a) $16^{1/4}$ b) $121^{1/2}$ c) $(-121)^{1/2}$ d) $(-27)^{1/3}$ e) $-81^{1/4}$

---

2 Evaluate Expressions of the Form $a^{m/n}$

We can add another relationship between rational exponents and radicals.

Definition

If m and n are positive integers and $\dfrac{m}{n}$ is in lowest terms, then

$$a^{m/n} = (a^{1/n})^m = (\sqrt[n]{a})^m$$

provided that $a^{1/n}$ is a real number.

(The denominator of the fractional exponent is the index of the radical, and the numerator is the power to which we raise the radical expression.) We can also think of $a^{m/n}$ this way: $a^{m/n} = (a^m)^{1/n} = \sqrt[n]{a^m}$.

EXAMPLE 2

Write in radical form, and evaluate.

a) $25^{3/2}$ b) $-64^{2/3}$ c) $(-81)^{3/2}$ d) $-81^{3/2}$ e) $(-1000)^{2/3}$

Solution

a) The *denominator* of the fractional exponent is the *index* of the radical, and the *numerator* is the *power* to which we raise the radical expression.

$$25^{3/2} = (25^{1/2})^3 \quad \text{Use the definition to rewrite the exponent.}$$
$$= (\sqrt{25})^3 \quad \text{Rewrite as a radical.}$$
$$= 5^3 \quad \sqrt{25} = 5$$
$$= 125$$

664   CHAPTER 10   **Radicals and Rational Exponents**

b) To evaluate $-64^{2/3}$, *first* evaluate $64^{2/3}$, *then* take the negative of that result.

$$-64^{2/3} = -(64^{2/3}) = -(64^{1/3})^2 \quad \text{Use the definition to rewrite the exponent.}$$
$$= -(\sqrt[3]{64})^2 \quad \text{Rewrite as a radical.}$$
$$= -(4)^2 \quad \sqrt[3]{64} = 4$$
$$= -16$$

c) $(-81)^{3/2} = [(-81)^{1/2}]^3$
$$= (\sqrt{-81})^3 \quad \text{Not a real number.} \quad \text{The even root of a negative number is not a real number.}$$

d) $-81^{3/2} = -(81^{1/2})^3 = -(\sqrt{81})^3 = -(9)^3 = -729$

e) $(-1000)^{2/3} = [(-1000)^{1/3}]^2 = (\sqrt[3]{-1000})^2 = (-10)^2 = 100$

[YOU TRY 2]   Write in radical form, and evaluate.

a) $32^{2/5}$ b) $-100^{3/2}$ c) $(-100)^{3/2}$ d) $(-1)^{4/5}$ e) $-1^{5/3}$

BE CAREFUL

In Example 2, notice how the parentheses affect how we evaluate an expression. The base of the expression $(-81)^{3/2}$ is -81, while the base of $-81^{3/2}$ is 81.

---

3 Evaluate Expressions of the Form $a^{-m/n}$

Recall that if n is any integer and $a \neq 0$, then $a^{-n} = \left(\dfrac{1}{a}\right)^n = \dfrac{1}{a^n}$.

That is, to rewrite the expression with a *positive* exponent, take the reciprocal of the base. For example,

$$2^{-4} = \left(\frac{1}{2}\right)^4 = \frac{1}{16}$$

We can extend this idea to rational exponents.

> ## Definition
>
> If $a^{m/n}$ is a nonzero real number, then
>
> $$a^{-m/n} = \left(\frac{1}{a}\right)^{m/n} = \frac{1}{a^{m/n}}$$
>
> (To rewrite the expression with a *positive* exponent, take the reciprocal of the base.)

EXAMPLE 3   Rewrite with a positive exponent, and evaluate.

a) $36^{-1/2}$ b) $32^{-2/5}$ c) $\left(\dfrac{125}{64}\right)^{-2/3}$

Solution

a) To write $36^{-1/2}$ with a positive exponent, take the reciprocal of the base.

$$36^{-1/2} = \left(\frac{1}{36}\right)^{1/2}$$ The reciprocal of 36 is $\frac{1}{36}$.

$$= \sqrt{\frac{1}{36}}$$ The denominator of the fractional exponent is the index of the radical.

$$= \frac{1}{6}$$

b)

$$32^{-2/5} = \left(\frac{1}{32}\right)^{2/5}$$ The reciprocal of 32 is $\frac{1}{32}$.

$$= \left(\sqrt[5]{\frac{1}{32}}\right)^2$$ The denominator of the fractional exponent is the index of the radical.

$$= \left(\frac{1}{2}\right)^2$$ $\sqrt[5]{\frac{1}{32}} = \frac{1}{2}$

$$= \frac{1}{4}$$

c)

$$\left(\frac{125}{64}\right)^{-2/3} = \left(\frac{64}{125}\right)^{2/3}$$ The reciprocal of $\frac{125}{64}$ is $\frac{64}{125}$.

$$= \left(\sqrt[3]{\frac{64}{125}}\right)^2$$ The denominator of the fractional exponent is the index of the radical.

$$= \left(\frac{4}{5}\right)^2$$ $\sqrt[3]{\frac{64}{125}} = \frac{4}{5}$

$$= \frac{16}{25}$$

[YOU TRY 3]   Rewrite with a positive exponent, and evaluate.

a) $144^{-1/2}$ b) $16^{-3/4}$ c) $\left(\frac{8}{27}\right)^{-2/3}$

---

4 Combine the Rules of Exponents

We can combine the rules presented in this section with the rules of exponents we learned in Chapter 2 to simplify expressions containing numbers or variables.

EXAMPLE 4   Simplify completely. The answer should contain only positive exponents.

a) $(6^{1/5})^2$ b) $25^{3/4} \cdot 25^{-1/4}$ c) $\dfrac{8^{2/9}}{8^{11/9}}$

Solution

a) $(6^{1/5})^2 = 6^{2/5}$ Multiply exponents.

b) $25^{3/4} \cdot 25^{-1/4} = 25^{\frac{3}{4}+\left(-\frac{1}{4}\right)}$ Add exponents.

$$= 25^{2/4} = 25^{1/2} = 5$$

c) $\dfrac{8^{2/9}}{8^{11/9}} = 8^{\frac{2}{9}-\frac{11}{9}}$ Subtract exponents.

$$= 8^{-9/9} \qquad \text{Subtract } \frac{2}{9} - \frac{11}{9}.$$

$$= 8^{-1} \qquad \text{Reduce } -\frac{9}{9}.$$

$$= \left(\frac{1}{8}\right)^1 = \frac{1}{8}$$

[YOU TRY 4] Simplify completely. The answer should contain only positive exponents.

a) $49^{3/8} \cdot 49^{1/8}$ b) $(16^{1/12})^3$ c) $\dfrac{7^{2/5}}{7^{4/5}}$

EXAMPLE 5   Simplify completely. Assume the variables represent positive real numbers. The answer should contain only positive exponents.

a) $r^{1/8} \cdot r^{3/8}$ b) $\left(\dfrac{x^{2/3}}{y^{1/4}}\right)^6$ c) $\dfrac{n^{-5/6} \cdot n^{1/3}}{n^{-1/6}}$ d) $\left(\dfrac{a^{-7}b^{1/2}}{a^5 c^{1/3}}\right)^{-3/4}$

Solution

a) $r^{1/8} \cdot r^{3/8} = r^{\frac{1}{8}+\frac{3}{8}}$ Add exponents.

$$= r^{4/8} = r^{1/2}$$

b) $\left(\dfrac{x^{2/3}}{y^{1/4}}\right)^6 = \dfrac{x^{\frac{2}{3}\cdot 6}}{y^{\frac{1}{4}\cdot 6}}$ Multiply exponents.

$$= \dfrac{x^4}{y^{3/2}} \qquad \text{Reduce.}$$

c) $\dfrac{n^{-5/6} \cdot n^{1/3}}{n^{-1/6}} = \dfrac{n^{-\frac{5}{6}+\frac{1}{3}}}{n^{-1/6}} = \dfrac{n^{-\frac{5}{6}+\frac{2}{6}}}{n^{-1/6}} = \dfrac{n^{-3/6}}{n^{-1/6}}$ Add exponents.

$$= n^{-\frac{3}{6}-\left(-\frac{1}{6}\right)} = n^{-2/6} = n^{-1/3} = \dfrac{1}{n^{1/3}} \qquad \text{Subtract exponents.}$$

d) $\left(\dfrac{a^{-7}b^{1/2}}{a^5 b^{1/3}}\right)^{-3/4} = \left(\dfrac{a^5 b^{1/3}}{a^{-7} b^{1/2}}\right)^{3/4}$ Eliminate the negative from the outermost exponent by taking the reciprocal of the base.

Simplify the expression inside the parentheses by subtracting the exponents.

$$= (a^{5-(-7)} b^{1/3-1/2})^{3/4} = (a^{5+7} b^{2/6-3/6})^{3/4} = (a^{12} b^{-1/6})^{3/4}$$

Apply the power rule, and simplify.

$$= (a^{12})^{3/4} (b^{-1/6})^{3/4} = a^9 b^{-1/8} = \dfrac{a^9}{b^{1/8}}$$

[YOU TRY 5] Simplify completely. Assume the variables represent positive real numbers. The answer should contain only positive exponents.

a) $(a^3 b^{1/5})^{10}$ b) $\dfrac{t^{3/10}}{t^{7/10}}$ c) $\dfrac{s^{3/4}}{s^{1/2} \cdot s^{-5/4}}$ d) $\left(\dfrac{x^4 y^{3/8}}{x^9 y^{1/4}}\right)^{-2/5}$

5 Convert a Radical Expression to Exponential Form and Simplify

Some radicals can be simplified by first putting them into rational exponent form and then converting them back to radicals.

EXAMPLE 6

Rewrite each radical in exponential form, then simplify. Write the answer in simplest (or radical) form. Assume the variable represents a nonnegative real number.

a) $\sqrt[8]{9^4}$ b) $\sqrt[6]{s^4}$

Solution

a) Because the index of the radical is the denominator of the exponent and the power is the numerator, we can write

$$\sqrt[8]{9^4} = 9^{4/8} \qquad \text{Write with a rational exponent.}$$
$$= 9^{1/2} = 3$$

b) $\sqrt[6]{s^4} = s^{4/6} \qquad \text{Write with a rational exponent.}$
$$= s^{2/3} = \sqrt[3]{s^2}$$

The expression $\sqrt[6]{s^4}$ is not in simplest form because the 4 and the 6 contain a common factor of 2, but $\sqrt[3]{s^2}$ *is* in simplest form because 2 and 3 do not have any common factors besides 1.

YOU TRY 6

Rewrite each radical in exponential form, then simplify. Write the answer in simplest (or radical) form. Assume the variable represents a nonnegative real number.

a) $\sqrt[6]{125^2}$ b) $\sqrt[10]{p^4}$

In Section 10.1 we said that if a is negative and n is a positive, even number, then $\sqrt[n]{a^n} = |a|$. For example, if we are *not* told that k is positive, then $\sqrt{k^2} = |k|$. However, if we assume that k is positive, then $\sqrt{k^2} = k$. **In the rest of this chapter, we will assume that all variables represent positive, real numbers unless otherwise stated.** When we make this assumption, we do not need to use absolute values when simplifying even roots. And if we consider this together with the relationship between radicals and rational exponents we have another way to explain why $\sqrt[n]{a^n} = a$.

EXAMPLE 7

Simplify.

a) $\sqrt[3]{5^3}$ b) $(\sqrt[4]{9})^4$ c) $\sqrt{k^2}$

Solution

a) $\sqrt[3]{5^3} = (5^3)^{1/3} = 5^{3 \cdot \frac{1}{3}} = 5^1 = 5$

b) $(\sqrt[4]{9})^4 = (9^{1/4})^4 = 9^{\frac{1}{4} \cdot 4} = 9^1 = 9$

c) $\sqrt{k^2} = (k^2)^{1/2} = k^{2 \cdot \frac{1}{2}} = k^1 = k$

YOU TRY 7

Simplify.

a) $(\sqrt{10})^2$ b) $\sqrt[3]{7^3}$ c) $\sqrt[4]{t^4}$

We can evaluate square roots, cube roots, or even higher roots by first rewriting the radical in exponential form and then using a graphing calculator.

For example, to evaluate $\sqrt{49}$, first rewrite the radical as $49^{1/2}$, then enter [4] [9], press [^] [(], enter [1] [÷] [2], and press [)] [ENTER]. The result is 7, as shown on the screen on the left below.

To approximate $\sqrt[3]{12^2}$ rounded to the nearest tenth, first rewrite the radical as $12^{2/3}$, then enter [1] [2], press [^] [(], enter [2] [÷] [3], and press [)] [ENTER]. The result is 5.241482788 as shown on the screen on the right below. The result rounded to the nearest tenth is then 5.2.

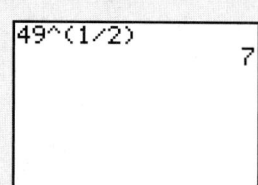
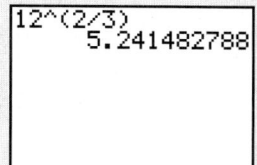

To evaluate radicals with an index greater than 3, follow the same procedure explained above. Evaluate by rewriting in exponential form if necessary and then using a graphing calculator. If necessary, approximate to the nearest tenth.

1) $16^{1/2}$ 2) $\sqrt[3]{512}$ 3) $\sqrt{37}$ 4) $361^{1/2}$ 5) $4096^{2/3}$ 6) $2401^{1/4}$

ANSWERS TO [YOU TRY] EXERCISES

1) a) 2 b) 11 c) not a real number d) -3 e) -3 2) a) 4 b) -1000
c) not a real number d) 1 e) -1 3) a) $\dfrac{1}{12}$ b) $\dfrac{1}{8}$ c) $\dfrac{9}{4}$ 4) a) 7 b) 2 c) $\dfrac{1}{7^{2/5}}$
5) a) $a^{30}b^2$ b) $\dfrac{1}{t^{2/5}}$ c) $s^{3/2}$ d) $\dfrac{x^2}{y^{1/20}}$ 6) a) 5 b) $\sqrt[5]{p^2}$ 7) a) 10 b) 7 c) t

---

ANSWERS TO TECHNOLOGY EXERCISES

1) 4   2) 8   3) 6.1   4) 19   5) 256   6) 7

[E] Evaluate   10.2   Exercises

Do the exercises, and check your work.

Objective 1: Evaluate Expressions of the Form $a^{1/n}$

1) Explain how to write $25^{1/2}$ in radical form.

2) Explain how to write $1^{1/3}$ in radical form.

Write in radical form, and evaluate.

3) $9^{1/2}$

4) $64^{1/2}$

5) $1000^{1/3}$

6) $27^{1/3}$

7) $32^{1/5}$

8) $81^{1/4}$

9) $-64^{1/6}$

10) $-125^{1/3}$

11) $\left(\dfrac{4}{121}\right)^{1/2}$

12) $\left(\dfrac{4}{9}\right)^{1/2}$

13) $\left(\dfrac{125}{64}\right)^{1/3}$

14) $\left(\dfrac{16}{81}\right)^{1/4}$

15) $-\left(\dfrac{36}{169}\right)^{1/2}$

16) $-\left(\dfrac{1000}{27}\right)^{1/3}$

17) $(-81)^{1/4}$

18) $(-169)^{1/2}$

19) $(-1)^{1/7}$

20) $(-8)^{1/3}$

Objective 2: Evaluate Expressions of the Form $a^{m/n}$

21) Explain how to write $16^{3/4}$ in radical form.

22) Explain how to write $100^{3/2}$ in radical form.

Write in radical form, and evaluate.

23) $8^{4/3}$

24) $81^{3/4}$

25) $64^{5/6}$

26) $32^{3/5}$

27) $(-125)^{2/3}$

28) $(-1000)^{2/3}$

29) $-36^{3/2}$

30) $-27^{4/3}$

31) $(-81)^{3/4}$

32) $(-25)^{3/2}$

33) $\left(\dfrac{16}{81}\right)^{3/4}$

34) $-16^{5/4}$

35) $-\left(\dfrac{1000}{27}\right)^{2/3}$

36) $-\left(\dfrac{8}{27}\right)^{4/3}$

Objective 3: Evaluate Expressions of the Form $a^{-m/n}$

Decide whether each statement is true or false. Explain your answer.

37) $81^{-1/2} = -9$

38) $\left(\dfrac{1}{100}\right)^{-3/2} = \left(\dfrac{1}{100}\right)^{2/3}$

Rewrite with a positive exponent, and evaluate.

Fill It In

Fill in the blanks with either the missing mathematical step or reason for the given step.

39) $64^{-1/2} = \left(\right)^{1/2}$ The reciprocal of 64 is ___ .

$= \sqrt{\dfrac{1}{64}}$ _____

$= \underline{}$ Simplify.

40) $\left(\dfrac{1}{1000}\right)^{-1/3} = (\underline{})^{1/3}$ The reciprocal of $\dfrac{1}{1000}$

is ___ .

$= \sqrt[3]{1000}$ _____

$= \underline{}$ Simplify.

41) $49^{-1/2}$

42) $100^{-1/2}$

43) $1000^{-1/3}$

44) $27^{-1/3}$

45) $\left(\dfrac{1}{81}\right)^{-1/4}$

46) $\left(\dfrac{1}{32}\right)^{-1/5}$

47) $-\left(\dfrac{1}{64}\right)^{-1/3}$

48) $-\left(\dfrac{1}{125}\right)^{-1/3}$

49) $64^{-5/6}$

50) $81^{-3/4}$

51) $125^{-2/3}$

52) $64^{-2/3}$

53) $\left(\dfrac{25}{4}\right)^{-3/2}$

54) $\left(\dfrac{9}{100}\right)^{-3/2}$

55) $\left(\dfrac{64}{125}\right)^{-2/3}$

56) $\left(\dfrac{81}{16}\right)^{-3/4}$

Objective 4: Combine the Rules of Exponents

Simplify completely. The answer should contain only positive exponents.

57) $2^{2/3} \cdot 2^{7/3}$

58) $5^{3/4} \cdot 5^{5/4}$

59) $(9^{1/4})^2$

60) $(7^{2/3})^3$

61) $8^{7/5} \cdot 8^{-3/5}$

62) $6^{-4/3} \cdot 6^{5/3}$

63) $\dfrac{2^{23/4}}{2^{3/4}}$

64) $\dfrac{5^{3/2}}{5^{9/2}}$

65) $\dfrac{4^{2/5}}{4^{6/5} \cdot 4^{3/5}}$

66) $\dfrac{6^{-1}}{6^{1/2} \cdot 6^{-5/2}}$

Simplify completely. The answer should contain only positive exponents.

67) $z^{1/6} \cdot z^{5/6}$

68) $h^{1/6} \cdot h^{-3/4}$

69) $(-9v^{5/8})(8v^{3/4})$

70) $(-3x^{-1/3})(8x^{4/9})$

71) $\dfrac{a^{5/9}}{a^{4/9}}$

72) $\dfrac{x^{1/6}}{x^{5/6}}$

73) $\dfrac{48w^{3/10}}{10w^{2/5}}$

74) $\dfrac{20c^{-2/3}}{72c^{5/6}}$

75) $(x^{-2/9})^3$

76) $(n^{-2/7})^3$

77) $(z^{1/5})^{2/3}$

78) $(r^{4/3})^{5/2}$

79) $(81u^{8/3}v^4)^{3/4}$

80) $(64x^6y^{12/5})^{5/6}$

81) $(32r^{1/3}s^{4/9})^{3/5}$

82) $(125a^9b^{1/4})^{2/3}$

83) $\left(\dfrac{f^{6/7}}{27g^{-5/3}}\right)^{1/3}$

84) $\left(\dfrac{16c^{-8}}{b^{-11/3}}\right)^{3/4}$

85) $\left(\dfrac{x^{-5/3}}{w^{3/2}}\right)^{-6}$

86) $\left(\dfrac{t^{-3/2}}{u^{1/4}}\right)^{-4}$

87) $\dfrac{y^{1/2} \cdot y^{-1/3}}{y^{5/6}}$

88) $\dfrac{t^5}{t^{1/2} \cdot t^{3/4}}$

89) $\left(\dfrac{a^4b^3}{32a^{-2}b^4}\right)^{2/5}$

90) $\left(\dfrac{16c^{-8}d^3}{c^4d^5}\right)^{3/2}$

91) $\left(\dfrac{r^{4/5}t^{-2}}{r^{2/3}t^5}\right)^{-3/2}$

92) $\left(\dfrac{x^{10}y^{1/6}}{x^{-8}y^{2/3}}\right)^{-2/3}$

93) $\left(\dfrac{h^{-2}k^{5/2}}{h^{-8}k^{5/6}}\right)^{-5/6}$

94) $\left(\dfrac{c^{1/8}d^{-4}}{c^{3/4}d}\right)^{-8/5}$

95) $p^{1/2}(p^{2/3} + p^{1/2})$

96) $w^{4/3}(w^{1/2} - w^3)$

Objective 5: Convert a Radical Expression to Exponential Form and Simplify

Rewrite each radical in exponential form, then simplify. Write the answer in simplest (or radical) form.

Fill It In

Fill in the blanks with either the missing mathematical step or reason for the given step.

97) $\sqrt[12]{25^6} =$ _____   Write with a rational exponent.

   $=$ ____   Reduce the exponent.

   $= 5$ _____

98) $\sqrt[10]{c^4} =$ ____   Write with a rational exponent.

   $= c^{2/5}$ _____

   $=$ ____   Write in radical form.

99) $\sqrt[6]{49^3}$

100) $\sqrt[9]{8^3}$

101) $\sqrt[4]{81^2}$

102) $\sqrt{3^2}$

103) $(\sqrt{5})^2$

104) $(\sqrt[3]{10})^3$

105) $(\sqrt[3]{12})^3$

106) $(\sqrt[4]{15})^4$

107) $\sqrt[3]{x^{12}}$

108) $\sqrt[4]{t^8}$

109) $\sqrt[6]{k^2}$

110) $\sqrt[9]{w^6}$

111) $\sqrt[4]{z^2}$

112) $\sqrt[8]{m^4}$

113) $\sqrt{d^4}$

114) $\sqrt{s^6}$

The wind chill temperature, WC, measures how cold it feels outside (for temperatures under 50 degrees F) when the velocity of the wind, V, is considered along with the air temperature, T. The stronger the wind at a given air temperature, the colder it feels.

The formula for calculating wind chill is

$$WC = 35.74 + 0.6215T - 35.75V^{4/25} + 0.4275TV^{4/25}$$

where WC and T are in degrees Fahrenheit and V is in miles per hour. (http://www.nws.noaa.gov/om/windchill/windchillglossary.shtml)

Use this information for Exercises 115 and 116, and round all answers to the nearest degree.

115) Determine the wind chill when the air temperature is 20 degrees and the wind is blowing at the given speed.

image100/Alamy

   a) 5 mph

   b) 15 mph

116) Determine the wind chill when the air temperature is 10 degrees and the wind is blowing at the given speed. Round your answer to the nearest degree.

   a) 12 mph

   b) 20 mph

R Rethink

R1) Describe how having a good understanding of the rules of exponents has helped you move quickly through the exercises.

R2) Where could you spend more time mastering an objective?

10.3 Simplifying Expressions Containing Square Roots

P Prepare

O Organize

What are your objectives for Section 10.3?	How can you accomplish each objective?
1 Multiply Square Roots	• Write the definition of the *product rule for square roots* in your own words. • Complete the given example on your own. • Complete You Try 1.
2 Simplify the Square Root of a Whole Number	• Write the property for **When Is a Square Root Simplified?** • Complete the given example on your own. • Complete You Try 2.
3 Use the Quotient Rule for Square Roots	• Write the definition of the *quotient rule for square roots* in your own words. • Complete the given example on your own. • Complete You Try 3.
4 Simplify Square Root Expressions Containing Variables with Even Exponents	• Write the property for $\sqrt{a^m}$ in your notes. • Complete the given example on your own. • Complete You Try 4.
5 Simplify Square Root Expressions Containing Variables with Odd Exponents	• Write the procedure for **Simplifying a Radical Containing Variables** in your own words. • Complete the given examples on your own. • Complete You Trys 5–7.
6 Simplify More Square Root Expressions Containing Variables	• Apply the rules from previous objectives, and follow the example. • Complete You Try 8.

Read the explanations, follow the examples, take notes, and complete the You Trys.

In this section, we will introduce rules for finding the product and quotient of square roots as well as for simplifying expressions containing square roots.

1 Multiply Square Roots

Let's begin with the product $\sqrt{4} \cdot \sqrt{9}$. We can find the product like this:
$\sqrt{4} \cdot \sqrt{9} = 2 \cdot 3 = 6$. Also notice that $\sqrt{4} \cdot \sqrt{9} = \sqrt{4 \cdot 9} = \sqrt{36} = 6$.

We obtain the same result. This leads us to the product rule for multiplying expressions containing square roots.

Definition Product Rule for Square Roots

Let a and b be nonnegative real numbers. Then,

$$\sqrt{a} \cdot \sqrt{b} = \sqrt{a \cdot b}$$

In other words, the product of two square roots equals the square root of the product.

EXAMPLE 1 Multiply. a) $\sqrt{5} \cdot \sqrt{2}$ b) $\sqrt{3} \cdot \sqrt{x}$

Solution

a) $\sqrt{5} \cdot \sqrt{2} = \sqrt{5 \cdot 2} = \sqrt{10}$ b) $\sqrt{3} \cdot \sqrt{x} = \sqrt{3 \cdot x} = \sqrt{3x}$

 **BE CAREFUL** We can multiply radicals this way *only if* the indices are the same. We will see later how to multiply radicals with different indices such as $\sqrt{5} \cdot \sqrt[3]{t}$.

[**YOU TRY 1**] Multiply.

a) $\sqrt{6} \cdot \sqrt{5}$ b) $\sqrt{10} \cdot \sqrt{r}$

2 Simplify the Square Root of a Whole Number

Knowing how to simplify radicals is very important in the study of algebra. We begin by discussing how to simplify expressions containing square roots.

How do we know when a square root is simplified?

Property When Is a Square Root Simplified?

An expression containing a square root is simplified when all of the following conditions are met:

1) The radicand does not contain any factors (other than 1) that are perfect squares.

2) The radicand does not contain any fractions.

3) There are no radicals in the denominator of a fraction.

Note: Condition 1) implies that the radical cannot contain variables with exponents greater than or equal to 2, the index of the square root.

We will discuss higher roots in Section 10.4.

To simplify expressions containing square roots we reverse the process of multiplying. That is, we use the product rule that says $\sqrt{a \cdot b} = \sqrt{a} \cdot \sqrt{b}$ where a or b are perfect squares.

EXAMPLE 2 Simplify completely.

a) $\sqrt{18}$ b) $\sqrt{500}$ c) $\sqrt{21}$ d) $\sqrt{48}$

Solution

a) The radical $\sqrt{18}$ is not in simplest form since 18 contains a factor (other than 1) that is a perfect square. Think of two numbers that multiply to 18 so that at least one of the numbers is a perfect square: $18 = 9 \cdot 2$.

(While it is true that $18 = 6 \cdot 3$, neither 6 nor 3 is a perfect square.)
Rewrite $\sqrt{18}$:

$$\begin{aligned} \sqrt{18} &= \sqrt{9 \cdot 2} &&\text{9 is a perfect square.} \\ &= \sqrt{9} \cdot \sqrt{2} &&\text{Product rule} \\ &= 3\sqrt{2} &&\sqrt{9} = 3 \end{aligned}$$

$3\sqrt{2}$ is completely simplified because 2 does not have any factors that are perfect squares.

W Hint

What is the first question you should ask yourself when simplifying?

b) Does 500 have a factor that is a perfect square? *Yes!* $500 = 100 \cdot 5$. To simplify $\sqrt{500}$, rewrite it as

$$\begin{aligned} \sqrt{500} &= \sqrt{100 \cdot 5} &&\text{100 is a perfect square.} \\ &= \sqrt{100} \cdot \sqrt{5} &&\text{Product rule} \\ &= 10\sqrt{5} &&\sqrt{100} = 10 \end{aligned}$$

$10\sqrt{5}$ is completely simplified because 5 does not have any factors that are perfect squares.

c) $21 = 3 \cdot 7$ Neither 3 nor 7 is a perfect square.
$21 = 1 \cdot 21$ Although 1 is a perfect square, it will not help us simplify $\sqrt{21}$.

$\sqrt{21}$ is in simplest form.

d) There are different ways to simplify $\sqrt{48}$. We will look at two of them.

i) Two numbers that multiply to 48 are 16 and 3 with 16 being a perfect square. We can write

$$\sqrt{48} = \sqrt{16 \cdot 3} = \sqrt{16} \cdot \sqrt{3} = 4\sqrt{3}$$

ii) We can also think of 48 as $4 \cdot 12$ since 4 is a perfect square. We can write

$$\sqrt{48} = \sqrt{4 \cdot 12} = \sqrt{4} \cdot \sqrt{12} = 2\sqrt{12}$$

Therefore, $\sqrt{48} = 2\sqrt{12}$. Is $\sqrt{12}$ in simplest form? *No, because $12 = 4 \cdot 3$ and 4 is a perfect square.* We must continue to simplify.

$$\begin{aligned} \sqrt{48} &= 2\sqrt{12} \\ &= 2\sqrt{4 \cdot 3} = 2\sqrt{4} \cdot \sqrt{3} = 2 \cdot 2 \cdot \sqrt{3} = 4\sqrt{3} \end{aligned}$$

$4\sqrt{3}$ is completely simplified because 3 does not have any factors that are perfect squares.

Example 2d) shows that using either $\sqrt{48} = \sqrt{16 \cdot 3}$ or $\sqrt{48} = \sqrt{4 \cdot 12}$ leads us to the same result. Furthermore, this example illustrates that a radical is not always *completely* simplified after just one iteration of the simplification process. It is necessary to always examine the radical to determine whether or not it can be simplified more.

After simplifying a radical, look at the result and ask yourself, "*Is the radical in simplest form?*" If it is not, simplify again. Asking yourself this question will help you to be sure that the radical *is* completely simplified.

[YOU TRY 2] Simplify completely.

a) $\sqrt{28}$ b) $\sqrt{75}$ c) $\sqrt{35}$ d) $\sqrt{72}$

---

3 Use the Quotient Rule for Square Roots

Let's simplify $\dfrac{\sqrt{36}}{\sqrt{9}}$. We can say $\dfrac{\sqrt{36}}{\sqrt{9}} = \dfrac{6}{3} = 2$. It is also true that $\dfrac{\sqrt{36}}{\sqrt{9}} = \sqrt{\dfrac{36}{9}} = \sqrt{4} = 2$.

This leads us to the quotient rule for dividing expressions containing square roots.

Definition Quotient Rule for Square Roots

Let a and b be nonnegative real numbers such that $b \neq 0$. Then,

$$\sqrt{\dfrac{a}{b}} = \dfrac{\sqrt{a}}{\sqrt{b}}$$

The square root of a quotient equals the quotient of the square roots.

EXAMPLE 3   Simplify completely.

a) $\sqrt{\dfrac{9}{49}}$ b) $\sqrt{\dfrac{200}{2}}$ c) $\dfrac{\sqrt{72}}{\sqrt{6}}$ d) $\sqrt{\dfrac{5}{81}}$

Solution

a) Because 9 and 49 are each perfect squares, find the square root of each separately.

$$\sqrt{\dfrac{9}{49}} = \dfrac{\sqrt{9}}{\sqrt{49}} \qquad \text{Quotient rule}$$

$$= \dfrac{3}{7} \qquad \sqrt{9} = 3 \text{ and } \sqrt{49} = 7$$

b) Neither 200 nor 2 is a perfect square, but if we simplify $\dfrac{200}{2}$ we get 100, which *is* a perfect square.

$$\sqrt{\dfrac{200}{2}} = \sqrt{100} \qquad \text{Simplify } \dfrac{200}{2}.$$

$$= 10$$

c) We can simplify $\dfrac{\sqrt{72}}{\sqrt{6}}$ using two different methods.

i) Begin by applying the quotient rule to obtain a fraction under *one* radical and simplify the fraction.

$$\frac{\sqrt{72}}{\sqrt{6}} = \sqrt{\frac{72}{6}} \qquad \text{Quotient rule}$$

$$= \sqrt{12} \; = \sqrt{4 \cdot 3} = \sqrt{4} \cdot \sqrt{3} = 2\sqrt{3}$$

ii) We can apply the product rule to rewrite $\sqrt{72}$ then simplify the fraction.

$$\frac{\sqrt{72}}{\sqrt{6}} = \frac{\sqrt{6} \cdot \sqrt{12}}{\sqrt{6}} \qquad \text{Product rule}$$

$$= \frac{\overset{1}{\cancel{\sqrt{6}}} \cdot \sqrt{12}}{\underset{1}{\cancel{\sqrt{6}}}} \qquad \text{Divide out the common factor.}$$

$$= \sqrt{12} = \sqrt{4 \cdot 3} = \sqrt{4} \cdot \sqrt{3} = 2\sqrt{3}$$

Either method will produce the same result.

d) The fraction $\dfrac{5}{81}$ is in simplest form, and 81 *is* a perfect square. Begin by applying the quotient rule.

$$\sqrt{\frac{5}{81}} = \frac{\sqrt{5}}{\sqrt{81}} \qquad \text{Quotient rule}$$

$$= \frac{\sqrt{5}}{9} \qquad \sqrt{81} = 9$$

[YOU TRY 3] Simplify completely.

a) $\sqrt{\dfrac{100}{169}}$ b) $\sqrt{\dfrac{27}{3}}$ c) $\dfrac{\sqrt{250}}{\sqrt{5}}$ d) $\sqrt{\dfrac{11}{36}}$

4   Simplify Square Root Expressions Containing Variables with Even Exponents

Recall that a square root is not simplified if it contains any factors that are perfect squares.

This means that a square root containing variables is simplified if the power on each variable is less than 2. For example, $\sqrt{r^6}$ is not in simplified form. If r represents a nonnegative real number, then we can use rational exponents to simplify $\sqrt{r^6}$.

$$\sqrt{r^6} = (r^6)^{1/2} = r^{6 \cdot \frac{1}{2}} = r^{6/2} = r^3$$

Multiplying $6 \cdot \frac{1}{2}$ is the same as dividing 6 by 2. We can generalize this result with the following statement.

Property $\sqrt{a^m}$

If a is a nonnegative real number and m is an integer, then

$$\sqrt{a^m} = a^{m/2}$$

We can combine this property with the product and quotient rules to simplify radical expressions.

EXAMPLE 4

Simplify completely.

a) $\sqrt{z^2}$ b) $\sqrt{49t^2}$ c) $\sqrt{18b^{14}}$ d) $\sqrt{\dfrac{32}{n^{20}}}$

Solution

a) $\sqrt{z^2} = z^{2/2} = z^1 = z$

b) $\sqrt{49t^2} = \sqrt{49} \cdot \sqrt{t^2} = 7 \cdot t^{2/2} = 7t$

c) $\sqrt{18b^{14}} = \sqrt{18} \cdot \sqrt{b^{14}}$ Product rule

           $= \sqrt{9} \cdot \sqrt{2} \cdot b^{14/2}$ 9 is a perfect square.

           $= 3\sqrt{2} \cdot b^7$ Simplify.

           $= 3b^7\sqrt{2}$ Rewrite using the commutative property.

d) We begin by using the quotient rule.

$$\sqrt{\frac{32}{n^{20}}} = \frac{\sqrt{32}}{\sqrt{n^{20}}} = \frac{\sqrt{16} \cdot \sqrt{2}}{n^{20/2}} = \frac{4\sqrt{2}}{n^{10}}$$

W Hint

Are you writing out the examples as you are reading them?

[YOU TRY 4]

Simplify completely.

a) $\sqrt{y^{10}}$ b) $\sqrt{144p^{16}}$ c) $\sqrt{54c^{10}}$ d) $\sqrt{\dfrac{45}{w^4}}$

---

5   Simplify Square Root Expressions Containing Variables with Odd Exponents

How do we simplify an expression containing a square root if the power under the square root is odd? We can use the product rule for radicals and fractional exponents to help us understand how to simplify such expressions.

EXAMPLE 5

Simplify completely.

a) $\sqrt{x^7}$ b) $\sqrt{c^{11}}$

Solution

a) To simplify $\sqrt{x^7}$, write x^7 as the product of two factors so that the exponent of one of the factors is the *largest* number less than 7 that is divisible by 2 (the index of the radical).

$\sqrt{x^7} = \sqrt{x^6 \cdot x^1}$ 6 is the largest number less than 7 that is divisible by 2.

      $= \sqrt{x^6} \cdot \sqrt{x}$ Product rule

      $= x^{6/2} \cdot \sqrt{x}$ Use a fractional exponent to simplify.

      $= x^3\sqrt{x}$ $6 \div 2 = 3$

b) To simplify $\sqrt{c^{11}}$, write c^{11} as the product of two factors so that the exponent of one of the factors is the *largest* number less than 11 that is divisible by 2 (the index of the radical).

$\sqrt{c^{11}} = \sqrt{c^{10} \cdot c^1}$ 10 is the largest number less than 11 that is divisible by 2.

      $= \sqrt{c^{10}} \cdot \sqrt{c}$ Product rule

      $= c^{10/2} \cdot \sqrt{c}$ Use a fractional exponent to simplify.

      $= c^5\sqrt{c}$ $10 \div 2 = 5$

[YOU TRY 5]   Simplify completely.

a) $\sqrt{m^5}$ b) $\sqrt{z^{19}}$

We used the product rule to simplify each radical in Example 5. During the simplification, however, we always divided an exponent by 2. This idea of division gives us another way to simplify radical expressions. Once again, let's look at the radicals and their simplified forms in Example 5 to see how we can simplify radical expressions using division.

$$\sqrt{x^7} = x^3\sqrt{x^1} = x^3\sqrt{x}$$

$$\begin{array}{r} 3 \to \text{Quotient} \\ \text{Index} \\ \text{of} \to 2\overline{)\ 7} \\ \text{radical} \quad \underline{-6} \\ 1 \to \text{Remainder} \end{array}$$

$$\sqrt{c^{11}} = c^5\sqrt{c^1} = c^5\sqrt{c}$$

$$\begin{array}{r} 5 \to \text{Quotient} \\ \text{Index} \\ \text{of} \to 2\overline{)\ 11} \\ \text{radical} \quad \underline{-10} \\ 1 \to \text{Remainder} \end{array}$$

W Hint

Write a sample expression that follows these steps in your notes.

Procedure Simplifying a Radical Containing Variables

To simplify a radical expression containing variables:

1) Divide the original exponent in the radicand by the index of the radical.

2) The exponent on the variable *outside* of the radical will be the *quotient* of the division problem.

3) The exponent on the variable *inside* of the radical will be the *remainder* of the division problem.

EXAMPLE 6   Simplify completely.

a) $\sqrt{t^9}$ b) $\sqrt{16b^5}$ c) $\sqrt{45y^{21}}$

Solution

a) To simplify $\sqrt{t^9}$, divide:

$$\begin{array}{r} 4 \to \text{Quotient} \\ 2\overline{)\ 9} \\ \underline{-8} \\ 1 \to \text{Remainder} \end{array}$$

$$\sqrt{t^9} = t^4\sqrt{t^1} = t^4\sqrt{t}$$

b) $\sqrt{16b^5} = \sqrt{16} \cdot \sqrt{b^5}$ Product rule

$= 4 \cdot b^2\sqrt{b^1}$ $5 \div 2$ gives a quotient of 2 and a remainder of 1.

$= 4b^2\sqrt{b}$

c) $\sqrt{45y^{21}} = \sqrt{45} \cdot \sqrt{y^{21}}$ Product rule

$= \sqrt{9} \cdot \sqrt{5} \cdot y^{10}\sqrt{y^1}$

　　　↑　　　　↑
　　Product　　21 ÷ 2 gives a quotient of 10
　　rule　　　and a remainder of 1.

$= 3\sqrt{5} \cdot y^{10}\sqrt{y}$ $\sqrt{9} = 3$

$= 3y^{10} \cdot \sqrt{5} \cdot \sqrt{y}$ Use the commutative property to rewrite the expression.

$= 3y^{10}\sqrt{5y}$ Use the product rule to write the expression with one radical.

[YOU TRY 6]   Simplify completely.

a) $\sqrt{m^{13}}$ b) $\sqrt{100v^7}$ c) $\sqrt{32a^3}$

If a radical contains more than one variable, apply the product or quotient rule.

EXAMPLE 7

Simplify completely.

a) $\sqrt{8a^{15}b^3}$ b) $\sqrt{\dfrac{5r^{27}}{s^8}}$

Solution

a) $\sqrt{8a^{15}b^3} = \sqrt{8} \cdot \sqrt{a^{15}} \cdot \sqrt{b^3}$

$= \sqrt{4} \cdot \sqrt{2} \cdot a^7\sqrt{a^1} \cdot b^1\sqrt{b^1}$

     ↗   ↑   ↑

Product rule   15 ÷ 2 gives a quotient   3 ÷ 2 gives a quotient
of 7 and a remainder of 1. of 1 and a remainder of 1.

$= 2\sqrt{2} \cdot a^7\sqrt{a} \cdot b\sqrt{b}$ $\sqrt{4} = 2$

$= 2a^7b \cdot \sqrt{2} \cdot \sqrt{a} \cdot \sqrt{b}$ Use the commutative property to rewrite the expression.

$= 2a^7b\sqrt{2ab}$ Use the product rule to write the expression with one radical.

b) $\sqrt{\dfrac{5r^{27}}{s^8}} = \dfrac{\sqrt{5r^{27}}}{\sqrt{s^8}}$ Quotient rule

$= \dfrac{\sqrt{5} \cdot \sqrt{r^{27}}}{s^4}$ ⟵ Product rule
                       ⟵ 8 ÷ 2 = 4

$= \dfrac{\sqrt{5} \cdot r^{13}\sqrt{r^1}}{s^4}$ 27 ÷ 2 gives a quotient of 13 and a remainder of 1.

$= \dfrac{r^{13} \cdot \sqrt{5} \cdot \sqrt{r}}{s^4}$ Use the commutative property to rewrite the expression.

$= \dfrac{r^{13}\sqrt{5r}}{s^4}$ Use the product rule to write the expression with one radical.

[YOU TRY 7]

Simplify completely.

a) $\sqrt{c^5d^{12}}$ b) $\sqrt{27x^{10}y^9}$ c) $\sqrt{\dfrac{40u^{13}}{v^{20}}}$

6 Simplify More Square Root Expressions Containing Variables

Next we will look at some examples of multiplying and dividing radical expressions that contain variables. Remember to always look at the result and ask yourself, "*Is the radical in simplest form?*" If it is not, simplify completely.

EXAMPLE 8

Perform the indicated operation, and simplify completely.

a) $\sqrt{6t} \cdot \sqrt{3t}$ b) $\sqrt{2a^3b} \cdot \sqrt{8a^2b^5}$ c) $\dfrac{\sqrt{20x^5}}{\sqrt{5x}}$

Solution

a) $\sqrt{6t} \cdot \sqrt{3t} = \sqrt{6t \cdot 3t}$ Product rule

$= \sqrt{18t^2}$

$= \sqrt{18} \cdot \sqrt{t^2}$ Product rule

$= \sqrt{9 \cdot 2} \cdot t = \sqrt{9} \cdot \sqrt{2} \cdot t = 3\sqrt{2} \cdot t = 3t\sqrt{2}$

W Hint

In parts b) and c), think about which method you prefer and why.

b) $\sqrt{2a^3b} \cdot \sqrt{8a^2b^5}$

There are two good methods for multiplying these radicals.

i) Multiply the radicands to obtain one radical.

$$\sqrt{2a^3b} \cdot \sqrt{8a^2b^5} = \sqrt{2a^3b \cdot 8a^2b^5} \qquad \text{Product rule}$$
$$= \sqrt{16a^5b^6} \qquad \text{Multiply.}$$

Is the radical in simplest form? *No.*

$$= \sqrt{16} \cdot \sqrt{a^5} \cdot \sqrt{b^6} \qquad \text{Product rule}$$
$$= 4 \cdot a^2\sqrt{a} \cdot b^3 \qquad \text{Evaluate.}$$
$$= 4a^2b^3\sqrt{a} \qquad \text{Commutative property}$$

ii) Simplify each radical, then multiply.

$$\sqrt{2a^3b} = \sqrt{2} \cdot \sqrt{a^3} \cdot \sqrt{b} \qquad \qquad \sqrt{8a^2b^5} = \sqrt{8} \cdot \sqrt{a^2} \cdot \sqrt{b^5}$$
$$= \sqrt{2} \cdot a\sqrt{a} \cdot \sqrt{b} \qquad \qquad \qquad = 2\sqrt{2} \cdot a \cdot b^2\sqrt{b}$$
$$= a\sqrt{2ab} \qquad \qquad \qquad \qquad = 2ab^2\sqrt{2b}$$

Then, $\sqrt{2a^3b} \cdot \sqrt{8a^2b^5} = a\sqrt{2ab} \cdot 2ab^2\sqrt{2b}$

$$= a \cdot 2ab^2 \cdot \sqrt{2ab} \cdot \sqrt{2b} \qquad \text{Commutative property}$$
$$= 2a^2b^2\sqrt{4ab^2} \qquad \text{Multiply.}$$
$$= 2a^2b^2 \cdot 2 \cdot b \cdot \sqrt{a} \qquad \sqrt{4ab^2} = 2b\sqrt{a}$$
$$= 4a^2b^3\sqrt{a} \qquad \text{Multiply.}$$

Both methods give the same result.

c) We can use the quotient rule first or simplify first.

i) $\dfrac{\sqrt{20x^5}}{\sqrt{5x}} = \sqrt{\dfrac{20x^5}{5x}} \qquad \text{Use the quotient rule first.}$

$$= \sqrt{4x^4} = \sqrt{4} \cdot \sqrt{x^4} = 2x^2$$

ii) $\dfrac{\sqrt{20x^5}}{\sqrt{5x}} = \dfrac{\sqrt{20} \cdot \sqrt{x^5}}{\sqrt{5x}} \qquad \text{Simplify first by using the product rule.}$

$$= \dfrac{\sqrt{4} \cdot \sqrt{5} \cdot x^2\sqrt{x}}{\sqrt{5x}} \qquad \text{Product rule; simplify } \sqrt{x^5}.$$

$$= \dfrac{2\sqrt{5} \cdot x^2\sqrt{x}}{\sqrt{5x}} \qquad \sqrt{4} = 2$$

$$= \dfrac{2x^2\sqrt{5x}}{\sqrt{5x}} \qquad \text{Product rule}$$

$$= 2x^2 \qquad \text{Divide out the common factor.}$$

Both methods give the same result. In this case, the second method was longer. Sometimes, however, this method *can* be more efficient.

[YOU TRY 8] Perform the indicated operation, and simplify completely.

a) $\sqrt{2n^3} \cdot \sqrt{6n}$ 　　　 b) $\sqrt{15cd^5} \cdot \sqrt{3c^2d}$ 　　　 c) $\dfrac{\sqrt{128k^9}}{\sqrt{2k}}$

E Evaluate 10.3 Exercises

Do the exercises, and check your work.

Unless otherwise stated, assume all variables represent nonnegative real numbers.

Objective 1: Multiply Square Roots
Multiply and simplify.

1) $\sqrt{3} \cdot \sqrt{7}$ 2) $\sqrt{11} \cdot \sqrt{5}$

3) $\sqrt{10} \cdot \sqrt{3}$ 4) $\sqrt{7} \cdot \sqrt{2}$

5) $\sqrt{6} \cdot \sqrt{y}$ 6) $\sqrt{5} \cdot \sqrt{p}$

Objective 2: Simplify the Square Root of a Whole Number
Label each statement as true or false. Give a reason for your answer.

7) $\sqrt{20}$ is in simplest form.

8) $\sqrt{35}$ is in simplest form.

9) $\sqrt{42}$ is in simplest form.

10) $\sqrt{63}$ is in simplest form.

Simplify completely.

Fill It In
Fill in the blanks with either the missing mathematical step or reason for the given step.

11) $\sqrt{60} = \sqrt{4 \cdot 15}$ _____

    $= $ _____   Product rule

    $= $ _____   Simplify.

12) $\sqrt{200} = $ _____   Factor.

    $= \sqrt{100} \cdot \sqrt{2}$ _____

    $= $ ____   Simplify.

Simplify completely. If the radical is already simplified, then say so.

 13) $\sqrt{20}$ 14) $\sqrt{12}$

15) $\sqrt{54}$ 16) $\sqrt{63}$

17) $\sqrt{33}$ 18) $\sqrt{15}$

19) $\sqrt{108}$ 20) $\sqrt{80}$

21) $\sqrt{98}$ 22) $\sqrt{96}$

23) $\sqrt{38}$ 24) $\sqrt{46}$

25) $\sqrt{400}$ 26) $\sqrt{900}$

27) $\sqrt{750}$ 28) $\sqrt{420}$

Objective 3: Use the Quotient Rule for Square Roots
Simplify completely.

29) $\sqrt{\dfrac{144}{25}}$ 30) $\sqrt{\dfrac{16}{81}}$

31) $\dfrac{\sqrt{4}}{\sqrt{49}}$ 32) $\dfrac{\sqrt{64}}{\sqrt{121}}$

33) $\dfrac{\sqrt{54}}{\sqrt{6}}$ 34) $\dfrac{\sqrt{48}}{\sqrt{3}}$

35) $\sqrt{\dfrac{60}{5}}$ 36) $\sqrt{\dfrac{40}{5}}$

37) $\dfrac{\sqrt{120}}{\sqrt{6}}$ 38) $\dfrac{\sqrt{54}}{\sqrt{3}}$

39) $\dfrac{\sqrt{35}}{\sqrt{5}}$ 40) $\dfrac{\sqrt{30}}{\sqrt{2}}$

41) $\sqrt{\dfrac{6}{49}}$ 42) $\sqrt{\dfrac{2}{81}}$

43) $\sqrt{\dfrac{45}{16}}$ 44) $\sqrt{\dfrac{60}{49}}$

Objective 4: Simplify Square Root Expressions Containing Variables with Even Exponents
Simplify completely.

45) $\sqrt{x^8}$ 46) $\sqrt{q^6}$

47) $\sqrt{w^{14}}$ 48) $\sqrt{t^{16}}$

49) $\sqrt{100c^2}$ 50) $\sqrt{9z^8}$

51) $\sqrt{64k^6m^{10}}$

52) $\sqrt{25p^{20}q^{14}}$

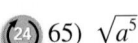 53) $\sqrt{28r^4}$

54) $\sqrt{27z^{12}}$

55) $\sqrt{300q^{22}t^{16}}$

56) $\sqrt{50n^4y^4}$

57) $\sqrt{\dfrac{81}{c^6}}$

58) $\sqrt{\dfrac{h^2}{169}}$

59) $\dfrac{\sqrt{40}}{\sqrt{t^8}}$

60) $\dfrac{\sqrt{18}}{\sqrt{m^{30}}}$

61) $\sqrt{\dfrac{75x^2}{y^{12}}}$

62) $\sqrt{\dfrac{44}{w^2z^{18}}}$

Objective 5: Simplify Square Root Expressions Containing Variables with Odd Exponents

Simplify completely.

Fill It In

Fill in the blanks with either the missing mathematical step or reason for the given step.

63) $\sqrt{w^9} = \sqrt{w^8 \cdot w^1}$ _____

$\phantom{\sqrt{w^9}} = $ _____ Product rule

$\phantom{\sqrt{w^9}} = w^4\sqrt{w}$ _____

64) $\sqrt{z^{19}} = \sqrt{z^{18} \cdot z^1}$ _____

$\phantom{\sqrt{z^{19}}} = \sqrt{z^{18}} \cdot \sqrt{z^1}$ _____

$\phantom{\sqrt{z^{19}}} = $ _____ Simplify.

65) $\sqrt{a^5}$

66) $\sqrt{c^7}$

67) $\sqrt{g^{13}}$

68) $\sqrt{k^{15}}$

69) $\sqrt{h^{31}}$

70) $\sqrt{b^{25}}$

71) $\sqrt{72x^3}$

72) $\sqrt{100a^5}$

73) $\sqrt{13q^7}$

74) $\sqrt{20c^9}$

75) $\sqrt{75t^{11}}$

76) $\sqrt{45p^{17}}$

77) $\sqrt{c^8d^2}$

78) $\sqrt{r^4s^{12}}$

79) $\sqrt{a^4b^3}$

80) $\sqrt{x^2y^9}$

81) $\sqrt{u^5v^7}$

82) $\sqrt{f^3g^9}$

83) $\sqrt{36m^9n^4}$

84) $\sqrt{4t^6u^5}$

85) $\sqrt{44x^{12}y^5}$

86) $\sqrt{63c^7d^4}$

87) $\sqrt{32t^5u^7}$

88) $\sqrt{125k^3l^9}$

89) $\sqrt{\dfrac{a^7}{81b^6}}$

90) $\sqrt{\dfrac{x^5}{49y^6}}$

91) $\sqrt{\dfrac{3r^9}{s^2}}$

92) $\sqrt{\dfrac{17h^{11}}{k^8}}$

Objective 6: Simplify More Square Root Expressions Containing Variables

Perform the indicated operation, and simplify. Assume all variables represent positive real numbers.

93) $\sqrt{5} \cdot \sqrt{10}$

94) $\sqrt{8} \cdot \sqrt{6}$

95) $\sqrt{21} \cdot \sqrt{3}$

96) $\sqrt{2} \cdot \sqrt{14}$

97) $\sqrt{w} \cdot \sqrt{w^5}$

98) $\sqrt{d^3} \cdot \sqrt{d^{11}}$

99) $\sqrt{n^3} \cdot \sqrt{n^4}$

100) $\sqrt{a^{10}} \cdot \sqrt{a^3}$

101) $\sqrt{2k} \cdot \sqrt{8k^5}$

102) $\sqrt{5z^9} \cdot \sqrt{5z^3}$

103) $\sqrt{5a^6b^5} \cdot \sqrt{10ab^4}$

104) $\sqrt{6x^4y^3} \cdot \sqrt{2x^5y^2}$

105) $\sqrt{8c^9d^2} \cdot \sqrt{5cd^7}$

106) $\sqrt{6t^3u^3} \cdot \sqrt{3t^7u^4}$

107) $\dfrac{\sqrt{18k^{11}}}{\sqrt{2k^3}}$

108) $\dfrac{\sqrt{48m^{15}}}{\sqrt{3m^9}}$

109) $\dfrac{\sqrt{120h^8}}{\sqrt{3h^2}}$

110) $\dfrac{\sqrt{72c^{10}}}{\sqrt{6c^2}}$

111) $\dfrac{\sqrt{50a^{16}b^9}}{\sqrt{5a^7b^4}}$

112) $\dfrac{\sqrt{21y^8z^{18}}}{\sqrt{3yz^{13}}}$

113) The velocity v of a moving object can be determined from its mass m and its kinetic energy KE using the formula $v = \sqrt{\dfrac{2KE}{m}}$, where the velocity is in meters/second, the mass is in kilograms, and the KE is measured in joules. A 600-kg roller coaster car is moving along a track and has kinetic energy of 120,000 joules. What is the velocity of the car?

114) The length of a side s of an equilateral triangle is a function of its area A and can be described by $s(A) = \sqrt{\dfrac{4\sqrt{3}A}{3}}$. If an equilateral triangle has an area of $6\sqrt{3}$ cm^2, how long is each side of the triangle?

R Rethink

R1) Describe the steps you take to manually check your answers.

R2) After completing the exercises, do you feel that you have memorized the definitions, properties, and procedures by practicing them? Explain.

10.4 Simplifying Expressions Containing Higher Roots

P Prepare   **O Organize**

What are your objectives for Section 10.4?	How can you accomplish each objective?
1 Multiply Higher Roots	• Write the *product rule for higher roots* in your own words. • Complete the given example on your own. • Complete You Try 1.
2 Simplify Higher Roots of Integers	• Write the property for **When Is a Radical Simplified?** in your own words. • Know the two different methods for simplifying higher roots. • Complete the given example on your own. • Complete You Try 2.
3 Use the Quotient Rule for Higher Roots	• Write the *quotient rule for higher roots* in your own words. • Complete the given example on your own. • Complete You Try 3.
4 Simplify Radicals Containing Variables	• Write the property for $\sqrt[n]{a^m}$ in your notes. • Complete the given examples on your own. • Complete You Trys 4–6.
5 Multiply and Divide Radicals with Different Indices	• Create a procedure by following Example 7, and write it in your notes. • Complete You Try 7.

W Work   **Read the explanations, follow the examples, take notes, and complete the You Trys.**

In Section 10.1 we first discussed finding higher roots like $\sqrt[4]{16} = 2$ and $\sqrt[3]{-27} = -3$. In this section, we will extend what we learned about multiplying, dividing, and simplifying *square* roots to doing the same with higher roots.

1   Multiply Higher Roots

> **Definition** Product Rule for Higher Roots
>
> If $\sqrt[n]{a}$ and $\sqrt[n]{b}$ are real numbers, then
>
> $$\sqrt[n]{a} \cdot \sqrt[n]{b} = \sqrt[n]{a \cdot b}$$

This rule enables us to multiply and simplify radicals with any index in a way that is similar to multiplying and simplifying square roots.

Multiply.

a) $\sqrt[3]{2} \cdot \sqrt[3]{7}$ b) $\sqrt[4]{t} \cdot \sqrt[4]{10}$

Solution

a) $\sqrt[3]{2} \cdot \sqrt[3]{7} = \sqrt[3]{2 \cdot 7} = \sqrt[3]{14}$ b) $\sqrt[4]{t} \cdot \sqrt[4]{10} = \sqrt[4]{t \cdot 10} = \sqrt[4]{10t}$

$\begin{bmatrix} \text{YOU TRY 1} \end{bmatrix}$ Multiply.

a) $\sqrt[4]{6} \cdot \sqrt[4]{5}$ b) $\sqrt[5]{8} \cdot \sqrt[5]{k^2}$

BE CAREFUL   Remember that we can apply the product rule in this way *only* if the indices of the radicals are the same. Later in this section we will discuss how to multiply radicals with different indices.

2 Simplify Higher Roots of Integers

In Section 10.3 we said that a simplified *square root* cannot contain any *perfect squares*. Next we list the conditions that determine when a radical with *any* index is in simplest form.

> **Property** When Is a Radical Simplified?
>
> Let P be an expression and let n be an integer greater than 1. Then $\sqrt[n]{P}$ is completely simplified when all of the following conditions are met:
>
> 1) The radicand does not contain any factors (other than 1) that are perfect nth powers.
> 2) The exponents in the radicand and the index of the radical do not have any common factors (other than 1).
> 3) The radicand does not contain any fractions.
> 4) There are no radicals in the denominator of a fraction.

Note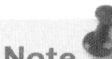
Condition 1) implies that the radical cannot contain variables with exponents greater than or equal to n, the index of the radical.

To simplify radicals with any index, use the product rule $\sqrt[n]{a \cdot b} = \sqrt[n]{a} \cdot \sqrt[n]{b}$, where a or b is an nth power.

Remember, to be certain that a radical is simplified completely, always look at the radical carefully and ask yourself, "*Is the radical in simplest form?*"

EXAMPLE 2

Simplify completely.

a) $\sqrt[3]{250}$ b) $\sqrt[4]{48}$

Solution

a)   We will look at two methods for simplifying $\sqrt[3]{250}$.

i)   Since we must simplify the *cube* root of 250, think of two numbers that multiply to 250 so that at least one of the numbers is a *perfect cube*.

$$250 = 125 \cdot 2$$

$\sqrt[3]{250} = \sqrt[3]{125 \cdot 2}$ 125 is a perfect cube.

$\quad\quad = \sqrt[3]{125} \cdot \sqrt[3]{2}$ Product rule

$\quad\quad = 5\sqrt[3]{2}$ $\sqrt[3]{125} = 5$

Is $5\sqrt[3]{2}$ in simplest form? Yes, because 2 does not have any factors that are perfect cubes.

ii)   Use a factor tree to find the prime factorization of 250: $250 = 2 \cdot 5^3$.

$\sqrt[3]{250} = \sqrt[3]{2 \cdot 5^3}$ $2 \cdot 5^3$ is the prime factorization of 250.

$\quad\quad = \sqrt[3]{2} \cdot \sqrt[3]{5^3}$ Product rule

$\quad\quad = \sqrt[3]{2} \cdot 5$ $\sqrt[3]{5^3} = 5$

$\quad\quad = 5\sqrt[3]{2}$ Commutative property

We obtain the same result using either method.

W Hint

In your notes and in your own words, explain how you know that a radical with any index is in simplest form.

b)   We will use two methods for simplifying $\sqrt[4]{48}$.

i)   Since we must simplify the *fourth* root of 48, think of two numbers that multiply to 48 so that at least one of the numbers is a *perfect fourth power*.

$$48 = 16 \cdot 3$$

$\sqrt[4]{48} = \sqrt[4]{16 \cdot 3}$ 16 is a perfect fourth power.

$\quad\quad = \sqrt[4]{16} \cdot \sqrt[4]{3}$ Product rule

$\quad\quad = 2\sqrt[4]{3}$ $\sqrt[4]{16} = 2$

Is $2\sqrt[4]{3}$ in simplest form? Yes, because 3 does not have any factors that are perfect fourth powers.

ii)   Use a factor tree to find the prime factorization of 48: $48 = 2^4 \cdot 3$.

$\sqrt[4]{48} = \sqrt[4]{2^4 \cdot 3}$ $2^4 \cdot 3$ is the prime factorization of 48.

$\quad\quad = \sqrt[4]{2^4} \cdot \sqrt[4]{3}$ Product rule

$\quad\quad = 2\sqrt[4]{3}$ $\sqrt[4]{2^4} = 2$

Once again, both methods give us the same result.

[YOU TRY 2]

Simplify completely.

a) $\sqrt[3]{40}$ b) $\sqrt[4]{96}$

3   Use the Quotient Rule for Higher Roots

Definition Quotient Rule for Higher Roots

If $\sqrt[n]{a}$ and $\sqrt[n]{b}$ are real numbers, $b \neq 0$, and n is a natural number then

$$\sqrt[n]{\frac{a}{b}} = \frac{\sqrt[n]{a}}{\sqrt[n]{b}}$$

We apply the quotient rule when working with *n*th roots the same way we apply it when working with square roots.

EXAMPLE 3

Simplify completely.

a) $\sqrt[3]{-\dfrac{81}{3}}$ b) $\dfrac{\sqrt[3]{96}}{\sqrt[3]{2}}$

Solution

a) We can think of $-\dfrac{81}{3}$ as $\dfrac{-81}{3}$ or $\dfrac{81}{-3}$. Let's think of it as $\dfrac{-81}{3}$.

Neither -81 nor 3 is a perfect cube, but if we simplify $\dfrac{-81}{3}$ we get -27, which *is* a perfect cube.

$$\sqrt[3]{-\dfrac{81}{3}} = \sqrt[3]{-27} = -3$$

W Hint

Make a list of perfect cubes and numbers raised to fourth and fifth powers.

b) Let's begin by applying the quotient rule to obtain a fraction under *one* radical, then simplify the fraction.

$$\dfrac{\sqrt[3]{96}}{\sqrt[3]{2}} = \sqrt[3]{\dfrac{96}{2}}$$ Quotient rule

$$= \sqrt[3]{48}$$ Simplify $\dfrac{96}{2}$.

$$= \sqrt[3]{8 \cdot 6}$$ 8 is a perfect cube.

$$= \sqrt[3]{8} \cdot \sqrt[3]{6}$$ Product rule

$$= 2\sqrt[3]{6}$$ $\sqrt[3]{8} = 2$

Is $2\sqrt[3]{6}$ in simplest form? Yes, because 6 does not have any factors that are perfect cubes.

YOU TRY 3

Simplify completely.

a) $\sqrt[5]{-\dfrac{5}{160}}$ b) $\dfrac{\sqrt[3]{162}}{\sqrt[3]{3}}$

4 Simplify Radicals Containing Variables

In Section 10.2 we discussed the relationship between radical notation and fractional exponents. Let's review that relationship.

Property $\sqrt[n]{a^m}$

If *a* is a nonnegative number and *m* and *n* are integers such that $n > 1$, then

$$\sqrt[n]{a^m} = a^{m/n}$$

That is, the index of the radical becomes the denominator of the fractional exponent, and the power in the radicand becomes the numerator of the fractional exponent.

This is the principle we use to simplify radicals with indices greater than 2.

EXAMPLE 4

Simplify completely.

a) $\sqrt[3]{y^{15}}$　　　b) $\sqrt[4]{16t^{24}u^{8}}$　　　c) $\sqrt[5]{\dfrac{c^{10}}{d^{30}}}$

Solution

a) $\sqrt[3]{y^{15}} = y^{15/3} = y^{5}$

b) $\sqrt[4]{16t^{24}u^{8}} = \sqrt[4]{16} \cdot \sqrt[4]{t^{24}} \cdot \sqrt[4]{u^{8}}$　　　Product rule

$\qquad = 2 \cdot t^{24/4} \cdot u^{8/4}$　　　Write with rational exponents.

$\qquad = 2t^{6}u^{2}$　　　Simplify exponents.

c) $\sqrt[5]{\dfrac{c^{10}}{d^{30}}} = \dfrac{\sqrt[5]{c^{10}}}{\sqrt[5]{d^{30}}} = \dfrac{c^{10/5}}{d^{30/5}} = \dfrac{c^{2}}{d^{6}}$　　　Quotient rule

---

[YOU TRY 4]　　Simplify completely.

a) $\sqrt[5]{p^{30}}$　　　b) $\sqrt[3]{a^{3}b^{21}}$　　　c) $\sqrt[4]{\dfrac{m^{12}}{16n^{20}}}$

To simplify a radical expression if the power in the radicand does not divide evenly by the index, we use the same methods we used in Section 10.3 for simplifying similar expressions with square roots. We can use the product rule or we can use the idea of quotient and remainder in a division problem.

EXAMPLE 5

Simplify $\sqrt[4]{x^{23}}$ completely in two ways: i) use the product rule and ii) divide the exponent by the index and use the quotient and remainder.

Solution

i) Using the product rule:
   To simplify $\sqrt[4]{x^{23}}$, write x^{23} as the product of two factors so that the exponent of one of the factors is the *largest* number less than 23 that is divisible by 4 (the index).

$\sqrt[4]{x^{23}} = \sqrt[4]{x^{20} \cdot x^{3}}$　　　20 is the largest number less than 23 that is divisible by 4.

$\qquad - \sqrt[4]{x^{20}} \cdot \sqrt[4]{x^{3}}$　　　Product rule

$\qquad = x^{20/4} \cdot \sqrt[4]{x^{3}}$　　　Use a fractional exponent to simplify.

$\qquad = x^{5}\sqrt[4]{x^{3}}$　　　$20 \div 4 = 5$

Hint

Be sure to do this example both ways. Which method do you prefer and why?

ii) Using the quotient and remainder:

$$
\begin{array}{r}
5 \leftarrow \text{Quotient} \\
4\overline{)\ 23} \\
-20 \\ \hline
3 \leftarrow \text{Remainder}
\end{array}
$$

To simplify $\sqrt[4]{x^{23}}$, divide

Recall from our work with square roots in Section 10.3 that

i) the exponent on the variable *outside* of the radical will be the *quotient* of the division problem,

and

ii) the exponent on the variable *inside* of the radical will be the *remainder* of the division problem.

$$\sqrt[4]{x^{23}} = x^{5}\sqrt[4]{x^{3}}$$

Is $x^{5}\sqrt[4]{x^{3}}$ in simplest form? Yes, because the exponent inside of the radical is less than the index, and they contain no common factors other than 1.

We can apply the product and quotient rules together with the methods in Example 5 to simplify certain radical expressions.

EXAMPLE 6

Completely simplify $\sqrt[3]{56a^{16}b^8}$.

Solution

$$\sqrt[3]{56a^{16}b^8} = \sqrt[3]{56} \cdot \sqrt[3]{a^{16}} \cdot \sqrt[3]{b^8} \qquad \text{Product rule}$$
$$= \sqrt[3]{8} \cdot \sqrt[3]{7} \cdot a^5\sqrt[3]{a^1} \cdot b^2\sqrt[3]{b^2}$$

Product rule   16 ÷ 3 gives a quotient of 5 and a remainder of 1.   8 ÷ 3 gives a quotient of 2 and a remainder of 2.

$$= 2\sqrt[3]{7} \cdot a^5\sqrt[3]{a} \cdot b^2\sqrt[3]{b^2} \qquad \text{Simplify } \sqrt[3]{8}.$$
$$= 2a^5b^2 \cdot \sqrt[3]{7} \cdot \sqrt[3]{a} \cdot \sqrt[3]{b^2} \qquad \text{Use the commutative property to rewrite the expression.}$$
$$= 2a^5b^2\sqrt[3]{7ab^2} \qquad \text{Product rule}$$

[YOU TRY 6] Simplify completely.

a) $\sqrt[4]{48x^{15}y^{22}}$ b) $\sqrt[3]{\dfrac{r^{19}}{27s^{12}}}$

5 Multiply and Divide Radicals with Different Indices

The product and quotient rules for radicals apply only when the radicals have the *same* indices. To multiply or divide radicals with *different* indices, we first change the radical expressions to rational exponent form.

EXAMPLE 7

Multiply the expressions, and write the answer in simplest radical form.

$$\sqrt[3]{x^2} \cdot \sqrt{x}$$

Solution

The indices of $\sqrt[3]{x^2}$ and $\sqrt{x}$ are different, so we *cannot* use the product rule right now. Rewrite each radical as a fractional exponent, use the product rule for *exponents*, then convert the answer back to radical form.

$$\sqrt[3]{x^2} \cdot \sqrt{x} = x^{2/3} \cdot x^{1/2} \qquad \text{Change radicals to fractional exponents.}$$
$$= x^{4/6} \cdot x^{3/6} \qquad \text{Get a common denominator to add exponents.}$$
$$= x^{\frac{4}{6}+\frac{3}{6}} = x^{7/6} \qquad \text{Add exponents.}$$
$$= \sqrt[6]{x^7} = x\sqrt[6]{x} \qquad \text{Rewrite in radical form, and simplify.}$$

Ⓦ Hint
Write your own procedure in your notes.

[YOU TRY 7] Perform the indicated operation, and write the answer in simplest radical form.

a) $\sqrt[4]{y} \cdot \sqrt[6]{y}$ b) $\dfrac{\sqrt[3]{c^2}}{\sqrt{c}}$

E Evaluate **10.4** Exercises   Do the exercises, and check your work.

Mixed Exercises: Objectives 1–3

1) In your own words, explain the product rule for radicals.

2) In your own words, explain the quotient rule for radicals.

3) How do you know that a radical expression containing a cube root is completely simplified?

4) How do you know that a radical expression containing a fourth root is completely simplified?

For the remainder of the exercises, assume all variables represent positive real numbers.

Objective 1: Multiply Higher Roots
Multiply.

5) $\sqrt[5]{6} \cdot \sqrt[5]{2}$

6) $\sqrt[3]{5} \cdot \sqrt[3]{4}$

7) $\sqrt[5]{9} \cdot \sqrt[5]{m^2}$

8) $\sqrt[4]{11} \cdot \sqrt[4]{h^3}$

9) $\sqrt[3]{a^2} \cdot \sqrt[3]{b}$

10) $\sqrt[5]{t^2} \cdot \sqrt[5]{u^4}$

Mixed Exercises: Objectives 2 and 3
Simplify completely.

Fill It In

Fill in the blanks with either the missing mathematical step or reason for the given step.

11) $\sqrt[3]{56} = \sqrt[3]{8 \cdot 7}$ _____

$=$ _____   Product rule

$=$ _____   Simplify.

12) $\sqrt[4]{80} = \sqrt[4]{16 \cdot 5}$ _____

$= \sqrt[4]{16} \cdot \sqrt[4]{5}$ _____

$=$ _____   Simplify.

13) $\sqrt[3]{24}$

14) $\sqrt[3]{48}$

15) $\sqrt[4]{64}$

16) $\sqrt[4]{32}$

17) $\sqrt[3]{54}$

18) $\sqrt[3]{88}$

19) $\sqrt[3]{2000}$

20) $\sqrt[3]{108}$

21) $\sqrt[5]{64}$

22) $\sqrt[4]{162}$

23) $\sqrt[3]{\dfrac{1}{125}}$

24) $\sqrt[4]{\dfrac{1}{16}}$

25) $\sqrt[3]{-\dfrac{54}{2}}$

26) $\sqrt[4]{\dfrac{48}{3}}$

27) $\dfrac{\sqrt[3]{48}}{\sqrt[3]{2}}$

28) $\dfrac{\sqrt[3]{500}}{\sqrt[3]{2}}$

29) $\dfrac{\sqrt[4]{240}}{\sqrt[4]{3}}$

30) $\dfrac{\sqrt[3]{8000}}{\sqrt[3]{4}}$

Objective 4: Simplify Radicals Containing Variables
Simplify completely.

31) $\sqrt[3]{d^6}$

32) $\sqrt[3]{g^9}$

33) $\sqrt[4]{n^{20}}$

34) $\sqrt[4]{t^{36}}$

35) $\sqrt[5]{x^5y^{15}}$

36) $\sqrt[6]{a^{12}b^6}$

37) $\sqrt[3]{w^{14}}$

38) $\sqrt[3]{b^{19}}$

39) $\sqrt[4]{y^9}$

40) $\sqrt[4]{m^7}$

41) $\sqrt[3]{d^5}$

42) $\sqrt[3]{c^{29}}$

43) $\sqrt[3]{u^{10}v^{15}}$

44) $\sqrt[3]{x^9y^{16}}$

45) $\sqrt[3]{b^{16}c^5}$

46) $\sqrt[4]{r^{15}s^9}$

47) $\sqrt[4]{m^3n^{18}}$

48) $\sqrt[3]{a^{11}b}$

49) $\sqrt[3]{24x^{10}y^{12}}$

50) $\sqrt[3]{54y^{10}z^{24}}$

51) $\sqrt[3]{72t^{17}u^7}$

52) $\sqrt[3]{250w^4x^{16}}$

53) $\sqrt[4]{\dfrac{m^8}{81}}$

54) $\sqrt[4]{\dfrac{16^8}{x^{12}}}$

55) $\sqrt[5]{\dfrac{32a^{23}}{b^{15}}}$

56) $\sqrt[3]{\dfrac{h^{17}}{125k^{21}}}$

57) $\sqrt[4]{\dfrac{t^9}{81s^{24}}}$

58) $\sqrt[5]{\dfrac{32c^9}{d^{20}}}$

59) $\sqrt[3]{\dfrac{u^{28}}{v^3}}$

60) $\sqrt[4]{\dfrac{m^{13}}{n^8}}$

Perform the indicated operation, and simplify.

61) $\sqrt[3]{6} \cdot \sqrt[3]{4}$

62) $\sqrt[3]{4} \cdot \sqrt[3]{10}$

63) $\sqrt[3]{9} \cdot \sqrt[3]{12}$

64) $\sqrt[3]{9} \cdot \sqrt[3]{6}$

65) $\sqrt[3]{20} \cdot \sqrt[3]{4}$

66) $\sqrt[3]{28} \cdot \sqrt[3]{2}$

67) $\sqrt[3]{m^4} \cdot \sqrt[3]{m^5}$

68) $\sqrt[3]{t^5} \cdot \sqrt[3]{t}$

69) $\sqrt[4]{k^7} \cdot \sqrt[4]{k^9}$

70) $\sqrt[4]{a^9} \cdot \sqrt[4]{a^{11}}$

71) $\sqrt[3]{r^7} \cdot \sqrt[3]{r^4}$

72) $\sqrt[3]{y^2} \cdot \sqrt[3]{y^{17}}$

73) $\sqrt[5]{p^{14}} \cdot \sqrt[5]{p^9}$

74) $\sqrt[5]{c^{17}} \cdot \sqrt[5]{c^9}$

75) $\sqrt[3]{9z^{11}} \cdot \sqrt[3]{3z^8}$

76) $\sqrt[3]{2h^4} \cdot \sqrt[3]{4h^{16}}$

77) $\sqrt[3]{\dfrac{h^{14}}{h^2}}$

78) $\sqrt[3]{\dfrac{a^{20}}{a^{14}}}$

79) $\sqrt[3]{\dfrac{c^{11}}{c^4}}$

80) $\sqrt[3]{\dfrac{z^{16}}{z^5}}$

81) $\sqrt[4]{\dfrac{162d^{21}}{2d^2}}$

82) $\sqrt[4]{\dfrac{48t^{11}}{3t^6}}$

Objective 5: Multiply and Divide Radicals with Different Indices

The following radical expressions do not have the same indices. Perform the indicated operation, and write the answer in simplest radical form.

Fill It In

Fill in the blanks with either the missing mathematical step or the reason for the given step.

83) $\sqrt{a} \cdot \sqrt[4]{a^3} = a^{1/2} \cdot a^{3/4}$ _____

$= a^{2/4} \cdot a^{3/4}$ _____

$=$ _____ Add exponents.

$= \sqrt[4]{a^5}$ _____

$=$ _____ Simplify.

84) $\sqrt[5]{r^4} \cdot \sqrt[3]{r^2} =$ _____ Change radicals to fractional exponents.

$=$ _____ Rewrite exponents with a common denominator.

$= r^{22/15}$ _____

$=$ _____ Rewrite in radical form.

$=$ _____ Simplify.

85) $\sqrt{p} \cdot \sqrt[3]{p}$

86) $\sqrt[3]{y^2} \cdot \sqrt[4]{y}$

87) $\sqrt[4]{n^3} \cdot \sqrt{n}$

88) $\sqrt[5]{k^4} \cdot \sqrt{k}$

89) $\sqrt[5]{c^3} \cdot \sqrt[3]{c^2}$

90) $\sqrt[3]{a^2} \cdot \sqrt[5]{a^2}$

91) $\dfrac{\sqrt{w}}{\sqrt[4]{w}}$

92) $\dfrac{\sqrt[4]{m^3}}{\sqrt{m}}$

93) $\dfrac{\sqrt[4]{h^3}}{\sqrt[3]{h^2}}$

94) $\dfrac{\sqrt[5]{t^4}}{\sqrt[3]{t^2}}$

95) A block of candle wax in the shape of a cube has a volume of 64 in³. The length of a side of the block, s, is given by $s = \sqrt[3]{V}$, where V is the volume of the block of candle wax. How long is each side of the block?

96) The radius $r(V)$ of a sphere is a function of its volume V and can be described by the function $r(V) = \sqrt[3]{\dfrac{3V}{4\pi}}$. If a spherical water tank has a volume of $\dfrac{256\pi}{3}$ ft³, what is the radius of the tank?

R Rethink

R1) Which method did you use to simplify in Objective 2? Why?

R2) How did it help you to learn these same rules, but only apply them to square roots before applying them to higher roots in this section?

10.5 Adding, Subtracting, and Multiplying Radicals

P Prepare

What are your objectives for Section 10.5?	## O Organize How can you accomplish each objective?
1 Add and Subtract Radical Expressions	• Write the definition of *like radicals* in your notes. • Write the property for **Adding and Subtracting Radicals** in your notes. • Complete the given examples on your own. • Complete You Trys 1 and 2.
2 Simplify Before Adding and Subtracting	• Write the procedure for **Adding and Subtracting Radicals** in your own words. • Complete the given example on your own. • Complete You Try 3.
3 Multiply a Binomial Containing Radical Expressions by a Monomial	• Write a procedure that outlines how to multiply in this objective. • Complete the given example on your own. • Complete You Try 4.
4 Multiply Radical Expressions Using FOIL	• Compare using FOIL in Chapter 6 to using it for binomials containing radicals, and note what is similar and what is different. • Complete the given example on your own. • Complete You Try 5.
5 Square a Binomial Containing Radical Expressions	• Review the formulas for squaring a binomial in Chapter 6, if necessary. • Complete the given example on your own. • Complete You Try 6.
6 Multiply Radical Expressions of the Form $(a + b)(a - b)$	• Use the same formula derived in Chapter 6. • Complete the given example on your own. • Complete You Try 7.

 **W Work**   **Read the explanations, follow the examples, take notes, and complete the You Trys.**

Just as we can add and subtract like terms such as $4x + 6x = 10x$, we can add and subtract *like radicals* such as $4\sqrt{3} + 6\sqrt{3}$.

Definition

Like radicals have the same index and the same radicand.

Some examples of like radicals are

$$4\sqrt{3} \text{ and } 6\sqrt{3}, \qquad -\sqrt[3]{5} \text{ and } 8\sqrt[3]{5}, \qquad \sqrt{x} \text{ and } 7\sqrt{x}, \qquad 2\sqrt[3]{a^2 b} \text{ and } \sqrt[3]{a^2 b}$$

In this section, assume all variables represent nonnegative real numbers.

1 Add and Subtract Radical Expressions

W Hint
Where have you seen similar definitions and procedures before?

Property Adding and Subtracting Radicals

In order to add or subtract radicals, they must be *like* radicals.

We add and subtract like radicals in the same way we add and subtract like terms—add or subtract the "coefficients" of the radicals and multiply that result by the radical. We are using the distributive property when we combine like terms in this way.

EXAMPLE 1

Perform the operations, and simplify.

a) $4x + 6x$ b) $4\sqrt{3} + 6\sqrt{3}$ c) $\sqrt[4]{5} - 9\sqrt[4]{5}$ d) $7\sqrt{2} + 4\sqrt{3}$

Solution

a) First notice that $4x$ and $6x$ are like terms. Therefore, they can be added.

$$4x + 6x = (4 + 6)x \qquad \text{Distributive property}$$
$$= 10x \qquad \text{Simplify.}$$

Or, we can say that by just adding the coefficients, $4x + 6x = 10x$.

W Hint
When adding or subtracting "like radicals," the radical does not change.

b) Before attempting to add $4\sqrt{3}$ and $6\sqrt{3}$, we must be certain that they are like radicals. Since they *are* like, they can be added.

$$4\sqrt{3} + 6\sqrt{3} = (4 + 6)\sqrt{3} \qquad \text{Distributive property}$$
$$= 10\sqrt{3} \qquad \text{Simplify.}$$

Or, we can say that by just adding the coefficients of $\sqrt{3}$, we get $4\sqrt{3} + 6\sqrt{3} = 10\sqrt{3}$.

c) $\sqrt[4]{5} - 9\sqrt[4]{5} = 1\sqrt[4]{5} - 9\sqrt[4]{5} = (1 - 9)\sqrt[4]{5} = -8\sqrt[4]{5}$

d) The radicands in $7\sqrt{2} + 4\sqrt{3}$ are different, so these expressions cannot be combined.

[YOU TRY 1]

Perform the operations, and simplify.

a) $9c + 7c$ b) $9\sqrt{10} + 7\sqrt{10}$ c) $\sqrt[3]{4} - 6\sqrt[3]{4}$ d) $5\sqrt{6} - 2\sqrt{3}$

EXAMPLE 2

Perform the operations, and simplify. $6\sqrt{x} + 11\sqrt[3]{x} + 2\sqrt{x} - 6\sqrt[3]{x}$

Solution

Begin by noticing that there are *two* different types of radicals: $\sqrt{x}$ and $\sqrt[3]{x}$. Write the like radicals together.

$$6\sqrt{x} + 11\sqrt[3]{x} + 2\sqrt{x} - 6\sqrt[3]{x} = 6\sqrt{x} + 2\sqrt{x} + 11\sqrt[3]{x} - 6\sqrt[3]{x} \qquad \text{Commutative property}$$
$$= (6 + 2)\sqrt{x} + (11 - 6)\sqrt[3]{x} \qquad \text{Distributive property}$$
$$= 8\sqrt{x} + 5\sqrt[3]{x}$$

Is $8\sqrt{x} + 5\sqrt[3]{x}$ in simplest form? *Yes.* The radicals are not like (they have different indices) so they cannot be combined further. Also, each radical, $\sqrt{x}$ and $\sqrt[3]{x}$, is in simplest form.

[YOU TRY 2]

Perform the operations, and simplify. $8\sqrt[3]{2n} - 3\sqrt{2n} + 5\sqrt{2n} + 5\sqrt[3]{2n}$

2 Simplify Before Adding and Subtracting

Sometimes it looks like two radicals cannot be added or subtracted. But if the radicals can be *simplified* and they turn out to be *like* radicals, then we can add or subtract them.

> **Procedure** Adding and Subtracting Radicals
>
> 1) Write each radical expression in simplest form.
>
> 2) Combine like radicals.

EXAMPLE 3

Perform the operations, and simplify.

a) $8\sqrt{2} + 3\sqrt{50} - \sqrt{45}$ b) $-7\sqrt[3]{40} + \sqrt[3]{5}$

c) $10\sqrt{8t} - 9\sqrt{2t}$ d) $\sqrt[3]{xy^6} + \sqrt[3]{x^7}$

Solution

a) The radicals $8\sqrt{2}$, $3\sqrt{50}$, and $\sqrt{45}$ are not like. The first radical is in simplest form, but $3\sqrt{50}$ and $\sqrt{45}$ should be simplified to determine if any of the radicals can be combined.

$$\begin{aligned}
8\sqrt{2} + 3\sqrt{50} - \sqrt{45} &= 8\sqrt{2} + 3\sqrt{25 \cdot 2} - \sqrt{9 \cdot 5} && \text{Factor.}\\
&= 8\sqrt{2} + 3\sqrt{25} \cdot \sqrt{2} - \sqrt{9} \cdot \sqrt{5} && \text{Product rule}\\
&= 8\sqrt{2} + 3 \cdot 5 \cdot \sqrt{2} - 3\sqrt{5} && \text{Simplify radicals.}\\
&= 8\sqrt{2} + 15\sqrt{2} - 3\sqrt{5} && \text{Multiply.}\\
&= 23\sqrt{2} - 3\sqrt{5} && \text{Add like radicals.}
\end{aligned}$$

W Hint

Write out each step very carefully to avoid making mistakes.

b)
$$\begin{aligned}
-7\sqrt[3]{40} + \sqrt[3]{5} &= -7\sqrt[3]{8 \cdot 5} + \sqrt[3]{5} && \text{8 is a perfect cube.}\\
&= -7\sqrt[3]{8} \cdot \sqrt[3]{5} + \sqrt[3]{5} && \text{Product rule}\\
&= -7 \cdot 2 \cdot \sqrt[3]{5} + \sqrt[3]{5} && \sqrt[3]{8} = 2\\
&= -14\sqrt[3]{5} + \sqrt[3]{5} && \text{Multiply.}\\
&= -13\sqrt[3]{5} && \text{Add like radicals.}
\end{aligned}$$

c) The radical $\sqrt{2t}$ is simplified, but $\sqrt{8t}$ is not. We must simplify $\sqrt{8t}$:

$$\sqrt{8t} = \sqrt{8} \cdot \sqrt{t} = \sqrt{4} \cdot \sqrt{2} \cdot \sqrt{t} = 2\sqrt{2} \cdot \sqrt{t} = 2\sqrt{2t}$$

Substitute $2\sqrt{2t}$ for $\sqrt{8t}$ in the original expression.

$$\begin{aligned}
10\sqrt{8t} - 9\sqrt{2t} &= 10(2\sqrt{2t}) - 9\sqrt{2t} && \text{Substitute } 2\sqrt{2t} \text{ for } \sqrt{8t}.\\
&= 20\sqrt{2t} - 9\sqrt{2t} && \text{Multiply.}\\
&= 11\sqrt{2t} && \text{Subtract.}
\end{aligned}$$

d) Each radical in the expression $\sqrt[3]{xy^6} + \sqrt[3]{x^7}$ must be simplified.

$$\sqrt[3]{xy^6} = \sqrt[3]{x} \cdot \sqrt[3]{y^6} = \sqrt[3]{x} \cdot y^2 = y^2\sqrt[3]{x} \qquad \Big| \qquad \sqrt[3]{x^7} = x^2\sqrt[3]{x^1}$$

7 ÷ 3 gives a quotient of 2 and a remainder of 1.

$$\begin{aligned}
\sqrt[3]{xy^6} + \sqrt[3]{x^7} &= y^2\sqrt[3]{x} + x^2\sqrt[3]{x} && \text{Substitute the simplified radicals in the original expression.}\\
&= (y^2 + x^2)\sqrt[3]{x} && \text{Factor out } \sqrt[3]{x} \text{ from each term.}
\end{aligned}$$

In this problem we cannot *add* $y^2\sqrt[3]{x} + x^2\sqrt[3]{x}$ like we added radicals in previous examples, but we *can* factor out $\sqrt[3]{x}$.

$(y^2 + x^2)\sqrt[3]{x}$ is the completely simplified form of the sum.

[YOU TRY 3]

Perform the operations, and simplify.

a) $7\sqrt{3} - \sqrt{12}$ b) $2\sqrt{63} - 11\sqrt{28} + 2\sqrt{21}$ c) $\sqrt[3]{54} + 5\sqrt[3]{16}$

d) $2\sqrt{6k} + 4\sqrt{54k}$ e) $\sqrt[4]{mn^{11}} + \sqrt[4]{81mn^3}$

In the rest of this section, we will learn how to simplify expressions that combine multiplication, addition, and subtraction of radicals.

3 Multiply a Binomial Containing Radical Expressions by a Monomial

EXAMPLE 4

Multiply and simplify.

a) $4(\sqrt{5} - \sqrt{20})$ b) $\sqrt{2}(\sqrt{10} + \sqrt{15})$ c) $\sqrt{x}(\sqrt{x} + \sqrt{32y})$

Solution

a) Because $\sqrt{20}$ can be simplified, we will do that first.

$$\sqrt{20} = \sqrt{4 \cdot 5} = \sqrt{4} \cdot \sqrt{5} = 2\sqrt{5}$$

Substitute $2\sqrt{5}$ for $\sqrt{20}$ in the original expression.

$$\begin{aligned}
4(\sqrt{5} - \sqrt{20}) &= 4(\sqrt{5} - 2\sqrt{5}) &&\text{Substitute } 2\sqrt{5} \text{ for } \sqrt{20}. \\
&= 4(-\sqrt{5}) &&\text{Subtract.} \\
&= -4\sqrt{5} &&\text{Multiply.}
\end{aligned}$$

W Hint

Use a pen with multiple colors to perform the steps in different colors as it is done in the examples.

b) Neither $\sqrt{10}$ nor $\sqrt{15}$ can be simplified. Begin by applying the distributive property.

$$\begin{aligned}
\sqrt{2}(\sqrt{10} + \sqrt{15}) &= \sqrt{2} \cdot \sqrt{10} + \sqrt{2} \cdot \sqrt{15} &&\text{Distribute.} \\
&= \sqrt{20} + \sqrt{30} &&\text{Product rule}
\end{aligned}$$

Is $\sqrt{20} + \sqrt{30}$ in simplest form? *No.* $\sqrt{20}$ can be simplified.

$$= \sqrt{4 \cdot 5} + \sqrt{30} = \sqrt{4} \cdot \sqrt{5} + \sqrt{30} = 2\sqrt{5} + \sqrt{30}$$

c) Since $\sqrt{32y}$ can be simplified, we will do that first.

$$\sqrt{32y} = \sqrt{32} \cdot \sqrt{y} = \sqrt{16 \cdot 2} \cdot \sqrt{y} = \sqrt{16} \cdot \sqrt{2} \cdot \sqrt{y} = 4\sqrt{2y}$$

Substitute $4\sqrt{2y}$ for $\sqrt{32y}$ in the original expression.

$$\begin{aligned}
\sqrt{x}(\sqrt{x} + \sqrt{32y}) &= \sqrt{x}(\sqrt{x} + 4\sqrt{2y}) &&\text{Substitute } 4\sqrt{2y} \text{ for } \sqrt{32y}. \\
&= \sqrt{x} \cdot \sqrt{x} + \sqrt{x} \cdot 4\sqrt{2y} &&\text{Distribute.} \\
&= x + 4\sqrt{2xy} &&\text{Multiply.}
\end{aligned}$$

[YOU TRY 4]

Multiply and simplify.

a) $6(\sqrt{75} + 2\sqrt{3})$ b) $\sqrt{3}(\sqrt{3} + \sqrt{21})$ c) $\sqrt{c}(\sqrt{c^3} - \sqrt{100d})$

4 Multiply Radical Expressions Using FOIL

In Chapter 6, we first multiplied binomials using **FOIL** (**F**irst **O**uter **I**nner **L**ast).

$$(2x + 3)(x + 4) = 2x \cdot x + 2x \cdot 4 + 3 \cdot x + 3 \cdot 4$$

$$\quad \text{F}\qquad \text{O}\qquad\;\; \text{I}\qquad\;\; \text{L}$$

$$= 2x^2 + 8x + 3x + 12$$
$$= 2x^2 + 11x + 12$$

We can multiply binomials containing radicals the same way.

EXAMPLE 5

Multiply and simplify.

a) $(2 + \sqrt{5})(4 + \sqrt{5})$

b) $(2\sqrt{3} + \sqrt{2})(\sqrt{3} - 5\sqrt{2})$

c) $(\sqrt{r} + \sqrt{3s})(\sqrt{r} + 8\sqrt{3s})$

Solution

a) Since we must multiply two binomials, we will use FOIL.

$$(2 + \sqrt{5})(4 + \sqrt{5}) = 2 \cdot 4 + 2 \cdot \sqrt{5} + 4 \cdot \sqrt{5} + \sqrt{5} \cdot \sqrt{5}$$

$$\phantom{(2 + \sqrt{5})(4 + \sqrt{5}) =}\quad \text{F}\qquad\;\; \text{O}\qquad\;\; \text{I}\qquad\;\; \text{L}$$

$$= 8 + 2\sqrt{5} + 4\sqrt{5} + 5 \qquad \text{Multiply.}$$
$$= 13 + 6\sqrt{5} \qquad\qquad\qquad\;\; \text{Combine like terms.}$$

W Hint

Multiplication of binomials containing radicals uses the same procedure as multiplying binomials in Chapter 6!

b) $(2\sqrt{3} + \sqrt{2})(\sqrt{3} - 5\sqrt{2})$

$$\phantom{= 2\sqrt{3}}\quad \text{F}\qquad\;\; \text{O}\qquad\qquad\;\; \text{I}\qquad\;\; \text{L}$$

$$= 2\sqrt{3} \cdot \sqrt{3} + 2\sqrt{3} \cdot (-5\sqrt{2}) + \sqrt{2} \cdot \sqrt{3} + \sqrt{2} \cdot (-5\sqrt{2})$$
$$= 2 \cdot 3 + (-10\sqrt{6}) + \sqrt{6} + (-5 \cdot 2) \qquad \text{Multiply.}$$
$$= 6 - 10\sqrt{6} + \sqrt{6} - 10 \qquad\qquad\; \text{Multiply.}$$
$$= -4 - 9\sqrt{6} \qquad\qquad\qquad\qquad\qquad\quad\; \text{Combine like terms.}$$

c) $(\sqrt{r} + \sqrt{3s})(\sqrt{r} + 8\sqrt{3s})$

$$\quad \text{F}\qquad\;\; \text{O}\qquad\qquad\;\; \text{I}\qquad\qquad\;\; \text{L}$$

$$= \sqrt{r} \cdot \sqrt{r} + \sqrt{r} \cdot 8\sqrt{3s} + \sqrt{3s} \cdot \sqrt{r} + \sqrt{3s} \cdot 8\sqrt{3s}$$
$$= r + 8\sqrt{3rs} + \sqrt{3rs} + 8 \cdot 3s \qquad \text{Multiply.}$$
$$= r + 8\sqrt{3rs} + \sqrt{3rs} + 24s \qquad\;\; \text{Multiply.}$$
$$= r + 9\sqrt{3rs} + 24s \qquad\qquad\qquad\; \text{Combine like terms.}$$

[YOU TRY 5]

Multiply and simplify.

a) $(6 - \sqrt{7})(5 + \sqrt{7})$

b) $(\sqrt{2} + 4\sqrt{5})(3\sqrt{2} + \sqrt{5})$

c) $(\sqrt{6p} - \sqrt{2q})(\sqrt{6p} - 3\sqrt{2q})$

5 Square a Binomial Containing Radical Expressions

Recall, again, from Chapter 6, that we can use FOIL to square a binomial or we can use these special formulas:

$$(a + b)^2 = a^2 + 2ab + b^2 \qquad (a - b)^2 = a^2 - 2ab + b^2$$

For example,

$$(k + 7)^2 = (k)^2 + 2(k)(7) + (7)^2 \qquad \text{and} \qquad (2p - 5)^2 = (2p)^2 - 2(2p)(5) + (5)^2$$
$$= k^2 + 14k + 49 \qquad\qquad\qquad\qquad\qquad = 4p^2 - 20p + 25$$

To square a binomial containing radicals, we can either use FOIL or we can use the formulas above. Understanding how to use the formulas to square a binomial will make it easier to solve radical equations in Section 10.7.

EXAMPLE 6

Multiply and simplify.

a) $(\sqrt{10} + 3)^2$ b) $(2\sqrt{x} - 6)^2$

Solution

a) Use $(a + b)^2 = a^2 + 2ab + b^2$.

$$(\sqrt{10} + 3)^2 = (\sqrt{10})^2 + 2(\sqrt{10})(3) + (3)^2 \qquad \text{Substitute } \sqrt{10} \text{ for } a \text{ and } 3 \text{ for } b.$$
$$= 10 + 6\sqrt{10} + 9 \qquad\qquad\qquad\qquad \text{Multiply.}$$
$$= 19 + 6\sqrt{10} \qquad\qquad\qquad\qquad\quad \text{Combine like terms.}$$

b) Use $(a - b)^2 = a^2 - 2ab + b^2$.

$$(2\sqrt{x} - 6)^2 = (2\sqrt{x})^2 - 2(2\sqrt{x})(6) + (6)^2 \qquad \text{Substitute } 2\sqrt{x} \text{ for } a \text{ and } 6 \text{ for } b.$$
$$= (4 \cdot x) - (4\sqrt{x})(6) + 36 \qquad\qquad \text{Multiply.}$$
$$= 4x - 24\sqrt{x} + 36 \qquad\qquad\qquad \text{Multiply.}$$

[YOU TRY 6]

Multiply and simplify.

a) $(\sqrt{6} + 5)^2$ b) $(3\sqrt{2} - 4)^2$ c) $(\sqrt{w} + \sqrt{11})^2$

---

6   Multiply Radical Expressions of the Form $(a + b)(a - b)$

We will review one last rule from Chapter 6 on multiplying binomials. We will use this in Section 10.6 when we divide radicals.

$$(a + b)(a - b) = a^2 - b^2$$

For example, $(t + 8)(t - 8) = (t)^2 - (8)^2 = t^2 - 64$.

The same rule applies when we multiply binomials containing radicals.

EXAMPLE 7

Multiply and simplify $(2\sqrt{x} + \sqrt{y})(2\sqrt{x} - \sqrt{y})$.

Solution

Use $(a + b)(a - b) = a^2 - b^2$.

$$(2\sqrt{x} + \sqrt{y})(2\sqrt{x} - \sqrt{y}) = (2\sqrt{x})^2 - (\sqrt{y})^2 \qquad \text{Substitute } 2\sqrt{x} \text{ for } a \text{ and } \sqrt{y} \text{ for } b.$$
$$= 4(x) - y \qquad\qquad\qquad \text{Square each term.}$$
$$= 4x - y \qquad\qquad\qquad\quad \text{Simplify.}$$

$$\left[\text{YOU TRY 7} \right]$$ Multiply and simplify.

    a) $(4 + \sqrt{10})(4 - \sqrt{10})$ b) $(\sqrt{5h} + \sqrt{k})(\sqrt{5h} - \sqrt{k})$

ANSWERS TO $\left[\text{YOU TRY} \right]$ **EXERCISES**

1) a) $16c$ b) $16\sqrt{10}$ c) $-5\sqrt[3]{4}$ d) $5\sqrt{6} - 2\sqrt{3}$ 2) $13\sqrt[3]{2n} + 2\sqrt{2n}$ 3) a) $5\sqrt{3}$
   b) $-16\sqrt{7} + 2\sqrt{21}$ c) $13\sqrt[3]{2}$ d) $14\sqrt{6k}$ e) $(n^2 + 3)\sqrt[4]{mn^3}$ 4) a) $42\sqrt{3}$ b) $3 + 3\sqrt{7}$
   c) $c^2 - 10\sqrt{cd}$ 5) a) $23 + \sqrt{7}$ b) $26 + 13\sqrt{10}$ c) $6p - 8\sqrt{3pq} + 6q$
   6) a) $31 + 10\sqrt{6}$ b) $34 - 24\sqrt{2}$ c) $w + 2\sqrt{11w} + 11$ 7) a) 6 b) $5h - k$

E Evaluate **10.5** Exercises   Do the exercises, and check your work.

Assume all variables represent nonnegative real numbers.

Objective 1: Add and Subtract Radical Expressions

1) How do you know if two radicals are *like* radicals?

2) Are $5\sqrt{3}$ and $7\sqrt[3]{3}$ like radicals? Why or why not?

Perform the operations, and simplify.

3) $5\sqrt{2} + 9\sqrt{2}$ 4) $11\sqrt{7} + 7\sqrt{7}$

5) $7\sqrt[3]{4} + 8\sqrt[3]{4}$ 6) $10\sqrt[3]{5} - 2\sqrt[3]{5}$

 7) $6 - \sqrt{13} + 5 - 2\sqrt{13}$

8) $-8 + 3\sqrt{6} - 4\sqrt{6} + 9$

9) $15\sqrt[3]{z^2} - 20\sqrt[3]{z^2}$

10) $7\sqrt[3]{p} - 4\sqrt[3]{p}$

11) $2\sqrt[3]{n^2} + 9\sqrt[5]{n^2} - 11\sqrt[3]{n^2} + \sqrt[5]{n^2}$

12) $5\sqrt[4]{s} - 3\sqrt[3]{s} + 2\sqrt[3]{s} + 4\sqrt[4]{s}$

13) $\sqrt{5c} - 8\sqrt{6c} + \sqrt{5c} + 6\sqrt{6c}$

14) $10\sqrt{2m} + 6\sqrt{3m} - \sqrt{2m} + 8\sqrt{3m}$

Objective 2: Simplify Before Adding and Subtracting

15) What are the steps for adding or subtracting radicals?

16) Is $6\sqrt{2} + \sqrt{10}$ in simplified form? Explain.

Perform the operations, and simplify.

Fill It In

Fill in the blanks with either the missing mathematical step or reason for the given step.

17) $\sqrt{48} + \sqrt{3}$
    $= \sqrt{16 \cdot 3} + \sqrt{3}$ _____
    $=$ _____   Product rule
    $= 4\sqrt{3} + \sqrt{3}$ _____
    $=$ ___   Add like radicals.

18) $\sqrt{44} - 8\sqrt{11}$
    $= \sqrt{4 \cdot 11} - 8\sqrt{11}$ _____
    $= \sqrt{4} \cdot \sqrt{11} - 8\sqrt{11}$ _____
    $=$ _____   Simplify.
    $=$ _____   Subtract like radicals.

19) $6\sqrt{3} - \sqrt{12}$ 20) $\sqrt{45} + 4\sqrt{5}$

21) $\sqrt{32} - 3\sqrt{8}$ 22) $3\sqrt{24} + \sqrt{96}$

23) $\sqrt{12} + \sqrt{75} - \sqrt{3}$ 24) $\sqrt{96} + \sqrt{24} - 5\sqrt{54}$

25) $8\sqrt[3]{9} + \sqrt[3]{72}$ 26) $5\sqrt[3]{88} + 2\sqrt[3]{11}$

27) $\sqrt[3]{6} - \sqrt[3]{48}$ 28) $11\sqrt[3]{16} + 7\sqrt[3]{2}$

29) $6q\sqrt{q} + 7\sqrt{q^3}$ 30) $11\sqrt{m^3} + 8m\sqrt{m}$

31) $4d^2\sqrt{d} - 24\sqrt{d^5}$ 32) $16k^4\sqrt{k} - 13\sqrt{k^9}$

33) $9t^3\sqrt[3]{t} - 5\sqrt[3]{t^{10}}$ 34) $8r^4\sqrt[3]{r} - 16\sqrt[3]{r^{13}}$

35) $5a\sqrt[4]{a^7} + \sqrt[4]{a^{11}}$ 36) $-3\sqrt[4]{c^{11}} + 6c^2\sqrt[4]{c^3}$

37) $2\sqrt{8p} - 6\sqrt{2p}$ 38) $4\sqrt{63t} + 6\sqrt{7t}$

39) $7\sqrt[3]{81a^5} + 4a\sqrt[3]{3a^2}$

40) $3\sqrt[3]{40x} - 12\sqrt[3]{5x}$

41) $\sqrt{xy^3} + 3y\sqrt{xy}$

42) $5a\sqrt{ab} + 2\sqrt{a^3b}$

43) $6c^2\sqrt{8d^3} - 9d\sqrt{2c^4d}$

44) $11v\sqrt{5u^3} - 2u\sqrt{45uv^2}$

45) $8p^2q\sqrt[3]{11pq^2} + 3p^2\sqrt[3]{88pq^5}$

46) $18a^5\sqrt[3]{7a^2b} + 2a^3\sqrt[3]{7a^8b}$

47) $15cd\sqrt[4]{9cd} - \sqrt[4]{9c^5d^5}$

48) $7yz^2\sqrt[4]{11y^4z} + 3z\sqrt[4]{11y^8z^5}$

49) $\sqrt[3]{a^9b} - \sqrt[3]{b^7}$

50) $\sqrt[3]{c^8} + \sqrt[3]{c^2d^3}$

Objective 3: Multiply a Binomial Containing Radical Expressions by a Monomial

Multiply and simplify.

51) $3(x + 5)$

52) $8(k + 3)$

53) $7(\sqrt{6} + 2)$

54) $5(4 - \sqrt{7})$

55) $\sqrt{10}(\sqrt{3} - 1)$

56) $\sqrt{2}(9 + \sqrt{11})$

57) $-6(\sqrt{32} + \sqrt{2})$

58) $10(\sqrt{12} - \sqrt{3})$

59) $4(\sqrt{45} - \sqrt{20})$

60) $-3(\sqrt{18} + \sqrt{50})$

61) $\sqrt{5}(\sqrt{24} - \sqrt{54})$

62) $\sqrt{2}(\sqrt{20} + \sqrt{45})$

63) $\sqrt[4]{3}(5 - \sqrt[4]{27})$

64) $\sqrt[3]{4}(2\sqrt[3]{5} + 7\sqrt[3]{4})$

65) $\sqrt{t}(\sqrt{t} - \sqrt{81u})$

66) $\sqrt{s}(\sqrt{12r} + \sqrt{7s})$

67) $\sqrt{2xy}(\sqrt{2y} - y\sqrt{x})$

68) $\sqrt{ab}(\sqrt{5a} + \sqrt{27b})$

69) $\sqrt[3]{c^2}(\sqrt[3]{c^2} + \sqrt[3]{125cd})$

70) $\sqrt[5]{mn^3}(\sqrt[5]{2m^2n} - n\sqrt[5]{mn^2})$

Mixed Exercises: Objectives 4–6

71) How are the problems *Multiply* $(x + 8)(x + 3)$ and *Multiply* $(3 + \sqrt{2})(1 + \sqrt{2})$ similar? What method can be used to multiply each of them?

72) How are the problems *Multiply* $(y - 5)^2$ and *Multiply* $(\sqrt{7} - 2)^2$ similar? What method can be used to multiply each of them?

73) What formula can be used to multiply $(5 + \sqrt{6})(5 - \sqrt{6})$?

74) What happens to the radical terms whenever we multiply $(a + b)(a - b)$ where the binomials contain square roots?

Objective 4: Multiply Radical Expressions Using FOIL

Multiply and simplify.

75) $(p + 7)(p + 6)$

76) $(z - 8)(z + 2)$

Fill It In

Fill in the blanks with either the missing mathematical step or reason for the given step.

77) $(6 + \sqrt{7})(2 + \sqrt{7})$

= _____   Use FOIL.

$= 12 + 6\sqrt{7} + 2\sqrt{7} + 7$ _____

= _____   Combine like terms.

78) $(3 + \sqrt{5})(1 + \sqrt{5})$

$= 3 \cdot 1 + 3\sqrt{5} + 1\sqrt{5} + \sqrt{5} \cdot \sqrt{5}$ _____

= _____   Multiply.

= _____   Combine like terms.

79) $(\sqrt{2} + 8)(\sqrt{2} - 3)$

80) $(\sqrt{6} - 7)(\sqrt{6} + 2)$

81) $(4 + \sqrt{x})(10 + \sqrt{x})$

82) $(\sqrt{y} + 9)(\sqrt{y} + 8)$

83) $(5\sqrt{2} - \sqrt{3})(2\sqrt{3} - \sqrt{2})$

84) $(\sqrt{5} - 4\sqrt{3})(2\sqrt{5} - \sqrt{3})$

85) $(5 + 2\sqrt{3})(\sqrt{7} + \sqrt{2})$

86) $(\sqrt{5} + 4)(\sqrt{3} - 6\sqrt{2})$

87) $(\sqrt{7n} + 1)(\sqrt{3n} - 8)$

88) $(\sqrt{2c} - 5)(\sqrt{11c} - 4)$

89) $(\sqrt[3]{25} - 3)(\sqrt[3]{5} - \sqrt[3]{6})$

90) $(\sqrt[4]{8} - \sqrt[4]{3})(\sqrt[4]{6} + \sqrt[4]{2})$

91) $(\sqrt{6p} - 2\sqrt{q})(8\sqrt{q} + 5\sqrt{6p})$

92) $(4\sqrt{3r} + \sqrt{s})(3\sqrt{s} - 2\sqrt{3r})$

Objective 5: Square a Binomial Containing Radical Expressions

93) $(\sqrt{3} + 1)^2$

94) $(2 + \sqrt{5})^2$

95) $(\sqrt{k} - 6)^2$

96) $(8 + \sqrt{t})^2$

97) $(1 + 4\sqrt{3v})^2$

98) $(5\sqrt{2z} - 3)^2$

99) $(\sqrt{11} - \sqrt{5})^2$

100) $(\sqrt{3} + \sqrt{13})^2$

101) $(\sqrt{h} + \sqrt{7})^2$

102) $(\sqrt{m} + \sqrt{3})^2$

103) $(\sqrt{x} - \sqrt{y})^2$

104) $(\sqrt{b} - \sqrt{a})^2$

Objective 6: Multiply Radical Expressions of the Form $(a + b)(a - b)$

105) $(c + 9)(c - 9)$

106) $(g - 7)(g + 7)$

(24) 107) $(6 - \sqrt{5})(6 + \sqrt{5})$

108) $(4 - \sqrt{7})(4 + \sqrt{7})$

109) $(\sqrt{p} + 7)(\sqrt{p} - 7)$

110) $(\sqrt{w} + 10)(\sqrt{w} - 10)$

111) $(4\sqrt{3} + \sqrt{2})(4\sqrt{3} - \sqrt{2})$

112) $(2\sqrt{2} - 2\sqrt{7})(2\sqrt{2} + 2\sqrt{7})$

113) $(\sqrt[3]{2} - 3)(\sqrt[3]{2} + 3)$

114) $(1 + \sqrt[3]{6})(1 - \sqrt[3]{6})$

(24) 115) $(\sqrt{c} + \sqrt{d})(\sqrt{c} - \sqrt{d})$

116) $(\sqrt{2y} + \sqrt{z})(\sqrt{2y} - \sqrt{z})$

117) $(8\sqrt{f} - \sqrt{g})(8\sqrt{f} + \sqrt{g})$

118) $(\sqrt{a} + 3\sqrt{4b})(\sqrt{a} - 3\sqrt{4b})$

Extension

Multiply and simplify.

119) $(1 + 2\sqrt[3]{5})(1 - 2\sqrt[3]{5} + 4\sqrt[3]{25})$

120) $(3 + \sqrt[3]{2})(9 - 3\sqrt[3]{2} + \sqrt[3]{4})$

Let $f(x) = x^2$. Find each function value.

121) $f(\sqrt{7} + 2)$

122) $f(5 - \sqrt{6})$

123) $f(1 - 2\sqrt{3})$

124) $f(3\sqrt{2} + 4)$

R Rethink

R1) Explain the difference between mastering the concepts from Chapter 6 and completing these exercises. How did they compare?

R2) Explain how you could substitute "x" for the radical $\sqrt{3}$ and "y" for the radical $\sqrt{2}$ in Exercise 111 to help you multiply.

10.6 Dividing Radicals

P Prepare

O Organize

What are your objectives for Section 10.6?	How can you accomplish each objective?
1 Rationalize a Denominator: One Square Root	• Understand what *rationalizing a denominator* means. • Complete the given examples on your own, and develop a procedure for rationalizing a denominator that contains one square root. • Complete You Trys 1–3.
2 Rationalize a Denominator: One Higher Root	• Follow the explanation to understand the logic behind rationalizing a higher root. • Complete the given examples on your own, and develop a procedure for rationalizing a denominator that contains one higher root. • Complete You Trys 4 and 5.
3 Rationalize a Denominator Containing Two Terms	• Understand the definition of a *conjugate*, and write a few examples in your notes. • Recall the formula for $(a + b)(a - b)$, and know how it can help you multiply conjugates. • Learn the procedure for **Rationalizing a Denominator That Contains Two Terms.** • Complete the given examples on your own. • Complete You Trys 6 and 7.

What are your objectives for Section 10.6?	How can you accomplish each objective?
4 Rationalize a Numerator	• Follow the same procedures developed for rationalizing the denominator, but focus instead on the numerator. • Complete the given example on your own. • Complete You Try 8.
5 Divide Out Common Factors from the Numerator and Denominator	• Know how to apply the properties of real numbers, especially factoring and distributing, to simplify. • Complete the given example on your own. • Complete You Try 9.

W Work

Read the explanations, follow the examples, take notes, and complete the You Trys.

It is generally agreed that a radical expression is *not* in simplest form if its denominator contains a radical. For example, $\dfrac{1}{\sqrt{3}}$ is not simplified, but an equivalent form, $\dfrac{\sqrt{3}}{3}$, is simplified.

Later we will show that $\dfrac{1}{\sqrt{3}} = \dfrac{\sqrt{3}}{3}$. The process of eliminating radicals from the denominator of an expression is called **rationalizing the denominator.** We will look at two types of rationalizing problems.

      1)   Rationalizing a denominator containing one term

      2)   Rationalizing a denominator containing two terms

To rationalize a denominator, we will use the fact that multiplying the numerator and denominator of a fraction by the same quantity results in an equivalent fraction:

$$\frac{2}{3} \cdot \frac{4}{4} = \frac{8}{12} \qquad \frac{2}{3} \text{ and } \frac{8}{12} \text{ are equivalent because } \frac{4}{4} = 1.$$

We use the same idea to rationalize the denominator of a radical expression.

1 Rationalize a Denominator: One Square Root

The goal of rationalizing a denominator is to eliminate the radical from the denominator. With regard to square roots, recall that $\sqrt{a} \cdot \sqrt{a} = \sqrt{a^2} = a$ for $a \geq 0$. For example,

$$\sqrt{19} \cdot \sqrt{19} = \sqrt{(19)^2} = 19 \quad \text{and} \quad \sqrt{t} \cdot \sqrt{t} = \sqrt{t^2} = t \; (t \geq 0)$$

We will use this property to rationalize the denominators of the following expressions.

EXAMPLE 1

Rationalize the denominator of each expression.

a) $\dfrac{1}{\sqrt{3}}$ b) $\dfrac{36}{\sqrt{18}}$ c) $\dfrac{5\sqrt{3}}{\sqrt{2}}$

Solution

a) To eliminate the square root from the denominator of $\dfrac{1}{\sqrt{3}}$, ask yourself, "By what do I multiply $\sqrt{3}$ to get a *perfect square* under the square root?" The answer is $\sqrt{3}$ because $\sqrt{3} \cdot \sqrt{3} = \sqrt{3^2} = \sqrt{9} = 3$. Multiply by $\sqrt{3}$ in the numerator *and* denominator. (We are actually multiplying by 1.)

$$\frac{1}{\sqrt{3}} = \frac{1}{\sqrt{3}} \cdot \frac{\sqrt{3}}{\sqrt{3}} = \frac{\sqrt{3}}{\sqrt{3^2}} = \frac{\sqrt{3}}{\sqrt{9}} = \frac{\sqrt{3}}{3}$$

Rationalize the denominator.

BE CAREFUL

$\dfrac{\sqrt{3}}{3}$ is in simplest form. We cannot "simplify" terms inside and outside of the radical.

$$\frac{\sqrt{3}}{3} = \frac{\sqrt{3}^{\,1}}{3_{1}} = \sqrt{1} = 1 \qquad \textbf{Incorrect!}$$

W Hint

Be sure you understand *why* you multiply by a particular radical to rationalize the denominator!

b) First, simplify the denominator of $\dfrac{36}{\sqrt{18}}$.

$$\frac{36}{\sqrt{18}} = \frac{36}{3\sqrt{2}} = \frac{12}{\sqrt{2}} = \frac{12}{\sqrt{2}} \cdot \frac{\sqrt{2}}{\sqrt{2}} = \frac{12\sqrt{2}}{2} = 6\sqrt{2}$$

Simplify $\sqrt{18}$.   Simplify.   Rationalize the denominator.

c) To rationalize $\dfrac{5\sqrt{3}}{\sqrt{2}}$, multiply the numerator and denominator by $\sqrt{2}$.

$$\frac{5\sqrt{3}}{\sqrt{2}} = \frac{5\sqrt{3}}{\sqrt{2}} \cdot \frac{\sqrt{2}}{\sqrt{2}} = \frac{5\sqrt{6}}{2}$$

[YOU TRY 1] Rationalize the denominator of each expression.

a) $\dfrac{1}{\sqrt{7}}$ b) $\dfrac{15}{\sqrt{27}}$ c) $\dfrac{9\sqrt{6}}{\sqrt{5}}$

Sometimes we will apply the quotient or product rule before rationalizing.

EXAMPLE 2   Simplify completely.

a) $\sqrt{\dfrac{3}{24}}$ b) $\sqrt{\dfrac{5}{14}} \cdot \sqrt{\dfrac{7}{3}}$

Solution

a) Begin by simplifying the fraction $\dfrac{3}{24}$ under the radical.

$$\sqrt{\frac{3}{24}} = \sqrt{\frac{1}{8}} \qquad \text{Simplify.}$$

$$= \frac{\sqrt{1}}{\sqrt{8}} = \frac{1}{\sqrt{4} \cdot \sqrt{2}} = \frac{1}{2\sqrt{2}} = \frac{1}{2\sqrt{2}} \cdot \frac{\sqrt{2}}{\sqrt{2}} = \frac{\sqrt{2}}{2 \cdot 2} = \frac{\sqrt{2}}{4}$$

b) Begin by using the product rule to multiply the radicands.

$$\sqrt{\frac{5}{14}} \cdot \sqrt{\frac{7}{3}} = \sqrt{\frac{5}{14} \cdot \frac{7}{3}} \qquad \text{Product rule}$$

Multiply the fractions under the radical.

$$= \sqrt{\frac{5}{\underset{2}{\cancel{14}}} \cdot \frac{\cancel{7}^{\,1}}{3}} = \sqrt{\frac{5}{6}} \qquad \text{Multiply.}$$

$$= \frac{\sqrt{5}}{\sqrt{6}} = \frac{\sqrt{5}}{\sqrt{6}} \cdot \frac{\sqrt{6}}{\sqrt{6}} = \frac{\sqrt{30}}{6}$$

$\left[\text{YOU TRY 2}\right]$ Simplify completely.

a) $\sqrt{\dfrac{10}{35}}$ b) $\sqrt{\dfrac{21}{10}} \cdot \sqrt{\dfrac{2}{7}}$

We work with radical expressions containing variables the same way. **In the rest of this section, we will assume that all variables represent positive real numbers.**

EXAMPLE 3   Simplify completely.

a) $\dfrac{2}{\sqrt{x}}$ b) $\sqrt{\dfrac{12m^3}{7n}}$ c) $\sqrt{\dfrac{6cd^2}{cd^3}}$

Solution

a) Ask yourself, "By what do I multiply $\sqrt{x}$ to get a *perfect square* under the square root?" The perfect square we want to get is $\sqrt{x^2}$.

$$\sqrt{x} \cdot \sqrt{?} = \sqrt{x^2} = x$$
$$\sqrt{x} \cdot \sqrt{x} = \sqrt{x^2} = x$$

$$\frac{2}{\sqrt{x}} = \frac{2}{\sqrt{x}} \cdot \frac{\sqrt{x}}{\sqrt{x}} = \frac{2\sqrt{x}}{\sqrt{x^2}} = \frac{2\sqrt{x}}{x}$$

$\uparrow$
Rationalize the denominator.

W Hint

Ask yourself the questions found in parts a) and b). This will help you understand the next objective too.

b) Before rationalizing, apply the quotient rule and simplify the numerator.

$$\sqrt{\frac{12m^3}{7n}} = \frac{\sqrt{12m^3}}{\sqrt{7n}} = \frac{2m\sqrt{3m}}{\sqrt{7n}}$$

Rationalize the denominator. "By what do I multiply $\sqrt{7n}$ to get a *perfect square* under the square root?" The perfect square we want to get is $\sqrt{7^2n^2}$ or $\sqrt{49n^2}$.

$$\sqrt{7n} \cdot \sqrt{?} = \sqrt{7^2n^2} = 7n$$
$$\sqrt{7n} \cdot \sqrt{7n} = \sqrt{7^2n^2} = 7n$$

$$\sqrt{\frac{12m^3}{7n}} = \frac{2m\sqrt{3m}}{\sqrt{7n}}$$

$$= \frac{2m\sqrt{3m}}{\sqrt{7n}} \cdot \frac{\sqrt{7n}}{\sqrt{7n}} = \frac{2m\sqrt{21mn}}{7n}$$

$\uparrow$
Rationalize the denominator.

c) $\sqrt{\dfrac{6cd^2}{cd^3}} = \sqrt{\dfrac{6}{d}}$ Simplify the radicand using the quotient rule for exponents.

$$= \dfrac{\sqrt{6}}{\sqrt{d}} = \dfrac{\sqrt{6}}{\sqrt{d}} \cdot \dfrac{\sqrt{d}}{\sqrt{d}} = \dfrac{\sqrt{6d}}{d}$$

[YOU TRY 3]   Simplify completely.

a) $\dfrac{5}{\sqrt{p}}$ b) $\dfrac{\sqrt{18k^5}}{\sqrt{10m}}$ c) $\sqrt{\dfrac{20r^3s}{s^2}}$

---

2 Rationalize a Denominator: One Higher Root

Many students assume that to rationalize *all* denominators we simply multiply the numerator and denominator of the expression by the denominator as in

$\dfrac{4}{\sqrt{3}} = \dfrac{4}{\sqrt{3}} \cdot \dfrac{\sqrt{3}}{\sqrt{3}} = \dfrac{4\sqrt{3}}{3}$. *We will see, however, why this reasoning is incorrect.*

To rationalize an expression like $\dfrac{4}{\sqrt{3}}$ we asked ourselves, "By what do I multiply $\sqrt{3}$ to get a *perfect square* under the *square root*?"

To rationalize an expression like $\dfrac{5}{\sqrt[3]{2}}$ we must ask ourselves, "By what do I multiply $\sqrt[3]{2}$ to get a *perfect cube* under the *cube root*?" The perfect cube we want is 2^3 (since we began with 2) so that $\sqrt[3]{2} \cdot \sqrt[3]{2^2} = \sqrt[3]{2^3} = 2$.

We will practice some fill-in-the-blank problems to eliminate radicals before we move on to rationalizing.

EXAMPLE 4   Fill in the blank.

a) $\sqrt[3]{5} \cdot \sqrt[3]{?} = \sqrt[3]{5^3} = 5$ b) $\sqrt[3]{3} \cdot \sqrt[3]{?} = \sqrt[3]{3^3} = 3$

c) $\sqrt[3]{x^2} \cdot \sqrt[3]{?} = \sqrt[3]{x^3} = x$ d) $\sqrt[5]{8} \cdot \sqrt[5]{?} = \sqrt[5]{2^5} = 2$

e) $\sqrt[4]{27} \cdot \sqrt[4]{?} = \sqrt[4]{3^4} = 3$

Solution

a)   Ask yourself, "By what do I multiply $\sqrt[3]{5}$ to get $\sqrt[3]{5^3}$?" The answer is $\sqrt[3]{5^2}$.

$$\sqrt[3]{5} \cdot \sqrt[3]{?} = \sqrt[3]{5^3} = 5$$
$$\sqrt[3]{5} \cdot \sqrt[3]{5^2} = \sqrt[3]{5^3} = 5$$

b)   "By what do I multiply $\sqrt[3]{3}$ to get $\sqrt[3]{3^3}$?" $\sqrt[3]{3^2}$

$$\sqrt[3]{3} \cdot \sqrt[3]{?} = \sqrt[3]{3^3} = 3$$
$$\sqrt[3]{3} \cdot \sqrt[3]{3^2} = \sqrt[3]{3^3} = 3$$

c)   "By what do I multiply $\sqrt[3]{x^2}$ to get $\sqrt[3]{x^3}$?" $\sqrt[3]{x}$

$$\sqrt[3]{x^2} \cdot \sqrt[3]{?} = \sqrt[3]{x^3} = x$$
$$\sqrt[3]{x^2} \cdot \sqrt[3]{x} = \sqrt[3]{x^3} = x$$

d) In this example, $\sqrt[5]{8} \cdot \sqrt[5]{?} = \sqrt[5]{2^5} = 2$, why are we trying to obtain $\sqrt[5]{2^5}$ instead of $\sqrt[5]{8^5}$? Because in the first radical, $\sqrt[5]{8}$, 8 *is a power of* 2. Before attempting to fill in the blank, rewrite 8 as 2^3.

$$\sqrt[5]{8} \cdot \sqrt[5]{?} = \sqrt[5]{2^5} = 2$$
$$\sqrt[5]{2^3} \cdot \sqrt[5]{?} = \sqrt[5]{2^5} = 2$$
$$\sqrt[5]{2^3} \cdot \sqrt[5]{2^2} = \sqrt[5]{2^5} = 2$$

W Hint

Don't move to the next example unless you have fully grasped this concept first!

e)
$$\sqrt[4]{27} \cdot \sqrt[4]{?} = \sqrt[4]{3^4} = 3$$
$$\sqrt[4]{3^3} \cdot \sqrt[4]{?} = \sqrt[4]{3^4} = 3 \qquad \text{Since 27 is a power of 3, rewrite } \sqrt[4]{27} \text{ as } \sqrt[4]{3^3}.$$
$$\sqrt[4]{3^3} \cdot \sqrt[4]{3} = \sqrt[4]{3^4} = 3$$

[**YOU TRY 4**] Fill in the blank.

a) $\sqrt[3]{2} \cdot \sqrt[3]{?} = \sqrt[3]{2^3} = 2$ b) $\sqrt[5]{t^2} \cdot \sqrt[5]{?} = \sqrt[5]{t^5} = t$

c) $\sqrt[4]{125} \cdot \sqrt[4]{?} = \sqrt[4]{5^4} = 5$

We will use the technique presented in Example 4 to rationalize denominators with indices higher than 2.

EXAMPLE 5   Rationalize the denominator.

a) $\dfrac{7}{\sqrt[3]{3}}$ b) $\sqrt[5]{\dfrac{3}{4}}$ c) $\dfrac{7}{\sqrt[4]{n}}$

Solution

a) *First* identify what we want the denominator to be *after* multiplying. **We want to obtain $\sqrt[3]{3^3}$ since $\sqrt[3]{3^3} = 3$.**

$$\frac{7}{\sqrt[3]{3}} \cdot \frac{}{\underset{\uparrow}{}} = \frac{}{\sqrt[3]{3^3}} \quad \longleftarrow \text{ This is what we want to get.}$$
$$\text{What is needed here?}$$

Ask yourself, "By what do I multiply $\sqrt[3]{3}$ to get $\sqrt[3]{3^3}$?" $\sqrt[3]{3^2}$

W Hint

Be sure to ask yourself this question so that you multiply by the correct radical!

$$\frac{7}{\sqrt[3]{3}} \cdot \frac{\sqrt[3]{3^2}}{\sqrt[3]{3^2}} = \frac{7\sqrt[3]{3^2}}{\sqrt[3]{3^3}} \qquad \text{Multiply.}$$
$$= \frac{7\sqrt[3]{9}}{3} \qquad \text{Simplify.}$$

b) Use the quotient rule for radicals to rewrite $\sqrt[5]{\dfrac{3}{4}}$ as $\dfrac{\sqrt[5]{3}}{\sqrt[5]{4}}$. Then, write 4 as 2^2 to get

$$\frac{\sqrt[5]{3}}{\sqrt[5]{4}} = \frac{\sqrt[5]{3}}{\sqrt[5]{2^2}}$$

What denominator do we want to get *after* multiplying? **We want to obtain $\sqrt[5]{2^5}$ since $\sqrt[5]{2^5} = 2$.**

$$\frac{\sqrt[5]{3}}{\sqrt[5]{2^2}} \cdot \frac{}{\underset{\uparrow}{}} = \frac{}{\sqrt[5]{2^5}} \quad \longleftarrow \text{ This is what we want to get.}$$
$$\text{What is needed here?}$$

"By what do I multiply $\sqrt[5]{2^2}$ to get $\sqrt[5]{2^5}$?" $\sqrt[5]{2^3}$

$$\frac{\sqrt[5]{3}}{\sqrt[5]{2^2}} \cdot \frac{\sqrt[5]{2^3}}{\sqrt[5]{2^3}} = \frac{\sqrt[5]{3} \cdot \sqrt[5]{2^3}}{\sqrt[5]{2^5}} \qquad \text{Multiply.}$$

$$= \frac{\sqrt[5]{3} \cdot \sqrt[5]{8}}{2} = \frac{\sqrt[5]{24}}{2} \qquad \text{Multiply.}$$

W Hint

In your own words, summarize how to rationalize denominators in Objective 2.

c) What denominator do we want to get *after* multiplying? **We want to obtain** $\sqrt[4]{n^4}$ **since** $\sqrt[4]{n^4} = n.$

$$\frac{7}{\sqrt[4]{n}} \cdot \frac{}{} = \frac{}{\sqrt[4]{n^4}} \longleftarrow \text{This is what we want to get.}$$
$$\uparrow$$
$$\text{What is needed here?}$$

Ask yourself, "By what do I multiply $\sqrt[4]{n}$ to get $\sqrt[4]{n^4}$?" $\sqrt[4]{n^3}$

$$\frac{7}{\sqrt[4]{n}} \cdot \frac{\sqrt[4]{n^3}}{\sqrt[4]{n^3}} = \frac{7\sqrt[4]{n^3}}{\sqrt[4]{n^4}} \qquad \text{Multiply.}$$

$$= \frac{7\sqrt[4]{n^3}}{n} \qquad \text{Simplify.}$$

[YOU TRY 5] Rationalize the denominator.

a) $\dfrac{4}{\sqrt[3]{7}}$ 　　 b) $\sqrt[4]{\dfrac{2}{27}}$ 　　 c) $\sqrt[5]{\dfrac{8}{w^3}}$

---

3 Rationalize a Denominator Containing Two Terms

To rationalize the denominator of an expression like $\dfrac{1}{5 + \sqrt{3}}$, we multiply the numerator and the denominator of the expression by the *conjugate* of $5 + \sqrt{3}$.

Definition

The **conjugate** of a binomial is the binomial obtained by changing the sign between the two terms.

Expression	Conjugate
$\sqrt{7} - 2\sqrt{5}$	$\sqrt{7} + 2\sqrt{5}$
$\sqrt{a} + \sqrt{b}$	$\sqrt{a} - \sqrt{b}$

In Section 10.5 we applied the formula $(a + b)(a - b) = a^2 - b^2$ to multiply binomials containing square roots. Recall that the terms containing the square roots were eliminated.

EXAMPLE 6

Multiply $8 - \sqrt{6}$ by its conjugate.

Solution

The conjugate of $8 - \sqrt{6}$ is $8 + \sqrt{6}$. We will first multiply using FOIL to show *why* the radical drops out, then we will multiply using the formula

$$(a + b)(a - b) = a^2 - b^2$$

i) Use FOIL to multiply.

$$(8 - \sqrt{6})(8 + \sqrt{6}) = 8 \cdot 8 + 8 \cdot \sqrt{6} - 8 \cdot \sqrt{6} - \sqrt{6} \cdot \sqrt{6}$$
$$\phantom{(8 - \sqrt{6})(8 + \sqrt{6}) = } \text{F} \qquad \text{O} \qquad \text{I} \qquad \text{L}$$
$$= 64 - 6$$
$$= 58$$

ii) Use $(a + b)(a - b) = a^2 - b^2$.

$$(8 - \sqrt{6})(8 + \sqrt{6}) = (8)^2 - (\sqrt{6})^2 \qquad \text{Substitute 8 for } a \text{ and } \sqrt{6} \text{ for } b.$$
$$= 64 - 6 = 58$$

Each method gives the same result.

[YOU TRY 6] Multiply $2 + \sqrt{11}$ by its conjugate.

Procedure Rationalize a Denominator Containing Two Terms

If the denominator of an expression contains two terms, including one or two square roots, then to rationalize the denominator we multiply the numerator and denominator of the expression by the *conjugate* of the denominator.

EXAMPLE 7

Rationalize the denominator, and simplify completely.

a) $\dfrac{3}{5 + \sqrt{3}}$
b) $\dfrac{\sqrt{a} + b}{\sqrt{b} - a}$

Solution

a) The denominator of $\dfrac{3}{5 + \sqrt{3}}$ has two terms, so we multiply the numerator and denominator by $5 - \sqrt{3}$, the conjugate of the denominator.

$$\frac{3}{5 + \sqrt{3}} \cdot \frac{5 - \sqrt{3}}{5 - \sqrt{3}} \qquad \text{Multiply by the conjugate.}$$

$$= \frac{3(5 - \sqrt{3})}{(5)^2 - (\sqrt{3})^2} \qquad (a + b)(a - b) = a^2 - b^2$$

$$= \frac{15 - 3\sqrt{3}}{25 - 3} \qquad \text{Simplify.}$$

$$= \frac{15 - 3\sqrt{3}}{22} \qquad \text{Subtract.}$$

b)

$$\frac{\sqrt{a} + b}{\sqrt{b} - a} \cdot \frac{\sqrt{b} + a}{\sqrt{b} + a}$$ Multiply by the conjugate.

In the numerator we must multiply $(\sqrt{a} + b)(\sqrt{b} + a)$. We will use FOIL.

$$\frac{\sqrt{a} + b}{\sqrt{b} - a} \cdot \frac{\sqrt{b} + a}{\sqrt{b} + a} = \frac{\sqrt{ab} + a\sqrt{a} + b\sqrt{b} + ab}{(\sqrt{b})^2 - (a)^2}$$ ← Use FOIL in the numerator.
$$\qquad\qquad\qquad\qquad\qquad\qquad\qquad\qquad\quad ← (a+b)(a-b) = a^2 - b^2$$
$$= \frac{\sqrt{ab} + a\sqrt{a} + b\sqrt{b} + ab}{b - a^2}$$ Square the terms.

[YOU TRY 7] Rationalize the denominator, and simplify completely.

a) $\dfrac{1}{\sqrt{7} - 2}$ b) $\dfrac{c - \sqrt{d}}{c + \sqrt{d}}$

---

4 Rationalize a Numerator

In higher-level math courses, sometimes it is necessary to rationalize the *numerator* of a radical expression so that the numerator does not contain a radical.

EXAMPLE 8

Rationalize the numerator, and simplify completely.

a) $\dfrac{\sqrt{7}}{\sqrt{2}}$ b) $\dfrac{8 - \sqrt{5}}{3}$

Solution

a) Rationalizing the numerator of $\dfrac{\sqrt{7}}{\sqrt{2}}$ means eliminating the square root from the *numerator*. Multiply the numerator and denominator by $\sqrt{7}$.

$$\frac{\sqrt{7}}{\sqrt{2}} = \frac{\sqrt{7}}{\sqrt{2}} \cdot \frac{\sqrt{7}}{\sqrt{7}} = \frac{7}{\sqrt{14}}$$

W Hint

Follow the same process you did for rationalizing the denominator, but focus on the numerator.

b) To rationalize the numerator we must multiply the numerator and denominator by $8 + \sqrt{5}$, the conjugate of the numerator.

$$\frac{8 - \sqrt{5}}{3} \cdot \frac{8 + \sqrt{5}}{8 + \sqrt{5}}$$ Multiply by the conjugate.

$$= \frac{8^2 - (\sqrt{5})^2}{3(8 + \sqrt{5})}$$ $\longleftarrow$ $(a+b)(a-b) = a^2 - b^2$
$$\qquad\qquad\qquad\qquad \longleftarrow$ Multiply.

$$= \frac{64 - 5}{24 + 3\sqrt{5}} = \frac{59}{24 + 3\sqrt{5}}$$

[YOU TRY 8] Rationalize the numerator, and simplify completely.

a) $\dfrac{\sqrt{3}}{\sqrt{5}}$ b) $\dfrac{6 + \sqrt{7}}{4}$

---

5 Divide Out Common Factors from the Numerator and Denominator

Sometimes it is necessary to simplify a radical expression by dividing out common factors from the numerator and denominator. This is a skill we will need in Chapter 11 to solve quadratic equations, so we will look at an example here.

EXAMPLE 9

Simplify completely: $\dfrac{4\sqrt{5} + 12}{4}$

Solution

 **BE CAREFUL**

It is tempting to do one of the following:

$$\frac{\cancel{4}\sqrt{5} + 12}{\cancel{4}} = \sqrt{5} + 12 \qquad \text{Incorrect!}$$

or

$$\frac{4\sqrt{5} + \overset{3}{\cancel{12}}}{\cancel{4}} = 4\sqrt{5} + 3 \qquad \text{Incorrect!}$$

Each is incorrect because $4\sqrt{5}$ is a *term* in a sum and 12 is a *term* in a sum.

The correct way to simplify $\dfrac{4\sqrt{5} + 12}{4}$ is to begin by factoring out a 4 in the numerator and *then* divide the numerator and denominator by any common factors.

$$\frac{4\sqrt{5} + 12}{4} = \frac{4(\sqrt{5} + 3)}{4} \qquad \text{Factor out 4 from the numerator.}$$

$$= \frac{\overset{1}{\cancel{4}}(\sqrt{5} + 3)}{\underset{1}{\cancel{4}}} \qquad \text{Divide by 4.}$$

$$= \sqrt{5} + 3 \qquad \text{Simplify.}$$

We can divide the numerator and denominator by 4 in $\dfrac{4(\sqrt{5} + 3)}{4}$ because the 4 in the numerator is part of a *product*, not a sum or difference.

[YOU TRY 9]

Simplify completely.

a) $\dfrac{5\sqrt{7} - 40}{5}$ b) $\dfrac{20 + 6\sqrt{2}}{4}$

ANSWERS TO [YOU TRY] EXERCISES

1) a) $\dfrac{\sqrt{7}}{7}$ b) $\dfrac{5\sqrt{3}}{3}$ c) $\dfrac{9\sqrt{30}}{5}$ 2) a) $\dfrac{\sqrt{14}}{7}$ b) $\dfrac{\sqrt{15}}{5}$ 3) a) $\dfrac{5\sqrt{p}}{p}$ b) $\dfrac{3k^2\sqrt{5km}}{5m}$
c) $\dfrac{2r\sqrt{5rs}}{s}$ 4) a) 2^2 or 4   b) t^3 c) 5   5) a) $\dfrac{4\sqrt[3]{49}}{7}$ b) $\dfrac{\sqrt[4]{6}}{3}$ c) $\dfrac{\sqrt[5]{8w^2}}{w}$ 6) -7
7) a) $\dfrac{\sqrt{7} + 2}{3}$ b) $\dfrac{c^2 - 2c\sqrt{d} + d}{c^2 - d}$ 8) a) $\dfrac{3}{\sqrt{15}}$ b) $\dfrac{29}{24 - 4\sqrt{7}}$ 9) a) $\sqrt{7} - 8$ b) $\dfrac{10 + 3\sqrt{2}}{2}$

Assume all variables represent positive real numbers.

Objective 1: Rationalize a Denominator: One Square Root

1) What does it mean to rationalize the denominator of a radical expression?

2) In your own words, explain how to rationalize the denominator of an expression containing one term in the denominator.

Rationalize the denominator of each expression.

3) $\dfrac{1}{\sqrt{5}}$

4) $\dfrac{1}{\sqrt{6}}$

5) $\dfrac{9}{\sqrt{6}}$

6) $\dfrac{25}{\sqrt{10}}$

7) $-\dfrac{20}{\sqrt{8}}$

8) $-\dfrac{18}{\sqrt{45}}$

9) $\dfrac{\sqrt{3}}{\sqrt{28}}$

10) $\dfrac{\sqrt{8}}{\sqrt{27}}$

11) $\sqrt{\dfrac{20}{60}}$

12) $\sqrt{\dfrac{12}{80}}$

13) $\dfrac{\sqrt{56}}{\sqrt{48}}$

14) $\dfrac{\sqrt{66}}{\sqrt{12}}$

Multiply and simplify.

15) $\sqrt{\dfrac{10}{7}} \cdot \sqrt{\dfrac{7}{3}}$

16) $\sqrt{\dfrac{11}{5}} \cdot \sqrt{\dfrac{5}{2}}$

17) $\sqrt{\dfrac{6}{5}} \cdot \sqrt{\dfrac{1}{8}}$

18) $\sqrt{\dfrac{11}{10}} \cdot \sqrt{\dfrac{8}{11}}$

Simplify completely.

19) $\dfrac{8}{\sqrt{y}}$

20) $\dfrac{4}{\sqrt{w}}$

21) $\dfrac{\sqrt{5}}{\sqrt{t}}$

22) $\dfrac{\sqrt{2}}{\sqrt{m}}$

23) $\sqrt{\dfrac{64v^7}{5w}}$

24) $\sqrt{\dfrac{81c^5}{2d}}$

25) $\sqrt{\dfrac{a^3 b^3}{3ab^4}}$

26) $\sqrt{\dfrac{m^2 n^5}{7m^3 n}}$

27) $-\dfrac{\sqrt{75}}{\sqrt{b^3}}$

28) $-\dfrac{\sqrt{24}}{\sqrt{v^3}}$

29) $\dfrac{\sqrt{13}}{\sqrt{j^5}}$

30) $\dfrac{\sqrt{22}}{\sqrt{w^7}}$

Objective 2: Rationalize a Denominator: One Higher Root

Fill in the blank.

31) $\sqrt[3]{2} \cdot \sqrt[3]{?} = \sqrt[3]{2^3} = 2$

32) $\sqrt[3]{5} \cdot \sqrt[3]{?} = \sqrt[3]{5^3} = 5$

33) $\sqrt[3]{9} \cdot \sqrt[3]{?} = \sqrt[3]{3^3} = 3$

34) $\sqrt[3]{4} \cdot \sqrt[3]{?} = \sqrt[3]{2^3} = 2$

35) $\sqrt[3]{c} \cdot \sqrt[3]{?} = \sqrt[3]{c^3} = c$

36) $\sqrt[3]{p} \cdot \sqrt[3]{?} = \sqrt[3]{p^3} = p$

37) $\sqrt[5]{4} \cdot \sqrt[5]{?} = \sqrt[5]{2^5} = 2$

38) $\sqrt[5]{16} \cdot \sqrt[5]{?} = \sqrt[5]{2^5} = 2$

39) $\sqrt[4]{m^3} \cdot \sqrt[4]{?} = \sqrt[4]{m^4} = m$

40) $\sqrt[4]{k} \cdot \sqrt[4]{?} = \sqrt[4]{k^4} = k$

41) Inez simplifies $\dfrac{8}{\sqrt[3]{3}}$ like this: $\dfrac{8}{\sqrt[3]{3}} \cdot \dfrac{\sqrt[3]{3}}{\sqrt[3]{3}} = \dfrac{8\sqrt[3]{3}}{3}$. Is she right or wrong? If her answer is wrong, explain her mistake and do the problem correctly.

42) Stavros simplifies $\dfrac{5}{\sqrt[8]{x^3}}$ like this: $\dfrac{5}{\sqrt[8]{x^3}} \cdot \dfrac{\sqrt[8]{x^5}}{\sqrt[8]{x^5}} = \dfrac{5\sqrt[8]{x^5}}{x}$. Is he right or wrong? If his answer is wrong, explain his mistake and do the problem correctly.

Rationalize the denominator of each expression.

43) $\dfrac{4}{\sqrt[3]{3}}$

44) $\dfrac{26}{\sqrt[3]{5}}$

45) $\dfrac{12}{\sqrt[3]{2}}$

46) $\dfrac{21}{\sqrt[3]{3}}$

47) $\dfrac{9}{\sqrt[3]{25}}$

48) $\dfrac{6}{\sqrt[3]{4}}$

49) $\sqrt[4]{\dfrac{5}{9}}$

50) $\sqrt[4]{\dfrac{2}{25}}$

51) $\sqrt[5]{\dfrac{3}{8}}$

52) $\sqrt[5]{\dfrac{7}{4}}$

53) $\dfrac{10}{\sqrt[3]{z}}$

54) $\dfrac{6}{\sqrt[3]{u}}$

55) $\sqrt[3]{\dfrac{3}{n^2}}$

56) $\sqrt[3]{\dfrac{5}{x^2}}$

57) $\dfrac{\sqrt[3]{7}}{\sqrt[3]{2k^2}}$

58) $\dfrac{\sqrt[3]{2}}{\sqrt[3]{25t}}$

59) $\dfrac{9}{\sqrt[5]{a^3}}$

60) $\dfrac{8}{\sqrt[5]{h^2}}$

61) $\sqrt[4]{\dfrac{5}{2m}}$

62) $\sqrt[4]{\dfrac{2}{3t^2}}$

Objective 3: Rationalize a Denominator Containing Two Terms

63) How do you find the conjugate of an expression with two radical terms?

64) When you multiply a binomial containing a square root by its conjugate, what happens to the radical?

Find the conjugate of each expression. Then, multiply the expression by its conjugate.

65) $(5 + \sqrt{2})$

66) $(\sqrt{5} - 4)$

67) $(\sqrt{2} + \sqrt{6})$

68) $(\sqrt{3} - \sqrt{10})$

69) $(\sqrt{t} - 8)$

70) $(\sqrt{p} + 5)$

Rationalize the denominator, and simplify completely.

Fill It In

Fill in the blanks with either the missing mathematical step or reason for the given step.

71) $\dfrac{6}{4 - \sqrt{5}} = \dfrac{6}{4 - \sqrt{5}} \cdot \dfrac{4 + \sqrt{5}}{4 + \sqrt{5}}$ _____

$= \dfrac{6(4 + \sqrt{5})}{(4)^2 - (\sqrt{5})^2}$ _____

$=$ _____   Multiply terms in numerator; square terms in denominator.

$=$ _____   Simplify.

72) $\dfrac{\sqrt{6}}{\sqrt{7} + \sqrt{2}} = \dfrac{\sqrt{6}}{\sqrt{7} + \sqrt{2}} \cdot \dfrac{\sqrt{7} - \sqrt{2}}{\sqrt{7} - \sqrt{2}}$ _____

$= \dfrac{\sqrt{6}(\sqrt{7} - \sqrt{2})}{(\sqrt{7})^2 - (\sqrt{2})^2}$ _____

$=$ _____   Multiply terms in numerator; square terms in denominator.

$=$ _____   Simplify.

73) $\dfrac{3}{2 + \sqrt{3}}$

74) $\dfrac{8}{6 - \sqrt{5}}$

75) $\dfrac{10}{9 - \sqrt{2}}$

76) $\dfrac{5}{4 + \sqrt{6}}$

77) $\dfrac{\sqrt{8}}{\sqrt{3} + \sqrt{2}}$

78) $\dfrac{\sqrt{32}}{\sqrt{5} - \sqrt{7}}$

79) $\dfrac{\sqrt{3} - \sqrt{5}}{\sqrt{10} - \sqrt{3}}$

80) $\dfrac{\sqrt{3} + \sqrt{6}}{\sqrt{2} + \sqrt{5}}$

81) $\dfrac{\sqrt{m}}{\sqrt{m} + \sqrt{n}}$

82) $\dfrac{\sqrt{u}}{\sqrt{u} - \sqrt{v}}$

83) $\dfrac{b - 25}{\sqrt{b} - 5}$

84) $\dfrac{d - 9}{\sqrt{d} + 3}$

85) $\dfrac{\sqrt{x} + \sqrt{y}}{\sqrt{x} - \sqrt{y}}$

86) $\dfrac{\sqrt{f} - \sqrt{g}}{\sqrt{f} + \sqrt{g}}$

Objective 4: Rationalize a Numerator

Rationalize the numerator of each expression, and simplify.

87) $\dfrac{\sqrt{5}}{3}$

88) $\dfrac{\sqrt{2}}{9}$

89) $\dfrac{\sqrt{x}}{\sqrt{7}}$

90) $\dfrac{\sqrt{8a}}{\sqrt{b}}$

91) $\dfrac{2 + \sqrt{3}}{6}$

92) $\dfrac{1 + \sqrt{7}}{3}$

93) $\dfrac{\sqrt{x} - 2}{x - 4}$

94) $\dfrac{3 - \sqrt{n}}{n - 9}$

95) $\dfrac{4 - \sqrt{c + 11}}{c - 5}$

96) $\dfrac{\sqrt{x + h} - \sqrt{x}}{h}$

97) Does rationalizing the denominator of an expression change the value of the original expression? Explain your answer.

98) Does rationalizing the numerator of an expression change the value of the original expression? Explain your answer.

Objective 5: Divide Out Common Factors from the Numerator and Denominator

Simplify completely.

99) $\dfrac{5 + 10\sqrt{3}}{5}$

100) $\dfrac{18 - 6\sqrt{7}}{6}$

101) $\dfrac{30 - 18\sqrt{5}}{4}$

102) $\dfrac{36 + 20\sqrt{2}}{12}$

103) $\dfrac{\sqrt{45} + 6}{9}$

104) $\dfrac{\sqrt{48} + 28}{4}$

105) $\dfrac{-10 - \sqrt{50}}{5}$

106) $\dfrac{-35 + \sqrt{200}}{15}$

107) The function $r(A) = \sqrt{\dfrac{A}{\pi}}$ describes the radius of a circle, $r(A)$, in terms of its area, A.

    a) If the area of a circle is measured in square inches, find $r(8\pi)$ and explain what it means in the context of the problem.

b) If the area of a circle is measured in square inches, find $r(7)$ and rationalize the denominator. Explain the meaning of $r(7)$ in the context of the problem.

c) Obtain an equivalent form of the function by rationalizing the denominator.

108) The function $r(V) = \sqrt[3]{\dfrac{3V}{4\pi}}$ describes the radius of a sphere, $r(V)$, in terms of its volume, V.

a) If the volume of a sphere is measured in cubic centimeters, find $r(36\pi)$ and explain what it means in the context of the problem.

b) If the volume of a sphere is measured in cubic centimeters, find $r(11)$ and rationalize the denominator. Explain the meaning of $r(11)$ in the context of the problem.

c) Obtain an equivalent form of the function by rationalizing the denominator.

R Rethink

R1) Why is it important to always read the directions closely?

R2) Which section of the book could you go to for help or review if you struggled with this section?

R3) Did you circle any problems you struggled with? Ask about them in class!

Putting It All Together

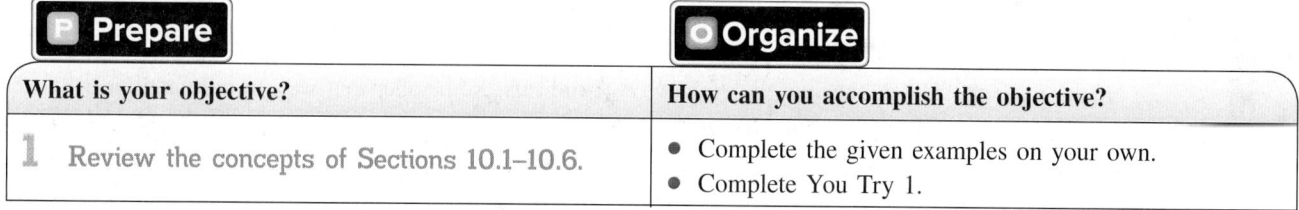

What is your objective?	How can you accomplish the objective?
1 Review the concepts of Sections 10.1–10.6.	• Complete the given examples on your own. • Complete You Try 1.

 **Read the explanations, follow the examples, take notes, and complete the You Trys.**

1   Review the Concepts Presented in 10.1–10.6

In Section 10.1, we learned how to find roots of numbers. For example, $\sqrt[3]{-64} = -4$ because $(-4)^3 = -64$. In Section 10.2, we learned about the relationship between rational exponents and radicals. Recall that if m and n are positive integers and $\dfrac{m}{n}$ is in lowest terms, then $a^{m/n} = (a^{1/n})^m = (\sqrt[n]{a})^m$ provided that $a^{1/n}$ is a real number. **For the rest of this section we will assume that all variables represent positive real numbers.**

EXAMPLE 1   Simplify completely. The answer should contain only positive exponents.

a) $(32)^{4/5}$ b) $\left(\dfrac{a^7 b^{9/8}}{25 a^9 b^{3/4}}\right)^{-3/2}$

Solution

a)   The denominator of the fractional exponent is the index of the radical, and the numerator is the power to which we raise the radical expression.

$$32^{4/5} = (\sqrt[5]{32})^4 \qquad \text{Write in radical form.}$$
$$= (2)^4 \qquad \sqrt[5]{32} = 2$$
$$= 16$$

b) $\left(\dfrac{a^7 b^{9/8}}{25 a^9 b^{3/4}}\right)^{-3/2} = \left(\dfrac{25 a^9 b^{3/4}}{a^7 b^{9/8}}\right)^{3/2}$ Eliminate the negative from the outermost exponent by taking the reciprocal of the base.

Simplify the expression inside the parentheses by subtracting the exponents.

$$= (25 a^{9-7} b^{3/4 - 9/8})^{3/2} = (25 a^2 b^{6/8 - 9/8})^{3/2} = (25 a^2 b^{-3/8})^{3/2}$$

Apply the power rule, and simplify.

$$= (25)^{3/2}(a^2)^{3/2}(b^{-3/8})^{3/2} = (\sqrt{25})^3 a^3 b^{-9/16} = 5^3 a^3 b^{-9/16} = \frac{125 a^3}{b^{9/16}}$$

In Sections 10.3–10.6 we learned how to simplify, multiply, divide, add, and subtract radicals. Let's look at these operations together so that we will learn to recognize the techniques needed to perform these operations.

EXAMPLE 2   Perform the operations, and simplify.

a) $\sqrt{3} + 10\sqrt{6} - 4\sqrt{3}$ b) $\sqrt{3}(10\sqrt{6} - 4\sqrt{3})$

Solution

a)   This is the *sum and difference* of radicals. Remember that we can only add and subtract radicals that are like radicals.

$$\sqrt{3} + 10\sqrt{6} - 4\sqrt{3} = \sqrt{3} - 4\sqrt{3} + 10\sqrt{6} \qquad \text{Write like radicals together.}$$
$$= -3\sqrt{3} + 10\sqrt{6} \qquad \text{Subtract.}$$

b)   This is the *product* of radical expressions. We must multiply the binomial $10\sqrt{6} - 4\sqrt{3}$ by $\sqrt{3}$ using the distributive property.

$$\sqrt{3}(10\sqrt{6} - 4\sqrt{3}) = \sqrt{3} \cdot 10\sqrt{6} - \sqrt{3} \cdot 4\sqrt{3} \qquad \text{Distribute.}$$
$$= 10\sqrt{18} - 4 \cdot 3 \qquad \text{Product rule; } \sqrt{3} \cdot \sqrt{3} = 3.$$
$$= 10\sqrt{18} - 12 \qquad \text{Multiply.}$$

Ask yourself, "Is $10\sqrt{18} - 12$ in simplest form?" *No.* $\sqrt{18}$ can be simplified.

$$= 10\sqrt{9 \cdot 2} - 12 \qquad \text{9 is a perfect square.}$$
$$= 10\sqrt{9} \cdot \sqrt{2} - 12 \qquad \text{Product rule}$$
$$= 10 \cdot 3\sqrt{2} - 12 \qquad \sqrt{9} = 3$$
$$= 30\sqrt{2} - 12 \qquad \text{Multiply.}$$

The expression is now in simplest form.

Next we will look at multiplication problems involving binomials. Remember that the rules we used to multiply binomials like $(x + 4)(x - 9)$ are the same rules we use to multiply binomials containing radicals.

EXAMPLE 3

Multiply and simplify.

a) $(8 + \sqrt{2})(9 - \sqrt{11})$ b) $(\sqrt{n} + \sqrt{7})(\sqrt{n} - \sqrt{7})$

c) $(2\sqrt{5} - 3)^2$

Solution

a) Since we must multiply two binomials, we will use FOIL.

$$\overset{}{(8 + \sqrt{2})(9 - \sqrt{11})} = \overset{F}{8 \cdot 9} - \overset{O}{8 \cdot \sqrt{11}} + \overset{I}{9 \cdot \sqrt{2}} - \overset{L}{\sqrt{2} \cdot \sqrt{11}} \qquad \text{Use FOIL.}$$
$$= 72 - 8\sqrt{11} + 9\sqrt{2} - \sqrt{22} \qquad \text{Multiply.}$$

All radicals are simplified and none of them are like radicals, so this expression is in simplest form.

> **W Hint**
>
> Notice the connection between what you learned in Chapter 6 and multiplying binomials containing radicals.

b) We can multiply $(\sqrt{n} + \sqrt{7})(\sqrt{n} - \sqrt{7})$ using FOIL or, if we notice that this product is in the form $(a + b)(a - b)$ we can apply the rule $(a + b)(a - b) = a^2 - b^2$. Either method will give us the correct answer. We will use the second method.

$$(a + b)(a - b) = a^2 - b^2$$
$$(\sqrt{n} + \sqrt{7})(\sqrt{n} - \sqrt{7}) = (\sqrt{n})^2 - (\sqrt{7})^2 = n - 7 \qquad \text{Substitute } \sqrt{n} \text{ for } a \text{ and } \sqrt{7} \text{ for } b.$$

c) Once again we can either use FOIL to expand $(2\sqrt{5} - 3)^2$, or we can use the special formula we learned for squaring a binomial.

We will use $(a - b)^2 = a^2 - 2ab + b^2$.

$$(2\sqrt{5} - 3)^2 = (2\sqrt{5})^2 - 2(2\sqrt{5})(3) + (3)^2 \qquad \text{Substitute } 2\sqrt{5} \text{ for } a \text{ and } 3 \text{ for } b.$$
$$= (4 \cdot 5) - 4\sqrt{5}(3) + 9 \qquad \text{Multiply.}$$
$$= 20 - 12\sqrt{5} + 9 \qquad \text{Multiply.}$$
$$= 29 - 12\sqrt{5} \qquad \text{Combine like terms.}$$

Remember that an expression is not considered to be in simplest form if it contains a radical in its denominator. To rationalize the denominator of a radical expression, we must keep in mind the index on the radical and the number of terms in the denominator.

EXAMPLE 4

Rationalize the denominator of each expression.

a) $\dfrac{10}{\sqrt{2x}}$ b) $\dfrac{10}{\sqrt[3]{2x}}$ c) $\dfrac{\sqrt{10}}{\sqrt{2} - 1}$

Solution

a) First, notice that the denominator of $\dfrac{10}{\sqrt{2x}}$ contains only one term and it is a *square* root.

Ask yourself, "By what do I multiply $\sqrt{2x}$ to get a perfect *square* under the radical?" The answer is $\sqrt{2x}$ since $\sqrt{2x} \cdot \sqrt{2x} = \sqrt{4x^2} = 2x$. Multiply the numerator and denominator by $\sqrt{2x}$, and simplify.

$$\frac{10}{\sqrt{2x}} = \frac{10}{\sqrt{2x}} \cdot \frac{\sqrt{2x}}{\sqrt{2x}} \qquad \text{Rationalize the denominator.}$$

$$= \frac{10\sqrt{2x}}{\sqrt{4x^2}} = \frac{10\sqrt{2x}}{2x} = \frac{5\sqrt{2x}}{x}$$

b) The denominator of $\dfrac{10}{\sqrt[3]{2x}}$ contains only one term, but it is a *cube* root. Ask yourself, "By what do I multiply $\sqrt[3]{2x}$ to get a radicand that is a perfect *cube*?" The answer is $\sqrt[3]{4x^2}$ since $\sqrt[3]{2x} \cdot \sqrt[3]{4x^2} = \sqrt[3]{8x^3} = 2x$. Multiply the numerator and denominator by $\sqrt[3]{4x^2}$, and simplify.

$$\frac{10}{\sqrt[3]{2x}} = \frac{10}{\sqrt[3]{2x}} \cdot \frac{\sqrt[3]{4x^2}}{\sqrt[3]{4x^2}} \qquad \text{Rationalize the denominator.}$$

$$= \frac{10\sqrt[3]{4x^2}}{\sqrt[3]{8x^3}} = \frac{10\sqrt[3]{4x^2}}{2x} = \frac{5\sqrt[3]{4x^2}}{x}$$

c) The denominator of $\dfrac{\sqrt{10}}{\sqrt{2}-1}$ contains two terms, so how do we rationalize the denominator of this expression? We multiply the numerator and denominator by the *conjugate* of the denominator.

$$\frac{\sqrt{10}}{\sqrt{2}-1} = \frac{\sqrt{10}}{\sqrt{2}-1} \cdot \frac{\sqrt{2}+1}{\sqrt{2}+1} \qquad \text{Multiply by the conjugate.}$$

$$= \frac{\sqrt{10}(\sqrt{2}+1)}{(\sqrt{2})^2 - (1)^2} \qquad \begin{array}{l}\text{Multiply.}\\ (a+b)(a-b) = a^2 - b^2\end{array}$$

$$= \frac{\sqrt{20} + \sqrt{10}}{1} \qquad \begin{array}{l}\text{Distribute.}\\ \text{Simplify.}\end{array}$$

$$= 2\sqrt{5} + \sqrt{10} \qquad \sqrt{20} = 2\sqrt{5};\ \text{simplify.}$$

[YOU TRY 1]

a) Perform the operations, and simplify.

     i) $(\sqrt{w} + 8)^2$ ii) $(3 - \sqrt{5a})(4 + \sqrt{5a})$ iii) $\sqrt{2}(9\sqrt{10} - \sqrt{2})$

     iv) $\sqrt{2} + 9\sqrt{10} - 5\sqrt{2}$ v) $(2\sqrt{3} + y)(2\sqrt{3} - y)$

b) Find each root.

     i) $-\sqrt{\dfrac{121}{16}}$ ii) $\sqrt[3]{-1000}$ iii) $\sqrt{0.09}$ iv) $\sqrt{-49}$

c) Simplify completely. The answer should contain only positive exponents.

     i) $(-64)^{2/3}$ ii) $\left(\dfrac{81x^3y^{1/2}}{x^{-5}y^6}\right)^{-3/4}$

d) Rationalize the denominator of each expression.

     i) $\dfrac{24}{\sqrt[3]{9h}}$ ii) $\dfrac{7 + \sqrt{6}}{4 + \sqrt{6}}$ iii) $\dfrac{56}{\sqrt{7}}$

ANSWERS TO [YOU TRY] EXERCISES

1) a) i) $w + 16\sqrt{w} + 64$ ii) $12 - \sqrt{5a} - 5a$ iii) $18\sqrt{5} - 2$ iv) $-4\sqrt{2} + 9\sqrt{10}$ v) $12 - y^2$

    b) i) $-\dfrac{11}{4}$ ii) -10 iii) 0.3 iv) not a real number

    c) i) 16 ii) $\dfrac{y^{33/8}}{27x^6}$ d) i) $\dfrac{8\sqrt[3]{3h^2}}{h}$ ii) $\dfrac{22 - 3\sqrt{6}}{10}$ iii) $8\sqrt{7}$

Putting It All Together Exercises

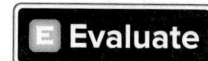
Objective 1: Review the Concepts Presented in 10.1–10.6

Fill in the blank with *always*, *sometimes*, or *never* to make the statement true.

1) The odd root of a real number is _____ a real number.

2) The even root of a negative number is _____ a real number.

Assume all variables represent positive real numbers.

Find each root, if possible.

(24) 3) $\sqrt[4]{81}$

4) $\sqrt[3]{-1000}$

5) $-\sqrt[6]{64}$

6) $\sqrt{121}$

7) $\sqrt{-169}$

8) $\sqrt{\dfrac{144}{49}}$

Simplify completely. The answer should contain only positive exponents.

9) $(144)^{1/2}$

10) $(-32)^{4/5}$

(24) 11) $-1000^{2/3}$

12) $\left(-\dfrac{16}{81}\right)^{3/4}$

13) $125^{-1/3}$

14) $\left(\dfrac{100}{9}\right)^{-3/2}$

15) $k^{-3/5} \cdot k^{3/10}$

16) $\left(t^{3/8}\right)^{16}$

17) $\left(\dfrac{27u^{-8}}{b^9}\right)^{2/3}$

18) $\left(\dfrac{18x^9 y^{4/3}}{2x^3 y}\right)^{-3/2}$

Simplify completely.

19) $\sqrt{24}$

20) $\sqrt[4]{32}$

(24) 21) $\sqrt[3]{72}$

22) $\sqrt[3]{\dfrac{500}{2}}$

23) $\sqrt[4]{243}$

24) $\sqrt{45c^{11}}$

25) $\sqrt[3]{96m^7 n^{15}}$

26) $\sqrt[5]{\dfrac{64x^{19}}{y^{20}}}$

Perform the operations, and simplify.

27) $\sqrt[3]{12} \cdot \sqrt[3]{2}$

28) $\sqrt[4]{\dfrac{96k^{11}}{2k^3}}$

29) $(6 + \sqrt{7})(2 + \sqrt{7})$

30) $4c^2 \sqrt[3]{108c} - 15\sqrt[3]{32c^7}$

31) $\dfrac{18}{\sqrt{6}}$

32) $\dfrac{5}{\sqrt{3} - \sqrt{2}}$

(24) 33) $3\sqrt{75m^3 n} + m\sqrt{12mn}$

34) $\sqrt{6p^7 q^3} \cdot \sqrt{15pq^2}$

35) $\dfrac{9}{\sqrt[3]{2}}$

36) $\dfrac{\sqrt{60t^8 u^3}}{\sqrt{5t^2 u}}$

37) $(2\sqrt{3} + 10)^2$

38) $(\sqrt{2} + 3)(\sqrt{2} - 3)$

(24) 39) $\dfrac{\sqrt{2}}{4 + \sqrt{10}}$

40) $\sqrt[3]{r^2} \cdot \sqrt{r}$

41) $\sqrt[3]{\dfrac{b^2}{9c}}$

42) $\dfrac{\sqrt[4]{32}}{\sqrt[4]{w^{11}}}$

43) $(8 - \sqrt{5w})^2$

44) $(2\sqrt{x} - \sqrt{y})(5\sqrt{r} - 6\sqrt{y})$

Rationalize the numerator of each expression, and simplify.

45) $\dfrac{\sqrt{11}}{4}$

46) $\dfrac{\sqrt{20k}}{\sqrt{h}}$

47) $\dfrac{\sqrt{a} - 5}{a - 25}$

48) $\dfrac{7 + \sqrt{2}}{8}$

R Rethink

R1) Which types of problems did you struggle with the most?

R2) Would you be able to complete exercises like these without any help?

10.7 Solving Radical Equations

P Prepare	**O** Organize
What are your objectives for Section 10.7?	**How can you accomplish each objective?**
1 Understand the Steps for Solving a Radical Equation	• Write the procedure for **Solving Radical Equations Containing Square Roots** in your own words, and be aware of *extraneous solutions*.
2 Solve an Equation Containing One Square Root	• Use the steps for **Solving Radical Equations Containing Square Roots.** • Be aware that you will, often, have to square a binomial. • Complete the given examples on your own. • Complete You Trys 1 and 2.
3 Solve an Equation Containing Two Square Roots	• Use the steps for **Solving Radical Equations Containing Square Roots.** • Recognize when you need to square both sides of the equation twice. • Complete the given examples on your own. • Complete You Trys 3–5.
4 Solve an Equation Containing a Cube Root	• Follow the explanation to create your own procedure for solving radical equations containing a cube root. • Complete the given example on your own. • Complete You Try 6.

 **Work** **Read the explanations, follow the examples, take notes, and complete the You Trys.**

In this section, we will learn how to solve *radical equations.*

An equation containing a variable in the radicand is a **radical equation.** Some examples of radical equations are

$$\sqrt{p} = 7, \qquad \sqrt[3]{n} = 2, \qquad \sqrt{2x + 1} + 1 = x, \qquad \sqrt{5w + 6} - \sqrt{4w + 1} = 1$$

1 Understand the Steps for Solving a Radical Equation

Let's review what happens when we square a square root expression: If $x \geq 0$, then $(\sqrt{x})^2 = x$. That is, to eliminate the radical from $\sqrt{x}$, we *square* the expression. Therefore to solve equations like those above containing *square roots,* we *square* both sides of the equation to obtain new equations. The solutions of the new equations contain all of the solutions of the original equation and may also contain *extraneous solutions.*

An **extraneous solution** is a value that satisfies one of the new equations but does not satisfy the original equation. Extraneous solutions occur frequently when solving radical equations, so we *must* check all possible solutions in the original equation and discard any that are extraneous.

> **Procedure** Solving Radical Equations Containing Square Roots
>
> **Step 1:** Get a radical on a side by itself.
>
> **Step 2:** Square both sides of the equation to eliminate a radical.
>
> **Step 3:** Combine like terms on each side of the equation.
>
> **Step 4:** If the equation still contains a radical, repeat Steps 1–3.
>
> **Step 5:** Solve the equation.
>
> **Step 6:** Check the proposed solutions *in the original equation,* and discard extraneous solutions.

2 Solve an Equation Containing One Square Root

EXAMPLE 1

Solve.

a) $\sqrt{c-2} = 3$ b) $\sqrt{t+5} + 6 = 0$

Solution

a) **Step 1:** The radical *is* on a side by itself: $\sqrt{c-2} = 3$

Step 2: *Square* both sides to eliminate the *square root.*

$$(\sqrt{c-2})^2 = 3^2 \quad \text{Square both sides.}$$
$$c - 2 = 9$$

Steps 3 and 4 do not apply because there are no like terms to combine and no radicals remain.

> **W Hint**
>
> Write out the steps as you are reading them. Be sure to write neatly and in a very orderly way.

Step 5: Solve the equation.

$$c = 11 \quad \text{Add 2 to each side.}$$

Step 6: Check $c = 11$ in the *original* equation.

$$\sqrt{c-2} = 3$$
$$\sqrt{11-2} \overset{?}{=} 3$$
$$\sqrt{9} = 3 \quad \checkmark$$

The solution set is $\{11\}$.

b) The first step is to get the radical on a side by itself.

$$\sqrt{t+5} + 6 = 0$$
$$\sqrt{t+5} = -6 \quad \text{Subtract 6 from each side.}$$
$$(\sqrt{t+5})^2 = (-6)^2 \quad \text{Square both sides to eliminate the radical.}$$
$$t + 5 = 36 \quad \text{The square root has been eliminated.}$$
$$t = 31 \quad \text{Solve the equation.}$$

Check $t = 31$ in the *original* equation.

$$\sqrt{t+5} + 6 = 0$$
$$\sqrt{31+5} + 6 \overset{?}{=} 0$$
$$6 + 6 \overset{?}{=} 0 \quad \text{FALSE}$$

Because $t = 31$ gives us a false statement, it is an *extraneous solution.* The equation has no real solution. The solution set is $\emptyset$.

[YOU TRY 1]
Solve.

a) $\sqrt{a+4}=7$ b) $\sqrt{m-7}+12=9$

Sometimes, we have to square a binomial in order to solve a radical equation. Don't forget that when we square a binomial we can either use FOIL or one of the following formulas: $(a+b)^2 = a^2 + 2ab + b^2$ or $(a-b)^2 = a^2 - 2ab + b^2$.

EXAMPLE 2

Solve $\sqrt{2x+1}+1=x$.

Solution

Start by getting the radical on a side by itself.

$\sqrt{2x+1} = x - 1$	Subtract 1 from each side.
$(\sqrt{2x+1})^2 = (x-1)^2$	Square both sides to eliminate the radical.
$2x + 1 = x^2 - 2x + 1$	Simplify; square the binomial.
$0 = x^2 - 4x$	Subtract $2x$; subtract 1.
$0 = x(x-4)$	Factor.

$x = 0$	or	$x - 4 = 0$	Set each factor equal to zero.
$x = 0$	or	$x = 4$	Solve.

W Hint

Review Section 6.3 if you need help squaring a binomial.

Check $x = 0$ and $x = 4$ in the *original* equation.

$x = 0$: $\sqrt{2x+1}+1=x$ $x = 4$: $\sqrt{2x+1}+1=x$

$\sqrt{2(0)+1}+1 \overset{?}{=} 0$ $\sqrt{2(4)+1}+1 \overset{?}{=} 4$

$\sqrt{1}+1 \overset{?}{=} 0$ $\sqrt{9}+1 \overset{?}{=} 4$

$2 \overset{?}{=} 1$ FALSE $3 + 1 = 4$ TRUE

$x = 4$ *is* a solution but $x = 0$ is **not** because $x = 0$ does not satisfy the original equation. The solution set is $\{4\}$.

[YOU TRY 2]
Solve.

a) $\sqrt{3p+10}-4=p$ b) $\sqrt{4h-3}-h=-2$

3 Solve an Equation Containing Two Square Roots

Next, we will take our first look at solving an equation containing two square roots.

EXAMPLE 3

Solve $\sqrt{2a+4}-3\sqrt{a-5}=0$.

Solution

Begin by getting a radical on a side by itself.

$\sqrt{2a+4} = 3\sqrt{a-5}$	Add $3\sqrt{a-5}$ to each side.
$(\sqrt{2a+4})^2 = (3\sqrt{a-5})^2$	Square both sides to eliminate the radicals.
$2a + 4 = 9(a-5)$	$3^2 = 9$
$2a + 4 = 9a - 45$	Distribute.
$-7a = -49$	
$a = 7$	Solve.

Check $a = 7$ in the original equation.

$$\sqrt{2a + 4} - 3\sqrt{a - 5} = 0$$
$$\sqrt{2(7) + 4} - 3\sqrt{7 - 5} = 0$$
$$\sqrt{14 + 4} - 3\sqrt{2} \stackrel{?}{=} 0$$
$$\sqrt{18} - 3\sqrt{2} \stackrel{?}{=} 0$$
$$3\sqrt{2} - 3\sqrt{2} = 0 \checkmark$$

The solution set is $\{7\}$.

[YOU TRY 3] Solve $4\sqrt{r - 3} - \sqrt{6r + 2} = 0$.

Recall from Section 10.5 that we can square binomials containing radical expressions just like we squared $(x - 1)^2$ in Example 2. We can use FOIL or the formulas

$$(a + b)^2 = a^2 + 2ab + b^2 \qquad \text{or} \qquad (a - b)^2 = a^2 - 2ab + b^2$$

EXAMPLE 4 Square and simplify $(3 - \sqrt{m + 2})^2$.

Solution

Use the formula $(a - b)^2 = a^2 - 2ab + b^2$.

$$(3 - \sqrt{m + 2})^2 = (3)^2 - 2(3)(\sqrt{m + 2}) + (\sqrt{m + 2})^2 \qquad \text{Substitute 3 for } a \text{ and } \sqrt{m + 2} \text{ for } b.$$

$$= 9 - 6\sqrt{m + 2} + (m + 2)$$
$$= m + 11 - 6\sqrt{m + 2} \qquad \text{Combine like terms.}$$

W Hint
Make sure you understand this example before going on to the next.

[YOU TRY 4] Square and simplify each expression.

a) $(\sqrt{z} - 4)^2$ b) $(5 + \sqrt{3d - 1})^2$

To solve the next two equations, we will have to square both sides of the equation twice to eliminate the radicals. Be very careful when you are squaring the binomials that contain a radical.

EXAMPLE 5 Solve each equation.

a) $\sqrt{x + 5} + \sqrt{x} = 5$ b) $\sqrt{5w + 6} - \sqrt{4w + 1} = 1$

Solution

a) This equation contains two radicals *and* a constant. Get one of the radicals on a side by itself, then square both sides.

W Hint
Always use parentheses to organize your work when squaring both sides.

$$\sqrt{x + 5} = 5 - \sqrt{x} \qquad \text{Subtract } \sqrt{x} \text{ from each side.}$$
$$(\sqrt{x + 5})^2 = (5 - \sqrt{x})^2 \qquad \text{Square both sides.}$$
$$x + 5 = (5)^2 - 2(5)(\sqrt{x}) + (\sqrt{x})^2 \qquad \text{Use the formula } (a - b)^2 = a^2 - 2ab + b^2.$$
$$x + 5 = 25 - 10\sqrt{x} + x \qquad \text{Simplify.}$$

The equation still contains a radical. Therefore, repeat Steps 1–3. Begin by getting the radical on a side by itself.

$$5 = 25 - 10\sqrt{x} \qquad \text{Subtract } x \text{ from each side.}$$
$$-20 = -10\sqrt{x} \qquad \text{Subtract 25 from each side.}$$
$$2 = \sqrt{x} \qquad \text{Divide by } -10.$$
$$2^2 = (\sqrt{x})^2 \qquad \text{Square both sides.}$$
$$4 = x \qquad \text{Solve.}$$

The check is left to the student. The solution set is {4}.

W Hint

These problems can be very long, so write out each step *carefully* and *neatly* to minimize mistakes.

b) **Step 1:** Get a radical on a side by itself.

$$\sqrt{5w + 6} - \sqrt{4w + 1} = 1$$
$$\sqrt{5w + 6} = 1 + \sqrt{4w + 1} \qquad \text{Add } \sqrt{4w + 1} \text{ to each side.}$$

Step 2: Square both sides of the equation to eliminate a radical.

$$(\sqrt{5w + 6})^2 = (1 + \sqrt{4w + 1})^2 \qquad \text{Square both sides.}$$
$$5w + 6 = (1)^2 + 2(1)(\sqrt{4w + 1}) + (\sqrt{4w + 1})^2 \qquad \text{Use the formula}$$
$$\qquad\qquad (a+b)^2 = a^2 + 2ab + b^2.$$
$$5w + 6 = 1 + 2\sqrt{4w + 1} + 4w + 1$$

Step 3: Combine like terms on the right side.

$$5w + 6 = 4w + 2 + 2\sqrt{4w + 1} \qquad \text{Combine like terms.}$$

Step 4: **The equation still contains a radical, so repeat Steps 1–3.**

Step 1: Get the radical on a side by itself.

$$5w + 6 = 4w + 2 + 2\sqrt{4w + 1}$$
$$w + 4 = 2\sqrt{4w + 1} \qquad \text{Subtract } 4w \text{ and subtract 2.}$$

We do not need to eliminate the 2 from in front of the radical before squaring both sides. The radical must not be a part of a *sum* or *difference* when we square.

Step 2: Square both sides of the equation to eliminate the radical.

$$(w + 4)^2 = (2\sqrt{4w + 1})^2 \qquad \text{Square both sides.}$$
$$w^2 + 8w + 16 = 4(4w + 1) \qquad \text{Square the binomial; } 2^2 = 4.$$

Steps 3 and 4 no longer apply.

Step 5: Solve the equation.

$$w^2 + 8w + 16 = 16w + 4 \qquad \text{Distribute.}$$
$$w^2 - 8w + 12 = 0 \qquad \text{Subtract } 16w \text{ and subtract 4.}$$
$$(w - 2)(w - 6) = 0 \qquad \text{Factor.}$$
$$w - 2 = 0 \quad \text{or} \quad w - 6 = 0 \qquad \text{Set each factor equal to zero.}$$
$$w = 2 \quad \text{or} \quad w = 6 \qquad \text{Solve.}$$

Step 6: The check is left to the student. Verify that $w = 2$ and $w = 6$ each satisfy the original equation. The solution set is {2, 6}.

[YOU TRY 5] Solve each equation.

a) $\sqrt{2y + 1} - \sqrt{y} = 1$ 　　　　 b) $\sqrt{3t + 4} + \sqrt{t + 2} = 2$

 **BE CAREFUL** Watch out for two common mistakes that students make when solving an equation like the one in Example 5b).

1) Do not square both sides before getting a radical on a side by itself.

This is incorrect: $(\sqrt{5w+6} - \sqrt{4w+1})^2 = 1^2$
$$5w + 6 - (4w + 1) = 1$$

2) The *second* time we perform Step 2, watch out for this common error.

This is incorrect: $(w+4)^2 = (2\sqrt{4w+1})^2$
$$w^2 + 16 = 2(4w + 1)$$

On the left we must multiply using FOIL or the formula $(a + b)^2 = a^2 + 2ab + b^2$, and on the right we must remember to square the 2.

4 Solve an Equation Containing a Cube Root

We can solve many equations containing cube roots the same way we solve equations containing square roots except, to eliminate a *cube root,* we *cube* both sides of the equation.

EXAMPLE 6

Solve $\sqrt[3]{7a+1} - 2\sqrt[3]{a-1} = 0$.

Solution

Begin by getting a radical on a side by itself.

$$\sqrt[3]{7a+1} = 2\sqrt[3]{a-1} \qquad \text{Add } 2\sqrt[3]{a-1} \text{ to each side.}$$
$$(\sqrt[3]{7a+1})^3 = (2\sqrt[3]{a-1})^3 \qquad \text{Cube both sides to eliminate the radicals.}$$
$$7a + 1 = 8(a-1) \qquad \text{Simplify; } 2^3 = 8.$$
$$7a + 1 = 8a - 8 \qquad \text{Distribute.}$$
$$9 = a \qquad \text{Subtract } 7a; \text{ add } 8.$$

> **W Hint**
> Write a procedure to help you solve equations containing a cube root.

Check $a = 9$ in the original equation.

$$\sqrt[3]{7a+1} - 2\sqrt[3]{a-1} = 0$$
$$\sqrt[3]{7(9)+1} - 2\sqrt[3]{9-1} \overset{?}{=} 0$$
$$\sqrt[3]{64} - 2\sqrt[3]{8} \overset{?}{=} 0$$
$$4 - 2(2) \overset{?}{=} 0$$
$$4 - 4 = 0 \checkmark$$

The solution set is $\{9\}$.

[YOU TRY 6]

Solve $3\sqrt[3]{r-4} - \sqrt[3]{5r+2} = 0$.

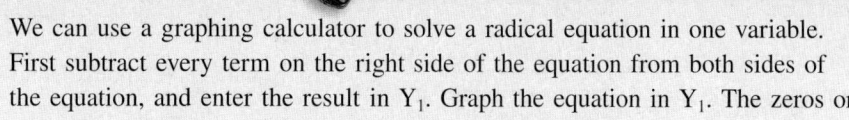

Using Technology

We can use a graphing calculator to solve a radical equation in one variable. First subtract every term on the right side of the equation from both sides of the equation, and enter the result in Y_1. Graph the equation in Y_1. The zeros or x-intercepts of the graph are the solutions to the equation.

We will solve $\sqrt{x + 3} = 2$ using a graphing calculator.

1) Enter $\sqrt{x + 3} - 2$ in Y_1.

2) Press $\boxed{\text{ZOOM}}$ $\boxed{6}$ to graph the function in Y_1 as shown.

3) Press $\boxed{\text{2nd}}$ $\boxed{\text{TRACE}}$ 2:zero, move the cursor to the left of the zero and press $\boxed{\text{ENTER}}$, move the cursor to the right of the zero and press $\boxed{\text{ENTER}}$, and move the cursor close to the zero and press $\boxed{\text{ENTER}}$ to display the zero. The solution to the equation is $x = 1$.

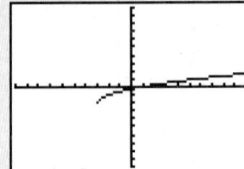 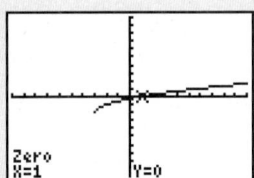

Solve each equation using a graphing calculator.

1) $\sqrt{x - 2} = 1$ 2) $\sqrt{3x - 2} = 5$ 3) $\sqrt{3x - 2} = \sqrt{x} + 2$

4) $\sqrt{4x - 5} = \sqrt{x + 4}$ 5) $\sqrt{2x - 7} = \sqrt{x} - 1$ 6) $\sqrt{\sqrt{x} - 1} = 1$

ANSWERS TO $\boxed{\text{YOU TRY}}$ EXERCISES

1) a) $\{45\}$ b) $\varnothing$ 2) a) $\{-3, -2\}$ b) $\{7\}$ 3) $\{5\}$ 4) a) $z - 8\sqrt{z} + 16$
b) $3d + 24 + 10\sqrt{3d - 1}$ 5) a) $\{0, 4\}$ b) $\{-1\}$ 6) $\{5\}$

ANSWERS TO TECHNOLOGY EXERCISES

1) $\{3\}$ 2) $\{9\}$ 3) $\{9\}$ 4) $\{3\}$ 5) $\{4\}$ 6) $\{4\}$

$\boxed{\text{E Evaluate}}$ **10.7** Exercises   Do the exercises, and check your work.

Objective 1: Understand the Steps for Solving a Radical Equation

1) Why is it necessary to check the proposed solutions to a radical equation in the original equation?

2) How do you know, without actually solving and checking the solution, that $\sqrt{y} = -3$ has no solution?

Objective 2: Solve an Equation Containing One Square Root

Fill in the blank with *always, sometimes,* or *never* to make the statement true.

3) A negative number can _____ be a solution to an equation containing a square root expression.

4) An equation containing a square root expression will _____ have extraneous solutions.

Solve.

5) $\sqrt{q} = 7$ 6) $\sqrt{z} = 10$

7) $\sqrt{w} - \dfrac{2}{3} = 0$ 8) $\sqrt{r} - \dfrac{3}{5} = 0$

9) $\sqrt{a} + 5 = 3$ 10) $\sqrt{k} + 8 = 2$

11) $\sqrt{b - 11} - 3 = 0$ 12) $\sqrt{d + 3} - 5 = 0$

13) $\sqrt{4g - 1} + 7 = 1$ 14) $\sqrt{3v + 4} + 10 = 6$

15) $\sqrt{3f + 2} + 9 = 11$ 16) $\sqrt{5u - 4} + 12 = 17$

17) $m = \sqrt{m^2 - 3m + 6}$ 18) $b = \sqrt{b^2 + 4b - 24}$

19) $\sqrt{9r^2 - 2r + 10} = 3r$ 20) $\sqrt{4p^2 - 3p + 6} = 2p$

Square each binomial, and simplify.

21) $(n + 5)^2$ 22) $(z - 3)^2$

23) $(c - 6)^2$ 24) $(2k + 1)^2$

Solve.

25) $c - 7 = \sqrt{2c + 1}$ 26) $p + 6 = \sqrt{12 + p}$

27) $6 + \sqrt{c^2 + 3c - 9} = c$

28) $-4 + \sqrt{z^2 + 5z - 8} = z$

24) 29) $w - \sqrt{10w + 6} = -3$ 30) $3 - \sqrt{8t + 9} = -t$

31) $3v = 8 + \sqrt{3v + 4}$ 32) $4k = 3 + \sqrt{10k + 5}$

33) $m + 4 = 5\sqrt{m}$ 34) $b + 5 = 6\sqrt{b}$

35) $y + 2\sqrt{6 - y} = 3$

36) $r - 3\sqrt{r + 2} = 2$

37) $\sqrt{r^2 - 8r - 19} = r - 9$

38) $\sqrt{x^2 + x + 4} = x + 8$

Objective 3: Solve an Equation Containing Two Square Roots

Solve.

24) 39) $5\sqrt{1 - 5h} = 4\sqrt{1 - 8h}$

40) $3\sqrt{6a - 2} - 4\sqrt{3a + 3} = 0$

41) $3\sqrt{3x + 6} - 2\sqrt{9x - 9} = 0$

42) $5\sqrt{q + 11} = 2\sqrt{8q + 25}$

43) $\sqrt{m} = 3\sqrt{7}$

44) $4\sqrt{3} = \sqrt{p}$

45) $2\sqrt{3t + 4} + \sqrt{t - 6} = 0$

46) $\sqrt{2w - 1} + 2\sqrt{w + 4} = 0$

Square each expression, and simplify.

24) 47) $(\sqrt{x} + 5)^2$ 48) $(\sqrt{y} - 8)^2$

49) $(9 - \sqrt{a + 4})^2$ 50) $(4 + \sqrt{p + 5})^2$

51) $(2\sqrt{3n - 1} + 7)^2$ 52) $(5 - 3\sqrt{2k - 3})^2$

Solve.

53) $\sqrt{2y - 1} = 2 + \sqrt{y - 4}$

54) $\sqrt{3n + 4} = \sqrt{2n + 1} + 1$

24) 55) $1 + \sqrt{3s - 2} = \sqrt{2s + 5}$

56) $\sqrt{4p + 12} - 1 = \sqrt{6p - 11}$

57) $\sqrt{5a + 19} - \sqrt{a + 12} = 1$

58) $\sqrt{2u + 3} - \sqrt{5u + 1} = -1$

59) $\sqrt{3k + 1} - \sqrt{k - 1} = 2$

60) $\sqrt{4z - 3} - \sqrt{5z + 1} = -1$

61) $\sqrt{3x + 4} - 5 = \sqrt{3x - 11}$

62) $\sqrt{4c - 7} = \sqrt{4c + 1} - 4$

63) $\sqrt{3v + 3} - \sqrt{v - 2} = 3$

64) $\sqrt{2y + 1} - \sqrt{y} = 1$

Objective 4: Solve an Equation Containing a Cube Root

65) How do you eliminate the radical from an equation like $\sqrt[3]{x} = 2$?

66) Give a reason why $\sqrt[3]{h} = -3$ has no extraneous solutions.

Solve.

67) $\sqrt[3]{y} = 5$ 68) $\sqrt[3]{c} = 3$

69) $\sqrt[3]{m} = -4$ 70) $\sqrt[3]{t} = -2$

24) 71) $\sqrt[3]{2x - 5} + 3 = 1$

72) $\sqrt[3]{4a + 1} + 7 = 4$

73) $\sqrt[3]{6j - 2} = \sqrt[3]{j - 7}$

74) $\sqrt[3]{w + 3} = \sqrt[3]{2w - 11}$

75) $\sqrt[3]{3y - 1} - \sqrt[3]{2y - 3} = 0$

76) $\sqrt[3]{2 - 2b} + \sqrt[3]{b - 5} = 0$

77) $\sqrt[3]{2n^2} = \sqrt[3]{7n + 4}$

78) $\sqrt[3]{4c^2 - 5c + 11} = \sqrt[3]{c^2 + 9}$

Extension

Solve.

79) $p^{1/2} = 6$ 80) $\dfrac{2}{3} = t^{1/2}$

81) $7 = (2z - 3)^{1/2}$ 82) $(3k + 1)^{1/2} = 4$

83) $(y + 4)^{1/3} = 3$ 84) $-5 = (a - 2)^{1/3}$

85) $\sqrt[4]{n + 7} = 2$ 86) $\sqrt[4]{x - 3} = -1$

87) $\sqrt{13 + \sqrt{r}} = \sqrt{r + 7}$

88) $\sqrt{m - 1} = \sqrt{m} - \sqrt{m - 4}$

89) $\sqrt{y + \sqrt{y + 5}} = \sqrt{y + 2}$

90) $\sqrt{2d - \sqrt{d + 6}} = \sqrt{d + 6}$

Mixed Exercises: Objectives 2 and 4

Solve for the indicated variable.

24) 91) $v = \sqrt{\dfrac{2E}{m}}$ for E 92) $V = \sqrt{\dfrac{300VP}{m}}$ for P

93) $c = \sqrt{a^2 + b^2}$ for b^2 94) $r = \sqrt{\dfrac{A}{\pi}}$ for A

95) $T = \sqrt[4]{\dfrac{E}{\sigma}}$ for σ

96) $r = \sqrt[3]{\dfrac{3V}{4\pi}}$ for V

97) The speed of sound is proportional to the square root of the air temperature in still air. The speed of sound is given by the formula.

$$V_S = 20\sqrt{T + 273}$$

where V_S is the speed of sound in meters/second and T is the temperature of the air in °Celsius.

a) What is the speed of sound when the temperature is $-17°C$ (about 1°F)?

b) What is the speed of sound when the temperature is 16°C (about 61°F)?

c) What happens to the speed of sound as the temperature increases?

d) Solve the equation for T.

98) If the area of a square is A and each side has length l, then the length of a side is given by

$$l = \sqrt{A}$$

A square rug has an area of 25 ft².

a) Find the dimensions of the rug.

b) Solve the equation for A.

99) Let V represent the volume of a cylinder, h represent its height, and r represent its radius. V, h, and r are related according to the formula

$$r = \sqrt{\dfrac{V}{\pi h}}$$

a) A cylindrical soup can has a volume of 28π in³. It is 7 in. high. What is the radius of the can?

b) Solve the equation for V.

100) For shallow water waves, the wave velocity is given by
$$c = \sqrt{gH}$$

Goodshoot/Alamy Stock Photo

where g is the acceleration due to gravity (32 ft/sec²) and H is the depth of the water (in feet).

a) Find the velocity of a wave in 8 ft of water.

b) Solve the equation for H.

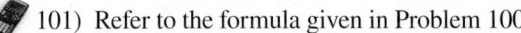

101) Refer to the formula given in Problem 100.

The catastrophic Indian Ocean tsunami that hit Banda Aceh, Sumatra, Indonesia, on December 26, 2004 was caused by an earthquake whose epicenter was off the coast of northern Sumatra. The tsunami originated in about 14,400 ft of water.

a) Find the velocity of the wave near the epicenter, in miles per hour. Round the answer to the nearest unit. (Hint: 1 mile = 5280 ft.)

b) Banda Aceh, the area hardest hit by the tsunami, was about 60 mi from the tsunami's origin. Approximately how many minutes after the earthquake occurred did the tsunami hit Banda Aceh? (*Exploring Geology*, McGraw-Hill, 2008.)

102) The radius r of a cone with height h and volume V is given by $r = \sqrt{\dfrac{3V}{\pi h}}$.
A hanging glass vase in the shape of a cone is 8 in. tall, and the radius of the top of the cone is 2 in. How much water will the vase hold? Give an exact answer and an approximation to the tenths place.

Use the following information for Exercises 103 and 104.

The distance a person can see to the horizon is approximated by the function $D(h) = 1.2\sqrt{h}$, where D is the number of miles a person can see to the horizon from a height of h ft.

103) Sig is the captain of an Alaskan crab fishing boat and can see 4.8 mi to the horizon when he is sailing his ship. Find his height above the sea.

104) Phil is standing on the deck of a boat and can see 3.6 mi to the horizon. What is his height above the water?

Use the following information for Exercises 105 and 106.

When the air temperature is 0°F, the wind chill temperature, W, in degrees Fahrenheit is a function of the velocity of the wind, V, in miles per hour and is given by the formula

$$W(V) = 35.74 - 35.75V^{4/25}$$

105) Calculate the wind speed when the wind chill temperature is $-10°F$. Round to the nearest whole number.

106) Find V so that $W(V) = -20$. Round to the nearest whole number. Explain your result in the context of the problem. (http://www.nws.noaa.gov/om/windchill/windchillglossary.shtml)

R1) How has a good understanding of binomials helped you complete these exercises?

R2) Could you explain to a friend how to solve these problems?

10.8 Complex Numbers

P Prepare

O Organize

What are your objectives for Section 10.8?	How can you accomplish each objective?
1 Find the Square Root of a Negative Number	• Understand the definition of the *imaginary number, i.* • Learn the definition of a *complex number.* • Understand the property of **Real Numbers and Complex Numbers.** • Complete the given example on your own. • Complete You Try 1.
2 Multiply and Divide Square Roots Containing Negative Numbers	• Follow the explanation and example to create a procedure for this objective. • Complete the given example on your own. • Complete You Try 2.
3 Add and Subtract Complex Numbers	• Write the procedure for **Adding and Subtracting Complex Numbers** in your own words. • Complete the given example on your own. • Complete You Try 3.
4 Multiply Complex Numbers	• Compare the steps for multiplying complex numbers to the notes you took on multiplying polynomials. • Complete the given example on your own. • Complete You Try 4.
5 Multiply a Complex Number by Its Conjugate	• Understand the definition of a *complex conjugate.* • Follow the explanation, and write the summary of **Complex Conjugates** in your notes. • Complete the given example on your own. • Complete You Try 5.
6 Divide Complex Numbers	• Write the procedure for **Dividing Complex Numbers** in your own words. • Complete the given example on your own. • Complete You Try 6.
7 Simplify Powers of i	• Follow the explanation, and summarize how to simplify powers of i in your notes. • Complete the given example on your own. • Complete You Try 7.

Read the explanations, follow the examples, take notes, and complete the You Trys.

1 Find the Square Root of a Negative Number

We have seen throughout this chapter that the square root of a negative number does not exist in the real number system because there is no real number that, when squared, will result in a negative number. For example, $\sqrt{-4}$ is not a real number because there is no real number whose square is -4.

The square roots of negative numbers do exist, however, under another system of numbers called *complex numbers*. Before we define a complex number, we must define the number i. The number i is called an *imaginary number*.

Definition

The **imaginary number** i is defined as

$$i = \sqrt{-1}$$

Therefore, squaring both sides gives us

$$i^2 = -1$$

Note

$i = \sqrt{-1}$ and $i^2 = -1$ are two *very* important facts to remember. We will be using them often!

Definition

A **complex number** is a number of the form $a + bi$, where a and b are real numbers; a is called the **real part,** and b is called the **imaginary part.**

The following table lists some examples of complex numbers and their real and imaginary parts.

Complex Number	Real Part	Imaginary Part
$-5 + 2i$	-5	2
$\dfrac{1}{3} - 7i$	$\dfrac{1}{3}$	-7
$8i$	0	8
4	4	0

Note

The complex number $8i$ can be written in the form $a + bi$ as $0 + 8i$.
Likewise, besides being a real number, 4 is a complex number because it can be written as $4 + 0i$.

Because all real numbers, a, can be written in the form $a + 0i$, all real numbers are also complex numbers.

> **Property** Real Numbers and Complex Numbers
>
> The set of real numbers is a subset of the set of complex numbers.

Since we defined i as $i = \sqrt{-1}$, we can now evaluate square roots of negative numbers.

EXAMPLE 1

Simplify.

a) $\sqrt{-9}$ b) $\sqrt{-7}$ c) $\sqrt{-12}$

Solution

a) $\sqrt{-9} = \sqrt{-1 \cdot 9} = \sqrt{-1} \cdot \sqrt{9} = i \cdot 3 = 3i$

b) $\sqrt{-7} = \sqrt{-1 \cdot 7} = \sqrt{-1} \cdot \sqrt{7} = i\sqrt{7}$

c) $\sqrt{-12} = \sqrt{-1 \cdot 12} = \sqrt{-1} \cdot \sqrt{12} = i\sqrt{4}\sqrt{3} = i \cdot 2\sqrt{3} = 2i\sqrt{3}$

 **Hint**

Create a procedure for simplifying square roots of negative numbers.

 **Note**

In Example 1b) we wrote $i\sqrt{7}$ instead of $\sqrt{7}i$, and in Example 1c) we wrote $2i\sqrt{3}$ instead of $2\sqrt{3}i$. We do this to be clear that the i is not under the radical. It is good practice to write the i *before* the radical.

YOU TRY 1

Simplify.

a) $\sqrt{-36}$ b) $\sqrt{-13}$ c) $\sqrt{-20}$

2 Multiply and Divide Square Roots Containing Negative Numbers

When multiplying or dividing square roots with negative radicands, write each radical in terms of i first. Remember, also, that since $i = \sqrt{-1}$ it follows that $i^2 = -1$. We must keep this in mind when simplifying expressions.

 **Note**

Whenever an i^2 appears in an expression, replace it with -1.

EXAMPLE 2

Multiply and simplify. $\sqrt{-8} \cdot \sqrt{-2}$

Solution

$$\sqrt{-8} \cdot \sqrt{-2} = i\sqrt{8} \cdot i\sqrt{2} \qquad \text{Write each radical in terms of } i \text{ before multiplying.}$$
$$= i^2\sqrt{16} \qquad \text{Multiply.}$$
$$= (-1)(4) \qquad \text{Replace } i^2 \text{ with } -1.$$
$$= -4$$

[**YOU TRY 2**]
Perform the operation, and simplify.

a) $\sqrt{-6} \cdot \sqrt{-3}$ b) $\dfrac{\sqrt{-72}}{\sqrt{-2}}$

3 Add and Subtract Complex Numbers

Just as we can add, subtract, multiply, and divide real numbers, we can perform all of these operations with complex numbers.

Procedure Adding and Subtracting Complex Numbers

1) To add complex numbers, add the real parts and add the imaginary parts.

2) To subtract complex numbers, apply the distributive property and combine the real parts and combine the imaginary parts.

EXAMPLE 3

Add or subtract.

a) $(8 + 3i) + (4 + 2i)$ b) $(7 + i) - (3 - 4i)$

Solution

a) $(8 + 3i) + (4 + 2i) = (8 + 4) + (3 + 2)i$ Add real parts; add imaginary parts.
$= 12 + 5i$

b) $(7 + i) - (3 - 4i) = 7 + i - 3 + 4i$ Distributive property
$= (7 - 3) + (1 + 4)i$ Add real parts; add imaginary parts.
$= 4 + 5i$

[**YOU TRY 3**]
Add or subtract.

a) $(-10 + 6i) + (1 + 8i)$ b) $(2 - 5i) - (-1 + 6i)$

4 Multiply Complex Numbers

We multiply complex numbers just like we would multiply polynomials. There may be an additional step, however. Remember to replace i^2 with -1.

EXAMPLE 4

Multiply and simplify.

a) $5(-2 + 3i)$ b) $(8 + 3i)(-1 + 4i)$ c) $(6 + 2i)(6 - 2i)$

Solution

a) $5(-2 + 3i) = -10 + 15i$ Distributive property

W **Hint**

How does this compare to the techniques you learned in Section 6.3?

b) Look carefully at $(8 + 3i)(-1 + 4i)$. Each complex number has two terms, similar to, say, $(x + 3)(x + 4)$. How can we multiply these two binomials? We can use FOIL.

$$\begin{array}{cccc} & \text{F} & \text{O} & \text{I} & \text{L} \\ (8 + 3i)(-1 + 4i) = & (8)(-1) + (8)(4i) + (3i)(-1) + (3i)(4i) \\ = & -8 \;\;+\;\; 32i \;\;-\;\; 3i \;\;+\;\; 12i^2 \end{array}$$
$$= -8 + 29i + 12(-1) \qquad \text{Replace } i^2 \text{ with } -1.$$
$$= -8 + 29i - 12$$
$$= -20 + 29i$$

c) Use FOIL to find the product $(6 + 2i)(6 - 2i)$.

$$\begin{array}{cccc} & \text{F} & \text{O} & \text{I} & \text{L} \\ (6 + 2i)(6 - 2i) = & (6)(6) + (6)(-2i) + (2i)(6) + (2i)(-2i) \\ = & 36 \;\;-\;\; 12i \;\;+\;\; 12i \;\;-\;\; 4i^2 \end{array}$$
$$= 36 - 4(-1) \qquad \text{Replace } i^2 \text{ with } -1.$$
$$= 36 + 4$$
$$= 40$$

[**YOU TRY 4**]

Multiply and simplify.

a) $-3(6 - 7i)$ b) $(5 + i)(4 + 8i)$ c) $(-2 - 9i)(-2 + 9i)$

5   Multiply a Complex Number by Its Conjugate

In Section 10.6, we learned about conjugates of radical expressions. For example, the conjugate of $3 + \sqrt{5}$ is $3 - \sqrt{5}$.

The complex numbers in Example 4c), $6 + 2i$ and $6 - 2i$, are **complex conjugates.**

Definition

The **conjugate** of $a + bi$ is $a - bi$.

We found that $(6 + 2i)(6 - 2i) = 40$, which is a real number. **The product of a complex number and its conjugate is *always* a real number,** as illustrated next.

$$\begin{array}{cccc} & \text{F} & \text{O} & \text{I} & \text{L} \\ (a + bi)(a - bi) = & (a)(a) + (a)(-bi) + (bi)(a) + (bi)(-bi) \\ = & a^2 \;\;-\;\; abi \;\;+\;\; abi \;\;-\;\; b^2i^2 \end{array}$$
$$= a^2 - b^2(-1) \qquad \text{Replace } i^2 \text{ with } -1.$$
$$= a^2 + b^2$$

We can summarize these facts about complex numbers and their conjugates as follows:

Summary   Complex Conjugates

1) The conjugate of $a + bi$ is $a - bi$.

2) The product of $a + bi$ and $a - bi$ is a real number.

3) We can find the product $(a + bi)(a - bi)$ by using FOIL or by using
$(a + bi)(a - bi) = a^2 + b^2$.

EXAMPLE 5

Multiply $-3 + 4i$ by its conjugate using the formula $(a + bi)(a - bi) = a^2 + b^2$.

Solution

The conjugate of $-3 + 4i$ is $-3 - 4i$.

$$(-3 + 4i)(-3 - 4i) = (-3)^2 + (4)^2 \qquad a = -3, b = 4$$
$$= 9 + 16$$
$$= 25$$

W Hint
Always use parentheses!

[YOU TRY 5]   Multiply $2 - 9i$ by its conjugate using the formula $(a + bi)(a - bi) = a^2 + b^2$.

---

6 Divide Complex Numbers

To rationalize the denominator of a radical expression like $\dfrac{2}{3 + \sqrt{5}}$, we multiply the numerator and denominator by $3 - \sqrt{5}$, the conjugate of the denominator. We divide complex numbers in the same way.

> **Procedure** Dividing Complex Numbers
>
> To divide complex numbers, multiply the numerator and denominator by the *conjugate of the denominator*. Write the quotient in the form $a + bi$.

EXAMPLE 6

Divide. Write the quotient in the form $a + bi$.

a) $\dfrac{3}{4 - 5i}$ b) $\dfrac{6 - 2i}{-7 + i}$

Solution

a) $\dfrac{3}{4 - 5i} = \dfrac{3}{(4 - 5i)} \cdot \dfrac{(4 + 5i)}{(4 + 5i)}$ Multiply the numerator and denominator by the conjugate of the denominator.

$= \dfrac{12 + 15i}{(4)^2 + (5)^2}$ Multiply numerators.

$\qquad\qquad\qquad$ $(a + bi)(a - bi) = a^2 + b^2$

$= \dfrac{12 + 15i}{16 + 25}$

$= \dfrac{12 + 15i}{41}$

$= \dfrac{12}{41} + \dfrac{15}{41}i$ Write the quotient in the form $a + bi$.

Recall that we can find the product $(4 - 5i)(4 + 5i)$ using FOIL *or* by using the formula $(a + bi)(a - bi) = a^2 + b^2$.

b) $\dfrac{6 - 2i}{-7 + i} = \dfrac{(6 - 2i)}{(-7 + i)} \cdot \dfrac{(-7 - i)}{(-7 - i)}$ Multiply the numerator and denominator by the conjugate of the denominator.

$= \dfrac{-42 - 6i + 14i + 2i^2}{(-7)^2 + (1)^2}$ Multiply using FOIL.

$\qquad\qquad\qquad$ $(a + bi)(a - bi) = a^2 + b^2$

$= \dfrac{-42 + 8i - 2}{49 + 1} = \dfrac{-44 + 8i}{50} = -\dfrac{44}{50} + \dfrac{8}{50}i = -\dfrac{22}{25} + \dfrac{4}{25}i$

[YOU TRY 6]

Divide. Write the result in the form $a + bi$.

a) $\dfrac{9}{2 + i}$ b) $\dfrac{5 + 3i}{-6 - 4i}$

7 Simplify Powers of i

All powers of i larger than i^1 (or just i) can be simplified. We use the fact that $i^2 = -1$ to simplify powers of i.

Let's write i through i^4 in their simplest forms.

i is in simplest form.
$i^2 = -1$
$i^3 = i^2 \cdot i = -1 \cdot i = -i$
$i^4 = (i^2)^2 = (-1)^2 = 1$

Let's continue by simplifying i^5 and i^6.

$$i^5 = i^4 \cdot i \qquad\qquad i^6 = (i^2)^3$$
$$= (i^2)^2 \cdot i \qquad\qquad = (-1)^3$$
$$= (-1)^2 \cdot i \qquad\qquad = -1$$
$$= 1i$$
$$= i$$

The pattern repeats so that all powers of i can be simplified to i, -1, $-i$, or 1.

EXAMPLE 7

Simplify each power of i.

a) i^8 b) i^{14} c) i^{11} d) i^{37}

Solution

a) Use the power rule for exponents to simplify i^8. Since the exponent is even, we can rewrite it in terms of i^2.

$$i^8 = (i^2)^4 \qquad \text{Power rule}$$
$$= (-1)^4 \qquad i^2 = -1$$
$$= 1 \qquad \text{Simplify.}$$

b) As in Example 7a), the exponent is even. Rewrite i^{14} in terms of i^2.

$$i^{14} = (i^2)^7 \qquad \text{Power rule}$$
$$= (-1)^7 \qquad i^2 = -1$$
$$= -1 \qquad \text{Simplify.}$$

c) The exponent of i^{11} is odd, so first use the product rule to write i^{11} as a product of i and i^{11-1} or i^{10}.

$$i^{11} = i^{10} \cdot i \qquad \text{Product rule}$$
$$= (i^2)^5 \cdot i \qquad \begin{array}{l}\text{10 is even; write}\\ i^{10} \text{ in terms of } i^2.\end{array}$$
$$= (-1)^5 \cdot i \qquad i^2 = -1$$
$$= -1 \cdot i \qquad \text{Simplify.}$$
$$= -i \qquad \text{Multiply.}$$

d) The exponent of i^{37} is odd. Use the product rule to write i^{37} as a product of i and i^{37-1} or i^{36}.

$$i^{37} = i^{36} \cdot i \qquad \text{Product rule}$$
$$= (i^2)^{18} \cdot i \qquad \begin{array}{l}\text{36 is even; write}\\ i^{36} \text{ in terms of } i^2.\end{array}$$
$$= (-1)^{18} \cdot i \qquad i^2 = -1$$
$$= 1 \cdot i \qquad \text{Simplify.}$$
$$= i \qquad \text{Multiply.}$$

[YOU TRY 7]

Simplify each power of i.

a) i^{18} b) i^{32} c) i^7 d) i^{25}

We can use a graphing calculator to perform operations on complex numbers or to evaluate square roots of negative numbers.

If the calculator is in the default REAL mode the result is an error message "ERR: NONREAL ANS," which indicates that $\sqrt{-4}$ is a complex number rather than a real number. Before evaluating $\sqrt{-4}$ on the home screen of your calculator, check the mode by pressing MODE and looking at row 7. Change the mode to complex numbers by selecting $a + bi$, as shown at the left below.

Now evaluating $\sqrt{-4}$ on the home screen results in the correct answer $2i$, as shown on the right below.

 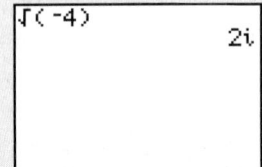

Operations can be performed on complex numbers with the calculator in either REAL or $a + bi$ mode. Simply use the arithmetic operators on the right column on your calculator. To enter the imaginary number i, press 2nd . . To add $2 - 5i$ and $4 + 3i$, enter $(2 - 5i) + (4 + 3i)$ on the home screen and press ENTER as shown on the left screen below. To subtract $8 + 6i$ from $7 - 2i$, enter $(7 - 2i) - (8 + 6i)$ on the home screen and press ENTER as shown.

To multiply $3 - 5i$ and $7 + 4i$, enter $(3 - 5i) \cdot (7 + 4i)$ on the home screen and press ENTER as shown on the middle screen below. To divide $2 + 9i$ by $2 - i$, enter $(2 + 9i)/(2 - i)$ on the home screen and press ENTER as shown.

To raise $3 - 4i$ to the fifth power, enter $(3 - 4i)^5$ on the home screen and press ENTER as shown.

Consider the quotient $(5 + 3i)/(4 - 7i)$. The exact answer is $-\dfrac{1}{65} + \dfrac{47}{65}i$. The calculator automatically displays the decimal result. Press MATH 1 ENTER to convert the decimal result to the exact fractional result, as shown on the right screen below.

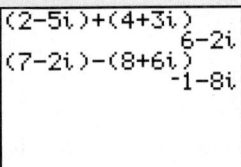 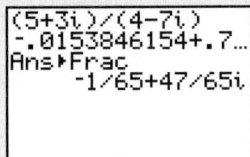

Perform the indicated operation using a graphing calculator.

1) Simplify $\sqrt{-36}$ 2) $(3 + 7i) + (5 - 8i)$ 3) $(10 - 3i) - (4 + 8i)$

4) $(3 + 2i)(6 - 3i)$ 5) $(4 + 3i) \div (1 - i)$ 6) $(5 - 3i)^3$

ANSWERS TO [YOU TRY] **EXERCISES**

1) a) $6i$ b) $i\sqrt{13}$ c) $2i\sqrt{5}$ 2) a) $-3\sqrt{2}$ b) 6 3) a) $-9 + 14i$ b) $3 - 11i$

4) a) $-18 + 21i$ b) $12 + 44i$ c) 85 5) 85 6) a) $\dfrac{18}{5} - \dfrac{9}{5}i$ b) $-\dfrac{21}{26} + \dfrac{1}{26}i$

7) a) -1 b) 1 c) $-i$ d) i

E Evaluate **10.8** Exercises   Do the exercises, and check your work.

Objective 1: Find the Square Root of a Negative Number

Determine if each statement is true or false.

1) Every complex number is a real number.

2) Every real number is a complex number.

3) Since $i = \sqrt{-1}$, it follows that $i^2 = -1$.

4) In the complex number $-6 + 5i$, -6 is the real part and $5i$ is the imaginary part.

Simplify.

 5) $\sqrt{-81}$ 6) $\sqrt{-16}$

7) $\sqrt{-25}$ 8) $\sqrt{-169}$

9) $\sqrt{-6}$ 10) $\sqrt{-30}$

11) $\sqrt{-27}$ 12) $\sqrt{-75}$

13) $\sqrt{-60}$ 14) $\sqrt{-28}$

Objective 2: Multiply and Divide Square Roots Containing Negative Numbers

Find the error in each of the following exercises, then find the correct answer.

15) $\sqrt{-5} \cdot \sqrt{-10} = \sqrt{-5 \cdot (-10)}$
$= \sqrt{50}$
$= \sqrt{25} \cdot \sqrt{2}$
$= 5\sqrt{2}$

16) $(\sqrt{-7})^2 = \sqrt{(-7)^2}$
$= \sqrt{49}$
$= 7$

Perform the indicated operation, and simplify.

17) $\sqrt{-1} \cdot \sqrt{-5}$ 18) $\sqrt{-5} \cdot \sqrt{-15}$

19) $\sqrt{-20} \cdot \sqrt{-5}$ 20) $\sqrt{-12} \cdot \sqrt{-3}$

21) $\dfrac{\sqrt{-60}}{\sqrt{-15}}$ 22) $\dfrac{\sqrt{-2}}{\sqrt{-128}}$

23) $(\sqrt{-13})^2$ 24) $(\sqrt{-1})^2$

Mixed Exercises: Objectives 3–6

25) Explain how to add complex numbers.

26) How is multiplying $(1 + 3i)(2 - 7i)$ similar to multiplying $(x + 3)(2x - 7)$?

27) When i^2 appears in an expression, it should be replaced with what?

28) Explain how to divide complex numbers.

Objective 3: Add and Subtract Complex Numbers

Perform the indicated operations.

 29) $(-4 + 9i) + (7 + 2i)$ 30) $(6 + i) + (8 - 5i)$

31) $(13 - 8i) - (9 + i)$ 32) $(-12 + 3i) - (-7 - 6i)$

33) $\left(-\dfrac{3}{4} - \dfrac{1}{6}i\right) - \left(-\dfrac{1}{2} + \dfrac{2}{3}i\right)$

34) $\left(\dfrac{1}{2} + \dfrac{7}{9}i\right) - \left(\dfrac{7}{8} - \dfrac{1}{6}i\right)$

35) $16i - (3 + 10i) + (3 + i)$

36) $(-6 - 5i) + (2 + 6i) - (-4 + i)$

Objective 4: Multiply Complex Numbers

Multiply and simplify.

37) $3(8 - 5i)$ 38) $-6(8 - i)$

39) $\dfrac{2}{3}(-9 + 2i)$ 40) $\dfrac{1}{2}(18 + 7i)$

41) $-4i(6 + 11i)$ 42) $6i(5 + 6i)$

43) $(2 + 5i)(1 + 6i)$ 44) $(2 + i)(10 + 5i)$

45) $(-1 + 3i)(4 - 6i)$ 46) $(-4 - 9i)(3 - i)$

47) $(5 - 3i)(9 - 3i)$ 48) $(3 - 4i)(6 + 7i)$

49) $\left(\dfrac{3}{4} + \dfrac{3}{4}i\right)\left(\dfrac{2}{5} + \dfrac{1}{5}i\right)$ 50) $\left(\dfrac{1}{3} - \dfrac{4}{3}i\right)\left(\dfrac{3}{4} + \dfrac{2}{3}i\right)$

Objective 5: Multiply a Complex Number by Its Conjugate

Identify the conjugate of each complex number, then multiply the number and its conjugate.

51) $11 + 4i$ 52) $-1 - 2i$

53) $-3 - 7i$ 54) $4 + 9i$

55) $-6 + 4i$ 56) $6 - 5i$

57) How are conjugates of complex numbers like conjugates of expressions containing real numbers and radicals?

58) Is the product of two complex numbers always a complex number? Explain your answer.

Objective 6: Divide Complex Numbers
Divide. Write the result in the form $a + bi$.

 59) $\dfrac{4}{2 - 3i}$

60) $\dfrac{-10}{8 - 9i}$

61) $\dfrac{8i}{4 + i}$

62) $\dfrac{i}{6 - 5i}$

63) $\dfrac{2i}{-3 + 7i}$

64) $\dfrac{9i}{-4 + 10i}$

65) $\dfrac{3 - 8i}{-6 + 7i}$

66) $\dfrac{-5 + 2i}{4 - i}$

67) $\dfrac{2 + 3i}{5 - 6i}$

68) $\dfrac{1 + 6i}{5 + 2i}$

69) $\dfrac{9}{i}$

70) $\dfrac{16 + 3i}{-i}$

Objective 7: Simplify Powers of i
Simplify each power of i.

Fill It In
Fill in the blanks with either the missing mathematical step or reason for the given step.

71) $i^{24} = $ _____   Rewrite i^{24} in terms of i^2 using the power rule.

$= (-1)^{12}$ _____

$= $ _   Simplify.

72) $i^{31} = i^{30} \cdot i$ _____

$= (i^2)^{15} \cdot i$

$= $ _____ $i^2 = -1$

$= $ _   Simplify.

$= $ _   Multiply.

73) i^{24}

74) i^{16}

75) i^{28}

76) i^{30}

77) i^9

78) i^{19}

79) i^{35}

80) i^{29}

81) i^{23}

82) i^{40}

83) i^{42}

84) i^{33}

85) $(2i)^5$

86) $(2i)^6$

87) $(-i)^{14}$

88) $(-i)^{15}$

Expand.

89) $(-2 + 5i)^3$

90) $(3 - 4i)^3$

Simplify each expression. Write the result in the form $a + bi$.

91) $1 + \sqrt{-8}$

92) $-7 - \sqrt{-48}$

93) $8 - \sqrt{-45}$

94) $3 + \sqrt{-20}$

95) $\dfrac{-12 + \sqrt{-32}}{4}$

96) $\dfrac{21 - \sqrt{-18}}{3}$

Used in the field of electronics, the **impedance, Z,** is the total opposition to the current flow of an alternating current (AC) within an electronic component, circuit, or system. It is expressed as a complex number $Z = R + Xj$, where the i used to represent an imaginary number in most areas of mathematics is replaced by j in electronics. R represents the resistance of a substance, and X represents the reactance.

The **total impedance, Z,** of components connected in series is the *sum* of the individual impedances of each component.

Each exercise contains the impedance of individual circuits. Find the total impedance of a system formed by connecting the circuits in series by finding the sum of the individual impedances.

97) $Z_1 = 3 + 2j$
$Z_2 = 7 + 4j$

98) $Z_1 = 5 + 3j$
$Z_2 = 9 + 6j$

99) $Z_1 = 5 - 2j$
$Z_2 = 11 + 6j$

100) $Z_1 = 4 - 1.5j$
$Z_2 = 3 + 0.5j$

R Rethink

R1) Did you check your answers manually before looking up the answers?

R2) Which learning objectives from previous chapters helped you master the concepts of this exercise set?

Group Activity – Radical Equations

The science of forensics is used in accident investigations. The speed of a car before brakes are applied can be approximated from the length of its skid marks by using the following equation:

$$v = 5.47\sqrt{\mu d} \quad \text{where} \quad \begin{aligned} v &= \text{speed of the car (in mph)} \\ d &= \text{length of skid marks (in feet)} \\ \mu &= \text{coefficient of friction determined by road conditions} \end{aligned}$$

Graph the equation $v = 5.47\sqrt{\mu d}$ by filling in the table and plotting points if the pavement is dry (the coefficient of friction is 0.7).

d	$v = 5.47\sqrt{\mu d}$
0	
50	
100	
150	
200	
250	
300	
400	
500	

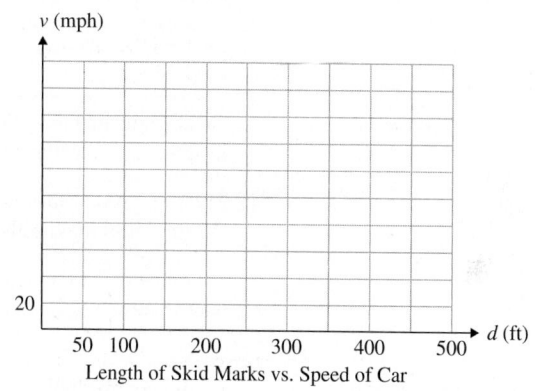

Length of Skid Marks vs. Speed of Car

Graph the equation if the coefficient of friction on wet pavement is 0.4.

As the road conditions become more slippery, how does the graph change?

Think about a time you have worked in a study group. Did it work well for you? Why or why not? Complete this survey to see how effective it was and to help you think about the different elements involved when working or studying in a group.

Statement	Strongly Agree	Agree	Disagree	Strongly Disagree
1. I knew the material better after meeting with the study group than before I walked in.				
2. I helped the members of my group to succeed.				
3. Other group members offered me new insights that helped me to succeed.				
4. My group stayed focused throughout our time working together.				
5. Everyone in the group tried to contribute as much as he or she could.				
6. By working as a team, we ensured that we did not overlook any major topics.				
7. Studying with my classmates was more fun than studying alone.				
8. We had the right number of people in our study group.				
9. I enjoyed the experience of helping members of my group to learn.				
10. I was motivated to help the others in my group to do well.				

Based on the results of this survey, try to identify what worked in your study group and what did not. Consider whether the issues apply to a particular study group or to your group work in general. Apply your conclusions to your future group work, both in academic contexts and beyond.

Chapter 10: Summary

Definition/Procedure	Example

10.1 Finding Roots

If the radicand is a perfect square, then the square root is a *rational* number.	$\sqrt{49} = 7$ since $7^2 = 49$.
If the radicand is a negative number, then the square root is *not* a real number.	$\sqrt{-36}$ is not a real number.
If the radicand is positive and not a perfect square, then the square root is an *irrational* number.	$\sqrt{7}$ is irrational because 7 is not a perfect square.
The symbol $\sqrt[n]{a}$ is read as "the *n*th root of *a*." If $\sqrt[n]{a} = b$, then $b^n = a$. We call *n* the **index** of the radical.	$\sqrt[5]{32} = 2$ since $2^5 = 32$.
For any *positive* number *a* and any *even* index *n*, the **principal *n*th root** of *a* is $\sqrt[n]{a}$, and the **negative *n*th root** of *a* is $-\sqrt[n]{a}$.	$\sqrt[4]{16} = 2$ $-\sqrt[4]{16} = -2$
The *odd root* of a negative number is a negative number.	$\sqrt[3]{-125} = -5$ since $(-5)^3 = -125$.
The *even root* of a negative number is not a real number.	$\sqrt[4]{-16}$ is not a real number.
If *n* is a positive, *even* integer, then $\sqrt[n]{a^n} = \lvert a \rvert$. If *n* is a positive, *odd* integer, then $\sqrt[n]{a^n} = a$.	$\sqrt[4]{(-2)^4} = \lvert -2 \rvert = 2$ $\sqrt[3]{(-2)^3} = -2$

10.2 Rational Exponents

If *n* is a positive integer greater than 1 and $\sqrt[n]{a}$ is a real number, then $a^{1/n} = \sqrt[n]{a}$.	$8^{1/3} = \sqrt[3]{8} = 2$
If *m* and *n* are positive integers and $\frac{m}{n}$ is in lowest terms, then $a^{m/n} = (a^{1/n})^m = (\sqrt[n]{a})^m$ provided that $a^{1/n}$ is a real number.	$16^{3/4} = (\sqrt[4]{16})^3 = 2^3 = 8$
If $a^{m/n}$ is a nonzero real number, then $$a^{-m/n} = \left(\frac{1}{a}\right)^{m/n} = \frac{1}{a^{m/n}}.$$	$25^{-3/2} = \left(\frac{1}{25}\right)^{3/2} = \left(\sqrt{\frac{1}{25}}\right)^3 = \left(\frac{1}{5}\right)^3 = \frac{1}{125}$ **BE CAREFUL** The negative exponent does not make the expression negative.

Definition/Procedure	Example

10.3 Simplifying Expressions Containing Square Roots

Product Rule for Square Roots

Let a and b be nonnegative real numbers. Then,
$\sqrt{a} \cdot \sqrt{b} = \sqrt{a \cdot b}$.

$\sqrt{5} \cdot \sqrt{7} = \sqrt{5 \cdot 7} = \sqrt{35}$

An expression containing a square root is simplified when all of the following conditions are met:

1) The radicand does not contain any factors (other than 1) that are perfect squares.
2) The radicand does not contain any fractions.
3) There are no radicals in the denominator of a fraction.

To *simplify square roots,* rewrite using the product rule as $\sqrt{a \cdot b} = \sqrt{a} \cdot \sqrt{b}$, where a or b is a perfect square.

After simplifying a radical, look at the result and ask yourself, "*Is the radical in simplest form?*" If it is not, simplify again.

Simplify $\sqrt{24}$.

$$\begin{aligned}
\sqrt{24} &= \sqrt{4 \cdot 6} && \text{4 is a perfect square.}\\
&= \sqrt{4} \cdot \sqrt{6} && \text{Product rule}\\
&= 2\sqrt{6} && \sqrt{4} = 2
\end{aligned}$$

Quotient Rule for Square Roots

Let a and b be nonnegative real numbers such that $b \neq 0$.
Then, $\sqrt{\dfrac{a}{b}} = \dfrac{\sqrt{a}}{\sqrt{b}}$.

$$\begin{aligned}
\sqrt{\frac{72}{25}} &= \frac{\sqrt{72}}{\sqrt{25}} && \text{Quotient rule}\\
&= \frac{\sqrt{36} \cdot \sqrt{2}}{5} && \text{Product rule; } \sqrt{25} = 5\\
&= \frac{6\sqrt{2}}{5} && \sqrt{36} = 6
\end{aligned}$$

If a is a nonnegative real number and m is an integer, then $\sqrt{a^m} = a^{m/2}$.

$\sqrt{k^{18}} = k^{18/2} = k^9$
(provided k represents a nonnegative real number)

Two Approaches to Simplifying Radical Expressions Containing Variables

Let a represent a nonnegative real number. To simplify $\sqrt{a^n}$ where n is odd and positive,

i) Method 1:
Write a^n as the product of two factors so that the exponent of one of the factors is the *largest* number less than n that is divisible by 2 (the index of the radical).

ii) Method 2:
1) Divide the exponent in the radicand by the index of the radical.
2) The exponent on the variable *outside* of the radical will be the *quotient* of the division problem.
3) The exponent on the variable *inside* of the radical will be the *remainder* of the division problem.

i) Simplify $\sqrt{x^9}$.

$$\begin{aligned}
\sqrt{x^9} &= \sqrt{x^8 \cdot x^1} && \text{8 is the largest number less than 9 that is divisible by 2.}\\
&= \sqrt{x^8} \cdot \sqrt{x} && \text{Product rule}\\
&= x^{8/2}\sqrt{x}\\
&= x^4\sqrt{x} && 8 \div 2 = 4
\end{aligned}$$

ii) Simplify $\sqrt{p^{15}}$.

$$\begin{aligned}
\sqrt{p^{15}} &= p^7\sqrt{p^1} && \text{15} \div \text{2 gives a quotient of}\\
&= p^7\sqrt{p} && \text{7 and a remainder of 1.}
\end{aligned}$$

Definition/Procedure	Example

10.4 Simplifying Expressions Containing Higher Roots

Product Rule for Higher Roots

If $\sqrt[n]{a}$ and $\sqrt[n]{b}$ are real numbers such that the roots exist, then $\sqrt[n]{a} \cdot \sqrt[n]{b} = \sqrt[n]{a \cdot b}$.

$\sqrt[3]{3} \cdot \sqrt[3]{5} = \sqrt[3]{15}$

Let P be an expression and let n be a positive integer greater than 1. Then $\sqrt[n]{P}$ **is completely simplified when all of the following conditions are met:**

1) The radicand does not contain any factors (other than 1) that are perfect nth powers.
2) The exponents in the radicand and the index of the radical do not have any common factors (other than 1).
3) The radicand does not contain any fractions.
4) There are no radicals in the denominator of a fraction.

To *simplify radicals with any index,* reverse the process of multiplying radicals, where a or b is an nth power.

$$\sqrt[n]{a \cdot b} = \sqrt[n]{a} \cdot \sqrt[n]{b}$$

Simplify $\sqrt[3]{40}$.

Method 1:

Think of two numbers that multiply to 40 so that one of them is a *perfect cube.*

$$40 = 8 \cdot 5 \quad \text{8 is a perfect cube.}$$

Then, $\quad \sqrt[3]{40} = \sqrt[3]{8 \cdot 5}$
$\qquad\qquad = \sqrt[3]{8} \cdot \sqrt[3]{5} \quad$ Product rule
$\qquad\qquad = 2\sqrt[3]{5} \qquad \sqrt[3]{8} = 2$

Method 2:

Begin by using a factor tree to find the prime factorization of 40.

$$40 = 2^3 \cdot 5$$
$$\sqrt[3]{40} = \sqrt[3]{2^3 \cdot 5}$$
$$\qquad = \sqrt[3]{2^3} \cdot \sqrt[3]{5} \quad \text{Product rule}$$
$$\qquad = 2\sqrt[3]{5} \qquad \sqrt[3]{2^3} = 2$$

Quotient Rule for Higher Roots

If $\sqrt[n]{a}$ and $\sqrt[n]{b}$ are real numbers, $b \neq 0$ and n is a natural number, then $\sqrt[n]{\dfrac{a}{b}} = \dfrac{\sqrt[n]{a}}{\sqrt[n]{b}}$.

$\sqrt[4]{\dfrac{32}{81}} = \dfrac{\sqrt[4]{32}}{\sqrt[4]{81}} = \dfrac{\sqrt[4]{16} \cdot \sqrt[4]{2}}{3} = \dfrac{2\sqrt[4]{2}}{3}$

Simplifying Higher Roots with Variables in the Radicand

If a is a nonnegative number and m and n are integers such that $n > 1$, then $\sqrt[n]{a^m} = a^{m/n}$.

Simplify $\sqrt[4]{a^{12}}$.

$\sqrt[4]{a^{12}} = a^{12/4} = a^3$

If the exponent does not divide evenly by the index, we can use two methods for simplifying the radical expression. If a is a nonnegative number and m and n are integers such that $n > 1$, then

i) **Method 1:** Use the product rule.
   To simplify $\sqrt[n]{a^m}$, write a^m as the product of two factors so that the exponent of one of the factors is the *largest* number less than m that is divisible by n (the index).

ii) **Method 2:** Use the quotient and remainder (presented in Section 10.3).

i) Simplify $\sqrt[5]{c^{17}}$.

$\sqrt[5]{c^{17}} = \sqrt[5]{c^{15} \cdot c^2} \qquad$ 15 is the largest number less than 17 that is divisible by 5.
$\qquad = \sqrt[5]{c^{15}} \cdot \sqrt[5]{c^2} \qquad$ Product rule
$\qquad = c^{15/5} \cdot \sqrt[5]{c^2}$
$\qquad = c^3\sqrt[5]{c^2} \qquad\qquad 15 \div 5 = 3$

ii) Simplify $\sqrt[4]{m^{11}}$.

$\sqrt[4]{m^{11}} = m^2\sqrt[4]{m^3} \qquad$ $11 \div 4$ gives a quotient of 2 and a remainder of 3.

10.5 Adding, Subtracting, and Multiplying Radicals

Like radicals have the same index and the same radicand. In order to add or subtract radicals, they must be like radicals.

Steps for Adding and Subtracting Radicals

1) Write each radical expression in simplest form.
2) Combine like radicals.

Perform the operations, and simplify.

a) $5\sqrt{2} + 9\sqrt{7} - 3\sqrt{2} + 4\sqrt{7}$
$= 2\sqrt{2} + 13\sqrt{7}$
b) $\sqrt{72} + \sqrt{18} - \sqrt{45}$
$\qquad = \sqrt{36} \cdot \sqrt{2} + \sqrt{9} \cdot \sqrt{2} - \sqrt{9} \cdot \sqrt{5}$
$\qquad = 6\sqrt{2} + 3\sqrt{2} - 3\sqrt{5}$
$\qquad = 9\sqrt{2} - 3\sqrt{5}$

Definition/Procedure	Example

Combining Multiplication, Addition, and Subtraction of Radicals

Multiply expressions containing radicals using the same techniques that are used for multiplying polynomials.

Multiply and simplify.

a) $\sqrt{m}(\sqrt{2m} + \sqrt{n})$
$= \sqrt{m} \cdot \sqrt{2m} + \sqrt{m} \cdot \sqrt{n}$ Distribute.
$= \sqrt{2m^2} + \sqrt{mn}$ Multiply.
$= m\sqrt{2} + \sqrt{mn}$ Simplify.

b) $(\sqrt{k} + \sqrt{6})(\sqrt{k} - \sqrt{2})$

Since we are multiplying two binomials, multiply using FOIL.

$(\sqrt{k} + \sqrt{6})(\sqrt{k} - \sqrt{2})$
$\qquad\quad\text{F}\qquad\text{O}\qquad\text{I}\qquad\text{L}$
$= \sqrt{k} \cdot \sqrt{k} - \sqrt{2} \cdot \sqrt{k} + \sqrt{6} \cdot \sqrt{k} - \sqrt{6} \cdot \sqrt{2}$
$= k^2 - \sqrt{2k} + \sqrt{6k} - \sqrt{12}$ Product rule
$= k^2 - \sqrt{2k} + \sqrt{6k} - 2\sqrt{3}$ $\sqrt{12} = 2\sqrt{3}$

Squaring a Radical Expression with Two Terms

To square a binomial we can either use FOIL or one of the special formulas from Chapter 6:

$$(a + b)^2 = a^2 + 2ab + b^2$$
$$(a - b)^2 = a^2 - 2ab + b^2$$

$(\sqrt{7} + 5)^2 = (\sqrt{7})^2 + 2(\sqrt{7})(5) + (5)^2$
$\qquad\qquad = 7 + 10\sqrt{7} + 25$
$\qquad\qquad = 32 + 10\sqrt{7}$

Multiply $(a + b)(a - b)$

To multiply binomials of the form $(a + b)(a - b)$ use the formula $(a + b)(a - b) = a^2 - b^2$.

$(3 + \sqrt{10})(3 - \sqrt{10}) = (3)^2 - (\sqrt{10})^2$
$\qquad\qquad\qquad\qquad\quad = 9 - 10$
$\qquad\qquad\qquad\qquad\quad = -1$

10.6 Dividing Radicals

The process of eliminating radicals from the denominator of an expression is called **rationalizing the denominator.**

First, we give examples of rationalizing denominators containing one term.

Rationalize the denominator of each expression.

a) $\dfrac{9}{\sqrt{2}} = \dfrac{9}{\sqrt{2}} \cdot \dfrac{\sqrt{2}}{\sqrt{2}} = \dfrac{9\sqrt{2}}{2}$

b) $\dfrac{5}{\sqrt[3]{2}} = \dfrac{5}{\sqrt[3]{2}} \cdot \dfrac{\sqrt[3]{2^2}}{\sqrt[3]{2^2}} = \dfrac{5\sqrt[3]{2^2}}{\sqrt[3]{2^3}} = \dfrac{5\sqrt[3]{4}}{2}$

The **conjugate** of an expression of the form $a + b$ is $a - b$.

$\sqrt{11} - 4$ conjugate: $\sqrt{11} + 4$
$-8 + \sqrt{5}$ conjugate: $-8 - \sqrt{5}$

Rationalizing a Denominator with Two Terms

If the denominator of an expression contains two terms, including one or two square roots, then to rationalize the denominator, we multiply the numerator and denominator of the expression by the conjugate of the denominator.

Rationalize the denominator of $\dfrac{4}{\sqrt{2} - 3}$.

$\dfrac{4}{\sqrt{2} - 3} = \dfrac{4}{\sqrt{2} - 3} \cdot \dfrac{\sqrt{2} + 3}{\sqrt{2} + 3}$ Multiply by the conjugate of the denominator.

$= \dfrac{4(\sqrt{2} + 3)}{(\sqrt{2})^2 - (3)^2}$ $(a + b)(a - b) = a^2 - b^2$

$= \dfrac{4(\sqrt{2} + 3)}{2 - 9}$ Square the terms.

$= \dfrac{4(\sqrt{2} + 3)}{-7} = -\dfrac{4\sqrt{2} + 12}{7}$

Definition/Procedure	Example

10.7 Solving Radical Equations

Solving Radical Equations Containing Square Roots

Step 1: Get a radical on a side by itself.

Step 2: Square both sides of the equation to eliminate a radical.

Step 3: Combine like terms on each side of the equation.

Step 4: If the equation still contains a radical, repeat Steps 1–3.

Step 5: Solve the equation.

Step 6: Check the proposed solutions *in the original equation,* and discard extraneous solutions.

Solve $t = 2 + \sqrt{2t - 1}$.

$$t - 2 = \sqrt{2t - 1} \qquad \text{Get the radical by itself.}$$
$$(t - 2)^2 = (\sqrt{2t - 1})^2 \qquad \text{Square both sides.}$$
$$t^2 - 4t + 4 = 2t - 1$$
$$t^2 - 6t + 5 = 0 \qquad \text{Get all terms on the same side.}$$
$$(t - 5)(t - 1) = 0 \qquad \text{Factor.}$$
$$t - 5 = 0 \quad \text{or} \quad t - 1 = 0$$
$$t = 5 \quad \text{or} \qquad t = 1$$

Check $t = 5$ and $t = 1$ in the *original* equation.

$t = 5$: $\quad t = 2 + \sqrt{2t - 1}$
$\qquad 5 \stackrel{?}{=} 2 + \sqrt{2(5) - 1}$
$\qquad 5 \stackrel{?}{=} 2 + \sqrt{9}$
$\qquad 5 = 2 + 3$
$\qquad\qquad$ True

$t = 1$: $\quad t = 2 + \sqrt{2t - 1}$
$\qquad 1 \stackrel{?}{=} 2 + \sqrt{2(1) - 1}$
$\qquad 1 \stackrel{?}{=} 2 + 1$
$\qquad 1 = 3$
$\qquad\qquad$ False

$t = 5$ *is* a solution, but $t = 1$ is *not* because $t = 1$ does not satisfy the original equation.

The solution set is $\{5\}$.

10.8 Complex Numbers

Definition of *i*: $\quad i = \sqrt{-1}$

Therefore, $i^2 = -1$.

A **complex number** is a number of the form $a + bi$, where a and b are real numbers. a is called the **real part** and b is called the **imaginary part.** The set of real numbers is a subset of the set of complex numbers.

Examples of complex numbers:

$\quad -2 + 7i$
$\quad 5 \qquad$ (since it can be written $5 + 0i$)
$\quad 4i \qquad$ (since it can be written $0 + 4i$)

Simplifying Square Roots with Negative Radicands

Use the product rule and $i = \sqrt{-1}$.

Simplify $\sqrt{-25}$.

$$\sqrt{-25} = \sqrt{-1} \cdot \sqrt{25}$$
$$= i \cdot 5$$
$$= 5i$$

When multiplying or dividing square roots with negative radicands, write each radical in terms of i first.

Multiply $\sqrt{-12} \cdot \sqrt{-3}$.

$$\sqrt{-12} \cdot \sqrt{-3} = i\sqrt{12} \cdot i\sqrt{3} = i^2\sqrt{36}$$
$$= -1 \cdot 6 = -6$$

Adding and Subtracting Complex Numbers

To add and subtract complex numbers, combine the real parts and combine the imaginary parts.

Subtract $(10 + 7i) - (-2 + 4i)$.

$$(10 + 7i) - (-2 + 4i) = 10 + 7i + 2 - 4i$$
$$= 12 + 3i$$

Multiply complex numbers like we multiply polynomials. Remember to replace i^2 with -1.

Multiply and simplify.

a) $4(9 + 5i) = 36 + 20i$

$\qquad\qquad\qquad$ F $\quad$ O $\quad$ I $\quad$ L
b) $(-3 + i)(2 - 7i) = -6 + 21i + 2i - 7i^2$
$\qquad\qquad\qquad\quad = -6 + 23i - 7(-1)$
$\qquad\qquad\qquad\quad = -6 + 23i + 7$
$\qquad\qquad\qquad\quad = 1 + 23i$

Definition/Procedure	Example
Complex Conjugates 1) The conjugate of $a + bi$ is $a - bi$. 2) The product of $a + bi$ and $a - bi$ is a real number. 3) Find the product $(a + bi)(a - bi)$ using FOIL or recall that $(a + bi)(a - bi) = a^2 + b^2$.	Multiply $-5 - 3i$ by its conjugate. The conjugate of $-5 - 3i$ is $-5 + 3i$. Use $(a + bi)(a - bi) = a^2 + b^2$. $$(-5 - 3i)(-5 + 3i) = (-5)^2 + (3)^2$$ $$= 25 + 9$$ $$= 34$$
Dividing Complex Numbers To divide complex numbers, multiply the numerator and denominator by the *conjugate of the denominator*. Write the quotient in the form $a + bi$.	Divide $\dfrac{6i}{2 + 5i}$. Write the result in the form $a + bi$. $$\frac{6i}{2 + 5i} = \frac{6i}{2 + 5i} \cdot \frac{(2 - 5i)}{(2 - 5i)}$$ $$= \frac{12i - 30i^2}{(2)^2 + (5)^2}$$ $$= \frac{12i - 30(-1)}{29}$$ $$= \frac{30}{29} + \frac{12}{29}i$$
Simplify Powers of i We can simplify powers of i using $i^2 = -1$.	Simplify i^{14}. $$i^{14} = (i^2)^7 \quad \text{Power rule}$$ $$= (-1)^7 \quad i^2 = -1$$ $$= -1 \quad \text{Simplify.}$$

Chapter 10:   **Review Exercises**

(10.1) Find each root, if possible.

1) $\sqrt{\dfrac{169}{4}}$

2) $\sqrt{-16}$

3) $-\sqrt{81}$

4) $\sqrt[5]{32}$

5) $\sqrt[3]{-1}$

6) $-\sqrt[4]{81}$

7) $\sqrt[6]{-64}$

8) $\sqrt{9 - 16}$

Simplify. Use absolute values when necessary.

9) $\sqrt{(-13)^2}$

10) $\sqrt[5]{(-8)^5}$

11) $\sqrt{p^2}$

12) $\sqrt[6]{c^6}$

13) $\sqrt[3]{h^3}$

14) $\sqrt[4]{(y + 7)^4}$

15) Approximate $\sqrt{34}$ to the nearest tenth and plot it on a number line.

(10.2)

16) Explain how to write $8^{2/3}$ in radical form.

Evaluate.

17) $36^{1/2}$

18) $32^{1/5}$

19) $\left(\dfrac{27}{125}\right)^{1/3}$

20) $-16^{1/4}$

21) $32^{3/5}$

22) $\left(\dfrac{64}{27}\right)^{2/3}$

23) $81^{-1/2}$

24) $\left(\dfrac{1}{27}\right)^{-1/3}$

25) $81^{-3/4}$

26) $1000^{-2/3}$

27) $\left(\dfrac{27}{1000}\right)^{-2/3}$

28) $\left(\dfrac{25}{16}\right)^{-3/2}$

From this point forward, assume all variables represent positive real numbers.

Simplify completely. The answer should contain only positive exponents.

29) $3^{6/7} \cdot 3^{8/7}$

30) $(169^4)^{1/8}$

31) $(8^{1/5})^{10}$

32) $\dfrac{8^2}{8^{11/3}}$

33) $\dfrac{7^2}{7^{5/3} \cdot 7^{1/3}}$

34) $(2k^{-5/6})(3k^{1/2})$

35) $(64a^4b^{12})^{5/6}$

36) $\left(\dfrac{t^4u^3}{7t^7u^5}\right)^{-2}$

37) $\left(\dfrac{81c^{-5}d^9}{16c^{-1}d^2}\right)^{-1/4}$

Rewrite each radical in exponential form, then simplify. Write the answer in simplest (or radical) form.

38) $\sqrt[4]{36^2}$

39) $\sqrt[12]{27^4}$

40) $(\sqrt{17})^2$

41) $\sqrt[3]{7^3}$

42) $\sqrt[5]{t^{20}}$

43) $\sqrt[4]{k^{28}}$

44) $\sqrt{x^{10}}$

45) $\sqrt{w^6}$

(10.3) Simplify completely.

46) $\sqrt{28}$

47) $\sqrt{1000}$

48) $\dfrac{\sqrt{63}}{\sqrt{7}}$

49) $\sqrt{\dfrac{18}{49}}$

50) $\dfrac{\sqrt{48}}{\sqrt{121}}$

51) $\sqrt{k^{12}}$

52) $\sqrt{\dfrac{40}{m^4}}$

53) $\sqrt{x^9}$

54) $\sqrt{y^5}$

55) $\sqrt{45t^2}$

56) $\sqrt{80n^{21}}$

57) $\sqrt{72x^7y^{13}}$

58) $\sqrt{\dfrac{m^{11}}{36n^2}}$

Perform the indicated operation, and simplify.

59) $\sqrt{5} \cdot \sqrt{3}$

60) $\sqrt{6} \cdot \sqrt{15}$

61) $\sqrt{2} \cdot \sqrt{12}$

62) $\sqrt{b^7} \cdot \sqrt{b^3}$

63) $\sqrt{11x^5} \cdot \sqrt{11x^8}$

64) $\sqrt{5a^2b} \cdot \sqrt{15a^6b^4}$

65) $\dfrac{\sqrt{200k^{21}}}{\sqrt{2k^5}}$

66) $\dfrac{\sqrt{63c^{17}}}{\sqrt{7c^9}}$

(10.4) Simplify completely.

67) $\sqrt[3]{16}$

68) $\sqrt[3]{250}$

69) $\sqrt[4]{48}$

70) $\sqrt[3]{\dfrac{81}{3}}$

71) $\sqrt[4]{z^{24}}$

72) $\sqrt[5]{p^{40}}$

73) $\sqrt[3]{a^{20}}$

74) $\sqrt[5]{x^{14}y^7}$

75) $\sqrt[3]{16z^{15}}$

76) $\sqrt[3]{80m^{17}n^{10}}$

77) $\sqrt[4]{\dfrac{h^{12}}{81}}$

78) $\sqrt[5]{\dfrac{c^{22}}{32d^{10}}}$

Perform the indicated operation, and simplify.

79) $\sqrt[3]{3} \cdot \sqrt[3]{7}$

80) $\sqrt[3]{25} \cdot \sqrt[3]{10}$

81) $\sqrt[4]{4t^7} \cdot \sqrt[4]{8t^{10}}$

82) $\sqrt[5]{\dfrac{x^{21}}{x^{16}}}$

83) $\sqrt[3]{n} \cdot \sqrt{n}$

84) $\dfrac{\sqrt[4]{a^3}}{\sqrt[3]{a}}$

(10.5) Perform the operations, and simplify.

85) $8\sqrt{5} + 3\sqrt{5}$

86) $\sqrt{125} + \sqrt{80}$

87) $\sqrt{80} - \sqrt{48} + \sqrt{20}$

88) $9\sqrt[3]{72} - 8\sqrt[3]{9}$

89) $3p\sqrt{p} - 7\sqrt{p^3}$

90) $9n\sqrt{n} - 4\sqrt{n^3}$

91) $10d^2\sqrt{8d} - 32d\sqrt{2d^3}$

92) $\sqrt{6}(\sqrt{7} - \sqrt{6})$

93) $3\sqrt{k}(\sqrt{20k} + \sqrt{2})$

94) $(5 - \sqrt{3})(2 + \sqrt{3})$

95) $(\sqrt{6a} - 9)(\sqrt{2a} - 7)$

96) $(8 + 3\sqrt{c})(8 - 3\sqrt{c})$

97) $\sqrt[4]{32a^7b^{13}} + 7ab\sqrt[4]{2a^3b^9}$

98) $4\sqrt[3]{w^2}\left(\sqrt[3]{54w} - \sqrt[3]{4w^2}\right)$

99) $(\sqrt{2r} + 5\sqrt{s})(3\sqrt{s} + 4\sqrt{2r})$

100) $(2\sqrt{5} - 4)^2$

101) $(1 + \sqrt{y + 1})^2$

102) $(\sqrt{6} - \sqrt{5})(\sqrt{6} + \sqrt{5})$

(10.6) Rationalize the denominator of each expression.

103) $\dfrac{14}{\sqrt{3}}$

104) $\dfrac{20}{\sqrt{6}}$

105) $\dfrac{\sqrt{18k}}{\sqrt{n}}$

106) $\dfrac{\sqrt{45}}{\sqrt{m^5}}$

107) $\dfrac{7}{\sqrt[3]{2}}$

108) $-\dfrac{15}{\sqrt[3]{9}}$

109) $\dfrac{\sqrt[3]{x^2}}{\sqrt[3]{y}}$

110) $\sqrt[4]{\dfrac{3}{4k^2}}$

111) $\dfrac{2}{3 + \sqrt{3}}$

112) $\dfrac{z - 4}{\sqrt{z} + 2}$

Rationalize the numerator of each expression, and simplify.

113) $\dfrac{\sqrt{15}}{6}$

114) $\dfrac{4 - \sqrt{c}}{c - 16}$

Simplify completely.

115) $\dfrac{8 - 24\sqrt{2}}{8}$

116) $\dfrac{-\sqrt{48} - 6}{10}$

(10.7) Solve.

117) $\sqrt{x + 8} = 3$

118) $10 - \sqrt{3r - 5} - 2$

119) $\sqrt{3j + 4} = -\sqrt{4j - 1}$

120) $\sqrt[3]{6d - 14} = -2$

121) $a = \sqrt{a + 8} - 6$

122) $1 + \sqrt{6m + 7} = 2m$

123) $\sqrt{4a + 1} - \sqrt{a - 2} = 3$

124) $\sqrt{6x + 9} - \sqrt{2x + 1} = 4$

125) Solve for V: $r = \sqrt{\dfrac{3V}{\pi h}}$

126) The velocity of a wave in shallow water is given by $c = \sqrt{gH}$, where g is the acceleration due to gravity (32 ft/sec^2), and H is the depth of the water (in feet). Find the velocity of a wave in 10 ft of water.

(10.8) Simplify.

127) $\sqrt{-49}$

128) $\sqrt{-8}$

129) $\sqrt{-2} \cdot \sqrt{-8}$

130) $\sqrt{-6} \cdot \sqrt{-3}$

Perform the indicated operations.

131) $(2 + i) + (10 - 4i)$

132) $(4 + 3i) - (11 - 4i)$

133) $\left(\dfrac{4}{5} - \dfrac{1}{3}i\right) - \left(\dfrac{1}{2} + i\right)$

134) $\left(-\dfrac{3}{8} - 2i\right) + \left(\dfrac{5}{8} + \dfrac{3}{2}i\right) - \left(\dfrac{1}{4} - \dfrac{1}{2}i\right)$

Multiply and simplify.

135) $5(-6 + 7i)$

136) $-8i(4 + 3i)$

137) $3i(-7 + 12i)$

138) $(3 - 4i)(2 + i)$

139) $(4 - 6i)(3 - 6i)$

140) $\left(\dfrac{1}{5} - \dfrac{2}{3}i\right)\left(\dfrac{3}{2} - \dfrac{2}{3}i\right)$

Identify the conjugate of each complex number, then multiply the number and its conjugate.

141) $2 - 7i$

142) $-2 + 3i$

Divide. Write the quotient in the form $a + bi$.

143) $\dfrac{6}{2 + 5i}$

144) $\dfrac{-12}{4 - 3i}$

145) $\dfrac{8}{i}$

146) $\dfrac{4i}{1 - 3i}$

147) $\dfrac{9 - 4i}{6 - i}$

148) $\dfrac{5 - i}{-2 + 6i}$

Simplify.

149) i^{10}

150) i^{51}

Chapter 10:   Test

1) **Fill in the blank with *always*, *sometimes*, or *never* to make the statement true.**

   a) The even root of a negative number is _____ a real number.

   b) The product of a complex number and its conjugate is _____ a real number.

   c) A negative number can _____ be a solution to an equation containing a square root.

2) **Find each real root, if possible.**

   a) $\sqrt{144}$

   b) $\sqrt[5]{-32}$

   c) $\sqrt{-16}$

3) **Approximate $\sqrt{46}$ to the nearest tenth, and plot it on a number line.**

Simplify. Use absolute values when necessary.

4) $\sqrt[4]{w^4}$

5) $\sqrt[5]{(-19)^5}$

Evaluate.

6) $16^{1/4}$

7) $27^{4/3}$

8) $(49)^{-1/2}$

9) $\left(\dfrac{8}{125}\right)^{-2/3}$

From this point forward, assume all variables represent positive real numbers.

Simplify completely. The answer should contain only positive exponents.

10) $m^{3/8} \cdot m^{1/4}$

11) $\dfrac{35a^{1/6}}{14a^{5/6}}$

12) $(2x^{3/10}y^{-2/5})^{-5}$

Simplify completely.

13) $\sqrt{75}$

14) $\sqrt[3]{48}$

15) $\sqrt{\dfrac{24}{2}}$

Simplify completely.

16) $\sqrt{y^6}$

17) $\sqrt[4]{p^{24}}$

18) $\sqrt{t^9}$

19) $\sqrt{63m^5n^8}$

20) $\sqrt[3]{c^{23}}$

21) $\sqrt[3]{\dfrac{40a^{14}b^7}{27}}$

Perform the operations, and simplify.

22) $\sqrt{3} \cdot \sqrt{12}$

23) $\sqrt[3]{z^4} \cdot \sqrt[3]{z^6}$

24) $\dfrac{\sqrt{120w^{15}}}{\sqrt{2w^4}}$

25) $9\sqrt{7} - 3\sqrt{7}$

26) $\sqrt{12} - \sqrt{108} + \sqrt{18}$

27) $2h^3\sqrt[4]{h} - 16\sqrt[4]{h^{13}}$

Multiply and simplify.

28) $\sqrt{6}(\sqrt{2} - 5)$

29) $(3 - 2\sqrt{5})(\sqrt{2} + 1)$

30) $(5\sqrt{x} + \sqrt{11})(5\sqrt{x} - \sqrt{11})$

31) $(\sqrt{2p + 1} + 2)^2$

32) $2\sqrt{t}(\sqrt{t} - \sqrt{3u})$

Rationalize the denominator of each expression.

33) $\dfrac{2}{\sqrt{5}}$

34) $\dfrac{8}{\sqrt{7} + 3}$

35) $\dfrac{\sqrt{6}}{\sqrt{a}}$

36) $\dfrac{10n}{\sqrt[3]{25n}}$

37) Simplify completely: $\dfrac{2 - \sqrt{48}}{2}$

Solve.

38) $\sqrt{5h + 4} = 3$

39) $z = \sqrt{1 - 4z} - 5$

40) $\sqrt[3]{n - 5} - \sqrt[3]{2n - 18} = 0$

41) $\sqrt{3k + 1} - \sqrt{2k - 1} = 1$

42) In the formula $r = \sqrt{\dfrac{V}{\pi h}}$, V represents the volume of a cylinder, h represents the height of the cylinder, and r represents the radius.

   a) A cylindrical container has a volume of 72π in^3. It is 8 in. high. What is the radius of the container?

   b) Solve the formula for V.

Simplify.

43) $\sqrt{-64}$ 44) $\sqrt{-45}$ 45) i^{19}

Perform the indicated operation, and simplify. Write the answer in the form $a + bi$.

46) $(-10 + 3i) - (6 + i)$ 47) $(2 - 7i)(-1 + 3i)$

48) $\dfrac{8 + i}{2 - 3i}$

Chapter 10:   Cumulative Review for Chapters 1–10

1) Combine like terms.

$$4x - 3y + 9 - \frac{2}{3}x + y - 1$$

2) Write in scientific notation.

8,723,000

3) Solve $3(2c - 1) + 7 = 9c + 5(c + 2)$.

4) Graph $3x + 2y = 12$.

5) Write the equation of the line containing the points $(5, 3)$ and $(1, -2)$. Write the equation in slope-intercept form.

6) Solve by substitution.

$$\begin{aligned} 2x + 7y &= -12 \\ x - 4y &= -6 \end{aligned}$$

7) Multiply.

$$(5p^2 - 2)(3p^2 - 4p - 1)$$

8) Divide.

$$\frac{8n^3 - 1}{2n - 1}$$

Factor completely.

9) $9w^2 + 6w + 1$

10) $8 - 18t^2$

11) Solve $6y^2 - 4 = 5y$.

12) Solve $3(k^2 + 20) - 4k = 2k^2 + 11k + 6$.

13) *Write an equation, and solve.* The width of a rectangle is 5 in. less than its length. The area is 84 in^2. Find the dimensions of the rectangle.

Perform the operations, and simplify.

14) $\dfrac{5a^2 + 3}{a^2 + 4a} - \dfrac{3a - 2}{a + 4}$

15) $\dfrac{10m^2}{9n} \cdot \dfrac{6n^2}{35m^5}$

16) Solve $\dfrac{3}{r^2 + 8r + 15} - \dfrac{4}{r + 3} = 1$.

17) Solve $|6g + 1| \geq 11$. Write the answer in interval notation.

18) Solve using Gaussian elimination.

$$\begin{aligned} x + 3y + z &= 3 \\ 2x - y - 5z &= -1 \\ -x + 2y + 3z &= 0 \end{aligned}$$

19) Simplify. Assume all variables represent nonnegative real numbers.

   a) $\sqrt{500}$ b) $\sqrt[3]{56}$

   c) $\sqrt{p^{10}q^7}$ d) $\sqrt[4]{32a^{15}}$

20) Evaluate.

   a) $81^{1/2}$ b) $8^{4/3}$

   c) $(27)^{-1/3}$

21) Multiply and simplify $2\sqrt{3}(5 - \sqrt{3})$.

22) Rationalize the denominator. Assume the variables represent positive real numbers.

   a) $\sqrt{\dfrac{20}{50}}$ b) $\dfrac{6}{\sqrt[3]{2}}$

   c) $\dfrac{x}{\sqrt[3]{y^2}}$ d) $\dfrac{\sqrt{a} - 2}{1 - \sqrt{a}}$

23) Solve.

   a) $\sqrt{2b - 1} + 7 = 6$

   b) $\sqrt{3z + 10} = 2 - \sqrt{z + 4}$

24) Simplify.

   a) $\sqrt{-49}$ b) $\sqrt{-56}$

   c) i^8

25) Perform the indicated operation, and simplify. Write the answer in the form $a + bi$.

   a) $(-3 + 4i) + (5 + 3i)$

   b) $(3 + 6i)(-2 + 7i)$

   c) $\dfrac{2 - i}{-4 + 3i}$

Design Icon Credits: Notes box icon (push pin): ©Shutterstock/ksevgi; Using Technology icon (graphing calculator): ©mbbirdy/Getty Images.

Quadratic Equations

Get Ready

When we learn different methods for solving quadratic equations in this chapter, we will have to simplify radicals. Here's a quick review.

1) We use the **product and quotients rules for square roots** to simplify radicals.

   Examples: Simplify each radical.

   a) $\sqrt{45}$ $\qquad$ $\sqrt{45} = \sqrt{9 \cdot 5} = \sqrt{9} \cdot \sqrt{5} = 3\sqrt{5}$

   b) $\sqrt{\dfrac{48}{121}}$ $\qquad$ $\sqrt{\dfrac{48}{121}} = \dfrac{\sqrt{48}}{\sqrt{121}} = \dfrac{\sqrt{16 \cdot 3}}{11} = \dfrac{\sqrt{16} \cdot \sqrt{3}}{11} = \dfrac{4\sqrt{3}}{11}$

2) Recall that the **imaginary number, i,** is defined as $i = \sqrt{-1}$. We use this fact to simplify square roots of negative numbers.

   Example: Simplify $\sqrt{-24}$.

   $$\sqrt{-24} = \sqrt{-1 \cdot 24} = \sqrt{-1} \cdot \sqrt{24} = i\sqrt{4} \cdot \sqrt{6} = 2i\sqrt{6}$$

Get Ready Exercises

Simplify each radical.

1) $\sqrt{28}$ $\qquad$ 2) $\sqrt{27}$ $\qquad$ 3) $\sqrt{75}$ $\qquad$ 4) $\sqrt{72}$

5) $\sqrt{\dfrac{1}{144}}$ $\qquad$ 6) $\sqrt{\dfrac{4}{49}}$ $\qquad$ 7) $\sqrt{\dfrac{80}{81}}$ $\qquad$ 8) $\sqrt{\dfrac{50}{9}}$

9) $\sqrt{-25}$ $\qquad$ 10) $\sqrt{-36}$ $\qquad$ 11) $\sqrt{-63}$ $\qquad$ 12) $\sqrt{-40}$

Answers

1) $2\sqrt{7}$ 2) $3\sqrt{3}$ 3) $5\sqrt{3}$ 4) $6\sqrt{2}$ 5) $\dfrac{1}{12}$ 6) $\dfrac{2}{7}$ 7) $\dfrac{4\sqrt{5}}{9}$ 8) $\dfrac{5\sqrt{2}}{3}$ 9) $5i$ 10) $6i$ 11) $3i\sqrt{7}$ 12) $2i\sqrt{10}$

Use the following *Basic Skills Worksheets* to prepare students for this, and future, chapters: **WS2 Powers** (to practice quick recall of powers of integers) and **WS5 Roots** (to practice quick recall of roots of integers).

 **Study Strategies** Developing Grit

Angela Duckworth, professor of psychology at the University of Pennsylvania, defines **grit** as "passion and perseverance for long-term goals." Gritty people keep working toward their goals even when there are setbacks or their progress is slow. They keep their eyes on the prize. Duckworth's research has found that, more than IQ and other factors, *grit* "predicts success in the military, education, and business."* Let's learn how we can use the P.O.W.E.R. framework to help us develop the grit we need to succeed in this course.

- **I will work on developing grit to help me succeed in this course.**

- Complete the emPOWERme survey that appears before the Chapter Summary.
- Create a vision for yourself. *Where do you want to be in 5 years, 10 years, or 15 years? What do you need to do to get there?*
- Create a plan for your vision. Ask yourself, *"What long-term academic goal do I need to do what I want to do in the future?"* Realize that succeeding in this course is a step toward achieving that goal.
- Believe that you *can* achieve your goal.
- Understand that effort and achievement are linked. The harder, and smarter, you work, the more likely you are to achieve your goal. Your efforts can improve your future.
- Realize that there may be times you encounter difficulties in the course. With a plan, you can overcome them and get back on the track to success.

- Write down some goals for yourself specific to this course: Decide on the grade you want to earn. Attend every class. See your instructor when you need help.
- Work *smart*. Develop good note-taking skills, ask questions in class, learn where and how you do homework and study best, and use good test-taking strategies.
- Do your homework after each class. Practice *more* problems, if necessary. The more you practice, the better you will get.
- Identify someone you see as gritty and talk to that person about how he or she has achieved a major goal.
- Find a mentor, someone who will be encouraging, supportive, and helpful to you in achieving your goals. This can be an instructor, another student, or a family member.
- Surround yourself with supportive and gritty people, inside and outside of class. Organize a group with students in your class so that you can study together or communicate with someone when the math gets difficult.
- Find other people who share your interests. Join a club or extracurricular activity, and stick with it. The lessons learned there carry over into the classroom.
- If the going gets tough, don't get discouraged. Remind yourself of your long-term goal to motivate you to work hard in the course, and come up with a plan to get back on track. Talk with your instructor about what you can do differently. Tell yourself that you *can* succeed, and keep on working!

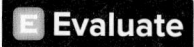

- As the term goes along ask yourself, *"Am I making the grades that I want to make on the quizzes and tests?"* If not, have you come up with a plan to get back on track?
- Did you use your long-term goal as motivation to keep working hard when you encountered difficulties in the course? Do you feel like you became grittier as the term went on?
- Evaluate how you have done at the *end of the term*. Did you earn the grade you had hoped to earn?

(continued on next page)

*Duckworth, Angela (2016). *Grit: The Power of Passion and Perseverance*. New York: Scribner.

Rethink

- First, *rethink* throughout the term. If you are doing well on quizzes and tests, ask yourself, "*What have I been doing right so far?*" so that you can continue to do those things. If not, ask yourself, "*Why not?*" Did you skip class? Did you do your homework at the last minute? Did you give up when the material became difficult? Ask yourself, "*What can I do differently to improve before the end of the term?*" Make adjustments so that you still have time to learn the material and improve your grade.

  - *Rethink* at the end of the term. If you have succeeded in the course, think about what you did to get there so that you can apply those strategies in the future. Did you come up with a plan that helped you to work through difficult parts of the course? Did you work smarter and with more persistence than in the past? Did you seek help from your instructor or other students in the class?

    - If you did not succeed in the class, ask yourself, "*Why not?*" Ask yourself the questions in the first bullet as well. This way, you can do things differently in the future.

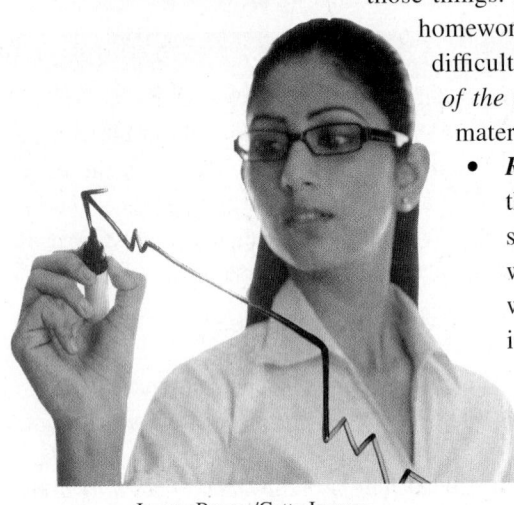
ImagesBazaar/Getty Images

11.1 Review of Solving Equations by Factoring

P Prepare

O Organize

What is your objective for Section 11.1?	How can you accomplish the objective?
1 Review How to Solve Quadratic Equations by Factoring	- Understand when and how to apply the *zero product rule*. - Recognize different forms of quadratic equations, and be able to rewrite them in standard form. - Go back to Section 7.5 if you need a more detailed review. - Complete the given example on your own. - Complete You Try 1.

W Work

Read the explanations, follow the examples, take notes, and complete the You Trys.

We defined a quadratic equation in Chapter 7. Let's restate the definition:

Definition

A **quadratic equation** can be written in the form $ax^2 + bx + c = 0$, where a, b, and c are real numbers and $a \neq 0$.

In Section 7.5, we learned how to solve quadratic equations by factoring. We will not be able to solve all quadratic equations by factoring, however. Therefore, we need to learn other methods. In this chapter, we will discuss the following four methods for solving quadratic equations.

Four Methods for Solving Quadratic Equations

1) Factoring

2) Square root property

3) Completing the square

4) Quadratic formula

1 Review How to Solve Quadratic Equations by Factoring

We begin by reviewing how to solve an equation by factoring.

Procedure Solving a Quadratic Equation by Factoring

1) Write the equation in the form $ax^2 + bx + c = 0$ so that all terms are on one side of the equal sign and zero is on the other side.

2) Factor the expression.

3) Set each factor equal to zero, and solve for the variable.

4) Check the answer(s).

EXAMPLE 1

Solve by factoring.

a) $8t^2 + 3 = -14t$ b) $(a - 9)(a + 7) = -15$ c) $3x^3 + 10x^2 = 8x$

Solution

a) Begin by writing $8t^2 + 3 = -14t$ in standard form.

$$8t^2 + 14t + 3 = 0 \qquad \text{Standard form}$$
$$(4t + 1)(2t + 3) = 0 \qquad \text{Factor.}$$

$$4t + 1 = 0 \quad \text{or} \quad 2t + 3 = 0 \qquad \text{Set each factor equal to zero.}$$
$$4t = -1 \qquad\qquad 2t = -3$$
$$t = -\frac{1}{4} \quad \text{or} \quad t = -\frac{3}{2} \qquad \text{Solve.}$$

The check is left to the student. The solution set is $\left\{ -\frac{3}{2}, -\frac{1}{4} \right\}$.

> **W Hint**
> Look back at Section 7.5 if you need a more detailed review of this topic.

b) You may want to solve $(a - 9)(a + 7) = -15$ like this:

$$(a - 9)(a + 7) = -15$$

$$a - 9 = -15 \quad \text{or} \quad a + 7 = -15 \qquad \textbf{This is incorrect!}$$
$$a = -6 \quad \text{or} \qquad a = -22$$

To solve the equation correctly, one side of the equation must equal *zero* and the other side must be factored so that we may apply the zero product rule and set each factor equal to zero.

Begin by multiplying the binomials using FOIL.

$$(a - 9)(a + 7) = -15$$
$$a^2 - 2a - 63 = -15 \qquad \text{Multiply using FOIL.}$$
$$a^2 - 2a - 48 = 0 \qquad \text{Write in standard form.}$$
$$(a + 6)(a - 8) = 0 \qquad \text{Factor.}$$

$$a + 6 = 0 \quad \text{or} \quad a - 8 = 0 \qquad \text{Set each factor equal to zero.}$$
$$a = -6 \qquad\qquad a = 8 \qquad \text{Solve.}$$

The check is left to the student. The solution set is $\{-6, 8\}$.

c)   Although this is a cubic equation and not quadratic, we *can* solve it by factoring.

$$3x^3 + 10x^2 = 8x$$
$$3x^3 + 10x^2 - 8x = 0 \qquad \text{Get zero on one side of the equal sign.}$$
$$x(3x^2 + 10x - 8) = 0 \qquad \text{Factor out } x.$$
$$x(3x - 2)(x + 4) = 0 \qquad \text{Factor } 3x^2 + 10x - 8.$$

$$x = 0 \quad \text{or} \quad 3x - 2 = 0 \quad \text{or} \quad x + 4 = 0 \qquad \text{Set each factor equal to zero.}$$
$$3x = 2$$

$$x = 0 \quad \text{or} \qquad x = \frac{2}{3} \quad \text{or} \qquad x = -4 \qquad \text{Solve.}$$

The check is left to the student. The solution set is $\left\{-4, 0, \dfrac{2}{3}\right\}$.

YOU TRY 1

Solve by factoring.

a) $c^2 - 12 = c$

b) $p(7p + 18) + 8 = 0$

c) $(k - 7)(k - 5) = -1$

d) $2x^3 + 30x = 16x^2$

ANSWERS TO YOU TRY **EXERCISE**

1) a) $\{-3, 4\}$ b) $\left\{-2, -\dfrac{4}{7}\right\}$ c) $\{6\}$ d) $\{0, 3, 5\}$

E Evaluate **11.1** Exercises   Do the exercises, and check your work.

Objective 1: Review How to Solve Quadratic Equations by Factoring

Solve each equation.

1) $(t + 7)(t - 6) = 0$

2) $3z(2z - 9) = 0$

3) $u^2 + 15u + 44 = 0$

4) $n^2 + 10n - 24 = 0$

5) $x^2 = x + 56$

6) $c^2 + 3c = 54$

7) $1 - 100w^2 = 0$

8) $9j^2 = 49$

9) $5m^2 + 8 = 22m$

10) $19a + 20 = -3a^2$

11) $23d = -10 - 6d^2$

12) $8h^2 + 12 = 35h$

13) $2r = 7r^2$

14) $5n^2 = -6n$

15) $(z + 3)(z + 1) = 15$ 16) $(c - 10)(c - 1) = -14$

Identify each equation as linear or quadratic.

17) $9m^2 - 2m + 1 = 0$ 18) $17 = 3z - z^2$

19) $13 - 4x = 19$

20) $10 - 2(3d + 1) = 5d + 19$

21) $y(2y - 5) = 3y + 1$ 22) $3(4y - 3) = y(y + 1)$

23) $-4(b + 7) + 5b = 2b + 9$

24) $6 + 2k(k - 1) = 5k$

In this section, there is a mix of linear and quadratic equations as well as equations of higher degree. Solve each equation.

25) $13c = 2c^2 + 6$ 26) $12x - 1 = 2x + 9$

27) $2p(p + 4) = p^2 + 5p + 10$

28) $z^2 - 20 = 22 - z$

 29) $5(3n - 2) - 11n = 2n - 1$

30) $5a^2 = 45a$ 31) $3t^3 + 5t = -8t^2$

32) $6(2k - 3) + 10 = 3(2k - 5)$

33) $2(r + 5) = 10 - 4r^2$

34) $3d - 4 = d(d + 8)$

35) $9y - 6(y + 1) = 12 - 5y$

36) $3m(2m + 5) - 8 = 2m(3m + 5) + 2$

37) $\frac{1}{16}w^2 + \frac{1}{8}w = \frac{1}{2}$ 38) $6h = 4h^3 + 5h^2$

39) $(c - 8)(c - 6) = -1$

40) $(4k - 1)(k + 1) = 9$

41) $12n + 3 = -12n^2$ 42) $u = u^2$

 43) $3b^2 - b - 7 = 4b(2b + 3) - 1$

44) $\frac{1}{2}q^2 + \frac{3}{4} = \frac{5}{4}q$

45) $t^3 + 7t^2 - 4t - 28 = 0$

46) $5m^3 + 2m^2 - 5m - 2 = 0$

Write an equation and solve.

47) The length of a rectangle is 5 in. more than its width. Find the dimensions of the rectangle if its area is 14 in^2.

48) The width of a rectangle is 3 cm shorter than its length. If the area is 70 cm^2, what are the dimensions of the rectangle?

49) The length of a rectangle is 1 cm less than twice its width. The area is 45 cm^2. What are the dimensions of the rectangle?

50) A rectangle has an area of 32 in^2. Its length is 4 in. less than three times its width. Find the length and width.

Find the base and height of each triangle.

51)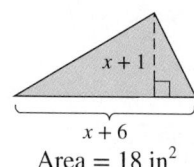
$x + 1$
$x + 6$
Area = 18 in^2

52)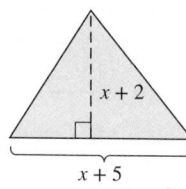
$x + 2$
$x + 5$
Area = 27 in^2

53)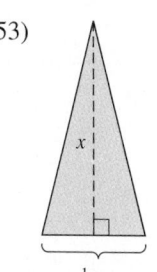
x
$\frac{1}{2}x$
Area = 36 cm^2

54)
$2x$
x
Area = 16 cm^2

Find the lengths of the sides of the following right triangles. (Hint: Use the Pythagorean theorem.)

 55)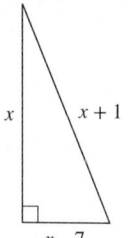
x $x + 1$
$x - 7$

56)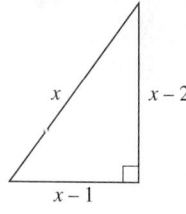
x $x - 2$
$x - 1$

57)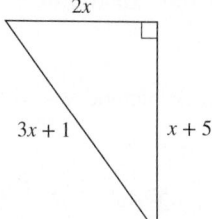
$2x$
$3x + 1$ $x + 5$

58)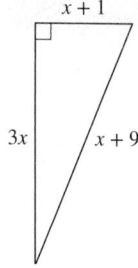
$x + 1$
$3x$ $x + 9$

R Rethink

R1) Thinking about Exercises 17–46, how would you distinguish among a linear equation, a quadratic equation, and a cubic equation? Compare the methods for solving them.

R2) Which exercises could you do easily and which were more difficult? Is there a pattern? What could you do to get better at solving equations?

11.2 The Square Root Property and Completing the Square

What are your objectives for Section 11.2?	How can you accomplish each objective?
1 Solve an Equation of the Form $x^2 = k$	• Understand the *square root property*. • Complete the given example on your own. • Complete You Try 1.
2 Solve an Equation of the Form $(ax + b)^2 = k$	• Be able to recognize a binomial that is being squared. • Understand the *square root property*. • Review the procedure for solving equations. • Complete the given examples on your own. • Complete You Trys 2 and 3.
3 Use the Distance Formula	• Understand the definition of *the distance formula*. • Review the **Pythagorean theorem**. • Complete the given example on your own. • Complete You Try 4.
4 Complete the Square for an Expression of the Form $x^2 + bx$	• Learn the procedure for **Completing the Square for $x^2 + bx$.** • Review factoring of *perfect square trinomials*. • Complete the given examples on your own. • Complete You Trys 5 and 6.
5 Solve a Quadratic Equation by Completing the Square	• Learn the procedure for **Solving a Quadratic Equation by Completing the Square.** • Review the *square root property*. • Complete the given example on your own. • Complete You Try 7.

 **Work**   **Read the explanations, follow the examples, take notes, and complete the You Trys.**

The next method we will learn for solving quadratic equations is the square root property.

1 Solve an Equation of the Form $x^2 = k$

Look at the equation $x^2 = 9$. We can solve this equation by factoring, like this:

$$x^2 = 9$$
$$x^2 - 9 = 0 \qquad \text{Get all terms on the same side.}$$
$$(x + 3)(x - 3) = 0 \qquad \text{Factor.}$$

$$x + 3 = 0 \quad \text{or} \quad x - 3 = 0 \qquad \text{Set each factor equal to zero.}$$
$$x = -3 \quad \text{or} \quad x = 3 \qquad \text{Solve.}$$

The solution set is $\{-3, 3\}$.

Or, we can solve an equation like $x^2 = 9$ using the **square root property** as we will see in Example 1a).

Definition The Square Root Property

Let k be a constant. If $x^2 = k$, then $x = \sqrt{k}$ or $x = -\sqrt{k}$.

(The solution is often written as $x = \pm\sqrt{k}$, read as "x equals plus or minus the square root of k.")

Note

We can use the square root property to solve an equation containing a squared quantity and a constant. To do so we will get the squared quantity containing the variable on one side of the equal sign and the constant on the other side.

EXAMPLE 1

Solve using the square root property.

a) $x^2 = 9$ b) $t^2 - 20 = 0$ c) $2a^2 + 21 = 3$

Solution

a)
$$x^2 = 9$$

$x = \sqrt{9}$ or $x = -\sqrt{9}$ Square root property
$x = 3$ or $x = -3$

The solution set is $\{-3, 3\}$. The check is left to the student.

An equivalent way to solve $x^2 = 9$ is to write it as

$$x^2 = 9$$
$$x = \pm\sqrt{9} \quad \text{Square root property}$$
$$x = \pm 3$$

The solution set is $\{-3, 3\}$. We will use this approach when solving equations using the square root property.

W Hint

Be sure you are writing out each step as you are reading the example.

b) To solve $t^2 - 20 = 0$, begin by getting t^2 on a side by itself.

$$t^2 - 20 = 0$$
$$t^2 = 20 \qquad \text{Add 20 to each side.}$$
$$t = \pm\sqrt{20} \qquad \text{Square root property}$$
$$t = \pm\sqrt{4} \cdot \sqrt{5} \qquad \text{Product rule for radicals}$$
$$t = \pm 2\sqrt{5} \qquad \sqrt{4} = 2$$

Check:

$t = 2\sqrt{5}$: $\quad t^2 - 20 = 0$
$\qquad (2\sqrt{5})^2 - 20 \overset{?}{=} 0$
$\qquad (4 \cdot 5) - 20 \overset{?}{=} 0$
$\qquad 20 - 20 = 0$ ✓

$t = -2\sqrt{5}$: $\quad t^2 - 20 = 0$
$\qquad (-2\sqrt{5})^2 - 20 \overset{?}{=} 0$
$\qquad (4 \cdot 5) - 20 \overset{?}{=} 0$
$\qquad 20 - 20 = 0$ ✓

The solution set is $\{-2\sqrt{5}, 2\sqrt{5}\}$.

c) $2a^2 + 21 = 3$

$\qquad 2a^2 = -18$ Subtract 21.

$\qquad a^2 = -9$ Divide by 2.

$\qquad a = \pm\sqrt{-9}$ Square root property

$\qquad a = \pm 3i$

Check:

$a = 3i$: $2a^2 + 21 = 3$ $a = -3i$: $2a^2 + 21 = 3$

$\qquad\qquad 2(3i)^2 + 21 \overset{?}{=} 3$ $2(-3i)^2 + 21 \overset{?}{=} 3$

$\qquad\qquad 2(9i^2) + 21 \overset{?}{=} 3$ $2(9i^2) + 21 \overset{?}{=} 3$

$\qquad\qquad 2(9)(-1) + 21 \overset{?}{=} 3$ $2(9)(-1) + 21 \overset{?}{=} 3$

$\qquad\qquad -18 + 21 = 3 \; \checkmark$ $-18 + 21 = 3 \; \checkmark$

The solution set is $\{-3i, 3i\}$.

[YOU TRY 1]

Solve using the square root property.

a) $p^2 = 100$ b) $w^2 - 32 = 0$ c) $3m^2 + 19 = 7$

Can we solve $(w - 4)^2 = 25$ using the square root property? Yes. The equation has a *squared quantity* and a *constant*.

2   Solve an Equation of the Form $(ax + b)^2 = k$

EXAMPLE 2

Solve $x^2 = 25$ and $(w - 4)^2 = 25$ using the square root property.

Solution

While the equation $(w - 4)^2 = 25$ has a *binomial* that is being squared, the two equations are actually in the same form.

$$x^2 = 25 \qquad\qquad\qquad (w - 4)^2 = 25$$

$\qquad\qquad\quad \uparrow \quad\quad \uparrow \qquad\qquad\qquad\qquad \uparrow \qquad\quad \uparrow$

$\qquad\qquad$ *x* squared = constant $\qquad\quad$ (*w* − 4) squared = constant

Solve $x^2 = 25$:

$$x^2 = 25$$

$\qquad\qquad x = \pm\sqrt{25}$ Square root property

$\qquad\qquad x = \pm 5$

The solution set is $\{-5, 5\}$.

We solve $(w - 4)^2 = 25$ in the same way with some additional steps.

$$(w - 4)^2 = 25$$

$\qquad\qquad w - 4 = \pm\sqrt{25}$ Square root property

$\qquad\qquad w - 4 = \pm 5$

This means $w - 4 = 5$ or $w - 4 = -5$. Solve both equations.

$\qquad\qquad w - 4 = 5$ or $w - 4 = -5$

$\qquad\qquad\quad w = 9$ or $w = -1$ Add 4 to each side.

Check:

$w = 9$: $(w-4)^2 = 25$
$\quad\quad\quad (9-4)^2 \overset{?}{=} 25$
$\quad\quad\quad\quad\quad 5^2 = 25$ ✓

$w = -1$: $(w-4)^2 = 25$
$\quad\quad\quad\quad (-1-4)^2 \overset{?}{=} 25$
$\quad\quad\quad\quad\quad (-5)^2 = 25$ ✓

The solution set is $\{-1, 9\}$.

[YOU TRY 2] Solve $(c + 6)^2 = 81$ using the square root property.

EXAMPLE 3 Solve.

a) $(3t + 4)^2 = 9$ b) $(2m - 5)^2 = 12$ c) $(z + 8)^2 + 11 = 7$

d) $(6k - 5)^2 + 20 = 0$

Solution

a) $(3t + 4)^2 = 9$
$\quad\quad 3t + 4 = \pm\sqrt{9}$ Square root property
$\quad\quad 3t + 4 = \pm 3$

This means $3t + 4 = 3$ or $3t + 4 = -3$. Solve both equations.

$\quad\quad 3t + 4 = 3$ $\quad$ or $\quad 3t + 4 = -3$
$\quad\quad\quad\quad 3t = -1$ $\quad\quad\quad\quad\quad 3t = -7$ Subtract 4 from each side.
$\quad\quad\quad\quad\quad t = -\dfrac{1}{3}$ $\quad$ or $\quad\quad t = -\dfrac{7}{3}$ Divide by 3.

The solution set is $\left\{-\dfrac{7}{3}, -\dfrac{1}{3}\right\}$.

W Hint

Look back at Section 10.3 if you need to review how to simplify square roots.

b) $(2m - 5)^2 = 12$
$\quad\quad 2m - 5 = \pm\sqrt{12}$ $\quad\quad$ Square root property
$\quad\quad 2m - 5 = \pm 2\sqrt{3}$ $\quad\quad$ Simplify $\sqrt{12}$.
$\quad\quad\quad 2m = 5 \pm 2\sqrt{3}$ $\quad\quad$ Add 5 to each side.
$\quad\quad\quad\quad m = \dfrac{5 \pm 2\sqrt{3}}{2}$ $\quad\quad$ Divide by 2.

One solution is $\dfrac{5 + 2\sqrt{3}}{2}$, and the other is $\dfrac{5 - 2\sqrt{3}}{2}$.

The solution set, $\left\{\dfrac{5 - 2\sqrt{3}}{2}, \dfrac{5 + 2\sqrt{3}}{2}\right\}$, can also be written as $\left\{\dfrac{5 \pm 2\sqrt{3}}{2}\right\}$.

c) $(z + 8)^2 + 11 = 7$
$\quad\quad (z + 8)^2 = -4$ $\quad\quad$ Subtract 11 from each side.
$\quad\quad z + 8 = \pm\sqrt{-4}$ $\quad\quad$ Square root property
$\quad\quad z + 8 = \pm 2i$ $\quad\quad$ Simplify $\sqrt{-4}$.
$\quad\quad\quad\quad z = -8 \pm 2i$ $\quad\quad$ Subtract 8 from each side.

The check is left to the student. The solution set is $\{-8 - 2i, -8 + 2i\}$.

d) $(6k - 5)^2 + 20 = 0$

$$(6k - 5)^2 = -20 \qquad \text{Subtract 20 from each side.}$$
$$6k - 5 = \pm\sqrt{-20} \qquad \text{Square root property}$$
$$6k - 5 = \pm 2i\sqrt{5} \qquad \text{Simplify } \sqrt{-20}.$$
$$6k = 5 \pm 2i\sqrt{5} \qquad \text{Add 5 to each side.}$$
$$k = \frac{5 \pm 2i\sqrt{5}}{6} \qquad \text{Divide by 6.}$$

The check is left to the student. The solution set is $\left\{ \dfrac{5 - 2i\sqrt{5}}{6}, \dfrac{5 + 2i\sqrt{5}}{6} \right\}$.

[YOU TRY 3] Solve.

a) $(7q + 1)^2 = 36$ b) $(5a - 3)^2 = 24$

c) $(c - 7)^2 + 100 = 0$ d) $(2y + 3)^2 - 5 = -23$

Did you notice in Examples 1c), 3c), and 3d) that a complex number *and* its conjugate were the solutions to the equations? This will always be true provided that the variables in the equation have real number coefficients.

Note

If $a + bi$ is a solution of a quadratic equation having only real coefficients, then $a - bi$ is also a solution.

3 Use the Distance Formula

In mathematics, we sometimes need to find the distance between two points in a plane. The **distance formula** enables us to do that. We can use the Pythagorean theorem and the square root property to develop the distance formula.

Suppose we want to find the distance between any two points with coordinates (x_1, y_1) and (x_2, y_2) as pictured here. [We also include the point (x_2, y_1) in our drawing so that we get a right triangle.]

The lengths of the legs are a and b. The length of the hypotenuse is c. Our goal is to find the *distance* between (x_1, y_1) and (x_2, y_2), *which is the same as* finding the length of c.

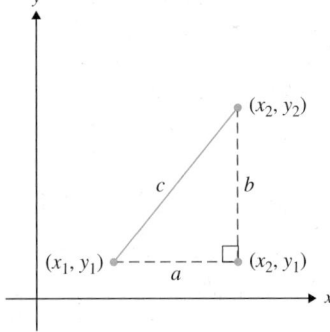

How long is side a? $|x_2 - x_1|$
How long is side b? $|y_2 - y_1|$

The Pythagorean theorem states that $a^2 + b^2 = c^2$. Substitute $|x_2 - x_1|$ for a and $|y_2 - y_1|$ for b, then solve for c.

$$a^2 + b^2 = c^2 \qquad \text{Pythagorean theorem}$$
$$|x_2 - x_1|^2 + |y_2 - y_1|^2 = c^2 \qquad \text{Substitute values.}$$
$$\pm\sqrt{(x_2 - x_1)^2 + (y_2 - y_1)^2} = c \qquad \text{Solve for } c \text{ using the square root property.}$$

W Hint

Distances will always be a positive number or zero.

The distance between the points (x_1, y_1) and (x_2, y_2) is $c = \sqrt{(x_2 - x_1)^2 + (y_2 - y_1)^2}$. We want only the positive square root since c is a length.

Because this formula represents the *distance* between two points, we usually use the letter *d* instead of *c*.

Definition The Distance Formula

The distance, *d*, between two points with coordinates (x_1, y_1) and (x_2, y_2) is given by

$$d = \sqrt{(x_2 - x_1)^2 + (y_2 - y_1)^2}.$$

EXAMPLE 4

Find the distance between the points $(-4, 1)$ and $(2, 5)$.

Solution

Begin by labeling the points: $(\overset{x_1, y_1}{-4, 1}), (\overset{x_2, y_2}{2, 5})$.

Substitute the values into the distance formula.

$$\begin{aligned} d &= \sqrt{(x_2 - x_1)^2 + (y_2 - y_1)^2} \\ &= \sqrt{[2 - (-4)]^2 + (5 - 1)^2} \quad \text{Substitute values.} \\ &= \sqrt{(2 + 4)^2 + (4)^2} \\ &= \sqrt{(6)^2 + (4)^2} = \sqrt{36 + 16} = \sqrt{52} = 2\sqrt{13} \end{aligned}$$

[YOU TRY 4]   Find the distance between the points $(1, 2)$ and $(7, -3)$.

The next method we will learn for solving a quadratic equation is *completing the square*. We need to review an idea first presented in Section 7.4.

A **perfect square trinomial** is a trinomial whose factored form is the square of a binomial. Some examples of perfect square trinomials are

Perfect Square Trinomials	Factored Form
$x^2 + 10x + 25$	$(x + 5)^2$
$d^2 - 8d + 16$	$(d - 4)^2$

In the trinomial $x^2 + 10x + 25$, x^2 is called the *quadratic term*, $10x$ is called the *linear term*, and 25 is called the *constant*.

4 Complete the Square for an Expression of the Form $x^2 + bx$

In a perfect square trinomial where the coefficient of the quadratic term is 1, the constant term is related to the coefficient of the linear term in the following way: *If you find half of the linear coefficient and square the result, you will get the constant term.*

$x^2 + 10x + 25$:   The constant, 25, is obtained by

1) finding half of the coefficient of *x*;   then   2) squaring the result.

$$\frac{1}{2}(10) = 5 \qquad\qquad 5^2 = 25 \text{ (the constant)}$$

$d^2 - 8d + 16$: The constant, 16, is obtained by

1) finding half of the coefficient of d; then 2) squaring the result.

$$\frac{1}{2}(-8) = -4$$

$$(-4)^2 = 16 \text{ (the constant)}$$

We can generalize this procedure so that we can find the constant needed to obtain the perfect square trinomial for any quadratic expression of the form $x^2 + bx$. Finding this perfect square trinomial is called **completing the square** because the trinomial will factor to the square of a binomial.

Procedure Completing the Square for $x^2 + bx$

To find the constant needed to complete the square for $x^2 + bx$:

Step 1: Find half of the coefficient of x: $\frac{1}{2}b$.

Step 2: Square the result: $\left(\frac{1}{2}b\right)^2$.

Step 3: Then add it to $x^2 + bx$ to get $x^2 + bx + \left(\frac{1}{2}b\right)^2$. The factored form is $\left(x + \frac{1}{2}b\right)^2$.

 **BE CAREFUL** The coefficient of the squared term *must* be 1 before you complete the square!

EXAMPLE 5 Complete the square for each expression to obtain a perfect square trinomial. Then, factor.

a) $y^2 + 6y$ b) $t^2 - 14t$

Solution

a) Find the constant needed to complete the square for $y^2 + 6y$.

Step 1: Find half of the coefficient of y:

$$\frac{1}{2}(6) = 3$$

Step 2: Square the result:

$$3^2 = 9$$

Step 3: Add 9 to $y^2 + 6y$:

$$y^2 + 6y + 9$$

The perfect square trinomial is $y^2 + 6y + 9$. The factored form is $(y + 3)^2$.

b) Find the constant needed to complete the square for $t^2 - 14t$.

Step 1: Find half of the coefficient of t:

$$\frac{1}{2}(-14) = -7$$

Step 2: Square the result:

$$(-7)^2 = 49$$

Step 3: Add 49 to $t^2 - 14t$:

$$t^2 - 14t + 49$$

The perfect square trinomial is $t^2 - 14t + 49$. The factored form is $(t - 7)^2$.

Complete the square for each expression to obtain a perfect square trinomial. Then, factor.

a) $w^2 + 2w$ b) $z^2 - 16z$

We've seen the following perfect square trinomials and their factored forms. Let's look at the relationship between the constant in the factored form and the coefficient of the linear term.

W Hint

Be sure you understand these relationships before reading Example 6.

Perfect Square Trinomial		Factored Form
$x^2 + 10x + 25$	5 is $\frac{1}{2}(10)$.	$(x + 5)^2$
$d^2 - 8d + 16$	-4 is $\frac{1}{2}(-8)$.	$(d - 4)^2$
$y^2 + 6y + 9$	3 is $\frac{1}{2}(6)$.	$(d + 3)^2$
$t^2 - 14t + 49$	-7 is $\frac{1}{2}(-14)$.	$(t - 7)^2$

This pattern will always hold true and can be helpful in factoring some perfect square trinomials.

EXAMPLE 6

Complete the square for $n^2 + 5n$ to obtain a perfect square trinomial. Then, factor.

Solution

Find the constant needed to complete the square for $n^2 + 5n$.

Step 1: Find half of the coefficient of n: $\frac{1}{2}(5) = \frac{5}{2}$

Step 2: Square the result: $\left(\frac{5}{2}\right)^2 = \frac{25}{4}$

Step 3: Add $\frac{25}{4}$ to $n^2 + 5n$. The perfect square trinomial is $n^2 + 5n + \frac{25}{4}$.

The factored form is $\left(n + \frac{5}{2}\right)^2$.

$\frac{5}{2}$ is $\frac{1}{2}(5)$, the coefficient of n.

Check: $\left(n + \frac{5}{2}\right)^2 = n^2 + 2n\left(\frac{5}{2}\right) + \left(\frac{5}{2}\right)^2 = n^2 + 5n + \frac{25}{4}$

Complete the square for $p^2 - 3p$ to obtain a perfect square trinomial. Then, factor.

5 Solve a Quadratic Equation by Completing the Square

Any quadratic equation of the form $ax^2 + bx + c = 0$ $(a \neq 0)$ can be written in the form $(x - h)^2 = k$ by completing the square. Once an equation is in this form, we can use the square root property to solve for the variable.

> **Procedure** Solve a Quadratic Equation $(ax^2 + bx + c = 0)$ by Completing the Square
>
> **Step 1:** **The coefficient of the squared term must be 1.** If it is not 1, divide both sides of the equation by a to obtain a leading coefficient of 1.
>
> **Step 2:** **Get the variables on one side of the equal sign and the constant on the other side.**
>
> **Step 3:** **Complete the square.** Find half of the linear coefficient, then square the result. Add that quantity to *both* sides of the equation.
>
> **Step 4:** **Factor.**
>
> **Step 5:** **Solve using the square root property.**

EXAMPLE 7

Solve by completing the square.

a) $x^2 + 6x + 8 = 0$ b) $12h + 4h^2 = -24$

Solution

a) $x^2 + 6x + 8 = 0$

Step 1: The coefficient of x^2 is already 1.

Step 2: Get the variables on one side of the equal sign and the constant on the other side: $x^2 + 6x = -8$

Step 3: Complete the square: $\dfrac{1}{2}(6) = 3$

$3^2 = 9$

Add 9 to both sides of the equation: $x^2 + 6x + 9 = -8 + 9$
$$x^2 + 6x + 9 = 1$$

Step 4: Factor: $(x + 3)^2 = 1$

Step 5: Solve using the square root property.

$$(x + 3)^2 = 1$$
$$x + 3 = \pm\sqrt{1}$$
$$x + 3 = \pm 1$$

$x + 3 = 1$ or $x + 3 = -1$
$x = -2$ or $x = -4$

The check is left to the student. The solution set is $\{-4, -2\}$.

Note

We would have obtained the same result if we had solved the equation by factoring.

$$x^2 + 6x + 8 = 0$$
$$(x + 4)(x + 2) = 0$$
$$x + 4 = 0 \quad \text{or} \quad x + 2 = 0$$
$$x = -4 \quad \text{or} \quad x = -2$$

b) $12h + 4h^2 = -24$

Step 1: Because the coefficient of h^2 is *not* 1, divide the whole equation by 4.

$$\frac{12h}{4} + \frac{4h^2}{4} = \frac{-24}{4}$$
$$3h + h^2 = -6$$

Step 2: The constant is on a side by itself. Rewrite the left side of the equation.

$$h^2 + 3h = -6$$

Step 3: Complete the square: $\dfrac{1}{2}(3) = \dfrac{3}{2}$

$$\left(\frac{3}{2}\right)^2 = \frac{9}{4}$$

Add $\dfrac{9}{4}$ to both sides of the equation.

$$h^2 + 3h + \frac{9}{4} = -6 + \frac{9}{4}$$
$$h^2 + 3h + \frac{9}{4} = -\frac{24}{4} + \frac{9}{4} \qquad \text{Get a common denominator.}$$
$$h^2 + 3h + \frac{9}{4} = -\frac{15}{4}$$

Step 4: Factor.

$$\left(h + \frac{3}{2}\right)^2 = -\frac{15}{4}$$

$$\uparrow$$

$\dfrac{3}{2}$ is $\dfrac{1}{2}(3)$, the coefficient of h.

Step 5: Solve using the square root property.

$$\left(h + \frac{3}{2}\right)^2 = -\frac{15}{4}$$
$$h + \frac{3}{2} = \pm\sqrt{-\frac{15}{4}}$$
$$h + \frac{3}{2} = \pm\frac{\sqrt{15}}{2}i \qquad \text{Simplify the radical.}$$
$$h = -\frac{3}{2} \pm \frac{\sqrt{15}}{2}i \qquad \text{Subtract } \frac{3}{2}.$$

The check is left to the student. The solution set is

$$\left\{-\frac{3}{2} - \frac{\sqrt{15}}{2}i, \ -\frac{3}{2} + \frac{\sqrt{15}}{2}i\right\}.$$

W Hint

This quadratic equation has nonreal, complex solutions.

ANSWERS TO [YOU TRY] EXERCISES

1) a) $\{-10, 10\}$ b) $\{-4\sqrt{2}, 4\sqrt{2}\}$ c) $\{-2i, 2i\}$ 2) $\{-15, 3\}$

3) a) $\left\{-1, \dfrac{5}{7}\right\}$ b) $\left\{\dfrac{3 - 2\sqrt{6}}{5}, \dfrac{3 + 2\sqrt{6}}{5}\right\}$ c) $\{7 - 10i, 7 + 10i\}$

d) $\left\{-\dfrac{3}{2} - \dfrac{3\sqrt{2}}{2}i, \dfrac{3}{2} + \dfrac{3\sqrt{2}}{2}i\right\}$ 4) $\sqrt{61}$

5) a) $w^2 + 2w + 1; (w + 1)^2$ b) $z^2 - 16z + 64; (z - 8)^2$

6) $p^2 - 3p + \dfrac{9}{4}; \left(p - \dfrac{3}{2}\right)^2$ 7) a) $\{-12, 2\}$ b) $\left\{\dfrac{5}{2} - \dfrac{\sqrt{7}}{2}i, \dfrac{5}{2} + \dfrac{\sqrt{7}}{2}i\right\}$

E Evaluate **11.2** **Exercises**   Do the exercises, and check your work.

Objective 1: Solve an Equation of the Form $x^2 = k$

1) Choose two methods to solve $y^2 - 16 = 0$. Solve the equation using both methods.

2) If k is a negative number and $x^2 = k$, what can you conclude about the solution to the equation?

Solve using the square root property.

3) $b^2 = 36$

4) $h^2 = 64$

5) $r^2 - 27 = 0$

6) $a^2 - 30 = 0$

7) $n^2 = \dfrac{4}{9}$

8) $v^2 = \dfrac{121}{16}$

9) $q^2 = -4$

10) $w^2 = -121$

11) $z^2 + 3 = 0$

12) $h^2 + 14 = -23$

13) $z^2 + 5 = 19$

14) $q^2 - 3 = 15$

15) $4m^2 + 1 = 37$

16) $2d^2 + 5 = 55$

17) $5f^2 + 39 = -21$

18) $2y^2 + 56 = 0$

Objective 2: Solve an Equation of the Form $(ax + b)^2 = k$

Solve using the square root property.

19) $(r + 10)^2 = 4$

20) $(x - 5)^2 = 81$

21) $(q - 7)^2 = 1$

22) $(c + 12)^2 = 25$

23) $(p + 4)^2 - 18 = 0$

24) $(d + 2)^2 - 7 = 13$

 25) $(c + 3)^2 - 4 = -29$

26) $(u - 15)^2 - 4 = -8$

27) $1 = 15 + (k - 2)^2$

28) $2 = 14 + (g + 4)^2$

29) $20 = (2w + 1)^2$

30) $(5b - 6)^2 = 11$

31) $8 = (3q - 10)^2 - 6$

32) $22 = (6x + 11)^2 + 4$

33) $36 + (4p - 5)^2 = 6$

34) $(3k - 1)^2 + 20 = 4$

35) $(6g + 11)^2 + 50 = 1$

36) $9 = 38 + (9s - 4)^2$

37) $\left(\dfrac{3}{4}n - 8\right)^2 = 4$

38) $\left(\dfrac{2}{3}j + 10\right)^2 = 16$

39) $(5y - 2)^2 + 6 = 22$

40) $-6 = 3 - (2q - 9)^2$

Objective 3: Use the Distance Formula

Find the distance between the given points.

41) $(7, -1)$ and $(3, 2)$

42) $(3, 10)$ and $(12, 6)$

43) $(-5, -6)$ and $(-2, -8)$

44) $(5, -2)$ and $(-3, 4)$

45) $(0, 3)$ and $(3, -1)$

46) $(-8, 3)$ and $(2, 1)$

47) $(-4, 11)$ and $(2, 6)$

48) $(0, 13)$ and $(0, 7)$

49) $(3, -3)$ and $(5, -7)$

50) $(-5, -6)$ and $(-1, 2)$

Objective 4: Complete the Square for an Expression of the Form $x^2 + bx$

51) What is a perfect square trinomial? Give an example.

52) Can you complete the square on $3y^2 + 15y$ as it is given? Why or why not?

Complete the square for each expression to obtain a perfect square trinomial. Then, factor.

Fill It In

Fill in the blanks with either the missing mathematical step or reason for the given step.

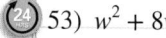 53) $w^2 + 8w$ _____

     _____ Find half of the coefficient of w.

     _____ Square the result.

     _____ Add the constant to the expression.

The perfect square trinomial is _____

The factored form of the trinomial is _____

54) $n^2 - n$

$\frac{1}{2}(-1) = -\frac{1}{2}$ _____

$\left(-\frac{1}{2}\right)^2 = \frac{1}{4}$ _____

$n^2 - n + \frac{1}{4}$ _____

The perfect square trinomial is

     _____

The factored form of the trinomial is

     _____

55) $a^2 + 12a$ 56) $g^2 + 4g$

57) $c^2 - 18c$ 58) $k^2 - 16k$

59) $t^2 + 5t$ 60) $z^2 - 7z$

61) $b^2 - 9b$ 62) $r^2 + 3r$

63) $x^2 + \frac{1}{3}x$ 64) $y^2 - \frac{3}{5}y$

Objective 5: Solve a Quadratic Equation by Completing the Square

65) What is the first thing you should do if you want to solve $2p^2 - 7p = 8$ by completing the square?

66) Can $x^3 + 10x - 3 = 0$ be solved by completing the square? Give a reason for your answer.

Solve by completing the square.

67) $x^2 + 6x + 8 = 0$ 68) $t^2 + 12t - 13 = 0$

69) $k^2 - 8k + 15 = 0$ 70) $v^2 - 6v - 27 = 0$

71) $u^2 - 9 = 2u$ 72) $s^2 + 10 = -10s$

73) $p^2 = -10p - 26$ 74) $t^2 = 2t - 9$

75) $a^2 + 19 = 8a$ 76) $v^2 + 4v + 8 = 0$

77) $m^2 + 3m - 40 = 0$ 78) $p^2 + 5p + 4 = 0$

79) $x^2 - 7x + 12 = 0$ 80) $d^2 + d - 72 = 0$

81) $r^2 - r = 3$ 82) $y^2 - 3y = 7$

83) $c^2 + 5c + 7 = 0$ 84) $b^2 + 14 = 7b$

85) $3k^2 - 6k + 12 = 0$ 86) $4f^2 + 16f + 48 = 0$

87) $4r^2 + 24r = 8$ 88) $3h^2 + 6h = 15$

89) $10d = 2d^2 + 12$ 90) $54x - 6x^2 = 48$

91) $2n^2 + 8 = 5n$ 92) $2t^2 + 3t + 4 = 0$

93) $4a^2 - 7a + 3 = 0$ 94) $n + 2 = 3n^2$

95) $(y + 5)(y - 3) = 5$ 96) $(b - 4)(b + 10) = -17$

97) $(2m + 1)(m - 3) = -7$ 98) $(3c + 4)(c + 2) = 3$

Use the Pythagorean theorem and the square root property to find the length of the missing side.

99)
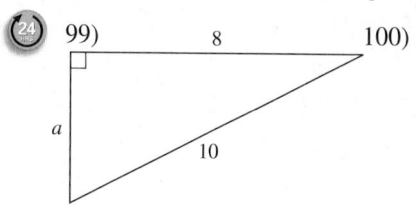

100)
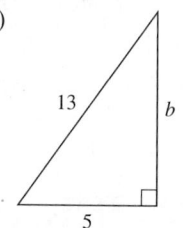

101)   102)
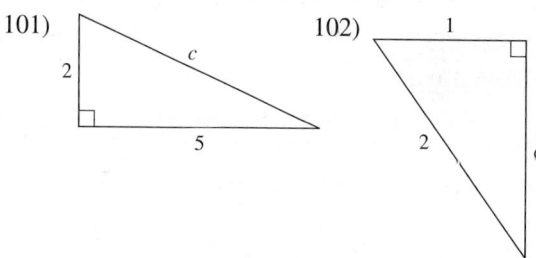

Write an equation, and solve. (Hint: Draw a picture.)

103) The width of a rectangle is 4 in., and its diagonal is $2\sqrt{13}$ in. long. What is the length of the rectangle?

104) Find the length of the diagonal of a rectangle if it has a width of 5 cm and a length of $4\sqrt{2}$ cm.

Write an equation, and solve.

105) A 13-ft ladder is leaning against a wall so that the base of the ladder is 5 ft away from the wall. How high on the wall does the ladder reach?

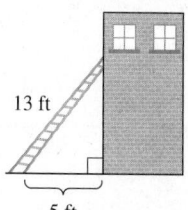

106) Salma is flying a kite. It is 30 ft from her horizontally, and it is 40 ft above her hand. How long is the kite string?

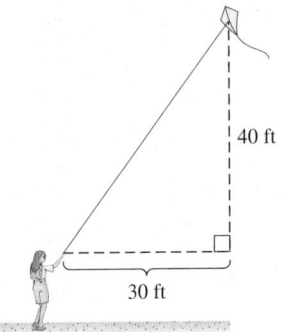

40 ft

30 ft

107) Let $f(x) = (x + 3)^2$. Find x so that $f(x) = 49$.

108) Let $g(t) = (t - 5)^2$. Find t so that $g(t) = 12$.

Solve each problem by writing an equation and solving it by completing the square.

109) The length of a rectangular garden is 8 ft more than its width. Find the dimensions of the garden if it has an area of 153 ft^2.

110) The rectangular screen on a laptop has an area of 375 cm^2. Its width is 10 cm less than its length. What are the dimensions of the screen?

R Rethink

R1) Why must the coefficient be "1" before you use the square root property?

R2) What is a real-life situation where you may need to use the distance formula?

R3) What do you find most difficult about completing the square? After you complete the square, which method do you use next?

11.3 The Quadratic Formula

P Prepare

What are your objectives for Section 11.3?

O Organize

How can you accomplish each objective?

What are your objectives for Section 11.3?	How can you accomplish each objective?
1 Derive the Quadratic Formula	• Understand where the *quadratic formula* comes from. • Learn the *quadratic formula*. Practice writing it several times in your notes.
2 Solve a Quadratic Equation Using the Quadratic Formula	• Know how to write a quadratic equation in standard form. • Be able to identify a, b, and c in the quadratic formula. • Complete the given examples on your own. • Complete You Trys 1 and 2.
3 Determine the Number and Type of Solutions of a Quadratic Equation Using the Discriminant	• Write out all the possible types/number of solutions for a *discriminant* in your notes. • Complete the given example on your own. • Complete You Try 3.
4 Solve an Applied Problem Using the Quadratic Formula	• Read the problem carefully. • In a motion equation, be able to determine if you are solving for height (h) or solving for time (t), if applicable. • Review the **Pythagorean theorem.** • Complete the given example on your own. • Complete You Try 4.

Read the explanations, follow the examples, take notes, and complete the You Trys.

1 Derive the Quadratic Formula

In Section 11.2, we saw that any quadratic equation of the form $ax^2 + bx + c = 0$ $(a \neq 0)$ can be solved by completing the square. Therefore, we can solve equations like $x^2 - 8x + 5 = 0$ and $2x^2 + 3x - 1 = 0$ using this method.

We can develop another method for solving quadratic equations by completing the square on the general quadratic equation $ax^2 + bx + c = 0$ $(a \neq 0)$. This will let us derive the *quadratic formula*.

The steps we use to complete the square on $ax^2 + bx + c = 0$ are *exactly* the same steps we use to solve an equation like $2x^2 + 3x - 1 = 0$. We will do these steps side by side so that you can more easily understand how we are solving $ax^2 + bx + c = 0$ for x by completing the square.

Solve for x by Completing the Square.

$$2x^2 + 3x - 1 = 0 \qquad\qquad ax^2 + bx + c = 0$$

Step 1: The coefficient of the squared term must be 1.

$$2x^2 + 3x - 1 = 0 \qquad\qquad ax^2 + bx + c = 0$$

$$\frac{2x^2}{2} + \frac{3x}{2} - \frac{1}{2} = \frac{0}{2} \quad \text{Divide by 2.} \qquad \frac{ax^2}{a} + \frac{bx}{a} + \frac{c}{a} = \frac{0}{a} \quad \text{Divide by } a.$$

$$x^2 + \frac{3}{2}x - \frac{1}{2} = 0 \quad \text{Simplify.} \qquad x^2 + \frac{b}{a}x + \frac{c}{a} = 0 \quad \text{Simplify.}$$

W Hint

Notice that we are using the same steps that we used in Section 11.2 to solve *both* equations.

Step 2: Get the constant on the other side of the equal sign.

$$x^2 + \frac{3}{2}x = \frac{1}{2} \quad \text{Add } \frac{1}{2}. \qquad\qquad x^2 + \frac{b}{a}x = -\frac{c}{a} \quad \text{Subtract } \frac{c}{a}.$$

Step 3: Complete the square.

$$\frac{1}{2}\left(\frac{3}{2}\right) = \frac{3}{4} \quad \tfrac{1}{2} \text{ of } x\text{-coefficient} \qquad \frac{1}{2}\left(\frac{b}{a}\right) = \frac{b}{2a} \quad \tfrac{1}{2} \text{ of } x\text{-coefficient}$$

$$\left(\frac{3}{4}\right)^2 = \frac{9}{16} \quad \text{Square the result.} \qquad \left(\frac{b}{2a}\right)^2 = \frac{b^2}{4a^2} \quad \text{Square the result.}$$

Add $\dfrac{9}{16}$ to both sides of the equation. $\qquad$ Add $\dfrac{b^2}{4a^2}$ to both sides of the equation.

$$x^2 + \frac{3}{2}x + \frac{9}{16} = \frac{1}{2} + \frac{9}{16} \qquad\qquad x^2 + \frac{b}{a}x + \frac{b^2}{4a^2} = -\frac{c}{a} + \frac{b^2}{4a^2}$$

$$x^2 + \frac{3}{2}x + \frac{9}{16} = \frac{8}{16} + \frac{9}{16} \quad \begin{array}{l}\text{Get a}\\ \text{common}\\ \text{denominator.}\end{array} \qquad x^2 + \frac{b}{a}x + \frac{b^2}{4a^2} = -\frac{4ac}{4a^2} + \frac{b^2}{4a^2} \quad \begin{array}{l}\text{Get a}\\ \text{common}\\ \text{denominator.}\end{array}$$

$$x^2 + \frac{3}{2}x + \frac{9}{16} = \frac{17}{16} \quad \text{Add.} \qquad x^2 + \frac{b}{a}x + \frac{b^2}{4a^2} = \frac{b^2 - 4ac}{4a^2} \quad \text{Add.}$$

Step 4: **Factor.**

$$\left(x + \frac{3}{4}\right)^2 = \frac{17}{16}$$

$\uparrow$

$\frac{3}{4}$ is $\frac{1}{2}\left(\frac{3}{2}\right)$, the coefficient of x.

$$\left(x + \frac{b}{2a}\right)^2 = \frac{b^2 - 4ac}{4a^2}$$

$\uparrow$

$\frac{b}{2a}$ is $\frac{1}{2}\left(\frac{b}{a}\right)$, the coefficient of x.

Step 5: **Solve using the square root property.**

$$\left(x + \frac{3}{4}\right)^2 = \frac{17}{16}$$

$$x + \frac{3}{4} = \pm\sqrt{\frac{17}{16}}$$

$$x + \frac{3}{4} = \frac{\pm\sqrt{17}}{4} \qquad \sqrt{16} = 4$$

$$x = -\frac{3}{4} \pm \frac{\sqrt{17}}{4} \qquad \text{Subtract } \frac{3}{4}.$$

$$x = \frac{-3 \pm \sqrt{17}}{4} \qquad \begin{array}{l}\text{Same denomi-}\\ \text{nators, add}\\ \text{numerators.}\end{array}$$

$$\left(x + \frac{b}{2a}\right)^2 = \frac{b^2 - 4ac}{4a^2}$$

$$x + \frac{b}{2a} = \pm\sqrt{\frac{b^2 - 4ac}{4a^2}}$$

$$x + \frac{b}{2a} = \frac{\pm\sqrt{b^2 - 4ac}}{2a} \qquad \sqrt{4a^2} = 2a$$

$$x = -\frac{b}{2a} \pm \frac{\sqrt{b^2 - 4ac}}{2a} \qquad \text{Subtract } \frac{b}{2a}.$$

$$x = \frac{-b \pm \sqrt{b^2 - 4ac}}{2a} \qquad \begin{array}{l}\text{Same denomi-}\\ \text{nators, add}\\ \text{numerators.}\end{array}$$

The result on the right is called the *quadratic formula*.

W Hint

Memorize this formula!

Definition The Quadratic Formula

The solutions of any quadratic equation of the form $ax^2 + bx + c = 0$ $(a \neq 0)$ are

$$x = \frac{-b \pm \sqrt{b^2 - 4ac}}{2a}$$

This formula is called the **quadratic formula.**

Note

1) To use the quadratic formula, write the equation to be solved in the form $ax^2 + bx + c = 0$ so that a, b, and c can be identified correctly.

2) $x = \dfrac{-b \pm \sqrt{b^2 - 4ac}}{2a}$ represents the two solutions

$$x = \frac{-b + \sqrt{b^2 - 4ac}}{2a} \quad \text{and} \quad x = \frac{-b - \sqrt{b^2 - 4ac}}{2a}.$$

3) Notice that the fraction bar runs under $-b$ and under the radical.

$$x = \frac{-b \pm \sqrt{b^2 - 4ac}}{2a} \qquad\qquad x = -b \pm \frac{\sqrt{b^2 - 4ac}}{2a}$$

Correct   Incorrect

4) When deriving the quadratic formula, using the $\pm$ allows us to say that $\sqrt{4a^2} = 2a$.

5) The quadratic formula is a *very* important result, and we will use it often. *It should be memorized!*

2 Solve a Quadratic Equation Using the Quadratic Formula

EXAMPLE 1

Solve using the quadratic formula.

a) $2x^2 + 3x - 1 = 0$ b) $k^2 = 10k - 29$

Solution

a) Is $2x^2 + 3x - 1 = 0$ in the form $ax^2 + bx + c = 0$? Yes. Identify the values of a, b, and c, and substitute them into the quadratic formula.

$$a = 2 \qquad b = 3 \qquad c = -1$$

<table>
<tr><td>

$$x = \dfrac{-b \pm \sqrt{b^2 - 4ac}}{2a}$$

</td><td>Quadratic formula</td></tr>
<tr><td>

$$= \dfrac{-(3) \pm \sqrt{(3)^2 - 4(2)(-1)}}{2(2)}$$

</td><td>Substitute $a = 2$, $b = 3$, and $c = -1$.</td></tr>
<tr><td>

$$= \dfrac{-3 \pm \sqrt{9 - (-8)}}{4}$$

</td><td>Perform the operations.</td></tr>
<tr><td>

$$= \dfrac{-3 \pm \sqrt{17}}{4}$$

</td><td></td></tr>
</table>

> **W Hint**
>
> Write down the quadratic formula and the values of a, b, and c on your paper.

The solution set is $\left\{ \dfrac{-3 - \sqrt{17}}{4}, \dfrac{-3 + \sqrt{17}}{4} \right\}$. This is the same result we obtained when we solved this equation by completing the square at the beginning of the section.

b) Is $k^2 = 10k - 29$ in the form $ax^2 + bx + c = 0$? *No*. Begin by writing the equation in the correct form.

<table>
<tr><td>

$$k^2 - 10k + 29 = 0$$

</td><td>Subtract $10k$, and add 29 to both sides.</td></tr>
<tr><td>

$$a = 1 \qquad b = -10 \qquad c = 29$$

</td><td>Identify a, b, and c.</td></tr>
<tr><td>

$$k = \dfrac{-b \pm \sqrt{b^2 - 4ac}}{2a}$$

</td><td>Quadratic formula</td></tr>
<tr><td>

$$= \dfrac{-(-10) \pm \sqrt{(-10)^2 - 4(1)(29)}}{2(1)}$$

</td><td>Substitute $a = 1$, $b = -10$, and $c = 29$.</td></tr>
<tr><td>

$$= \dfrac{10 \pm \sqrt{100 - 116}}{2}$$

</td><td>Perform the operations.</td></tr>
<tr><td>

$$= \dfrac{10 \pm \sqrt{-16}}{2}$$

</td><td>$100 - 116 = -16$</td></tr>
<tr><td>

$$= \dfrac{10 \pm 4i}{2}$$

</td><td>$\sqrt{-16} = 4i$</td></tr>
<tr><td>

$$= \dfrac{10}{2} \pm \dfrac{4}{2}i = 5 \pm 2i$$

</td><td></td></tr>
</table>

The solution set is $\{5 - 2i, 5 + 2i\}$.

[**YOU TRY 1**]

Solve using the quadratic formula.

a) $2x^2 + 3x - 7 = 0$ b) $y^2 = 6y - 25$

Equations in various forms may be solved using the quadratic formula.

EXAMPLE 2   Solve using the quadratic formula.

$$(3p - 1)(3p + 4) = 3p - 5$$

Solution

Is $(3p - 1)(3p + 4) = 3p - 5$ in the form $ax^2 + bx + c = 0$? *No*. Before we can apply the quadratic formula, we must write it in that form.

$$(3p - 1)(3p + 4) = 3p - 5$$
$$9p^2 + 9p - 4 = 3p - 5 \quad \text{Multiply using FOIL.}$$
$$9p^2 + 6p + 1 = 0 \quad \text{Subtract } 3p, \text{ and add 5 to both sides.}$$

The equation is in the correct form. Identify a, b, and c: $a = 9$ $b = 6$ $c = 1$

$$p = \frac{-b \pm \sqrt{b^2 - 4ac}}{2a} \quad \text{Quadratic formula}$$

$$= \frac{-(6) \pm \sqrt{(6)^2 - 4(9)(1)}}{2(9)} \quad \text{Substitute } a = 9, b = 6, \text{ and } c = 1.$$

$$= \frac{-6 \pm \sqrt{36 - 36}}{18} \quad \text{Perform the operations.}$$

$$= \frac{-6 \pm \sqrt{0}}{18}$$

$$= \frac{-6 \pm 0}{18} = \frac{-6}{18} = -\frac{1}{3}$$

The solution set is $\left\{ -\dfrac{1}{3} \right\}$.

[YOU TRY 2]   Solve $(d + 6)(d - 2) = -10$ using the quadratic formula.

To solve a quadratic equation containing fractions, first multiply by the LCD to eliminate the fractions. Then, solve using the quadratic formula.

3 Determine the Number and Type of Solutions of a Quadratic Equation Using the Discriminant

We can find the solutions of any quadratic equation of the form $ax^2 + bx + c = 0$ $(a \neq 0)$ using the quadratic formula.

$$x = \frac{-b \pm \sqrt{b^2 - 4ac}}{2a}$$

The radicand in the quadratic formula determines the type of solution a quadratic equation has.

Property The Discriminant and Solutions

The expression under the radical, $b^2 - 4ac$, is called the **discriminant.** The discriminant tells us what kind of solution a quadratic equation has. If a, b, and c are integers, then

1) if $b^2 - 4ac$ is *positive and the square of an integer*, the equation has *two rational solutions*.

2) if $b^2 - 4ac$ is *positive but not a perfect square*, the equation has *two irrational solutions*.

3) if $b^2 - 4ac$ is *negative*, the equation has *two nonreal, complex solutions of the form $a + bi$ and $a - bi$*.

4) if $b^2 - 4ac = 0$, the equation has *one rational solution*.

EXAMPLE 3   Find the value of the discriminant. Then, determine the number and type of solutions of each equation.

a) $z^2 + 6z - 4 = 0$ b) $5h^2 = 6h - 2$

Solution

a) Is $z^2 + 6z - 4 = 0$ in the form $ax^2 + bx + c = 0$? *Yes.* Identify a, b, and c.

$$a = 1 \qquad b = 6 \qquad c = -4$$

$$\text{Discriminant} = b^2 - 4ac = (6)^2 - 4(1)(-4) = 36 + 16 = 52$$

Because 52 is positive but *not* a perfect square, the equation will have *two irrational solutions*. ($\sqrt{52}$, or $2\sqrt{13}$, will appear in the solution, and $2\sqrt{13}$ is irrational.)

b) Is $5h^2 = 6h - 2$ in the form $ax^2 + bx + c = 0$? *No.* Rewrite the equation in that form, and identify a, b, and c.

$$5h^2 - 6h + 2 = 0$$

$$a = 5 \qquad b = -6 \qquad c = 2$$

$$\text{Discriminant} = b^2 - 4ac = (-6)^2 - 4(5)(2) = 36 - 40 = -4$$

The discriminant is -4, so the equation will have *two nonreal, complex solutions of the form $a + bi$ and $a - bi$*, where $b \neq 0$.

The discriminant is $b^2 - 4ac$ *not* $\sqrt{b^2 - 4ac}$.

[YOU TRY 3] Find the value of the discriminant. Then, determine the number and type of solutions of each equation.

a) $2x^2 + x + 5 = 0$ b) $m^2 + 5m = 24$ c) $-3v^2 = 4v - 1$

d) $4r(2r - 3) = -1 - 6r - r^2$

4 Solve an Applied Problem Using the Quadratic Formula

EXAMPLE 4

A ball is thrown upward from a height of 20 ft. The height h of the ball (in feet) t sec after the ball is released is given by

$$h = -16t^2 + 16t + 20$$

a) How long does it take the ball to reach a height of 8 ft?

b) How long does it take the ball to hit the ground?

Solution

a) Find the *time* it takes for the ball to reach a height of 8 ft.

Find t when $h = 8$.

$$h = -16t^2 + 16t + 20$$
$$8 = -16t^2 + 16t + 20 \qquad \text{Substitute 8 for } h.$$
$$0 = -16t^2 + 16t + 12 \qquad \text{Write in standard form.}$$
$$0 = 4t^2 - 4t - 3 \qquad \text{Divide by } -4.$$

$$t = \frac{-b \pm \sqrt{b^2 - 4ac}}{2a} \qquad \text{Quadratic formula}$$

$$= \frac{-(-4) \pm \sqrt{(-4)^2 - 4(4)(-3)}}{2(4)} \qquad \begin{array}{l}\text{Substitute } a = 4, b = -4, \\ \text{and } c = -3.\end{array}$$

$$= \frac{4 \pm \sqrt{16 + 48}}{8} \qquad \text{Perform the operations.}$$

$$= \frac{4 \pm \sqrt{64}}{8} = \frac{4 \pm 8}{8}$$

$$t = \frac{4 + 8}{8} \qquad \text{or} \qquad t = \frac{4 - 8}{8} \qquad \text{The equation has two rational solutions.}$$

$$t = \frac{12}{8} = \frac{3}{2} \qquad \text{or} \qquad t = \frac{-4}{8} = -\frac{1}{2}$$

Because t represents time, t cannot equal $-\frac{1}{2}$. We reject that as a solution.

Therefore, $t = \frac{3}{2}$ sec or 1.5 sec. The ball will be 8 ft above the ground after 1.5 sec.

W Hint

Think *carefully* about the problem you are asked to solve. What does it mean in terms of the formula?

b) When the ball hits the ground, it is 0 ft above the ground.

Find t when $h = 0$.

$$h = -16t^2 + 16t + 20$$
$$0 = -16t^2 + 16t + 20 \qquad \text{Substitute 0 for } h.$$
$$0 = 4t^2 - 4t - 5 \qquad \text{Divide by } -4.$$

$$t = \frac{-(-4) \pm \sqrt{(-4)^2 - 4(4)(-5)}}{2(4)} \qquad \text{Substitute } a = 4, b = -4, \text{ and } c = -5.$$

$$= \frac{4 \pm \sqrt{16 + 80}}{8} \qquad \text{Perform the operations.}$$

$$= \frac{4 \pm \sqrt{96}}{8}$$

$$= \frac{4 \pm 4\sqrt{6}}{8} \qquad \sqrt{96} = \sqrt{16} \cdot \sqrt{6} = 4\sqrt{6}$$

$$= \frac{4(1 \pm \sqrt{6})}{8} \qquad \text{Factor out 4 in the numerator.}$$

$$t = \frac{1 \pm \sqrt{6}}{2} \qquad \text{Divide numerator and denominator by 4 to simplify.}$$

$$t = \frac{1 + \sqrt{6}}{2} \qquad \text{or} \qquad t = \frac{1 - \sqrt{6}}{2} \qquad \text{The equation has two irrational solutions.}$$

$$t \approx \frac{1 + 2.4}{2} \qquad \text{or} \qquad t \approx \frac{1 - 2.4}{2} \qquad \sqrt{6} \approx 2.4$$

$$t \approx \frac{3.4}{2} = 1.7 \qquad \text{or} \qquad t \approx -0.7$$

Because t represents time, t cannot equal $\dfrac{1 - \sqrt{6}}{2}$. We reject this as a solution.

Therefore, $t = \dfrac{1 + \sqrt{6}}{2}$ sec or $t \approx 1.7$ sec. The ball will hit the ground after about 1.7 sec.

[YOU TRY 4]

An object is thrown upward from a height of 12 ft. The height h of the object (in feet) t sec after the object is thrown is given by

$$h = -16t^2 + 56t + 12$$

a) How long does it take the object to reach a height of 36 ft?

b) How long does it take the object to hit the ground?

ANSWERS TO [YOU TRY] EXERCISES

1) a) $\left\{ \dfrac{-3 - \sqrt{65}}{4}, \dfrac{-3 + \sqrt{65}}{4} \right\}$ b) $\{3 - 4i, 3 + 4i\}$

2) $\{-2 - \sqrt{6}, -2 + \sqrt{6}\}$ 3) a) -39; two nonreal, complex solutions
b) 121; two rational solutions   c) 28; two irrational solutions   d) 0; one rational solution

4) a) It takes $\dfrac{1}{2}$ sec to reach 36 ft on its way up and 3 sec to reach 36 ft on its way down.

b) $\dfrac{7 + \sqrt{61}}{4}$ sec or approximately 3.7 sec

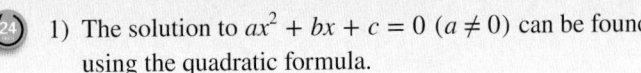

Mixed Exercises: Objectives 2 and 3

Find the error in each, and correct the mistake.

1) The solution to $ax^2 + bx + c = 0$ $(a \neq 0)$ can be found using the quadratic formula.

$$x = -b \pm \frac{\sqrt{b^2 - 4ac}}{2a}$$

2) In order to solve $5n^2 - 3n = 1$ using the quadratic formula, a student substitutes a, b, and c into the formula in this way: $a = 5$, $b = -3$, $c = 1$.

$$n = \frac{-(-3) \pm \sqrt{(-3)^2 - 4(5)(1)}}{2(5)}$$

3) $\dfrac{-2 \pm 6\sqrt{11}}{2} = -1 \pm 6\sqrt{11}$

4) The discriminant of $3z^2 - 4z + 1 = 0$ is

$$\sqrt{b^2 - 4ac} = \sqrt{(-4)^2 - 4(3)(1)}$$
$$= \sqrt{16 - 12}$$
$$= \sqrt{4}$$
$$= 2$$

Objective 2: Solve a Quadratic Equation Using the Quadratic Formula

Solve using the quadratic formula.

5) $x^2 + 4x + 3 = 0$

6) $v^2 - 8v + 7 = 0$

7) $3t^2 + t - 10 = 0$

8) $6q^2 + 11q + 3 = 0$

9) $k^2 + 2 = 5k$

10) $n^2 = 5 - 3n$

11) $y^2 = 8y - 25$

12) $-4x + 5 = -x^2$

13) $3 - 2w = -5w^2$

14) $2d^2 = -4 - 5d$

15) $r^2 + 7r = 0$

16) $p^2 - 10p = 0$

17) $2k(k - 3) = -3$

18) $3v(v + 3) = 7v + 4$

19) $(2c - 5)(c - 5) = -3$

20) $-11 = (3z - 1)(z - 5)$

21) $\dfrac{1}{6}u^2 + \dfrac{4}{3}u = \dfrac{5}{2}$

22) $\dfrac{1}{6}h + \dfrac{1}{2} = \dfrac{3}{4}h^2$

23) $2(p + 10) = (p + 10)(p - 2)$

24) $(t - 8)(t - 3) = 3(3 - t)$

25) $4g^2 + 9 = 0$

26) $25q^2 - 1 = 0$

27) $x(x + 6) = -34$

28) $c(c - 4) = -22$

29) $(2s + 3)(s - 1) = s^2 - s + 6$

30) $(3m + 1)(m - 2) = (2m - 3)(m + 2)$

31) $3(3 - 4y) = -4y^2$

32) $5a(5a + 2) = -1$

33) $-\dfrac{1}{6} = \dfrac{2}{3}p^2 + \dfrac{1}{2}p$

34) $\dfrac{1}{2}n = \dfrac{3}{4}n^2 + 2$

35) $4q^2 + 6 = 20q$

36) $4w^2 = 6w + 16$

37) Let $f(x) = x^2 + 6x - 2$. Find x so that $f(x) = 0$.

38) Let $g(x) = 3x^2 - 4x - 1$. Find x so that $g(x) = 0$.

39) Let $h(t) = 2t^2 - t + 7$. Find t so that $h(t) = 12$.

40) Let $P(a) = a^2 + 8a + 9$. Find a so that $P(a) = -3$.

41) Let $f(x) = 5x^2 + 21x - 1$ and $g(x) = 2x + 3$. Find all values of x such that $f(x) = g(x)$.

42) Let $F(x) = -x^2 + 3x - 2$ and $G(x) = x^2 + 12x + 6$. Find all values of x such that $F(x) = G(x)$.

Objective 3: Determine the Number and Type of Solutions of a Quadratic Equation Using the Discriminant

43) If the discriminant of a quadratic equation is zero, what do you know about the solutions of the equation?

44) If the discriminant of a quadratic equation is negative, what do you know about the solutions of the equation?

Find the value of the discriminant. Then, determine the number and type of solutions of each equation. *Do not solve.*

45) $10d^2 - 9d + 3 = 0$

46) $3j^2 + 8j + 2 = 0$

47) $4y^2 + 49 = -28y$

48) $3q = 1 + 5q^2$

49) $-5 = u(u + 6)$

50) $g^2 + 4 = 4g$

51) $2w^2 - 4w - 5 = 0$

52) $3 + 2p^2 - 7p = 0$

Find the value of a, b, or c so that each equation has only one rational solution.

53) $z^2 + bz + 16 = 0$

54) $k^2 + bk + 49 = 0$

55) $4y^2 - 12y + c = 0$

56) $25t^2 - 20t + c = 0$

57) $ap^2 + 12p + 9 = 0$

58) $ax^2 - 6x + 1 = 0$

Objective 4: Solve an Applied Problem Using the Quadratic Formula

Write an equation, and solve.

59) One leg of a right triangle is 1 in. more than twice the other leg. The hypotenuse is $\sqrt{29}$ in. long. Find the lengths of the legs.

60) The hypotenuse of a right triangle is $\sqrt{34}$ in. long. The length of one leg is 1 in. less than twice the other leg. Find the lengths of the legs.

Solve.

 61) An object is thrown upward from a height of 24 ft. The height h of the object (in feet) t sec after the object is released is given by $h = -16t^2 + 24t + 24$.

 a) How long does it take the object to reach a height of 8 ft?

 b) How long does it take the object to hit the ground?

62) A ball is thrown upward from a height of 6 ft. The height h of the ball (in feet) t sec after the ball is released is given by $h = -16t^2 + 44t + 6$.

 a) How long does it take the ball to reach a height of 16 ft?

 b) How long does it take the object to hit the ground?

R Rethink

R1) How many different types of solutions are possible for a quadratic equation?

R2) Do you know any songs that might help you remember the quadratic formula?

R3) What do you find most difficult about solving applied problems?

Putting It All Together

P Prepare

O Organize

What is your objective for Putting It All Together?	How can you accomplish the objective?
1 Decide Which Method to Use to Solve a Quadratic Equation	• Be able to write out all the different methods in your notes. • Review characteristics of each method, and be able to identify the most efficient method for each problem. • Try solving some problems using more than one method, if time permits, and check your answers. • Complete the given example on your own. • Complete You Try 1.

 **Read the explanations, follow the examples, take notes, and complete the You Try.**

We have learned four methods for solving quadratic equations.

Methods for Solving Quadratic Equations

 1) Factoring

 2) Square root property

 3) Completing the square

 4) Quadratic formula

While it is true that the quadratic formula can be used to solve *every* quadratic equation of the form $ax^2 + bx + c = 0$ $(a \neq 0)$, it is not always the most *efficient* method. In this section we will discuss how to decide which method to use to solve a quadratic equation.

1 Decide Which Method to Use to Solve a Quadratic Equation

Solve.

a) $p^2 - 6p = 16$

b) $m^2 - 8m + 13 = 0$

c) $3t^2 + 8t + 7 = 0$

d) $(2z - 7)^2 - 6 = 0$

Solution

a) Write $p^2 - 6p = 16$ in standard form: $p^2 - 6p - 16 = 0$

Does $p^2 - 6p - 16$ factor? Yes. *Solve by factoring.*

$$(p - 8)(p + 2) = 0$$

$p - 8 = 0$ or $p + 2 = 0$ Set each factor equal to 0.

$p = 8$ or $p = -2$ Solve.

The solution set is $\{-2, 8\}$.

> **W Hint**
>
> Write the solutions to the equations in this example in your notes. In your own words, explain why each method is chosen to solve each equation.

b) To solve $m^2 - 8m + 13 = 0$ ask yourself, "Can I factor $m^2 - 8m + 13$?" No, it does not factor. We could solve this using the quadratic formula, but *completing the square* is also a good method for solving this equation. Why?

Note

Completing the square is a good method for solving a quadratic equation when the coefficient of the squared term is 1 or –1 and when the coefficient of the linear term is even.

We will solve $m^2 - 8m + 13 = 0$ by completing the square.

Step 1: The coefficient of m^2 is 1.

Step 2: Get the variables on one side of the equal sign and the constant on the other side.

$$m^2 - 8m = -13$$

Step 3: Complete the square: $\dfrac{1}{2}(-8) = -4$

$$(-4)^2 = 16$$

Add 16 to both sides of the equation.

$$m^2 - 8m + 16 = -13 + 16$$
$$m^2 - 8m + 16 = 3$$

Step 4: Factor: $(m - 4)^2 = 3$

Step 5: Solve using the square root property:

$$(m - 4)^2 = 3$$
$$m - 4 = \pm\sqrt{3}$$
$$m = 4 \pm \sqrt{3}$$

The solution set is $\{4 - \sqrt{3}, 4 + \sqrt{3}\}$.

c) Ask yourself, "Can I solve $3t^2 + 8t + 7 = 0$ by factoring?" No, $3t^2 + 8t + 7$ does not factor. Completing the square would not be a very efficient way to solve the equation because the coefficient of t^2 is 3, and dividing the equation by 3 would give us $t^2 + \dfrac{8}{3}t + \dfrac{7}{3} = 0$.

We will solve $3t^2 + 8t + 7 = 0$ using the quadratic formula.

Identify a, b, and c: $a = 3$ $b = 8$ $c = 7$

$$t = \frac{-b \pm \sqrt{b^2 - 4ac}}{2a} \qquad \text{Quadratic formula}$$

$$= \frac{-(8) \pm \sqrt{(8)^2 - 4(3)(7)}}{2(3)} \qquad \text{Substitute } a = 3, b = 8, \text{ and } c = 7.$$

$$= \frac{-8 \pm \sqrt{64 - 84}}{6} \qquad \text{Perform the operations.}$$

$$= \frac{-8 \pm \sqrt{-20}}{6}$$

$$= \frac{-8 \pm 2i\sqrt{5}}{6} \qquad \sqrt{-20} = i\sqrt{4}\sqrt{5} = 2i\sqrt{5}$$

$$= \frac{2(-4 \pm i\sqrt{5})}{6} \qquad \text{Factor out 2 in the numerator.}$$

$$= \frac{-4 \pm i\sqrt{5}}{3} \qquad \begin{array}{l}\text{Divide numerator and} \\ \text{denominator by 2 to simplify.}\end{array}$$

$$= -\frac{4}{3} \pm \frac{\sqrt{5}}{3}i \qquad \text{Write in the form } a + bi.$$

The solution set is $\left\{ -\dfrac{4}{3} - \dfrac{\sqrt{5}}{3}i, -\dfrac{4}{3} + \dfrac{\sqrt{5}}{3}i \right\}$.

d) Which method should we use to solve $(2z - 7)^2 - 6 = 0$?

We *could* square the binomial, combine like terms, then solve, possibly, by factoring or using the quadratic formula. However, this would be very inefficient.

Notice that this equation contains a squared quantity and a constant.

We will solve $(2z - 7)^2 - 6 = 0$ using the square root property.

$$(2z - 7)^2 - 6 = 0$$
$$(2z - 7)^2 = 6 \qquad \text{Add 6 to each side.}$$
$$2z - 7 = \pm\sqrt{6} \qquad \text{Square root property}$$
$$2z = 7 \pm \sqrt{6} \qquad \text{Add 7 to each side.}$$
$$z = \frac{7 \pm \sqrt{6}}{2} \qquad \text{Divide by 2.}$$

The solution set is $\left\{ \dfrac{7 - \sqrt{6}}{2}, \dfrac{7 + \sqrt{6}}{2} \right\}$.

Putting It All Together Exercises

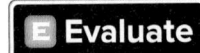
Do the exercises, and check your work.

Objective 1: Decide Which Method to Use to Solve a Quadratic Equation

Keep in mind the four methods we have learned for solving quadratic equations: *factoring, the square root property, completing the square, and the quadratic formula.* Solve the equations using one of these methods.

1) $z^2 - 50 = 0$

2) $j^2 - 6j = 8$

3) $a(a + 1) = 20$

4) $2x^2 + 6 = 3x$

5) $u^2 + 7u + 9 = 0$

6) $3p^2 - p - 4 = 0$

7) $2k(2k + 7) = 3(k + 1)$

8) $2 = (w + 3)^2 + 8$

9) $m^2 + 14m + 60 = 0$

10) $\dfrac{1}{2}y^2 = \dfrac{3}{4} - \dfrac{1}{2}y$

11) $10 + (3b - 1)^2 = 4$

12) $c^2 + 8c + 25 = 0$

13) $\dfrac{9}{2a^2} = \dfrac{1}{6} + \dfrac{1}{a}$

14) $100 = 4d^2$

15) $r^2 - 4r = 3$

16) $2t^3 + 108t = -30t^2$

17) $p(p + 8) = 3(p^2 + 2) + p$

18) $h^2 = h$

19) $\dfrac{10}{z} = 1 + \dfrac{21}{z^2}$

20) $2s(2s + 3) = 4s + 5$

21) $(3v + 4)(v - 2) = -9$

22) $34 = 6y - y^2$

23) $(c - 5)^2 + 16 = 0$

24) $(2b + 1)(b + 5) = -7$

25) $3g = g^2$

26) $5z^2 + 15z + 30 = 0$

27) $4m^3 = 9m$

28) $1 = \dfrac{x^2}{12} - \dfrac{x}{3}$

29) $\dfrac{1}{3}q^2 + \dfrac{5}{6}q + \dfrac{4}{3} = 0$

30) $-3 = (12d + 5)^2 + 6$

R Rethink

R1) Which method is easiest to use? Which one is the most difficult?

R2) Why would you need to complete the square? Why must the coefficient be equal to 1 before you complete the square?

11.4 Equations in Quadratic Form

What are your objectives for Section 11.4?	How can you accomplish each objective?
1 Solve Quadratic Equations Resulting from Equations Containing Fractions or Radicals	• Review the procedure for eliminating fractions in equations by using the LCD. • Review the procedures for solving equations containing radicals. • Complete the given examples on your own. • Complete You Trys 1 and 2.
2 Solve an Equation in Quadratic Form by Factoring	• Be able to recognize when an equation is in *quadratic form.* • Apply the same methods for solving equations in *quadratic form* as you would use with *quadratic equations.* • Check all solutions in the original equation. • Complete the given example on your own. • Complete You Try 3.
3 Solve an Equation in Quadratic Form Using Substitution	• Understand why and how we can use substitution. • Be familiar with *all* methods for solving quadratic equations. • Substitute the original variables back into the equation to finish solving. • Complete the given example on your own. • Complete You Trys 4 and 5.
4 Use Substitution for a Binomial to Solve a Quadratic Equation	• Choose a variable for the binomial quantity and substitute this into the equation. • Be familiar with *all* methods for solving quadratic equations. • Substitute the original binomial back into the equation to finish solving. • Complete the given example on your own. • Complete You Try 6.

W Work

Read the explanations, follow the examples, take notes, and complete the You Trys.

In Chapters 8 and 10, we solved some equations that were *not* quadratic but could be rewritten in the form of a quadratic equation, $ax^2 + bx + c = 0$. Two such examples are:

$$\frac{10}{x} - \frac{7}{x + 1} = \frac{2}{3} \qquad \text{and} \qquad r + \sqrt{r} = 12$$

Rational equation (Ch. 8)   Radical equation (Ch. 10)

We will review how to solve each type of equation.

1 Solve Quadratic Equations Resulting from Equations Containing Fractions or Radicals

---

EXAMPLE 1

Solve $\dfrac{10}{x} - \dfrac{7}{x+1} = \dfrac{2}{3}$.

Solution

To solve an equation containing rational expressions, *multiply the equation by the LCD of all of the fractions to eliminate the denominators,* then solve.

$$\text{LCD} = 3x(x+1)$$

<table>
<tr><td></td><td></td></tr>
</table>

$$3x(x+1)\left(\dfrac{10}{x} - \dfrac{7}{x+1}\right) = 3x(x+1)\left(\dfrac{2}{3}\right)$$

Multiply both sides of the equation by the LCD of the fractions.

$$3\cancel{x}(x+1) \cdot \dfrac{10}{\cancel{x}} - 3x\cancel{(x+1)} \cdot \dfrac{7}{\cancel{x+1}} = \cancel{3}x(x+1) \cdot \left(\dfrac{2}{\cancel{3}}\right)$$

Distribute, and divide out common factors.

$$30(x+1) - 3x(7) = 2x(x+1)$$

$$30x + 30 - 21x = 2x^2 + 2x \qquad \text{Distribute.}$$

$$9x + 30 = 2x^2 + 2x \qquad \text{Combine like terms.}$$

$$0 = 2x^2 - 7x - 30 \qquad \text{Write in the form } ax^2 + bx + c = 0.$$

$$0 = (2x + 5)(x - 6) \qquad \text{Factor.}$$

$$2x + 5 = 0 \quad \text{or} \quad x - 6 = 0 \qquad \text{Set each factor equal to zero.}$$

$$2x = -5$$

$$x = -\dfrac{5}{2} \quad \text{or} \quad x = 6 \qquad \text{Solve.}$$

Recall that you *must* check the proposed solutions in the original equation to be certain they do not make a denominator equal zero. The solution set is $\left\{-\dfrac{5}{2}, 6\right\}$.

> **Ⓦ Hint**
>
> Why do you have to check your solutions in the original equations?

[YOU TRY 1]

Solve $\dfrac{1}{m} = \dfrac{1}{2} + \dfrac{m}{m+4}$.

EXAMPLE 2

Solve $r + \sqrt{r} = 12$.

Solution

The first step in solving a radical equation is getting a radical on a side by itself.

$$r + \sqrt{r} = 12$$

$$\sqrt{r} = 12 - r \qquad \text{Subtract } r \text{ from each side.}$$

$$(\sqrt{r})^2 = (12 - r)^2 \qquad \text{Square both sides.}$$

$$r = 144 - 24r + r^2$$

$$0 = r^2 - 25r + 144 \qquad \text{Write in the form } ax^2 + bx + c = 0.$$

$$0 = (r - 16)(r - 9) \qquad \text{Factor.}$$

$$r - 16 = 0 \quad \text{or} \quad r - 9 = 0 \qquad \text{Set each factor equal to zero.}$$

$$r = 16 \quad \text{or} \quad r = 9 \qquad \text{Solve.}$$

Recall that you *must* check the proposed solutions in the *original* equation.

Check $r = 16$:
$$r + \sqrt{r} = 12$$
$$16 + \sqrt{16} \stackrel{?}{=} 12$$
$$16 + 4 = 12 \qquad \text{False}$$

Check $r = 9$:
$$r + \sqrt{r} = 12$$
$$9 + \sqrt{9} \stackrel{?}{=} 12$$
$$9 + 3 = 12 \qquad \text{True}$$

16 is an extraneous solution. The solution set is $\{9\}$.

$\left[\text{YOU TRY 2}\right]$ Solve $y + 3\sqrt{y} = 10$.

2 Solve an Equation in Quadratic Form by Factoring

Some equations that are not quadratic can be solved using the same methods that can be used to solve quadratic equations. These are called **equations in quadratic form.** Some examples of equations in quadratic form are:

$$x^4 - 10x^2 + 9 = 0, \qquad t^{2/3} + t^{1/3} - 6 = 0, \qquad 2n^4 - 5n^2 = -1$$

Let's compare the equations above to *quadratic equations* to understand why they are said to be in quadratic form.

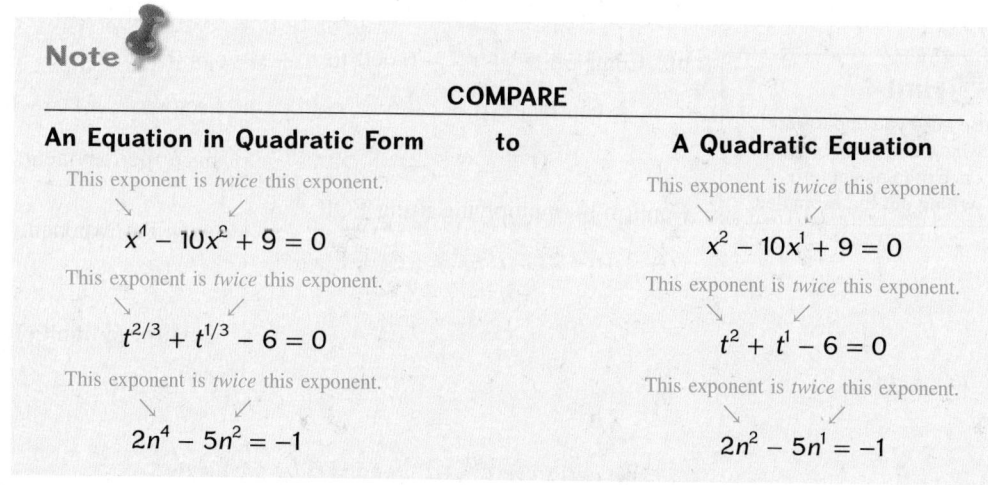

Note

COMPARE

An Equation in Quadratic Form	to	A Quadratic Equation

This exponent is *twice* this exponent.
$$x^4 - 10x^2 + 9 = 0$$

This exponent is *twice* this exponent.
$$x^2 - 10x^1 + 9 = 0$$

This exponent is *twice* this exponent.
$$t^{2/3} + t^{1/3} - 6 = 0$$

This exponent is *twice* this exponent.
$$t^2 + t^1 - 6 = 0$$

This exponent is *twice* this exponent.
$$2n^4 - 5n^2 = -1$$

This exponent is *twice* this exponent.
$$2n^2 - 5n^1 = -1$$

This pattern enables us to work with equations in quadratic form like we can work with quadratic equations.

EXAMPLE 3

Solve.

a) $x^4 - 10x^2 + 9 = 0$ b) $t^{2/3} + t^{1/3} - 6 = 0$

Solution

a) Let's compare $x^4 - 10x^2 + 9 = 0$ to $x^2 - 10x + 9 = 0$.

We can factor $x^2 - 10x + 9$:

$$(x - 9)(x - 1)$$

Confirm by multiplying using FOIL:

$$(x - 9)(x - 1) = x^2 - x - 9x + 9$$
$$= x^2 - 10x + 9$$

Factor $x^4 - 10x^2 + 9$ in a similar way since the exponent, 4, of the first term is twice the exponent, 2, of the second term:

$$x^4 - 10x^2 + 9 = (x^2 - 9)(x^2 - 1)$$

Confirm by multiplying using FOIL:

$$(x^2 - 9)(x^2 - 1) = x^4 - x^2 - 9x^2 + 9$$
$$= x^4 - 10x^2 + 9$$

We can solve $x^4 - 10x^2 + 9 = 0$ by factoring.

$$x^4 - 10x^2 + 9 = 0$$
$$(x^2 - 9)(x^2 - 1) = 0 \qquad \text{Factor.}$$

$x^2 - 9 = 0$ or $x^2 - 1 = 0$ Set each factor equal to 0.
$x^2 = 9$ $\qquad\qquad x^2 = 1$ Square root property
$x = \pm 3$ $\qquad\qquad x = \pm 1$

The check is left to the student. Check the answers in the *original* equation. The solution set is $\{-3, -1, 1, 3\}$.

**Hint**

Be sure you understand *why* you are performing each step as you are writing out the examples.

b) Compare $t^{2/3} + t^{1/3} - 6 = 0$ to $t^2 + t - 6 = 0$.

We can factor $t^2 + t - 6$:

$$(t + 3)(t - 2)$$

Confirm by multiplying using FOIL:

$$(t + 3)(t - 2) = t^2 - 2t + 3t - 6$$
$$= t^2 + t - 6$$

Factor $t^{2/3} + t^{1/3} - 6$ in a similar way since the exponent, $\dfrac{2}{3}$, of the first term is twice the exponent, $\dfrac{1}{3}$, of the second term:

$$t^{2/3} + t^{1/3} - 6 = (t^{1/3} + 3)(t^{1/3} - 2)$$

Confirm by multiplying using FOIL:

$$(t^{1/3} + 3)(t^{1/3} - 2) = t^{2/3} - 2t^{1/3} + 3t^{1/3} - 6$$
$$= t^{2/3} + t^{1/3} - 6$$

We can solve $t^{2/3} + t^{1/3} - 6 = 0$ by factoring.

$$t^{2/3} + t^{1/3} - 6 = 0$$
$$(t^{1/3} + 3)(t^{1/3} - 2) = 0 \qquad \text{Factor.}$$

$t^{1/3} + 3 = 0$ or $t^{1/3} - 2 = 0$ Set each factor equal to 0.
$t^{1/3} = -3$ $\qquad\qquad t^{1/3} = 2$ Isolate the constant.
$\sqrt[3]{t} = -3$ $\qquad\qquad \sqrt[3]{t} = 2$ $t^{1/3} = \sqrt[3]{t}$
$(\sqrt[3]{t})^3 = (-3)^3$ $\qquad (\sqrt[3]{t})^3 = 2^3$ Cube both sides.
$t = -27$ or $t = 8$ Solve.

The check is left to the student. The solution set is $\{-27, 8\}$.

Solve.

a) $r^4 - 13r^2 + 36 = 0$ b) $c^{2/3} + 4c^{1/3} - 5 = 0$

3 Solve an Equation in Quadratic Form Using Substitution

The equations in Example 3 can also be solved using a method called **substitution.** We will illustrate the method in Example 4.

EXAMPLE 4

Solve $x^4 - 10x^2 + 9 = 0$ using substitution.

Solution

$$x^4 - 10x^2 + 9 = 0$$
$$\downarrow$$
$$x^4 = (x^2)^2$$

W Hint

When using substitution, make special notations so that you do not forget to substitute back to the original variables.

To rewrite $x^4 - 10x^2 + 9 = 0$ in quadratic form, let $u = x^2$.

$$\text{If} \quad u = x^2, \text{ then}$$
$$u^2 = x^4.$$

$$x^4 - 10x^2 + 9 = 0$$
$$u^2 - 10u + 9 = 0 \qquad \text{Substitute } u^2 \text{ for } x^4 \text{ and } u \text{ for } x^2.$$
$$(u - 9)(u - 1) = 0 \qquad \text{Solve by factoring.}$$
$$\swarrow \qquad \searrow$$
$$u - 9 = 0 \quad \text{or} \quad u - 1 = 0 \qquad \text{Set each factor equal to 0.}$$
$$u = 9 \quad \text{or} \qquad u = 1 \qquad \text{Solve for } u.$$

Be careful! $u = 9$ and $u = 1$ are *not* the solutions to $x^4 - 10x^2 + 9 = 0$. **We still need to solve for x.** Above we let $u = x^2$. *To solve for x, substitute 9 for u and solve for x, and then substitute 1 for u and solve for x.*

$$\begin{array}{cc} u = x^2 & u = x^2 \\ \text{Substitute 9 for } u. \qquad 9 = x^2 & 1 = x^2 \qquad \text{Substitute 1 for } u. \\ \text{Square root property} \qquad \pm 3 = x & \pm 1 = x \qquad \text{Square root property} \end{array}$$

The solution set is $\{-3, -1, 1, 3\}$. This is the same as the result we obtained in Example 3a).

[YOU TRY 4]

Solve by substitution.

a) $r^4 - 13r^2 + 36 = 0$ b) $c^{2/3} + 4c^{1/3} - 5 = 0$

If, after substitution, an equation cannot be solved by factoring, we can use the quadratic formula.

EXAMPLE 5 Solve $2n^4 - 5n^2 = -1$.

Solution

Write the equation in standard form: $2n^4 - 5n^2 + 1 = 0$.

Can we solve the equation by factoring? *No.*

We will solve $2n^4 - 5n^2 + 1 = 0$ using the quadratic formula. Begin with substitution.

$$\text{If } u = n^2, \text{ then}$$
$$u^2 = n^4.$$

$$2n^4 - 5n^2 + 1 = 0$$
$$2u^2 - 5u + 1 = 0 \qquad \text{Substitute } u^2 \text{ for } n^4 \text{ and } u \text{ for } n^2.$$

$$u = \frac{-(-5) \pm \sqrt{(-5)^2 - 4(2)(1)}}{2(2)} \qquad a = 2, b = -5, c = 1$$

$$u = \frac{5 \pm \sqrt{25 - 8}}{4} = \frac{5 \pm \sqrt{17}}{4}$$

Note that $u = \dfrac{5 \pm \sqrt{17}}{4}$ does not solve the *original* equation. We must solve for x using the

fact that $u = x^2$. Since $u = \dfrac{5 \pm \sqrt{17}}{4}$ means $u = \dfrac{5 + \sqrt{17}}{4}$ or $u = \dfrac{5 - \sqrt{17}}{4}$, we get

$$\begin{array}{cc}
u = x^2 & u = x^2 \\[4pt]
\dfrac{5 + \sqrt{17}}{4} = x^2 & \dfrac{5 - \sqrt{17}}{4} = x^2 \\[10pt]
\pm\sqrt{\dfrac{5 + \sqrt{17}}{4}} = x & \pm\sqrt{\dfrac{5 - \sqrt{17}}{4}} = x \qquad \text{Square root property} \\[10pt]
\dfrac{\pm\sqrt{5 + \sqrt{17}}}{2} = x & \dfrac{\pm\sqrt{5 - \sqrt{17}}}{2} = x \qquad \sqrt{4} = 2
\end{array}$$

The solution set is $\left\{ \dfrac{\sqrt{5 + \sqrt{17}}}{2}, -\dfrac{\sqrt{5 + \sqrt{17}}}{2}, \dfrac{\sqrt{5 - \sqrt{17}}}{2}, -\dfrac{\sqrt{5 - \sqrt{17}}}{2} \right\}$.

[YOU TRY 5] Solve $2k^4 + 3 = 9k^2$.

---

4 Use Substitution for a Binomial to Solve a Quadratic Equation

We can use substitution to solve an equation like the one in Example 6.

EXAMPLE 6

Solve $2(3a + 1)^2 - 7(3a + 1) - 4 = 0$.

Solution

The binomial $3a + 1$ appears as a *squared quantity* and as a *linear quantity*. Begin by using substitution.

$$\text{Let } u = 3a + 1. \qquad \text{Then, } u^2 = (3a + 1)^2.$$

Substitute: $\begin{aligned} 2(3a + 1)^2 - 7(3a + 1) - 4 &= 0 \\ 2u^2 \qquad - 7u \qquad - 4 &= 0 \end{aligned}$

Does $2u^2 - 7u - 4$ factor? *Yes.* Solve by factoring.

$$(2u + 1)(u - 4) = 0 \qquad \text{\small Factor } 2u^2 - 7u - 4.$$

$$2u + 1 = 0 \quad \text{or} \quad u - 4 = 0 \qquad \text{\small Set each factor equal to 0.}$$

$$u = -\frac{1}{2} \quad \text{or} \qquad u = 4 \qquad \text{\small Solve for } u.$$

Solve for a using $u = 3a + 1$.

W Hint

Don't forget to solve for the variable in the *original* equation.

When $u = -\dfrac{1}{2}$:

$$u = 3a + 1$$

$$-\frac{1}{2} = 3a + 1$$

$$\text{\small Subtract 1.} \quad -\frac{3}{2} = 3a$$

$$\text{\small Multiply by } \frac{1}{3}. \quad -\frac{1}{2} = a$$

When $u = 4$:

$$u = 3a + 1$$

$$4 = 3a + 1$$

$$3 = 3a \quad \text{\small Subtract 1.}$$

$$1 = a \quad \text{\small Divide by 3.}$$

The solution set is $\left\{ -\dfrac{1}{2}, 1 \right\}$. Check these values in the original equation.

YOU TRY 6

Solve $3(2p - 1)^2 - 11(2p - 1) + 10 = 0$.

ANSWERS TO $\lceil$ **YOU TRY** $\rceil$ **EXERCISES**

1) $\left\{ -2, \dfrac{4}{3} \right\}$ 2) $\{4\}$ 3) a) $\{-3, -2, 2, 3\}$ b) $\{-125, 1\}$ 4) a) $\{-3, -2, 2, 3\}$

b) $\{-125, 1\}$ 5) $\left\{ \dfrac{\sqrt{9 + \sqrt{57}}}{2}, -\dfrac{\sqrt{9 + \sqrt{57}}}{2}, \dfrac{\sqrt{9 - \sqrt{57}}}{2}, -\dfrac{\sqrt{9 - \sqrt{57}}}{2} \right\}$ 6) $\left\{ \dfrac{4}{3}, \dfrac{3}{2} \right\}$

Objective 1: Solve Quadratic Equations Resulting from Equations Containing Fractions or Radicals

Solve.

1) $t - \dfrac{48}{t} = 8$

2) $z + 11 = -\dfrac{24}{z}$

3) $\dfrac{2}{x} + \dfrac{6}{x - 2} = -\dfrac{5}{2}$

4) $\dfrac{3}{y} - \dfrac{6}{y - 1} = \dfrac{1}{2}$

5) $1 = \dfrac{2}{c} + \dfrac{1}{c - 5}$

6) $\dfrac{2}{g} = 1 + \dfrac{g}{g + 5}$

7) $\dfrac{3}{2v + 2} + \dfrac{1}{v} = \dfrac{3}{2}$

8) $\dfrac{1}{b + 3} + \dfrac{1}{b} = \dfrac{1}{3}$

9) $\dfrac{9}{n^2} = 5 + \dfrac{4}{n}$

10) $3 - \dfrac{16}{a^2} = \dfrac{8}{a}$

11) $\dfrac{5}{6r} = 1 - \dfrac{r}{6r - 6}$

12) $\dfrac{7}{4} - \dfrac{x}{4x + 4} = \dfrac{1}{x}$

13) $g = \sqrt{g + 20}$

14) $c = \sqrt{7c - 6}$

15) $a = \sqrt{\dfrac{14a - 8}{5}}$

16) $k = \sqrt{\dfrac{6 - 11k}{2}}$

17) $v + \sqrt{v} = 2$

18) $p - \sqrt{p} = 6$

19) $x = 5\sqrt{x} - 4$

20) $10 = m - 3\sqrt{m}$

21) $2 + \sqrt{2y - 1} = y$

22) $1 - \sqrt{5t + 1} = -t$

23) $2 = \sqrt{6k + 4} - k$

24) $\sqrt{10 - 3q} - 6 = q$

Mixed Exercises: Objectives 2–3

Determine whether each is an equation in quadratic form. Do *not* solve.

25) $n^4 - 12n^2 + 32 = 0$

26) $p^6 + 8p^3 - 9 = 0$

27) $2t^6 + 3t^3 - 5 = 0$

28) $a^4 - 4a - 3 = 0$

29) $c^{2/3} - 4c - 6 = 0$

30) $3z^{2/3} + 2z^{1/3} + 1 = 0$

31) $m + 9m^{1/2} = 4$

32) $2x^{1/2} - 5x^{1/4} = 2$

33) $5k^4 + 6k - 7 = 0$

34) $r^{-2} = 10 - 4r^{-1}$

Solve.

35) $x^4 - 10x^2 + 9 = 0$

36) $d^4 - 29d^2 + 100 = 0$

37) $p^4 - 11p^2 + 28 = 0$

38) $k^4 - 9k^2 + 8 = 0$

39) $c^4 + 9c^2 = -18$

40) $a^4 + 12a^2 = -35$

41) $b^{2/3} + 3b^{1/3} + 2 = 0$

42) $z^{2/3} + z^{1/3} - 12 = 0$

43) $p^{2/3} - p^{1/3} = 6$

44) $t^{2/3} - 6t^{1/3} = 40$

45) $4h^{1/2} + 21 = h$

46) $s + 12 = -7s^{1/2}$

47) $2a - 5a^{1/2} - 12 = 0$

48) $2w = 9w^{1/2} + 18$

49) $9n^4 = -15n^2 - 4$

50) $4h^4 + 19h^2 + 12 = 0$

51) $z^4 - 2z^2 = 15$

52) $a^4 + 2a^2 = 24$

53) $w^4 - 6w^2 + 2 = 0$

54) $p^4 - 8p^2 + 3 = 0$

55) $2m^4 + 1 = 7m^2$

56) $8x^4 + 2 = 9x^2$

57) $t^{-2} - 4t^{-1} - 12 = 0$

58) $d^{-2} + d^{-1} - 6 = 0$

59) $4 = 13y^{-1} - 3y^{-2}$

60) $14h^{-1} + 3 = 5h^{-2}$

Objective 4: Use Substitution for a Binomial to Solve a Quadratic Equation

Solve.

61) $(x - 2)^2 + 11(x - 2) + 24 = 0$

62) $(r + 1)^2 - 3(r + 1) - 10 = 0$

63) $2(3q + 4)^2 - 13(3q + 4) + 20 = 0$

64) $4(2b - 3)^2 - 9(2b - 3) - 9 = 0$

65) $(5a - 3)^2 + 6(5a - 3) = -5$

66) $(3z - 2)^2 - 8(3z - 2) = 20$

67) $3(k + 8)^2 + 5(k + 8) = 12$

68) $5(t + 9)^2 + 37(t + 9) + 14 = 0$

69) $1 - \dfrac{8}{2w + 1} = -\dfrac{16}{(2w + 1)^2}$

70) $1 - \dfrac{8}{4p + 3} = -\dfrac{12}{(4p + 3)^2}$

71) $1 + \dfrac{2}{h - 3} = \dfrac{1}{(h - 3)^2}$

72) $\dfrac{2}{(c + 6)^2} + \dfrac{2}{(c + 6)} = 1$

Mixed Exercises: Objectives 1–4

Solve.

73) $2n^{2/3} = 7n^{1/3} + 15$

74) $j - 6\sqrt{j} = -5$

75) $y^4 + 3y^2 = 108$

76) $\dfrac{1}{w-3} - \dfrac{2}{w} = \dfrac{3}{4}$

77) $5(2x-1)^2 + 9(2x-1) + 4 = 0$

78) $3k^{2/3} + 10k^{1/3} + 8 = 0$

79) $v - 8\sqrt{v} + 12 = 0$

80) $2a^4 - 20 = 3a^2$

81) $\dfrac{1}{6} = \dfrac{2}{m+4} - \dfrac{3}{m}$

82) $14 = 4(3c+2)^2 + (3c+2)$

Write an equation and solve.

83) It takes Kevin 3 hr longer than Walter to build a tree house. Together they can do the job in 2 hr. How long would it take each man to build the tree house on his own?

Corbis/SuperStock

84) It takes one pipe 4 hours more to empty a pool than it takes another pipe to fill a pool. If both pipes are accidentally left open, it takes 24 hr to fill the pool. How long does it take the single pipe to fill the pool?

85) A boat can travel 9 mi downstream and then 6 mi back upstream in 1 hr. If the speed of the current is 3 mph, what is the speed of the boat in still water?

86) A plane can travel 800 mi with the wind and then 650 mi back against the wind in 5 hr. If the wind blows at 30 mph, what is the speed of the plane?

87) A large fish tank at an aquarium needs to be emptied so that it can be cleaned. When its large and small drains are opened together, the tank can be emptied in 2 hr. By itself, it takes the small drain 3 hr longer to empty the tank than it takes the large drain to empty the tank on its own. How much time would it take for each drain to empty the pool on its own?

Dinodia Photo Library/
Brand X Pictures/Jupiter Images

88) Working together, a professor and her teaching assistant can grade a set of exams in 1.2 hr. On her own, the professor can grade the tests 1 hr faster than the teaching assistant can grade them on her own. How long would it take for each person to grade the test by herself?

89) Miguel took his son to college in Boulder, Colorado, 600 mi from their hometown. On his way home, he was slowed by a snowstorm so that his speed was 10 mph less than when he was driving to Boulder. His total driving time was 22 hr. How fast did Miguel drive on each leg of the trip?

90) Nariko was training for a race and went out for a run. Her speed was 2 mph faster during the first 6 mi than it was for the last 3 mi. If her total running time was $1\dfrac{3}{4}$ hr, what was her speed on each part of the run?

R Rethink

R1) What does it mean for an equation to be in quadratic form?

R2) What is the hardest part of using substitution?

R3) Can you solve any of the problems in this section by using two different methods?

11.5 Formulas and Applications

What are your objectives for Section 11.5?	How can you accomplish each objective?
1 Solve a Formula for a Variable	• Use the procedures for solving quadratic equations or equations containing radicals. • Complete the given examples on your own. • Complete You Trys 1 and 2.
2 Solve an Applied Problem Involving Volume	• Use the **Steps for Solving Applied Problems.** • Complete the given example on your own. • Complete You Try 3.
3 Solve an Applied Problem Involving Area	• Use the **Steps for Solving Applied Problems.** • Complete the given example on your own. • Complete You Try 4.
4 Solve an Applied Problem Using a Quadratic Equation	• Read the problem carefully, and understand how the given information relates to the given equation. • Complete the given example on your own.

W Work   Read the explanations, follow the examples, take notes, and complete the You Trys.

Sometimes, solving a formula for a variable involves using one of the techniques we've learned for solving a quadratic equation or for solving an equation containing a radical.

1 Solve a Formula for a Variable

EXAMPLE 1

Solve $v = \sqrt{\dfrac{300VP}{m}}$ for m.

Solution

Put a box around the m. The goal is to get m on a side by itself.

$$v = \sqrt{\frac{300VP}{\boxed{m}}}$$

$$v^2 = \frac{300VP}{\boxed{m}} \qquad \text{Square both sides.}$$

Since we are solving for m and it is in the denominator, multiply both sides by m to eliminate the denominator.

$$\boxed{m}v^2 = 300VP \qquad \text{Multiply both sides by } m.$$
$$m = \frac{300VP}{v^2} \qquad \text{Divide both sides by } v^2.$$

YOU TRY 1

Solve $v = \sqrt{\dfrac{2E}{m}}$ for m.

We may need to use the quadratic formula to solve a formula for a variable. Compare the following equations. Each equation is *quadratic in x* because each is written in the form $ax^2 + bx + c = 0$.

$$8x^2 + 3x - 2 = 0 \qquad \text{and} \qquad 8x^2 + tx - z = 0$$
$$a = 8 \quad b = 3 \quad c = -2 \qquad\qquad a = 8 \quad b = t \quad c = -z$$

To solve the equations for x, we can use the quadratic formula.

EXAMPLE 2

Solve for x.

a) $8x^2 + 3x - 2 = 0$ b) $8x^2 + tx - z = 0$

Solution

a) $8x^2 + 3x - 2$ does not factor, so we will solve using the quadratic formula.

$$8x^2 + 3x - 2 = 0$$
$$a = 8 \quad b = 3 \quad c = -2$$
$$x = \frac{-3 \pm \sqrt{(3)^2 - 4(8)(-2)}}{2(8)} = \frac{-3 \pm \sqrt{9 + 64}}{16} = \frac{-3 \pm \sqrt{73}}{16}$$

The solution set is $\left\{ \dfrac{-3 - \sqrt{73}}{16}, \dfrac{-3 + \sqrt{73}}{16} \right\}$.

b) Solve $8x^2 + tx - z = 0$ for x using the quadratic formula.

$$a = 8 \quad b = t \quad c = -z$$
$$x = \frac{-t \pm \sqrt{t^2 - 4(8)(-z)}}{2(8)} \qquad x = \frac{-b \pm \sqrt{b^2 - 4ac}}{2a}$$
$$= \frac{-t \pm \sqrt{t^2 + 32z}}{16} \qquad \text{Perform the operations.}$$

The solution set is $\left\{ \dfrac{-t - \sqrt{t^2 + 32z}}{16}, \dfrac{-t + \sqrt{t^2 + 32z}}{16} \right\}$.

YOU TRY 2

Solve for n.

a) $3n^2 + 5n - 1 = 0$ b) $3n^2 + pn - r = 0$

2 Solve an Applied Problem Involving Volume

EXAMPLE 3

A rectangular piece of cardboard is 5 in. longer than it is wide. A square piece that measures 2 in. on each side is cut from each corner, then the sides are turned up to make an uncovered box with volume 252 in³. Find the length and width of the original piece of cardboard.

Solution

Step 1: **Read** the problem carefully. Draw a picture.

Step 2: **Choose a variable** to represent the unknown, and define the other unknown in terms of this variable.

$$\text{Let} \quad x = \text{the width of the cardboard}$$
$$x + 5 = \text{the length of the cardboard}$$

 **Hint**

Draw a picture to help you understand the problem.

Step 3: **Translate** the information that appears in English into an algebraic equation.

The volume of a box is (length)(width)(height). We will use the formula (length)(width)(height) = 252.

Original Cardboard   **Box**

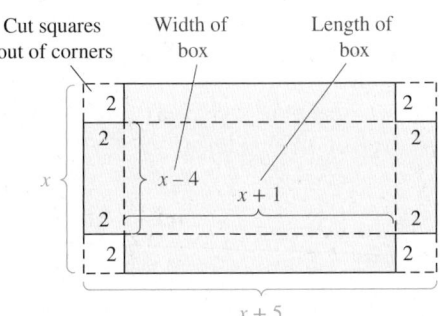

The figure on the left shows the original piece of cardboard with the sides labeled. The figure on the right illustrates how to label the box when the squares are cut out of the corners. When the sides are folded along the dotted lines, we must label the length, width, and height of the box.

Length of box = Length of original cardboard − Length of side cut out on the left − Length of side cut out on the right

$$= \quad x + 5 \quad - \quad 2 \quad - \quad 2$$
$$= \quad x + 1$$

Width of box = Width of original cardboard − Length of side cut out on top − Length of side cut out on bottom

$$= \quad x \quad - \quad 2 \quad - \quad 2$$
$$= \quad x - 4$$

Height of box = Length of side cut out

$$= \quad 2$$

Statement: Volume of box = (length)(width)(height)

Equation: $252 = (x + 1)(x - 4)(2)$

Step 4: **Solve** the equation.

$$252 = (x + 1)(x - 4)(2)$$
$$126 = (x + 1)(x - 4) \qquad \text{Divide both sides by 2.}$$
$$126 = x^2 - 3x - 4 \qquad \text{Multiply.}$$
$$0 = x^2 - 3x - 130 \qquad \text{Write in standard form.}$$
$$0 = (x + 10)(x - 13) \qquad \text{Factor.}$$
$$x + 10 = 0 \quad \text{or} \quad x - 13 = 0 \qquad \text{Set each factor equal to zero.}$$
$$x = -10 \quad \text{or} \qquad x = 13 \qquad \text{Solve.}$$

Step 5: **Check** the answer, and **interpret** the solution as it relates to the problem.

Because x represents the width, it cannot be negative. Therefore, the width of the original piece of cardboard is 13 in.

The length of the cardboard is $x + 5$, so $13 + 5 = 18$ in.

Width of cardboard = 13 in.   Length of cardboard = 18 in.

Check:

Width of box = $13 - 4 = 9$ in.; Length of box = $13 + 1 = 14$ in.;
Height of box = 2 in.
Volume of box = $9(14)(2) = 252$ in^3.

[**YOU TRY 3**]

The width of a rectangular piece of cardboard is 2 in. less than its length. A square piece that measures 3 in. on each side is cut from each corner, then the sides are turned up to make a box with volume 504 in^3. Find the length and width of the original piece of cardboard.

---

3 Solve an Applied Problem Involving Area

EXAMPLE 4

A rectangular pond is 20 ft long and 12 ft wide. The pond is bordered by a strip of grass of uniform (the same) width. The area of the grass is 320 ft^2. How wide is the border of grass around the pond?

Solution

Step 1: **Read** the problem carefully. Draw a picture.

Step 2: **Choose a variable** to represent the unknown, and define the other unknowns in terms of this variable.

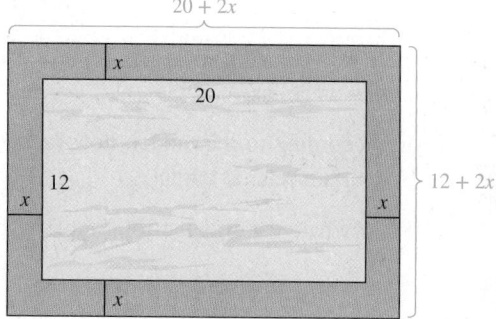

x = width of the strip of grass
$20 + 2x$ = length of pond plus two strips of grass
$12 + 2x$ = width of pond plus two strips of grass

Step 3: **Translate** from English into an algebraic equation.

We know that the area of the grass border is 320 ft^2. We can calculate the area of the pond since we know its length and width. The pond plus grass border forms a large rectangle of length $20 + 2x$ and width $12 + 2x$. The equation will come from the following relationship:

Statement: Area of pond plus grass $-$ Area of pond $=$ Area of grass border

Equation: $(20 + 2x)(12 + 2x) - 20(12) = 320$

Step 4: **Solve** the equation.

$$(20 + 2x)(12 + 2x) - 20(12) = 320$$
$$240 + 64x + 4x^2 - 240 = 320 \qquad \text{Multiply.}$$
$$4x^2 + 64x = 320 \qquad \text{Combine like terms.}$$
$$x^2 + 16x = 80 \qquad \text{Divide by 4.}$$
$$x^2 + 16x - 80 = 0 \qquad \text{Write in standard form.}$$
$$(x + 20)(x - 4) = 0 \qquad \text{Factor.}$$
$$x + 20 = 0 \quad \text{or} \quad x - 4 = 0 \qquad \text{Set each factor equal to 0.}$$
$$x = -20 \quad \text{or} \quad x = 4 \qquad \text{Solve.}$$

Step 5: **Check** the answer, and **interpret** the solution as it relates to the problem.

x represents the width of the strip of grass, so x cannot equal -20.

The width of the strip of grass is 4 ft.

Check: Substitute $x = 4$ into the equation written in Step 3.

$$[20 + 2(4)][12 + 2(4)] - 20(12) \overset{?}{=} 320$$
$$(28)(20) - 240 \overset{?}{=} 320$$
$$560 - 240 = 320 \quad \checkmark$$

[YOU TRY 4] A rectangular pond is 6 ft wide and 10 ft long and is surrounded by a concrete border of uniform width. The area of the border is 80 ft^2. Find the width of the border.

---

4 Solve an Applied Problem Using a Quadratic Equation

Steve Mason/Getty Images

EXAMPLE 5

The total tourism-related output in a country from 2015 to 2019 can be modeled by

$$y = 16.4x^2 - 50.6x + 896$$

where x is the number of years since 2015 and y is the total tourism-related output in billions of dollars.

a)   According to the model, how much money was generated in 2017 due to tourism-related output?

b)   In what year was the total tourism-related output about $955 billion?

Solution

a) Because x is the number of years *after* 2015, the year 2017 corresponds to $x = 2$.

$$y = 16.4x^2 - 50.6x + 896$$
$$y = 16.4(2)^2 - 50.6(2) + 896 \quad \text{Substitute 2 for } x.$$
$$y = 860.4$$

The total tourism-related output in 2017 was approximately $860.4 billion.

b) Because y represents the total tourism-related output (in billions), substitute 955 for y and solve for x.

$$y = 16.4x^2 - 50.6x + 896$$
$$955 = 16.4x^2 - 50.6x + 896 \quad \text{Substitute 955 for } y.$$
$$0 = 16.4x^2 - 50.6x - 59 \quad \text{Write in standard form.}$$

Use the quadratic formula to solve for x.

$$a = 16.4 \qquad b = -50.6 \qquad c = -59$$
$$x = \frac{50.6 \pm \sqrt{(-50.6)^2 - 4(16.4)(-59)}}{2(16.4)} \quad \begin{array}{l}\text{Substitute the values into}\\ \text{the quadratic formula.}\end{array}$$
$$x \approx 3.99 \approx 4 \text{ or } x \approx -0.90$$

The negative value of x does not make sense in the context of the problem. Use $x \approx 4$, which corresponds to the year 2019. The total tourism-related output was about $955 billion in 2019.

ANSWERS TO **YOU TRY** **EXERCISES**

1) $m = \dfrac{2E}{v^2}$
2) a) $\left\{ \dfrac{-5 - \sqrt{37}}{6}, \dfrac{-5 + \sqrt{37}}{6} \right\}$
b) $\left\{ \dfrac{-p - \sqrt{p^2 + 12r}}{6}, \dfrac{-p + \sqrt{p^2 + 12r}}{6} \right\}$

3) length = 20 in., width = 18 in.
4) 2 ft

E Evaluate **11.5** Exercises *Do the exercises, and check your work.*

Objective 1: Solve a Formula for a Variable
Solve for the indicated variable.

1) $A = \pi r^2$ for r

2) $V = \dfrac{1}{3}\pi r^2 h$ for r

3) $a = \dfrac{v^2}{r}$ for v

4) $K = \dfrac{1}{2}Iw^2$ for w

5) $E = \dfrac{I}{d^2}$ for d

6) $L = \dfrac{2U}{I^2}$ for I

7) $F = \dfrac{kq_1q_2}{r^2}$ for r

8) $E = \dfrac{kq}{r^2}$ for r

9) $d = \sqrt{\dfrac{4A}{\pi}}$ for A

10) $d = \sqrt{\dfrac{12V}{\pi h}}$ for V

11) $T_p = 2\pi\sqrt{\dfrac{l}{g}}$ for l

12) $V = \sqrt{\dfrac{3RT}{M}}$ for T

13) $T_p = 2\pi\sqrt{\dfrac{l}{g}}$ for g

14) $V = \sqrt{\dfrac{3RT}{M}}$ for M

15) Compare the equations $3x^2 - 5x + 4 = 0$ and $rx^2 - 5x + s = 0$.

a) How are the equations alike?

b) How can both equations be solved for x?

16) What method could be used to solve $2t^2 + 7t + 1 = 0$ and $kt^2 + mt + n = 0$ for t? Why?

Solve for the indicated variable.

17) $rx^2 - 5x + s = 0$ for x

18) $cx^2 + dx - 3 = 0$ for x

19) $pz^2 + rz - q = 0$ for z

20) $hr^2 - kr + j = 0$ for r

21) $da^2 - ha = k$ for a

22) $kt^2 + mt = -n$ for t

23) $s = \dfrac{1}{2}gt^2 + vt$ for t

24) $s = 2\pi rh + \pi r^2$ for r

Mixed Exercises: Objectives 2 and 3

Write an equation, and solve.

25) The length of a rectangular piece of sheet metal is 3 in. longer than its width. A square piece that measures 1 in. on each side is cut from each corner, then the sides are turned up to make a box with volume 70 in³. Find the length and width of the original piece of sheet metal.

26) The width of a rectangular piece of cardboard is 8 in. less than its length. A square piece that measures 2 in. on each side is cut from each corner, then the sides are turned up to make a box with volume 480 in³. Find the length and width of the original piece of cardboard.

27) A rectangular swimming pool is 60 ft wide and 80 ft long. A nonskid surface of uniform width is to be installed around the pool. If there is 576 ft² of the nonskid material, how wide can the strip of the nonskid surface be?

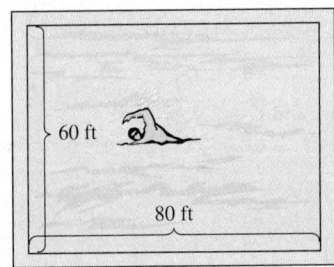

60 ft

80 ft

28) A picture measures 10 in. by 12 in. Emilio will get it framed with a border around it so that the total area of the picture plus the frame of uniform width is 168 in². How wide is the border?

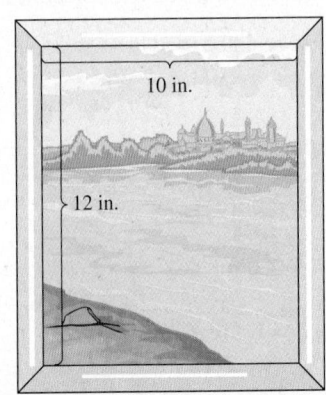

10 in.

12 in.

29) The height of a triangular sail is 1 ft less than twice the base of the sail. Find its height and the length of its base if the area of the sail is 60 ft².

30) Chandra cuts fabric into isosceles triangles for a quilt. The height of each triangle is 1 in. less than the length of the base. The area of each triangle is 15 in². Find the height and base of each triangle.

31) Valerie makes a bike ramp in the shape of a right triangle. The base of the ramp is 4 in. more than twice its height, and the length of the incline is 4 in. less than three times its height. How high is the ramp?

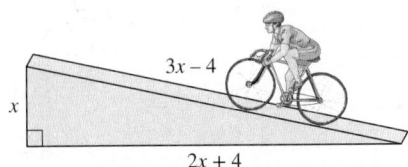

$3x - 4$

x

$2x + 4$

32) The width of a widescreen TV is 10 in. less than its length. The diagonal of the rectangular screen is 10 in. more than the length. Find the length and width of the screen.

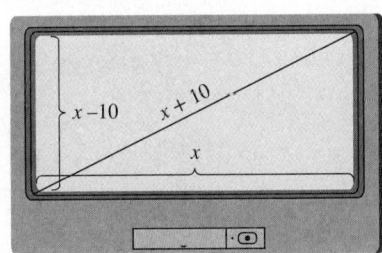

$x - 10$ $x + 10$

x

Objective 4: Solve an Applied Problem Using a Quadratic Equation

Solve.

33) An object is propelled upward from a height of 4 ft. The height h of the object (in feet) t sec after the object is released is given by

$$h = -16t^2 + 60t + 4$$

a) How long does it take the object to reach a height of 40 ft?

b) How long does it take the object to hit the ground?

34) An object is launched from the ground. The height h of the object (in feet) t sec after the object is released is given by

$$h = -16t^2 + 64t$$

When will the object be 48 ft in the air?

35) Attendance at sporting events in a large city from 2012 to 2019 can be modeled by

$$y = -0.25x^2 + 1.5x + 9.5$$

where x represents the number of years after 2012 and y represents the number of people who attended a sporting event (in millions).

   a) Approximately how many people saw a sporting event in 2012?

   b) In what year did approximately 11.75 million people see a sporting event?

36) The illuminance E (measure of the light emitted, in lux) of a light source is given by

$$E = \frac{1}{d^2}$$

where E is the luminous intensity (measured in candela) and d is the distance, in meters, from the light source. The luminous intensity of a lamp is 2700 candela at a distance of 3 m from the lamp. Find the illuminance, E, in lux.

37) A sandwich shop has determined that the demand for its turkey sandwich is $\frac{65}{P}$ per day, where P is the price of the sandwich in dollars. The daily supply is given by $10P + 3$. Find the price at which the demand for the sandwich equals the supply.

38) A hardware store determined that the demand for shovels one winter was $\frac{2800}{P}$, where P is the price of the shovel in dollars. The supply was given by

$12P + 32$. Find the price at which demand for the shovels equals the supply.

Use the following formula for Exercises 39 and 40.

A wire is stretched between two poles separated by a distance d, and a weight is in the center of the wire of length L so that the wire is pulled taut as pictured here. The vertical distance, D, between the weight on the wire and the top of the poles is given by $D = \dfrac{\sqrt{L^2 - d^2}}{2}$.

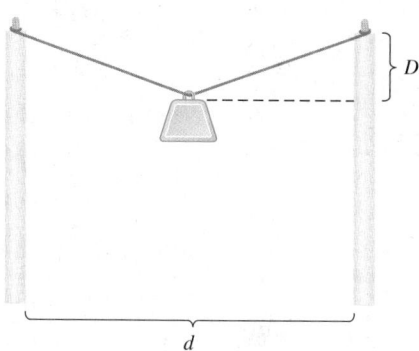

39) A 12.5-ft clothesline is attached to the top of two poles that are 12 ft apart. A shirt is hanging in the middle of the clothesline. Find the distance, D, that the shirt is hanging down.

40) An 11-ft wire is attached to a ceiling in a loft apartment by hooks that are 10 ft apart. A light fixture is hanging in the middle of the wire. Find the distance, D, between the ceiling and the top of the light fixture. Round the answer to the nearest tenths place.

R Rethink

R1) Have you ever had to calculate volume or area before studying this section?

R2) Are you allowed to have negative answers for volume or area?

R3) What do you find most difficult about doing applied problems?

Group Activity – Heat Index (*www.weather.gov*)

Why does 80°F in Arizona feel cooler than 80°F in Florida? The answer: relative humidity. During the summer, the relative humidity in Arizona is below 40%, while in Florida the relative humidity is generally between 50% and 70%. When the relative humidity is factored with the air temperature, the result is known as the heat index.

Mathematically, the heat index T_{HI} (°F) is a function of the relative humidity R ($40\% \leq R \leq 100\%$) and the air temperature T ($T \geq 80°F$).

$$T_{HI} = -42.379 + 2.049T + 10.143R - 0.225RT - 0.00684T^2 - 0.0548R^2$$
$$+ 0.001228RT^2 + 0.000853R^2T - 0.000002R^2T^2$$

We can group the terms as follows:

$$T_{HI} = (-0.00684T^2 + 0.001228RT^2 - 0.000002R^2T^2)$$
$$+ (2.049T - 0.225RT + 0.000853R^2T) + (-42.379 + 10.143R - 0.0548R^2)$$

Factoring T^2 from the first group and T from the second group yields:

$$T_{HI} = \underbrace{(-0.00684 + 0.001228R - 0.000002R^2)}_{a} T^2 + \underbrace{(2.049 - 0.225R + 0.000853R^2)}_{b} T$$
$$+ \underbrace{(-42.379 + 10.143R - 0.0548R^2)}_{c}$$

For a specific value of relative humidity, R, this becomes a quadratic equation in the variable T.

1) a) Find the values of a, b, and c when $R = 70\%$.

    b) Write the equation for $R = 70\%$, and use the equation to fill in the following table.

T (air temperature)	T_{HI} (when $R = 70\%$)
80°F	
84°F	
88°F	
92°F	
96°F	
100°F	
104°F	
108°F	

☐ Caution
☐ Extreme Caution
☐ Danger
☐ Extreme Danger

2) a) Find the values of a, b, and c when $R = 40\%$.

   b) Write the equation for $R = 40\%$, and use the equation to fill in the following table.

T (air temperature)	T_{HI} (when $R = 40\%$)	
80°F		
84°F		☐ Caution
88°F		
92°F		☐ Extreme Caution
96°F		
100°F		☐ Danger
104°F		
108°F		■ Extreme Danger

3) Look up a Heat Index Chart, and compare your findings.

4) What was the highest heat index registered in your city?

One essential component to success—in the classroom and beyond—is taking responsibility for your results. Yes, there will always be things that happen that are beyond your control. But if you blame the alarm that didn't go off or the mosquito that wouldn't stop buzzing during the test for your poor results on a math test, you will never improve. You have the power, and the responsibility, to achieve your goals.

To get a sense of your ideas of why things happen to you, circle the statement from each of the pairs below that best describes your views.

1) a) In the long run, people get the respect they deserve in this world.
   b) Unfortunately, an individual's value often goes unrecognized no matter how hard he or she tries.

2) a) The idea that teachers are unfair to students is nonsense.
   b) Most students don't realize the extent to which their exam results are influenced by random events.

3) a) I have found that much of what happens will happen no matter what I do.
   b) Trusting fate has never turned out as well for me as making a decision to take a definite course of action.

4) a) For a well prepared student, there is rarely, if ever, such a thing as an unfair exam.
   b) Many times, exam questions are unrelated to coursework, and studying is often useless.

5) a) Becoming a success is a matter of hard work; luck has little or nothing to do with it.
   b) Getting a good job depends mainly on being in the right place at the right time.

6) a) It is not always wise to plan too far ahead because you can never predict what's going to happen to you.
   b) When I make plans, I am almost certain that I can make them work.

7) a) In my case, getting what I want has little or nothing to do with luck.
   b) I often feel like I might as well decide what to do by flipping a coin.

8) a) In general, I feel that I have little influence over the things that happen to me.
   b) It is impossible for me to believe that chance or luck plays an important role in my life.

9) a) What happens to me is my own doing.
   b) Sometimes I feel that I don't have enough control over the direction my life is taking.

10) a) Sometimes I can't understand how teachers arrive at the grades they give.
    b) There is a direct connection between how hard a person studies and the grades he or she gets.

Scoring: Give yourself one point for each of the following answers and then add up your score:

1) a, 2) a, 3) b, 4) a, 5) a, 6) b, 7) a, 8) b, 9) a, 10) b

Your total score can range from 0 to 10. The higher your score, the more you believe that you have a strong influence over what happens to you and that you are in control of your life and your own behavior. The lower your score, the more you believe that your life is outside of your control and what happens to you is caused by luck or fate.

If you score below 5 on this questionnaire, consider how rethinking your views of the causes of what happens to you might lead to greater success.

Adapted from "Do you control what happens to you?" in Nathenson, M. (1985). *The Book of Tests*. New York: Penguin.

Chapter 11: Summary

Definition/Procedure	Example

11.1 Review of Solving Equations by Factoring

Steps for Solving a Quadratic Equation by Factoring

1) Write the equation in the form $ax^2 + bx + c = 0$.

2) Factor the expression.

3) Use the zero product rule: Set each factor equal to zero, and solve for the variable.

4) Check the answer(s).

Solve $5z^2 - 7z = 6$.

$$5z^2 - 7z - 6 = 0$$
$$(5z + 3)(z - 2) = 0$$

$5z + 3 = 0$ or $z - 2 = 0$
$5z = -3$

$z = -\dfrac{3}{5}$ or $z = 2$

The check is left to the student.

The solution set is $\left\{ -\dfrac{3}{5}, 2 \right\}$.

11.2 The Square Root Property and Completing the Square

The Square Root Property

Let k be a constant. If $x^2 = k$, then $x = \sqrt{k}$ or $x = -\sqrt{k}$.

Solve $6p^2 = 54$.

$p^2 = 9$ Divide by 6.
$p = \pm\sqrt{9}$ Square root property
$p = \pm 3$ $\sqrt{9} = 3$

The solution set is $\{-3, 3\}$.

The Distance Formula

The **distance,** d, between two points with coordinates (x_1, y_1) and (x_2, y_2) is given by $d = \sqrt{(x_2 - x_1)^2 + (y_2 - y_1)^2}$.

Find the distance between the points $(6, -2)$ and $(0, 2)$.

Label the points: $(\overset{x_1}{6}, \overset{y_1}{-2})\ (\overset{x_2}{0}, \overset{y_2}{2})$

Substitute the values into the distance formula.

$$d = \sqrt{(0 - 6)^2 + (2 - (-2))^2}$$
$$= \sqrt{(-6)^2 + (4)^2}$$
$$= \sqrt{36 + 16} = \sqrt{52} = 2\sqrt{13}$$

A **perfect square trinomial** is a trinomial whose factored form is the square of a binomial.

Perfect Square Trinomial	Factored Form
$y^2 + 8y + 16$	$(y + 4)^2$
$9t^2 - 30t + 25$	$(3t - 5)^2$

Complete the Square for $x^2 + bx$

To find the constant needed to complete the square for $x^2 + bx$,

Step 1: Find half of the coefficient of x: $\dfrac{1}{2}b$

Step 2: Square the result: $\left(\dfrac{1}{2}b\right)^2$

Step 3: Add it to $x^2 + bx$: $x^2 + bx + \left(\dfrac{1}{2}b\right)^2$. The factored form is $\left(x + \dfrac{1}{2}b\right)^2$.

Complete the square for $x^2 + 12x$ to obtain a perfect square trinomial. Then, factor.

Step 1: Find half of the coefficient of x: $\dfrac{1}{2}(12) = 6$

Step 2: Square the result: $6^2 = 36$

Step 3: Add 36 to $x^2 + 12x$: $x^2 + 12x + 36$

The perfect square trinomial is $x^2 + 12x + 36$.
The factored form is $(x + 6)^2$.

Definition/Procedure	Example

Solve a Quadratic Equation ($ax^2 + bx + c = 0$) by Completing the Square

Step 1: **The coefficient of the squared term must be 1.** If it is not 1, divide both sides of the equation by a to obtain a leading coefficient of 1.

Step 2: **Get the variables on one side of the equal sign and the constant on the other side.**

Step 3: **Complete the square.** Find half of the linear coefficient, then square the result. Add that quantity to both sides of the equation.

Step 4: **Factor.**

Step 5: **Solve using the square root property.**

Solve $x^2 + 6x + 7 = 0$ by completing the square.

$$x^2 + 6x + 7 = 0 \qquad \text{The coefficient of } x^2 \text{ is 1.}$$
$$x^2 + 6x = -7 \qquad \text{Get the constant on the other side of the equal sign.}$$

Complete the square: $\dfrac{1}{2}(6) = 3$

$$(3)^2 = 9$$

Add 9 to both sides of the equation.

$$x^2 + 6x + 9 = -7 + 9$$
$$(x + 3)^2 = 2 \qquad \text{Factor.}$$
$$x + 3 = \pm\sqrt{2} \qquad \text{Square root property}$$
$$x = -3 \pm \sqrt{2}$$

The solution set is $\{-3 - \sqrt{2}, -3 + \sqrt{2}\}$.

11.3 The Quadratic Formula

The Quadratic Formula

The solutions of any quadratic equation of the form $ax^2 + bx + c = 0$ $(a \neq 0)$ are

$$x = \frac{-b \pm \sqrt{b^2 - 4ac}}{2a}$$

This formula is called the **quadratic formula.**

Solve $2x^2 - 5x - 2 = 0$ using the quadratic formula.

$$a = 2 \qquad b = -5 \qquad c = -2$$

Substitute the values into the quadratic formula, and simplify.

$$x = \frac{-(-5) \pm \sqrt{(-5)^2 - 4(2)(-2)}}{2(2)}$$

$$x = \frac{5 \pm \sqrt{25 + 16}}{4} = \frac{5 \pm \sqrt{41}}{4}$$

The solution set is $\left\{\dfrac{5 - \sqrt{41}}{4}, \dfrac{5 + \sqrt{41}}{4}\right\}$.

The expression under the radical, $b^2 - 4ac$, is called the **discriminant.**

1) If $b^2 - 4ac$ is **positive and the square of an integer,** the equation has **two rational solutions.**

2) If $b^2 - 4ac$ is **positive but not a perfect square,** the equation has **two irrational solutions.**

3) If $b^2 - 4ac$ is **negative,** the equation has **two nonreal, complex solutions of the form** $a + bi$ and $a - bi$.

4) If $b^2 - 4ac = 0$, the equation has **one rational solution.**

Find the value of the discriminant for $3m^2 + 4m + 5 = 0$, and determine the number and type of solutions of the equation.

$$a = 3 \qquad b = 4 \qquad c = 5$$

$$b^2 - 4ac = (4)^2 - 4(3)(5) = 16 - 60 = -44$$

Discriminant $= -44$. The equation has two nonreal, complex solutions of the form $a + bi$ and $a - bi$.

11.4 Equations in Quadratic Form

Some equations that are not quadratic can be solved using the same methods that can be used to solve quadratic equations. These are called **equations in quadratic form.**

Solve $\qquad r^4 + 2r^2 - 24 = 0$.

$$(r^2 - 4)(r^2 + 6) = 0 \qquad \text{Factor.}$$

$$r^2 - 4 = 0 \qquad \text{or} \qquad r^2 + 6 = 0$$
$$r^2 = 4 \qquad\qquad\qquad r^2 = -6$$
$$r = \pm\sqrt{4} \qquad\qquad\quad r = \pm\sqrt{-6}$$
$$r = \pm 2 \qquad\qquad\quad\ r = \pm i\sqrt{6}$$

The solution set is $\{-i\sqrt{6}, i\sqrt{6}, -2, 2\}$.

Definition/Procedure	Example

11.5 Formulas and Applications

Solve a Formula for a Variable.

Solve for s: $\quad g = \dfrac{10}{s^2}$

$$s^2 g = 10 \qquad \text{Multiply both sides by } s^2.$$
$$s^2 = \dfrac{10}{g} \qquad \text{Divide both sides by } g.$$
$$s = \pm\sqrt{\dfrac{10}{g}} \qquad \text{Square root property}$$
$$s = \dfrac{\pm\sqrt{10}}{\sqrt{g}} \cdot \dfrac{\sqrt{g}}{\sqrt{g}} \qquad \text{Rationalize the denominator.}$$
$$s = \dfrac{\pm\sqrt{10g}}{g}$$

Solving Application Problems Using a Quadratic Equation.

A woman dives off of a cliff 49 m above the ocean. Her height, $h(t)$, in meters, above the water is given by

$$h(t) = -9.8t^2 + 49$$

where t is the time, in seconds, after she leaves the cliff. When will she hit the water?

Let $h(t) = 0$, and solve for t.

$$\begin{aligned} h(t) &= -9.8t^2 + 49 \\ 0 &= -9.8t^2 + 49 \qquad \text{Substitute 0 for } h. \\ 9.8t^2 &= 49 \qquad \text{Add } 9.8t^2 \text{ to each side.} \\ t^2 &= 5 \qquad \text{Divide by 9.8.} \\ t &= \pm\sqrt{5} \qquad \text{Square root property} \end{aligned}$$

Since t represents time, we discard $-\sqrt{5}$. She will hit the water in $\sqrt{5}$, or about 2.2, sec.

Chapter 11:   Review Exercises

(11.1) Solve by factoring.

1) $a^2 - 3a - 54 = 0$

2) $2t^2 + 9t + 10 = 0$

3) $\dfrac{2}{3}c^2 = \dfrac{2}{3}c + \dfrac{1}{2}$

4) $4k = 12k^2$

5) $x^3 + 3x^2 - 16x - 48 = 0$

6) $3p - 16 = p(p - 7)$

Write an equation and solve.

7) A rectangle has an area of 96 cm². Its width is 4 cm less than its length. Find the length and width.

8) Find the base and height of the triangle if its area is 30 in².

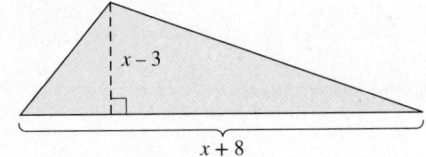

(11.2) Solve using the square root property.

9) $d^2 = 144$

10) $m^2 = 75$

11) $v^2 + 4 = 0$

12) $2c^2 - 11 = 25$

13) $(b - 3)^2 = 49$

14) $(6y + 7)^2 - 15 = 0$

15) $27k^2 - 30 = 0$

16) $(j - 14)^2 + 5 = 0$

17) Find the distance between the points $(-8, 3)$ and $(-12, 5)$.

18) A rectangle has a length of $5\sqrt{2}$ in. and a width of 4 in. How long is its diagonal?

Complete the square for each expression to obtain a perfect square trinomial. Then, factor.

19) $r^2 + 10r$

20) $z^2 - 12z$

21) $c^2 - 5c$

22) $x^2 + x$

23) $a^2 + \dfrac{2}{3}a$

24) $d^2 - \dfrac{5}{2}d$

Solve by completing the square.

25) $p^2 - 6p - 16 = 0$

26) $w^2 - 2w - 35 = 0$

27) $n^2 + 10n = 6$

28) $t^2 + 9 = -4t$

29) $f^2 + 3f + 1 = 0$

30) $j^2 - 7j = 4$

31) $-3q^2 + 7q = 12$

32) $6v^2 - 15v + 3 = 0$

(11.3) Solve using the quadratic formula.

33) $m^2 + 4m - 12 = 0$

34) $3y^2 = 10y - 8$

35) $10g - 5 = 2g^2$

36) $20 = 4x - 5x^2$

37) $\dfrac{1}{6}t^2 - \dfrac{1}{3}t + \dfrac{2}{3} = 0$

38) $(s - 3)(s - 5) = 9$

39) $(6r + 1)(r - 4) = -2(12r + 1)$

40) $z^2 - \dfrac{3}{2}z + \dfrac{13}{16} = 0$

Find the value of the discriminant. Then, determine the number and type of solutions of each equation. Do not solve.

41) $3n^2 - 2n - 5 = 0$

42) $t^2 = -3(t + 2)$

43) Find the value of b so that $4k^2 + bk + 9 = 0$ has only one rational solution.

44) A ball is thrown upward from a height of 4 ft. The height, h, of the ball (in feet) t sec after the ball is released is given by $h = -16t^2 + 52t + 4$.

a) How long does it take the ball to reach a height of 16 ft?

b) How long does it take the ball to hit the ground?

(11.1–11.3) Keep in mind the four methods we have learned for solving quadratic equations: *factoring, the square root property, completing the square, and the quadratic formula*. Solve the equations using one of these methods.

45) $3k^2 + 4 = 7k$

46) $n^2 - 6n + 11 = 0$

47) $15 = 3 + (y + 8)^2$

48) $(2a + 1)(a + 2) = 14$

49) $\dfrac{1}{3}w^2 + w = -\dfrac{5}{6}$

50) $4t^2 + 5 = 7$

51) $6 + p(p - 10) = 2(4p - 15)$

52) $6 = 2m - 3m^2$

53) $x^3 = x$

54) $\dfrac{1}{12}b^2 - \dfrac{9}{2} = \dfrac{1}{4}b$

55) Let $f(x) = (2x - 1)^2$. Find all values of x so that $f(x) = 25$.

56) Let $f(x) = \dfrac{1}{10}x^2 + 3x$ and $g(x) = 4x - \dfrac{11}{5}$. Find all values of x such that $f(x) = g(x)$.

(11.4) Solve.

57) $\dfrac{5k}{k + 1} = 3k - 4$

58) $\dfrac{10}{m} = 3 + \dfrac{8}{m^2}$

59) $f = \sqrt{7f - 12}$

60) $x - 4\sqrt{x} = 5$

61) $n^4 - 17n^2 + 16 = 0$

62) $b^4 + 5b^2 - 14 = 0$

63) $q^{2/3} + 2q^{1/3} - 3 = 0$

64) $y + 2 = 3y^{1/2}$

65) $2r^4 = 7r^2 - 2$

66) $2(v + 2)^2 + (v + 2) - 3 = 0$

67) $(2k - 5)^2 - 5(2k - 5) - 6 = 0$

Write an equation, and solve.

68) At the end of the day, the employees at Forever Young have to put all clothes left in the dressing room back to their proper places. Working together, Lorena and Erica can put away the clothes in 1 hr 12 min. On her own, it takes Lorena 1 hr longer to put away the clothes than it takes Erica to do it by herself. How long does it take each girl to put away the clothes by herself?

(11.5) Solve for the indicated variable.

69) $F = \dfrac{mv^2}{r}$ for v

70) $U = \dfrac{1}{2}kx^2$ for x

71) $r = \sqrt{\dfrac{A}{\pi}}$ for A

72) $r = \sqrt{\dfrac{V}{\pi l}}$ for V

73) $kn^2 - ln - m = 0$ for n

74) $2p^2 + t = rp$ for p

Write an equation, and solve.

75) Ayesha is making a pillow sham by sewing a border onto an old pillowcase. The rectangular pillowcase measures 18 in. by 27 in. When she sews a border of uniform width around the pillowcase, the total area of the surface of the pillow sham will be 792 in². How wide is the border?

76) The width of a rectangular piece of cardboard is 4 in. less than its length. A square piece that measures 2 in. on each side is cut from each corner, then the sides are turned up to make a box with volume 280 in^3. Find the length and width of the original piece of cardboard.

77) A flower shop determined that the demand, $D(P)$, for its tulip bouquet is $D(P) = \dfrac{240}{P}$ per week, where P is the price of the bouquet in dollars. The weekly supply, $S(P)$, is given by $S(P) = 4p - 2$. Find the price at which demand for the tulips equals the supply.

78) U.S. sales of a certain brand of wine can be modeled by
$$y = -0.20x^2 + 4.0x + 8.4$$
for the years 2004 to 2019. x is the number of years after 2004, and y is the number of bottles sold, in millions.

   a) How many bottles were sold in 2004?

   b) How many bottles were sold in 2017?

   c) In what year did sales reach 28.4 million bottles?

Chapter 11:   Test

Solve the equations in Exercises 1 and 2 by factoring.

1) $k^2 - 8k = 48$

2) $16 - 9w^2 = 0$

3) Solve $t^2 + 7 = 25$ using the square root property.

4) If k is a negative number and $x^2 = k$, what can you conclude about the solution set of the equation?

Solve the equations in Exercises 5 and 6 by completing the square.

5) $b^2 + 4b - 7 = 0$

6) $2x^2 - 6x + 14 = 0$

7) Solve $x^2 - 8x + 17 = 0$ using the quadratic formula.

Solve using any method.

8) $(c + 5)^2 + 8 = 2$

9) $3q^2 + 2q = 8$

10) $y^2 - \dfrac{4}{25} = 0$

11) $(4n + 1)^2 + 9(4n + 1) + 18 = 0$

12) $(2t - 3)(t - 2) = 2$

13) $45a = 54a^2$

14) $\dfrac{3}{10x} = \dfrac{x}{x - 1} - \dfrac{4}{5}$

15) $p^4 + p^2 - 72 = 0$

16) Find the value of the discriminant. Then, determine the number and type of solutions of the equation. *Do not solve.*
$$5z^2 - 6z - 1 = 0$$

17) Find the length of the missing side.

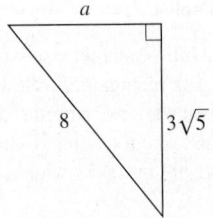

18) Let $P(x) = 5x^2$ and $Q(x) = 2x$. Find all values of x so that $P(x) = Q(x)$.

19) Find the distance between the points $(7, -4)$ and $(5, 6)$.

20) Solve for V. $r = \sqrt{\dfrac{3V}{\pi h}}$

21) Solve for t. $rt^2 - st = 6$

22) A ball is projected upward from the top of a 200-ft tall building. The height h of the ball above the ground (in feet) t sec after the ball is released is given by
$$h = -16t^2 + 24t + 200.$$

   a) When will the ball be 40 ft above the ground?

   b) When will the ball hit the ground?

Write an equation, and solve.

23) It takes Kimora 20 min more to type a report than it takes Justine. It would take them 24 min to type the report together. How long would it take each woman to type the report alone?

Westend61/Getty Images

24) The length of a rectangular garden is 2 ft more than the width. The diagonal of the garden is 10 ft. Find the length and width of the garden.

25) A rectangular piece of sheet metal is 6 in. longer than it is wide. A square piece that measures 3 in. on each side is cut from each corner, then the sides are turned up to make a box with volume 273 in^3. Find the length and width of the original piece of sheet metal.

Chapter 11:   Cumulative Review for Chapters 1–11

1) Simplify $\dfrac{\dfrac{12}{35}}{\dfrac{24}{49}}$.

Simplify. The final answer should contain only positive exponents.

2) $(5x^4y^{-10})(3xy^3)^2$

3) Solve for m. $y = mx + b$

4) Given the relation
$\{(4, 0), (3, 1), (3, -1), (0, 2)\}$,

a) what is the domain?

b) what is the range?

c) is the relation a function?

5) In the second quarter of 2018, Facebook reported that its average revenue per user in the United States and Canada was \$25.91. This can be modeled by the function $R(u) = 25.91u$, where u is the number of users in the United States and Canada and $R(u)$ is the revenue, or amount of money earned, from those users. (techcrunch.com)

a) What is Facebook's revenue from 20,000 users?

b) How much does Facebook earn from 1,500,000 users?

c) How many users generate \$1,295,500,000 in revenue?

Dolphfyn/Alamy Stock Photo

6) *Write a system of two equations in two variables, and solve.* Two bags of chips and three cans of soda cost \$3.85, while one bag of chips and two cans of soda cost \$2.30. Find the cost of a bag of chips and a can of soda.

7) Solve this system: $4x - 2y + z = -7$
$-3x + y - 2z = 5$
$2x + 3y + 5z = 4$

8) Subtract
$(4x^2y^2 - 11x^2y + xy + 2) - (x^2y^2 - 6x^2y + 3xy^2 + 10xy - 6)$

9) Multiply and simplify $3(r - 5)^2$.

Factor completely.

10) $4p^3 + 14p^2 - 8p$

11) $a^3 + 125$

12) Let $f(x) = \dfrac{x + 1}{2x^2 - x - 10}$.

a) Find $f(5)$.

b) Find x so that $f(x) = 0$.

c) Determine the domain of f.

13) Add $\dfrac{z - 8}{z + 4} + \dfrac{3}{z}$

14) Simplify $\dfrac{2 + \dfrac{6}{c}}{\dfrac{2}{c^2} - \dfrac{8}{c}}$

Solve.

15) $|3k + 7| = 19$

16) $|3k + 7| \geq 19$

Simplify. Assume all variables represent nonnegative real numbers.

17) $\sqrt{75}$ 18) $\sqrt[3]{10}$

19) $\sqrt{63x^7y^4}$

20) Simplify $64^{2/3}$.

21) Rationalize the denominator: $\dfrac{5}{2 + \sqrt{3}}$.

22) Multiply and simplify $(10 + 3i)(1 - 8i)$.

Solve.

23) $1 - \dfrac{1}{3h - 2} = \dfrac{20}{(3h - 2)^2}$

24) $p^2 + 6p = 27$

25) Solve for V: $r = \sqrt{\dfrac{V}{\pi h}}$.

Functions and Their Graphs

Get Ready

In this chapter we will learn how to graph many different types of equations, so now we will review how to find x- and y-intercepts.

Recall that the **x-intercept** of the graph of an equation is the point where the graph intersects the x-axis. The **y-intercept** of the graph of an equation is the point where the graph intersects the y-axis.

1) To **find the x-intercept of the graph of an equation,** let $y = 0$ and solve for x.
2) To **find the y-intercept of the graph of an equation,** let $x = 0$ and solve for y.

Remember to write the intercepts as ordered pairs.

  Example: Find the x- and y-intercept of the graph of $y = -4x + 8$.

  x-*intercept*: Let $y = 0$ and solve for x. $0 = -4x + 8$
$$-8 = -4x$$
$$2 = x$$

  The x-intercept is $(2, 0)$.

  y-*intercept*: Let $x = 0$ and solve for y. $y = -4(0) + 8$
$$y = 0 + 8$$
$$y = 8$$

  The y-intercept is $(0, 8)$.

Use the following *Basic Skills Worksheets* to prepare students for this, and future, chapters: WS2 Powers (to practice quick recall of powers of integers) and WS5 Roots (to practice quick recall of roots of integers).

Find the x- and y-intercept(s) of the graph of each equation. **Do not graph.** (You may not know how to graph some of these equations yet, but you can still find the intercepts.)

1) $y = 3x + 12$

2) $y = \frac{2}{7}x - 6$

3) $4x - 3y = 10$

4) $-x - 8y = -12$

5) $f(x) = -\frac{8}{5}x - 4$

6) $g(x) = \frac{1}{4}x + 5$

7) $y = x^2 - 9$

8) $y = -x^2 + 4$

9) $y = -x^3 + 8$

10) $y = x^3 + 1$

Answers

1) x-int: $(-4, 0)$ 2) x-int: $(21, 0)$ 3) x-int: $\left(\frac{5}{2}, 0\right)$ 4) x-int: $(12, 0)$ 5) x-int: $\left(-\frac{5}{2}, 0\right)$ 6) x-int: $(-20, 0)$
y-int: $(0, 12)$ y-int: $(0, -6)$ y-int: $\left(0, -\frac{10}{3}\right)$ y-int: $(0, -4)$ y-int: $(0, 5)$

7) x-int: $(-3, 0)$ and $(3, 0)$ 8) x-int: $(-2, 0)$ and $(2, 0)$ 9) x-int: $(2, 0)$ 10) x-int: $(-1, 0)$
y-int: $(0, -9)$ y-int: $(0, 4)$ y-int: $(0, 8)$ y-int: $(0, 1)$

Study Strategies

Developing Financial Literacy

Would you go on vacation without first thinking about, and planning for, where you would go? Would you start cooking dinner by putting ingredients in a bowl without any idea of what you would make? For any project to succeed, you need a plan. When it comes to personal finances, that plan is a budget. A budget will help you keep your spending under control and remain on track to meet your larger financial goals. Use the following strategies to create an effective budget.

- **I will create and use a budget that will allow me to achieve my long-term and short-term goals.**

- Complete the emPOWERme survey that appears before the Chapter Summary to learn about your saving style.
- Write down your financial goals. These might include paying tuition, paying off a car loan, saving money for a large purchase, or putting money in a savings account every month.
- Figure out how you spend your money by making a list of expenses. For a week, write down or use a phone app to keep track of *every* purchase you make, large and small.
- Make a list of everything you will need to spend money on in the coming year. This list should include things like tuition, rent or mortgage, food, gas money, and entertainment.
- Make a list of long-term goals as well, such as saving a certain amount of money every year.
- Make a list of all of your income for the coming year, such as the amount of money you earn at your job and any financial aid you might receive.
- Add up all of your sources of income as well as everything you spend money on, including small, daily purchases and your larger expenses.

- Create your budget. It could be a weekly budget or a monthly budget, whichever works best for you. Look at your income and list of expenses.
- Prioritize your spending. Assign money in your budget to the most important things first such as rent, food, and transportation. Give lower priority to things

like entertainment. If possible, try to save a little bit of money every week or month.

- If you find that your spending is greater than your income, look at your app or what you wrote down when you were keeping track of how you spend your money. Think about where you could cut some costs, and adjust your budget accordingly.
- If possible, look for ways to increase your income such as by finding a part-time job, if your schedule allows.
- When you feel like you have created a reasonable budget, start using it right away!

- After using your budget for a couple of weeks or months, ask yourself, "Was I able to stick to my budget and pay all expenses? Did I spend less money on things I really didn't need compared to *before* I made a budget? Was I able to save some money every month?"

- If you were able to stick to your budget, continue using it. No matter what, however, think about ways you may want to or need to be flexible. In the future, for example, you may need to adjust the budget if your rent increases.
- Were you able to save a little bit of money while on your budget? Look at your income and expenses and, if possible, try to save some money every week or month.
- If you did not stick to the budget, think about the reason why. Was the budget unrealistic? Did you spend more on extra things that you didn't really need that took money away from more important expenses? Think about what you could do differently, and adjust your budget accordingly.

12.1 Relations and Functions

P Prepare	**O Organize**
What are your objectives for Section 12.1?	**How can you accomplish each objective?**
1 Identify Relation, Function, Domain, and Range	• Learn the definitions of a *relation, function, domain,* and *range*. • Know how to use the **vertical line test.** • Understand when and how to use function notation. • Complete the given examples on your own. • Complete You Trys 1 and 2.
2 Review Linear, Polynomial, and Rational Functions	• Know the definitions of *linear, polynomial,* and *rational functions,* and make up your own example of each. • Be able to identify the domains of these functions. • Complete the given examples on your own. • Complete You Trys 3–5.
3 Define and Evaluate Quadratic Functions	• Write the definition of a *quadratic function* in your own words and write down an example. • Understand how to evaluate a function for a real number, a variable, and an expression. • Complete the given example on your own. • Complete You Try 6.

What are your objectives for Section 12.1?	How can you accomplish each objective?
4 Determine the Domains of Square Root and Cube Root Functions	• Know how to identify a *radical expression, radical function,* and a *square root function.* • Write the definitions of the *domain of a square root function* and *domain of a cube root function* in your own words. • Complete the given example on your own. • Complete You Try 7.
5 Graph Square Root and Cube Root Functions	• Follow Example 8 to create your own procedure for **Graphing a Square Root Function.** • Follow Example 9 to create your own procedure for **Graphing a Cube Root Function.** • Complete You Trys 8 and 9.

 **W Work**   **Read the explanations, follow the examples, take notes, and complete the You Trys.**

1 Identify Relation, Function, Domain, and Range

We first studied functions in Chapter 4. Let's review some of the concepts discussed in previous chapters, and we will extend what we have learned to include new functions. Recall the following definitions.

Definition

A **relation** is any set of ordered pairs. The **domain** of a relation is the set of all values of the independent variable (the first coordinates in the set of ordered pairs). The **range** of a relation is the set of all values of the dependent variable (the second coordinates in the set of ordered pairs).

Definition

A **function** is a special type of relation. If each element of the domain corresponds to *exactly one* element of the range, then the relation is a function.

EXAMPLE 1   Identify the domain and range of each relation, and determine whether each relation is a function.

a) $\{(-2, -7), (0, -1), (1, 2), (5, 14)\}$ b)

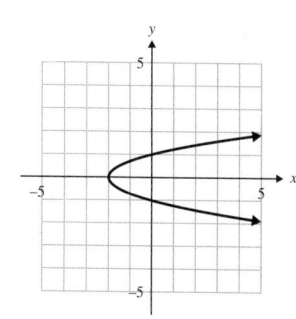

Solution

a) The *domain* is the set of first coordinates, $\{-2, 0, 1, 5\}$.

   The *range* is the set of second coordinates, $\{-7, -1, 2, 14\}$.

   Ask yourself, "Does every first coordinate correspond to *exactly one* second coordinate?" *Yes.* This relation *is* a function.

b) The domain is $[-2, \infty)$. The range is $(-\infty, \infty)$.

   To determine whether this graph represents a function, recall that we can use the *vertical line test.*

   The **vertical line test** says that if there is no vertical line that can be drawn through a graph so that it intersects the graph more than once, then the graph represents a function.

   This graph fails the vertical line test because we can draw a vertical line through the graph that intersects it more than once. *This graph does* not *represent a function.*

[YOU TRY 1] Identify the domain and range of each relation, and determine whether each relation is a function.

a) $\{(-8, 1), (-5, 2), (-5, -2), (7, -4)\}$

b)

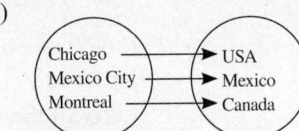

c)

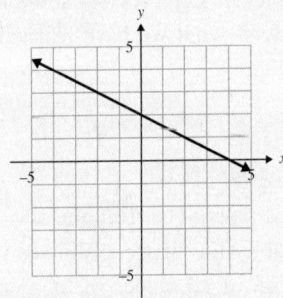

If a relation is written *as an equation* so that y is in terms of x, then we can think of the **domain** as the set of all real numbers that can be substituted for the independent variable, x. The resulting set of real numbers that are obtained for y, the dependent variable, is the **range.** To determine the domain, sometimes it is helpful to ask yourself, *"Is there any number that cannot be substituted for x?"*

EXAMPLE 2   Determine whether each relation describes y as a function of x, and determine the domain of the relation.

a) $y = -3x + 4$ b) $y^2 = x$

Solution

a) *Every* value substituted for x will have *exactly one* corresponding value of y. For example, if we substitute 5 for x, the only value of y is -11. Therefore, $y = -3x + 4$ *is a function.*

   To determine the domain, ask yourself, *"Is there any number that cannot be substituted for x in $y = -3x + 4$?"* No. Any real number can be substituted for x, and $y = -3x + 4$ will be defined.

   The domain consists of all real numbers. This can be written as $(-\infty, \infty)$.

b) If we substitute a number such as 4 for x and solve for y, we get

$$y^2 = 4$$
$$y = \pm\sqrt{4}$$
$$y = \pm 2$$

The ordered pairs $(4, 2)$ and $(4, -2)$ satisfy $y^2 = x$. Since $x = 4$ corresponds to two *different* y-values, $y^2 = x$ is *not* a function.

To determine the domain of this relation, ask yourself, *"Is there any number that cannot be substituted for x in $y^2 = x$?"* In this case, let's first look at y. Since y is squared, any real number substituted for y will produce a number that is greater than or equal to zero. Therefore, in the equation, x will equal a number that is greater than or equal to zero. The domain is $[0, \infty)$.

[**YOU TRY 2**] Determine whether each relation describes y as a function of x, and determine the domain of the relation.

a) $y = x^2 + 5$ b) $y = x^3$

Recall that if a function describes the relationship between x and y so that x is the independent variable and y is the dependent variable, then y *is a function of* x. That is, *the value of y depends on the value of x.* We can use *function notation* to represent this relationship.

Definition

$y = f(x)$ is called **function notation,** and it is read as, "y equals f of x." $y = f(x)$ means that y is a function of x (that is, y depends on x).

In Example 2, we concluded that $y = -3x + 4$ is a function. Using function notation, we can write $y = -3x + 4$ as $f(x) = -3x + 4$. *They mean the same thing.* $f(x) = -3x + 4$ is a *linear function*, and its graph is a line.

2   Review Linear, Polynomial, and Rational Functions

Let's organize what we have already learned about linear, polynomial, and rational functions.

Function	Definition	Example
Linear	A **linear function** has the form $f(x) = mx + b$, where m and b are real numbers, m is the slope, and $(0, b)$ is the y-intercept. The **domain of a linear function** is all real numbers, $(-\infty, \infty)$.	$f(x) = \dfrac{1}{2}x - 3$
Polynomial	A **polynomial function of degree n** has the form $f(x) = a_n x^n + a_{n-1} x^{n-1} + \cdots + a_1 x + a_0$, where $a_n, a_{n-1}, \ldots, a_1,$ and a_0 are real numbers, $a_n \neq 0$, and n is a whole number. The **domain of a polynomial function** is $(-\infty, \infty)$.	$f(x) = x^3 + 2x^2 - 9x + 5$
Rational	A **rational function** has the form $f(x) = \dfrac{P(x)}{Q(x)}$, where P and Q are polynomials and $Q(x) \neq 0$. The **domain of a rational function** consists of all real numbers *except* the value(s) of the variable that make the denominator equal zero.	$f(x) = \dfrac{x + 4}{2x + 1}$

EXAMPLE 3

$$f(x) = \frac{1}{2}x - 3$$

a) What is the domain of f?   b) Graph the function.

c) What is the range of f?

Solution

a) The domain is the set of all real numbers that can be substituted for x. Ask yourself,

"*Is there any number that cannot be substituted for x in $f(x) = \frac{1}{2}x - 3$?*"

No. Any real number can be substituted for x, and $f(x) = \frac{1}{2}x - 3$ will be defined.

The domain consists of all real numbers. This can be written as $(-\infty, \infty)$.

b) This is a *linear* function. The y-intercept is $(0, -3)$, and

the slope of the line is $\frac{1}{2}$. Use this information to graph

the line.

c) Look at the graph. If we could extend the line infinitely in both directions, all real number values of y would be included on the line. The range is $(-\infty, \infty)$.

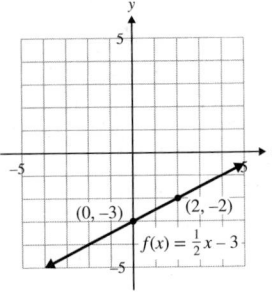

YOU TRY 3 $g(x) = -x + 2$

a) What is the domain of g?   b) Graph the function.

c) What is the range of g?

EXAMPLE 4   Let $f(x) = x^3 + 2x^2 - 9x + 5$.

a) Is $f(x)$ a linear, polynomial, or rational function?

b) What is the domain of f?

c) Find $f(-3)$.

Solution

a) It is a **polynomial function** of degree 3 because, in the expression $x^3 + 2x^2 - 9x + 5$, each exponent is a whole number, all coefficients are real numbers, and the degree of the polynomial is 3.

b) The **domain** is $(-\infty, \infty)$ because any real number can be substituted for x and the function is defined.

c) Finding $f(-3)$ means that we are evaluating the function for $x = -3$.

$$f(x) = x^3 + 2x^2 - 9x + 5$$
$$f(-3) = (-3)^3 + 2(-3)^2 - 9(-3) + 5 \qquad \text{Substitute } -3 \text{ for } x.$$
$$= -27 + 18 + 27 + 5 \qquad \text{Evaluate.}$$
$$= 23$$

Therefore, $f(-3) = 23$. We can also say that the ordered pair $(-3, 23)$ satisfies $f(x) = x^3 + 2x^2 - 9x + 5$.

YOU TRY 4

Let $f(x) = 3x^4 - 10x^3 + x - 1$.

a) Is $f(x)$ a linear, polynomial, or rational function?

b) What is the domain of f?

c) Find $f(2)$.

When trying to find the domain of a rational function, sometimes it is helpful to ask yourself, *"Is there any number that cannot be substituted for the variable?"*

Note

To determine the domain of a rational function, we set the denominator equal to zero and solve for the variable. Any values that make the denominator equal to zero are *not* in the domain of the function.

EXAMPLE 5

Determine the domain of each rational function.

a) $g(x) = \dfrac{x + 4}{2x + 1}$ b) $r(t) = \dfrac{2t^2 + 9}{6}$

Solution

a) Ask yourself, *"Is there any number that cannot be substituted for x in $g(x) = \dfrac{x + 4}{2x + 1}$?"*

Look at the denominator. When will it equal 0? Set the denominator equal to 0 and solve for x.

$$2x + 1 = 0 \qquad \text{Set the denominator} = 0.$$
$$2x = -1$$
$$x = -\frac{1}{2} \qquad \text{Solve.}$$

 **Hint**

Get in the habit of asking yourself the same questions that are asked in the examples.

When $x = -\dfrac{1}{2}$, the denominator of $g(x) = \dfrac{x + 4}{2x + 1}$ equals zero. The domain contains all real numbers *except* $-\dfrac{1}{2}$. Write the domain in interval notation as

$$\left(-\infty, -\frac{1}{2}\right) \cup \left(-\frac{1}{2}, \infty\right).$$

b) *Is there any number that cannot be substituted for t in $r(t) = \dfrac{2t^2 + 9}{6}$?* The denominator is a constant, 6, and it can never equal zero. Therefore, any real number can be substituted for t and the function will be defined.

The domain consists of all real numbers, which can be written as $(-\infty, \infty)$.

YOU TRY 5

Determine the domain of each rational function.

a) $f(x) = \dfrac{9}{x}$ b) $k(c) = \dfrac{c}{3c + 4}$ c) $g(n) = \dfrac{n - 2}{n^2 - 9}$

3 Define and Evaluate Quadratic Functions

A *quadratic function* is a special type of polynomial function.

<table>
</table>

W Hint

In your notes, make a table that summarizes the four types of functions we have learned about so far. Include an example of each.

Definition

A **quadratic function** is a function that can be written in the form

$$f(x) = ax^2 + bx + c$$

where a, b, and c are real numbers and $a \neq 0$. The domain of a quadratic function is $(-\infty, \infty)$.

An example of a **quadratic function** is $f(x) = 3x^2 + x + 8$. (Notice that this is similar to a quadratic equation, an equation of the form $ax^2 + bx + c = 0$.)

In Example 4, we evaluated $f(x) = x^3 + 2x^2 - 9x + 5$ for a *number*, $x = -3$, and found that $f(-3) = 23$. We can also evaluate functions for variables and expressions.

EXAMPLE 6

Given the quadratic function $h(x) = x^2 + 2x - 11$, find each of the following and simplify.

a) $h(p)$ b) $h(w - 4)$

Solution

a) To find $h(p)$ (read as *h of p*) means to substitute p for x in the function h, and simplify the expression as much as possible.

$$h(x) = x^2 + 2x - 11$$
$$h(p) = p^2 + 2p - 11 \qquad \text{Substitute } p \text{ for } x.$$

b) To find $h(w - 4)$ (read as *h of w minus* 4) means to substitute $w - 4$ for x in function h, and simplify the expression as much as possible. *When we substitute, we must put $w - 4$ in parentheses because $w - 4$ consists of more than one term.*

W Hint

When you substitute an expression for a variable, remember to put the expression in parentheses!

$$h(x) = x^2 + 2x - 11$$
$$h(w - 4) = (w - 4)^2 + 2(w - 4) - 11 \qquad \text{Substitute } w - 4 \text{ for } x.$$
$$h(w - 4) = w^2 - 8w + 16 + 2w - 8 - 11 \qquad \text{Multiply.}$$
$$h(w - 4) = w^2 - 6w - 3 \qquad \text{Combine like terms.}$$

[YOU TRY 6]

Let $f(x) = -9x + 2$ and $k(x) = x^2 + 5x + 8$. Find each of the following and simplify.

a) $f(c)$ b) $f(r - 5)$ c) $k(z)$ d) $k(m + 2)$

In Sections 12.2 and 12.3, we will learn much more about quadratic functions, including how to graph them.

4 Determine the Domains of Square Root and Cube Root Functions

Radical functions are functions of the form $f(x) = \sqrt[n]{x}$. The domain and range of a radical function consist of all *real numbers* that can be substituted for the independent variable so that the function values are *real numbers*. (Complex numbers are *not* included in the domain or range.) Let's begin with square root functions.

When the radicand of a square root function is negative, the function value is not a real number. For example, if $f(x) = \sqrt{x}$, then $f(9) = \sqrt{9} = 3$, but $f(-25) = \sqrt{-25}$, which is not a real number. Therefore, 9 *is* in the domain of f but -25 is *not* in the domain of f. *Any number that makes the radicand negative is not in the domain of a square root function.*

Definition

The **domain of a square root function** consists of all of the real numbers that can be substituted for the variable so that the radicand is nonnegative.

Procedure Finding the Domain of a Square Root Function

To determine the domain of a square root function, set up an inequality so that the radicand ≥ 0. Solve for the variable. These are the real numbers in the domain of the function.

EXAMPLE 7

Determine the domain of each square root function.

a) $f(x) = \sqrt{x}$　　b) $g(r) = \sqrt{2r - 9}$

Solution

a) The radicand, x, must be a real number that is greater than or equal to zero. We write that as the inequality $x \geq 0$. In interval notation, we write the domain as $[0, \infty)$.

b) In the square root function $g(r) = \sqrt{2r - 9}$, the radicand, $2r - 9$, must be nonnegative. We write this as $2r - 9 \geq 0$. To determine the domain of the function, solve the inequality $2r - 9 \geq 0$.

$$2r - 9 \geq 0 \quad \text{The value of the radicand must be} \geq 0.$$
$$2r \geq 9$$
$$r \geq \frac{9}{2} \quad \text{Solve.}$$

Any value of r that satisfies $r \geq \frac{9}{2}$ will make the radicand greater than or equal to zero. The domain of $g(r) = \sqrt{2r - 9}$ is $\left[\frac{9}{2}, \infty\right)$.

YOU TRY 7

Determine the domain of each square root function.

a) $h(x) = \sqrt{x - 9}$　　b) $k(t) = \sqrt{7t + 2}$

Unlike a square root function, when the radicand of a **cube root function** is negative, the function value *is* a real number. For example, if $f(x) = \sqrt[3]{x}$, then $f(125) = \sqrt[3]{125} = 5$ and $f(-8) = \sqrt[3]{-8} = -2$. Therefore, *any real number may be substituted into a cube root function and the function will be defined.*

Definition

The **domain of a cube root function** is the set of all real numbers. We can write this in interval notation as $(-\infty, \infty)$.

In fact, we can say that when n is an odd, positive number, the domain of $f(x) = \sqrt[n]{x}$ is all real numbers, or $(-\infty, \infty)$. This is because the odd root of any real number is, itself, a real number.

5 Graph Square Root and Cube Root Functions

We need to know the domain of a square root function in order to sketch its graph.

EXAMPLE 8

Graph each function.

a) $f(x) = \sqrt{x}$ b) $g(x) = \sqrt{x + 4}$

Solution

a) In Example 7 we found that the domain of $f(x) = \sqrt{x}$ is $[0, \infty)$. When we make a table of values, we will start by letting $x = 0$, the smallest number in the domain, and then choose real numbers greater than 0. Usually it is easiest to choose values for x that are perfect squares so that it will be easier to plot the points. We will also plot the point $(6, \sqrt{6})$ so that you can see where it lies on the graph. Connect the points with a smooth curve.

$f(x) = \sqrt{x}$	
x	$f(x)$
0	0
1	1
4	2
6	$\sqrt{6} \approx 2.4$
9	3

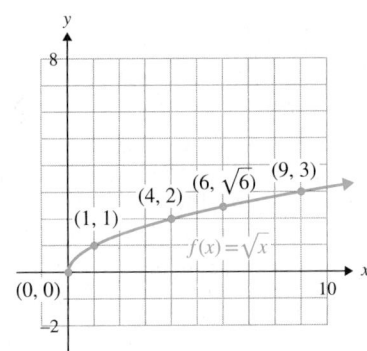

The graph reinforces the fact that this is a function. It passes the vertical line test.

W Hint

Create a procedure for graphing a square root function.

b) To graph $g(x) = \sqrt{x + 4}$ we will begin by determining its domain. Solve $x + 4 \geq 0$.

$$x + 4 \geq 0 \quad \text{The value of the radicand must be } \geq 0.$$
$$x \geq -4 \quad \text{Solve.}$$

The domain of $g(x)$ is $[-4, \infty)$. When we make a table of values, *we will start by letting $x = -4$, the smallest number in the domain, and then choose real numbers greater than -4.* We will choose values for x so that the radicand will be a perfect square. This will make it easier to plot the points. We will also plot the point $(1, \sqrt{5})$ so that you can see where it lies on the graph. Connect the points with a smooth curve.

$g(x) = \sqrt{x + 4}$	
x	$g(x)$
-4	0
-3	1
0	2
1	$\sqrt{5} \approx 2.2$
5	3

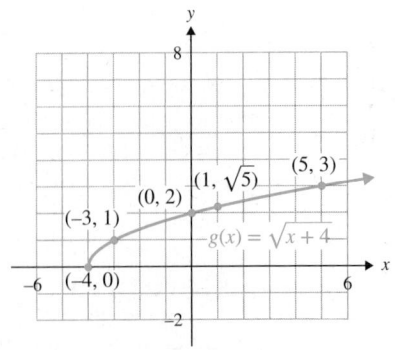

Since this graph represents a function, it passes the vertical line test.

[YOU TRY 8]   Graph $f(x) = \sqrt{x + 2}$.

The domain of a cube root function consists of all real numbers. Therefore, we can substitute any real number into the function and it will be defined. However, we want to choose our numbers carefully. *To make the table of values, pick values in the domain so that the radicand will be a perfect cube, and choose values for the variable that will give us positive numbers, negative numbers, and zero for the value of the radicand.* This will help us to graph the function correctly.

EXAMPLE 9   Graph each function.

a) $f(x) = \sqrt[3]{x}$ b) $g(x) = \sqrt[3]{x - 1}$

Solution

a)   Make a table of values. Choose x-values that are perfect cubes. Also, remember to choose x-values that are positive, negative, and zero. Plot the points, and connect them with a smooth curve.

$f(x) = \sqrt[3]{x}$	
x	**f(x)**
0	0
1	1
8	2
−1	−1
−8	−2

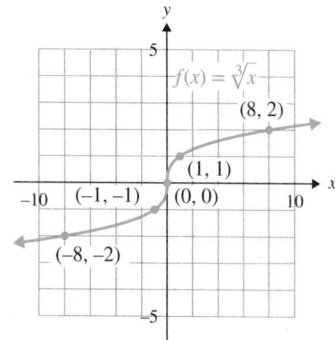

The graph passes the vertical line test for functions.

b)   Remember, for the table of values we want to choose values for x that will give us positive numbers, negative numbers, and zero *in the radicand.* First we will determine which value of x will make the radicand in $g(x) = \sqrt[3]{x - 1}$ equal to zero.

$$x - 1 = 0$$
$$x = 1$$

If $x = 1$, the radicand equals zero. Therefore, the first value we will put in the table of values is $x = 1$. Then, choose a couple of numbers *greater than* 1 and a couple that are *less than* 1 so that we get positive and negative numbers in the radicand. Also, we will choose our x-values so that the radicand will be a perfect cube. Plot the points, and connect them with a smooth curve.

x	$g(x) = \sqrt[3]{x - 1}$
1	$\sqrt[3]{1 - 1} = \sqrt[3]{0} = 0$
2	$\sqrt[3]{2 - 1} = \sqrt[3]{1} = 1$
9	$\sqrt[3]{9 - 1} = \sqrt[3]{8} = 2$
0	$\sqrt[3]{0 - 1} = \sqrt[3]{-1} = -1$
−7	$\sqrt[3]{-7 - 1} = \sqrt[3]{-8} = -2$

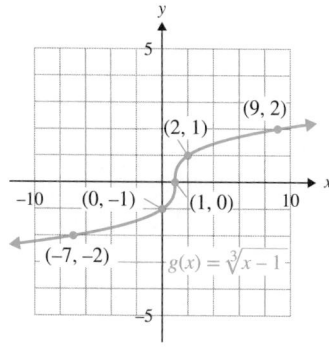

Since this graph represents a function, it passes the vertical line test.

Using Technology

We can use a graphing calculator to find the domain of a square root function or cube root function visually. The domain consists of the x-values of the points shown on the graph.

We first consider the basic shape of a square root function. To graph the equation $f(x) = \sqrt{x}$, press **2nd** **x^2** **X,T,Θ,n** **)** to the right of \Y_1 =. Press **ZOOM** and select 6:ZStandard to graph the equation.

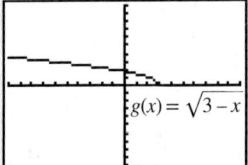

The left side of the graph begins at the point $(0, 0)$, and the right side of the graph continues up and to the right forever. The x-values of the graph consist of all x-values greater than or equal to 0. In interval notation, the domain is $[0, \infty)$.

The domain of any square root function can be found using a similar approach. First graph the function, and then look at the x-values of the points on the graph. The graph of a square root function will always start at a number and extend to positive or negative infinity.

For example, consider the graph of the function $g(x) = \sqrt{3 - x}$ as shown.

The largest x-value on the graph is 3. The x-values of the graph consist of all x-values less than or equal to 3. In interval notation, the domain is $(-\infty, 3]$.

Next consider the basic shape of a cube root function.

To graph the equation $f(x) = \sqrt[3]{x}$ press **MATH**, select 4: $\sqrt[3]{\ }$ (, and press **X,T,Θ,n** **)** to the right of \Y_1 =. Press **ZOOM** and select 6:ZStandard to graph the equation as shown on the graph at right.

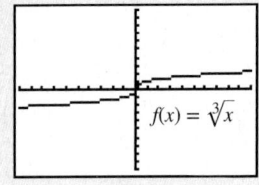

The left side of the graph extends down and to the left forever, and the right side of the graph extends up and to the right forever. In interval notation, the domain is $(-\infty, \infty)$. This is true for any cube root function, so the domain is always $(-\infty, \infty)$.

Determine the domain using a graphing calculator. Use interval notation in your answer.

1) $f(x) = \sqrt{x - 2}$ 2) $g(x) = \sqrt{x + 3}$ 3) $h(x) = \sqrt{2 - x}$

4) $f(x) = -\sqrt{x + 1}$ 5) $f(x) = \sqrt[3]{x + 5}$ 6) $g(x) = \sqrt[3]{4 - x}$

ANSWERS TO [**YOU TRY**] **EXERCISES**

1) a) domain: $\{-8, -5, 7\}$; range: $\{-4, -2, 1, 2\}$; no   b) domain: {Chicago, Mexico City, Montreal};
   range: {USA, Mexico, Canada}; yes   c) domain: $(-\infty, \infty)$; range: $(-\infty, \infty)$; yes
2) a) is a function; domain: $(-\infty, \infty)$ b) is a function; domain: $(-\infty, \infty)$
3) a) $(-\infty, \infty)$ b)   c) $(-\infty, \infty)$

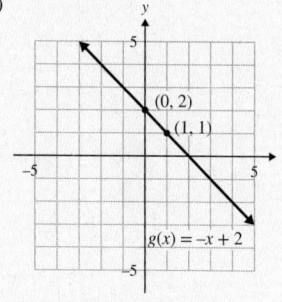

4) a) polynomial function   b) $(-\infty, \infty)$ c) -31
5) a) $(-\infty, 0) \cup (0, \infty)$ b) $\left(-\infty, -\dfrac{4}{3}\right) \cup \left(-\dfrac{4}{3}, \infty\right)$ c) $(-\infty, -3) \cup (-3, 3) \cup (3, \infty)$
6) a) $f(c) = -9c + 2$ b) $f(r - 5) = -9r + 47$ c) $k(z) = z^2 + 5z + 8$ d) $k(m + 2) = m^2 + 9m + 22$
7) a) $[9, \infty)$ b) $\left[-\dfrac{2}{7}, \infty\right)$

8)   9)

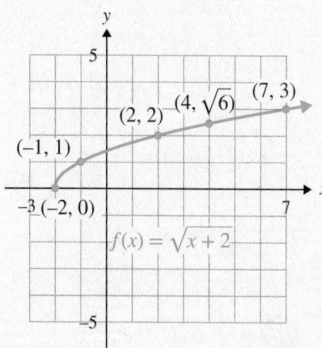   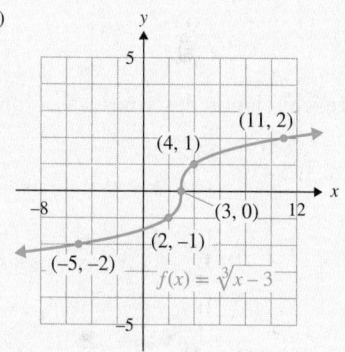

ANSWERS TO TECHNOLOGY EXERCISES

1) $[2, \infty)$ 2) $[-3, \infty)$ 3) $(-\infty, 2]$ 4) $[-1, \infty)$ 5) $(-\infty, \infty)$ 6) $(-\infty, \infty)$

E Evaluate **12.1** Exercises   Do the exercises, and check your work.

Objective 1: Identify Relation, Function, Domain, and Range

1) What is a function?

2) What is the domain of a relation?

Identify the domain and range of each relation, and determine whether each relation is a function.

 3) $\{(5, 0), (6, 1), (14, 3), (14, -3)\}$

4) $\{(-5, 7), (-4, 5), (0, -3), (0.5, -4), (3, -9)\}$

5)

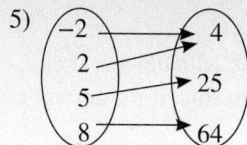

6)

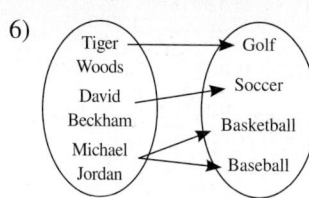

7)

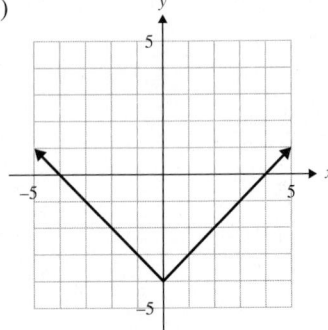

8)

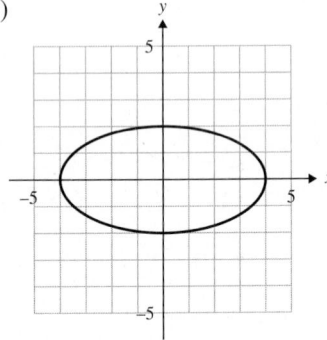

Determine whether each relation describes y as a function of x.

9) $y = 5x + 17$

10) $y = 4x^2 - 10x + 3$

 11) $y = \dfrac{x}{x + 6}$

12) $y^2 = x + 2$

13) $y^2 = x - 8$

14) $y = \sqrt{x + 3}$

15) $x = |y|$

16) $y = |x|$

Objective 2: Review Linear, Polynomial, and Rational Functions

17) Give an example of a

   a) linear function, f, in terms of x.

   b) polynomial function, p, in terms of a.

   c) rational function, R, in terms of w.

18) If you were reading $f(x)$ and $f(4)$ out loud, how would you read them? Write out the answer, in words.

For Exercises 19–22, fill in the blank with *always, sometimes*, or *never* to make the statement true. If the answer is *sometimes*, explain why.

19) The graph of a linear function is _____ a line.

20) The domain of a linear function is _____ $(-\infty, \infty)$.

21) The domain of a rational function is _____ $(-\infty, \infty)$.

22) The domain of a polynomial function is _____ $[0, \infty)$.

23) Let $f(x) = 7x + 4$. Find $f(6), f(-2)$, and $f(0)$.

24) Let $g(n) = n^3 - 6n^2 + 2n + 15$. Find $g(5), g(-3)$, and $g(0)$.

Graph each linear function.

25) $f(x) = x - 5$

26) $g(x) = -x + 3$

27) $g(x) = -\dfrac{3}{2}x - 1$

28) $f(x) = \dfrac{1}{4}x + 2$

29) $k(c) = c$

30) $h(x) = -3$

31) How do you find the domain of a rational function?

32) Is $h(x) = \dfrac{x^2 + 1}{3x + 5}$ defined for $x = -\dfrac{5}{3}$? Explain your answer.

Determine the domain of each function.

33) $f(x) = x + 10$

34) $h(x) = -8x - 2$

35) $p(a) = 8a^2 + 4a - 9$

36) $r(t) = t^3 - 7t^2 + t + 4$

37) $f(x) = \dfrac{6}{x + 8}$

38) $k(x) = \dfrac{2x}{x - 9}$

39) $h(x) = \dfrac{10}{x}$

40) $Q(r) = \dfrac{7}{2r}$

41) $R(t) = -\dfrac{t - 4}{7t + 3}$

42) $k(n) = \dfrac{8}{1 - 3n}$

43) $h(x) = \dfrac{9x + 2}{4}$

44) $p(c) = \dfrac{c - 2}{7}$

45) $k(x) = \dfrac{1}{x^2 + 11x + 24}$

46) $f(t) = \dfrac{5}{t^2 - 7t + 6}$

47) $r(c) = \dfrac{c + 3}{c^2 - 5c - 36}$

48) $g(a) = \dfrac{4}{2a^2 + 3a}$

49) Write a rational function with a domain of $(-\infty, 10) \cup (10, \infty)$.

50) Write a rational function with a domain of $(-\infty, -1) \cup (-1, 4) \cup (4, \infty)$.

Objective 3: Define and Evaluate Quadratic Functions

For Exercises 51–54, identify each as a linear, quadratic, polynomial, or rational function.

51) $h(x) = x^3 + 9x - 4$

52) $f(a) = \dfrac{a + 10}{a^2 - 8a - 20}$

53) $g(w) = -\dfrac{3}{4}w^2 + w + \dfrac{1}{2}$

54) $k(t) = \dfrac{2}{3}t + 5$

55) What is the domain of a quadratic function?

56) Is every quadratic function also a polynomial function? Explain your answer.

Let $f(x) = 3x - 7$ and $g(x) = x^2 - 4x - 9$. Find each of the following and simplify.

57) $f(6)$

58) $f(0)$

59) $g(3)$

60) $g(-2)$

61) $f(a)$

62) $f(z)$

63) $g(d)$

64) $g(r)$

65) $f(c + 4)$

66) $f(w + 9)$

67) $g(t + 2)$

68) $g(a + 3)$

69) $g(h - 1)$

70) $g(p - 5)$

71) $p(x) = x^2 - 6x - 16$. Find x so that $p(x) = 0$.

72) $g(x) = 2x^2 - 5x - 9$. Find x so that $g(x) = -6$.

Objective 4: Determine the Domains of Square Root and Cube Root Functions

73) Is -1 in the domain of $f(x) = \sqrt{x}$? Explain your answer.

74) Is -1 in the domain of $f(x) = \sqrt[3]{x}$? Explain your answer.

75) Let $f(x) = \sqrt{x}$ and $g(t) = \sqrt{11 - 3t}$.

   a) Find $f(1)$ and $g(1)$.

   b) Is 25 in the domain of both f and g? Explain your answer.

   c) Find $f(c)$ and $g(c)$.

   d) Find $f(a + 4)$ and $g(a + 4)$.

   e) Find $f(-2k + 1)$ and $g(-2k + 1)$.

76) Let $f(a) = \sqrt[3]{a}$ and $g(n) = \sqrt[3]{4n - 5}$.

   a) Find $f(8)$ and $g(8)$.

   b) Is 0 in the domain of both f and g? Explain your answer.

   c) Find $f(p)$ and $g(p)$.

   d) Find $f(w - 3)$ and $g(w - 3)$.

   e) Find $f(2 - 5t)$ and $g(2 - 5t)$.

77) How do you find the domain of a square root function?

78) What is the domain of a cube root function?

Determine the domain of each function.

79) $h(n) = \sqrt{n + 2}$

80) $g(c) = \sqrt{c + 10}$

81) $p(a) = \sqrt{a - 8}$

82) $f(a) = \sqrt{a - 1}$

83) $f(a) = \sqrt[3]{a - 7}$

84) $h(t) = \sqrt[3]{t}$

85) $r(k) = \sqrt{3k + 7}$

86) $k(x) = \sqrt{2x - 5}$

87) $g(x) = \sqrt[3]{2x - 5}$

88) $h(c) = \sqrt[3]{-c}$

89) $g(t) = \sqrt{-t}$

90) $h(x) = \sqrt{3 - x}$

91) $r(a) = \sqrt{9 - 7a}$

92) $g(c) = \sqrt{8 - 5c}$

Objective 5: Graph Square Root and Cube Root Functions

Determine the domain, and then graph each function.

93) $f(x) = \sqrt{x - 1}$

94) $g(x) = \sqrt{x - 4}$

95) $g(x) = \sqrt{x + 3}$

96) $h(x) = \sqrt{x + 1}$

97) $h(x) = \sqrt{x} - 2$

98) $f(x) = \sqrt{x} + 2$

99) $f(x) = \sqrt{-x}$

100) $g(x) = \sqrt{-x} - 3$

101) $f(x) = \sqrt[3]{x + 1}$

102) $g(x) = \sqrt[3]{x + 2}$

103) $h(x) = \sqrt[3]{x - 2}$

104) $g(x) = \sqrt[3]{-x}$

105) $g(x) = \sqrt[3]{x} + 1$

106) $h(x) = \sqrt[3]{x} - 1$

Use the following information for Exercises 107–110.

The period of a pendulum is the time it takes for the pendulum to make one complete swing back and forth. The period, $T(L)$ in seconds, can be described by the function $T(L) = 2\pi\sqrt{\dfrac{L}{32}}$, where L is the length of the pendulum, in feet. For Exercises 107 and 108, give an *exact answer and an answer rounded to two decimal places*. Use 3.14 for π.

107) Find $T\left(\dfrac{1}{2}\right)$, and explain what it means in the context of the problem.

108) Find $T(1.5)$, and explain what it means in the context of the problem.

109) Find the length of the pendulum with a period of $\dfrac{\pi}{2}$ sec.

110) Find the length of the pendulum with a period of π sec.

111) The area, A, of a circle is a function of its radius, r.

   a) Write an equation using function notation to describe this relationship between A and r.

   b) If the radius is given in centimeters, find $A(3)$ and explain what this means in the context of the problem.

   c) If the radius is given in inches, find $A(5)$ and explain what this means in the context of the problem.

   d) What is the radius of a circle with an area of 64π in^2?

112) The perimeter, P, of a square is a function of the length of its side, s.

   a) Write an equation using function notation to describe this relationship between P and s.

   b) If the length of a side is given in feet, find $P(2)$ and explain what this means in the context of the problem.

   c) If the length of a side is given in centimeters, find $P(11)$ and explain what this means in the context of the problem.

   d) What is the length of each side of a square that has a perimeter of 18 inches?

R Rethink

R1) How can you distinguish among linear, polynomial, and rational functions?

R2) What are some similarities and differences between square root and cube root functions?

12.2 Graphs of Functions and Transformations

P Prepare

O Organize

What are your objectives for Section 12.2?	How can you accomplish each objective?
1 Learn About Vertical Shifts Using Absolute Value Functions	• Understand how to graph $f(x) = |x|$. • Write the property for **vertical shifts** in your own words, and include an example. • Complete the given example on your own. • Complete You Try 1.
2 Learn About Horizontal Shifts Using Quadratic Functions	• Know that the graph of a quadratic function is a *parabola*, and write the definition of *vertex* in your own words. • Understand how to graph $f(x) = x^2$. • Write the property for **horizontal shifts** in your own words, and include an example. • Complete the given example on your own. • Complete You Try 2.

What are your objectives for Section 12.2?	How can you accomplish each objective?
3 Learn About Reflecting a Graph About the x-Axis Using Square Root Functions	• Understand how to graph $f(x) = \sqrt{x}$. • Write the property for **reflecting a graph about the x-axis** in your own words, and include an example. • Complete the given example on your own. • Complete You Try 3.
4 Graph Functions Using a Combination of the Transformations	• Summarize the transformation properties learned in this section. • Complete the given example on your own. • Complete You Try 4.
5 Graph a Piecewise Function	• Learn the definition of a *piecewise function*. • Be careful when choosing values for the domain. • Complete the given example on your own. • Complete You Try 5.
6 Define, Graph, and Apply the Greatest Integer Function	• Learn the definition of the *greatest integer function*. • Understand how to make a table of values. • For applied problems, include the appropriate values for the domain. • Complete the given examples on your own. • Complete You Trys 6–8.

Read the explanations, follow the examples, take notes, and complete the You Trys.

Some functions and their graphs appear often when studying algebra. We will look at the basic graphs of

1) the absolute value function, $f(x) = |x|$.

2) the quadratic function, $f(x) = x^2$.

3) the square root function, $f(x) = \sqrt{x}$.

It is possible to obtain the graph of any function by plotting points. But we will also see how we can graph other, similar functions by *transforming* the graphs of the functions above.

First, we will graph two absolute value functions. We will begin by plotting points so that we can observe the pattern that develops.

1 Learn About Vertical Shifts Using Absolute Value Functions

EXAMPLE 1

Graph $f(x) = |x|$ and $g(x) = |x| + 2$ on the same axes. Identify the domain and range.

Solution

$f(x) = |x|$	
x	$f(x)$
0	0
1	1
2	2
−1	1
−2	2

$g(x) = |x| + 2$	
x	$g(x)$
0	2
1	3
2	4
−1	3
−2	4

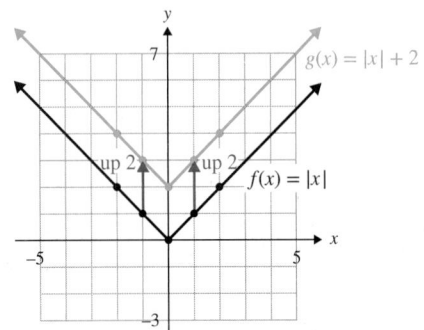

The domain of $f(x) = |x|$ is $(-\infty, \infty)$. The range is $[0, \infty)$.

The domain of $g(x) = |x| + 2$ is $(-\infty, \infty)$. The range is $[2, \infty)$.

Absolute value functions like these have V-shaped graphs. We can see from the tables of values that although the *x*-values are the same in each table, the corresponding *y*-values in the table for $g(x)$ are 2 *more than* the *y*-values in the first table.

$$f(x) = |x| \qquad g(x) = |x| + 2$$
$$g(x) = f(x) + 2 \qquad \text{Substitute } f(x) \text{ for } |x|.$$

The *y*-coordinates of the ordered pairs of $g(x)$ are 2 *more than* the *y*-coordinates of the ordered pairs of $f(x)$ when the ordered pairs of *f* and *g* have the same *x*-coordinates. This means that **the graph of *g* is the same shape as the graph of *f*, but *g* is shifted up 2 units.**

W Hint

In your own words, describe how you know when to shift a graph vertically.

Property Vertical Shifts

Given the graph of $f(x)$, if $g(x) = f(x) + k$, where *k* is a constant, then the graph of $g(x)$ is the same shape as the graph of $f(x)$, but *g* is shifted **vertically** *k* units.

In Example 1, $k = 2$. $\qquad f(x) = |x|$ and $\qquad g(x) = |x| + 2$
$$\text{or} \qquad g(x) = f(x) + 2$$

The graph of *g* is the same shape as the graph of *f*, but the graph of *g* is shifted *up* 2 units. We say that we can graph $g(x) = |x| + 2$ by *transforming* the graph of $f(x) = |x|$. This vertical shifting works not only for absolute value functions but for any function.

[YOU TRY 1] Graph $g(x) = |x| - 1$.

2 Learn About Horizontal Shifts Using Quadratic Functions

In the previous section, we said that a quadratic function can be written in the form $f(x) = ax^2 + bx + c$, where a, b, and c are real numbers and $a \neq 0$. Here we begin our discussion of graphing quadratic functions.

The graph of a quadratic function is called a **parabola.** Let's look at the simplest form of a quadratic function, $f(x) = x^2$, and a variation of it. In Section 12.3, we will discuss graphing quadratic functions in much greater detail.

EXAMPLE 2

Graph $f(x) = x^2$ and $g(x) = (x + 3)^2$ on the same axes. Identify the domain and range.

Solution

$f(x) = x^2$	
x	**f(x)**
0	0
1	1
2	4
−1	1
−2	4

$g(x) = (x + 3)^2$	
x	**g(x)**
−3	0
−2	1
−1	4
−4	1
−5	4

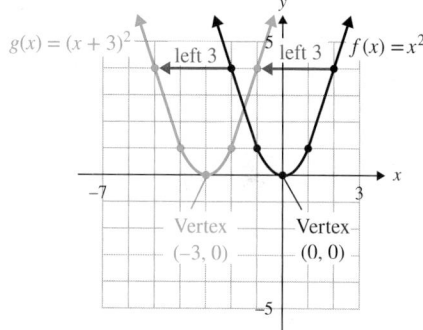

Notice that the graphs of $f(x)$ and $g(x)$ open upward. The lowest point on a parabola that opens upward or the highest point on a parabola that opens downward is called the **vertex.** The vertex of the graph of $f(x)$ is $(0, 0)$, and the vertex of the graph of $g(x)$ is $(-3, 0)$. When graphing a quadratic function by plotting points, it is important to locate the vertex. *The x-coordinate of the vertex is the value of x that makes the expression that is being squared equal to zero.*

The domain of $f(x) = x^2$ is $(-\infty, \infty)$. The range is $[0, \infty)$.

The domain of $g(x) = (x + 3)^2$ is $(-\infty, \infty)$. The range is $[0, \infty)$.

We can see from the tables of values that although the y-values are the same in each table, the corresponding x-values in the table for $g(x)$ are 3 *less than* the x-values in the first table.

The x-coordinates of the ordered pairs of $g(x)$ are 3 *less than* the x-coordinates of the ordered pairs of $f(x)$ when the ordered pairs of f and g have the same y-coordinates. This means that **the graph of g is the same shape as the graph of f, but the graph of g is shifted left 3 units.**

W Hint

In your own words, describe how you know when to shift a graph horizontally.

Property Horizontal Shifts

Given the graph of $f(x)$, if $g(x) = f(x - h)$, where h is a constant, then the graph of $g(x)$ is the same shape as the graph of $f(x)$, but g is shifted **horizontally** h units.

In Example 2, $h = -3$.

$$f(x) = x^2 \quad \text{and} \quad g(x) = (x + 3)^2$$
$$\text{or} \quad g(x) = f(x - (-3))$$

The graph of g is the same shape as the graph of f, but the graph of g is shifted -3 units horizontally or 3 units to the *left*. This horizontal shifting works for any function, not just quadratic functions.

$$\left[\text{YOU TRY 2}\right]\quad \text{Graph } g(x) = (x + 4)^2.$$

BE CAREFUL It is important to distinguish between the graph of an absolute value function and the graph of a quadratic function. The absolute value functions we will study have V-shaped graphs. The graph of a quadratic function is *not* shaped like a V. It is a parabola.

The next type of transformation we will discuss is reflecting the graph of a function about the *x*-axis.

3 Learn About Reflecting a Graph About the *x*-Axis Using Square Root Functions

EXAMPLE 3 Graph $f(x) = \sqrt{x}$ and $g(x) = -\sqrt{x}$ on the same axes. Identify the domain and range.

Solution

The domain of each function is $[0, \infty)$. From the graphs, we can see that the range of $f(x) = \sqrt{x}$ is $[0, \infty)$, while the range of $g(x) = -\sqrt{x}$ is $(-\infty, 0]$.

$f(x) = \sqrt{x}$	
x	**f(x)**
0	0
1	1
4	2
9	3

$g(x) = -\sqrt{x}$	
x	g(x)
0	0
1	−1
4	−2
9	−3

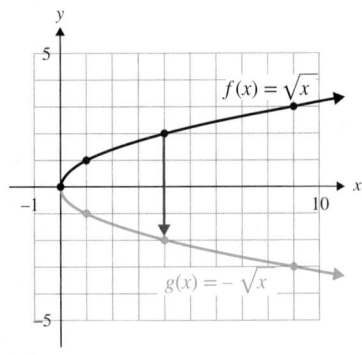

The tables of values show us that although the *x*-values are the same in each table, the corresponding *y*-values in the table for $g(x)$ are the *negatives* of the *y*-values in the first table.

We say that *the graph of g is the reflection of the graph of f about the x-axis.* (*g* is the mirror image of *f* with respect to the *x*-axis.)

W Hint

In your own words, describe how you know when to reflect a graph about the *x*-axis.

Property Reflection About the *x*-Axis

Given the graph of any function $f(x)$, if $g(x) = -f(x)$ then the graph of $g(x)$ will be the **reflection of the graph of *f* about the *x*-axis.** That is, obtain the graph of *g* by keeping the *x*-coordinate of each point on *f* the same, but take the negative of the *y*-coordinate.

In Example 3,

$$f(x) = \sqrt{x} \quad \text{and} \quad g(x) = -\sqrt{x}$$
$$\text{or} \quad g(x) = -f(x)$$

The graph of g is the mirror image of the graph of f with respect to the x-axis. This is true for any function where $g(x) = -f(x)$.

> **Note**
> The properties of vertical and horizontal shifting as well as reflection about the x-axis apply to *all* functions.

[**YOU TRY 3**] Graph $g(x) = -x^2$.

We can combine the techniques used in the transformation of the graphs of functions to help us graph more complicated functions.

4 Graph Functions Using a Combination of the Transformations

EXAMPLE 4 Graph $h(x) = |x + 2| - 3$.

Solution

The graph of $h(x)$ will be the same shape as the graph of $f(x) = |x|$. So, let's see what the constants in $h(x)$ tell us about transforming the graph of $f(x) = |x|$.

$$h(x) = |x + 2| - 3$$
$$\qquad\quad \uparrow \qquad\quad \uparrow$$
$$\qquad \text{Shift } f(x) \quad \text{Shift } f(x)$$
$$\qquad \text{left 2.} \qquad \text{down 3.}$$

Sketch the graph of $f(x) = |x|$, including some key points, then *move every point on the graph of f left 2 and down 3 to obtain the graph of h.*

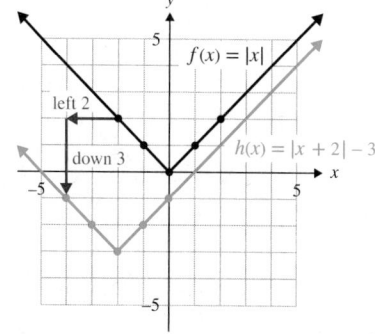

W Hint

Do you understand how to put vertical and horizontal shifting together?

[**YOU TRY 4**] Graph $h(x) = |x - 2| - 4$.

5 Graph a Piecewise Function

Definition

A **piecewise function** is a single function defined by two or more different rules.

EXAMPLE 5

Graph the piecewise function

$$f(x) = \begin{cases} 2x - 4, & x \geq 3 \\ -x + 2, & x < 3 \end{cases}$$

Solution

This is a piecewise function because $f(x)$ is defined by two different rules. *The rule we use to find $f(x)$ depends on which value is substituted for x.*

Graph $f(x)$ by making two separate tables of values, one for each rule.

When $x \geq 3$, use the rule

$$f(x) = 2x - 4$$

The first x-value we will put in the table of values is 3 because it is the smallest number (lower bound) of the domain of $f(x) = 2x - 4$. *The other values we choose for x must be greater than 3 because this is when we use the rule $f(x) = 2x - 4$.* **This part of the graph will not extend to the left of (3, 2).**

$$f(x) = 2x - 4$$
$$(x \geq 3)$$

x	$f(x) = 2x - 4, x \geq 3$
3	2
4	4
5	6
6	8

When $x < 3$, use the rule

$$f(x) = -x + 2$$

The first x-value we will put in the table of values is 3 because it is the upper bound of the domain. *Notice that 3 is not included in the domain (the inequality is $<$, **not** $\leq$) so the point $(3, f(3))$ will be represented as an open circle on the graph.* The other values we choose for x must be less than 3 because this is when we use the rule $f(x) = -x + 2$. **This part of the graph will not extend to the right of (3, −1).**

$$f(x) = -x + 2$$
$$(x < 3)$$

x	$f(x) = -x + 2, x < 3$
3	−1
2	0
1	1
0	2

$(3, -1)$ is an open circle.

$$f(x) = \begin{cases} 2x - 4, x \geq 3 \\ -x + 2, x < 3 \end{cases}$$

The graph of $f(x)$ is at the left.

[YOU TRY 5]

Graph the piecewise function

$$f(x) = \begin{cases} -2x + 3, & x \leq -2 \\ \dfrac{3}{2}x - 1, & x > -2 \end{cases}$$

---

6 Define, Graph, and Apply the Greatest Integer Function

Another function that has many practical applications is the greatest integer function.

Definition

The **greatest integer function**

$$f(x) = [\![x]\!]$$

represents the largest integer less than or equal to x.

EXAMPLE 6

Let $f(x) = [\![x]\!]$. Find the following function values.

a) $f\left(9\frac{1}{2}\right)$ b) $f(6)$ c) $f(-2.3)$

Solution

a) $f\left(9\frac{1}{2}\right) = \left[\!\!\left[9\frac{1}{2}\right]\!\!\right]$. This is the largest integer *less than or equal to* $9\frac{1}{2}$. That number is 9. So $f\left(9\frac{1}{2}\right) = \left[\!\!\left[9\frac{1}{2}\right]\!\!\right] = 9$.

b) $f(x) = [\![6]\!] = 6$. The largest integer *less than or equal to* 6 is 6.

c) To help us understand how to find this function value, we will locate -2.3 on a number line.

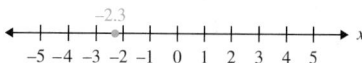

The largest integer *less than or equal to* -2.3 is -3, so $f(-2.3) = [\![-2.3]\!] = -3$.

YOU TRY 6

Let $f(x) = [\![x]\!]$. Find the following function values.

a) $f(5.1)$ b) $f(0)$ c) $f\left(-5\frac{1}{4}\right)$

EXAMPLE 7

Graph $f(x) = [\![x]\!]$.

Solution

First, let's look at the part of this function between $x = 0$ and $x = 1$ (when $0 \le x \le 1$).

x	$f(x) = [\![x]\!]$
0	0
$\frac{1}{4}$	0
$\frac{1}{2}$	0
$\frac{3}{4}$	0
$\vdots$	0
1	1

For all values of x *greater than or equal to* 0 and *less than* 1, the function value, $f(x)$, equals zero.

⟶ When $x = 1$ the function value changes to 1.

The graph has an open circle at (1, 0) because if $x < 1, f(x) = 0$. That means that x can get *very close to* 1 and the function value will be zero, but $f(1) \neq 0$.

This pattern continues so that for the x-values in the interval [1, 2), the function values are 1. The graph has an open circle at (2, 1).

For the x-values in the interval [2, 3), $f(x) = 2$. The graph has an open circle at (3, 2).

Continuing in this way, we get the graph to the right.

The domain of the function is $(-\infty, \infty)$.
The range is the set of all integers
$\{\ldots, -3, -2, -1, 0, 1, 2, 3, \ldots\}$.

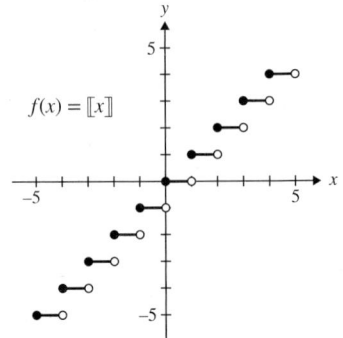

$f(x) = [\![x]\!]$

W Hint

In your notes, summarize the different types of functions we have learned to graph in this section.

Because of the appearance of the graph, $f(x) = [\![x]\!]$ is also called a **step function.**

[YOU TRY 7] Graph $f(x) = [\![x]\!] - 3$.

EXAMPLE 8

To mail a large envelope within the United States in 2018, the U.S. Postal Service charged \$1.00 for the first ounce and \$0.21 for each additional ounce or fraction of an ounce. Let $C(x)$ represent the cost of mailing a large envelope within the United States, and let x represent the weight of the envelope, in ounces. Graph $C(x)$ for any large envelope weighing up to (and including) 5 oz.

(www.usps.com)

C Squared Studios/Getty Images

Solution

If a large envelope weighs between 0 and 1 oz $(0 < x \leq 1)$, the cost, $C(x)$, is \$1.00.

If it weighs more than 1 oz but less than or equal to 2 oz $(1 < x \leq 2)$, the cost, $C(x)$, is \$1.00 + \$0.21 = \$1.21.

The pattern continues, and we get the graph to the right.

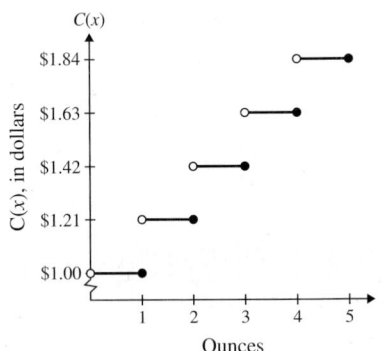

[YOU TRY 8] To mail a letter outside the United States in 2018, the U.S. Postal Service charged \$1.15 for the first ounce and \$0.21 for each additional ounce or fraction of an ounce. Let $C(x)$ represent the cost of mailing an international letter, and let x represent the weight of the letter, in ounces. Graph $C(x)$ for any letter weighing up to (and including) 5 oz. (www.usps.com)

We can graph piecewise functions using a graphing calculator by entering each piece separately. Suppose that we wish to graph the piecewise function

$f(x) = \begin{cases} 2x - 4, & x \geq 3 \\ -x + 2, & x < 3 \end{cases}$. Enter $2x - 4$ in Y_1 and $-x + 2$ in Y_2. First put

parentheses around each function, and then put parentheses around the interval of x-values for which that part of the function is defined. In order to enter the inequality symbols, press 2nd $\boxed{\text{MATH}}$ and use the arrow keys to scroll to the desired symbol before pressing $\boxed{\text{ENTER}}$.

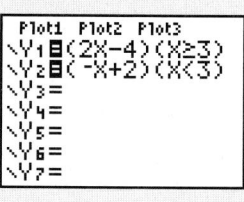

Display the piecewise function using the standard view.

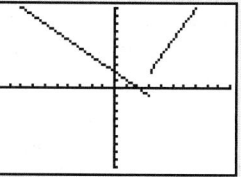

Graph the following functions using a graphing calculator.

1) $f(x) = \begin{cases} 3x - 2, & x \geq 1 \\ -x - 5, & x < 1 \end{cases}$

2) $f(x) = \begin{cases} x + 3, & x \geq 2 \\ -2x + 1, & x < 2 \end{cases}$

3) $f(x) = \begin{cases} -\dfrac{1}{2}x - 1, & x \leq -3 \\ x - 3, & x > -3 \end{cases}$

4) $f(x) = \begin{cases} 4, & x < 4 \\ -x + 2, & x \geq 4 \end{cases}$

5) $f(x) = \begin{cases} -\dfrac{2}{3}x + 1, & x \geq -1 \\ x + 3, & x < -1 \end{cases}$

6) $f(x) = \begin{cases} x, & x \geq 0 \\ 5x - 2, & x < 0 \end{cases}$

ANSWERS TO $\boxed{\text{YOU TRY}}$ **EXERCISES**

1)

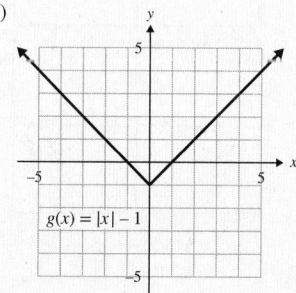

$g(x) = |x| - 1$

2)

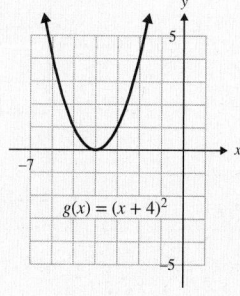

$g(x) = (x + 4)^2$

3)

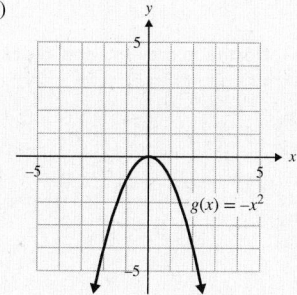

$g(x) = -x^2$

4)

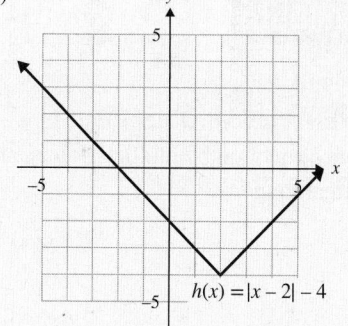

$h(x) = |x - 2| - 4$

5)

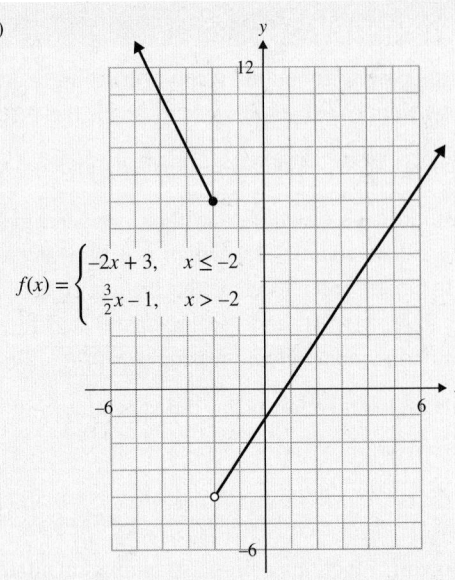

$$f(x) = \begin{cases} -2x + 3, & x \le -2 \\ \frac{3}{2}x - 1, & x > -2 \end{cases}$$

6) a) 5 b) 0 c) −6

7)

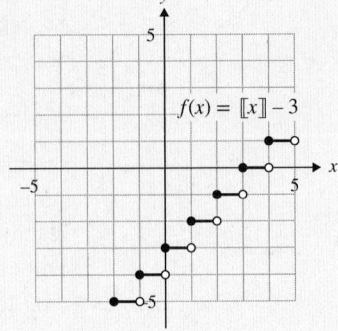

$f(x) = [\![x]\!] - 3$

8)

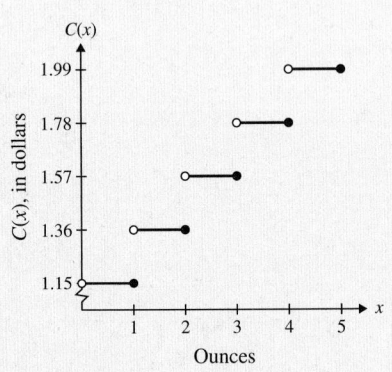

Ounces

ANSWERS TO TECHNOLOGY EXERCISES

1)
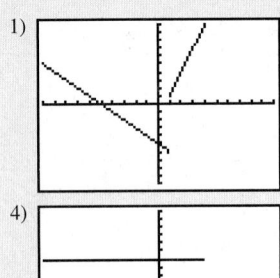

2)

3)

4)

5)

6)

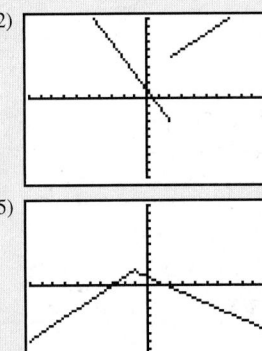

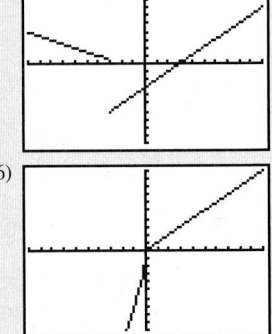

E Evaluate **12.2** Exercises   Do the exercises, and check your work.

Mixed Exercises: Objectives 1–4

Graph each function by plotting points, and identify the domain and range.

1) $f(x) = |x| + 3$

2) $g(x) = |x - 2|$

3) $k(x) = \frac{1}{2}|x|$

4) $g(x) = 2|x|$

(24) 5) $g(x) = x^2 - 4$

6) $h(x) = (x - 2)^2$

7) $f(x) = -x^2 - 1$

8) $f(x) = (x - 2)^2 - 5$

(24) 9) $f(x) = \sqrt{x + 3}$

10) $g(x) = \sqrt{x} + 2$

11) $f(x) = 2\sqrt{x}$

12) $h(x) = -\frac{1}{2}\sqrt{x}$

Given the following pairs of functions, explain how the graph of $g(x)$ can be obtained from the graph of $f(x)$ using the transformation techniques discussed in this section.

13) $f(x) = |x|$, $g(x) = |x| - 2$

14) $f(x) = |x|$, $g(x) = |x| + 1$

15) $f(x) = x^2$, $g(x) = (x + 2)^2$

16) $f(x) = x^2$, $g(x) = (x - 3)^2$

17) $f(x) = x^2$, $g(x) = -x^2$

18) $f(x) = \sqrt{x}$, $g(x) = -\sqrt{x}$

Sketch the graph of $f(x)$. Then, graph $g(x)$ on the same axes using the transformation techniques discussed in this section.

19) $f(x) = |x|$
    $g(x) = |x| - 2$

20) $f(x) = |x|$
    $g(x) = |x| + 1$

21) $f(x) = |x|$
    $g(x) = |x| + 3$

22) $f(x) = |x|$
    $g(x) = |x| - 4$

23) $f(x) = x^2$
    $g(x) = (x + 2)^2$

24) $f(x) = x^2$
    $g(x) = (x - 3)^2$

25) $f(x) = x^2$
    $g(x) = (x - 4)^2$

26) $f(x) = x^2$
    $g(x) = (x + 1)^2$

27) $f(x) = x^2$
    $g(x) = -x^2$

28) $f(x) = \sqrt{x}$
    $g(x) = -\sqrt{x}$

29) $f(x) = \sqrt{x + 1}$
    $g(x) = -\sqrt{x + 1}$

30) $f(x) = \sqrt{x - 2}$
    $g(x) = -\sqrt{x - 2}$

31) $f(x) = |x - 3|$
    $g(x) = -|x - 3|$

32) $f(x) = |x + 4|$
    $g(x) = -|x + 4|$

Use the transformation techniques discussed in this section to graph each of the following functions.

33) $f(x) = |x| - 5$

34) $f(x) = \sqrt{x} + 3$

35) $y = \sqrt{x - 4}$

36) $y = (x - 2)^2$

37) $g(x) = |x + 2| + 3$

38) $h(x) = |x + 1| - 5$

39) $y = (x - 3)^2 + 1$

40) $f(x) = (x + 2)^2 - 3$

41) $f(x) = \sqrt{x + 4} - 2$

42) $y = \sqrt{x - 3} + 2$

43) $h(x) = -x^2 + 6$

44) $y = -(x - 1)^2$

45) $g(x) = -|x - 1| + 3$

46) $h(x) = -|x + 3| - 2$

47) $f(x) = -\sqrt{x + 5}$

48) $y = -\sqrt{x + 2}$

Match each function to its graph.

49) $f(x) = x^2 - 3$, $g(x) = (x - 3)^2$,
    $h(x) = -(x + 3)^2$, $k(x) = -x^2 + 3$

a)

b)

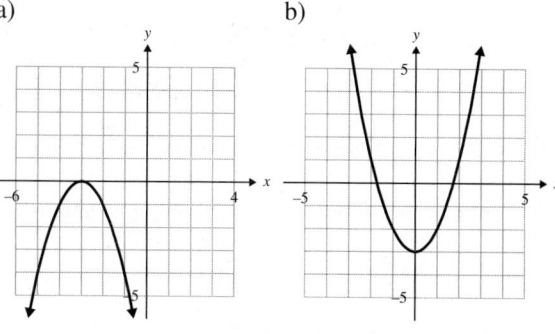

c)

d)

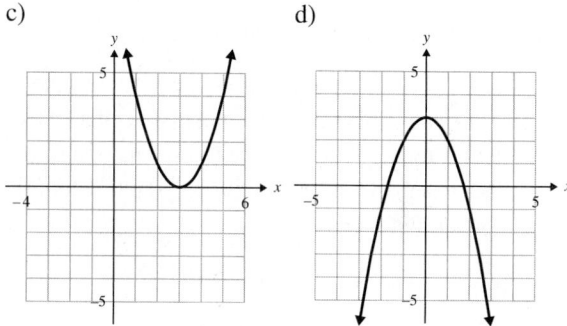

50) $f(x) = -|x - 2|$, $g(x) = |x + 2|$,
    $h(x) = -|x| - 2$, $k(x) = |x| + 2$

a)

b)

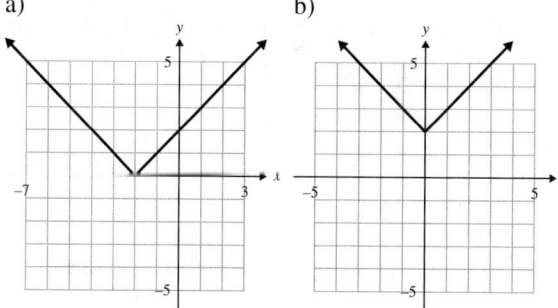

c)

d)

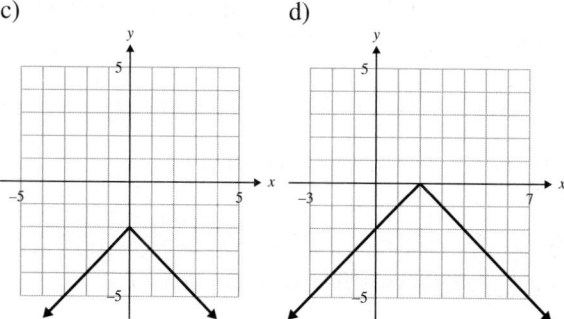

If the following transformations are performed on the graph of $f(x)$ to obtain the graph of $g(x)$, write the equation of $g(x)$.

51) $f(x) = \sqrt{x}$ is shifted 5 units to the left.

52) $f(x) = \sqrt{x}$ is shifted down 6 units.

53) $f(x) = |x|$ is shifted left 2 units and down 1 unit.

54) $f(x) = |x|$ is shifted right 1 unit and up 4 units.

(24) 55) $f(x) = x^2$ is shifted left 3 units and up $\dfrac{1}{2}$ unit.

56) $f(x) = x^2$ is shifted right 5 units and down 1.5 units.

57) $f(x) = x^2$ is reflected about the x-axis.

58) $f(x) = |x|$ is reflected about the x-axis.

59) Graph $f(x) = x^3$ by plotting points. (Hint: Make a table of values and choose 0, positive, and negative numbers for x.) Then, use the transformation techniques discussed in this section to graph each of the following functions.

a) $g(x) = (x + 2)^3$ b) $h(x) = x^3 - 3$

c) $k(x) = -x^3$ d) $r(x) = (x - 1)^3 - 2$

60) Graph $f(x) = \sqrt[3]{x}$ by plotting points. (Hint: Make a table of values and choose 0, positive, and negative numbers for x.) Then, use the transformation techniques discussed in this section to graph each of the following functions.

a) $g(x) = \sqrt[3]{x} + 4$ b) $h(x) = -\sqrt[3]{x}$

c) $k(x) = \sqrt[3]{x - 2}$ d) $r(x) = -\sqrt[3]{x} - 3$

If the following transformations are performed on the graph of $f(x)$ to obtain the graph of $g(x)$, write the equation of $g(x)$.

61) $f(x) = x^3$ is shifted left 2 units and down 1 unit.

62) $f(x) = \sqrt[3]{x}$ is shifted down 6 units.

63) $f(x) = \sqrt[3]{x}$ is reflected about the x-axis.

64) $f(x) = x^3$ is reflected about the x-axis.

Objective 5: Graph a Piecewise Function
Graph the following piecewise functions.

(24) 65) $f(x) = \begin{cases} -x - 3, & x \le -1 \\ 2x + 2, & x > -1 \end{cases}$

66) $g(x) = \begin{cases} x - 1, & x \ge 2 \\ -3x + 3, & x < 2 \end{cases}$

67) $h(x) = \begin{cases} -x + 5, & x \ge 3 \\ \dfrac{1}{2}x + 1, & x < 3 \end{cases}$

68) $f(x) = \begin{cases} 2x + 13, & x \le -4 \\ -\dfrac{1}{2}x + 1, & x > -4 \end{cases}$

69) $g(x) = \begin{cases} -\dfrac{3}{2}x - 3, & x < 0 \\ 1, & x \ge 0 \end{cases}$

70) $h(x) = \begin{cases} -\dfrac{2}{3}x - \dfrac{7}{3}, & x \ge -1 \\ 2, & x < -1 \end{cases}$

71) $k(x) = \begin{cases} x + 1, & x \ge -2 \\ 2x + 8, & x < -2 \end{cases}$

72) $g(x) = \begin{cases} x, & x \le 0 \\ 2x + 3, & x > 0 \end{cases}$

73) $f(x) = \begin{cases} 2x - 4, & x > 1 \\ -\dfrac{1}{3}x - \dfrac{5}{3}, & x \le 1 \end{cases}$

74) $k(x) = \begin{cases} \dfrac{1}{2}x + \dfrac{5}{2}, & x < 3 \\ -x + 7, & x \ge 3 \end{cases}$

Objective 6: Define, Graph, and Apply the Greatest Integer Function
Let $f(x) = [\![x]\!]$. Find the following function values.

(24) 75) $f\left(3\dfrac{1}{4}\right)$ 76) $f\left(10\dfrac{3}{8}\right)$

77) $f(7.8)$ 78) $f(9.2)$

79) $f(8)$ 80) $f\left(\dfrac{4}{5}\right)$

81) $f\left(-6\dfrac{2}{5}\right)$ 82) $f\left(-1\dfrac{3}{4}\right)$

83) $f(-8.1)$ 84) $f(-3.6)$

Graph the following greatest integer functions.

(24) 85) $f(x) = [\![x]\!] + 1$ 86) $g(x) = [\![x]\!] - 2$

87) $h(x) = [\![x]\!] - 4$ 88) $k(x) = [\![x]\!] + 3$

89) $g(x) = [\![x + 2]\!]$ 90) $h(x) = [\![x - 1]\!]$

91) $k(x) = \left[\!\left[\dfrac{1}{2}x\right]\!\right]$ 92) $f(x) = [\![2x]\!]$

93) To ship small packages within the United States, a shipping company charges $3.75 for the first pound and $1.10 for each additional pound or fraction of a pound. Let $C(x)$ represent the cost of shipping a package, and let x represent the weight of the package. Graph $C(x)$ for any package weighing up to (and including) 6 lb.

94) To deliver small packages overnight, an express delivery service charges $15.40 for the first pound and $4.50 for each additional pound or fraction of a pound. Let $C(x)$ represent the cost of shipping a package overnight, and let x represent the weight of the package. Graph $C(x)$ for any package weighing up to (and including) 6 lb.

95) The Saffir-Simpson Hurricane Wind Scale rates a hurricane's strength on a scale from 1 to 5 based upon its sustained wind speed. Each scale number is called a *category*. For example, the table tells us that a Category 4 hurricane has sustained winds from 130 mph through but not including 157 mph. Let $C(w)$

represent the category of a hurricane when its wind speed, w, is in mph. Graph $C(w)$. (www.nhc.noaa.gov)

Hurricane Strength	
Sustained Wind Speed, w	Category, C(w)
74 mph through but not including 96 mph	1
96 mph through but not including 111 mph	2
111 mph through but not including 130 mph	3
130 mph through but not including 157 mph	4
157 mph and greater	5

96) To consult with an attorney costs $35 for every 10 min or fraction of this time. Let $C(t)$ represent the cost of meeting an attorney, and let t represent the length of the meeting, in minutes. Graph $C(t)$ for meeting with the attorney for up to (and including) 1 hr.

R Rethink

R1) Give an example of an applied problem that would produce a piecewise function.

R2) What makes the graph of a function open down?

R3) Which of the graphs is the hardest to graph?

12.3 Quadratic Functions and Their Graphs

P Prepare

O Organize

What are your objectives for Section 12.3?	How can you accomplish each objective?
1 Graph a Quadratic Function by Shifting the Graph of $f(x) = x^2$	• Be able to write the definition of a *quadratic function* in your own words. • Be familiar with the graph of $f(x) = x^2$. • Know the definition of a *parabola* and a *vertex*. • Understand how to apply **vertical** and **horizontal shifting** and **reflecting about the x-axis** to graph functions. • Complete the given example on your own.
2 Graph $f(x) = a(x - h)^2 + k$ Using Characteristics of a Parabola	• Follow the procedure for **Graphing a Quadratic Function of the Form $f(x) = a(x - h)^2 + k$.** • Review procedures for finding intercepts. • Follow the procedure for **Graphing Parabolas from the Form $f(x) = ax^2 + bx + c$.** • Complete the given example on your own. • Complete You Try 1.

(continued)

What are your objectives for Section 12.3?	How can you accomplish each objective?
3 Graph $f(x) = ax^2 + bx + c$ by Completing the Square	• Follow the procedure for **Rewriting** $f(x) = ax^2 + bx + c$ **in the Form** $f(x) = a(x - h)^2 + k$ **by Completing the Square.** • Complete the given example on your own. • Complete You Try 2.
4 Graph $f(x) = ax^2 + bx + c$ Using $\left(-\dfrac{b}{2a}, f\left(-\dfrac{b}{2a}\right)\right)$	• Learn the **Vertex Formula.** • Complete the given example on your own. • Complete You Try 3.

Read the explanations, follow the examples, take notes, and complete the You Trys.

1 Graph a Quadratic Function by Shifting the Graph of $f(x) = x^2$

In this section, we will study quadratic functions in more detail and see how the rules we learned in Section 12.2 apply specifically to these functions. We restate the definition of a quadratic function here.

Definition

A **quadratic function** is a function that can be written in the form

$$f(x) = ax^2 + bx + c$$

where a, b, and c are real numbers and $a \neq 0$. An example is $f(x) = x^2 + 6x + 10$. The graph of a quadratic function is called a **parabola**. The lowest point on a parabola that opens upward or the highest point on a parabola that opens downward is called the **vertex.**

Quadratic functions can be written in other forms, as well. One common form is $f(x) = a(x - h)^2 + k$. An example is $f(x) = 2(x - 3)^2 + 1$. We will study the form $f(x) = a(x - h)^2 + k$ first because graphing parabolas from this form comes directly from the transformation techniques we learned earlier.

EXAMPLE 1   Graph $g(x) = (x - 2)^2 - 1$.

Solution

If we compare $g(x)$ to $f(x) = x^2$, what do the constants in $g(x)$ tell us about transforming the graph of $f(x)$?

$$g(x) = (x - 2)^2 - 1$$

Shift $f(x)$ Shift $f(x)$
right 2.   down 1.

Sketch the graph of $f(x) = x^2$, then move every point on the graph of f right 2 and down 1 to obtain the graph of $g(x)$. This moves the vertex from $(0, 0)$ to $(2, -1)$. The domain of $g(x)$ is $(-\infty, \infty)$; the range is $[-1, \infty)$.

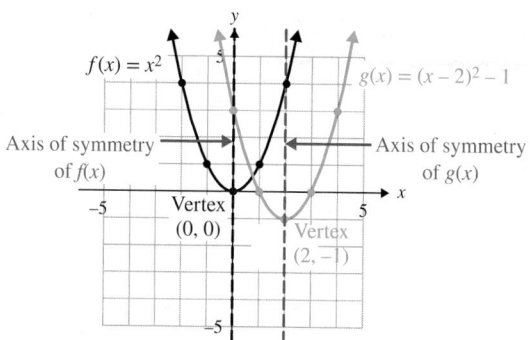

 Hint

In your own words, define the **vertex** of a parabola and the **axis of symmetry**.

Every parabola has symmetry. Let's look at the graph of $f(x) = x^2$ in Example 1. If we were to fold the paper along the y-axis, one half of the graph of $f(x) = x^2$ would fall exactly on the other half. The y-axis, or the line $x = 0$, is the **axis of symmetry** of $f(x) = x^2$. Now look at the graph of $g(x) = (x - 2)^2 - 1$ in Example 1. This parabola is symmetric with respect to the vertical line $x = 2$ through its vertex $(2, -1)$. If we were to fold the paper along the line $x = 2$, half of the graph of $g(x)$ would fall exactly on the other half. The line $x = 2$ is the *axis of symmetry* of $g(x) = (x - 2)^2 - 1$.

2 Graph $f(x) = a(x - h)^2 + k$ Using Characteristics of a Parabola

When a quadratic function is in the form $f(x) = a(x - h)^2 + k$, we can read the vertex directly from the equation. Furthermore, the value of a tells us whether the parabola opens upward or downward and whether the graph is narrower, wider, or the same width as $y = x^2$.

W Hint

Summarize this procedure in your notes.

Procedure Graphing a Quadratic Function of the Form $f(x) = a(x - h)^2 + k$

1) The vertex of the parabola is (h, k).

2) The axis of symmetry is the vertical line with equation $x = h$.

3) If a is positive, the parabola opens upward.

   If a is negative, the parabola opens downward.

4) If $|a| < 1$, then the graph of $f(x) = a(x - h)^2 + k$ is *wider* than the graph of $y = x^2$.

   If $|a| > 1$, then the graph of $f(x) = a(x - h)^2 + k$ is *narrower* than the graph of $y = x^2$.

   If $a = 1$ or $a = -1$, the graph is the *same* width as $y = x^2$.

EXAMPLE 2 Graph $f(x) = 2(x + 1)^2 - 4$. Also find the x- and y-intercepts.

Solution

Here is the information we can get from the equation:

1) $h = -1$ and $k = -4$. The vertex is $(-1, -4)$.

2) The axis of symmetry is $x = -1$.

3) $a = 2$. Because a is positive, the parabola opens upward.

4) Since $|a| > 1$, the graph of $f(x) = 2(x + 1)^2 - 4$ is *narrower* than the graph of $f(x) = x^2$.

To graph the function, start by putting the vertex on the axes. Then, choose a couple of values of x to the left or right of the vertex to plot more points. Use the axis of symmetry to find the points $(-2, -2)$ and $(-3, 4)$ on the graph of $f(x) = 2(x + 1)^2 - 4$.

x	y
0	-2
1	4

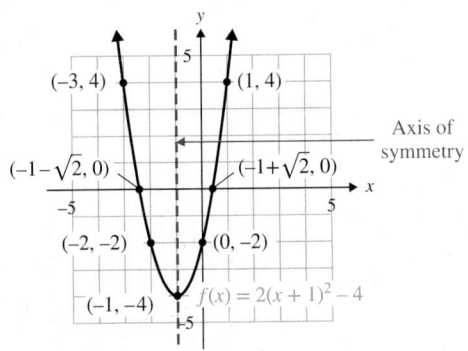

We can read the y-intercept from the graph: $(0, -2)$. To find the x-intercepts, let $f(x) = 0$ and solve for x.

$$f(x) = 2(x + 1)^2 - 4$$
$$0 = 2(x + 1)^2 - 4 \qquad \text{Substitute 0 for } f(x).$$
$$4 = 2(x + 1)^2 \qquad \text{Add 4.}$$
$$2 = (x + 1)^2 \qquad \text{Divide by 2.}$$
$$\pm\sqrt{2} = x + 1 \qquad \text{Square root property}$$
$$-1 \pm \sqrt{2} = x \qquad \text{Add } -1.$$

The x-intercepts are $(-1 - \sqrt{2}, 0)$ and $(-1 + \sqrt{2}, 0)$. The domain is $(-\infty, \infty)$; the range is $[-4, \infty)$.

[**YOU TRY 1**] Graph $f(x) = 2(x - 1)^2 - 2$. Also find the x- and y-intercepts.

When a quadratic function is written in the form $f(x) = ax^2 + bx + c$, there are two methods we can use to graph the function.

Procedure Graphing Parabolas from the Form $f(x) = ax^2 + bx + c$

There are two methods we can use to graph the function $f(x) = ax^2 + bx + c$.

Method 1: Rewrite $f(x) = ax^2 + bx + c$ in the form $f(x) = a(x - h)^2 + k$ by *completing the square.*

Method 2: Use the formula $x = -\dfrac{b}{2a}$ to find the x-coordinate of the vertex.

Then, the vertex has coordinates $\left(-\dfrac{b}{2a},\ f\left(-\dfrac{b}{2a}\right)\right)$.

We will begin with Method 1. We will modify the steps we used in Section 11.2 to solve quadratic equations by completing the square.

3 Graph $f(x) = ax^2 + bx + c$ by Completing the Square

W Hint

Summarize this procedure in your notes.

Procedure Rewriting $f(x) = ax^2 + bx + c$ in the Form $f(x) = a(x - h)^2 + k$ by Completing the Square

Step 1: The coefficient of the square term must be 1. If it is not 1, multiply or divide both sides of the equation (*including* $f(x)$) by the appropriate value to obtain a leading coefficient of 1.

Step 2: Separate the constant from the terms containing the variables by grouping the variable terms with parentheses.

Step 3: Complete the square for the quantity in the parentheses. Find half of the linear coefficient, then square the result. *Add* that quantity inside the parentheses, and *subtract* the quantity from the constant. (Adding and subtracting the same number on the same side of an equation is like adding 0 to the equation.)

Step 4: Factor the expression inside the parentheses.

Step 5: Solve for $f(x)$.

EXAMPLE 3

Graph each function. Begin by completing the square to rewrite each function in the form $f(x) = a(x - h)^2 + k$. Include the intercepts.

a) $f(x) = x^2 + 6x + 10$

b) $g(x) = -\dfrac{1}{2}x^2 + 4x - 6$

Solution

a) **Step 1:** The coefficient of x^2 is 1.

Step 2: Separate the constant from the variable terms using parentheses.

$$f(x) = (x^2 + 6x) + 10$$

Step 3: Complete the square for the quantity in the parentheses.

$$\frac{1}{2}(6) = 3$$

$$3^2 = 9$$

Add 9 inside the parentheses, and subtract 9 from the 10. This is like adding 0 to the equation.

$$f(x) = (x^2 + 6x + 9) + 10 - 9$$
$$f(x) = (x^2 + 6x + 9) + 1$$

Step 4: Factor the expression inside the parentheses.

$$f(x) = (x + 3)^2 + 1$$

Step 5: The equation *is* solved for $f(x)$.

From the equation $f(x) = (x + 3)^2 + 1$ we can see that

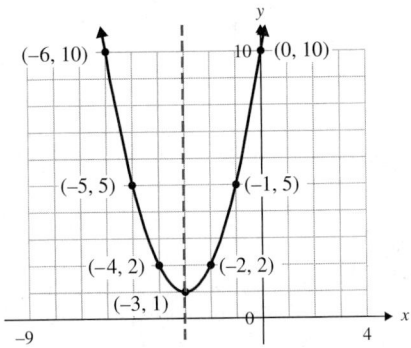

i) The vertex is $(-3, 1)$.
ii) The axis of symmetry is $x = -3$.
iii) $a = 1$ so the parabola opens upward.
iv) Since $a = 1$, the graph is the same width as $y = x^2$.

Find some other points on the parabola. Use the axis of symmetry.

To find the x-intercepts, let $f(x) = 0$ and solve for x. Use *either* form of the equation. We will use $f(x) = (x + 3)^2 + 1$.

x	$f(x)$
-2	2
-1	5

$$0 = (x + 3)^2 + 1 \qquad \text{Let } f(x) = 0.$$
$$-1 = (x + 3)^2 \qquad \text{Subtract 1.}$$
$$\pm\sqrt{-1} = x + 3 \qquad \text{Square root property}$$
$$-3 \pm i = x \qquad \sqrt{-1} = i; \text{ subtract 3.}$$

Because the solutions to $f(x) = 0$ are *not* real numbers, *there are no x-intercepts.* To find the y-intercept, let $x = 0$ and solve for $f(0)$.

$$f(x) = (x + 3)^2 + 1$$
$$f(0) = (0 + 3)^2 + 1$$
$$f(0) = 9 + 1 = 10$$

The y-intercept is $(0, 10)$. The domain is $(-\infty, \infty)$, and the range is $[1, \infty)$.

b) **Step 1:** The coefficient of x^2 is $-\frac{1}{2}$. Multiply both sides of the equation (including the $g(x)$) by -2 so that the coefficient of x^2 will be 1.

$$g(x) = -\frac{1}{2}x^2 + 4x - 6$$

$$-2g(x) = -2\left(-\frac{1}{2}x^2 + 4x - 6\right) \qquad \text{Multiply by } -2.$$

$$-2g(x) = x^2 - 8x + 12 \qquad \text{Distribute.}$$

Step 2: Separate the constant from the variable terms using parentheses.

$$-2g(x) = (x^2 - 8x) + 12$$

Step 3: Complete the square for the quantity in parentheses.

$$\frac{1}{2}(-8) = -4$$
$$(-4)^2 = 16$$

Add 16 inside the parentheses, and subtract 16 from the 12.

$$-2g(x) = (x^2 - 8x + 16) + 12 - 16$$
$$-2g(x) = (x^2 - 8x + 16) - 4$$

Step 4: Factor the expression inside the parentheses.

$$-2g(x) = (x - 4)^2 - 4$$

Step 5: Solve the equation for $g(x)$ by dividing by -2.

$$\frac{-2g(x)}{-2} = \frac{(x - 4)^2}{-2} - \frac{4}{-2}$$
$$g(x) = -\frac{1}{2}(x - 4)^2 + 2$$

From $g(x) = -\frac{1}{2}(x - 4)^2 + 2$ we can see that

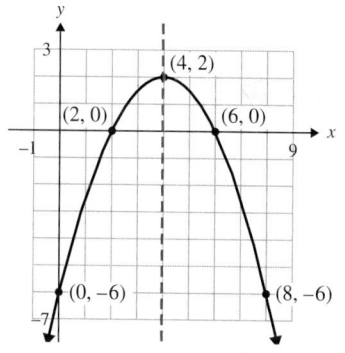

i) The vertex is (4, 2).
ii) The axis of symmetry is $x = 4$.
iii) $a = -\frac{1}{2}$ (the same as in the form

$g(x) = -\frac{1}{2}x^2 + 4x - 6$) so the parabola

opens downward.

iv) Since $a = -\frac{1}{2}$, the graph of $g(x)$ will be
wider than $y = x^2$.

Find some other points on the parabola.
Use the axis of symmetry.

x	g(x)
6	0
8	-6

Using the axis of symmetry, we can see that the x-intercepts are (6, 0) and (2, 0) and that the y-intercept is (0, −6). The domain is $(-\infty, \infty)$; the range is $(-\infty, 2]$.

[YOU TRY 2]

Graph each function. Begin by completing the square to rewrite each function in the form $f(x) = a(x - h)^2 + k$. Include the intercepts.

a) $f(x) = x^2 + 4x + 3$ b) $g(x) = -2x^2 + 12x - 8$

---

4 Graph $f(x) = ax^2 + bx + c$ Using $\left(-\frac{b}{2a}, f\left(-\frac{b}{2a}\right)\right)$

We can also graph quadratic functions of the form $f(x) = ax^2 + bx + c$ by using the formula $h = -\frac{b}{2a}$ to find the x-coordinate of the vertex. This formula comes from completing the square on $f(x) = ax^2 + bx + c$.

Although there is a formula for k, it is only necessary to remember the formula for h. The y-coordinate of the vertex, then, is $k = f\left(-\dfrac{b}{2a}\right)$. The axis of symmetry is $x = h$.

W Hint

Write the Vertex Formula in your notes. In your own words, explain how to use it.

Property The Vertex Formula

The **vertex** of the graph of $f(x) = ax^2 + bx + c$ $(a \neq 0)$ has

coordinates $\left(-\dfrac{b}{2a},\, f\left(-\dfrac{b}{2a}\right)\right)$.

EXAMPLE 4

Graph $f(x) = x^2 - 6x + 3$ using the vertex formula. Include the intercepts.

Solution

$a = 1$, $b = -6$, $c = 3$. Since $a = +1$, the graph opens upward. The x-coordinate, h, of the vertex is

$$h = -\frac{b}{2a} = -\frac{(-6)}{2(1)} = \frac{6}{2} = 3$$

$h = 3$. Then, the y-coordinate, k, of the vertex is $k = f(3)$.

$$f(x) = x^2 - 6x + 3$$
$$f(3) = (3)^2 - 6(3) + 3$$
$$= 9 - 18 + 3 = -6$$

The vertex is $(3, -6)$. The axis of symmetry is $x = 3$.
Find more points on the graph of $f(x) = x^2 - 6x + 3$, then use the axis of symmetry to find other points on the parabola.

x	$f(x)$
4	-5
5	-2
6	3

To find the x-intercepts, let $f(x) = 0$ and solve for x.

$$0 = x^2 - 6x + 3$$
$$x = \frac{-(-6) \pm \sqrt{(-6)^2 - 4(1)(3)}}{2(1)}$$ Solve using the quadratic formula.
$$x = \frac{6 \pm \sqrt{24}}{2} = \frac{6 \pm 2\sqrt{6}}{2}$$ Simplify.
$$x = 3 \pm \sqrt{6}$$

The x-intercepts are $(3 + \sqrt{6}, 0)$ and $(3 - \sqrt{6}, 0)$.
We can see from the graph that the y-intercept is $(0, 3)$. The domain is $(-\infty, \infty)$; the range is $[-6, \infty)$.

Graph $f(x) = -x^2 - 8x - 13$ using the vertex formula. Include the intercepts.

Using Technology

In Section 7.5 we said that the solutions of the equation $x^2 - x - 6 = 0$ are the
x-intercepts of the graph of $y = x^2 - x - 6$.

The x-intercepts are also called the zeros of the
equation since they are the values of x that make $y = 0$.
Enter $x^2 - x - 6$ in Y_1 then find the x-intercepts shown
on the graph by pressing 2nd TRACE and then selecting
2:zero. Move the cursor to the left of an x-intercept using
the right arrow key and press ENTER. Move the cursor
to the right of the x-intercept using the right arrow key
and press ENTER. Move the cursor close to the
x-intercept using the left arrow key and press ENTER.
Repeat these steps for each x-intercept. The x-intercepts
are $(-2, 0)$ and $(3, 0)$ as shown in the graphs to the right.

The y-intercept is found by graphing the function and
pressing TRACE 0 ENTER. As shown on the graph,
the y-intercept for $y = x^2 - x - 6$ is $(0, -6)$.

The x-value of the vertex can be found using the vertex
formula. In this case, $a = 1$ and $b = -1$, so $-\dfrac{b}{2a} = \dfrac{1}{2}$.

To find the vertex on the graph, press TRACE, type 1/2,
and press ENTER. The vertex is shown as $(0.5, -6.25)$
on the graph.

Remember, you can convert the coordinates of the vertex
to fractions. Go to the home screen by pressing 2nd MODE.
To display the x-value of the vertex, press X, T, Θ, n MATH ENTER ENTER.
To display the y-value of the vertex, press ALPHA I MATH ENTER ENTER.
The vertex is then $\left(\dfrac{1}{2}, -\dfrac{25}{4}\right)$.

Find the x-intercepts, y-intercept, and vertex using a graphing calculator.

1) $y = x^2 - 2x + 2$ 2) $y = x^2 - 4x - 5$ 3) $y = -(x + 1)^2 + 4$

4) $y = x^2 - 4$ 5) $y = x^2 - 6x + 9$ 6) $y = -x^2 - 8x - 19$

ANSWERS TO [YOU TRY] EXERCISES

1) x-ints: (0, 0), (2, 0); y-int: (0, 0)

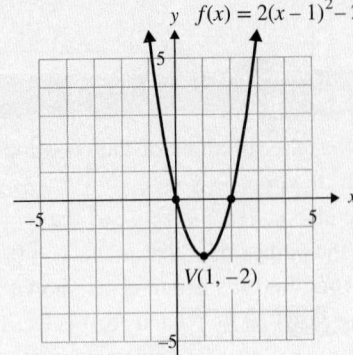

2) a) x-ints: (−3, 0), (−1, 0); y-int: (0, 3)

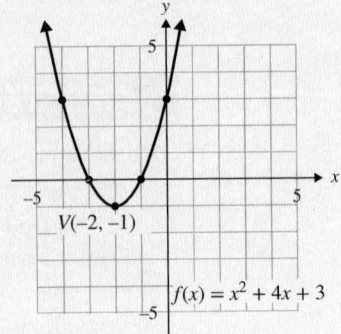

b) x-ints: $(3 + \sqrt{5}, 0)$, $(3 − \sqrt{5}, 0)$;
   y-int: (0, −8)

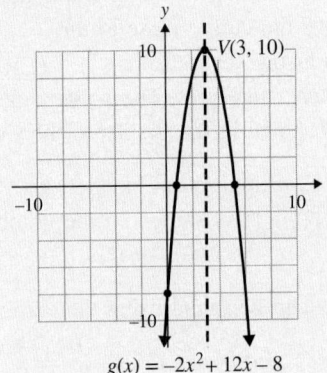

3) x-ints: $(−4 + \sqrt{3}, 0)$, $(−4 − \sqrt{3}, 0)$;
   y-int: (0, −13)

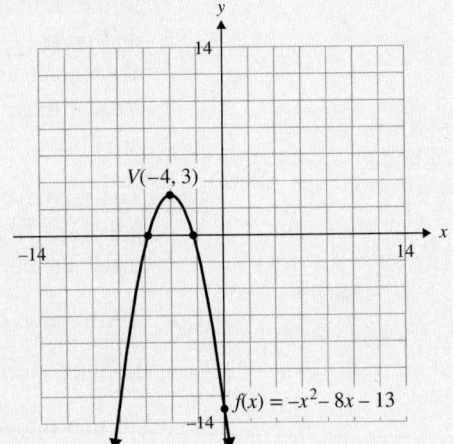

ANSWERS TO TECHNOLOGY EXERCISES

1) no x-intercepts; y-intercept: (0, 2); vertex: (1, 1)
2) x-intercepts: (−1, 0), (5, 0); y-intercept: (0, −5); vertex: (2, −9)
3) x-intercepts: (−3, 0), (1, 0); y-intercept: (0, 3); vertex: (−1, 4)
4) x-intercepts: (−2, 0), (2, 0); y-intercept: (0, −4); vertex: (0, −4)
5) x-intercept: (3, 0); y-intercept: (0, 9); vertex: (3, 0)
6) no x-intercepts; y-intercept: (0, −19); vertex: (−4, −3)

E Evaluate 12.3 Exercises

Do the exercises, and check your work.

Objective 1: Graph a Quadratic Function by Shifting the Graph of $f(x) = x^2$

1) How does the graph of $g(x) = x^2 + 6$ compare to the graph of $f(x) = x^2$?

2) How does the graph of $h(x) = x^2 − 5$ compare to the graph of $f(x) = x^2$?

3) How does the graph of $h(x) = (x + 5)^2$ compare to the graph of $f(x) = x^2$?

4) How does the graph of $g(x) = (x − 4)^2$ compare to the graph of $f(x) = x^2$?

For Exercises 5–18, sketch the graph of $f(x) = x^2$. Then graph $g(x)$ on the same axes by shifting the graph of $f(x)$.

5) $g(x) = x^2 + 3$　　　　6) $g(x) = x^2 + 5$

7) $g(x) = x^2 − 4$　　　　8) $g(x) = x^2 − 1$

9) $g(x) = (x + 2)^2$　　　10) $g(x) = (x + 1)^2$

11) $g(x) = (x-3)^2$

12) $g(x) = (x-4)^2$

13) $g(x) = -x^2$

14) $g(x) = -(x-1)^2$

15) $g(x) = \dfrac{1}{2}x^2$

16) $g(x) = 2x^2$

17) $g(x) = (x-1)^2 - 3$

18) $g(x) = (x+2)^2 + 1$

Objective 2: Graph $f(x) = a(x-h)^2 + k$ Using Characteristics of a Parabola

19) Given a quadratic function of the form
$f(x) = a(x-h)^2 + k,$

a) what is the vertex?

b) what is the equation of the axis of symmetry?

c) how do you know if the parabola opens upward?

d) how do you know if the parabola opens downward?

e) how do you know if the parabola is narrower than the graph of $y = x^2$?

f) how do you know if the parabola is wider than the graph of $y = x^2$?

For each quadratic function, identify the vertex, axis of symmetry, and x- and y-intercepts. Then, graph the function. Determine the domain and range.

20) $g(x) = (x-3)^2 - 1$

21) $f(x) = (x+1)^2 - 4$

22) $h(x) = (x+2)^2 + 7$

23) $g(x) = (x-2)^2 + 3$

24) $y = (x+1)^2 - 5$

25) $y = (x-4)^2 - 2$

26) $g(x) = -(x-3)^2 + 2$

27) $f(x) = -(x+3)^2 + 6$

28) $f(x) = -(x-2)^2 - 4$

29) $y - -(x+1)^2 - 5$

30) $y = 2(x+1)^2 - 2$

31) $f(x) = 2(x-1)^2 - 8$

32) $h(x) = \dfrac{1}{2}(x+4)^2$

33) $g(x) = \dfrac{1}{4}x^2 - 1$

34) $y = -x^2 + 5$

35) $f(x) = -\dfrac{1}{3}(x+4)^2 + 3$

36) $y = -\dfrac{1}{2}(x-4)^2 + 2$

37) $g(x) = 3(x+2)^2 + 5$

38) $f(x) = 2(x-3)^2 + 3$

In Exercises 39 and 40, match each function to its graph.

39) $f(x) = x^2 - 3$, $g(x) = (x-3)^2$,
$h(x) = -(x+3)^2$, $k(x) = -x^2 + 3$

a)

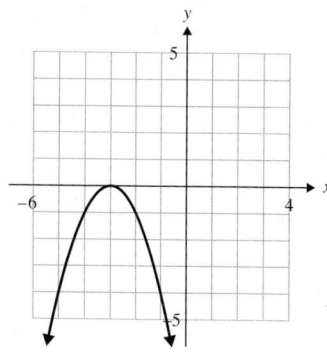

b)

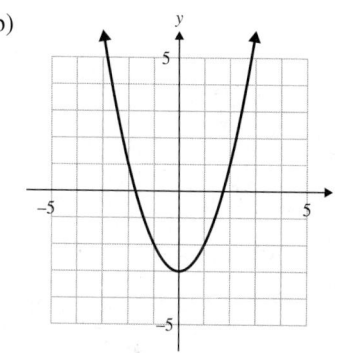

c)

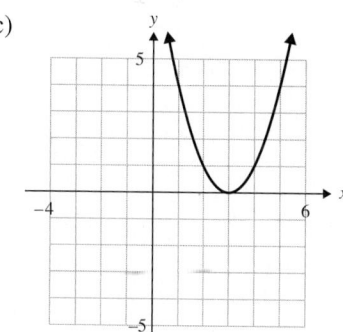

d)

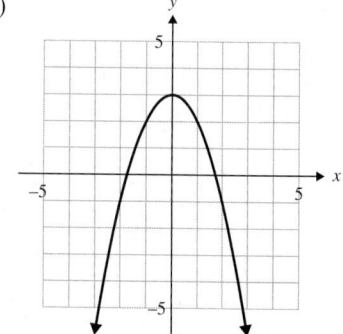

40) $f(x) = (x + 1)^2$, $g(x) = x^2 + 1$,
    $h(x) = -(x + 1)^2$, $k(x) = -x^2 + 1$

a)

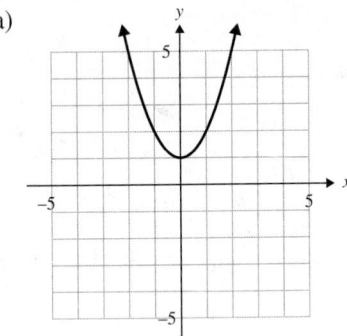

b)

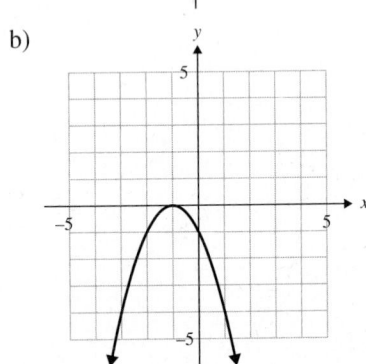

c)

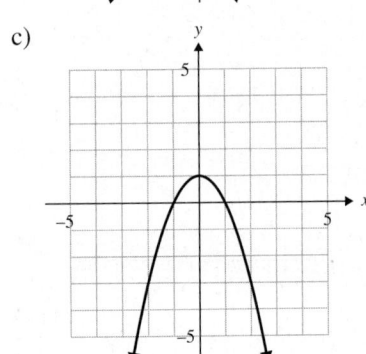

d)

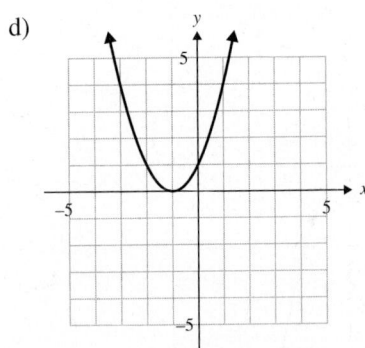

Exercises 41–46 each present a shift that is performed on the graph of $f(x) = x^2$ to obtain the graph of $g(x)$. Write the equation of $g(x)$.

41) $f(x)$ is shifted 8 units to the right.

42) $f(x)$ is shifted down 9 units.

43) $f(x)$ is shifted up 3.5 units.

44) $f(x)$ is shifted left 1.2 units.

45) $f(x)$ is shifted left 4 units and down 7 units.

46) $f(x)$ is shifted up 2 units and right 5 units.

Objective 3: Graph $f(x) = ax^2 + bx + c$ by Completing the Square

Rewrite each function in the form $f(x) = a(x - h)^2 + k$.

Fill It In

Fill in the blanks with either the missing mathematical step or the reason for the given step.

47) $f(x) = x^2 + 8x + 11$

_____ Group the variable terms together using parentheses.

_____ Find the number that completes the square in the parentheses.

$f(x) = (x^2 + 8x + 16) + 11 - 16$

_____
_____
_____ Factor and simplify.

48) $f(x) = x^2 - 4x - 7$ _____
    $f(x) = (x^2 - 4x) - 7$ _____

_____ Find the number that completes the square in the parentheses.

_____ Add and subtract the number above to the same side of the equation.

$f(x) = (x - 2)^2 - 11$ _____

Rewrite each function in the form $f(x) = a(x - h)^2 + k$ by completing the square. Then, graph the function. Include the intercepts. Determine the domain and range.

49) $f(x) = x^2 - 2x - 3$ 50) $g(x) = x^2 + 6x + 8$

51) $y = x^2 + 6x + 7$ 52) $h(x) = x^2 - 4x + 1$

53) $g(x) = x^2 + 4x$ 54) $y = x^2 - 8x + 18$

55) $h(x) = -x^2 - 4x + 5$ 56) $f(x) = -x^2 - 2x + 3$

57) $y = -x^2 + 6x - 10$ 58) $g(x) = -x^2 - 4x - 6$

59) $y = 2x^2 - 8x + 2$ 60) $f(x) = 2x^2 - 8x + 4$

61) $g(x) = -\dfrac{1}{3}x^2 - 2x - 9$

62) $h(x) = -\dfrac{1}{2}x^2 - 3x - \dfrac{19}{2}$

63) $y = x^2 - 3x + 2$

64) $f(x) = x^2 + 5x + \dfrac{21}{4}$

Objective 4: Graph $f(x) = ax^2 + bx + c$ Using $\left(-\dfrac{b}{2a}, f\left(-\dfrac{b}{2a}\right)\right)$

Graph each function using the vertex formula. Include the intercepts. Determine the domain and range.

65) $y = x^2 + 2x - 3$

66) $g(x) = x^2 - 6x + 8$

67) $f(x) = -x^2 - 8x - 13$

68) $y = -x^2 + 2x + 2$

69) $g(x) = 2x^2 - 4x + 4$

70) $f(x) = -4x^2 - 8x - 6$

71) $y = -3x^2 + 6x + 1$

72) $h(x) = 2x^2 - 12x + 9$

73) $f(x) = \dfrac{1}{2}x^2 - 4x + 5$

74) $y = \dfrac{1}{2}x^2 + 2x - 3$

75) $h(x) = -\dfrac{1}{3}x^2 - 2x - 5$

76) $g(x) = \dfrac{1}{5}x^2 - 2x + 8$

R Rethink

R1) What shape is a quadratic function? What causes the graph to open up?

R2) What are the different ways to find the vertex of a parabola? Which do you prefer?

R3) What do you find most difficult about completing the square?

12.4 Applications of Quadratic Functions and Graphing Other Parabolas

P Prepare

O Organize

What are your objectives for Section 12.4?	How can you accomplish each objective?
1 Find the Maximum or Minimum Value of a Quadratic Function	• Learn how to find the **Maximum or Minimum Value of a Quadratic Function.** • Complete the given example on your own. • Complete You Try 1.
2 Given a Quadratic Function, Solve an Applied Problem Involving a Maximum or Minimum Value	• Be able to determine whether the problem involves a maximum or minimum. • Complete the given example on your own. • Complete You Try 2.
3 Write a Quadratic Function to Solve an Applied Problem Involving a Maximum or Minimum Value	• Follow the procedure for **Solving a Max/Min Problem.** Write this procedure in your notes. • Review methods for solving quadratic equations. • Complete the given example on your own. • Complete You Try 3.

(continued)

What are your objectives for Section 12.4?	How can you accomplish each objective?
4 Graph Parabolas of the Form $x = a(y - k)^2 + h$	• Follow the procedure for **Graphing an Equation of the Form $x = a(y - k)^2 + h$.** • Complete the given example on your own. • Complete You Try 4.
5 Rewrite $x = ay^2 + by + c$ as $x = a(y - k)^2 + h$ by Completing the Square	• Review the procedure for **Completing the Square.** Write the procedure in your own words. • Complete the given example on your own. • Complete You Try 5.
6 Find the Vertex of the Graph of $x = ay^2 + by + c$ Using $y = -\dfrac{b}{2a}$, and Graph the Equation	• Review the vertex formula in Section 12.3, and notice the slight differences when using the formula for these graphs. • Know the procedures for **Graphing Parabolas from the Form $x = ay^2 + by + c$.** • Complete the given example on your own. • Complete You Try 6.

 **Work**   **Read the explanations, follow the examples, take notes, and complete the You Trys.**

1 Find the Maximum or Minimum Value of a Quadratic Function

From our work with quadratic functions, we have seen that the vertex is either the lowest point or the highest point on the graph depending on whether the parabola opens upward or downward.

If the parabola opens upward, the vertex is the *lowest* point on the parabola.

If the parabola opens downward, the vertex is the *highest* point on the parabola.

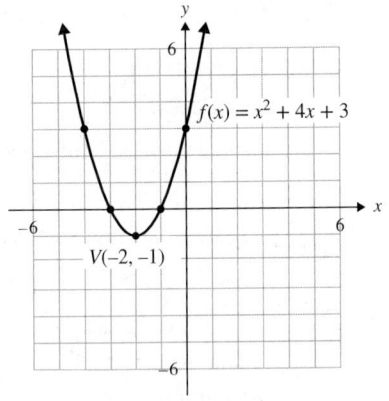

$$f(x) = x^2 + 4x + 3$$

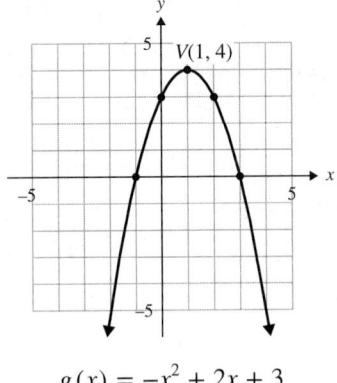

$$g(x) = -x^2 + 2x + 3$$

 **Hint**
In your notes and in your own words, summarize this information about the minimum and maximum values of a quadratic function.

The y-coordinate of the vertex, -1, is the *smallest* y-value the function will have. We say that **-1 is the minimum value of the function.** $f(x)$ has no maximum because the graph continues upward indefinitely—the y-values get larger without bound.

The y-coordinate of the vertex, 4, is the *largest* y-value the function will have. We say that **4 is the maximum value of the function.** $g(x)$ has no minimum because the graph continues downward indefinitely—the y-values get smaller without bound.

Property Maximum and Minimum Values of a Quadratic Function

Let $f(x) = ax^2 + bx + c$.

1) If a is **positive,** the graph of $f(x)$ opens upward, so the vertex is the lowest point on the parabola. The y-coordinate of the vertex is the **minimum** value of the function $f(x)$.

2) If a is **negative,** the graph of $f(x)$ opens downward, so the vertex is the highest point on the parabola. The y-coordinate of the vertex is the **maximum** value of the function $f(x)$.

We can use this information about the vertex to help us solve problems.

EXAMPLE 1

Let $f(x) = -x^2 + 4x + 2$.

a) Does the function attain a minimum or maximum value at its vertex?

b) Find the vertex of the graph of $f(x)$.

c) What is the minimum or maximum value of the function?

d) Graph the function to verify parts a)–c).

Solution

a) Because $a = -1$, the graph of $f(x)$ will open downward. Therefore, the vertex will be the *highest* point on the parabola. The function will attain its *maximum* value at the vertex.

b) Use $x = -\dfrac{b}{2a}$ to find the x-coordinate of the vertex. For $f(x) = -x^2 + 4x + 2$,

$$x = -\frac{b}{2a} = -\frac{(4)}{2(-1)} = 2$$

The y-coordinate of the vertex is $f(2)$.

$$f(2) = -(2)^2 + 4(2) + 2$$
$$= -4 + 8 + 2 = 6$$

The vertex is (2, 6).

c) $f(x)$ has no minimum value. The *maximum* value of the function is 6, the y-coordinate of the vertex. (The largest y-value of the function is 6.)

   We say that the maximum value of the function is 6 and that it occurs at $x = 2$ (the x-coordinate of the vertex).

d) From the graph of $f(x)$, we can see that our conclusions in parts a)–c) make sense.

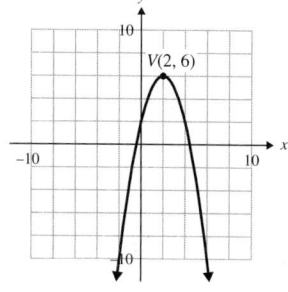

YOU TRY 1

Let $f(x) = x^2 + 6x + 7$. Repeat parts a)–d) from Example 1.

2 Given a Quadratic Function, Solve an Applied Problem Involving a Maximum or Minimum Value

EXAMPLE 2

A ball is thrown upward from a height of 24 ft. The height h of the ball (in feet) t sec after the ball is released is given by

$$h(t) = -16t^2 + 16t + 24.$$

a) How long does it take the ball to reach its maximum height?

b) What is the maximum height attained by the ball?

Solution

a) Begin by understanding what the function $h(t)$ tells us: $a = -16$, so the graph of h opens downward. Therefore, the vertex is the highest point on the parabola. The maximum value of the function occurs at the vertex. The ordered pairs that satisfy $h(t)$ are of the form $(t, h(t))$.

To determine how long it takes the ball to reach its maximum height, we must find the t-coordinate of the vertex.

$$t = -\frac{b}{2a} = -\frac{16}{2(-16)} = \frac{1}{2}$$

The ball will reach its maximum height after $\frac{1}{2}$ sec.

b) The maximum height the ball reaches is the y-coordinate (or $h(t)$-coordinate) of the vertex. Since the ball attains its maximum height when $t = \frac{1}{2}$, find $h\left(\frac{1}{2}\right)$.

$$h\left(\frac{1}{2}\right) = -16\left(\frac{1}{2}\right)^2 + 16\left(\frac{1}{2}\right) + 24$$

$$= -16\left(\frac{1}{4}\right) + 8 + 24$$

$$= -4 + 32 = 28$$

The ball reaches a maximum height of 28 ft.

YOU TRY 2

An object is propelled upward from a height of 10 ft. The height h of the object (in feet) t sec after the ball is released is given by

$$h(t) = -16t^2 + 32t + 10$$

a) How long does it take the object to reach its maximum height?

b) What is the maximum height attained by the object?

3 Write a Quadratic Function to Solve an Applied Problem Involving a Maximum or Minimum Value

Ayesha plans to put a fence around her rectangular garden. If she has 32 ft of fencing, what is the maximum area she can enclose?

Solution

Begin by drawing a picture.

Let $x =$ the width of the garden

Let $y =$ the length of the garden

Label the picture.

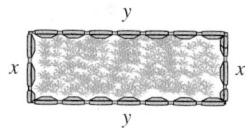

We will write two equations for a problem like this:

1) *The maximize or minimize equation;* this equation describes what we are trying to maximize or minimize.

2) *The constraint equation;* this equation describes the restrictions on the variables or the conditions the variables must meet.

Here is how we will get the equations.

1) We will write a *maximize* equation because we are trying to find the *maximum area* of the garden.

$$\text{Let } A = \text{area of the garden}$$

The area of the rectangle above is xy. Our equation is

$$\text{Maximize:} \quad A = xy$$

2) To write the *constraint* equation, think about the restriction put on the variables. We cannot choose *any* two numbers for x and y. Since Ayesha has 32 ft of fencing, the distance around the garden is 32 ft. This is the *perimeter* of the rectangular garden. The perimeter of the rectangle drawn above is $2x + 2y$, and it must equal 32 ft.
    The constraint equation is

$$\text{Constraint:} \quad 2x + 2y = 32$$

Set up this maximization problem as

$$\text{Maximize:} \quad A = xy$$
$$\text{Constraint:} \quad 2x + 2y = 32$$

Solve the constraint for a variable, and then substitute the expression into the maximize equation.

$$2x + 2y = 32$$
$$2y = 32 - 2x$$
$$y = 16 - x \qquad \text{Solve the constraint for } y.$$

Substitute $y = 16 - x$ into $A = xy$.

$$A = x(16 - x)$$
$$A = 16x - x^2 \qquad \text{Distribute.}$$
$$A = -x^2 + 16x \qquad \text{Write in descending powers.}$$

W Hint

Do you understand how the solution of this problem is related to the graph of a quadratic equation?

Look carefully at $A = -x^2 + 16x$. This is a quadratic function! Its graph is a parabola that opens downward (since $a = -1$). At the vertex, the function attains its maximum. The ordered pairs that satisfy this function are of the form $(x, A(x))$, where x represents the width and $A(x)$ represents the area of the rectangular garden. *The second coordinate of the vertex is the maximum area we are looking for.*

$$A = -x^2 + 16x$$

Use $x = -\dfrac{b}{2a}$ with $a = -1$ and $b = 16$ to find the x-coordinate of the vertex (the width of the rectangle that produces the maximum area).

$$x = -\frac{16}{2(-1)} = 8$$

Substitute $x = 8$ into $A = -x^2 + 16x$ to find the maximum area.

$$A = -(8)^2 + 16(8)$$
$$A = -64 + 128$$
$$A = 64$$

The graph of $A = -x^2 + 16x$ is a parabola that opens downward with vertex $(8, 64)$.
The maximum area of the garden is 64 ft^2, and this will occur when the width of the garden is 8 ft. (The length will be 8 ft as well.)

Let's summarize the steps we can use to solve a max/min problem.

Procedure Steps for Solving a Max/Min Problem Like Example 3

1) Draw a picture, if applicable.
2) Define the unknowns. Label the picture.
3) Write the max/min equation.
4) Write the constraint equation.
5) Solve the constraint for a variable. Substitute the expression into the max/min equation to obtain a quadratic function.
6) Find the vertex of the parabola using the vertex formula, $x = -\dfrac{b}{2a}$.
7) Answer the question being asked.

$\left[\text{ YOU TRY 3 }\right]$ Find the maximum area of a rectangle that has a perimeter of 28 in.

4 Graph Parabolas of the Form $x = a(y - k)^2 + h$

Not all parabolas are functions. Parabolas can open in the x-direction as illustrated below. Clearly, these fail the vertical line test for functions.

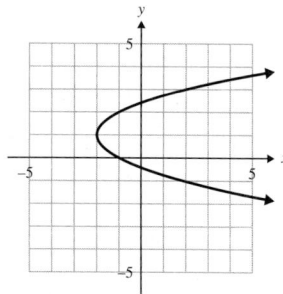 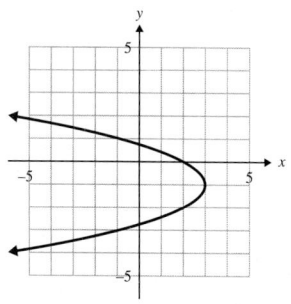

Hint

When graphing equations that begin with $x =$, many earlier processes are switched when graphing.

Parabolas that open in the y-direction, or vertically, result from the functions

$$y = a(x - h)^2 + k \qquad \text{or} \qquad y = ax^2 + bx + c.$$

If we interchange the x and y, we obtain the equations

$$x = a(y - k)^2 + h \qquad \text{or} \qquad x = ay^2 + by + c.$$

The graphs of these equations are parabolas that open in the x-direction, or horizontally.

Procedure Graphing an Equation of the Form $x = a(y - k)^2 + h$

1) The vertex of the parabola is (h, k). (Notice, however, that h and k have changed their positions in the equation when compared to a quadratic function.)

2) The axis of symmetry is the horizontal line $y = k$.

3) If a is positive, the graph opens to the right.
   If a is negative, the graph opens to the left.

EXAMPLE 4

Graph each equation. Find the x- and y-intercepts and the domain and range.

a) $x = (y + 2)^2 - 1$ b) $x = -2(y - 2)^2 + 4$

Solution

a) 1) $h = -1$ and $k = -2$. The vertex is $(-1, -2)$.

     2) The axis of symmetry is $y = -2$.

     3) $a = +1$, so the parabola opens to the right. It is the same width as $y = x^2$.

To find the x-intercept, let $y = 0$ and solve for x.

$$x = (y + 2)^2 - 1$$
$$x = (0 + 2)^2 - 1$$
$$x = 4 - 1 = 3$$

The x-intercept is $(3, 0)$.

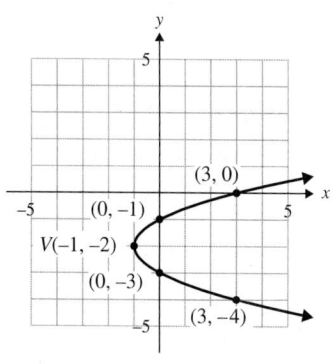

SECTION 12.4   Applications of Quadratic Functions and Graphing Other Parabolas   851

Find the y-intercepts by substituting 0 for x and solving for y.

$$x = (y + 2)^2 - 1$$
$$0 = (y + 2)^2 - 1 \qquad \text{Substitute 0 for } x.$$
$$1 = (y + 2)^2 \qquad \text{Add 1.}$$
$$\pm 1 = y + 2 \qquad \text{Square root property}$$

$$1 = y + 2 \qquad \text{or} \qquad -1 = y + 2$$
$$-1 = y \qquad\qquad -3 = y \qquad \text{Solve.}$$

The y-intercepts are $(0, -3)$ and $(0, -1)$. Use the axis of symmetry to locate the point $(3, -4)$ on the graph. The domain is $[-1, \infty)$, and the range is $(-\infty, \infty)$.

b) $x = -2(y - 2)^2 + 4$

W Hint

Do you notice the similarities and differences between graphing these equations and graphing quadratic functions?

1) $h = 4$ and $k = 2$. The vertex is $(4, 2)$.

2) The axis of symmetry is $y = 2$.

3) $a = -2$, so the parabola opens to the left. It is narrower than $y = x^2$.

To find the x-intercept, let $y = 0$ and solve for x.

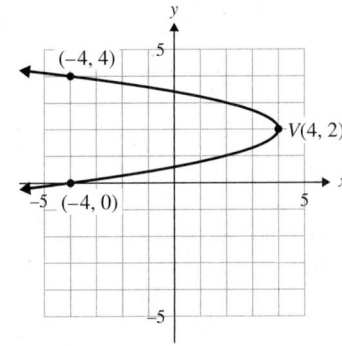

$$x = -2(y - 2)^2 + 4$$
$$x = -2(0 - 2)^2 + 4$$
$$x = -2(4) + 4 = -4$$

The x-intercept is $(-4, 0)$.

Find the y-intercepts by substituting 0 for x and solving for y.

$$x = -2(y - 2)^2 + 4$$
$$0 = -2(y - 2)^2 + 4 \qquad \text{Substitute 0 for } x.$$
$$-4 = -2(y - 2)^2 \qquad \text{Subtract 4.}$$
$$2 = (y - 2)^2 \qquad \text{Divide by } -2.$$
$$\pm \sqrt{2} = y - 2 \qquad \text{Square root property}$$
$$2 \pm \sqrt{2} = y \qquad \text{Add 2.}$$

The y-intercepts are $(0, 2 - \sqrt{2})$ and $(0, 2 + \sqrt{2})$. Use the axis of symmetry to locate the point $(-4, 4)$ on the graph. The domain is $(-\infty, 4]$; the range is $(-\infty, \infty)$.

$\left[\text{YOU TRY 4}\right]$ Graph $x = -(y + 1)^2 - 3$. Find the x- and y-intercepts and the domain and range.

Procedure Graphing Parabolas from the Form $x = ay^2 + by + c$

We can use two methods to graph $x = ay^2 + by + c$.

W Hint

Notice how this compares to the procedure for graphing $f(x) = ax^2 + bx + c$.

Method 1: Rewrite $x = ay^2 + by + c$ in the form $x = a(y - k)^2 + h$ by *completing the square*.

Method 2: Use the formula $y = -\dfrac{b}{2a}$ to find the y-*coordinate* of the vertex.

Find the x-coordinate by substituting the y-value into the equation $x = ay^2 + by + c$.

5 Rewrite $x = ay^2 + by + c$ as $x = a(y - k)^2 + h$ by Completing the Square

EXAMPLE 5

Rewrite $x = 2y^2 - 4y + 8$ in the form $x = a(y - k)^2 + h$ by completing the square.

Solution

To complete the square, follow the same procedure used for quadratic functions. (This is outlined in Section 12.3.)

Step 1: Divide the equation by 2 so that the coefficient of y^2 is 1.

$$\frac{x}{2} = y^2 - 2y + 4$$

Step 2: Separate the constant from the variable terms using parentheses.

$$\frac{x}{2} = (y^2 - 2y) + 4$$

Step 3: Complete the square for the quantity in parentheses. Add 1 *inside* the parentheses, and *subtract* 1 from the 4.

$$\frac{x}{2} = (y^2 - 2y + 1) + 4 - 1$$

$$\frac{x}{2} = (y^2 - 2y + 1) + 3$$

Step 4: Factor the expression inside the parentheses.

$$\frac{x}{2} = (y - 1)^2 + 3$$

Step 5: Solve the equation for x by multiplying by 2.

$$2\left(\frac{x}{2}\right) = 2[(y - 1)^2 + 3]$$
$$x = 2(y - 1)^2 + 6$$

YOU TRY 5

Rewrite $x = -y^2 - 6y - 1$ in the form $x = a(y - k)^2 + h$ by completing the square.

6 Find the Vertex of the Graph of $x = ay^2 + by + c$ Using $y = -\dfrac{b}{2a}$, and Graph the Equation

EXAMPLE 6

Graph $x = y^2 - 2y + 5$. Find the vertex using the vertex formula. Find the x- and y-intercepts and the domain and range.

Solution

Since this equation is solved for x and is quadratic in y, it opens in the x-direction. $a = 1$, so it opens to the right. Use the vertex formula to find the y-*coordinate* of the vertex.

$$y = -\frac{b}{2a}$$

$$y = -\frac{-2}{2(1)} = 1 \qquad a = 1, b = -2$$

Substitute $y = 1$ into $x = y^2 - 2y + 5$ to find the x-coordinate of the vertex.

$$x = (1)^2 - 2(1) + 5$$
$$x = 1 - 2 + 5 = 4$$

The vertex is (4, 1). Because the vertex is (4, 1) and the parabola opens to the right, the graph has *no y-intercepts*.

To find the x-intercept, let $y = 0$ and solve for x.

$$x = y^2 - 2y + 5$$
$$x = (0)^2 - 2(0) + 5$$
$$x = 5$$

The x-intercept is (5, 0).

Find another point on the parabola by choosing a value for y that is close to the y-coordinate of the vertex. Let $y = -1$. Find x.

$$x = (-1)^2 - 2(-1) + 5$$
$$x = 1 + 2 + 5 = 8$$

Another point on the parabola is (8, −1). Use the axis of symmetry to locate the additional points (5, 2) and (8, 3). The domain is [4, ∞), and the range is (−∞, ∞)

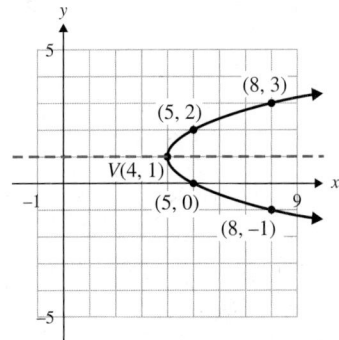

YOU TRY 6

Graph $x = y^2 + 6y + 3$. Find the vertex using the vertex formula. Find the x- and y-intercepts and the domain and range.

To graph a parabola that is a function, just enter the equation and press GRAPH.

Example 1: Graph $f(x) = -x^2 + 2$.

Enter $Y_1 = -x^2 + 2$ to graph the function on a calculator.

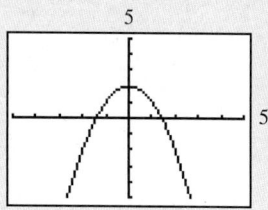

To graph an equation on a calculator, it must be entered so that y is a function of x. Since a parabola that opens horizontally is not a function, we must solve for y in terms of x so that the equation is represented by two different functions.

Example 2: Graph $x = y^2 - 4$ on a calculator.

Solve for y.

$$x = y^2 - 4$$
$$x + 4 = y^2$$
$$\pm\sqrt{x + 4} = y$$

Now the equation $x = y^2 - 4$ is rewritten so that y is in terms of x. In the graphing calculator, enter $y = \sqrt{x + 4}$ as Y_1. This represents the top half of the parabola since the y-values are positive above the x-axis. Enter $y = -\sqrt{x + 4}$ as Y_2. This represents the bottom half of the parabola since the y-values are negative below the x-axis. Set an appropriate window and press GRAPH.

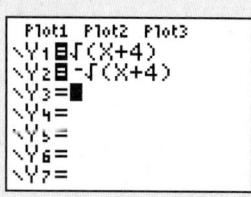 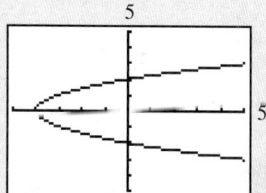

Graph each parabola on a graphing calculator. Where appropriate, rewrite the equation for y in terms of x. These problems come from the homework exercises so that the graphs can be found in the Answers to Exercises appendix.

1) $f(x) = x^2 + 6x + 9$; Exercise 9

2) $x = y^2 + 2$; Exercise 33

3) $x = \dfrac{1}{4}(y + 2)^2$; Exercise 39

4) $f(x) = -\dfrac{1}{2}x^2 + 4x - 6$; Exercise 11

5) $x = -(y - 4)^2 + 5$; Exercise 35

6) $x = y^2 - 4y + 5$; Exercise 41

ANSWERS TO [YOU TRY] EXERCISES

1) a) minimum value b) vertex $(-3, -2)$
   c) The minimum value of the function is -2.
   d)

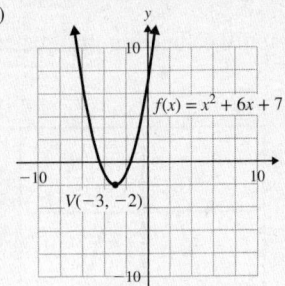

2) a) 1 sec b) 26 ft   3) 49 in^2
4) $V(-3, -1)$; x-int: $(-4, 0)$; y-int: none; domain:
   $(-\infty, -1]$; range: $(-\infty, \infty)$

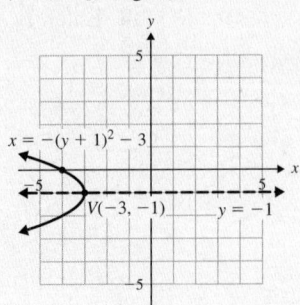

5) $x = -(y + 3)^2 + 8$
6) $V(-6, -3)$; x-int: $(3, 0)$; y-ints: $(0, -3 - \sqrt{6})$, $(0, -3 + \sqrt{6})$; domain: $[-6, \infty)$; range: $(-\infty, \infty)$

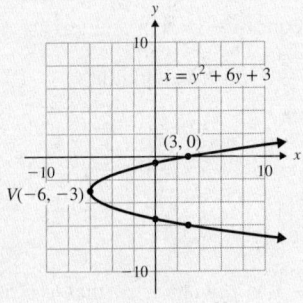

ANSWERS TO TECHNOLOGY EXERCISES

1) The equation can be entered as it is.
2) $Y_1 = \sqrt{x - 2}$; $Y_2 = -\sqrt{x - 2}$
3) $Y_1 = -2 + \sqrt{4x}$; $Y_2 = -2 - \sqrt{4x}$

4) The equation can be entered as it is.
5) $Y_1 = 4 + \sqrt{5 - x}$; $Y_2 = 4 - \sqrt{5 - x}$
6) $Y_1 = 2 + \sqrt{x - 1}$; $Y_2 = 2 - \sqrt{x - 1}$

E Evaluate **12.4** Exercises   Do the exercises, and check your work.

Objective 1: Find the Maximum or Minimum Value of a Quadratic Function

For Exercises 1–6, determine whether the function has a maximum value, minimum value, or neither.

1)

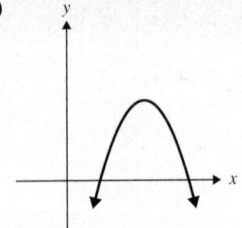

2)

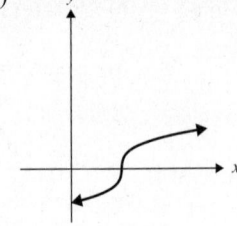

3)

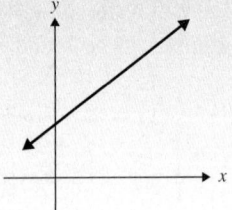

4)

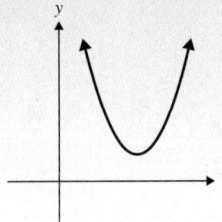

5)

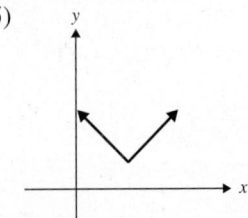

6)

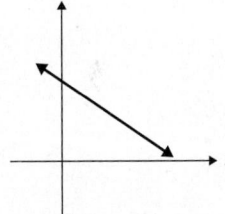

7) Let $f(x) = ax^2 + bx + c$. How do you know whether the function has a maximum or minimum value at the vertex?

8) Is there a maximum value of the function $y = 2x^2 + 12x + 11$? Explain your answer.

For Problems 9–12, answer parts a)–d) for each function, $f(x)$.

a) Does the function attain a minimum or maximum value at its vertex?

b) Find the vertex of the graph of $f(x)$.

c) What is the minimum or maximum value of the function?

d) Graph the function to verify parts a)–c).

9) $f(x) = x^2 + 6x + 9$

10) $f(x) = -x^2 + 2x + 4$

11) $f(x) = -\dfrac{1}{2}x^2 + 4x - 6$

12) $f(x) = 2x^2 + 4x$

Objective 2: Given a Quadratic Function, Solve an Applied Problem Involving a Maximum or Minimum Value

Solve.

 13) An object is fired upward from the ground so that its height h (in feet) t sec after being fired is given by

$$h(t) = -16t^2 + 320t$$

a) How long does it take the object to reach its maximum height?

b) What is the maximum height attained by the object?

c) How long does it take the object to hit the ground?

14) An object is thrown upward from a height of 64 ft so that its height h (in feet) t sec after being thrown is given by

$$h(t) = -16t^2 + 48t + 64$$

a) How long does it take the object to reach its maximum height?

b) What is the maximum height attained by the object?

c) How long does it take the object to hit the ground?

15) The average number of traffic tickets issued in a city on any given day Sunday–Saturday can be approximated by

$$T(x) = -7x^2 + 70x + 43$$

where x represents the number of days after Sunday ($x = 0$ represents Sunday, $x = 1$ represents Monday, etc.), and $T(x)$ represents the number of traffic tickets issued. On which day are the most tickets written? How many tickets are issued on that day?

16) The number of guests staying at the Toasty Inn from January to December 2019 can be approximated by

Emma Lee/Life File/Getty Images

$$N(x) = -10x^2 + 120x + 120$$

where x represents the number of months after January 2019 ($x = 0$ represents January, $x = 1$ represents February, etc.), and $N(x)$ represents the number of guests who stayed at the inn. During which month did the inn have the greatest number of guests? How many people stayed at the inn during that month?

17) The number of triplet and higher order births from 1990 to 2015 can be approximated by $N(t) = -0.266t^2 + 6.974t + 27.056$ where t represents the number of years after 1990, and $N(t)$ represents the number of triplet and higher order births born (in hundreds). According to this model, in what year was the number of triplet and higher order births the greatest? How many triplet and higher order births occurred that year? (Round t and $N(t)$ to the nearest whole number.) (www.cdc.gov)

18) The average audience for an NFL football game each year from 2012 to 2017 can be approximated by $A(x) = -0.59x^2 + 3.5x + 12.71$ where x represents the number of years after 2012, and $A(x)$ represents the average number of viewers per game (in millions). During which year did the greatest average viewership occur, and how many were watching each game? (www.businessinsider.com)

Objective 3: Write a Quadratic Function to Solve an Applied Problem Involving a Maximum or Minimum Value

Solve.

19) Every winter Rich makes a rectangular ice rink in his backyard. He has 100 ft of material to use as the border. What is the maximum area of the ice rink?

20) Find the dimensions of the rectangular garden of greatest area that can be enclosed with 40 ft of fencing.

21) The Soo family wants to fence in a rectangular area to hold their dogs. One side of the pen will be their barn. Find the dimensions of the pen of greatest area that can be enclosed with 48 ft of fencing.

22) A farmer wants to enclose a rectangular area with 120 ft of fencing. One side is a river and will not require a fence. What is the maximum area that can be enclosed?

23) Find two integers whose sum is 18 and whose product is a maximum.

24) Find two integers whose sum is 26 and whose product is a maximum.

25) Find two integers whose difference is 12 and whose product is a minimum.

26) Find two integers whose difference is 30 and whose product is a minimum.

Objective 4: Graph Parabolas of the Form $x = a(y - k)^2 + h$

Given a quadratic equation of the form $x = a(y - k)^2 + h$, answer the following.

27) What is the vertex?

28) What is the equation of the axis of symmetry?

29) If a is negative, which way does the parabola open?

30) If a is positive, which way does the parabola open?

For each equation, identify the vertex, axis of symmetry, and x- and y-intercepts. Then, graph the equation. Determine the domain and range.

31) $x = (y - 1)^2 - 4$ 32) $x = (y + 3)^2 - 1$

33) $x = y^2 + 2$ 34) $x = (y - 4)^2$

(24) 35) $x = -(y - 4)^2 + 5$ 36) $x = -(y + 1)^2 - 7$

37) $x = -2(y - 2)^2 - 9$ 38) $x = -\dfrac{1}{2}(y - 4)^2 + 7$

39) $x = \dfrac{1}{4}(y + 2)^2$ 40) $x = 2y^2 + 3$

Objective 5: Rewrite $x = ay^2 + by + c$ as $x = a(y - k)^2 + h$ by Completing the Square

Rewrite each equation in the form $x = a(y - k)^2 + h$ by completing the square and graph it. Determine the domain and range.

(24) 41) $x = y^2 + 4y - 6$ 42) $x = y^2 - 4y + 5$

43) $x = -y^2 - 2y - 5$ 44) $x = -y^2 + 6y + 6$

45) $x = \dfrac{1}{3}y^2 + \dfrac{8}{3}y - \dfrac{5}{3}$ 46) $x = 2y^2 - 4y + 5$

47) $x = -4y^2 - 8y - 10$ 48) $x = \dfrac{1}{2}y^2 + 4y - 1$

Objective 6: Find the Vertex of the Graph of $x = ay^2 + by + c$ Using $y = -\dfrac{b}{2a}$, and Graph the Equation

Graph each equation using the vertex formula. Find the x- and y-intercepts. Determine the domain and range.

49) $x = y^2 - 4y + 3$ 50) $x = -y^2 + 2y + 2$

51) $x = -y^2 + 4y$ 52) $x = y^2 + 6y - 4$

(24) 53) $x = -2y^2 + 4y - 6$ 54) $x = 3y^2 + 6y - 1$

55) $x = 4y^2 - 16y + 13$ 56) $x = 2y^2 + 4y + 8$

57) $x = \dfrac{1}{4}y^2 - \dfrac{1}{2}y + \dfrac{25}{4}$ 58) $x = -\dfrac{3}{4}y^2 + \dfrac{3}{2}y - \dfrac{11}{4}$

Mixed Exercises

Exercises 59–68 contain parabolas that open either horizontally or vertically. Graph each equation. Determine the domain and range.

59) $h(x) = -x^2 + 6$ 60) $y = x^2 - 6x - 1$

61) $x = y^2$ 62) $f(x) = -3x^2 + 12x - 8$

63) $x = -\dfrac{1}{2}y^2 - 4y - 5$ 64) $x = (y - 4)^2 + 3$

65) $y = x^2 + 2x - 3$ 66) $x = -3(y + 2)^2 + 11$

67) $f(x) = -2(x - 4)^2 + 3$

68) $g(x) = \dfrac{3}{2}x^2 - 12x + 20$

R Rethink

R1) What does it mean to have a maximum or a minimum?

R2) How do you determine whether a parabola opens left or right?

R3) What do you find most difficult about solving applied problems? Which type is the hardest for you?

12.5 Quadratic and Rational Inequalities

P Prepare

O Organize

What are your objectives for Section 12.5?	How can you accomplish each objective?
1 Solve a Quadratic Inequality by Graphing	• Know the definition of a *quadratic inequality*. • Review the procedure for **Graphing Quadratic Functions Using the Vertex Formula.** • Review interval notation, and understand the meanings of $<$, $\leq$, $>$, and $\geq$ on a graph. • Complete the given example on your own. • Complete You Try 1.
2 Solve a Quadratic Inequality Using Test Points	• Learn the procedure for **How to Solve a Quadratic Inequality.** • Review methods for solving quadratic equations. • Write solutions in interval notation and review that procedure, if needed. • Complete the given example on your own. • Complete You Try 2.
3 Solve Quadratic Inequalities with Special Solutions	• Analyze the inequality carefully to determine whether picking test points or graphing is necessary. • Complete the given example on your own. • Complete You Try 3.
4 Solve an Inequality of Higher Degree	• Use the procedure for **How to Solve a Quadratic Inequality.** You may use more intervals and more test points. • Complete the given example on your own. • Complete You Try 4.
5 Solve a Rational Inequality	• Know the definition of a *rational inequality*. • Learn the procedure for **How to Solve a Rational Inequality.** • Write the answer in interval notation. • Complete the given examples on your own. • Complete You Trys 5 and 6.

 **W Work**

Read the explanations, follow the examples, take notes, and complete the You Trys.

In Chapter 3, we learned how to solve *linear* inequalities such as $3x - 5 \leq 16$. In this section, we will discuss how to solve *quadratic* and *rational* inequalities.

Definition

A **quadratic inequality** can be written in the form

$$ax^2 + bx + c \leq 0 \qquad \text{or} \qquad ax^2 + bx + c \geq 0$$

where a, b, and c are real numbers and $a \neq 0$. ($<$ and $>$ may be substituted for $\leq$ and $\geq$.)

1 Solve a Quadratic Inequality by Graphing

To understand how to solve a quadratic inequality, let's look at the graph of a quadratic function.

EXAMPLE 1

a) Graph $y = x^2 - 2x - 3$.

b) Solve $x^2 - 2x - 3 < 0$.

c) Solve $x^2 - 2x - 3 \geq 0$.

Solution

a) The graph of the quadratic function $y = x^2 - 2x - 3$ is a parabola that opens upward. Use the vertex formula to confirm that the vertex is $(1, -4)$.

To find the y-intercept, let $x = 0$ and solve for y.

$$y = 0^2 - 2(0) - 3$$
$$y = -3$$

The y-intercept is $(0, -3)$.
To find the x-intercepts, let $y = 0$ and solve for x.

$$
\begin{aligned}
0 &= x^2 - 2x - 3 & \\
0 &= (x - 3)(x + 1) & \text{Factor.} \\
x - 3 = 0 \quad &\text{or} \quad x + 1 = 0 & \text{Set each factor equal to 0.} \\
x = 3 \quad &\text{or} \qquad\quad x = -1 & \text{Solve.}
\end{aligned}
$$

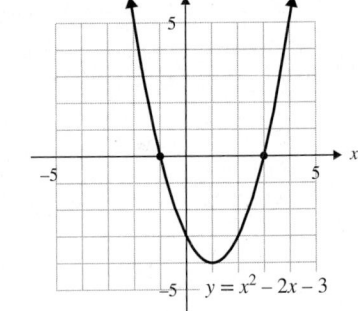

W Hint

Be sure to notice the difference between $<$ and $\leq$ as well as $>$ and $\geq$.

b) We will use the graph of $y = x^2 - 2x - 3$ to solve the inequality $x^2 - 2x - 3 < 0$. That is, to solve $x^2 - 2x - 3 < 0$ we must ask ourselves, "Where are the y-*values* of the function *less than* zero?"

The y-values of the function are less than zero when the x-values are greater than -1 and less than 3, as shown to the right.

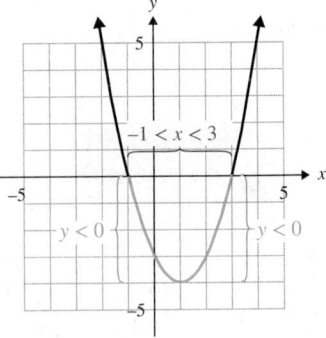

The solution set of $x^2 - 2x - 3 < 0$ (in interval notation) is $(-1, 3)$.

c) To solve $x^2 - 2x - 3 \geq 0$ means to find the x-values for which the y-*values* of the function $y = x^2 - 2x - 3$ are *greater than or equal to* zero. (Recall that the x-intercepts are where the function equals zero.)

The y-values of the function are greater than or equal to zero when $x \leq -1$ or when $x \geq 3$. The solution set of $x^2 - 2x - 3 \geq 0$ is $(-\infty, -1] \cup [3, \infty)$.

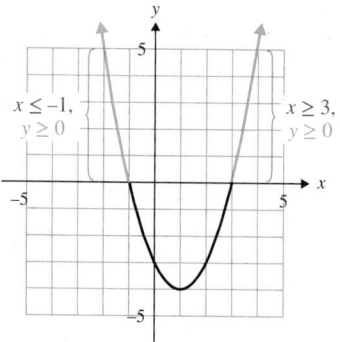

$x \leq -1,$
$y \geq 0$

$x \geq 3,$
$y \geq 0$

When $x \leq -1$ or $x \geq 3$, the y-values are greater than or equal to 0.

[**YOU TRY 1**]

a) Graph $y = x^2 + 6x + 5$. b) Solve $x^2 + 6x + 5 \leq 0$.

c) Solve $x^2 + 6x + 5 > 0$.

2 Solve a Quadratic Inequality Using Test Points

Example 1 illustrates how the x-intercepts of $y = x^2 - 2x - 3$ break up the x-axis into the three separate intervals: $x < -1$, $-1 < x < 3$, and $x > 3$. We can use this idea of intervals to solve a quadratic inequality without graphing.

EXAMPLE 2

Solve $x^2 - 2x - 3 < 0$.

Solution

Begin by solving the equation $x^2 - 2x - 3 = 0$.

 **Hint**

Be sure that you *understand* what is being done in each step.

$$x^2 - 2x - 3 = 0$$
$$(x - 3)(x + 1) = 0 \qquad \text{Factor.}$$
$$x - 3 = 0 \quad \text{or} \quad x + 1 = 0 \qquad \text{Set each factor equal to 0.}$$
$$x = 3 \quad \text{or} \qquad x = -1 \qquad \text{Solve.}$$

(These are the x-intercepts of $y = x^2 - 2x - 3$.)

 **Note**

The $<$ indicates that we want to find the values of x that will make $x^2 - 2x - 3 < 0$; that is, find the values of x that make $x^2 - 2x - 3$ a *negative* number.

Put $x = 3$ and $x = -1$ on a number line with the smaller number on the left. This breaks up the number line into three intervals: $x < -1$, $-1 < x < 3$, and $x > 3$.

Choose a test number in each interval and substitute it into $x^2 - 2x - 3$ to determine whether that value makes $x^2 - 2x - 3$ positive or negative. (If one number in the interval makes $x^2 - 2x - 3$ positive, then *all* numbers in that interval will make $x^2 - 2x - 3$ positive.) Indicate the result on the number line.

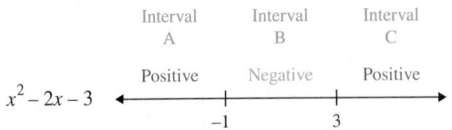

W Hint

Read this example *very* carefully.

Interval A: $(x < -1)$ As a test number, choose any number less than -1. We will choose -2. Evaluate $x^2 - 2x - 3$ for $x = -2$.

$$x^2 - 2x - 3 = (-2)^2 - 2(-2) - 3 \qquad \text{Substitute } -2 \text{ for } x.$$
$$= 4 + 4 - 3$$
$$= 8 - 3 = 5$$

When $x = -2$, $x^2 - 2x - 3$ is *positive*. Therefore, $x^2 - 2x - 3$ will be positive for all values of x in this interval. Indicate this on the number line as seen above.

Interval B: $(-1 < x < 3)$ As a test number, choose any number between -1 and 3. We will choose 0. Evaluate $x^2 - 2x - 3$ for $x = 0$.

$$x^2 - 2x - 3 = (0)^2 - 2(0) - 3 \qquad \text{Substitute } 0 \text{ for } x.$$
$$= 0 - 0 - 3 = -3$$

When $x = 0$, $x^2 - 2x - 3$ is *negative*. Therefore, $x^2 - 2x - 3$ will be negative for all values of x in this interval. Indicate this on the number line above.

Interval C: $(x > 3)$ As a test number, choose any number greater than 3. We will choose 4. Evaluate $x^2 - 2x - 3$ for $x = 4$.

$$x^2 - 2x - 3 = (4)^2 - 2(4) - 3 \qquad \text{Substitute } 4 \text{ for } x.$$
$$= 16 - 8 - 3$$
$$= 8 - 3 = 5$$

When $x = 4$, $x^2 - 2x - 3$ is *positive*. Therefore, $x^2 - 2x - 3$ will be positive for all values of x in this interval. Indicate this on the number line.

Look at the number line above. The solution set of $x^2 - 2x - 3 < 0$ consists of the interval(s) where $x^2 - 2x - 3$ is *negative*. This is in **Interval B,** $(-1, 3)$.

The graph of the solution set is

$$\longleftarrow \!\!+\!\!+\!\!+\!\!+\!\!\diamond\!\!+\!\!+\!\!+\!\!\diamond\!\!+\!\!+\!\!\longrightarrow$$
$$\text{-5 -4 -3 -2 -1 0 1 2 3 4 5}$$

The solution set is $(-1, 3)$. This is the same as the result we obtained in Example 1 by graphing.

[YOU TRY 2] Solve $x^2 + 5x + 4 \le 0$. Graph the solution set, and write the solution in interval notation.

Next we will summarize how to solve a quadratic inequality.

W Hint

In your own words,
summarize this procedure.

Step 1: **Write the inequality in the form $ax^2 + bx + c \leq 0$ or $ax^2 + bx + c \geq 0$.**
($<$ and $>$ may be substituted for $\leq$ and ≥ 0.) If the inequality symbol is $<$ or $\leq$, we are looking for a *negative* quantity in the interval on the number line. If the inequality symbol is $>$ or $\geq$, we are looking for a *positive* quantity in the interval.

Step 2: **Solve** the equation $ax^2 + bx + c = 0$.

Step 3: **Put the solutions of $ax^2 + bx + c = 0$ on a number line.** These values break up the number line into intervals.

Step 4: **Choose a test number in each interval** to determine whether $ax^2 + bx + c$ is positive or negative in each interval. Indicate this on the number line.

Step 5: **If the inequality is in the form $ax^2 + bx + c \leq 0$ or $ax^2 + bx + c < 0$, then the solution set contains the numbers in the interval where $ax^2 + bx + c$ is *negative*.**

If the inequality is in the form $ax^2 + bx + c \geq 0$ or $ax^2 + bx + c > 0$, then the solution set contains the numbers in the interval where $ax^2 + bx + c$ is *positive*.

Step 6: **If the inequality symbol is $\leq$ or $\geq$, then the endpoints of the interval(s) (the numbers found in Step 3) are included in the solution set.** Indicate this with brackets in the interval notation.

If the inequality symbol is $<$ or $>$, then the endpoints of the interval(s) are not included in the solution set. Indicate this with parentheses in interval notation.

3 Solve Quadratic Inequalities with Special Solutions

We should look carefully at the inequality before trying to solve it. Sometimes, it is not necessary to go through all of the steps.

EXAMPLE 3

Solve.

a) $(y + 4)^2 \geq -5$ b) $(t - 8)^2 < -3$

Solution

a) The inequality $(y + 4)^2 \geq -5$ says that a squared quantity, $(y + 4)^2$, is greater than or equal to a *negative* number, -5. *This is always true.* (A squared quantity will *always* be greater than or equal to zero.) Any real number, y, will satisfy the inequality.

The solution set is (∞, ∞).

b) The inequality $(t - 8)^2 < -3$ says that a squared quantity, $(t - 8)^2$, is less than a *negative* number, -3. *There is no real number value for t so that $(t - 8)^2 < -3$.*

The solution set is $\varnothing$.

[YOU TRY 3]

Solve.

a) $(k + 2)^2 \leq -4$ b) $(z - 9)^2 > -1$

4 Solve an Inequality of Higher Degree

Other polynomial inequalities in factored form can be solved in the same way that we solve quadratic inequalities.

EXAMPLE 4

Solve $(c - 2)(c + 5)(c - 4) < 0$.

Solution

This is the factored form of a third-degree polynomial. Since the inequality is $<$, the solution set will contain the intervals where $(c - 2)(c + 5)(c - 4)$ is *negative*.

$$\text{Solve } (c - 2)(c + 5)(c - 4) = 0.$$

$c - 2 = 0$	or	$c + 5 = 0$	or	$c - 4 = 0$	Set each factor equal to 0.
$c = 2$	or	$c = -5$	or	$c = 4$	Solve.

Put $c = 2$, $c = -5$, and $c = 4$ on a number line, and test a number in each interval.

Interval	$c < -5$	$-5 < c < 2$	$2 < c < 4$	$c > 4$
Test number	$c = -6$	$c = 0$	$c = 3$	$c = 5$
Evaluate	$(-6 - 2)(-6 + 5)(-6 - 4)$	$(0 - 2)(0 + 5)(0 - 4)$	$(3 - 2)(3 + 5)(3 - 4)$	$(5 - 2)(5 + 5)(5 - 4)$
$(c - 2)(c + 5)(c - 4)$	$= (-8)(-1)(-10)$	$= (-2)(5)(-4)$	$= (1)(8)(-1)$	$= (3)(10)(1)$
	$= -80$	$= 40$	$= -8$	$= 30$
Sign	Negative	Positive	Negative	Positive

$$(c - 2)(c + 5)(c - 4)$$

We can see that the intervals where $(c - 2)(c + 5)(c - 4)$ is negative are $(-\infty, -5)$ and $(2, 4)$. The endpoints are not included because the inequality is $<$.

The graph of the solution set is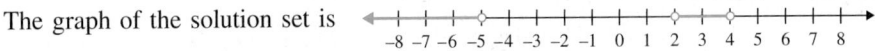

The solution set of $(c - 2)(c + 5)(c - 4) < 0$ is $(-\infty, -5) \cup (2, 4)$.

[**YOU TRY 4**]

Solve $(y + 3)(y - 1)(y + 1) \geq 0$. Graph the solution set, and write the solution in interval notation.

5 Solve a Rational Inequality

An inequality containing a rational expression, $\dfrac{p}{q}$, where p and q are polynomials, is called a **rational inequality.** The way we solve rational inequalities is very similar to the way we solve quadratic inequalities.

Step 1: **Write the inequality so that there is a 0 on one side and only one rational expression on the other side.** If the inequality symbol is $<$ or $\leq$, we are looking for a *negative* quantity in the interval on the number line. If the inequality symbol is $>$ or $\geq$, we are looking for a *positive* quantity in the interval.

Step 2: **Find the numbers that make the numerator equal 0 and any numbers that make the denominator equal 0.**

Step 3: **Put the numbers found in Step 2 on a number line.** These values break up the number line into intervals.

Step 4: **Choose a test number in each interval** to determine whether the rational inequality is positive or negative in each interval. Indicate this on the number line.

Step 5: **If the inequality is in the form $\dfrac{p}{q} \leq 0$ or $\dfrac{p}{q} < 0$, then the solution set contains the numbers in the interval where $\dfrac{p}{q}$ is *negative*.**

If the inequality is in the form $\dfrac{p}{q} \geq 0$ or $\dfrac{p}{q} > 0$, then the solution set contains the numbers in the interval where $\dfrac{p}{q}$ is *positive*.

Step 6: **Determine whether the endpoints of the intervals are included in or excluded from the solution set.** Do not include any values that make the denominator equal 0.

EXAMPLE 5

Solve $\dfrac{5}{x+3} > 0$. Graph the solution set, and write the solution in interval notation.

Solution

Step 1: The inequality is in the correct form—zero on one side and only one rational expression on the other side. Since the inequality symbol is > 0, the solution set will contain the interval(s) where $\dfrac{5}{x+3}$ is *positive*.

Step 2: Find the numbers that make the numerator equal 0 and any numbers that make the denominator equal 0.

Numerator: 5	Denominator: $x + 3$
The numerator is a constant, 5, so it cannot equal 0.	Set $x + 3 = 0$ and solve for x. $x + 3 = 0$ $x = -3$

Step 3: Put -3 on a number line to break it up into intervals.

Step 4: Choose a test number in each interval to determine whether $\dfrac{5}{x+3}$ is positive or negative in each interval.

Interval	$x < -3$	$x > -3$
Test number	$x = -4$	$x = 0$
Evaluate $\dfrac{5}{x+3}$	$\dfrac{5}{-4+3} = \dfrac{5}{-1} = -5$	$\dfrac{5}{0+3} = \dfrac{5}{3}$
Sign	Negative	Positive

Step 5: The solution set of $\dfrac{5}{x+3} > 0$ contains the numbers in the interval where $\dfrac{5}{x+3}$ is *positive*. This interval is $(-3, \infty)$.

$$\dfrac{5}{x+3} \xleftarrow{\hspace{1cm}}\underset{-3}{\overset{\text{Negative} \qquad\qquad \text{Positive}}{\Big|}}\xrightarrow{\hspace{1cm}}$$

Step 6: Since the inequality symbol is $>$, the endpoint of the interval, -3, is not included in the solution set.

The graph of the solution set is

$$\xleftarrow{\hspace{0.3cm}}\underset{\substack{-5\;-4\;-3\;-2\;-1\;\;0\;\;1\;\;2\;\;3\;\;4\;\;5}}{+\;\;+\;\;+\;\;\diamond\!\!-\!\!+\!\!-\!\!+\!\!-\!\!+\!\!-\!\!+\!\!-\!\!+\!\!-\!\!+\!\!-\!\!+\!\!-\!\!+}\xrightarrow{\hspace{0.3cm}}$$

The solution set is $(-3, \infty)$.

[YOU TRY 5] Solve $\dfrac{2}{y-6} < 0$. Graph the solution set, and write the solution in interval notation.

EXAMPLE 6 Solve $\dfrac{7}{a+2} \le 3$. Graph the solution set, and write the solution in interval notation.

Solution

Step 1: Get a zero on one side of the inequality symbol and only one rational expression on the other side.

$$\dfrac{7}{a+2} \le 3$$

$$\dfrac{7}{a+2} - 3 \le 0 \qquad \text{Subtract 3.}$$

$$\dfrac{7}{a+2} - \dfrac{3(a+2)}{a+2} \le 0 \qquad \text{Get a common denominator.}$$

$$\dfrac{7}{a+2} - \dfrac{3a+6}{a+2} \le 0 \qquad \text{Distribute.}$$

$$\dfrac{1-3a}{a+2} \le 0 \qquad \text{Combine numerators, and combine like terms.}$$

From this point forward, we will work with the inequality $\dfrac{1-3a}{a+2} \le 0$. It is equivalent to the original inequality. Because the inequality symbol is $\le$, the solution set contains the interval(s) where $\dfrac{1-3a}{a+2}$ is negative.

Step 2: Find the numbers that make the numerator equal 0 and any numbers that make the denominator equal 0.

Numerator	Denominator
$1 - 3a = 0$	$a + 2 = 0$
$-3a = -1$	$a = -2$
$a = \dfrac{1}{3}$	

Step 3: Put $\dfrac{1}{3}$ and -2 on a number line to break it up into intervals.

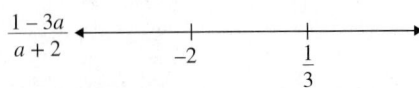

$$\dfrac{1-3a}{a+2} \xleftarrow{\hspace{1cm}}\underset{\substack{-2 \qquad\qquad \frac{1}{3}}}{\Big|\qquad\qquad\Big|}\xrightarrow{\hspace{1cm}}$$

Step 4: Choose a test number in each interval.

Interval	$a < -2$	$-2 < a < \dfrac{1}{3}$	$a > \dfrac{1}{3}$
Test number	$a = -3$	$a = 0$	$a = 1$
Evaluate $\dfrac{1 - 3a}{a + 2}$	$\dfrac{1 - 3(-3)}{-3 + 2} = \dfrac{10}{-1} = -10$	$\dfrac{1 - 3(0)}{0 + 2} = \dfrac{1}{2}$	$\dfrac{1 - 3(1)}{1 + 2} = -\dfrac{2}{3}$
Sign	Negative	Positive	Negative

Step 5: The solution set of $\dfrac{1 - 3a}{a + 2} \le 0$ $\left(\text{and therefore } \dfrac{7}{a + 2} \le 3\right)$ will contain the numbers in the intervals where $\dfrac{1 - 3a}{a + 2}$ is *negative*. These are the first and last intervals.

Step 6: Determine whether the endpoints of the intervals, -2 and $\dfrac{1}{3}$, are included in the solution set. The endpoint $\dfrac{1}{3}$ is included because it does not make the denominator equal 0. *But -2 is not included because it makes the denominator equal 0.*

The graph of the solution set of $\dfrac{7}{a + 2} \le 3$ is

The solution set is $(-\infty, -2) \cup \left[\dfrac{1}{3}, \infty\right)$.

> **BE CAREFUL** Although an inequality symbol may be $\le$ or $\ge$, an endpoint cannot be included in the solution set if it makes the denominator equal 0.

[YOU TRY 6] Solve $\dfrac{3}{z + 4} \ge 2$. Graph the solution set, and write the solution in interval notation.

ANSWERS TO [YOU TRY] EXERCISES

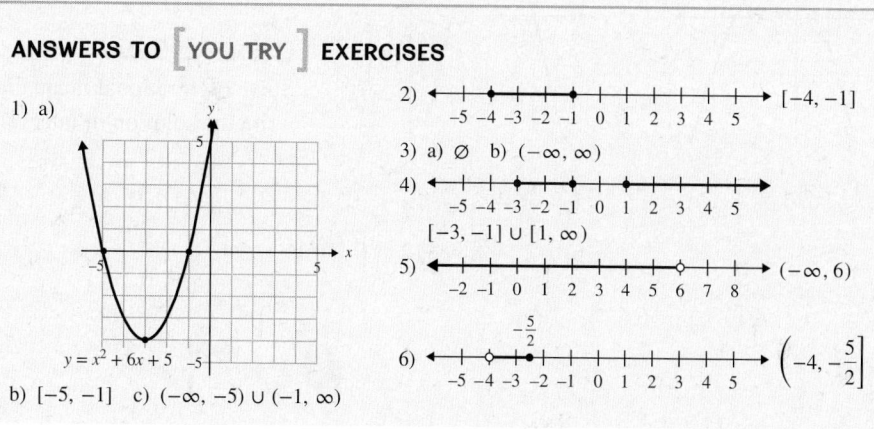

1) a)

$y = x^2 + 6x + 5$

b) $[-5, -1]$ c) $(-\infty, -5) \cup (-1, \infty)$

2) $[-4, -1]$

3) a) $\varnothing$ b) $(-\infty, \infty)$

4) $[-3, -1] \cup [1, \infty)$

5) $(-\infty, 6)$

6) $\left(-4, -\dfrac{5}{2}\right]$

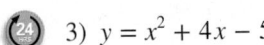

1) When solving a quadratic inequality, how do you know when to include and when to exclude the endpoints in the solution set?

2) If a rational inequality contains a ≤ or ≥ symbol, will the endpoints of the solution set always be included? Explain your answer.

Objective 1: Solve a Quadratic Inequality by Graphing

For Exercises 3–6, use the graph of the function to solve each inequality.

3) $y = x^2 + 4x - 5$

4) $y = x^2 - 6x + 8$

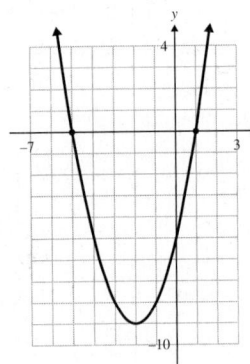

a) $x^2 + 4x - 5 \le 0$

b) $x^2 + 4x - 5 > 0$

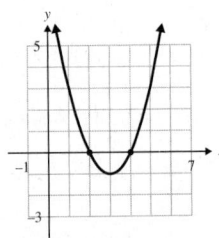

a) $x^2 - 6x + 8 > 0$

b) $x^2 - 6x + 8 \le 0$

5) $y = -\dfrac{1}{2}x^2 + x + \dfrac{3}{2}$

6) $y = -x^2 - 8x - 12$

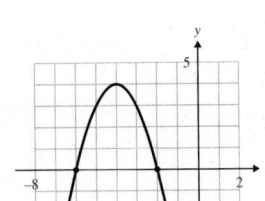

a) $-\dfrac{1}{2}x^2 + x + \dfrac{3}{2} \ge 0$

b) $-\dfrac{1}{2}x^2 + x + \dfrac{3}{2} < 0$

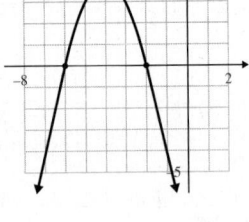

a) $-x^2 - 8x - 12 < 0$

b) $-x^2 - 8x - 12 \ge 0$

Objective 2: Solve a Quadratic Inequality Using Test Points

Solve each quadratic inequality. Graph the solution set, and write the solution in interval notation.

7) $x^2 + 6x - 7 \ge 0$

8) $m^2 - 2m - 24 > 0$

9) $c^2 + 5c < 36$

10) $t^2 + 36 \le 15t$

11) $3z^2 + 14z - 24 \le 0$

12) $5k^2 + 36k + 7 \ge 0$

13) $7p^2 - 4 > 12p$

14) $4w^2 - 19w < 30$

15) $b^2 - 9b > 0$

16) $c^2 + 12c \le 0$

17) $m^2 - 64 < 0$

18) $p^2 - 144 > 0$

19) $121 - h^2 \le 0$

20) $1 - d^2 > 0$

Objective 3: Solve Quadratic Inequalities with Special Solutions

Solve each inequality.

21) $(h + 5)^2 \ge -2$

22) $(3v - 11)^2 > -20$

23) $(2y - 1)^2 < -8$

24) $(r + 4)^2 < -3$

25) $(4d - 3)^2 > -1$

26) $(5s - 2)^2 \le -9$

Objective 4: Solve an Inequality of Higher Degree

Solve each inequality. Graph the solution set, and write the solution in interval notation.

27) $(r + 2)(r - 5)(r - 1) \le 0$

28) $(b + 2)(b - 3)(b - 12) > 0$

29) $(6c + 1)(c + 7)(4c - 3) < 0$

30) $(t + 2)(4t - 7)(5t - 1) \ge 0$

Objective 5: Solve a Rational Inequality

Solve each rational inequality. Graph the solution set, and write the solution in interval notation.

31) $\dfrac{7}{p + 6} > 0$

32) $\dfrac{3}{v - 2} < 0$

33) $\dfrac{5}{z + 3} \le 0$

34) $\dfrac{9}{m - 4} \ge 0$

35) $\dfrac{x - 4}{x - 3} > 0$

36) $\dfrac{a - 2}{a + 1} < 0$

37) $\dfrac{h - 9}{3h + 1} \le 0$

38) $\dfrac{2c + 1}{c + 4} \ge 0$

39) $\dfrac{k}{k + 3} \le 0$

40) $\dfrac{r}{r - 7} \ge 0$

41) $\dfrac{7}{t + 6} < 3$

42) $\dfrac{3}{x + 7} < -2$

43) $\dfrac{3}{a + 7} \ge 1$

44) $\dfrac{5}{w - 3} \le 1$

45) $\dfrac{2y}{y - 6} \le -3$

46) $\dfrac{3z}{z + 4} \ge 2$

47) $\dfrac{3w}{w + 2} > -4$

48) $\dfrac{4h}{h + 3} < 1$

49) $\dfrac{(4t - 3)^2}{t - 5} > 0$

50) $\dfrac{(2y + 3)^2}{y + 3} < 0$

51) $\dfrac{m + 1}{m^2 + 3} \ge 0$

52) $\dfrac{w - 7}{w^2 + 8} \le 0$

53) $\dfrac{s^2 + 2}{s - 4} \le 0$

54) $\dfrac{z^2 + 10}{z + 6} \le 0$

Mixed Exercises: Objectives 2 and 5
Write an inequality, and solve.

55) Compu Corp. estimates that its total profit function, $P(x)$, for producing x thousand units is given by $P(x) = -2x^2 + 32x - 96$.

a) At what level of production does the company make a profit?

b) At what level of production does the company lose money?

56) A model rocket is launched from the ground with an initial velocity of 128 ft/s. The height $s(t)$, in ft, of the rocket t seconds after liftoff is given by the function $s(t) = -16t^2 + 128t$.

a) When is the rocket more than 192 ft above the ground?

b) When does the rocket hit the ground?

57) A designer purse company has found that the average cost, $\overline{C}(x)$, of producing x purses per month can be described by the function $\overline{C}(x) = \dfrac{10x + 100{,}000}{x}$. How many purses must the company produce each month so that the average cost of producing each purse is no more than $20?

58) A company that produces clay pigeons for target shooting has determined that the average cost, $\overline{C}(x)$, of producing x cases of clay pigeons per month can be described by the function $\overline{C}(x) = \dfrac{2x + 15{,}000}{x}$. How many cases of clay pigeons must the company produce each month so that the average cost of producing each case is no more than $3?

R Rethink

R1) What is a rational inequality? How is this different from a rational equation?

R2) Are there any similarities between a quadratic inequality and a rational inequality?

R3) Which type of inequality do you find easiest to solve?

R4) Why do you need to pick test points?

12.6 The Algebra of Functions

What are your objectives for Section 12.6?	How can you accomplish each objective?
1 Add, Subtract, Multiply, and Divide Functions	• Learn the rules for performing operations with functions. • Understand that there are two ways to find a function value for a sum, difference, product, or quotient of functions. • Complete the given examples on your own. • Complete You Trys 1 and 2.
2 Find the Composition of Functions	• Write the definition of a *composition function* in your own words. • Understand how to evaluate the *composition of functions* for a given value. • Understand what it means to find the *decomposition of functions*. • Complete the given examples on your own. • Complete You Trys 3–5.
3 Use Function Composition	• Be able to use the composition of functions to solve an applied problem. • Complete the given example on your own. • Complete You Try 6.

W Work   **Read the explanations, follow the examples, take notes, and complete the You Trys.**

1 Add, Subtract, Multiply, and Divide Functions

We have learned that we can add, subtract, multiply, and divide polynomials. These same operations can be performed with functions.

Definition Operations with Functions

Given the functions $f(x)$ and $g(x)$, the **sum, difference, product,** and **quotient** of f and g are defined by

1) $(f + g)(x) = f(x) + g(x)$

2) $(f - g)(x) = f(x) - g(x)$

3) $(fg)(x) = f(x) \cdot g(x)$

4) $\left(\dfrac{f}{g}\right)(x) = \dfrac{f(x)}{g(x)}$, where $g(x) \neq 0$

The domain of $(f + g)(x)$, $(f - g)(x)$, $(fg)(x)$, and $\left(\dfrac{f}{g}\right)(x)$ is the *intersection* of the domains of $f(x)$ and $g(x)$.

EXAMPLE 1

Let $f(x) = x^2 - 2x + 7$ and $g(x) = 4x - 3$. Find each of the following.

a)　$(f + g)(x)$　　b)　$(f - g)(x)$ and $(f - g)(-1)$　　c)　$(fg)(x)$　　d)　$\left(\dfrac{f}{g}\right)(x)$

Solution

a)　$(f + g)(x) = f(x) + g(x)$

$\qquad = (x^2 - 2x + 7) + (4x - 3)$　　Substitute the functions.

$\qquad = x^2 + 2x + 4$　　Combine like terms.

b)　$(f - g)(x) = f(x) - g(x)$

$\qquad = (x^2 - 2x + 7) - (4x - 3)$　　Substitute the functions.

$\qquad = x^2 - 2x + 7 - 4x + 3$　　Distribute.

$\qquad = x^2 - 6x + 10$　　Combine like terms.

> **W Hint**
>
> Remember to use parentheses when you are substituting values!

Use the result above to find $(f - g)(-1)$.

$$(f - g)(x) = x^2 - 6x + 10$$
$$(f - g)(-1) = (-1)^2 - 6(-1) + 10 \quad \text{Substitute } -1 \text{ for } x.$$
$$= 1 + 6 + 10$$
$$= 17$$

We can also find $(f - g)(-1)$ using the rule this way:

$$(f - g)(-1) = f(-1) - g(-1)$$
$$= [(-1)^2 - 2(-1) + 7] - [4(-1) - 3] \quad \text{Substitute } -1 \text{ for } x \text{ in}$$
$$= (1 + 2 + 7) - (-4 - 3) \qquad\qquad\qquad f(x) \text{ and } g(x).$$
$$= 10 - (-7)$$
$$= 17$$

c)　$(fg)(x) = f(x) \cdot g(x)$

$\qquad = (x^2 - 2x + 7)(4x - 3)$　　Substitute the functions.

$\qquad = 4x^3 - 3x^2 - 8x^2 + 6x + 28x - 21$　　Multiply.

$\qquad = 4x^3 - 11x^2 + 34x - 21$　　Combine like terms.

d)　$\left(\dfrac{f}{g}\right)(x) = \dfrac{f(x)}{g(x)}$, where $g(x) \neq 0$

$\qquad = \dfrac{x^2 - 2x + 7}{4x - 3}$, where $x \neq \dfrac{3}{4}$　　Substitute the functions.

Notice that $x \neq \dfrac{3}{4}$ because this is the value that makes the denominator equal 0.

[YOU TRY 1]

Let $f(x) = 3x^2 - 8$ and $g(x) = 2x + 1$. Find each of the following.

a)　$(f + g)(x)$ and $(f + g)(-2)$　　b)　$(f - g)(x)$　　c)　$(fg)(x)$　　d)　$\left(\dfrac{f}{g}\right)(3)$

We can use functions to solve real-world problems.

EXAMPLE 2

A publisher sells paperback romance novels to a large bookstore chain for $4.00 per book. Therefore, the publisher's revenue, in dollars, is defined by the function

$$R(x) = 4x$$

where x is the number of books sold to the retailer. The publisher's cost, in dollars, to produce x books is

$$C(x) = 2.5x + 1200$$

In business, profit is defined as revenue − cost. In terms of functions, this is written as $P(x) = R(x) - C(x)$, where $P(x)$ is the profit function.

a) Find the profit function, $P(x)$, that describes the publisher's profit from the sale of x books.

b) If the publisher sells 10,000 books to this chain of bookstores, what is the publisher's profit?

Solution

Creatas Images/Getty Images

a) $P(x) = R(x) - C(x)$
    $\quad\quad = 4x - (2.5x + 1200)$ Substitute the functions.
    $\quad\quad = 1.5x - 1200$
    $\quad P(x) = 1.5x - 1200$

b) Find $P(10,000)$.

$$P(10,000) = 1.5(10,000) - 1200$$
$$= 15,000 - 1200$$
$$= 13,800$$

The publisher's profit is $13,800.

[**YOU TRY 2**] A candy company sells its Valentine's Day candy to a grocery store retailer for $6.00 per box. The candy company's revenue, in dollars, is defined by $R(x) = 6x$, where x is the number of boxes sold to the retailer. The company's cost, in dollars, to produce x boxes of candy is $C(x) = 4x + 900$.

a) Find the profit function, $P(x)$, that defines the company's profit from the sale of x boxes of candy.

b) Find the candy company's profit from the sale of 2000 boxes of candy.

2 Find the Composition of Functions

W Hint

Read this explanation carefully so that you understand *how* the composition of functions works.

We have learned how to add, subtract, multiply, and divide functions. Now we will combine functions in a new way, using *function composition*. We use these *composite functions* when we are given certain two-step processes and want to combine them into a single step.

For example, if you work x hours per week earning $16 per hour, your earnings before taxes and other deductions can be described by the function $f(x) = 16x$. Your take-home pay is different, however, because of taxes and other deductions. So, if your take-home pay is 75% of your earnings before taxes, then $g(x) = 0.75x$ can be used to compute your take-home pay when x is your earnings before taxes.

We can describe what is happening with two tables of values.

| $f(x) = 16x$ | | | $g(x) = 0.75x$ | |
Hours Worked x	Earnings Before Deductions $f(x)$		Earnings Before Deductions x	Take-Home Pay $g(x)$
6	96		96	72
10	160		160	120
20	320		320	240
40	640		640	480

$x \rightarrow f(x)$ $x \rightarrow g(x)$

One function, $f(x)$, describes total earnings before deductions in terms of the number of hours worked. Another function, $g(x)$, describes take-home pay in terms of the total earnings before deductions. It would be convenient to have a function that would allow us to compute, directly, the take-home pay in terms of the number of hours worked.

| | $f(x) = 16x$ | $g(x) = 0.75x$ | |
Hours Worked	Earnings Before Deductions	Take-Home Pay
6	96	72
10	160	120
20	320	240
40	640	480

$x \rightarrow f(x) \rightarrow h(x) = g(f(x))$

f g

$$h(x) = (g \circ f)(x) = g(f(x))$$

If we substitute the function $f(x)$ for x in the function $g(x)$, we will get a new function, $h(x)$, where $h(x) = g(f(x))$. The take-home pay in terms of the number of hours worked, $h(x)$, is given by the composition function $g(f(x))$, read as "g of f of x" and is given by

$$h(x) = g(f(x)) = g(16x)$$
$$= 0.75(16x)$$
$$= 12x$$

Therefore, $h(x) = 12x$ allows us to directly compute the take-home pay from the number of hours worked. To find out your take-home pay when you work 20 hr in a week, find $h(20)$.

$$h(x) = 12x$$
$$h(20) = 12(20) = 240$$

Working 20 hr will result in take-home pay of $240. Notice that this is the same as the take-home pay computed in the tables.

Another way to write $g(f(x))$ is $(g \circ f)(x)$, and both can be read as "g of f of x," or "g composed with f," or "the composition of g and f." Likewise, $f(g(x)) = (f \circ g)(x)$, and these can be read as "f of g of x," or "f composed with g," or "the composition of f and g."

Definition

Given the functions $f(x)$ and $g(x)$, the **composition function** $f \circ g$ (read "f of g") is defined as

$$(f \circ g)(x) = f(g(x))$$

where $g(x)$ is in the domain of f.

EXAMPLE 3

Let $f(x) = 2x - 5$ and $g(x) = x + 8$. Find

a) $g(3)$ b) $(f \circ g)(3)$ c) $(f \circ g)(x)$

Solution

a) $g(x) = x + 8$
    $g(3) = 3 + 8 = 11$

b) $(f \circ g)(3) = f(g(3))$ In part a) we found $g(3) = 11$.
              $= f(11)$
              $= 2(11) - 5$ Substitute 11 for x in $f(x) = 2x - 5$.
              $= 17$

c) $(f \circ g)(x) = f(g(x))$
              $= f(x + 8)$ Substitute $x + 8$ for $g(x)$.
              $= 2(x + 8) - 5$ Substitute $x + 8$ for x in $f(x)$.
              $= 2x + 11$

We can also find $(f \circ g)(3)$, the question for part b), by substituting 3 for x in $(f \circ g)(x)$ found in part c).

$$(f \circ g)(x) = 2x + 11$$
$$(f \circ g)(3) = 2(3) + 11 = 17$$

Notice that this is the same as the result we obtained in b).

YOU TRY 3

Let $f(x) = 3x + 4$ and $g(x) = x - 10$. Find

a) $g(-2)$ b) $(f \circ g)(-2)$ c) $(f \circ g)(x)$

BE CAREFUL   The notation $(f \circ g)(x)$ represents the *composition* of functions, $f(g(x))$; the notation $(f \cdot g)(x)$ represents the *product* of functions, $f(x) \cdot g(x)$.

EXAMPLE 4

Let $f(x) = 4x - 1$, $g(x) = x^2$, and $h(x) = x^2 + 5x - 2$. Find

a) $(f \circ g)(x)$ b) $(g \circ f)(x)$ c) $(h \circ f)(x)$

Solution

a) $(f \circ g)(x) = f(g(x))$
              $= f(x^2)$ Substitute x^2 for $g(x)$.
              $= 4(x^2) - 1$ Substitute x^2 for x in $f(x)$.
              $= 4x^2 - 1$

b) $(g \circ f)(x) = g(f(x))$

$\qquad\qquad = g(4x - 1)$ Substitute $4x - 1$ for $f(x)$.

$\qquad\qquad = (4x - 1)^2$ Substitute $4x - 1$ for x in $g(x)$.

$\qquad\qquad = 16x^2 - 8x + 1$ Expand the binomial.

W Hint

In general,
$(f \circ g)(x) \neq (g \circ f)(x)$.

c) $(h \circ f)(x) = h(f(x))$

$\qquad\qquad = h(4x - 1)$ Substitute $4x - 1$ for $f(x)$.

$\qquad\qquad = (4x - 1)^2 + 5(4x - 1) - 2$ Substitute $4x - 1$ for x in $h(x)$.

$\qquad\qquad = 16x^2 - 8x + 1 + 20x - 5 - 2$ Distribute.

$\qquad\qquad = 16x^2 + 12x - 6$ Combine like terms.

[YOU TRY 4]

Let $f(x) = x^2 + 6$, $g(x) = 2x - 3$, and $h(x) = x^2 - 4x + 9$. Find

a) $(g \circ f)(x)$ b) $(f \circ g)(x)$ c) $(h \circ g)(x)$

Sometimes, it is necessary to rewrite a single function in terms of the composition of two other functions. This is called the **decomposition** of functions.

EXAMPLE 5

Let $h(x) = \sqrt{x^2 + 5}$. Find f and g such that $h(x) = (f \circ g)(x)$.

Solution

Think about what is happening in the function $h(x)$. We can "build" $h(x)$ in the following way: first find $x^2 + 5$, then take the square root of that quantity. So if we let $g(x) = x^2 + 5$ and $f(x) = \sqrt{x}$, we will get $h(x) = (f \circ g)(x)$. Let's check by finding the composition function, $(f \circ g)(x)$.

$$g(x) = x^2 + 5 \qquad f(x) = \sqrt{x}$$

$(f \circ g)(x) = f(g(x))$

$\qquad\qquad = f(x^2 + 5)$ Substitute $x^2 + 5$ for $g(x)$.

$\qquad\qquad = \sqrt{x^2 + 5}$ Substitute $x^2 + 5$ for x in $f(x) = \sqrt{x}$.

Our result is $h(x) = \sqrt{x^2 + 5}$.

W Hint

In your notes and in your own words, explain how to decompose a function. Give an example.

In Example 5, forming $h(x)$ using $f(x) = \sqrt{x}$ and $g(x) = x^2 + 5$ is probably the easiest decomposition to "see." However, there is more than one way to decompose a function, $h(x)$, into two functions $f(x)$ and $g(x)$ so that $h(x) = (f \circ g)(x)$.

For example, if $f(x) = \sqrt{x + 5}$ and $g(x) = x^2$, we get

$(f \circ g)(x) = f(g(x))$

$\qquad\qquad = f(x^2)$ Substitute x^2 for $g(x)$.

$\qquad\qquad = \sqrt{x^2 + 5}$ Substitute x^2 for x in $f(x) = \sqrt{x + 5}$.

This is another way to obtain $h(x) = \sqrt{x^2 + 5}$.

[YOU TRY 5]

Let $h(x) = \sqrt{2x^2 + 1}$. Find f and g such that $h(x) = (f \circ g)(x)$.

3 Use Function Composition

EXAMPLE 6

The area, A, of a square expressed in terms of its perimeter, P, is defined by the function

$$A(P) = \frac{1}{16}P^2$$

The perimeter of a square with a side of length x is defined by the function

$$P(x) = 4x$$

a) Find $(A \circ P)(x)$, and explain what it represents.

b) Find $(A \circ P)(3)$, and explain what it represents.

Solution

a)
$$
\begin{aligned}
(A \circ P)(x) &= A(P(x)) \\
&= A(4x) \qquad \text{Substitute } 4x \text{ for } P(x). \\
&= \frac{1}{16}(4x)^2 \qquad \text{Substitute } 4x \text{ for } P \text{ in } A(P) = \frac{1}{16}P^2. \\
&= \frac{1}{16}(16x^2) \\
&= x^2
\end{aligned}
$$

$(A \circ P)(x) = x^2$. This is the formula for the area of a square in terms of the length of a side, x.

b) To find $(A \circ P)(3)$, use the result obtained in a).

$$(A \circ P)(x) = x^2$$
$$(A \circ P)(3) = 3^2 = 9$$

A square that has a side of length 3 units has an area of 9 square units.

[YOU TRY 6]

Let $f(x) = 100x$ represent the number of centimeters in x meters. Let $g(y) = 1000y$ represent the number of meters in y kilometers.

a) Find $(f \circ g)(y)$, and explain what it represents.

b) Find $(f \circ g)(4)$, and explain what it represents.

Using Technology

The composition of two functions can be evaluated analytically, numerically, and graphically using a graphing calculator.

Consider the composition $h(x) = f(g(x))$ given the functions $f(x) = x^2$ and $g(x) = 3x - 2$. The function $h(x)$ is determined analytically by substituting $g(x)$ into the function f.

We can evaluate $h(2) = f(g(2))$ by substituting 2 in for x.

$$h(x) = f(g(x))$$
$$= f(3x - 2)$$
$$= (3x - 2)^2$$
$$= 9x^2 - 12x + 4$$

$$9(2)^2 - 12(2) + 4 = 16$$

Using a graphing calculator, enter $f(x) = x^2$ into Y_1, enter $g(x) = 3x - 2$ into Y_2, and enter $Y_1(Y_2)$ into Y_3 to represent the composition h, as shown. Recall that Y_1 and Y_2 are found by pressing **VARS**, pressing the right arrow key, and pressing ENTER.

To evaluate $h(2) = f(g(2))$ using the calculator, press 2nd **MODE** to return to the home screen and $Y_3(2)$ as shown.

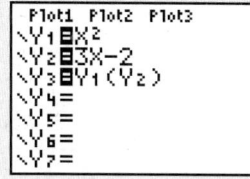   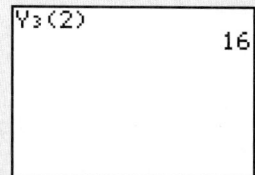

Start with the function Y_1, Y_2 as shown above. $h(2) = f(g(2))$ can be evaluated numerically by setting up a table showing x, Y_1, and Y_2 near $x = 2$. Press 2nd WINDOW and enter 0 after TblStart =. Then press 2nd ENTER to display the table. First evaluate $g(2) = Y_2(2)$ by moving the cursor down to $x = 2$ and then across to the column under Y_2 as shown.

Then substitute the result $g(2) = Y_2(2) = 4$ into the function f to evaluate $f(4) = Y_1(4) = 16$ by moving the cursor down to $x = 4$ and then across to the column under Y_1 as shown. The result is 16 as desired.

To illustrate this composition graphically, change the window by increasing Ymax to 20. $h(2) = f(g(2))$ can be evaluated graphically using the following approach.

First evaluate $g(2) = Y_2(2)$ by pressing TRACE, pressing the down arrow key to switch to Y_2, and pressing 2 ENTER resulting in the point $(2, 4)$ on the graph of g, as shown.

Next evaluate $f(4) = Y_1(4)$ by pressing the down arrow key to switch to Y_1, and pressing 4 ENTER resulting in the point $(4, 16)$ on the graph of f as shown on the graph. The result is 16 as desired.

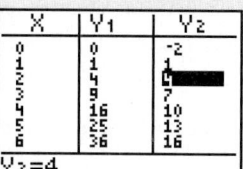

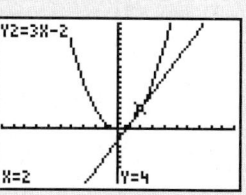

Given the functions $f(x) = x^2 - 5x$ and $g(x) = 2x + 3$, evaluate the following function values using a graphing calculator.

1) $(f \circ g)(3)$ 2) $(f \circ g)(-2)$ 3) $(g \circ f)(2)$

4) $(g \circ f)(-2)$ 5) $(f \circ f)(1)$ 6) $(g \circ g)(2)$

$\boxed{\text{E } \textbf{Evaluate}}$ **12.6** Exercises   Do the exercises, and check your work.

Objective 1: Add, Subtract, Multiply, and Divide Functions

For each pair of functions, find a) $(f + g)(x)$,
b) $(f + g)(5)$, c) $(f - g)(x)$, and d) $(f - g)(2)$.

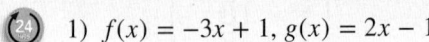 1) $f(x) = -3x + 1, g(x) = 2x - 11$

2) $f(x) = 5x - 9, g(x) = x + 4$

3) $f(x) = 4x^2 - 7x - 1, g(x) = x^2 + 3x - 6$

4) $f(x) = -2x^2 + x + 8, g(x) = 3x^2 - 4x - 6$

For each pair of functions, find a) $(fg)(x)$ and
b) $(fg)(-3)$.

5) $f(x) = x, g(x) = -x + 5$

6) $f(x) = -2x, g(x) = 3x + 1$

7) $f(x) = 2x + 3, g(x) = 3x + 1$

8) $f(x) = 4x + 7, g(x) = x - 5$

For each pair of functions, find a) $\left(\dfrac{f}{g}\right)(x)$ and b) $\left(\dfrac{f}{g}\right)(-2)$.

Identify any values that are not in the domain of $\left(\dfrac{f}{g}\right)(x)$.

9) $f(x) = 6x + 9, g(x) = x + 4$

10) $f(x) = 3x - 8, g(x) = x - 1$

11) $f(x) = x^2 - 5x - 24, g(x) = x - 8$

12) $f(x) = x^2 - 15x + 54, g(x) = x - 9$

13) $f(x) = 3x^2 + 14x + 8, g(x) = 3x + 2$

14) $f(x) = 2x^2 + x - 15, g(x) = 2x - 5$

15) Find two polynomial functions $f(x)$ and $g(x)$ so that
$(f + g)(x) = 5x^2 + 8x - 2$.

16) Let $f(x) = 6x^3 - 9x^2 - 4x + 10$. Find $g(x)$ so that
$(f - g)(x) = x^3 + 3x^2 + 8$.

17) Let $f(x) = 4x - 5$. Find $g(x)$ so that
$(fg)(x) = 8x^2 - 22x + 15$.

18) Let $f(x) = 12x^3 - 18x^2 + 2x$. Find $g(x)$ so that
$\left(\dfrac{f}{g}\right)(x) = 6x^2 - 9x + 1$.

19) A manufacturer's revenue, $R(x)$ in dollars, from the
sale of x calculators is given by $R(x) = 12x$. The
company's cost, $C(x)$ in dollars, to produce x
calculators is $C(x) = 8x + 2000$.

   a) Find the profit function, $P(x)$, that defines the
   manufacturer's profit from the sale of x
   calculators.

   b) What is the profit from the sale of 1500
   calculators?

20) $R(x) = 80x$ is the revenue function for the sale of
x bicycles, in dollars. The cost to manufacture
x bikes, in dollars, is $C(x) = 60x + 7000$.

   a) Find the profit function, $P(x)$, that describes the
   manufacturer's profit from the sale of x bicycles.

   b) What is the profit from the sale of 500 bicycles?

21) $R(x) = 18x$ is the revenue function for the sale of x toasters, in dollars. The cost to manufacture x toasters, in dollars, is $C(x) = 15x + 2400$.

   a) Find the profit function, $P(x)$, that describes the profit from the sale of x toasters.

   b) What is the profit from the sale of 800 toasters?

22) A company's revenue, $R(x)$ in dollars, from the sale of x dog houses is given by $R(x) = 60x$. The company's cost, $C(x)$ in dollars, to produce x dog houses is $C(x) = 45x + 6000$.

   a) Find the profit function, $P(x)$, that describes the company's profit from the sale of x dog houses.

   b) What is the profit from the sale of 300 dog houses?

For Exercises 23 and 24, let x be the number of items sold (in hundreds), and let $R(x)$ and $C(x)$ be in thousands of dollars.

23) A manufacturer's revenue, $R(x)$, from the sale of flat-screen TVs is given by $R(x) = -0.2x^2 + 23x$, while the cost, $C(x)$, is given by $C(x) = 4x + 9$.

   a) Find the profit function, $P(x)$, that describes the company's profit from the sale of x hundred flat-screen TVs.

   b) What is the profit from the sale of 2000 TVs?

24) A manufacturer's revenue, $R(x)$, from the sale of laptop computers is given by $R(x) = -0.4x^2 + 30x$, while the cost, $C(x)$, is given by $C(x) = 3x + 11$.

   a) Find the profit function, $P(x)$, that describes the company's profit from the sale of x hundred laptop computers.

   b) What is the profit from the sale of 1500 computers?

Objective 2: Find the Composition of Functions

25) Given two functions $f(x)$ and $g(x)$, explain how to find $(f \circ g)(x)$.

26) Given two functions $f(x)$ and $g(x)$, explain the difference between $(f \circ g)(x)$ and $(f \cdot g)(x)$.

For Exercises 27–30, find

   a) $g(4)$ b) $(f \circ g)(4)$ using the result in part a)

   c) $(f \circ g)(x)$ d) $(f \circ g)(4)$ using the result in part c)

27) $f(x) = 3x + 1$, $g(x) = 2x - 9$

28) $f(x) = -x + 5$, $g(x) = x + 7$

29) $f(x) = x^2 - 5$, $g(x) = x + 3$

30) $f(x) = x^2 + 2$, $g(x) = x - 1$

31) Let $f(x) = 5x - 4$ and $g(x) = x + 7$. Find

   a) $(f \circ g)(x)$ b) $(g \circ f)(x)$

   c) $(f \circ g)(3)$

32) Let $f(x) = x - 10$ and $g(x) = 4x + 3$. Find

   a) $(f \circ g)(x)$ b) $(g \circ f)(x)$

   c) $(f \circ g)(-6)$

33) Let $h(x) = -2x + 9$ and $k(x) = 3x - 1$. Find

   a) $(k \circ h)(x)$ b) $(h \circ k)(x)$

   c) $(k \circ h)(-1)$

34) Let $r(x) = 6x + 2$ and $v(x) = -7x - 5$. Find

   a) $(v \circ r)(x)$ b) $(r \circ v)(x)$

   c) $(r \circ v)(2)$

35) Let $g(x) = x^2 - 6x + 11$ and $h(x) = x - 4$. Find

   a) $(h \circ g)(x)$ b) $(g \circ h)(x)$

   c) $(g \circ h)(4)$

36) Let $f(x) = x^2 + 7x - 9$ and $g(x) = x + 2$. Find

   a) $(g \circ f)(x)$ b) $(f \circ g)(x)$

   c) $(g \circ f)(3)$

37) Let $m(x) = x + 8$ and $n(x) = -x^2 + 3x - 8$. Find

   a) $(n \circ m)(x)$ b) $(m \circ n)(x)$

   c) $(m \circ n)(0)$

38) Let $f(x) = -x^2 + 10x + 4$ and $g(x) = x + 1$. Find

   a) $(g \circ f)(x)$ b) $(f \circ g)(x)$

   c) $(f \circ g)(-2)$

39) Let $f(x) = \sqrt{x + 10}$, $g(x) = x^2 - 6$. Find

   a) $(f \circ g)(x)$ b) $(g \circ f)(x)$

   c) $(f \circ g)(-3)$

40) Let $h(x) = x^2 + 7$, $k(x) = \sqrt{x - 1}$. Find

   a) $(h \circ k)(x)$ b) $(k \circ h)(x)$

   c) $(k \circ h)(0)$

41) Let $P(t) = \dfrac{1}{t + 8}$, $Q(t) = t^2$. Find

   a) $(P \circ Q)(t)$ b) $(Q \circ P)(t)$

   c) $(Q \circ P)(-5)$

42) Let $F(a) = \dfrac{1}{5a}$, $G(a) = a^2$. Find

    a) $(G \circ F)(a)$ b) $(F \circ G)(a)$ c) $(G \circ F)(-2)$

For Exercises 43–48, find $f(x)$ and $g(x)$ such that $h(x) = (f \circ g)(x)$.

43) $h(x) = \sqrt{x^2 + 13}$

44) $h(x) = \sqrt{2x^2 + 7}$

45) $h(x) = (8x - 3)^2$

46) $h(x) = (4x + 9)^2$

47) $h(x) = \dfrac{1}{6x + 5}$

48) $h(x) = \dfrac{2}{x - 10}$

Objective 3: Use Function Composition

49) Oil spilled from a ship off the coast of Alaska with the oil spreading out in a circle across the surface of the water. The radius of the oil spill is given by $r(t) = 4t$ where t is the number of minutes after the leak began and $r(t)$ is in feet. The area of the spill is given by $A(r) = \pi r^2$ where r represents the radius of the oil slick. Find each of the following, and explain their meanings.

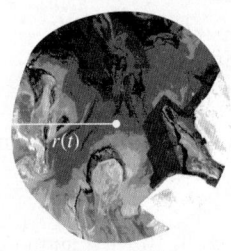

    a) $r(5)$ b) $A(20)$ c) $A(r(t))$ d) $A(r(5))$

50) The sales tax on goods in a major metropolitan area is 7% so that the final cost of an item, $f(x)$, is given by $f(x) = 1.07x$, where x is the cost of the item. A women's clothing store is having a sale so that all of its merchandise is 20% off. If the regular price of an item is x dollars then the sale price, $s(x)$, is given by $s(x) = 0.80x$. Find each of the following and explain their meanings.

    a) $s(40)$

    b) $f(32)$

    c) $(f \circ s)(x)$

    d) $(f \circ s)(40)$

51) The radius of a circle is half its diameter. We can express this with the function $r(d) = \dfrac{1}{2}d$, where d is the diameter of a circle and r is the radius. The area of a circle in terms of its radius is $A(r) = \pi r^2$. Find each of the following and explain their meanings.

    a) $r(6)$ b) $A(3)$ c) $A(r(d))$ d) $A(r(6))$

52) The function $C(F) = \dfrac{5}{9}(F - 32)$ can be used to convert a temperature from degrees Fahrenheit, F, to degrees Celsius, C. The relationship between the Celsius scale, C, and the Kelvin scale, K, is given by $K(C) = C + 273$. Find each of the following and explain their meanings.

    a) $C(59)$ b) $K(15)$ c) $K(C(F))$ d) $K(C(59))$

R Rethink

R1) Think of a real-life situation where you could use an operation with functions.

R2) Can using composite functions make some calculations faster and easier? How?

R3) Were there certain exercises that were more difficult than others? What made them more difficult? What might you be able to do to better understand them?

Group Activity – Functions

1) Given the functions $p(x) = x^2$ and $q(x) = \sqrt{19 - 3x}$,

   a) Evaluate the following:

       i) $p(-2)$ ii) $p(2)$

       iii) $p(4)$ iv) $p(6)$

       v) $q(-10)$ vi) $q(-2)$

       vii) $q(5)$

   b) Using the results of part a), evaluate the following expressions. Remember the order of operations!

       i) $2 \cdot p(4) + q(5)$

       ii) $2[p(4) + q(5)]$

       iii) $4 - 5[q(-2) + q(5)]^2$

       iv) $\sqrt{\dfrac{q(5)}{5 - p(-2)}}$

       v) $[q(-2) - q(-10)]^2 + \dfrac{p(6) + 6}{p(2) + 6}$

2) Given the functions $f(x) = x^3 - 5x^2 - 9x + 45$ and $g(x) = x^2 - 8x + 15$, simplify the following expressions.

   a) $\dfrac{f(x)}{g(x)}$

   b) $\dfrac{1}{f(x)} + \dfrac{1}{g(x)}$

   c) $f(x) - g(x)$

3) Given $f(x) = 2x + 5$, $g(x) = 3x - 2$, and $k(x) = \sqrt{2x + 10}$, solve each equation for x.

   a) $k(x) = f(x)$

   b) $f(g(x)) + 5 \cdot g(x) = k(45)$

 me Determine Your Saving Style

There are as many ways to save money as there are people looking to save it. To help you save money and keep your budget in order, identify your saving style. Read each of the following statements, and rate how well it describes you, using this scale:

1 = That's me

2 = Sometimes

3 = That's not me

	1	2	3
1. I count the change I'm given by cashiers in stores and restaurants.			
2. I always pick up all the change I receive from a transaction in a store, even if it's only a few cents.			
3. I don't buy something right away if I'm pretty sure it will go on sale soon.			
4. I feel a real sense of accomplishment if I buy something on sale.			
5. I always remember how much I paid for something.			
6. If something goes on sale soon after I've bought it, I feel cheated.			
7. I have money in at least one interest-bearing account.			
8. I rarely lend people money.			
9. If I lend money to someone repeatedly without getting it back, I stop lending it to that person.			
10. I share resources (e.g., cell phone plan, carpool, an apartment) with other people to save money.			
11. I'm good at denying myself small purchases when I know I am low on cash.			
12. I believe most generic or off-brand items are just as good as name brands.			

Add up your ratings. Interpret your total score according to this informal guide:

12–15: Very aggressive saving style

16–20: Careful saving style

21–27: Fairly loose saving style

28–32: Loose saving style

33–36: Nonexistent saving style

What are the advantages and disadvantages of your saving style? How do you think your saving style affects your ability to keep a healthy budget? If you are dissatisfied with your saving style, how might you be able to change it?

Chapter 12:   Summary

Definition/Procedure	Example

12.1 Relations and Functions

A **relation** is any set of ordered pairs. A relation can also be represented as a correspondence or mapping from one set to another or as an equation.

a) $\{(-6, 2), (-3, 1), (0, 0), (9, -3)\}$
b) $y^2 = x$

The **domain** of a relation is the set of values of the independent variable (the first coordinates in the set of ordered pairs). The **range** of a relation is the set of all values of the dependent variable (the second coordinates in the set of ordered pairs).

    If a relation is written as an equation so that y is in terms of x, then the **domain** is the set of all real numbers that can be substituted for the independent variable, x. The resulting set of real numbers that are obtained for y, the dependent variable, is the **range.**

In a) above, the domain is $\{-6, -3, 0, 9\}$, and the range is $\{-3, 0, 1, 2\}$.
In b) above, the domain is $[0, \infty)$, and the range is $(-\infty, \infty)$.

A **function** is a relation in which each element of the domain corresponds to *exactly one* element of the range.

The relation in a) *is* a function.
The relation in b) *is not* a function since there are elements of the domain that correspond to more than one element in the range.
    For example, if $x = 4$, then $y = 2$ or $y = -2$.

$y = f(x)$ is called **function notation,** and it is read as, "y equals f of x."

Let $f(x) = 7x + 2$. Find $f(-2)$.

$$\begin{aligned} f(-2) &= 7(-2) + 2 && \text{Substitute } -2 \text{ for } x. \\ &= -14 + 2 && \text{Multiply.} \\ &= -12 && \text{Add.} \end{aligned}$$

Linear Functions

A **linear function** has the form $f(x) = mx + b$, where m and b are real numbers, m is the *slope,* and $(0, b)$ is the *y-intercept.* The **domain** of a linear function is $(-\infty, \infty)$.

Graph $f(x) = -3x + 1$.

The slope is -3, and the y-intercept is $(0, 1)$.

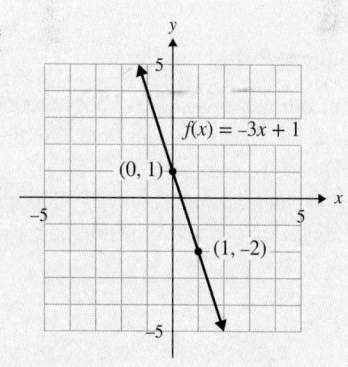

Polynomial Functions

A **polynomial function of degree n** has the form $f(x) = a_n x^n + a_{n-1} x^{n-1} + \cdots + a_1 x + a_0$, where $a_n, a_{n-1}, ..., a_1,$ and a_0 are real numbers, $a_n \neq 0$, and n is a whole number. The **domain of a polynomial function** is $(-\infty, \infty)$.

$f(x) = 4x^3 - 10x^2 - x + 7$ is an example of a polynomial function of degree 3 because $4x^3 - 10x^2 - x + 7$ is a polynomial of degree 3.

Definition/Procedure	Example

Rational Functions

A **rational function** has the form $f(x) = \dfrac{P(x)}{Q(x)}$, where P and Q are polynomials and $Q(x) \neq 0$. The **domain of a rational function** consists of all real numbers *except* the value(s) of the variable that make the denominator equal zero.

Find the domain of the rational function $f(x) = \dfrac{x-2}{6x+1}$.

Ask yourself, *"Is there any number that cannot be substituted for x?"* A rational expression is undefined when its denominator equals zero, so the value or values that make the denominator equal zero are *not* in the domain of the function.

$$6x + 1 = 0 \quad \text{Set the denominator} = 0.$$
$$6x = -1 \quad \text{Subtract 1 from each side.}$$
$$x = -\frac{1}{6} \quad \text{Divide by 6.}$$

When $x = -\dfrac{1}{6}$, the denominator equals zero. The domain of $f(x)$ contains all real numbers *except* $-\dfrac{1}{6}$. Write the domain in interval notation as $\left(-\infty, -\dfrac{1}{6}\right) \cup \left(-\dfrac{1}{6}, \infty\right)$.

Quadratic Functions

A **quadratic function** can be written in the form $f(x) = ax^2 + bx + c$, where a, b, and c are real numbers and $a \neq 0$. (It is a special type of polynomial function.) The **domain of a quadratic function** is $(-\infty, \infty)$.

Given the quadratic function $f(x) = x^2 + 5x - 8$, find
a) $f(-2)$ and b) $f(k+3)$.

a) Substitute -2 for x in $f(x)$ and simplify.
$$f(-2) = (-2)^2 + 5(-2) - 8 = 4 - 10 - 8 = -14$$

b) Substitute $k + 3$ for x in $f(x)$ and simplify. *We must put $k + 3$ in parentheses because the expression contains more than one term.*

$$f(x) = x^2 + 5x - 8$$
$$f(k+3) = (k+3)^2 + 5(k+3) - 8 \quad \text{Substitute } k + 3 \text{ for } x.$$
$$f(k+3) = k^2 + 6k + 9 + 5k + 15 - 8 \quad \text{Multiply.}$$
$$f(k+3) = k^2 + 11k + 16 \quad \text{Combine like terms.}$$

Square Root Functions

The **domain of a square root function** consists of all of the real numbers that can be substituted for the variable so that the radicand is nonnegative.

To find the domain, set up an inequality so that the radicand is ≥ 0 and solve for the variable.

Determine the domain of the square root function.

$$f(x) = \sqrt{6x - 7}$$
$$6x - 7 \geq 0 \quad \text{The value of the radicand must be} \geq 0.$$
$$6x \geq 7$$
$$x \geq \frac{7}{6} \quad \text{Solve.}$$

The domain of $f(x) = \sqrt{6x-7}$ is $\left[\dfrac{7}{6}, \infty\right)$.

Definition/Procedure	Example
Cube Root Functions The **domain of a cube root function** is the set of all real numbers. We can write this in interval notation as $(-\infty, \infty)$.	The domain of $g(x) = \sqrt[3]{x}$ is $(-\infty, \infty)$.
To **graph a square root function,** find the domain, make a table of values, and graph.	Graph $g(x) = \sqrt{x}$. 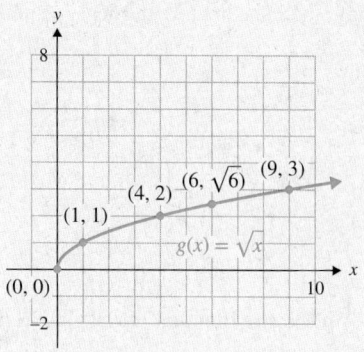

x	$g(x)$
0	0
1	1
4	2
6	$\sqrt{6}$
9	3

12.2 Graphs of Functions and Transformations

Vertical Shifts Given the graph of $f(x)$, if $g(x) = f(x) + k$, where k is a constant, then the graph of $g(x)$ is the same shape as the graph of $f(x)$, but the graph of g is shifted *vertically* k units.	The graph of $g(x) = |x| + 2$ is the same shape as the graph of $f(x) = |x|$, but $g(x)$ is shifted *up* 2 units. 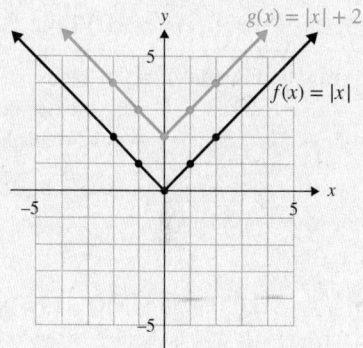 Functions f and g are **absolute value functions.**
Horizontal Shifts Given the graph of $f(x)$, if $g(x) = f(x - h)$, where h is a constant, the graph of $g(x)$ is the same shape as the graph of $f(x)$, but the graph of g is shifted *horizontally* h units.	The graph of $g(x) = (x + 3)^2$ is the same shape as the graph of $f(x) = x^2$, but g is shifted *left* 3 units. 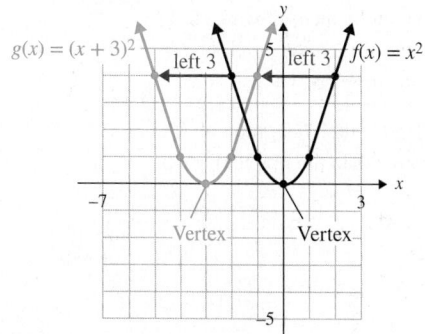 Functions f and g are **quadratic functions,** and their graphs are called **parabolas.**

Definition/Procedure	Example

Reflection About the x-Axis

Given the graph of $f(x)$, if $g(x) = -f(x)$, then the graph of $g(x)$ is the *reflection of the graph of f about the x-axis*. That is, obtain the graph of g by keeping the x-coordinate of each point on the graph of f the same but take the negative of the y-coordinate.

Let $f(x) = \sqrt{x}$ and $g(x) = -\sqrt{x}$. The graph of $g(x)$ is the reflection of $f(x)$ about the x-axis.

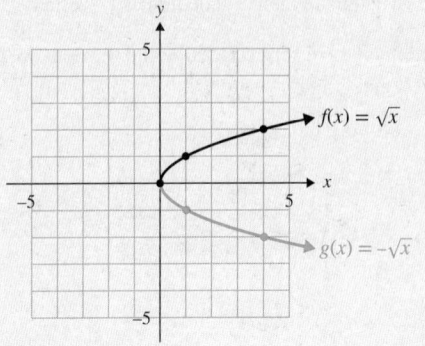

A **piecewise function** is a single function defined by two or more different rules.

$$f(x) = \begin{cases} x + 3, & x > -2 \\ -\dfrac{1}{2}x + 2, & x \le -2 \end{cases}$$

The **greatest integer function,** $f(x) = [\![x]\!]$, represents the largest integer less than or equal to x.

$[\![8.3]\!] = 8, \quad \left[\!\left[-4\dfrac{3}{8}\right]\!\right] = -5$

12.3 Quadratic Functions and Their Graphs

A **quadratic function** is a function that can be written in the form $f(x) = ax^2 + bx + c$, where a, b, and c are real numbers and $a \ne 0$. The graph of a quadratic function is called a **parabola.**

The lowest point on an upward-opening parabola or the highest point on a downward-opening parabola is called the **vertex.**

$f(x) = 5x^2 + 7x - 9$ is a quadratic function.

A quadratic function can also be written in the form $f(x) = a(x - h)^2 + k$:

1) The vertex of the parabola is (h, k).
2) The axis of symmetry is the vertical line with equation $x = h$.
3) If a is positive, the parabola opens upward. If a is negative, the parabola opens downward.
4) If $|a| < 1$, then the graph of $f(x) = a(x - h)^2 + k$ is *wider* than the graph of $y = x^2$.

   If $|a| > 1$, then the graph of $f(x) = a(x - h)^2 + k$ is *narrower* than the graph of $y = x^2$.

Graph $f(x) = -(x + 3)^2 + 4$.

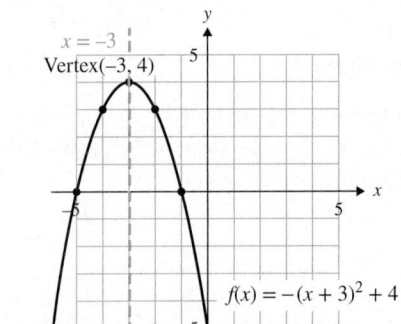

Vertex: $(-3, 4)$

Axis of symmetry: $x = -3$

$a = -1$, so the graph opens downward. The domain is $(-\infty, \infty)$; the range is $(-\infty, 4]$.

Definition/Procedure	Example

When a quadratic function is written in the form $f(x) = ax^2 + bx + c$, there are two methods we can use to graph the function.

Method 1: Rewrite $f(x) = ax^2 + bx + c$ in the form $f(x) = a(x - h)^2 + k$ by *completing the square.*

Method 2: Use the formula $h = -\dfrac{b}{2a}$ to find the x-coordinate of the vertex. The vertex has coordinates
$$\left(-\frac{b}{2a}, f\left(-\frac{b}{2a}\right)\right).$$

Graph $f(x) = x^2 + 4x + 5$.

Method 1: Complete the square.

$$f(x) = x^2 + 4x + 5$$
$$f(x) = (x^2 + 4x + 2^2) + 5 - 2^2$$
$$f(x) = (x^2 + 4x + 4) + 5 - 4$$
$$f(x) = (x + 2)^2 + 1$$

The vertex of the parabola is $(-2, 1)$. The axis of symmetry is $x = -2$. The parabola opens upward and has the same shape as $f(x) = x^2$. The graph is shown.

Method 2: Use the formula $h = -\dfrac{b}{2a}$.

$h = -\dfrac{4}{2(1)} = -2$. Then, $f(-2) = 1$.

The vertex of the parabola is $(-2, 1)$. The axis of symmetry is $x = -2$. The domain is $(-\infty, \infty)$; the range is $[1, \infty)$.

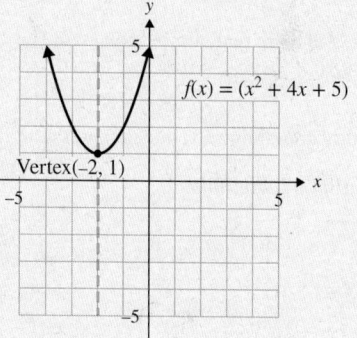

12.4 Applications of Quadratic Functions and Graphing Other Parabolas

Maximum and Minimum Values of a Quadratic Function	

Maximum and Minimum Values of a Quadratic Function

Let $f(x) = ax^2 + bx + c$.

1) If a is *positive,* the y-coordinate of the vertex is the **minimum** value of the function $f(x)$.

2) If a is *negative,* the y-coordinate of the vertex is the **maximum** value of the function $f(x)$.

Find the minimum value of the function
$$f(x) = 2x^2 + 12x + 7.$$

Because a is positive ($a = 2$), the function's *minimum value* is at the vertex.

The x-coordinate of the vertex is $h = -\dfrac{b}{2a} = -\dfrac{12}{2(2)} = -3$.

The y-coordinate of the vertex is

$$f(-3) = 2(-3)^2 + 12(-3) + 7$$
$$= 18 - 36 + 7 = -11$$

The minimum value of the function is -11.

Definition/Procedure	Example

The graph of the quadratic equation $x = ay^2 + by + c$ is a parabola that opens in the x-direction, or horizontally.

The quadratic equation $x = ay^2 + by + c$ can also be written in the form $x = a(y - k)^2 + h$. When it is written in this form we can find the following.

1) The vertex of the parabola is (h, k).

2) The axis of symmetry is the horizontal line $y = k$.

3) If a is positive, the graph opens to the right. If a is negative, the graph opens to the left.

Graph $x = \dfrac{1}{2}(y + 4)^2 - 2$.

Vertex: $(-2, -4)$

Axis of symmetry: $y = -4$

$a = \dfrac{1}{2}$, so the graph opens to the right. The domain is $[-2, \infty)$; the range is $(-\infty, \infty)$.

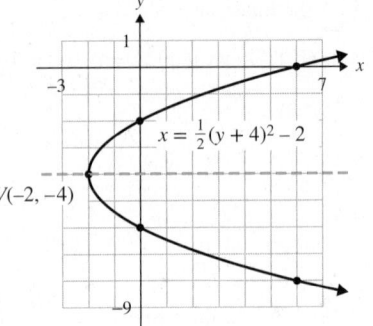

12.5 Quadratic and Rational Inequalities

A **quadratic inequality** can be written in the form

$$ax^2 + bx + c \le 0 \qquad \text{or} \qquad ax^2 + bx + c \ge 0$$

where a, b, and c are real numbers and $a \ne 0$. ($<$ and $>$ may be substituted for $\le$ and $\ge$.)

An inequality containing a rational expression, like $\dfrac{c - 5}{c + 1} \le 0$, is called a **rational inequality.**

Solve $r^2 - 4r \ge 12$.

Step 1: $r^2 - 4r - 12 \ge 0$ Subtract 12.

Because the inequality symbol is $\ge$, the solution set contains the interval(s) where the quantity $r^2 - 4r - 12$ is *positive*.

Step 2: Solve $r^2 - 4r - 12 = 0$.

$$(r - 6)(r + 2) = 0 \qquad \text{Factor.}$$

$$r - 6 = 0 \qquad \text{or} \qquad r + 2 = 0$$
$$r = 6 \qquad \text{or} \qquad r = -2$$

Step 3: Put $r = 6$ and $r = -2$ on a number line.

$r^2 - 4r - 12$ Positive   Negative   Positive

-2 6

Step 4: Choose a test number in each interval to determine the sign of $r^2 - 4r - 12$.

Step 5: The solution set will contain the numbers in the intervals where $r^2 - 4r - 12$ is *positive*.

Step 6: The endpoints of the intervals are included because the inequality is $\ge$. The graph of the solution set is

$-5\ -4\ -3\ -2\ -1\ \ 0\ \ 1\ \ 2\ \ 3\ \ 4\ \ 5\ \ 6\ \ 7\ \ 8\ \ 9\ \ 10$

The solution set of $r^2 - 4r - 12$ is $(-\infty, -2] \cup [6, \infty)$.

12.6 The Algebra of Functions

Operations with Functions

Given the functions $f(x)$ and $g(x)$, we can find their sum, difference, product, and quotient. The domain of $(f + g)(x)$, $(f - g)(x)$, $(fg)(x)$, and $\left(\dfrac{f}{g}\right)(x)$ is the intersection of the domains of f and g.

Let $f(x) = 5x - 1$ and $g(x) = x + 4$. Find $(f + g)(x)$.

$$(f + g)(x) = f(x) + g(x)$$
$$= (5x - 1) + (x + 4)$$
$$= 6x + 3$$

Definition/Procedure	Example
Composition of Functions Given the functions $f(x)$ and $g(x)$, the **composition function** $f \circ g$ (read "f of g") is defined as $$(f \circ g)(x) = f(g(x))$$ where $g(x)$ is in the domain of f.	$f(x) = 4x - 10$ and $g(x) = -3x + 2$. Find $(f \circ g)(x)$. $\begin{aligned}(f \circ g)(x) &= f(g(x)) \\ &= f(-3x + 2) \\ &= 4(-3x + 2) - 10 \quad \text{Substitute } -3x + 2 \\ &= -12x + 8 - 10 \quad \text{for } x \text{ in } f(x). \\ &= -12x - 2\end{aligned}$

Chapter 12: Review Exercises

(12.1) Identify the domain and range of each relation, and determine whether each relation is a function.

1) $\{(-7, -4), (-5, -1), (2, 3), (2, 5), (4, 9)\}$

2)
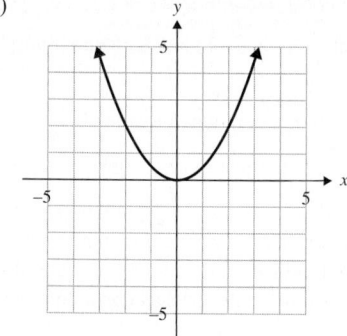

Determine whether each relation describes y as a function of x.

3) $x = |y|$

4) $y = \sqrt{-x}$

5) How do you find the domain of a square root function?

6) How do you find the domain of a rational function?

Identify each function as a linear, quadratic, polynomial, rational, square root, or cube root function, and determine the domain.

7) $h(a) = \dfrac{a + 6}{a^2 - 7a - 8}$

8) $k(c) = c^2 + 11c + 2$

9) $C(n) = -4n - 9$

10) $f(p) = \sqrt{3 - 4p}$

11) $g(x) = \sqrt[3]{x - 12}$

12) $P(a) = 3a^4 - 7a^3 + 12a^2 + a - 8$

13) $h(t) = \sqrt{5t - 7}$

14) $R(c) = \dfrac{c + 8}{2c + 3}$

Graph each function, and identify the domain and range.

15) $f(x) = \sqrt{x}$

16) $h(x) = -\dfrac{1}{3}x + 4$

17) $f(x) = 2x - 5$

18) $g(x) = \sqrt[3]{-x}$

19) Let $f(x) = -8x + 3$ and $g(x) = x^2 + 7x - 12$. Find each of the following and simplify.

    a) $f(5)$ b) $f(-4)$

    c) $g(-2)$ d) $g(3)$

    e) $f(c)$ f) $g(r)$

    g) $f(p - 3)$ h) $g(t + 4)$

20) $g(x) = x^2 + 3x - 28$. Find x so that $g(x) = 0$.

21) To rent a car for one day, a company charges its customers a flat fee of $26 plus $0.20 per mile. This can be described by the function $C(m) = 0.20m + 26$, where m is the number of miles driven and C is the cost of renting a car, in dollars.

    a) What is the cost of renting a car that is driven 30 mi?

    b) What is the cost of renting a car that is driven 100 mi?

    c) If a customer paid $56 to rent a car, how many miles did she drive?

    d) If a customer paid $42 to rent a car, how many miles did he drive?

22) The area, A, of a square is a function of the length of its side, s.

    a) Write an equation using function notation to describe this relationship between A and s.

    b) If the length of a side is given in inches, find $A(4)$ and explain what this means in the context of the problem.

    c) If the length of a side is given in feet, find $A(1)$ and explain what this means in the context of the problem.

    d) What is the length of each side of a square that has an area of 49 cm²?

(12.2) Graph each function, and identify the domain and range.

23) $h(x) = (x + 4)^2$

24) $g(x) = |x|$

25) $k(x) = -|x| + 5$

26) $f(x) = \sqrt{x} - 3$

27) $g(x) = \sqrt{x - 2} - 1$

28) $h(x) = |x - 1| + 2$

Graph each piecewise function.

29) $f(x) = \begin{cases} -\dfrac{1}{2}x - 2, & x \le 2 \\ x - 3, & x > 2 \end{cases}$

30) $g(x) = \begin{cases} 1, & x < -3 \\ x + 4, & x \ge -3 \end{cases}$

Let $f(x) = [\![x]\!]$. Find the following function values.

31) $f\left(7\dfrac{2}{3}\right)$

32) $f\left(\dfrac{3}{8}\right)$

33) $f(-5.8)$

Graph each greatest integer function.

34) $f(x) = [\![x]\!]$

35) $g(x) = \left[\!\left[\dfrac{1}{2}x\right]\!\right]$

36) Visitors to downtown Hinsdale must pay the parking meters to park their cars. The cost of parking is 5¢ for the first 12 min and 5¢ for each additional 12 min or fraction of this time. Let $P(t)$ represent the cost of parking, and let t represent the number of minutes the car is parked at the meter. Graph $P(t)$ for parking a car for up to (and including) 1 hr.

If the following transformations are performed on the graph of $f(x)$ to obtain the graph of $g(x)$, write the equation of $g(x)$.

37) $f(x) = |x|$ is shifted right 5 units.

38) $f(x) = \sqrt{x}$ is shifted left 2 units and up 1 unit.

(12.3 and 12.4)

39) Given a quadratic function in the form $f(x) = a(x - h)^2 + k$, answer the following.

   a) What is the vertex?

   b) What is the equation of the axis of symmetry?

   c) What does the sign of a tell us about the graph of f?

40) What are two ways to find the vertex of the graph of $f(x) = ax^2 + bx + c$?

41) Given a quadratic equation of the form $x = a(y - k)^2 + h$, answer the following.

   a) What is the vertex?

   b) What is the equation of the axis of symmetry?

   c) What does the sign of a tell us about the graph of the equation?

42) What are two ways to find the vertex of the graph of $x = ay^2 + by + c$?

For each quadratic equation, identify the vertex, axis of symmetry, and x- and y-intercepts. Then, graph the equation.

43) $f(x) = (x + 2)^2 - 1$

44) $g(x) = -\dfrac{1}{2}(x - 3)^2 - 2$

45) $x = -y^2 - 1$

46) $y = 2x^2$

47) $x = -(y - 3)^2 + 11$

48) $x = (y + 1)^2 - 5$

Rewrite each equation in the form $f(x) = a(x - h)^2 + k$ or $x = a(y - k)^2 + h$ by completing the square. Then graph the equation. Include the intercepts.

49) $x = y^2 + 8y + 7$

50) $f(x) = -2x^2 - 8x + 2$

51) $y = \dfrac{1}{2}x^2 - 4x + 9$

52) $x = -y^2 + 4y - 4$

Graph each equation using the vertex formula. Include the intercepts.

53) $f(x) = x^2 - 2x - 4$

54) $x = 3y^2 - 12y$

55) $x = -\dfrac{1}{2}y^2 - 3y - \dfrac{5}{2}$

56) $y = -x^2 - 6x - 10$

Solve.

57) An object is thrown upward from a height of 240 ft so that its height h (in feet) t sec after being thrown is given by

$$h(t) = -16t^2 + 32t + 240$$

   a) How long does it take the object to reach its maximum height?

   b) What is the maximum height attained by the object?

   c) How long does it take the object to hit the ground?

58) A restaurant wants to add outdoor seating to its inside service. It has 56 ft of fencing to enclose a rectangular, outdoor café. Find the dimensions of the outdoor café of maximum area if the building will serve as one side of the café.

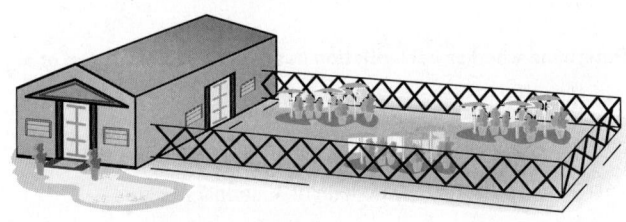

(12.5) Solve each inequality. Graph the solution set, and write the solution in interval notation.

59) $a^2 + 2a - 3 < 0$

60) $4m^2 + 8m \ge 21$

61) $64v^2 \ge 25$

62) $36 - r^2 > 0$

63) $(5c + 2)(c - 4)(3c + 1) < 0$

64) $(p - 6)^2 \le -5$

65) $\dfrac{t + 7}{2t - 3} > 0$

66) $\dfrac{6}{g - 7} \le 0$

67) $\dfrac{z}{z - 2} \le 3$

68) $\dfrac{1}{n - 4} > -3$

69) $\dfrac{r^2 + 4}{r - 7} \ge 0$

Solve.

70) Custom Bikes, Inc., estimates that its total profit function, $P(x)$, for producing x thousand units is given by $P(x) = -2x^2 + 32x - 110$. At what level of production does the company make a profit?

(12.6) Let $f(x) = 5x + 2$, $g(x) = -x + 4$, $h(x) = 3x^2 - 7$, and $k(x) = x^2 - 7x - 8$. Find each of the following.

71) $(f + g)(x)$

72) $(h - k)(x)$

73) $(g - h)(2)$

74) $(f + k)(-3)$

75) $(fg)(x)$

76) $(gk)(1)$

For Exercises 77 and 78, find a) $\left(\dfrac{f}{g}\right)(x)$ and b) $\left(\dfrac{f}{g}\right)(3)$.

Identify any values that are not in the domain of $\left(\dfrac{f}{g}\right)(x)$.

77) $f(x) = 6x - 5$, $g(x) = x + 4$

78) $f(x) = 3x^2 - 5x + 2$, $g(x) = 3x - 2$

79) $R(x) = 20x$ is the revenue function for the sale of x children's soccer uniforms, in dollars. The cost to produce x soccer uniforms, in dollars, is

$$C(x) = 14x + 400$$

a) Find the profit function, $P(x)$, that describes the profit from the sale of x uniforms.

b) What is the profit from the sale of 200 uniforms?

Stockbyte/Getty Images

80) Write out, in words, how to read "$(f \circ g)(x)$."

81) Let $f(x) = x + 6$ and $g(x) = 2x - 9$. Find

a) $(g \circ f)(x)$

b) $(f \circ g)(x)$

c) $(f \circ g)(5)$

82) Let $g(x) = x^2 + 10$ and $h(x) = \sqrt{x - 7}$. Find

a) $(g \circ h)(x)$

b) $(h \circ g)(x)$

c) $(h \circ g)(6)$

83) Let $h(x) = 2x - 1$ and $k(x) = x^2 + 5x - 4$. Find

a) $(k \circ h)(x)$

b) $(h \circ k)(x)$

c) $(h \circ k)(-3)$

84) Antoine's gross weekly pay, G, in terms of the number of hours, h, he worked is given by $G(h) = 12h$. His net weekly pay, N, in terms of his gross pay is given by $N(G) = 0.8G$.

a) Find $(N \circ G)(h)$ and explain what it represents.

b) Find $(N \circ G)(30)$ and explain what it represents.

c) What is his net pay if he works 40 hr in 1 week?

Given $h(x)$, find f and g such that $h(x) = (f \circ g)(x)$.

85) a) $h(x) = (3x + 10)^2$ b) $h(x) = \dfrac{1}{2 - 5x}$

86) a) $h(x) = (8x - 7)^3$ b) $h(x) = \sqrt{x^2 + 6}$

Chapter 12: Test

1) What is a function?

2) Given the relation $\{(-8, -1), (2, 3), (5, 3), (7, 10)\}$,

a) determine the domain.

b) determine the range.

c) is this a function?

3) For $y = \sqrt{3x + 7}$,

a) determine the domain.

b) is y a function of x?

4) Explain how to find the domain of a rational function.

5) Determine the domain of $g(t) = \dfrac{t + 10}{7t - 8}$.

Let $f(x) = 4x + 3$ and $g(x) = x^2 - 6x + 10$. Find each of the following and simplify.

6) $g(4)$

7) $f(c)$

8) $f(n - 7)$

9) $g(k + 5)$

10) Let $h(x) = -2x + 6$. Find x so that $h(x) = 9$.

11) A garden supply store charges $50 per cubic yard plus a $60 delivery fee to deliver cedar mulch. This can be described by the function $C(m) = 50m + 60$, where m is the amount of cedar mulch delivered, in cubic yards, and C is the cost, in dollars.

a) Find $C(3)$ and explain what it means in the context of the problem.

b) If a customer paid $360 to have cedar mulch delivered to his home, how much did he order?

Graph each function and identify the domain and range.

12) $h(x) = -|x - 1|$

13) $g(x) = \sqrt{x + 3}$

14) $h(x) = -\dfrac{1}{3}x - 1$

15) $g(x) = \sqrt[3]{x} + 1$

16) If the graph of $f(x) = |x|$ is shifted up 5 units and left 2 units to obtain the graph of $g(x)$, what is the equation of $g(x)$?

17) Graph $f(x) = \begin{cases} x + 3, & x > -1 \\ -2x - 5, & x \le -1 \end{cases}$

Graph each equation. Identify the vertex, axis of symmetry, and intercepts. Determine the domain and range.

18) $f(x) = -(x + 2)^2 + 4$

19) $x = y^2 - 3$

20) $x = 3y^2 - 6y + 5$

21) $g(x) = x^2 - 6x + 8$

22) A ball is projected upward from the top of a 200-ft tall building. The height $h(t)$ of the ball above the ground (in feet) t sec after the ball is released is given by

$$h(t) = -16t^2 + 24t + 200.$$

a) What is the maximum height attained by the ball?

b) When will the ball be 40 ft above the ground?

c) When will the ball hit the ground?

Let $f(x) = 2x + 7$ and $g(x) = x^2 + 5x - 3$. Find each of the following.

23) $(g - f)(x)$

24) $(f + g)(-1)$

25) $(fg)(x)$

26) $(f \circ g)(x)$

27) $(g \circ f)(x)$

28) $(g \circ f)(-3)$

29) Let $h(x) = \sqrt[3]{5x + 8}$. Find f and g such that $h(x) = (g \circ f)(x)$.

Solve each inequality. Graph the solution set, and write the solution in interval notation.

30) $y^2 + 4y - 45 \ge 0$

31) $\dfrac{m - 5}{m + 3} \ge 0$

32) A company has determined that the average cost, $\overline{C}(x)$, of producing x backpacks per month can be described by the function $\overline{C}(x) = \dfrac{5x + 80{,}000}{x}$. How many backpacks must the company produce each month so that the average cost of producing each backpack is no more than \$15?

Chapter 12: Cumulative Review for Chapters 1–12

1) Write each number as a power of 10.

   a) 10,000   b) 0.001   c) $\dfrac{1}{10}$

2) Evaluate.

   a) 7^2 b) 10^3 c) 3^4

   d) 5^{-3} e) $\left(\dfrac{1}{2}\right)^6$ f) $\left(\dfrac{1}{10}\right)^{-2}$

3) Write the equation of the line parallel to $4x + 3y = 15$ containing the point $(-5, 6)$. Express it in standard form.

4) Multiply and simplify $(p - 8)(p + 7)$.

5) Factor completely.

   a) $k^2 - 15k + 54$ b) $100 - 9m^2$

6) Divide.

   a) $\dfrac{12r - 40r^2 + 6r^3 + 4}{4r^2}$

   b) $\dfrac{c - 8}{2c^2 - 5c - 12} \div \dfrac{3c - 24}{c^2 - 16}$

   c) $\dfrac{4 - 2i}{2 + 3i}$

Solve.

7) $4(y^2 + 2y) = 5$

8) $\dfrac{4x + 2}{x + 5} = 10$

9) $|7y + 6| \le -8$

10) $q - 8q^{1/2} + 7 = 0$

11) $6(2y + 1) - 4y = 5(y + 2)$

12) $(3n - 4)^2 + 9 = 0$

13) $x = -6 + \sqrt{x + 8}$

14) Solve this system using any method.

$$x - \dfrac{1}{4}y = \dfrac{5}{2}$$
$$\dfrac{1}{2}x + \dfrac{1}{3}y = \dfrac{13}{6}$$

15) Graph the compound inequality

$$y \le -\dfrac{1}{2}x + 4 \text{ and } 2x - y \le 2$$

16) Solve the compound inequality

$$x + 8 \leq 6 \text{ or } 1 - 2x \leq -5$$

Graph the solution set and write the answer in interval notation.

17) Suppose y varies inversely as the square of x. If $y = 12$ when $x = 2$, find y when $x = 4$.

Simplify. Assume all variables represent nonnegative real numbers.

18) $\sqrt{60}$

19) $\sqrt[4]{16}$

20) $\sqrt{18c^6 d^{11}}$

21) $(100)^{-3/2}$

22) Add $\sqrt{12} + \sqrt{3} + \sqrt{48}$.

23) Let $g(x) = x + 1$ and $h(x) = x^2 + 4x + 3$.

a) Find $g(7)$.

b) Find $\left(\dfrac{h}{g}\right)(x)$. Identify any values that are not in the domain of $\left(\dfrac{h}{g}\right)(x)$.

c) Find x so that $g(x) = 5$.

d) Find $(g \circ h)(x)$.

24) Graph $f(x) = -x^2 + 4$ and identify the domain and range.

25) Graph $x = -y^2 - 2y - 3$.

Exponential and Logarithmic Functions

Get Ready

In this chapter, we will learn about logarithms. The properties of logarithms and the rules of exponents are related, so let's review some of the rules here.

1) Recall that an **exponent** (or **power**) is used to represent repeated multiplication. For example, $2 \cdot 2 \cdot 2 = 2^3$ where the *exponent*, 3, tells us that the *base*, 2, is multiplied by itself three times. (It is necessary to know the powers of integers listed in Section 1.2.)

2) Here are some of the **rules of exponents** that help us simplify expressions.

   Examples: Simplify using the rules of exponents. Assume all variables represent nonzero real numbers. The answer should contain only positive exponents.

   a) $p^7 \cdot p^3 = p^{7+3} = p^{10}$ b) $(k^2)^6 = k^{2 \cdot 6} = k^{12}$

   c) $(3c)^4 = 3^4 \cdot c^4 = 81c^4$ d) $\dfrac{n^8}{n^3} = n^{8-3} = n^5$

   e) $7^0 = 1$ f) $t^{-2} = \dfrac{1}{t^2}$

3) Remember the **relationship between radical and fractional exponent notations.**

   If n is a positive integer greater than 1 and $\sqrt[n]{a}$ is a real number, then $\sqrt[n]{a} = a^{1/n}$.

   (*The denominator of the fractional exponent is the index of the radical.*)

   Examples: Write each expression using a fractional exponent.

   a) $\sqrt[4]{7} = 7^{1/4}$ b) $\sqrt{3} = 3^{1/2}$

4) If an equation contains more than one variable, we can **solve for a specific variable.**

   Examples: Solve $x = 2y - 9$ for y.

   $$x = 2y - 9$$
   $$x + 9 = 2y \qquad \text{Add 9 to each side.}$$
   $$\frac{x+9}{2} = \frac{2y}{2} \qquad \text{Divide by 2.}$$
   $$\frac{1}{2}x + \frac{9}{2} = y \qquad \text{Simplify.}$$

In this chapter, use the following *Basic Skills Worksheets* to prepare students for this, and future, chapters: **WS2 Powers** and **WS5 Roots.**

1) Write each product in exponential form.
   a) $8 \cdot 8 \cdot 8 \cdot 8 \cdot 8 \cdot 8$ b) $5 \cdot 5 \cdot 5 \cdot 5$
2) Evaluate.
   a) 9^2 b) 2^5 c) 4^3 d) 10^4 e) 3^3

For Exercises 3–8, simplify using the rules of exponents. Assume all variables represent nonzero real numbers. The answer should contain only positive exponents.

3) $(2n^4)^5$ 4) $\dfrac{h^5}{h}$ 5) w^{-2} 6) 10^{-3} 7) 9^0 8) $k^7 \cdot k^2$

9) Write 64 as a power of 2.   10) Write 1000 as a power of 10.

11) Write $\dfrac{1}{125}$ as a power of $\dfrac{1}{5}$.   12) Write $\dfrac{1}{49}$ as a power of 7.

Write each expression using a fractional exponent.

13) $\sqrt[3]{5}$ 14) $\sqrt{10}$ 15) $\sqrt{6}$ 16) $\sqrt[5]{32}$

Solve each equation for y.

17) $x = 8y + 5$ 18) $x = -\dfrac{2}{3}y - 4$

1) a) 8^6 b) 5^4 2) a) 81 b) 32 c) 64 d) 10,000 e) 27   3) $32n^{20}$ 4) h^4 5) $\dfrac{1}{w^2}$ 6) $\dfrac{1}{1000}$ 7) 1   8) k^9 9) 2^6
10) 10^3 11) $\left(\dfrac{1}{5}\right)^3$ 12) 7^{-2} 13) $5^{1/3}$ 14) $10^{1/2}$ 15) $6^{1/2}$ 16) $32^{1/5}$ 17) $y = \dfrac{1}{8}x - \dfrac{8}{5}$ 18) $y = -\dfrac{3}{2}x - 6$

Answers

Study Strategies   Using the Internet for Learning

The Internet provides access to millions of resources to supplement your learning. Using these resources can help you master difficult math concepts. But just as you may have a particular learning style in the classroom, it is the same for online resources. It is important for you to find websites that fit your style of learning, and it is *equally* important that you learn how to use those resources in a way that will be most helpful to you.

- **I will investigate and use online resources to help me learn math.**

- Complete the emPOWERme survey that appears before the Chapter Summary to help you understand how much you know about online learning resources.
- Gather the materials you will need such as a computer, your book, paper, and pencils.
- Find a workspace that is quiet and that has enough room to work on the computer *and* to take notes.
- Decide which math skills you want to learn or review.
- Be aware of any online resources that are available with your textbook or other course materials. Familiarize yourself with the types of educational materials contained within this resource. If you are using an interactive e-book, you should realize that videos are embedded in the book so that you can watch them as you read.
- Ask your instructor for recommendations, or search the Internet to find appropriate learning materials.

(continued on next page)

- Think about your own learning style and decide which type of resource is best for you. Do you want a website that will solve an equation for you? Or do you prefer to watch a video where a teacher explains a topic? Would you like a discussion board where you can pose questions to other students?
- Consider the source of the resource before using it. Well-known sources are more likely to be accurate and of high quality.
- Some well-known resources are: the many online resources that accompany this book (see the front of the book or ask your instructor for information), www.youtube.com, and www.khanacademy.org.

- Explore the online resource. Learn which types of educational tools it contains such as explanations that you read, problem-solvers, and videos.
- Work with the resource thoroughly. Watch one or two complete videos, or work through a couple of problems step by step.
- **Be an active user!** Take notes as you go along. Don't just watch or read passively!
- Keep a list of the websites or resources that are most helpful to you. Bookmark them so that you can find them easily in the future.
- Be sure that the methods used in the resources are consistent with what your instructor has taught you and what is shown in your book. If not, ask your instructor about the differences. Sometimes, it can be confusing if instructors solve a problem one way but an outside resource uses a different method.

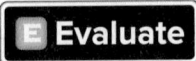

- Did you find any websites or online resources that were helpful to you? Did you understand the material better after using that resource for learning?

- If you identified some helpful resources, be sure to note what they were so that you can use them in the future. Next time, explore them more; you may find other types of learning resources that are even better!
- If you did not find anything that was helpful, think about why. Did you use the wrong content within that resource? Should you have watched a video instead of just having a website solve problems for you? Or, should you try a different website or resource next time?

13.1 Inverse Functions

P Prepare

O Organize

What are your objectives for Section 13.1?	How can you accomplish each objective?
1 Determine Whether a Function Is One-to-One	• Write the definition of a *one-to-one function* in your own words. • Use the *horizontal line test* to determine whether a function is one-to-one. • Complete the given examples on your own. • Complete You Trys 1 and 2.
2 Find the Inverse of a Function	• Write the definition of an *inverse function* in your own words. • Know how to write the inverse of a function. • Know how to graph a function and its inverse on the same axes. • Complete the given examples on your own. • Complete You Trys 3–5.

What are your objectives for Section 13.1?	How can you accomplish each objective?
3 Given the Graph of $f(x)$, Graph $f^{-1}(x)$	• Review how to determine the domain and range of a function. • Complete the given example on your own. • Complete You Try 6.
4 Show That $(f^{-1} \circ f)(x) = x$ and $(f \circ f^{-1})(x) = x.$	• Complete the given example on your own. • Complete You Try 7.

W Work

Read the explanations, follow the examples, take notes, and complete the You Trys.

In this chapter, we will study inverse functions and two very useful types of functions in mathematics: exponential and logarithmic functions. But first, we must learn about one-to-one and inverse functions. This is because exponential and logarithmic functions are related in a special way: They are *inverses* of one another.

1 Determine Whether a Function Is One-to-One

Recall from Sections 4.6 and 12.1 that a relation is a *function* if each x-value corresponds to exactly one y-value. Let's look at two functions, f and g.

$$f = \{(1, -3), (2, -1), (4, 3), (7, 9)\} \quad g = \{(0, 3), (1, 4), (-1, 4), (2, 7)\}$$

In functions f and g, each x-value corresponds to exactly one y-value. That is why they are functions. In function f, each *y-value also corresponds to exactly one x-value.* Therefore, f is a *one-to-one function.* In function g, however, each y-value does *not* correspond to exactly one x-value. (The y-value of 4 corresponds to $x = 1$ and $x = -1$.) Therefore, g is *not* a one-to-one function.

Definition

In order for a function to be a **one-to-one function,** each x value corresponds to exactly one y-value, and each y-value corresponds to exactly one x-value.

Alternatively, we can say that a function is one-to-one if each value in its domain corresponds to exactly one value in its range *and* if each value in its range corresponds to exactly one value in its domain.

EXAMPLE 1

Determine whether each function is one-to-one.

a) $f = \{(-1, 9), (1, -3), (2, -6), (4, -6)\}$

b) $g = \{(-3, 13), (-1, 5), (5, -19), (8, -31)\}$

c)

State	Number of Representatives in U.S. House of Representatives (2019)
Alaska	1
California	53
Connecticut	5
Delaware	1
Ohio	16

d)

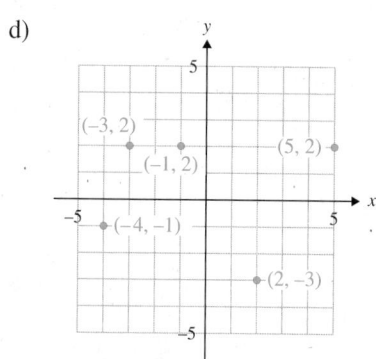

Solution

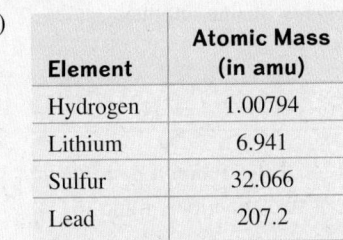

Hint

In your notes, describe the difference between a function that is one-to-one and a function that is not.

a) *f* is *not* a one-to-one function because the *y*-value −6 corresponds to two different *x*-values: (2, −6) and (4, −6).

b) *g* *is* a one-to-one function because each *y*-value corresponds to exactly one *x*-value.

c) The information in the table does *not* represent a one-to-one function because the value 1 in the range corresponds to two different values in the domain, Alaska and Delaware.

d) The graph does *not* represent a one-to-one function because three points have the same *y*-value: (−3, 2), (−1, 2), and (5, 2).

[**YOU TRY 1**] Determine whether each function is one-to-one.

a) $f = \{(-2, -13), (0, -7), (4, 5), (5, 8)\}$ b) $g = \{(-4, 2), (-1, 1), (0, 2), (3, 5)\}$

c)

Element	Atomic Mass (in amu)
Hydrogen	1.00794
Lithium	6.941
Sulfur	32.066
Lead	207.2

d)

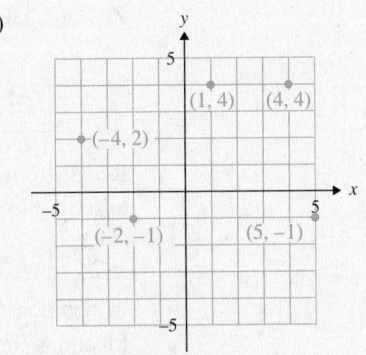

Just as we can use the vertical line test to determine whether a graph represents a function, we can use the *horizontal line test* to determine whether a function is one-to-one.

Definition

Horizontal Line Test: If every horizontal line that could be drawn through a function would intersect the graph at most once, then the function is one-to-one.

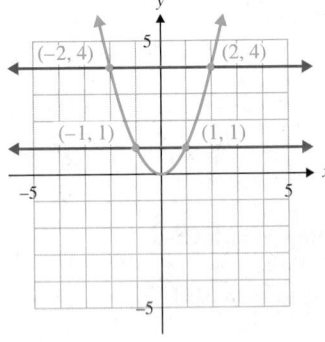

Look at the graph of the given function, in blue. We can see that if a horizontal line intersects the graph more than once, then one *y*-value corresponds to more than one *x*-value. This means that the function is *not* one-to-one. For example, the *y*-value of 1 corresponds to $x = 1$ and $x = -1$.

EXAMPLE 2

Determine whether each graph represents a one-to-one function.

a)
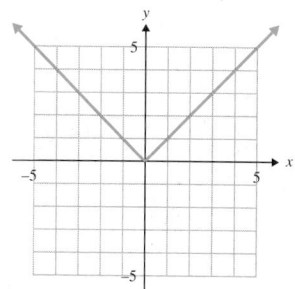

b)
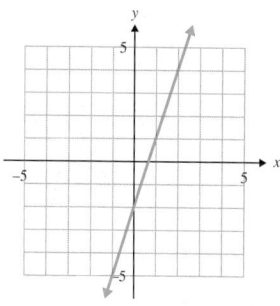

Solution

a) *Not* one-to-one. It is possible to draw a horizontal line through the graph so that it intersects the graph more than once.

b) *Is* one-to-one. Every horizontal line that could be drawn through the graph would intersect the graph at most once.

[YOU TRY 2]

Determine whether each graph represents a one-to-one function.

a)
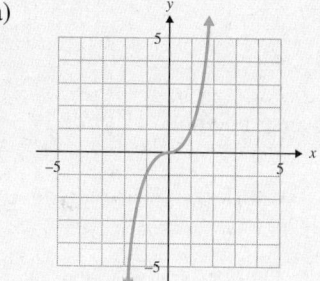

b)
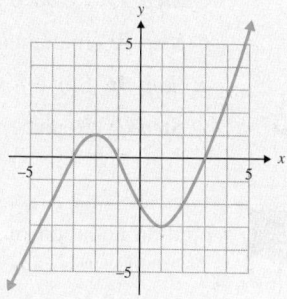

---

2 Find the Inverse of a Function

One-to-one functions lead to other special functions—inverse functions. *A one-to-one function has an inverse function.*

To find the inverse of a one-to-one function, we interchange the coordinates of the ordered pairs.

EXAMPLE 3

Find the inverse function of $f = \{(4, 2), (9, 3), (36, 6)\}$.

Solution

To find the inverse of f, switch the x- and y-coordinates of each ordered pair. The inverse of f is $\{(2, 4), (3, 9), (6, 36)\}$.

[YOU TRY 3]

Find the inverse function of $f = \{(-5, -1), (-3, 2), (0, 7), (4, 13)\}$.

We use special notation to represent the inverse of a function. If f is a one-to-one function, then f^{-1} (read "f inverse") represents the inverse of f. For Example 3, we can write the inverse as $f^{-1} = \{(2, 4), (3, 9), (6, 36)\}$.

Definition

Inverse Function: Let f be a one-to-one function. The **inverse** of f, denoted by f^{-1}, is a one-to-one function that contains the set of all ordered pairs (y, x), where (x, y) belongs to f.

1) f^{-1} is read "f inverse" *not* "f to the negative one."

2) f^{-1} does *not* mean $\dfrac{1}{f}$.

3) If a function is not one-to-one, it does not have an inverse.

We said that if (x, y) belongs to the one-to-one function $f(x)$, then (y, x) belongs to its inverse, $f^{-1}(x)$ (read as f *inverse of* x). We use this idea to find the equation for the inverse of $f(x)$.

Procedure How to Find an Equation of the Inverse of $y = f(x)$

Step 1: Replace $f(x)$ with y.

Step 2: Interchange x and y.

Step 3: Solve for y.

Step 4: Replace y with the inverse notation, $f^{-1}(x)$.

EXAMPLE 4

Find an equation of the inverse of $f(x) = 3x + 4$.

Solution

$$f(x) = 3x + 4$$

Step 1: $\qquad\qquad y = 3x + 4$ $\qquad$ Replace $f(x)$ with y.

Step 2: $\qquad\qquad x = 3y + 4$ $\qquad$ Interchange x and y.

Step 3: Solve for y.

$$x - 4 = 3y \qquad \text{Subtract 4.}$$

$$\frac{x - 4}{3} = y \qquad \text{Divide by 3.}$$

$$\frac{1}{3}x - \frac{4}{3} = y \qquad \text{Simplify.}$$

Step 4: $\qquad\qquad f^{-1}(x) = \dfrac{1}{3}x - \dfrac{4}{3}$ $\qquad$ Replace y with $f^{-1}(x)$.

[YOU TRY 4] Find an equation of the inverse of $f(x) = -5x + 10$.

In Example 5, we look more closely at the relationship between a function and its inverse.

EXAMPLE 5

Find the equation of the inverse of $f(x) = 2x - 4$. Then, graph $f(x)$ and $f^{-1}(x)$ on the same axes.

Solution

$$f(x) = 2x - 4$$

Step 1: $\quad\quad\quad\quad y = 2x - 4 \quad$ Replace $f(x)$ with y.

Step 2: $\quad\quad\quad\quad x = 2y - 4 \quad$ Interchange x and y.

Step 3: Solve for y.

$$x + 4 = 2y \quad\quad \text{Add 4.}$$

$$\frac{x + 4}{2} = y \quad\quad \text{Divide by 2.}$$

$$\frac{1}{2}x + 2 = y \quad\quad \text{Simplify.}$$

Step 4: $\quad\quad\quad\quad f^{-1}(x) = \frac{1}{2}x + 2 \quad$ Replace y with $f^{-1}(x)$.

We will graph $f(x)$ and $f^{-1}(x)$ by making a table of values for each. Then we can see another relationship between the two functions.

$f(x) = 2x - 4$	
x	$y = f(x)$
0	−4
1	−2
2	0
5	6

$f^{-1}(x) = \frac{1}{2}x + 2$	
x	$y = f^{-1}(x)$
−4	0
−2	1
0	2
6	5

Notice that the x- and y-coordinates have switched when we compare the tables of values. Graph $f(x)$ and $f^{-1}(x)$.

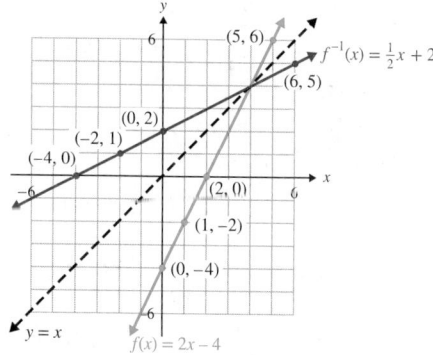

YOU TRY 5

Find the equation of the inverse of $f(x) = -3x + 1$. Then, graph $f(x)$ and $f^{-1}(x)$ on the same axes.

3 Given the Graph of $f(x)$, Graph $f^{-1}(x)$

Look again at the tables in Example 5. The x-values for $f(x)$ become the y-values of $f^{-1}(x)$, and the y-values of $f(x)$ become the x-values of $f^{-1}(x)$. This is true not only for the values in the tables but for *all* values of x and y. That is, for all ordered pairs (x, y) that belong to $f(x)$, (y, x) belongs to $f^{-1}(x)$. Another way to say this is *the domain of f becomes the range of f^{-1}, and the range of f becomes the domain of f^{-1}.*

Let's turn our attention to the graph in Example 5. The graphs of $f(x)$ and $f^{-1}(x)$ are mirror images of one another with respect to the line $y = x$. We say that *the graphs of $f(x)$ and $f^{-1}(x)$ are symmetric with respect to the line $y = x$.* This is true for every function $f(x)$ and its inverse, $f^{-1}(x)$.

W Hint

Write down the relationships between a function and its inverse that you have learned so far.

EXAMPLE 6

Given the graph of $f(x)$, graph $f^{-1}(x)$.

Solution

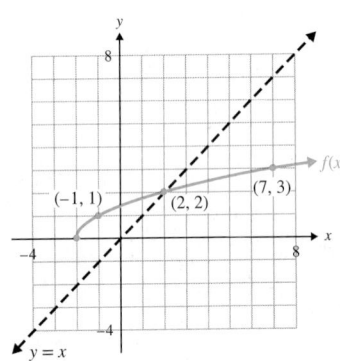
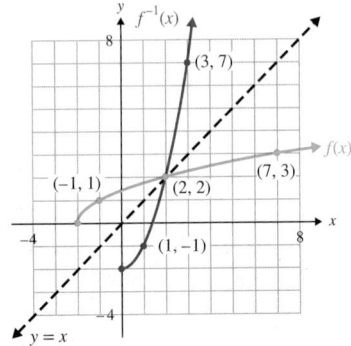

Some points on the graph of $f(x)$ are $(-2, 0)$, $(-1, 1)$, $(2, 2)$, and $(7, 3)$. We can obtain points on the graph of $f^{-1}(x)$ by interchanging the x- and y-values.

Some points on the graph of $f^{-1}(x)$ are $(0, -2)$, $(1, -1)$, $(2, 2)$, and $(3, 7)$. Plot these points to get the graph of $f^{-1}(x)$. Notice that the graphs are symmetric with respect to the line $y = x$.

[YOU TRY 6]

Given the graph of $f(x)$, graph $f^{-1}(x)$.

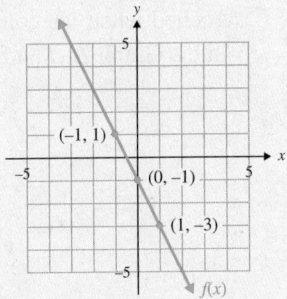

4 Show That $(f^{-1} \circ f)(x) = x$ and $(f \circ f^{-1})(x) = x$

Going back to the tables in Example 5, we see from the first table that $f(0) = -4$ and from the second table that $f^{-1}(-4) = 0$. The first table also shows that $f(1) = -2$ while the second table shows that $f^{-1}(-2) = 1$. That is, putting x into the function f produces $f(x)$. And putting $f(x)$ into $f^{-1}(x)$ produces x.

$$f(0) = -4 \quad \text{and} \quad f^{-1}(-4) = 0$$
$$f(1) = -2 \quad \text{and} \quad f^{-1}(-2) = 1$$
$$\uparrow \quad \uparrow \qquad\qquad \uparrow \quad \uparrow$$
$$x \quad f(x) \qquad f^{-1}(f(x)) = x$$

This leads us to another fact about functions and their inverses.

Note

Let f be a one-to-one function. Then f^{-1} is the inverse of f such that $(f^{-1} \circ f)(x) = x$ and $(f \circ f^{-1})(x) = x$.

EXAMPLE 7

If $f(x) = 4x + 3$, show that $f^{-1}(x) = \dfrac{1}{4}x - \dfrac{3}{4}$.

Solution

Show that $(f^{-1} \circ f)(x) = x$ and $(f \circ f^{-1})(x) = x$.

$$(f^{-1} \circ f)(x) = f^{-1}(f(x))$$

$$= f^{-1}(4x + 3) \qquad \text{Substitute } 4x + 3 \text{ for } f(x).$$

$$= \frac{1}{4}(4x + 3) - \frac{3}{4} \qquad \text{Evaluate.}$$

$$= x + \frac{3}{4} - \frac{3}{4} \qquad \text{Distribute.}$$

$$= x$$

$$(f \circ f^{-1})(x) = f(f^{-1}(x))$$

$$= f\left(\frac{1}{4}x - \frac{3}{4}\right) \qquad \text{Substitute } \frac{1}{4}x - \frac{3}{4} \text{ for } f^{-1}(x).$$

$$= 4\left(\frac{1}{4}x - \frac{3}{4}\right) + 3 \qquad \text{Evaluate.}$$

$$= x - 3 + 3 \qquad \text{Distribute.}$$

$$= x$$

[YOU TRY 7]

If $f(x) = -6x + 2$, show that $f^{-1}(x) = -\dfrac{1}{6}x + \dfrac{1}{3}$.

Using Technology

A graphing calculator can list tables of values on one screen for more than one equation. The graphing calculator screen shown here is the table of values generated when the equation of one line is entered as Y_1 and the equation of another line is entered as Y_2.

X	Y₁	Y₂
0	4	-2
2	8	-1
4	12	0
6	16	1
8	20	2
10	24	3
12	28	4

X=0

We read the table as follows:

- The points (0, 4), (2, 8), (4, 12), (6, 16), (8, 20), (10, 24), and (12, 28) are points on the line entered as Y_1.
- The points (0, −2), (2, −1) (4, 0), (6, 1), (8, 2), (10, 3), and (12, 4) are points on the line entered as Y_2.

Equations Y_1 and Y_2 are linear functions, and they are inverses.

1) Looking at the table of values, what evidence is there that the functions Y_1 and Y_2 are inverses of each other?

2) Find the equations of the lines Y_1 and Y_2.

3) Graph Y_1 and Y_2. Is there evidence from their graphs that they are inverses?

4) Using the methods of this chapter, show that Y_1 and Y_2 are inverses.

ANSWERS TO [YOU TRY] EXERCISES

1) a) yes b) no c) yes d) no   2) a) yes b) no   3) {(−1, −5), (2, −3), (7, 0), (13, 4)}

4) $f^{-1}(x) = -\frac{1}{5}x + 2$ 5) $f^{-1}(x) = -\frac{1}{3}x + \frac{1}{3}$

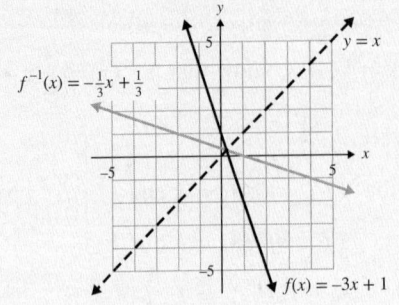

6)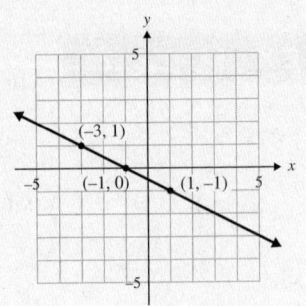

7) Show that $(f^{-1} \circ f)(x) = x$ and $(f \circ f^{-1})(x) = x$.

ANSWERS TO TECHNOLOGY EXERCISES

1) If Y_1 and Y_2 are inverses, then if (x, y) is a point on Y_1, (y, x) is a point on Y_2. We see this is true with (0, 4) on Y_1 and (4, 0) on Y_2, with (2, 8) on Y_1 and (8, 2) on Y_2, and with (4, 12) on Y_1 and (12, 4) on Y_2.
2) $Y_1 = 2x + 4$, $Y_2 = 0.5x − 2$
3) Yes. They appear to be symmetric with respect to the line $y = x$.
4) Let $f(x) = Y_1$ and $f^{-1}(x) = Y_2$. We can show that $(f \circ f^{-1})(x) = x$ and $(f^{-1} \circ f)(x) = x$.

E Evaluate **13.1** Exercises   Do the exercises, and check your work.

Objective 1: Determine Whether a Function Is One-to-One

Determine whether each function is one-to-one. If it is one-to-one, find its inverse.

1) $f = \{(-6, 3), (-1, 8), (4, 3)\}$

2) $g = \{(0, -7), (1, -6), (4, -5), (25, -2)\}$

(24) 3) $h = \{(-5, -16), (-1, -4), (3, 8)\}$

4) $f = \{(-4, 3), (-2, -3), (2, -3), (6, 13)\}$

Determine whether each function is one-to-one.

5) The table shows the average temperature during selected months in Tulsa, Oklahoma. The function matches each month with the average temperature, in °F. Is it one-to-one? (www.noaa.gov)

Month	Average Temp. (°F)
Jan.	36.4
Apr.	60.8
July	83.5
Oct.	62.6

6) The table shows some NCAA college football conferences and the number of schools in the conference in 2019. The function matches each conference with the number of schools it contains. Is it one-to-one?

Conference	Number of Member Schools
ACC	14
Big 10	14
Big 12	10
SEC	14
Pac-12	12

Mixed Exercises: Objectives 1–3

7) Do all functions have inverses? Explain your answer.

8) What test can be used to determine whether the graph of a function has an inverse?

Determine whether each statement is true or false. If it is false, rewrite the statement so that it is true.

9) $f^{-1}(x)$ is read as "f to the negative one of x."

10) If f^{-1} is the inverse of f, then $(f^{-1} \circ f)(x) = x$ and $(f \circ f^{-1})(x) = x$.

11) The domain of f is the range of f^{-1}.

12) If f is one-to-one and $(5, 9)$ is on the graph of f, then $(-5, -9)$ is on the graph of f^{-1}.

13) The graphs of $f(x)$ and $f^{-1}(x)$ are symmetric with respect to the x-axis.

14) Let $f(x)$ be one-to-one. If $f(7) = 2$, then $f^{-1}(2) = 7$.

For each function graphed here, answer the following.

   a) Determine whether it is one-to-one.

   b) If it is one-to-one, graph its inverse.

 15)

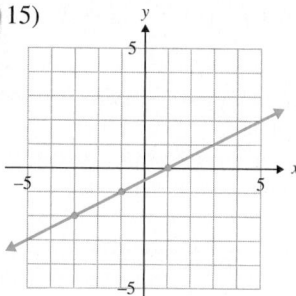

16)

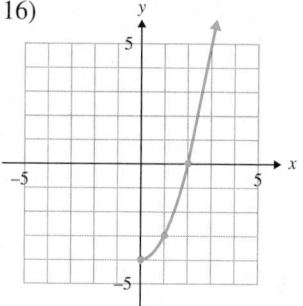

17)

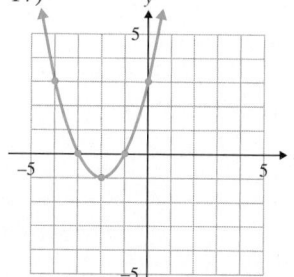

18)

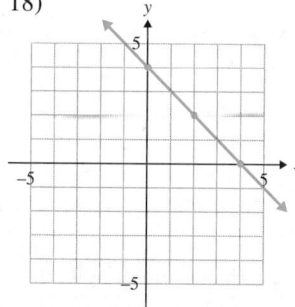

19)

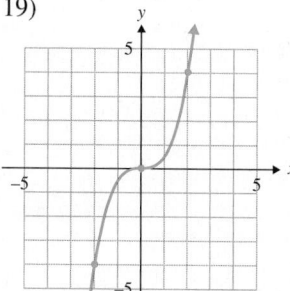

20)

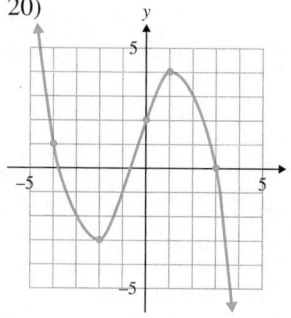

Objective 2: Find the Inverse of a Function

Find the inverse of each one-to-one function.

Fill It In

Fill in the blanks with either the missing mathematical step or the reason for the given step.

21) $f(x) = 2x - 10$

    $y = 2x - 10$ _____

    _____   Interchange x and y.

    Solve for y.

    $x + 10 = 2y$

    _____   Divide by 2, and simplify.

    $f^{-1}(x) = \dfrac{1}{2}x + 5$ _____

22) $g(x) = \dfrac{1}{3}x + 4$

    _____   Replace $g(x)$ with y.

    $x = \dfrac{1}{3}y + 4$ _____

    Solve for y.

    _____   Subtract 4.

    $3x - 12 = y$

    _____   Replace y with $g^{-1}(x)$.

Find the inverse of each one-to-one function. Then, graph the function and its inverse on the same axes.

23) $g(x) = x - 6$

24) $h(x) = x + 3$

25) $f(x) = -2x + 5$

26) $g(x) = 4x - 9$

27) $g(x) = \dfrac{1}{2}x$

28) $h(x) = -\dfrac{1}{3}x$

29) $f(x) = x^3$

30) $g(x) = \sqrt[3]{x} + 4$

Find the inverse of each one-to-one function.

31) $f(x) = 2x - 6$

32) $g(x) = -4x + 8$

33) $h(x) = -\frac{3}{2}x + 4$

34) $f(x) = \frac{2}{5}x + 1$

35) $h(x) = \sqrt[3]{x - 7}$

36) $g(x) = \sqrt[3]{x + 2}$

37) $f(x) = \sqrt{x}, x \geq 0$

38) $g(x) = \sqrt{x + 3}, x \geq -3$

Objective 4: Show That $(f^{-1} \circ f)(x) = x$ and $(f \circ f^{-1})(x) = x$

Given the one-to-one function $f(x)$, find the function values *without* finding the equation of $f^{-1}(x)$. Find the value in a) before b).

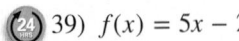 39) $f(x) = 5x - 2$

    a) $f(1)$ b) $f^{-1}(3)$

40) $f(x) = 3x + 7$

    a) $f(-4)$ b) $f^{-1}(-5)$

41) $f(x) = -\frac{1}{3}x + 5$

    a) $f(9)$ b) $f^{-1}(2)$

42) $f(x) = \frac{1}{2}x - 1$

    a) $f(6)$ b) $f^{-1}(2)$

43) $f(x) = -x + 3$

    a) $f(-7)$ b) $f^{-1}(10)$

44) $f(x) = -\frac{5}{4}x + 2$

    a) $f(8)$ b) $f^{-1}(-8)$

45) $f(x) = 2^x$

    a) $f(3)$ b) $f^{-1}(8)$

46) $f(x) = 3^x$

    a) $f(-2)$ b) $f^{-1}\left(\frac{1}{9}\right)$

47) If $f(x) = x + 9$, show that $f^{-1}(x) = x - 9$.

48) If $f(x) = x - 12$, show that $f^{-1}(x) = x + 12$.

49) If $f(x) = -6x + 4$, show that $f^{-1}(x) = -\frac{1}{6}x + \frac{2}{3}$.

50) If $f(x) = -\frac{1}{7}x + \frac{2}{7}$, show that $f^{-1}(x) = -7x + 2$.

51) If $f(x) = \frac{3}{2}x - 9$, show that $f^{-1}(x) = \frac{2}{3}x + 6$.

52) If $f(x) = -\frac{5}{8}x + 10$, show that

$$f^{-1}(x) = -\frac{8}{5}x + 16.$$

53) If $f(x) = \sqrt[3]{x - 10}$, show that $f^{-1}(x) = x^3 + 10$.

54) If $f(x) = x^3 - 1$, show that $f^{-1}(x) = \sqrt[3]{x + 1}$.

R **Rethink**

R1) Explain, in your own words, how to tell whether a function is one-to-one.

R2) How can you tell whether the graphs of two functions are inverses of each other?

13.2 Exponential Functions

P Prepare

O Organize

What are your objectives for Section 13.2?	How can you accomplish each objective?
1 Define an Exponential Function	• Learn the definition of an *exponential function*. • Understand why $a > 0$ and $a \neq 1$ in the exponential function $f(x) = a^x$.
2 Graph $f(x) = a^x$	• Understand why we choose values of x that are negative, positive, and zero. • Know what the graph should look like based on the value of a. • Complete the given examples on your own. • Complete You Trys 1 and 2.
3 Graph $f(x) = a^{x+c}$	• Understand why we choose values of x that will make the exponent negative, positive, and zero. • Complete the given example on your own. • Complete You Try 3.
4 Define the Number e and Graph $f(x) = e^x$	• Learn the definition of e and its numerical approximation. • Be able to use a calculator to generate a table of values for $f(x) = e^x$. • Complete the given example on your own.
5 Solve an Exponential Equation	• Review the rules of exponents. • Learn the procedure for **Solving an Exponential Equation.** • Complete the given example on your own. • Complete You Try 4.
6 Solve an Applied Problem Using a Given Exponential Function	• Read the given example carefully. • Complete the given example on your own. • Complete You Try 5.

Read the explanations, follow the examples, take notes, and complete the You Trys.

We have already studied the following types of functions:

W Hint

Do you remember what the graph of each of these functions looks like?

Linear functions like $f(x) = 2x + 5$

Quadratic functions like $g(x) = x^2 - 6x + 8$

Absolute value functions like $h(x) = |x|$

Square root functions like $k(x) = \sqrt{x - 3}$

1 Define an Exponential Function

In this section, we will learn about *exponential functions*.

Definition

An **exponential function** is a function of the form

$$f(x) = a^x$$

where $a > 0$, $a \neq 1$, and x is a real number.

Note

1) We stipulate that a is a positive number ($a > 0$) because if a were a negative number, some expressions would not be real numbers.

   Example: If $a = -2$ and $x = \dfrac{1}{2}$, we get $f(x) = (-2)^{1/2} = \sqrt{-2}$ (not real).

   Therefore, a *must* be a positive number.

2) We add the condition that $a \neq 1$ because if $a = 1$, the function would be linear, not exponential.

   Example: If $a = 1$, then $f(x) = 1^x$. This is equivalent to $f(x) = 1$, which is a linear function.

2 Graph $f(x) = a^x$

We can graph exponential functions by plotting points. *It is important to choose many values for the variable so that we obtain positive numbers, negative numbers, and zero in the exponent.*

EXAMPLE 1 Graph $f(x) = 2^x$ and $g(x) = 3^x$ on the same axes. Determine the domain and range.

Solution

Make a table of values for each function. Be sure to choose values for x that will give us *positive numbers, negative numbers, and zero* in the exponent.

W Hint
Be sure to choose values of x that will make the exponent *positive*, *negative*, and *zero* so that we can see the complete graph.

$f(x) = 2^x$	
x	$f(x)$
0	1
1	2
2	4
3	8
-1	$\dfrac{1}{2}$
-2	$\dfrac{1}{4}$

$g(x) = 3^x$	
x	$g(x)$
0	1
1	3
2	9
3	27
-1	$\dfrac{1}{3}$
-2	$\dfrac{1}{9}$

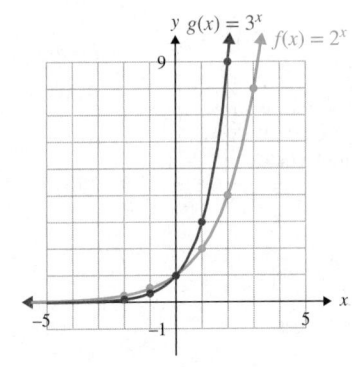

Plot each set of points, and connect them with a smooth curve. Note that the larger the value of a, the more rapidly the y-values increase. Additionally, as x increases, the value of y also increases. Here are some other interesting facts to note about the graphs of these functions.

1) Each graph passes the vertical line test so the graphs *do* represent functions.

2) Each graph passes the horizontal line test, so the functions are one-to-one.

3) The y-intercept of each function is $(0, 1)$.

4) The domain of each function is $(-\infty, \infty)$, and the range is $(0, \infty)$.

[YOU TRY 1] Graph $f(x) = 4^x$. Determine the domain and range.

EXAMPLE 2

Graph $f(x) = \left(\dfrac{1}{2}\right)^x$. Determine the domain and range.

Solution

Make a table of values and be sure to choose values for x that will give us *positive numbers, negative numbers, and zero* in the exponent.

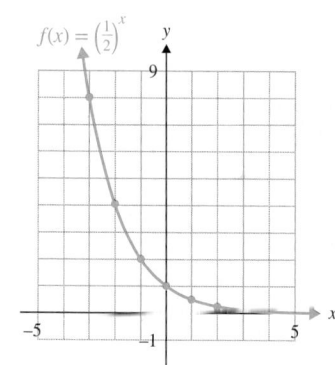

$f(x) = \left(\dfrac{1}{2}\right)^x$	
x	$f(x)$
0	1
1	$\dfrac{1}{2}$
2	$\dfrac{1}{4}$
-1	2
-2	4
-3	8

Like the graphs of $f(x) = 2^x$ and $g(x) = 3^x$ in Example 1, the graph of $f(x) = \left(\dfrac{1}{2}\right)^x$ passes both the vertical and horizontal line tests, making it a one-to-one function. The y-intercept is $(0, 1)$. The domain is $(-\infty, \infty)$, and the range is $(0, \infty)$.

In the case of $f(x) = \left(\dfrac{1}{2}\right)^x$, however, as the value of x increases, the value of y *decreases*. This is because $0 < a < 1$.

[YOU TRY 2] Graph $g(x) = \left(\dfrac{1}{3}\right)^x$. Determine the domain and range.

We can summarize what we have learned so far about exponential functions:

Summary Characteristics of $f(x) = a^x$, where $a > 0$ and $a \neq 1$

1) If $f(x) = a^x$ where $a > 1$, the value of y increases as the value of x increases.

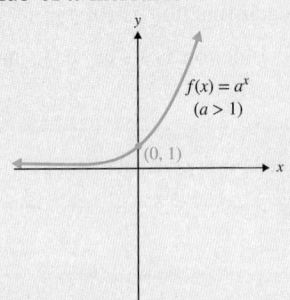

2) If $f(x) = a^x$, where $0 < a < 1$, the value of y decreases as the value of x increases.

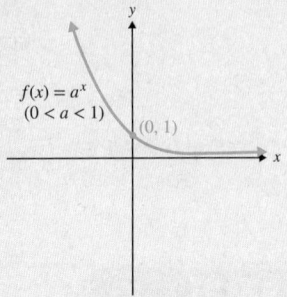

3) The function is one-to-one.

4) The *y*-intercept is $(0, 1)$.

5) The domain is $(-\infty, \infty)$, and the range is $(0, \infty)$.

3 Graph $f(x) = a^{x+c}$

Next we will graph an exponential function with an expression other than *x* as its exponent.

EXAMPLE 3

Graph $f(x) = 3^{x-2}$. Determine the domain and range.

Solution

Remember, for the table of values we want to choose values of *x* that will give us positive numbers, negative numbers, and zero *in the exponent*. First we will determine which value of *x* will make the exponent equal zero.

$$x - 2 = 0$$
$$x = 2$$

If $x = 2$, the exponent equals zero. Choose a couple of numbers *greater than* 2 and a couple that are *less than* 2 to get positive and negative numbers in the exponent.

	x	$x - 2$	$f(x) = 3^{x-2}$	Plot
	2	0	$3^0 = 1$	$(2, 1)$
Values greater than 2	3	1	$3^1 = 3$	$(3, 3)$
	4	2	$3^2 = 9$	$(4, 9)$
Values less than 2	1	−1	$3^{-1} = \dfrac{1}{3}$	$\left(1, \dfrac{1}{3}\right)$
	0	−2	$3^{-2} = \dfrac{1}{9}$	$\left(0, \dfrac{1}{9}\right)$

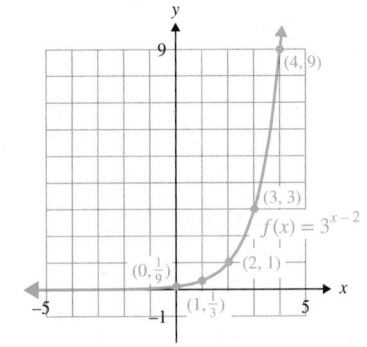

Note that the *y*-intercept is not $(0, 1)$ because the exponent is $x - 2$, *not x*, as in $f(x) = a^x$. The graph of $f(x) = 3^{x-2}$ is the same shape as the graph of $g(x) = 3^x$ except that the graph of *f* is shifted 2 units to the right. This is because $f(x) = g(x - 2)$. The domain of *f* is $(-\infty, \infty)$, and the range is $(0, \infty)$.

$$\left[\text{YOU TRY 3}\right]$$

Graph $f(x) = 2^{x+4}$. Determine the domain and range.

4 Define the Number e and Graph $f(x) = e^x$

Next we will introduce a special exponential function, one with a base of e.

Like the number π, e is an irrational number that has many uses in mathematics. In the 1700s, the work of Swiss mathematician Leonhard Euler led him to the approximation of e.

Definition

Approximation of e

$$e \approx 2.718281828459045235$$

One of the questions Euler set out to answer was, what happens to the value of $\left(1 + \dfrac{1}{n}\right)^n$ as n gets larger and larger? He found that as n gets larger, $\left(1 + \dfrac{1}{n}\right)^n$ gets closer to a fixed number. This number is e. Euler approximated e to the 18 decimal places in the definition, and the letter e was chosen to represent this number in his honor. It should be noted that there are other ways to generate e. Finding the value that $\left(1 + \dfrac{1}{n}\right)^n$ approaches as n gets larger and larger is just one way. Also, since e is irrational, it is a nonterminating, nonrepeating decimal.

EXAMPLE 4

Graph $f(x) = e^x$. Determine the domain and range.

Solution

A calculator is needed to generate a table of values. We will use either the $\boxed{e^x}$ key or the two keys $\boxed{\text{INV}}$ (or $\boxed{\text{2ND}}$) and $\boxed{\ln x}$ to find powers of e. (Calculators will approximate powers of e to a few decimal places.)

For example, if a calculator has an $\boxed{e^x}$ key, find e^2 by pressing the following keys:

$$\boxed{2}\ \boxed{e^x} \quad \text{or} \quad \boxed{e^x}\ \boxed{2}\ \boxed{\text{ENTER}}$$

To four decimal places, $e^2 \approx 7.3891$.

If a calculator has an $\boxed{\ln x}$ key with e^x written above it, find e^2 by pressing the following keys:

$$\boxed{2}\ \boxed{\text{INV}}\ \boxed{\ln x} \quad \text{or} \quad \boxed{\text{INV}}\ \boxed{\ln x}\ \boxed{2}\ \boxed{\text{ENTER}}$$

The same approximation for e^2 is obtained.

Remember to choose positive numbers, negative numbers, and zero for x when making the table of values. We will approximate the values of e^x to four decimal places.

$f(x) = e^x$	
x	$f(x)$
0	1
1	2.7183
2	7.3891
3	20.0855
−1	0.3679
−2	0.1353

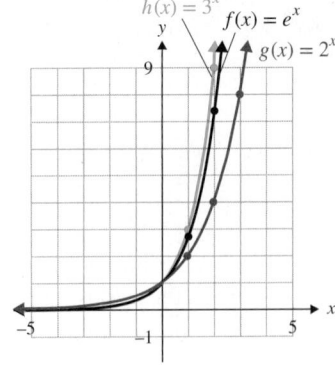

Notice that the graph of $f(x) = e^x$ is between the graphs of $g(x) = 2^x$ and $h(x) = 3^x$. This is because $2 < e < 3$, so e^x grows more quickly than 2^x, but e^x grows more slowly than 3^x. The domain of $f(x) = e^x$ is $(-\infty, \infty)$, and the range is $(0, \infty)$.

We will study e^x and its special properties in more detail later in the chapter.

5 Solve an Exponential Equation

An **exponential equation** is an equation that has a variable in the exponent. Some examples of exponential equations are

$$2^x = 8, \qquad 3^{a-5} = \frac{1}{9}, \qquad e^t = 14, \qquad 5^{2y-1} = 6^{y+4}$$

In this section, we will learn how to solve exponential equations like the first two examples. We can solve those equations by getting the same base.

We know that the exponential function $f(x) = a^x \ (a > 0, a \neq 1)$ is one-to-one. This leads to the following property that enables us to solve many exponential equations.

$$\text{If } a^x = a^y, \text{ then } x = y. \quad (a > 0, a \neq 1)$$

This property says that if two sides of an equation have the same base, set the exponents equal and solve for the unknown variable.

Procedure Solving an Exponential Equation

Step 1: **If possible, express each side of the equation with the same base.** If it is *not* possible to get the same base, a different method must be used. (This is presented in Section 13.6.)

Step 2: **Use the rules of exponents to simplify the exponents.**

Step 3: **Set the exponents equal, and solve for the variable.**

EXAMPLE 5

Solve each equation.

a) $2^x = 8$ b) $49^{c+3} = 7^{3c}$ c) $9^{6n} = 27^{n-4}$ d) $3^{a-5} = \dfrac{1}{9}$

Solution

a) **Step 1:** Express each side of the equation with the same base.

$$2^x = 8$$
$$2^x = 2^3 \qquad \text{Rewrite 8 with a base of 2: } 8 = 2^3.$$

W Hint

In your notes, write out each example as you are reading it.

Step 2: The exponents are simplified.

Step 3: Since the bases are the same, set the exponents equal and solve.

$$x = 3$$

The solution set is $\{3\}$.

b) *Step 1:* Express each side of the equation with the same base.

$$49^{c+3} = 7^{3c}$$
$$(7^2)^{c+3} = 7^{3c} \qquad \text{Both sides are powers of 7; } 49 = 7^2.$$

Step 2: Use the rules of exponents to simplify the exponents.

$$7^{2(c+3)} = 7^{3c} \qquad \text{Power rule for exponents}$$
$$7^{2c+6} = 7^{3c} \qquad \text{Distribute.}$$

Step 3: Since the bases are the same, set the exponents equal and solve.

$$2c + 6 = 3c \qquad \text{Set the exponents equal.}$$
$$6 = c \qquad \text{Subtract } 2c.$$

The solution set is $\{6\}$. Check the answer in the original equation.

W Hint

Do you see why it is important to know the powers of numbers? Review them, if necessary.

c) *Step 1:* Express each side of the equation with the same base. 9 *and* 27 *are each powers of* 3.

$$9^{6n} = 27^{n-4}$$
$$(3^2)^{6n} = (3^3)^{n-4} \qquad 9 = 3^2; \quad 27 = 3^3$$

Step 2: Use the rules of exponents to simplify the exponents.

$$3^{2(6n)} = 3^{3(n-4)} \qquad \text{Power rule for exponents}$$
$$3^{12n} = 3^{3n-12} \qquad \text{Multiply.}$$

Step 3: Since the bases are the same, set the exponents equal and solve.

$$12n = 3n - 12 \qquad \text{Set the exponents equal.}$$
$$9n = -12 \qquad \text{Subtract } 3n.$$
$$n = -\frac{12}{9} = -\frac{4}{3} \qquad \text{Divide by 9; simplify.}$$

The solution set is $\left\{-\dfrac{4}{3}\right\}$.

d) *Step 1:* Express each side of the equation $3^{a-5} = \dfrac{1}{9}$ with the same base. $\dfrac{1}{9}$ *can be expressed with a base of* 3: $\dfrac{1}{9} = \left(\dfrac{1}{3}\right)^2 = 3^{-2}.$

$$3^{a-5} = \frac{1}{9}$$

$$3^{a-5} = 3^{-2} \qquad \text{Rewrite } \frac{1}{9} \text{ with a base of 3.}$$

Step 2: The exponents are simplified.

Step 3: Set the exponents equal and solve.

$$a - 5 = -2 \qquad \text{Set the exponents equal.}$$
$$a = 3 \qquad \text{Add 5.}$$

The solution set is $\{3\}$.

[YOU TRY 4] Solve each equation.

a) $(12)^x = 144$ b) $6^{t-5} = 36^{t+4}$ c) $32^{2w} = 8^{4w-1}$ d) $8^k = \dfrac{1}{64}$

6 Solve an Applied Problem Using a Given Exponential Function

EXAMPLE 6

The value of a car depreciates (decreases) over time. The value, $V(t)$, in dollars, of a sedan t yr after it is purchased is given by

$$V(t) = 18{,}200(0.794)^t$$

a) What was the purchase price of the car?

b) What will the car be worth 5 yr after purchase?

Solution

a) To find the purchase price of the car, let $t = 0$. Evaluate $V(0)$ given that $V(t) = 18{,}200(0.794)^t$.

$$V(0) = 18{,}200(0.794)^0$$
$$= 18{,}200(1)$$
$$= 18{,}200$$

The purchase price of the car was $18,200.

b) To find the value of the car after 5 yr, let $t = 5$. Use a calculator to find $V(5)$.

$$V(5) = 18{,}200(0.794)^5$$
$$= 5743.46$$

The car will be worth about $5743.46.

[YOU TRY 5]

The value, $V(t)$, in dollars, of a pickup truck t yr after it is purchased is given by

$$V(t) = 23{,}500(0.785)^t$$

a) What was the purchase price of the pickup?

b) What will the pickup truck be worth 4 yr after purchase?

ANSWERS TO [YOU TRY] EXERCISES

1)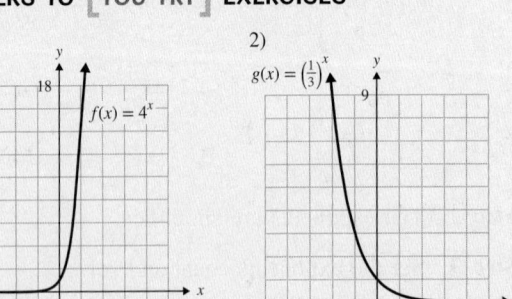

domain: $(-\infty, \infty)$; range: $(0, \infty)$

2)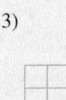

domain: $(-\infty, \infty)$; range: $(0, \infty)$

3)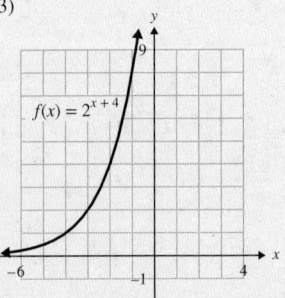

domain: $(-\infty, \infty)$; range: $(0, \infty)$

4) a) $\{2\}$ b) $\{-13\}$ c) $\left\{\dfrac{3}{2}\right\}$ d) $\{-2\}$ 5) a) $23,500 b) $8923.73

Mixed Exercises: Objectives 1 and 2

1) When making a table of values to graph an exponential function, what kinds of values should be chosen for the variable?

2) What is the y-intercept of the graph of $f(x) = a^x$ where $a > 0$ and $a \neq 1$?

Graph each exponential function. Determine the domain and range.

3) $y = 2^x$

4) $g(x) = 4^x$

5) $f(x) = 5^x$

6) $f(x) = 3^x$

7) $h(x) = \left(\dfrac{1}{3}\right)^x$

8) $y = \left(\dfrac{1}{4}\right)^x$

For an exponential function of the form $f(x) = a^x$ $(a > 0, a \neq 1)$, answer the following.

9) What is the domain?

10) What is the range?

Objective 3: Graph $f(x) = a^{x+c}$

Graph each exponential function. State the domain and range.

11) $g(x) = 2^{x+1}$

12) $y = 3^{x+2}$

13) $f(x) = 3^{x-4}$

14) $h(x) = 2^{x-3}$

15) $f(x) = 2^{2x}$

16) $h(x) = 3^{\frac{1}{2}x}$

17) $y = 2^x + 1$

18) $f(x) = 2^x - 3$

19) $g(x) = 3^x - 2$

20) $h(x) = 3^x + 1$

21) $y = -2^x$

22) $f(x) = -\left(\dfrac{1}{3}\right)^x$

23) As the value of x gets larger, would you expect $f(x) = 2x$ or $g(x) = 2^x$ to grow faster? Why?

24) Let $f(x) = \left(\dfrac{1}{5}\right)^x$. The graph of $f(x)$ gets very close to the line $y = 0$ (the x-axis) as the value of x gets larger. Why?

25) If you are given the graph of $f(x) = a^x$, where $a > 0$ and $a \neq 1$, how would you obtain the graph of $g(x) = a^x - 2$?

26) If you are given the graph of $f(x) = a^x$, where $a > 0$ and $a \neq 1$, how would you obtain the graph of $g(x) = a^{x-3}$?

Objective 4: Define the Number e and Graph $f(x) = e^x$

27) What is the approximate value of e to four decimal places?

28) Is e a rational or an irrational number? Explain your answer.

For Exercises 29–32, match each exponential function with its graph.

A)

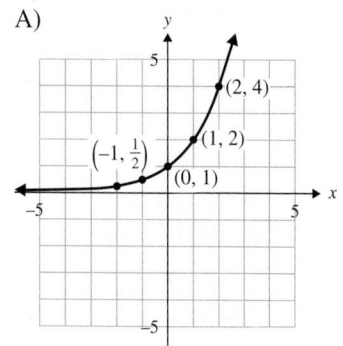

B)

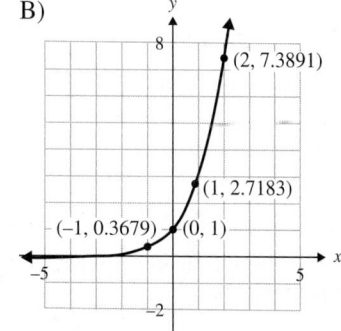

C)

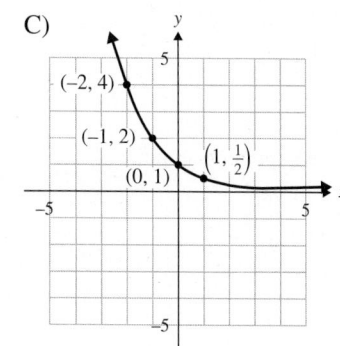

D)

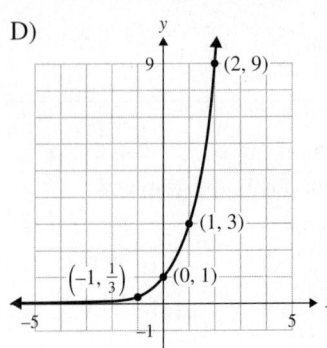

29) $f(x) = e^x$

30) $g(x) = 2^x$

31) $h(x) = 3^x$

32) $k(x) = \left(\dfrac{1}{2}\right)^x$

Graph each function. State the domain and range.

33) $f(x) = e^x - 2$

34) $g(x) = e^x + 1$

35) $y = e^{x+1}$

36) $h(x) = e^{x-3}$

37) $g(x) = \dfrac{1}{2}e^x$

38) $y = 2e^x$

39) $h(x) = -e^x$

40) $f(x) = e^{-x}$

41) Graph $y = e^x$, and compare it with the graph of $h(x) = -e^x$ in Exercise 39. What can you say about these graphs?

42) Graph $y = e^x$, and compare it with the graph of $f(x) = e^{-x}$ in Exercise 40. What can you say about these graphs?

Objective 5: Solve an Exponential Equation

Solve each exponential equation.

Fill It In

Fill in the blanks with either the missing mathematical step or the reason for the given step.

43) $6^{3n} = 36^{n-4}$

_____   Express each side with the same base.

$6^{3n} = 6^{2(n-4)}$ _____

$6^{3n} = 6^{2n-8}$ _____

_____   Set the exponents equal.

_____   Solve for n.

The solution set is _____.

44) $125^{2w} = 5^{w+2}$

$(5^3)^{2w} = 5^{w+2}$ _____

_____   Power rule for exponents

$5^{6w} = 5^{w+2}$ _____

$6w = w + 2$ _____

_____   Solve for w.

The solution set is _____.

45) $9^x = 81$

46) $4^y = 16$

47) $5^{4d} = 125$

48) $4^{3a} = 64$

49) $3^{5t} = 9^{t+4}$

50) $16^{m-2} = 2^{3m}$

51) $(1000)^{2p-3} = 10^{4p+1}$

52) $7^{2k-6} = 49^{3k+1}$

53) $32^{3c} = 8^{c+4}$

54) $(125)^{2x-9} = 25^{x-3}$

55) $81^{3n+9} = 27^{2n+6}$

56) $100^{5z-1} = (1000)^{2z+7}$

57) $27^{5v} = 9^{v+4}$

58) $32^{y+1} = 64^{y+2}$

59) $6^x = \dfrac{1}{36}$

60) $11^t = \dfrac{1}{121}$

61) $2^a = \dfrac{1}{8}$

62) $3^z = \dfrac{1}{81}$

63) $9^r = \dfrac{1}{27}$

64) $16^c = \dfrac{1}{8}$

65) $\left(\dfrac{5}{6}\right)^{3x+7} = \left(\dfrac{36}{25}\right)^{2x}$

66) $\left(\dfrac{7}{2}\right)^{5w} = \left(\dfrac{4}{49}\right)^{4w+3}$

67) $\left(\dfrac{3}{4}\right)^{5k} = \left(\dfrac{27}{64}\right)^{k+1}$

68) $\left(\dfrac{3}{2}\right)^{y+4} = \left(\dfrac{81}{16}\right)^{y-2}$

Objective 6: Solve an Applied Problem Using a Given Exponential Function

Solve each application.

69) The value of a car depreciates (decreases) over time. The value, $V(t)$, in dollars, of an SUV t yr after it is purchased is given by

$$V(t) = 32{,}700(0.812)^t$$

a) What was the purchase price of the SUV?

b) What will the SUV be worth 3 yr after purchase?

70) The value, $V(t)$, in dollars, of a sports car t yr after it is purchased is given by

$$V(t) = 48{,}600(0.820)^t$$

a) What was the purchase price of the sports car?

b) What will the sports car be worth 4 yr after purchase?

71) From 2009 to 2019, the value of homes in a suburb increased by 3% per year. The value, $V(t)$, in dollars, of a particular house t yr after 2009 is given by

$$V(t) = 185,200(1.03)^t$$

a) How much was the house worth in 2009?

b) How much was the house worth in 2016?

72) From 2009 to 2019, the value of condominiums in a big city high-rise building increased by 2% per year. The value, $V(t)$, in dollars, of a particular condo t yr after 2009 is given by

$$V(t) = 420,000(1.02)^t$$

Mike Tauber/Blend Images LLC

a) How much was the condominium worth in 2009?

b) How much was the condominium worth in 2019?

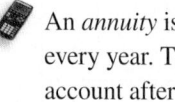 An *annuity* is an account into which money is deposited every year. The amount of money, A in dollars, in the account after t yr of depositing c dollars at the beginning of every year earning an interest rate r (as a decimal) is

$$A = c\left[\frac{(1 + r)^t - 1}{r}\right](1 + r)$$

Use the formula for Exercises 73 and 74.

73) After Fernando's daughter is born, he decides to begin saving for her college education. He will deposit $2000 every year in an annuity for 18 yr at a rate of 9%. How much will be in the account after 18 yr?

74) To save for retirement, Susan plans to deposit $6000 per year in an annuity for 30 yr at a rate of 8.5%. How much will be in the account after 30 yr?

 **R Rethink**

R1) What are some ways that graphing exponential functions is similar to and different from graphing a linear function?

R2) How do the graphs of exponential functions differ when the base is greater than 1

75) After taking a certain antibiotic, the amount of amoxicillin $A(t)$, in milligrams, remaining in the patient's system t hr after taking 1000 mg of amoxicillin is

$$A(t) = 1000e^{-0.5332t}$$

How much amoxicillin is in the patient's system 6 hr after taking the medication?

76) Some cockroaches can reproduce according to the formula

$$y = 2(1.65)^t$$

where y is the number of cockroaches resulting from the mating of two cockroaches and their offspring t months after the first two cockroaches mate.

If Morris finds two cockroaches in his kitchen (assuming one is male and one is female) how large can the cockroach population become after 12 months?

77) Jameson Irish whiskey is one of the largest single-distillery whiskey producers in the world. After distillation, the liquid is poured into wooden maturation casks to age. Each cask is filled with 200 L of the liquid, and each year, 2% of the liquid is lost to evaporation. What remains in the cask is the whiskey. (www.jamesonwhiskey.com)

a) Write an exponential function to determine the amount of whiskey, $A(t)$, that remains in a cask t years after it has been filled with the liquid.

b) Evaluate the function in part a) for $t = 12$, and explain the meaning of the answer. Round your answer to the nearest whole number.

c) If the whiskey has aged for 18 years, how much is in the cask? Round your answer to the nearest whole number.

d) In general, the longer that a whiskey has been aged, the more expensive it is to buy. Thinking about the answers to parts b) and c), why might this be true?

compared to a nonzero base that is between -1 and 1?

R3) Make your own example of an exponential equation like those in this section. Solve it and, in your own words, explain how you solved it.

13.3 Logarithmic Functions

P Prepare **O Organize**

What are your objectives for Section 13.3?	How can you accomplish each objective?
1 Define a Logarithm	• Learn the definition of a *logarithm*. • Know that $a > 0$, $x > 0$, and $a \neq 1$. • In your notes, write the relationship between the logarithmic form of an equation and the exponential form of an equation.
2 Convert from Logarithmic Form to Exponential Form	• Review the conversion from log form to exponential form. • Complete the given example on your own. • Complete You Try 1.
3 Convert from Exponential Form to Logarithmic Form	• Review the conversion from exponential form to log form. • Complete the given example on your own. • Complete You Try 2.
4 Solve an Equation of the Form $\log_a b = c$	• Be able to identify a *logarithmic equation*. • Learn the procedure for **Solving an Equation of the Form $\log_a b = c$.** • Complete the given example on your own. • Complete You Try 3.
5 Evaluate a Logarithm	• Know what it means to evaluate a logarithmic expression. • Complete the given example on your own. • Complete You Try 4.
6 Evaluate Common Logarithms, and Solve Equations of the Form $\log b = c$	• Learn the definition of a *common logarithm*. • Complete the given examples on your own. • Complete You Trys 5 and 6.
7 Use the Properties $\log_a a = 1$ and $\log_a 1 = 0$	• Learn the **Properties of Logarithms** in this section. • Complete the given example on your own. • Complete You Try 7.
8 Define and Graph a Logarithmic Function	• Learn the definition for a *logarithmic function*. • When graphing a logarithmic function, choose values for y in the table of values. • Review the **Summary of Characteristics of Logarithmic Functions.** • Complete the given examples on your own. • Complete You Trys 8–10.
9 Solve an Applied Problem Using a Logarithmic Equation	• Read the given example carefully. • Choose the correct value for t, and be sure you understand what this value represents. • Complete the given example on your own. • Complete You Try 11.

 **Work**   Read the explanations, follow the examples, take notes, and complete the You Trys.

1 Define a Logarithm

In Section 13.2, we graphed $f(x) = 2^x$ by making a table of values and plotting the points. The graph passes the horizontal line test, making the function one-to-one. Recall that if (x, y) is on the graph of a function, then (y, x) is on the graph of its inverse. We can graph the inverse of $f(x) = 2^x$, $f^{-1}(x)$, by switching the x- and y-coordinates in the table of values and plotting the points.

$f(x) = 2^x$	
x	$y = f(x)$
0	1
1	2
2	4
3	8
-1	$\dfrac{1}{2}$
-2	$\dfrac{1}{4}$

$f^{-1}(x)$	
x	$y = f^{-1}(x)$
1	0
2	1
4	2
8	3
$\dfrac{1}{2}$	-1
$\dfrac{1}{4}$	-2

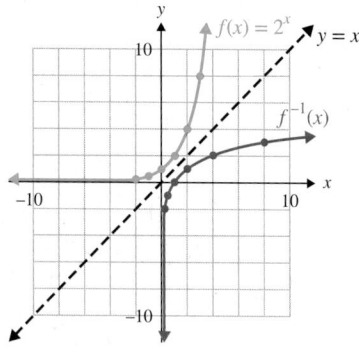

Above is the graph of $f(x) = 2^x$ and its inverse. Notice that, like the graphs of all functions and their inverses, they are symmetric with respect to the line $y = x$.

What is the equation of $f^{-1}(x)$ if $f(x) = 2^x$? We will use the procedure outlined in Section 13.1 to find the equation of $f^{-1}(x)$.

If $f(x) = 2^x$, then find the equation of $f^{-1}(x)$ as follows:

Step 1: Replace $f(x)$ with y.

$$y = 2^x$$

Step 2: Interchange x and y.

$$x = 2^y$$

Step 3: Solve for y.

How do we solve $x = 2^y$ for y? To answer this question, we must introduce another concept called *logarithms*.

Definition

Definition of Logarithm: If $a > 0$, $a \neq 1$, and $x > 0$, then for every real number y,

$$y = \log_a x \text{ means } x = a^y.$$

The word **log** is an abbreviation for **logarithm.** We read $\log_a x$ as "log of x to the base a" or "log to the base a of x." *This definition of a logarithm should be memorized!*

Note

It is very important to note that the base of the logarithm must be positive and not equal to 1, and that x must be positive as well.

The relationship between the logarithmic form of an equation ($y = \log_a x$) and the exponential form of an equation ($x = a^y$) is one that has many uses. Notice the relationship between the two forms.

Logarithmic Form	Exponential Form
Value of the logarithm	Exponent
$\downarrow$	$\downarrow$
$y = \log_a x$	$x = a^y$
$\uparrow$	$\uparrow$
Base	Base

From the above, you can see that *a logarithm is an exponent*. $\log_a x$ is the power to which we raise a to get x.

2 Convert from Logarithmic Form to Exponential Form

Much of our work with logarithms involves converting between logarithmic and exponential notation. After working with logs and exponential form, we will come back to the question of how to solve $x = 2^y$ for y.

EXAMPLE 1

Write in exponential form.

a) $\log_6 36 = 2$ b) $\log_4 \dfrac{1}{64} = -3$ c) $\log_7 1 = 0$

Solution

a) $\log_6 36 = 2$ means that 2 is the power to which we raise 6 to get 36. The exponential form is $6^2 = 36$.

$$\log_6 36 = 2 \text{ means } 6^2 = 36.$$

W Hint

You will need to understand this example before you can do Example 3.

b) $\log_4 \dfrac{1}{64} = -3$ means $4^{-3} = \dfrac{1}{64}$.

c) $\log_7 1 = 0$ means $7^0 = 1$.

$\left[\text{YOU TRY 1}\right]$ Write in exponential form.

a) $\log_3 81 = 4$ b) $\log_5 \dfrac{1}{25} = -2$ c) $\log_{64} 8 = \dfrac{1}{2}$ d) $\log_{13} 13 = 1$ e) $\log_9 1 = 0$

3 Convert from Exponential Form to Logarithmic Form

EXAMPLE 2

Write in logarithmic form.

a) $10^4 = 10{,}000$ b) $9^{-2} = \dfrac{1}{81}$ c) $8^1 = 8$

d) $\sqrt{25} = 5$

Solution

a) $10^4 = 10,000$ means $\log_{10} 10,000 = 4$.

b) $9^{-2} = \dfrac{1}{81}$ means $\log_9 \dfrac{1}{81} = -2$.

c) $8^1 = 8$ means $\log_8 8 = 1$.

d) To write $\sqrt{25} = 5$ in logarithmic form, rewrite $\sqrt{25}$ as $25^{1/2}$.

$25^{1/2} = 5$ means $\log_{25} 5 = \dfrac{1}{2}$.

Note

When working with logarithms, we will often change radical notation to the equivalent fractional exponent. This is because a logarithm *is* an exponent.

[**YOU TRY 2**] Write in logarithmic form.

a) $7^2 = 49$ b) $5^{-4} = \dfrac{1}{625}$ c) $19^0 = 1$ d) $\sqrt{144} = 12$

---

4 Solve an Equation of the Form $\log_a b = c$

A **logarithmic equation** is an equation in which at least one term contains a logarithm. In this section, we will learn how to solve a logarithmic equation of the form $\log_a b = c$. We will learn how to solve other types of logarithmic equations in Sections 13.5 and 13.6.

Procedure Solve an Equation of the Form $\log_a b = c$

To solve a logarithmic equation of the form $\log_a b = c$, write the equation in exponential form ($a^c = b$) and solve for the variable. Check the answer in the original equation.

EXAMPLE 3   Solve each logarithmic equation.

a) $\log_{10} r = 3$ b) $\log_w 25 = 2$ c) $\log_3 (7a + 18) = 4$ d) $\log_2 16 = c$

e) $\log_{36} \sqrt[4]{6} = x$

Solution

a) Write the equation in exponential form, and solve for r.

$$\log_{10} r = 3 \quad \text{means} \quad 10^3 = r$$
$$1000 = r$$

The solution set is $\{1000\}$.

b) Write $\log_w 25 = 2$ in exponential form and solve for w.

$$\log_w 25 = 2 \quad \text{means} \quad w^2 = 25$$
$$w = \pm 5 \qquad \text{Square root property}$$

Although we get $w = 5$ or $w = -5$ when we solve $w^2 = 25$, recall that the base of a logarithm must be a positive number. Therefore, $w = -5$ is *not* a solution of the original equation.

The solution set is $\{5\}$.

c) Write $\log_3(7a + 18) = 4$ in exponential form and solve for a.

$$\log_3(7a + 18) = 4 \quad \text{means} \quad 3^4 = 7a + 18$$
$$81 = 7a + 18$$
$$63 = 7a \qquad \text{Subtract 18.}$$
$$9 = a \qquad \text{Divide by 7.}$$

The solution set is $\{9\}$. The check is left to the student.

d) Write $\log_2 16 = c$ in exponential form, and solve for c.

$$\log_2 16 = c \quad \text{means} \quad 2^c = 16$$
$$c = 4$$

Verify that the solution set is $\{4\}$.

e) $\log_{36} \sqrt[4]{6} = x \quad \text{means} \quad 36^x = \sqrt[4]{6}$

$$(6^2)^x = 6^{1/4} \qquad \text{Express each side with the same base; rewrite the radical as a fractional exponent.}$$
$$6^{2x} = 6^{1/4} \qquad \text{Power rule for exponents}$$
$$2x = \frac{1}{4} \qquad \text{Set the exponents equal.}$$
$$x = \frac{1}{8} \qquad \text{Divide by 2.}$$

The solution set is $\left\{\dfrac{1}{8}\right\}$. The check is left to the student.

[YOU TRY 3] Solve each logarithmic equation.

a) $\log_2 y = 5$ b) $\log_x 169 = 2$ c) $\log_6 36 = n$

d) $\log_{64} \sqrt[5]{8} = k$ e) $\log_5(3p + 11) = 3$

---

5 Evaluate a Logarithm

Often when working with logarithms, we are asked to *evaluate* them or to find the value of a log.

EXAMPLE 4

Evaluate.

a) $\log_3 9$ b) $\log_{10} \dfrac{1}{10}$ c) $\log_{25} 5$

Solution

a) To *evaluate* (or *find the value of*) $\log_3 9$ means to find the power to which we raise 3 to get 9. That power is **2.**

$$\log_3 9 = 2 \qquad \text{because} \qquad 3^2 = 9$$

b) To evaluate $\log_{10} \dfrac{1}{10}$ means to find the power to which we raise 10 to get $\dfrac{1}{10}$.

That power is **−1.**

If you don't see that this is the answer, set the expression $\log_{10} \dfrac{1}{10}$ equal to x, write the equation in exponential form, and solve for x as in Example 3.

$$\log_{10} \dfrac{1}{10} = x \quad \text{means} \quad 10^x = \dfrac{1}{10}$$
$$10^x = 10^{-1} \qquad \dfrac{1}{10} = 10^{-1}$$
$$x = -1$$

Then, $\log_{10} \dfrac{1}{10} = -1$.

c) To evaluate $\log_{25} 5$ means to find the power to which we raise 25 to get 5. That power is $\dfrac{1}{2}$.

Once again, we can also find the value of $\log_{25} 5$ by setting it equal to x, writing the equation in exponential form, and solving for x.

$$\log_{25} 5 = x \quad \text{means} \quad 25^x = 5$$
$$(5^2)^x = 5 \qquad \text{Express each side with the same base.}$$
$$5^{2x} = 5^1 \qquad \text{Power rule; } 5 = 5^1$$
$$2x = 1 \qquad \text{Set the exponents equal.}$$
$$x = \dfrac{1}{2} \qquad \text{Divide by 2.}$$

Therefore, $\log_{25} 5 = \dfrac{1}{2}$.

[YOU TRY 4]

Evaluate.

a) $\log_{10} 100$ 　　　b) $\log_8 \dfrac{1}{8}$ 　　　c) $\log_{144} 12$

---

6 Evaluate Common Logarithms, and Solve Equations of the Form log b = c

Logarithms have many applications not only in mathematics but also in other areas such as chemistry, biology, engineering, and economics.

Since our number system is a base 10 system, logarithms to the base 10 are very widely used and are called **common logarithms** or **common logs.** A base 10 log has a special notation—$\log_{10} x$ is written as $\log x$. *When a log is written in this way, the base is assumed to be* 10.

W Hint

Write the definition of a common logarithm in your notes.

$$\log x \text{ means } \log_{10} x$$

We must keep this in mind when evaluating logarithms and when solving logarithmic equations.

EXAMPLE 5

Evaluate log 100.

Solution

log 100 is equivalent to $\log_{10} 100$. To evaluate log 100 means to find the power to which we raise 10 to get 100. That power is **2.**

$$\log 100 = 2$$

[YOU TRY 5]

Evaluate log 1000.

EXAMPLE 6

Solve $\log(3x - 8) = 1$.

Solution

$\log(3x - 8) = 1$ is equivalent to $\log_{10}(3x - 8) = 1$. Write the equation in exponential form, and solve for x.

$$\log(3x - 8) = 1 \quad \text{means} \quad 10^1 = 3x - 8$$
$$10 = 3x - 8$$
$$18 = 3x \qquad \text{Add 8.}$$
$$6 = x \qquad \text{Divide by 3.}$$

Check $x = 6$ in the original equation. The solution set is $\{6\}$.

[YOU TRY 6]

Solve $\log(12q + 16) = 2$.

We will study common logs in more depth in Section 13.5.

7 Use the Properties $\log_a a = 1$ and $\log_a 1 = 0$

There are a couple of properties of logarithms that can simplify our work.

If a is any real number, then $a^1 = a$. Furthermore, if $a \neq 0$, then $a^0 = 1$. Write $a^1 = a$ and $a^0 = 1$ in logarithmic form to obtain these two properties of logarithms:

Properties of Logarithms

If $a > 0$ and $a \neq 1$,

1) $\log_a a = 1$
2) $\log_a 1 = 0$

EXAMPLE 7

Use the properties of logarithms to evaluate each.

a) $\log_{12} 1$ b) $\log 10$

Solution

a) By Property 2, $\log_{12} 1 = 0$.

b) The base of log 10 is 10. Therefore, $\log 10 = \log_{10} 10$. By Property 1, $\log 10 = 1$.

[**YOU TRY 7**] Use the properties of logarithms to evaluate each.

a) $\log_{1/3} 1$ b) $\log_{\sqrt{11}} \sqrt{11}$

8 Define and Graph a Logarithmic Function

Next we define a logarithmic function.

Definition

For $a > 0$, $a \neq 1$, and $x > 0$, $f(x) = \log_a x$ is the **logarithmic function with base a.**

Note

$f(x) = \log_a x$ can also be written as $y = \log_a x$. Changing $y = \log_a x$ to exponential form, we get $a^y = x$. Remembering that a is a *positive number not equal to* 1, it follows that
1) any real number may be substituted for y. Therefore, **the range of $y = \log_a x$ is $(-\infty, \infty)$.**
2) x must be a positive number. So, **the domain of $y = \log_a x$ is $(0, \infty)$.**

Let's return to the problem of finding the equation of the inverse of $f(x) = 2^x$ that was first introduced at the beginning of this section.

EXAMPLE 8

Find the equation of the inverse of $f(x) = 2^x$.

Solution

Step 1: Replace $f(x)$ with y: $y = 2^x$

Step 2: Interchange x and y: $x = 2^y$

Step 3: Solve for y.

To solve $x = 2^y$ for y, write the equation in logarithmic form.

$$x = 2^y \quad \text{means} \quad y = \log_2 x$$

Step 4: Replace y with $f^{-1}(x)$.

$$f^{-1}(x) = \log_2 x$$

The inverse of the exponential function $f(x) = 2^x$ is $f^{-1}(x) = \log_2 x$.

YOU TRY 8   Find the equation of the inverse of $f(x) = 6^x$.

W Hint

Summarize this Note box in your notes.

Note

The inverse of the exponential function $f(x) = a^x$ (where $a > 0$, $a \neq 1$, and x is any real number) is $f^{-1}(x) = \log_a x$. Furthermore,

1) the domain of $f(x)$ is the range of $f^{-1}x$.
2) the range of $f(x)$ is the domain of $f^{-1}x$.

Their graphs are symmetric with respect to $y = x$.

To graph a logarithmic function, write it in exponential form first. Then make a table of values, plot the points, and draw the curve through the points.

EXAMPLE 9   Graph $f(x) = \log_2 x$.

Solution

Substitute y for $f(x)$ and write the equation in exponential form.

$$y = \log_2 x \quad \text{means} \quad 2^y = x$$

To make a table of values, it will be easier to *choose values for y* and compute the corresponding values of x. Remember to choose values of y that will give positive numbers, negative numbers, and zero in the exponent.

W Hint

This is the result from Example 8!

$2^y = x$	
x	**y**
1	0
2	1
4	2
8	3
$\dfrac{1}{2}$	-1
$\dfrac{1}{4}$	-2

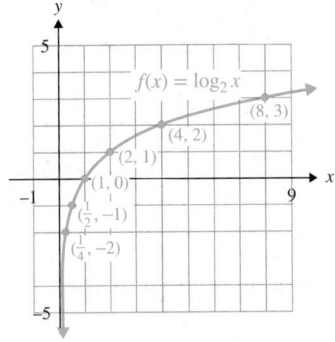

From the graph, we can see that the domain of f is $(0, \infty)$, and the range of f is $(-\infty, \infty)$.

YOU TRY 9   Graph $f(x) = \log_4 x$.

EXAMPLE 10   Graph $f(x) = \log_{1/3} x$.

Solution

Substitute y for $f(x)$ and write the equation in exponential form.

$$y = \log_{1/3} x \quad \text{means} \quad \left(\frac{1}{3}\right)^y = x$$

For the table of values, *choose values for* y and compute the corresponding values of x.

$\left(\dfrac{1}{3}\right)^y = x$	
x	**y**
1	0
$\dfrac{1}{3}$	1
$\dfrac{1}{9}$	2
3	−1
9	−2

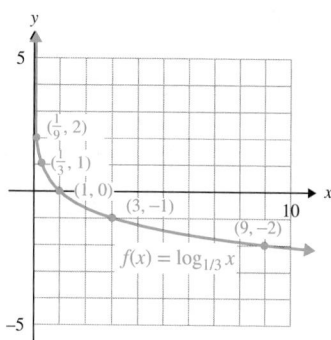

The domain of f is $(0, \infty)$, and the range is $(-\infty, \infty)$.

YOU TRY 10   Graph $f(x) = \log_{1/4} x$.

The graphs in Examples 9 and 10 are typical of the graphs of logarithmic functions—Example 9 for functions where $a > 1$ and Example 10 for functions where $0 < a < 1$. Next is a summary of some characteristics of logarithmic functions.

Summary Characteristics of a Logarithmic Function $f(x) = \log_a x$, where $a > 0$ and $a \neq 1$

1) If $f(x) = \log_a x$ where $a > 1$, the value of y increases as the value of x increases.

2) If $f(x) = \log_a x$ where $0 < a < 1$, the value of y decreases as the value of x increases.

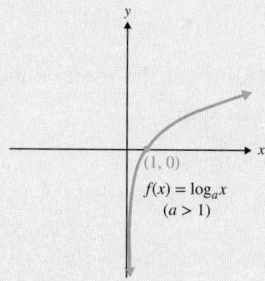
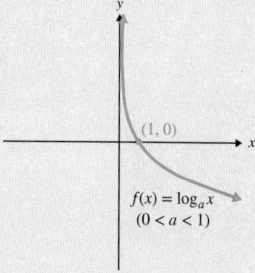

3) The function is one-to-one.

4) The x-intercept is $(1, 0)$.

5) The domain is $(0, \infty)$, and the range is $(-\infty, \infty)$.

6) The inverse of $f(x) = \log_a x$ is $f^{-1}(x) = a^x$.

Compare these characteristics of logarithmic functions to the characteristics of exponential functions on p. 910 in Section 13.2. The domain and range of logarithmic and exponential functions are interchanged because they are inverse functions.

9 Solve an Applied Problem Using a Logarithmic Equation

EXAMPLE 11

image100/PunchStock

A hospital has found that the function $A(t) = 50 + 8 \log_2(t + 2)$ approximates the number of people treated each year since 2005 for severe allergic reactions to peanuts. If $t = 0$ represents the year 2005, answer the following.

a) How many people were treated in 2005?

b) How many people were treated in 2011?

c) In what year were approximately 82 people treated for allergic reactions to peanuts?

Solution

a) The year 2005 corresponds to $t = 0$. Let $t = 0$, and find $A(0)$.

$$
\begin{aligned}
A(0) &= 50 + 8 \log_2(0 + 2) && \text{Substitute 0 for } t. \\
&= 50 + 8 \log_2 2 \\
&= 50 + 8(1) && \log_2 2 = 1 \\
&= 58
\end{aligned}
$$

In 2005, 58 people were treated for peanut allergies.

b) The year 2011 corresponds to $t = 6$. Let $t = 6$, and find $A(6)$.

$$
\begin{aligned}
A(6) &= 50 + 8 \log_2(6 + 2) && \text{Substitute 6 for } t. \\
&= 50 + 8 \log_2 8 \\
&= 50 + 8(3) && \log_2 8 = 3 \\
&= 50 + 24 \\
&= 74
\end{aligned}
$$

In 2011, 74 people were treated for peanut allergies.

c) To determine in what year 82 people were treated, let $A(t) = 82$ and solve for t.

$$82 = 50 + 8 \log_2(t + 2) \qquad \text{Substitute 82 for } A(t).$$

To solve for t, we first need to get the term containing the logarithm on a side by itself. Subtract 50 from each side.

$$
\begin{aligned}
32 &= 8 \log_2(t + 2) && \text{Subtract 50.} \\
4 &= \log_2(t + 2) && \text{Divide by 8.} \\
2^4 &= t + 2 && \text{Write in exponential form.} \\
16 &= t + 2 \\
14 &= t
\end{aligned}
$$

$t = 14$ corresponds to the year 2019. (Add 14 to the year 2005.)
82 people were treated for peanut allergies in 2019.

[YOU TRY 11]

The amount of garbage (in millions of pounds) collected in a certain town each year since 2010 can be approximated by $G(t) = 6 + \log_2(t + 1)$, where $t = 0$ represents the year 2010.

a) How much garbage was collected in 2010?

b) How much garbage was collected in 2017?

c) In what year would it be expected that 11,000,000 pounds of garbage will be collected? [Hint: Let $G(t) = 11$.]

ANSWERS TO [YOU TRY] EXERCISES

1) a) $3^4 = 81$ b) $5^{-2} = \dfrac{1}{25}$ c) $64^{1/2} = 8$ d) $13^1 = 13$ e) $9^0 = 1$

2) a) $\log_7 49 = 2$ b) $\log_5 \dfrac{1}{625} = -4$ c) $\log_{19} 1 = 0$ d) $\log_{144} 12 = \dfrac{1}{2}$

3) a) {32}   b) {13}   c) {2}   d) $\left\{\dfrac{1}{10}\right\}$ e) {38}

4) a) 2   b) -1 c) $\dfrac{1}{2}$ 5) 3   6) {7}   7) a) 0   b) 1   8) $f^{-1}(x) = \log_6 x$

9)

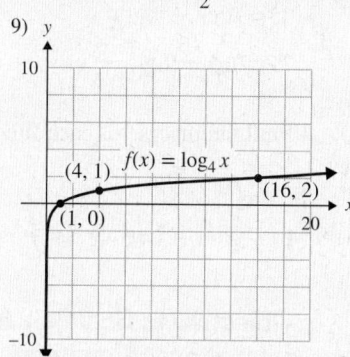

10)

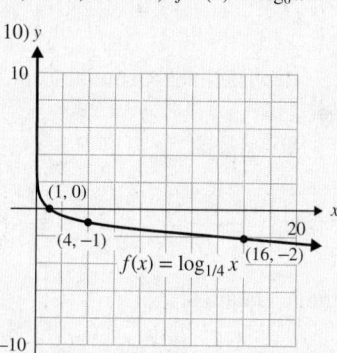

11) a) 6,000,000 lb   b) 9,000,000 lb   c) 2041

E Evaluate **13.3** Exercises

Do the exercises, and check your work.

Mixed Exercises: Objectives 1 and 2

1) In the equation $y = \log_a x$, a must be what kind of number?

2) In the equation $y = \log_a x$, x must be what kind of number?

3) What is the base of $y = \log x$?

4) A base 10 logarithm is called a _____ logarithm.

Write in exponential form.

5) $\log_7 49 = 2$

6) $\log_{11} 121 = 2$

7) $\log 1,000,000 = 6$

8) $\log 10,000 = 4$

9) $\log_9 \dfrac{1}{81} = -2$

10) $\log_8 \dfrac{1}{64} = -2$

11) $\log_{25} 5 = \dfrac{1}{2}$

12) $\log_{64} 4 = \dfrac{1}{3}$

13) $\log_9 1 = 0$

14) $\log_{13} 13 = 1$

Objective 3: Convert from Exponential Form to Logarithmic Form

Write in logarithmic form.

15) $9^2 = 81$

16) $12^2 = 144$

17) $10^2 = 100$

18) $10^3 = 1000$

19) $2^{-5} = \dfrac{1}{32}$

20) $3^{-4} = \dfrac{1}{81}$

21) $10^1 = 10$

22) $10^0 = 1$

23) $169^{1/2} = 13$

24) $27^{1/3} = 3$

25) $\sqrt{9} = 3$

24) $\sqrt{64} = 8$

27) $\sqrt[3]{64} = 4$

28) $\sqrt[4]{81} = 3$

SECTION 13.3   **Logarithmic Functions**   **929**

Mixed Exercises: Objectives 4 and 6

29) Explain how to solve a logarithmic equation of the form $\log_a b = c$.

30) A student solves $\log_x 9 = 2$ and gets the solution set $\{-3, 3\}$. Is this correct? Why or why not?

Solve each logarithmic equation.

Fill It In

Fill in the blanks with either the missing mathematical step or the reason for the given step.

31) $\log_2 x = 6$

$2^6 = x$ _____

_____ Solve for x.

The solution set is ____.

32) $\log_5 t = -3$

_____ Rewrite in exponential form.

_____ Solve for t.

The solution set is _____.

Solve each logarithmic equation.

33) $\log_5 k = 3$

34) $\log_{11} x = 2$

35) $\log_4 r = 3$

36) $\log_2 y = 4$

37) $\log w = 2$

38) $\log p = 5$

39) $\log_m 49 = 2$

40) $\log_x 4 = 2$

41) $\log_6 h = -2$

42) $\log_4 b = -3$

43) $\log_2(a + 2) = 4$

44) $\log_6(5y + 1) = 2$

45) $\log_3(4t - 3) = 3$

46) $\log_2(3n + 7) = 5$

47) $\log_{125} \sqrt{5} = c$

48) $\log_{16} \sqrt[5]{4} = k$

49) $\log_{81} \sqrt[4]{9} = x$

50) $\log_{49} \sqrt[3]{7} = d$

51) $\log_{144} w = \dfrac{1}{2}$

52) $\log_{64} p = \dfrac{1}{3}$

53) $\log_8 x = \dfrac{2}{3}$

54) $\log_{16} t = \dfrac{3}{4}$

55) $\log_{(3m-1)} 25 = 2$

56) $\log_{(y-1)} 4 = 2$

Mixed Exercises: Objectives 5–7

Evaluate each logarithm.

57) $\log_2 32$

58) $\log_4 64$

59) $\log 100$

60) $\log 1000$

61) $\log_{49} 7$

62) $\log_{36} 6$

63) $\log_8 \dfrac{1}{8}$

64) $\log_3 \dfrac{1}{3}$

65) $\log_5 5$

66) $\log_2 1$

67) $\log_{1/4} 16$

68) $\log_{1/3} 27$

Objective 8: Define and Graph a Logarithmic Function

69) Explain how to graph a logarithmic function of the form $f(x) = \log_a x$.

70) What are the domain and range of $f(x) = \log_a x$?

Graph each logarithmic function.

71) $f(x) = \log_2 x$

72) $f(x) = \log_5 x$

73) $f(x) = \log_3 x$

74) $f(x) = \log_4 x$

75) $f(x) = \log_{1/2} x$

76) $f(x) = \log_{1/3} x$

77) $f(x) = \log_{1/4} x$

78) $f(x) = \log_{1/5} x$

Find the inverse of each function.

79) $f(x) = 3^x$

80) $f(x) = 4^x$

81) $f(x) = \log_2 x$

82) $f(x) = \log_5 x$

Objective 9: Solve an Applied Problem Using a Logarithmic Equation

Solve each problem.

83) The function $L(t) = 1800 + 68 \log_3(t + 3)$ approximates the number of dog licenses issued by a city each year since 1995. If $t = 0$ represents the year 1995, answer the following.

a) How many dog licenses were issued in 1995?

b) How many were issued in 2019?

c) In what year would it be expected that 2072 dog licenses will be issued?

84) Until the 2000s, Rock Glen was a rural community outside of a large city. In 2004, subdivisions of homes began to be built. The number of houses in Rock Glen t years after 2004 can be approximated by

$$H(t) = 142 + 58 \log_2(t + 1)$$

where $t = 0$ represents 2004.

a) Determine the number of homes in Rock Glen in 2004.

b) Determine the number of homes in Rock Glen in 2007.

c) In what year were there approximately 374 homes?

85) A company plans to introduce a new type of cookie to the market. The company predicts that its sales over the next 24 months can be approximated by

$$S(t) = 14 \log_3 (2t + 1)$$

where t is the number of months after the product is introduced, and $S(t)$ is in thousands of boxes of cookies.

Stockbyte/PictureQuest

a) How many boxes of cookies were sold after they were on the market for 1 month?

b) How many boxes were sold after they were on the market for 4 months?

c) After 13 months, sales were approximately 43,000. Does this number fall short of, meet, or exceed the number of sales predicted by the formula?

86) Based on previous data, city planners have calculated that the number of tourists (in millions) to their city each year can be approximated by

$$N(t) = 10 + 1.2 \log_2 (t + 2)$$

where t is the number of years after 2005.

a) How many tourists visited the city in 2005?

b) How many tourists visited the city in 2011?

c) In 2019, actual data put the number of tourists at 14,720,000. How does this number compare to the number predicted by the formula?

R Rethink

R1) Why are logarithms useful?

R2) In your own words, explain to a classmate how to convert from logarithmic form to exponential form.

R3) Are log functions and exponential functions inverses of each other? How do you know?

R4) What is the most difficult part of learning logarithmic functions?

13.4 Properties of Logarithms

P Prepare

O Organize

What are your objectives for Section 13.4?	How can you accomplish each objective?
1 Use the Product Rule for Logarithms	• Learn the **Product Rule for Logarithms.** • Review the rules for exponents and understand how they relate to logarithms. • Complete the given examples on your own. • Complete You Trys 1 and 2.
2 Use the Quotient Rule for Logarithms	• Learn the **Quotient Rule for Logarithms.** • Complete the given examples on your own. • Complete You Trys 3 and 4.

(continued)

What are your objectives for Section 13.4?	How can you accomplish each objective?
3 Use the Power Rule for Logarithms	• Learn the **Power Rule for Logarithms.** • Review the rules for rewriting radicals as fractional exponents. • Complete the given example on your own. • Complete You Try 5.
4 Use the Properties $\log_a a^x = x$ and $a^{\log_a x} = x$	• Learn the **Other Properties of Logarithms.** • Complete the given example on your own. • Complete You Try 6.
5 Combine the Properties of Logarithms	• Review the **Summary of Properties of Logarithms,** and write the summary in your notes. • Review the list of common errors. • Write an explanation next to each step to help you remember the rules. • Complete the given examples on your own. • Complete You Trys 7–9.

 Work   **Read the explanations, follow the examples, take notes, and complete the You Trys.**

Logarithms have properties that are very useful in applications and in higher mathematics.

In this section, we will learn more properties of logarithms, and we will practice using them because they can make some very difficult mathematical calculations much easier. *The properties of logarithms come from the properties of exponents.*

1   Use the Product Rule for Logarithms

The product rule for logarithms can be derived from the product rule for exponents.

 **Hint**

Write this in your notes, and include an example.

Property The Product Rule for Logarithms

Let x, y, and a be positive real numbers where $a \neq 1$. Then,

$$\log_a xy = \log_a x + \log_a y$$

The logarithm of a product, xy, is the same as the sum of the logarithms of each factor, x and y.

 $\log_a xy \neq (\log_a x)(\log_a y)$

EXAMPLE 1

Rewrite as the sum of logarithms and simplify, if possible. Assume the variables represent positive real numbers.

a) $\log_6(4 \cdot 7)$ b) $\log_4 16t$ c) $\log_8 y^3$ d) $\log 10pq$

Solution

a) The logarithm of a product equals the *sum* of the logs of the factors. Therefore,

$$\log_6(4 \cdot 7) = \log_6 4 + \log_6 7 \qquad \text{Product rule}$$

b) $\log_4 16t = \log_4 16 + \log_4 t \qquad$ Product rule
   $$= 2 + \log_4 t \qquad \log_4 16 = 2$$

Evaluate logarithms, like $\log_4 16$, when possible.

W Hint

Write out each example as you are reading it!

c) $\log_8 y^3 = \log_8(y \cdot y \cdot y) \qquad$ Write y^3 as $y \cdot y \cdot y$.
   $$= \log_8 y + \log_8 y + \log_8 y \qquad \text{Product rule}$$
   $$= 3 \log_8 y$$

d) Recall that if no base is written, then it is assumed to be 10.

$$\log 10pq = \log 10 + \log p + \log q \qquad \text{Product rule}$$
$$= 1 + \log p + \log q \qquad \log 10 = 1$$

YOU TRY 1

Rewrite as the sum of logarithms and simplify, if possible. Assume the variables represent positive real numbers.

a) $\log_9(2 \cdot 5)$ b) $\log_2 32k$ c) $\log_6 c^4$ d) $\log 100yz$

We can use the product rule for exponents in the "opposite" direction, too. That is, given the sum of logarithms we can write a single logarithm.

EXAMPLE 2

Write as a single logarithm. Assume the variables represent positive real numbers.

a) $\log_8 5 + \log_8 3$ b) $\log 7 + \log r$ c) $\log_3 x + \log_3(x + 4)$

Solution

a) $\log_8 5 + \log_8 3 = \log_8(5 \cdot 3) \qquad$ Product rule
   $$= \log_8 15 \qquad 5 \cdot 3 = 15$$

b) $\log 7 + \log r = \log 7r \qquad$ Product rule

c) $\log_3 x + \log_3(x + 4) = \log_3 x(x + 4) \qquad$ Product rule
   $$= \log_3(x^2 + 4x) \qquad \text{Distribute.}$$

BE CAREFUL

$\log_a(x + y) \neq \log_a x + \log_a y$. Therefore, $\log_3(x^2 + 4x)$ does *not* equal $\log_3 x^2 + \log_3 4x$.

YOU TRY 2

Write as a single logarithm. Assume the variables represent positive real numbers.

a) $\log_5 9 + \log_5 4$ b) $\log_6 13 + \log_6 c$ c) $\log y + \log(y + 6)$

2 Use the Quotient Rule for Logarithms

The quotient rule for logarithms can be derived from the quotient rule for exponents.

Property The Quotient Rule for Logarithms

Let x, y, and a be positive real numbers where $a \neq 1$. Then,

$$\log_a \frac{x}{y} = \log_a x - \log_a y$$

The logarithm of a quotient, $\frac{x}{y}$, is the same as the logarithm of the numerator *minus* the logarithm of the denominator.

BE CAREFUL

$$\log_a \frac{x}{y} \neq \frac{\log_a x}{\log_a y}.$$

EXAMPLE 3

Write as the difference of logarithms and simplify, if possible. Assume $w > 0$.

a) $\log_7 \dfrac{3}{10}$ b) $\log_3 \dfrac{81}{w}$

Solution

a) $\log_7 \dfrac{3}{10} = \log_7 3 - \log_7 10$ Quotient rule

b) $\log_3 \dfrac{81}{w} = \log_3 81 - \log_3 w$ Quotient rule

$\qquad = 4 - \log_3 w$ $\log_3 81 = 4$

[YOU TRY 3]

Write as the difference of logarithms and simplify, if possible. Assume $n > 0$.

a) $\log_6 \dfrac{2}{9}$ b) $\log_5 \dfrac{n}{25}$

EXAMPLE 4

Write as a single logarithm. Assume the variable is defined so that the expressions are positive.

a) $\log_2 18 - \log_2 6$ b) $\log_4(z - 5) - \log_4(z^2 + 9)$

Solution

a) $\log_2 18 - \log_2 6 = \log_2 \dfrac{18}{6}$ Quotient rule

$\qquad = \log_2 3$ $\dfrac{18}{6} = 3$

b) $\log_4(z - 5) - \log_4(z^2 + 9) = \log_4 \dfrac{z - 5}{z^2 + 9}$ Quotient rule

 $\log_a(x - y) \neq \log_a x - \log_a y$

[YOU TRY 4] Write as a single logarithm. Assume the variable is defined so that the expressions are positive.

a) $\log_4 36 - \log_4 3$ b) $\log_5(c^2 - 2) - \log_5(c + 1)$

3 Use the Power Rule for Logarithms

In Example 1c), we saw that $\log_8 y^3 = 3 \log_8 y$ because

$$\log_8 y^3 = \log_8(y \cdot y \cdot y)$$
$$= \log_8 y + \log_8 y + \log_8 y$$
$$= 3 \log_8 y$$

This result can be generalized as the next property and comes from the power rule for exponents.

W Hint
Write this in your notes, and include an example.

Property The Power Rule for Logarithms

Let x and a be positive real numbers, where $a \neq 1$, and let r be any real number. Then,

$$\log_a x^r = r \log_a x$$

 The rule applies to $\log_a x^r$ *not* $(\log_a x)^r$. Be sure you can distinguish between the two expressions.

EXAMPLE 5 Rewrite each expression using the power rule and simplify, if possible. Assume the variables represent positive real numbers and that the variable bases are positive real numbers not equal to 1.

a) $\log_9 y^4$ b) $\log_2 8^5$ c) $\log_a \sqrt{3}$ d) $\log_w \dfrac{1}{w}$

Solution

a) $\log_9 y^4 = 4 \log_9 y$ Power rule

b) $\log_2 8^5 = 5 \log_2 8$ Power rule
    $\quad\quad\quad = 5(3)$ $\log_2 8 = 3$
    $\quad\quad\quad = 15$ Multiply.

W Hint
Are you writing out each step as you read the example?

c) *It is common practice to rewrite radicals as fractional exponents when applying the properties of logarithms. This will be our first step.*

$$\log_a \sqrt{3} = \log_a 3^{1/2}   \text{Rewrite as a fractional exponent.}$$
$$= \frac{1}{2} \log_a 3   \text{Power rule}$$

d) Rewrite $\frac{1}{w}$ as w^{-1}: $\quad \log_w \frac{1}{w} = \log_w w^{-1} \qquad \frac{1}{w} = w^{-1}$

$$= -1 \log_w w \qquad \text{Power rule}$$
$$= -1(1) \qquad \log_w w = 1$$
$$= -1 \qquad \text{Multiply.}$$

[YOU TRY 5] Rewrite each expression using the power rule and simplify, if possible. Assume the variables represent positive real numbers and that the variable bases are positive real numbers not equal to 1.

a) $\log_8 t^9$ b) $\log_3 9^7$ c) $\log_a \sqrt[3]{5}$ d) $\log_m \frac{1}{m^8}$

The next properties we will look at can be derived from the power rule and from the fact that $f(x) = a^x$ and $g(x) = \log_a x$ are inverse functions.

4 Use the Properties $\log_a a^x = x$ and $a^{\log_a x} = x$

Other Properties of Logarithms

Let a be a positive real number such that $a \neq 1$. Then,

1) $\log_a a^x = x$ for any real number x.
2) $a^{\log_a x} = x$ for $x > 0$.

EXAMPLE 6 Evaluate each expression.

a) $\log_6 6^7$ b) $\log 10^8$ c) $5^{\log_5 3}$

Solution

a) $\log_6 6^7 = 7 \qquad \log_a a^x = x$

b) $\log 10^8 = 8 \qquad \text{The base of the log is 10.}$

c) $5^{\log_5 3} = 3 \qquad a^{\log_a x} = x$

[YOU TRY 6] Evaluate each expression.

a) $\log_3 3^{10}$ b) $\log 10^{-6}$ c) $7^{\log_7 9}$

Next is a summary of the properties of logarithms. The properties presented in Section 13.3 are included as well.

Summary Properties of Logarithms

Let x, y, and a be positive real numbers where $a \neq 1$, and let r be any real number. Then,

1) $\log_a a = 1$

2) $\log_a 1 = 0$

3) $\log_a xy = \log_a x + \log_a y$ Product rule

4) $\log_a \dfrac{x}{y} = \log_a x - \log_a y$ Quotient rule

5) $\log_a x^r = r \log_a x$ Power rule

6) $\log_a a^x = x$ for any real number x

7) $a^{\log_a x} = x$

Many students make the same mistakes when working with logarithms. Keep in mind the following to avoid these common errors.

1) $\log_a xy \neq (\log_a x)(\log_a y)$

2) $\log_a(x + y) \neq \log_a x + \log_a y$

3) $\log_a \dfrac{x}{y} \neq \dfrac{\log_a x}{\log_a y}$

4) $\log_a(x - y) \neq \log_a x - \log_a y$

5) $(\log_a x)^r \neq r \log_a x$

5 Combine the Properties of Logarithms

Not only can the properties of logarithms simplify some very complicated computations, they are also needed for solving some types of logarithmic equations. The properties of logarithms are also used in calculus and many areas of science.

Next, we will see how to use different properties of logarithms together to rewrite logarithmic expressions.

EXAMPLE 7

Write each expression as the sum or difference of logarithms in simplest form. Assume all variables represent positive real numbers and that the variable bases are positive real numbers not equal to 1.

a) $\log_8 r^5 t$ b) $\log_3 \dfrac{27}{ab^2}$ c) $\log_7 \sqrt[4]{7p}$ d) $\log_a(4a + 5)$

Solution

a) $\log_8 r^5 t = \log_8 r^5 + \log_8 t$ Product rule

              $= 5 \log_8 r + \log_8 t$ Power rule

b) $\log_3 \dfrac{27}{ab^2} = \log_3 27 - \log_3 ab^2$ Quotient rule

            $= 3 - (\log_3 a + \log_3 b^2)$ $\log_3 27 = 3$; product rule

            $= 3 - (\log_3 a + 2 \log_3 b)$ Power rule

            $= 3 - \log_3 a - 2 \log_3 b$ Distribute.

c) $\log_7 \sqrt[4]{7p} = \log_7 (7p)^{1/4}$ Rewrite the radical as a fractional exponent.

$$= \frac{1}{4} \log_7 (7p)$$ Power rule

$$= \frac{1}{4} (\log_7 7 + \log_7 p)$$ Product rule

$$= \frac{1}{4} (1 + \log_7 p)$$ $\log_7 7 = 1$

$$= \frac{1}{4} + \frac{1}{4} \log_7 p$$ Distribute.

d) $\log_a(4a + 5)$ is in simplest form and cannot be rewritten using any properties of logarithms. [Recall that $\log_a(x + y) \neq \log_a x + \log_a y$.]

[YOU TRY 7] Write each expression as the sum or difference of logarithms in simplest form. Assume all variables represent positive real numbers and that the variable bases are positive real numbers not equal to 1.

a) $\log_2 8s^2 t^5$ b) $\log_a \dfrac{4c^2}{b^3}$ c) $\log_5 \sqrt[3]{\dfrac{25}{n}}$ d) $\dfrac{\log_4 k}{\log_4 m}$

EXAMPLE 8 Write each as a single logarithm in simplest form. Assume the variable represents a positive real number.

a) $2 \log_7 5 + 3 \log_7 2$ b) $\dfrac{1}{2} \log_6 s - 3 \log_6(s^2 + 1)$

Solution

a) $2 \log_7 5 + 3 \log_7 2 = \log_7 5^2 + \log_7 2^3$ Power rule

$$= \log_7 25 + \log_7 8$$ $5^2 = 25;\ 2^3 = 8$

$$= \log_7 (25 \cdot 8)$$ Product rule

$$= \log_7 200$$ Multiply.

b) $\dfrac{1}{2} \log_6 s - 3 \log_6(s^2 + 1) = \log_6 s^{1/2} - \log_6(s^2 + 1)^3$ Power rule

$$= \log_6 \sqrt{s} - \log_6(s^2 + 1)^3$$ Write in radical form.

$$= \log_6 \frac{\sqrt{s}}{(s^2 + 1)^3}$$ Quotient rule

[YOU TRY 8] Write each as a single logarithm in simplest form. Assume the variables are defined so that the expressions are positive.

a) $2 \log 4 + \log 5$ b) $\dfrac{2}{3} \log_5 c + \dfrac{1}{3} \log_5 d - 2 \log_5(c - 6)$

Given the values of logarithms, we can compute the values of other logarithms using the properties we have learned in this section.

EXAMPLE 9

Given that $\log 6 \approx 0.7782$ and $\log 4 \approx 0.6021$, use the properties of logarithms to approximate the following.

a) $\log 24$ b) $\log\sqrt{6}$

Solution

a) To find the value of $\log 24$, we must determine how to write 24 in terms of 6 or 4 or some combination of the two. Because $24 = 6 \cdot 4$, we can write

$$\begin{aligned}
\log 24 &= \log(6 \cdot 4) & \scriptstyle 24 = 6 \cdot 4 \\
&= \log 6 + \log 4 & \scriptstyle \text{Product rule} \\
&\approx 0.7782 + 0.6021 & \scriptstyle \text{Substitute.} \\
&= 1.3803 & \scriptstyle \text{Add.}
\end{aligned}$$

b) We can write $\sqrt{6}$ as $6^{1/2}$.

$$\begin{aligned}
\log \sqrt{6} &= \log 6^{1/2} & \scriptstyle \sqrt{6} = 6^{1/2} \\
&= \frac{1}{2}\log 6 & \scriptstyle \text{Power rule} \\
&\approx \frac{1}{2}(0.7782) & \scriptstyle \log 6 \approx 0.7782 \\
&= 0.3891 & \scriptstyle \text{Multiply.}
\end{aligned}$$

[YOU TRY 9] Using the values given in Example 9, use the properties of logarithms to approximate the following.

a) $\log 16$ b) $\log\dfrac{6}{4}$ c) $\log\sqrt[3]{4}$ d) $\log\dfrac{1}{6}$

ANSWERS TO [YOU TRY] EXERCISES

1) a) $\log_9 2 + \log_9 5$ b) $5 + \log_2 k$ c) $4\log_6 c$ d) $2 + \log y + \log z$ 2) a) $\log_5 36$
b) $\log_6 13c$ c) $\log(y^2 + 6y)$ 3) a) $\log_6 2 - \log_6 9$ b) $\log_5 n - 2$ 4) a) $\log_4 12$
b) $\log_5 \dfrac{c^2 - 2}{c + 1}$ 5) a) $9\log_8 t$ b) 14 c) $\dfrac{1}{3}\log_u 5$ d) 8 6) a) 10 b) -6 c) 9
7) a) $3 + 2\log_2 s + 5\log_2 t$ b) $\log_a 4 + 2\log_a c - 3\log_a b$ c) $\dfrac{2}{3} - \dfrac{1}{3}\log_5 n$

d) cannot be simplified   8) a) $\log 80$ b) $\log_5 \dfrac{\sqrt[3]{c^2 d}}{(c - 6)^2}$ 9) a) 1.2042 b) 0.1761

c) 0.2007 d) -0.7782

E Evaluate **13.4** Exercises   Do the exercises, and check your work.

Mixed Exercises: Objectives 1–5

Decide whether each statement is true or false.

1) $\log_6 8c = \log_6 8 + \log_6 c$

2) $\log_5 \dfrac{m}{3} = \log_5 m - \log_5 3$

3) $\log_9 \dfrac{7}{2} = \dfrac{\log_9 7}{\log_9 2}$

4) $\log 1000 = 3$

5) $(\log_4 k)^2 = 2\log_4 k$

6) $\log_2(x^2 + 8) = \log_2 x^2 + \log_2 8$

7) $5^{\log_5 4} = 4$

8) $\log_3 4^5 = 5\log_3 4$

Write as the sum or difference of logarithms and simplify, if possible. Assume all variables represent positive real numbers.

Fill It In

Fill in the blanks with either the missing mathematical step or the reason for the given step.

9) $\log_5 25y$

$\log_5 25y = \log_5 25 + \log_5 y$ _____

$= \underline{\hspace{1cm}}$ Evaluate $\log_5 25$.

10) $\log_3 \dfrac{81}{n^2}$

$\log_3 \dfrac{81}{n^2} = \log_3 81 - \log_3 n^2$ _____

$= \underline{\hspace{1cm}}$ Evaluate $\log_3 81$; use power rule.

11) $\log_8(3 \cdot 10)$

13) $\log_7 5d$

15) $\log_5 \dfrac{20}{17}$

17) $\log_8 10^4$

19) $\log p^8$

21) $\log_3 \sqrt{7}$

23) $\log_2 16p$

25) $\log_2 \dfrac{8}{k}$

27) $\log_7 49^3$

29) $\log 1000b$

31) $\log_2 2^9$

33) $\log_5 \sqrt{5}$

35) $\log \sqrt[3]{100}$

37) $\log_6 w^4 z^3$

39) $\log_7 \dfrac{a^2}{b^5}$

41) $\log \dfrac{\sqrt[5]{11}}{y^2}$

43) $\log_2 \dfrac{4\sqrt{n}}{m^3}$

45) $\log_4 \dfrac{x^3}{yz^2}$

12) $\log_2(6 \cdot 5)$

14) $\log_4 6w$

16) $\log_9 \dfrac{4}{7}$

18) $\log_5 2^3$

20) $\log_3 z^5$

22) $\log_7 \sqrt[3]{4}$

24) $\log_5 25t$

26) $\log_3 \dfrac{x}{9}$

28) $\log_8 64^{12}$

30) $\log_3 27m$

32) $\log_2 32^7$

34) $\log \sqrt[3]{10}$

36) $\log_2 \sqrt{8}$

38) $\log_5 x^2 y$

40) $\log_4 \dfrac{s^4}{t^6}$

42) $\log_3 \dfrac{\sqrt{x}}{y^4}$

44) $\log_9 \dfrac{gf^2}{h^3}$

46) $\log \dfrac{3}{ab^2}$

47) $\log_5 \sqrt{5c}$

48) $\log_8 \sqrt[3]{\dfrac{z}{8}}$

49) $\log k(k - 6)$

50) $\log_2 \dfrac{m^5}{m^2 + 3}$

Write as a single logarithm. Assume the variables are defined so that the variable expressions are positive and so that the bases are positive real numbers not equal to 1.

Fill It In

Fill in the blanks with either the missing mathematical step or the reason for the given step.

51) $2 \log_6 x + \log_6 y$

$2 \log_6 x + \log_6 y = \log_6 x^2 + \log_6 y$ _____

$= \underline{\hspace{1cm}}$ Product rule

52) $5 \log 2 + \log c - 3 \log d$

$5 \log 2 + \log c - 3 \log d$

$= \underline{\hspace{1cm}}$ Power rule

$= \log 32 + \log c - \log d^3$ _____

$= \underline{\hspace{1cm}}$ Product rule

$= \log \dfrac{32c}{d^3}$ _____

53) $\log_a m + \log_a n$

55) $\log_7 d - \log_7 3$

57) $4 \log_3 f + \log_3 g$

59) $\log_8 t + 2 \log_8 u - 3 \log_8 v$

60) $3 \log a + 4 \log c - 6 \log b$

61) $\log(r^2 + 3) - 2 \log(r^2 - 3)$

62) $2 \log_2 t - 3 \log_2(5t + 1)$

63) $3 \log_n 2 + \dfrac{1}{2} \log_n k$

65) $\dfrac{1}{3} \log_d 5 - 2 \log_d z$

67) $\log_6 y - \log_6 3 - 3 \log_6 z$

68) $\log_7 8 - 4 \log_7 x - \log_7 y$

69) $4 \log_3 t - 2 \log_3 6 - 2 \log_3 u$

70) $2 \log_9 m - 4 \log_9 2 - 4 \log_9 n$

71) $\dfrac{1}{2} \log_b (c + 4) - 2 \log_b (c + 3)$

72) $\dfrac{1}{2} \log_a r + \dfrac{1}{2} \log_a (r - 2) - \log_a (r + 2)$

54) $\log_4 7 + \log_4 x$

56) $\log_p r - \log_p s$

58) $5 \log_y m + 2 \log_y n$

64) $2 \log_z 9 + \dfrac{1}{3} \log_z w$

66) $\dfrac{1}{2} \log_5 a - 4 \log_5 b$

73) $\log(a^2 + b^2) - \log(a^4 - b^4)$

74) $\log_n(x^3 - y^3) - \log_n(x - y)$

Given that $\log 5 \approx 0.6990$ and $\log 9 \approx 0.9542$, use the properties of logarithms to approximate the following.
Do not use a calculator.

75) $\log 45$

76) $\log 25$

77) $\log 81$

78) $\log \dfrac{9}{5}$

 79) $\log \dfrac{5}{9}$

80) $\log \sqrt{5}$

81) $\log 3$

82) $\log \dfrac{1}{9}$

83) $\log \dfrac{1}{5}$

84) $\log 5^8$

85) $\log \dfrac{1}{81}$

86) $\log 90$

87) $\log 50$

88) $\log \dfrac{25}{9}$

89) Since $8 = (-4)(-2)$, can we use the properties of logarithms in the following way? Explain.

$$\log_2 8 = \log_2(-4)(-2)$$
$$= \log_2(-4) + \log_2(-2)$$

90) Derive the product rule for logarithms from the product rule for exponents. Assume a, x, and y are positive real numbers with $a \neq 1$. Let $a^m = x$ so that $\log_a x = m$, and let $a^n = y$ so that $\log_a y = n$. Since $a^m \cdot a^n = xy$, show that $\log_a xy = \log_a x + \log_a y$.

R Rethink

R1) How are the rules of exponents and the properties of logarithms similar? How are they different?

R2) Could you write all the rules without looking at your notes?

R3) Which rule is the hardest to remember?

13.5 Common and Natural Logarithms and Change of Base

P Prepare

O Organize

What are your objectives for Section 13.5?	How can you accomplish each objective?
1 Evaluate Common Logarithms Using a Calculator	• Know the definition of a *common logarithm*. • Complete the given example on your own. • Complete You Try 1.
2 Solve an Equation Containing a Common Logarithm	• Review converting from the logarithmic form of an equation to the exponential form of an equation. • Know what it means to find an exact solution and an approximate solution. • Complete the given example on your own. • Complete You Try 2.
3 Solve an Applied Problem Given an Equation Containing a Common Logarithm	• Read the problem carefully. • Complete the given example on your own. • Complete You Try 3.

(continued)

What are your objectives for Section 13.5?	How can you accomplish each objective?
4 Define and Evaluate a Natural Logarithm	• Learn the definition of a *natural logarithm,* and write it in your notes. • Complete the given example on your own. • Complete You Try 4.
5 Graph a Natural Logarithm Function	• Use a table of values to find ordered pairs. • Review how to find the domain and range of a function. • Complete the given example on your own. • Complete You Try 5.
6 Solve an Equation Containing a Natural Logarithm	• Review converting from the logarithmic form of an equation to the exponential form of an equation. • Know what it means to find an exact solution and an approximate solution. • Complete the given example on your own. • Complete You Try 6.
7 Solve Applied Problems Using Exponential Functions	• Understand how to use the *compound interest* formula. • Understand *continuous compounding,* and learn the formula. • Complete the given examples on your own. • Complete You Trys 7 and 8.
8 Use the Change-of-Base Formula	• Learn the **Change-of-Base Formula,** and write it in your own words. • Practice using either base 10 or base *e*. • Complete the given example on your own. • Complete You Try 9.

 **Work**   **Read the explanations, follow the examples, take notes, and complete the You Trys.**

In this section, we will focus our attention on two widely used logarithmic bases—base 10 and base *e*.

1 Evaluate Common Logarithms Using a Calculator

In Section 13.3, we said that a base 10 logarithm is called a **common logarithm.** It is often written as log *x*.

$$\log x \text{ means } \log_{10} x$$

We can evaluate many logarithms without the use of a calculator because we can write them in terms of a base of 10.

For example, $\log 1000 = 3$ because $10^3 = 1000$, and $\log \dfrac{1}{100} = -2$.

W Hint
Expressions are entered into different calculators in different ways. Be sure you know how to use yours.

Common logarithms are used throughout mathematics and other fields to make calculations easier to solve in applications. Often, however, we need a calculator to evaluate the logarithms. Next we will learn how to use a calculator to find the value of a base 10 logarithm. **We will approximate the value to four decimal places.**

EXAMPLE 1

Find log 12.

Solution

Enter 12 $\boxed{\text{LOG}}$ or $\boxed{\text{LOG}}$ 12 $\boxed{\text{ENTER}}$ into your calculator.

$$\log 12 \approx 1.0792$$

(Note that $10^{1.0792} \approx 12$. Press 10 $\boxed{y^x}$ 1.0792 $\boxed{=}$ to evaluate $10^{1.0792}$.)

[**YOU TRY 1**] Find log 3.

We can solve logarithmic equations with or without the use of a calculator.

2 Solve an Equation Containing a Common Logarithm

For the equation in Example 2, we will give an exact solution *and* a solution that is approximated to four decimal places. This will give us an idea of the size of the exact solution.

EXAMPLE 2

Solve log $x = 2.4$. Give an exact solution and a solution that is approximated to four decimal places.

Solution

Change to exponential form, and solve for x.

$$\log x = 2.4 \quad \text{means} \quad \log_{10} x = 2.4$$
$$10^{2.4} = x \qquad \text{Exponential form}$$
$$251.1886 \approx x \qquad \text{Approximation}$$

The exact solution is $\{10^{2.4}\}$. This is approximately $\{251.1886\}$.

[**YOU TRY 2**] Solve log $x = 0.7$. Give an exact solution and a solution that is approximated to four decimal places.

3 Solve an Applied Problem Given an Equation Containing a Common Logarithm

EXAMPLE 3

The loudness of sound, $L(I)$ in decibels (dB), is given by

$$L(I) = 10 \log \frac{I}{10^{-12}}$$

where I is the intensity of sound in watts per square meter (W/m²). Fifty meters from the stage at a concert, the intensity of sound is 0.01 W/m². Find the loudness of the music at the concert 50 m from the stage.

Solution

Substitute 0.01 for I and find $L(0.01)$.

$$L(0.01) = 10 \log \frac{0.01}{10^{-12}}$$

$$= 10 \log \frac{10^{-2}}{10^{-12}} \qquad 0.01 = 10^{-2}$$

$$= 10 \log 10^{10} \qquad \text{Quotient rule for exponents}$$

$$= 10(10) \qquad \log 10^{10} = 10$$

$$= 100$$

The sound level of the music 50 m from the stage is 100 dB. (To put this in perspective, a normal conversation has a loudness of about 50 dB.)

[YOU TRY 3]

The intensity of sound from a thunderstorm is about 0.001 W/m². Find the loudness of the storm, in decibels.

4 Define and Evaluate a Natural Logarithm

Another base that is often used for logarithms is the number e. In Section 13.2, we said that e, like π, is an irrational number. To four decimal places, $e \approx 2.7183$.

A base e logarithm is called a **natural logarithm** or **natural log.** The notation used for a base e logarithm is $\ln x$ (read as "*the natural log of x*" or "*ln of x*"). Since it is a base e logarithm, it is important to remember that

$$\ln x \text{ means } \log_e x$$

> **W Hint**
>
> In your notes, summarize this information about the natural logarithm, ln x.

Using the properties $\log_a a^x = x$ and $\log_a a = 1$, we can find the value of some natural logarithms without using a calculator. For some natural logs, we will use a calculator to approximate the value to four decimal places.

EXAMPLE 4

Evaluate.

a) $\ln e$ b) $\ln e^2$ c) $\ln 5$

Solution

a)   To evaluate $\ln e$, remember that $\ln e = \log_e e = 1$ since $\log_a a = 1$.

$$\ln e = 1$$

> **W Hint**
>
> Remember that ln e = 1. We will use it when solving some equations.

This is a value you should remember. We will use this in Section 13.6 to solve exponential equations with base e.

b) $\ln e^2 = \log_e e^2 = 2 \qquad \log_a a^x = x$

c)   We can use a calculator to *approximate natural logarithms to four decimal places* if the properties do not give us an exact value.

Enter $\boxed{5}\ \boxed{\text{LN}}$ or $\boxed{\text{LN}}\ \boxed{5}\ \boxed{\text{ENTER}}$ into your calculator.

$$\ln 5 \approx 1.6094$$

[YOU TRY 4]

Evaluate.

a) $5 \ln e$ b) $\ln e^8$ c) $\ln 9$

5 Graph a Natural Logarithm Function

We can graph $y = \ln x$ by substituting values for x and using a calculator to approximate the values of y.

EXAMPLE 5

Graph $y = \ln x$. Determine the domain and range.

Solution

Choose values for x, and use a calculator to approximate the corresponding values of y. Remember that $\ln e = 1$, so e is a good choice for x.

x	y
1	0
$e \approx 2.72$	1
6	1.79
0.5	−0.69
0.25	−1.39

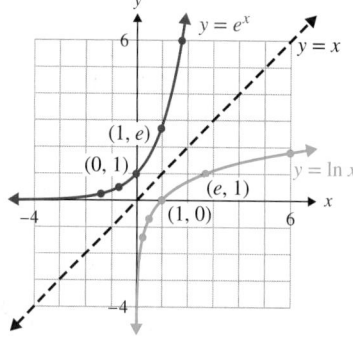

The domain of $y = \ln x$ is $(0, \infty)$, and the range is $(-\infty, \infty)$.

The graph of the inverse of $y = \ln x$ is also shown. We can obtain the graph of the inverse of $y = \ln x$ by reflecting the graph about the line $y = x$. The inverse of $y = \ln x$ is $y = e^x$.

Notice that the domain of $y = e^x$ is $(-\infty, \infty)$, while the range is $(0, \infty)$, the opposite of the domain and range of $y = \ln x$. This is a direct result of the relationship between a function and its inverse.

[YOU TRY 5] Graph $y = \ln(x + 4)$. Determine the domain and range.

W Hint

Remember that $y = \ln x$ is equivalent to $y = \log_e x$.

It is important to remember that $y = \ln x$ means $y = \log_e x$. Understanding this relationship allows us to make the following connections:

$$y = \ln x \text{ is equivalent to } y = \log_e x$$

and

$$y = \log_e x \text{ can be written in exponential form as } e^y = x.$$

Therefore, $y = \ln x$ can be written in exponential form as $e^y = x$.

We can use this relationship to show that the inverse of $y = \ln x$ is $y = e^x$. (You will be asked to verify this relationship in Exercise 97.) Also, in Example 5, notice that the graphs of $y = \ln x$ and $y = e^x$ are symmetric with respect to $y = x$.

6 Solve an Equation Containing a Natural Logarithm

Note

To solve an equation containing a natural logarithm, like ln $x = 4$, we change to exponential form and solve for the variable. We can give an exact solution and a solution that is approximated to four decimal places.

EXAMPLE 6

Solve each equation. Give an exact solution and a solution that is approximated to four decimal places.

a) ln $x = 4$ b) ln$(2x + 5) = 3.8$

Solution

a) ln $x = 4$ means $\log_e x = 4$

$$e^4 = x \qquad \text{Exponential form}$$
$$54.5982 \approx x \qquad \text{Approximation}$$

The exact solution is $\{e^4\}$. This is approximately $\{54.5982\}$.

b) ln$(2x + 5) = 3.8$ means $\log_e(2x + 5) = 3.8$

$$e^{3.8} = 2x + 5 \qquad \text{Exponential form}$$
$$e^{3.8} - 5 = 2x \qquad \text{Subtract 5.}$$
$$\frac{e^{3.8} - 5}{2} = x \qquad \text{Divide by 2.}$$
$$19.8506 \approx x \qquad \text{Approximation}$$

The exact solution is $\left\{ \dfrac{e^{3.8} - 5}{2} \right\}$. This is approximately $\{19.8506\}$.

[YOU TRY 6]

Solve each equation. Give an exact solution and a solution that is approximated to four decimal places.

a) ln $y = 2.7$ b) ln$(3a - 1) = 0.5$

7 Solve Applied Problems Using Exponential Functions

One of the most practical applications of exponential functions is for compound interest.

Definition

Compound Interest: The amount of money, A, in dollars, in an account after t years is given by

$$A = P\left(1 + \frac{r}{n}\right)^{nt}$$

where P (the principal) is the amount of money (in dollars) deposited in the account, r is the annual interest rate, and n is the number of times the interest is compounded (paid) per year.

Note

We can also think of this formula in terms of the amount of money owed, A, after t yr when P is the amount of money loaned.

EXAMPLE 7

If $2000 is deposited in an account paying 4% per year, find the total amount in the account after 5 yr if the interest is compounded

a) quarterly.   b) monthly.

(We assume no withdrawals or additional deposits are made.)

Solution

a) If interest compounds quarterly, then interest is paid four times per year. Use

$$A = P\left(1 + \frac{r}{n}\right)^{nt}$$

W Hint

To use the interest rate in the formula, you must change the percent to a decimal.

with $P = 2000$, $r = 0.04$, $t = 5$, $n = 4$.

$$A = 2000\left(1 + \frac{0.04}{4}\right)^{4(5)}$$
$$= 2000(1.01)^{20}$$
$$\approx 2440.3801$$

Because A is an amount of money, round to the nearest cent. The account will contain $2440.38 after 5 yr.

b) If interest is compounded monthly, then interest is paid 12 times per year. Use

$$A = P\left(1 + \frac{r}{n}\right)^{nt}$$

with $P = 2000$, $r = 0.04$, $t = 5$, $n = 12$.

$$A = 2000\left(1 + \frac{0.04}{12}\right)^{12(5)}$$
$$\approx 2441.9932$$

Round A to the nearest cent. The account will contain $2441.99 after 5 yr.

[YOU TRY 7]

If $1500 is deposited in an account paying 5% per year, find the total amount in the account after 8 yr if the interest is compounded

a) monthly.   b) weekly.

In Example 7 we saw that the account contained more money after 5 yr when the interest compounded monthly (12 times per year) versus quarterly (four times per year). This will always be true. The more often interest is compounded each year, the more money that accumulates in the account.

If interest *compounds continuously,* we obtain the formula for *continuous compounding,* $A = Pe^{rt}$.

Definition

Continuous Compounding: If P dollars is deposited in an account earning interest rate r compounded continuously, then the amount of money, A (in dollars), in the account after t years is given by

$$A = Pe^{rt}$$

EXAMPLE 8

Determine the amount of money in an account after 5 yr if $2000 was initially invested at 4% compounded continuously.

Solution

Use $A = Pe^{rt}$ with $P = 2000$, $r = 0.04$, and $t = 5$.

$$
\begin{aligned}
A &= 2000e^{0.04(5)} && \text{Substitute values.} \\
&= 2000e^{0.20} && \text{Multiply } (0.04)(5). \\
&\approx 2442.8055 && \text{Evaluate using a calculator.}
\end{aligned}
$$

Round A to the nearest cent.

The account will contain $2442.81 after 5 yr. Note that, as expected, this is more than the amounts obtained in Example 7 when the same amount was deposited for 5 yr at 4% but the interest was compounded quarterly and monthly.

[YOU TRY 8]

Determine the amount of money in an account after 8 yr if $1500 was initially invested at 5% compounded continuously.

8 Use the Change-of-Base Formula

Sometimes we need to find the value of a logarithm with a base other than 10 or e—like $\log_3 7$. Some calculators, however, do not calculate logarithms other than common logs (base 10) and natural logs (base e). In such cases, we can use the change-of-base formula to evaluate logarithms with bases other than 10 or e.

Definition

Change-of-Base Formula: If a, b, and x are positive real numbers and $a \neq 1$ and $b \neq 1$, then

$$\log_a x = \frac{\log_b x}{\log_b a}$$

Note

We can choose any positive real number not equal to 1 for b, but it is most convenient to choose 10 or e since these will give us common logarithms and natural logarithms, respectively.

EXAMPLE 9

Find the value of $\log_3 7$ to four decimal places using

a) common logarithms.   b) natural logarithms.

Solution

a) The base we will use to evaluate $\log_3 7$ is 10; this is the base of a common logarithm. Then

$$\log_3 7 = \frac{\log_{10} 7}{\log_{10} 3} \qquad \text{Change-of-base formula}$$

$$\approx 1.7712 \qquad \text{Use a calculator.}$$

b) The base of a natural logarithm is e. Then

$$\log_3 7 = \frac{\log_e 7}{\log_e 3}$$

$$= \frac{\ln 7}{\ln 3}$$

$$\approx 1.7712 \qquad \text{Use a calculator.}$$

Using either base 10 or base e gives us the same result.

W Hint

Decide whether you prefer to use a common log or a natural log in the change-of-base formula.

[**YOU TRY 9**]

Find the value of $\log_5 38$ to four decimal places using

a) common logarithms.   b) natural logarithms.

Using Technology

Graphing calculators will graph common logarithmic functions and natural logarithmic functions directly using the **log** or **LN** keys.

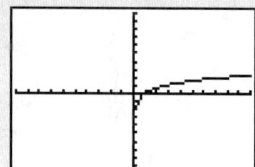

For example, let's graph $f(x) = \ln x$.

To graph a logarithmic function with a base other than 10 or e, it is necessary to use the change-of-base formula. For example, to graph the function $f(x) = \log_2 x$, first rewrite the function as a quotient of natural logarithms or common logarithms: $f(x) = \log_2 x = \dfrac{\ln x}{\ln 2}$ or $\dfrac{\log x}{\log 2}$. Enter one of these quotients in Y_1 and press $\boxed{\text{GRAPH}}$ to graph as shown below. To illustrate that the same graph results in either case, trace to the point where $x = 3$.

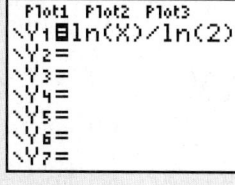
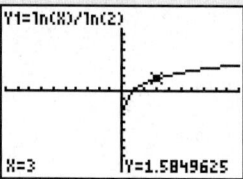
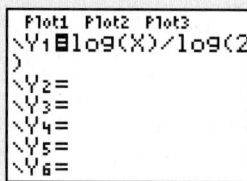
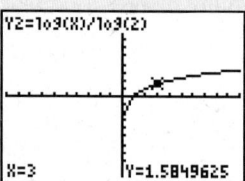

Graph the following functions using a graphing calculator.

1) $f(x) = \log_3 x$
2) $f(x) = \log_5 x$
3) $f(x) = 4 \log_2 x + 1$
4) $f(x) = \log_2 (x - 3)$
5) $f(x) = 2 - \log_4 x$
6) $f(x) = 3 - \log_2 (x + 1)$

ANSWERS TO [YOU TRY] **EXERCISES**

1) 0.4771   2) $\{10^{0.7}\}$; $\{5.0119\}$ 3) 90 dB   4) a) 5   b) 8   c) 2.1972
5) domain: $(-4, \infty)$; range: $(-\infty, \infty)$

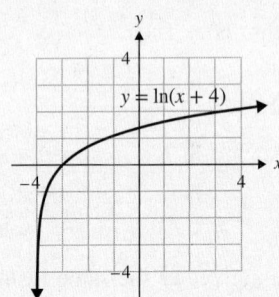

6) a) $\{e^{2.7}\}$; $\{14.8797\}$ b) $\left\{\dfrac{e^{0.5} + 1}{3}\right\}$; $\{0.8829\}$

7) a) \$2235.88   b) \$2237.31   8) \$2237.74
9) a) 2.2602   b) 2.2602

ANSWERS TO TECHNOLOGY EXERCISES

1)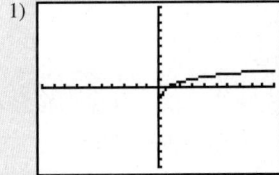
2)
3)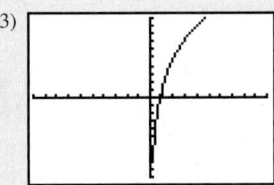

4)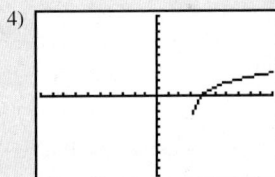
5)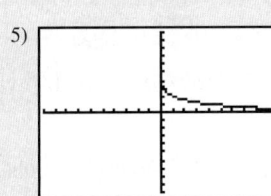
6)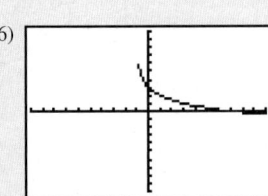

E Evaluate **13.5** Exercises   Do the exercises, and check your work.

Mixed Exercises: Objectives 1 and 4

1) What is the base of $\ln x$?

2) What is the base of $\log x$?

Evaluate each logarithm. Do *not* use a calculator.

3) $\log 100$

4) $\log 10,000$

5) $\log \dfrac{1}{1000}$

6) $\log \dfrac{1}{100,000}$

7) $\log 0.1$

8) $\log 0.01$

9) $\log 10^9$

10) $\log 10^7$

11) $\log \sqrt[4]{10}$

12) $\log \sqrt[5]{10}$

13) $\ln e^{10}$

14) $\ln e^6$

15) $\ln \sqrt{e}$

16) $\ln \sqrt[3]{e}$

17) $\ln \dfrac{1}{e^5}$

18) $\ln \dfrac{1}{e^2}$

19) $\ln 1$

20) $\log 1$

Mixed Exercises: Objectives 1 and 4

Use a calculator to find the approximate value of each logarithm to four decimal places.

21) $\log 16$

22) $\log 23$

23) $\log 0.5$

24) $\log 627$

25) $\ln 3$

26) $\ln 6$

27) $\ln 1.31$

28) $\ln 0.218$

Objective 2: Solve an Equation Containing a Common Logarithm

Solve each equation. Do *not* use a calculator.

29) $\log x = 3$

30) $\log z = 5$

31) $\log k = -1$

32) $\log c = -2$

33) $\log(4a) = 2$

34) $\log(5w) = 1$

35) $\log(3t + 4) = 1$

36) $\log(2p + 12) = 2$

Mixed Exercises: Objectives 2 and 6

Solve each equation. Give an exact solution and a solution that is approximated to four decimal places.

37) $\log a = 1.5$

38) $\log y = 1.8$

39) $\log r = 0.8$

40) $\log c = 0.3$

41) $\ln x = 1.6$

42) $\ln p = 1.1$

43) $\ln t = -2$

44) $\ln z = 0.25$

45) $\ln(3q) = 2.1$

46) $\ln\left(\dfrac{1}{4}m\right) = 3$

47) $\log\left(\dfrac{1}{2}c\right) = 0.47$

48) $\log(6k) = -1$

49) $\log(5y - 3) = 3.8$

50) $\log(8x + 15) = 2.7$

51) $\ln(10w + 19) = 1.85$

52) $\ln(7a - 4) = 0.6$

53) $\ln(2d - 5) = 0$

54) $\log(3t + 14) = 2.4$

Objective 5: Graph a Natural Logarithm Function

Graph each function. State the domain and range.

55) $f(x) = \ln x - 3$

56) $y = \ln x + 2$

57) $h(x) = \ln x - 1$

58) $g(x) = \ln x + 1$

59) $g(x) = \ln(x + 3)$

60) $h(x) = \ln(x + 2)$

61) $f(x) = \ln(x - 2)$

62) $y = \ln(x - 1)$

63) $y = -\ln x$

64) $f(x) = \ln(-x)$

65) $h(x) = \log x$

66) $k(x) = \log(x + 4)$

67) If you are given the graph of $f(x) = \ln x$, how could you obtain the graph of $g(x) = \ln(x + 5)$ without making a table of values and plotting points?

68) If you are given the graph of $f(x) = \ln x$, how could you obtain the graph of $h(x) = \ln x + 4$ without making a table of values and plotting points?

Objective 8: Use the Change-of-Base Formula

Use the change-of-base formula with either base 10 or base e to approximate each logarithm to four decimal places.

69) $\log_2 13$

70) $\log_6 25$

71) $\log_9 70$

72) $\log_3 52$

73) $\log_{1/3} 16$

74) $\log_{1/5} 13$

75) $\log_5 3$

76) $\log_7 4$

Mixed Exercises: Objectives 3 and 7

For Exercises 77–80, use the formula

$$L(I) = 10 \log \frac{I}{10^{-12}}$$

where I is the intensity of sound, in watts per square meter, and $L(I)$ is the loudness of sound in decibels. Do *not* use a calculator.

77) The intensity of sound from a dishwasher is about 0.000001 W/m^2. Find the loudness of the dishwasher, in decibels.

78) The intensity of sound from a refrigerator is about 0.00000001 W/m^2. Find the loudness of the refrigerator, in decibels.

79) The intensity of sound from fireworks is about 0.1 W/m^2. Find the loudness of the fireworks, in decibels.

Thinkstock

80) The intensity of sound from the takeoff of a space shuttle is 1,000,000 W/m². Find the loudness of the sound made by the space shuttle at takeoff, in decibels.

Use the formula $A = P\left(1 + \dfrac{r}{n}\right)^{nt}$ to solve each problem. See Example 7.

81) Isabel deposits $3000 in an account earning 5% per year compounded monthly. How much will be in the account after 3 yr?

82) How much money will Pavel have in his account after 8 yr if he initially deposited $6000 at 4% interest compounded quarterly?

83) Find the amount Christopher owes at the end of 5 yr if he borrows $4000 at a rate of 6.5% compounded quarterly.

84) How much will Anna owe at the end of 4 yr if she borrows $5000 at a rate of 7.2% compounded weekly?

Use the formula $A = Pe^{rt}$ for Exercises 85–88. See Example 8.

85) If $3000 is deposited in an account earning 5% compounded continuously, how much will be in the account after 3 yr?

86) If $6000 is deposited in an account earning 4% compounded continuously, how much will be in the account after 8 yr?

87) How much will Cyrus owe at the end of 6 yr if he borrows $10,000 at a rate of 7.5% compounded continuously?

88) Find the amount Nadia owes at the end of 5 yr if she borrows $4500 at a rate of 6.8% compounded continuously.

89) The number of bacteria, $N(t)$, in a culture t hr after the bacteria are placed in a dish is given by

$$N(t) = 5000e^{0.0617t}$$

a) How many bacteria were originally in the culture?

b) How many bacteria are present after 8 hr?

90) The number of bacteria, $N(t)$, in a culture t hr after the bacteria are placed in a dish is given by

$$N(t) = 8000e^{0.0342t}$$

a) How many bacteria were originally in the culture?

b) How many bacteria are present after 10 hr?

91) The function $N(t) = 10,000e^{0.0492t}$ describes the number of bacteria in a culture t hr after 10,000 bacteria were placed in the culture. How many bacteria are in the culture after 1 day?

92) How many bacteria are present 2 days after 6000 bacteria are placed in a culture if the number of bacteria in the culture is

$$N(t) = 6000e^{0.0285t}$$

t hr after the bacteria are placed in a dish?

In chemistry, the pH of a solution is given by

$$pH = -\log[H^+]$$

where $[H^+]$ is the molar concentration of the hydronium ion. A neutral solution has pH = 7. *Acidic solutions* have pH < 7, and *basic solutions* have pH > 7.

For Exercises 93–96, the hydronium ion concentrations, $[H^+]$, are given for some common substances. Find the pH of each substance (to the tenths place), and determine whether each substance is acidic or basic.

93) Cola: $[H^+] = 2 \times 10^{-3}$

94) Tomatoes: $[H^+] = 1 \times 10^{-4}$

95) Ammonia: $[H^+] = 6 \times 10^{-12}$

96) Egg white: $[H^+] = 2 \times 10^{-8}$

Extension

97) Show that the inverse of $y = \ln x$ is $y = e^x$.

R Rethink

R1) What is the difference between a common logarithm and a natural logarithm?

R2) When using the change-of-base formula, can you use either base 10 or base e?

R3) Which part of using the calculator is hardest for you when evaluating logarithms?

13.6 Solving Exponential and Logarithmic Equations

What are your objectives for Section 13.6?	How can you accomplish each objective?
1 Solve an Exponential Equation	• Review the properties of logarithms. • Learn the procedure for **Solving an Exponential Equation,** and summarize it in your notes. • Complete the given examples on your own. • Complete You Trys 1 and 2.
2 Solve Logarithmic Equations Using the Properties of Logarithms	• Learn the procedure for **How to Solve an Equation Where Each Term Contains a Logarithm.** • Learn the procedure for **How to Solve an Equation Where One Term Does Not Contain a Logarithm.** • Complete the given examples on your own. • Complete You Trys 3 and 4.
3 Solve Applied Problems Involving Exponential Functions Using a Calculator	• Review the formula for **Continuous Compounding.** • Complete the given examples on your own. • Complete You Trys 5 and 6.
4 Solve an Applied Problem Involving Exponential Growth or Decay	• Understand and learn the formula for **Exponential Growth or Decay.** • Understand the meaning of *half-life*. • Complete the given example on your own. • Complete You Try 7.

Read the explanations, follow the examples, take notes, and complete the You Trys.

In this section, we will learn another property of logarithms that will allow us to solve additional types of exponential and logarithmic equations.

Properties for Solving Exponential and Logarithmic Equations

Let a, x, and y be positive real numbers, where $a \neq 1$.

1) If $x = y$, then $\log_a x = \log_a y$.
2) If $\log_a x = \log_a y$, then $x = y$.

For example, 1) tells us that if $x = 3$, then $\log_a x = \log_a 3$. Likewise, 2) tells us that if $\log_a 5 = \log_a y$, then $5 = y$. We can use the properties above to solve exponential and logarithmic equations that we could not solve previously.

1 Solve an Exponential Equation

We will look at two types of exponential equations—equations where both sides *can* be expressed with the same base and equations where both sides *cannot* be expressed with the same base. If the two sides of an exponential equation *cannot* be expressed with the same base, we will use logarithms to solve the equation.

EXAMPLE 1

Solve.

a) $2^x = 8$ b) $2^x = 12$

Solution

a) Because 8 is a power of 2, we can solve $2^x = 8$ by expressing each side of the equation with the same base and setting the exponents equal to each other.

$$2^x = 8$$
$$2^x = 2^3 \qquad 8 = 2^3$$
$$x = 3 \qquad \text{Set the exponents equal.}$$

The solution set is $\{3\}$.

**Hint**

Write the solutions to these equations in your notes. Explain, in your own words, why they are solved differently.

b) Can we express both sides of $2^x = 12$ with the same base? *No. We will use property 1) to solve $2^x = 12$ by taking the logarithm of each side.*

We can use a logarithm of *any* base. It is most convenient to use base 10 (common logarithm) or base e (natural logarithm) because this is what we can find most easily on our calculators. *We will take the natural log of both sides.*

$$2^x = 12$$
$$\ln 2^x = \ln 12 \qquad \text{Take the natural log of each side.}$$
$$x \ln 2 = \ln 12 \qquad \log_a x^r = r \log_a x$$
$$x = \frac{\ln 12}{\ln 2} \qquad \text{Divide by } \ln 2.$$

The exact solution is $\left\{ \dfrac{\ln 12}{\ln 2} \right\}$. Use a calculator to get an approximation to four decimal places: $x \approx 3.5850$.

The approximation is $\{3.5850\}$. We can verify the solution by substituting it for x in $2^x = 12$: $2^{3.5850} \approx 12$.

BE CAREFUL

The exact solution is written as $\left\{ \dfrac{\ln 12}{\ln 2} \right\}$. Recall that $\dfrac{\ln 12}{\ln 2} \neq \ln 6$.

Procedure Solving an Exponential Equation

Begin by asking yourself, "*Can I express each side with the same base?*"

Hint

Get in the habit of asking yourself this question.

1) If the answer is **yes,** then write each side of the equation with the same base, set the exponents equal, and solve for the variable.

2) If the answer is **no,** then take the natural logarithm of each side, use the properties of logarithms, and solve for the variable.

Solve.

a) $3^t = 9$ b) $3^t = 24$

EXAMPLE 2

Solve.

a) $5^{x-2} = 16$ b) $e^{5n} = 4$

Solution

a) Ask yourself, *"Can I express each side with the same base?"* **No.** Therefore, take the natural log of each side.

$$5^{x-2} = 16$$
$$\ln 5^{x-2} = \ln 16 \qquad \text{Take the natural log of each side.}$$
$$(x - 2)\ln 5 = \ln 16 \qquad \log_a x^r = r \log_a x$$

$(x - 2)$ *must* be in parentheses because it contains two terms.

$$x \ln 5 - 2 \ln 5 = \ln 16 \qquad \text{Distribute.}$$
$$x \ln 5 = \ln 16 + 2 \ln 5 \qquad \text{Add } 2\ln 5 \text{ to get the } x\text{-term by itself.}$$
$$x = \frac{\ln 16 + 2 \ln 5}{\ln 5} \qquad \text{Divide by } \ln 5.$$

The exact solution is $\left\{ \dfrac{\ln 16 + 2 \ln 5}{\ln 5} \right\}$. This is approximately $\{3.7227\}$.

 **Hint**

Remember that $\ln e = 1$.

b) Begin by taking the natural log of each side.

$$e^{5n} = 4$$
$$\ln e^{5n} = \ln 4 \qquad \text{Take the natural log of each side.}$$
$$5n \ln e = \ln 4 \qquad \log_a x^r = r \log_a x$$
$$5n(1) = \ln 4 \qquad \ln e = 1$$
$$5n = \ln 4$$
$$n = \frac{\ln 4}{5} \qquad \text{Divide by } 5.$$

The exact solution is $\left\{ \dfrac{\ln 4}{5} \right\}$. The approximation is $\{0.2773\}$.

YOU TRY 2

Solve.

a) $9^{k+4} = 2$ b) $e^{6c} = 2$

---

2 Solve Logarithmic Equations Using the Properties of Logarithms

We learned earlier that to solve a logarithmic equation like $\log_2(t + 5) = 4$, we write the equation in exponential form and solve for the variable.

$$\log_2(t + 5) = 4$$
$$2^4 = t + 5 \qquad \text{Write in exponential form.}$$
$$16 = t + 5 \qquad 2^4 = 16$$
$$11 = t \qquad \text{Subtract 5.}$$

In this section, we will learn how to solve other types of logarithmic equations as well. We will look at equations where

1) each term in the equation contains a logarithm.

2) one term in the equation does *not* contain a logarithm.

Let's begin with the first case.

Procedure How to Solve an Equation Where Each Term Contains a Logarithm

1) Use the properties of logarithms to write the equation in the form $\log_a x = \log_a y$.

2) Set $x = y$, and solve for the variable.

3) Check the proposed solution(s) in the original equation to be sure the values satisfy the equation.

EXAMPLE 3

Solve.

a) $\log_5(m - 4) = \log_5 9$ b) $\log x + \log(x + 6) = \log 16$

Solution

a) To solve $\log_5(m - 4) = \log_5 9$, use the property that states if $\log_a x = \log_a y$, then $x = y$.

$$\log_5(m - 4) = \log_5 9$$
$$m - 4 = 9$$
$$m = 13 \qquad \text{Add 4.}$$

Check to be sure that $m = 13$ satisfies the original equation.

$$\log_5(13 - 4) \overset{?}{=} \log_5 9$$
$$\log_5 9 = \log_5 9 \quad \checkmark$$

The solution set is $\{13\}$.

b) To solve $\log x + \log(x + 6) = \log 16$, we must begin by using the product rule for logarithms to obtain one logarithm on the left side.

Hint

Review the properties of logarithms in Section 13.4, if necessary.

$$\log x + \log(x + 6) = \log 16$$
$$\log x(x + 6) = \log 16 \qquad \text{Product rule}$$
$$x(x + 6) = 16 \qquad \text{If } \log_a x = \log_a y, \text{ then } x = y.$$
$$x^2 + 6x = 16 \qquad \text{Distribute.}$$
$$x^2 + 6x - 16 = 0 \qquad \text{Subtract 16.}$$
$$(x + 8)(x - 2) = 0 \qquad \text{Factor.}$$
$$x + 8 = 0 \quad \text{or} \quad x - 2 = 0 \qquad \text{Set each factor equal to 0.}$$
$$x = -8 \quad \text{or} \quad x = 2 \qquad \text{Solve.}$$

Check to be sure that $x = -8$ and $x = 2$ satisfy the original equation.

Check $x = -8$:

$$\log x + \log(x + 6) = \log 16$$
$$\log(-8) + \log(-8 + 6) \overset{?}{=} \log 16$$
$$\text{FALSE}$$

We reject $x = -8$ as a solution because it leads to $\log(-8)$, which is undefined.

The solution set is $\{2\}$.

Check $x = 2$:

$$\log x + \log(x + 6) = \log 16$$
$$\log 2 + \log(2 + 6) \overset{?}{=} \log 16$$
$$\log 2 + \log 8 \overset{?}{=} \log 16$$
$$\log(2 \cdot 8) \overset{?}{=} \log 16$$
$$\log 16 = \log 16 \quad \checkmark$$

$x = 2$ satisfies the original equation.

BE CAREFUL Just because a proposed solution is a negative number does *not* mean it should be rejected. You *must* check it in the original equation; it may satisfy the equation.

[**YOU TRY 3**]

Solve.

a) $\log_8(z + 3) = \log_8 5$

b) $\log_3 c + \log_3(c - 1) = \log_3 12$

Procedure How to Solve an Equation Where One Term Does *Not* Contain a Logarithm

1) Use the properties of logarithms to get one logarithm on one side of the equation and a constant on the other side. That is, write the equation in the form $\log_a x = y$.

2) Write $\log_a x = y$ in exponential form, $a^y = x$, and solve for the variable.

3) Check the proposed solution(s) in the original equation to be sure the values satisfy the equation.

EXAMPLE 4

Solve $\log_2 3w - \log_2(w - 5) = 3$.

Solution

Notice that one term in the equation $\log_2 3w - \log_2(w - 5) = 3$ does *not* contain a logarithm. Therefore, we want to use the properties of logarithms to get *one* logarithm on the left. Then, write the equation in exponential form, and solve.

W Hint

Are you writing down the example as you are reading it?

$$\log_2 3w - \log_2(w - 5) = 3$$

$$\log_2 \frac{3w}{w - 5} = 3 \qquad \text{Quotient rule}$$

$$2^3 = \frac{3w}{w - 5} \qquad \text{Write in exponential form.}$$

$$8 = \frac{3w}{w - 5} \qquad 2^3 = 8$$

$$8(w - 5) = 3w \qquad \text{Multiply by } w - 5.$$

$$8w - 40 = 3w \qquad \text{Distribute.}$$

$$-40 = -5w \qquad \text{Subtract } 8w.$$

$$8 = w \qquad \text{Divide by } -5.$$

Verify that $w = 8$ satisfies the original equation. The solution set is $\{8\}$.

Solve.

a) $\log_4(7p + 1) = 3$ b) $\log_3 2x - \log_3(x - 14) = 2$

Let's look at the two types of equations we have discussed side by side. Notice the difference between them.

Solve each equation

1) $\log_3 x + \log_3(2x + 5) = \log_3 12$

Use the properties of logarithms to get one log on the left.

$$\log_3 x(2x + 5) = \log_3 12$$

Since *both terms contain logarithms,* use the property that states if $\log_a x = \log_a y$, then $x = y$.

$$x(2x + 5) = 12$$
$$2x^2 + 5x = 12$$
$$2x^2 + 5x - 12 = 0$$
$$(2x - 3)(x + 4) = 0$$
$$2x - 3 = 0 \quad \text{or} \quad x + 4 = 0$$
$$x = \frac{3}{2} \quad \text{or} \quad x = -4$$

Check. Reject -4 as a solution. The solution set is $\left\{\dfrac{3}{2}\right\}$.

2) $\log_3 x + \log_3(2x + 5) = 1$

Use the properties of logarithms to get one log on the left.

$$\log_3 x(2x + 5) = 1$$

The term on the right does *not* contain a logarithm. Write the equation in exponential form, and solve.

$$3^1 = x(2x + 5)$$
$$3 = 2x^2 + 5x$$
$$0 = 2x^2 + 5x - 3$$
$$0 = (2x - 1)(x + 3)$$
$$2x - 1 = 0 \quad \text{or} \quad x + 3 = 0$$
$$x = \frac{1}{2} \quad \text{or} \quad x = -3$$

Check. Reject $x = -3$ as a solution. The solution set is $\left\{\dfrac{1}{2}\right\}$.

3 Solve Applied Problems Involving Exponential Functions Using a Calculator

Recall that $A = Pe^{rt}$ is the formula for **continuous compound interest** where P (the principal) is the amount invested, r is the interest rate, and A is the amount (in dollars) in the account after t yr. Here we will look at how we can use the formula to solve a different problem from the type we solved in Section 13.5.

EXAMPLE 5

If $3000 is invested at 5% interest compounded continuously, how long would it take for the investment to grow to $4000?

Solution

In this problem, we are asked to find t, the amount of *time* it will take for $3000 to grow to $4000 when invested at 5% compounded continuously.

Use $A = Pe^{rt}$ with $P = 3000$, $A = 4000$, and $r = 0.05$.

$$A = Pe^{rt}$$

$$4000 = 3000e^{0.05t} \quad \text{Substitute the values.}$$

$$\frac{4}{3} = e^{0.05t} \quad \text{Divide by 3000.}$$

$$\ln \frac{4}{3} = \ln e^{0.05t} \quad \text{Take the natural log of both sides.}$$

$$\ln \frac{4}{3} = 0.05t \ln e \quad \log_a x^r = r \log_a x$$

$$\ln \frac{4}{3} = 0.05t(1) \quad \ln e = 1$$

$$\ln \frac{4}{3} = 0.05t$$

$$\frac{\ln \frac{4}{3}}{0.05} = t \quad \text{Divide by 0.05.}$$

$$5.75 \approx t \quad \text{Use a calculator to get the approximation.}$$

It would take about 5.75 yr for $3000 to grow to $4000.

[**YOU TRY 5**] If $4500 is invested at 6% interest compounded continuously, how long would it take for the investment to grow to $5000?

The amount of time it takes for a quantity to double in size is called the *doubling time*. We can use this in many types of applications.

EXAMPLE 6

The number of bacteria, $N(t)$, in a culture t hr after the bacteria are placed in a dish is given by

$$N(t) = 5000e^{0.0462t}$$

where 5000 bacteria are initially present. How long will it take for the number of bacteria to double?

Solution

If there are 5000 bacteria present initially, there will be $2(5000) = 10,000$ bacteria when the number doubles. This is $N(t)$.

Find t when $N(t) = 10,000$.

$$N(t) = 5000e^{0.0462t}$$

$$10,000 = 5000e^{0.0462t} \quad \text{Substitute 10,000 for } N(t).$$

$$2 = e^{0.0462t} \quad \text{Divide by 5000.}$$

$$\ln 2 = \ln e^{0.0462t} \quad \text{Take the natural log of both sides.}$$

$$\ln 2 = 0.0462t \ln e \quad \log_a x^r = r \log_a x$$

$$\ln 2 = 0.0462t(1) \quad \ln e = 1$$

$$\ln 2 = 0.0462t$$

$$\frac{\ln 2}{0.0462} = t \quad \text{Divide by 0.0462.}$$

$$15 \approx t$$

W Hint

Don't forget: $\ln e = 1$.

It will take about 15 hr for the number of bacteria to double.

[YOU TRY 6] The number of bacteria, $N(t)$, in a culture t hr after the bacteria are placed in a dish is given by

$$N(t) = 12{,}000e^{0.0385t}$$

where 12,000 bacteria are initially present. How long will it take for the number of bacteria to double?

4 Solve an Applied Problem Involving Exponential Growth or Decay

W Hint

Learn this formula, and understand what each variable represents.

We can generalize the formulas used in Examples 5 and 6 with a formula widely used to model situations that grow or decay exponentially. That formula is

$$y = y_0 e^{kt}$$

where y_0 is the initial amount or quantity at time $t = 0$, y is the amount present after time t, and k is a constant. If k is positive, it is called a *growth constant* because the quantity will *increase* over time. If k is negative, it is called a *decay constant* because the quantity will *decrease* over time.

EXAMPLE 7

In April 1986, an accident at the Chernobyl nuclear power plant released many radioactive substances into the environment. One such substance was cesium-137. Cesium-137 decays according to the equation

$$y = y_0 e^{-0.0230t}$$

where y_0 is the initial amount present at time $t = 0$ and y is the amount present after t yr. If a sample of soil contains 10 g of cesium-137 immediately after the accident,

a) how many grams will remain after 15 yr?

b) how long would it take for the initial amount of cesium-137 to decay to 2 g?

c) the **half-life** of a substance is the amount of time it takes for a substance to decay to half its original amount. What is the half-life of cesium-137?

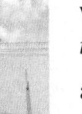

Creatas/PunchStock

Solution

a) The initial amount of cesium-137 is 10 g, so $y_0 = 10$. We must find y when $y_0 = 10$ and $t = 15$.

$$y = y_0 e^{-0.0230t}$$
$$= 10e^{-0.0230(15)} \qquad \text{Substitute the values.}$$
$$\approx 7.08 \qquad \text{Use a calculator to get the approximation.}$$

There will be about 7.08 g of cesium-137 remaining after 15 yr.

b) The initial amount of cesium-137 is $y_0 = 10$. To determine how long it will take to decay to 2 g, let $y = 2$ and solve for t.

$$y = y_0 e^{-0.0230t}$$
$$2 = 10e^{-0.0230t} \qquad \text{Substitute 2 for } y \text{ and 10 for } y.$$
$$0.2 = e^{-0.0230t} \qquad \text{Divide by 10.}$$
$$\ln 0.2 = \ln e^{-0.0230t} \qquad \text{Take the natural log of both sides.}$$
$$\ln 0.2 = -0.0230t \ln e \qquad \log_a x^r = r \log_a x$$
$$\ln 0.2 = -0.0230t \qquad \ln e = 1$$
$$\frac{\ln 0.2}{-0.0230} = t \qquad \text{Divide by } -0.0230.$$
$$69.98 \approx t \qquad \text{Use a calculator to get the approximation.}$$

It will take about 69.98 yr for 10 g of cesium-137 to decay to 2 g.

W **Hint**

Write a definition of **half-life** in your notes, in your own words.

c) Since there are 10 g of cesium-137 in the original sample, to determine the half-life we will determine how long it will take for the 10 g to decay to 5 g because $\frac{1}{2}(10) = 5$.

Let $y_0 = 10$, $y = 5$, and solve for t.

$$y = y_0 e^{-0.0230t}$$
$$5 = 10e^{-0.0230t} \qquad \text{Substitute the values.}$$
$$0.5 = e^{-0.0230t} \qquad \text{Divide by 10.}$$
$$\ln 0.5 = \ln e^{-0.0230t} \qquad \text{Take the natural log of both sides.}$$
$$\ln 0.5 = -0.0230t \ln e \qquad \log_a x^r = r \log_a x$$
$$\ln 0.5 = -0.0230t \qquad \ln e = 1$$
$$\frac{\ln 0.5}{-0.0230} = t \qquad \text{Divide by } -0.0230.$$
$$30.14 \approx t \qquad \text{Use a calculator to get the approximation.}$$

The half-life of cesium-137 is about 30.14 yr. This means that it would take about 30.14 yr for any quantity of cesium-137 to decay to half of its original amount.

$\left[\text{YOU TRY 7}\right]$ Radioactive strontium-90 decays according to the equation

$$y = y_0 e^{-0.0244t}$$

where t is in years. If a sample contains 40 g of strontium-90,

a) how many grams will remain after 8 yr?

b) how long would it take for the initial amount of strontium-90 to decay to 30 g?

c) what is the half-life of strontium-90?

Using Technology

We can solve exponential and logarithmic equations in the same way that we solved other equations—by graphing both sides of the equation and finding where the graphs intersect.

In Example 2 of this section, we learned how to solve $5^{x-2} = 16$. Because the right side of the equation is 16, the graph will have to go at least as high as 16. So set the $Y_{\max}$ to be 20, enter the left side of the equation as Y_1 and the right side as Y_2, and press $\boxed{\text{GRAPH}}$:

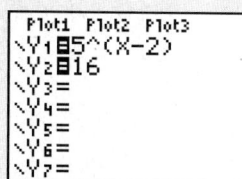
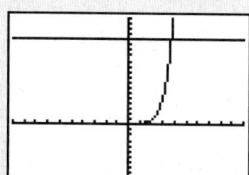

Recall that the x-coordinate of the point of intersection is the solution to the equation. To find the point of intersection, press $\boxed{2^{\text{nd}}}$ $\boxed{\text{TRACE}}$ and then highlight 5:intersect and press $\boxed{\text{ENTER}}$. Press $\boxed{\text{ENTER}}$ three more times to see that the x-coordinate of the point of intersection is approximately 3.723.

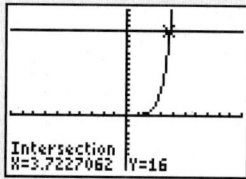

Remember, while the calculator can sometimes save you time, it will often give an approximate answer and not an exact solution.

Use a graphing calculator to solve each equation. Round your answer to the nearest thousandth.

1) $7^x = 49$

2) $6^{2b+1} = 13$

3) $5^{4a+7} = 8^{2a}$

4) $\ln x = 1.2$

5) $\log(k + 9) = \log 11$

6) $\ln(x + 3) = \ln(x - 2)$

E Evaluate **13.6** Exercises   Do the exercises, and check your work.

Objective 1: Solve an Exponential Equation

Solve each equation. Give the exact solution. If the answer contains a logarithm, approximate the solution to four decimal places.

1) $7^x = 49$

2) $5^c = 125$

3) $7^n = 15$

4) $5^a = 38$

5) $8^z = 3$

6) $4^y = 9$

7) $6^{5p} = 36$

8) $2^{3t} = 32$

9) $4^{6k} = 2.7$

10) $3^{2x} = 7.8$

11) $2^{4n+1} = 5$

12) $6^{2b+1} = 13$

13) $5^{3a-2} = 8$

14) $3^{2x-3} = 14$

15) $4^{2c+7} = 64^{3c-1}$

16) $27^{5m-2} = 3^{m+6}$

17) $9^{5d-2} = 4^{3d}$

18) $5^{4a+7} = 8^{2a}$

Solve each equation. Give the exact solution and the approximation to four decimal places.

19) $e^y = 12.5$

20) $e^t = 0.36$

21) $e^{-4x} = 9$

22) $e^{3p} = 4$

23) $e^{0.01r} = 2$

24) $e^{-0.08k} = 10$

25) $e^{0.006t} = 3$

26) $e^{0.04a} = 12$

27) $e^{-0.4y} = 5$

28) $e^{-0.005c} = 16$

Objective 2: Solve Logarithmic Equations Using the Properties of Logarithms

Solve each equation.

29) $\log_6(k + 9) = \log_6 11$

30) $\log_5(d - 4) = \log_5 2$

31) $\log_7(3p - 1) = \log_7 9$

32) $\log_4(5y + 2) = \log_4 10$

33) $\log x + \log(x - 2) = \log 15$

34) $\log_9 r + \log_9(r + 7) = \log_9 18$

35) $\log_3 n + \log_3(12 - n) = \log_3 20$

36) $\log m + \log(11 - m) = \log 24$

37) $\log_2(-z) + \log_2(z - 8) = \log_2 15$

38) $\log_5 8y - \log_5(3y - 4) = \log_5 2$

39) $\log_3(4c + 5) = 3$

40) $\log_6(5b - 4) = 2$

41) $\log(3p + 4) = 1$

42) $\log(7n - 11) = 1$

43) $\log_3 y + \log_3(y - 8) = 2$

44) $\log_4 k + \log_4(k - 6) = 2$

45) $\log_2 r + \log_2(r + 2) = 3$

46) $\log_9(z + 8) + \log_9 z = 1$

47) $\log_4 20c - \log_4(c + 1) = 2$

48) $\log_6 40x - \log_6(1 + x) = 2$

49) $\log_2 8d - \log_2(2d - 1) = 4$

50) $\log_6(13 - x) + \log_6 x = 2$

Mixed Exercises: Objectives 3 and 4

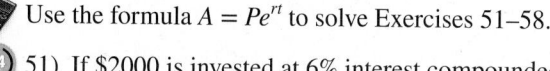
Use the formula $A = Pe^{rt}$ to solve Exercises 51–58.

51) If $2000 is invested at 6% interest compounded continuously, how long would it take

   a) for the investment to grow to $2500?

   b) for the initial investment to double?

52) If $5000 is invested at 7% interest compounded continuously, how long would it take

   a) for the investment to grow to $6000?

   b) for the initial investment to double?

53) How long would it take for an investment of $7000 to earn $800 in interest if it is invested at 7.5% compounded continuously?

54) How long would it take for an investment of $4000 to earn $600 in interest if it is invested at 6.8% compounded continuously?

55) Cynthia wants to invest some money now so that she will have $5000 in the account in 10 yr. How much should she invest in an account earning 8% compounded continuously?

56) How much should Leroy invest now at 7.2% compounded continuously so that the account contains $8000 in 12 yr?

57) Raj wants to invest $3000 now so that it grows to $4000 in 4 yr. What interest rate should he look for? (Round to the nearest tenth of a percent.)

58) Marisol wants to invest $12,000 now so that it grows to $20,000 in 7 yr. What interest rate should she look for? (Round to the nearest tenth of a percent.)

59) The number of bacteria, $N(t)$, in a culture t hr after the bacteria are placed in a dish is given by

$$N(t) = 4000e^{0.0374t}$$

where 4000 bacteria are initially present.

   a) After how many hours will there be 5000 bacteria in the culture?

   b) How long will it take for the number of bacteria to double?

60) The number of bacteria, $N(t)$, in a culture t hr after the bacteria are placed in a dish is given by

$$N(t) = 10,000e^{0.0418t}$$

where 10,000 bacteria are initially present.

   a) After how many hours will there be 15,000 bacteria in the culture?

   b) How long will it take for the number of bacteria to double?

61) The population of an Atlanta suburb is growing at a rate of 3.6% per year. If 21,000 people lived in the suburb in 2015, determine how many people will live in the town in 2023. Use $y = y_0 e^{0.036t}$.

62) The population of a Seattle suburb is growing at a rate of 3.2% per year. If 30,000 people lived in the suburb in 2017, determine how many people will live in the town in 2024. Use $y = y_0 e^{0.032t}$.

63) A rural town in South Dakota is losing residents at a rate of 1.3% per year. The population of the town was 2470 in 2000. Use $y = y_0 e^{-0.013t}$ to answer the following questions.

   a) What was the population of the town in 2015?

   b) In what year would it be expected that the population of the town is 1600?

64) In 2015, the population of a rural town in Kansas was 1682. The population is decreasing at a rate of 0.8% per year. Use $y = y_0 e^{-0.008t}$ to answer the following questions.

   a) What was the population of the town in 2019?

   b) In what year would it be expected that the population of the town is 1000?

65) Radioactive carbon-14 is a substance found in all living organisms. After the organism dies, the carbon-14 decays according to the equation

$$y = y_0 e^{-0.000121t}$$

where t is in years, y_0 is the initial amount present at time $t = 0$, and y is the amount present after t yr.

a) If a sample initially contains 15 g of carbon-14, how many grams will be present after 2000 yr?

b) How long would it take for the initial amount to decay to 10 g?

c) What is the half-life of carbon-14?

66) Plutonium-239 decays according to the equation

$$y = y_0 e^{-0.0000287t}$$

where t is in years, y_0 is the initial amount present at time $t = 0$, and y is the amount present after t yr.

a) If a sample initially contains 8 g of plutonium-239, how many grams will be present after 5000 yr?

b) How long would it take for the initial amount to decay to 5 g?

c) What is the half-life of plutonium-239?

67) Radioactive iodine-131 is used in the diagnosis and treatment of some thyroid-related illnesses. The concentration of the iodine in a patient's system is given by

$$y = 0.4e^{-0.086t}$$

where t is in days, and y is in the appropriate units.

a) How much iodine-131 is given to the patient?

b) How much iodine-131 remains in the patient's system 7 days after treatment?

68) The amount of cobalt-60 in a sample is given by

$$y = 30e^{-0.131t}$$

where t is in years, and y is in grams.

a) How much cobalt-60 is originally in the sample?

b) How long would it take for the initial amount to decay to 10 g?

Extension

Solve. Where appropriate, give the exact solution and the approximation to four decimal places.

69) $\log_2 (\log_2 x) = 2$

70) $\log_3 (\log y) = 1$

71) $\log_3 \sqrt{n^2 + 5} = 1$

72) $\log (p - 7)^2 = 4$

73) $e^{|t|} = 13$

74) $e^{r^2 - 25} = 1$

75) $e^{2y} + 3e^y - 4 = 0$

76) $e^{2x} - 9e^x + 8 = 0$

77) $5^{2c} - 4 \cdot 5^c - 21 = 0$

78) $9^{2a} + 5 \cdot 9^a - 24 = 0$

79) $(\log x)^2 = \log x^3$

80) $\log 6^y = y^2$

R Rethink

R1) How are exponential equations and logarithmic equations the same? How are they different?

R2) Can you think of a real-life situation that you have encountered that involves decay or growth?

R3) What does continuous compounding mean?

Group Activity – Exponential Equations

Blood alcohol concentration (BAC) is the amount of alcohol present in your blood as you drink. It is determined by the number of milligrams of alcohol present in 100 milliliters of blood. The chart below shows approximate values of the BAC based on the number of drinks per hour. A drink is defined as 0.5 oz. of pure ethyl alcohol. The following are considered as one drink: 1.25 oz. of 80 proof liquor, 12 oz. of beer, or 5 oz. of table wine.

Table 1 Blood Alcohol Concentration (BAC) Table for Specific Body Weights

		Number of Drinks per Hour									
	Weight	1	2	3	4	5	6	7	8	9	10
Females	100 lb	0.050	0.100	0.150	0.200	0.250	0.300	0.350	0.400	0.450	0.500
	120 lb	0.042	0.084	0.126	0.168	0.210	0.252	0.294	0.336	0.378	0.420
	140 lb	0.036	0.072	0.108	0.144	0.180	0.216	0.252	0.288	0.324	0.360
	160 lb	0.032	0.064	0.096	0.128	0.160	0.192	0.224	0.256	0.288	0.320
	180 lb	0.028	0.056	0.084	0.112	0.140	0.168	0.196	0.224	0.252	0.280
	200 lb	0.025	0.050	0.075	0.100	0.125	0.150	0.175	0.200	0.225	0.250
Males	100 lb	0.039	0.078	0.117	0.156	0.195	0.234	0.273	0.312	0.351	0.390
	120 lb	0.035	0.070	0.105	0.140	0.175	0.210	0.245	0.280	0.315	0.350
	140 lb	0.029	0.058	0.087	0.116	0.145	0.174	0.203	0.232	0.261	0.290
	160 lb	0.026	0.052	0.078	0.104	0.130	0.156	0.182	0.208	0.234	0.260
	180 lb	0.023	0.046	0.069	0.092	0.115	0.138	0.161	0.184	0.207	0.230
	200 lb	0.020	0.040	0.060	0.080	0.100	0.120	0.140	0.160	0.180	0.200

Table 2 Symptoms for Values of BAC

BAC	Symptoms
0.02–0.04	lightheaded
0.05–0.10	warm and relaxed; slurred speech, impaired motor skills
0.10–0.15	lack of coordination and balance; markedly impaired motor skills; memory loss
0.15–0.20	difficulty talking, standing; vomiting; possible blackout
0.25	emotionally and physically numb
0.30	drunken stupor; difficult to awaken if passed out
0.35	may stop breathing
0.40	coma; possible death

1) Select your gender and approximate weight from Table 1.

   a) Using the appropriate data, graph the number of drinks per hour (x) vs. the blood alcohol concentration (c) for your gender and weight. What kind of function fits the data?

   b) What is the slope of the line? What does the slope of the line represent?

   c) Find the linear function $c(x)$.

   d) What is your BAC if you have $1\frac{1}{2}$ drinks in one hour?

   e) Is this function the same for everyone?

2) The risk of being involved in a car accident increases exponentially as the concentration of alcohol in the blood increases. This can be modeled by the equation:

$$R(c) = 6e^{12.8c}, \text{ where } R = \text{percent of risk, } c = \text{blood alcohol concentration}$$

a) Graph the function $R(c)$ using the BAC for your weight and gender.

b) Is this function the same for everyone?

3) A body metabolizes alcohol at the rate of 1.5% per hour. For example, a 100-lb female who has two drinks in three hours will have a BAC of 0.055 [0.100 − 3(0.015) = 0.055]. If a 160-lb male consumes 6 drinks in 4 hours, what is his BAC? What is his risk of being involved in a car accident?

4) What blood alcohol concentration is considered legally drunk in your state?

Read each statement, and check T for *true* or F for *false*.

	T	F
1) I am aware of and know how to access all of the online resources available with this textbook.	☐	☐
2) I have used some of the online resources that accompany this textbook.	☐	☐
3) I have asked my instructor whether he or she can recommend any websites that are a good fit for the text we are using.	☐	☐
4) I frequently search the Internet for supplemental learning resources.	☐	☐
5) I have a good understanding of which Web sources are trustworthy, or I am willing to put in the time to determine which sources can be trusted.	☐	☐
6) I have decided which type of Internet resource best fits my learning style.	☐	☐
7) I understand that Internet resources are presented at different difficulty levels, and I am willing to determine whether a given level of a lesson is appropriate for my class.	☐	☐
8) I feel I learn more when I am watching a video because I can rewind it when I do not understand the material.	☐	☐
9) I understand that some video formats will not fit my particular learning style.	☐	☐
10) I understand that I may learn techniques that are different from what my instructor used in class. In these cases, I will talk with my instructor before I use the technique on the exam.	☐	☐
11) I understand how to take into account user statistics regarding the number of viewers who have watched a video.	☐	☐
12) I know that there is value in taking into account the user comments relating to videos posted on the Internet.	☐	☐

Scoring:

Scoring of this emPOWERme is simple: The more statements you agree are true about yourself, the greater your ability and readiness to use online resources as a way to supplement your instruction. Each of these statements involves important considerations in the use of instructional resources you find on the Web.

Chapter 13: Summary

Definition/Procedure	Example

13.1 Inverse Functions

One-to-One Function

In order for a function to be a **one-to-one function,** each x-value corresponds to exactly one y-value, and each y-value corresponds to exactly one x-value.

The **horizontal line test** tells us how we can determine whether a graph represents a one-to-one function:

If every horizontal line that could be drawn through a function would intersect the graph at most once, then the function is one-to-one.

Determine whether each function is one-to-one.

a) $f = \{(-2, 9), (1, 3), (3, -1), (7, -9)\}$ *is* one-to-one.

b) $g = \{(0, 9), (2, 1), (4, 1), (5, 4)\}$ is *not* one-to-one because the y-value 1 corresponds to two different x-values.

c)

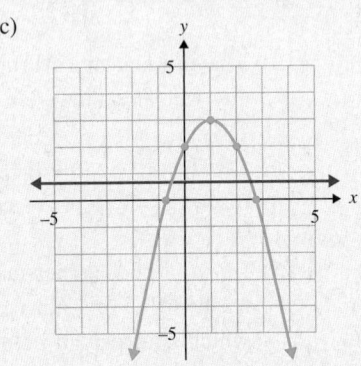

No. It fails the horizontal line test.

Inverse Function

Let f be a one-to-one function. The **inverse** of f, denoted by f^{-1}, is a one-to-one function that contains the set of all ordered pairs (y, x) where (x, y) belongs to f.

How to Find an Equation of the Inverse of $y = f(x)$

Step 1: Replace $f(x)$ with y.

Step 2: Interchange x and y.

Step 3: Solve for y.

Step 4: Replace y with the inverse notation, $f^{-1}(x)$.

The graphs of $f(x)$ and $f^{-1}(x)$ are symmetric with respect to the line $y = x$.

Find an equation of the inverse of $f(x) = 2x - 4$.

Step 1: $y = 2x - 4$ Replace $f(x)$ with y.

Step 2: $x = 2y - 4$ Interchange x and y.

Step 3: Solve for y.

$$x + 4 = 2y \quad \text{Add 4.}$$

$$\frac{x + 4}{2} = y \quad \text{Divide by 2.}$$

$$\frac{1}{2}x + 2 = y \quad \text{Simplify.}$$

Step 4: $f^{-1}(x) = \dfrac{1}{2}x + 2$ Replace y with $f^{-1}(x)$.

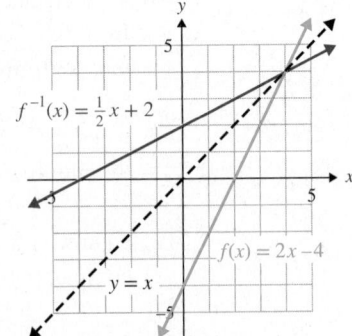

Definition/Procedure	Example

13.2 Exponential Functions

An **exponential function** is a function of the form

$$f(x) = a^x$$

where $a > 0$, $a \neq 1$, and x is a real number.

$f(x) = 3^x$

Characteristics of an Exponential Function

$$f(x) = a^x$$

1) If $f(x) = a^x$, where $a > 1$, the value of y increases as the value of x increases.
2) If $f(x) = a^x$ where $0 < a < 1$, the value of y decreases as the value of x increases.
3) The function is one-to-one.
4) The y-intercept is $(0, 1)$.
5) The domain is $(-\infty, \infty)$, and the range is $(0, \infty)$.

$f(x) = e^x$ is a special exponential function that has many uses in mathematics. Like the number π, e is an irrational number.

$$e \approx 2.7183$$

1)

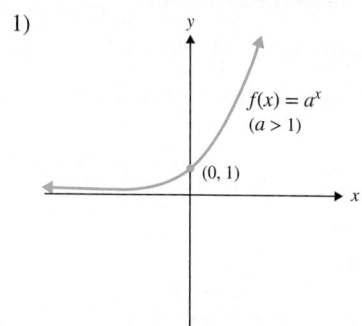

2)

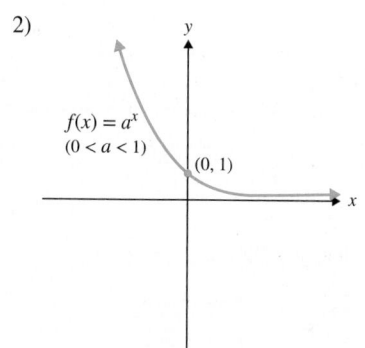

Solving an Exponential Equation

Step 1: **If possible, express each side of the equation with the same base.** If it is not possible to get the same base, a method in Section 13.6 can be used.

Step 2: **Use the rules of exponents to simplify the exponents.**

Step 3: **Set the exponents equal, and solve for the variable.**

Solve $5^{4x-1} = 25^{3x+4}$.

Step 1: $5^{4x-1} = (5^2)^{(3x+4)}$ Both sides are powers of 5.

Step 2: $5^{4x-1} = 5^{2(3x+4)}$ Power rule for exponents

$5^{4x-1} = 5^{6x+8}$ Distribute.

Step 3: $4x - 1 = 6x + 8$ The bases are the same. Set the exponents equal.

$-2x = 9$ Subtract $6x$; add 1.

$x = -\dfrac{9}{2}$ Divide by -2.

The solution set is $\left\{ -\dfrac{9}{2} \right\}$.

13.3 Logarithmic Functions

Definition of Logarithm

If $a > 0$, $a \neq 1$, and $x > 0$, then for every real number y,
$y = \log_a x$ means $x = a^y$.

Write $\log_5 125 = 3$ in exponential form.

$$\log_5 125 = 3 \text{ means } 5^3 = 125$$

A **logarithmic equation** is an equation in which at least one term contains a logarithm.

To solve a logarithmic equation of the form

$$\log_a b = c$$

write the equation in exponential form ($a^c = b$) and solve for the variable.

Solve $\log_2 k = 3$.

Write the equation in exponential form, and solve for k.

$$\log_2 k = 3 \text{ means } 2^3 = k.$$
$$8 = k$$

The solution set is $\{8\}$.

Definition/Procedure	Example
To evaluate $\log_a b$ means *to find the power to which we raise a to get b*.	Evaluate $\log_7 49$. $\log_7 49 = 2$ because $7^2 = 49$
A base 10 logarithm is called a **common logarithm.** A base 10 logarithm is often written without the base.	$\log x$ means $\log_{10} x$.

Characteristics of a Logarithmic Function

$$f(x) = \log_a x, \text{ where } a > 0 \text{ and } a \neq 1$$

1) If $f(x) = \log_a x$, where $a > 1$, the value of y increases as the value of x increases.
2) If $f(x) = \log_a x$, where $0 < a < 1$, the value of y decreases as the value of x increases.
3) The function is one-to-one.
4) The x-intercept is $(1, 0)$.
5) The domain is $(0, \infty)$, and the range is $(-\infty, \infty)$.
6) The inverse of $f(x) = \log_a x$ is $f^{-1}(x) = a^x$.

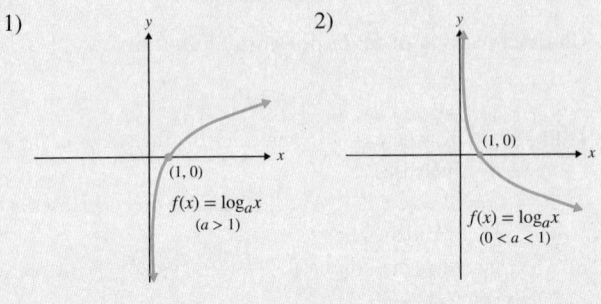

13.4 Properties of Logarithms

Let x, y, and a be positive real numbers where $a \neq 1$, and let r be any real number. Then,

1) $\log_a a = 1$
2) $\log_a 1 = 0$
3) $\log_a xy = \log_a x + \log_a y$ Product rule
4) $\log_a \dfrac{x}{y} = \log_a x - \log_a y$ Quotient rule
5) $\log_a x^r = r \log_a x$ Power rule
6) $\log_a a^x = x$ for any real number x
7) $a^{\log_a x} = x$

Write $\log_4 \dfrac{c^5}{d^2}$ as the sum or difference of logarithms in simplest form. Assume c and d represent positive real numbers.

$$\log_4 \frac{c^5}{d^2} = \log_4 c^5 - \log_4 d^2 \quad \text{Quotient rule}$$
$$= 5 \log_4 c - 2 \log_4 d \quad \text{Power rule}$$

13.5 Common and Natural Logarithms and Change of Base

We can evaluate common logarithms with or without a calculator.	Find the value of each. a) $\log 100$ b) $\log 53$ a) $\log 100 = \log_{10} 100 = \log_{10} 10^2 = 2$ b) Using a calculator, we get $\log 53 \approx 1.7243$.
The number e is approximately equal to 2.7183. A base e logarithm is called a **natural logarithm.** The notation used for a natural logarithm is $\ln x$. $$f(x) = \ln x \quad \text{means} \quad f(x) = \log_e x$$ The domain of $f(x) = \ln x$ is $(0, \infty)$, and the range is $(-\infty, \infty)$.	The graph of $f(x) = \ln x$ looks like this: 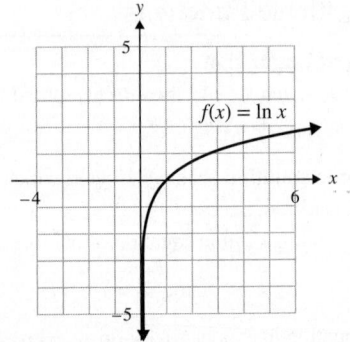

Definition/Procedure	Example

$\ln e = 1$ because $\ln e = 1$ means $\log_e e = 1$.

We can find the values of some natural logarithms using the properties of logarithms. We can approximate the values of other natural logarithms using a calculator.

Find the value of each.

a) $\ln e^{12}$ b) $\ln 18$

a) $\ln e^{12} = 12 \ln e$ Power rule
   $= 12(1)$ $\ln e = 1$
   $= 12$

b) Using a calculator, we get $\ln 18 \approx 2.8904$.

To solve an equation such as $\ln x = 1.6$, change to exponential form and solve for the variable.

Solve $\ln x = 1.6$.

$\ln x = 1.6$ means $\log_e x = 1.6$.

$$\log_e x = 1.6$$
$$e^{1.6} = x \quad \text{Exponential form}$$
$$4.9530 \approx x \quad \text{Approximation}$$

The exact solution is $\{e^{1.6}\}$. The approximation is $\{4.9530\}$.

Applications of Exponential Functions

Continuous Compounding

If P dollars are deposited in an account earning interest rate r *compounded continuously*, then the amount of money, A (in dollars), in the account after t years is given by $A = Pe^{rt}$.

Determine the amount of money in an account after 6 yr if $3000 was initially invested at 5% compounded continuously.

$$A = Pe^{rt}$$
$$= 3000e^{0.05(6)} \quad \text{Substitute the values.}$$
$$= 3000e^{0.30} \quad \text{Multiply } (0.05)(6).$$
$$\approx 4049.5764 \quad \text{Evaluate using a calculator.}$$
$$\approx \$4049.58 \quad \text{Round to the nearest cent.}$$

Change-of-Base Formula

If a, b, and x are positive real numbers and $a \neq 1$ and $b \neq 1$, then

$$\log_a x = \frac{\log_b x}{\log_b a}$$

Find $\log_2 75$ to four decimal places.

$$\log_2 75 = \frac{\log_{10} 75}{\log_{10} 2} \approx 6.2288$$

13.6 Solving Exponential and Logarithmic Equations

Let a, x, and y be positive real numbers, where $a \neq 1$.

1) If $x = y$, then $\log_a x = \log_a y$.
2) If $\log_a x = \log_a y$, then $x = y$.

Solve each equation.

a) $4^x = 64$

Ask yourself, "*Can I express both sides with the same base?*" **Yes.**

$$4^x = 64$$
$$4^x = 4^3$$
$$x = 3 \quad \text{Set the exponents equal.}$$

The solution set is $\{3\}$.

b) $4^x = 9$

How to Solve an Exponential Equation

Begin by asking yourself, "*Can I express each side with the same base?*"

1) If the answer is **yes,** then write each side of the equation with the same base, set the exponents equal, and solve for the variable.
2) If the answer is **no,** then take the natural logarithm of each side, use the properties of logarithms, and solve for the variable.

Ask yourself, "*Can I express both sides with the same base?*" **No.** Take the natural logarithm of each side.

$$4^x = 9$$
$$\ln 4^x = \ln 9 \quad \text{Take the natural log of each side.}$$
$$x \ln 4 = \ln 9 \quad \log_a x^r = r \log_a x$$
$$x = \frac{\ln 9}{\ln 4} \quad \text{Divide by } \ln 4.$$
$$x \approx 1.5850 \quad \text{Use a calculator to get the approximation.}$$

The exact solution is $\left\{\dfrac{\ln 9}{\ln 4}\right\}$. The approximation is $\{1.5850\}$.

Definition/Procedure	Example
Solve an exponential equation with base e by taking the natural logarithm of each side.	Solve $e^y = 35.8$. $\ln e^y = \ln 35.8$ Take the natural log of each side. $y \ln e = \ln 35.8$ $\log_a x^r = r \log_a x$ $y(1) = \ln 35.8$ $\ln e = 1$ $y = \ln 35.8$ $y \approx 3.5779$ Approximation The exact solution is $\{\ln 35.8\}$. The approximation is $\{3.5779\}$.
Solving Logarithmic Equations Sometimes we must use the properties of logarithms to solve logarithmic equations.	Solve $\log x + \log(x - 3) = \log 28$. $\log x + \log(x - 3) = \log 28$ $\log x(x - 3) = \log 28$ Product rule $x(x - 3) = 28$ If $\log_a x = \log_a y$, then $x = y$. $x^2 - 3x = 28$ Distribute. $x^2 - 3x - 28 = 0$ Subtract 28. $(x - 7)(x + 4) = 0$ Factor. $x - 7 = 0$ or $x + 4 = 0$ Set each factor equal to 0. $x = 7$ or $x = -4$ Solve. Verify that only 7 satisfies the original equation. The solution set is $\{7\}$.

Chapter 13: Review Exercises

(13.1) Determine whether each function is one-to-one. If it is one-to-one, find its inverse.

1) $f = \{(-7, -4), (-2, 1), (1, 5), (6, 11)\}$

2) $g = \{(1, 4), (3, 7), (6, 4), (10, 9)\}$

Determine whether each function is one-to-one. If it is one-to-one, graph its inverse.

3)

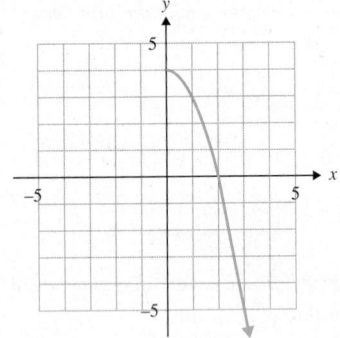

4)

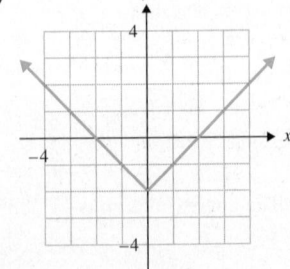

Find the inverse of each one-to-one function. Graph each function and its inverse on the same axes.

5) $f(x) = x + 4$

6) $g(x) = 2x - 10$

7) $h(x) = \dfrac{1}{3}x - 1$

8) $f(x) = \sqrt[3]{x} + 2$

Given each one-to-one function $f(x)$, find the following function values *without* finding an equation of $f^{-1}(x)$. Find the value in a) before b).

9) $f(x) = 6x - 1$

a) $f(2)$

b) $f^{-1}(11)$

10) $f(x) = \sqrt[3]{x + 5}$

a) $f(-13)$

b) $f^{-1}(-2)$

11) If $f(x) = -9x + 2$, show that $f^{-1}(x) = -\dfrac{1}{9}x + \dfrac{2}{9}$.

12) Does $f(x) = x^2$ have an inverse for all values in its domain? Explain.

In Exercises 13 and 14, determine whether each statement is *true* or *false*. If it is false, explain why.

13) Every function has an inverse.

14) The domain of $f(x)$ is the range of $f^{-1}(x)$.

(13.2) Graph each exponential function. State the domain and range.

15) $f(x) = 2x$

16) $h(x) = \left(\dfrac{1}{3}\right)^x$

17) $y = 2^x - 4$

18) $f(x) = 3^{x-2}$

19) $f(x) = e^x$

20) $g(x) = e^x + 2$

Solve each exponential equation.

21) $2^c = 64$

22) $7^{m+5} = 49$

23) $16^{3z} = 32^{2z-1}$

24) $9^y = \dfrac{1}{81}$

25) $\left(\dfrac{3}{2}\right)^{x+4} = \left(\dfrac{4}{9}\right)^{x-3}$

26) The value, $V(t)$, in dollars, of a luxury car t yr after it is purchased is given by $V(t) = 67,300(0.816)^t$.

   a) What was the purchase price of the car?

   b) What will the car be worth 4 yr after purchase?

(13.3)

27) What is the domain of $y = \log_a x$?

28) In the equation $y = \log_a x$, a must be what kind of number?

Write in exponential form.

29) $\log_5 125 = 3$

30) $\log_{16} \dfrac{1}{4} = -\dfrac{1}{2}$

31) $\log 100 = 2$

32) $\log 1 = 0$

Write in logarithmic form.

33) $3^4 = 81$

34) $\left(\dfrac{2}{3}\right)^{-2} = \dfrac{9}{4}$

35) $10^3 = 1000$

36) $\sqrt{121} = 11$

Solve.

37) $\log_2 x = 3$

38) $\log_9(4x + 1) = 2$

39) $\log_{32} 16 = x$

40) $\log(2x + 5) = 1$

Evaluate.

41) $\log_8 64$

42) $\log_3 27$

43) $\log 1000$

44) $\log 1$

45) $\log_{1/2} 16$

46) $\log_{1/5} \dfrac{1}{25}$

Graph each logarithmic function.

47) $f(x) = \log_2 x$

48) $f(x) = \log_{1/4} x$

49) $h(x) = \log_{1/3} x$

50) $g(x) = \log_3 x$

Find the inverse of each function.

51) $f(x) = 5^x$

52) $h(x) = \log_6 x$

Solve.

53) A company plans to test market its new dog food in a large metropolitan area before taking it nationwide. The company predicts that its sales over the next 12 months can be approximated by

$$S(t) = 10 \log_3(2t + 1)$$

where t is the number of months after the dog food is introduced, and $S(t)$ is in thousands of bags of dog food.

   a) How many bags of dog food were sold after 1 month on the market?

   b) How many bags of dog food were sold after 4 months on the market?

(13.4) Decide whether each statement is true or false.

54) $\log_5(x + 4) = \log_5 x + \log_5 4$

55) $\log_2 \dfrac{k}{6} = \log_2 k - \log_2 6$

Write as the sum or difference of logarithms and simplify, if possible. Assume all variables represent positive real numbers.

56) $\log_8 3z$

57) $\log_7 \dfrac{49}{t}$

58) $\log_4 \sqrt{64}$

59) $\log \dfrac{1}{100}$

60) $\log_5 c^4 d^3$

61) $\log_4 m\sqrt{n}$

62) $\log_a \dfrac{xy}{z^3}$

63) $\log_4 \dfrac{a^2}{bc^4}$

64) $\log p(p + 8)$

65) $\log_6 \dfrac{r^3}{r^2 - 5}$

Write as a single logarithm. Assume the variables are defined so that the variable expressions are positive and so that the bases are positive real numbers not equal to 1.

66) $\log c + \log d$

67) $9 \log_2 a + 3 \log_2 b$

68) $\log_5 r - 2 \log_5 t$

69) $\log_3 5 + 4 \log_3 m - 2 \log_3 n$

70) $\dfrac{1}{2} \log_z a - \log_z b$

Given that $\log 7 \approx 0.8451$ and $\log 9 \approx 0.9542$, use the properties of logarithms to approximate the following. Do NOT use a calculator.

71) $\log 49$

72) $\log \dfrac{1}{7}$

(13.5)

73) What is the base of $\ln x$?   74) Evaluate $\ln e$.

Evaluate each logarithm. Do not use a calculator.

75) $\log 100$

76) $\log \sqrt{10}$

77) $\log \dfrac{1}{100}$

78) $\log 0.001$

79) $\ln 1$

80) $\ln \sqrt[3]{e}$

Use a calculator to find the approximate value of each logarithm to four decimal places.

81) $\log 8$

82) $\log 0.3$

83) $\ln 1.75$

84) $\ln 0.924$

Solve each equation. Do not use a calculator.

85) $\log p = 2$

86) $\log(5n) = 3$

87) $\log\left(\dfrac{1}{2}c\right) = -1$

88) $\log(6z - 5) = 1$

Solve each equation. Give an exact solution and a solution that is approximated to four decimal places.

89) $\log x = 2.1$

90) $\log k = -1.4$

91) $\ln y = 2$

92) $\ln c = -0.5$

93) $\log(4t) = 1.75$

94) $\ln(2a - 3) = 1$

Graph each function. State the domain and range.

95) $f(x) = \ln(x - 3)$

96) $g(x) = \ln x - 2$

Use the change-of-base formula with either base 10 or base e to approximate each logarithm to four decimal places.

97) $\log_4 19$

98) $\log_9 42$

99) $\log_{1/2} 38$

100) $\log_6 0.82$

For Exercises 101 and 102, use the formula $L(I) = 10 \log \dfrac{I}{10^{-12}}$, where I is the intensity of sound, in watts per square meter, and $L(I)$ is the loudness of sound in decibels. Do *not* use a calculator.

101) The intensity of sound from the crowd at a college basketball game reached 0.1 W/m². Find the loudness of the crowd, in decibels.

Vetta/Getty Images

102) Find the intensity of the sound of a jet taking off if the noise level can reach 140 dB 25 m from the jet.

Use the formula $A = P\left(1 + \dfrac{r}{n}\right)^{nt}$ and a calculator to solve.

103) Pedro deposits $2500 in an account earning 6% interest compounded quarterly. How much will be in the account after 5 yr?

Use the formula $A = Pe^{rt}$ and a calculator to solve.

104) Find the amount Liang will owe at the end of 4 yr if he borrows $9000 at a rate of 6.2% compounded continuously.

105) The number of bacteria, $N(t)$, in a culture t hr after the bacteria are placed in a dish is given by

$$N(t) = 6000e^{0.0514t}$$

a) How many bacteria were originally in the culture?

b) How many bacteria are present after 12 hr?

106) The pH of a solution is given by pH $= -\log[H^+]$, where $[H^+]$ is the molar concentration of the hydronium ion. Find the ideal pH of blood if $[H^+] = 3.98 \times 10^{-8}$.

(13.6) Solve each equation. Give the exact solution. If the answer contains a logarithm, approximate the solution to four decimal places. *Some of these exercises require the use of a calculator to obtain a decimal approximation.*

107) $2^y = 16$

108) $3^n = 7$

109) $125^{m-4} = 25^{1-m}$

110) $e^z = 22$

111) $e^{0.03t} = 19$

Solve each logarithmic equation.

112) $\log(3n - 5) = 3$

113) $\log_2 x + \log_2(x + 2) = \log_2 24$

114) $\log_7 10p - \log_7(p - 8) = \log_7 6$

115) $\log_4 k + \log_4(k - 12) = 3$

116) $\log_3 12m - \log_3(1 + m) = 2$

Use the formula $A = Pe^{rt}$ to solve Exercises 117 and 118.

117) Jamar wants to invest some money now so that he will have $10,000 in the account in 6 yr. How much should he invest in an account earning 6.5% compounded continuously?

118) Samira wants to invest $6000 now so that it grows to $9000 in 5 yr. What interest rate (compounded continuously) should she look for? (Round to the nearest tenth of a percent.)

119) The population of a suburb is growing at a rate of 1.6% per year. The population of the suburb was 16,410 in 2015. Use $y = y_0 e^{0.016t}$ to answer the following questions.

 a) What was the population of the town in 2020?

 b) In what year would it be expected that the population of the town is 23,000?

120) Radium-226 decays according to the equation

$$y = y_0 e^{-0.000436t}$$

where t is in years, y_0 is the initial amount present at time $t = 0$, and y is the amount present after t yr.

a) If a sample initially contains 80 g of radium-226, how many grams will be present after 500 yr?

b) How long would it take for the initial amount to decay to 20 g?

c) What is the half-life of radium-226?

Mixed Exercises

Fill in the blank with *always*, *sometimes*, or *never* to make the statement true.

121) A function _____ has an inverse.

122) The domain of $f(x)$ is _____ the range of its inverse.

123) The solution of a logarithmic equation can _____ be a negative number.

124) The inverse of an exponential function $f(x) = a^x$ is _____ a logarithmic function.

125) The graphs of a function and its inverse are _____ symmetric with respect to the line $y = x$.

126) The range of $f(x) = e^x$ is _____ $(-\infty, \infty)$.

Chapter 13:   Test

Use a calculator only where indicated.

Determine whether each function is one-to-one. If it is one-to-one, find its inverse.

1) $f = \{(-4, 5), (-2, 7), (0, 3), (6, 5)\}$

2) $g = \left\{ (2, 4), (6, 6), \left(9, \dfrac{15}{2}\right), (14, 10) \right\}$

3) Is this function one-to-one? If it is one-to-one, graph its inverse.

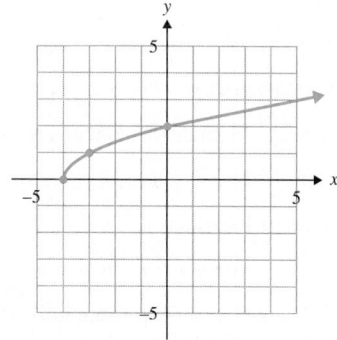

4) If $k(x) = \dfrac{5}{3}x - \dfrac{1}{2}$, show that $k^{-1}(x) = \dfrac{3}{5}x + \dfrac{3}{10}$.

5) Find an equation of the inverse of $f(x) = -3x + 12$.

Use $f(x) = 2^x$ and $g(x) = \log_2 x$ for Exercises 6–9.

6) Graph $f(x)$.

7) Graph $g(x)$.

8) a) What is the domain of $g(x)$?

 b) What is the range of $g(x)$?

9) How are the functions $f(x)$ and $g(x)$ related?

10) Write $3^{-2} = \dfrac{1}{9}$ in logarithmic form.

Solve each equation.

11) $9^{4x} = 81$

12) $125^{2c} = 25^{c-4}$

13) $\log_{12} y = -2$

14) $\log(3r + 13) = 2$

15) $\log_6(2m) + \log_6(2m - 3) = \log_6 40$

16) $\log_5 x + \log_5(3x - 10) = 2$

17) Evaluate.

 a) $\log_2 16$ b) $\log_7 \sqrt{7}$

18) Find $\ln e$.

Write as the sum or difference of logarithms and simplify, if possible. Assume all variables represent positive real numbers.

19) $\log_8 5n$

20) $\log_3 \dfrac{9a^4}{b^5 c}$

21) Write as a single logarithm.

$$2 \log x - 3 \log (x + 1)$$

Use a calculator for the rest of the exercises.

Solve each equation. Give an exact solution and a solution that is approximated to four decimal places.

22) $\log w = 0.8$

23) $e^{0.3t} = 5$

24) $\ln x = -0.25$

25) $4^{4a+3} = 9$

Graph the functions in Exercises 26 and 27. State the domain and range.

26) $y = e^x - 4$

27) $f(x) = \ln(x + 1)$

28) Approximate $\log_5 17$ to four decimal places.

29) If $6000 is deposited in an account earning 7.4% interest compounded continuously, how much will be in the account after 5 yr? Use $A = Pe^{rt}$.

30) Polonium-210 decays according to the equation
$$y = y_0 e^{-0.00495t}$$
where t is in days, y_0 is the initial amount present at time $t = 0$, and y is the amount present after t days.

a) If a sample initially contains 100 g of polonium-210, how many grams will be present after 30 days?

b) How long would it take for the initial amount to decay to 20 g?

c) What is the half-life of polonium-210?

Chapter 13:   Cumulative Review for Chapters 1–13

Simplify. The answer should not contain any negative exponents.

1) $(-5a^2)(3a^4)$

2) $\left(\dfrac{2c^{10}}{d^3}\right)^{-3}$

3) Write 0.00009231 in scientific notation.

4) *Write an equation, and solve.*
A watch is on sale for $38.40. This is 20% off of the regular price. What was the regular price of the watch?

5) Solve $-4x + 7 < 13$. Graph the solution set, and write the answer in interval notation.

6) Solve the system.
$$6x + 5y = -8$$
$$3x - y = 3$$

7) Divide $(6c^3 - 7c^2 - 22c + 5) \div (2c - 5)$.

8) Factor $w^2 + 8w + 16$.

9) Solve $x^2 + 14x = -48$.

10) Solve $\dfrac{9}{y + 6} + \dfrac{4}{y - 6} = \dfrac{-4}{y^2 - 36}$.

11) Graph the compound inequality $x + 2y \geq 6$ and $y - x \leq -2$.

Simplify. Assume all variables represent positive real numbers.

12) $\sqrt{45t^9}$

13) $\sqrt{\dfrac{36a^5}{a^3}}$

14) Solve $\sqrt{h^2 + 2h - 7} = h - 3$.

15) Solve $k^2 - 8k + 4 = 0$ by completing the square.

Solve each equation in Exercises 16–18.

16) $r^2 + 5r = -2$

17) $t^2 = 10t - 41$

18) $4m^4 + 4 = 17m^2$

19) Find the distance between the points (3, –4) and (5, 1).

20) Graph $y = -x^2 + 2x + 2$. Identify the vertex and intercepts. Determine the domain and range.

21) Let $f(x) = x^2 - 6x + 2$ and $g(x) = x - 3$.

a) Find $f(-1)$.

b) Find $(f \circ g)(x)$.

c) Find x so that $g(x) = -7$.

22) Graph $f(x) = 2^x - 3$. State the domain and range.

23) Solve $\log_4(5x + 1) = 2$.

24) Write as a single logarithm.
$$\log a + 2 \log b - 5 \log c$$

25) Solve $e^{-0.04t} = 6$. Give an exact solution and an approximation to four decimal places.

Conic Sections, Nonlinear Systems, and Nonlinear Inequalities

Get Ready

To prepare for this chapter, we will review how to **complete the square** for an expression of the form $x^2 + bx$ to obtain a **perfect square trinomial**. Then, we will **factor** the trinomial.

1) To **complete the square** for an expression of the form $x^2 + bx$:

Step 1: Find half of the coefficient of x: $\dfrac{1}{2}b$

Step 2: Square the result: $\left(\dfrac{1}{2}b\right)^2$

Step 3: Add the quantity in Step 2 to $x^2 + bx$ to obtain $x^2 + bx + \left(\dfrac{1}{2}b\right)^2$. The *factored form* is $\left(x + \dfrac{1}{2}b\right)^2$.

*The coefficient of the squared term **must** be 1 before completing the square!*

Examples: Complete the square for each expression to obtain a perfect square trinomial. Then, factor.

a) $x^2 + 8x$ **Step 1:** Find half of the x coefficient: $\dfrac{1}{2}(8) = 4$

　　　　　Step 2: Square the result: $4^2 = 16$

　　　　　Step 3: Add 16 to $x^2 + 8x$: $x^2 + 8x + 16$

The perfect square trinomial is $x^2 + 8x + 16$. The factored form is $(x + 4)^2$.

b) $y^2 - 14y$ **Step 1:** Find half of the y coefficient: $\dfrac{1}{2}(-14) = -7$

　　　　　Step 2: Square the result: $(-7)^2 = 49$

　　　　　Step 3: Add 49 to $y^2 - 14y$: $y^2 - 14y + 49$

The perfect square trinomial is $y^2 - 14y + 49$. The factored form is $(y - 7)^2$.

In this chapter, use the following *Basic Skills Worksheets* to prepare students for Chapter 15: WS2 Powers.

Complete the square for each expression to obtain a perfect square trinomial. Then, factor.

1) $x^2 + 6x$ 2) $y^2 + 10y$ 3) $n^2 - 2n$ 4) $p^2 - 4p$ 5) $k^2 - 20k$

6) $z^2 - 18z$ 7) $w^2 + 3w$ 8) $t^2 + 5t$ 9) $c^2 - c$ 10) $m^2 - 7m$

Answers

1) $x^2 + 6x + 9$; $(x+3)^2$ 2) $y^2 + 10y + 25$; $(y+5)^2$ 3) $n^2 - 2n + 1$; $(n-1)^2$ 4) $p^2 - 4p + 4$; $(p-2)^2$

5) $k^2 - 20k + 100$; $(k-10)^2$ 6) $z^2 - 18z + 81$; $(z-9)^2$ 7) $w^2 + 3w + \dfrac{9}{4}$; $\left(w + \dfrac{3}{2}\right)^2$ 8) $t^2 + 5t + \dfrac{25}{4}$; $\left(t + \dfrac{5}{2}\right)^2$

9) $c^2 - c + \dfrac{1}{4}$; $\left(c - \dfrac{1}{2}\right)^2$ 10) $m^2 - 7m + \dfrac{49}{4}$; $\left(m - \dfrac{7}{2}\right)^2$

 **Study Strategies**   Improving Your Memory

Whatever field you choose to pursue, you will need to recall a great deal of information. You will need to remember concepts you learned in college as well as new ideas and practices related to your specific job. The following strategies will help you get the most out of your memory.

P Prepare

- **I will learn and apply strategies for improving my memory.**

O Organize

- Complete the emPOWERme survey that appears before the Chapter Summary to learn how you *best* remember information as well as how *easily* you remember information.
- Identify the specific information you need to memorize. For example, is it facts in a history book? A procedure for solving an equation? Policies at work?
- Plan ahead and allow enough time for remembering material that you have to read or learn from a video. Do not leave memorization for the last minute; your mind needs time to form long-lasting memories.
- Think about strategies that have worked for you in the past, and use them again. Do you have different strategies for remembering or learning different topics?

 **W Work**

- Start far enough in advance so that your learning has time to sink in and so that you do not feel rushed or stressed.
- Apply previously successful strategies to remembering new information.
- Try relating new information to what you already know. For example, try to remember new math procedures by relating them to something you have already learned.
- Rehearse new material, repeating it again and again, until you can do so automatically.
- Create an *acronym* to help you remember a list or order of steps. An **acronym** is a word or phrase formed by the first letters of a list of terms. You know at least one acronym already: **P.O.W.E.R.** Each letter in the word P.O.W.E.R. stands for a step in the framework.
- Create an *acrostic* to help you remember a list or order of steps. An **acrostic** is a sentence in which the first letters spell out something that needs to be recalled. The acrostic **P**lease **E**xcuse **M**y **D**ear **A**unt **S**ally is a way to remember the order of operations: **P**arentheses, **E**xponents, **M**ultiplication, **D**ivision, **A**ddition, **S**ubtraction.
- Be an active learner. When you are in class, *write down* what your instructor is saying or writing on the board. If you are learning from a textbook, *write out* the examples in your math book as you are reading them.
- Some people remember information when they say it out loud. If you are learning from a textbook, try reading to yourself out loud.
- Try drawing diagrams or outlines.

 **Evaluate**

- Test your recall of the information you have tried to memorize. Flashcards are a great way of testing yourself, as are Chapter Reviews and Tests at the ends of chapters in your books. Use these resources *without looking at your book or notes* to assess, honestly, how well you remembered the material.
- Form a study group and quiz each other over what you set out to memorize.

 **Rethink**

- If you remembered everything you set out to memorize, that is great! Recall which strategies worked for you so that you can use them in the future.
- If you did not remember as much as you would have liked, think about what you could do differently next time. Should you start working earlier? Are there other strategies you didn't use that you could try next time?
- Return to the material you've tried to memorize a few days later. This will both further fix the information in your mind and help you identify the areas where you still need work.

14.1 The Circle

P Prepare

O Organize

What are your objectives for Section 14.1?	How can you accomplish each objective?
1 Use the Midpoint Formula	<ul><li>Learn the **midpoint formula.**</li><li>Understand what the *midpoint* is and where it is located.</li><li>Complete the given example on your own.</li><li>Complete You Try 1.</li></ul>
2 Graph a Circle Given in the Form $(x - h)^2 + (y - k)^2 = r^2$	<ul><li>Understand what a *conic section* is and how it is formed. Write all the different conic sections in your notes with an example of each.</li><li>Learn the definitions of a *circle*, the *center* of a circle, and the *radius* of a circle.</li><li>Draw a circle in your notes and label its parts.</li><li>Know the definition of the *standard form for the equation of a circle.*</li><li>Complete the given examples on your own.</li><li>Complete You Trys 2–4.</li></ul>
3 Graph a Circle of the Form $Ax^2 + Ay^2 + Cx + Dy + E = 0$	<ul><li>Know the definition of the *general form for the equation of a circle.*</li><li>Review the procedure for **Completing the Square.**</li><li>Complete the given example on your own.</li><li>Complete You Try 5.</li></ul>

 **Work**

Read the explanations, follow the examples, take notes, and complete the You Trys.

In this chapter, we will study the *conic sections*. When a right circular cone is intersected by a plane, the result is a **conic section.** The conic sections are parabolas, circles, ellipses, and hyperbolas. The following figures show how each conic section is obtained from the intersection of a cone and a plane.

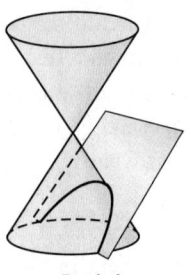
Parabola

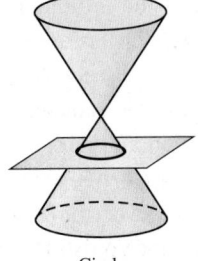
Circle

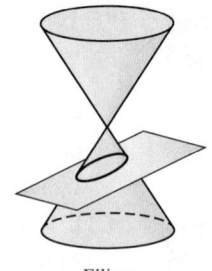
Ellipse

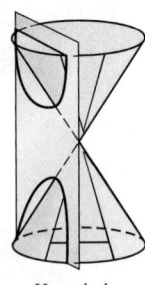
Hyperbola

In Chapter 12 we learned how to graph parabolas. The graph of a quadratic function, $f(x) = ax^2 + bx + c$, is a *parabola* that opens vertically. Another form this function may take is $f(x) = a(x - h)^2 + k$. The graph of a quadratic equation of the form $x = ay^2 + by + c$, or $x = a(y - k)^2 + h$, is a *parabola* that opens horizontally. The next conic section we will discuss is the circle.

We will use the distance formula, presented in Section 11.2, to derive the equation of a circle. But first, let's learn the *midpoint formula*.

1 Use the Midpoint Formula

The **midpoint** of a line segment is the point that is exactly halfway between the endpoints of a line segment. We use the midpoint formula to find the midpoint.

W Hint

Notice that the midpoint formula calculates the arithmetic average of the given *x*- and *y*-values!

Definition The Midpoint Formula

If (x_1, y_1) and (x_2, y_2) are the endpoints of a line segment, then the **midpoint** of the segment has coordinates

$$\left(\frac{x_1 + x_2}{2}, \frac{y_1 + y_2}{2} \right)$$

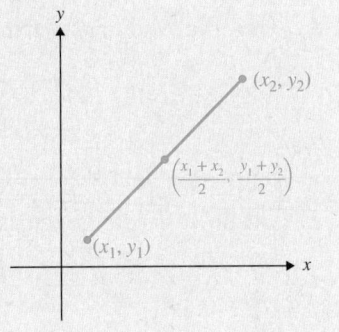

EXAMPLE 1

Find the midpoint of the line segment with endpoints $(-3, 4)$ and $(1, -2)$.

Solution

Begin by labeling the points: $\overset{(x_1, y_1)}{(-3, 4)}, \overset{(x_2, y_2)}{(1, -2)}$.
Substitute the values into the midpoint formula.

$$\text{Midpoint} = \left(\frac{x_1 + x_2}{2}, \frac{y_1 + y_2}{2} \right)$$

$$= \left(\frac{-3 + 1}{2}, \frac{4 + (-2)}{2} \right) \qquad \text{Substitute values.}$$

$$= \left(\frac{-2}{2}, \frac{2}{2} \right)$$

$$= (-1, 1) \qquad \text{Simplify.}$$

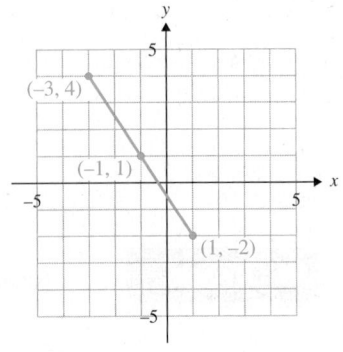

[YOU TRY 1] Find the midpoint of the line segment with endpoints $(5, 2)$ and $(1, -3)$.

The midpoint of a diameter of a circle is the **center** of the circle.

2 Graph a Circle Given in the Form $(x - h)^2 + (y - k)^2 = r^2$

A **circle** is defined as the set of all points in a plane equidistant (the same distance) from a fixed point. The fixed point is the **center** of the circle. The distance from the center to a point on the circle is the **radius** of the circle.

Let the center of a circle have coordinates (h, k) and let (x, y) represent any point on the circle. Let r represent the distance between these two points. r is the radius of the circle.

We will use the distance formula to find the distance between the center, (h, k), and the point (x, y) on the circle.

$$d = \sqrt{(x_2 - x_1)^2 + (y_2 - y_1)^2} \qquad \text{Distance formula}$$

Substitute (x, y) for (x_2, y_2), (h, k) for (x_1, y_1), and r for d.

$$r = \sqrt{(x - h)^2 + (y - k)^2}$$
$$r^2 = (x - h)^2 + (y - k)^2 \qquad \text{Square both sides.}$$

This is the **standard form** for the equation of a circle.

Hint

Write this formula in your notes. Include an example.

Definition

Standard Form for the Equation of a Circle: The standard form for the equation of a circle with center (h, k) and radius r is

$$(x - h)^2 + (y - k)^2 = r^2$$

Hint

In your notes, write down these relationships between the radius and diameter of a circle.

Note

Recall that a **diameter**, d, of a circle is a line segment that passes through the center of the circle and has its endpoints on the circle. The *length* of this segment is also called the diameter. Therefore, the diameter of a circle is twice its radius, or $d = 2r$. It is also true that the radius of a circle is half its diameter, or $r = \dfrac{1}{2}d$. *You should remember these important relationships.*

EXAMPLE 2   Graph $(x - 2)^2 + (y + 1)^2 = 9$.

Solution

Standard form is $(x - h)^2 + (y - k)^2 = r^2$.
Our equation is $(x - 2)^2 + (y + 1)^2 = 9$.

$$h = 2 \qquad k = -1 \qquad r = \sqrt{9} = 3$$

The center is $(2, -1)$. The radius is 3.

To graph the circle, first plot the center $(2, -1)$. Use the radius to locate four points on the circle. From the center, move 3 units up, down, left, and right. Draw a circle through the four points.

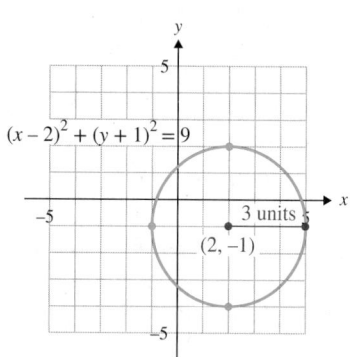

[YOU TRY 2] Graph $(x + 3)^2 + (y - 1)^2 = 16$.

EXAMPLE 3 Graph $x^2 + y^2 = 1$.

Solution

Standard form is $(x - h)^2 + (y - k)^2 = r^2$.
Our equation is $\quad x^2 \quad + \quad y^2 \quad = 1$.

$$h = 0 \qquad k = 0 \qquad r = \sqrt{1} = 1$$

The center is $(0, 0)$. The radius is 1. Plot $(0, 0)$, then use the radius to locate four points on the circle. From the center, move 1 unit up, down, left, and right. Draw a circle through the four points.
    The circle $x^2 + y^2 = 1$ is used often in other areas of mathematics such as trigonometry. $x^2 + y^2 = 1$ is called the **unit circle.**

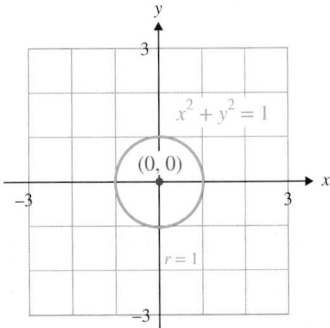

[YOU TRY 3] Graph $x^2 + y^2 = 25$.

If we are told the center and radius of a circle, we can write its equation.

EXAMPLE 4 Find an equation of the circle with center $(-5, 0)$ and radius $\sqrt{7}$.

Solution

The x-coordinate of the center is h: $h = -5$
The y-coordinate of the center is k: $k = 0$

$$r = \sqrt{7}$$

Substitute these values into $(x - h)^2 + (y - k)^2 = r^2$.

$$[x - (-5)]^2 + (y - 0)^2 = (\sqrt{7})^2 \qquad \text{Substitute } -5 \text{ for } x, 0 \text{ for } k, \text{ and } \sqrt{7} \text{ for } r.$$
$$(x + 5)^2 + y^2 = 7$$

[YOU TRY 4] Find an equation of the circle with center $(4, 7)$ and radius 5.

---

3 Graph a Circle of the Form $Ax^2 + Ay^2 + Cx + Dy + E = 0$

The equation of a circle can take another form—general form.

Hint

Write this formula in your notes along with an example.

Definition

General Form for the Equation of a Circle: An equation of the form $Ax^2 + Ay^2 + Cx + Dy + E = 0$, where A, C, D, and E are real numbers, is the **general form** for the equation of a circle.

(The coefficients of x^2 and y^2 must be the *same* in order for this to be the equation of a circle.)

To graph a circle given in this form, we complete the square on x and on y to put it into standard form.

After we learn *all* of the conic sections, it is very important that we understand how to identify each one. To do this we will usually look at the coefficients of the square terms.

EXAMPLE 5

Graph $x^2 + y^2 + 6x + 2y + 6 = 0$.

Solution

The coefficients of x^2 and y^2 are each 1. Therefore, this is the equation of a circle.

Our goal is to write the given equation in standard form, $(x - h)^2 + (y - k)^2 = r^2$, so that we can identify its center and radius. To do this we will group x^2 and $6x$ together, group y^2 and $2y$ together, then complete the square on each group of terms.

$$x^2 + y^2 + 6x + 2y + 6 = 0$$
$$(x^2 + 6x) + (y^2 + 2y) = -6$$

Group x^2 and $6x$ together.
Group y^2 and $2y$ together.
Move the constant to the other side.

Complete the square for each group of terms.

$$(x^2 + 6x + 9) + (y^2 + 2y + 1) = -6 + 9 + 1$$
$$(x + 3)^2 + (y + 1)^2 = 4$$

Because 9 and 1 are added on the left, they must also be added on the right.
Factor; add.

Hint

See Section 11.2 if you need to review how to complete the square.

The center of the circle is $(-3, -1)$. The radius is 2.

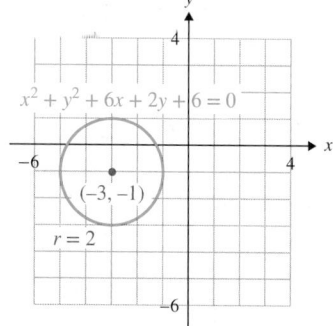

$x^2 + y^2 + 6x + 2y + 6 = 0$
$(-3, -1)$
$r = 2$

YOU TRY 5

Graph $x^2 + y^2 + 10x - 4y + 20 = 0$.

Note

If we rewrite $Ax^2 + Ay^2 + Cx + Dy + E = 0$ in standard form and get $(x - h)^2 + (y - k)^2 = 0$, then the graph is just the point (h, k). If the constant on the right side of the standard form equation is a negative number, then the equation has no graph.

Recall that the equation of a circle is not a function. However, if we want to graph an equation on a graphing calculator, it must be entered as a function or a pair of functions. Therefore, to graph a circle we must solve the equation for y in terms of x.

Let's discuss how to graph $x^2 + y^2 = 4$ on a graphing calculator.

We must solve the equation for y.

$$x^2 + y^2 = 4$$
$$y^2 = 4 - x^2$$
$$y = \pm\sqrt{4 - x^2}$$

Now the equation of the circle $x^2 + y^2 = 4$ is rewritten so that y is in terms of x. In the graphing calculator, enter $y = \sqrt{4 - x^2}$ as Y_1. This represents the top half of the circle since the y-values are positive above the x-axis. Enter $y = -\sqrt{4 - x^2}$ as Y_2. This represents the bottom half of the circle since the y-values are negative below the x-axis. Here we have the window set from -3 to 3 in both the x- and y-directions. Press GRAPH .

The graph is distorted and does not actually look like a circle! This is because the screen is rectangular, and the graph is longer in the x-direction. We can "fix" this by squaring the window.

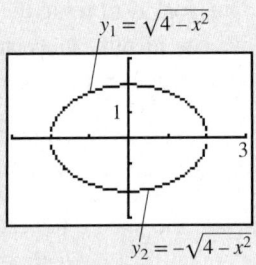

To square the window and get a better representation of the graph of $x^2 + y^2 = 4$, press ZOOM and choose 5:ZSquare. The graph reappears on a "squared" window and now looks like a circle.

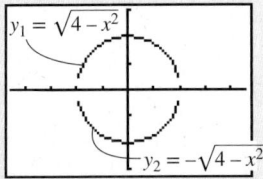

Identify the center and radius of each circle. Then, rewrite each equation for y in terms of x, and graph each circle on a graphing calculator. These problems come from the homework exercises.

1) $x^2 + y^2 = 36$; Exercise 23

2) $x^2 + y^2 = 9$; Exercise 25

3) $(x + 3)^2 + y^2 = 4$; Exercise 19

4) $x^2 + (y - 1)^2 = 25$; Exercise 27

ANSWERS TO [YOU TRY] EXERCISES

1) $\left(3, -\dfrac{1}{2}\right)$

2)

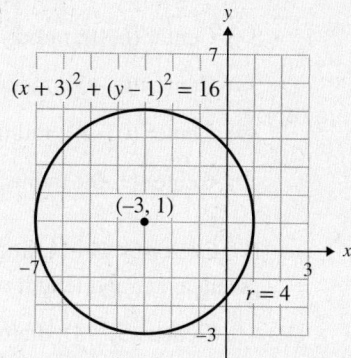

3)

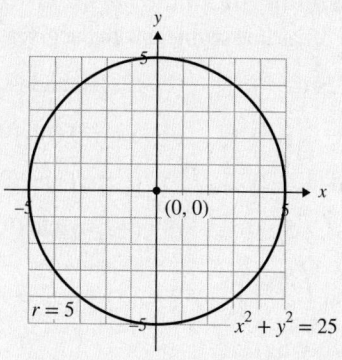

4) $(x - 4)^2 + (y - 7)^2 = 25$ 5)

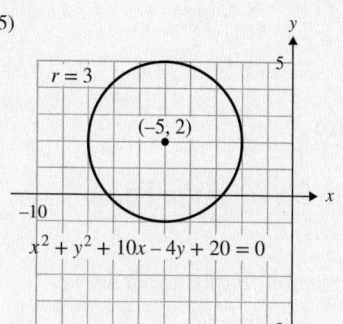

ANSWERS TO TECHNOLOGY EXERCISES

1) Center $(0, 0)$; radius $= 6$; $Y_1 = \sqrt{36 - x^2}$, $Y_2 = -\sqrt{36 - x^2}$

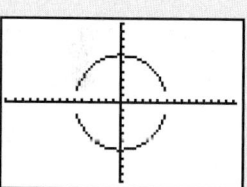

2) Center $(0, 0)$; radius $= 3$; $Y_1 = \sqrt{9 - x^2}$, $Y_2 = -\sqrt{9 - x^2}$

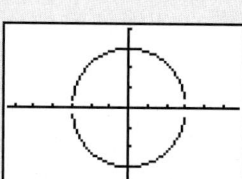

3) Center $(-3, 0)$; radius $= 2$; $Y_1 = \sqrt{4 - (x + 3)^2}$, $Y_2 = -\sqrt{4 - (x + 3)^2}$

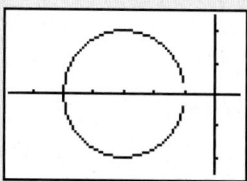

4) Center $(0, 1)$; radius $= 5$; $Y_1 = 1 + \sqrt{25 - x^2}$, $Y_2 = 1 - \sqrt{25 - x^2}$

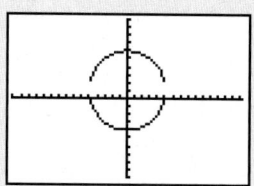

Objective 1: Use the Midpoint Formula

Find the midpoint of the line segment with the given endpoints.

 1) (1, 3) and (7, 9)

2) (2, 10) and (8, 4)

3) (−5, 2) and (−1, −8)

4) (6, −3) and (0, 5)

5) (−3, −7) and (1, −2)

6) (−1, 3) and (2, −9)

7) (4, 0) and (−3, −5)

8) (−2, 4) and (9, 3)

9) $\left(\frac{3}{2}, -1\right)$ and $\left(\frac{5}{2}, \frac{7}{2}\right)$

10) $\left(\frac{9}{2}, \frac{3}{2}\right)$ and $\left(-\frac{7}{2}, -5\right)$

11) (−6.2, 1.5) and (4.8, 5.7)

12) (−3.7, −1.8) and (3.7, −3.6)

Objective 2: Graph a Circle Given in the Form $(x − h)^2 + (y − k)^2 = r^2$

13) Is the equation of a circle a function? Explain your answer.

14) The standard form for the equation of a circle is

$$(x − h)^2 + (y − k)^2 = r^2$$

Identify the center and the radius.

Identify the center and radius of each circle and graph.

15) $(x + 2)^2 + (y − 4)^2 = 9$

16) $(x + 1)^2 + (y + 3)^2 = 25$

17) $(x − 5)^2 + (y − 3)^2 = 1$

18) $x^2 + (y − 5)^2 = 9$

19) $(x + 3)^2 + y^2 = 4$

20) $(x − 2)^2 + (y − 2)^2 = 36$

21) $(x − 6)^2 + (y + 3)^2 = 16$

22) $(x + 8)^2 + (y − 4)^2 = 4$

23) $x^2 + y^2 = 36$

24) $x^2 + y^2 = 16$

25) $x^2 + y^2 = 9$

26) $x^2 + y^2 = 25$

27) $x^2 + (y − 1)^2 = 25$

28) $(x + 3)^2 + y^2 = 1$

For Exercises 29–40, find an equation of the circle with the given center and radius.

29) Center (4, 1); radius = 5

30) Center (3, 5); radius = 2

31) Center (−3, 2); radius = 1

32) Center (4, −6); radius = 3

33) Center (−1, −5); radius = $\sqrt{3}$

34) Center (−2, −1); radius = $\sqrt{5}$

35) Center (0, 0); radius = $\sqrt{10}$

36) Center (0, 0); radius = $\sqrt{6}$

37) Center (6, 0); radius = 4

38) Center (0, −3); radius = 5

39) Center (0, −4); radius = $2\sqrt{2}$

40) Center (1, 0); radius = $3\sqrt{2}$

For Exercises 41–44, find the equation of the circle with the given center and length of its diameter.

41) Center (5, −3); diameter = 6

42) Center (−4, 1); diameter = 10

43) Center (2, 7); diameter = $4\sqrt{5}$

44) Center (0, −6); diameter = $8\sqrt{3}$

45) A circle contains the point (3, −4) and has a center of (6, 2). Find the standard form for the equation of this circle.

46) A circle contains the point (−5, 2) and has a center of (−2, −1). Find the standard form for the equation of this circle.

47) The endpoints of a diameter of a circle are (−5, 1) and (−1, 3). Find the standard form for the equation of this circle.

48) The endpoints of a diameter of a circle are (3, −7) and (6, −2). Find the standard form for the equation of this circle.

Objective 3: Graph a Circle of the Form $Ax^2 + Ay^2 + Cx + Dy + E = 0$

Write the equation of the circle in standard form.

Fill It In

Fill in the blanks with either the missing mathematical step or the reason for the given step.

49) $x^2 + y^2 − 8x + 2y + 8 = 0$
$(x^2 − 8x) + (y^2 + 2y) = −8$ _____

_____

_____   Complete the square.
_____   Factor.

50) $x^2 + y^2 + 2x + 10y + 10 = 0$
$(x^2 + 2x) + (y^2 + 10y) = −10$ _____

_____

_____   Complete the square.
_____   Factor.

Put the equation of each circle in the form $(x - h)^2 + (y - k)^2 = r^2$, identify the center and the radius, and graph.

51) $x^2 + y^2 + 2x + 10y + 17 = 0$

52) $x^2 + y^2 - 4x - 6y + 9 = 0$

53) $x^2 + y^2 + 8x - 2y - 8 = 0$

54) $x^2 + y^2 - 6x + 8y + 24 = 0$

(24) 55) $x^2 + y^2 - 10x - 14y + 73 = 0$

56) $x^2 + y^2 + 12x + 12y + 63 = 0$

57) $x^2 + y^2 + 6y + 5 = 0$

58) $x^2 + y^2 + 2x - 24 = 0$

59) $x^2 + y^2 - 4x - 1 = 0$

60) $x^2 + y^2 - 10y + 22 = 0$

61) $x^2 + y^2 - 8x + 8y - 4 = 0$

62) $x^2 + y^2 - 6x + 2y - 6 = 0$

63) $4x^2 + 4y^2 - 12x - 4y - 6 = 0$
(Hint: Begin by dividing the equation by 4.)

64) $16x^2 + 16y^2 + 16x - 24y - 3 = 0$
(Hint: Begin by dividing the equation by 16.)

Mixed Exercises: Objectives 1–3

(24) 65) The London Eye is a Ferris wheel that opened in London in March 2000. It is 135 m high, and the bottom of the wheel is 7 m off the ground.

Ingram Publishing/AGE Fotostock

a) What is the diameter of the wheel?

b) What is the radius of the wheel?

c) Using the axes in the illustration, what are the coordinates of the center of the wheel?

d) Write the equation of the wheel.
(www.aviewoncities.com/london/londoneye.htm)

66) The first Ferris wheel was designed and built by George W. Ferris in 1893 for the Chicago World's

Fair. It was 264 ft tall, and the wheel had a diameter of 250 ft.

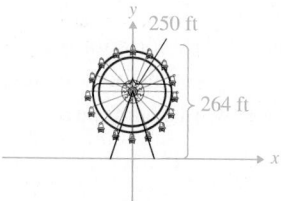

a) What is the radius of the wheel?

b) Using the axes in the illustration, what are the coordinates of the center of the wheel?

c) Write the equation of the wheel.

67) A CD is placed on axes as shown in the figure where the units of measurement for x and y are millimeters. Using $\pi = 3.14$, what is the surface area of a CD (to the nearest square millimeter)?

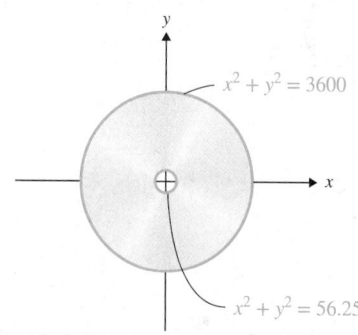

68) A storage container is in the shape of a right circular cylinder. The top of the container may be described by the equation $x^2 + y^2 = 5.76$, as shown in the figure (x and y are in feet). If the container is 3.2 ft tall, what is the storage capacity of the container (to the nearest ft³)?

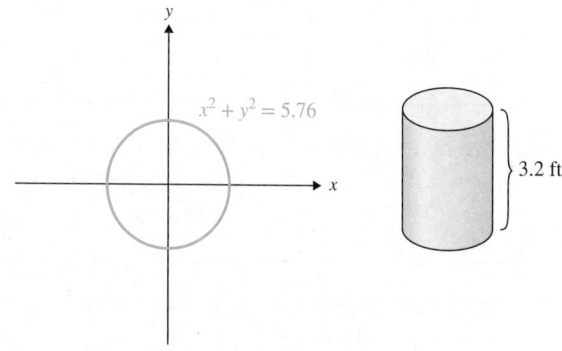

Use a graphing calculator to graph each equation. Square the viewing window.

69) $(x - 5)^2 + (y - 3)^2 = 1$

70) $x^2 + y^2 = 16$

71) $(x + 2)^2 + (y - 4)^2 = 9$

72) $(x + 3)^2 + y^2 = 1$

R1) Write the standard form for the equation of a circle without looking at the book or your notes.

R2) In your own words, summarize how to graph a circle.

R3) What profession might deal with circles on a daily basis?

14.2 The Ellipse and the Hyperbola

P Prepare   **O** Organize

What are your objectives for Section 14.2?	How can you accomplish each objective?
1 Graph an Ellipse	• Know the definition of the *standard form for the equation of an ellipse*. • Know how to identify the *center* and *foci* of the ellipse. Draw an ellipse in your notes, and label its parts. • Understand the properties of an ellipse with center at the origin. • Review the procedure for **Completing the Square.** • Complete the given examples on your own. • Complete You Trys 1–4.
2 Graph a Hyperbola in Standard Form	• Learn the definition of a *hyperbola*. Draw an example in your notes, and label the diagram. • Understand what *center, vertex, focus, transverse axis,* and *asymptotes* are. • Learn the definition of the *standard form for the equation of a hyperbola*. • Understand the properties of an equation of a hyperbola with center at the origin. • Complete the given examples on your own. • Complete You Trys 5–7.
3 Graph a Hyperbola in Nonstandard Form	• Learn the definition of a *nonstandard form for the equation of a hyperbola*. • Complete the given example on your own. • Complete You Try 8.
4 Graph Other Square Root Functions	• Know the standard forms of equations of conic sections. • Review the squaring procedure when rearranging equations. • Complete the given example on your own. • Complete You Try 9.

 **W** Work   **Read the explanations, follow the examples, take notes, and complete the You Trys.**

1 Graph an Ellipse

The next conic section we will study is the *ellipse*. An **ellipse** is the set of all points in a plane such that the *sum* of the distances from a point on the ellipse to two fixed points is constant. Each fixed point is called a **focus** (plural: **foci**). The point halfway between the foci is the **center** of the ellipse.

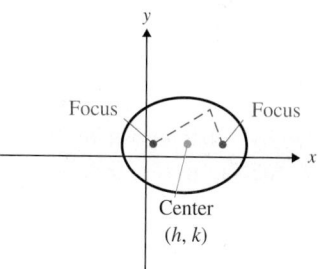

The orbits of planets around the sun as well as satellites around the Earth are elliptical. Statuary Hall in the U.S. Capitol building is an ellipse. If a person stands at one focus of this ellipse and whispers, a person standing across the room on the other focus can clearly hear what was said. Properties of the ellipse are used in medicine as well. One procedure for treating kidney stones involves immersing the patient in an elliptical tub of water. The kidney stone is at one focus, while at the other focus, high energy shock waves are produced, which destroy the kidney stone.

W Hint

Write this equation in your notes along with the information above, in bold. Include an example.

Definition

Standard Form for the Equation of an Ellipse: The standard form for the equation of an ellipse is

$$\frac{(x - h)^2}{a^2} + \frac{(y - k)^2}{b^2} = 1$$

The center of the ellipse is (h, k).

It is important to remember that the terms on the left are *both* positive quantities.

EXAMPLE 1

Graph $\dfrac{(x - 3)^2}{16} + \dfrac{(y - 1)^2}{4} = 1.$

Solution

Standard form is $\dfrac{(x - h)^2}{a^2} + \dfrac{(y - k)^2}{b^2} = 1.$

Our equation is $\dfrac{(x - 3)^2}{16} + \dfrac{(y - 1)^2}{4} = 1.$

$$h = 3 \qquad k = 1$$
$$a = \sqrt{16} = 4 \qquad b = \sqrt{4} = 2$$

The center is $(3, 1)$.

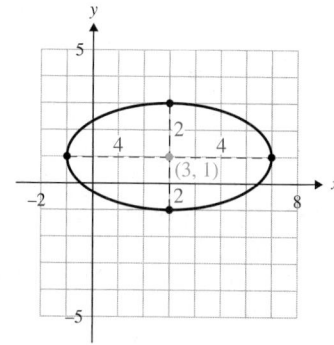

To graph the ellipse, first plot the center $(3, 1)$. Since $a = 4$ and a^2 is under the squared quantity containing the **x**, move 4 units each way in the *x*-direction from the center. These are two points on the ellipse.

Since $b = 2$ and b^2 is under the squared quantity containing the **y**, move 2 units each way in the *y*-direction from the center. These are two more points on the ellipse. Sketch the ellipse through the four points.

Graph $\dfrac{(x+2)^2}{25} + \dfrac{(y-3)^2}{16} = 1$.

EXAMPLE 2

Graph $\dfrac{x^2}{9} + \dfrac{y^2}{25} = 1$.

Solution

Standard form is $\dfrac{(x-h)^2}{a^2} + \dfrac{(y-k)^2}{b^2} = 1$.

Our equation is $\dfrac{x^2}{9} + \dfrac{y^2}{25} = 1$.

$$h = 0 \qquad k = 0$$
$$a = \sqrt{9} = 3 \qquad b = \sqrt{25} = 5$$

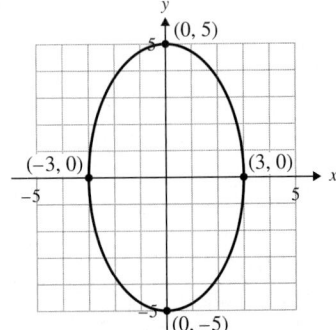

The center is (0, 0).

Plot the center (0, 0). Since $a = 3$ and a^2 is under the x^2, move 3 units each way in the x-direction from the center. These are two points on the ellipse.

Since $b = 5$ and b^2 is under the y^2, move 5 units each way in the y-direction from the center. These are two more points on the ellipse. Sketch the ellipse through the four points.

YOU TRY 2

Graph $\dfrac{x^2}{36} + \dfrac{y^2}{9} = 1$.

In Example 2, note that the *origin*, (0, 0), is the center of the ellipse. Notice that $a = 3$ and the x-intercepts are (3, 0) and $(-3, 0)$; $b = 5$ and the y-intercepts are (0, 5) and (0, −5). We can generalize these relationships as follows.

Property Equation of an Ellipse with Center at the Origin

The graph of $\dfrac{x^2}{a^2} + \dfrac{y^2}{b^2} = 1$ is an ellipse with

center at the origin, x-intercepts $(a, 0)$ and $(-a, 0)$, and y-intercepts $(0, b)$ and $(0, -b)$.

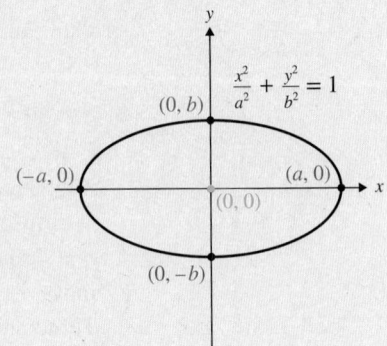

Looking at Examples 1 and 2, we can make another interesting observation.

Example 1

$$\frac{(x-3)^2}{16} + \frac{(y-1)^2}{4} = 1$$
$$a^2 = 16 \quad b^2 = 4$$
$$a^2 > b^2$$

The number under $(x-3)^2$ is greater than the number under $(y-1)^2$. The ellipse is longer in the x-direction.

Example 2

$$\frac{x^2}{9} + \frac{y^2}{25} = 1$$
$$a^2 = 9 \quad b^2 = 25$$
$$b^2 > a^2$$

The number under y^2 is greater than the number under x^2. The ellipse is longer in the y-direction.

This relationship between a^2 and b^2 will always produce the same result. The equation of an ellipse can take other forms.

EXAMPLE 3

Graph $4x^2 + 25y^2 = 100$.

Solution

How can we tell if this is a circle or an ellipse? We look at the coefficients of x^2 and y^2. Both of the coefficients are positive, *and* they are different. *This is an ellipse.* (If this were a circle, the coefficients would be the same.)

Since the standard form for the equation of an ellipse has a 1 on one side of the = sign, divide both sides of $4x^2 + 25y^2 = 100$ by 100 to obtain a 1 on the right.

$$4x^2 + 25y^2 = 100$$

$$\frac{4x^2}{100} + \frac{25y^2}{100} = \frac{100}{100} \qquad \text{Divide both sides by 100.}$$

$$\frac{x^2}{25} + \frac{y^2}{4} = 1 \qquad \text{Simplify.}$$

W Hint

Use the coefficients of x^2 and y^2 to help you identify the type of conic section you are given.

The center is $(0, 0)$. $a = \sqrt{25} = 5$ and $b = \sqrt{4} = 2$. Plot $(0, 0)$. Move 5 units each way from the center in the x-direction. Move 2 units each way from the center in the y-direction

Notice that the x-intercepts are $(5, 0)$ and $(-5, 0)$. The y-intercepts are $(0, 2)$ and $(0, -2)$.

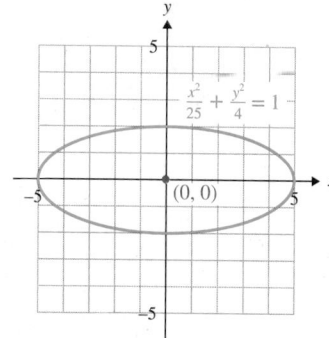

YOU TRY 3

Graph $x^2 + 4y^2 = 4$.

Just like with the equation of a circle, sometimes we must complete the square to put the equation of an ellipse into standard form.

EXAMPLE 4

Graph $9x^2 + 4y^2 + 18x - 16y - 11 = 0$.

Solution

Our goal is to write the given equation in standard form, $\dfrac{(x-h)^2}{a^2} + \dfrac{(y-k)^2}{b^2} = 1$. To do this we will group the *x*-terms together, group the *y*-terms together, then complete the square on each group of terms.

$$9x^2 + 4y^2 + 18x - 16y - 11 = 0$$

Group the *x*-terms together and the *y*-terms together. Move the constant to the other side.

$$(9x^2 + 18x) + (4y^2 - 16y) = 11$$

$$9(x^2 + 2x) + 4(y^2 - 4y) = 11$$

Factor out the coefficients of the squared terms.

Complete the square, inside the parentheses, for each group of terms.

$$9 \cdot 1 = 9 \qquad 4 \cdot 4 = 16$$

$$9(x^2 + 2x + 1) + 4(y^2 - 4y + 4) = 11 + 9 + 16$$

Since 9 and 16 are added on the left, they must also be added on the right.

$$9(x + 1)^2 + 4(y - 2)^2 = 36$$

Factor; add.

$$\frac{9(x+1)^2}{36} + \frac{4(y-2)^2}{36} = \frac{36}{36}$$

Divide by 36 to get 1 on the right.

$$\frac{(x+1)^2}{4} + \frac{(y-2)^2}{9} = 1$$

Standard form

The center of the ellipse is $(-1, 2)$. The graph is at right.

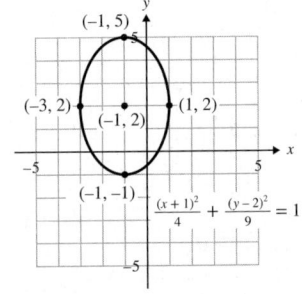

Graph $4x^2 + 25y^2 - 24x - 50y - 39 = 0$.

2 Graph a Hyperbola in Standard Form

The last of the conic sections is the *hyperbola*. A **hyperbola** is the set of all points, *P*, in a plane such that the absolute value of the *difference* of the distances $|d_1 - d_2|$, from two fixed points is constant. Each fixed point is called a **focus.** The point halfway between the foci is the **center** of the hyperbola.

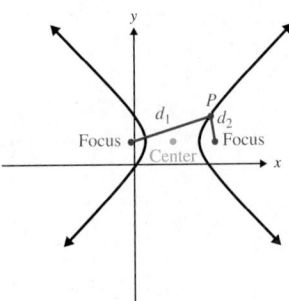

The graph of a hyperbola consists of two **branches,** and each branch has a **vertex** that is an endpoint of the *transverse axis* of the hyperbola. The **transverse axis** is the line segment whose endpoints are the vertices of the hyperbola.

Some navigation systems used by ships are based on the properties of hyperbolas. A lamp casts a hyperbolic shadow on a wall, and many telescopes use hyperbolic lenses.

Definition Standard Form for the Equation of a Hyperbola

1) A hyperbola with center (h, k) and branches that open in the *x-direction* has equation

$$\frac{(x-h)^2}{a^2} - \frac{(y-k)^2}{b^2} = 1.$$

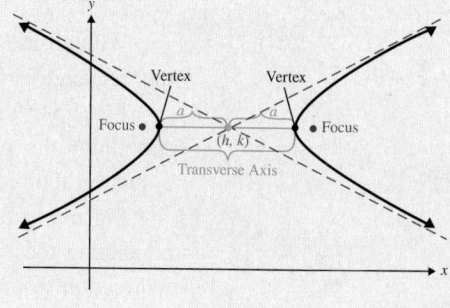

Its graph is to the right.

The distance from the center, (h, k), to a vertex is a. Therefore, the length of the transverse axis is $2a$.

2) A hyperbola with center (h, k) and branches that open in the *y-direction* has equation

$$\frac{(y-k)^2}{b^2} - \frac{(x-h)^2}{a^2} = 1.$$

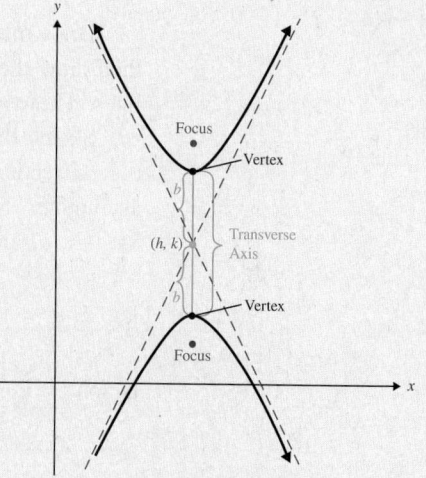

Its graph is to the right.
The distance from the center, (h, k), to a vertex is b. Therefore, the length of the transverse axis is $2b$.

Notice in 1) that $\dfrac{(x-h)^2}{a^2}$ is the positive quantity, and the branches open in the *x*-direction. In 2), the positive quantity is $\dfrac{(y-k)^2}{b^2}$, and the branches open in the *y*-direction.

In 1) and 2) notice how the branches of the hyperbola get closer to the dotted lines as the branches continue indefinitely. These dotted lines are called **asymptotes.** They are not an actual part of the graph of the hyperbola, but we can use them to help us obtain the hyperbola.

EXAMPLE 5

Graph $\dfrac{(x+2)^2}{9} - \dfrac{(y-1)^2}{4} = 1.$

Solution

How do we know that this is a hyperbola and not an ellipse? *It is a hyperbola because there is a subtraction sign between the two quantities on the left. If it was addition, it would be an ellipse.*

Standard form is $\dfrac{(x-h)^2}{a^2} - \dfrac{(y-k)^2}{b^2} = 1.$

Our equation is $\dfrac{(x+2)^2}{9} - \dfrac{(y-1)^2}{4} = 1.$

$$h = -2 \qquad k = 1$$
$$a = \sqrt{9} = 3 \qquad b = \sqrt{4} = 2$$

The center is $(-2, 1)$ *Because the quantity* $\dfrac{(x-h)^2}{a^2}$ *is the* positive *quantity, the branches of the hyperbola will open in the x-direction.*

We will use the center, $a = 3$, and $b = 2$ to draw a *reference rectangle*. The diagonals of this rectangle are the asymptotes of the hyperbola.

First, plot the center $(-2, 1)$ Since $a = 3$ and a^2 is under the squared quantity containing the x, move 3 units each way in the x-direction from the center. These are two points on the rectangle.

Since $b = 2$ and b^2 is under the squared quantity containing the y, move 2 units each way in the y-direction from the center. These are two more points on the rectangle.

Draw the rectangle containing these four points, then draw the diagonals of the rectangle as dotted lines. These are the asymptotes of the hyperbola.

Sketch the branches of the hyperbola opening in the x-direction with the branches approaching the asymptotes. Note that the vertices of the hyperbola are $(-5, 1)$ and $(1, 1)$.

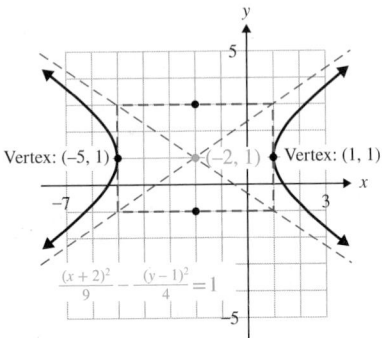

[YOU TRY 5]

Graph $\dfrac{(x+1)^2}{9} - \dfrac{(y+1)^2}{16} = 1.$

EXAMPLE 6

Graph $\dfrac{y^2}{4} - \dfrac{x^2}{25} = 1.$

Solution

Standard form is $\dfrac{(y-k)^2}{b^2} - \dfrac{(x-h)^2}{a^2} = 1.$ Our equation is $\dfrac{y^2}{4} - \dfrac{x^2}{25} = 1.$

$$k = 0 \qquad h = 0$$
$$b = \sqrt{4} = 2 \qquad a = \sqrt{25} = 5$$

The center is $(0, 0)$. *Because the quantity* $\dfrac{y^2}{4}$ *is the* positive *quantity, the branches of the hyperbola will open in the y-direction.*

Use the center, $a = 5$, and $b = 2$ to draw the reference rectangle and its diagonals.

Plot the center $(0, 0)$. Because $a = 5$ and a^2 is under the x^2, move 5 units each way in the x-direction from the center to get two points on the rectangle.

Because $b = 2$ and b^2 is under the y^2, move 2 units each way in the y-direction from the center to get two more points on the rectangle.

Draw the rectangle and its diagonals as dotted lines. These are the asymptotes of the hyperbola.

Sketch the branches of the hyperbola opening in the y-direction approaching the asymptotes. The vertices of the hyperbola are $(0, 2)$ and $(0, -2)$.

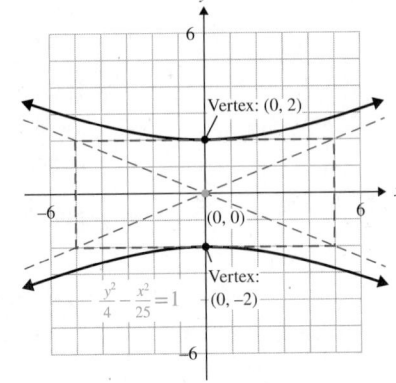

W Hint

After reading Examples 5 and 6, create a procedure for graphing a hyperbola.

Graph $\dfrac{y^2}{16} - \dfrac{x^2}{16} = 1$.

W Hint

Summarize these properties in your notes. Include examples.

Property Equation of a Hyperbola with Center at the Origin

1) The graph of $\dfrac{x^2}{a^2} - \dfrac{y^2}{b^2} = 1$ is a hyperbola with center $(0, 0)$ and vertices and x-intercepts $(a, 0)$ and $(-a, 0)$ as shown below.

2) The graph $\dfrac{y^2}{b^2} - \dfrac{x^2}{a^2} = 1$ is a hyperbola with center $(0, 0)$ and vertices and y-intercepts $(0, b)$ and $(0, -b)$ as shown below.

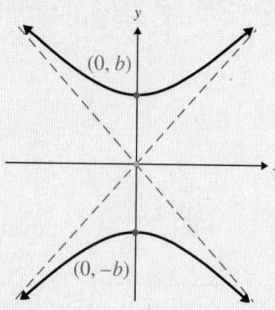

The equations of the asymptotes are $y = \dfrac{b}{a}x$ and $y = -\dfrac{b}{a}x$.

Let's look at another example.

EXAMPLE 7

Graph $y^2 - 9x^2 = 9$.

Solution

This is a hyperbola because there is a subtraction sign between the two terms.

Since the standard form for the equation of the hyperbola has a 1 on one side of the = sign, divide both sides of $y^2 - 9x^2 = 9$ by 9 to obtain a 1 on the right.

$$y^2 - 9x^2 = 9$$

$$\dfrac{y^2}{9} - \dfrac{9x^2}{9} = \dfrac{9}{9} \qquad \text{Divide both sides by 9.}$$

$$\dfrac{y^2}{9} - x^2 = 1 \qquad \text{Simplify.}$$

The center is $(0, 0)$. *The branches of the hyperbola will open in the y-direction since $\dfrac{y^2}{9}$ is a positive quantity.*

x^2 is the same as $\dfrac{x^2}{1}$, so $a = \sqrt{1} = 1$ and $b = \sqrt{9} = 3$.

Plot the center at the origin. Move 1 unit each way in the x-direction from the center and 3 units each way in the y-direction. Draw the rectangle and the asymptotes.

We can find the equations of the asymptotes using $a = 1$ and $b = 3$:

$$y = \frac{3}{1}x \text{ and } y = -\frac{3}{1}x \qquad y = \frac{b}{a}x \text{ and } y = -\frac{b}{a}x.$$

The equations of the asymptotes are $y = 3x$ and $y = -3x$.

Sketch the branches of the hyperbola opening in the y-direction approaching the asymptotes. The vertices are $(0, 3)$ and $(0, -3)$. These are also the y-intercepts. There are no x-intercepts.

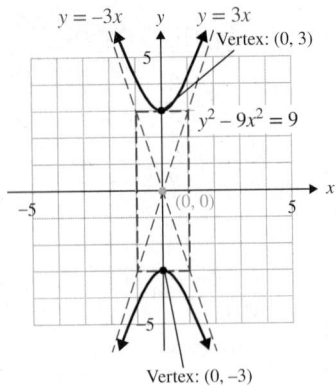

[**YOU TRY 7**] Graph $4x^2 - 9y^2 = 36$.

3 Graph a Hyperbola in Nonstandard Form

Equations of hyperbolas can take other forms. We look at one here.

 Hint

Write this definition in your notes. Include an example.

> **Definition** A Nonstandard Form for the Equation of a Hyperbola
>
> The graph of the equation $xy = c$, where c is a nonzero constant, is a hyperbola whose asymptotes are the x- and y-axes.

EXAMPLE 8 Graph $xy = 4$.

Solution

Solve for y.

$$y = \frac{4}{x} \qquad \text{Divide by } x.$$

Notice that we cannot substitute 0 for x in the equation $y = \frac{4}{x}$ because then the denominator would equal zero. Also notice that as $|x|$ gets larger, the value of y gets closer to 0. Likewise, we cannot substitute 0 for y. The x-axis is a horizontal asymptote, and the y-axis is a vertical asymptote.

Make a table of values, plot the points, and sketch the branches of the hyperbola so that they approach the asymptotes.

x	y
1	4
−1	−4
2	2
−2	−2
4	1
−4	−1
8	$\frac{1}{2}$
−8	$-\frac{1}{2}$

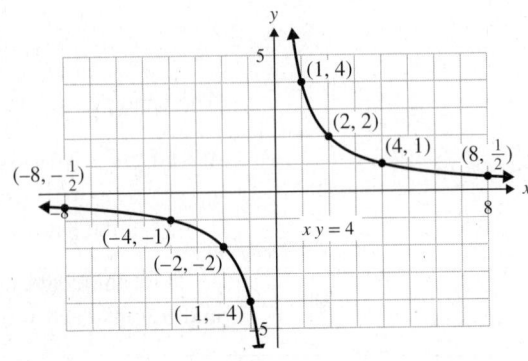

Graph $xy = 8$.

4 Graph Other Square Root Functions

We have already learned how to graph square root functions like $f(x) = \sqrt{x}$ and $g(x) = \sqrt{x - 3}$. Next, we will learn how to graph other square root functions by relating them to the graphs of conic sections.

The vertical line test shows that horizontal parabolas, circles, ellipses, and some hyperbolas are not the graphs of functions. What happens, however, if we look at a *portion* of the graph of a conic section? Let's start with a circle.

The graph of $x^2 + y^2 = 16$, at the left, is a circle with center (0, 0) and radius 4. If we solve this equation for y we get $y = \pm\sqrt{16 - x^2}$. This represents two equations, $y = \sqrt{16 - x^2}$ and $y = -\sqrt{16 - x^2}$.

The graph of $y = \sqrt{16 - x^2}$ is the **top half of the circle** since the y-coordinates of all points on the graph will be non-negative. The domain is $[-4, 4]$, and the range is $[0, 4]$.

Because of the negative sign out front of the radical, the graph of $y = -\sqrt{16 - x^2}$ is the **bottom half of the circle** since the y-coordinates of all points on the graph will be nonpositive. The domain is $[-4, 4]$, and the range is $[-4, 0]$.

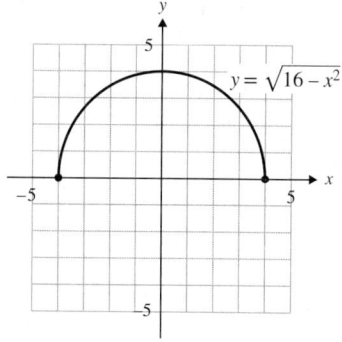

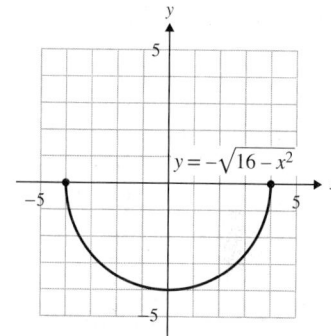

Therefore, to graph $y = \sqrt{16 - x^2}$ it is helpful if we recognize that it is the top half of the circle with equation $x^2 + y^2 = 16$. Likewise, if we are asked to graph $y = -\sqrt{16 - x^2}$, we should recognize that it is the bottom half of the graph of $x^2 + y^2 = 16$.

Let's graph another square root function by first relating it to a conic section.

EXAMPLE 9

Graph $f(x) = -5\sqrt{1 - \dfrac{x^2}{9}}$. Identify the domain and range.

Solution

The graph of this function is half of the graph of a conic section. First, notice that $f(x)$ is always a nonpositive quantity. Since all nonpositive values of y are on or below the x-axis, the graph of this function will only be on or below the x-axis.

Replace $f(x)$ with y, and rearrange the equation into a form we recognize as a conic section.

$$y = -5\sqrt{1 - \frac{x^2}{9}}$$ Replace $f(x)$ with y.

$$-\frac{y}{5} = \sqrt{1 - \frac{x^2}{9}}$$ Divide by -5.

$$\frac{y^2}{25} = 1 - \frac{x^2}{9}$$ Square both sides.

$$\frac{x^2}{9} + \frac{y^2}{25} = 1$$ Add $\frac{x^2}{9}$.

The equation $\frac{x^2}{9} + \frac{y^2}{25} = 1$ represents an ellipse centered at the origin. Graph it. The domain is $[-3, 3]$, and the range is $[-5, 5]$. It is not a function.

The graph of $f(x) = -5\sqrt{1 - \frac{x^2}{9}}$ is the *bottom half* of the ellipse, shown below, and it is a function. Its domain is $[-3, 3]$, and its range is $[-5, 0]$.

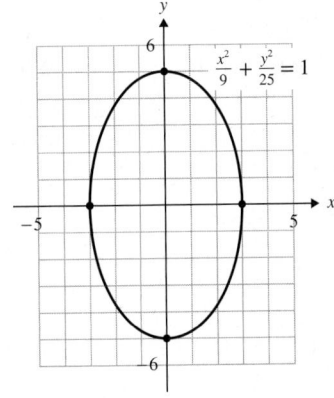

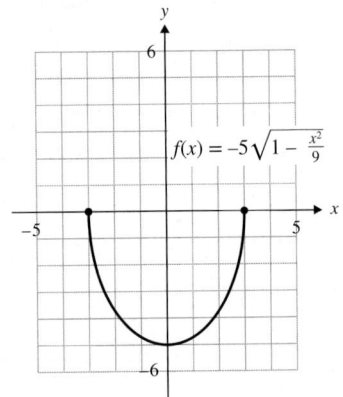

[**YOU TRY 9**]

Graph $f(x) = \sqrt{\frac{x^2}{4} - 1}$. Identify the domain and range.

More about the conic sections and their characteristics are studied in later mathematics courses.

Using Technology

We graph ellipses and hyperbolas on a graphing calculator in the same way that we graphed circles: Solve the equation for y in terms of x, enter both values of y, and graph both equations.

Let's graph the ellipse given in Example 3: $4x^2 + 25y^2 = 100$.

Solve for y.

$$25y^2 = 100 - 4x^2$$

$$y^2 = 4 - \frac{4x^2}{25}$$

$$y = \pm\sqrt{4 - \frac{4x^2}{25}}$$

Enter $Y_1 = \sqrt{4 - \dfrac{4x^2}{25}}$, the top half of the ellipse
(the y-values are positive above the x-axis). Enter
$Y_2 = -\sqrt{4 - \dfrac{4x^2}{25}}$, the bottom half of the ellipse
(the y-values are negative below the x-axis). Set an
appropriate window, and press [GRAPH]. We use the
same technique to graph a hyperbola.

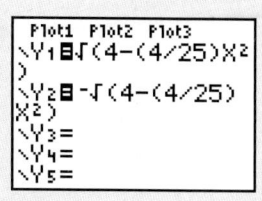

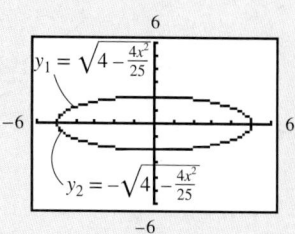

Identify each conic section as either an ellipse
or a hyperbola. Identify the center of each, rewrite each
equation for y in terms of x, and graph each equation
on a graphing calculator. These problems come from
the homework exercises so that the graphs can be found in the Answers to Exercises
appendix at the back of the book.

1) $4x^2 + 9y^2 = 36$; Exercise 19

2) $x^2 + \dfrac{y^2}{4} = 1$; Exercise 13

3) $\dfrac{x^2}{25} + (y + 4)^2 = 1$; Exercise 15

4) $9x^2 - y^2 = 36$; Exercise 59

5) $\dfrac{y^2}{16} - \dfrac{x^2}{4} = 1$; Exercise 49

6) $y^2 - \dfrac{(x - 1)^2}{9} = 1$; Exercise 55

ANSWERS TO [YOU TRY] **EXERCISES**

1)

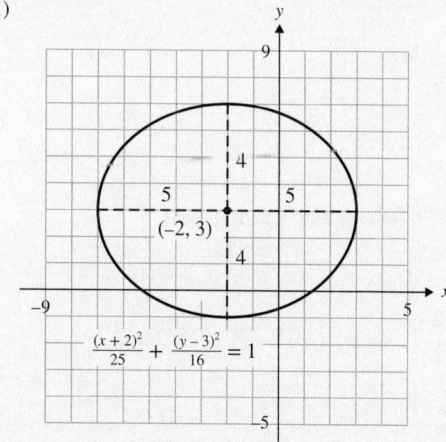

2)

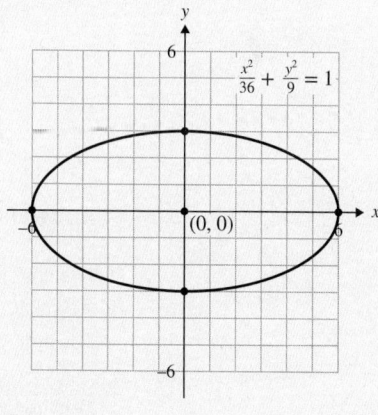

3)

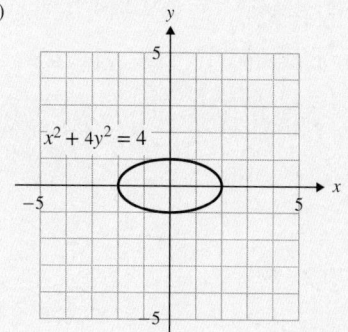

4)

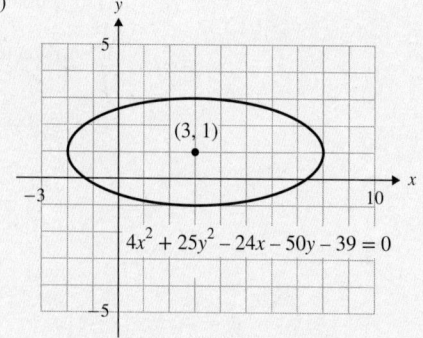

5)

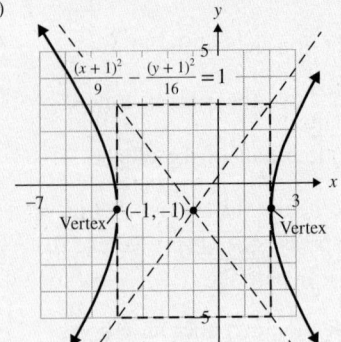

$$\frac{(x+1)^2}{9} - \frac{(y+1)^2}{16} = 1$$

Vertex $(-1,-1)$ Vertex

6)

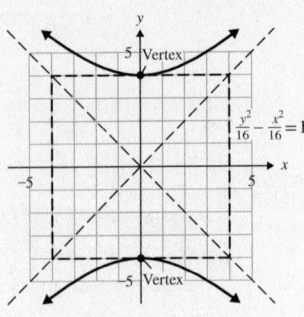

Vertex

$$\frac{y^2}{16} - \frac{x^2}{16} = 1$$

Vertex

7)

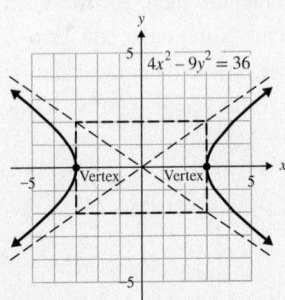

$4x^2 - 9y^2 = 36$

Vertex Vertex

8)

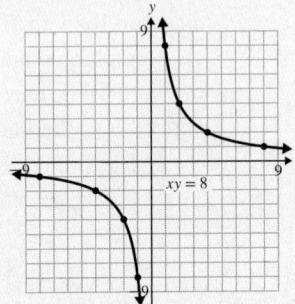

$xy = 8$

9) domain: $(-\infty, -2] \cup [2, \infty)$; range: $[0, \infty)$

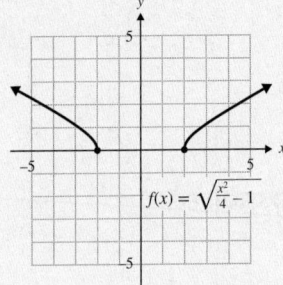

$$f(x) = \sqrt{\frac{x^2}{4} - 1}$$

ANSWERS TO TECHNOLOGY EXERCISES

1) ellipse with center $(0, 0)$; $Y_1 = \sqrt{4 - \frac{4x^2}{9}}$, $Y_2 = -\sqrt{4 - \frac{4x^2}{9}}$

2) ellipse with center $(0, 0)$; $Y_1 = \sqrt{4 - 4x^2}$, $Y_2 = -\sqrt{4 - 4x^2}$

3) ellipse with center $(0, -4)$; $Y_1 = -4 + \sqrt{1 - \frac{x^2}{25}}$, $Y_2 = -4 - \sqrt{1 - \frac{x^2}{25}}$

4) hyperbola with center $(0, 0)$; $Y_1 = \sqrt{9x^2 - 36}$, $Y_2 = -\sqrt{9x^2 - 36}$

5) hyperbola with center $(0, 0)$; $Y_1 = \sqrt{16 + 4x^2}$, $Y_2 = -\sqrt{16 + 4x^2}$

6) hyperbola with center $(1, 0)$; $Y_1 = \sqrt{1 + \frac{(x-1)^2}{9}}$, $Y_2 = -\sqrt{1 + \frac{(x-1)^2}{9}}$

Objective 1: Graph an Ellipse

Determine whether each statement is *true* or *false*.

 1) The graph of an ellipse is a function.

2) The center of the ellipse with equation

$$\frac{(x-1)^2}{16} + \frac{(y-5)^2}{9} = 1 \text{ is } (-1, -5).$$

3) The center of the ellipse with equation

$$\frac{(x-8)^2}{9} + \frac{(y+3)^2}{25} = 1 \text{ is } (8, -3).$$

4) The center of the ellipse with equation

$$\frac{x^2}{7} + \frac{y^2}{4} = 1 \text{ is } (0, 0).$$

5) The graph of $4x^2 + 25y^2 = 100$ is an ellipse.

6) The graph of $\dfrac{(x+3)^2}{4} - \dfrac{(y+2)^2}{9} = 1$ is an ellipse.

7) The graph of $9y^2 - x^2 = 9$ is an ellipse.

8) The equation of $4x^2 + y^2 + 8x - 10y + 8 = 0$ can be put into the standard form for the equation of an ellipse by completing the square.

Identify the center of each ellipse, and graph the equation.

 9) $\dfrac{(x+2)^2}{9} + \dfrac{(y-1)^2}{4} = 1$

10) $\dfrac{(x \quad 4)^2}{4} + \dfrac{(y-3)^2}{16} = 1$

11) $\dfrac{(x-3)^2}{9} + \dfrac{(y+2)^2}{16} = 1$

12) $\dfrac{(x+4)^2}{25} + \dfrac{(y-5)^2}{16} = 1$

13) $x^2 + \dfrac{y^2}{4} = 1$

14) $\dfrac{x^2}{9} + y^2 = 1$

15) $\dfrac{x^2}{25} + (y+4)^2 = 1$

16) $(x+3)^2 + \dfrac{(y+4)^2}{9} = 1$

17) $\dfrac{(x+1)^2}{4} + \dfrac{(y+3)^2}{9} = 1$

18) $\dfrac{(x-2)^2}{16} + \dfrac{y^2}{25} = 1$

19) $4x^2 + 9y^2 = 36$ 20) $x^2 + 4y^2 = 16$

21) $25x^2 + y^2 = 25$ 22) $9x^2 + y^2 = 36$

Write the equation of the ellipse in standard form.

Fill It In

Fill in the blanks with either the missing mathematical step or the reason for the given step.

23) $3x^2 + 2y^2 - 6x + 4y - 7 = 0$
    $(3x^2 - 6x) + (2y^2 + 4y) = 7$

    _____

    _____   Factor out the
                       coefficients of the
                       squared terms.

    $3(x^2 - 2x + 1) + 2(y^2 + 2y + 1)$
    $= 7 + 3(1) + 2(1)$

    _____   Factor.

    _____   Divide both sides
                       by 12.

24) $4x^2 + 9y^2 + 16x + 54y + 61 = 0$
    $(4x^2 + 16x) + (9y^2 + 54y) = -61$

    _____

    _____   Factor out the
                       coefficients of the
                       squared terms.

    _____   Complete the
                       square.

    $4(x+2)^2 + 9(y+3)^2 = 36$
    _____

    _____   Divide both sides
                       by 36.

Put each equation into the standard form for the equation of an ellipse, and graph.

25) $x^2 + 4y^2 - 2x - 24y + 21 = 0$

26) $9x^2 + 4y^2 + 36x - 8y + 4 = 0$

27) $9x^2 + y^2 + 72x + 2y + 136 = 0$

28) $x^2 + 4y^2 - 6x - 40y + 105 = 0$

29) $4x^2 + 9y^2 - 16x - 54y + 61 = 0$

30) $4x^2 + 25y^2 + 8x + 200y + 304 = 0$

31) $25x^2 + 4y^2 + 150x + 125 = 0$

32) $4x^2 + y^2 - 2y - 15 = 0$

Extension: The Ellipse

Write an equation of the ellipse containing the following points.

33) $(-3, 0)$, $(3, 0)$, $(0, -5)$, and $(0, 5)$

34) $(-6, 0)$, $(6, 0)$, $(0, -2)$, and $(0, 2)$

35) $(-7, 0)$, $(7, 0)$, $(0, -1)$, and $(0, 1)$

36) $(-1, 0)$, $(1, 0)$, $(0, -9)$, and $(0, 9)$

37) $(3, 5)$, $(3, -3)$, $(1, 1)$, and $(5, 1)$

38) $(-1, 1)$, $(-1, -5)$, $(4, -2)$, and $(-6, -2)$

39) Is a circle a special type of ellipse? Explain your answer.

40) The Oval Office in the White House is an ellipse about 36 ft long and 29 ft wide. If the center of the room is at the origin of a Cartesian coordinate system and the length of the room is along the x-axis, write an equation of the elliptical room. (www.whitehousehistory.org)

41) The fuselage of a Boeing 767 jet has an elliptical cross section that is 198 in. wide and 213 in. tall. If the center of this cross section is at the origin and the width is along the x-axis, write an equation of this ellipse. (Jan Roskam, *Airplane Design*, p. 89).

42) The arch of a bridge over a canal in Amsterdam is half of an ellipse. At water level the arch is 14 ft wide, and it is 6 ft tall at its highest point.

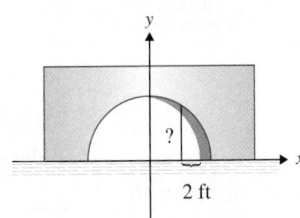

   2 ft

  a) Write an equation of the arch.

  b) What is the height of the arch (to the nearest foot) 2 ft from the bottom edge?

Objective 2: Graph a Hyperbola in Standard Form

Determine whether each statement is *true* or *false*.

43) The graph of $\dfrac{(x-7)^2}{9} + \dfrac{(y-2)^2}{16} = 1$ is a hyperbola with center $(7, 2)$.

44) The center of the hyperbola with equation $\dfrac{(x-5)^2}{16} - \dfrac{(y+3)^2}{25} = 1$ is $(5, -3)$.

45) The center of the hyperbola with equation $\dfrac{(y-6)^2}{25} - \dfrac{(x+4)^2}{4} = 1$ is $(-4, 6)$.

46) The graph of $9y^2 - x^2 = 9$ is a hyperbola with center at the origin.

Graph each hyperbola. Identify the center and vertices, and sketch the asymptotes.

47) $\dfrac{x^2}{9} - \dfrac{y^2}{25} = 1$

48) $\dfrac{x^2}{9} - \dfrac{y^2}{4} = 1$

49) $\dfrac{y^2}{16} - \dfrac{x^2}{4} = 1$

50) $\dfrac{y^2}{4} - \dfrac{x^2}{4} = 1$

51) $\dfrac{(x-2)^2}{9} - \dfrac{(y+3)^2}{16} = 1$

52) $\dfrac{(x+3)^2}{4} - \dfrac{(y+1)^2}{16} = 1$

53) $\dfrac{(y+1)^2}{25} - \dfrac{(x+4)^2}{4} = 1$

54) $\dfrac{(y-1)^2}{36} - \dfrac{(x+1)^2}{9} = 1$

55) $y^2 - \dfrac{(x-1)^2}{9} = 1$

56) $\dfrac{(y+4)^2}{4} - x^2 = 1$

57) $\dfrac{(x-1)^2}{25} - \dfrac{(y-2)^2}{25} = 1$

58) $\dfrac{(x-2)^2}{16} - \dfrac{(y-3)^2}{9} = 1$

59) $9x^2 - y^2 = 36$

60) $4y^2 - x^2 = 16$

61) $y^2 - x^2 = 1$

62) $x^2 - y^2 = 25$

Write the equations of the asymptotes of the graph in each of the following exercises. *Do not graph.*

63) Exercise 47

64) Exercise 48

65) Exercise 49

66) Exercise 50

67) Exercise 59

68) Exercise 60

69) Exercise 61

70) Exercise 62

Objective 3: Graph a Hyperbola in Nonstandard Form

Graph each equation.

71) $xy = 1$

72) $xy = 6$

73) $xy = 2$

74) $xy = 10$

75) $xy = -4$

76) $xy = -8$

77) $xy = -6$

78) $xy = -1$

Objective 4: Graph Other Square Root Functions

Graph each square root function. Identify the domain and range.

 79) $g(x) = \sqrt{25 - x^2}$

80) $f(x) = \sqrt{9 - x^2}$

81) $h(x) = -\sqrt{1 - x^2}$

82) $k(x) = -\sqrt{9 - x^2}$

83) $g(x) = -2\sqrt{1 - \dfrac{x^2}{9}}$

84) $f(x) = 3\sqrt{1 - \dfrac{x^2}{16}}$

85) $h(x) = -3\sqrt{\dfrac{x^2}{4} - 1}$

86) $k(x) = 2\sqrt{\dfrac{x^2}{16} - 1}$

Sketch the graph of each equation.

87) $x = \sqrt{16 - y^2}$

88) $x = -\sqrt{4 - y^2}$

89) $x = -3\sqrt{1 - \dfrac{y^2}{4}}$

90) $x = \sqrt{1 - \dfrac{y^2}{9}}$

Extension: The Hyperbola

We have learned that sometimes it is necessary to complete the square to put the equations of circles and ellipses into standard form. The same is true for hyperbolas. In Exercises 91 and 92, practice going through the steps of putting the equation of a hyperbola into standard form.

Fill It In

Fill in the blanks with either the missing mathematical step or the reason for the given step.

 91) $4x^2 - 9y^2 - 8x - 18y - 41 = 0$

$4x^2 - 8x - 9y^2 - 18y = 41$

_____

_____ Factor out the coefficients of the squared terms.

$4(x^2 - 2x + 1) - 9(y^2 + 2y + 1)$
$= 41 + 4(1) - 9(1)$

_____

_____ Factor.

_____ Divide both sides by 36.

_____

92) $-x^2 + 4y^2 + 6x - 16y + 3 = 0$

$-x^2 + 6x + 4y^2 - 16y = -3$

_____

_____ Factor out the coefficients of the squared terms.

_____ Complete the square.

_____

$-(x - 3)^2 + 4(y - 2)^2 = 4$

_____ Divide both sides by 4.

_____

For Exercises 93–96, put each equation into the standard form for the equation of a hyperbola and graph.

 93) $x^2 - 4y^2 - 2x - 24y - 51 = 0$

94) $9x^2 - 4y^2 + 90x - 16y + 173 = 0$

95) $16y^2 - 9x^2 + 18x - 64y - 89 = 0$

96) $y^2 - 4x^2 - 16x - 6y - 23 = 0$

We know that the standard form for the equation of a hyperbola is

$$\frac{(x - h)^2}{a^2} - \frac{(y - k)^2}{b^2} = 1 \text{ or } \frac{(y - k)^2}{b^2} - \frac{(x - h)^2}{a^2} = 1.$$

The equations for the asymptotes are

$$y - k = \frac{b}{a}(x - h) \text{ and } y - k = -\frac{b}{a}(x - h)$$

Use these formulas to write the equations of the asymptotes of the graph in each of the following exercises. *Do not graph.*

97) Exercise 51

98) Exercise 52

99) Exercise 53

100) Exercise 54

101) Exercise 55

102) Exercise 56

103) A hyperbola centered at the origin opens in the y-direction and has asymptotes with equations $y = \dfrac{1}{2}x$ and $y = -\dfrac{1}{2}x$. Write an equation of the hyperbola.

104) A hyperbola centered at the origin opens in the x-direction and has asymptotes with equations $y = \dfrac{3}{2}x$ and $y = -\dfrac{3}{2}x$. Write an equation of the hyperbola.

Use a graphing calculator to graph the following ellipses and hyperbolas. Square the viewing window.

105) $\dfrac{(x - 2)^2}{9} - \dfrac{(y + 3)^2}{16} = 1$

106) $\dfrac{(x + 3)^2}{4} - \dfrac{(y + 1)^2}{16} = 1$

107) $\dfrac{x^2}{36} + \dfrac{y^2}{16} = 1$

108) $9x^2 + y^2 = 36$

109) $\dfrac{(y + 1)^2}{25} - \dfrac{(x + 4)^2}{4} = 1$

110) $\dfrac{(y - 1)^2}{36} - \dfrac{(x + 1)^2}{9} = 1$

R1) If you are given an equation of a hyperbola, explain how to determine when the branches open in the *x*-direction and when they open in the *y*-direction.

R2) Draw a graph of a hyperbola and explain the asymptotes to a classmate.

R3) Explain how to differentiate between the equation of an ellipse and the equation of a hyperbola.

Putting It All Together

P Prepare

What is your objective for Putting It All Together?
1 Identify and Graph Different Types of Conic Sections

O Organize

How can you accomplish the objective?
• Review the equations for all the different conic sections, and summarize them in your notes. • Draw a graph of all the conic sections in your notes, and label the graphs. • Complete the given examples on your own. • Complete You Try 1.

W Work

Read the explanations, follow the examples, take notes, and complete the You Try.

1 Identify and Graph Different Types of Conic Sections

Sometimes the most difficult part of graphing a conic section is identifying which type of graph will result from the given equation. In this section, we will discuss how to look at an equation and determine which type of conic section it represents.

EXAMPLE 1

Graph $x^2 + y^2 + 4x - 6y + 9 = 0$.

Solution

First, notice that this equation has two squared terms. Therefore, its graph cannot be a parabola because the equation of a parabola contains only one squared term. Next, observe that the coefficients of x^2 and y^2 are each 1. Since the coefficients are the same, *this is the equation of a circle.*

Write the equation in the form $(x - h)^2 + (y - k)^2 = r^2$ by completing the square on the x-terms and on the y-terms.

$$x^2 + y^2 + 4x - 6y + 9 = 0$$

Group the x-terms together, and group the y-terms together. Move the constant to the other side.

$$(x^2 + 4x) + (y^2 - 6y) = -9$$

$$(x^2 + 4x + 4) + (y^2 - 6y + 9) = -9 + 4 + 9$$

Complete the square for each group of terms.

$$(x + 2)^2 + (y - 3)^2 = 4$$

Factor; add.

W Hint

Pay close attention to how you identify the different types of conic sections in Examples 1–4.

The center of the circle is $(-2, 3)$. The radius is 2.

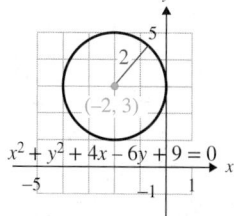

EXAMPLE 2

Graph $x = y^2 + 4y + 3$.

Solution

This equation contains only one squared term. Therefore, *this is the equation of a parabola.* Since the squared term is y^2 and $a = 1$, the parabola will open to the right.

Use the formula $y = -\dfrac{b}{2a}$ to find the y-coordinate of the vertex.

$$a = 1 \quad b = 4 \quad c = 3$$

$$y = -\frac{4}{2(1)} = -2$$

$$x = (-2)^2 + 4(-2) + 3 = -1$$

The vertex is $(-1, -2)$. Make a table of values to find other points on the parabola, and use the axis of symmetry to find more points.

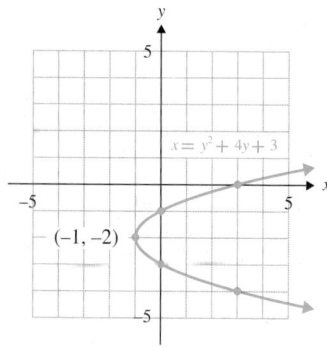

x	y
0	-1
3	0

Plot the points in the table. Locate the points $(0, -3)$ and $(3, -4)$ using the axis of symmetry, $y = -2$.

EXAMPLE 3

Graph $\dfrac{(y - 1)^2}{9} - \dfrac{(x - 3)^2}{4} = 1$.

Solution

In this equation we see the *difference* of two squares. *The graph of this equation is a hyperbola. The branches of the hyperbola will open in the y-direction because the quantity containing the variable y,* $\dfrac{(y - 1)^2}{9}$, *is the positive, squared quantity.*

The center is $(3, 1)$; $a = \sqrt{4} = 2$ and $b = \sqrt{9} = 3$.

Draw the reference rectangle and its diagonals, the asymptotes of the graph.

Draw the branches of the graph so that they approach the asymptotes. The vertices are $(3, 4)$ and $(3, -2)$.

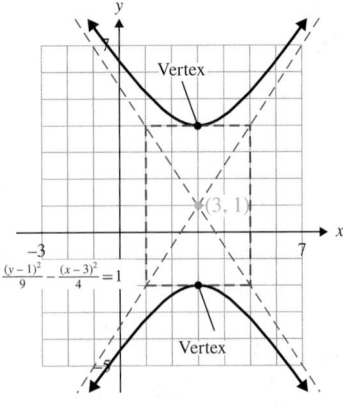

EXAMPLE 4

Graph $x^2 + 9y^2 = 36$.

Solution

This equation contains the *sum* of two squares with *different* coefficients. *This is the equation of an ellipse.* (If the coefficients were the same, the graph would be a circle.)

Divide both sides of the equation by 36 to get 1 on the right side of the $=$ sign.

$$x^2 + 9y^2 = 36$$

$$\frac{x^2}{36} + \frac{9y^2}{36} = \frac{36}{36} \qquad \text{Divide both sides by 36.}$$

$$\frac{x^2}{36} + \frac{y^2}{4} = 1 \qquad \text{Simplify.}$$

The center is $(0, 0)$, $a = 6$ and $b = 2$.

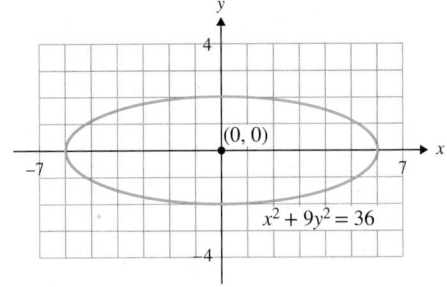

YOU TRY 1

Determine whether the graph of each equation is a parabola, circle, ellipse, or hyperbola. Then, graph each equation.

a) $4x^2 - 25y^2 = 100$

b) $x^2 + y^2 - 6x - 12y + 9 = 0$

c) $y = -x^2 - 2x + 4$

d) $x^2 + \frac{(y + 4)^2}{9} = 1$

W Hint

After completing Examples 1–4, summarize what you've learned about each of the conic sections. Explain, in your own words, how to determine which type of conic section an equation represents. Include examples.

ANSWERS TO YOU TRY **EXERCISES**

1) a) hyperbola

b) circle

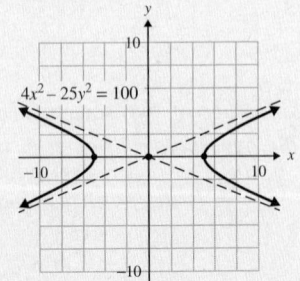

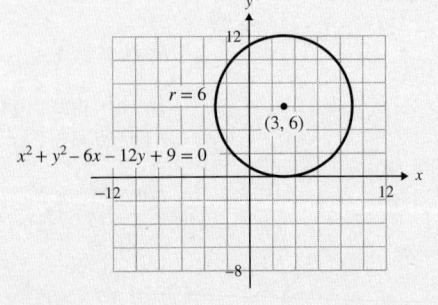

c) parabola

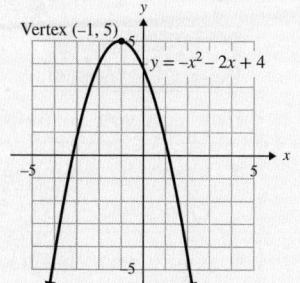

d) ellipse

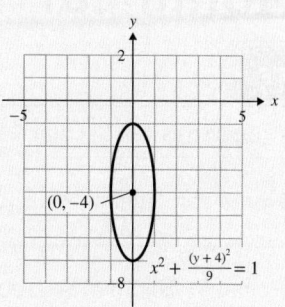

Putting it All Together Exercises

Do the exercises, and check your work.

Objective 1: Identify and Graph Different Types of Conic Sections

Determine whether the graph of each equation is a parabola, circle, ellipse, or hyperbola. Then, graph each equation.

1) $y = x^2 + 4x + 8$

2) $(x + 5)^2 + (y - 3)^2 = 25$

3) $\dfrac{(y + 4)^2}{9} - \dfrac{(x + 1)^2}{4} = 1$

4) $x = (y - 1)^2 + 8$

5) $16x^2 + 9y^2 = 144$

6) $x^2 - 4y^2 = 36$

7) $x^2 + y^2 + 8x - 6y - 11 = 0$

8) $\dfrac{(x - 2)^2}{25} + \dfrac{(y - 2)^2}{36} = 1$

9) $(x - 1)^2 + \dfrac{y^2}{16} = 1$

10) $x^2 + y^2 + 8y + 7 = 0$

11) $x = -(y + 4)^2 - 3$

12) $\dfrac{(y + 4)^2}{4} - (x + 2)^2 = 1$

13) $25x^2 - 4y^2 = 100$

14) $4x^2 + y^2 = 16$

15) $(x - 3)^2 + y^2 = 16$

16) $y = -x^2 + 6x - 7$

17) $x = \dfrac{1}{2}y^2 + 2y + 3$

18) $\dfrac{(x - 3)^2}{16} + y^2 = 1$

19) $(x - 2)^2 - (y + 1)^2 = 9$

20) $x^2 + y^2 + 6x - 8y + 9 = 0$

Where appropriate, write the equation in standard form. Then, graph each equation.

21) $xy = 5$

22) $25x^2 - 4y^2 + 150x + 125 = 0$

23) $9x^2 + y^2 - 54x + 4y + 76 = 0$

24) $xy = -2$

25) $9y^2 - 4x^2 - 18y + 16x - 43 = 0$

26) $4x^2 + 9y^2 - 8x - 54y + 49 = 0$

Use a graphing calculator to graph Exercises 27–30. Square the viewing window.

27) $x^2 + y^2 + 8x - 6y - 11 = 0$

28) $(x - 1)^2 + \dfrac{y^2}{16} = 1$

29) $x = -(y + 4)^2 - 3$

30) $25x^2 - 4y^2 = 100$

R Rethink

R1) Were you able to distinguish between the different conic sections without looking back at your notes or the book? Summarize how you recognize which type of conic section will result from a particular equation.

R2) Are all conic sections functions? Why or why not?

R3) Which of the conic sections is the hardest for you to graph? What can you do to get better at graphing it?

14.3 Nonlinear Systems of Equations

What are your objectives for Section 14.3?	How can you accomplish each objective?
1 Define a Nonlinear System of Equations	• Know the definition of a *nonlinear system of equations*.
2 Solve a Nonlinear System by Substitution	• Know what the graphs of each equation look like in order to know how many possible intersection points exist. Sketch a rough graph in your notes. • Review the substitution method for solving a system of equations. • Check the proposed solutions in the original equations. • Complete the given examples on your own. • Complete You Trys 1 and 2.
3 Solve a Nonlinear System Using the Elimination Method	• Be able to recognize when both equations are second-degree equations. • Review the elimination method for solving a system of equations. • Check the proposed solutions in the original equations. • Complete the given examples on your own. • Complete You Trys 3 and 4.

 **Work**   Read the explanations, follow the examples, take notes, and complete the You Trys.

1 Define a Nonlinear System of Equations

In Chapter 5, we learned to solve systems of linear equations by graphing, substitution, and the elimination method. We can use these same techniques for solving a *nonlinear system of equations* in two variables. A **nonlinear system of equations** is a system in which at least one of the equations is not linear.

Solving a nonlinear system by graphing is not practical since it would be very difficult (if not impossible) to accurately read the points of intersection. Therefore, we will solve the systems using substitution and the elimination method. We will graph the equations, however, so that we can visualize the solution(s) as the point(s) of intersection of the graphs.

We are interested only in real-number solutions. If a system has imaginary solutions, then the graphs of the equations do not intersect in the real-number plane.

2 Solve a Nonlinear System by Substitution

When one of the equations in a system is linear, it is often best to use the substitution method to solve the system.

EXAMPLE 1

Solve the system $x^2 - 2y = 2$ (1)
$-x + y = 3$ (2)

Solution

The graph of equation (1) is a parabola, and the graph of equation (2) is a line. Let's begin by thinking about the number of possible points of intersection the graphs can have.

W Hint

Make a rough sketch for yourself so that you know how many solutions the system *could* have.

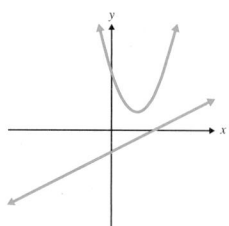

No points of intersection

The system has no solution.

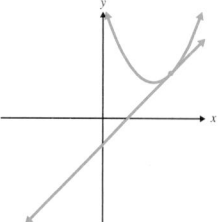

One point of intersection

The system has one solution.

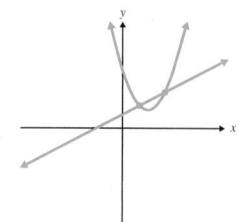

Two points of intersection

The system has two solutions.

Solve the linear equation for one of the variables.

$$-x + y = 3$$
$$y = x + 3 \quad (3) \qquad \text{Solve for } y.$$

Substitute $x + 3$ for y in equation (1).

$$x^2 - 2y = 2 \qquad \text{Equation (1)}$$
$$x^2 - 2(x + 3) = 2 \qquad \text{Substitute.}$$
$$x^2 - 2x - 6 = 2 \qquad \text{Distribute.}$$
$$x^2 - 2x - 8 = 0 \qquad \text{Subtract 2.}$$
$$(x - 4)(x + 2) = 0 \qquad \text{Factor.}$$
$$x = 4 \text{ or } x = -2 \qquad \text{Solve.}$$

To find the corresponding value of y for each value of x, we can substitute $x = 4$ and then $x = -2$ into *either* equation (1), (2), or (3). No matter which equation you choose, you should always check the solutions in *both* of the original equations. We will substitute the values into equation (3) because this is just an alternative form of equation (2), and it is already solved for y.

Substitute each value into equation (3) to find y.

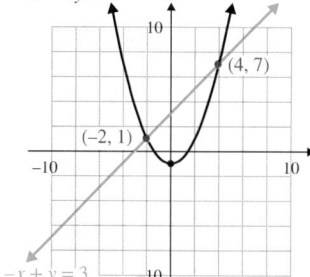

$x = 4$: $y = x + 3$ $x = -2$: $y = x + 3$
$y = 4 + 3$ $y = -2 + 3$
$y = 7$ $y = 1$

The proposed solutions are $(4, 7)$ and $(-2, 1)$. Verify that they solve the system by checking them in equation (1). (Remember, the ordered pair must satisfy *both* equations in the system.) The solution set is $\{(4, 7), (-2, 1)\}$. We can see on the graph to the left that these are the points of intersection of the graphs.

[YOU TRY 1]

Solve the system $x^2 + 3y = 6$
$x + y = 2$

EXAMPLE 2

Solve the system $x^2 + y^2 = 1$ (1)

                   $x + 2y = -1$ (2)

Solution

The graph of equation (1) is a circle, and the graph of equation (2) is a line. These graphs can intersect at zero, one, or two points. Therefore, this system will have zero, one, or two solutions.

We will not solve equation (1) for a variable because doing so would give us a radical in the expression. It will be easiest to solve equation (2) for x because its coefficient is 1.

W Hint

Write out each step as you are reading the example.

$$x + 2y = -1 \qquad (2)$$
$$x = -2y - 1 \qquad (3) \qquad \text{Solve for } x.$$

Substitute $-2y - 1$ for x in equation (1).

$$x^2 + y^2 = 1 \quad (1)$$
$$(-2y - 1)^2 + y^2 = 1 \qquad \text{Substitute.}$$
$$4y^2 + 4y + 1 + y^2 = 1 \qquad \text{Expand } (-2y - 1)^2.$$
$$5y^2 + 4y = 0 \qquad \text{Combine like terms; subtract 1.}$$
$$y(5y + 4) = 0 \qquad \text{Factor.}$$
$$y = 0 \quad \text{or} \quad 5y + 4 = 0 \qquad \text{Set each factor equal to zero.}$$
$$y = -\frac{4}{5} \qquad \text{Solve for y.}$$

Substitute $y = 0$ and then $y = -\dfrac{4}{5}$ into equation (3) to find their corresponding values of x.

$$
\begin{array}{l|l}
y = 0: \quad x = -2y - 1 & y = \dfrac{4}{5}: \quad x = -2y - 1 \\[2mm]
\qquad\quad\; x = -2(0) - 1 & \qquad\quad\; x = -2\left(-\dfrac{4}{5}\right) - 1 \\[2mm]
\qquad\quad\; x = -1 & \qquad\quad\; x = \dfrac{8}{5} - 1 = \dfrac{3}{5}
\end{array}
$$

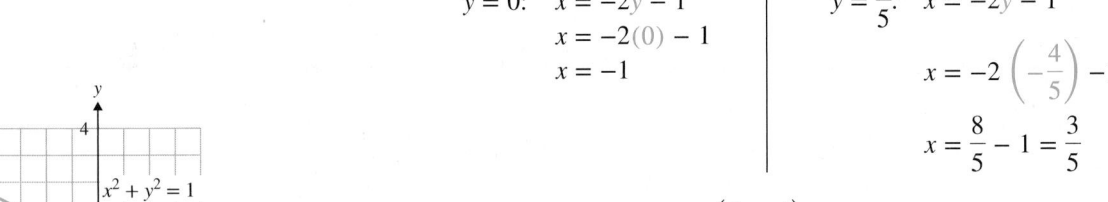

The proposed solutions are $(-1, 0)$ and $\left(\dfrac{3}{5}, -\dfrac{4}{5}\right)$. Check them in equations (1) and (2).

The solution set is $\left\{ (-1, 0), \left(\dfrac{3}{5}, -\dfrac{4}{5}\right) \right\}$. The graph at left shows that these are the points of intersection of the graphs.

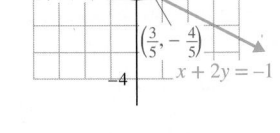

[YOU TRY 2] Solve the system $x^2 + y^2 = 25$

                     $x - y = 7$

Note

We must always check the proposed solutions in *each* equation in the system.

3 Solve a Nonlinear System Using the Elimination Method

The elimination method can be used to solve a system when both equations are second-degree equations.

EXAMPLE 3

Solve the system $5x^2 + 3y^2 = 21$ (1)
$$4x^2 - y^2 = 10 \quad (2)$$

Solution

Each equation is a second-degree equation. The first is an ellipse, and the second is a hyperbola. They can have zero, one, two, three, or four points of intersection. Multiply equation (2) by 3. Then adding the two equations will eliminate the y^2-terms.

<table>
<tr><td align="center">**Original System**</td><td></td><td align="center">**Rewrite the System**</td></tr>
<tr><td align="center">$5x^2 + 3y^2 = 21$</td><td>$\longrightarrow$</td><td align="center">$5x^2 + 3y^2 = 21$</td></tr>
<tr><td align="center">$4x^2 - y^2 = 10$</td><td></td><td align="center">$12x^2 - 3y^2 = 30$</td></tr>
</table>

$$\begin{array}{r} 5x^2 + 3y^2 = 21 \\ + \underline{12x^2 - 3y^2 = 30} \\ 17x^2 = 51 \end{array}$$ Add the equations to eliminate y^2.
$$x^2 = 3$$
$$x = \pm\sqrt{3}$$

> **W Hint**
>
> Make a rough sketch of the possible ways an ellipse and a hyperbola can (or might not) intersect.

Find the corresponding values of y for $x = \sqrt{3}$ and $x = -\sqrt{3}$.

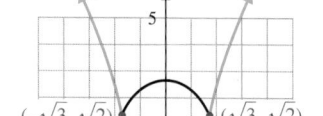

$x = \sqrt{3}$:
$$4x^2 - y^2 = 10 \quad (2)$$
$$4(\sqrt{3})^2 - y^2 = 10$$
$$12 - y^2 = 10$$
$$-y^2 = -2$$
$$y^2 = 2$$
$$y = \pm\sqrt{2}$$

This gives us $(\sqrt{3}, \sqrt{2})$ and $(\sqrt{3}, -\sqrt{2})$.

$x = -\sqrt{3}$:
$$4x^2 - y^2 = 10 \quad (2)$$
$$4(-\sqrt{3})^2 - y^2 = 10$$
$$12 - y^2 = 10$$
$$-y^2 = -2$$
$$y^2 = 2$$
$$y = \pm\sqrt{2}$$

This gives us $(-\sqrt{3}, \sqrt{2})$ and $(-\sqrt{3}, -\sqrt{2})$.

Check the proposed solutions in equation (1) to verify that they satisfy that equation as well.

The solution set is $\{(\sqrt{3}, \sqrt{2}), (\sqrt{3}, -\sqrt{2}), (-\sqrt{3}, \sqrt{2}), (-\sqrt{3}, -\sqrt{2})\}$.

[YOU TRY 3]

Solve the system $2x^2 - 13y^2 = 20$
$$-x^2 + 10y^2 = 4$$

For solving some systems, using *either* substitution or the elimination method works well. Look carefully at each system to decide which method to use.

We will see in Example 4 that not all systems have solutions.

EXAMPLE 4

Solve the system $y = \sqrt{x}$ (1)
$$y^2 - 4x^2 = 4 \quad (2)$$

Solution

The graph of the square root function $y = \sqrt{x}$ is half of a parabola. The graph of equation (2) is a hyperbola. Solve this system by substitution. Replace y in equation (2) with $\sqrt{x}$ from equation (1).

$$y^2 - 4x^2 = 4 \quad (2)$$
$$(\sqrt{x})^2 - 4x^2 = 4$$ Substitute $y = \sqrt{x}$ into equation (2).
$$x - 4x^2 = 4$$
$$0 = 4x^2 - x + 4$$

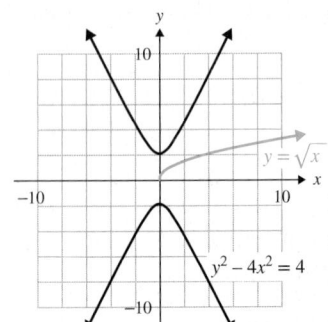

The right-hand side does not factor, so solve it using the quadratic formula.

$$4x^2 - x + 4 = 0 \qquad a = 4 \qquad b = -1 \qquad c = 4$$

$$x = \frac{-(-1) \pm \sqrt{(-1)^2 - 4(4)(4)}}{2(4)} = \frac{1 \pm \sqrt{1 - 64}}{8} = \frac{1 \pm \sqrt{-63}}{8}$$

Because $\sqrt{-63}$ is not a real number, there are no real-number values for x. The system has no solution, so the solution set is $\varnothing$. The graph is shown on the left.

$$\left[\textbf{YOU TRY 4} \right]$$ Solve the system $4x^2 + y^2 = 4$
$$\qquad\qquad\qquad\qquad\qquad x - y = 3$$

Using Technology

We can solve systems of nonlinear equations on the graphing calculator just like we solved systems of linear equations in Chapter 5—graph the equations and find their points of intersection.

Let's look at Example 3:

$$5x^2 + 3y^2 = 21$$
$$4x^2 - y^2 = 10$$

Solve each equation for y and enter them into the calculator.

Solve $5x^2 + 3y^2 = 21$ for y: | Solve $4x^2 - y^2 = 10$ for y:

$$y = \pm\sqrt{7 - \frac{5}{3}x^2}$$ | $$y = \pm\sqrt{4x^2 - 10}$$

Enter $\sqrt{7 - \frac{5}{3}x^2}$ as Y_1. | Enter $\sqrt{4x^2 - 10}$ as Y_3.

Enter $-\sqrt{7 - \frac{5}{3}x^2}$ as Y_2. | Enter $-\sqrt{4x^2 - 10}$ as Y_4.

After entering the equations, press $\boxed{\text{GRAPH}}$.

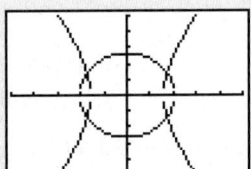

The system has four real solutions since the graphs have four points of intersection. We can use the INTERSECT option to find the solutions. Since we graphed four functions, we must tell the calculator which point of intersection we want to find. Note that the point where the graphs intersect in the first quadrant comes from the intersection of equations Y_1 and Y_3. Press $\boxed{2^{\text{nd}}}$ $\boxed{\text{TRACE}}$ and choose 5:intersect and you will see the screen to the right.

Notice that the top left of the screen to the right displays the function Y_1. Since we want to find the intersection of Y_1 and Y_3, press $\boxed{\text{ENTER}}$ when Y_1 is displayed. Now Y_2 appears at the top left, but we do not need this function. Press

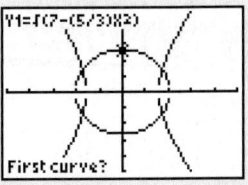

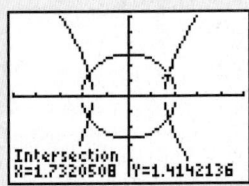

the down arrow to see the equation for Y_3 and be sure that the cursor is close to the intersection point in quadrant I. Press [ENTER] twice. You will see the approximate solution (1.732, 1.414), as shown to the right.

In Example 3 we found the exact solutions algebraically. The calculator solution, (1.732, 1.414), is an approximation of the exact solution $(\sqrt{3}, \sqrt{2})$.

The other solutions of the system can be found in the same way.

Use the graphing calculator to find all real-number solutions of each system. These are taken from the examples in the section and from the Chapter Summary.

1) $x^2 - 2y = 2$
   $-x + y = 3$

2) $x^2 + y^2 = 1$
   $x + 2y = -1$

3) $x - y^2 = 3$
   $x - 2y = 6$

4) $\quad y = \sqrt{x}$
   $y^2 - 4x^2 = 4$

ANSWERS TO [YOU TRY] EXERCISES

1) $\{(0, 2), (3, -1)\}$ 2) $\{(4, -3), (3, -4)\}$ 3) $\{(6, 2), (6, -2), (-6, 2), (-6, -2)\}$ 4) $\varnothing$

ANSWERS TO TECHNOLOGY EXERCISES

1) $\{(4, 7), (-2, 1)\}$ 2) $\{(-1, 0), (0.6, -0.8)\}$ 3) $\{(12, 3), (4, -1)\}$ 4) $\varnothing$

E Evaluate **14.3** Exercises   Do the exercises, and check your work.

Get Ready
Solve each system by graphing.

1) $y = \dfrac{1}{2}x + 1$
   $y = -x + 4$

2) $2x + y = 3$
   $-x + y = -3$

3) $y = -x$
   $5x + 5y = 10$

4) $-6x + 9y = -12$
   $4x - 6y = 8$

Solve using the substitution method.

5) $y = 4x - 5$
   $-2x + y = 1$

6) $-9x - y = 6$
   $6x - y = -4$

7) $2x - 9y = -28$
   $x + 6y = 7$

8) $5x - 10y = 3$
   $-x + 2y = -8$

Solve using the elimination method.

9) $7x + y = 11$
   $3x - y = 9$

10) $10x - 3y = -12$
    $-5x + 2y = 8$

11) $-6x + 4y = 5$
    $8x - 5y = -7$

12) $9y + 4 = -12x$
    $8x + 6y = -1$

Objective 1: Define a Nonlinear System of Equations

If a nonlinear system consists of equations with the following graphs,

a) sketch the different ways in which the graphs can intersect.

b) make a sketch in which the graphs do not intersect.

c) how many possible solutions can each system have?

13) circle and line

14) parabola and line

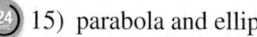 15) parabola and ellipse

16) ellipse and hyperbola

17) parabola and hyperbola

18) circle and ellipse

Mixed Exercises: Objectives 2 and 3

Solve each system using either substitution or the elimination method.

19) $x^2 + 4y = 8$
    $x + 2y = -8$

20) $x^2 + y = 1$
    $-x + y = -5$

21) $x + 2y = 5$
$x^2 + y^2 = 10$

22) $y = 2$
$x^2 + y^2 = 8$

23) $y = x^2 - 6x + 10$
$y = 2x - 6$

24) $y = x^2 - 10x + 22$
$y = 4x - 27$

25) $x^2 + 2y^2 = 11$
$x^2 - y^2 = 8$

26) $2x^2 - y^2 = 7$
$2y^2 - 3x^2 = 2$

27) $x^2 + y^2 = 6$
$2x^2 + 5y^2 = 18$

28) $5x^2 - y^2 = 16$
$x^2 + y^2 = 14$

29) $3x^2 + 4y = -1$
$x^2 + 3y = -12$

30) $2x^2 + y = 9$
$y = 3x^2 + 4$

31) $y = 6x^2 - 1$
$2x^2 + 5y = -5$

32) $x^2 + 2y = 5$
$-3x^2 + 2y = 5$

33) $x^2 + y^2 = 4$
$-2x^2 + 3y = 6$

34) $x^2 + y^2 = 49$
$x - 2y^2 = 7$

35) $x^2 + y^2 = 3$
$x + y = 4$

36) $y - x = 1$
$4y^2 - 16x^2 = 64$

37) $x = \sqrt{y}$
$x^2 - 9y^2 = 9$

38) $x = \sqrt{y}$
$x^2 - y^2 = 4$

39) $9x^2 + y^2 = 9$
$x^2 + y^2 = 5$

40) $x^2 + y^2 = 6$
$5x^2 + y^2 = 10$

41) $x^2 + y^2 = 1$
$y = x^2 + 1$

42) $y = -x^2 - 2$
$x^2 + y^2 = 4$

In Exercises 43 and 44, write a system of equations, and solve.

43) Find two numbers whose product is 40 and whose sum is 13.

44) Find two numbers whose product is 28 and whose sum is 11.

45) The perimeter of a rectangular computer screen is 38 in. Its area is 88 in^2. Find the dimensions of the screen.

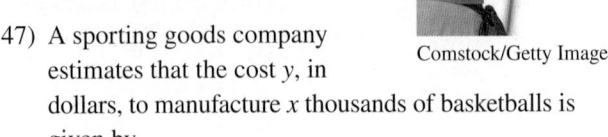

46) The area of a rectangular bulletin board is 180 in^2, and its perimeter is 54 in. Find the dimensions of the bulletin board.

47) A sporting goods company estimates that the cost y, in dollars, to manufacture x thousands of basketballs is given by

$$y = 6x^2 + 33x + 12$$

The revenue y, in dollars, from the sale of x thousands of basketballs is given by

$$y = 15x^2$$

The company breaks even on the sale of basketballs when revenue equals cost. The point, (x, y), at which this occurs is called the *break-even point*. Find the break-even point for the manufacture and sale of the basketballs.

48) A backpack manufacturer estimates that the cost y, in dollars, to make x thousands of backpacks is given by

$$y = 9x^2 + 30x + 18$$

The revenue y, in dollars, from the sale of x thousands of backpacks is given by

$$y = 21x^2$$

Find the break-even point for the manufacture and sale of the backpacks. (See Exercise 47 for an explanation.)

R Rethink

R1) Do all nonlinear systems of equations have solutions? Why or why not?

R2) Why do you need to check the solutions in the original equations?

R3) Why is it helpful to sketch a graph of the equations before solving them?

14.4 Second-Degree Inequalities and Systems of Inequalities

What are your objectives for Section 14.4?	How can you accomplish each objective?
1 Graph Second-Degree Inequalities	• Learn the definition of a *second-degree inequality*. • Know how to graph conic sections. • Recognize when to graph a conic section as a solid curve or as a dotted curve, depending on the sign of the inequality. • Review how to choose a test point to determine which region to shade. • Complete the given examples on your own. • Complete You Trys 1 and 2.
2 Graph Systems of Nonlinear Inequalities	• Understand the definition of a *solution set of a system of inequalities*. Know that the solution set is the intersection of the shaded regions. • Know how to graph the conic sections and other functions we have learned so far. • Complete the given examples on your own. • Complete You Trys 3 and 4.

W Work   **Read the explanations, follow the examples, take notes, and complete the You Trys.**

1 Graph Second-Degree Inequalities

In Section 9.3, we learned how to graph linear inequalities in two variables such as $2x + y \leq 3$. To graph this inequality, first graph the boundary line $2x + y = 3$. Then we can choose a test point on one side of the line, say $(0, 0)$. Since $(0, 0)$ satisfies the inequality $2x + y \leq 3$, we shade the side of the line containing $(0, 0)$. All points in the shaded region satisfy $2x + y \leq 3$. (If the test point had *not* satisfied the inequality, we would have shaded the other side of the line.)

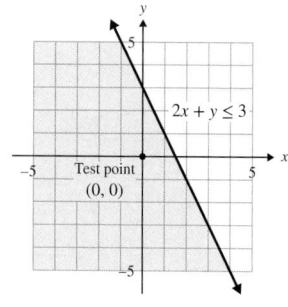

The graph of $2x + y \leq 3$ is shown at the right.

A **second-degree inequality** contains at least one squared term and no variable with degree greater than 2. We graph second-degree inequalities the same way we graph linear inequalities in two variables.

| EXAMPLE 1 | Graph $x^2 + y^2 < 25$. |

Solution

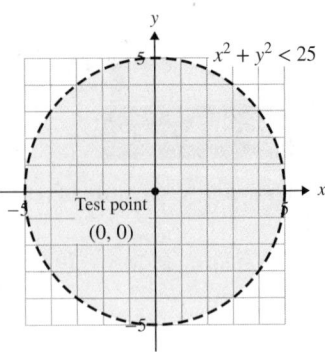

Begin by graphing the *circle*, $x^2 + y^2 = 25$, as a dotted curve since the inequality is $<$. (Points *on* the circle will not satisfy the inequality.)

Next, choose a test point not on the boundary curve: $(0, 0)$. Does the test point satisfy $x^2 + y^2 < 25$? *Yes:* $0^2 + 0^2 < 25$ is true. Shade the region inside the circle. All points in the shaded region satisfy $x^2 + y^2 < 25$.

| YOU TRY 1 | Graph $\dfrac{x^2}{25} + \dfrac{y^2}{9} < 1$. |

| EXAMPLE 2 | Graph $4x^2 - 9y^2 \geq 36$. |

Solution

First, graph the *hyperbola*, $4x^2 - 9y^2 = 36$, as a solid curve because the inequality is $\geq$.

$$4x^2 - 9y^2 = 36$$
$$\frac{4x^2}{36} - \frac{9y^2}{36} = \frac{36}{36} \qquad \text{Divide by 36.}$$
$$\frac{x^2}{9} - \frac{y^2}{4} = 1 \qquad \text{Simplify.}$$

The center is $(0, 0)$, $a = 3$, and $b = 2$.

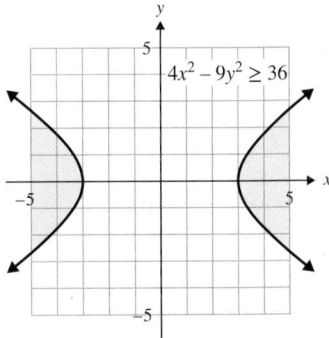

Next, choose a test point not on the boundary curve: $(0, 0)$. Does $(0, 0)$ satisfy $4x^2 - 9y^2 \geq 36$? *No:* $4(0)^2 - 9(0)^2 \geq 36$ is false.

Since $(0, 0)$ does not satisfy the inequality, we do not shade the region containing $(0, 0)$. Shade the other side of the branches of the hyperbola.

All points in the shaded region, as well as *on* the hyperbola, satisfy $4x^2 - 9y^2 \geq 36$.

| YOU TRY 2 | Graph $y \geq -x^2 + 3$. |

2 Graph Systems of Nonlinear Inequalities

The **solution set of a system of inequalities** consists of the set of points that satisfy *all* the inequalities in the system.

We first discussed this in Section 9.3 when we graphed the solution set of a system like

$$x \geq -2 \text{ and } y \leq x - 3$$

The solution set of such a system of linear inequalities is the intersection of their graphs. The solution set of a system of nonlinear inequalities is also the intersection of the graphs of the individual inequalities.

EXAMPLE 3

Graph the solution set of the system.

$$4x^2 + y^2 < 16$$
$$-x + 2y > 2$$

Solution

First, graph the *ellipse*, $4x^2 + y^2 = 16$, as a dotted curve because the inequality is $<$.

$$4x^2 + y^2 = 16$$
$$\frac{4x^2}{16} + \frac{y^2}{16} = \frac{16}{16} \qquad \text{Divide by 16.}$$
$$\frac{x^2}{4} + \frac{y^2}{16} = 1 \qquad \text{Simplify.}$$

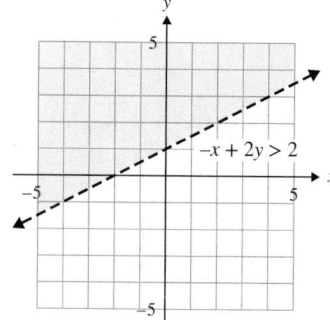

The test point $(0, 0)$ satisfies the inequality $4x^2 + y^2 < 16$, so shade inside the dotted curve of the ellipse, as shown to the right.

Graph the *line*, $-x + 2y = 2$, as a dotted line since the inequality is $>$.

$$-x + 2y > 2$$
$$2y > x + 2$$
$$y > \frac{1}{2}x + 1 \qquad \text{Solve for } y.$$

Shade above the line since the test point $(0, 0)$ does *not* satisfy $y > \frac{1}{2}x + 1$.

W Hint

Notice the similarities between this system and linear systems of inequalities.

The solution set of the system is the *intersection* of the two graphs. The shaded regions overlap as shown to the right. This is the solution set of the system.

All points in the shaded region satisfy *both* inequalities.

$$4x^2 + y^2 < 16$$
$$-x + 2y > 2$$

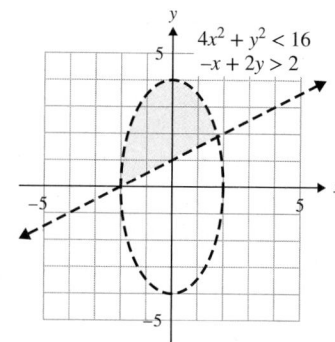

[YOU TRY 3]

Graph the solution set of the system.

$$y < x - 4$$
$$x^2 + y > 0$$

EXAMPLE 4 Graph the solution set of the system.

$$x \geq 0$$
$$y - x^2 \leq 1$$
$$x^2 + y^2 \leq 4$$

Solution

First, the graph of $x = 0$ is the y-axis. Therefore, the graph of $x \geq 0$ consists of quadrants I and IV. See the figure below left.

Graph the *parabola*, $y - x^2 = 1$, as a solid curve since the inequality is $\leq$. Rewrite the inequality as $y \leq x^2 + 1$ to determine that the vertex is $(0, 1)$. The test point $(0, 0)$ satisfies the inequality $y \leq x^2 + 1$, so shade outside of the parabola. See the figure below center.

The graph of $x^2 + y^2 \leq 4$ is the inside of the circle $x^2 + y^2 = 4$. See the figure below right.

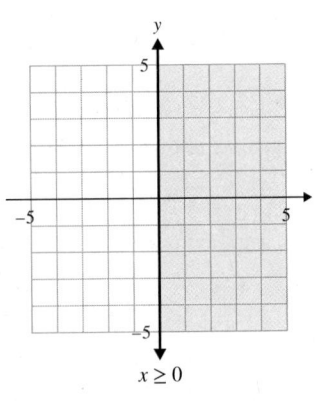

$x \geq 0$

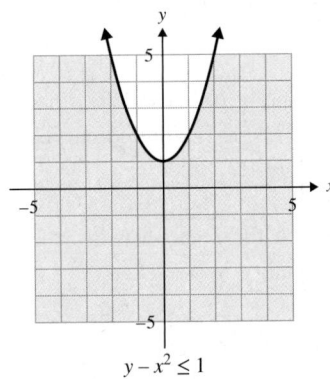

$y - x^2 \leq 1$

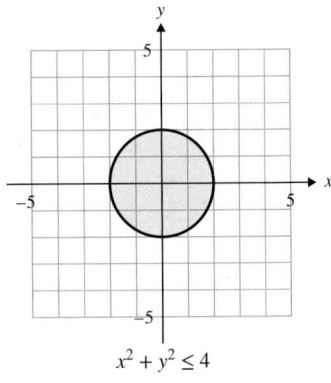

$x^2 + y^2 \leq 4$

Finally, the graph of the solution set of the system is the intersection, or overlap, of these three regions, as shown to the right.

All points in the shaded region satisfy each of the inequalities in the system

$$x \geq 0$$
$$y - x^2 \leq 1$$
$$x^2 + y^2 \leq 4$$

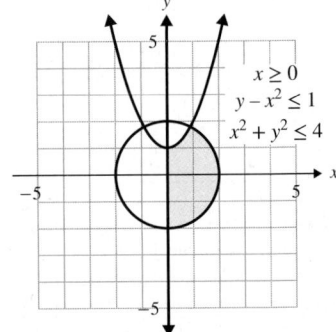

$x \geq 0$
$y - x^2 \leq 1$
$x^2 + y^2 \leq 4$

YOU TRY 4 Graph the solution set of the system.

$$y \geq 0$$
$$x^2 + y^2 \leq 9$$
$$y - x^2 \geq -2$$

1)
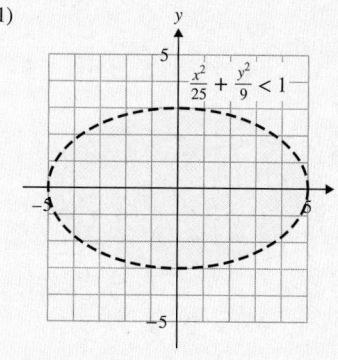

2)
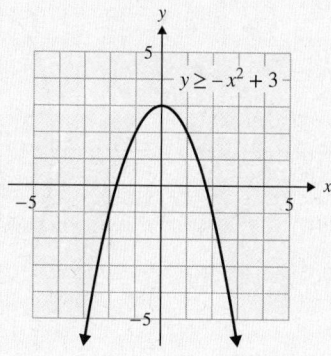

3)
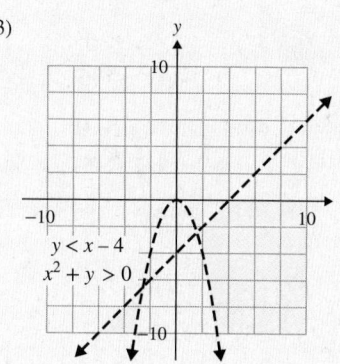

4)
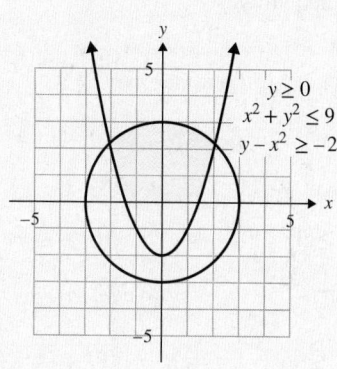

E Evaluate **14.4** Exercises   Do the exercises, and check your work.

Get Ready

Graph each inequality.

1) $y > -\dfrac{1}{3}x + 1$

2) $2x + 5y < 15$

3) $4x - 2y \geq 6$

4) $x \geq -2$

Graph each compound inequality.

5) $x \leq 3$ and $y \geq -\dfrac{4}{3}x + 2$

6) $y > -4$ and $y > 3x - 1$

7) $5x - 2y < 8$ and $2x + y < 2$

8) $x \geq 1$ and $y \leq -2$

Objective 1: Graph Second-Degree Inequalities

The graphs of second-degree inequalities are given below. For each, find three points which satisfy the inequality and three points that are not in the solution set.

 9) $x^2 + y^2 \geq 36$

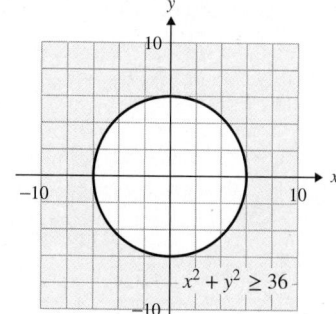

10) $x > y^2 + 2$

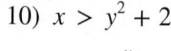

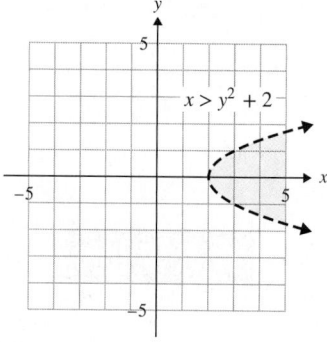

11) $4y^2 - x^2 < 4$

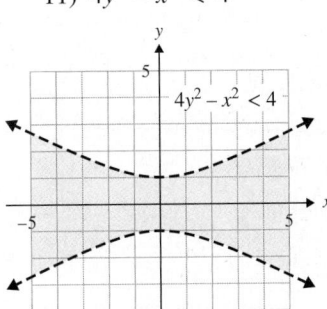

12) $25x^2 + 4y^2 \le 100$

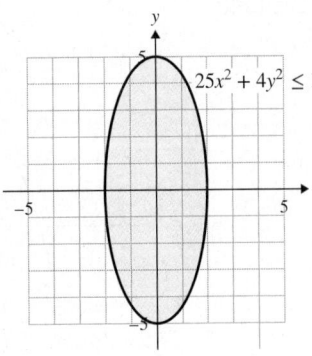

Graph each inequality.

13) $\dfrac{x^2}{9} + \dfrac{y^2}{16} < 1$

14) $y \ge (x - 2)^2 - 1$

15) $y < (x + 1)^2 + 3$

16) $x^2 + y^2 < 16$

17) $25y^2 - 4x^2 \ge 100$

18) $x^2 - 4y^2 \le 4$

19) $x^2 + y^2 \ge 3$

20) $y \ge -x^2 + 5$

21) $x > y^2 - 2$

22) $x^2 + 4y^2 \le 4$

23) $\dfrac{x^2}{4} - \dfrac{y^2}{9} \le 1$

24) $x \ge -y^2 - 2y + 1$

25) $x^2 + (y + 4)^2 < 9$

26) $y^2 - \dfrac{x^2}{9} > 1$

27) $x^2 + 9y^2 \ge 9$

28) $(x + 3)^2 + y^2 \ge 4$

29) $y \le -x^2 - 2x + 3$

30) $\dfrac{x^2}{36} + \dfrac{y^2}{25} > 1$

Objective 2: Graph Systems of Nonlinear Inequalities

Graph the solution set of each system.

31) $\begin{aligned} y &\ge x - 2 \\ x^2 + y &\le 1 \end{aligned}$

32) $\begin{aligned} y + x &< 3 \\ x^2 + y^2 &< 9 \end{aligned}$

33) $\begin{aligned} 2y - x &> 4 \\ 4x^2 + 9y^2 &> 36 \end{aligned}$

34) $\begin{aligned} y &< x - 3 \\ y - x^2 &< -5 \end{aligned}$

35) $\begin{aligned} x^2 + y^2 &\ge 16 \\ 25x^2 - 4y^2 &\le 100 \end{aligned}$

36) $\begin{aligned} x^2 - y^2 &\le 1 \\ 4x^2 + 9y^2 &\le 36 \end{aligned}$

37) $\begin{aligned} x^2 + y^2 &\le 9 \\ 9x^2 + 4y^2 &\ge 36 \end{aligned}$

38) $\begin{aligned} x^2 + y^2 &\le 16 \\ y^2 - 4x^2 &\ge 4 \end{aligned}$

39) $\begin{aligned} y^2 - x^2 &< 1 \\ 4x^2 + y^2 &> 16 \end{aligned}$

40) $\begin{aligned} x - y^2 &> -4 \\ 2y - 3x &> 4 \end{aligned}$

41) $\begin{aligned} x + y^2 &< 0 \\ x^2 + y^2 &< 16 \end{aligned}$

42) $\begin{aligned} x^2 + y^2 &\ge 4 \\ x + y &\ge 1 \end{aligned}$

43) $\begin{aligned} \dfrac{x^2}{16} + \dfrac{y^2}{9} &\le 1 \\ 4x^2 - y^2 &\ge 16 \end{aligned}$

44) $\begin{aligned} \dfrac{x^2}{25} + \dfrac{y^2}{9} &> 1 \\ y &> x^2 \end{aligned}$

45) $\begin{aligned} y &\le -x \\ y - 2x^2 &\le -2 \end{aligned}$

46) $\begin{aligned} x^2 + y^2 &\ge 1 \\ x^2 + 25y^2 &\le 25 \end{aligned}$

47) $\begin{aligned} y &\le 0 \\ x^2 + 4y^2 &\le 36 \end{aligned}$

48) $\begin{aligned} x &\ge 0 \\ x^2 + y^2 &\le 25 \end{aligned}$

49) $\begin{aligned} x &> 0 \\ y^2 - 4x^2 &< 4 \end{aligned}$

50) $\begin{aligned} y &< 0 \\ 4x^2 - 9y^2 &< 36 \end{aligned}$

51) $\begin{aligned} x &\ge 0 \\ y &\ge x^2 + 4 \\ x + 2y &\le 12 \end{aligned}$

52) $\begin{aligned} y &\ge 0 \\ 4x^2 + 25y^2 &\le 100 \\ 2x + 5y &\le 10 \end{aligned}$

53) $\begin{aligned} y &< 0 \\ x^2 + y^2 &< 9 \\ y &> x + 1 \end{aligned}$

54) $\begin{aligned} y &< 0 \\ y &< -x^2 + 4 \\ y &< \dfrac{1}{2}x - 1 \end{aligned}$

55) $\begin{aligned} y &\ge 0 \\ y &\le x^2 \\ \dfrac{x^2}{4} + \dfrac{y^2}{9} &\le 1 \end{aligned}$

56) $\begin{aligned} x &\ge 0 \\ y &\le x^2 - 1 \\ x^2 + y^2 &\le 16 \end{aligned}$

57) $\begin{aligned} y &< 0 \\ 4x^2 + 9y^2 &< 36 \\ x^2 - y^2 &> 1 \end{aligned}$

58) $\begin{aligned} x &< 0 \\ x^2 + y^2 &< 9 \\ y &< -x^2 \end{aligned}$

59) $\begin{aligned} x &\ge 0 \\ y &\ge 0 \\ x^2 + y^2 &\ge 4 \\ x &\ge y^2 \end{aligned}$

60) $\begin{aligned} x &\ge 0 \\ y &\ge 0 \\ y &\ge x^2 \\ x^2 + 4y^2 &\ge 16 \end{aligned}$

R Rethink

R1) When do you use a dotted line to graph, and when do you use a solid line to graph?

R2) What do the solutions of a nonlinear inequality represent?

Group Activity – The Ellipse and the Hyperbola

Recall that an ellipse is the set of all points in the plane, the *sum* of whose distances from two fixed points (the foci) is constant. The equation of an ellipse is given by $\dfrac{x^2}{a^2} + \dfrac{y^2}{b^2} = 1$.

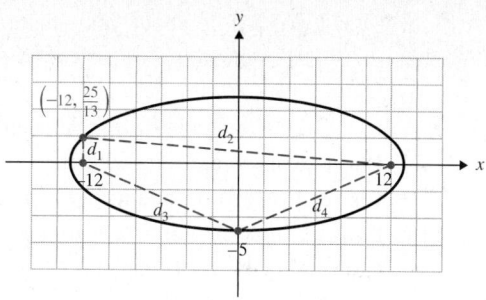

1) Verify the definition for the ellipse given above by finding $d_1 + d_2$ and $d_3 + d_4$.
2) What is the sum in each case?

3) The equation of the ellipse is $\dfrac{x^2}{13^2} + \dfrac{y^2}{5^2} = 1$. Define a and b. How is the sum found in Problem 2 related to a?

4) How can you find the x-values of the foci using a and b?

Recall that a hyperbola is the set of all points in the plane, the difference of whose distances from two fixed points (the foci) is constant. The equation of a hyperbola is given by $\dfrac{x^2}{a^2} - \dfrac{y^2}{b^2} = 1$.

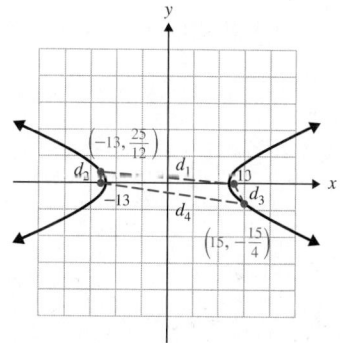

5) Verify the definition for the hyperbola given above by finding $|d_1 - d_2|$ and $|d_3 - d_4|$.
6) What is the difference in each case?

7) The equation of the hyperbola shown above is $\dfrac{x^2}{12^2} - \dfrac{y^2}{5^2} = 1$. Define a. How is the difference found in Problem 6 related to a?

8) How can you find the x-values of the foci using a and b?

9) Use the figure at the right to find

    a) the equations of the asymptotes of the hyperbola.

    b) the equation of the hyperbola.

    c) the equation of the circle passing through the foci and corners of the construction rectangle.

    d) the equation of the ellipse passing through the foci of the hyperbola.

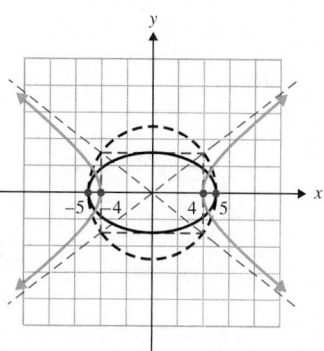

em **POWER** me How Is My Memory?

We've all heard people say, "I have a terrible memory!" or "I remember everything!" Do either of these statements apply to you, or are you somewhere in between? Take this survey to learn a little about your memory. Check all boxes that apply and fill in the blanks, where applicable.

☐ I feel like I remember *everything* in most areas of my life.

☐ I have no problem remembering things that are very important to me.

☐ I have a very good memory when it comes to learning all subjects in school.

☐ I have trouble remembering things that I learn in most of my classes.

☐ I remember most of what I learn in school *except* in these classes:_____.

☐ I remember best while listening to someone talk.

☐ I remember best while watching my instructor write information on the board.

☐ I remember best when I write down what my instructor says or writes on the board.

☐ I tend to forget things when I am rushed or under stress.

☐ I use memory devices to help me remember important information.

☐ I feel that I am aware of how I best remember things and use those methods, consistently, for learning.

Think about the items that you have, and have *not*, checked in this survey. In the Study Strategies at the beginning of this chapter, you will find some strategies for improving your memory.

Chapter 14: Summary

Definition/Procedure	Example

14.1 The Circle

The Midpoint Formula

If (x_1, y_1) and (x_2, y_2) are the endpoints of a line segment, then the **midpoint** of the segment has coordinates

$$\left(\frac{x_1 + x_2}{2}, \frac{y_1 + y_2}{2}\right)$$

Find the midpoint of the line segment with endpoints $(-2, 5)$ and $(6, 3)$.

$$\text{Midpoint} = \left(\frac{-2 + 6}{2}, \frac{5 + 3}{2}\right) = \left(\frac{4}{2}, \frac{8}{2}\right) = (2, 4)$$

Parabolas, circles, ellipses, and hyperbolas are called **conic sections.**

The **standard form for the equation of a circle** with center (h, k) and radius r is

$$(x - h)^2 + (y - k)^2 = r^2$$

Graph $(x + 3)^2 + y^2 = 4$.

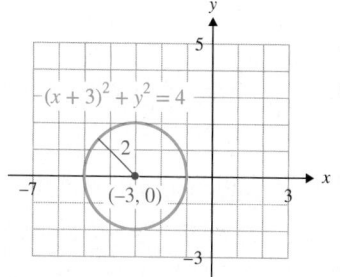

The center is $(-3, 0)$. The radius is $\sqrt{4} = 2$.

The **general form for the equation of a circle** is

$$Ax^2 + Ay^2 + Cx + Dy + E = 0$$

where A, C, D, and E are real numbers.

To rewrite the equation in the form $(x - h)^2 + (y - k)^2 = r^2$, divide the equation by A so that the coefficient of each squared term is 1, then complete the square on x and on y to put it into standard form.

Write $x^2 + y^2 - 16x + 4y + 67 = 0$ in the form $(x - h)^2 + (y - k)^2 = r^2$.

Group the x-terms together, and group the y-terms together.

$$(x^2 - 16x) + (y^2 + 4y) = -67$$

Complete the square for each group of terms.

$$(x^2 - 16x + 64) + (y^2 + 4y + 4) = -67 + 64 + 4$$
$$(x - 8)^2 + (y + 2)^2 = 1$$

14.2 The Ellipse and the Hyperbola

The **standard form for the equation of an ellipse** is

$$\frac{(x - h)^2}{a^2} + \frac{(y - k)^2}{b^2} = 1$$

The center of the ellipse is (h, k).

Graph $\dfrac{(x - 1)^2}{9} + \dfrac{(y - 2)^2}{4} = 1$.

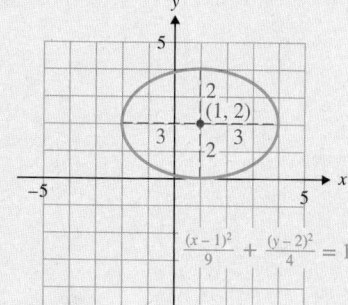

The center is $(1, 2)$.
$a = \sqrt{9} = 3$
$b = \sqrt{4} = 2$

Definition/Procedure	Example

Standard Form for the Equation of a Hyperbola

1) A hyperbola with center (h, k) with branches that open in the *x-direction* has equation

$$\frac{(x-h)^2}{a^2} - \frac{(y-k)^2}{b^2} = 1$$

2) A hyperbola with center (h, k) with branches that open in the *y-direction* has equation

$$\frac{(y-k)^2}{b^2} - \frac{(x-h)^2}{a^2} = 1$$

Notice in 1) that $\dfrac{(x-h)^2}{a^2}$ is the positive quantity, and the branches open in the *x*-direction.

In 2), the positive quantity is $\dfrac{(y-k)^2}{b^2}$, and the branches open in the *y*-direction.

Graph $\dfrac{(y-1)^2}{9} - \dfrac{(x-4)^2}{4} = 1$.

The center is $(4, 1)$, $a = \sqrt{4} = 2$, and $b = \sqrt{9} = 3$.

Use the center, $a = 2$, and $b = 3$ to draw the reference rectangle. The diagonals of the rectangle are the asymptotes of the hyperbola. The vertices of the hyperbola are $(4, 4)$ and $(4, -2)$.

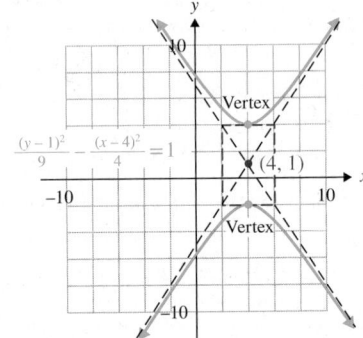

A Nonstandard Form for the Equation of a Hyperbola

The graph of the equation $xy = c$, where c is a nonzero constant, is a hyperbola whose asymptotes are the *x*- and *y*-axes.

Graph $xy = 4$.

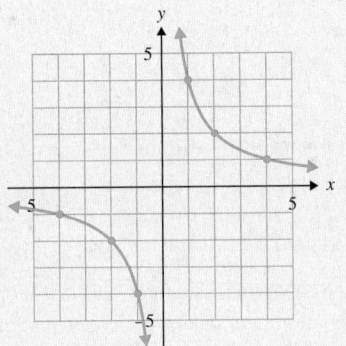

14.3 Nonlinear Systems of Equations

A **nonlinear system of equations** is a system in which at least one of the equations is not linear. We can solve nonlinear systems by substitution or the elimination method.

Solve
$$x - y^2 = 3 \qquad (1)$$
$$x - 2y = 6 \qquad (2)$$

$$x - y^2 = 3 \qquad (1) \quad \text{Solve equation (1) for } x.$$
$$x = y^2 + 3 \qquad (3)$$

Substitute $x = y^2 + 3$ into equation (2).

$$(y^2 + 3) - 2y = 6$$
$$y^2 - 2y - 3 = 0 \qquad \text{Subtract 6.}$$
$$(y - 3)(y + 1) = 0 \qquad \text{Factor.}$$
$$y - 3 = 0 \quad \text{or} \quad y + 1 = 0 \qquad \text{Set each factor equal to 0.}$$
$$y = 3 \quad \text{or} \quad y = -1 \qquad \text{Solve.}$$

Substitute each value into equation (3).

$y = 3$: $x = y^2 + 3$ $y = -1$: $x = y^2 + 3$
         $x = (3)^2 + 3$ $x = (-1)^2 + 3$
         $x = 12$ $x = 4$

The proposed solutions are $(12, 3)$ and $(4, -1)$. Verify that they also satisfy (2).
The solution set is $\{(12, 3), (4, -1)\}$.

Definition/Procedure	Example

14.4 Second-Degree Inequalities and Systems of Inequalities

To graph a **second-degree inequality,** graph the boundary curve, choose a test point, and shade the appropriate region.

Graph $x^2 + y^2 < 9$.

Graph the *circle* $x^2 + y^2 = 9$ as a dotted curve since the inequality is $<$. Choose a test point not on the boundary curve: $(0, 0)$. Since the test point $(0, 0)$ satisfies the inequality, shade the region inside the circle.

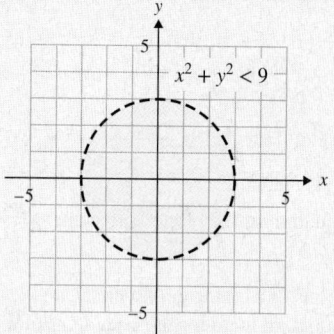

All points in the shaded region satisfy $x^2 + y^2 < 9$.

The **solution set of a system of inequalities** consists of the set of points that satisfy all inequalities in the system. The solution set is the intersection of the graphs of the individual inequalities.

Graph the solution set of the system

$$x^2 + y \leq 3$$
$$x + y \geq 1$$

First, graph the *parabola* $x^2 + y = 3$ as a solid curve since the inequality is $\leq$. The test point $(0, 0)$ satisfies the inequality $x^2 + y \leq 3$, so shade inside the parabola.

Graph the *line* $x + y = 1$ as a solid line since the inequality is $\geq$. Shade above the line since the test point $(0, 0)$ does not satisfy $x + y \geq 1$.

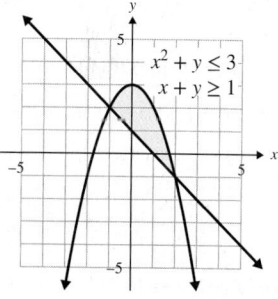

The solution set of the system is the *intersection* of the shaded regions of the two graphs.

All points in the shaded region satisfy both inequalities.

Chapter 14: Review Exercises

(14.1) Find the midpoint of the line segment with the given endpoints.

1) $(3, 8)$ and $(5, 2)$

2) $(-6, 1)$ and $(-2, -1)$

3) $(7, -3)$ and $(6, -4)$

4) $\left(\dfrac{2}{3}, \dfrac{1}{4}\right)$ and $\left(-\dfrac{1}{6}, \dfrac{5}{8}\right)$

Identify the center and radius of each circle and graph.

5) $(x + 3)^2 + (y - 5)^2 = 36$

6) $x^2 + (y + 4)^2 = 9$

7) $x^2 + y^2 - 10x - 4y + 13 = 0$

8) $x^2 + y^2 + 4x + 16y + 52 = 0$

9) Find an equation of the circle with center $(1, -5)$ and radius $= \sqrt{7}$.

10) The endpoints of a diameter of a circle are $(4, 3)$ and $(-1, -5)$. Find the standard form for the equation of this circle.

(14.2)

11) When is an ellipse also a circle?

Identify the center of the ellipse, and graph the equation.

12) $\dfrac{x^2}{25} + \dfrac{y^2}{36} = 1$

13) $\dfrac{(x + 3)^2}{9} + \dfrac{(y - 3)^2}{4} = 1$

14) $(x - 4)^2 + \dfrac{(y - 2)^2}{16} = 1$

15) $25x^2 + 4y^2 = 100$

Write each equation in standard form, and graph.

16) $4x^2 + 9y^2 + 8x + 36y + 4 = 0$

17) $25x^2 + 4y^2 - 100x = 0$

18) How can you distinguish between the equation of an ellipse and the equation of a hyperbola?

For Exercises 19–22, identify the center of the hyperbola, and graph the equation. Also, identify the vertices.

19) $\dfrac{y^2}{9} - \dfrac{x^2}{25} = 1$

20) $\dfrac{(y - 3)^2}{4} - \dfrac{(x + 2)^2}{9} = 1$

21) $\dfrac{(x + 1)^2}{4} - \dfrac{(y + 2)^2}{4} = 1$

22) $16x^2 - y^2 = 16$

23) Graph $xy = 6$.

24) Write the equations of the asymptotes of the graph in Exercise 19.

Write each equation in standard form, and graph.

25) $16y^2 - x^2 + 2x + 96y + 127 = 0$

26) $x^2 - y^2 - 4x + 6y - 9 = 0$

Graph each function. Identify the domain and range.

27) $h(x) = 2\sqrt{1 - \dfrac{x^2}{9}}$

28) $f(x) = -\sqrt{4 - x^2}$

(14.1–14.2) Determine whether the graph of each equation is a parabola, circle, ellipse, or hyperbola, then graph each equation.

29) $x^2 + 9y^2 = 9$

30) $x^2 + y^2 = 25$

31) $x = -y^2 + 6y - 5$

32) $x^2 - y = 3$

33) $\dfrac{(x - 3)^2}{16} - \dfrac{(y - 4)^2}{25} = 1$

34) $\dfrac{(x + 3)^2}{16} + \dfrac{(y + 1)^2}{25} = 1$

35) $x^2 + y^2 - 2x + 2y - 2 = 0$

36) $4y^2 - 9x^2 = 36$

37) $y = \dfrac{1}{2}(x + 2)^2 + 1$

38) $x^2 + y^2 - 6x - 8y + 16 = 0$

(14.3)

39) If a nonlinear system of equations consists of an ellipse and a hyperbola, how many possible solutions can the system have?

40) If a nonlinear system of equations consists of a line and a circle, how many possible solutions can the system have?

Solve each system.

41) $-4x^2 + 3y^2 = 3$
$7x^2 - 5y^2 = 7$

42) $y - x^2 = 7$
$3x^2 + 4y = 28$

43) $y = 3 - x^2$
$x - y = -1$

44) $x^2 + y^2 = 9$
$8x + y^2 = 21$

45) $4x^2 + 9y^2 = 36$
$y = \dfrac{1}{3}x - 5$

46) $4x + 3y = 0$
$4x^2 + 4y^2 = 25$

Write a system of equations, and solve.

47) Find two numbers whose product is 36 and whose sum is 13.

48) The perimeter of a rectangular window is 78 in., and its area is 378 in^2. Find the dimensions of the window.

(14.4) Graph each inequality.

49) $x^2 + y^2 \leq 4$

50) $y > -(x - 3)^2 + 2$

51) $\dfrac{x^2}{9} + \dfrac{y^2}{4} > 1$

52) $\dfrac{x^2}{9} - \dfrac{y^2}{4} \geq 1$

53) $x < -y^2 + 2y - 12$

54) $4x^2 + 25y^2 < 100$

Graph the solution set of each system.

55) $y - x^2 > 2$
$x + y > 5$

56) $x^2 + y^2 \le 25$
$y - x \le 2$

57) $\dfrac{x^2}{16} + \dfrac{y^2}{25} < 1$
$y + 3 > x^2$

58) $x^2 - y^2 \le 4$
$x^2 + y^2 \ge 16$

59) $y \ge 0$
$x^2 + y^2 \le 36$
$x + y^2 \le 0$

60) $x \le 0$
$4x^2 + 25y^2 \ge 100$
$-x + y \le 1$

Chapter 14: Test

1) Find the midpoint of the line segment with endpoints (2, 1) and (10, −6).

Determine whether each statement is true or false.

2) The equation of an ellipse represents a function.

3) A circle is a special type of ellipse.

Determine whether the graph of each equation is a parabola, circle, ellipse, or hyperbola, then graph each equation.

4) $\dfrac{(x-2)^2}{25} + \dfrac{(y+3)^2}{4} = 1$

5) $y = -2x^2 + 6$

6) $y^2 - 4x^2 = 16$

7) $x^2 + (y-1)^2 = 9$

8) $xy = 2$

9) Write $x^2 + y^2 + 2x - 6y - 6 = 0$ in the form $(x-h)^2 + (y-k)^2 = r^2$. Identify the center and radius, and graph the equation.

10) Write an equation of the circle with center (5, 2) and radius $\sqrt{11}$.

11) The Colosseum in Rome is an ellipse measuring 188 m long and 156 m wide. If the Colosseum is represented on a Cartesian coordinate system with the center of the ellipse at the origin and the longer axis along the x-axis, write an equation of this elliptical structure.
(www.romaviva.com/Colosseo/colosseum.htm)

12) Graph $f(x) = -\sqrt{25 - x^2}$. State the domain and range.

13) Suppose a nonlinear system consists of the equation of a parabola and a circle.

a) Sketch the different ways in which the graphs can intersect.

b) Make a sketch in which the graphs do not intersect.

c) How many possible solutions can the system have?

Solve each system.

14) $x - 2y^2 = -1$
$x + 4y = -1$

15) $2x^2 + 3y^2 = 21$
$-x^2 + 12y^2 = 3$

16) The perimeter of a rectangular picture frame is 44 in. The area is 112 in². Find the dimensions of the frame.

Graph each inequality.

17) $y \ge x^2 - 2$

18) $x^2 - 4y^2 < 36$

Graph the solution set of each system.

19) $x^2 + y^2 > 9$
$y < 2x - 1$

20) $x \ge 0$
$y \le x^2$
$4x^2 + 9y^2 \le 36$

Chapter 14: Cumulative Review for Chapters 1–14

Perform the indicated operations, and simplify.

1) $\dfrac{1}{6} - \dfrac{11}{12}$

2) $16 + 20 \div 4 - (5-2)^2$

Find the area and perimeter of each figure.

3)
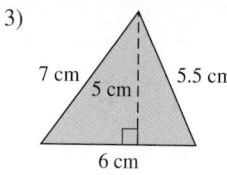

4)
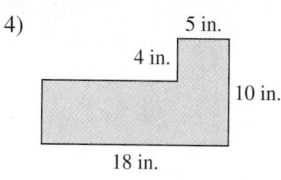

Evaluate.

5) $(-1)^5$

6) 2^4

7) Simplify $\left(\dfrac{2a^8 b}{a^2 b^{-4}}\right)^{-3}$.

8) Solve $\dfrac{3}{8}k + 11 = -4$.

9) Solve for n. $an + z = c$

10) Solve $8 - 5p \le 28$.

11) *Write an equation, and solve.*
The sum of three consecutive odd integers is 13 more than twice the largest integer. Find the numbers.

12) Find the slope of the line containing the points (−6, 4) and (2, −4).

13) What is the slope of the line with equation $y = 3$?

14) Graph $y = -2x + 5$.

15) Write the slope-intercept form of the line containing the points $(-4, 7)$ and $(4, 1)$.

16) Solve the system $3x + 4y = 3$
$5x + 6y = 4$

17) *Write a system of equations, and solve.*
How many milliliters of an 8% alcohol solution and how many milliliters of a 16% alcohol solution must be mixed to make 20 mL of a 14% alcohol solution?

18) Subtract $5p^2 - 8p + 4$ from $2p^2 - p + 10$.

19) Multiply and simplify. $(4w - 3)(2w^2 + 9w - 5)$

20) Divide. $(x^3 - 7x - 36) \div (x - 4)$

Factor completely.

21) $6c^2 - 14c + 8$

22) $m^3 - 8$

23) Solve $(x + 1)(x + 2) = 2(x + 7) + 5x$.

24) Multiply and simplify. $\dfrac{a^2 + 3a - 54}{4a + 36} \cdot \dfrac{10}{36 - a^2}$

25) Simplify $\dfrac{\dfrac{t^2 - 9}{4}}{\dfrac{t - 3}{24}}$.

26) Solve $|3n + 11| = 7$.

27) Solve $|5r + 3| > 12$.

Simplify. Assume all variables represent nonnegative real numbers.

28) $\sqrt{75}$

29) $\sqrt[3]{48}$

30) $\sqrt[3]{27a^5b^{13}}$

31) $(16)^{-3/4}$

32) $\dfrac{18}{\sqrt{12}}$

33) Rationalize the denominator of $\dfrac{5}{\sqrt{3} + 4}$.

Solve.

34) $(2p - 1)^2 + 16 = 0$

35) $y^2 = -7y - 3$

36) Solve $\dfrac{t - 3}{2t + 5} > 0$.

37) Given the relation $\{(4, 0), (3, 1), (3, -1), (0, 2)\}$,

a) what is the domain?

b) what is the range?

c) is the relation a function?

38) Graph $f(x) = \sqrt{x}$ and $g(x) = \sqrt{x + 3}$ on the same axes.

39) Graph $x = -y^2 + 2y + 2$. Identify the vertex, the x- and y-intercepts, and determine the domain and range.

40) $f(x) = 3x - 7$ and $g(x) = 2x + 5$.

a) Find $g(-4)$.

b) Find $(g \circ f)(x)$.

c) Find x so that $f(x) = 11$.

41) Given the function $f = \{(-5, 9), (-2, 11), (3, 14), (7, 9)\}$,

a) is f one-to-one?

b) does f have an inverse?

42) Find an equation of the inverse of $f(x) = \dfrac{1}{3}x + 4$.

Solve the equations in Exercises 43 and 44.

43) $8^{5t} = 4^{t-3}$

44) $\log_3 (4n - 11) = 2$

45) Evaluate log 100.

46) Graph $f(x) = \log_2 x$.

47) Solve $e^{3k} = 8$. Give an exact solution and an approximation to four decimal places.

48) Graph $\dfrac{y^2}{4} - \dfrac{x^2}{9} = 1$.

49) Graph $x^2 + y^2 - 2x + 6y - 6 = 0$.

50) Solve the system. $y - 5x^2 = 3$
$x^2 + 2y = 6$

Answers to Selected Exercises

Chapter 1
Section 1.1

1) a) $\dfrac{2}{5}$ b) $\dfrac{2}{3}$ c) 1   3) $\dfrac{1}{2}$ 5) a) 1, 2, 3, 6, 9, 18

  b) 1, 2, 4, 5, 8, 10, 20, 40 c) 1, 23

7) a) composite b) composite c) prime

9) Composite. It is divisible by 2 and has other factors as well.

11) a) $2 \cdot 3 \cdot 3$ b) $2 \cdot 3 \cdot 3 \cdot 3$ c) $2 \cdot 3 \cdot 7$ d) $2 \cdot 3 \cdot 5 \cdot 5$

13) a) $\dfrac{3}{4}$ b) $\dfrac{3}{4}$ c) $\dfrac{12}{5}$ or $2\dfrac{2}{5}$ d) $\dfrac{3}{7}$

15) a) $\dfrac{6}{35}$ b) $\dfrac{10}{39}$ c) $\dfrac{7}{15}$ d) $\dfrac{12}{25}$ e) $\dfrac{1}{2}$ f) $\dfrac{7}{4}$ or $1\dfrac{3}{4}$

17) She multiplied the whole numbers and multiplied the fractions. She should have converted the mixed numbers to improper fractions before multiplying. Correct answer: $\dfrac{77}{6}$ or $12\dfrac{5}{6}$.

19) a) $\dfrac{1}{12}$ b) $\dfrac{15}{44}$ c) $\dfrac{4}{7}$ d) 7 e) $\dfrac{24}{7}$ or $3\dfrac{3}{7}$ f) $\dfrac{1}{14}$

21) 30   23) a) 30 b) 24 c) 36

25) a) $\dfrac{8}{11}$ b) $\dfrac{3}{5}$ c) $\dfrac{3}{5}$ d) $\dfrac{7}{18}$ e) $\dfrac{29}{30}$ f) $\dfrac{1}{18}$

  g) $\dfrac{71}{63}$ or $1\dfrac{8}{63}$ h) $\dfrac{7}{12}$ i) $\dfrac{7}{5}$ or $1\dfrac{2}{5}$ j) $\dfrac{41}{54}$

27) a) $14\dfrac{7}{11}$ b) $11\dfrac{2}{5}$ c) $6\dfrac{1}{2}$ d) $5\dfrac{9}{20}$ e) $1\dfrac{2}{5}$ f) $4\dfrac{13}{40}$

  g) $11\dfrac{5}{28}$ h) $8\dfrac{9}{20}$

29) Four bears: $\dfrac{1}{3}$ yd remaining   31) 128

33) $16\dfrac{1}{2}$ in. by $22\dfrac{5}{8}$ in.   35) $3\dfrac{5}{12}$ cups   37) $5\dfrac{3}{20}$ gal

39) 35   41) $7\dfrac{23}{24}$ in.

Section 1.2

1) a) base: 6; exponent: 4 b) base: 2; exponent: 3

  c) base: $\dfrac{9}{8}$; exponent: 5

3) a) 9^4 b) 2^8 c) $\left(\dfrac{1}{4}\right)^3$

5) a) 64 b) 121 c) 16 d) 125 e) 81 f) 144 g) 1

  h) $\dfrac{9}{100}$ i) $\dfrac{1}{64}$ j) 0.09   7) 0.25   9) always

11) never   13) Answers may vary.   15) 19   17) 38

19) 23   21) 17   23) $\dfrac{13}{20}$ 25) $\dfrac{19}{18}$ or $1\dfrac{1}{18}$

27) 4   29) 15   31) 19   33) 37   35) 11

37) $\dfrac{5}{6}$ 39) $\dfrac{27}{7}$ or $3\dfrac{6}{7}$

Section 1.3

1) acute   3) straight   5) supplementary; complementary

7) 31°   9) 78°   11) 37°   13) 142°

15) $m\angle A = m\angle C = 149°$; $m\angle B = 31°$

17) 180   19) 39°; obtuse   21) 39°; right

23) equilateral   25) isosceles   27) true

29) $A = 80$ ft^2; $P = 36$ ft   31) $A = 42$ cm^2; $P = 29.25$ cm

33) $A = 42.25$ mi^2; $P = 26$ mi   35) $A = 162$ in^2; $P = 52$ in.

37) a) $A = 25\pi$ in^2; $A \approx 78.5$ in^2 b) $C = 10\pi$ in.; $C \approx 31.4$ in.

39) a) $A = 6.25\pi$ m^2; $A \approx 19.625$ m^2 b) $C = 5\pi$ m; $C \approx 15.7$ m

41) $A = \dfrac{1}{4}\pi$ m^2; $C = \pi$ m   43) $A = 49\pi$ ft^2; $C = 14\pi$ ft

45) $A = 376$ m^2; $P = 86$ m   47) $A = 201.16$ in^2; $P = 67.4$ in.

49) 88 in^2 51) 25.75 ft^2 53) 177.5 cm^2 55) 70 m^3

57) 288π in^3 59) $\dfrac{500}{3}\pi$ ft^3 61) 136π cm^3

63) a) 58.5 ft^2 b) No, it would cost $1170 to use this glass.

65) a) 226.08 ft^3 b) 1691 gal

67) a) 62.8 in. b) 314 in^2 69) 6395.4 gal

71) No. This granite countertop would cost $2970.00.

73) a) 140.4 in^2 b) 54 in.   75) a) 44 ft^2 b) $752

77) 1205.76 ft^3

Section 1.4

1) Answers may vary.

3) a) 17 b) 17, 0 c) 17, 0, -25 d) 17, 3.8, $\dfrac{4}{5}$, 0, -25, $6.\overline{7}$, $-2\dfrac{1}{8}$ e) $\sqrt{10}$, 9.721983... f) all numbers in the set

5) always   7) sometimes   9) never

11)

13)

15) >   17) <   19) >   21) >   23) >   25) $11 > 6$

27) $-8 < 4$ 29) the distance of the number from zero

31) -8 33) 15   35) $\dfrac{3}{4}$ 37) 10   39) $\dfrac{9}{4}$

41) -14 43) 13   45) $-4\dfrac{1}{7}$ 47) true

49) $-10, -2, 0, \dfrac{9}{10}, 3.8, 7$

51) $-6.51, -6.5, -5, 2, 7\dfrac{1}{3}, 7\dfrac{5}{6}$ 53) true   55) false

57) false   59) false   61) -22 shots on goal

63) 9 million active Twitter users   65) $-58{,}600$ active duty
    military members

Section 1.5

1) Answers may vary.   3) Answers may vary.

5)
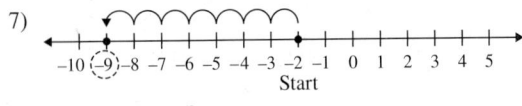

$$6 - 11 = -5$$

7)

$$-2 + (-7) = -9$$

9) -7 11) -14 13) 23   15) -11

17) -850 19) $\dfrac{1}{6}$ 21) $-\dfrac{25}{24}$ or $-1\dfrac{1}{24}$ 23) $-\dfrac{8}{45}$

25) 2.7   27) -9.23 29) 15   31) -11

33) -2 35) -19 37) 23   39) $-\dfrac{19}{18}$ or $-1\dfrac{1}{18}$

41) $\dfrac{5}{24}$ 43) 12   45) 11   47) -7

49) sometimes   51) false   53) false   55) true

57) $-18 + 5 = -13$. His score in the 2019 Masters was -13.

59) $10.43 - 5.01 = 5.42$. The carbon emissions of China
    were 5.42 billion metric tons more than those of the
    United States.

61) $867{,}049 + 586 = 867{,}635$. In 2016, 867,635 flights went
    through O'Hare.

63) a) -3 b) -134 c) -214 d) -194

65) a) -0.7 b) 0.4 c) -0.5 d) -0.8 67) $5 + 7$; 12

69) $10 - 16$; -6 71) $9 - (-8)$; 17   73) $-21 + 13$; -8

75) $-20 + 30$; 10   77) $23 - 19$; 4

79) $(-5 + 11) - 18$; -12

81) Answers may vary. One example is *"The sum of 19 and -4."*

83) Answers may vary. One example is *"6 less than the sum of
    2 and 7."*

1) negative   3) -56 5) 45   7) 84

9) $-\dfrac{2}{15}$ 11) 1.4   13) 135   15) -84

17) when k is negative   19) when $k \neq 0$ 21) 36

23) -125 25) 9   27) -49 29) -32

31) positive   33) 10   35) -4 37) -8

39) $\dfrac{10}{13}$ 41) 0   43) $-\dfrac{3}{2}$ or $-1\dfrac{1}{2}$ 45) -33

47) 43   49) 16   51) 16   53) $\dfrac{1}{4}$

55) $-12 \cdot 6$; -72 57) $(-7)(-5) + 9$; 44

59) $\dfrac{63}{-9} + 7$; 0   61) $(-4)(-8) - 19$; 13

63) $\dfrac{-100}{4} - (-7 + 2)$; -20 65) $2[18 + (-31)]$; -26

67) $\dfrac{2}{3}(-27)$; -18 69) $12(-5) + \dfrac{1}{2}(36)$; -42

71) Answers may vary. One example is *"The quotient of 34 and 17."*

73) Answers may vary. One example is *"3 less than the product
    of 6 and -10."*

Section 1.7

1)

Term	Coeff.
$7p^2$	7
$-6p$	-6
4	4

The constant is 4.

3)

Term	Coeff.
x^2y^2	1
$2xy$	2
$-y$	-1
11	11

The constant is 11.

5)

Term	Coeff.
$-2g^5$	-2
$\dfrac{g^4}{5}$	$\dfrac{1}{5}$
$3.8g^2$	3.8
g	1
-1	-1

The constant is -1.

7) It is a collection of numbers, variables, and grouping symbols
   connected by operation symbols such as $+$, $-$, $\times$, and $\div$.

9) 23   11) -21 13) -1 15) 7   17) -9.4

19) -13 21) $\frac{1}{2}$ 23) -11 25) $\frac{9}{7}$

27) No. The exponents are different.

29) Yes. Both are a^3b-terms.   31) 1

33) -5 35) distributive   37) identity

39) associative   41) distributive

43) $19 + p$ 45) $(8 + 1) + 9$

47) y 49) 4   51) $\frac{11}{35}$

53) No. Subtraction is not commutative.

55) $2 \cdot 1 + 2 \cdot 9 = 2 + 18 = 20$

57) $9(a + b)$ 59) $8(3 - 10) = 8(-7) = -56$

61) $-10 + 4 = -6$ 63) $8y + 8 \cdot 3 = 8y + 24$

65) $-\frac{2}{3}z + \left(-\frac{2}{3}\right) \cdot 6 = -\frac{2}{3}z - 4$

67) $-3x - (-3) \cdot (4y) - (-3) \cdot (6) = -3x + 12y + 18$

69) $8c - 9d + 14$ 71) $24p + 7$

73) $-19y^2 + 30$ 75) $\frac{16}{5}r - \frac{2}{9}$ 77) $8h^2 - 12h + 14$

79) $c^3 + 7c^2 - 6$ 81) $7w + 10$

83) $26x + 37$ 85) $k^2 + 5k + 6$ 87) $18n^2 - 10$

89) $\frac{11}{10}z + \frac{13}{2}$ 91) $\frac{65}{8}t - \frac{19}{16}$ 93) $-1.1x - 19.6$

95) $x + 18$ 97) $x - 6$ 99) $8x$ 101) $\frac{x}{7}$

103) $x - 3$ 105) $12 + 2x$

107) $\frac{1}{2}x - 1$ 109) $(3 + 2x) - 7; 2x - 4$

111) Answers may vary. One example is, *"The product of 10 and a number."*

113) Answers may vary. One example is, *"13 decreased by a number."*

115) Answers may vary. One example is, *"Eight more than half a number."*

Chapter 1: Group Activity

1)

8	1	6
3	5	7
4	9	2

2)

$\frac{1}{9}$	$\frac{1}{2}$	$\frac{2}{9}$
$\frac{7}{18}$	$\frac{5}{18}$	$\frac{1}{6}$
$\frac{1}{3}$	$\frac{1}{18}$	$\frac{4}{9}$

3)

38	8	16	21
19	18	2	44
9	43	20	11
17	14	45	7

Chapter 1: Review Exercises

1) a) 1, 2, 4, 8, 16 b) 1, 37   3) a) $\frac{2}{5}$ b) $\frac{23}{39}$

5) $\frac{3}{10}$ 7) 40   9) $\frac{3}{14}$ 11) $\frac{11}{12}$ 13) $\frac{7}{10}$

15) $\frac{19}{56}$ 17) $6\frac{13}{24}$ 19) 81   21) $\frac{27}{64}$ 23) 10

25) $\frac{1}{4}$ 27) $102°$ 29) $A = 6\frac{9}{16}$ mi^2; $P = 10\frac{3}{4}$ mi

31) $A = 100$ in^2; $P = 40$ in.

33) a) $A = 9\pi$ in^2; $A \approx 28.26$ in^2 b) $C = 6\pi$ in.; $C \approx 18.84$ in.

35) 124 ft^2 37) 1.3π ft^3 39) $\frac{125}{8}$ in^3 or $15\frac{5}{8}$ in^3

41) always   43) a) $\{-16, 0, 4\}$

b) $\left\{\frac{7}{15}, -16, 0, 3.\overline{2}, 8.5, 4\right\}$ c) $\{4\}$ d) $\{0, 4\}$

e) $\{\sqrt{31}, 6.01832...\}$ 45) a) 18 b) -7

47) 19   49) $-\frac{5}{24}$ 51) -12 53) 72   55) -12

57) $\frac{15}{4}$ or $3\frac{3}{4}$ 59) sometimes   61) always

63) always   65) -36 67) 64   69) 27   71) -9

73) -31 75) $\frac{-120}{-3}$; 40   77) $(-4) \cdot 7 - 15$; -43

79) -55 81) associative   83) identity

85) commutative   87) $10 \cdot 5 + 4 \cdot 5 = 50 + 20 = 70$

89) $(-6) \cdot 9p - (-6)(4q) + (-6) \cdot 1 = -54p + 24q - 6$

91) $12m - 10$ 93) $17y^2 - 3y - 1$

95) $\frac{31}{4}n - \frac{9}{2}$ 97) $x + 15$ 99) $8x$

Chapter 1: Test

1) $2 \cdot 3 \cdot 5 \cdot 7$ 2) a) $\frac{5}{8}$ b) $\frac{3}{4}$ 3) $\frac{5}{24}$ 4) $\frac{23}{36}$

5) $7\frac{5}{12}$ 6) $\frac{1}{27}$ 7) $-\frac{1}{4}$ 8) -17 9) 20

10) $-\frac{1}{12}$ 11) 60 12) -3.7 13) -49 14) $\frac{2}{3}$

15) 15,211 ft   16) a) 125   b) 43   c) -37

17) $(-2)^4 = -2 \cdot (-2) \cdot (-2) \cdot (-2) = 16$;
$-2^4 = -1 \cdot 2^4 = -1 \cdot 2 \cdot 2 \cdot 2 \cdot 2 = -16$

18) $149°$ 19) $49°$; acute

20) a) $A = 9 \text{ mm}^2; P = 14.6 \text{ mm}$
   b) $A = 105 \text{ cm}^2; P = 44 \text{ cm}$
   c) $A = 200 \text{ in}^2; P = 68 \text{ in.}$

21) 9 ft^3

22) a) $22, 0$ b) 22 c) $\sqrt{43}, 8.0934...$ d) $22, -7, 0$
   e) $3\frac{1}{5}, 22, -7, 0, 6.2, 1.\overline{5}$

23)

24) a) $-4 + 27; 23$ b) $17 - 5(-6); 47$

25)

Term	Coeff.
$4p^3$	4
$-p^2$	-1
$\frac{1}{3}p$	$\frac{1}{3}$
-10	-10

26) $\frac{1}{3}$

27) a) commutative   b) associative
   c) inverse   d) distributive

28) a) $(-4) \cdot 2 + (-4) \cdot 7 = -8 + (-28) = -36$
   b) $3 \cdot 8m - 3 \cdot 3n + 3 \cdot 11 = 24m - 9n + 33$

29) a) $-6k^2 + 4k - 14$ b) $6c - \frac{49}{6}$ 30) $2x - 9$

Chapter 2

Section 2.1A

1) 9^6 3) $\left(\frac{1}{7}\right)^4$ 5) $(-5)^7$ 7) $(-3y)^8$

9) base: 6; exponent: 8   11) base: 0.05; exponent: 7

13) base: -8; exponent: 5   15) base: $9x$; exponent: 8

17) base: $-11a$; exponent: 2   19) base: p; exponent: 4

21) base: y; exponent: 2

23) $(3 + 4)^2 = 49, 3^2 + 4^2 = 25$. They are not equivalent
   because when evaluating $(3 + 4)^2$, first add $3 + 4$ to get 7,
   then square the 7.

25) Answers may vary.

27) No, $3t^4 = 3 \cdot t^4$; $(3t)^4 = 3^4 \cdot t^4 = 81t^4$

29) 32 31) 121 33) 16 35) -81 37) -8

39) $\frac{1}{125}$ 41) never   43) sometimes   45) 32

47) 81 49) 200 51) $\frac{1}{64}$ 53) 8^{12} 55) 5^{11}

57) $(-7)^8$ 59) b^6 61) k^6 63) $8y^5$ 65) $54m^{15}$

67) $-42r^5$ 69) $28t^{16}$ 71) $-40x^6$ 73) $8b^{15}$

75) y^{12} 77) w^{77} 79) 729 81) $(-5)^6$ 83) $\frac{1}{81}$

85) $\frac{36}{a^2}$ 87) $\frac{m^5}{n^5}$ 89) $10,000y^4$ 91) $81p^4$

93) $-64a^3b^3$ 95) $6x^3y^3$ 97) $-9t^4u^4$

99) a) $A = 3w^2$ sq units; $P = 8w$ units
   b) $A = 5k^5$ sq units; $P = 10k^3 + 2k^2$ units

101) $\frac{3}{8}x^2$ sq units

Section 2.1B

1) operations   3) k^{24} 5) $200z^{26}$ 7) $-6a^{31}b^7$

9) 121 11) $-64t^{18}u^{26}$ 13) $288k^{14}t^4$ 15) $\frac{3}{4g^{15}}$

17) $\frac{49}{4}n^{22}$ 19) $900h^{28}$ 21) $-147w^{45}$ 23) $\frac{36x^6}{25y^{10}}$

25) $\frac{d^{18}}{4c^{30}}$ 27) $\frac{2a^{36}b^{63}}{9c^2}$ 29) $\frac{r^{39}}{242t^5}$ 31) $\frac{2}{3}x^{24}y^{14}$

33) $-\frac{1}{10}c^{29}d^{18}$ 35) $\frac{125x^{15}y^6}{z^{12}}$ 37) $\frac{81t^{16}u^{36}}{16v^{28}}$

39) $\frac{9w^{10}}{x^6y^{12}}$ 41) false   43) never

45) a) $20l^2$ units   b) $25l^4$ sq units

47) a) $\frac{3}{8}x^2$ sq units   b) $\frac{11}{4}x$ units

49) Answers may vary. One example is $k^{10} \cdot k^2$.

51) Answers may vary. One example is $\left(\frac{n^5}{m^8}\right)^3$.

Section 2.2A

1) false   3) true   5) 1 7) -1 9) 0 11) 2

13) $\frac{1}{36}$ 15) $\frac{1}{16}$ 17) $\frac{1}{125}$ 19) 64 21) 32

23) $\frac{27}{64}$ 25) $\frac{49}{81}$ 27) -64 29) $\frac{64}{9}$ 31) $-\frac{1}{64}$

33) -1 35) $\frac{1}{16}$ 37) $\frac{13}{36}$ 39) $\frac{83}{81}$

41) 13^{-1} 43) 10^{-2} 45) $\left(\frac{3}{5}\right)^{-3}$ 47) $-\left(\frac{7}{2}\right)^{-2}$

1) a) w b) n c) $2p$ d) c 3) 1   5) −2   7) 2

9) $\dfrac{1}{d^3}$ 11) $\dfrac{1}{p}$ 13) $\dfrac{b^3}{a^{10}}$ 15) $\dfrac{x^5}{y^8}$ 17) $\dfrac{t^5 u^3}{8}$

19) $\dfrac{5m^6}{n^2}$ 21) $2t^{11}u^5$ 23) $\dfrac{8a^6 c^{10}}{5bd}$ 25) $2x^7 y^6 z^4$

27) $\dfrac{36}{a^2}$ 29) $\dfrac{q^5}{32n^5}$ 31) $\dfrac{c^2 d^2}{144b^2}$ 33) $-\dfrac{9}{k^2}$ 35) $\dfrac{3}{t^3}$

37) $-\dfrac{1}{m^9}$ 39) z^{10} 41) j 43) $5n^2$ 45) cd^3

47) always   49) always   51) true   53) false

Section 2.3

1) −5   3) −3   5) 7

7) You must subtract the denominator's exponent from the numerator's exponent; a^2.

9) false   11) false   13) true   15) d^5 17) m^4

19) $8t^7$ 21) 36   23) 81   25) $\dfrac{1}{16}$ 27) $\dfrac{1}{125}$

29) $10d^2$ 31) $\dfrac{2}{3}c^5$ 33) $\dfrac{1}{y^5}$ 35) $\dfrac{1}{x^9}$ 37) $\dfrac{1}{t^3}$

39) $\dfrac{1}{a^{10}}$ 41) t^3 43) $\dfrac{15}{w^8}$ 45) $-\dfrac{6}{k^3}$ 47) $a^3 b^7$

49) $\dfrac{2k^3}{3l^8}$ 51) $\dfrac{10}{x^5 y^5}$ 53) $\dfrac{w^6}{9v^3}$ 55) $\dfrac{3}{8}c^4 d$

57) $(x+y)^7$ 59) $(c+d)^6$

61) Answers may vary. One example is $\dfrac{n^{10}}{n^4}$.

63) Answers may vary. One example is $\dfrac{a^9 b^3}{a^5 b^2}$.

65) Answers may vary. One example is $\dfrac{4k^2}{k^7}$.

Chapter 2: Putting It All Together

1) $\dfrac{16}{81}$ 2) 64   3) 1   4) −125   5) $\dfrac{9}{100}$ 6) $\dfrac{49}{9}$

7) 9   8) −125   9) $\dfrac{1}{100}$ 10) $\dfrac{1}{8}$ 11) $\dfrac{1}{32}$

12) 81   13) $-\dfrac{27}{125}$ 14) 64   15) $\dfrac{1}{36}$ 16) $\dfrac{13}{36}$

17) $270g^{12}$ 18) $56d^9$ 19) $\dfrac{33}{s^{11}}$ 20) $\dfrac{1}{c^5}$

21) $\dfrac{16}{81}x^{40}y^{24}$ 22) $\dfrac{a^9 b^{15}}{1000}$ 23) $\dfrac{n^6}{81m^{16}}$ 24) $\dfrac{r^8 s^{24}}{81}$

25) $-b^{15}$ 26) h^{88} 27) $-27m^{15}n^6$ 28) $169a^{12}b^2$

29) $-6z^3$ 30) $-9w^9$ 31) $\dfrac{t^{18}}{s^{42}}$ 32) $\dfrac{1}{m^3 n^{14}}$

33) $a^{14}b^3 c^7$ 34) $\dfrac{4}{9v^{10}}$ 35) $\dfrac{27u^{30}}{64v^{21}}$ 36) $\dfrac{81}{x^6 y^8}$

37) $-27t^6 u^{15}$ 38) $\dfrac{1}{144}k^{16}m^2$ 39) $\dfrac{1}{h^{18}}$ 40) $-\dfrac{1}{d^{20}}$

41) $\dfrac{h^4}{16}$ 42) $\dfrac{13}{f^2}$ 43) $56c^{10}$ 44) $80p^{15}$ 45) $\dfrac{3}{a^5}$

46) $\dfrac{1}{9r^2 s^2}$ 47) $\dfrac{6}{55}r^{10}$ 48) $\dfrac{1}{f^{36}}$ 49) $\dfrac{a^2 b^9}{c}$

50) $\dfrac{x^9 y^{24}}{z^3}$ 51) $\dfrac{72n^3}{5m^5}$ 52) $\dfrac{t^7}{100s^7}$ 53) 1   54) 1

55) $\dfrac{9}{49d^6}$ 56) $\dfrac{1}{100x^{10}y^8}$ 57) p^{12c} 58) $25d^{8t}$

59) y^{4m} 60) x^{4c} 61) $\dfrac{1}{t^{3b}}$ 62) $\dfrac{1}{a^{7y}}$ 63) $\dfrac{5}{8c^{7x}}$

64) $-\dfrac{3}{8y^{8a}}$

Section 2.4

1) yes   3) no   5) no   7) yes

9) Answers may vary.   11) 9,802,000

13) 15,020,000   15) 0.00674   17) 1,920,000

19) 2034.49   21) −0.0007   23) −0.0095

25) 60,000   27) −0.09815   29) 300,000,000 receptors

31) 0.000000000025 m   33) 2.1105×10^3

35) 9.6×10^{-5} 37) -7×10^6 39) 3.4×10^3

41) 8×10^{-4} 43) -7.6×10^{-2} 45) 6×10^3

47) 3.808×10^8 kg   49) 1×10^{-8} cm   51) 30,000

53) 690,000   55) −1200   57) −0.06

59) −0.0005   61) 160,000   63) 0.0001239

65) 5,256,000,000 particles   67) 22,872 lb/cow

69) 26,400,000 droplets   71) $4687.50

73) 0.94 lb per person

Chapter 2: Review Exercises

1) a) 8^6 b) $(-7)^4$ 3) a) 32 b) $\dfrac{1}{27}$ c) 7^{12} d) k^{30}

5) a) $125y^3$ b) $-14m^{16}$ c) $\dfrac{a^6}{b^6}$ d) $6x^2 y^2$ e) $\dfrac{25}{3}c^8$

7) a) z^{22} b) $-18c^{10}d^{16}$ c) 125 d) $\dfrac{25t^6}{2u^{21}}$

9) a) 1 b) −1 c) $\dfrac{1}{9}$ d) $-\dfrac{5}{36}$ e) $\dfrac{125}{64}$

11) a) $\dfrac{1}{v^9}$ b) $\dfrac{c^2}{81}$ c) y^8 d) $-\dfrac{7}{k^9}$

 e) $\dfrac{19a}{z^4}$ f) $\dfrac{20n^5}{m^6}$ g) $\dfrac{k^5}{32j^5}$

13) a) 9 b) r^8 c) $\dfrac{3}{2t^5}$ d) $\dfrac{3x^7}{5y}$

15) a) $81s^{16}t^{20}$ b) $2a^{16}$ c) $\dfrac{y^{18}}{z^{24}}$ d) $-36x^{11}y^{11}$

 e) $\dfrac{d^{25}}{c^{35}}$ f) $8m^3n^{12}$ g) $\dfrac{125t^9}{27k^{18}}$ h) 14

17) y^{10k} 19) 938,000 21) 9000 23) 0.00000105

25) 5.75×10^{-5} 27) 3.2×10^7 29) 9.315×10^{-4}

31) 70,000,000 33) -0.0002 35) 0.00019

37) $-1,302,000$ 39) 937.5 acres

41) 178,000,000

Chapter 2: Test

1) $(-3)^3$ 2) x^5 3) 125 4) $\dfrac{1}{x^7}$ 5) 8^{36}

6) p^5 7) 81 8) 1 9) $\dfrac{1}{32}$ 10) $\dfrac{3}{16}$

11) $-\dfrac{27}{64}$ 12) $\dfrac{49}{100}$ 13) $125n^{18}$ 14) $-30p^{12}$

15) m^6 16) $\dfrac{a^4}{b^6}$ 17) $-\dfrac{t^{33}}{27u^{27}}$ 18) $\dfrac{8}{y^9}$ 19) 1

20) $2m + n$ 21) $\dfrac{3a^4c^2}{5b^3d^3}$ 22) y^{18} 23) t^{13k}

24) 728,300 25) 1.65×10^{-4} 26) $-50,000$

27) 28,300,000 28) 0.00000000000000000182 g

Chapter 2: Cumulative Review for Chapters 1–2

1) $\dfrac{3}{5}$ 2) $\dfrac{7}{12}$ 3) $\dfrac{7}{25}$ 4) 12 5) -28

6) -81 7) -1 8) 42 9) a) $346\dfrac{2}{3}$ yd

 b) $11,520 10) 62 11) $V = \dfrac{4}{3}\pi r^3$

12) a) $-4, 3$ b) $\sqrt{11}$ c) 3 d) $3, -4, -2.1\overline{3}, 2\dfrac{2}{3}$ e) 3

13) 261 14) $\dfrac{9}{2}m - 15n + \dfrac{21}{4}$ 15) $-3t^2 + 37t - 25$

16) $\dfrac{1}{2}x - 13$ 17) 4^{10} 18) $\dfrac{y^3}{x^3}$ 19) $\dfrac{1}{4x^5}$

20) $-\dfrac{81r^4}{t^{12}}$ 21) $-28z^8$ 22) $\dfrac{1}{n^7}$ 23) $-\dfrac{32b^5}{a^{30}}$

24) 7.29×10^{-4} 25) 58,280

Chapter 3

Section 3.1

1) 7.5 3) 3.9 5) 5.4 7) 15 9) $-\dfrac{5}{12}$

11) expression 13) equation 15) No, it is an expression.

17) b, c 19) no 21) yes 23) yes 25) $\{17\}$

27) $\{-6\}$ 29) $\{-4\}$ 31) $\{-0.7\}$ 33) $\left\{\dfrac{17}{20}\right\}$

35) Answers may vary. 37) $\{4\}$ 39) $\{-7\}$

41) $\{12\}$ 43) $\{14\}$ 45) $\{-1\}$ 47) $\{48\}$

49) $\left\{-\dfrac{9}{2}\right\}$ 51) $\{-45\}$ 53) $\left\{\dfrac{5}{3}\right\}$

55) $\{18\}$ 57) $\{7\}$ 59) $\{0\}$

61) $\left\{-\dfrac{2}{5}\right\}$ 63) $\{-1\}$ 65) $\{10\}$ 67) $\{-5\}$

69) $\left\{-\dfrac{33}{5}\right\}$ 71) $\left\{\dfrac{24}{5}\right\}$ 73) $\left\{-\dfrac{3}{2}\right\}$

75) $\{-18\}$ 77) $\{-6.2\}$

79) *Step 1:* Clear parentheses and combine like terms on each side of the equation. *Step 2:* Isolate the variable. *Step 3:* Solve for the variable. *Step 4:* Check the solution.

81) Combine like terms; $8x + 11 - 11 = 27 - 11$; Combine like terms; $\dfrac{8x}{8} = \dfrac{16}{8}$; $x = 2$; $\{2\}$

83) sometimes 85) $\{4\}$ 87) $\left\{-\dfrac{5}{2}\right\}$

89) $\{3\}$ 91) $\{-2\}$ 93) $\{5\}$ 95) $\{0\}$

97) $\left\{-\dfrac{4}{7}\right\}$ 99) $\{6\}$ 101) $\{-1\}$ 103) $\left\{\dfrac{1}{4}\right\}$

Section 3.2

1) Answers may vary.

3) $\{1\}$ 5) $\left\{-\dfrac{5}{2}\right\}$ 7) $\{-6\}$ 9) $\{1\}$

11) $\{0\}$ 13) $\left\{\dfrac{1}{2}\right\}$ 15) $\{8\}$

17) Eliminate the fractions by multiplying both sides of the equation by the LCD of all the fractions in the equation.

19) Multiply both sides of the equation by 8.

21) $\{5\}$ 23) $\left\{-\dfrac{2}{3}\right\}$ 25) $\{-8\}$

27) $\left\{\dfrac{20}{9}\right\}$ 29) $\{-1\}$ 31) $\{3\}$ 33) $\{-20\}$

35) $\{2\}$ 37) $\{-0.15\}$ 39) $\{600\}$

41) sometimes 43) The variable is eliminated, and you get a false statement like $5 = -12$.

45) $\varnothing$ 47) {all real numbers} 49) $\varnothing$ 51) $\{100\}$

53) $\{25\}$ 55) $\{6\}$ 57) $\{16\}$ 59) $\left\{-\dfrac{2}{3}\right\}$

61) $\{8\}$ 63) {all real numbers}   65) $\left\{-\dfrac{41}{10}\right\}$

67) $\{0\}$ 69) $\{6000\}$ 71) $\varnothing$ 73) $\left\{\dfrac{3}{4}\right\}$

75) **Step 1.** Read the problem until you understand it. **Step 2.** Choose a variable to represent an unknown quantity. **Step 3.** Translate the problem from English into an equation. **Step 4.** Solve the equation. **Step 5.** Check the answer in the original problem, and interpret the solution as it relates to the problem.

77) $x + 12 = 5; -7$ 79) $x - 9 = 12; 21$

81) $7x = 63; 9$ 83) $2x + 5 = 17; 6$

85) $3x - 8 = 40; 16$ 87) $\dfrac{3}{4}x = 33; 44$

89) $\dfrac{1}{2}x - 9 = 3; 24$ 91) $2x - 3 = x + 8; 11$

93) $x - 24 = \dfrac{x}{9}; 27$ 95) $x + \dfrac{2}{3}x = 25; 15$

97) Answers may vary. One example is *"Nine less than five times a number is sixteen."*

99) Answers may vary. One example is *"Eight more than a number is the same as one more than twice the number."*

Section 3.3

1) $c + 14$ 3) $c - 37$ 5) $\dfrac{1}{2}s$ 7) $14 - x$

9) The number of children must be a whole number.

11) It is an even number.   13) 1905: 4.2 in.; 2004: 3.0 in.

15) Capitals: 33, Golden Knights: 31

17) regular: 260 mg, decaf: 20 mg

19) Spanish: 186; French: 124   21) 11 in., 25 in.

23) bracelet: 9.5 in.; necklace: 19 in.

25) Derek: 2 ft; Cory: 3 ft; Tamara: 1 ft

27) 41, 42, 43   29) 18, 20   31) $-15, -13, -11$

33) 107, 108   35) Jimmy: 13; Kelly: 7   37) 5 ft, 11 ft

39) Bonnaroo: 80,000; Lollapalooza: 300,000   41) 57, 58, 59

43) Ileana: 946, Charlie: 728, Deepa: 501

45) 12 in., 24 in., 36 in.

47) Ed Sheeran: 2,764,000; Taylor Swift: 2,336,000; Drake: 2,227,000

49) 72, 74, 76

Section 3.4

1) 0.40 or 0.4   3) 0.027   5) 0.05   7) 2

9) 7.28   11) 400   13) 150   15) 60%   17) 150%

19) a) \$13.80   b) \$55.20   21) \$157.62   23) \$17.60

25) 20%   27) 4.8%   29) \$239.20   31) \$104.00

33) \$30.00   35) 1800 acres   37) \$38,500

39) \$32   41) \$380   43) \$7000 at 3% and \$5000 at 4%

45) \$4500 at 6.5% and \$3000 at 9.5%

47) \$2100 at 5% and \$1650 at 6%   49) 3 oz   51) 8.25 mL

53) 12 oz of the 2% solution, 24 oz of the 8% solution   55) 3 L

57) 20 mL of pure alcohol, 460 mL of the 4% solution

59) 3 lb   61) \$56,000   63) \$3400 at 5% and \$1900 at 6%

65) 36 mL of the 4% solution, 18 mL of the 10% solution

67) \$8.75   69) 16 oz of orange juice, 60 oz of fruit drink

Section 3.5

1) $V = lwh$ 3) Add the lengths of the three sides.

5) No. The height of a triangle cannot be a negative number.

7) cubic centimeters   9) true   11) $\dfrac{11}{4}$ 13) 3000

15) 2.5   17) 9.2π 19) 4   21) 4   23) 9

25) 4   27) 10   29) 78 ft   31) 8 in.   33) 314 yd^2

35) 67 mph   37) 24 in.   39) 3 ft   41) 18 in. $\times$ 28 in.

43) 12 ft $\times$ 19 ft   45) 2 in., 8 in.   47) 1.5 ft, 1.5 ft, 2.5 ft

49) $m\angle A = 35°, m\angle C = 62°$

51) $m\angle A = 26°, m\angle B = 52°$

53) $m\angle A = 44°, m\angle B = m\angle C = 68°$ 55) 43°, 43°

57) 172°, 172°   59) 38°, 38°   61) 144°, 36°

63) 120°, 60°   65) 73°, 107°   67) $180 - x$ 69) 63°

71) 24°   73) angle: 20°; comp: 70°; supp: 160°   75) 72°

77) 45°   79) a) $x = 21$ b) $x = y - h$ c) $x = c - r$

81) a) $c = 7$ b) $c = \dfrac{d}{a}$ c) $c = \dfrac{v}{m}$

83) a) $a = 44$ b) $a = ry$ c) $a = dw$

85) a) $d = 3$ b) $d = \dfrac{z + a}{k}$

87) a) $h = -\dfrac{2}{3}$ b) $h = \dfrac{n - v}{q}$ 89) $m = \dfrac{F}{a}$

91) $c = nv$ 93) $\sigma = \dfrac{E}{T^4}$ 95) $h = \dfrac{3V}{\pi r^2}$

97) $E = IR$ 99) $R = \dfrac{I}{PT}$

101) $l = \dfrac{P - 2w}{2}$ or $l = \dfrac{P}{2} - w$ 103) $N = \dfrac{2.5H}{D^2}$

105) $b_2 = \dfrac{2A}{h} - b_1$ or $b_2 = \dfrac{2A - hb_1}{h}$

107) a) $w = \dfrac{P - 2l}{2}$ or $w = \dfrac{P}{2} - l$ b) 3 cm

109) a) $F = \dfrac{9}{5}C + 32$ b) 68°F

Section 3.6

1) Answers may vary, but some possible answers are
$\dfrac{6}{8}, \dfrac{9}{12}$, and $\dfrac{12}{16}$.

3) Yes, a percent can be written as a fraction with a denominator of 100. For example, 25% can be written as $\dfrac{25}{100}$ or $\dfrac{1}{4}$.

5) $\dfrac{4}{3}$ 7) $\dfrac{2}{25}$ 9) $\dfrac{1}{4}$ 11) $\dfrac{2}{3}$ 13) $\dfrac{3}{8}$

15) package of 8: $0.786 per battery

17) 48-oz jar: $0.177 per oz   19) 24-oz box: $0.262 per oz

21) A ratio is a quotient of two quantities. A proportion is a statement that two ratios are equal.

23) true   25) false   27) true   29) {2}

31) {40}   33) {18}   35) $\left\{\dfrac{8}{3}\right\}$ 37) {−2}

39) {11}   41) {−1}   43) $\left\{\dfrac{5}{2}\right\}$ 45) $3.54

47) $\dfrac{1}{2}$ cup   49) 82.5 mg   51) 360

53) 8 lb   55) 42.75 Euros   57) $x = 10$

59) $x = 13$ 61) $x = 63$

63) a) $0.80 b) 80¢   65) a) $2.17 b) 217¢

67) a) $2.95 b) 295¢   69) a) $0.25q$ dollars b) $25q$ cents

71) a) $0.10d$ dollars b) $10d$ cents

73) a) $0.01p + 0.25q$ dollars b) $p + 25q$ cents

75) 9 nickels, 17 quarters   77) 11 $5 bills, 14 $1 bills

79) 38 adult tickets, 19 children's tickets

81) eastbound: 65 mph; westbound: 73 mph   83) $\dfrac{5}{6}$ hr

85) passenger train: 50 mph; freight train: 30 mph

87) 36 min   89) 4.30 P.M.   91) 48 mph

93) 23 dimes, 16 quarters   95) $\frac{1}{4}$ hr or 15 min

97) 30,550,000   99) jet: 400 mph, small plane: 200 mph

Section 3.7

1) You use parentheses when there is a $<$ or $>$ symbol or when you use ∞ or $-\infty$.

3) $(-\infty, 4)$ 5) $[-3, \infty)$

7)
a) $\{x \mid x \geq 3\}$ b) $[3, \infty)$

9)
a) $\{c \mid c < -1\}$ b) $(-\infty, -1)$

11)
a) $\left\{w \mid w > -\dfrac{11}{3}\right\}$ b) $\left(-\dfrac{11}{3}, \infty\right)$

13)
a) $\{r \mid r \leq 4\}$ b) $(-\infty, 4]$

15)
a) $\{y \mid y \geq -4\}$ b) $[-4, \infty)$

17)
a) $\{c \mid c > 4\}$ b) $(4, \infty)$

19)
a) $\left\{k \mid k < -\dfrac{11}{3}\right\}$ b) $\left(-\infty, -\dfrac{11}{3}\right)$

21)
a) $\{b \mid b \geq -8\}$ b) $[-8, \infty)$

23)
a) $\{w \mid w < 3\}$ b) $(-\infty, 3)$

25)
a) $\{z \mid z \geq -15\}$ b) $[-15, \infty)$

27)
a) $\{y \mid y > 8\}$ b) $(8, \infty)$

29)
$(-1, \infty)$

31)
$\left(-\infty, -\dfrac{3}{7}\right]$

33)
$(-3, \infty)$

35)
$\left(\dfrac{6}{11}, \infty\right)$

37)
$(-\infty, 4)$

39)
$(0, \infty)$

41)
$(-\infty, 5]$

43)
a) $\{n \mid 1 \le n \le 4\}$ b) $[1, 4]$

45)
a) $\{a \mid -2 < a < 1\}$ b) $(-2, 1)$

47)
a) $\left\{z \,\middle|\, \dfrac{1}{2} < z \le 3\right\}$ b) $\left(\dfrac{1}{2}, 3\right]$

49)
$[-3, 1]$

51)
$\left(\dfrac{3}{2}, 3\right)$

53)
$[-4, -1]$

55)
$\left[\dfrac{7}{4}, 3\right)$

57)
$(-8, 4)$

59)
$\left[-1, -\dfrac{2}{5}\right]$

61)
$[5, 8)$

63)
$\left[-1, \dfrac{4}{3}\right]$

65) $(-7, \infty)$ 67) $\left(-\infty, \dfrac{4}{3}\right]$ 69) $(-\infty, -12)$

71) $\left(-15, -\dfrac{15}{4}\right]$ 73) $[-9, \infty)$ 75) $[-2, 0)$

77) at most $5\dfrac{1}{2}$ hr   79) at most 8 mi   81) 89 or higher

Section 3.8

1) $A \cap B$ means "A intersect B." $A \cap B$ is the set of all elements which are in set A *and* in set B.

3) $\{8, 10\}$ 5) $\{2, 4, 5, 6, 7, 8, 9, 10\}$ 7) $\varnothing$

9) $\{1, 2, 3, 4, 5, 6, 8, 10\}$

11)
$[-3, 2]$

13)
$(-1, 3)$

15)
$[3, \infty)$

17)
$\varnothing$

19)
$[2, 5]$

21)
$(-2, 3)$

23)
$(-3, 4]$

25)
$\varnothing$

27)
$(3, \infty)$

29)
$[-4, 1]$

31)
$(-\infty, -1) \cup (5, \infty)$

33)
$\left(-\infty, \dfrac{5}{3}\right] \cup (4, \infty)$

35)
$(1, \infty)$

37)
$(-\infty, \infty)$

39)
$(-\infty, -1) \cup (3, \infty)$

41)
$\left(-\infty, \dfrac{7}{2}\right] \cup (6, \infty)$

43)
$(-5, \infty)$

45)
$(-\infty, -6) \cup [-3, \infty)$

47)
$(-\infty, \infty)$

49)

$(-\infty, -2]$

51) $\left[-5, \dfrac{1}{2}\right]$ 53) $\left(-\infty, -\dfrac{9}{4}\right) \cup [5, \infty)$

55) $(-\infty, \infty)$ 57) $(-\infty, 0)$ 59) $[-8, -4]$

61) {Susanne Klatten, Jacqueline Mars, Alice Walton}

63) {Susanne Klatten, Jacqueline Mars, Sandra Ortega Mera, Alice Walton}

Chapter 3: Group Activity

1) shirt: $28.20; jeans: $42.00   2) $105.20

3) $78.90   4) $82.85

5)

Month	Balance	Payment	Remaining Balance	Interest	New Balance
1	$82.85	$10.00	$72.85	$1.46	$74.31
2	$74.31	$10.00	$64.31	$1.29	$65.60
3	$65.60	$10.00	$55.60	$1.11	$56.71
4	$56.71	$10.00	$46.71	$0.93	$47.64
5	$47.64	$10.00	$37.64	$0.75	$38.39
6	$38.39	$10.00	$28.39	$0.57	$28.96
7	$28.96	$10.00	$18.96	$0.38	$19.34
8	$19.34	$10.00	$9.34	$0.19	$9.53
9	$9.53	$9.53			
10					

6) $89.53

Chapter 3: Review Exercises

1) no

3) The variables are eliminated, and you get a false statement like $5 = 13$.

5) $\{-19\}$ 7) $\{-8\}$ 9) $\{36\}$ 11) $\left\{\dfrac{15}{2}\right\}$

13) $\{5\}$ 15) $\left\{-\dfrac{2}{3}\right\}$ 17) $\{0\}$ 19) $\{10\}$

21) {all real numbers}   23) $2x - 9 = 25$; 17

25) Thursday: 75; Friday: 51   27) 14 in., 22 in.

29) 6 lb   31) $1500 at 2%, $4500 at 4%

33) 7   35) 7 in.

37) $m\angle A = 55°$, $m\angle B = 55°$, $m\angle C = 70°$

39) $61°, 61°$ 41) $p = z + n$ 43) $b = \dfrac{2A}{h}$

45) Yes. It can be written as $\dfrac{15}{100}$ or $\dfrac{3}{20}$.   47) $\dfrac{4}{5}$

49) $\{12\}$ 51) 720   53) 45 min

55)

$[8, \infty)$

57)

$(-3, \infty)$

59)

$(-\infty, 4]$

61)

$(-2, 3]$

63)

$(-\infty, 4)$

65)

$(4, 8]$

67) 87 or higher

69)

$[1, 3]$

71)

$(-1, \infty)$

73) {Toyota}   75) $\left\{\dfrac{5}{2}\right\}$

77) $\{-42\}$ 79) $\left(-\infty, -\dfrac{1}{2}\right)$

81) $\varnothing$ 83) $(-\infty, -4)$

85) $\{13\}$ 87) $[3, 6]$

89) $\{-1\}$ 91) $57°$ 93) 17, 19   95) 360

Chapter 3: Test

1) $\{6\}$ 2) $\{-3\}$ 3) $\{15\}$ 4) $\left\{\dfrac{2}{3}\right\}$

5) $\varnothing$ 6) $\{1\}$

7) A ratio is a quotient of two quantities. A proportion is a statement that two ratios are equal.

8) Fall: $1305, Spring: $1479   9) 36, 38, 40

10) $31.90   11) 10 in. by 15 in.   12) 4.5

13) 3.5 qt of regular oil, 1.5 qt of synthetic oil   14) 50

15) eastbound: 66 mph; westbound: 72 mph   16) $1\dfrac{1}{5}$ gal

17) $a = \dfrac{4B}{n}$ 18) $h = \dfrac{S - 2\pi r^2}{2\pi r}$ or $h = \dfrac{S}{2\pi r} - r$

19) $m\angle A = 26°$, $m\angle B = 115°$, $m\angle C = 39°$

20) $(-\infty, -2]$

21) $\left(\frac{3}{4}, \infty\right)$

22) [0, 6)

23)

(−3, 0)

24) at most 6 hr   25) a) {1, 2, 3, 6, 9, 12}   b) {1, 2, 12}

26) $(-\infty, -8) \cup \left(\frac{7}{3}, \infty\right)$ 27) [0, 3]

28) $(-\infty, \infty)$

Chapter 3: Cumulative Review for Chapters 1–3

1) $-\frac{11}{24}$ 2) $\frac{15}{2}$ 3) 54   4) −87   5) −47

6) 27 cm^2 7) {−5, 0, 9}   8) $\left\{\frac{3}{4}, -5, 2.5, 0, 0.\overline{4}, 9\right\}$

9) {0, 9}   10) distributive

11) No. For example, $10 - 3 \neq 3 - 10$.   12) $17y^2 - 18y$

13) $\frac{5}{4}r^{12}$ 14) $-72\, m^{33}$ 15) $-\frac{9}{2z^6}$

16) $\frac{1}{4c^6d^{10}}$ 17) 8.95×10^{-6} 18) $\left\{-\frac{23}{2}\right\}$

19) {4}   20) {all real numbers}   21) $\left\{\frac{11}{8}\right\}$

22) {18}   23) car: 60 mph, train: 70 mph

24) $(-\infty, -6]$ 25) $(-\infty, -3] \cup \left[\frac{11}{4}, \infty\right)$

Chapter 4
Section 4.1

1) 7.2 gal   3) 2012, 2014, and 2015; 6.6 gal

5) Consumption was decreasing.

7) New Jersey; 90.1%

9) Florida's graduation rate is about 9.4% less than New Jersey's.

11) Answers may vary.   13) yes   15) yes   17) no

19) yes   21) 5   23) $-\frac{7}{2}$ 25) 5

27)

x	y
0	−4
1	−2
−1	−6
−2	−8

29)

x	y
0	0
$\frac{1}{2}$	2
3	12
−5	−20

31)

x	y
0	−2
$-\frac{8}{5}$	0
1	$-\frac{13}{4}$
$-\frac{12}{5}$	1

33)

x	y
0	−2
−3	−2
8	−2
17	−2

35) Answers may vary.

37) A: (−2, 1); quadrant II
B: (5, 0); no quadrant
C: (−2, −1); quadrant III
D: (0, −1); no quadrant
E: (2, −2); quadrant IV
F: (3, 4); quadrant I

39, 41)

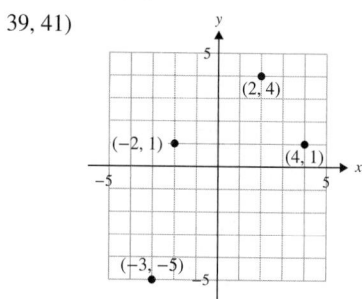

43, 45)

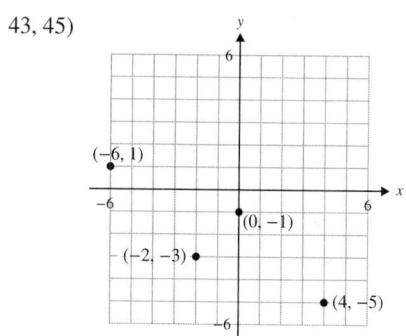

47, 49)

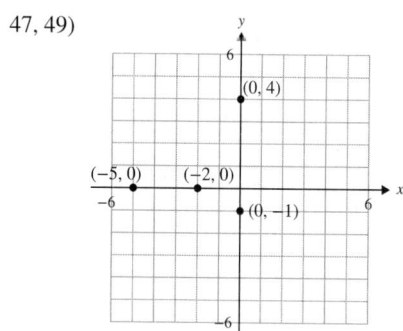

51, 53)

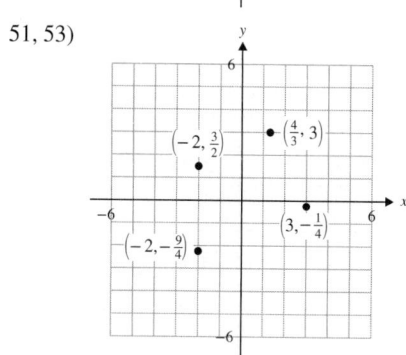

55)

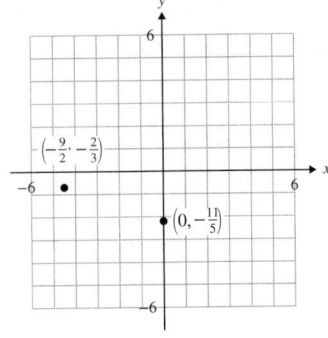

57) never   **59)** sometimes

61)

x	y
0	3
$\frac{3}{4}$	0
2	-5
-1	7

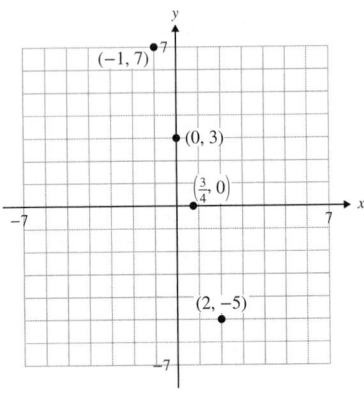

63)

x	y
0	0
-1	-1
3	3
-5	-5

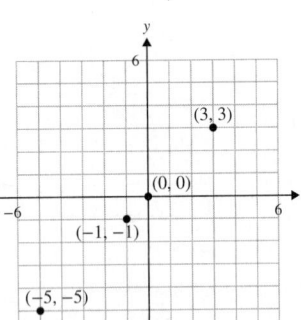

65)

x	y
0	3
4	0
1	$\frac{9}{4}$
-4	6

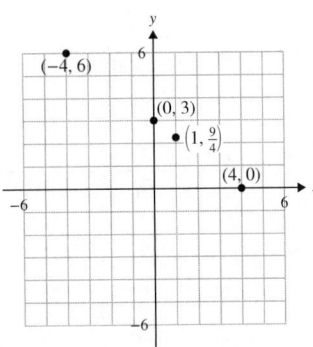

67)

x	y
0	-1
1	-1
-3	-1
-1	-1

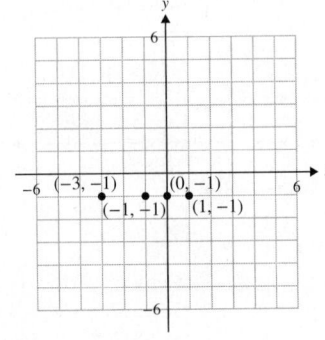

69)

x	y
0	2
-2	$\frac{3}{2}$
4	3
-1	$\frac{7}{4}$

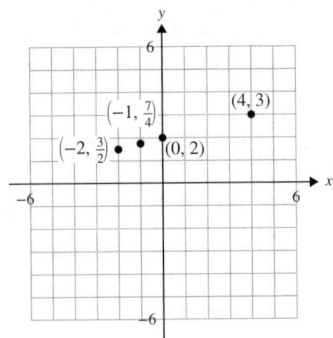

71) a) $(3, -5), (6, -3), (-3, -9)$

b) $\left(1, -\dfrac{19}{3}\right), \left(5, -\dfrac{11}{3}\right), \left(-2, -\dfrac{25}{3}\right)$

c) The x-values in part a) are multiples of the denominator of $\dfrac{2}{3}$. So, when you multiply $\dfrac{2}{3}$ by a multiple of 3 the fraction is eliminated.

73) negative   **75)** positive

77) positive   **79)** zero

81) a) x represents the year; y represents the number of visitors in millions

b) In 2015, there were 42.3 million visitors to Las Vegas.

c) 42.2 million   d) In 2012 and 2013, there were approximately 39.7 million visitors.

e) 4.0 million   f) (2016, 42.9)

83) a) (2011, 37), (2012, 33), (2013, 30), (2014, 31), (2015, 33), (2016, 31)

b)

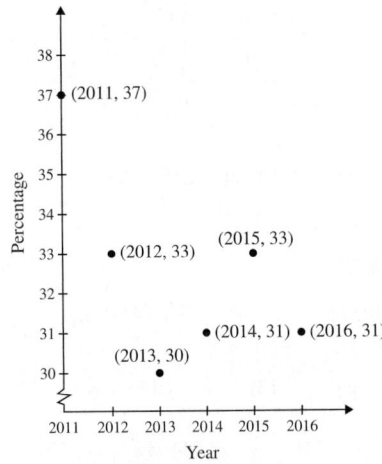

c) In the year 2016, 31% of all fatal accidents in Minnesota involved alcohol.

85) a)

x	y
100.00	10.10
140.00	14.14
210.70	21.2807
250.00	25.25

b)

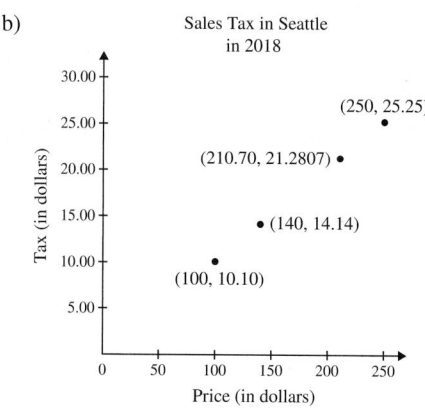

Sales Tax in Seattle in 2018

(250, 25.25)
(210.70, 21.2807)
(140, 14.14)
(100, 10.10)

c) If a bill totals $140.00, the sales tax will be $14.14.
d) $21.28   e) They lie on a straight line.   f) $200.00

Section 4.2

1) line

3)
x	y
0	4
-1	6
2	0
3	-2

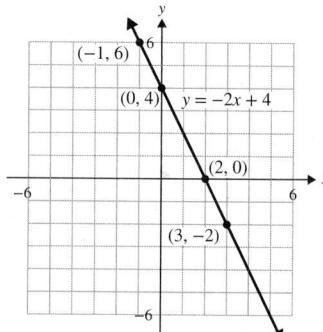

5)
x	y
0	7
2	10
-2	4
-4	1

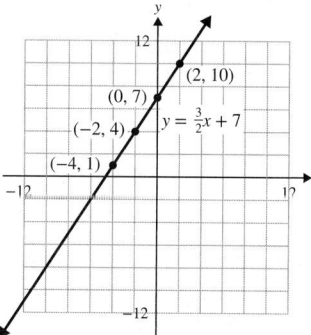

7)
x	y
$\frac{3}{2}$	0
0	3
$\frac{1}{2}$	2
-1	5

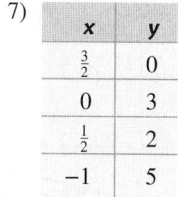
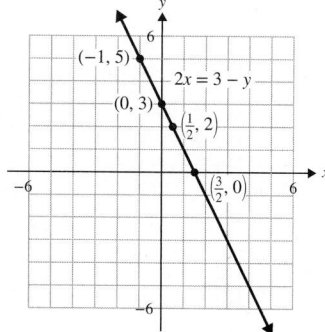

9)
x	y
$-\frac{4}{9}$	5
$-\frac{4}{9}$	0
$-\frac{4}{9}$	-1
$-\frac{4}{9}$	-2

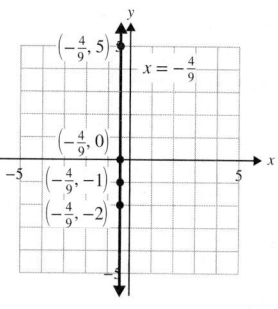
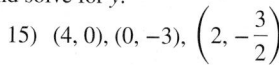

11) It is the point where the graph intersects the y-axis.
Let $x = 0$ in the equation, and solve for y.

13) (1, 0), (0, -1), (2, 1)   15) (4, 0), (0, -3), $\left(2, -\frac{3}{2}\right)$

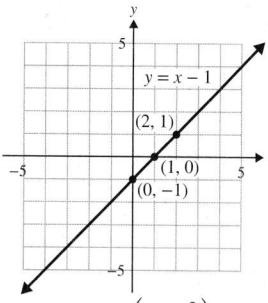
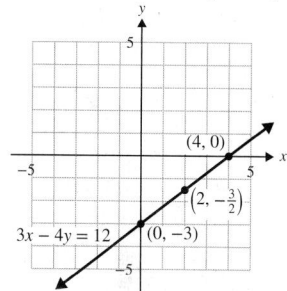

17) (-2, 0), $\left(0, -\frac{3}{2}\right)$, (2, -3)   19) (4, 0), (0, -8), (2, -4)

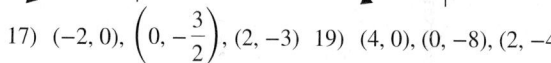
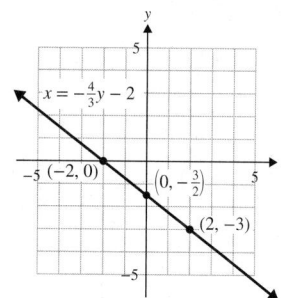
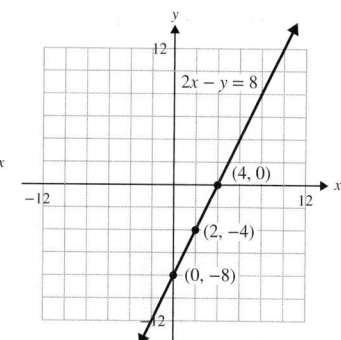

21) (0, 0), (1, -1), (-1, 1)   23) (0, 0), (3, 4), (-3, -4)

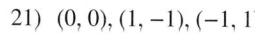
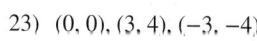
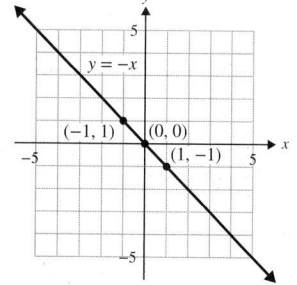
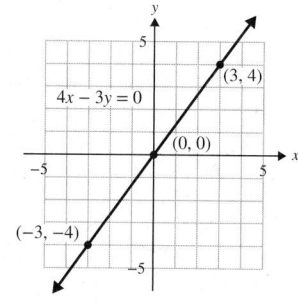

25) (5, 0), (5, 2), (5, -1)   27) (0, 0), (1, 0), (-2, 0)

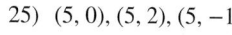
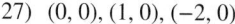
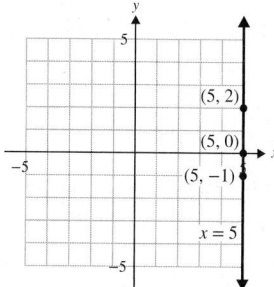
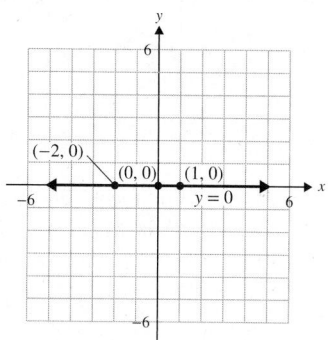

29) $\left(\frac{4}{3}, 0\right), \left(\frac{4}{3}, 1\right), \left(\frac{4}{3}, -2\right)$ 31) $\left(\frac{9}{4}, 0\right), (0, -9), (3, 3)$

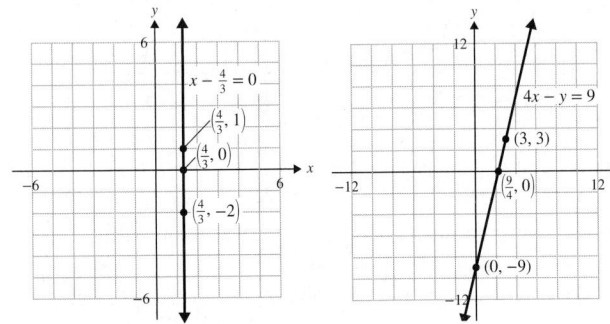

33) always   35) never   37) $(0, 0)$

39) a)

x	y
0	0
4	5.16
7	9.03
12	15.48

$(0, 0), (4, 5.16), (7, 9.03),$ $(12, 15.48)$

b) $(0, 0)$: If no songs are purchased, the cost is $0. $(4, 5.16)$; The cost of downloading 4 songs is $5.16. $(7, 9.03)$; The cost of downloading 7 songs is $9.03. $(12, 15.48)$; The cost of downloading 12 songs is $15.48.

c)

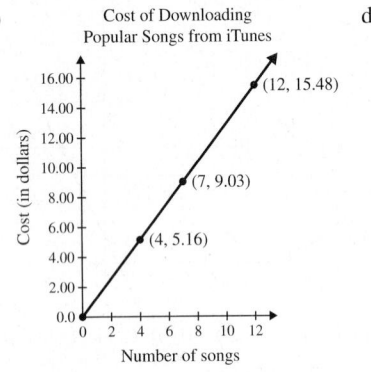

d) 9

41) a) 2005: 1550; 2015: 2005
   b) 2005: 1552; 2015: 2002; yes, they are close.

c)

d) $(0, 1330)$; It looks like it is within 10 units of the number given by the equation.
e) 2317

Section 4.3

1) $\frac{5}{9}$ 3) $-\frac{1}{3}$ 5) 7

7) The slope of a line is the ratio of vertical change to horizontal change. It is $\frac{\text{Change in } y}{\text{Change in } x}$ or $\frac{\text{Rise}}{\text{Run}}$ or $\frac{y_2 - y_1}{x_2 - x_1}$, where (x_1, y_1) and (x_2, y_2) are points on the line.

9) It slants upward from left to right.

11) undefined   13) $m = \frac{3}{4}$

15) $m = -\frac{2}{3}$ 17) $m = -3$

19) Slope is undefined.

21)

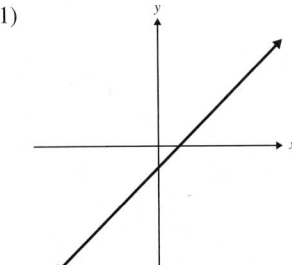

23) 2   25) -4 27) $-\frac{11}{5}$ 29) undefined

31) 0   33) $\frac{3}{7}$ 35) 0.75   37) $\frac{12}{79}$

39) No. The slope of the slide is $0.\overline{6}$. This is more than the recommended slope.

41) Yes. The slope of the driveway is 0.0375. This is less than the maximum slope allowed.

43) $\frac{6}{13}$

45) a) 1182; 1109
   b) negative
   c) The number of injuries is decreasing.
   d) $m = -18.25$; the number of injuries is decreasing by about 18.25 per year.

47) 49)

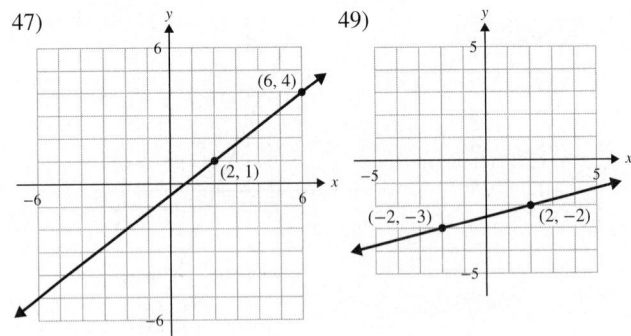

51)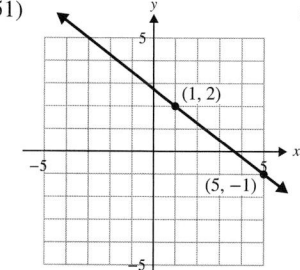

53)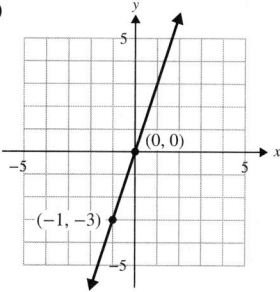

5) $m = -\frac{3}{2}$, y-int: $(0, 3)$

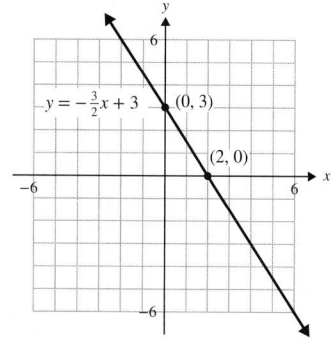

55)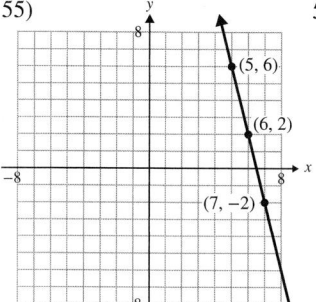

57)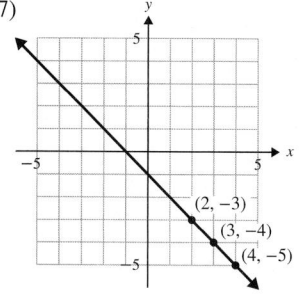

7) $m = \frac{3}{4}$, y-int: $(0, 2)$

9) $m = -2$, y-int: $(0, -3)$

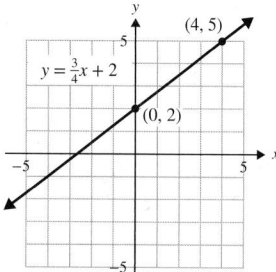

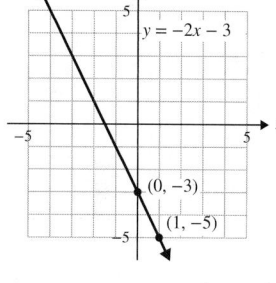

59)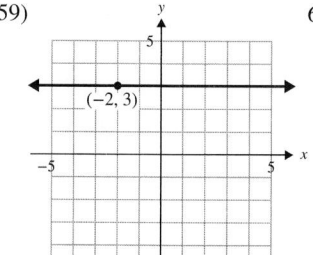

61)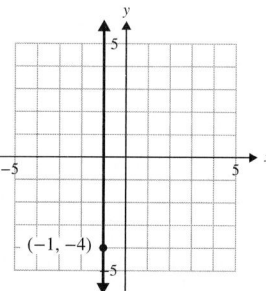

11) $m = 5$, y-int: $(0, 0)$

13) $m = -\frac{3}{2}$, y-int: $\left(0, -\frac{7}{2}\right)$

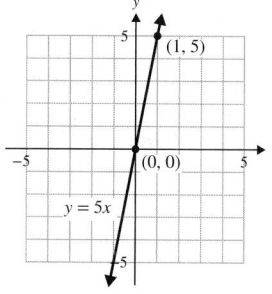

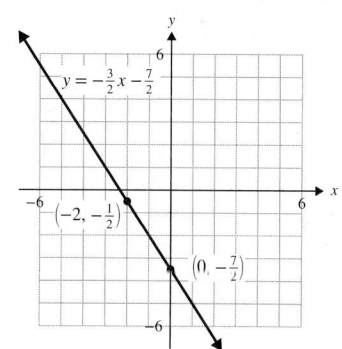

63)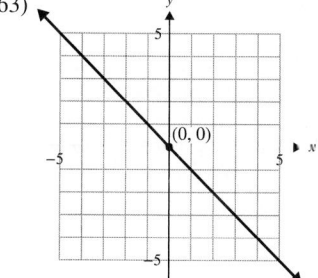

Section 4.4

1) The slope is m, and the y-intercept is $(0, b)$.

3) $m = \frac{2}{5}$, y-int: $(0, -6)$

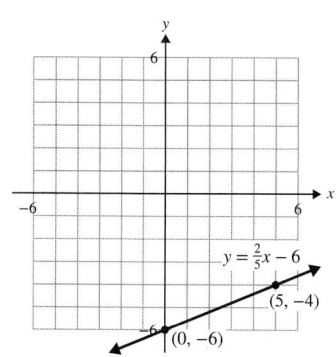

15) $m = 0$, y-int: $(0, 6)$

17) $y = -\frac{1}{3}x - 2$

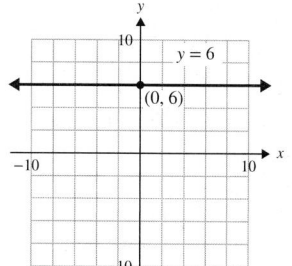

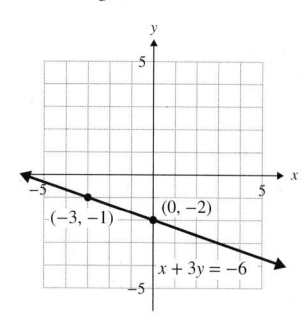

19) $y = -\frac{4}{3}x + 7$

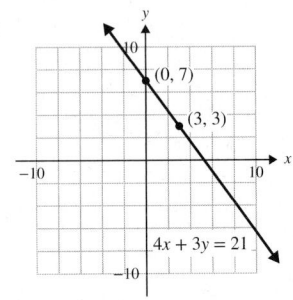

21) This cannot be written in slope-intercept form.

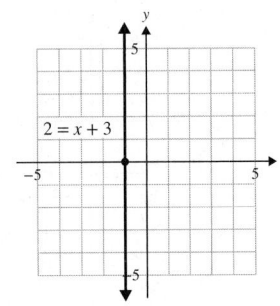

23) $y = -\frac{2}{3}x + 6$

25) $y = -5$

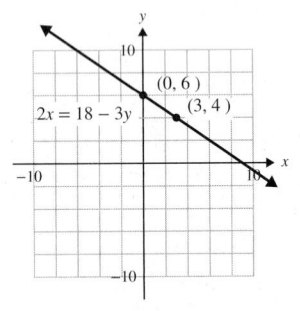

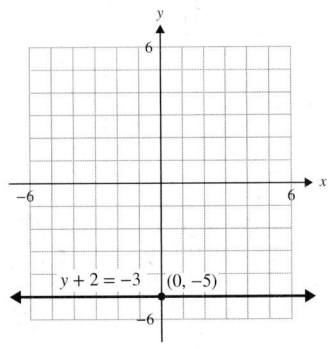

27) a) (0, 0); if Kolya works 0 hr, he earns $0. b) $m = 12.50$; Kolya earns $12.50 per hour. c) $150.00

29) a) (0, 18); when the joey comes out of the pouch, it weighs 18 oz. b) 24 oz c) A joey gains 2 oz per week after coming out of its mother's pouch. d) 7 weeks

31) a) (0, 0); $0 = 0$ rupees b) 72.00; each American dollar is worth 72.00 rupees. c) 5040 rupees d) $45.00

33) $y = -4x + 7$ 35) $y = \frac{9}{5}x - 3$

37) $y = -\frac{5}{2}x - 1$ 39) $y = x + 2$ 41) $y = 0$

43) Their slopes are negative reciprocals, or one line is vertical and one is horizontal.

45) true   47) false   49) perpendicular   51) parallel

53) neither   55) neither   57) perpendicular

59) parallel   61) perpendicular   63) parallel

65) perpendicular   67) neither   69) parallel

71) parallel   73) perpendicular

Section 4.5

1) $2x + y = -4$ 3) $x - y = 1$

5) $4x - 5y = -5$ 7) $4x + 12y = -15$

9) Substitute the slope and y-intercept into $y = mx + b$.

11) $y = -7x + 2$ 13) $4x + y = 6$ 15) $2x - 7y = 21$

17) $y = -x$ 19) a) $y - y_1 = m(x - x_1)$ b) Substitute the slope and point into the point-slope formula.

21) $y = x + 2$ 23) $y = -5x + 19$

25) $4x - y = -7$ 27) $2x - 5y = -50$

29) $y = -\frac{5}{4}x + \frac{29}{4}$ 31) $5x - 6y = -15$

33) Find the slope, and use it and one of the points in the point-slope formula.

35) $y = -3x + 4$ 37) $y = 2x - 3$ 39) $y = -\frac{1}{3}x + \frac{10}{3}$

41) $x + 3y = -2$ 43) $5x - 3y = 18$

45) $y = -3.0x + 1.4$ 47) $y = \frac{3}{4}x - 1$

49) $y = -3x - 4$ 51) $y = 3$ 53) $y = -\frac{4}{3}x + \frac{5}{3}$

55) $y = x + 2$ 57) $y = 7x + 6$

59) $x = 3$ 61) $y = 3$ 63) $y = -4x - 4$

65) $y = -3x + 20$ 67) $y = \frac{1}{2}x - 2$

69) They have the same slopes and different y-intercepts.

71) $y = 4x + 2$ 73) $4x - y = 0$ 75) $x + 2y = 10$

77) $y = 5x - 2$ 79) $y = \frac{3}{2}x - 4$ 81) $x - 5y = 10$

83) $y = -x - 5$ 85) $3x - y = 10$ 87) $y = -3x - 12$

89) $y = -x + 11$ 91) $y = 4$ 93) $y = 2$

95) $y = -\frac{2}{7}x + \frac{1}{7}$ 97) $y = -\frac{3}{2}$

99) a) $y = 0.276x + 17.22$ b) The number of employees in health care and social assistance in the United States is increasing by 276,000 per year.   c) 20.532 mil

101) a) $y = -15,000x + 500,000$ b) The budget is being cut by $15,000 per year.   c) $455,000   d) 2026

103) a) $y = 8x + 100$ b) A kitten gains about 8 g per day.   c) 140 g; 212 g   d) 23 days

105) a) $E = 1.6A + 28.4$ b) 40

Section 4.6

1) a) any set of ordered pairs
   b) Answers may vary.
   c) Answers may vary.

3) domain: $\{-8, -2, 1, 5\}$; range: $\{-3, 4, 6, 13\}$; function

5) domain: $\{1, 9, 25\}$; range: $\{-3, -1, 1, 5, 7\}$; not a function

7) domain: $\{-1, 2, 5, 8\}$; range: $\{-7, -3, 12, 19\}$; not a function

9) domain: $(-\infty, \infty)$; range: $(-\infty, \infty)$; function

11) domain: $[-5, 1]$; range: $[-6, 0]$; not a function

13) domain: $(-\infty, \infty)$; range: $(-\infty, 6]$; function

15) yes   17) yes   19) no   21) no

23) $(-\infty, \infty)$; function   25) $(-\infty, \infty)$; function

27) $[0, \infty)$; not a function   29) $(-\infty, 0) \cup (0, \infty)$; function

31) $(-\infty, -4) \cup (-4, \infty)$; function

33) $(-\infty, 5) \cup (5, \infty)$; function

35) $\left(-\infty, \dfrac{3}{5}\right) \cup \left(\dfrac{3}{5}, \infty\right)$; function

37) $\left(-\infty, -\dfrac{4}{3}\right) \cup \left(-\dfrac{4}{3}, \infty\right)$; function

39) $(-\infty, 3) \cup (3, \infty)$; function   41) $(-\infty, \infty)$; function

43) y is a function, and y is a function of x.

45) a) $y = 7$ b) $f(3) = 7$ 47) -13 49) 7

51) 50   53) -10 55) $-\dfrac{25}{4}$ 57) -105

59) $f(-1) = \dfrac{5}{2}, f(4) = 5$ 61) $f(-1) = 6, f(4) = 2$

63) $f(-1) = 7, f(4) = 3$ 65) -4 67) 6

69) $f(n - 3) = -9(n - 3) + 2$; Distribute.; $= -9n + 29$

71) a) $f(c) = -7c + 2$ b) $f(t) = -7t + 2$
    c) $f(a + 4) = -7a - 26$ d) $f(z - 9) = -7z + 65$
    e) $g(k) = k^2 - 5k + 12$ f) $g(m) = m^2 - 5m + 12$
    g) $-7x - 7h + 2$ h) $-7h$

73)
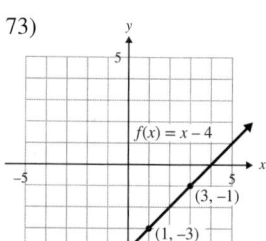

75)
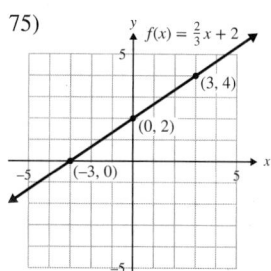

77)
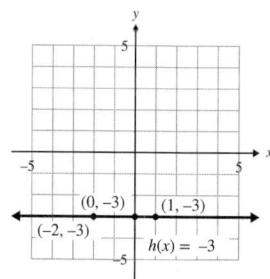

79) x-int: $(-1, 0)$; y-int: $(0, 3)$
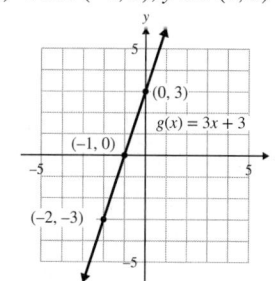

81) x-int: $(4, 0)$; y-int: $(0, 2)$
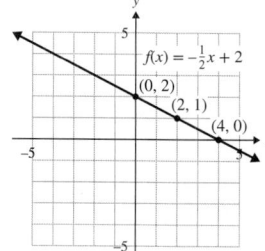

83) intercept: $(0, 0)$
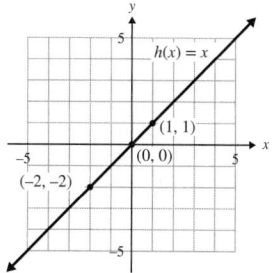

85) $m = -4$; y-int: $(0, -1)$
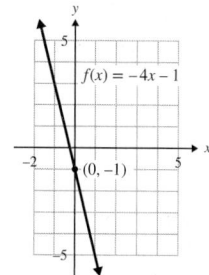

87) $m = -\dfrac{1}{4}$; y-int: $(0, -2)$
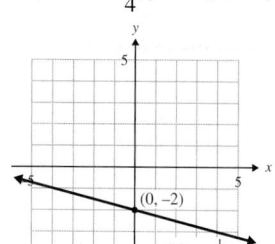

89) $m = 2$; y-int: $\left(0, \dfrac{1}{2}\right)$
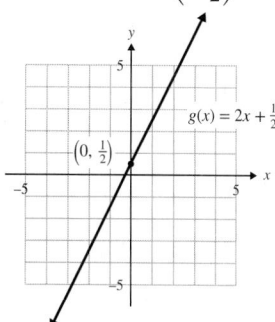

91)
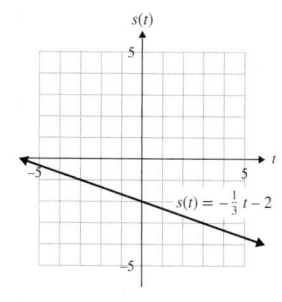

93)
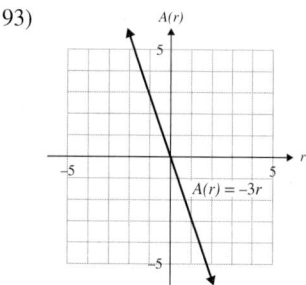

95) a) 108 mi   b) 216 mi   c) 2.5 hr
    d)
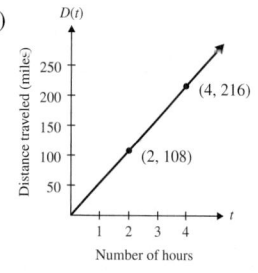

97) a) $C(8) = 28$; 8 gal of gas cost \$28.00.
    b) $C(15) = 52.5$; 15 gal of gas cost \$52.50.
    c) $g = 12$; 12 gal of gas can be purchased for \$42.00.

99) a) 253.56 MB   b) 1267.80 MB   c) 20 sec
    d)
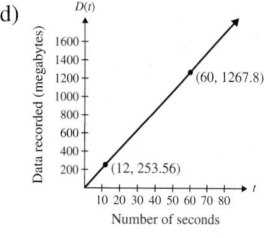

101) a) 60,000 b) 3.5 sec

103) a) $L(15) = 43.5$; in 15 min, there are 43.5 million likes on Instagram.
   b) $L(180) = 522$; in 180 min, or 3 hr, there are 522 million likes on Instagram.
   c) $t = 480$; there are 1392 million likes on Instagram in 480 min, or 8 hr.

105) a) 2 hr; 400 mg b) after about 30 min and after 6 hr
   c) 200 mg d) $A(8) = 50$. After 8 hr there are 50 mg of ibuprofen in Sasha's bloodstream.

Chapter 4: Group Activity

1) $T_{WC} = [(0.4275 \cdot 15^{0.16}) + 0.6215]T + [35.74 - (35.75 \cdot 15^{0.16})]$

a)

T (air temperature)	T_{WC} (when V = 15 mph)
−40°F	−70.6°F
−30°F	−57.8°F
−20°F	−45.0°F
−10°F	−32.2°F
0°F	−19.4°F
10°F	−6.6°F
20°F	6.2°F
30°F	19.0°F
40°F	31.8°F
50°F	44.6°F

b) The ratio of change of wind chill temperature to the change of air temperature.

2) a)

T (air temperature)	T_{WC} (when V = 5 mph)	T_{WC} (when V = 50 mph)
−40°F	−59°F	−86°F
−30°F	−47°F	−72°F
−20°F	−35°F	−58°F
−10°F	−23°F	−44°F
0°F	−11°F	−30°F
10°F	1°F	−16°F
20°F	13°F	−2°F
30°F	25°F	12°F
40°F	37°F	26°F
50°F	49°F	40°F

b) Round-off error

3) Answers may vary.

4) Answers may vary.

Chapter 4: Review Exercises

1) yes   3) yes   5) 14   7) −9

9)

x	y
0	−14
6	−8
−3	−17
−8	−22

11)
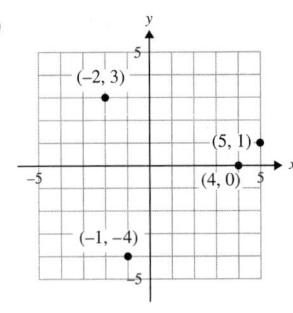

13) a)

x	y
10	50
18	54
29	59.50
36	63

b)
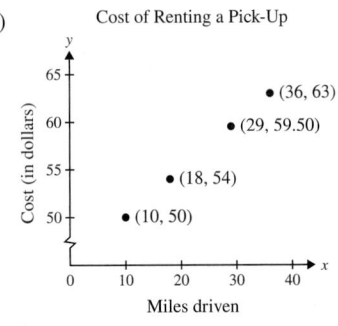

c) The cost of renting the pick-up is $74.00 if it is driven 58 miles.

15)

x	y
0	4
1	2
2	0
3	−2

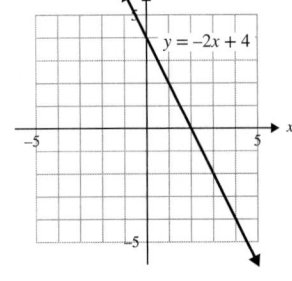

17) (2, 0), (0, −1); (4, 1) may vary.

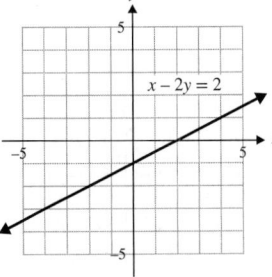

19) (2, 0), (0, 1); (−2, 2) may vary.

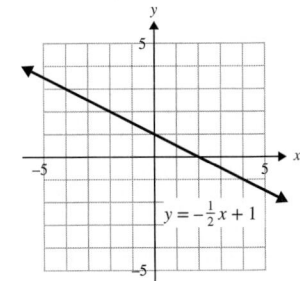

21) (0, 4); (2, 4), (−1, 4) may vary.

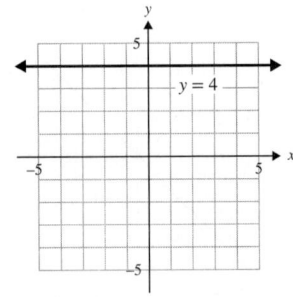

23) $\dfrac{3}{2}$ 25) 5   27) $\dfrac{2}{3}$ 29) -7 31) 0

33) a) $4.00   b) The slope is positive, so the value of the album is increasing over time.   c) $m = 1$; the value of the album is increasing by $1.00 per year.

35) 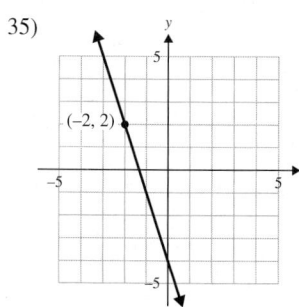   37)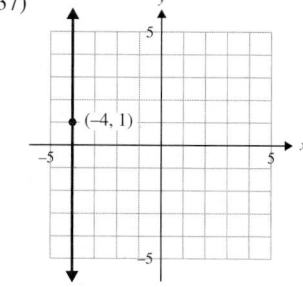

39) $m = -1$, y-int: $(0, 5)$ 41) $m = \dfrac{2}{5}$, y-int: $(0, -6)$

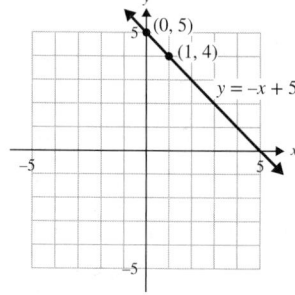   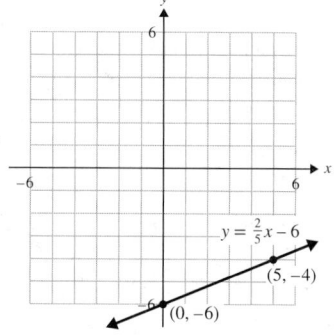

43) $m = -\dfrac{1}{3}$, y-int: $(0, -2)$ 45) $m = -1$, y-int: $(0, 0)$

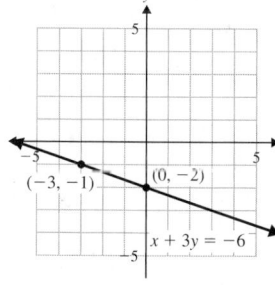   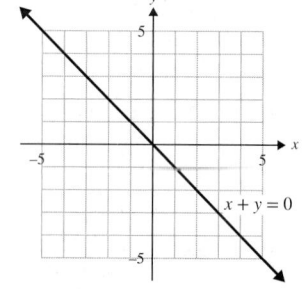

47) a) $(0, 50.00)$; when Sirena received the gift card, it was worth $50.00.   b) $m = -2.95$; each time Sirena buys her favorite drink, the value of the gift card decreases by $2.95.
c) $35.25   d) 14

49) parallel   51) perpendicular   53) $y = 6x + 10$

55) $y = -\dfrac{3}{4}x + 7$ 57) $y = -2x + 9$ 59) $y = 7$

61) $3x - y = 7$ 63) $5x - 2y = 8$ 65) $4x + y = 0$

67) $x + y = 7$ 69) a) $y = 3500x + 62{,}000$
b) Mr. Romanski's salary is increasing by $3500 per year.
c) $72,500   d) 2023

71) $y = -8x + 6$ 73) $x - 2y = -18$

75) $y = -\dfrac{1}{5}x + 10$ 77) $y = x - 12$

79) $y = \dfrac{3}{2}x + 2$ 81) $y = 4$

83) domain: $\{-3, 5, 12\}$; range: $\{1, 3, -3, 4\}$; not a function

85) domain: $\{$Beagle, Siamese, Parrot$\}$;
range: $\{$Dog, Cat, Bird$\}$; function

87) domain: $[0, 4]$; range: $[0, 2]$; not a function

89) $(-\infty, -3) \cup (-3, \infty)$; function

91) $[0, \infty)$; not a function

93) $\left(-\infty, \dfrac{2}{7}\right) \cup \left(\dfrac{2}{7}, \infty\right)$; function

95) $f(3) = 27, f(-2) = -8$

97) a) 8   b) -27 c) 32   d) 5   e) $5a - 12$
f) $t^2 + 6t + 5$ g) $5k + 28$ h) $5c - 22$
i) $5x + 5h - 12$ j) $5h$

99) $\dfrac{1}{3}$

101) a) 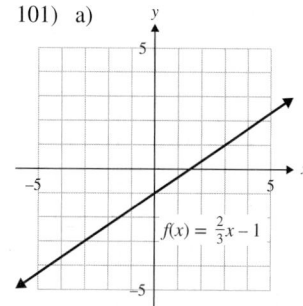   b)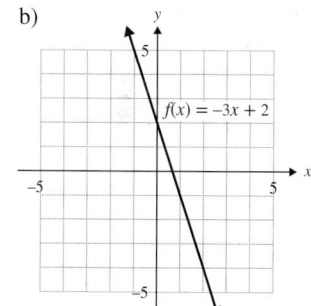

103)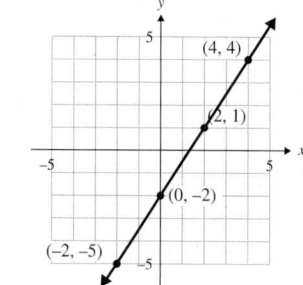

105) a) $8406 (Including the bride and groom, there are 53 people.)
b) They can invite 66 people. (With the bride and groom, there will be 68 people attending the reception for a total cost of $9936.)

Chapter 4: Test

1) yes

2)

x	y
0	-2
-2	-5
4	4
2	1

3) positive; negative

4) a) $(2, 0)$ b) $\left(0, -\dfrac{3}{2}\right)$ c) Answers may vary.

d)

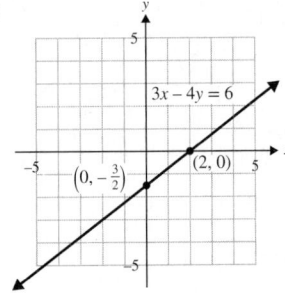

5)

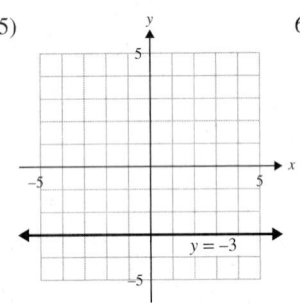

6)

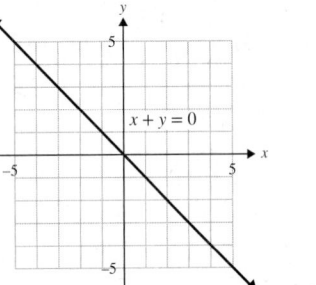

7) $-\dfrac{5}{4}$ 8) 0

9)

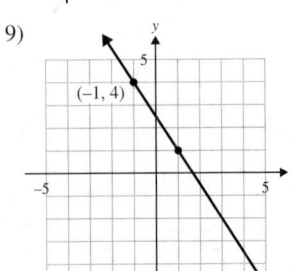

10)

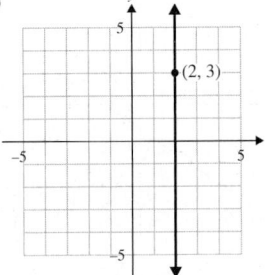

11) $y = \dfrac{3}{2}x - 5$

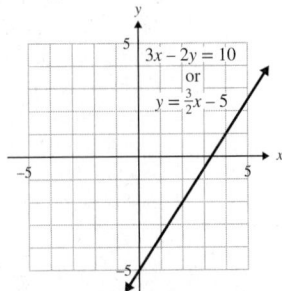

12) $y = 7x - 10$ 13) $x + 3y = 12$

14) $y = \dfrac{5}{2}x + \dfrac{3}{2}$ 15) perpendicular

16) a) $y = -\dfrac{1}{2}x + 7$ b) $y = \dfrac{3}{4}x - \dfrac{1}{4}$

17) 399   18) $y = -9x + 419$

19) According to the equation, 401 students attended the school in 2016. The actual number was 399.

20) The school is losing 9 students per year.

21) $(0, 419)$; in 2014, 419 students attended this school.

22) 347

23) domain: $\{-2, 1, 3, 8\}$; range: $\{-5, -1, 1, 4\}$; function

24) domain: $[-3, \infty)$; range: $(-\infty, \infty)$; not a function

25) a) $(-\infty, \infty)$ b) yes

26) a) $\left(-\infty, \dfrac{5}{2}\right) \cup \left(\dfrac{5}{2}, \infty\right)$ b) yes

27) a) -3 b) 5

28) a) -22 b) 5   c) $t^2 - 3t + 7$ d) $-4h + 30$

29)

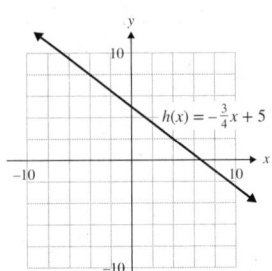

30) a) 30 GB
    b) $t = 13.2$; in 13.2 sec, 132 GB of data can be transferred.

Chapter 4: Cumulative Review for Chapters 1–4

1) $1, 2, 3, 5, 6, 9, 10, 15, 18, 30, 45, 90$ 2) $\dfrac{14}{33}$

3) 39 in.   4) -81 5) $\dfrac{14}{25}$

6) a) $18k^{14}$ b) $\dfrac{b^{10}}{6a^9}$ c) $-\dfrac{27}{125x^{24}y^{18}}$ 7) $6.73 \cdot 10^{-6}$

8) $61°$ 9) -14 10) commutative

11) $14b + 21$ 12) $\{-15\}$

13) $\{1\}$ 14) $\varnothing$ 15) $\left[\dfrac{5}{2}, \infty\right)$ 16) \$340,000

17) $m\angle A = 29°$, $m\angle B = 131°$

18) yes   19) a) negative   b) zero   20) $m = \dfrac{8}{3}$

21)

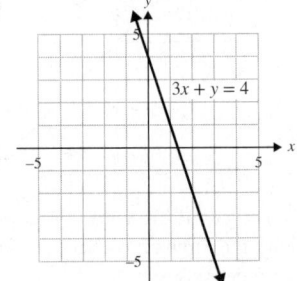

22) $3x - 8y = 8$ 23) $y = \dfrac{3}{4}x$

24) a) -37 b) $8a + 3$ c) $8t + 19$

25) a)   b)

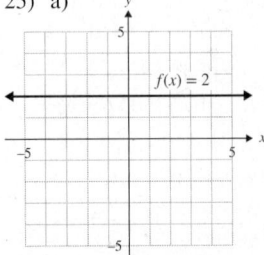

Chapter 5

Section 5.1

1) yes   3) no   5) yes   7) no

9) The lines are parallel.

11) (3, 1)

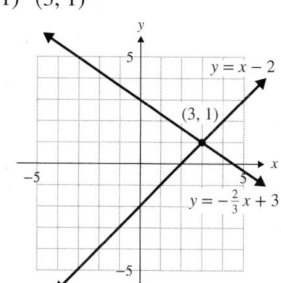

13) (2, 3)

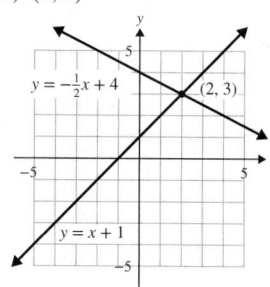

15) (4, −5)

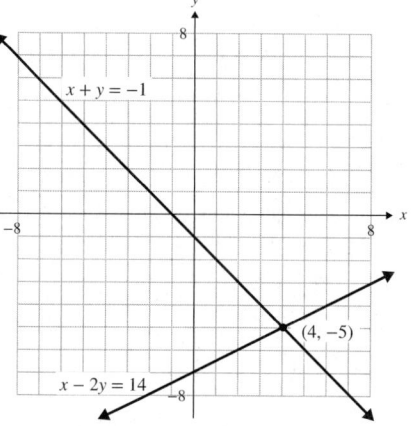

17) (−1, −4)

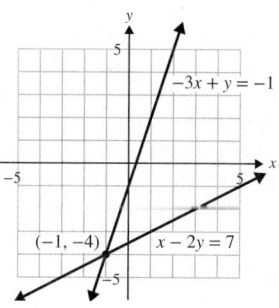

19) ∅; inconsistent system

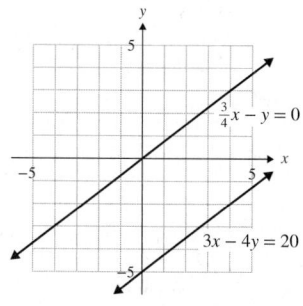

21) infinite number of solutions of the form
$$\left\{(x, y) \,\middle|\, y = \frac{1}{3}x - 2\right\};\ \text{dependent equations}$$

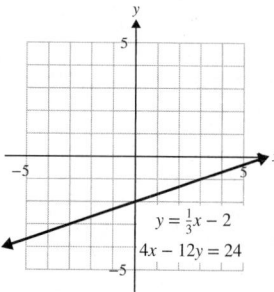

23) (0, 2)

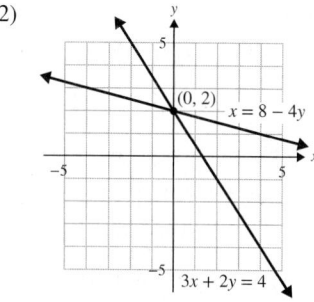

25) infinite number of solutions of the form
$\{(x, y) \,|\, y = -3x + 1\};$ dependent equations

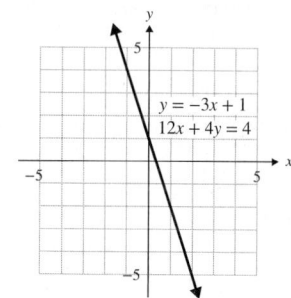

27) (−2, 2)

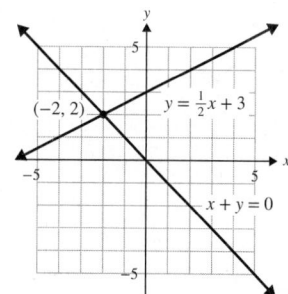

29) (1, −1)

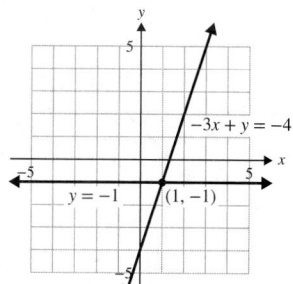

31) ∅; inconsistent system

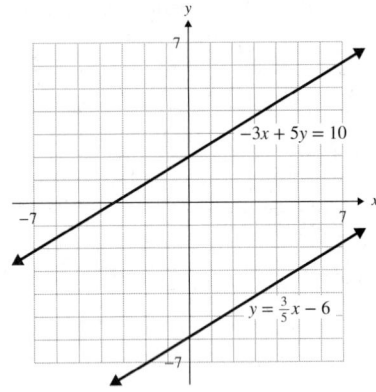

33) Answers may vary.   35) Answers may vary.

37) Answers may vary.

39) C; (−3, 4) is in quadrant II.

41) B; (4.1, 0) is the only point on the positive *x*-axis.

43) The slopes are different.   45) one solution

47) no solution   49) infinite number of solutions

51) one solution   53) no solution

55) a) No
    b) (2017, 2.3); in 2017, the percentage of electric and hybrid cars that were registered was the same: 2.3%.
    c) 2017–2018
    d) 2017–2018; this line segment has the steepest slope of any of the other line segments for hybrid cars.

57) 3

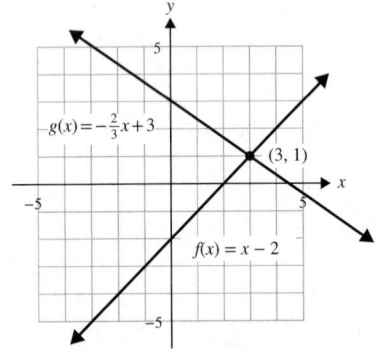

59) −1

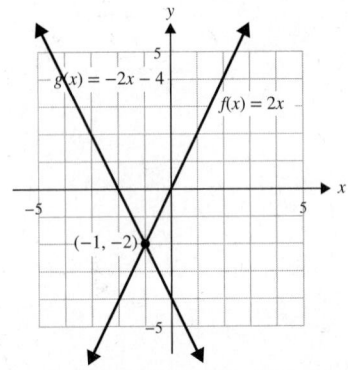

61) (3, −4)   63) (4, 1)   65) (−2.25, −1.6)

Section 5.2

1) It is the only variable with a coefficient of 1.

3) The variables are eliminated, and you get a false statement.

5) (2, 5)   7) (−3, −2)   9) (1, −2)

11) (0, −7)   13) ∅

15) infinite number of solutions of the form
$\{(x, y) | x − 2y = 10\}$

17) $\left(-\dfrac{4}{5}, 3\right)$ 19) (4, 5)

21) infinite number of solutions of the form
$\{(x, y) | -x + 2y = 2\}$

23) (−3, 4)   25) $\left(\dfrac{5}{3}, 2\right)$ 27) ∅

29) Multiply the equation by the LCD of the fractions to eliminate the fractions.

31) (6, 1)   33) (−6, 4)   35) (3, −2)

37) infinite number of solutions of the form
$\left\{(x, y) \,\middle|\, y − \dfrac{5}{2}x = -2\right\}$

39) (8, 0)   41) (3, 5)

43) (1.5, −1)   45) (−16, −12)   47) ∅

49) (6, 4)   51) (−4, −9)   53) (0, −8)

55) a) A+: $96.00; Rock Bottom: $110.00
    b) A+: $180.00; Rock Bottom: $145.00
    c) (200, 120); If the cargo trailer is driven 200 miles, the cost would be the same from each company: $120.00.
    d) If it is driven less than 200 miles, it is cheaper to rent from A+. If it is driven more than 200 miles, it is cheaper to rent from Rock Bottom Rental. If the trailer is driven exactly 200 miles, the cost is the same from each company.

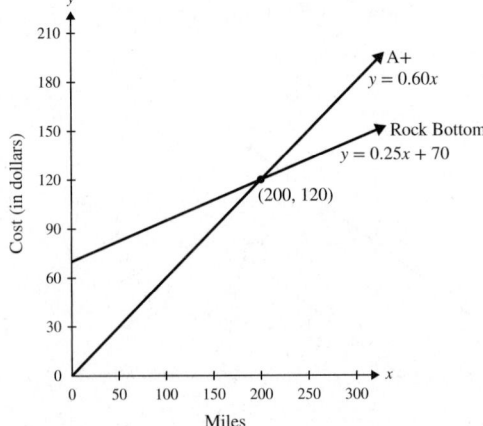

57) Answers may vary. One system is $x + y = 5$
                                      $x + 2y = 13$

59) Answers may vary. One system is $4x − 2y = -4$
                                     $4x + 4y = 11$

Section 5.3

1) Add the equations.   3) always   5) (5, 8)

7) $(-6, -2)$ 9) $(-7, 2)$ 11) $(2, 0)$

13) $(7, -1)$ 15) $(0, 2)$ 17) $\left(-\dfrac{2}{3}, -5\right)$

19) infinite number of solutions of the form $\{(x, y)\,|\,9x - y = 2\}$

21) $(8, 1)$ 23) $\left(5, -\dfrac{3}{2}\right)$ 25) $\varnothing$ 27) $(9, 5)$ 29) $\varnothing$

31) Eliminate the fractions. Multiply the first equation by 4, and multiply the second equation by 24.

33) $\left(\dfrac{5}{8}, 4\right)$ 35) $\left(-\dfrac{9}{2}, -13\right)$ 37) $(-6, 1)$

39) infinite number of solutions of the form
$$\left\{(x, y)\,\middle|\,y = \dfrac{2}{3}x - 7\right\}$$

41) $(1, 1)$ 43) $(-7, -4)$ 45) $(12, -1)$

47) $\varnothing$ 49) $(0.25, 5)$ 51) $(4, 3)$

53) $\left(-\dfrac{3}{2}, 4\right)$ 55) $(1, 1)$ 57) $\left(-\dfrac{123}{17}, \dfrac{78}{17}\right)$

59) $\left(-\dfrac{203}{10}, \dfrac{49}{5}\right)$ 61) 3   63) -8

65) a) 5   b) c can be any real number except 5.

67) a) 3   b) a can be any real number except 3.

69) $\left(\dfrac{2}{5}, \dfrac{2}{b}\right)$ 71) $\left(-\dfrac{1}{4a}, \dfrac{19}{4b}\right)$

Chapter 5: Putting It All Together

1) Elimination method; none of the coefficients is 1 or -1; $(5, 6)$.

2) Substitution; the first equation is solved for x and does not contain any fractions; $(3, 5)$.

3) Since the coefficient of y in the second equation is 1, you can solve for y and use substitution. Or, multiply the second equation by 5 and use the elimination method. Either method will work well; $\left(\dfrac{1}{4}, -3\right)$.

4) Elimination method; none of the coefficients is 1 or -1; $\left(0, -\dfrac{2}{5}\right)$.

5) Substitution; the second equation is solved for x and does not contain any fractions; $(1, -7)$.

6) The second equation is solved for y, but it contains two fractions. Multiply this equation by 4 to eliminate the fractions, then write it in the form $Ax + By = C$. Use the elimination method to solve the system; $(6, 4)$.

7) $(6, 0)$ 8) $(-2, 2)$ 9) $\left(-\dfrac{2}{3}, \dfrac{1}{5}\right)$

10) $(1, 4)$ 11) $\varnothing$

12) infinite number of solutions of the form
$$\{(x, y)\,|\,y = -6x + 5\}$$

13) $\left(-\dfrac{1}{2}, 1\right)$ 14) $\left(\dfrac{5}{6}, -\dfrac{3}{2}\right)$ 15) $(4, 3)$ 16) $\left(\dfrac{1}{6}, 6\right)$

17) $(9, -7)$ 18) $(10, -9)$ 19) $(0, 4)$ 20) $(-3, -1)$

21) infinite number of solutions of the form
$$\{(x, y)\,|\,3x - y = 5\}$$

22) $\varnothing$ 23) $(-9, -14)$ 24) $\left(2, \dfrac{4}{3}\right)$ 25) $\left(-\dfrac{68}{41}, \dfrac{64}{41}\right)$

26) $\left(-\dfrac{53}{34}, \dfrac{24}{17}\right)$ 27) $\left(\dfrac{3}{4}, 0\right)$ 28) $\left(5, \dfrac{3}{4}\right)$ 29) $(1, 1)$

30) $(-2, 4)$ 31) $\left(-\dfrac{5}{2}, 10\right)$ 32) $(-4, -6)$

33) Answers may vary.   34) Answers may vary.

35) Answers may vary.   36) Answers may vary.

37) $(2, 2)$

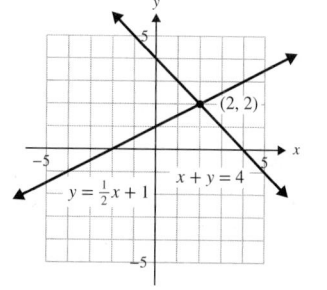

38) $(-1, -2)$

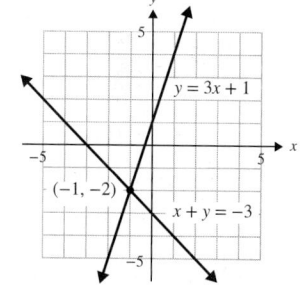

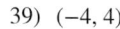

39) $(-4, 4)$

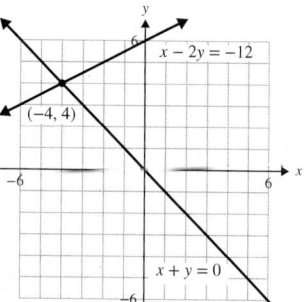

40) $(2, 4)$

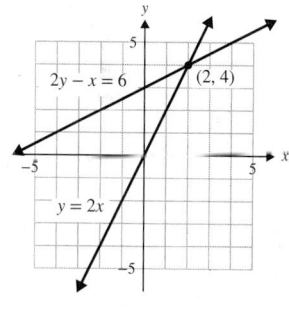

41) $\varnothing$

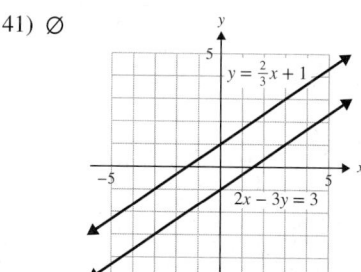

42) infinite number of
solutions of the form
$\left\{ (x, y) \middle| y = -\dfrac{5}{2}x - 3 \right\}$

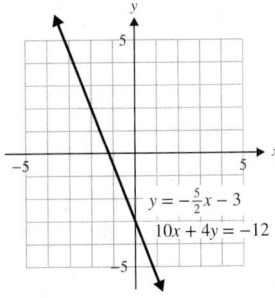

$y = -\frac{5}{2}x - 3$

$10x + 4y = -12$

43) $(-1.25, -0.5)$ 44) $(7.5, -13)$

Section 5.4

1) 38 and 49

3) *Star Wars*: $248.0 million; *Jurassic World*: $208.8 million

5) Beyoncé: 49; Jay-Z: 35

7) Vietnamese: 1,560,000; Chinese: 3,120,000

9) Gagarin: 108 min; Shepard: 15 min

11) width: 30 in.; height: 80 in.

13) length: 110 mm; width: 61.8 mm

15) width: 34 ft; length: 51 ft

17) $m\angle x = 67.5°$; $m\angle y = 112.5°$

19) T-shirt: $20.00; hockey puck: $8.00

21) Hodor mug: $14.99; Daenerys figurine: $9.99

23) hamburger: $0.82; medium fries: $2.15

25) wrapping paper: $7.00; gift bags: $8.00

27) 9%: 3 oz; 17%: 9 oz   29) pure acid: 2 L; 25%: 8 L

31) Asian Treasure: 24 oz; Pearadise: 36 oz

33) taco: 310 mg; chalupa: 560 mg

35) 2%: $2500; 4%: $3500   37) $0.49: 12; $0.34: 4

39) Sheldon: 9 mph; Amy: 8 mph

41) small plane: 240 mph; jet: 400 mph

43) Lori: 8 mph; Rick: 10 mph

45) speed of boat in still water: 6 mph; speed of the current: 1 mph

47) speed of boat in still water: 13 mph; speed of the current: 3 mph

49) speed of jet in still air: 450 mph; speed of the wind: 50 mph

Section 5.5

1) yes   3) no   5) Answers may vary.   7) $(-2, 0, 5)$

9) $(1, -1, 4)$ 11) $\left(2, -\dfrac{1}{2}, \dfrac{5}{2}\right)$ 13) $\varnothing$; inconsistent

15) $\{(x, y, z) | 5x + y - 3z = -1\}$; dependent equations

17) $\{(a, b, c) | -a + 4b - 3c = -1\}$; dependent equations

19) $(2, 5, -5)$ 21) $\left(-4, \dfrac{3}{5}, 4\right)$ 23) $(0, -7, 6)$

25) $(1, 5, 2)$ 27) $\left(-\dfrac{1}{4}, -5, 3\right)$ 29) $\varnothing$; inconsistent

31) $\left(4, -\dfrac{3}{2}, 0\right)$ 33) $(4, 4, 4)$

35) $\{(x, y, z) | -4x + 6y + 3z = 3\}$; dependent equations

37) $\left(1, -7, \dfrac{1}{3}\right)$ 39) $(0, -1, 2)$ 41) $(-3, -1, 1)$

43) Answers may vary.

45) hot dog: $2.00; fries: $1.50; soda: $2.00

47) Clif Bar: 10 g; Balance Bar: 14 g; PowerBar: 20 g

49) Knicks: $3.6 billion; Lakers: $3.3 billion; Warriors: $3.1 billion

51) bronze: $19; silver: $26; gold: $38   53) $104°, 52°, 24°$

55) $80°, 64°, 36°$ 57) 12 cm, 10 cm, 7 cm

59) $(3, 1, 2, 1)$ 61) $(0, -3, 1, -4)$

Section 5.6

1) $\begin{bmatrix} 1 & -7 & | & 15 \\ 4 & 3 & | & -1 \end{bmatrix}$ 3) $\begin{bmatrix} 1 & 6 & -1 & | & -2 \\ 3 & 1 & 4 & | & 7 \\ -1 & -2 & 3 & | & 8 \end{bmatrix}$

5) $3x + 10y = -4$
$x - 2y = 5$

7) $x - 6y = 8$
$y = -2$

9) $x - 3y + 2z = 7$
$4x - y + 3z = 0$
$-2x + 2y - 3z = -9$

11) $x + 5y + 2z = 14$
$y - 8z = 2$
$z = -3$

13) $(3, -1)$ 15) $(10, -4)$ 17) $(0, -2)$

19) $(-1, 4, 8)$ 21) $(10, 1, -4)$ 23) $(0, 1, 8)$

25) $\varnothing$; inconsistent   27) $(-5, 2, 1, -1)$ 29) $(3, 0, -2, 1)$

Chapter 5: Group Activity

1) $a = 12, b = 13$ 2) $\dfrac{x}{y} = -\dfrac{4}{17}$ 3) yes; $\left(\dfrac{4}{9}, \dfrac{47}{9}\right)$

Chapter 5: Review Exercises

1) no   3) The lines are parallel.

5) $\varnothing$ 7) $(-3, -1)$

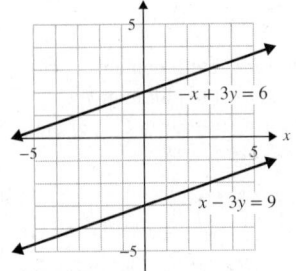

$-x + 3y = 6$

$x - 3y = 9$

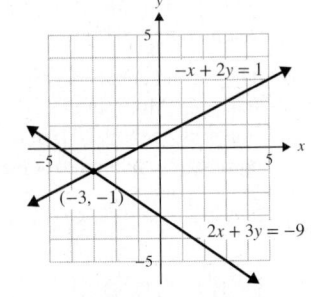

$-x + 2y = 1$

$(-3, -1)$

$2x + 3y = -9$

9) infinite number of solutions   11) $(2, 5)$

13) $(5, 3)$ 15) $(-4, -1)$

17) infinite number of solutions of the form $\{(x, y) \mid 5x - 2y = 4\}$

19) $\left(\dfrac{33}{43}, -\dfrac{36}{43}\right)$

21) when one of the variables has a coefficient of 1 or -1

23) $\left(-\dfrac{5}{3}, 2\right)$ 25) $\varnothing$ 27) $(0, 3)$ 29) $\left(\dfrac{3}{4}, 0\right)$

31) infinite number of solutions of the form
$$\left\{(x, y) \,\middle|\, y = -\dfrac{9}{7}x + \dfrac{6}{7}\right\}$$

33) white: 94; chocolate: 47

35) Edwin: 8 mph; Camille: 6 mph

37) length: 12 cm; width: 7 cm   39) quarters: 35; dimes: 28

41) hand warmers: \$4.50; socks: \$18.50

43) no   45) $(3, -1, 4)$ 47) $\left(-1, 2, \dfrac{1}{2}\right)$

49) $\left(3, \dfrac{2}{3}, -\dfrac{1}{2}\right)$ 51) $\varnothing$; inconsistent

53) $\{(a, b, c) \mid 3a - 2b + c = 2\}$; dependent

55) $(1, 0, 3)$ 57) $\left(\dfrac{3}{4}, -2, 1\right)$

59) Propel: 160 mg; Powerade: 150 mg; Gatorade: 160 mg

61) Blair: 65; Serena: 50; Chuck: 25

63) ice cream cone: \$1.50; shake: \$2.50; sundae: \$3.00

65) $92°, 66°, 22°$ 67) $(-9, 2)$ 69) $(1, 0)$ 71) $(5, -2, 6)$

Chapter 5: Test

1) yes   2) $(4, -2)$

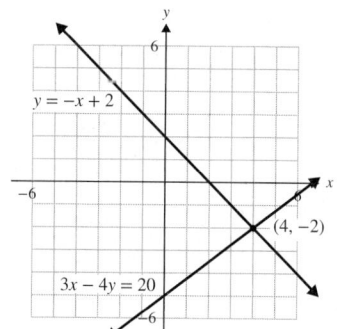

3) $\varnothing$

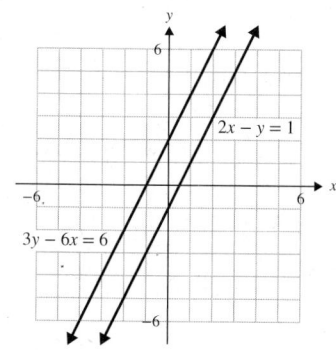

4) a) In 2017, approximately 3.5% of the people in Indiana were unemployed.

   b) (2015, 4.8); In 2015, 4.8% of the population of Indiana and Massachusetts were unemployed.

   c) 2014 and 2016

   d) Indiana from 2014 to 2015; During this time, Indiana had the largest decrease in its unemployment rate of all of the years reported on the graph.

5) $\left(-5, -\dfrac{1}{2}\right)$

6) infinite number of solutions of the form
$$\left\{(x, y) \,\middle|\, y = \dfrac{1}{2}x - 3\right\}$$

7) $(3, 1)$ 8) $(0, 6)$ 9) $\varnothing$ 10) $(-2, -8)$

11) $(2, -4)$ 12) $\left(-\dfrac{4}{3}, 0\right)$ 13) $(0, 3, -2)$

14) Answers may vary.

15) Yellowstone: 2.2 million acres; Death Valley: 3.3 million acres

16) adult: \$45.00; child: \$20.00

17) length: 38 cm; width: 19 cm

18) 12%: 40 mL; 30%: 32 mL

19) Rory: 42 mph; Lorelai: 38 mph

20) $105°, 42°, 33°$ 21) $(6, -2)$ 22) $(1, -1, 1)$

Chapter 5: Cumulative Review for Chapters 1–5

1) $\dfrac{41}{30}$ 2) $9\dfrac{1}{3}$ 3) -29 4) 30 in^2

5) $-12x^2 - 15x + 3$ 6) $80°, 100°$

7) $\{25\}$ 8) $\{48\}$ 9) $\left\{-\dfrac{13}{6}\right\}$ 10) $\varnothing$

11) $\left\{\dfrac{1}{2}\right\}$ 12) $(1, 6)$ 13) 21.8 mpg

14) a) $h = \dfrac{2A}{b_1 + b_2}$ b) 6 cm

15)

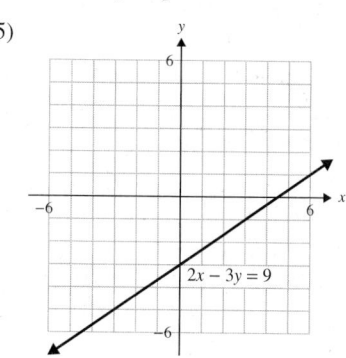

16) x-int: $(16, 0)$; y-int: $(0, -2)$ 17) $y = \dfrac{1}{4}x + \dfrac{5}{4}$

18) perpendicular   19) (4, 10)   20) (3, 1)   21) $\emptyset$

22) infinite number of solutions of the form
$\{(x, y) | 3x + 9y = -2\}$

23) $(-2, -7)$

24) 4-foot boards: 16; 6-foot boards: 32

25) kitty: 3.8 million; puppy: 2.2 million; pig: 1.1 million

Chapter 6

Section 6.1

1) quotient rule; k^6 3) power rule for a product; $16h^4$

5) 64   7) -64 9) $\dfrac{1}{6}$ 11) 81   13) $\dfrac{16}{81}$

15) -31 17) $\dfrac{1}{64}$ 19) t^{13} 21) $-16c^9$ 23) z^{24}

25) $125p^{30}$ 27) $-\dfrac{8}{27}a^{21}b^3$ 29) f^4 31) $7v$

33) $\dfrac{d^4}{6}$ 35) $\dfrac{1}{x^6}$ 37) $\dfrac{1}{m}$ 39) $\dfrac{3}{2k^4}$ 41) $20m^8n^{14}$

43) $24y^8$ 45) $\dfrac{1}{49a^8b^2}$ 47) $\dfrac{b^5}{a^3}$ 49) $\dfrac{x^3}{y^{20}}$ 51) $\dfrac{4a^2b^{15}}{3c^{17}}$

53) $\dfrac{y^{15}}{x^5}$ 55) $\dfrac{64c^6}{a^6b^3}$ 57) $\dfrac{9}{h^8k^8}$ 59) $\dfrac{c^6}{27d^{18}}$ 61) $\dfrac{u^{13}v^4}{32}$

63) $A = 10x^2$ sq units;   65) $A = \dfrac{3}{16}p^2$ sq units;
$P = 14x$ units $P = 2p$ units

67) k^{6a} 69) g^{8x} 71) x^{3b} 73) $\dfrac{1}{8r^{18m}}$

Section 6.2

1) Yes; the coefficients are real numbers and the exponents are whole numbers.

3) No; one of the exponents is a negative number.

5) No; two of the exponents are fractions.

7) binomial   9) trinomial

11) monomial   13) always

15) It is the same as the degree of the term in the polynomial with the highest degree.

17) Add the exponents on the variables.

19)

Term	Coeff.	Degree
$3y^4$	3	4
$7y^3$	7	3
$-2y$	-2	1
8	8	0

Degree of polynomial is 4.

21)

Term	Coeff.	Degree
$-4x^2y^3$	-4	5
$-x^2y^2$	-1	4
$\frac{2}{3}xy$	$\frac{2}{3}$	2
$5y$	5	1

Degree of polynomial is 5.

23) a) 1   b) 13   25) 37

27) 146   29) -23 31) never

33) a) $y = 680$; If he rents the equipment for 5 hours, the cost of building the road will be $680.00.
b) $920.00   c) 8 hours

35) $-11c^2 + 7c$ 37) $-4.3t^6 + 4.2t^2$ 39) $6a^2b^2 - 7ab^2$

41) $9n - 5$ 43) $12r^2 + 6r + 11$

45) $4b^2 - 3$ 47) $\dfrac{7}{18}w^4 - \dfrac{1}{2}w^2 - \dfrac{3}{8}w - \dfrac{3}{2}$

49) $2m^2 + 3m + 19$ 51) $10c^4 - \dfrac{1}{5}c^3 - \dfrac{1}{4}c + \dfrac{25}{9}$

53) $1.2d^3 + 7.7d^2 - 11.3d + 0.6$ 55) $12w - 4$

57) $-y + 2$ 59) $-2b^2 - 10b + 19$

61) $4f^4 - 14f^3 + 6f^2 - 8f + 9$ 63) $5.8r^2 + 6.5r + 11.8$

65) $7j^2 + 9j - 5$ 67) $8s^5 - 4s^4 - 4s^2 + 1$

69) $\dfrac{1}{16}r^2 + \dfrac{7}{9}r - \dfrac{5}{6}$ 71) Answers may vary.

73) No. If the coefficients of the like terms are opposite in sign, their sum will be zero.
Example: $(3x^2 + 4x + 5) + (2x^2 - 4x + 1) = 5x^2 + 6$

75) $-7a^4 - 12a^2 + 14$ 77) $4w^3 + 7w^2 - w - 2$

79) $\dfrac{4}{3}y^3 - \dfrac{7}{4}y^2 + 3y - \dfrac{1}{14}$ 81) $m^3 - 7m^2 - 4m - 7$

83) $9p^2 + 2p - 8$ 85) $5z^6 + 9z^2 - 4$ 87) $-7p^2 + 9p$

89) sometimes   91) $4w + 14z$ 93) $-5ac + 12a + 5c$

95) $4u^2v^2 + 31uv - 4$ 97) $17x^3y^2 - 11x^2y^2 - 20$

99) $6x + 6$ units   101) $10p^2 - 2p - 6$ units

103) a) 16   b) 4   105) a) 29   b) 5   107) 6

109) 40   111) a) 29,000   b) 20,880,000   c) 90 sec or 1.5 min

Section 6.3

1) Answers may vary.   3) $24m^8$ 5) $-32c^6$

7) $10a^2 - 35a$ 9) $6v^5 - 24v^4 - 12v^3$

11) $-36b^5 + 18b^4 + 54b^3 + 81b^2$

13) $3a^3b^3 + 18a^3b^2 - 39a^2b^2 + 21a^2b$

15) $-9k^6 - 12k^5 + \dfrac{9}{5}k^4$ 17) always

19) $6c^2 + 31c + 28$ 21) $3f^3 - 13f^2 - 14f + 20$

23) $8x^4 - 22x^3 + 17x^2 - 26x - 10$

25) $4y^4 + \frac{7}{3}y^3 + 45y^2 + 28y - 36$

27) $-24h^5 + 26h^4 - 53h^3 + 19h^2 - 18h$

29) $15y^3 - 22y^2 + 17y - 6$ 31) First, Outer, Inner, Last

33) $w^2 + 12w + 35$ 35) $r^2 + 6r - 27$ 37) $y^2 - 8y + 7$

39) $3p^2 + p - 14$ 41) $21n^3 + 7n^2 + 12n + 4$

43) $4w^2 - 17w + 15$ 45) $12a^2 + ab - 20b^2$

47) $v^2 + \frac{13}{12}v + \frac{1}{4}$ 49) $\frac{1}{3}a^2 + \frac{17}{6}ab^2 - 5b^4$

51) a) $4y + 4$ units   b) $y^2 + 2y - 15$ sq units

53) Both are correct.   55) $15y^2 + 54y - 24$

57) $-24g^4 - 36g^3 + 60g^2$ 59) $c^3 + 6c^2 + 5c - 12$

61) $5n^5 + 55n^3 + 150n$ 63) $2r^3 - 3r^2t - 3rt^2 + 2t^3$

65) $9m^2 - 4$ 67) $49a^2 - 64$ 69) $4p^2 - 49q^2$

71) $n^2 - \frac{1}{4}$ 73) $\frac{4}{9} - k^2$ 75) $0.09x^2 - 0.16y^2$

77) $25x^4 - 16$ 79) $y^2 + 16y + 64$ 81) $t^2 - 22t + 121$

83) $16w^2 + 8w + 1$ 85) $4d^2 - 20d + 25$

87) $36a^2 - 60ab + 25b^2$ 89) $c^4 - 18c^2 + 81$

91) $9m^2 + 6mn + n^2 + 12m + 4n + 4$

93) $x^2 - 8x + 16 - 2xy + 8y + y^2$

95) No. The order of operations tells us to perform exponents, $(t + 3)^2$, before multiplying by 4.

97) $6x^2 + 12x + 6$ 99) $2a^3 + 12a^2 + 18a$

101) $r^3 + 15r^2 + 75r + 125$ 103) $s^3 - 6s^2 + 12s - 8$

105) $y^4 + 8y^3 + 24y^2 + 32y + 16$

107) $v^2 - 10vw + 25w^2 - 16$ 109) $4a^2 + 4ab + b^2 - c^2$

111) No; $(x + 5)^2 = x^2 + 10x + 25$

113) $c^2 - 5c - 84$ 115) $10k^3 - 37k^2 - 38k + 9$

117) $\frac{1}{36} - h^2$ 119) $27c^3 + 27c^2 + 9c + 1$ 121) $\frac{9}{32}p^{11}$

123) $14y^2 - 35y + 18$ 125) $h^3 + 6h^2 + 12h + 8$ cubic units

127) $9x^2 + 33x + 14$ sq units

Section 6.4

1) dividend: $12c^3 + 20c^2 - 4c$; divisor: $4c$;
   quotient: $3c^2 + 5c - 1$

3) Answers may vary.   5) $7k^2 + 2k - 10$

7) $2u^5 + 2u^3 + 5u^2 - 8$ 9) $-5d^3 + 1$

11) $\frac{3}{2}w^2 + 7w - 1 + \frac{1}{2w}$

13) $\frac{5}{2}v^3 - 9v - \frac{11}{2} - \frac{5}{4v^2} + \frac{1}{4v^4}$

15) $9a^3b + 6a^2b - 4a^2 + 10a$

17) $-t^4u^2 + 7t^3u^2 + 12t^2u^2 - \frac{1}{9}t^2$

19) The answer is incorrect. When you divide $4t$ by $4t$, you get 1. The quotient should be $4t^2 - 9t + 1$.

21) 528   23) $158\frac{1}{6}$ 25) $437\frac{4}{9}$

27) $g + 4$ 29) $p + 6$ 31) $k - 5$ 33) $h + 8$

35) $2a^2 - 7a - 3$ 37) $3p^2 + 7p - 1$

39) $6t + 23 + \frac{119}{t - 5}$ 41) $4z^2 + 8z + 7 - \frac{72}{3z + 5}$

43) $w^2 - 4w + 16$ 45) $2r^2 + 8r + 3$

47) $3t^2 - 8t - 6 + \frac{2t - 4}{5t^2 - 1}$ 49) sometimes

51) $x^2y^2 + 5x^2y - \frac{1}{6} + \frac{1}{2xy}$ 53) $-2g^3 - 5g^2 + g - 4$

55) $6t + 5 + \frac{20}{t - 8}$ 57) $4n^2 + 10n + 25$ 59) $5x^2 - 7x + 3$

61) $4a^2 - 3a + 7$ 63) $5h^2 - 3h - 2 + \frac{1}{2h^2 - 9}$

65) $3d^2 - d - 8$ 67) $9c^3 + 8c + 3 + \frac{7c + 4}{c^2 - 10c + 4}$

69) $k^2 - 9$ 71) $-\frac{5}{7}a^3 - 7a + 2 + \frac{15}{7a}$

73) $4v^2 - 5v + 2 - \frac{4v + 5}{3v^2 - 1}$ 75) $\frac{1}{2}x + 3$

77) $\frac{2}{3}w + 2$ 79) $4y + 1$ 81) $2a^2 - 5a + 1$

83) $12h^2 + 6h + 2$

Chapter 6: Group Activity

1) a)

b) $n + 1$ c) an odd number; an even number

d) 2^n e) yes

2) a) 4; $n + 1$ b) They are in row $n = 3$ of Pascal's triangle.

c) a decreases by 1 from 3 to 0: b increases by 1 from 0 to 3.

d) Their sum equals n.

3) $(a + b)^9 = a^9 + 9a^8b + 36a^7b^2 + 84a^6b^3 + 126a^5b^4 + 126a^4b^5 + 84a^3b^6 + 36a^2b^7 + 9ab^8 + b^9$

4) a) $a^4 + 4a^3b + 6a^2b^2 + 4ab^3 + b^4$
   b) $x^4 + 8x^3 + 24x^2 + 32x + 16$
   c) $x^4 - 8x^3 + 24x^2 - 32x + 16$
   d) $81x^4 + 216x^3y + 216x^2y^2 + 96xy^3 + 16y^4$
   e) $x^8 + 8x^6y + 24x^4y^2 + 32x^2y^3 + 16y^4$

5) a) 125   b) x^{12} 6) a) 16   b) 9   c) 46   d) 8

Chapter 6: Review Exercises

1) 81   3) $\dfrac{64}{125}$ 5) z^{18} 7) $-54t^7$ 9) $\dfrac{1}{k^8}$

11) $-\dfrac{40b^4}{a^6}$ 13) $\dfrac{4q^{30}}{9p^6}$ 15) 14   17) y^{6p}

19) False. $-x^2 = -1 \cdot x \cdot x$. If $x \neq 0$, then $-x^2$ is a negative number. But $(-x)^2 = -x \cdot (-x)$, which is a positive number when $x \neq 0$.

21)

Term	Coeff.	Degree
$7s^3$	7	3
$-9s^2$	-9	2
s	1	1
6	6	0

Degree of polynomial is 3.

23) 31

25) a) 20   b) -6   27) $-2c^2 + c + 5$

29) $9.8j^3 + 4.3j^2 + 3.4j - 0.5$ 31) $\dfrac{1}{2}k^2 - k + 6$

33) $-3x^2y^2 + 9x^2y - 5xy + 6$ 35) a) \$150,000   b) \$60,800

37) $4d^2 + 6d + 6$ units   39) $24r^2 - 39r$

41) $-32w^4 - 24w^3 - 8w^2 - 2w + 3$

43) $y^2 - 12y + 27$ 45) $30d^3 + 25d^2 + 12d + 10$

47) $4p^4q^4 + 66p^3q^4 - 6p^2q^3 + 24pq^2$ 49) $4x^2 - 16xy - 9y^2$

51) $10x^6 + 50x^5 - 123x^4 - 15x^3 + 42x^2 + 30x - 72$

53) $-15u^2 - 27u + 132$ 55) $z^3 + 8z^2 + 19z + 12$

57) $\dfrac{1}{7}d^2 - \dfrac{11}{14}d - 24$ 59) $c^2 + 8c + 16$

61) $16p^2 - 24p + 9$ 63) $x^3 - 9x^2 + 27x - 27$

65) $8a^3 + 60a^2b + 150ab^2 + 125b^3$

67) $m^2 - 6m + 9 + 2mn - 6n + n^2$ 69) $p^2 - 169$

71) $\dfrac{81}{4} - \dfrac{25}{36}x^2$ 73) $9a^4 - \dfrac{1}{4}b^2$ 75) $3u^3 + 24u^2 + 48u$

77) a) $2n^2 + 7n - 22$ sq units   b) $6n + 18$ units

79) $8y^3$ 81) $4t^2 - 10t - 5$ 83) $w + 5$

85) $4r^2 + r - 3$ 87) $t + 2 - \dfrac{3}{2t} + \dfrac{10}{7t^2}$

89) $2v - 1 + \dfrac{6}{4v + 9}$ 91) $3v^2 - 7v + 8$

93) $c^2 + 2c + 4$ 95) $6k^2 - 4k + 7 - \dfrac{18}{3k + 2}$

97) $-\dfrac{5}{3}x^3y^2 - 4xy^2 + 1 - \dfrac{5}{4y^2}$

99) $2n^2 - n + 7 + \dfrac{4n + 1}{5n^2 + 3n - 2}$ 101) false

103) $8a + 2$ units   105) $20c^3 - 12c^2 - 11c + 1$

107) $144 - 49w^2$ 109) $-40r^{21}t^{27}$

111) $13a^3b^3 + 7ab^2 - \dfrac{5}{3}b + \dfrac{1}{3ab^2}$

113) $h^3 - 15h^2 + 75h - 125$ 115) $2c^2 - 8c + 9 + \dfrac{5}{c + 4}$

117) $\dfrac{y^{12}}{125}$ 119) $2p^2 + 3p - 5$

Chapter 6: Test

1) a) -81 b) $\dfrac{1}{32}$ c) -2 d) $\dfrac{1000}{27}$ e) $\dfrac{1}{64}$

2) $-30p^{12}$ 3) $\dfrac{1}{a^4b^6}$ 4) $\dfrac{8}{y^9}$ 5) $\dfrac{9x^4}{4y^{18}}$ 6) t^{13k}

7) a) -1 b) 3   8) 9   9) 3   10) Answers may vary.

11) a) $\dfrac{11}{6}$ b) -7 c) -44 12) $24h^5 - 12h^4 + 4h^3$

13) $12a^3b^2 - 3a^2b^2 - 3ab + 9$ 14) $9y^2 - 3y - 7$

15) $-8n^3 - 43n^2 + 12n - 9$ 16) $u^2 - 14u + 45$

17) $8g^2 + 10g + 3$ 18) $v^2 - \dfrac{4}{25}$ 19) $6x^2 - 11xy - 7y^2$

20) $36t^3 - 28t^2 + 9t - 7$ 21) $-12n^3 - 8n^2 + 63n - 40$

22) $2y^3 + 24y^2 + 72y$ 23) $9m^2 - 24m + 16$

24) $\dfrac{16}{9}x^2 + \dfrac{8}{3}xy + y^2$ 25) $25a^2 - 10ab + b^2 - 30a + 6b + 9$

26) $t^3 - 6t^2 + 12t - 8$ 27) $w + 3$

28) $3m^2 - 5m + 1 - \dfrac{3}{4m}$ 29) $6p^2 - p + 5 - \dfrac{15}{3p - 7}$

30) $y^2 + 3y + 9$ 31) $2r^2 + 3r - 4$

32) a) sometimes   b) sometimes   c) never

33) a) $3d^2 - 14d - 5$ sq units   b) $8d - 8$ units

34) a) \$121.80   b) 53

Chapter 6: Cumulative Review for Chapters 1–6

1) a) $\{41, 0\}$ b) $\{-15, 41, 0\}$

   c) $\left\{\dfrac{3}{8}, -15, 2.1, 41, 0.\overline{52}, 0\right\}$

2) -87 3) $\dfrac{75}{31}$ or $2\dfrac{13}{31}$ 4) $\left\{-\dfrac{35}{3}\right\}$ 5) $\varnothing$

6) $\left\{-\dfrac{1}{2}\right\}$ 7) $(-\infty, -5]$

8) 45 ml of 12% solution, 15 ml of 4% solution

9) x-int: (8, 0); y-int: (0, −3)

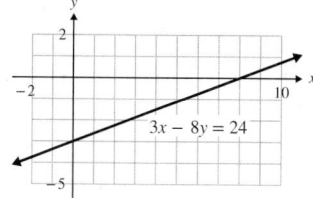

$3x − 8y = 24$

10)

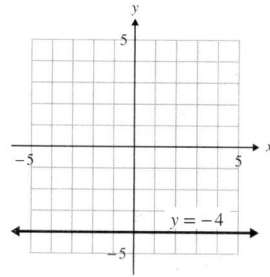

$y = −4$

11) $3x + y = −5$

12) $y = −\dfrac{1}{4}x + 3$ 13) $\left(-5, \dfrac{1}{2}\right)$

14) width: 12 cm; length: 35 cm

15) a) −13   b) 25   c) $−4a + 43$ d) $−\dfrac{5}{2}$

16) $−32a^{14}$ 17) c^{17} 18) $\dfrac{64}{p^{21}}$

19) $−2q^2 − 45$ 20) $n^2 + n − 56$

21) $9a^2 − 121$ 22) $ab^2 − \dfrac{3}{2b} + \dfrac{5}{a^2b} + \dfrac{1}{2a^3b}$

23) $5p^2 + p − 7 − \dfrac{16}{p − 3}$

24) $5c^3 − 40c^2 + 80c$ 25) $4z^2 − 2z + 1$

Chapter 7

Section 7.1

1) 7   3) $6p^2$ 5) $4n^6$ 7) $5a^2b$ 9) $21r^3s^2$

11) ab 13) $(k − 9)$ 15) Answers may vary.

17) yes   19) no   21) yes   23) $2(w + 5)$

25) $9(2z^2 − 1)$ 27) $10m(10m^2 − 3)$ 29) $r^2(r^7 + 1)$

31) $\dfrac{1}{5}y(y + 4)$ 33) does not factor

35) $5n^3(2n^2 − n + 8)$

37) $8p^3(5p^3 + 5p^2 − p + 1)$

39) $9a^2b(7ab^2 − 4ab + 1)$ 41) $−6(5n + 7)$

43) $−4w^3(3w^2 + 4)$ 45) $−1(k − 3)$

47) $(t − 5)(u + 6)$ 49) $(6x + 1)(y − z)$

51) $(q + 12)(p + 1)$ 53) $(9k + 8)(5h^2 − 1)$

55) $(b + 2)(a + 7)$ 57) $(3t + 4)(r − 9)$

59) $(2b + 5c)(4b + c^2)$ 61) $(g − 7)(f + 4)$

63) $(t − 10)(s − 6)$ 65) $(5u + 6)(t − 1)$

67) $(12g^3 + h)(3g − 8h)$ 69) Answers may vary.

71) $4(xy + 3x + 5y + 15)$; Group the terms and factor out the GCF from each group; $4(y + 3)(x + 5)$

73) $3(c + 7)(d + 2)$ 75) $2p(p − 4)(q − 5)$

77) $(5s − 6)(2t + 1)$ 79) $(3a^2 − 2b)(a − 7b)$

81) $2uv(4u + 5)(v + 2)$ 83) $(n + 7)(3m + 10)$

85) $8(2b − 3)$ 87) $(d + 6)(c − 4)$

89) $2a^3(3a − 4)(b + 2)$ 91) $(d + 4)(7c + 3)$

93) $(g − 1)(d + 1)$ 95) $x^3y^2(x + 12y)$

97) $4(m + 3)(n + 2)$ 99) $−2(3p^2 + 10p − 1)$

Section 7.2

1) a) 5, 2   b) −8, 7   c) 5, −1   d) −9, −4

3) They are negative.   5) Can I factor out a GCF?

7) Can I factor again?   9) $n + 2$ 11) $c − 10$

13) $x + 4$ 15) $(g + 6)(g + 2)$ 17) $(y + 8)(y + 2)$

19) $(w − 9)(w − 8)$ 21) $(b − 4)(b + 1)$ 23) prime

25) $(c − 9)(c − 4)$ 27) $(m + 10)(m − 6)$

29) $(r − 12)(r + 8)$ 31) prime

33) $(x + 8)(x + 8)$ or $(x + 8)^2$

35) $(n − 1)(n − 1)$ or $(n − 1)^2$

37) $(d + 12)(d + 2)$ 39) prime   41) $2(k − 3)(k − 8)$

43) $5h(h + 5)(h + 2)$ 45) $r^2(r + 12)(r − 11)$

47) $7q(q^2 − 7q − 6)$ 49) $3z^2(z + 4)(z + 4)$ or $3z^2(z + 4)^2$

51) $xy(y − 9)(y + 7)$ 53) $−(m + 5)(m + 7)$

55) $−(c + 7)(c − 4)$ 57) $−(z − 3)(z − 10)$

59) $−(p − 8)(p + 7)$ 61) $(r + 4y)(x + 3y)$

63) $(c − 8d)(c + d)$ 65) $(u − 5v)(u − 9v)$

67) $(m − 3n)(m + 7n)$

69) $(a + 12b)(a + 12b)$ or $(a + 12b)^2$

71) No; from $(3x + 6)$ you can factor out a 3. The correct answer is $3(x + 2)(x + 5)$.

73) yes   75) $2(x + 5)(x + 3)$

77) $(n − 4)(n − 2)$ 79) $(m − 4n)(m + 11n)$

81) prime   83) $4q(q − 3)(q − 4)$

85) $−(k + 9)(k + 9)$ or $−(k + 9)^2$ 87) $4h^3(h + 7)(h + 1)$

89) $(k + 12)(k + 9)$ 91) $pq(p − 7q)(p − 10q)$

93) prime   95) $(x − 12y)(x − y)$

97) $5v^3(v^2 + 11v − 9)$ 99) $6xy^2(x − 9)(x + 1)$

101) $(z − 9)(z − 4)$ 103) $(ab + 6)(ab + 7)$

105) $(x + y)(z + 10)(z − 3)$ 107) $(a − b)(c − 7)(c − 4)$

109) $(p + q)(r + 12)(r + 12)$ or $(p + q)(r + 12)^2$

Section 7.3

1) a) 10, −5   b) −27, −1   c) 6, 2   d) −12, 6

3) $(3c + 8)(c + 4)$ 5) $(6k − 7)(k − 1)$

7) $(2x − 9y)(3x + 4y)$ 9) Can I factor out a GCF?

11) $4k^2 + 17k + 18$ 13) $t + 2$

15) $3a + 2$ 17) $3x − y$

19) $(2h + 3)(h + 5)$ 21) $(7y − 4)(y − 1)$

23) $(5b − 6)(b + 3)$ 25) $(3p + 2)(2p − 1)$

27) $(2t + 3)(2t + 5)$ 29) $(9x − 4y)(x − y)$

31) because 2 can be factored out of $2x − 4$, but 2 cannot be factored out of $2x^2 + 13x − 24$

33) $(2r + 5)(r + 2)$ 35) $(3u − 5)(u − 6)$

37) $(7a − 4)(a + 5)$ 39) $(3y + 10)(2y + 1)$

41) $(9w − 7)(w + 3)$ 43) $(4c − 3)(2c − 9)$

45) $(2k + 11)(2k + 9)$ 47) $(10b + 9)(2b − 5)$

49) $(2r − 3t)(r + 8t)$ 51) $(6a − b)(a − 4b)$

53) $(4z − 3)(z + 2)$; the answer is the same.

55) never   57) $−(n + 12)(n − 4)$ 59) $−(5z − 2)(2z − 3)$

61) $−3z(z + 11)(z − 6)$ 63) $(n + 14)(n + 2)$

65) $(k − 4)(k − 11)$ 67) $(2w − 1)(w − 4)$

69) $(6y − 7)(4y − 1)$ 71) $(3p + 2)(p − 6)$

73) $(3k − 7)^2$ 75) $2(d + 5)(d − 4)$

77) $(7x − 3y)(x − 2y)$ 79) $4q(q − 3)(q − 4)$

81) prime   83) $−(5z − 2)(2z − 3)$

85) $(q − 1)^2(3p − 7)(4p − 7)$

87) $(3b + 25)(b + 3)$

89) $r^2t^2(6r + 1)(5r + 3)$

Section 7.4

1) a) 36   b) 100   c) 16   d) 121   e) 9

f) 64   g) 144   h) $\dfrac{1}{4}$ i) $\dfrac{9}{25}$

3) a) n^2 b) $5t$ c) $7k$ d) $4p^2$ e) $\dfrac{1}{3}$ f) $\dfrac{5}{2}$

5) $z^2 + 18z + 81$

7) The middle term does not equal $2(3c)(−4)$. It would have to equal $−24c$ to be a perfect square trinomial.

9) $(t + 8)^2$ 11) $(g − 9)^2$ 13) $(2y + 3)^2$

15) $(3k − 4)^2$ 17) prime

19) $\left(a + \dfrac{1}{3}\right)^2$ 21) $\left(v − \dfrac{3}{2}\right)^2$

23) $(x + 3y)^2$ 25) $(3a − 2b)^2$ 27) $4(f + 3)^2$

29) $2p^2(p − 6)^2$ 31) $−2(3d + 5)^2$ 33) $3c(4c^2 + c + 9)$

35) a) $x^2 − 16$ b) $16 − x^2$ 37) $(x + 3)(x − 3)$

39) $(n + 11)(n − 11)$ 41) prime

43) $\left(y + \dfrac{1}{5}\right)\left(y − \dfrac{1}{5}\right)$ 45) $\left(c + \dfrac{3}{4}\right)\left(c − \dfrac{3}{4}\right)$

47) $(6 + h)(6 − h)$ 49) $(13 + a)(13 − a)$

51) $\left(\dfrac{7}{8} + j\right)\left(\dfrac{7}{8} − j\right)$ 53) $(10m + 7)(10m − 7)$

55) $(4p + 9)(4p − 9)$ 57) prime

59) $\left(\dfrac{1}{2}k + \dfrac{2}{3}\right)\left(\dfrac{1}{2}k − \dfrac{2}{3}\right)$ 61) $(b^2 + 8)(b^2 − 8)$

63) $(12m + n^2)(12m − n^2)$ 65) $(r^2 + 1)(r + 1)(r − 1)$

67) $(4h^2 + g^2)(2h + g)(2h − g)$ 69) $4(a + 5)(a − 5)$

71) $2(m + 8)(m − 8)$ 73) $5r^2(3r + 1)(3r − 1)$

75) a) 64   b) 1   c) 1000   d) 27   e) 125   f) 8

77) a) y b) $2c$ c) $5r$ d) x^2 79) $x^2 − 3x + 9$

81) $(d + 1)(d^2 − d + 1)$ 83) $(p − 3)(p^2 + 3p + 9)$

85) $(k + 4)(k^2 − 4k + 16)$ 87) $\left(t + \dfrac{1}{2}\right)\left(t^2 − \dfrac{1}{2}t + \dfrac{1}{4}\right)$

89) $(3m − 5)(9m^2 + 15m + 25)$

91) $(5y − 2)(25y^2 + 10y + 4)$

93) $(10c − d)(100c^2 + 10cd + d^2)$

95) $(2j + 3k)(4j^2 − 6jk + 9k^2)$

97) $(4x + 5y)(16x^2 − 20xy + 25y^2)$

99) $6(c + 2)(c^2 − 2c + 4)$

101) $7(v − 10w^2)(v^2 + 10vw^2 + 100w^4)$

103) $(p + 1)(p − 1)(p^2 − p + 1)(p^2 + p + 1)$

105) $7(2x + 3)$ 107) $(3p + 7)(p − 1)$

109) $(t + 7)(t^2 + 8t + 19)$ 111) $(k − 10)(k^2 − 17k + 73)$

113) $(a + b + 7)(a − b + 7)$ 115) $(2x − y + 5)(2x − y − 5)$

Chapter 7: Putting It All Together

1) Can I factor out a greatest common factor?

2) Can I factor again?

3) $(m + 10)(m + 6)$ 4) $(h + 6)(h − 6)$

5) $(u + 9)(v + 6)$ 6) $(2y + 9)(y − 2)$

7) $(3k − 2)(k − 4)$ 8) $(n − 7)^2$ 9) $8d^4(2d^2 + d + 9)$

10) $(b + c)(b − 4c)$ 11) $10w(3w + 5)(2w − 1)$

12) $7(c − 1)(c^2 + c + 1)$ 13) $(t + 10)(t^2 − 10t + 100)$

14) $(p + 4)(q − 6)$ 15) $(7 + p)(7 − p)$

16) $(h − 7)(h − 8)$ 17) $(2x + y)^2$ 18) $9(3c − 2)$

19) $3z^2(z − 8)(z + 1)$ 20) $(3a − 2)(3a + 4)$

21) prime   22) $5c(a + 2)(b − 3)$

23) $5(2x - 3)(4x^2 + 6x + 9)$ 24) $(9z + 2)^2$

25) $\left(c + \dfrac{1}{2}\right)\left(c - \dfrac{1}{2}\right)$ 26) prime

27) $(3s + 1)(3s - 1)(5t - 4)$ 28) $3cd(2c^2 + 5d)(2c^2 - 5d)$

29) $(k + 3m)(k + 6m)$ 30) $8(2r + 1)(4r^2 - 2r + 1)$

31) $(z - 11)(z + 8)$ 32) $8fg^2(5f^3g^2 + f^2g + 2)$

33) $5(4y - 1)^2$ 34) $(4t - 5)(t + 1)$

35) $2(10c + 3d)(c + d)$ 36) $\left(x + \dfrac{3}{7}\right)\left(x - \dfrac{3}{7}\right)$

37) $(n^2 + 4m^2)(n + 2m)(n - 2m)$ 38) $(k - 12)(k - 9)$

39) $2(a - 9)(a + 4)$ 40) $(x + 2)(x - 2)(y + 7)$

41) $\left(r - \dfrac{1}{2}\right)^2$ 42) $(v - 5)(v^2 + 5v + 25)$

43) $(4g - 9)(7h + 4)$ 44) $-3x(2x - 1)(4x - 3)$

45) $(4b + 3)(2b - 5)$ 46) $2(5u + 3)^2$

47) $5a^2b(11a^4b^2 + 7a^3b^2 - 2a^2 - 4)$ 48) $(8 + u)(8 - u)$

49) prime   50) $2v^2w(v + w)(v + 6w)$ 51) $(3p - 4q)^2$

52) $(c^2 + 4)(c + 2)(c - 2)$ 53) $(6y - 1)(5y + 7)$

54) prime   55) $10(2a - 3b)(4a^2 + 6ab + 9b^2)$

56) $13n^3(2n^3 - 3n + 1)$ 57) $(r - 1)(t - 1)$

58) $(h + 5)^2$ 59) $4(g + 1)(g - 1)$

60) $(5a - 3b)(5a - 8b)$ 61) $3(c - 4)^2$

62) prime   63) $(12k + 11)(12k - 11)$

64) $(5p - 4q)(25p^2 + 20pq + 16q^2)$

65) $-4(6g + 1)(2g + 3)$ 66) $5(d + 11)(d + 1)$

67) $(q + 1)(q^2 - q + 1)$ 68) $(3x + 2)^2$

69) $(9u^2 + v^2)(3u + v)(3u - v)$ 70) $3(5v + w^2)(3v + 2w)$

71) $(11f + 3)(f + 3)$ 72) $4y(y - 5)(y + 4)$

73) $j^3(2j^8 - 1)$ 74) $\left(d + \dfrac{13}{10}\right)\left(d - \dfrac{13}{10}\right)$

75) $(w - 8)(w + 6)$ 76) $(4a - 5)^2$ 77) prime

78) $3(2y + 5)(4y^2 - 10y + 25)$ 79) $(m + 2)^2$

80) $(r - 9)(r - 6)$ 81) $4c^2(5c + 3)(5c - 3)$

82) $(3t + 8)(3t - 8)$ 83) $\left(x - \dfrac{2}{5}\right)\left(x^2 + \dfrac{2}{5}x + \dfrac{4}{25}\right)$

84) $\left(h + \dfrac{3}{4}\right)\left(h^2 - \dfrac{3}{4}h + \dfrac{19}{16}\right)$ 85) $(2z + 1)(y + 11)(y - 5)$

86) $(a + b)(c - 8)(c + 3)$ 87) $r(r + 3)$

88) $(n - 4)(n + 8)$ 89) $3p(3p - 13)$

90) $(5w - 8)(5w - 4)$ 91) $(7k + 3)(k - 1)$

92) $4(4z + 1)(z + 2)$ 93) $-3x(x - 2y)$

94) $5s(s - 2t)$ 95) $(n + p + 6)(n - p + 6)$

96) $(h + k - 5)(h - k - 5)$ 97) $(x - y + z)(x - y - z)$

98) $(a + b + c)(a + b - c)$

Section 7.5

1) It says that if the product of two quantities equals 0, then one or both of the quantities must be zero.

3) $\{-9, 8\}$ 5) $\{4, 7\}$ 7) $\left\{-\dfrac{3}{4}, 9\right\}$

9) $\{0, 8\}$ 11) $\left\{\dfrac{5}{6}\right\}$ 13) $\left\{-3, -\dfrac{7}{4}\right\}$

15) $\left\{-\dfrac{3}{2}, \dfrac{1}{4}\right\}$ 17) $\{0, 2.5\}$

19) No; the product of the factors must equal zero.

21) $\{-8, -7\}$ 23) $\{-15, 3\}$ 25) $\left\{-\dfrac{5}{3}, 2\right\}$

27) $\left\{-\dfrac{4}{7}, 0\right\}$ 29) $\{6, 9\}$ 31) $\{-7, 7\}$

33) $\left\{-\dfrac{6}{5}, \dfrac{6}{5}\right\}$ 35) $\{-12, 5\}$ 37) $\{8\}$

39) $\{-12, 5\}$ 41) $\left\{-2, -\dfrac{3}{4}\right\}$ 43) $\{0, 11\}$

45) $\left\{-\dfrac{1}{2}, 3\right\}$ 47) $\{-8, 12\}$ 49) $\left\{\dfrac{7}{2}, \dfrac{9}{2}\right\}$

51) $\{1, 10\}$ 53) $\left\{-\dfrac{5}{3}, \dfrac{1}{2}\right\}$ 55) $\{-9, 5\}$ 57) $\{1, 6\}$

59) $\left\{-3, -\dfrac{4}{5}\right\}$ 61) $\left\{\dfrac{3}{2}\right\}$ 63) $\left\{\dfrac{3}{2}\right\}$ 65) $\{-11, -3\}$

67) $\left\{-\dfrac{7}{2}, \dfrac{3}{4}\right\}$ 69) $\left\{\dfrac{2}{3}, 8\right\}$ 71) $\left\{-4, 0, \dfrac{1}{2}\right\}$

73) $\left\{-1, \dfrac{2}{9}, 11\right\}$ 75) $\left\{\dfrac{5}{2}, 3\right\}$ 77) $\{0, -8, 8\}$

79) $\{-4, 0, 9\}$ 81) $\{-12, 0, 5\}$ 83) $\left\{0, -\dfrac{3}{2}, \dfrac{3}{2}\right\}$

85) $\left\{-5, -\dfrac{2}{3}, \dfrac{7}{2}\right\}$ 87) $\left\{-\dfrac{3}{4}, \dfrac{2}{5}, \dfrac{1}{2}\right\}$ 89) $\{-6, -2, 2\}$

91) $\{-12, 7\}$ 93) $\left\{0, -\dfrac{7}{3}, \dfrac{7}{3}\right\}$ 95) $\{2, 9\}$

97) $\left\{-8, -\dfrac{1}{3}\right\}$ 99) $\left\{-6, -4, \dfrac{1}{2}\right\}$ 101) $-7, -3$

103) $\dfrac{5}{2}, 4$ 105) $-4, 4$ 107) $0, 1, 4$

Section 7.6

1) length = 12 in.; width = 3 in.

3) base = 3 cm; height = 8 cm

5) base = 6 in.; height = 3 in.

7) length = 10 in.; width = 6 in.

9) length = 9 ft; width = 5 ft

11) length = 9 in.; width = 6 in.

13) width = 12 in.; height = 6 in.

15) height = 10 cm; base = 7 cm

17) 5 and 6 or -4 and -3 19) 0, 2, 4 or 2, 4, 6

21) 6, 7, 8   23) Answers may vary.   25) 9

27) 15   29) 10   31) 8, 15, 17   33) 5, 12, 13

35) 8 in.   37) 5 ft   39) 5 mi

41) a) 144 ft   b) after 2 sec   c) 3 sec

43) a) 288 ft   b) 117 ft   c) 324 ft   d) 176 ft

45) a) \$3500   b) \$3375   c) \$12

47) a) 184 ft   b) 544 ft

    c) when $t = 2\frac{1}{2}$ sec and when $t = 10$ sec   d) $12\frac{1}{2}$ sec

Chapter 7: Group Activity

1) a) $R_1 = a(a - b)$; $R_2 = b(a - b)$
   b) $a^2 - b^2 = a(a - b) + b(a - b)$
   c) $a^2 - b^2 = (a - b)(a + b)$ d) Answers may vary.

2) a) Answers may vary; $(x - 5)(x + 5)$
   b) Answers may vary; $(x - 4)(x + 4)$
   c) Answers may vary; $(3x - 2)(3x + 2)$
   d) Answers may vary; $(5x - 1)(5x + 1)$

3) a) $S_1 = a^2(a - b)$; $S_2 = b^2(a - b)$; $S_3 = a(a - b)^2$
   b) $a^3 - b^3 = a^2(a - b) + b^2(a - b) + a(a - b)^2$
   c) $a^3 - b^3 = a^2(a - b) + b^2(a - b) + a(a - b)^2$
   d) Answers may vary.

4) a) Answers may vary; $(x - 2)(x^2 + 2x + 4)$
   b) Answers may vary; $(2x - 3)(4x^2 + 6x + 9)$

5) a) $a(a - b)(a + b)$
   b) $a(a - b)(a + b) + b^2(a + b)$
   c) $(a + b)(a^2 - ab + b^2)$

Chapter 7: Review Exercises

1) 8   3) $5h^3$ 5) $9(7t + 5)$ 7) $2p^4(p^2 - 10p + 1)$

9) $(m + 8)(n - 5)$ 11) $-5r(3r^2 + 8r - 1)$

13) $(a + 9)(b + 2)$ 15) $(x - 7)(4y - 3)$

17) $(q + 6)(q + 4)$ 19) $(z - 12)(z + 6)$

21) $(m - 3n)(m - 10n)$ 23) $4(v - 8)(v + 2)$

25) $-9w^2(w - 1)(w + 2)$ 27) $(a + 1)(b - 9)(b - 2)$

29) $(3r - 2)(r - 7)$ 31) $(2p - 5)(2p + 1)$

33) $2(3c + 2)(2c + 5)$ 35) $(5x - 3y)(2x + 9y)$

37) $(3c + 1)(3c - 1)$ 39) $(w + 7)(w - 7)$

41) $(8t + 5u)(8t - 5u)$ 43) prime   45) $(r + 6)^2$

47) $5(2k - 3)^2$ 49) $(h + 2)(h^2 - 2h + 4)$

51) $(3p - 4q)(9p^2 + 12pq + 16q^2)$ 53) never

55) $(7r - 6)(r + 2)$ 57) $\left(\dfrac{3}{5} + x\right)\left(\dfrac{3}{5} - x\right)$

59) $(s - 8)(t - 5)$ 61) $w^2(w - 1)(w^2 + w + 1)$

63) prime   65) $-4ab$ 67) $(3y - 14)(2y - 3)$

69) $\left\{0, \dfrac{1}{2}\right\}$ 71) $\left\{-1, \dfrac{2}{3}\right\}$ 73) $\{-3, 15\}$

75) $\left\{\dfrac{1}{2}\right\}$ 77) $\left\{-\dfrac{6}{7}, \dfrac{6}{7}\right\}$ 79) $\{-4, 8\}$ 81) $\left\{\dfrac{4}{5}, 1\right\}$

83) $\left\{0, -\dfrac{3}{2}, 2\right\}$ 85) $\{-7, 2\}$ 87) $\{-8, 9\}$

89) $\{-5, -3, 3\}$ 91) base = 5 in.; height = 6 in.

93) height = 3 in.; length = 8 in.

95) 12   97) length = 6 ft; width = 2.5 ft

99) $-1, 0, 1$ or 8, 9, 10   101) 5 in.

103) a) 0 ft   b) after 2 sec and after 4 sec
    c) 144 ft   d) after 6 sec

Chapter 7: Test

1) Determine whether you can factor out a GCF.

2) $(n - 6)(n - 5)$ 3) $(4 + b)(4 - b)$

4) $(5a + 2)(a - 3)$ 5) $7p^2q^3(8p^4q^3 - 11p^2q + 1)$

6) $(y - 2z)(y^2 + 2yz + 4z^2)$ 7) $2d(d + 9)(d - 2)$

8) prime   9) $(3h + 4)^2$ 10) $(2y - 3)(12x + 11)$

11) $(s - 7t)(s + 4t)$ 12) $(4s^2 + 9t^2)(2s + 3t)(2s - 3t)$

13) $(12p + 5)(3p + 7)$ 14) $(2b - 5)(6b - 7)$

15) $m^9(m + 1)(m^2 - m + 1)$

16) $\left(w + \dfrac{4}{3}\right)\left(w^2 - \dfrac{4}{3}w + \dfrac{16}{9}\right)$

17) $(a - b + x + y)(a - b - x - y)$

18) $\{-4, -3\}$ 19) $\{0, -5, 5\}$ 20) $\left\{-\dfrac{5}{12}, \dfrac{5}{12}\right\}$

21) $\{-4, 7\}$ 22) $\left\{-\dfrac{1}{4}, 8\right\}$ 23) $\left\{\dfrac{5}{3}, 2\right\}$

24) $\{3\}$ 25) $\left\{-\dfrac{7}{2}, -1, 1\right\}$

26) height = 10 ft; width = 4 ft   27) 5, 7, 9

28) 3 mi   29) length = 16 ft; width = 6 ft

30) a) $\dfrac{5}{4}$ sec and 3 sec   b) 60 ft   c) 132 ft   d) 5 sec

Chapter 7: Cumulative Review for Chapters 1–7

1) $\dfrac{1}{8}$ 2) $-\dfrac{9}{40}$ 3) $\dfrac{3t^4}{2u^6}$ 4) $-24k^{10}$

5) 481,300   6) $\{5\}$

7) $R = \dfrac{A - P}{PT}$ 8) California: 1295; Rhode Island: 25

9) $(-\infty, -4] \cup \left[\dfrac{38}{5}, \infty\right)$

10)

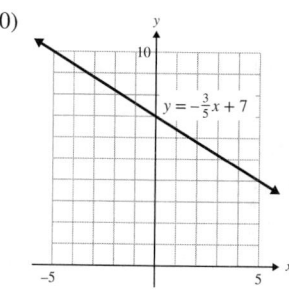

$y = -\dfrac{3}{5}x + 7$

11) $y = \dfrac{1}{3}x + 1$ 12) $(-2, -2)$ 13) $12y^2 - 8y - 15$

14) $8p^3 - 50p^2 + 95p - 56$ 15) $c^2 + 16c + 64$

16) $3a^2b^2 - 7a^2b - 5ab^2 + 19ab - 8$ 17) $6x^3 - 7x - 4$

18) $3r + 1 - \dfrac{5}{2r} + \dfrac{3}{4r^2}$ 19) $(b - 7)(c + 8)$

20) $6(3q - 7)(3q - 1)$ 21) prime

22) $(t^2 + 9)(t + 3)(t - 3)$ 23) $(x - 5)(x^2 + 5x + 25)$

24) $\{-8, 5\}$ 25) $\left\{-\dfrac{4}{3}, \dfrac{5}{2}\right\}$

Chapter 8

Section 8.1

1) when its denominator equals zero   3) a) $\dfrac{5}{17}$ b) $\dfrac{5}{8}$

5) a) $-\dfrac{4}{3}$ b) undefined   7) a) $\dfrac{4}{3}$ b) 0

9) Set the denominator equal to zero, and solve for the variable. That value cannot be substituted into the expression because it will make the denominator equal to zero.

11) a) -4 b) 0   13) a) $\dfrac{7}{2}$ b) $-\dfrac{1}{4}$

15) a) 0, 11 b) $\dfrac{9}{5}$ 17) a) never equals 0 b) 0

19) a) 0 b) $-4, -5$ 21) a) $-6, -3$ b) 0

23) a) 0 b) never undefined—any real number may be substituted for y

25) $\dfrac{7x}{3}$ 27) $\dfrac{3}{7g^2}$ 29) $\dfrac{4}{5}$ 31) $-\dfrac{13}{10}$

33) $g - 8$ 35) $\dfrac{1}{r - 2}$ 37) $\dfrac{q + 5}{2q + 3}$ 39) $\dfrac{w + 5}{5}$

41) $\dfrac{4(u - 5)}{13}$ 43) $\dfrac{x + y}{x^2 + xy + y^2}$ 45) -1 47) -1

49) $-m - 11$ 51) $-\dfrac{6}{x + 2}$ 53) $-4(b + 2)$

55) $-\dfrac{y^2 + 2}{7}$ 57) $-\dfrac{4t^2 + 6t + 9}{2t + 3}$ 59) $\dfrac{2a + 3}{2a - 3}$

61) $\dfrac{b - 6}{4(b + 1)}$ 63) $4h^3 - 8h^2 + 1$ 65) $\dfrac{9}{c + 3}$

67) $-\dfrac{5}{(v - 2)(v - 1)}$

69) possible answers:
$$\dfrac{-u - 7}{u - 2}, \dfrac{-(u + 7)}{u - 2}, \dfrac{u + 7}{2 - u}, \dfrac{u + 7}{-(u - 2)}, \dfrac{u + 7}{-u + 2}$$

71) possible answers:
$$\dfrac{-9 + 5t}{2t - 3}, \dfrac{5t - 9}{2t - 3}, \dfrac{-(9 - 5t)}{2t - 3}, \dfrac{9 - 5t}{-2t + 3}, \dfrac{9 - 5t}{3 - 2t}, \dfrac{9 - 5t}{-(2t - 3)}$$

73) possible answers:
$$-\dfrac{12m}{m^2 - 3}, \dfrac{12m}{-(m^2 - 3)}, \dfrac{12m}{-m^2 + 3}, \dfrac{12m}{3 - m^2}$$

75) $4y - 3$ 77) $4a^2 - 10a + 25$ 79) $3x + 2$

81) $2c + 4$ 83) $3k + 1$

85) a) $\dfrac{3}{2}$ b) -8 c) $(-\infty, -6) \cup (-6, \infty)$

87) a) undefined b) $\dfrac{3}{5}$ c) $(-\infty, -2) \cup (-2, \infty)$

89) a) -2 b) never equals zero
c) $(-\infty, -5) \cup (-5, -1) \cup (-1, \infty)$

91) $(-\infty, 7) \cup (7, \infty)$ 93) $\left(-\infty, \dfrac{7}{2}\right) \cup \left(\dfrac{7}{2}, \infty\right)$

95) $(-\infty, 1) \cup (1, 8) \cup (8, \infty)$

97) $(-\infty, -9) \cup (-9, 9) \cup (9, \infty)$

99) $(-\infty, \infty)$ 101) Answers may vary.

Section 8.2

1) $\dfrac{35}{54}$ 3) $\dfrac{5}{21}$ 5) $\dfrac{16b^4}{27}$ 7) $\dfrac{t^2}{6s^6}$

9) $\dfrac{5}{2}$ 11) $\dfrac{1}{2t(3t - 2)}$ 13) $\dfrac{2u^2}{3(4u - 5)^2}$

15) 6   17) $\dfrac{3(y - 5)}{2y^2}$ 19) $\dfrac{5}{c}$

21) $-7x(x + 11)$ 23) $\dfrac{(r - 2)(r^2 - 3r + 9)}{4(r - 3)}$ 25) 6

27) $\dfrac{28}{3}$ 29) $\dfrac{32m^7}{35}$ 31) $-\dfrac{20}{3g^3h^2}$ 33) $\dfrac{8}{3k^6(k - 2)}$

35) $8q(p + 7)$ 37) $\dfrac{5}{q(q - 7)}$

39) $\dfrac{z + 10}{(2z + 1)(z + 8)}$ 41) $\dfrac{3}{4(3a + 1)}$ 43) $-\dfrac{8}{2d + 5}$

45) $\dfrac{7}{2}$ 47) $\dfrac{c + 1}{6c}$ 49) $-\dfrac{t + 9}{5(t^2 + 2t + 4)}$

51) Answers may vary.   53) $h^2 + 3h - 10$

55) $7 - 2z$ 57) $\dfrac{25}{16}$ 59) $\dfrac{1}{18}$ 61) $\dfrac{1}{4}$ 63) $\dfrac{4}{3r^2}$

65) $\dfrac{a-5}{12a^8}$ 67) $\dfrac{2(4x+5)}{3x^4}$ 69) $\dfrac{(c+6)(c+1)}{9(c+5)}$

71) $\dfrac{5x+1}{4x}$ 73) $\dfrac{k(2k+3)}{2}$ 75) $-\dfrac{3}{4(x+y)}$

77) $-\dfrac{h^4}{3(h+8)}$ 79) $\dfrac{3x^6}{4y^6}$ 81) $-\dfrac{a}{6}$ 83) $\dfrac{7}{3}$

85) $\dfrac{a^2}{a-4}$ 87) $\dfrac{m+2}{16}$ 89) $\dfrac{4j-1}{3j+2}$ 91) $\dfrac{12x^5}{y^3}$

Section 8.3

1) 60   3) 120   5) n^{11} 7) $28r^7$

9) $36z^5$ 11) $110m^4$ 13) $24x^3y^2$

15) $11(z-3)$ 17) $w(2w+1)$

19) Factor the denominators.

21) $10(c-1)$ 23) $3p^5(3p-2)$

25) $(m-7)(m-3)$ 27) $(z+3)(z+8)(z-3)$

29) $(n+4)^2$ 31) $(t+6)(t-6)(t+3)$

33) $a-8$ or $8-a$ 35) $x-y$ or $y-x$

37) Answers may vary.

39) $\dfrac{28}{48}$ 41) $\dfrac{72}{9z}$ 43) $\dfrac{21k^3}{56k^4}$ 45) $\dfrac{12t^2u^3}{10t^7u^5}$

47) $\dfrac{7r}{r(3r+4)}$ 49) $\dfrac{4v^6}{16v^5(v-3)}$

51) $\dfrac{9x^2-45x}{(x+6)(x-5)}$ 53) $\dfrac{z^2+5z-24}{(2z-5)(z+8)}$

55) $-\dfrac{5}{p-3}$ 57) $\dfrac{8c}{7-6c}$

59) $\dfrac{8}{15} = \dfrac{16}{30}; \dfrac{1}{6} = \dfrac{5}{30}$ 61) $\dfrac{4}{u} = \dfrac{4u^2}{u^3}; \dfrac{8}{u^3} = \dfrac{8}{u^3}$

63) $\dfrac{9}{8n^6} = \dfrac{27}{24n^6}; \dfrac{2}{3n^2} = \dfrac{16n^4}{24n^6}$

65) $\dfrac{6}{4a^3b^5} = \dfrac{6a}{4a^4b^5}; \dfrac{6}{a^4b} = \dfrac{24b^4}{4a^4b^5}$

67) $\dfrac{r}{5} = \dfrac{r^2-4r}{5(r-4)}; \dfrac{2}{r-4} = \dfrac{10}{5(r-4)}$

69) $\dfrac{m}{m+7} = \dfrac{m^2}{m(m+7)}; \dfrac{3}{m} = \dfrac{3m+21}{m(m+7)}$

71) $\dfrac{a}{30a-15} = \dfrac{2a}{30(2a-1)}; \dfrac{1}{12a-6} = \dfrac{5}{30(2a-1)}$

73) $\dfrac{9}{k-9} = \dfrac{9k+27}{(k-9)(k+3)}; \dfrac{5k}{k+3} = \dfrac{5k^2-45k}{(k-9)(k+3)}$

75) $\dfrac{3}{a+2} = \dfrac{9a+12}{(a+2)(3a+4)}; \dfrac{2a}{3a+4} = \dfrac{2a^2+4a}{(a+2)(3a+4)}$

77) $\dfrac{9y}{y^2-y-42} = \dfrac{18y^2}{2y(y+6)(y-7)};$

$\dfrac{3}{2y^2+12y} = \dfrac{3y-21}{2y(y+6)(y-7)}$

79) $\dfrac{c}{c^2+9c+18} = \dfrac{c^2+6c}{(c+6)^2(c+3)};$

$\dfrac{11}{c^2+12c+36} = \dfrac{11c+33}{(c+6)^2(c+3)}$

81) $\dfrac{11}{g-3}$ already has the LCD. $\dfrac{4}{3-g} = -\dfrac{4}{g-3}$

83) $\dfrac{4}{3x-4} = \dfrac{12x+16}{(3x+4)(3x-4)};$

$\dfrac{7x}{16-9x^2} = -\dfrac{7x}{(3x+4)(3x-4)}$

85) $-\dfrac{9}{h^3+8} = -\dfrac{45}{5(h+2)(h^2-2h+4)};$

$\dfrac{2h}{5h^2-10h+20} = \dfrac{2h^2+4}{5(h+2)(h^2-2h+4)}$

87) $\dfrac{2}{z^2+3z} = \dfrac{6z+18}{3z(z+3)^2}; \dfrac{6}{3z^2+9z} = \dfrac{6z+18}{3z(z+3)^2};$

$\dfrac{8}{z^2+6z+9} = \dfrac{24z}{3z(z+3)^2}$

89) $\dfrac{t}{t^2-13t+30} = \dfrac{t^2+3t}{(t+3)(t-3)(t-10)};$

$\dfrac{6}{t-10} = \dfrac{6t^2-54}{(t+3)(t-3)(t-10)};$

$\dfrac{7}{t^2-9} = \dfrac{7t-70}{(t+3)(t-3)(t-10)}$

Section 8.4

1) $\dfrac{7}{8}$ 3) $\dfrac{4}{7}$ 5) $-\dfrac{18}{p}$ 7) $\dfrac{5}{c}$ 9) $\dfrac{z+6}{z-1}$

11) 2   13) $\dfrac{5}{t}$ 15) $\dfrac{d+6}{d+5}$

17) a) $18b^4$

b) Multiply the numerator and denominator of $\dfrac{4}{9b^2}$ by $2b^2$, and multiply the numerator and denominator of $\dfrac{5}{6b^4}$ by 3.

c) $\dfrac{4}{9b^2} = \dfrac{8b^2}{18b^4}; \dfrac{5}{6b^4} = \dfrac{15}{18b^4}$

19) $\dfrac{31}{40}$ 21) $\dfrac{8t+9}{6}$ 23) $\dfrac{6h^2+50}{15h^3}$ 25) $\dfrac{3-14f}{2f^2}$

27) $\dfrac{16y+9}{y(y+3)}$ 29) $\dfrac{11d+32}{d(d-8)}$ 31) $\dfrac{3(5c+16)}{(c-4)(c+8)}$

33) $\dfrac{m^2-16m-10}{(3m+5)(m-10)}$ 35) $\dfrac{3u+2}{u-1}$

37) $\dfrac{7g^2-53g+6}{(g+2)(g+8)(g-8)}$ 39) $\dfrac{3(a+1)}{(a-9)(a-1)}$

41) $\dfrac{2(x^2 - 5x + 8)}{(x - 4)(x + 5)(x - 3)}$ 43) $\dfrac{4b^2 + 28b + 3}{3(b - 4)(b + 3)}$

45) $\dfrac{-22n^2 + 25n - 42}{(2n + 3)(2n - 3)(4n^2 - 6n + 9)}$

47) No. If the sum is rewritten as $\dfrac{9}{x - 6} - \dfrac{4}{x - 6}$, then the

LCD $= x - 6$. If the sum is rewritten as $\dfrac{-9}{6 - x} + \dfrac{4}{6 - x}$,

then the LCD is $6 - x$.

49) $\dfrac{6}{q - 4}$ or $-\dfrac{6}{4 - q}$

51) $\dfrac{26}{f - 7}$ or $-\dfrac{26}{7 - f}$

53) $\dfrac{8 - x}{x - 4}$ or $\dfrac{x - 8}{4 - x}$ 55) 1

57) $\dfrac{3(1 + 2u)}{2u - 3v}$ or $-\dfrac{3(1 + 2u)}{3v - 2u}$

59) $-\dfrac{2(x - 1)}{(x + 3)(x - 3)}$ 61) $\dfrac{3(3a + 4)}{(2a + 3)(2a - 3)}$

63) $\dfrac{6c^2 + 5c + 5}{(1 - c)(c^2 + c + 1)}$ 65) $\dfrac{-10a^2 + 8a - 11}{a(a - 2)}$

67) $\dfrac{c^2 - 2c + 20}{(c - 4)^2(c + 3)}$ 69) $\dfrac{17a + 23b}{4(a + b)(a - b)}$

71) $\dfrac{2(9v + 2)}{(6v + 1)(3v + 2)(v - 5)}$ 73) $\dfrac{7g^2 - 45g + 205}{5g(g - 6)(2g - 5)}$

75) $\dfrac{2k^3 + k^2 + 7}{(2k + 1)(4k^2 - 2k + 1)}$ 77) $\dfrac{10b^2 + 2b + 15}{b(b + 8)(3b + 1)}$

79) $\dfrac{a^3 + a^2b + ab^2 + ab + b^2}{(a + b)(a - b)(a^2 + ab + b^2)}$

81) a) $\dfrac{2(k - 4)}{k + 1}$ b) $\dfrac{k^2 - 3k + 28}{2(k + 1)}$

83) a) $\dfrac{6h}{(h + 5)^2(h + 4)}$ b) $\dfrac{2(h^2 + 4h + 6)}{(h + 5)(h + 4)}$

85) $\dfrac{49x + 6}{4x^2}$

Chapter 8: Putting It All Together

1) a) 0   b) $\dfrac{1}{2}$ 2) a) $\dfrac{3}{5}$ b) undefined

3) a) undefined   b) $\dfrac{7}{45}$ 4) a) 3   b) $\dfrac{1}{2}$

5) a) $-6, 6$ b) 0   6) a) $-3, -\dfrac{5}{2}$ b) 4

7) a) $-4, 2$ b) $\dfrac{3}{5}$ 8) a) $-8, 8$ b) $\dfrac{8}{5}$

9) a) 0   b) never equals 0

10) a) never undefined—any real number may be substituted for t b) 15

11) $4w^{11}$ 12) $\dfrac{7}{3n^5}$ 13) $\dfrac{m + 9}{2(m + 4)}$

14) $\dfrac{1}{j - 5}$ 15) $-\dfrac{3}{n + 2}$ 16) $-\dfrac{1}{y + 5}$

17) $25t^2 - 10t + 4$ 18) $\dfrac{2(a^2 + ab + b^2)}{a + 3b}$

19) $\dfrac{32}{3}$ 20) $\dfrac{2(2f - 11)}{f(f + 11)}$

21) $\dfrac{4j^2 + 27j + 2}{(j + 9)(j - 9)(j + 6)}$

22) $\dfrac{5a^2b}{3}$ 23) $\dfrac{y}{3z^2}$ 24) $\dfrac{8q^2 - 37q + 21}{(q - 5)(q + 4)(q + 7)}$

25) $\dfrac{x^2 + 2x + 12}{(2x + 1)^2(x - 4)}$ 26) $-\dfrac{n - 4}{4}$ or $\dfrac{4 - n}{4}$

27) $-\dfrac{m + 7}{8}$ 28) $\dfrac{12}{r - 7}$

29) $\dfrac{9x}{(y + 8)(9x^2 + 15x + 25)}$

30) $\dfrac{6}{d^6}$ 31) $\dfrac{9d^2 + 8d + 24}{d^2(d + 3)}$ 32) $\dfrac{3}{5}$

33) $\dfrac{3k^3(3k + 1)}{2}$ 34) $\dfrac{35}{12z}$

35) $-\dfrac{(w - 15)(w - 1)}{(w + 5)(w - 5)(w - 7)}$ 36) $\dfrac{3a^3(a^2 + 5)}{10}$

37) $\dfrac{2(7x - 1)}{(x - 8)(x + 3)}$ 38) $\dfrac{3y^3 + 16}{12y^4}$

39) $\dfrac{h + 5}{8(2h + 1)}$ 40) $(b - 3)(b + 4)$

41) $\dfrac{3m + 20n}{7m - 4n}$ 42) -1 43) $\dfrac{-2p^2 - 8p + 11}{p(p + 7)(p - 8)}$

44) $\dfrac{5u^2 + 37u - 19}{u(3u - 2)(u + 1)}$ 45) $\dfrac{2(t + 1)^2}{t}$

46) $\dfrac{1}{5r^2(3r - 7)}$ 47) $\dfrac{35a^2 - 30ab + 16b^2}{4(5a + 2b)(25a^2 - 10ab + 4b^2)}$

48) $\dfrac{21t^2 - 17t - 6}{(3t + 1)(3t - 1)(9t^2 + 3t + 1)}$ 49) $\dfrac{c^2}{24}$

50) $\dfrac{9}{2}$ 51) 1   52) $\dfrac{12p^2 - 92p - 15}{(4p + 3)(p + 2)(p - 6)}$

53) $\dfrac{1}{12xy^7}$ 54) $\dfrac{1}{9r^4t^3}$ 55) $\dfrac{3m + 1}{7(m + 4)}$

56) $\dfrac{4(c^2 + 16)}{3(c - 2)}$ 57) $-\dfrac{1}{3k}$ 58) $-\dfrac{11}{5w}$

59) a) $\dfrac{6z}{(z + 5)(z + 2)}$ b) $\dfrac{2(z^2 + 8z + 30)}{(z + 5)(z + 2)}$

60) $\dfrac{111n + 8}{36n^2}$

Section 8.5

1) Method 1: Rewrite it as a division problem, then simplify.

$$\frac{2}{9} \div \frac{5}{18} = \frac{2}{\underset{1}{9}} \cdot \frac{\overset{2}{18}}{5} = \frac{4}{5}$$

Method 2: Multiply the numerator and denominator by 18, the LCD of $\frac{2}{9}$ and $\frac{5}{18}$. Then, simplify.

$$\frac{\overset{2}{18}\left(\dfrac{2}{9}\right)}{\underset{1}{18}\left(\dfrac{5}{18}\right)} = \frac{4}{5}$$

3) $\dfrac{14}{25}$ 5) ab^2 7) $\dfrac{1}{st^2}$ 9) $\dfrac{2m^4}{15n^2}$

11) $\dfrac{t}{5}$ 13) $\dfrac{4}{3(y-8)}$ 15) $\dfrac{5}{6w^4}$ 17) $x(x-3)$

19) $-\dfrac{1}{a+5b}$ 21) $\dfrac{3}{2}$ 23) $\dfrac{7d+2c}{d(c-5)}$ 25) $\dfrac{4z+7}{5z-7}$

27) Answers may vary.   29) $\dfrac{8}{y}$ 31) $\dfrac{x^2-7}{x^2-11}$

33) $-\dfrac{52}{15}$ 35) $\dfrac{4xy}{3(x+y)}$ 37) $\dfrac{r(r^2+s)}{s^2(sr+1)}$ 39) $\dfrac{t+3}{t+2}$

41) $\dfrac{b^2+1}{b^2-3}$ 43) $\dfrac{1}{m^3n}$ 45) $\dfrac{(h-1)(h+3)}{28}$

47) $\dfrac{2(x-9)(x+2)}{3(x+3)(x+1)}$ 49) $\dfrac{r}{20}$ 51) $\dfrac{a}{12}$ 53) $\dfrac{25}{18}$

55) $\dfrac{2(n+3)^2}{4n+7}$ 57) $\dfrac{2c}{(2c+3)^2}$ 59) $\dfrac{w(v-w)}{2v+w^2}$

61) $\dfrac{8y^2}{x(y^2-x)}$ 63) $\dfrac{a^3+b^2}{a^3(2-7b^2)}$ 65) $\dfrac{4n-m}{m(1+mn)}$

67) 0

x	f(x)
1	1
2	$\dfrac{1}{2}$
3	$\dfrac{1}{3}$
10	$\dfrac{1}{10}$
100	$\dfrac{1}{100}$
1000	$\dfrac{1}{1000}$

Section 8.6

1) Eliminate the denominators.

3) difference; $\dfrac{8r+15}{6}$ 5) equation; $\left\{-\dfrac{2}{5}\right\}$

7) sum; $\dfrac{a^2+3a+33}{a^2(a+11)}$ 9) equation; $\{12\}$

11) 0, 2   13) 0, 3, −3   15) −4, 9   17) {−1}

19) $\left\{\dfrac{16}{3}\right\}$ 21) {−4}   23) {−11}   25) {1}

27) {1, 12}   29) {−6}   31) $\left\{\dfrac{7}{3}\right\}$ 33) ∅

35) {2}   37) {−6, −2}   39) {3, 5}

41) {−8}   43) {−10, 12}   45) {3}

47) $\left\{\dfrac{21}{5}\right\}$ 49) ∅   51) {0, −12}

53) {8}   55) {3, 1}   57) ∅   59) {−5, −3}

61) {−3}   63) $\left\{-\dfrac{2}{3}\right\}$ 65) {−5, 4}   67) {5}

69) {−2, 8}   71) ∅   73) {−15}   75) {−1}

77) 0.4 m   79) 750 lb   81) $m = \dfrac{CA}{W}$

83) $b = \dfrac{rt}{2a}$ 85) $x = \dfrac{t+u}{3B}$ 87) $n = \dfrac{dz-t}{d}$

89) $s = \dfrac{3A-hr}{h}$ 91) $y = \dfrac{kx+raz}{r}$

93) $r = \dfrac{st}{s+t}$ 95) $z = \dfrac{4xy}{x-5y}$

Section 8.7

1) $7.10   3) 57 mg   5) 375 mL   7) 111

9) 3 cups of tapioca flour and 6 cups of potato-starch flour

11) length: 48 ft; width: 30 ft

13) stocks: $8000; bonds: $12,000

15) 1355   17) a) 7 mph   b) 13 mph

19) a) $x+30$ mph   b) $x-30$ mph

21) 20 mph   23) 4 mph   25) 260 mph   27) 2 mph

29) $\dfrac{1}{4}$ job/hr   31) $\dfrac{1}{t}$ job/hr   33) $1\dfrac{1}{5}$ hr   35) $3\dfrac{1}{13}$ hr

37) 20 min   39) 3 hr   41) 3 hr   43) $2\dfrac{1}{5}$ ft/sec

Section 8.8

1) increases   3) direct   5) inverse   7) combined

9) $M = kn$ 11) $h = \dfrac{k}{j}$ 13) $T = \dfrac{k}{c^2}$ 15) $s = krt$

17) $Q = \dfrac{k\sqrt{z}}{m}$ 19) a) 9   b) $z = 9x$ c) 54

21) a) 48   b) $N = \dfrac{48}{y}$ c) 16

23) a) 5   b) $Q = \dfrac{5r^2}{w}$ c) 45

25) 56   27) 18   29) 70   31) $500.00

33) $0.80   35) 180 watts   37) 48π cm^3

39) 200 cycles/sec   41) 3 ohms   43) 320 lb

SA-36 Answers to Selected Exercises

1)

Row A	x	-1	-2	-5	-10	-100	-1000
Row B	$y = \dfrac{1}{x}$	$y = -1$	$y = -\dfrac{1}{2}$	$y = -\dfrac{1}{5}$	$y = -\dfrac{1}{10}$	$y = -\dfrac{1}{100}$	$y = -\dfrac{1}{1000}$
	(x, y)	$(-1, -1)$	$\left(-2, -\dfrac{1}{2}\right)$	$\left(-5, -\dfrac{1}{5}\right)$	$\left(-10, -\dfrac{1}{10}\right)$	$\left(-100, -\dfrac{1}{100}\right)$	$\left(-1000, -\dfrac{1}{1000}\right)$

Row C	x	-0.5	-0.2	-0.1	-0.01	-0.001	
Row D	$y = \dfrac{1}{x}$	$y = -2$	$y = -5$	$y = -10$	$y = -100$	$y = -1000$	
	(x, y)	$(-0.5, -2)$	$(-0.2, -5)$	$(-0.1, -10)$	$(-0.01, -100)$	$(-0.001, -1000)$	

Row E	x	0.5	0.2	0.1	0.01	0.001	
Row F	$y = \dfrac{1}{x}$	$y = 2$	$y = 5$	$y = 10$	$y = 100$	$y = 1000$	
	(x, y)	$(0.5, 2)$	$(0.2, 5)$	$(0.1, 10)$	$(0.01, 100)$	$(0.001, 1000)$	

Row G	x	1	2	5	10	100	1000
Row H	$y = \dfrac{1}{x}$	$y = 1$	$y = \dfrac{1}{2}$	$y = \dfrac{1}{5}$	$y = \dfrac{1}{10}$	$y = \dfrac{1}{100}$	$y = \dfrac{1}{1000}$
	(x, y)	$(1, 1)$	$\left(2, \dfrac{1}{2}\right)$	$\left(5, \dfrac{1}{5}\right)$	$\left(10, \dfrac{1}{10}\right)$	$\left(100, \dfrac{1}{100}\right)$	$\left(1000, \dfrac{1}{1000}\right)$

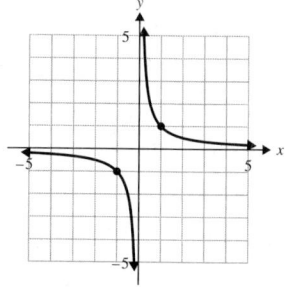

2) a) y gets larger; yes; 0 b) y gets smaller; $y \to -\infty$ c) $y \to +\infty$ d) $y \to 0$

3) Answers may vary. no 4) no; no 5) a) 1 b) 1 c) $+\infty$ d) $+\infty$

Chapter 8: Review Exercises

1) a) 0 b) 0   3) a) 0 b) $-\dfrac{11}{4}$

5) a) never equals 0 b) $-\dfrac{3}{2}, \dfrac{3}{2}$

7) a) $\dfrac{7}{12}$ b) -9 c) $\left(-\infty, \dfrac{1}{5}\right) \cup \left(\dfrac{1}{5}, \infty\right)$

9) $(-\infty, -4) \cup (-4, 6) \cup (6, \infty)$

11) $11k^6$ 13) $\dfrac{r - 8}{4r}$ 15) $-\dfrac{1}{x + 11}$

17) possible answers:
$$\dfrac{-4n - 1}{5 - 3n}, \dfrac{-(4n + 1)}{5 - 3n}, \dfrac{4n + 1}{3n - 5}, \dfrac{4n + 1}{-5 + 3n}, \dfrac{4n + 1}{-(5 - 3n)}$$

19) $b + 3$ 21) $\dfrac{24}{35}$ 23) $\dfrac{t + 2}{2(t + 6)}$ 25) $\dfrac{x - 3}{45}$

27) $\dfrac{1}{2r^2(r - 7)}$ 29) $-\dfrac{3k}{k^2 + 1}$ 31) $\dfrac{q}{35p^2}$ 33) $\dfrac{3}{10}$

35) $\dfrac{1}{3}$ 37) 30   39) k^5 41) $(4x + 9)(x - 7)$

43) $w - 5$ or $5 - w$ 45) $(c + 4)(c + 5)(c - 7)$

47) $\dfrac{12y^2}{20y^3}$ 49) $\dfrac{6z}{z(2z + 5)}$ 51) $\dfrac{t^2 + t - 12}{(3t + 1)(t + 4)}$

53) $\dfrac{8c}{c^2 + 5c - 24} = \dfrac{8c^2 - 24c}{(c - 3)^2(c + 8)}$;
$\dfrac{5}{c^2 - 6c + 9} = \dfrac{5c + 40}{(c - 3)^2(c + 8)}$

55) $\dfrac{7}{2q^2 - 12q} = \dfrac{7q + 42}{2q(q + 6)(q - 6)}$;
$\dfrac{3q}{36 - q^2} = -\dfrac{6q^2}{2q(q + 6)(q - 6)}$;
$\dfrac{q - 5}{2q^2 + 12q} = \dfrac{q^2 - 11q + 30}{2q(q + 6)(q - 6)}$

57) $\dfrac{4}{3c}$ 59) $\dfrac{36u - 5v}{40u^3v^2}$ 61) $\dfrac{n^2 - 12n + 20}{n(3n - 5)}$

63) $\dfrac{4y - 37}{(y - 3)(y + 2)}$ 65) $\dfrac{(k - 7)(k + 2)}{k(k + 7)^2}$ 67) $\dfrac{t + 20}{t - 18}$

69) $\dfrac{5w^2 - 51w + 8}{(2w - 7)(w + 8)(w + 3)}$ 71) $\dfrac{2b^2 + 7b + 2}{2b(3b + 2)(3b - 2)}$

73) a) $\dfrac{2}{x(x + 2)}$ b) $\dfrac{2x^3 + 4x + 8}{x^2(x + 2)}$ 75) $\dfrac{y}{x^2}$ 77) $\dfrac{1}{5}$

79) $\dfrac{(y + 4)(y - 9)}{(y - 8)(y + 6)}$ 81) $r + t$ 83) $\dfrac{y(x^2 + 2y)}{x(y^2 - x)}$

85) {4} 87) {-7, 5} 89) $\left\{-4, \dfrac{1}{2}\right\}$

91) {-3} 93) {-1, 5} 95) ∅

97) $D = \dfrac{s + T}{R}$ 99) $k = \dfrac{cw - N}{aw}$

101) $R_1 = \dfrac{R_2 R_3}{R_2 - R_3}$ 103) 2 mph 105) $3\dfrac{1}{13}$ hr

107) 21 109) 2 111) 216 cm² 113) $\dfrac{n^2 + 6n + 3}{(2n - 1)(n + 2)}$

115) $\dfrac{a + 2}{(4a - 7)(2a + 5)}$ 117) $-\dfrac{1}{r + 8}$ 119) $\dfrac{1}{p + q}$

121) {-10, 2} 123) {1, 3} 125) {-10}

Chapter 8: Test

1) $-\dfrac{3}{8}$ 2) a) -10 b) $\dfrac{9}{2}$

3) a) -4, 9 b) never equals zero

4) $\left(-\infty, -\dfrac{3}{2}\right) \cup \left(-\dfrac{3}{2}, \infty\right)$ 5) $\dfrac{1}{3t^4u^3}$

6) $\dfrac{6h}{h - 8}$ 7) $\dfrac{z + 3}{z^2 + 3z + 9}$

8) possible answers: $-\dfrac{m - 8}{4m - 5}, \dfrac{m - 8}{5 - 4m}, \dfrac{m - 8}{-4m + 5}$

9) $z(z + 6)$ 10) $\dfrac{2}{3r}$ 11) $\dfrac{7b}{5a^6}$ 12) $-\dfrac{13h}{36}$

13) $\dfrac{c^2 + 20c + 30}{(3c + 5)(c + 2)}$ 14) $-\dfrac{k^2 + 2}{4}$ 15) $\dfrac{2d}{5(d + 3)}$

16) $\dfrac{t - 14}{t - 7}$ or $\dfrac{14 - t}{7 - t}$ 17) $\dfrac{-2v^2 - 2v + 39}{(2v - 3)(v - 2)(v + 9)}$

18) $\dfrac{m}{m - 2}$ 19) $\dfrac{x + y}{4xy}$ 20) {-2} 21) {-5, -2}

22) ∅ 23) {-7, 3} 24) $b = \dfrac{ac}{a - c}$

25) $3k + 7$ 26) $0, \dfrac{1}{4}$ 27) Ricardo: 3 hr; Michael: 6 hr

28) 12 mph 29) 300 30) 10 L

Chapter 8: Cumulative Review for Chapters 1–8

1) 45 cm² 2) 63 3) 5 ft by 9 ft 4) $\left(-\infty, \dfrac{7}{4}\right)$

5) [0, 15] 6) x-int: $\left(\dfrac{3}{2}, 0\right)$; y-int: (0, -2) 7) $-\dfrac{4}{3}$

8) $\left(3, -\dfrac{5}{2}\right)$ 9) $32p^{15}$ 10) $\dfrac{1}{125y^6}$

11) $4n^2 - 12n + 9$ 12) $64a^2 - b^2$

13) $3h^2 - \dfrac{5}{3}h + 1 - \dfrac{2}{3h^2}$ 14) $5k^2 - 2k - 3 + \dfrac{4}{k + 4}$

15) $(2d + 5)(2d - 3)$ 16) $3(z^2 + 4)(z + 2)(z - 2)$

17) $(r + 8)(t - 1)$ 18) {-12, -3}

19) a) 0, 6 b) $-\dfrac{2}{7}$ 20) $\dfrac{3(c - 2)}{c - 6}$ 21) $\dfrac{2n}{4 - n}$

22) $\dfrac{3(y - 5)}{y(y + 5)}$ 23) $\dfrac{r - 6}{r - 11}$ 24) {6} 25) 3

Chapter 9
Section 9.1

1) Answers may vary. 3) {-6, 6} 5) {2, 8}

7) $\left\{-\dfrac{1}{2}, 3\right\}$ 9) $\left\{-\dfrac{1}{2}, -\dfrac{1}{3}\right\}$ 11) {-24, 15}

13) $\left\{-\dfrac{10}{3}, \dfrac{50}{3}\right\}$ 15) ∅ 17) {-10, 22}

19) {-5, 0} 21) {-14} 23) ∅

25) $\left\{-\dfrac{16}{5}, 2\right\}$ 27) ∅ 29) $\left\{-\dfrac{14}{3}, 4\right\}$ 31) $\left\{\dfrac{1}{2}, 4\right\}$

33) $\left\{\dfrac{2}{5}, 2\right\}$ 35) {10} 37) |x| = 9; answers may vary.

39) $|x| = \dfrac{1}{2}$; answers may vary. 41) {0, 14}

43) $\left\{-\dfrac{25}{3}, -\dfrac{5}{3}\right\}$ 45) $\left\{\dfrac{3}{10}\right\}$ 47) {-2, 2.4}

49) ∅ 51) $\left\{\dfrac{16}{3}, \dfrac{20}{3}\right\}$ 53) $\left\{-\dfrac{39}{8}, \dfrac{33}{20}\right\}$ 55) {-1.25}

Section 9.2

1) [-1, 5]

3) $(-\infty, 2) \cup (9, \infty)$

5) $\left(-\infty, -\dfrac{9}{2}\right] \cup \left[\dfrac{3}{5}, \infty\right)$

7) $[-7, 7]$

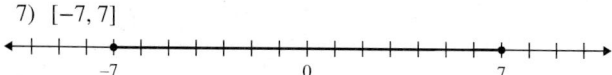

9) $(-4, 4)$

10) [number line from -5 to 5, open circles at -4 and 4]

11) $(-2, 6)$

[number line, open circles at -2 and 6]

13) $\left[-\dfrac{14}{3}, -2\right]$

[number line $-\frac{14}{3}$, closed circles]

15) $\left[\dfrac{2}{3}, \dfrac{5}{3}\right]$

[number line $\frac{2}{3}$ $\frac{5}{3}$, closed circles]

17) $\varnothing$

[number line from -5 to 5]

19) $\varnothing$

[number line from -5 to 5]

21) $(-8, 3)$

[number line, open circles at -8 and 3]

23) $[-12, 4]$

[number line -12 to 4, closed circles]

25) $(-\infty, -7] \cup [7, \infty)$

[number line, closed circles at -7 and 7]

27) $(-\infty, -14] \cup [-6, \infty)$

[number line, closed circles at -14 and -6]

29) $\left(-\infty, -\dfrac{3}{2}\right] \cup [3, \infty)$

[number line $-\frac{3}{2}$, closed circles]

31) $(-\infty, 2) \cup \left(\dfrac{11}{3}, \infty\right)$

[number line, open circles at 2 and $\frac{11}{3}$]

33) $(-\infty, \infty)$

[number line from -5 to 5]

35) $(-\infty, \infty)$

[number line from -5 to 5]

37) $(-\infty, -12] \cup [0, \infty)$

[number line, closed circles at -12 and 0]

39) $\left(-\infty, -\dfrac{27}{5}\right] \cup \left[\dfrac{21}{5}, \infty\right)$

[number line $-\frac{27}{5}$, $\frac{21}{5}$]

41) The absolute value of a quantity is always 0 or positive; it cannot be less than 0.

43) The absolute value of a quantity is always 0 or positive, so for any real number, x, the quantity $|2x + 1|$ will be greater than -3.

45) Answers may vary. One example is $|y| \le 8$.

47) $(-\infty, -6) \cup (-3, \infty)$

49) $\left\{-2, -\dfrac{1}{2}\right\}$ 51) $\left(-\infty, -\dfrac{1}{4}\right] \cup [2, \infty)$

53) $(-3, \infty)$ 55) $\varnothing$ 57) $\{-21, -3\}$

59) $\varnothing$ 61) $\left(-\infty, -\dfrac{1}{25}\right]$ 63) $(-\infty, \infty)$

65) $[-15, -1]$ 67) $\left(-\infty, \dfrac{1}{5}\right) \cup (3, \infty)$

69) $|a - 128| \le 0.75$; $127.25 \le a \le 128.75$; there is between 127.25 oz and 128.75 oz of milk in the container.

71) $|b - 38| \le 5$; $33 \le b \le 43$; he will spend between \$33 and \$43 on his daughter's gift.

Section 9.3

1) Answers may vary.   3) Answers may vary.

5) Answers may vary.   7) dotted

9)

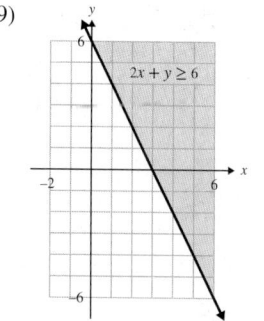

11)

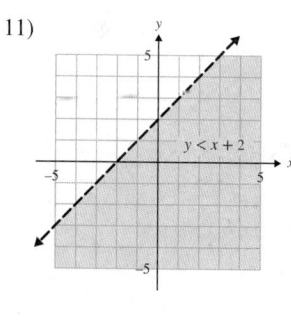

13)

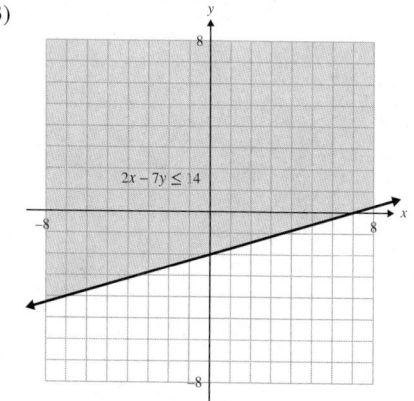

15)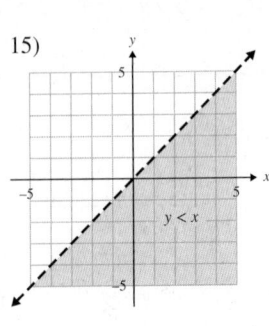

17)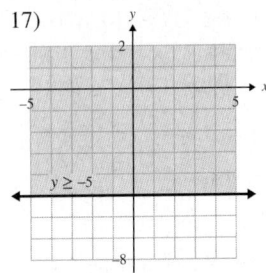

37)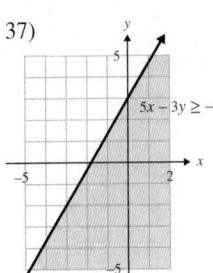

39)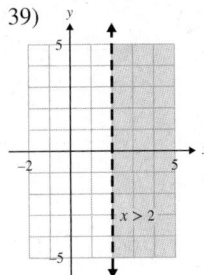

19) below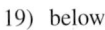

41)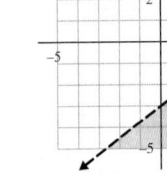

21)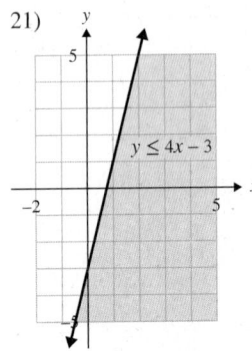

23)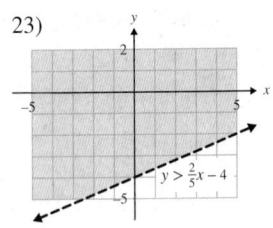

43) No; it does not satisfy $2x + y < 7$.

45)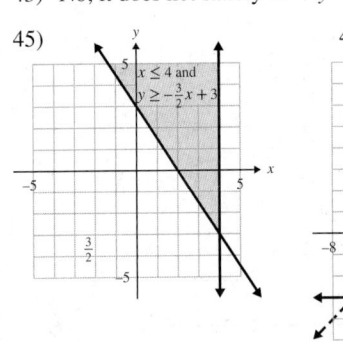

47)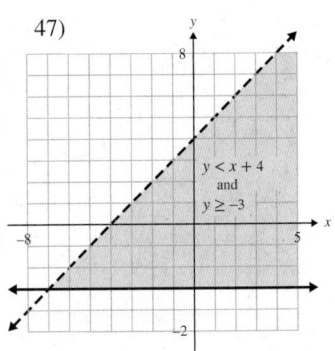

25)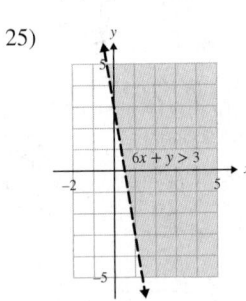

27)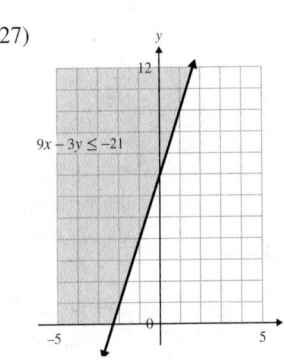

49)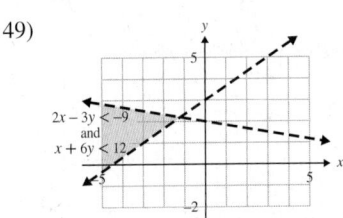

29)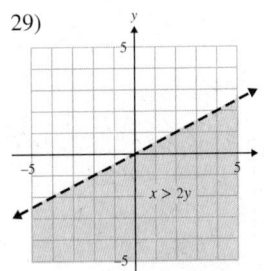

31) Answers may vary.

51)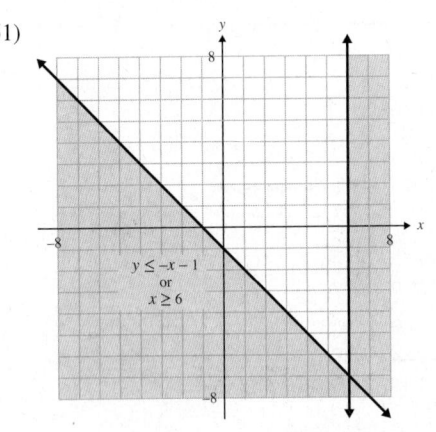

33)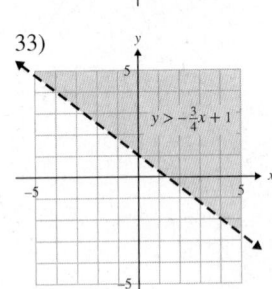

35)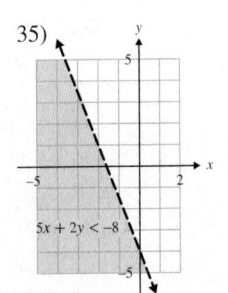

53)

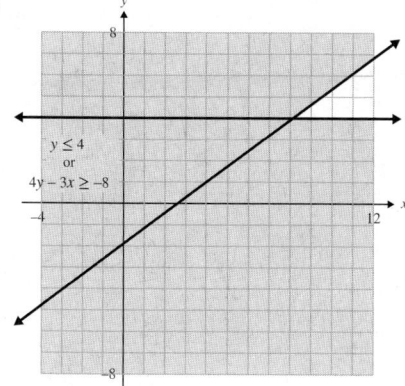

$y \leq 4$
or
$4y - 3x \geq -8$

55)

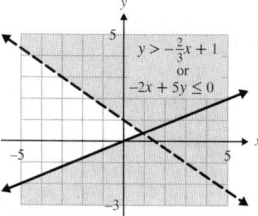

$y > -\frac{2}{3}x + 1$
or
$-2x + 5y \leq 0$

57)

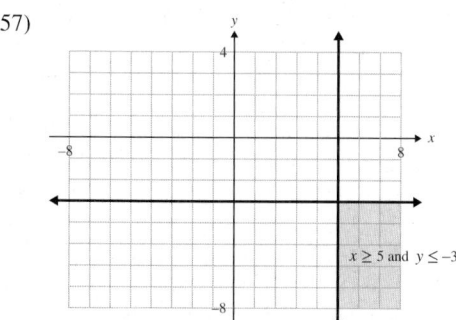

$x \geq 5$ and $y \leq -3$

59)

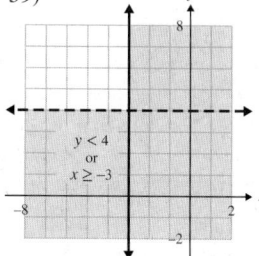

$y < 4$
or
$x \geq -3$

61)

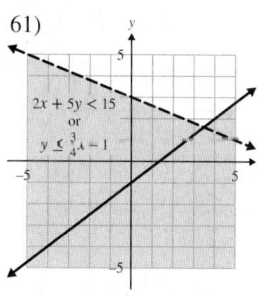

$2x + 5y < 15$
or
$y \leq \frac{3}{4}x - 1$

63)

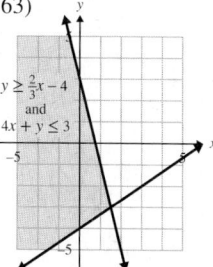

$y \geq \frac{2}{3}x - 4$
and
$4x + y \leq 3$

65) a) $x \geq 0$
        $y \geq 0$
        $x + y \leq 15$
    c) Answers may vary.
    d) Answers may vary.

b)

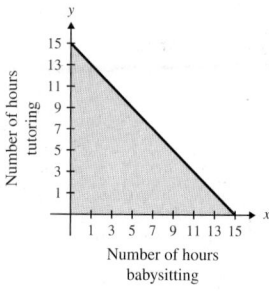

Number of hours tutoring

Number of hours babysitting

67) a) $150 \leq p \leq 250$ and $100 \leq r \leq 200$ and $p + r \geq 300$
    b)

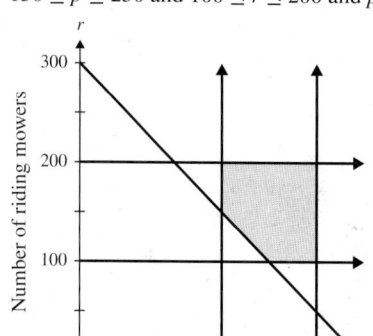

Number of riding mowers

Number of push mowers

c) It represents the production of 175 push mowers and 110 riding mowers per day. This does not meet the level of production needed because it is not a total of at least 300 mowers per day and is not in the feasible region.   d) Answers may vary.   e) Answers may vary.

Chapter 9: Group Activity

1)

×	Positive	Negative
Positive	Positive	Negative
Negative	Negative	Positive

÷	Positive	Negative
Positive	Positive	Negative
Negative	Negative	Positive

2) a) positive   b) positive; negative   c) negative

3) a) >   b) >   c) <   d) >   e) >   f) <   g) >   h) >

4) a) >   b) <   c) <   d) <   e) >   f) >

5) a) >   b) >   c) <   d) <

Chapter 9: Review Exercises

1) $\{-9, 9\}$

3) $\left\{-1, \frac{1}{7}\right\}$ 5) $\left\{-\frac{15}{8}, -\frac{7}{8}\right\}$ 7) $\left\{\frac{11}{5}, \frac{13}{5}\right\}$

9) $\left\{-8, \frac{4}{15}\right\}$ 11) $\varnothing$ 13) $\left\{-\frac{4}{9}\right\}$

15) $|a| = 4$; answers may vary.

17) $[-3, 3]$

19) $(-\infty, -2) \cup (2, \infty)$

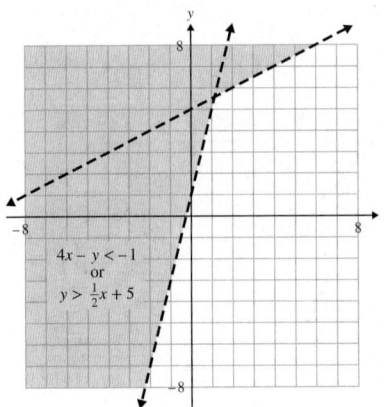

43)

21) $(-\infty, -1] \cup \left[\dfrac{1}{6}, \infty\right)$

23) $(-5, 13)$

25) $\left[-\dfrac{15}{4}, -\dfrac{3}{4}\right]$

27) $\left(-\infty, -\dfrac{19}{5}\right] \cup [-1, \infty)$

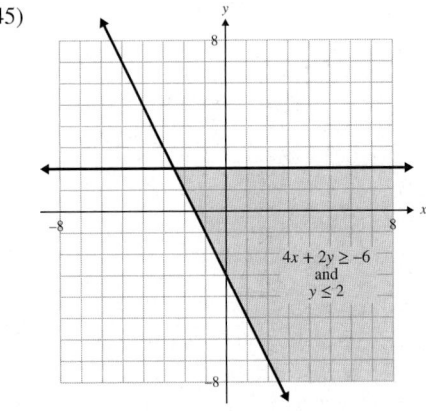

45)

29) $(-\infty, \infty)$

31) $\left\{-\dfrac{1}{12}\right\}$

33) Answers may vary. One example is $|x| < 9$.

35)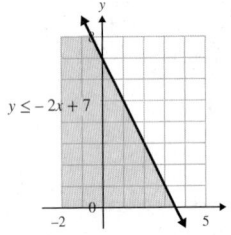

37)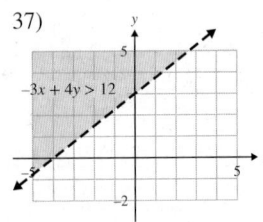

Chapter 9: Test

1) $\left\{-\dfrac{1}{2}, 5\right\}$ 2) $\{-16, 48\}$ 3) $\left\{-8, \dfrac{3}{2}\right\}$ 4) $\varnothing$

5) $|x| = 8$; answers may vary.   6) The absolute value of a quantity is always greater than or equal to zero, so for any real number a, the quantity $|0.8a + 1.3| \geq 0$.

7) $(-\infty, -4) \cup (4, \infty)$

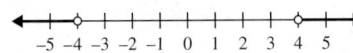

8) $[-1, 8]$

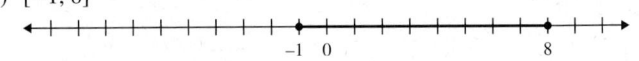

39)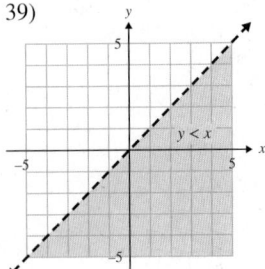

41)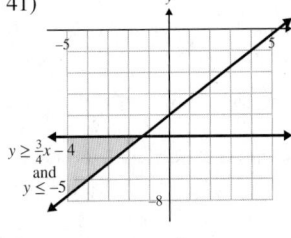

9) $\left(-\infty, -\dfrac{11}{2}\right] \cup [1, \infty)$

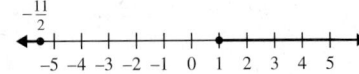

10) $\varnothing$

11) $|w - 168| \leq 0.75$; $167.25 \leq w \leq 168.75$; Thanh's weight is between 167.25 lb and 168.75 lb.

12) $|x| = 17$; answers may vary.

13) $|x| \leq 4$; answers may vary.

14)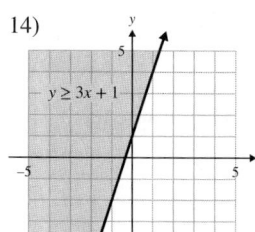

15)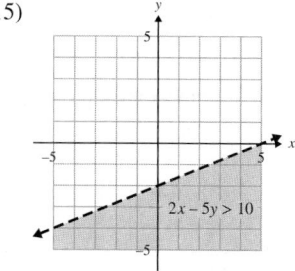

16)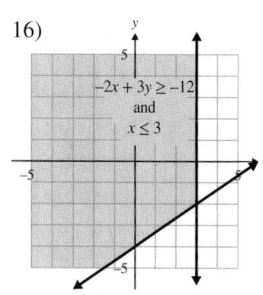

17)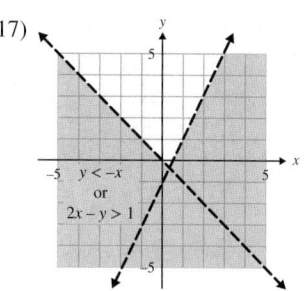

Chapter 9: Cumulative Review for Chapters 1–9

1) 26   2) $-\dfrac{11}{24}$ 3) 81   4) 32

5) $\dfrac{1}{64}$ 6) $\dfrac{1}{64}$ 7) 9.14×10^{-6} 8) $\left\{\dfrac{11}{5}\right\}$

9) $(-\infty, -21]$ 10) 4 oz   11) $y = \dfrac{1}{3}x - \dfrac{1}{3}$

12) $(1, -3)$ 13) $-12p^4 + 28p^3 + 4p^2$

14) $4k^2 - 25$ 15) $t^2 + 16t + 64$ 16) $2c^2 + 5c - 6$

17) $(3m + 11)(3m - 11)$ 18) $(z - 6)(z - 8)$ 19) $\{-3\}$

20) $\left\{-4, \dfrac{1}{2}\right\}$ 21) $\dfrac{-r^2 + 2r + 17}{2(r + 5)(r - 5)}$ 22) $\dfrac{w - 9}{w(w - 8)}$

23) $\{16, 40\}$

24) $(-\infty, -2) \cup \left(\dfrac{10}{9}, \infty\right)$

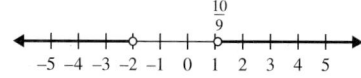

25)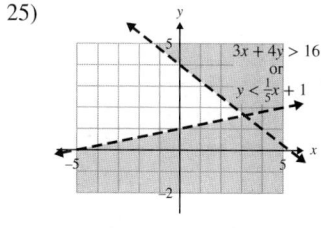

Chapter 10

Section 10.1

1) False; $\sqrt{121} = 11$ because the $\sqrt{}$ symbol means principal square root.

3) False; the square root of a negative number is not a real number.

---

5) 12 and -12 7) $\dfrac{6}{5}$ and $-\dfrac{6}{5}$ 9) 7

11) not real   13) $\dfrac{9}{5}$ 15) -6

17) 3.3

19) 1.4

21) 5.7

23) 7.4

25) True

27) False; the only even root of zero is zero.

29) $\sqrt[3]{64}$ is the number you cube to get 64. $\sqrt[3]{64} = 4$

31) No; the even root of a negative number is not a real number.

33) 5   35) -1 37) 3   39) not real   41) -2

43) -2 45) 10   47) not real   49) $\dfrac{2}{5}$ 51) 7

53) -3 55) 13

57) If a is negative and we didn't use the absolute values, the result would be negative. This is incorrect because if a is negative and n is even, then $a^n > 0$ so that $\sqrt[n]{a^n} > 0$. Using absolute values ensures a positive result.

59) 8   61) 6   63) $|y|$ 65) 5

67) z 69) $|h|$ 71) $|x + 7|$ 73) $2t - 1$

75) $|3n + 2|$ 77) $d - 8$

79) 2 ft   81) $\sqrt{10}$ sec; 3.16 sec

83) Yes. The car was traveling at 30 mph.

Section 10.2

1) The denominator of 2 becomes the index of the radical. $25^{1/2} = \sqrt{25}$

3) 3   5) 10   7) 2   9) -2 11) $\dfrac{2}{11}$ 13) $\dfrac{5}{4}$

15) $-\dfrac{6}{13}$ 17) not a real number   19) -1

21) The denominator of 4 becomes the index of the radical. The numerator of 3 is the power to which we raise the radical expression. $16^{3/4} = (\sqrt[4]{16})^3$

23) 16   25) 32   27) 25   29) -216

31) not a real number   33) $\dfrac{8}{27}$ 35) $-\dfrac{100}{9}$

37) False; the negative exponent does not make the result negative. $81^{-1/2} = \dfrac{1}{9}$

39) $\dfrac{1}{64}; \dfrac{1}{64};$ The denominator of the fractional exponent is the index of the radical.; $\dfrac{1}{8}$

41) $\dfrac{1}{7}$ 43) $\dfrac{1}{10}$ 45) 3   47) -4 49) $\dfrac{1}{32}$

51) $\dfrac{1}{25}$ 53) $\dfrac{8}{125}$ 55) $\dfrac{25}{16}$ 57) 8   59) 3

61) $8^{4/5}$ 63) 32   65) $\dfrac{1}{4^{7/5}}$ 67) z 69) $-72v^{11/8}$

71) $a^{1/9}$ 73) $\dfrac{24}{5w^{1/10}}$ 75) $\dfrac{1}{x^{2/3}}$ 77) $z^{2/15}$

79) $27u^2v^3$ 81) $8r^{1/5}s^{4/15}$ 83) $\dfrac{f^{2/7}g^{5/9}}{3}$ 85) $x^{10}w^9$

87) $\dfrac{1}{y^{2/3}}$ 89) $\dfrac{a^{12/5}}{4b^{2/5}}$ 91) $\dfrac{t^{21/2}}{r^{1/5}}$ 93) $\dfrac{1}{h^5k^{25/18}}$

95) $p^{7/6} + p$ 97) $25^{6/12}; 25^{1/2};$ Evaluate.   99) 7

101) 9   103) 5   105) 12   107) x^4 109) $\sqrt[3]{k}$

111) $\sqrt{z}$ 113) d^2 115) a) 13 degrees F   b) 6 degrees F

Section 10.3

1) $\sqrt{21}$ 3) $\sqrt{30}$ 5) $\sqrt{6y}$

7) False; 20 contains a factor of 4 which is a perfect square.

9) True; 42 does not have any factors (other than 1) that are perfect squares.

11) Factor; $\sqrt{4} \cdot \sqrt{15}; 2\sqrt{15}$

13) $2\sqrt{5}$ 15) $3\sqrt{6}$ 17) simplified   19) $6\sqrt{3}$

21) $7\sqrt{2}$ 23) simplified   25) 20   27) $5\sqrt{30}$

29) $\dfrac{12}{5}$ 31) $\dfrac{2}{7}$ 33) 3   35) $2\sqrt{3}$

37) $2\sqrt{5}$ 39) $\sqrt{7}$ 41) $\dfrac{\sqrt{6}}{7}$ 43) $\dfrac{3\sqrt{5}}{4}$

45) x^4 47) w^7 49) $10c$ 51) $8k^3m^5$

53) $2r^2\sqrt{7}$ 55) $10q^{11}t^8\sqrt{3}$ 57) $\dfrac{9}{c^3}$ 59) $\dfrac{2\sqrt{10}}{t^4}$

61) $\dfrac{5x\sqrt{3}}{y^6}$ 63) Factor; $\sqrt{w^8} \cdot \sqrt{w^1};$ Simplify.

65) $a^2\sqrt{a}$ 67) $g^6\sqrt{g}$ 69) $h^{15}\sqrt{h}$ 71) $6x\sqrt{2x}$

73) $q^3\sqrt{13q}$ 75) $5t^5\sqrt{3t}$ 77) c^4d 79) $a^2b\sqrt{b}$

81) $u^2v^3\sqrt{uv}$ 83) $6m^4n^2\sqrt{m}$ 85) $2x^6y^2\sqrt{11y}$

87) $4t^2u^3\sqrt{2tu}$ 89) $\dfrac{a^3\sqrt{a}}{9b^3}$ 91) $\dfrac{r^4\sqrt{3r}}{s}$ 93) $5\sqrt{2}$

95) $3\sqrt{7}$ 97) w^3 99) $n^3\sqrt{n}$ 101) $4k^3$

103) $5a^3b^4\sqrt{2ab}$ 105) $2c^5d^4\sqrt{10d}$ 107) $3k^4$

109) $2h^3\sqrt{10}$ 111) $a^4b^2\sqrt{10ab}$ 113) 20 m/s

Section 10.4

1) Answers may vary.

3) i) Its radicand will not contain any factors that are perfect cubes.
   ii) The radicand will not contain fractions.
   iii) There will be no radical in the denominator of a fraction.

5) $\sqrt[5]{12}$ 7) $\sqrt[5]{9m^2}$ 9) $\sqrt[3]{a^2b}$

11) Factor; $\sqrt[3]{8} \cdot \sqrt[3]{7}; 2\sqrt[3]{7}$

13) $2\sqrt[3]{3}$ 15) $2\sqrt[4]{4}$ 17) $3\sqrt[3]{2}$ 19) $10\sqrt[3]{2}$

21) $2\sqrt[5]{2}$ 23) $\dfrac{1}{5}$ 25) -3 27) $2\sqrt[3]{3}$ 29) $2\sqrt[4]{5}$

31) d^2 33) n^5 35) xy^3 37) $w^4\sqrt[3]{w^2}$ 39) $y^2\sqrt[4]{y}$

41) $d\sqrt[3]{d^2}$ 43) $u^3v^5\sqrt[3]{u}$ 45) $b^5c\sqrt[3]{bc^2}$

47) $n^4\sqrt[4]{m^3n^2}$ 49) $2x^3y^4\sqrt[3]{3x}$ 51) $2t^5u^2\sqrt[3]{9t^2u}$

53) $\dfrac{m^2}{3}$ 55) $\dfrac{2a^4\sqrt[5]{a^3}}{b^3}$ 57) $\dfrac{t^2\sqrt[4]{t}}{3s^6}$ 59) $\dfrac{u^9\sqrt[3]{u}}{v}$

61) $2\sqrt[3]{3}$ 63) $3\sqrt[3]{4}$ 65) $2\sqrt[3]{10}$ 67) m^3

69) k^4 71) $r^3\sqrt[3]{r^2}$ 73) $p^4\sqrt[5]{p^3}$ 75) $3z^6\sqrt[3]{z}$

77) h^4 79) $c^2\sqrt[3]{c}$ 81) $3d^4\sqrt[4]{d^3}$

83) Change radicals to fractional exponents.; Rewrite exponents with a common denominator.; $a^{5/4};$ Rewrite in radical form. $a\sqrt[4]{a}$

85) $\sqrt[6]{p^5}$ 87) $n\sqrt[4]{n}$ 89) $c\sqrt[15]{c^4}$ 91) $\sqrt[4]{w}$

93) $\sqrt[12]{h}$ 95) 4 in.

Section 10.5

1) They have the same index and the same radicand.

3) $14\sqrt{2}$ 5) $15\sqrt[3]{4}$ 7) $11 - 3\sqrt{13}$ 9) $-5\sqrt[3]{z^2}$

11) $-9\sqrt[3]{n^2} + 10\sqrt[5]{n^2}$ 13) $2\sqrt{5c} - 2\sqrt{6c}$

15) i) Write each radical expression in simplest form.
   ii) Combine like radicals.

17) Factor.; $\sqrt{16} \cdot \sqrt{3} + \sqrt{3};$ Simplify.; $5\sqrt{3}$

19) $4\sqrt{3}$ 21) $-2\sqrt{2}$ 23) $6\sqrt{3}$ 25) $10\sqrt[3]{9}$

27) $-\sqrt[3]{6}$ 29) $13q\sqrt{q}$ 31) $-20d^2\sqrt{d}$

33) $4t^3\sqrt[3]{t}$ 35) $6a^2\sqrt[4]{a^3}$ 37) $-2\sqrt{2p}$

39) $25a\sqrt[3]{3a^2}$ 41) $4y\sqrt{xy}$ 43) $3c^2d\sqrt{2d}$

45) $14p^2q\sqrt[3]{11pq^2}$ 47) $14cd\sqrt[4]{9cd}$ 49) $\sqrt[3]{b}(a^3 - b^2)$

51) $3x + 15$ 53) $7\sqrt{6} + 14$ 55) $\sqrt{30} - \sqrt{10}$

57) $-30\sqrt{2}$ 59) $4\sqrt{5}$ 61) $-\sqrt{30}$ 63) $5\sqrt[4]{3} - 3$

65) $t - 9\sqrt{tu}$ 67) $2y\sqrt{x} - xy\sqrt{2y}$

69) $c\sqrt[3]{c} + 5c\sqrt[3]{d}$

71) Both are examples of multiplication of two binomials. They can be multiplied using FOIL.

73) $(a + b)(a - b) = a^2 - b^2$ 75) $p^2 + 13p + 42$

77) $6 \cdot 2 + 6\sqrt{7} + 2\sqrt{7} + \sqrt{7} \cdot \sqrt{7};$ Multiply.; $19 + 8\sqrt{7}$

79) $-22 + 5\sqrt{2}$ 81) $x + 14\sqrt{x} + 40$

83) $-16 + 11\sqrt{6}$ 85) $5\sqrt{7} + 5\sqrt{2} + 2\sqrt{21} + 2\sqrt{6}$

87) $n\sqrt{21} - 8\sqrt{7n} + \sqrt{3n} - 8$

89) $5 - \sqrt[3]{150} - 3\sqrt[3]{5} + 3\sqrt[3]{6}$

91) $-2\sqrt{6pq} + 30p - 16q$ 93) $4 + 2\sqrt{3}$

95) $k - 12\sqrt{k} + 36$ 97) $48v + 8\sqrt{3v} + 1$

99) $16 - 2\sqrt{55}$ 101) $h + 2\sqrt{7h} + 7$

103) $x - 2\sqrt{xy} + y$ 105) $c^2 - 81$ 107) 31

109) $p - 49$ 111) 46   113) $\sqrt[3]{4} - 9$

115) $c - d$ 117) $64f - g$ 119) 41

121) $11 + 4\sqrt{7}$ 123) $13 - 4\sqrt{3}$

Section 10.6

1) Eliminate the radical from the denominator.

3) $\dfrac{\sqrt{5}}{5}$ 5) $\dfrac{3\sqrt{6}}{2}$ 7) $-5\sqrt{2}$ 9) $\dfrac{\sqrt{21}}{14}$

11) $\dfrac{\sqrt{3}}{3}$ 13) $\dfrac{\sqrt{42}}{6}$ 15) $\dfrac{\sqrt{30}}{3}$ 17) $\dfrac{\sqrt{15}}{10}$

19) $\dfrac{8\sqrt{y}}{y}$ 21) $\dfrac{\sqrt{5t}}{t}$ 23) $\dfrac{8v^3\sqrt{5vw}}{5w}$ 25) $\dfrac{a\sqrt{3b}}{3b}$

27) $-\dfrac{5\sqrt{3b}}{b^2}$ 29) $\dfrac{\sqrt{13j}}{j^3}$ 31) 2^2 or 4   33) 3

35) c^2 37) 2^3 or 8   39) m

41) Her answer is wrong because $\sqrt[3]{3} \cdot \sqrt[3]{3} = \sqrt[3]{9} \neq 3$. The right

way to simplify $\dfrac{8}{\sqrt[3]{3}}$ is $\dfrac{8}{\sqrt[3]{3}} \cdot \dfrac{\sqrt[3]{9}}{\sqrt[3]{9}} = \dfrac{8\sqrt[3]{9}}{\sqrt[3]{27}} = \dfrac{8\sqrt[3]{9}}{3}$.

43) $\dfrac{4\sqrt[3]{9}}{3}$ 45) $6\sqrt[3]{4}$

47) $\dfrac{9\sqrt[3]{5}}{5}$ 49) $\dfrac{\sqrt[4]{45}}{3}$ 51) $\dfrac{\sqrt[5]{12}}{2}$ 53) $\dfrac{10\sqrt[3]{z^2}}{z}$

55) $\dfrac{\sqrt[3]{3n}}{n}$ 57) $\dfrac{\sqrt[3]{28k}}{2k}$ 59) $\dfrac{9\sqrt[5]{a^2}}{a}$ 61) $\dfrac{\sqrt[4]{40m^3}}{2m}$

63) Change the sign between the two terms.

65) $(5 - \sqrt{2}); 23$ 67) $(\sqrt{2} - \sqrt{6}); -4$

69) $(\sqrt{t} + 8); t - 64$

71) Multiply by the conjugate.; $(a + b)(a - b) = a^2 - b^2$;

$\dfrac{24 + 6\sqrt{5}}{16 - 5}; \dfrac{24 + 6\sqrt{5}}{11}$

73) $6 - 3\sqrt{3}$ 75) $\dfrac{90 + 10\sqrt{2}}{79}$ 77) $2\sqrt{6} - 4$

79) $\dfrac{\sqrt{30} - 5\sqrt{2} + 3 - \sqrt{15}}{7}$ 81) $\dfrac{m - \sqrt{mn}}{m - n}$

83) $\sqrt{b} + 5$ 85) $\dfrac{x + 2\sqrt{xy} + y}{x - y}$ 87) $\dfrac{5}{3\sqrt{5}}$

89) $\dfrac{x}{\sqrt{7x}}$ 91) $\dfrac{1}{12 - 6\sqrt{3}}$

93) $\dfrac{1}{\sqrt{x} + 2}$ 95) $-\dfrac{1}{4 + \sqrt{c + 11}}$

97) No, because when we multiply the numerator and denominator by the conjugate of the denominator, we are multiplying the original expression by 1.

99) $1 + 2\sqrt{3}$ 101) $\dfrac{15 - 9\sqrt{5}}{2}$

103) $\dfrac{\sqrt{5} + 2}{3}$ 105) $-2 - \sqrt{2}$

107) a) $r(8\pi) = 2\sqrt{2}$; when the area of a circle is 8π in^2, its radius is $2\sqrt{2}$ in.

b) $r(7) = \dfrac{\sqrt{7\pi}}{\pi}$; when the area of a circle is 7 in^2, its

radius is $\dfrac{\sqrt{7\pi}}{\pi}$ in. (This is approximately 1.5 in.)

c) $r(A) = \dfrac{\sqrt{A\pi}}{\pi}$

Chapter 10: Putting It All Together

1) always   2) never

3) 3   4) -10 5) -2 6) 11   7) not a real number

8) $\dfrac{12}{7}$ 9) 12   10) 16   11) -100

12) not a real number   13) $\dfrac{1}{5}$ 14) $\dfrac{27}{1000}$ 15) $\dfrac{1}{k^{3/10}}$

16) t^6 17) $\dfrac{9}{a^{16/3}b^6}$ 18) $\dfrac{x^{30}}{243y^{5/6}}$ 19) $2\sqrt{6}$

20) $2\sqrt[4]{2}$ 21) $2\sqrt[3]{9}$ 22) $5\sqrt[3]{2}$ 23) $3\sqrt[4]{3}$

24) $3c^5\sqrt{5c}$ 25) $2m^2n^5\sqrt[3]{12m}$ 26) $\dfrac{2x^3\sqrt[5]{2x^4}}{y^4}$

27) $2\sqrt[3]{3}$ 28) $2k^2\sqrt[4]{3}$ 29) $19 + 8\sqrt{7}$

30) $-18c^2\sqrt[3]{4c}$ 31) $3\sqrt{6}$ 32) $5\sqrt{3} + 5\sqrt{2}$

33) $17m\sqrt{3mn}$ 34) $3p^4q^2\sqrt{10q}$ 35) $\dfrac{9\sqrt[3]{4}}{2}$

36) $2t^3u\sqrt{3}$ 37) $112 + 40\sqrt{3}$ 38) -7

39) $\dfrac{2\sqrt{2} - \sqrt{5}}{3}$ 40) $r\sqrt[6]{r}$ 41) $\dfrac{\sqrt[3]{3b^2c^2}}{3c}$

42) $\dfrac{2\sqrt[4]{2w}}{w^3}$ 43) $5w - 16\sqrt{5w} + 64$

44) $10x + 6y - 17\sqrt{xy}$ 45) $\dfrac{11}{4\sqrt{11}}$

46) $\dfrac{10k}{\sqrt{5hk}}$ 47) $\dfrac{1}{\sqrt{a} + 5}$ 48) $\dfrac{47}{56 - 8\sqrt{2}}$

Section 10.7

1) Sometimes there are extraneous solutions.   3) sometimes

5) $\{49\}$ 7) $\left\{\dfrac{4}{9}\right\}$ 9) $\varnothing$ 11) $\{20\}$ 13) $\varnothing$

15) $\left\{\dfrac{2}{3}\right\}$ 17) $\{2\}$ 19) $\{5\}$ 21) $n^2 + 10n + 25$

23) $c^2 - 12c + 36$ 25) {12}   27) ∅   29) {1, 3}

31) {4}   33) {1, 16}   35) {−3}   37) {10}

39) {−3}   41) {10}   43) {63}   45) ∅

47) $x + 10\sqrt{x} + 25$ 49) $85 - 18\sqrt{a + 4} + a$

51) $12n + 28\sqrt{3n - 1} + 45$ 53) {5, 13}   55) {2}

57) $\left\{\dfrac{1}{4}\right\}$ 59) {1, 5}   61) ∅   63) {2, 11}

65) Raise both sides of the equation to the third power.

67) {125}   69) {−64}   71) $\left\{-\dfrac{3}{2}\right\}$ 73) {−1}

75) {−2}   77) $\left\{-\dfrac{1}{2}, 4\right\}$ 79) {36}   81) {26}

83) {23}   85) {9}   87) {9}   89) {−1}

91) $E = \dfrac{mv^2}{2}$ 93) $b^2 = c^2 - a^2$ 95) $\sigma = \dfrac{E}{T^4}$

97) a) 320 m/sec   b) 340 m/sec
    c) The speed of sound increases.   d) $T = \dfrac{V_s^2}{400} - 273$

99) a) 2 in.   b) $V = \pi r^2 h$

101) a) 463 mph   b) about 8 min.

103) 16 ft   105) 5 mph

Section 10.8

1) False   3) True   5) $9i$ 7) $5i$

9) $i\sqrt{6}$ 11) $3i\sqrt{3}$ 13) $2i\sqrt{15}$

15) Write each radical in terms of i *before* multiplying.
$$\sqrt{-5} \cdot \sqrt{-10} = i\sqrt{5} \cdot i\sqrt{10}$$
$$= i^2\sqrt{50}$$
$$= (-1)\sqrt{25} \cdot \sqrt{2}$$
$$= -5\sqrt{2}$$

17) $-\sqrt{5}$ 19) -10 21) 2   23) -13

25) Add the real parts, and add the imaginary parts.

27) -1 29) $3 + 11i$ 31) $4 - 9i$ 33) $-\dfrac{1}{4} - \dfrac{5}{6}i$

35) $7i$ 37) $24 - 15i$ 39) $-6 + \dfrac{4}{3}i$

41) $44 - 24i$ 43) $-28 + 17i$ 45) $14 + 18i$

47) $36 - 42i$ 49) $\dfrac{3}{20} + \dfrac{9}{20}i$

51) conjugate: $11 - 4i$;   53) conjugate: $-3 + 7i$;
    product: 137   product: 58

55) conjugate: $-6 - 4i$;   57) Answers may vary.
    product: 52

59) $\dfrac{8}{13} + \dfrac{12}{13}i$ 61) $\dfrac{8}{17} + \dfrac{32}{17}i$ 63) $\dfrac{7}{29} - \dfrac{3}{29}i$

65) $-\dfrac{74}{85} + \dfrac{27}{85}i$ 67) $-\dfrac{8}{61} + \dfrac{27}{61}i$ 69) $-9i$

71) $(i^2)^{12}$; $i^2 = -1$; 1   73) 1   75) 1   77) i

79) $-i$ 81) $-i$ 83) -1 85) $32i$ 87) -1

89) $142 - 65i$ 91) $1 + 2i\sqrt{2}$ 93) $8 - 3i\sqrt{5}$

95) $-3 + i\sqrt{2}$ 97) $Z = 10 + 6j$ 99) $Z = 16 + 4j$

Chapter 10: Group Activity

d	$v = 5.47\sqrt{\mu d}$
0	0
50	32.4
100	45.8
150	56.1
200	64.7
250	72.4
300	79.3
400	91.5
500	102.3

Answers may vary.

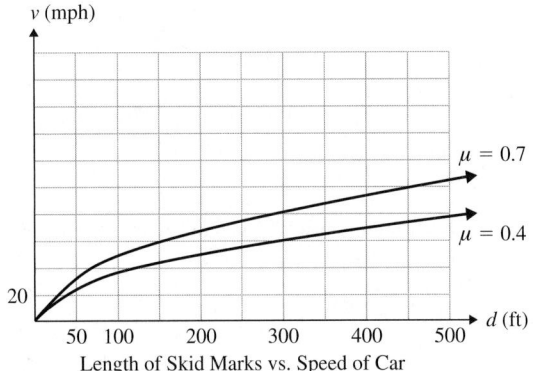

Length of Skid Marks vs. Speed of Car

Chapter 10: Review Exercises

1) $\dfrac{13}{2}$ 3) -9 5) -1

7) not real   9) 13   11) $|p|$ 13) h

15) 5.8

17) 6   19) $\dfrac{3}{5}$ 21) 8   23) $\dfrac{1}{9}$ 25) $\dfrac{1}{27}$

27) $\dfrac{100}{9}$ 29) 9   31) 64   33) 1

35) $32a^{10/3}b^{10}$ 37) $\dfrac{2c}{3d^{7/4}}$ 39) 3

41) 7   43) k^7 45) w^3 47) $10\sqrt{10}$

49) $\dfrac{3\sqrt{2}}{7}$ 51) k^6 53) $x^4\sqrt{x}$ 55) $3t\sqrt{5}$

57) $6x^3y^6\sqrt{2xy}$ 59) $\sqrt{15}$ 61) $2\sqrt{6}$ 63) $11x^6\sqrt{x}$

65) $10k^8$ 67) $2\sqrt[3]{2}$ 69) $2\sqrt[4]{3}$ 71) z^6

73) $a^6\sqrt[6]{a^2}$ 75) $2z^5\sqrt[3]{2}$ 77) $\dfrac{h^3}{3}$ 79) $\sqrt[3]{21}$

81) $2t^4\sqrt[4]{2t}$ 83) $\sqrt[6]{n^5}$ 85) $11\sqrt{5}$

87) $6\sqrt{5} - 4\sqrt{3}$ 89) $-4p\sqrt{p}$ 91) $-12d^2\sqrt{2d}$

93) $6k\sqrt{5} + 3\sqrt{2k}$ 95) $2a\sqrt{3} - 7\sqrt{6a} - 9\sqrt{2a} + 63$

97) $9ab^3\sqrt[4]{2a^3b}$ 99) $23\sqrt{2rs} + 8r + 15s$

101) $2 + 2\sqrt{y+1} + y$ 103) $\dfrac{14\sqrt{3}}{3}$ 105) $\dfrac{3\sqrt{2kn}}{n}$

107) $\dfrac{7\sqrt[3]{4}}{2}$ 109) $\dfrac{\sqrt[3]{x^2y^2}}{y}$ 111) $\dfrac{3 - \sqrt{3}}{3}$ 113) $\dfrac{5}{2\sqrt{15}}$

115) $1 - 3\sqrt{2}$ 117) $\{1\}$ 119) $\varnothing$ 121) $\{-4\}$

123) $\{2, 6\}$ 125) $V = \dfrac{1}{3}\pi r^2 h$ 127) $7i$ 129) -4

131) $12 - 3i$ 133) $\dfrac{3}{10} - \dfrac{4}{3}i$ 135) $-30 + 35i$

137) $-36 - 21i$ 139) $-24 - 42i$

141) conjugate: $2 + 7i$; product: 53 143) $\dfrac{12}{29} - \dfrac{30}{29}i$

145) $-8i$ 147) $\dfrac{58}{37} - \dfrac{15}{37}i$ 149) -1

Chapter 10: Test

1) a) never   b) always   c) sometimes

2) a) 12   b) -2 c) not real

3) 6.8

$$\underset{0\ 1\ 2\ 3\ 4\ 5\ 6\ 7\ 8}{\xleftrightarrow{\hspace{2cm}\bullet\hspace{1cm}}}\;\;\overset{\sqrt{46}}{}$$

4) $|w|$ 5) -19

6) 2   7) 81   8) $\dfrac{1}{7}$ 9) $\dfrac{25}{4}$ 10) $m^{5/8}$

11) $\dfrac{5}{2a^{2/3}}$ 12) $\dfrac{y^2}{32x^{3/2}}$ 13) $5\sqrt{3}$ 14) $2\sqrt[3]{6}$

15) $2\sqrt{3}$ 16) y^3 17) p^6 18) $t^4\sqrt{t}$

19) $3m^2n^4\sqrt{7m}$ 20) $c^7\sqrt[3]{c^2}$

21) $\dfrac{2a^4b^2\sqrt[3]{5a^2b}}{3}$ 22) 6   23) $z^3\sqrt[3]{z}$

24) $2w^5\sqrt{15w}$ 25) $6\sqrt{7}$ 26) $3\sqrt{2} - 4\sqrt{3}$

27) $-14h^3\sqrt[4]{h}$ 28) $2\sqrt{3} - 5\sqrt{6}$

29) $3\sqrt{2} + 3 - 2\sqrt{10} - 2\sqrt{5}$ 30) $25x - 11$

31) $2p + 5 + 4\sqrt{2p+1}$ 32) $2t - 2\sqrt{3tu}$ 33) $\dfrac{2\sqrt{5}}{5}$

34) $12 - 4\sqrt{7}$ 35) $\dfrac{\sqrt{6a}}{a}$ 36) $2\sqrt[3]{5n^2}$

37) $1 - 2\sqrt{3}$ 38) $\{1\}$ 39) $\{-2\}$ 40) $\{13\}$

41) $\{1, 5\}$ 42) a) 3 in.   b) $V = \pi r^2 h$

43) $8i$ 44) $3i\sqrt{5}$ 45) $-i$ 46) $-16 + 2i$

47) $19 + 13i$ 48) $1 + 2i$

1) $\dfrac{10}{3}x - 2y + 8$ 2) 8.723×10^6 3) $\left\{-\dfrac{3}{4}\right\}$

4)

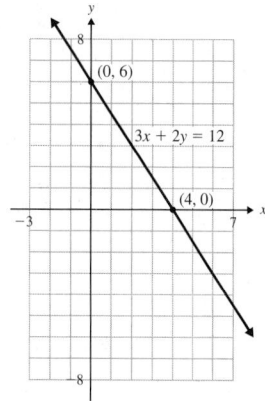

5) $y = \dfrac{5}{4}x - \dfrac{13}{4}$ 6) $(-6, 0)$

7) $15p^4 - 20p^3 - 11p^2 + 8p + 2$ 8) $4n^2 + 2n + 1$

9) $(3w + 1)^2$ 10) $2(2 + 3t)(2 - 3t)$

11) $\left\{-\dfrac{1}{2}, \dfrac{4}{3}\right\}$ 12) $\{6, 9\}$

13) length = 12 in., width = 7 in.

14) $\dfrac{2a^2 + 2a + 3}{a(a+4)}$ 15) $\dfrac{4n}{21m^3}$ 16) $\{-8, -4\}$

17) $(-\infty, -2] \cup \left[\dfrac{5}{3}, \infty\right)$ 18) $(4, -1, 2)$

19) a) $10\sqrt{5}$ b) $2\sqrt[3]{7}$ c) $p^5q^3\sqrt{q}$ d) $2a^3\sqrt[4]{2a^3}$

20) a) 9 b) 16 c) $\dfrac{1}{3}$ 21) $10\sqrt{3} - 6$

22) a) $\dfrac{\sqrt{10}}{5}$ b) $3\sqrt[3]{4}$ c) $\dfrac{x\sqrt[3]{y}}{y}$ d) $\dfrac{a - 2 - \sqrt{a}}{1 - a}$

23) a) $\varnothing$ b) $\{-3\}$ 24) a) $7i$ b) $2i\sqrt{14}$ c) 1

25) a) $2 + 7i$ b) $-48 + 9i$ c) $-\dfrac{11}{25} - \dfrac{2}{25}i$

Chapter 11

Section 11.1

1) $\{-7, 6\}$ 3) $\{-11, -4\}$ 5) $\{-7, 8\}$

7) $\left\{-\dfrac{1}{10}, \dfrac{1}{10}\right\}$ 9) $\left\{\dfrac{2}{5}, 4\right\}$ 11) $\left\{-\dfrac{10}{3}, -\dfrac{1}{2}\right\}$

13) $\left\{0, \dfrac{2}{7}\right\}$ 15) $\{-6, 2\}$ 17) quadratic

19) linear   21) quadratic   23) linear

25) $\left\{\dfrac{1}{2}, 6\right\}$ 27) $\{-5, 2\}$ 29) $\left\{\dfrac{9}{2}\right\}$

31) $\left\{-\dfrac{5}{3}, -1, 0\right\}$ 33) $\left\{-\dfrac{1}{2}, 0\right\}$ 35) $\left\{\dfrac{9}{4}\right\}$

37) $\{-4, 2\}$ 39) $\{7\}$ 41) $\left\{-\dfrac{1}{2}\right\}$ 43) $\left\{-2, -\dfrac{3}{5}\right\}$

45) $\{-7, -2, 2\}$ 47) width = 2 in., length = 7 in.

49) width = 5 cm, length = 9 cm

51) base = 9 in.; height = 4 in.

53) base = 6 cm; height = 12 cm

55) legs = 5, 12; hypotenuse = 13

57) legs = 6, 8; hypotenuse = 10

Section 11.2

1) Methods may vary; $\{-4, 4\}$ 3) $\{-6, 6\}$

5) $\{-3\sqrt{3}, 3\sqrt{3}\}$ 7) $\left\{-\dfrac{2}{3}, \dfrac{2}{3}\right\}$ 9) $\{-2i, 2i\}$

11) $\{-i\sqrt{3}, i\sqrt{3}\}$ 13) $\{-\sqrt{14}, \sqrt{14}\}$ 15) $\{-3, 3\}$

17) $\{-2i\sqrt{3}, 2i\sqrt{3}\}$ 19) $\{-12, -8\}$ 21) $\{6, 8\}$

23) $\{-4 - 3\sqrt{2}, -4 + 3\sqrt{2}\}$ 25) $\{-3 - 5i, -3 + 5i\}$

27) $\{2 - i\sqrt{14}, 2 + i\sqrt{14}\}$

29) $\left\{\dfrac{-1 - 2\sqrt{5}}{2}, \dfrac{-1 + 2\sqrt{5}}{2}\right\}$

31) $\left\{\dfrac{10 - \sqrt{14}}{3}, \dfrac{10 + \sqrt{14}}{3}\right\}$

33) $\left\{\dfrac{5}{4} - \dfrac{\sqrt{30}}{4}i, \dfrac{5}{4} + \dfrac{\sqrt{30}}{4}i\right\}$

35) $\left\{-\dfrac{11}{6} - \dfrac{7}{6}i, -\dfrac{11}{6} + \dfrac{7}{6}i\right\}$

37) $\left\{8, \dfrac{40}{3}\right\}$ 39) $\left\{-\dfrac{2}{5}, \dfrac{6}{5}\right\}$

41) 5   43) $\sqrt{13}$ 45) 5

47) $\sqrt{61}$ 49) $2\sqrt{5}$

51) A trinomial whose factored form is the square of a binomial; examples will vary.

53) $\dfrac{1}{2}(8) = 4$; $4^2 = 16$; $w^2 + 8w + 16$; $w^2 + 8w + 16$; $(w + 4)^2$

55) $a^2 + 12a + 36$; $(a + 6)^2$ 57) $c^2 - 18c + 81$; $(c - 9)^2$

59) $t^2 + 5t + \dfrac{25}{4}$; $\left(t + \dfrac{5}{2}\right)^2$ 61) $b^2 - 9b + \dfrac{81}{4}$; $\left(b - \dfrac{9}{2}\right)^2$

63) $x^2 + \dfrac{1}{3}x + \dfrac{1}{36}$; $\left(x + \dfrac{1}{6}\right)^2$

65) Divide both sides of the equation by 2.

67) $\{-4, -2\}$ 69) $\{3, 5\}$

71) $\{1 - \sqrt{10}, 1 + \sqrt{10}\}$ 73) $\{-5 - i, -5 + i\}$

75) $\{4 - i\sqrt{3}, 4 + i\sqrt{3}\}$ 77) $\{-8, 5\}$ 79) $\{3, 4\}$

81) $\left\{\dfrac{1}{2} - \dfrac{\sqrt{13}}{2}, \dfrac{1}{2} + \dfrac{\sqrt{13}}{2}\right\}$

83) $\left\{-\dfrac{5}{2} - \dfrac{\sqrt{3}}{2}i, -\dfrac{5}{2} + \dfrac{\sqrt{3}}{2}i\right\}$

85) $\{1 - i\sqrt{3}, 1 + i\sqrt{3}\}$ 87) $\{-3 - \sqrt{11}, -3 + \sqrt{11}\}$

89) $\{2, 3\}$ 91) $\left\{\dfrac{5}{4} - \dfrac{\sqrt{39}}{4}i, \dfrac{5}{4} + \dfrac{\sqrt{39}}{4}i\right\}$

93) $\left\{\dfrac{3}{4}, 1\right\}$ 95) $\{-1 - \sqrt{21}, -1 + \sqrt{21}\}$

97) $\left\{\dfrac{5}{4} - \dfrac{\sqrt{7}}{4}i, \dfrac{5}{4} + \dfrac{\sqrt{7}}{4}i\right\}$ 99) 6   101) $\sqrt{29}$

103) 6 in.   105) 12 ft   107) −10, 4

109) width = 9 ft, length = 17 ft

Section 11.3

1) The fraction bar should also be under $-b$:
$$x = \dfrac{-b \pm \sqrt{b^2 - 4ac}}{2a}$$

3) You cannot divide only the −2 by 2.
$$\dfrac{-2 \pm 6\sqrt{11}}{2} = \dfrac{2(-1 \pm 3\sqrt{11})}{2} = -1 \pm 3\sqrt{11}$$

5) $\{-3, -1\}$ 7) $\left\{-2, \dfrac{5}{3}\right\}$

9) $\left\{\dfrac{5 - \sqrt{17}}{2}, \dfrac{5 + \sqrt{17}}{2}\right\}$ 11) $\{4 - 3i, 4 + 3i\}$

13) $\left\{\dfrac{1}{5} - \dfrac{\sqrt{14}}{5}i, \dfrac{1}{5} + \dfrac{\sqrt{14}}{5}i\right\}$ 15) $\{-7, 0\}$

17) $\left\{\dfrac{3 - \sqrt{3}}{2}, \dfrac{3 + \sqrt{3}}{2}\right\}$ 19) $\left\{\dfrac{7}{2}, 4\right\}$

21) $\{-4 - \sqrt{31}, -4 + \sqrt{31}\}$ 23) $\{-10, 4\}$

25) $\left\{-\dfrac{3}{2}i, \dfrac{3}{2}i\right\}$ 27) $\{-3 - 5i, -3 + 5i\}$

29) $\{-1 - \sqrt{10}, -1 + \sqrt{10}\}$ 31) $\left\{\dfrac{3}{2}\right\}$

33) $\left\{-\dfrac{3}{8} - \dfrac{\sqrt{7}}{8}i, -\dfrac{3}{8} + \dfrac{\sqrt{7}}{8}i\right\}$

35) $\left\{\dfrac{5 - \sqrt{19}}{2}, \dfrac{5 + \sqrt{19}}{2}\right\}$

37) $-3 - \sqrt{11}, -3 + \sqrt{11}$

39) $\dfrac{1 - \sqrt{41}}{4}, \dfrac{1 + \sqrt{41}}{4}$ 41) $-4, \dfrac{1}{5}$

43) There is one rational solution.

45) −39; two nonreal, complex solutions

47) 0; one rational solution

49) 16; two rational solutions

51) 56; two irrational solutions

53) −8 or 8   55) 9   57) 4   59) 2 in., 5 in.

61) a) 2 sec   b) $\dfrac{3 + \sqrt{33}}{4}$ sec or about 2.2 sec

Chapter 11: Putting It All Together

1) $\{-5\sqrt{2}, 5\sqrt{2}\}$ 2) $\{3 - \sqrt{17}, 3 + \sqrt{17}\}$

3) $\{-5, 4\}$ 4) $\left\{\dfrac{3}{4} - \dfrac{\sqrt{39}}{4}i, \dfrac{3}{4} + \dfrac{\sqrt{39}}{4}i\right\}$

5) $\left\{\dfrac{-7 - \sqrt{13}}{2}, \dfrac{-7 + \sqrt{13}}{2}\right\}$ 6) $\left\{-1, \dfrac{4}{3}\right\}$

7) $\left\{-3, \dfrac{1}{4}\right\}$ 8) $\{-3 - i\sqrt{6}, -3 + i\sqrt{6}\}$

9) $\{-7 - i\sqrt{11}, -7 + i\sqrt{11}\}$

10) $\left\{\dfrac{-1 - \sqrt{7}}{2}, \dfrac{-1 + \sqrt{7}}{2}\right\}$

11) $\left\{\dfrac{1}{3} - \dfrac{\sqrt{6}}{3}i, \dfrac{1}{3} + \dfrac{\sqrt{6}}{3}i\right\}$ 12) $\{-4 - 3i, -4 + 3i\}$

13) $\{-9, 3\}$ 14) $\{-5, 5\}$ 15) $\{2 - \sqrt{7}, 2 + \sqrt{7}\}$

16) $\{-9, -6, 0\}$ 17) $\left\{\dfrac{3}{2}, 2\right\}$ 18) $\{0, 1\}$

19) $\{3, 7\}$ 20) $\left\{\dfrac{-1 - \sqrt{21}}{4}, \dfrac{-1 + \sqrt{21}}{4}\right\}$

21) $\left\{\dfrac{1}{3} - \dfrac{\sqrt{2}}{3}i, \dfrac{1}{3} + \dfrac{\sqrt{2}}{3}i\right\}$ 22) $\{3 - 5i, 3 + 5i\}$

23) $\{5 - 4i, 5 + 4i\}$ 24) $\left\{-4, -\dfrac{3}{2}\right\}$ 25) $\{0, 3\}$

26) $\left\{-\dfrac{3}{2} - \dfrac{\sqrt{15}}{2}i, -\dfrac{3}{2} + \dfrac{\sqrt{15}}{2}i\right\}$ 27) $\left\{-\dfrac{3}{2}, 0, \dfrac{3}{2}\right\}$

28) $\{-2, 6\}$ 29) $\left\{-\dfrac{5}{4} - \dfrac{\sqrt{39}}{4}i, -\dfrac{5}{4} + \dfrac{\sqrt{39}}{4}i\right\}$

30) $\left\{-\dfrac{5}{12} - \dfrac{1}{4}i, -\dfrac{5}{12} + \dfrac{1}{4}i\right\}$

Section 11.4

1) $\{-4, 12\}$ 3) $\left\{-2, \dfrac{4}{5}\right\}$

5) $\{4 - \sqrt{6}, 4 + \sqrt{6}\}$ 7) $\left\{\dfrac{1 - \sqrt{7}}{3}, \dfrac{1 + \sqrt{7}}{3}\right\}$

9) $\left\{-\dfrac{9}{5}, 1\right\}$ 11) $\left\{\dfrac{11 - \sqrt{21}}{10}, \dfrac{11 + \sqrt{21}}{10}\right\}$ 13) $\{5\}$

15) $\left\{\dfrac{4}{5}, 2\right\}$ 17) $\{1\}$ 19) $\{1, 16\}$ 21) $\{5\}$

23) $\{0, 2\}$ 25) yes   27) yes   29) no   31) yes

33) no   35) $\{-3, -1, 1, 3\}$ 37) $\{-\sqrt{7}, -2, 2, \sqrt{7}\}$

39) $\{-i\sqrt{6}, -i\sqrt{3}, i\sqrt{3}, i\sqrt{6}\}$ 41) $\{-8, -1\}$

43) $\{-8, 27\}$ 45) $\{49\}$ 47) $\{16\}$

49) $\left\{-\dfrac{2\sqrt{3}}{3}i, -\dfrac{\sqrt{3}}{3}i, \dfrac{\sqrt{3}}{3}i, \dfrac{2\sqrt{3}}{3}i\right\}$

51) $\{-\sqrt{5}, \sqrt{5}, -i\sqrt{3}, i\sqrt{3}\}$

53) $\{-\sqrt{3 + \sqrt{7}}, \sqrt{3 + \sqrt{7}}, -\sqrt{3 - \sqrt{7}}, \sqrt{3 - \sqrt{7}}\}$

55) $\left\{-\dfrac{\sqrt{7 + \sqrt{41}}}{2}, \dfrac{\sqrt{7 + \sqrt{41}}}{2}, -\dfrac{\sqrt{7 - \sqrt{41}}}{2}, \dfrac{\sqrt{7 - \sqrt{41}}}{2}\right\}$

57) $\left\{-\dfrac{1}{2}, \dfrac{1}{6}\right\}$ 59) $\left\{\dfrac{1}{4}, 3\right\}$ 61) $\{-6, -1\}$

63) $\left\{-\dfrac{1}{2}, 0\right\}$ 65) $\left\{-\dfrac{2}{5}, \dfrac{2}{5}\right\}$ 67) $\left\{-11, -\dfrac{20}{3}\right\}$

69) $\left\{\dfrac{3}{2}\right\}$ 71) $\{2 - \sqrt{2}, 2 + \sqrt{2}\}$ 73) $\left\{-\dfrac{27}{8}, 125\right\}$

75) $\{-3, 3, -2i\sqrt{3}, 2i\sqrt{3}\}$ 77) $\left\{0, \dfrac{1}{10}\right\}$

79) $\{4, 36\}$ 81) $\{-5 - i\sqrt{47}, -5 + i\sqrt{47}\}$

83) Walter: 3 hr; Kevin: 6 hr   85) 15 mph

87) large drain: 3 hr; small drain: 6 hr

89) to Boulder: 60 mph; going home: 50 mph

Section 11.5

1) $r = \dfrac{\pm\sqrt{A\pi}}{\pi}$ 3) $v = \pm\sqrt{ar}$

5) $d = \dfrac{\pm\sqrt{IE}}{E}$ 7) $r = \dfrac{\pm\sqrt{kq_1q_2F}}{F}$

9) $A = \dfrac{1}{4}\pi d^2$ 11) $l = \dfrac{gT_p^2}{4\pi^2}$ 13) $g = \dfrac{4\pi^2 l}{T_p^2}$

15) a) Both are written in the standard form for a quadratic equation, $ax^2 + bx + c = 0$.
    b) Use the quadratic formula.

17) $x = \dfrac{5 \pm \sqrt{25 - 4rs}}{2r}$ 19) $z = \dfrac{-r \pm \sqrt{r^2 + 4pq}}{2p}$

21) $a = \dfrac{h \pm \sqrt{h^2 + 4dk}}{2d}$ 23) $t = \dfrac{-v \pm \sqrt{v^2 + 2gs}}{g}$

25) length = 12 in., width = 9 in.   27) 2 ft

29) base = 8 ft, height = 15 ft   31) 10 in.

33) a) 0.75 sec on the way up, 3 sec on the way down
    b) $\dfrac{15 + \sqrt{241}}{8}$ sec or about 3.8 sec

35) a) 9.5 million   b) 2015   37) $2.40   39) 1.75 ft

Chapter 11: Group Activity

1) a) $a = 0.06932$; $b = -9.5213$; $c = 399.11$
   b) $T_{HI} = 0.06932T^2 - 9.5213T + 399.11$

T (air temperature)	T_{HI} (when $R = 70\%$)
80°F	81
84°F	88
88°F	98
92°F	110
96°F	124
100°F	140
104°F	159
108°F	179

2) a) $a = 0.03908$; $b = -5.5862$; $c = 275.661$
   b) $T_{HI} = 0.03908T^2 - 5.5862T + 275.661$

T (air temperature)	T_{HI} (when $R = 40\%$)
80°F	79
84°F	82
88°F	87
92°F	93
96°F	100
100°F	108
104°F	117
108°F	128

4) Answers may vary.

Chapter 11: Review Exercises

1) $\{-6, 9\}$ 3) $\left\{-\dfrac{1}{2}, \dfrac{3}{2}\right\}$ 5) $\{-4, -3, 4\}$

7) width = 8 cm, length = 12 cm

9) $\{-12, 12\}$ 11) $\{-2i, 2i\}$ 13) $\{-4, 10\}$

15) $\left\{-\dfrac{\sqrt{10}}{3}, \dfrac{\sqrt{10}}{3}\right\}$ 17) $2\sqrt{5}$

19) $r^2 + 10r + 25$; $(r + 5)^2$

21) $c^2 - 5c + \dfrac{25}{4}$; $\left(c - \dfrac{5}{2}\right)^2$

23) $a^2 + \dfrac{2}{3}a + \dfrac{1}{9}$; $\left(a + \dfrac{1}{3}\right)^2$ 25) $\{-2, 8\}$

27) $\{-5 - \sqrt{31}, -5 + \sqrt{31}\}$

29) $\left\{-\dfrac{3}{2} - \dfrac{\sqrt{5}}{2}, -\dfrac{3}{2} + \dfrac{\sqrt{5}}{2}\right\}$

31) $\left\{\dfrac{7}{6} - \dfrac{\sqrt{95}}{6}i, \dfrac{7}{6} + \dfrac{\sqrt{95}}{6}i\right\}$ 33) $\{-6, 2\}$

35) $\left\{\dfrac{5 - \sqrt{15}}{2}, \dfrac{5 + \sqrt{15}}{2}\right\}$ 37) $\{1 - i\sqrt{3}, 1 + i\sqrt{3}\}$

39) $\left\{-\dfrac{2}{3}, \dfrac{1}{2}\right\}$ 41) 64; two rational solutions

43) -12 or 12 45) $\left\{1, \dfrac{4}{3}\right\}$

47) $\{-8 - 2\sqrt{3}, -8 + 2\sqrt{3}\}$

49) $\left\{-\dfrac{3}{2} - \dfrac{1}{2}i, -\dfrac{3}{2} + \dfrac{1}{2}i\right\}$

51) $\{9 - 3\sqrt{5}, 9 + 3\sqrt{5}\}$ 53) $\{-1, 0, 1\}$

55) $-2, 3$ 57) $\left\{\dfrac{3 - \sqrt{21}}{3}, \dfrac{3 + \sqrt{21}}{3}\right\}$

59) $\{3, 4\}$ 61) $\{-4, -1, 1, 4\}$ 63) $\{-27, 1\}$

65) $\left\{-\dfrac{\sqrt{7 + \sqrt{33}}}{2}, \dfrac{\sqrt{7 - \sqrt{33}}}{2}, -\dfrac{\sqrt{7 - \sqrt{33}}}{2}, \dfrac{\sqrt{7 + \sqrt{33}}}{2}\right\}$

67) $\left\{2, \dfrac{11}{2}\right\}$ 69) $v = \dfrac{\pm\sqrt{Frm}}{m}$ 71) $A = \pi r^2$

73) $n = \dfrac{l \pm \sqrt{l^2 + 4km}}{2k}$ 75) 3 in.   77) \$8.00

Chapter 11: Test

1) $\{-4, 12\}$ 2) $\left\{-\dfrac{4}{3}, \dfrac{4}{3}\right\}$ 3) $\{-3\sqrt{2}, 3\sqrt{2}\}$

4) The solution set contains two nonreal, complex numbers.

5) $\{-2 - \sqrt{11}, -2 + \sqrt{11}\}$

6) $\left\{\dfrac{3}{2} - \dfrac{\sqrt{19}}{2}i, \dfrac{3}{2} + \dfrac{\sqrt{19}}{2}i\right\}$

7) $\{4 - i, 4 + i\}$ 8) $\{-5 - i\sqrt{6}, -5 + i\sqrt{6}\}$

9) $\left\{-2, \dfrac{4}{3}\right\}$ 10) $\left\{-\dfrac{2}{5}, \dfrac{2}{5}\right\}$ 11) $\left\{-\dfrac{7}{4}, -1\right\}$

12) $\left\{\dfrac{7 - \sqrt{17}}{4}, \dfrac{7 + \sqrt{17}}{4}\right\}$ 13) $\left\{0, \dfrac{5}{6}\right\}$

14) $\left\{-\dfrac{3}{2}, -1\right\}$ 15) $\{-2\sqrt{2}, 2\sqrt{2}, -3i, 3i\}$

16) 56; two irrational solutions   17) $\sqrt{19}$

18) $0, \dfrac{2}{5}$ 19) $2\sqrt{26}$ 20) $V = \dfrac{1}{3}\pi r^2 h$

21) $t = \dfrac{s \pm \sqrt{s^2 + 24r}}{2r}$

22) a) after 4 sec
    b) after $\dfrac{3 + \sqrt{209}}{4}$ sec or about 4.4 sec

23) Justine: 40 min; Kimora: 60 min

24) width = 6 ft; length = 8 ft

25) width = 13 in., length = 19 in.

1) $\dfrac{7}{10}$ 2) $\dfrac{45x^6}{y^4}$ 3) $m = \dfrac{y-b}{x}$

4) a) $\{0, 3, 4\}$ b) $\{-1, 0, 1, 2\}$ c) no

5) a) \$518,200   b) \$38,865,000   c) 50,000,000 users

6) chips: \$0.80, soda: \$0.75   7) $(0, 3, -1)$

8) $3x^2y^2 - 5x^2y - 3xy^2 - 9xy + 8$ 9) $3r^2 - 30r + 75$

10) $2p(2p-1)(p+4)$ 11) $(a+5)(a^2-5a+25)$

12) a) $\dfrac{6}{35}$ b) -1 c) $(-\infty, -2) \cup \left(-2, \dfrac{5}{2}\right) \cup \left(\dfrac{5}{2}, \infty\right)$

13) $\dfrac{z^2 - 5z + 12}{z(z+4)}$ 14) $\dfrac{c(c+3)}{1-4c}$ 15) $\left\{-\dfrac{26}{3}, 4\right\}$

16) $\left(-\infty, -\dfrac{26}{3}\right] \cup [4, \infty)$ 17) $5\sqrt{3}$ 18) $2\sqrt[3]{5}$

19) $3x^3y^2\sqrt{7x}$ 20) 16   21) $10 - 5\sqrt{3}$ 22) $34 - 77i$

23) $\left\{-\dfrac{2}{3}, \dfrac{7}{3}\right\}$ 24) $\{-9, 3\}$ 25) $V = \pi r^2 h$

Chapter 12

Section 12.1

1) It is a special type of relation in which each element of the domain corresponds to exactly one element in the range.

3) domain: $\{5, 6, 14\}$; range: $\{-3, 0, 1, 3\}$; not a function

5) domain: $\{-2, 2, 5, 8\}$; range: $\{4, 25, 64\}$; is a function

7) domain: $(-\infty, \infty)$; range: $[-4, \infty)$; is a function

9) yes   11) yes   13) no   15) no

17) a) Answers may vary. One example is $f(x) = 2x + 3$.
    b) Answers may vary. One example is
       $p(a) = a^3 + 6a^2 - 10a + 7$.
    c) Answers may vary. One example is $R(w) = \dfrac{w-3}{6w+1}$.

19) always

21) sometimes; The domain of a rational function like
    $f(x) = \dfrac{x-7}{x^2+2}$ is $(-\infty, \infty)$ because the function is defined
    for all real numbers, but the domain of a function like
    $g(x) = \dfrac{x+8}{x-3}$ is $(-\infty, 3) \cup (3, \infty)$ because if 3 is substituted
    for x, the denominator equals 0.

23) $f(6) = 46, f(-2) = -10, f(0) = 4$

25)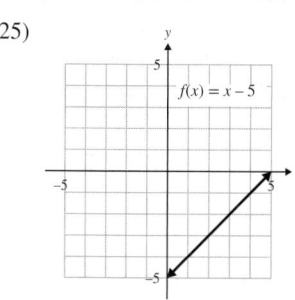

27)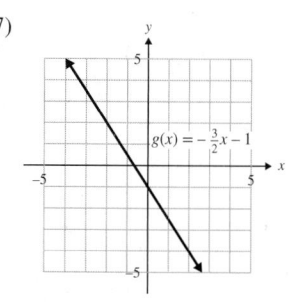

29)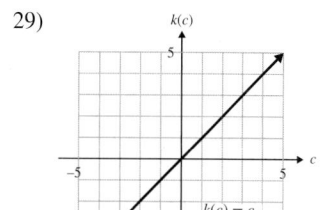

31) Set the denominator equal to 0 and solve for the variable. The domain consists of all real numbers *except* the values that make the denominator equal to 0.

33) $(-\infty, \infty)$ 35) $(-\infty, \infty)$

37) $(-\infty, -8) \cup (-8, \infty)$ 39) $(-\infty, 0) \cup (0, \infty)$

41) $\left(-\infty, -\dfrac{3}{7}\right) \cup \left(-\dfrac{3}{7}, \infty\right)$ 43) $(-\infty, \infty)$

45) $(-\infty, -8) \cup (-8, -3) \cup (-3, \infty)$

47) $(-\infty, -4) \cup (-4, 9) \cup (9, \infty)$

49) Answers may vary. One example is $f(x) = \dfrac{2x+9}{x-10}$.

51) polynomial function   53) quadratic function

55) $(-\infty, \infty)$ 57) 11   59) -12 61) $3a - 7$

63) $d^2 - 4d - 9$ 65) $3c + 5$ 67) $t^2 - 13$

69) $h^2 - 6h - 4$ 71) -2 or 8

73) No, because $\sqrt{-1}$ is not a real number.

75) a) $f(1) = 1, g(1) = 2\sqrt{2}$
    b) 25 is not in the domain of *both* f and g. It *is* in the domain of f because $f(25) = 5$, but it is not in the domain of g because $g(25) = \sqrt{-64}$, which is not a real number.
    c) $f(c) = \sqrt{c}, g(c) = \sqrt{11-3c}$
    d) $f(a+4) = \sqrt{a+4}, g(a+4) = \sqrt{-3a-1}$
    e) $f(2k+1) = \sqrt{2k+1}, g(-2k+1) = \sqrt{6k+8}$

77) Set up an inequality so that the radicand is greater than or equal to 0. Solve for the variable. These are the real numbers in the domain of the function.

79) $[-2, \infty)$ 81) $[8, \infty)$ 83) $(-\infty, \infty)$

85) $\left[-\dfrac{7}{3}, \infty\right)$ 87) $(-\infty, \infty)$

89) $(-\infty, 0]$ 91) $\left(-\infty, \dfrac{9}{7}\right]$

93) Domain: $[1, \infty)$ 95) Domain: $[-3, \infty)$

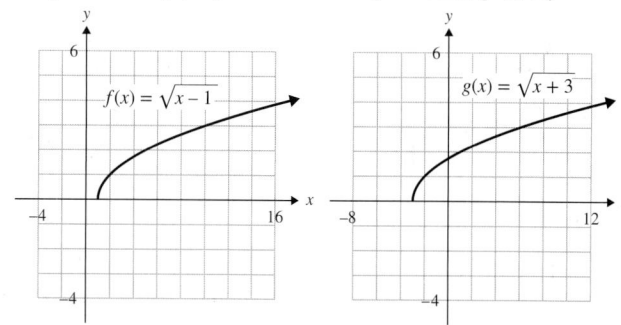

97) Domain: $[0, \infty)$

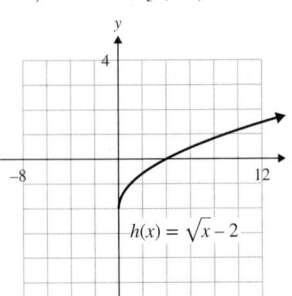
$h(x) = \sqrt{x} - 2$

99) Domain: $(-\infty, 0]$

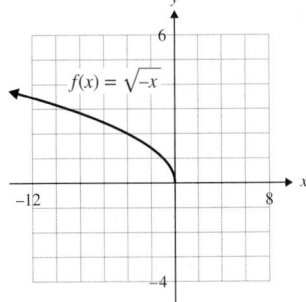
$f(x) = \sqrt{-x}$

101) Domain: $(-\infty, \infty)$

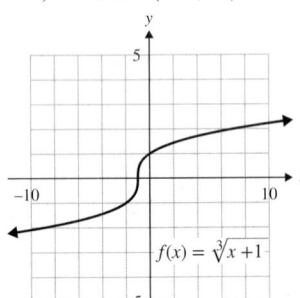
$f(x) = \sqrt[3]{x} + 1$

103) Domain: $(-\infty, \infty)$

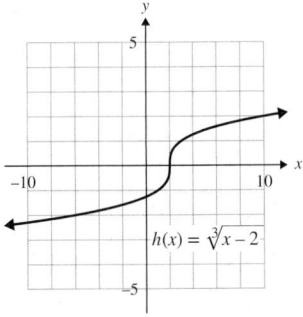
$h(x) = \sqrt[3]{x} - 2$

105) Domain: $(-\infty, \infty)$

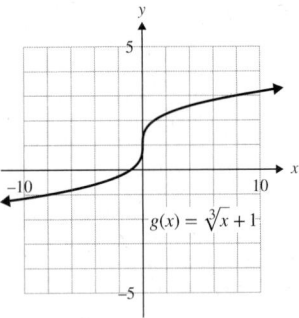
$g(x) = \sqrt[3]{x} + 1$

107) $T\left(\dfrac{1}{2}\right) = \dfrac{\pi}{4}$; A $\dfrac{1}{2}$-ft-long pendulum has a period of $\dfrac{\pi}{4}$ sec. This is approximately 0.79 sec.

109) 2 ft

111) a) $A(r) = \pi r^2$
b) $A(3) = 9\pi$. When the radius of the circle is 3 cm, the area of the circle is 9π cm^2.
c) $A(5) = 25\pi$. When the radius of the circle is 5 in., the area of the circle is 25π in^2.
d) $r = 8$ in.

Section 12.2

1) domain: $(-\infty, \infty)$;
range: $[3, \infty)$

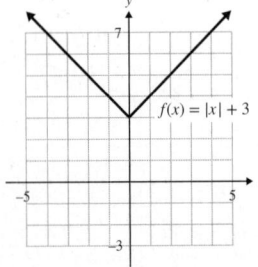
$f(x) = |x| + 3$

3) domain: $(-\infty, \infty)$;
range: $[0, \infty)$

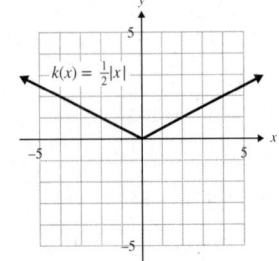
$k(x) = \frac{1}{2}|x|$

5) domain: $(-\infty, \infty)$;
range: $[-4, \infty)$

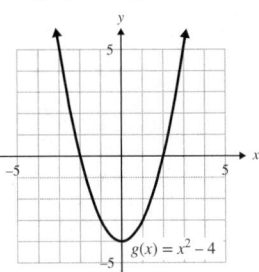
$g(x) = x^2 - 4$

7) domain: $(-\infty, \infty)$;
range: $(-\infty, -1]$

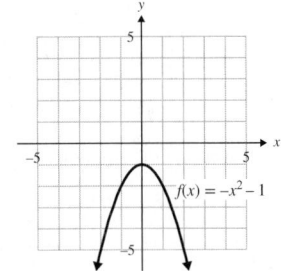
$f(x) = -x^2 - 1$

9) domain: $[-3, \infty)$;
range: $[0, \infty)$

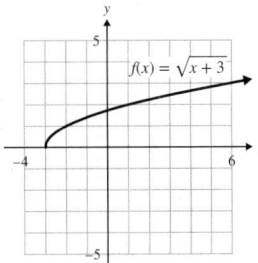
$f(x) = \sqrt{x + 3}$

11) domain: $[0, \infty)$;
range: $[0, \infty)$

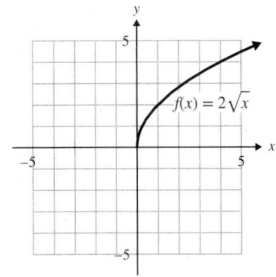
$f(x) = 2\sqrt{x}$

13) The graph of $g(x)$ is the same shape as $f(x)$, but g is shifted down 2 units.

15) The graph of $g(x)$ is the same shape as $f(x)$, but g is shifted left 2 units.

17) The graph of $g(x)$ is the reflection of $f(x)$ about the x-axis.

19)

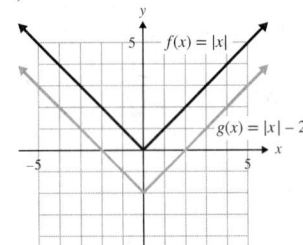
$f(x) = |x|$
$g(x) = |x| - 2$

21)

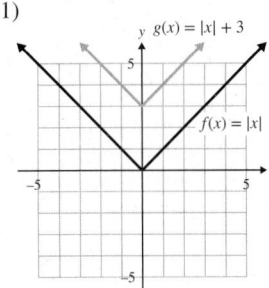
$g(x) = |x| + 3$
$f(x) = |x|$

23)

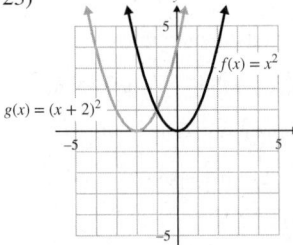
$g(x) = (x + 2)^2$
$f(x) = x^2$

25)

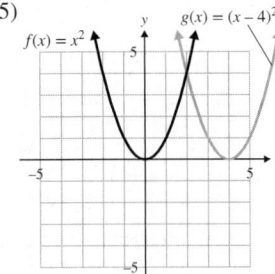
$f(x) = x^2$
$g(x) = (x - 4)^2$

27)

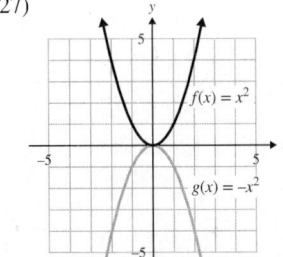
$f(x) = x^2$
$g(x) = -x^2$

29)

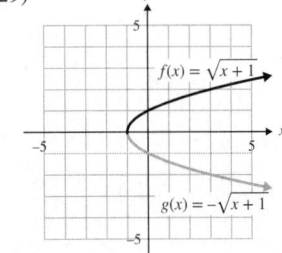
$f(x) = \sqrt{x + 1}$
$g(x) = -\sqrt{x + 1}$

31)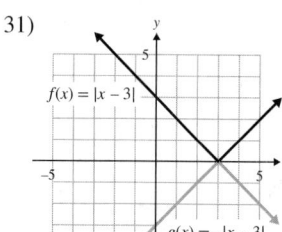
$f(x) = |x - 3|$
$g(x) = -|x - 3|$

33)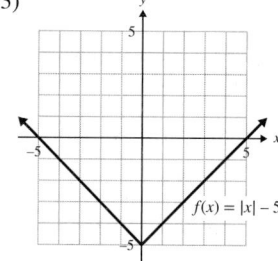
$f(x) = |x| - 5$

35)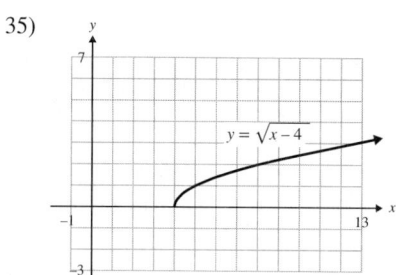
$y = \sqrt{x - 4}$

37)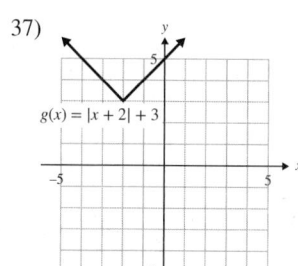
$g(x) = |x + 2| + 3$

39)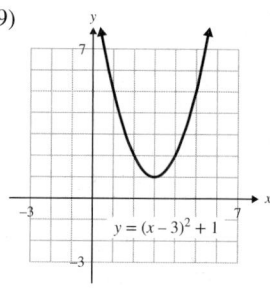
$y = (x - 3)^2 + 1$

41)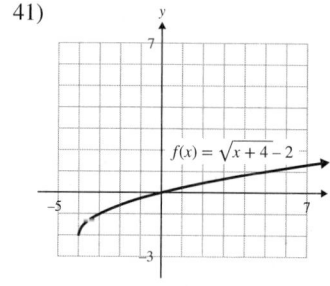
$f(x) = \sqrt{x + 4} - 2$

43)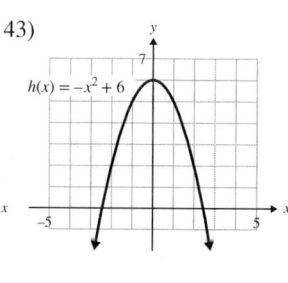
$h(x) = -x^2 + 6$

45)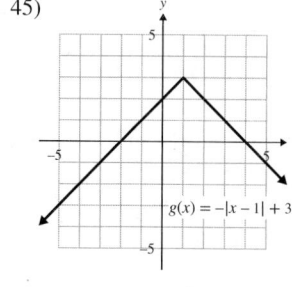
$g(x) = -|x - 1| + 3$

47)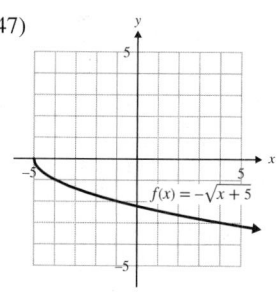
$f(x) = -\sqrt{x + 5}$

49) a) $h(x)$ b) $f(x)$ c) $g(x)$ d) $k(x)$

51) $g(x) = \sqrt{x + 5}$

53) $g(x) = |x + 2| - 1$ 55) $g(x) = (x + 3)^2 + \dfrac{1}{2}$

57) $g(x) = -x^2$

59)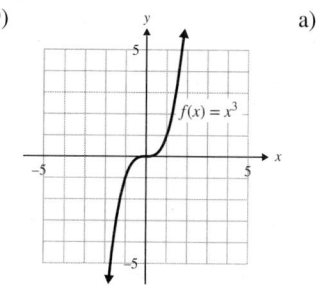
$f(x) = x^3$

a)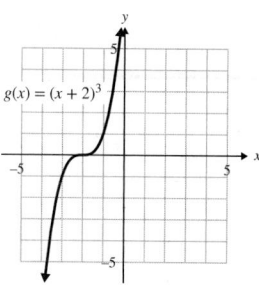
$g(x) = (x + 2)^3$

b)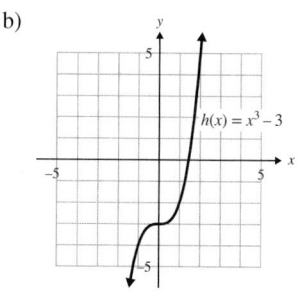
$h(x) = x^3 - 3$

c)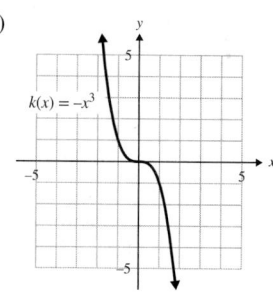
$k(x) = -x^3$

d)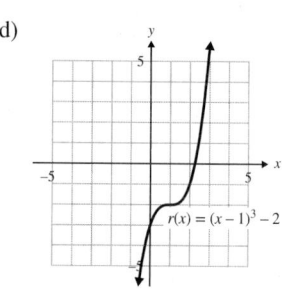
$r(x) = (x - 1)^3 - 2$

61) $g(x) = (x + 2)^3 - 1$

63) $g(x) = -\sqrt[3]{x}$

65)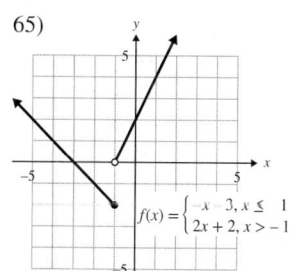
$f(x) = \begin{cases} -x - 3, & x \le -1 \\ 2x + 2, & x > -1 \end{cases}$

67)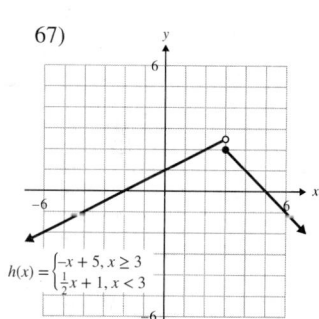
$h(x) = \begin{cases} -x + 5, & x \ge 3 \\ \frac{1}{2}x + 1, & x < 3 \end{cases}$

69)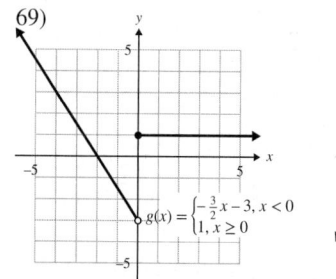
$g(x) = \begin{cases} -\frac{3}{2}x - 3, & x < 0 \\ 1, & x \ge 0 \end{cases}$

71)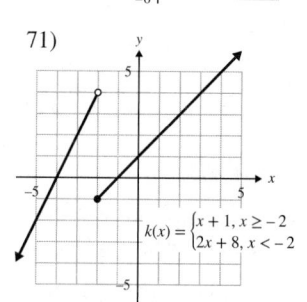
$k(x) = \begin{cases} x + 1, & x \ge -2 \\ 2x + 8, & x < -2 \end{cases}$

73)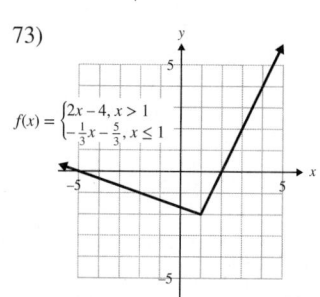
$f(x) = \begin{cases} 2x - 4, & x > 1 \\ -\frac{1}{3}x - \frac{5}{3}, & x \le 1 \end{cases}$

75) 3   77) 7   79) 8   81) −7   83) −9

85)
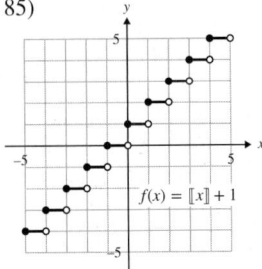
$f(x) = [\![x]\!] + 1$

87)
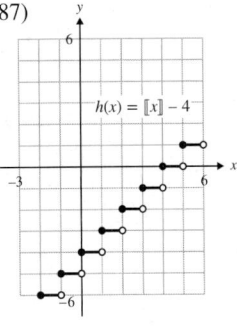
$h(x) = [\![x]\!] - 4$

89)
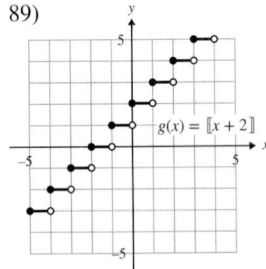
$g(x) = [\![x + 2]\!]$

91)
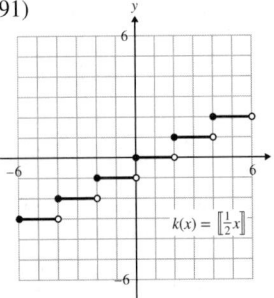
$k(x) = \left[\!\!\left[\tfrac{1}{2}x\right]\!\!\right]$

93)
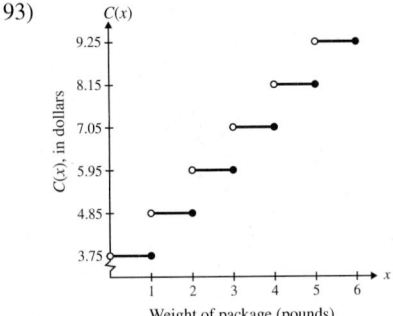

95) **Hurricane Strength**
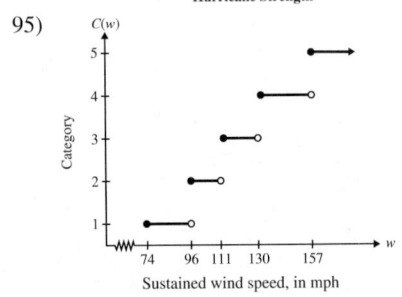

Section 12.3

1) The graph of $g(x)$ is the same shape as the graph of $f(x)$, but $g(x)$ is shifted up 6 units.

3) The graph of $h(x)$ is the same shape as the graph of $f(x)$, but $h(x)$ is shifted left 5 units.

5)
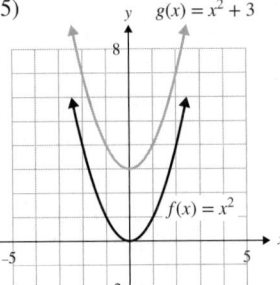

7)
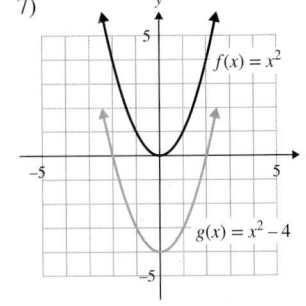

9)
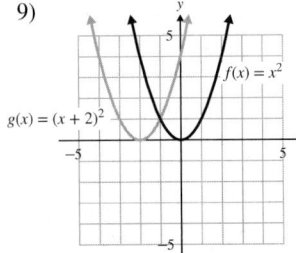

11)
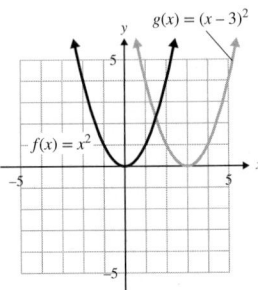

13)
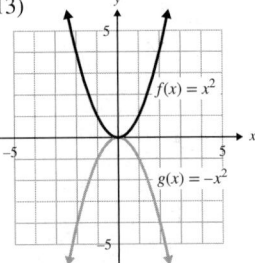

15)
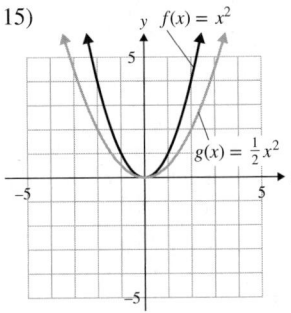

17)
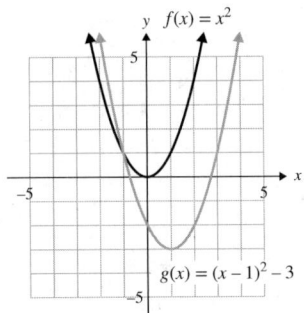

19) a) (h, k) b) $x = h$ c) a is positive.
   d) a is negative.
   e) $a > 1$ or $a < -1$
   f) $0 < a < 1$ or $-1 < a < 0$

21) $V(-1, -4)$; $x = -1$;
   x-ints: $(-3, 0)$, $(1, 0)$;
   y-int: $(0, -3)$; domain:
   $(-\infty, \infty)$; range: $[-4, \infty)$

23) $V(2, 3)$; $x = 2$; x-ints:
   none; y-int: $(0, 7)$;
   domain: $(-\infty, \infty)$;
   range: $[3, \infty)$

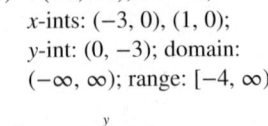

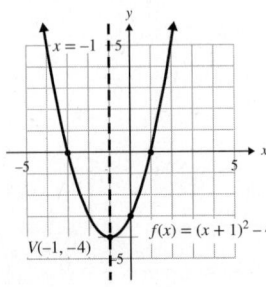

25) $V(4, -2)$; $x = 4$; x-ints:
$(4 - \sqrt{2}, 0)$,
$(4 + \sqrt{2}, 0)$; y-int:
$(0, 14)$; domain:
$(-\infty, \infty)$; range: $[-2, \infty)$

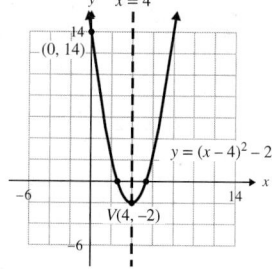

27) $V(-3, 6)$; $x = -3$;
x-ints: $(-3 - \sqrt{6}, 0)$,
$(-3 + \sqrt{6}, 0)$; y-int:
$(0, -3)$; domain:
$(-\infty, \infty)$; range: $(-\infty, 6]$

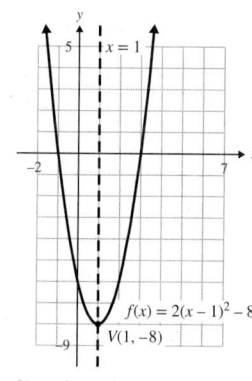

29) $V(-1, -5)$; $x = -1$:
x-ints: none; y-int:
$(0, -6)$; domain:
$(-\infty, \infty)$; range: $(-\infty, -5]$

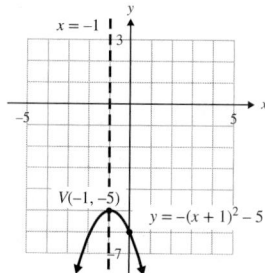

31) $V(1, -8)$; $x = 1$;
x-ints: $(-1, 0)$, $(3, 0)$;
y-int: $(0, -6)$; domain:
$(-\infty, \infty)$; range: $[-8, \infty)$

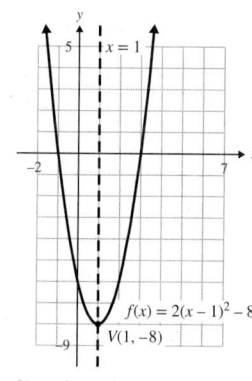

33) $V(0, -1)$; $x = 0$; x-ints: $(-2, 0)$, $(2, 0)$; y-int: $(0, -1)$;
domain: $(-\infty, \infty)$; range: $[-1, \infty)$

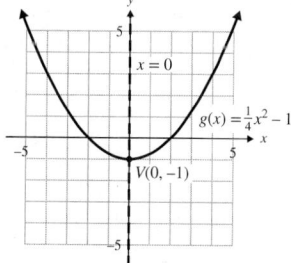

35) $V(-4, 3)$; $x = -4$;
x-ints: $(-7, 0)$, $(-1, 0)$;
y-int: $\left(0, -\dfrac{7}{3}\right)$; domain:
$(-\infty, \infty)$; range: $(-\infty, 3]$

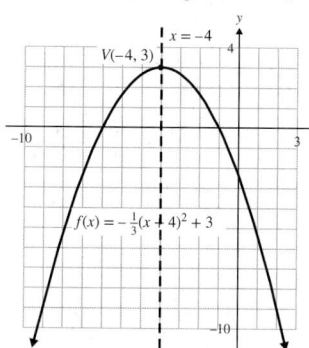

37) $V(-2, 5)$; $x = -2$;
x-int: none; y-int:
$(0, 17)$; domain:
$(-\infty, \infty)$; range: $[5, \infty)$

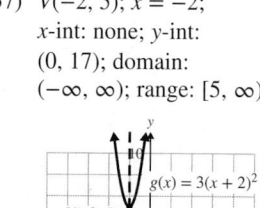

39) a) $h(x)$ b) $f(x)$ c) $g(x)$ d) $k(x)$

41) $g(x) = (x - 8)^2$ 43) $g(x) = x^2 + 3.5$

45) $g(x) = (x + 4)^2 - 7$

47) $f(x) = (x^2 + 8x) + 11$; $\left[\dfrac{1}{2}(8)\right]^2 = (4)^2 = 16$;

Add and subtract the number above to the same side of the
equation.; $f(x) = (x + 4)^2 - 5$

49) $f(x) = (x - 1)^2 - 4$;
x-ints: $(-1, 0)$, $(3, 0)$;
y-int: $(0, -3)$; domain:
$(-\infty, \infty)$; range: $[-4, \infty)$

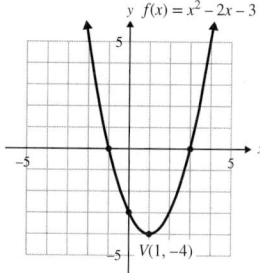

51) $y = (x + 3)^2 - 2$; x-ints:
$(-3 - \sqrt{2}, 0)$,
$(-3 + \sqrt{2}, 0)$; y-int:
$(0, 7)$; domain: $(-\infty, \infty)$;
range: $[-2, \infty)$

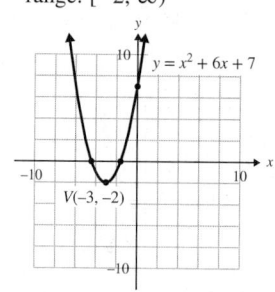

53) $g(x) = (x + 2)^2 - 4$;
x-ints: $(-4, 0)$, $(0, 0)$;
y-int: $(0, 0)$; domain:
$(-\infty, \infty)$; range: $[-4, \infty)$

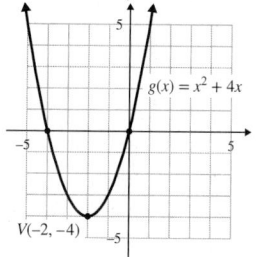

55) $h(x) = -(x + 2)^2 + 9$;
x-ints: $(-5, 0)$, $(1, 0)$;
y-int: $(0, 5)$; domain:
$(-\infty, \infty)$; range: $(-\infty, 9]$

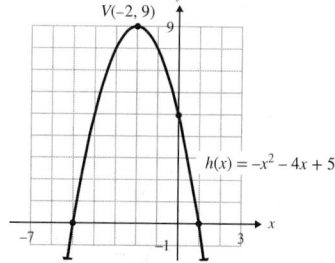

57) $y = -(x - 3)^2 - 1$;
x-ints: none; y-int: $(0, -10)$;
domain: $(-\infty, \infty)$;
range: $(-\infty, -1]$

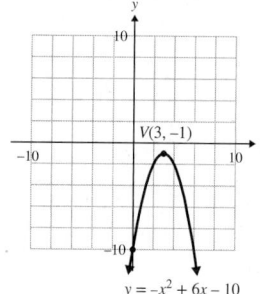

59) $y = 2(x - 2)^2 - 6$;
x-ints: $(2 - \sqrt{3}, 0)$,
$(2 + \sqrt{3}, 0)$; y-int:
$(0, 2)$; domain: $(-\infty, \infty)$;
range: $[-6, \infty)$

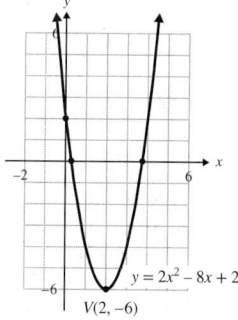

61) $g(x) = -\dfrac{1}{3}(x + 3)^2 - 6$;
x-ints: none; y-int: $(0, -9)$;
domain: $(-\infty, \infty)$;
range: $(-\infty, -6]$

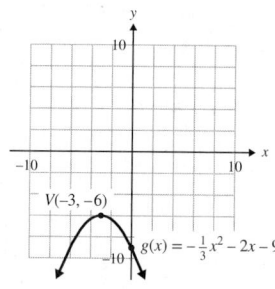

63) $y = \left(x - \dfrac{3}{2}\right)^2 - \dfrac{1}{4}$; x-ints: (1, 0), (2, 0); y-int: (0, 2);

domain: $(-\infty, \infty)$;

range: $\left[-\dfrac{1}{4}, \infty\right)$

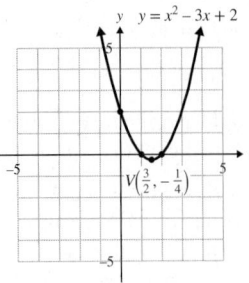

65) $V(-1, -4)$; x-ints: $(-3, 0)$, $(1, 0)$; y-int: $(0, -3)$; domain: $(-\infty, \infty)$; range: $[-4, \infty)$

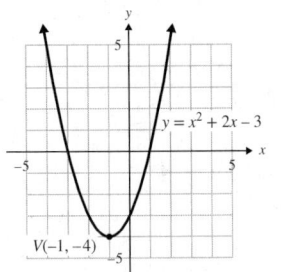

67) $V(-4, 3)$; x-ints: $(-4 - \sqrt{3}, 0)$, $(-4 + \sqrt{3}, 0)$; y-int: $(0, -13)$; domain: $(-\infty, \infty)$; range: $(-\infty, 3]$

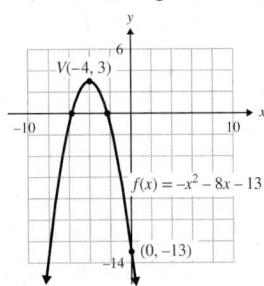

69) $V(1, 2)$; x-int: none; y-int: (0, 4); domain: $(-\infty, \infty)$; range: $[2, \infty)$

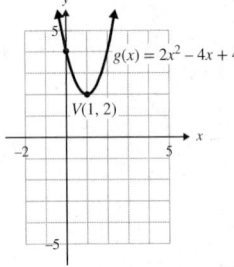

71) $V(1, 4)$; x-ints: $\left(1 + \dfrac{2\sqrt{3}}{3}, 0\right)$, $\left(1 - \dfrac{2\sqrt{3}}{3}, 0\right)$; y-int: (0, 1); domain: $(-\infty, \infty)$; range: $(-\infty, 4]$

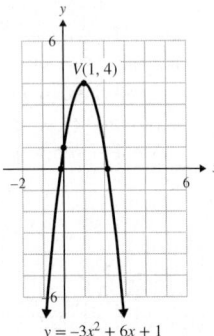

73) $V(4, -3)$; x-ints: $(4 - \sqrt{6}, 0)$, $(4 + \sqrt{6}, 0)$; y-int: (0, 5); domain: $(-\infty, \infty)$; range: $[-3, \infty)$

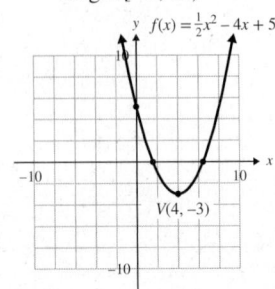

75) $V(-3, -2)$; x-int: none; y-int: $(0, -5)$; domain: $(-\infty, \infty)$; range: $(-\infty, -2]$

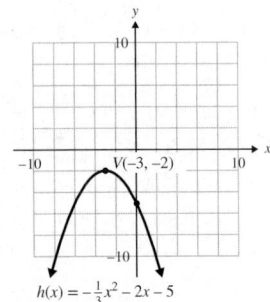

Section 12.4

1) maximum   3) neither   5) minimum

7) If a is positive the graph opens upward, so the y-coordinate of the vertex is the minimum value of the function. If a is negative the graph opens downward, so the y-coordinate of the vertex is the maximum value of the function.

9) a) minimum
   b) $(-3, 0)$ c) 0
   d)

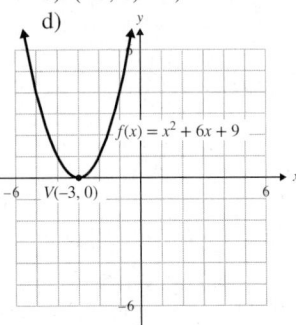

11) a) maximum
    b) $(4, 2)$ c) 2
    d)

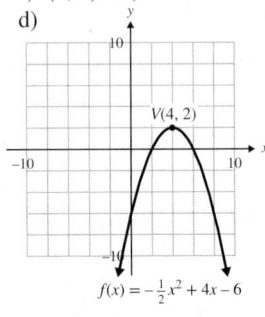

13) a) 10 sec   b) 1600 ft   c) 20 sec

15) Friday; 218 tickets   17) 2003; 7276

19) 625 ft²   21) 12 ft × 24 ft   23) 9 and 9

25) 6 and −6   27) (h, k) 29) to the left

31) $V(-4, 1)$; $y = 1$; x-int: $(-3, 0)$; y-ints: $(0, -1)$, $(0, 3)$; domain: $[-4, \infty)$; range: $(-\infty, \infty)$

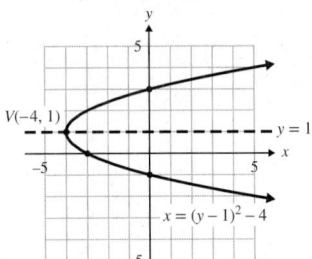

33) $V(2, 0)$; $y = 0$; x-int: (2, 0); y-int: none; domain: $[2, \infty)$; range: $(-\infty, \infty)$

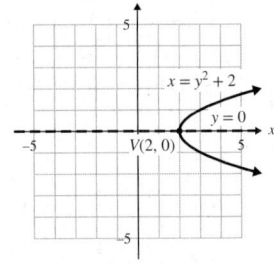

35) $V(5, 4)$; $y = 4$; x-int: $(-11, 0)$; y-ints: $(0, 4 - \sqrt{5})$, $(0, 4 + \sqrt{5})$; domain: $(-\infty, 5]$; range: $(-\infty, \infty)$

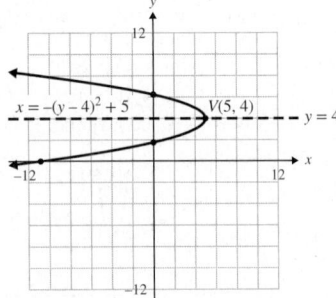

37) $V(-9, 2)$; $y = 2$; x-int: $(-17, 0)$; y-int: none; domain: $(-\infty, -9]$; range: $(-\infty, \infty)$

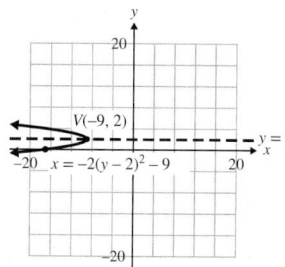

39) $V(0, -2)$; $y = -2$; x-int: $(1, 0)$; y-int: $(0, -2)$; domain: $[0, \infty)$; range: $(-\infty, \infty)$

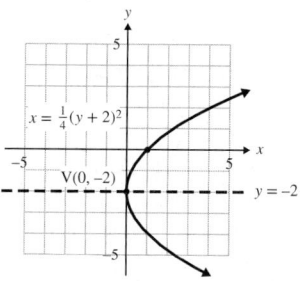

53) $V(-4, 1)$; x-int: $(-6, 0)$; y-int: none; domain: $(-\infty, -4]$; range: $(-\infty, \infty)$

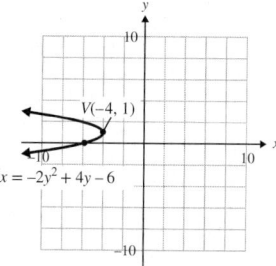

55) $V(-3, 2)$; x-int: $(13, 0)$; y-ints: $\left(0, 2 - \dfrac{\sqrt{3}}{2}\right)$, $\left(0, 2 + \dfrac{\sqrt{3}}{2}\right)$; domain: $[-3, \infty)$; range: $(-\infty, \infty)$

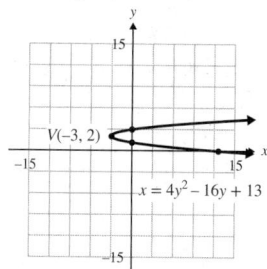

41) $x = (y + 2)^2 - 10$; domain: $[-10, \infty)$; range: $(-\infty, \infty)$

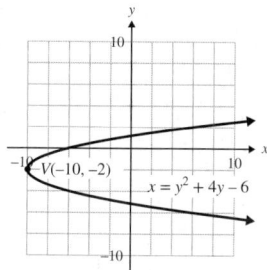

43) $x = -(y + 1)^2 - 4$; domain: $(-\infty, -4]$; range: $(-\infty, \infty)$

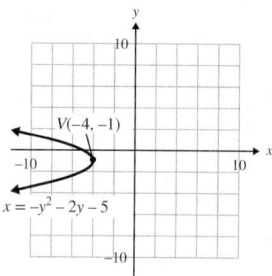

57) $V(6, 1)$; x-int: $\left(\dfrac{25}{4}, 0\right)$; y-int: none; domain: $[6, \infty)$; range: $(-\infty, \infty)$

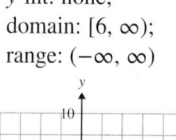
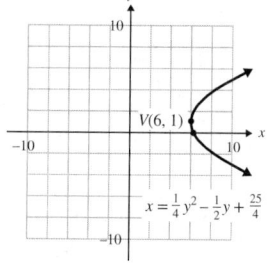

59) domain: $(-\infty, \infty)$; range: $(-\infty, 6]$

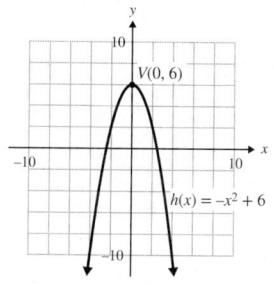

45) $x = \dfrac{1}{3}(y + 4)^2 - 7$; domain: $[-7, \infty)$; range: $(-\infty, \infty)$

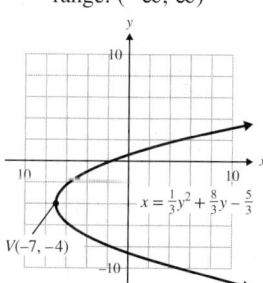

47) $x = -4(y + 1)^2 - 6$; domain: $(-\infty, -6]$; range: $(-\infty, \infty)$

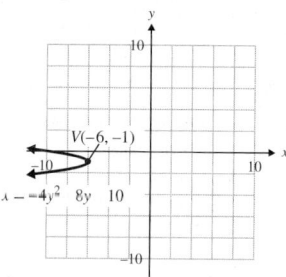

61) domain: $[0, \infty)$; range: $(-\infty, \infty)$

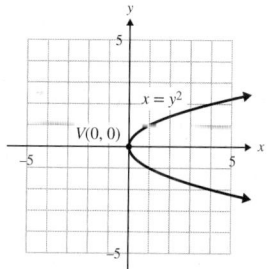

63) domain: $(-\infty, 3]$; range: $(-\infty, \infty)$

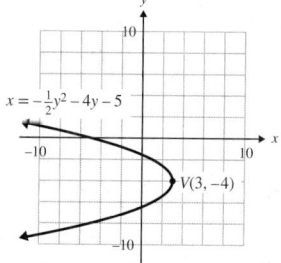

49) $V(-1, 2)$; x-int: $(3, 0)$; y-ints: $(0, 1)$, $(0, 3)$; domain: $[-1, \infty)$; range: $(-\infty, \infty)$

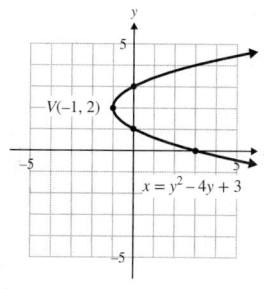

51) $V(4, 2)$; x-int: $(0, 0)$; y-ints: $(0, 0)$, $(0, 4)$; domain: $(-\infty, 4]$; range: $(-\infty, \infty)$

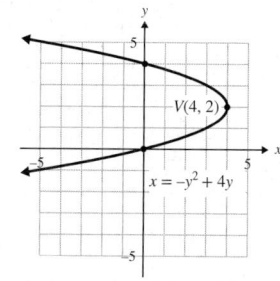

65) domain: $(-\infty, \infty)$; range: $[-4, \infty)$

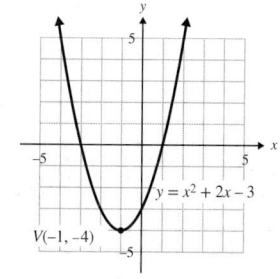

67) domain: $(-\infty, \infty)$; range: $(-\infty, 3]$

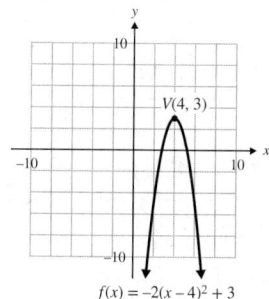

Section 12.5

1) The endpoints are included when the inequality symbol is $\leq$ or $\geq$. The endpoints are not included when the symbol is $<$ or $>$.

3) a) $[-5, 1]$ b) $(-\infty, -5) \cup (1, \infty)$

5) a) $[-1, 3]$ b) $(-\infty, -1) \cup (3, \infty)$

7) $(-\infty, -7] \cup [1, \infty)$

9) $(-9, 4)$

11) $\left[-6, \dfrac{4}{3}\right]$

13) $\left(-\infty, -\dfrac{2}{7}\right) \cup (2, \infty)$

15) $(-\infty, 0) \cup (9, \infty)$

17) $(-8, 8)$

19) $(-\infty, -11] \cup [11, \infty)$

21) $(-\infty, \infty)$ 23) $\varnothing$ 25) $(-\infty, \infty)$

27) $(-\infty, -2] \cup [1, 5]$

29) $(-\infty, -7) \cup \left(-\dfrac{1}{6}, \dfrac{3}{4}\right)$

31) $(-6, \infty)$

33) $(-\infty, -3)$

35) $(-\infty, 3) \cup (4, \infty)$

37) $\left(-\dfrac{1}{3}, 9\right]$

39) $(-3, 0]$

41) $(-\infty, -6) \cup \left(-\dfrac{11}{3}, \infty\right)$

43) $(-7, -4]$

45) $\left[\dfrac{18}{5}, 6\right)$

47) $(-\infty, -2) \cup \left(-\dfrac{8}{7}, \infty\right)$

49) $(5, \infty)$

51) $[-1, \infty)$

53) $(-\infty, 4)$

55) a) between 4000 and 12,000 units
    b) when it produces less than 4000 units or more than 12,000 units

57) 10,000 or more

Section 12.6

1) a) $-x - 10$ b) -15 c) $-5x + 12$ d) 2

3) a) $5x^2 - 4x - 7$ b) 98 c) $3x^2 - 10x + 5$ d) -3

5) a) $-x^2 + 5x$ b) -24 7) a) $6x^2 + 11x + 3$ b) 24

9) a) $\dfrac{6x + 9}{x + 4}, x \neq -4$ b) $-\dfrac{3}{2}$

11) a) $x + 3, x \neq 8$ b) 1

13) a) $x + 4, x \neq -\dfrac{2}{3}$ b) 2   15) Answers may vary.

17) $g(x) = 2x - 3$

19) a) $P(x) = 4x - 2000$ b) \$4000

21) a) $P(x) = 3x - 2400$ b) \$0

23) a) $P(x) = -0.2x^2 + 19x - 9$ b) \$291,000

25) $(f \circ g)(x) = f(g(x))$, so substitute the function $g(x)$ into the function $f(x)$ and simplify.

27) a) -1 b) -2 c) $6x - 26$ d) -2

29) a) 7 b) 44 c) $x^2 + 6x + 4$ d) 44

31) a) $5x + 31$ b) $5x + 3$ c) 46

33) a) $-6x + 26$ b) $-6x + 11$ c) 32

35) a) $x^2 - 6x + 7$ b) $x^2 - 14x + 51$ c) 11

37) a) $-x^2 - 13x - 48$ b) $-x^2 + 3x$ c) 0

39) a) $\sqrt{x^2 + 4}$ b) $x + 4$ c) $\sqrt{13}$

41) a) $\dfrac{1}{t^2 + 8}$ b) $\dfrac{1}{(t + 8)^2}$ c) $\dfrac{1}{9}$

43) $f(x) = \sqrt{x}, g(x) = x^2 + 13$; answers may vary.

45) $f(x) = x^2, g(x) = 8x - 3$; answers may vary.

47) $f(x) = \dfrac{1}{x}, g(x) = 6x + 5$; answers may vary.

49) a) $r(5) = 20$. The radius of the spill 5 min after the ship started leaking was 20 ft.
   b) $A(20) = 400\pi$. The area of the oil slick is 400π ft^2 when its radius is 20 ft.
   c) $A(r(t)) = 16\pi t^2$. This is the area of the oil slick in terms of t, the number of minutes after the leak began.
   d) $A(r(5)) = 400\pi$. The area of the oil slick 5 min after the ship began leaking was 400π ft^2.

51) a) $r(6) = 3$. When the diameter of a circle is 6 units, its radius is 3 units.
   b) $A(3) = 9\pi$. When the radius of a circle is 3 units, its area is 9π square units.
   c) $A(r(d)) = \dfrac{1}{4}\pi d^2$. This is the area of a circle in terms of its diameter.
   d) $A(r(6)) = 9\pi$. When the diameter of a circle is 6 units, its area is 9π square units.

Chapter 12: Group Activity

1) a) i) 4 ii) 4 iii) 16 iv) 36 v) 7 vi) 5 vii) 2
   b) i) 34 ii) 36 iii) -241 iv) $\sqrt{2}$ v) $\dfrac{41}{5}$

2) a) $x + 3$ b) $\dfrac{x + 4}{(x - 5)(x + 3)(x - 3)}$
   c) $x^3 - 6x^2 - x + 30$

3) a) $-\dfrac{9}{4} \pm \dfrac{\sqrt{21}}{4}$ b) $\dfrac{8}{7}$

Chapter 12 Review Exercises

1) domain: $\{-7, -5, 2, 4\}$; range: $\{-4, -1, 3, 5, 9\}$; not a function   3) no

5) Set up an inequality so that the radicand is greater than or equal to 0. Solve for the variable. These are the real numbers in the domain of the function.

7) rational; $(-\infty, -1) \cup (-1, 8) \cup (8, \infty)$

9) linear; $(-\infty, \infty)$ 11) cube root; $(-\infty, \infty)$

13) square root; $\left[\dfrac{7}{5}, \infty\right)$

15) domain: $[0, \infty)$; range: $[0, \infty)$

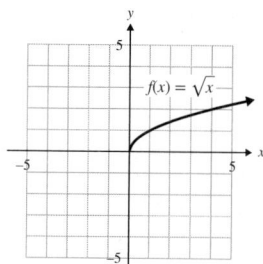

17) domain: $(-\infty, \infty)$; range: $(-\infty, \infty)$

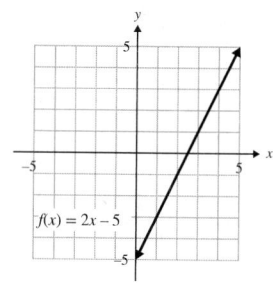

19) a) -37 b) 35 c) -22 d) 18 e) $-8c + 3$
   f) $r^2 + 7r - 12$ g) $-8p + 27$ h) $t^2 + 15t + 32$

21) a) \$32 b) \$46 c) 150 mi d) 80 mi

23) domain: $(-\infty, \infty)$; range: $[0, \infty)$

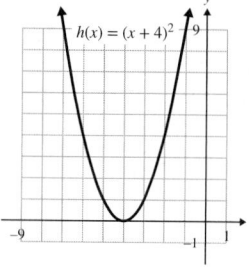

25) domain: $(-\infty, \infty)$; range: $(-\infty, 5]$

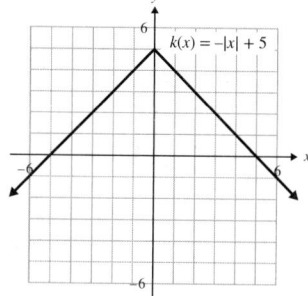

27) domain: $[2, \infty)$; range: $[-1, \infty)$

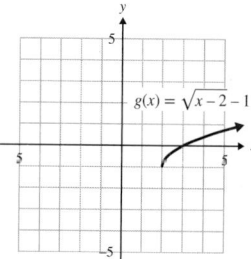

29)

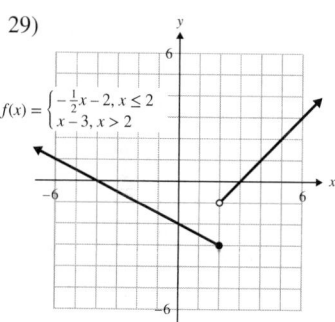

$f(x) = \begin{cases} -\frac{1}{2}x - 2, & x \le 2 \\ x - 3, & x > 2 \end{cases}$

31) 7   33) -6

35)

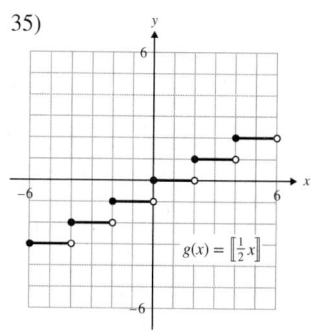

$g(x) = \left[\!\left[\frac{1}{2}x\right]\!\right]$

37) $g(x) = |x - 5|$

39) a) (h, k) b) $x = h$
   c) If a is positive, the parabola opens upward. If a is negative, the parabola opens downward.

41) a) (h, k) b) $y = k$
c) If a is positive, the parabola opens to the right. If a is negative, the parabola opens to the left.

43) $V(-2, -1)$; $x = -2$; x-ints: $(-3, 0), (-1, 0)$; y-int: $(0, 3)$

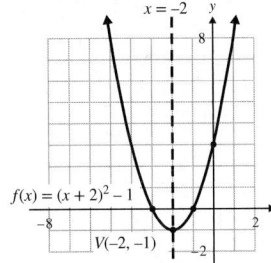

45) $V(-1, 0)$; $y = 0$; x-int: $(-1, 0)$; y-int: none

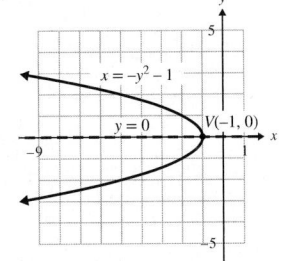

47) $V(11, 3)$; $y = 3$; x-int: $(2, 0)$; y-ints: $(0, 3 - \sqrt{11}), (0, 3 + \sqrt{11})$

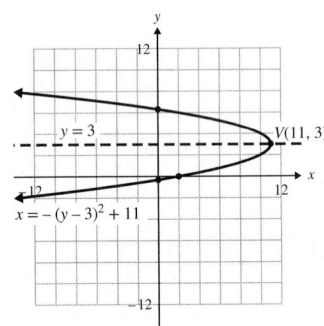

49) $x = (y + 4)^2 - 9$; x-int: $(7, 0)$; y-ints: $(0, -1), (0, -7)$

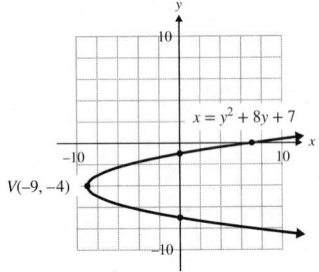

51) $y = \dfrac{1}{2}(x - 4)^2 + 1$; x-int: none; y-int: $(0, 9)$

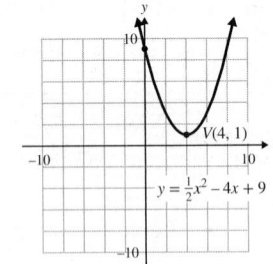

53) $V(1, -5)$; x-ints: $(1 - \sqrt{5}, 0)$, $(1 + \sqrt{5}, 0)$; y-int: $(0, -4)$

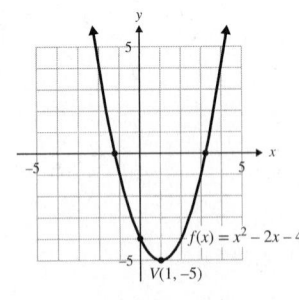

55) $V(2, -3)$; x-int: $\left(-\dfrac{5}{2}, 0\right)$; y-ints: $(0, -5), (0, -1)$

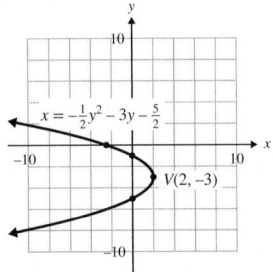

57) a) 1 sec b) 256 ft c) 5 sec

59) $(-3, 1)$

61) $\left(-\infty, -\dfrac{5}{8}\right] \cup \left[\dfrac{5}{8}, \infty\right)$

63) $\left(-\infty, -\dfrac{2}{5}\right) \cup \left(-\dfrac{1}{3}, 4\right)$

65) $(-\infty, -7) \cup \left(\dfrac{3}{2}, \infty\right)$

67) $(-\infty, 2) \cup [3, \infty)$

69) $(7, \infty)$

71) $4x + 6$ 73) -3 75) $-5x^2 + 18x + 8$

77) a) $\dfrac{6x - 5}{x + 4}, x \neq -4$ b) $\dfrac{13}{7}$

79) a) $P(x) = 6x - 400$ b) $\$800$

81) a) $2x + 3$ b) $2x - 3$ c) 7

83) a) $4x^2 + 6x - 8$ b) $2x^2 + 10x - 9$ c) -21

85) a) $f(x) = x^2$, $g(x) = 3x + 10$; answers may vary.
b) $f(x) = \dfrac{1}{x}$, $g(x) = 2 - 5x$; answers may vary.

Chapter 12 Test

1) It is a special type of relation in which each element of the domain corresponds to exactly one element of the range.

2) a) $\{-8, 2, 5, 7\}$ b) $\{-1, 3, 10\}$ c) yes

3) a) $\left[-\dfrac{7}{3}, \infty\right)$ b) yes

4) Set the expression in the denominator equal to zero, and solve for the variable. The domain of the rational function consists of all real numbers *except* those that make the denominator equal to zero.

5) $\left(-\infty, \dfrac{8}{7}\right) \cup \left(\dfrac{8}{7}, \infty\right)$ 6) 2   7) $4c + 3$

8) $4n - 25$ 9) $k^2 + 4k + 5$ 10) $-\dfrac{3}{2}$

11) a) $C(3) = 210$. The cost of delivering 3 yd^3 of cedar mulch is $210.   b) 6 yd^3

12) domain: $(-\infty, \infty)$;
range: $(-\infty, 0]$

13) domain: $[-3, \infty)$;
range: $[0, \infty)$

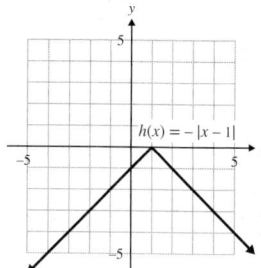

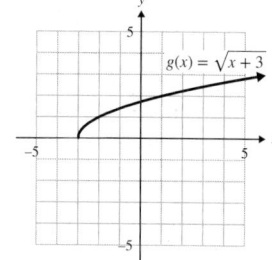

14) domain: $(-\infty, \infty)$;
range: $(-\infty, \infty)$

15) domain: $(-\infty, \infty)$;
range: $(-\infty, \infty)$

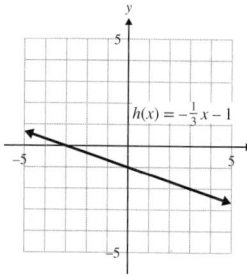

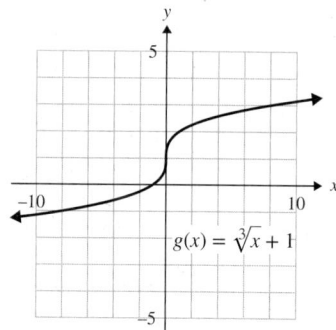

16) $g(x) = |x + 2| + 5$ 17)

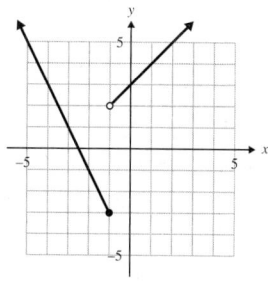

18) domain: $(-\infty, \infty)$;
range: $(-\infty, 4]$

19) domain: $[-3, \infty)$;
range: $(-\infty, \infty)$

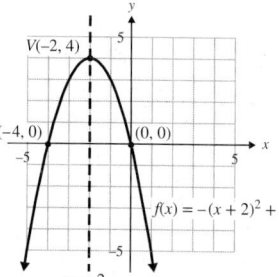

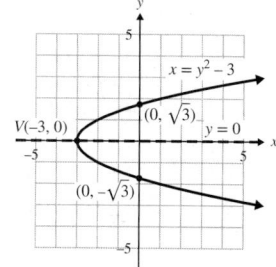

20) domain: $[2, \infty)$;
range: $(-\infty, \infty)$

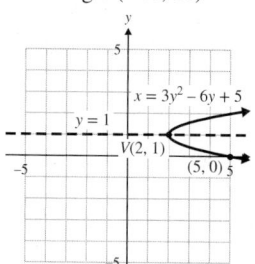

21) domain: $(-\infty, \infty)$;
range: $[-1, \infty)$

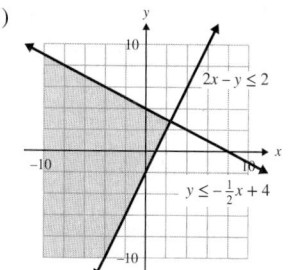

22) a) 209 ft   b) after 4 sec
c) after $\dfrac{3 + \sqrt{209}}{4}$ sec or about 4.4 sec

23) $x^2 + 3x - 10$ 24) -2

25) $2x^3 + 17x^2 + 29x - 21$ 26) $2x^2 + 10x + 1$

27) $4x^2 + 38x + 81$ 28) 3

29) Answers may vary. One possible answer is
$f(x) = 5x + 8$, $g(x) = \sqrt[3]{x}$.

30) $(-\infty, -9] \cup [5, \infty)$

31) $(-\infty, -3) \cup [5, \infty)$

32) 8000 or more

Chapter 12: Cumulative Review for Chapters 1–12

1) a) 10^4 b) 10^{-3} c) 10^{-1}

2) a) 49   b) 1000   c) 81   d) $\dfrac{1}{125}$ e) $\dfrac{1}{64}$ f) 100

3) $4x + 3y = -2$ 4) $p^2 - p - 56$

5) a) $(k - 9)(k - 6)$ b) $(10 + 3m)(10 - 3m)$

6) a) $\dfrac{3}{2}r - 10 + \dfrac{3}{r} + \dfrac{1}{r^2}$ b) $\dfrac{c + 4}{3(2c + 3)}$ c) $\dfrac{2}{13} - \dfrac{16}{13}i$

7) $\left\{-\dfrac{5}{2}, \dfrac{1}{2}\right\}$ 8) $\{-8\}$ 9) $\varnothing$ 10) $\{1, 49\}$

11) $\left\{\dfrac{4}{3}\right\}$ 12) $\left\{\dfrac{4}{3} - i, \dfrac{4}{3} + i\right\}$ 13) $\{-4\}$ 14) $(3, 2)$

15)

16) $(-\infty, -2] \cup [3, \infty)$

17) 3   18) $2\sqrt{15}$ 19) 2   20) $3c^3d^5\sqrt{2d}$

21) $\dfrac{1}{1000}$ 22) $7\sqrt{3}$

23) a) 8 b) $x + 3, x \neq -1$ c) 4 d) $x^2 + 4x + 4$

24) domain: $(-\infty, \infty)$;   25)
range: $(-\infty, 4]$

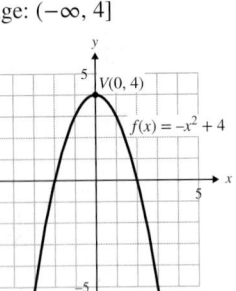

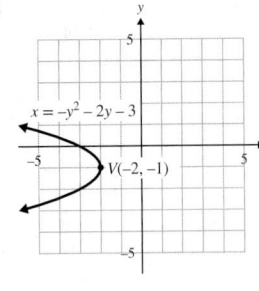

Chapter 13

Section 13.1

1) no   3) yes; $h^{-1} = \{(-16, -5), (-4, -1), (8, 3)\}$

5) yes   7) No; only one-to-one functions have inverses.

9) False; it is read "f inverse of x."   11) true

13) False; they are symmetric with respect to $y = x$.

15) a) yes b)   17) no

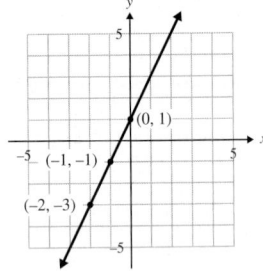

19) a) yes   b)

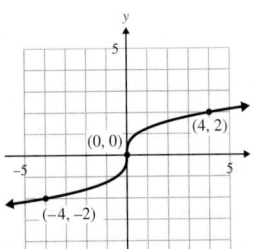

21) Replace $f(x)$ with y.; $x = 2y - 10$; Add 10.; $\dfrac{1}{2}x + 5 = y$;
Replace y with $f^{-1}(x)$.

23) $g^{-1}(x) = x + 6$

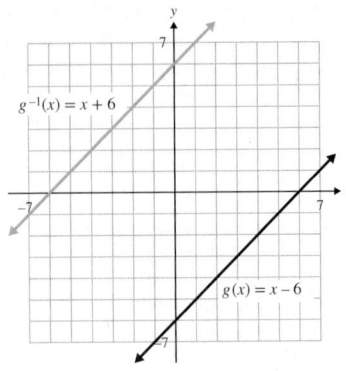

25) $f^{-1}(x) = -\dfrac{1}{2}x + \dfrac{5}{2}$ 27) $g^{-1}(x) = 2x$

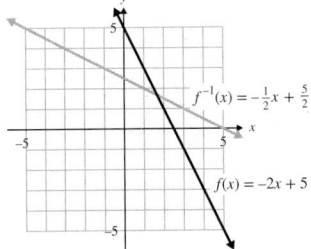

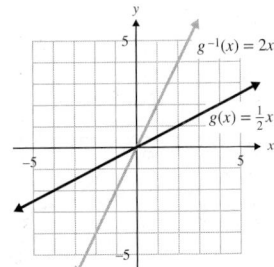

29) $f^{-1}(x) = \sqrt[3]{x}$

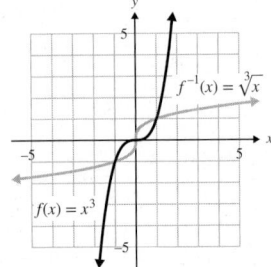

31) $f^{-1}(x) = \dfrac{1}{2}x + 3$ 33) $h^{-1}(x) = -\dfrac{2}{3}x + \dfrac{8}{3}$

35) $h^{-1}(x) = x^3 + 7$ 37) $f^{-1}(x) = x^2, x \geq 0$

39) a) 3 b) 1   41) a) 2 b) 9   43) a) 10 b) -7

45) a) 8 b) 3   47–53) Answers may vary.

Section 13.2

1) Choose values for the variable that will give positive numbers, negative numbers, and zero in the exponent.

3)   5)

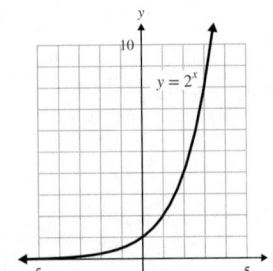

domain: $(-\infty, \infty)$;
range: $(0, \infty)$

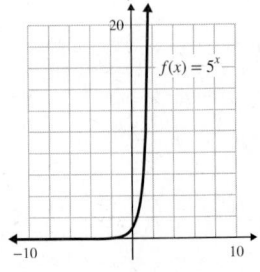

domain: $(-\infty, \infty)$;
range: $(0, \infty)$

7)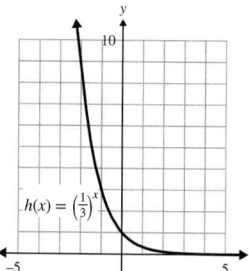
domain: $(-\infty, \infty)$;
range: $(0, \infty)$

9) $(-\infty, \infty)$

11)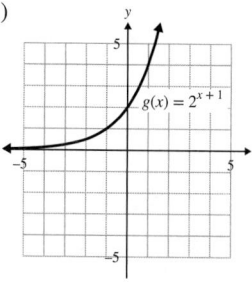
domain: $(-\infty, \infty)$;
range: $(0, \infty)$

13)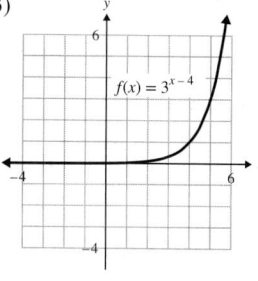
domain: $(-\infty, \infty)$;
range: $(0, \infty)$

15)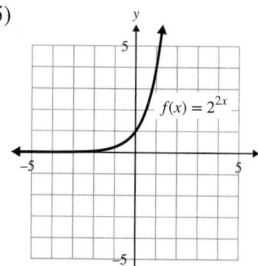
domain: $(-\infty, \infty)$;
range: $(0, \infty)$

17)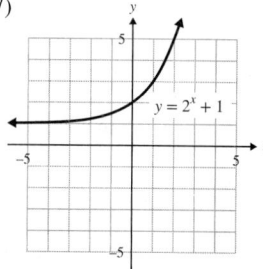
domain: $(-\infty, \infty)$;
range: $(1, \infty)$

19)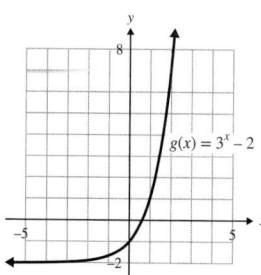
domain: $(-\infty, \infty)$;
range: $(-2, \infty)$

21)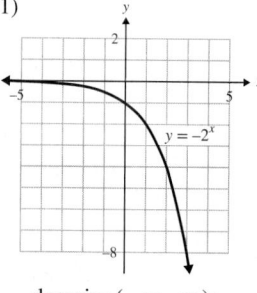
domain: $(-\infty, \infty)$;
range: $(-\infty, 0)$

23) $g(x) = 2^x$ would grow faster because for values of $x > 2$, $2^x > 2x$.

25) Shift the graph of $f(x)$ down 2 units.

27) 2.7183   29) B   31) D

33)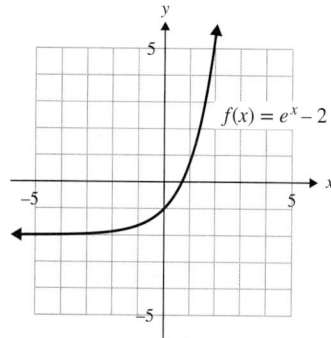
domain: $(-\infty, \infty)$; range: $(-2, \infty)$

35)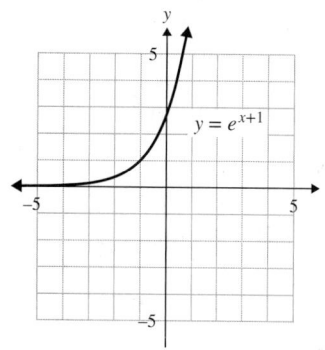
domain: $(-\infty, \infty)$; range: $(0, \infty)$

37)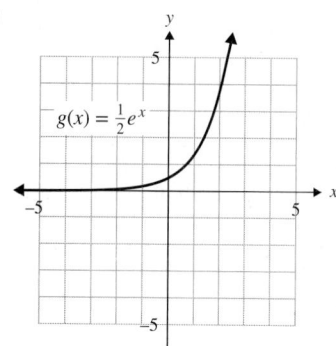
domain: $(-\infty, \infty)$; range: $(0, \infty)$

39)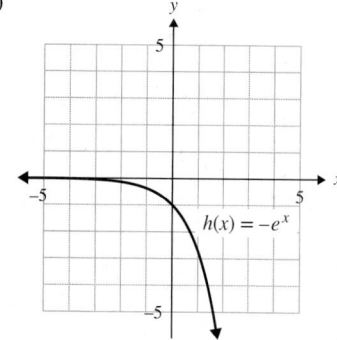
domain: $(-\infty, \infty)$; range: $(-\infty, 0)$

41) They are symmetric with respect to the x-axis.

43) $6^{3n} = (6^2)^{n-4}$; Power rule for exponents;
Distribute.; $3n = 2n - 8$; $n = -8$; $\{-8\}$

45) $\{2\}$ 47) $\left\{\dfrac{3}{4}\right\}$ 49) $\left\{\dfrac{8}{3}\right\}$ 51) $\{5\}$ 53) $\{1\}$

55) $\{-3\}$　　57) $\left\{\dfrac{8}{13}\right\}$　　59) $\{-2\}$　　61) $\{-3\}$

63) $\left\{-\dfrac{3}{2}\right\}$　　65) $\{-1\}$　　67) $\left\{\dfrac{3}{2}\right\}$

69) a) $32,700　b) $17,507.17

71) a) $185,200　b) $227,772.64

73) $90,036.92　　75) 40.8 mg

77) a) $A(t) = 200(0.98)^t$
   b) $A(12) = 157$; after 12 years, 157 L of whiskey is in
      the cask.
   c) After 18 years, there are 139 L of whiskey in the cask.
   d) The longer the whiskey has aged, the less there is
      remaining in the cask. The company needs to charge
      more for the older whiskey because there is less of it even
      though all ages of whiskey began with 200 L of
      liquid in the cask.

Section 13.3

1) a must be a positive real number that is not equal to 1.

3) 10　　5) $7^2 = 49$　　7) $10^6 = 1,000,000$　　9) $9^{-2} = \dfrac{1}{81}$

11) $25^{1/2} = 5$　　13) $9^0 = 1$　　15) $\log_9 81 = 2$

17) $\log_{10} 100 = 2$　　19) $\log_2 \dfrac{1}{32} = -5$

21) $\log_{10} 10 = 1$　　23) $\log_{169} 13 = \dfrac{1}{2}$　　25) $\log_9 3 = \dfrac{1}{2}$

27) $\log_{64} 4 = \dfrac{1}{3}$

29) Write the equation in exponential form, then solve for the
    variable.

31) Rewrite in exponential form.; $64 = x$; $\{64\}$

33) $\{125\}$　　35) $\{64\}$　　37) $\{100\}$　　39) $\{7\}$

41) $\left\{\dfrac{1}{36}\right\}$　　43) $\{14\}$　　45) $\left\{\dfrac{15}{2}\right\}$　　47) $\left\{\dfrac{1}{6}\right\}$

49) $\left\{\dfrac{1}{8}\right\}$　　51) $\{12\}$　　53) $\{4\}$　　55) $\{2\}$　　57) 5

59) 2　　61) $\dfrac{1}{2}$　　63) -1　　65) 1　　67) -2

69) Replace $f(x)$ with y, write $y = \log_a x$ in exponential form,
    make a table of values, then plot the points and draw the
    curve.

71)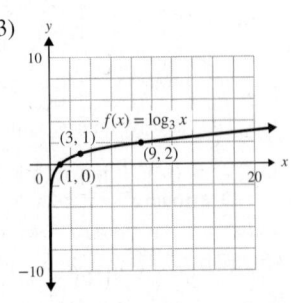

73)

75)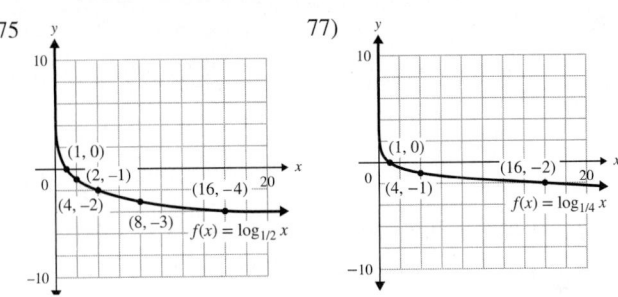

77)

79) $f^{-1}(x) = \log_3 x$　　81) $f^{-1}(x) = 2^x$

83) a) 1868　b) 2004　c) 2073

85) a) 14,000　b) 28,000
    c) It is 1000 more than what was predicted by the formula.

Section 13.4

1) true　　3) false　　5) false　　7) true

9) Product rule; $2 + \log_5 y$　　11) $\log_8 3 + \log_8 10$

13) $\log_7 5 + \log_7 d$　　15) $\log_5 20 - \log_5 17$　　17) $4 \log_8 10$

19) $8 \log p$　　21) $\dfrac{1}{2}\log_3 7$　　23) $4 + \log_2 p$

25) $3 - \log_2 k$　　27) 6　　29) $3 + \log b$　　31) 9

33) $\dfrac{1}{2}$　　35) $\dfrac{2}{3}$　　37) $4 \log_6 w + 3 \log_6 z$

39) $2 \log_7 a - 5 \log_7 b$　　41) $\dfrac{1}{5}\log 11 - 2 \log y$

43) $2 + \dfrac{1}{2}\log_2 n - 3 \log_2 m$

45) $3 \log_4 x - \log_4 y - 2 \log_4 z$　　47) $\dfrac{1}{2} + \dfrac{1}{2}\log_5 c$

49) $\log k + \log(k - 6)$　　51) Power rule; $\log_6 x^2 y$

53) $\log_a mn$　　55) $\log_7 \dfrac{d}{3}$　　57) $\log_3 f^4 g$　　59) $\log_8 \dfrac{tu^2}{v^3}$

61) $\log \dfrac{r^2 + 3}{(r^2 - 3)^2}$　　63) $\log_n 8\sqrt{k}$　　65) $\log_d \dfrac{\sqrt[3]{5}}{z^2}$

67) $\log_6 \dfrac{y}{3z^3}$　　69) $\log_3 \dfrac{t^4}{36u^2}$　　71) $\log_b \dfrac{\sqrt{c + 4}}{(c + 3)^2}$

73) $-\log(a^2 - b^2)$　　75) 1.6532　　77) 1.9084

79) -0.2552　　81) 0.4771　　83) -0.6990

85) -1.9084　　87) 1.6990

89) No. $\log_a xy$ is defined only if x and y are positive.

Section 13.5

1) e　　3) 2　　5) -3　　7) -1　　9) 9　　11) $\dfrac{1}{4}$

13) 10　　15) $\dfrac{1}{2}$　　17) -5　　19) 0　　21) 1.2041

23) -0.3010　　25) 1.0986　　27) 0.2700　　29) $\{1000\}$

31) $\left\{\dfrac{1}{10}\right\}$ 33) {25}   35) {2}   37) $\{10^{1.5}\}$; {31.6228}

39) $\{10^{0.8}\}$; {6.3096}   41) $\{e^{1.6}\}$; {4.9530}

43) $\left\{\dfrac{1}{e^2}\right\}$; {0.1353}   45) $\left\{\dfrac{e^{2.1}}{3}\right\}$; {2.7221}

47) $\{2 \cdot 10^{0.47}\}$; {5.9024}

49) $\left\{\dfrac{3 + 10^{3.8}}{5}\right\}$; {1262.5147}

51) $\left\{\dfrac{e^{1.85} - 19}{10}\right\}$; {-1.2640}   53) {3}

55)

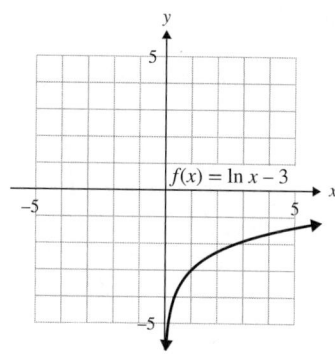

domain: $(0, \infty)$; range: $(-\infty, \infty)$

57)

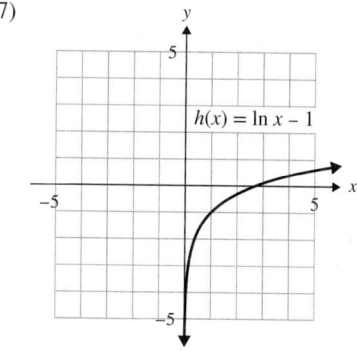

domain: $(0, \infty)$; range: $(-\infty, \infty)$

59)

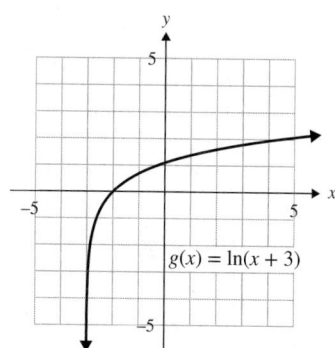

domain: $(-3, \infty)$; range: $(-\infty, \infty)$

61)

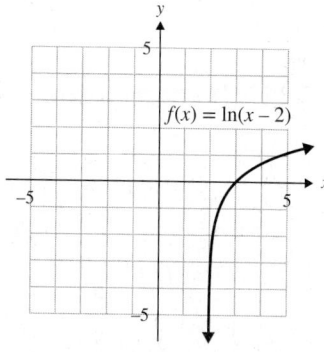

domain: $(2, \infty)$; range: $(-\infty, \infty)$

63)

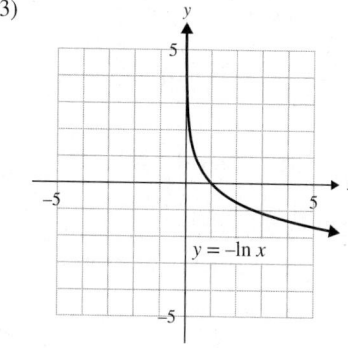

domain: $(0, \infty)$; range: $(-\infty, \infty)$

65)

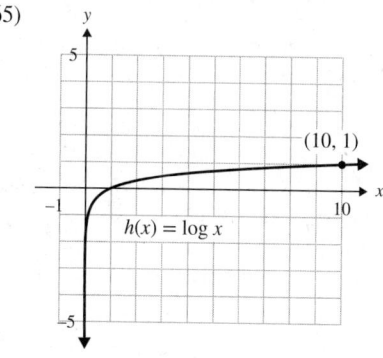

domain: $(0, \infty)$; range: $(-\infty, \infty)$

67) Shift the graph of $f(x)$ left 5 units.

69) 3.7004   71) 1.9336   73) −2.5237   75) 0.6826

77) 60 dB   79) 110 dB   81) $3484.42   83) $5521.68

85) $3485.50   87) $15,683.12   89) a) 5000 b) 8191

91) 32,570   93) 2.7; acidic   95) 11.2; basic

97) Answers may vary.

Section 13.6

1) {2}   3) $\left\{\dfrac{\ln 15}{\ln 7}\right\}$; {1.3917}   5) $\left\{\dfrac{\ln 3}{\ln 8}\right\}$; {0.5283}

7) $\left\{\dfrac{2}{5}\right\}$ 9) $\cdot \left\{\dfrac{\ln 2.7}{6 \ln 4}\right\}$; {0.1194}

11) $\left\{\dfrac{\ln 5 - \ln 2}{4 \ln 2}\right\}$; {0.3305}

13) $\left\{\dfrac{\ln 8 + 2 \ln 5}{3 \ln 5}\right\}$; {1.0973}

15) $\left\{\dfrac{10}{7}\right\}$ 17) $\left\{\dfrac{2 \ln 9}{5 \ln 9 - 3 \ln 4}\right\}$; {0.6437}

19) {ln 12.5}; {2.5257}   21) $\left\{-\dfrac{\ln 9}{4}\right\}$; {−0.5493}

23) $\left\{\dfrac{\ln 2}{0.01}\right\}$; {69.3147}   25) $\left\{\dfrac{\ln 3}{0.006}\right\}$; {183.1021}

27) $\left\{-\dfrac{\ln 5}{0.4}\right\}$; {−4.0236}   29) {2}   31) $\left\{\dfrac{10}{3}\right\}$

33) {5}   35) {2, 10}   37) ∅   39) $\left\{\dfrac{11}{2}\right\}$ 41) {2}

43) {9}   45) {2}   47) {4}   49) $\left\{\dfrac{2}{3}\right\}$

51) a) 3.72 yr   b) 11.55 yr   53) 1.44 yr

55) $2246.64   57) 7.2%   59) a) 6 hr   b) 18.5 hr

61) 28,009   63) a) 2032   b) 2033

65) a) 11.78 g   b) 3351 yr   c) 5728 yr

67) a) 0.4 units   b) 0.22 units   69) {16}   71) {−2, 2}

73) {−ln 13, ln 13}; {−2.5649, 2.5649}   75) {0}

77) $\left\{\dfrac{\ln 7}{\ln 5}\right\}$; {1.2091}   79) {1, 1000}

Chapter 13: Group Activity

1) a) Answers may vary; linear function
   b) Answers may vary; BAC
   c) Answers may vary.   d) Answers may vary.   e) No

2) a) Answers may vary.   b) No

3) 0.096; 20.5%   4) Answers may vary.

Chapter 13: Review Exercises

1) yes; {(−4, −7), (1, −2), (5, 1), (11, 6)}

3) yes   5) $f^{-1}(x) = x - 4$

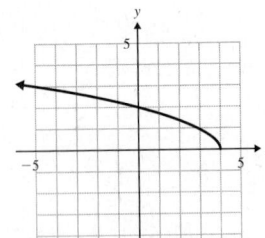

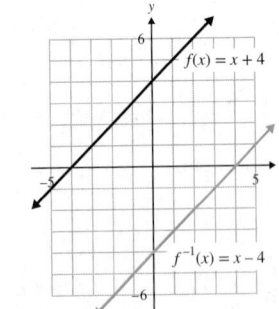

7) $h^{-1}(x) = 3x + 3$

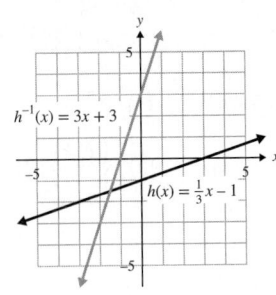

9) a) 11   b) 2

11) $(f \circ f^{-1})(x) = -9\left(-\dfrac{1}{9}x + \dfrac{2}{9}\right) + 2 = x - 2 + 2 = x$ and

$(f^{-1} \circ f)(x) = -\dfrac{1}{9}(-9x + 2) + 2 = x - 2 + 2 = x$

13) False. Only one-to-one functions have inverses.

15)

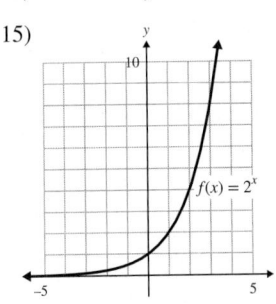

domain: (−∞, ∞);
range: (0, ∞)

17)

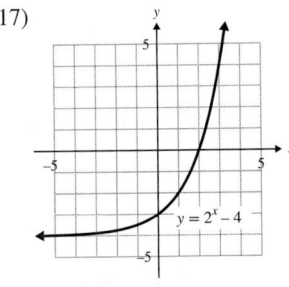

domain: (−∞, ∞);
range: (−4, ∞)

19)

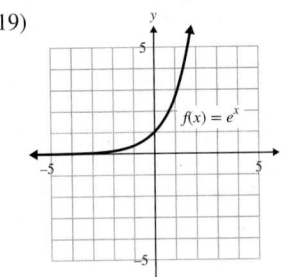

domain: (−∞, ∞);
range: (0, ∞)

21) {6}   23) $\left\{-\dfrac{5}{2}\right\}$ 25) $\left\{\dfrac{2}{3}\right\}$

27) (0, ∞)   29) $5^3 = 125$ 31) $10^2 = 100$

33) $\log_3 81 = 4$ 35) $\log 1000 = 3$ 37) {8}

39) $\left\{\dfrac{4}{5}\right\}$ 41) 2   43) 3   45) −4

47)

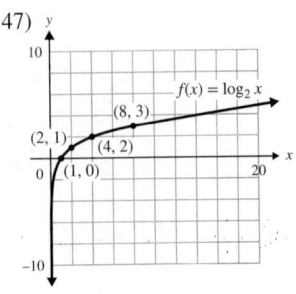

49)

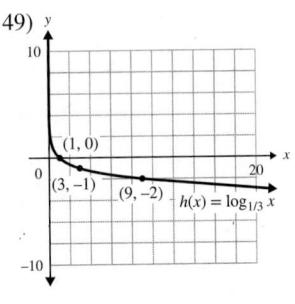

51) $f^{-1}(x) = \log_5 x$ 53) a) 10,000   b) 20,000

55) true   57) $2 - \log_7 t$ 59) -2

61) $\log_4 m + \dfrac{1}{2}\log_4 n$ 63) $2\log_4 a - \log_4 b - 4\log_4 c$

65) $3\log_6 r - \log_6(r^2 - 5)$ 67) $\log_2 a^9 b^3$ 69) $\log_3 \dfrac{5m^4}{n^2}$

71) 1.6902   73) e 75) 2   77) -2 79) 0

81) 0.9031   83) 0.5596   85) $\{100\}$ 87) $\left\{\dfrac{1}{5}\right\}$

89) $\{10^{2.1}\}$; $\{125.8925\}$ 91) $\{e^2\}$; $\{7.3891\}$

93) $\left\{\dfrac{10^{1.75}}{4}\right\}$; $\{14.0585\}$

95)

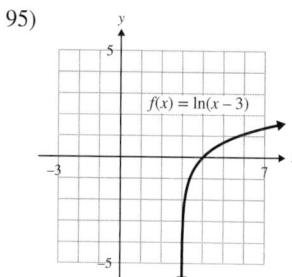

domain: $(3, \infty)$;
range: $(-\infty, \infty)$

97) 2.1240   99) -5.2479 101) 110 dB

103) $3367.14   105) a) 6000   b) 11,118   107) $\{4\}$

109) $\left\{\dfrac{14}{5}\right\}$ 110) $\left\{\dfrac{\ln 19}{0.03}\right\}$; $\{98.1480\}$ 113) $\{4\}$

115) $\{16\}$ 117) $6770.57   119) a) 17,777   b) 2036

121) sometimes   123) sometimes   125) always

Chapter 13: Test

1) no   2) yes; $g^{-1} = \left\{(4, 2), (6, 6), \left(\dfrac{15}{2}, 9\right), (10, 14)\right\}$

3) yes

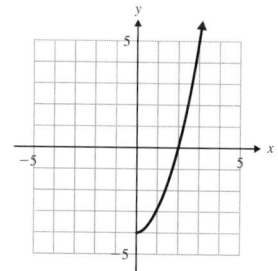

4) $(k \circ k^{-1})(x) = \dfrac{5}{3}\left(\dfrac{3}{5}x + \dfrac{3}{10}\right) - \dfrac{1}{2} = x - \dfrac{1}{2} + \dfrac{1}{2} = x$ and

$(k^{-1} \circ k)(x) = \dfrac{3}{5}\left(\dfrac{5}{3}x - \dfrac{1}{2}\right) + \dfrac{3}{10} = x - \dfrac{3}{10} + \dfrac{3}{10} = x$

5) $f^{-1}(x) = -\dfrac{1}{3}x + 4$

6)

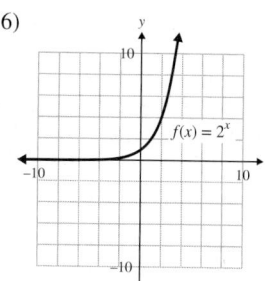

7)

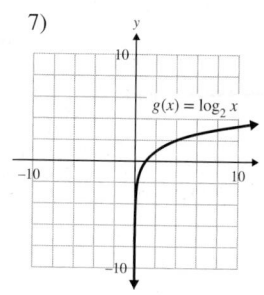

8) a) $(0, \infty)$ b) $(-\infty, \infty)$ 9) They are inverses.

10) $\log_3 \dfrac{1}{9} = -2$ 11) $\left\{\dfrac{1}{2}\right\}$ 12) $\{-2\}$ 13) $\left\{\dfrac{1}{144}\right\}$

14) $\{29\}$ 15) $\{4\}$ 16) $\{5\}$ 17) a) 4   b) $\dfrac{1}{2}$

18) 1   19) $\log_8 5 + \log_8 n$

20) $2 + 4\log_3 a - 5\log_3 b - \log_3 c$

21) $\log \dfrac{x^2}{(x+1)^3}$ 22) $\{10^{0.8}\}$; $\{6.3096\}$

23) $\left\{\dfrac{\ln 5}{0.3}\right\}$; $\{5.3648\}$ 24) $\{e^{-0.25}\}$; $\{0.7788\}$

25) $\left\{\dfrac{\ln 9 - 3\ln 4}{4\ln 4}\right\}$; $\{-0.3538\}$

26)

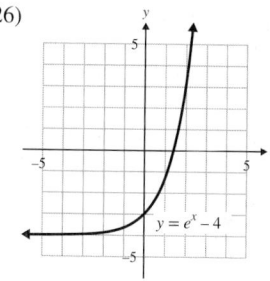

domain: $(-\infty, \infty)$;
range: $(-4, \infty)$

27)

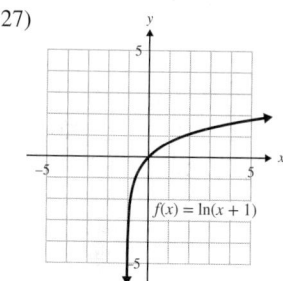

domain: $(-1, \infty)$;
range: $(-\infty, \infty)$

28) 1.7604   29) $8686.41

30) a) 86.2 g   b) 325.1 days   c) 140 days

Chapter 13: Cumulative Review for Chapters 1–13

1) $-15a^6$ 2) $\dfrac{d^9}{8c^{30}}$ 3) 9.231×10^{-5} 4) $48.00

5) $\left(-\dfrac{3}{2}, \infty\right)$

6) $\left(\dfrac{1}{3}, -2\right)$ 7) $3c^2 + 4c - 1$ 8) $(w + 4)^2$

9) $\{-8, -6\}$ 10) $\{2\}$

11)

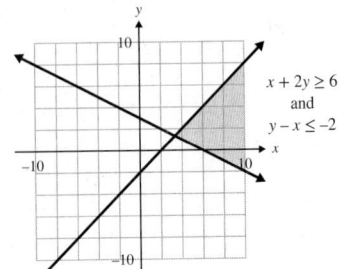

$$x + 2y \ge 6$$
and
$$y - x \le -2$$

12) $3t^4\sqrt{5t}$ 13) $6a$ 14) $\varnothing$

15) $\{4 - 2\sqrt{3}, 4 + 2\sqrt{3}\}$

16) $\left\{-\dfrac{5}{2} - \dfrac{\sqrt{17}}{2}, -\dfrac{5}{2} + \dfrac{\sqrt{17}}{2}\right\}$

17) $\{5 + 4i, 5 - 4i\}$

18) $\left\{-2, -\dfrac{1}{2}, \dfrac{1}{2}, 2\right\}$ 19) $\sqrt{29}$

20) $V(1, 3)$; x-ints:
$(1 - \sqrt{3}, 0)$,
$(1 + \sqrt{3}, 0)$; y-int:
$(0, 2)$; domain: $(-\infty, \infty)$;
range: $(-\infty, 3]$

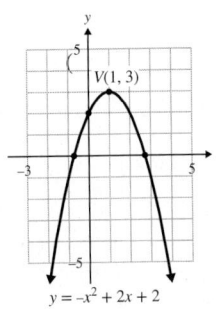

$$y = -x^2 + 2x + 2$$

21) a) 9 b) $x^2 - 12x + 29$ c) -4

22)

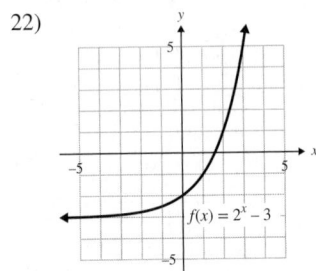

$$f(x) = 2^x - 3$$

domain: $(-\infty, \infty)$; range: $(-3, \infty)$

23) $\{3\}$ 24) $\log\dfrac{ab^2}{c^5}$

25) $\left\{-\dfrac{\ln 6}{0.04}\right\}$; $\{-44.7940\}$

Chapter 14

Section 14.1

1) $(4, 6)$ 3) $(-3, -3)$ 5) $\left(-1, -\dfrac{9}{2}\right)$

7) $\left(\dfrac{1}{2}, -\dfrac{5}{2}\right)$ 9) $\left(2, \dfrac{5}{4}\right)$ 11) $(-0.7, 3.6)$

13) No; there are values in the domain that give more than one value in the range. The graph fails the vertical line test.

15) center: $(-2, 4)$; $r = 3$

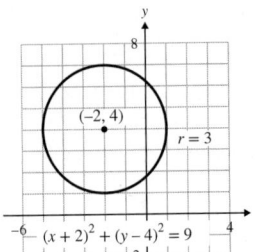

$$(x + 2)^2 + (y - 4)^2 = 9$$

17) center: $(5, 3)$; $r = 1$

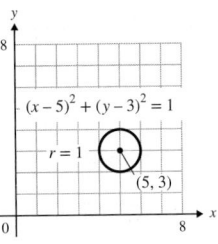

$$(x - 5)^2 + (y - 3)^2 = 1$$

19) center: $(-3, 0)$; $r = 2$

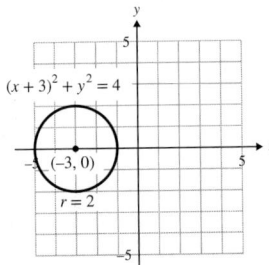

$$(x + 3)^2 + y^2 = 4$$

21) center: $(6, -3)$; $r = 4$

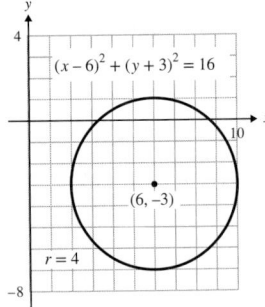

$$(x - 6)^2 + (y + 3)^2 = 16$$

23) center: $(0, 0)$; $r = 6$

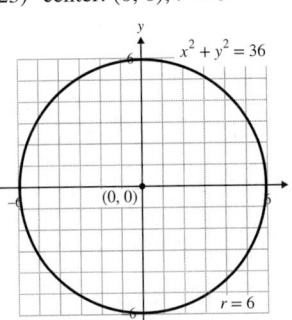

$$x^2 + y^2 = 36$$

25) center: $(0, 0)$; $r = 3$

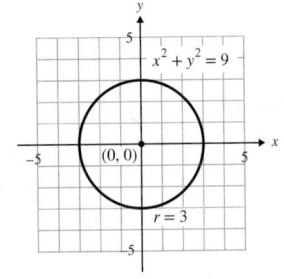

$$x^2 + y^2 = 9$$

27) center: $(0, 1)$; $r = 5$

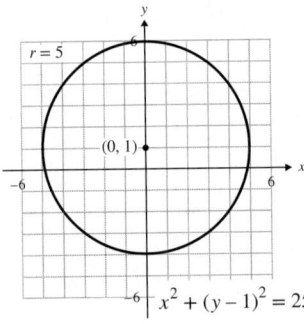

$$x^2 + (y - 1)^2 = 25$$

29) $(x - 4)^2 + (y - 1)^2 = 25$

31) $(x + 3)^2 + (y - 2)^2 = 1$

33) $(x + 1)^2 + (y + 5)^2 = 3$ 35) $x^2 + y^2 = 10$

37) $(x - 6)^2 + y^2 = 16$ 39) $x^2 + (y + 4)^2 = 8$

41) $(x - 5)^2 + (y + 3)^2 = 9$ 43) $(x - 2)^2 + (y - 7)^2 = 20$

45) $(x - 6)^2 + (y - 2)^2 = 45$ 47) $(x + 3)^2 + (y - 2)^2 = 5$

49) Group x- and y-terms separately;
$(x^2 - 8x + 16) + (y^2 + 2y + 1) = -8 + 16 + 1$;
$(x - 4)^2 + (y + 1)^2 = 9$

51) $(x + 1)^2 + (y + 5)^2 = 9$;
center: $(-1, -5)$; $r = 3$

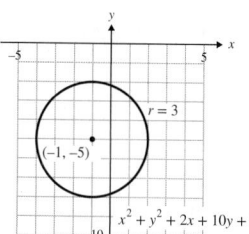

$x^2 + y^2 + 2x + 10y + 17 = 0$

53) $(x + 4)^2 + (y - 1)^2 = 25$;
center: $(-4, 1)$; $r = 5$

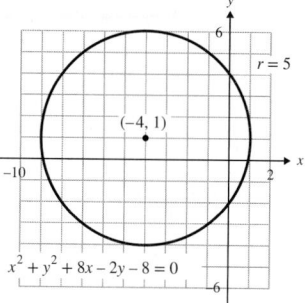

$x^2 + y^2 + 8x - 2y - 8 = 0$

55) $(x - 5)^2 + (y - 7)^2 = 1$;
center: $(5, 7)$; $r = 1$

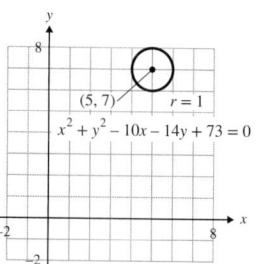

$x^2 + y^2 - 10x - 14y + 73 = 0$

57) $x^2 + (y + 3)^2 = 4$;
center: $(0, -3)$; $r = 2$

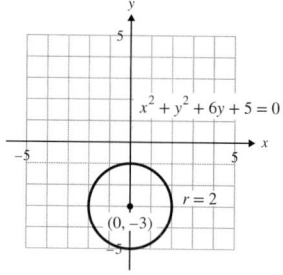

$x^2 + y^2 + 6y + 5 = 0$

59) $(x - 2)^2 + y^2 = 5$;
center: $(2, 0)$; $r = \sqrt{5}$

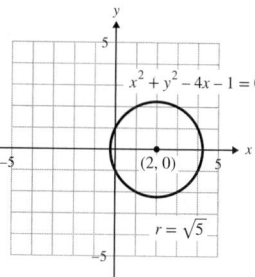

$x^2 + y^2 - 4x - 1 = 0$

61) $(x - 4)^2 + (y + 4)^2 = 36$;
center: $(4, -4)$; $r = 6$

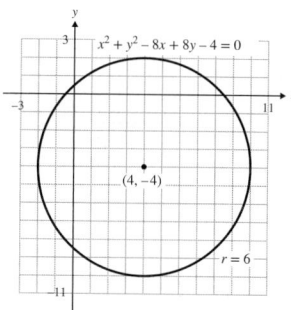

$x^2 + y^2 - 8x + 8y - 4 = 0$

63) $\left(x - \dfrac{3}{2}\right)^2 + \left(y - \dfrac{1}{2}\right)^2 = 4$; center: $\left(\dfrac{3}{2}, \dfrac{1}{2}\right)$; $r = 2$

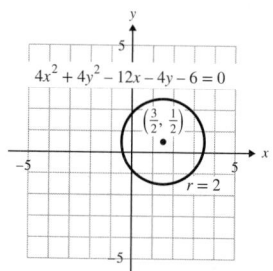

$4x^2 + 4y^2 - 12x - 4y - 6 = 0$

65) a) 128 m   b) 64 m   c) (0, 71)
d) $x^2 + (y - 71)^2 = 4096$

67) 11,127 mm²

69) $(x - 5)^2 + (y - 3)^2 = 1$

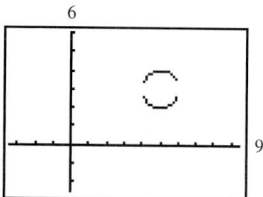

71) $(x + 2)^2 + (y - 4)^2 = 9$

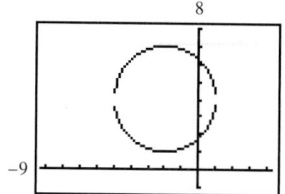

Section 14.2

1) false   3) true   5) true   7) false

9) center: $(-2, 1)$

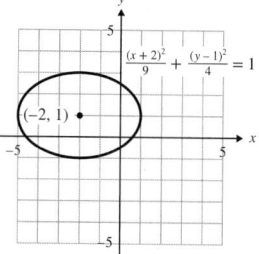

$\dfrac{(x + 2)^2}{9} + \dfrac{(y - 1)^2}{4} = 1$

11) center: $(3, -2)$

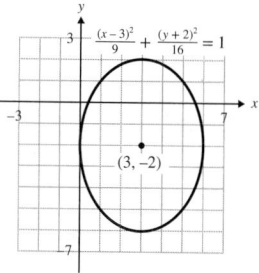

$\dfrac{(x - 3)^2}{9} + \dfrac{(y + 2)^2}{16} = 1$

13) center: $(0, 0)$

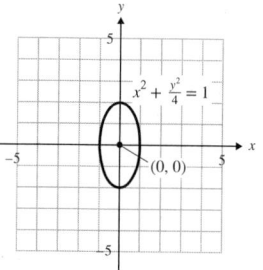

$x^2 + \dfrac{y^2}{4} = 1$

15) center: $(0, -4)$

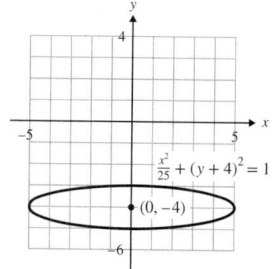

$\dfrac{x^2}{25} + (y + 4)^2 = 1$

17) center: $(-1, -3)$

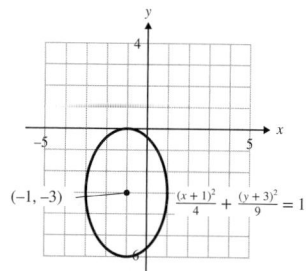

$\dfrac{(x + 1)^2}{4} + \dfrac{(y + 3)^2}{9} = 1$

19) center: $(0, 0)$

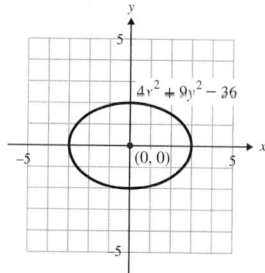

$4x^2 + 9y^2 = 36$

21) center: $(0, 0)$

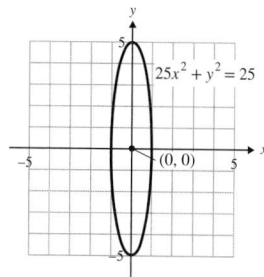

$25x^2 + y^2 = 25$

23) Group x- and y-terms separately;
$3(x^2 - 2x) + 2(y^2 + 2y) = 7$; Complete the square;
$3(x - 1)^2 + 2(y + 1)^2 = 12$; $\dfrac{(x - 1)^2}{4} + \dfrac{(y + 1)^2}{6} = 1$

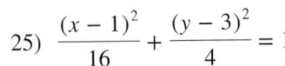

25) $\dfrac{(x - 1)^2}{16} + \dfrac{(y - 3)^2}{4} = 1$ 27) $(x + 4)^2 + \dfrac{(y + 1)^2}{9} = 1$

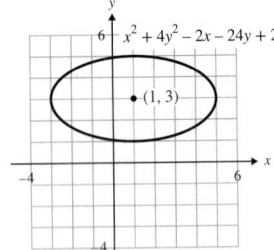

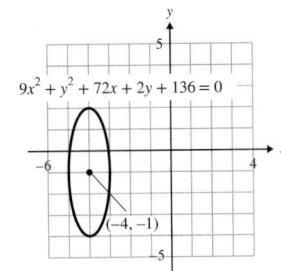

29) $\dfrac{(x - 2)^2}{9} + \dfrac{(y - 3)^2}{4} = 1$

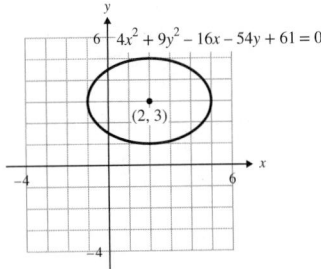

31) $\dfrac{(x + 3)^2}{4} + \dfrac{y^2}{25} = 1$

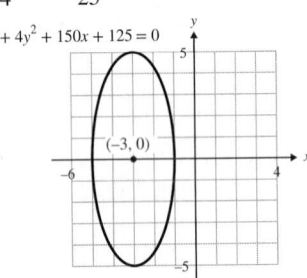

33) $\dfrac{x^2}{9} + \dfrac{y^2}{25} = 1$ 35) $\dfrac{x^2}{49} + y^2 = 1$

37) $\dfrac{(x - 3)^2}{4} + \dfrac{(y - 1)^2}{16} = 1$

39) Yes. If $a = b$ in the equation $\dfrac{(x - h)^2}{a^2} + \dfrac{(y - k)^2}{b^2} = 1$,
then the ellipse is a circle.

41) $\dfrac{x^2}{9801} + \dfrac{y^2}{11{,}342.25} = 1$

43) false   45) true

47) center: (0, 0);
vertices: (−3, 0), (3, 0)

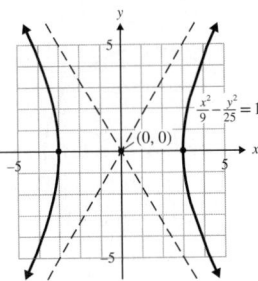

49) center: (0, 0);
vertices: (0, 4), (0, −4)

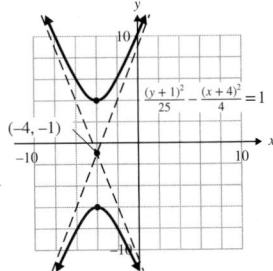

51) center: (2, −3);
vertices: (−1, −3), (5, −3)

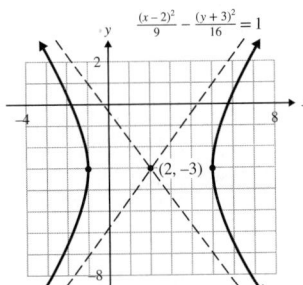

53) center: (−4, −1);
vertices: (−4, 4), (−4, −6)

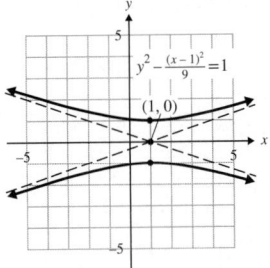

55) center: (1, 0);
vertices: (1, 1), (1, −1)

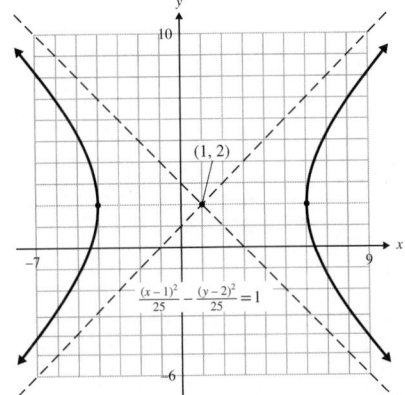

57) center: (1, 2); vertices: (−4, 2), (6, 2)

59) center: (0, 0);
   vertices: (−2, 0), (2, 0)

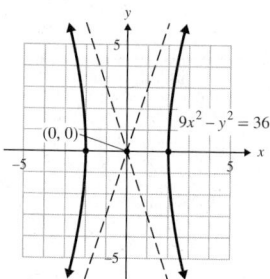

61) center: (0, 0);
   vertices: (0, 1), (0, −1)

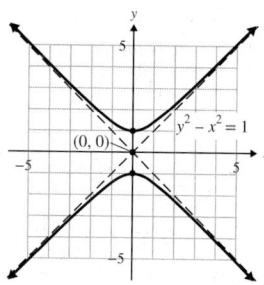

87)

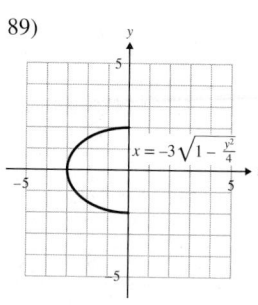

89)

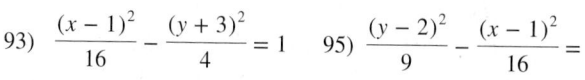

63) $y = \dfrac{5}{3}x$ and $y = -\dfrac{5}{3}x$

65) $y = 2x$ and $y = -2x$

67) $y = 3x$ and $y = -3x$

69) $y = x$ and $y = -x$

91) Group the x- and y-terms separately;
   $4(x^2 - 2x) - 9(y^2 + 2y) = 41$;
   Complete the square;
   $4(x - 1)^2 - 9(y + 1)^2 = 36;\ \dfrac{(x - 1)^2}{9} - \dfrac{(y + 1)^2}{4} = 1$

71)

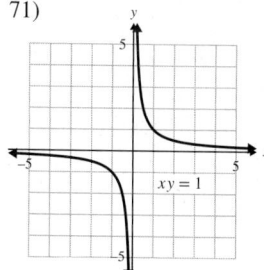

73)

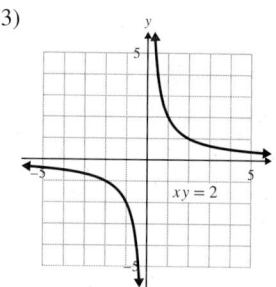

93) $\dfrac{(x - 1)^2}{16} - \dfrac{(y + 3)^2}{4} = 1$

95) $\dfrac{(y - 2)^2}{9} - \dfrac{(x - 1)^2}{16} = 1$

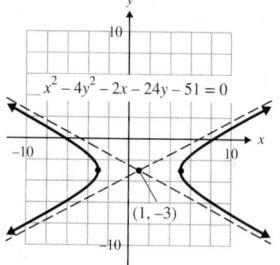

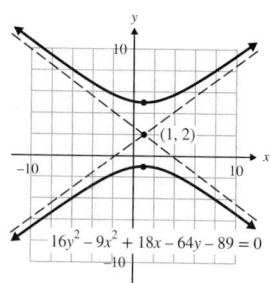

75)

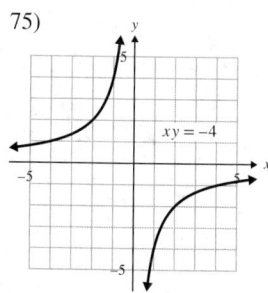

77)

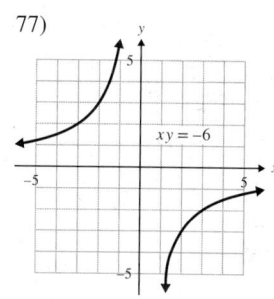

97) $y + 3 = \dfrac{4}{3}(x - 2)$ and $y + 3 = -\dfrac{4}{3}(x - 2)$

99) $y + 1 = \dfrac{5}{2}(x + 4)$ and $y + 1 = -\dfrac{5}{2}(x + 4)$

101) $y = \dfrac{1}{3}(x - 1)$ and $y = -\dfrac{1}{3}(x - 1)$

103) $y^2 - \dfrac{x^2}{4} = 1$

105) $\dfrac{(x - 2)^2}{9} - \dfrac{(y + 3)^2}{16} = 1$

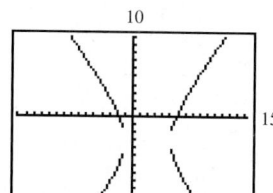

79) domain: [−5, 5];
   range: [0, 5]

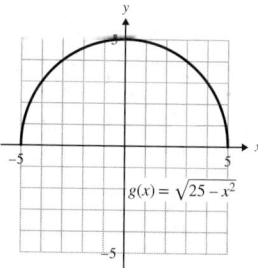

81) domain: [−1, 1];
   range: [−1, 0]

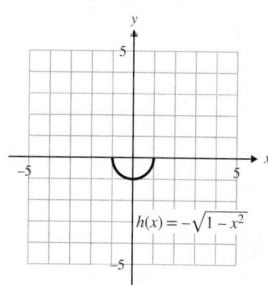

107) $\dfrac{x^2}{36} + \dfrac{y^2}{16} = 1$

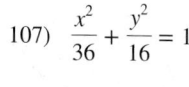

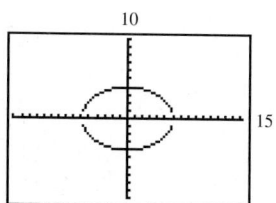

83) domain: [−3, 3];
   range: [−2, 0]

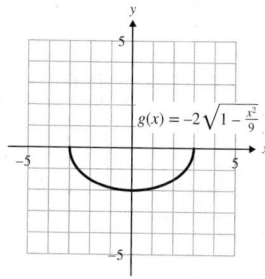

85) domain:
   $(-\infty, -2] \cup [2, \infty)$;
   range: $(-\infty, 0]$

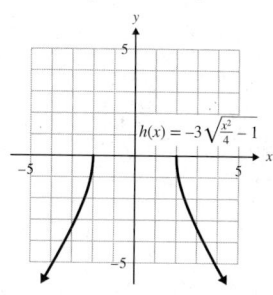

109) $\dfrac{(y+1)^2}{25} - \dfrac{(x+4)^2}{4} = 1$

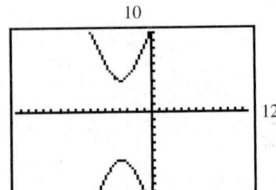

Chapter 14: Putting It All Together

1) parabola

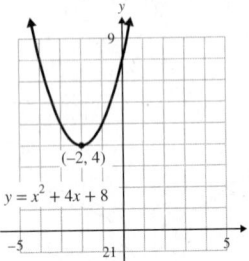

2) circle

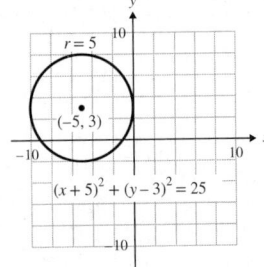

9) ellipse

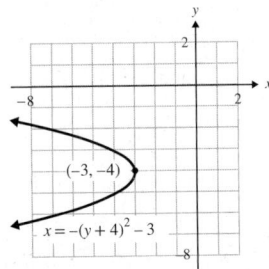

10) circle

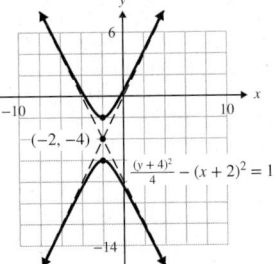

11) parabola

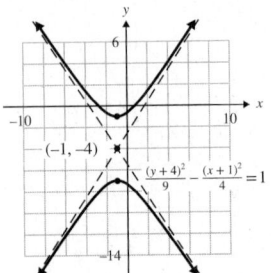

12) hyperbola

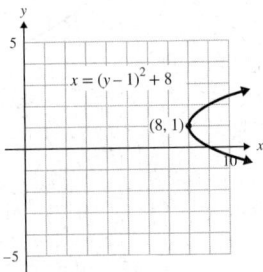

3) hyperbola

4) parabola

13) hyperbola

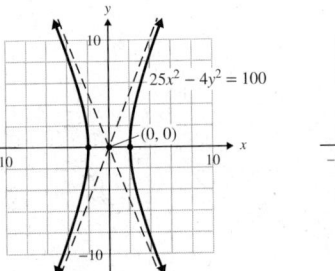

14) ellipse

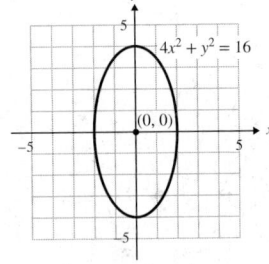

5) ellipse

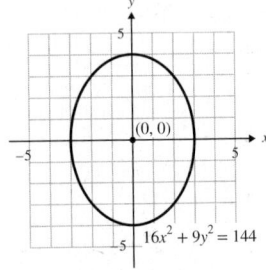

6) hyperbola

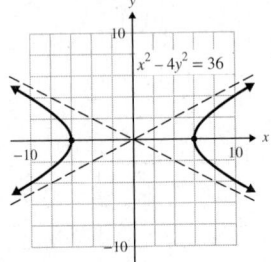

15) circle

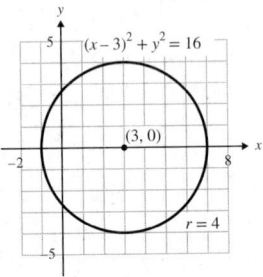

16) parabola

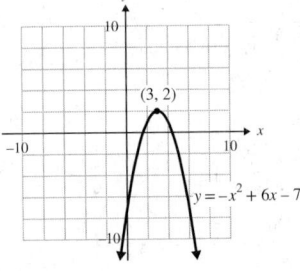

7) circle

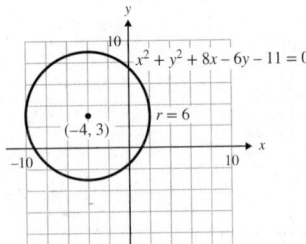

8) ellipse

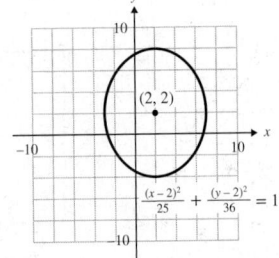

17) parabola

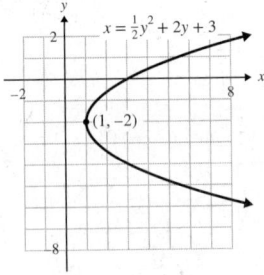

18) ellipse

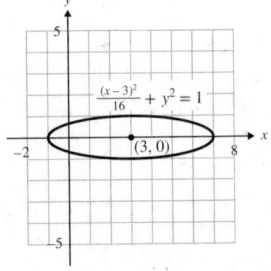

19) hyperbola

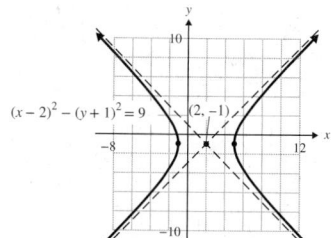

$(x-2)^2-(y+1)^2=9$ (2, −1)

20) circle

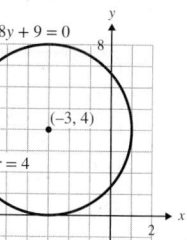

$x^2+y^2+6x-8y+9=0$ (−3, 4) $r=4$

29) $x=-(y+4)^2-3$

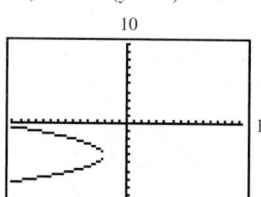

30) $25x^2-4y^2=100$

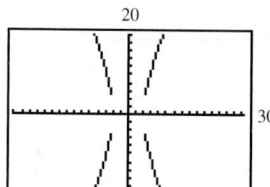

Section 14.3

1) (2, 2)

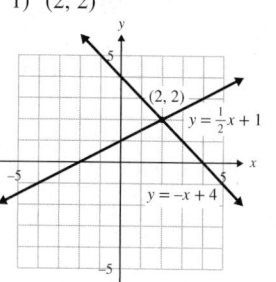

$y=\frac{1}{2}x+1$ (2, 2) $y=-x+4$

3) ∅

$y=-x$ $5x+5y=10$

21)

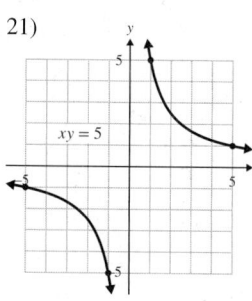

$xy=5$

22) $\dfrac{(x+3)^2}{4}-\dfrac{y^2}{25}=1$

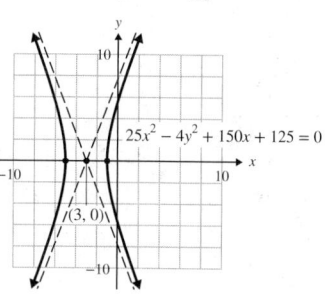

$25x^2-4y^2+150x+125=0$ (3, 0)

5) (3, 7)   7) (−5, 2)   9) (2, −3)   11) $\left(-\dfrac{3}{2}, -1\right)$

13) c) 0, 1, or 2

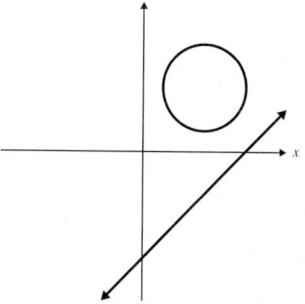

23) $(x-3)^2+\dfrac{(y+2)^2}{9}=1$

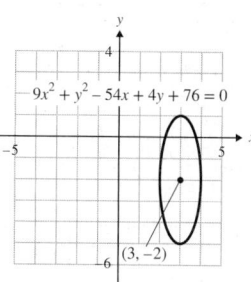

$-9x^2+y^2-54x+4y+76=0$ (3, −2)

24)

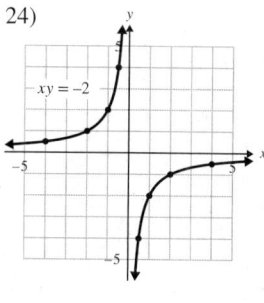

$xy=-2$

25) $\dfrac{(y-1)^2}{4}-\dfrac{(x-2)^2}{9}=1$

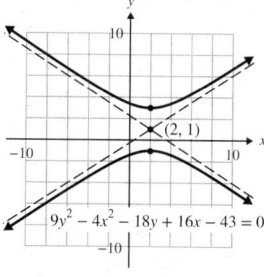

(2, 1) $9y^2-4x^2-18y+16x-43=0$

26) $\dfrac{(x-1)^2}{9}+\dfrac{(y-3)^2}{4}=1$

(1, 3) $-4x^2+9y^2-8x-54y+49=0$

15) c) 0, 1, 2, 3, or 4

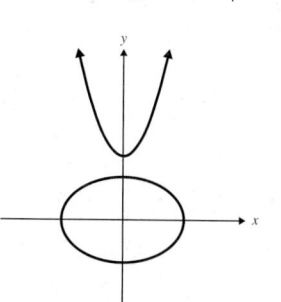

27) $x^2+y^2+8x-6y-11=0$

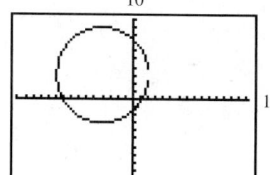

28) $(x-1)^2+\dfrac{y^2}{16}=1$

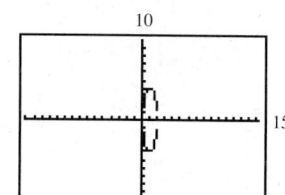

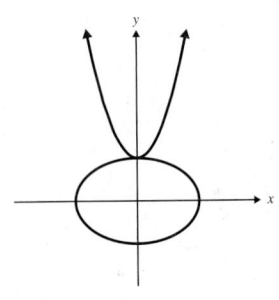
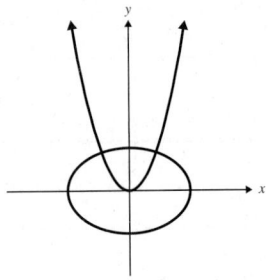

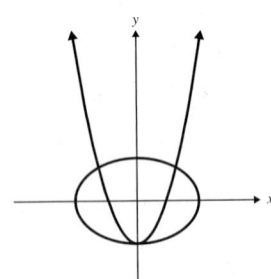
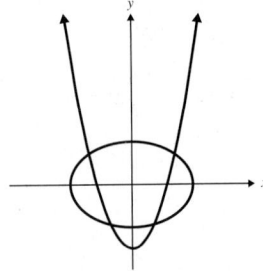

17) c) 0, 1, 2, 3, or 4

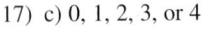
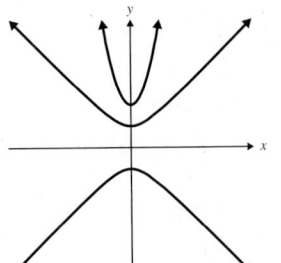

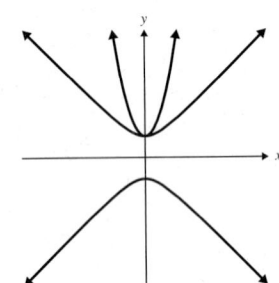
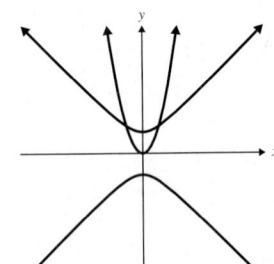

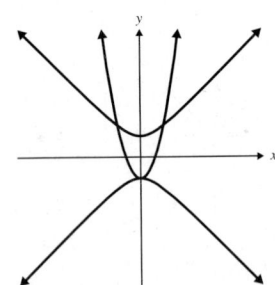
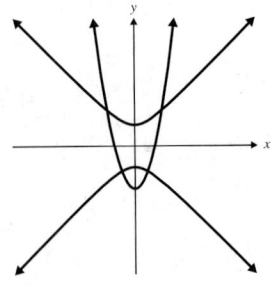

19) $\{(-4, -2), (6, -7)\}$ 21) $\{(-1, 3), (3, 1)\}$

23) $\{(4, 2)\}$ 25) $\{(3, 1), (3, -1), (-3, 1), (-3, -1)\}$

27) $\{(2, \sqrt{2}), (2, -\sqrt{2}), (-2, \sqrt{2}), (-2, -\sqrt{2})\}$

29) $\{(3, -7), (-3, -7)\}$ 31) $\{(0, -1)\}$

33) $\{(0, 2)\}$ 35) $\emptyset$ 37) $\emptyset$

39) $\left\{ \left(\dfrac{\sqrt{2}}{2}, \dfrac{3\sqrt{2}}{2} \right), \left(\dfrac{\sqrt{2}}{2}, -\dfrac{3\sqrt{2}}{2} \right), \right.$
$\left. \left(-\dfrac{\sqrt{2}}{2}, \dfrac{3\sqrt{2}}{2} \right), \left(-\dfrac{\sqrt{2}}{2}, -\dfrac{3\sqrt{2}}{2} \right) \right\}$

41) $\{(0, 1)\}$ 43) 8 and 5   45) 8 in. × 11 in.

47) 4000 basketballs; $240

Section 14.4

1)
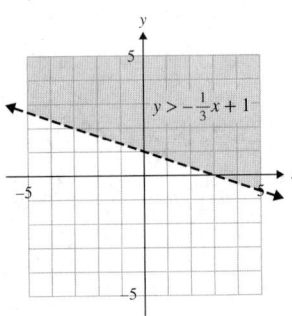

3)
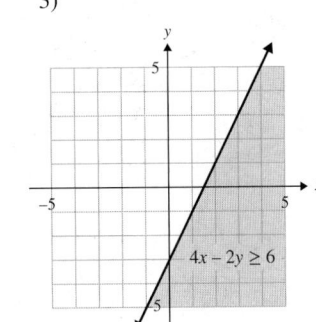

5)
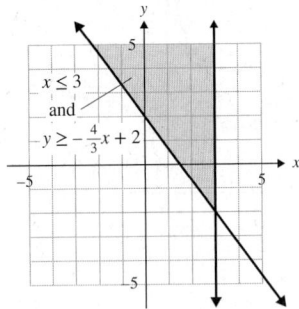

7)
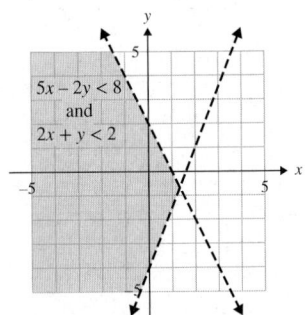

9) Three points that satisfy the inequality: (8, 0), (0, −8), and (6, 0); three that do not: (0, 0), (−3, −2), and (1, 1); answers may vary.

11) Three points that satisfy the inequality: (0, 0), (−4, −1), and (5, 2); three that do not: (0, 1), (0, −2), and (3, 3); answers may vary.

13)
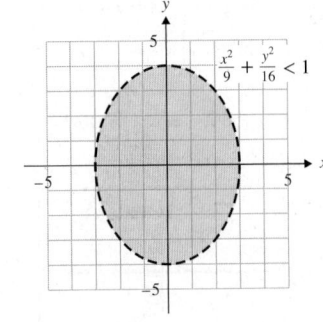

15)
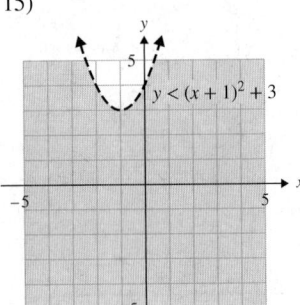

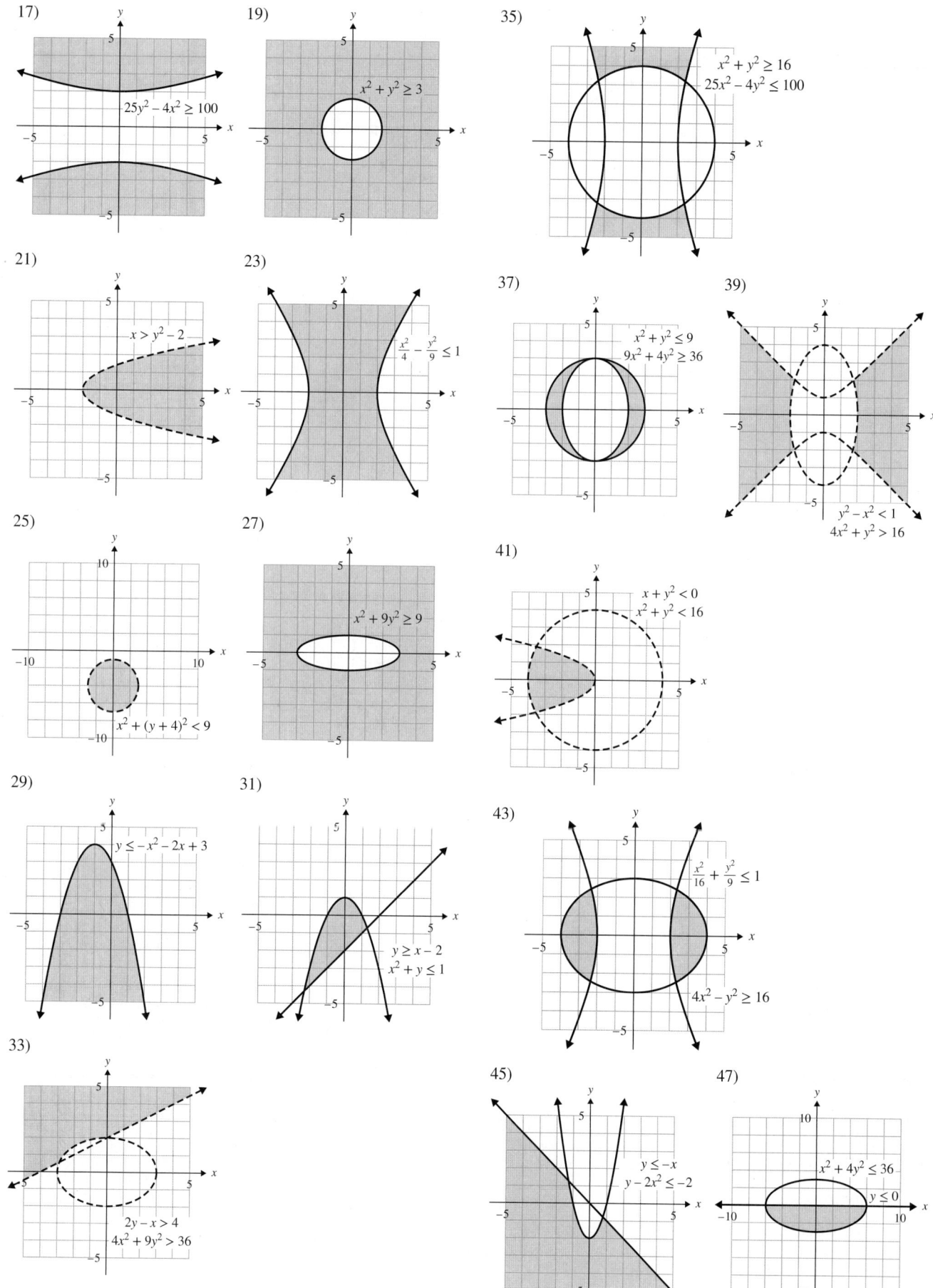

17) $25y^2 - 4x^2 \geq 100$

19) $x^2 + y^2 \geq 3$

35) $x^2 + y^2 \geq 16$
$25x^2 - 4y^2 \leq 100$

21) $-x > y^2 - 2$

23) $\dfrac{x^2}{4} - \dfrac{y^2}{9} \leq 1$

37) $x^2 + y^2 \leq 9$
$9x^2 + 4y^2 \geq 36$

39) $y^2 - x^2 < 1$
$4x^2 + y^2 > 16$

25) $x^2 + (y + 4)^2 < 9$

27) $x^2 + 9y^2 \geq 9$

41) $x + y^2 < 0$
$x^2 + y^2 < 16$

29) $y \leq -x^2 - 2x + 3$

31) $y \geq x - 2$
$x^2 + y \leq 1$

43) $\dfrac{x^2}{16} + \dfrac{y^2}{9} \leq 1$
$4x^2 - y^2 \geq 16$

33) $2y - x > 4$
$4x^2 + 9y^2 > 36$

45) $y \leq -x$
$y - 2x^2 \leq -2$

47) $x^2 + 4y^2 \leq 36$
$y \leq 0$

49)

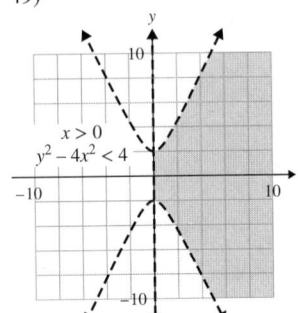

$x > 0$
$y^2 - 4x^2 < 4$

51)

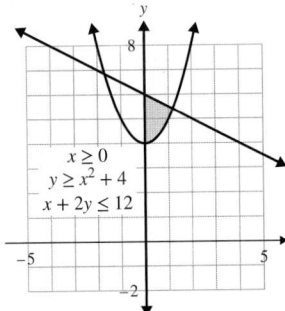

$x \geq 0$
$y \geq x^2 + 4$
$x + 2y \leq 12$

53)

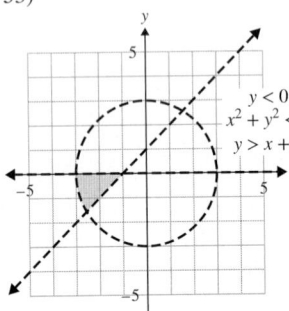

$y < 0$
$x^2 + y^2 < 9$
$y > x + 1$

55)

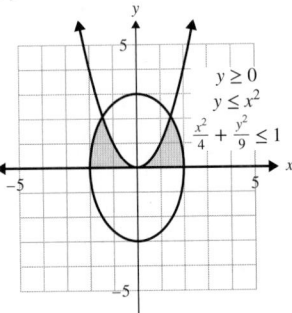

$y \geq 0$
$y \leq x^2$
$\dfrac{x^2}{4} + \dfrac{y^2}{9} \leq 1$

57)

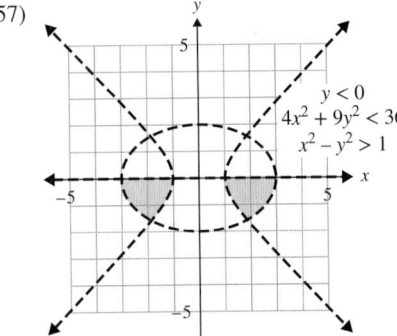

$y < 0$
$4x^2 + 9y^2 < 36$
$x^2 - y^2 > 1$

59)

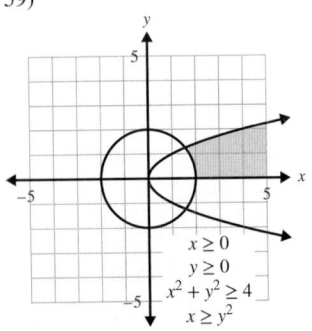

$x \geq 0$
$y \geq 0$
$x^2 + y^2 \geq 4$
$x \geq y^2$

Chapter 14: Group Activity

2) 26

3) a is the distance from the center to an x-intercept; b is the distance from the center to a y-intercept; $2a$

4) $x = \pm\sqrt{a^2 - b^2}$ 6) 24

7) a is the distance from the center to the x-intercepts; $2a$

8) $x = \pm\sqrt{a^2 + b^2}$

9) a) $y = \pm\dfrac{3}{4}x$ b) $\dfrac{x^2}{16} - \dfrac{y^2}{9} = 1$

c) $x^2 + y^2 = 25$ d) $\dfrac{x^2}{25} + \dfrac{y^2}{9} = 1$

Chapter 14: Review Exercises

1) $(4, 5)$ 3) $\left(\dfrac{13}{2}, -\dfrac{7}{2}\right)$

5) center: $(-3, 5)$; $r = 6$

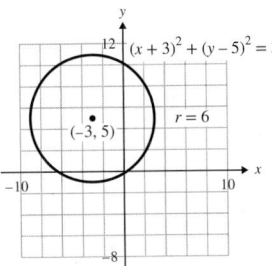

$(x + 3)^2 + (y - 5)^2 = 36$
$(-3, 5)$ $r = 6$

7) center: $(5, 2)$; $r = 4$

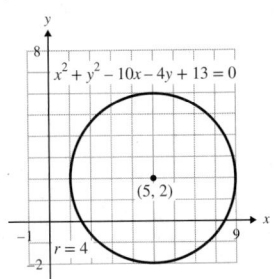

$x^2 + y^2 - 10x - 4y + 13 = 0$
$(5, 2)$
$r = 4$

9) $(x - 1)^2 + (y + 5)^2 = 7$

11) when $a = b$ in $\dfrac{(x - h)^2}{a^2} + \dfrac{(y - k)^2}{b^2} = 1$

13) center $(-3, 3)$

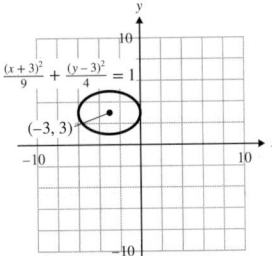

$\dfrac{(x + 3)^2}{9} + \dfrac{(y - 3)^2}{4} = 1$
$(-3, 3)$

15) center $(0, 0)$

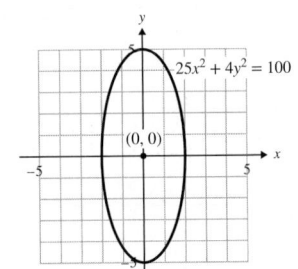

$25x^2 + 4y^2 = 100$
$(0, 0)$

17) $\dfrac{(x - 2)^2}{4} + \dfrac{y^2}{25} = 1$

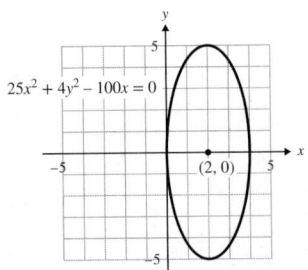

$25x^2 + 4y^2 - 100x = 0$
$(2, 0)$

19) center $(0, 0)$
vertices $(0, 3)$, $(0, -3)$

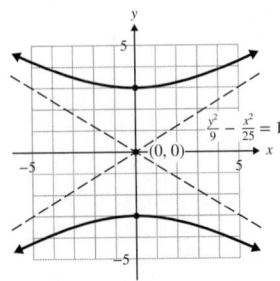

$\dfrac{y^2}{9} - \dfrac{x^2}{25} = 1$
$(0, 0)$

21) center $(-1, -2)$
vertices $(-3, -2)$, $(1, -2)$

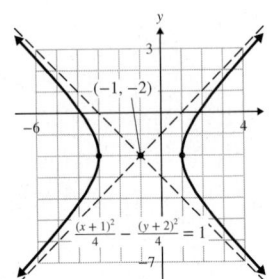

$(-1, -2)$
$\dfrac{(x + 1)^2}{4} - \dfrac{(y + 2)^2}{4} = 1$

23)

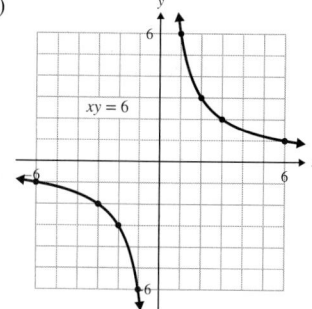

$xy = 6$

25) $(y + 3)^2 - \dfrac{(x - 1)^2}{16} = 1$

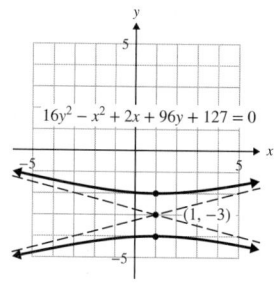

$16y^2 - x^2 + 2x + 96y + 127 = 0$

$(1, -3)$

27) domain: $[-3, 3]$; range: $[0, 2]$

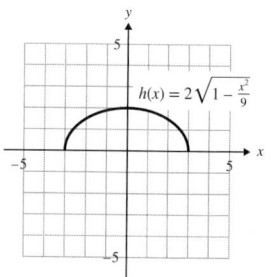

$h(x) = 2\sqrt{1 - \dfrac{x^2}{9}}$

29) ellipse

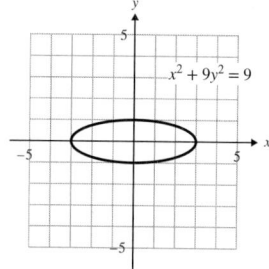

$x^2 + 9y^2 = 9$

31) parabola

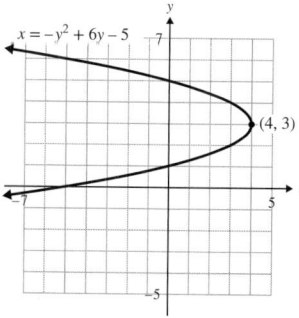

$x = -y^2 + 6y - 5$

$(4, 3)$

33) hyperbola

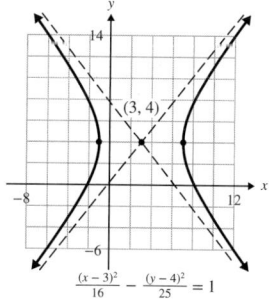

$(3, 4)$

$\dfrac{(x - 3)^2}{16} - \dfrac{(y - 4)^2}{25} = 1$

35) circle

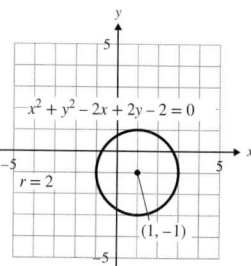

$x^2 + y^2 - 2x + 2y - 2 = 0$

$r = 2$

$(1, -1)$

37) parabola

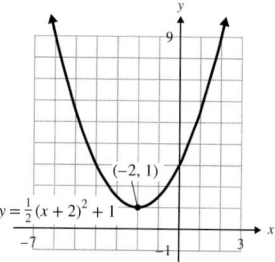

$y = \dfrac{1}{2}(x + 2)^2 + 1$

$(-2, 1)$

39) 0, 1, 2, 3, or 4   41) $\{(6, 7), (6, -7), (-6, 7), (-6, -7)\}$

43) $\{(1, 2), (-2, -1)\}$ 45) $\varnothing$ 47) 9 and 4

49)

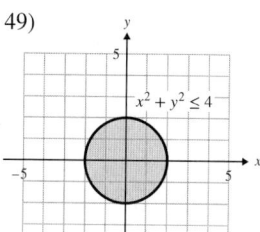

$x^2 + y^2 \le 4$

51)

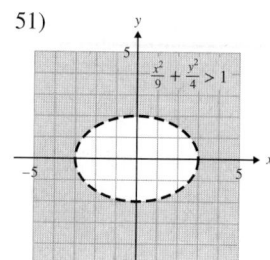

$\dfrac{x^2}{9} + \dfrac{y^2}{4} > 1$

53)

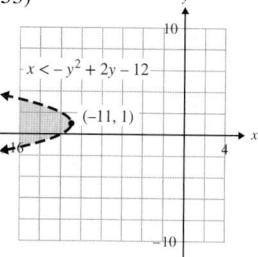

$x < -y^2 + 2y - 12$

$(-11, 1)$

55)

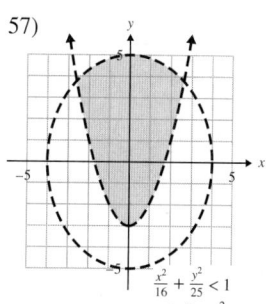

$y - x^2 > 2$
$x + y > 5$

57)

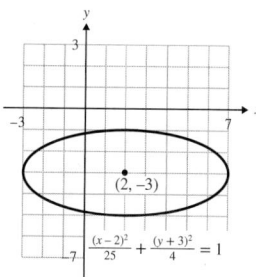

$\dfrac{x^2}{16} + \dfrac{y^2}{25} < 1$
$y + 3 > x^2$

59)

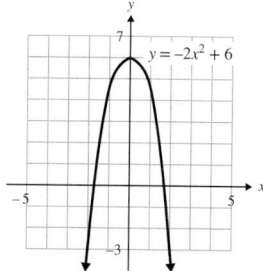

$y \ge 0$
$x^2 + y^2 \le 36$
$x + y \le 0$

Chapter 14: Test

1) $\left(6, -\dfrac{5}{2}\right)$ 2) false   3) true

4) ellipse

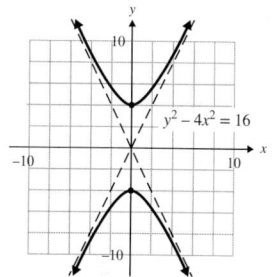

$(2, -3)$

$\dfrac{(x - 2)^2}{25} + \dfrac{(y + 3)^2}{4} = 1$

5) parabola

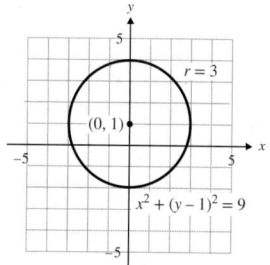

$y = -2x^2 + 6$

6) hyperbola

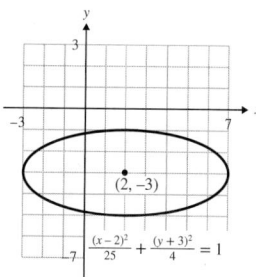

$y^2 - 4x^2 = 16$

7) circle

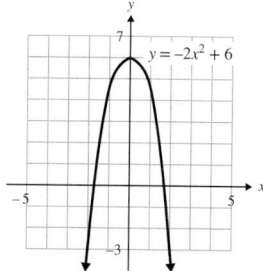

$r = 3$

$(0, 1)$

$x^2 + (y - 1)^2 = 9$

8) hyperbola

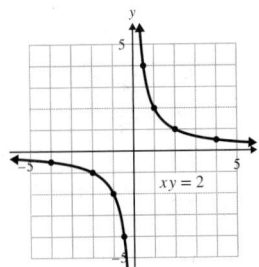

$xy = 2$

9) $(x + 1)^2 + (y - 3)^2 = 16$;
center $(-1, 3)$; $r = 4$

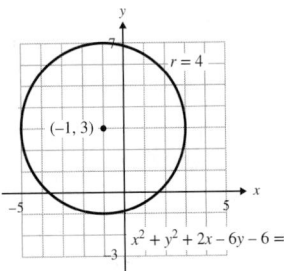

$r = 4$
$(-1, 3)$
$x^2 + y^2 + 2x - 6y - 6 = 0$

10) $(x - 5)^2 + (y - 2)^2 = 11$

11) $\dfrac{x^2}{8836} + \dfrac{y^2}{6084} = 1$

12) domain: $[-5, 5]$;
range: $[-5, 0]$

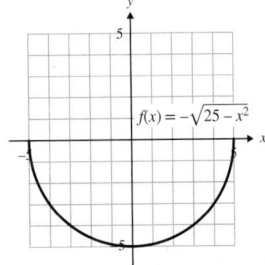

$f(x) = -\sqrt{25 - x^2}$

13) a)

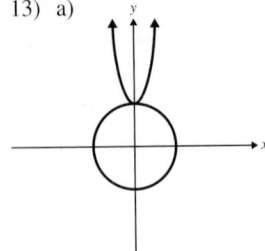

b)

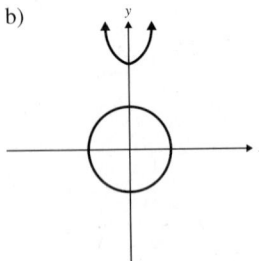

c) 0, 1, 2, 3, or 4

14) $\{(-1, 0), (7, -2)\}$

15) $\{(3, 1), (3, -1), (-3, 1), (-3, -1)\}$

16) 8 in. × 14 in.

17)

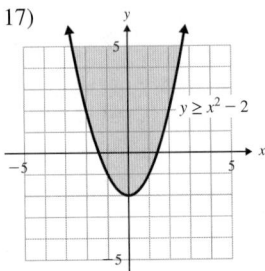

$y \geq x^2 - 2$

18)

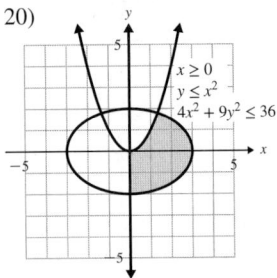

$x^2 - 4y^2 < 36$

19)

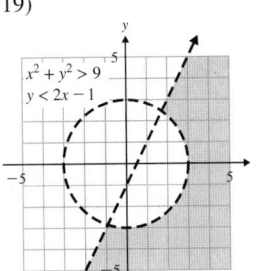

$x^2 + y^2 > 9$
$y < 2x - 1$

20)

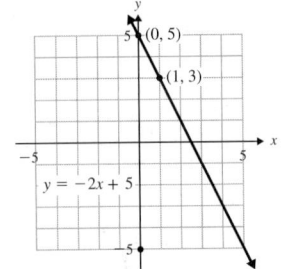

$x \geq 0$
$y \leq x^2$
$4x^2 + 9y^2 \leq 36$

Chapter 14: Cumulative Review for Chapters 1–14

1) $-\dfrac{3}{4}$ 2) 12   3) $A = 15$ cm^2; $P = 18.5$ cm

4) $A = 128$ in^2; $P = 56$ in.   5) -1 6) 16

7) $\dfrac{1}{8a^{18}b^{15}}$ 8) $\{-40\}$ 9) $n = \dfrac{c - z}{a}$ 10) $[-4, \infty)$

11) 15, 17, 19   12) -1 13) 0

14)

$(0, 5)$
$(1, 3)$
$y = -2x + 5$

15) $y = -\dfrac{3}{4}x + 4$ 16) $\left(-1, \dfrac{3}{2}\right)$

17) 5 mL of 8%, 15 mL of 16%   18) $-3p^2 + 7p + 6$

19) $8w^3 + 30w^2 - 47w + 15$ 20) $x^2 + 4x + 9$

21) $2(3c - 4)(c - 1)$ 22) $(m - 2)(m^2 + 2m + 4)$

23) $\{-2, 6\}$ 24) $-\dfrac{5}{2(a + 6)}$ 25) $6(t + 3)$

26) $\left\{-6, -\dfrac{4}{3}\right\}$ 27) $(-\infty, -3) \cup \left(\dfrac{9}{5}, \infty\right)$

28) $5\sqrt{3}$ 29) $2\sqrt[3]{6}$ 30) $3ab^4\sqrt[3]{a^2b}$ 31) $\dfrac{1}{8}$

32) $3\sqrt{3}$ 33) $\dfrac{20 - 5\sqrt{3}}{13}$ 34) $\left\{\dfrac{1}{2} - 2i, \dfrac{1}{2} + 2i\right\}$

35) $\left\{-\dfrac{7}{2} - \dfrac{\sqrt{37}}{2}, -\dfrac{7}{2} + \dfrac{\sqrt{37}}{2}\right\}$

36) $\left(-\infty, -\dfrac{5}{2}\right) \cup (3, \infty)$

37) a) $\{0, 3, 4\}$ b) $\{-1, 0, 1, 2\}$ c) no

38)

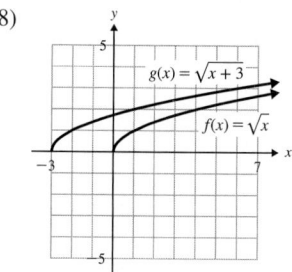

39) $V(3, 1)$; x-int: $(2, 0)$;
y-ints: $(0, 1 - \sqrt{3})$,
$(0, 1 + \sqrt{3})$;
domain: $(-\infty, 3]$;
range: $(-\infty, \infty)$

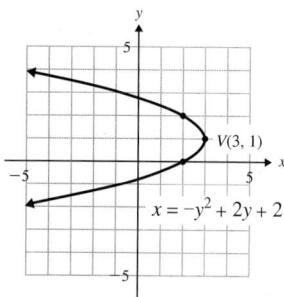

40) a) -3 b) $6x - 9$ c) 6 41) a) no   b) no

42) $f^{-1}(x) = 3x - 12$ 43) $\left\{-\dfrac{6}{13}\right\}$ 44) $\{5\}$ 45) 2

46)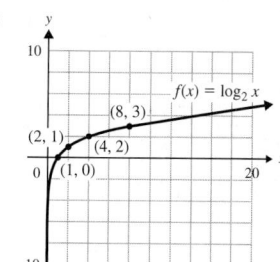

47) $\left\{\dfrac{\ln 8}{3}\right\}$; $\{0.6931\}$

48)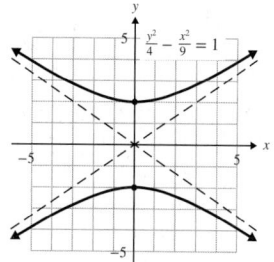

49)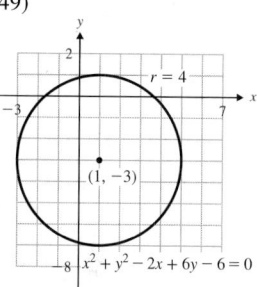

50) $(0, 3)$

Application Index

time to make doughnuts, 592
total fat in Caramel Frappuccino, 613
whiskey distillation, 917

FORENSIC SCIENCE

fingerprint identification, 310
speed at time of accident, 662

HEALTH AND MEDICINE

amoxicillin remaining in system, 917
dieter's weights, 291
doctor's weight scale range of values, 650
flu season school absences, 18
hair growth rate, 241–243
hip implant weights, 229
ibuprofen in bloodstream, 311
intravenous fluid drip rate, 592
iodine remaining in system, 964
mail-order drugs, 270
medication and body weight, 592
number of days in hospital for childbirth, 53
peanut reactions treated, 928
physicians practicing in a region, 278
prescription drug costs, 119
rhinovirus dimensions, 115
smokers beginning before 21st birthday, 199
sodium in each food item, 369–370
tuberculosis cases, 159

LANDSCAPE ARCHITECTURE

dimensions of garden fence, 849–850
flower box dimensions, 183
flower garden area and cost, 36
garden dimensions, 501, 615, 764, 802, 857
radius of flower garden, 662
reflecting pool volume, 36
width of border around pond, 789–790

SCIENCE AND TECHNOLOGY

CD surface area, 987
coin combinations, 192–194, 200, 201, 202, 230,
    B-20
current in circuit, 599, 602
data recorded on DVD, 310
distance between Earth and Sun, 115
droplets of ink in photo, 119
electron mass, 125
Fahrenheit/Celsius conversion, 186, 291
focal length of lens, 586
force exerted on object, 602
force needed to stretch spring, 258, 603
frequency of vibrating spring, 602, 614
half-life of radioactive substances, 960–961, 964,
    975, 976
illuminance of light source, 790
intensity of light source, 599
intensity of sound, 943–944, 951, 952, 974
kinetic energy, 602
laboratory water bath volume, 36
lowest and highest temperature differences, 50,
    53, 84
mass of water molecule, 124
motion of bouncing ball, 15-37
oceanic plates spreading, 586
oxygen-hydrogen bond length, 118
pendulum swing—distance, 15-33, 15-34, 15-37,
    15-53, 15-54

pendulum swing—period, 819
power in electrical system, 602
resistance of wire, 602
resistors in parallel, 561
size of atom, 118, B-11
solutions and mixtures, 167–168, 171–172, 231,
    364–366, 369, 400, 402, 445, 593, 650,
    15-55, B-20, B-36
speed of sound, 724
storage capacity of cylinder, 987
time spent in space by astronauts, 468
USB data transfer rate, 325
velocity, 671, 682, 724, 743
volume of gas, 615
weight of object above Earth, 603
wind chill temperature, 671, 724

SOCIAL STUDIES

age of each child, 380
albums on Billboard chart, 360–361
alcohol consumption by students, 230
annual salary of social workers, 247
attendance at festival, 159
attendance at sporting events, 793
attendance at tutoring service, 231
BET award nominations, 368
cellular phone subscribers, 19
children attending neighborhood school, 324–325
classroom whiteboard dimensions, 183
college applications, 158, B-37
college credits earned, 153
college tuition bills, 231
cruises leaving from the United States, 278
dimensions of computer screen, 996
doctorate degrees awarded in science, 258–259
emails received each day, 229
Facebook active users, 118
Facebook friends, 159
garbage amount collected each year, 929
health club membership, 265
housing starts, 45
immigration from selected countries, 368
math homework, 18
military officers on active duty, 45
news website visitors, 124
number of children in class, 157, 231
number of drive-in theaters in Wisconsin, 231
number of guests at inn, 857
numbers of students advised, B-15
numbers of students in each spinning class, 400
population changes, 45, 963, 975
population density, 119
population of selected cities, 361
public high school graduation rate, 244
residents wanting to secede from Quebec, 231
ribbon lengths, 159
salaries, 170, 171
Snapchat users, 124
students completing college by gender, 339
students enrolled in university, 229
test scores, 213, 230
text messages sent, 211, 400
tourism growth, 914
triplet and higher order births, 857
Twitter website visitors, 45
Vietnamese and Chinese speakers, 368
violent crime, 840

visitors to Las Vegas, 246–247
votes received by each candidate, 593

SPORTS

altitude and depth extremes, 85
area of tennis court, 183
attendance at football game, 199
baseball player's hits, 18
baseball ticket prices, 382
bench press improvement, 15-14
dimensions of lane on basketball court, 183
dimensions of triangular sail, 792
fish caught in derby, 158
football game audience, 857
goal shots for hockey teams, 158
golf tournament best and worst scores, 50
hockey shots on goal, 45
home plate perimeter, 36
human pyramid, 15-23, 15-24
ice rink dimensions, 857
Little League runs scored, 157
martial arts school students, 15-51
NCAA conferences and member schools, 904
Olympic medals won by swimmers, 151–152
participation in Olympics, 170
product endorsements by Tiger Woods, 124
ratio of male-to-female spectators at race, 592
running speed, 785
ski run slope, 269
soccer field center circle area, 183
soccer field dimensions, 174
soccer yellow cards, B-20
speed of baseball pitch, 649
Super Bowl viewers, 119
Tiger Woods' championship score, 52
TV ratings for World Series, 53
values of selected NBA teams, 382

TRANSPORTATION

caution flag dimensions, 183
cost of airline tickets, B-36
crashes involving alcohol, 247
flights through O'Hare Airport, 53
gas mileage, 402
motor vehicle accidents, 270
time spent commuting in Los Angeles, 247
traffic tickets issued, 857

WORK

annual salary, 247, 322, B-28
average hourly wage of embalmers, 310
gross salary, 278
home office dimensions, 36
hourly wages, 602
hours worked at home, 157
hours worked at two jobs by one person, 644
hours worked between two employees, 640
net pay, 891
number of employees, 170, 15-20, 15-52
rate to do a job alone, 594, 615
rate to paint a room, 590–592
salary increase, 15-14, 15-25, 15-51
time for one person to do work, 594, 785, 801,
    802
time for two persons working together, 590–592,
    594, 609–610, 613
unemployment rate in selected states, 401

Index

An "A–," "B–," or "C–," before page numbers indicates online appendix material. "15–" before page numbers indicated online Chapter 15 material.

rational equations *vs.,* 576–577
review of, 562–563
rewritten using their least common
denominator, 547–550
undefined or equal to zero, 527–529, 606,
B–51–B–52
Rational functions
domain of, 533–534, 606, 809–811, 884,
B–52–B–53
explanation of, 532, 606, 809, 884,
B–52–B–53
Rational inequalities
explanation of, 864, 888
method to solve, 864–867
Rational numbers, 39, 80, 526, 737, B–3–B–4
Rationalizing the denominator
containing two terms, 705–707, 740
explanation of, 700
for one higher root, 703–705
for one square root, 700–703
Rationalizing the numerator, 707
Ratios
applications with, 187–189, 227
explanation of, 188
Real numbers
addition of, 46–48, 81
as complex numbers, 727
division of, 58–59, 82
explanation of, 40, 80
geometry and, B–1–B–6
multiplication of, 55–56, 82
order of operations applied to, 50, B–4
product of complex number and its conjugate
as, 729
properties of, 64–70, 82, B–5
set of, 144, B–3–B–4
subtraction of, 49, 81
Real part, of complex numbers, 726, 741
Real-number bases, 122
Reciprocals
explanation of, 10, 68
multiplying by, 132
negative, 276
Rectangles, area and perimeter of, 28, B–2
Rectangular coordinate system. *See* Cartesian
coordinate system
Rectangular solids, 31
Reflection about the *x*-axis, 824–825, 886
Relations
domain of, 294, 295, 300, 319, 807, 808, 883,
B–25–B–26
explanation of, 294, 319, 807, 883, B–24
functions as type of, 294, 296, 319, 807, 809
range of, 294, 295, 807, 808, 883, B–25
written as equations, 298–300, 808, 883
Remainder, division with, 434
Remainder theorem, C–12–C–13
Right angles, 26
Right circular cones, 31
Right circular cylinders, 31
Right triangles, 27, 503, B–49–B–50
Rise, 262
Roots. *See also* Radicals; Square roots
cube, 657, 660, 721
evaluation of, 659–661
fifth, 658
fourth, 657
on graphing calculator, 660
method to find, 657–658, 737
n th, 658
product rule for higher, 683–684, 739
quotient rule for higher, 685–686, 739

Row echelon form, 385
Run, 262

S

Scalene triangles, 27, 80
Scientific notation
converting from, 123
explanation of, 114, 123
on graphing calculator, 117
operations with, 116
using, B–10
writing numbers in, 115–116
Second-degree equations. *See* Quadratic equations
Second-degree inequalities
explanation of, 1015, 1025
graphs of, 1015–1016, 1025
Sequence. *See also* Arithmetic sequence;
Geometric sequence
explained, 15–3
finite, 15–3–15–4, 15–48
general term of, 15–4, 15–6–15–6
on graphing calculator, 15–11–15–12
infinite, 15–3–15–4, 15–48, 15–49
solving applied problems using a, 15–8
writing terms of, 15–5–15–6, 15–16
Series
arithmetic, 15–20, 15–49
evaluating, 15–9–15–10
explained, 15–8–15–9, 15–48
writing using summation notation, 15–10
Set notation, 204
Sets
of integers, 38, B–3–B–4
union of, 214–215, 219, 228
Shifts
horizontal, 823, 828, 885
vertical, 822, 828, 885
Signed numbers. *See also* Real numbers
addition of, 47–48
applications involving, 42
division of, 58–59
explanation of, 38
multiplication of, 55–56
Similar triangles, 191
Simple interest, 165–167, B–15–B–16
Slope
applications involving, 265
explanation of, 260–261, 316
formula for, 262
on graphing calculator, 267–268
of horizontal and vertical lines, 266, 316
method to find, 262–264, B–22–B–23
and one point on line to graph line, 266–267
of parallel lines, 274–276
of perpendicular lines, 276
positive and negative, 264
writing equation of line given *y* -intercept and,
282
Slope-intercept form
equations in, 273–274, 317
explanation of, 272, 317
to graph linear inequalities in two variables,
636–637, 648, B–23
graphs of, 272–274
parallel and perpendicular lines and, 274–276,
317
Solution set
explanation of, 129
of inequalities, 208–210, 228, 626, 866, 867,
1016–1017
of system of inequalities, 1016–1019, 1025

Solutions
extraneous, 716, 717
infinite number of, 143, 144, 373, B–14, B–29,
B–31
of linear equations in two variables, 250,
251–252, B–29
no, 143, 144, B–29, B–31
of system of linear equations in three variables,
373–374
of system of linear equations in two variables,
330–331, 334–336, B–21
Spheres, 31
Square matrix, C–1
Square root functions
domain of, 813, 884
explanation of, 813
on graphing calculator, 816
graphs of, 816, 824–825, 997–998
reflecting graph about *x*-axis with, 824–825
Square root property
distance formula and, 757
explanation of, 753
to solve quadratic equations, 752–762, 775,
798
Square roots. *See also* Radicals
containing negative numbers, 727–728
containing variables with even exponents,
676–677
containing variables with odd exponents,
677–679
division of, 727–728
explanation of, 655, 657
on graphing calculator, 660
method to find, 654–656
method to simplify, 673–675, 676–681, 741
multiplication of, 672–673, 727–728
negative, 655
of negative number, 726–727
principal, 655
product rule for, 673, 738
quotient rule for, 675–676, 738
rationalizing the denominator and, 700–707
simplifying expressions containing, 738
solving equations containing, 717–721
symbol for, 655
Squares. *See also* Completing the square
area and perimeter of, 28, B–2
of binomials, 424–425, 475, 695–696, 718,
740
difference of two, 423, 478–480, 483, 1005,
B–47
perfect, 656
Standard form
of equation of circle, 981–985, 1023
of equation of ellipse, 989–992, 1023
of equation of hyperbola, 992–996, 1024
of equation of line, 273–274
of linear equations, 271, 281
of quadratic equations, 490, 492
Step functions, 828
Straight angles, 26
Substitution method
for binomial to solve quadratic equations,
782–783
explanation of, 340
to factor polynomials, 472
to solve equations in quadratic form,
781–782
to solve nonlinear system of equations,
1008–1010
to solve systems of linear equations, 340–345,
353–354, B–30–B–31

Figure		Perimeter	Area
Rectangle:		$P = 2l + 2w$	$A = lw$
Square:		$P = 4s$	$A = s^2$
Triangle: h = height		$P = a + b + c$	$A = \dfrac{1}{2}bh$
Parallelogram: h = height		$P = 2a + 2b$	$A = bh$
Trapezoid: h = height		$P = a + c + b_1 + b_2$	$A = \dfrac{1}{2}h(b_1 + b_2)$
Circle:		**Circumference** $C = 2\pi r$	**Area** $A = \pi r^2$

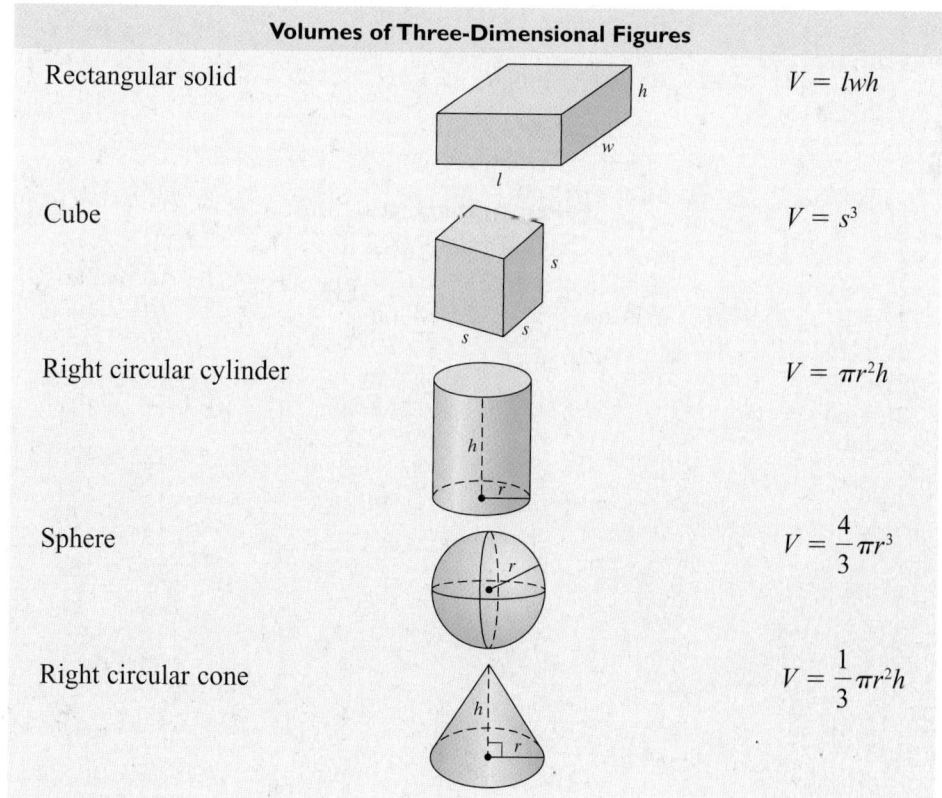

Volumes of Three-Dimensional Figures	
Rectangular solid	$V = lwh$
Cube	$V = s^3$
Right circular cylinder	$V = \pi r^2 h$
Sphere	$V = \dfrac{4}{3}\pi r^3$
Right circular cone	$V = \dfrac{1}{3}\pi r^2 h$

Divisibility Rules

A Number Is Divisible by	Example
. . . 2 if it is an even number.	7394 is divisible by 2 because it ends in 4, making it an even number.
. . . 3 if the sum of its digits is divisible by 3.	837—Add its digits: $8 + 3 + 7 = 18$. Since 18 is divisible by 3, the number 837 is divisible by 3.
. . . 4 if its last two digits form a number that is divisible by 4.	5932—The last two digits form the number 32. Since 32 is divisible by 4, the number 5932 is divisible by 4.
. . . 5 if the number ends in 0 or 5.	645 is divisible by 5 since it ends in 5.
. . . 6 if it is divisible by 2 and 3.	1248—The number is divisible by 2 since it is an even number. The number is divisible by 3 since the sum of its digits is divisible by 3: $1 + 2 + 4 + 8 = 15$. Therefore, the number 1248 is divisible by 6.
. . . 10 if it ends in a 0.	890 is divisible by 10 because it ends in 0.

Powers to Memorize

$2^1 = 2$	$3^1 = 3$	$4^1 = 4$	$5^1 = 5$	$6^1 = 6$	$8^1 = 8$	$10^1 = 10$
$2^2 = 4$	$3^2 = 9$	$4^2 = 16$	$5^2 = 25$	$6^2 = 36$	$8^2 = 64$	$10^2 = 100$
$2^3 = 8$	$3^3 = 27$	$4^3 = 64$	$5^3 = 125$			$10^3 = 1000$
$2^4 = 16$	$3^4 = 81$					
$2^5 = 32$				$7^1 = 7$	$9^1 = 9$	$11^1 = 11$
$2^6 = 64$				$7^2 = 49$	$9^2 = 81$	$11^2 = 121$
						$12^1 = 12$
						$12^2 = 144$
						$13^1 = 13$
						$13^2 = 169$